Statboard Videos

These brief, stepped-out whiteboard videos illustrate over 60 additional examples of difficult topics, and were created by a select group of statistics educators.

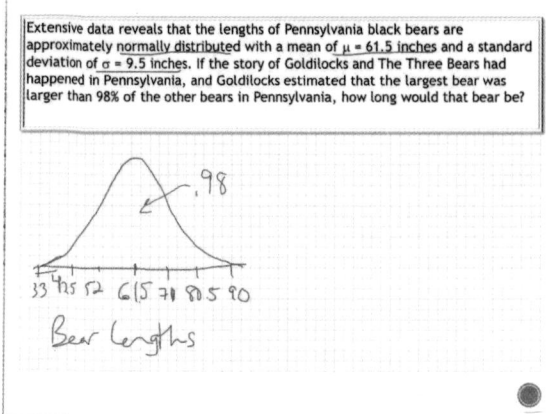

Statistics Videos

Easily assignable and assessable, these videos clarify the most important and most potentially difficult statistical concepts and practices. The collection includes:

- **StatClips:** Brief video tutorials by Alan Dabney of Texas A & M University that combine dynamic animation and interesting scenarios to explain essential concepts.

- **StatClips Step-By-Step Examples:** Animated whiteboard-style walkthroughs of statistical problems.

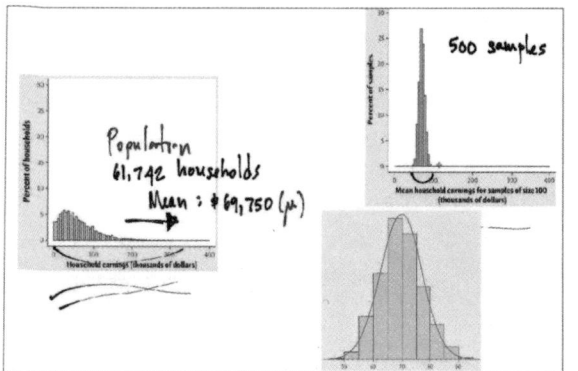

- **Statistically Speaking: Snapshots:** Documentary videos created by Coast Learning Systems and COMAP (Consortium for Mathematics and Its Applications), Inc., showing how statistics is used in a wide variety of fields and applications—from business to medicine, from the environment to the census.

StatTutors

These multimedia t... ...d procedures in a presenta... ...interac-
tive features. The r... ...assign-
able assessments.

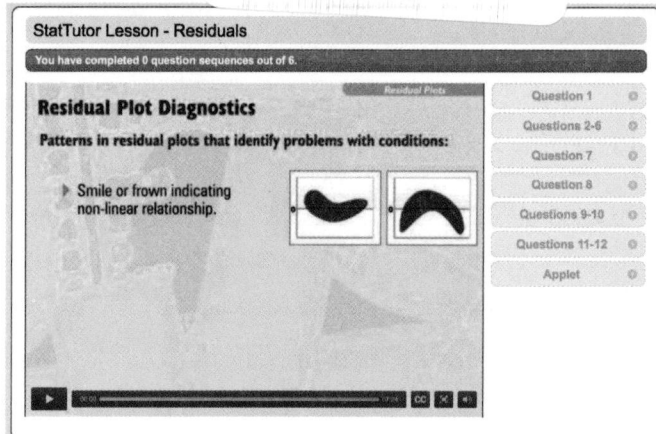

Statistical Applets

Now in HTML5 to be compatible across platforms, and including assignable assessment, Applet-based activities give students hands-on opportunities to familiarize themselves with important statistical concepts and procedures. The interactive setting allows them to manipulate variables and see the results graphically. Icons at specific points in the textbook indicate when there is a corresponding Applet online.

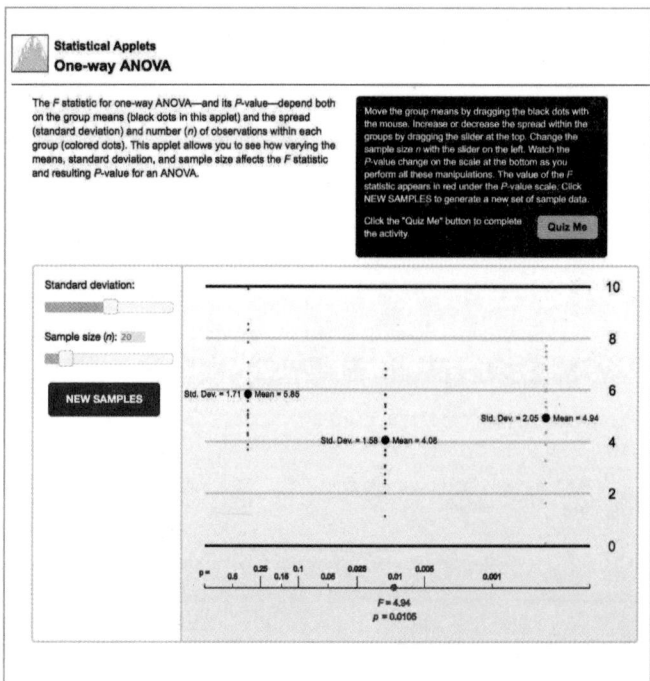

Stepped Tutorials

These new exercise tutorials (2–3 per chapter) are easily assignable and assessable. The tutorials are centered on algorithmically generated quizzing with step-by-step feedback to help students work their way toward the correct solution.

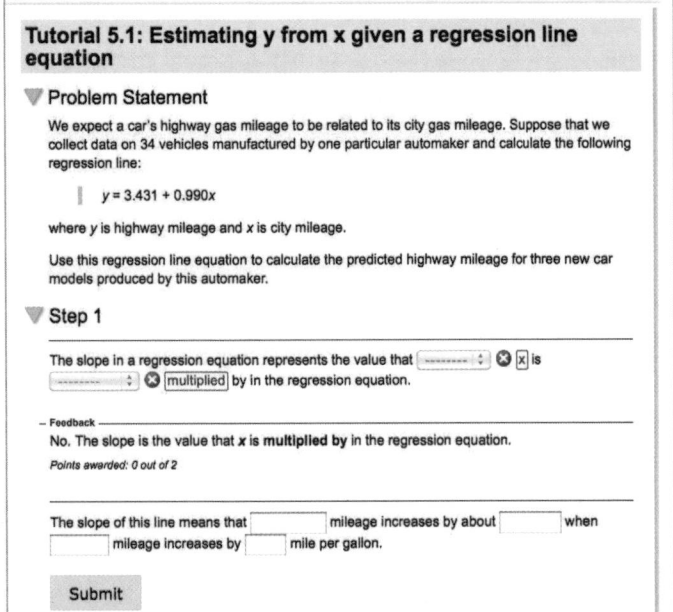

CRUNCH IT!

W. H. Freeman's online statistical software, powered by R, allows users to perform all of the statistical operations and graphing needed for an introductory statistics course and more. It saves users time by automatically loading data from W. H. Freeman's statistics textbooks, and it provides the flexibility to edit and import additional data. Help videos that show how to use CrunchIt! effectively are also available.

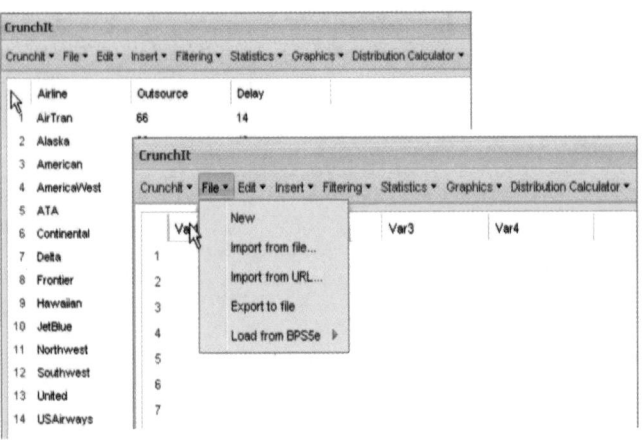

JMP Student Edition (developed by SAS) is easy to learn and contains all the capabilities required for introductory statistics. JMP is the commercial data analysis software of choice for scientists, engineers, and analysts at leading companies throughout the globe. Register inside LaunchPad at no additional cost.

Video Tech Manuals

These brief videos provide students with basic instruction for working with a variety of statistical software programs and technology. Video manuals are available for CrunchIT! and JMP, as well as Excel, SPSS, TI-83/84 calculators, Minitab, R, and RCmdr.

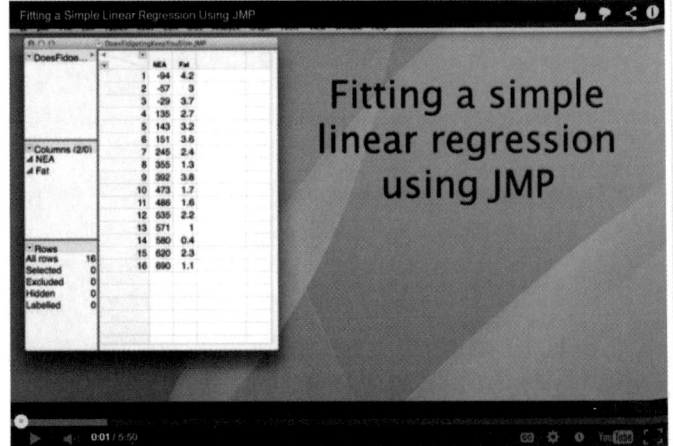

EESEE Case Studies

This newly updated series of cases, developed at The Ohio State University for the Electronic Encyclopedia of Statistical Examples and Exercises, involves students in applications of statistical practices in a variety of professional settings.

SolutionMaster

With this innovative digital tool, instructors can provide a secure solutions file for any set of exercises from the text. Solutions are easy to generate and can be exported in .pdf format for convenient downloading, posting, and printing.

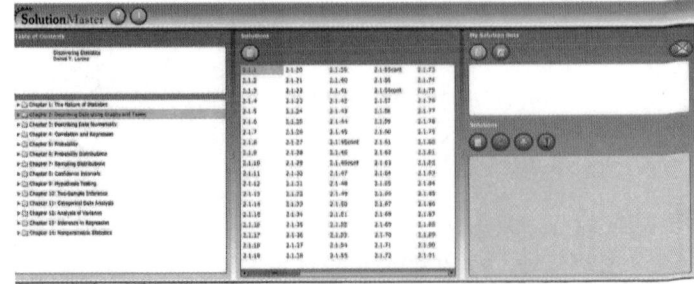

Instructor's Edition

Discovering Statistics

Third Edition

Instructor's Edition Material Written by
Daniel T. Larose and Chantal D. Larose

Daniel T. Larose

Central Connecticut State University

W. H. FREEMAN
& COMPANY

A Macmillan Education Imprint

Publisher: Terri Ward
Senior Acquisitions Editor: Karen Carson
Marketing Manager: Cara LeClair
Developmental Editor: Jorge Amaral
Executive Media Editor: Laura Judge
Associate Media Editor: Liam Ferguson
Associate Editor: Marie Dripchak
Editorial Assistant: Victoria Garvey
Marketing Assistant: Bailey James
Photo Editor: Robin Fadool
Cover Designer: Vicki Tomaselli
Text Designer: Patrice Sheridan
Art Director: Diana Blume
Project Editor: Jodi Isman
Illustrations: MPS Limited
Production Manager: Paul W. Rohloff
Composition: MPS Limited
Printing and Binding: RR Donnelley
Cover photo: Wavebreakmedia/Newscom

Library of Congress Control Number: 2015939181

Casebound ISBN-13: 978-1-4641-4200-0
ISBN-10: 1-4641-4200-9
Loose-leaf ISBN-13: 978-1-4641-8864-0
ISBN-10: 1-4641-8864-5
Instructor's Edition ISBN-13: 978-1-4641-9292-0
ISBN-10: 1-4641-9292-8

© 2016, 2013, 2010 by W. H. Freeman and Company

Printed in the United States of America

First printing

W. H. Freeman and Company
One New York Plaza
Suite 4500
New York, NY 10004-1562
www.whfreeman.com

BRIEF CONTENTS

CONTENTS

Mark Hooper/Getty Images

4. Correlation and Regression   186

Kris Hanke/Vetta/Getty Images

5. Probability   238

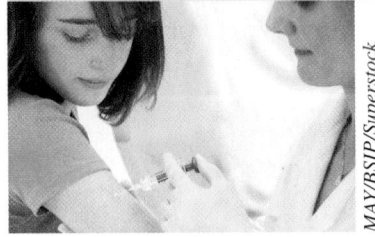

MAY/BSIP/Superstock

6. Probability Distributions   308

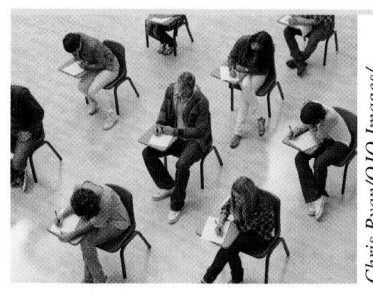

Chris Ryan/OJO Images/Getty Images

9. Hypothesis Testing **486**

Cultura/Henglein and Steets/Riser/Getty Images

Ariel Skelley/Blend Images/Getty Images

Susan Wides/Getty Images

Wavebreakmedia/Shutterstock

TO THE INSTRUCTOR

John Naisbitt wrote: "We are drowning in information, but we are starved for knowledge" (*Megatrends: Ten New Directions for Transforming Our Lives*, Warner Books, 1982). As we move deeper into the twenty-first century, the Information Age, the gap between the flood of information and the dearth of knowledge and meaning is ever widening. As statistics instructors, we stand on the front line in the struggle to close this gap, to help our citizen-students sort through the data deluge that daily confronts them, and to help them uncover nuggets of knowledge that have meaning in their lives. This is why I wrote *Discovering Statistics:* to provide a friendly and effective tool that will help instructors connect with their students and teach their students how to discover knowledge and meaning in data.

Discovering Statistics will help your students develop the quantitative and analytical tools needed to understand statistics in today's data-saturated world. The text incorporates unique pedagogical features, such as *Developing Your Statistical Sense, What Does This Mean?, What Results Might We Expect?,* and *What If Scenarios.* These features are aimed at anticipating students' questions and helping them understand the value of statistical intuition and interpretation, while at the same time also stressing computational skills. In the Information Age, it is important that students be able not only to find the right answer, but also to explain what the statistical results mean to those who have never taken a statistics course. Thus, the text stresses the importance of interpreting results to others.

In over 20 years of teaching introductory statistics, I have found that this approach works well at motivating students to learn statistics and often to enjoy the study of statistics. My hope is that instructors will find *Discovering Statistics* to be an effective tool for imparting to our twenty-first-century students an appreciation of the power of statistical analysis to extract knowledge from data.

The Introductory Statistics Course

Discovering Statistics is intended for an algebra-based, undergraduate, one- or two-semester course in general introductory statistics for non-majors. The only prerequisite is basic algebra. *Discovering Statistics* will prepare students to work with data in fields such as psychology, business, nursing, education, and liberal arts, to name a few.

The GAISE guidelines, endorsed by the American Statistical Association, include the following recommendations:

1. Emphasize statistical literacy and develop statistical thinking.
2. Use real data.
3. Stress conceptual understanding instead of mere knowledge of procedures.
4. Foster active learning in the classroom.
5. Use technology for developing conceptual understanding and analyzing data.
6. Use assessments to improve and evaluate student learning.

Discovering Statistics adopts these guidelines verbatim as the course pedagogical objectives, with the following single adjustment: (3) Stress conceptual understanding *in addition to* knowledge of procedures. To these, the text adds two course pedagogical objectives:

7. Use case studies to show how newly acquired analytic tools may be applied to a familiar problem.
8. Encourage student motivation.

Approach of *Discovering Statistics*, Third Edition

Balanced analytical and computational coverage. The text integrates data interpretation and discovery-based methods with complete computational coverage of introductory statistics topics. Through unique and careful use of pedagogy, the text helps students develop their "statistical sense"—understanding the meaning behind the numbers. Equally, the text includes integrated and comprehensive computational coverage, including step-by-step solutions within examples. Select examples include screenshots and computer output from TI-83/84, Excel, Minitab, SPSS, JMP, and CrunchIt!, with keystroke instructions located in the Step-by-Step Technology Guides at the ends of sections.

Communication of results. *Discovering Statistics,* Third Edition, emphasizes how, in the real world and in their future careers, students will need to explain statistical results to others who have never taken a statistics course.

Emphasis on variability. The importance of variability in the introductory statistics curriculum cannot be overstated. Without a solid appreciation of how statistics may vary, there is little chance that students will be able to understand the crucial topic of sampling distributions.

Use of powerful, current examples with real data. Video game sales, the use of cell-phone apps, and music videos available for download represent the variety of examples included in *Discovering Statistics,* Third Edition. Example and exercise topics reflect real-world problems and engage the interest of the reader in their solutions. Real data (with sources cited) are frequently used to further demonstrate relevance of topics.

Use of large data sets. The author leverages his expertise as an author and consultant in data mining and predictive analytics to provide the reader, in nearly every chapter, with large data sets to explore and analyze.

New to This Edition

*   With extensively revised examples and exercises, the Third Edition provides in-depth coverage of topics with increased clarity and comprehensiveness.
    *   60% of the examples are new or revised.
    *   50% of the exercises are new or revised.
*   The number and variety of exercises has been increased, to provide a greater opportunity for different learning styles and practice. Over 4,100 exercises are now included in the book.
*   Examples and exercises cover a wide range of applications and use current, real data.
*   **NEW JMP** and **SPSS** Statistical Software instructions and output screenshots are now included in the Step-by-Step Technology Guides at the end of most sections.
*   **NEW** Your Turn

    This new feature consists of a small set of exercises after approximately half of the examples. Students can use these exercises as a self-test, or instructors can use them in class to further reinforce concepts.

- **NEW** Check It Out

  This feature, located at the end of a section, ties the Practicing the Techniques exercises back to specific examples in the chapter, helping students learn more effectively and engage with the text, and helping instructors more easily organize and assign exercises.

- **NEW** Each Clarifying the Concepts exercise now provides the page number in the text where the answer may be found. Together with the Check It Out feature, this will assist students in completing their homework more easily.

- **NEW** Eight brand-new Chapter Case Studies, including

  - Chapter 1: Video Game Sales

  - Chapter 2: Criminal Justice in New York City

  - Chapter 4: Measuring the Human Body

  - Chapter 5: The Gardasil Vaccine

  - Chapter 6: SAT Scores and AP Exam Scores

  - Chapter 8: Motor Vehicle Fuel Efficiency

  - Chapter 9: Clothing Store Sales

  - Chapter 10: Bank Loans

- Simplified numbering of examples, tables, and figures

- In Chapter 4, examples for computational formulas for the correlation coefficient and the slope of the regression line have been eliminated and moved to the exercises.

- In Chapter 5, the use of contingency tables has been expanded throughout the chapter.

- In Chapter 6, the section on Continuous Random Variables and the Normal Probability Distribution and the section on Standard Normal Distribution have been combined. Also, the topic of Assessing Normality Using Normal Probability Plots has been moved to this chapter.

- In Chapter 7, the section on Introduction to Sampling Distributions and the section on Central Limit Theorem for Means have been combined.

- A new objective has been added in Chapter 9 (Calculate the test statistic Z_{data}; Find the critical regions and critical values for a hypothesis test).

Features of *Discovering Statistics*, Third Edition

The Third Edition retains many of the successful features from the Second Edition.

Case Studies. A Case Study begins each chapter and is developed throughout the section examples, using the new set of tools that the section provides. For the Third Edition, coverage of the Case Study in each chapter has been expanded to provide continuity throughout the chapter as students learn new topics and tools.

THE BIG PICTURE

Where we are coming from and where we are headed . . .

- In Chapter 1, we learned the basic concepts of statistics, such as population, sample, and types of variables, along with methods of collecting data.
- Here, in Chapter 2, we learn about graphs and tables for summarizing qualitative data and quantitative data, and we examine how to prevent our graphics from becoming misleading.
- Later, in Chapter 3, we will learn how to describe a data set using numerical measures instead of graphs and tables.

The Big Picture. Brief, bulleted lists at the beginning of each chapter look at "Where we are coming from and where we are headed . . .". (Chapter 2, page 39)

2.1   Graphs and Tables for Categorical Data

OBJECTIVES By the end of this section, I will be able to . . .

1   Construct and interpret a frequency distribution and a relative frequency distribution for qualitative data.
2   Build and interpret bar graphs and Pareto charts.
3   Construct and interpret pie charts.
4   Build crosstabulations to describe the relationship between two variables.
5   Work with tabular data to construct graphs and distributions.
6   Construct a clustered bar graph to describe the relationship between two variables.

Matched Objectives. Each section begins with a list of numbered objectives headed "By the end of this section, I will be able to . . .". The objective numbers are matched with the numbered topics within each section as well as the end-of-section summary. (Chapter 2, pages 40, 53)

Section 2.1 Summary

In this section, we learned about tabular and graphical methods for summarizing qualitative (categorical) data.
  1. Frequency distributions and relative frequency distributions list all the values that a qualitative variable can take, along with the frequencies (counts) or relative frequencies (percents) for each value.
  2. A bar graph is the graphical equivalent of a frequency distribution or a relative frequency distribution. When the rectangles are presented in decreasing order from left to right, the result is a Pareto chart.
  3. Pie charts are a common graphical device for displaying the relative frequencies of a categorical variable. A pie chart is a circle divided into sections (that is, slices or wedges),

with each section representing a particular category. The size of the section is proportional to the relative frequency of the category.
  4. Crosstabulation summarizes the relationship between two categorical variables. A crosstabulation is a table that gives the counts for each row-column combination, with totals for the rows and columns.
  5. Data often comes to us already summarized in a table. We can use this tabular data to construct graphs and distributions.
  6. Clustered bar graphs are useful for comparing two categorical variables and are often used in conjunction with crosstabulations.

NOW YOU CAN DO
Exercises 11–14 and
23–28.

Now You Can Do, found in the margin next to most examples, cues students to try related Practicing the Techniques exercises. These callouts are intended to prompt the student toward practicing the techniques shown in the example. For example, in Chapter 2, in the margin at the end of Example 2 on page 42, you will find "Now You Can Do Exercises 11–14 and 23–28." This callout lets the student know that they can use the example as a model when completing the exercise set. When working a particular exercise, the student can also easily look back through the section to find the callout to a related example.

Your Turn exercises follow about half of the examples. Students can use these exercises as a self-test or instructors can use them in class to further reinforce concepts. Answers to all Your Turn exercises are included in Appendix A. (Chapter 2, page 42)

YOUR TURN
#2

 Use Table 3 to construct relative frequency distributions for the following categorical variables.

1. *Borough*
2. *Violation type*

(The solutions are shown in Appendix A.)

BRINGING IT ALL TOGETHER

Shopping Enjoyment and Gender. Use the information in the crosstabulation for Exercises 105–119. The Pew Internet and American Life Project surveyed 4514 American men and women and asked them, "How much, if at all, do you enjoy shopping?" The results shown in the crosstabulation are missing some entries.

Bringing It All Together exercises within each section offer a culmination of everything students have learned in a particular section, using a related set of Applying the Concepts exercises to tie together the main concepts and techniques learned. (Chapter 2, page 58)

 Data sets, available in a variety of software formats, are each named and marked with an icon in the text. Students can locate the data sets online at **www.macmillanhighered.com/discostat3e**.

Developing Your Statistical Sense. This feature empowers students with some useful perspectives that real-world data analysts need to know. Students will learn to think like real-world statistical analysts. This feature implements the GAISE guideline "Develop statistical thinking." (Chapter 2, page 62)

Developing Your Statistical Sense

Choosing Which Distribution to Use

So which frequency distribution is the "right" one—Table 18 or Table 19? There is no absolute answer. It depends on the goals of the analysis, as well as other factors. For example, from Table 19, we can see that most of the artists in our sample had between one and three multi-platinum singles, a finding that was not immediately apparent from Table 18. Therefore, combining data values into classes can lead to interesting overall findings. However, whenever data values are combined into classes, some information is lost. For example, it is not possible, using Table 19 alone, to determine that the number of multi-platinums occurring with the greatest frequency in our sample is two.

What Does This Number Mean?

The Mean as the Balance Point of the Data

Let's explore our sample cell phone price data a bit further. Consider the dotplot of the cell phone prices in Figure 1. To find out where the mean price lies on this number line, imagine that the dots are little blocks on a ruler or a seesaw and that you must decide where to place the support (like the triangle in Figure 1) so that the ruler balances perfectly. *The place where the data set balances perfectly is the location of the mean.* Placing the fulcrum too far to the right or left would create an imbalance. This data set balances precisely at the sample mean, $\bar{x} = \$337.50$

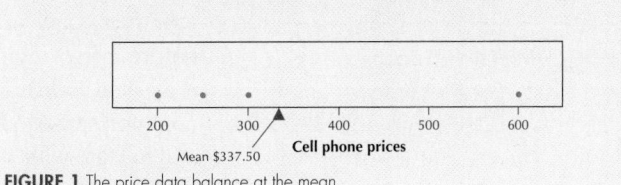

FIGURE 1 The price data balance at the mean.

What Does This Mean? feature boxes foster an intuitive approach and interpretation of results. Whenever a new formula or statistic is being introduced, the emphasis is on "What does this really mean?" Developing this understanding is just as important as getting the right answer, especially when the software can do the calculations. In the workplace, students may need to explain to their managers what the statistical results really mean. This feature helps to implement the GAISE guideline "Stress conceptual understanding." (Chapter 3, page 110)

What If Scenarios. These scenarios help students focus on statistical thinking instead of rote computation. Because of the availability of powerful statistical computer packages, statistical analysis is easy to do badly. The wrong analysis is worse than useless. It can cost companies lots of money, may convince lawmakers to pass bad legislation affecting millions of people, can incorrectly determine effects of pharmaceuticals or environmental pollution, as well as many other serious ramifications. The What If scenarios are extensions of examples or exercises aimed at honing students' critical-thinking skills. In

WHAT IF

What If Scenario

Consider Example 6 once again. Now imagine: *what if* there was an incorrect data entry, such as a typo, and the number of Michael Jackson's videos was greater than 31 by some unspecified amount?

Describe how and why this change would have affected the following, if at all:

a. The mean number of music videos

b. The median number of music videos

c. The mode number of music videos

Solution

a. Consider Figure 4, a dotplot of the number of music videos, with the triangle indicating the mean, or balance point, at 24.5. Recall that this represents the balance point of the data. As the number of Michael Jackson's videos increases (arrow), the point at which the data balance (the mean) also moves somewhat to the right. Thus, the mean number of followers will increase.

What If scenarios, the original problem setup is altered in a specific but nonquantifiable way. Students are then asked to think about how that change would percolate through the results, without recourse to calculations. The exercises, as well as the scenarios, are marked with the What If? icon. (Chapter 3, page 115)

Stepped Example Solutions. In selected examples, students are guided through the key steps needed to work through the calculations and find the solution. (Chapter 3, page 174)

Solution

From Example 31, the five-number summary for the state export data is Min = 1.6, Q1 = 3.4, Med = 4.65, Q3 = 5.4, Max = 7.7. The interquartile range for the state export data is IQR = Q3 − Q1 = 5.4 − 3.4 = 2.0.

Step 1   Determine the lower and upper fences:

a. Lower fence = Q1 − 1.5(IQR) = 3.4 − 1.5(2) = 0.4

b. Upper fence = Q3 + 1.5(IQR) = 5.4 + 1.5(2) = 8.4

Step 2   Draw a horizontal number line that encompasses the range of your data, including the fences. Above the number line, draw vertical lines at Q1 = 3.4, median = 4.65, and Q3 = 5.4. Connect the lines for Q1 and Q3 to each other so as to form a box, as shown in Figure 35a.

FIGURE 35A Constructing a boxplot by hand: Steps 1 and 2.

What Results Might We Expect?

Symmetric Data and Boxplots

So, can you now predict how a boxplot of *symmetric* data will look? The median will be about the same distance from Q1 (lower hinge) and Q3 (upper hinge). And the upper and lower whiskers will be about the same length. An example of a boxplot of symmetric data is shown in Figure 40.

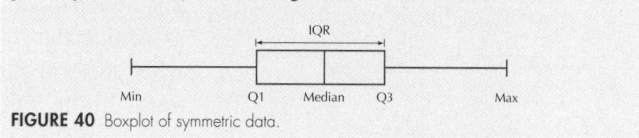

FIGURE 40 Boxplot of symmetric data.

What Results Might We Expect? This feature, located in example solutions, challenges students to predict what the result of a particular problem will be. Students are presented with a graphical view of the situation, and, before performing any calculations, are asked to bring their intuition and common sense to bear on the problem and to state what results we might expect once we do the number crunching. (Chapter 3, page 176)

Definitions and Formulas. Easily located in highlighted boxes, key definitions and formulas are important for students to understand when working examples and exercises. Important vocabulary and formulas are also listed (with page references) at the end of each chapter. (Chapter 3, page 177)

IQR Method to Detect Outliers

A data value is an outlier if

a. it is located 1.5(IQR) or more below Q1, or

b. it is located 1.5(IQR) or more above Q3.

Check It Out!, located at the end of a section, ties the Practicing the Techniques exercises back to specific examples in the chapter, helping instructors assign exercises and helping students learn. (Chapter 3, page 180)

✓ CHECK IT OUT!

To do	Check out	Topic
Exercises 7–8, 13–14, and 19–20.	Example 31	Five-number summary
Exercises 9–10, 15–16, and 21–22.	Example 34	Boxplots
Exercises 11–12, 17–18, and 23–24.	Example 37	IQR method for identifying outliers
Exercises 25 and 26	Examples 35 and 36	Boxplots and skewness
Exercises 27–30	Example 38	Comparison boxplots

Exercises. *Discovering Statistics,* Third Edition, contains a rich and varied collection of section and chapter exercises.

- Clarifying the Concepts (conceptual)
- Practicing the Techniques (skill-based)
- Applying the Concepts (real-world applications)
- Bringing It All Together

These exercises bring together everything students have learned in a particular section, using a related set of Applying the Concepts exercises to tie together the main concepts and techniques learned in the section.

At the end of each chapter, **Review Exercises** and a **Chapter Quiz** help to test students' overall understanding of each chapter's concepts and to practice for exams (Chapter 3, page 185). The student edition includes answers to odd-numbered exercises and chapter quiz exercises in the back of the book.

Chapter 3 QUIZ

TRUE OR FALSE

1. True or false: If two data sets have the same mean, median, and mode, then the two data sets are identical.

2. True or false: The variance is the square root of the standard deviation.

3. True or false: The Empirical Rule applies for any data set.

FILL IN THE BLANK

4. A(n) _____ is an extremely large or extremely small data value relative to the rest of the data set.

5. The mean can be viewed as the _____ point of the data.

6. The measure of center that is sensitive to the presence of extreme values is the _____.

SHORT ANSWER

7. What do we call summary descriptive measures that are not sensitive to the presence of outliers?

C+ in her two-credit physical education course. Calculate Angelita's grade point average for this semester.

12. A sample of 30 Americans yielded a sample mean consumption of carbonated beverages this year of 60 gallons, with a sample standard deviation of 40 gallons. Find the z-scores for the following amounts of carbonated beverage consumption.

   a. 120 gallons
   b. 20 gallons
   c. 100 gallons
   d. 0 gallons
   e. 60 gallons

13. Refer to the information in Exercise 12. Assume the distribution is bell-shaped. (*Hint:* Use your knowledge about the Empirical Rule to give a range for the proportions in parts (**b**) and (**d**)).

   a. Find the 50th percentile.
   b. Estimate the proportion of Americans who drink

STEP-BY-STEP TECHNOLOGY GUIDE: Boxplots

We will make boxplots for the exports data from Section 3.4, Example 28 on page 163.

TI-83/84

Step 1 Enter the data in list L1.
Step 2 Press 2nd Y =, and choose 1: Plot 1.
Step 3 Highlight On and press ENTER. Highlight the boxplot icon, as shown in Figure 44. Press ENTER.
Step 4 Press ZOOM, and choose 9: ZoomStat.

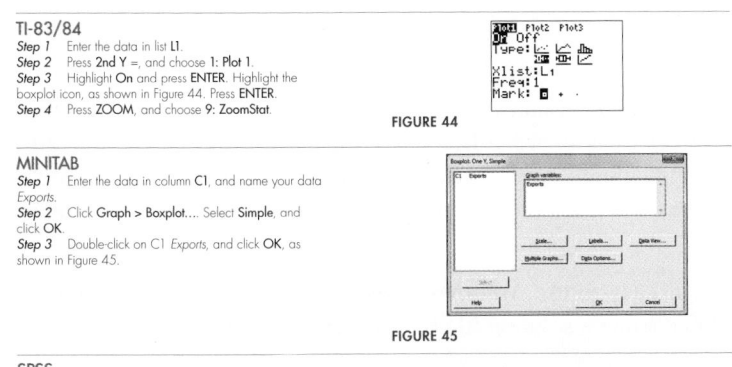

FIGURE 44

MINITAB

Step 1 Enter the data in column C1, and name your data *Exports*.
Step 2 Click Graph > Boxplot.... Select **Simple**, and click **OK**.
Step 3 Double-click on C1 *Exports*, and click **OK**, as shown in Figure 45.

FIGURE 45

SPSS

Step 1 Input the scores into the first column. Name the column *Exports*.
Step 2 Click Graphs > Chart Builder.... Click **OK**, then Scan Data.
Step 3 In the **Gallery** tab, find the **Choose from** menu and

Step 4 Click and drag the **Simple boxplot** to where it says "Drag a Gallery chart here..." Close the Element Properties box.
Step 5 Click and drag *Exports* to where it says "Y-Axis?" in the chart preview.
Step 6 Click **OK**.

Step-by-Step Technology Guide. This feature covers TI-83/84 calculators, Excel, Minitab, JMP, SPSS, and CrunchIt!, providing stepped keystroke instructions for working through selected examples in the text. Screenshots of the results are often provided, either within the Step-by-Step Technology Guide or in the corresponding example. (Chapter 3, page 179)

Applets. Applet icons in the text help to mark chapter material and exercises for which interactive statistical applets are available.

Caution notes. Signaled by the Caution icon, these warnings in the text help students avoid common errors and misconceptions.

Media and Supplements

The following electronic and print supplements are available with *Discovering Statistics*, Third Edition:

W. H. Freeman's new online homework system, **LaunchPad,** offers our quality content curated and organized for easy assignability in a simple yet powerful interface. We have taken what we have learned from thousands of instructors and hundreds of thousands of students to create a new generation of W. H. Freeman/Macmillan technology.

Curated units. Combining a curated collection of videos, homework sets, tutorials, applets, and e-Book content, LaunchPad's interactive units give instructors a building block to use as is or as a starting point for customized learning units. A majority of exercises from the text can be assigned as online homework, including an abundance of algorithmic exercises. An entire unit's worth of work can be assigned in seconds, drastically reducing the amount of time it takes for instructors to have their course up and running.

Easily customizable. Instructors can customize the LaunchPad units by adding quizzes and other activities from our vast wealth of resources. They can also add a discussion board, a drop box, and an RSS feed, with a few clicks. LaunchPad allows instructors to customize students' experiences as much or as little as desired.

Useful analytics. The gradebook quickly and easily allows instructors to look up performance metrics for classes, individual students, and individual assignments.

Intuitive interface and design. The student experience is simplified. Students' navigation options and expectations are clearly laid out at all times, ensuring they can never get lost in the system.

Assets integrated into LaunchPad include the following:

Interactive e-Book. Every LaunchPad e-Book comes with powerful study tools for students, video and multimedia content, and easy customization for instructors. Students can search, highlight, and bookmark, making it easier to study and access key content. Teachers can ensure that their classes get just the book they want to deliver by customizing and rearranging chapters, adding and sharing notes and discussions, and linking to quizzes, activities, and other resources.

LearningCurve provides students and instructors with powerful adaptive quizzing, a game-like format, direct links to the e-Book, and instant feedback. The quizzing system features questions tailored specifically to the text and adapts to students' responses, providing material at different difficulty levels and topics based on student performance.

SolutionMaster offers an easy-to-use, Web-based version of the instructor's solutions, allowing instructors to generate a solution file for any set of homework exercises.

Statistical Video Series consists of StatClips, StatClips Examples, and Statistically Speaking "Snapshots." View animated lecture videos, whiteboard lessons, and documentary-style footage that illustrate key statistical concepts and help students visualize statistics in real-world scenarios.

New Video Technology Manuals available for TI-83/84 calculators, Minitab, Excel, JMP, SPSS, R, Rcmdr, and CrunchIt! provide brief instructions for using specific statistical software.

Updated StatTutor Tutorials offer multimedia tutorials that explore important concepts and procedures in a presentation that combines video, audio, and interactive features. The newly revised format includes built-in, assignable assessments and a bright new interface.

Updated Statistical Applets give students hands-on opportunities to familiarize themselves with important statistical concepts and procedures, in an interactive setting that allows them to manipulate variables and see the results graphically. Icons in the textbook indicate when an applet is available for the material being covered.

 **CrunchIt!** is W. H. Freeman's Web-based statistical software that allows users to perform all the statistical operations and graphing needed for an introductory statistics course and more. It saves users time by automatically loading data from *Discovering Statistics,* and it provides the flexibility to edit and import additional data.

JMP Student Edition (developed by SAS) is easy to learn and contains all the capabilities required for introductory statistics. JMP is the leading commercial data analysis software of choice for scientists, engineers, and analysts at companies throughout the world (for Windows and Mac). Register inside LaunchPad at no additional cost.

Stats@Work Simulations put students in the role of the statistical consultant, helping them better understand statistics interactively within the context of real-life scenarios.

***EESEE* Case Studies** (*Electronic Encyclopedia of Statistical Examples and Exercises*), developed by The Ohio State University Statistics Department, teach students to apply their statistical skills by exploring actual case studies using real data.

Data files are available in CrunchIt!, JMP, ASCII, Excel, TI, Minitab, SPSS (an IBM Company)*, R, and CSV formats.

Student Solutions Manual provides solutions to the odd-numbered exercises in the text.

Instructor's Guide with Full Solutions includes teaching suggestions, chapter comments, and detailed solutions to all exercises, and is available electronically within LaunchPad.

Test Bank offers hundreds of multiple-choice questions and is available in LaunchPad.

Lecture Slides offer a customizable, detailed lecture presentation of statistical concepts covered in each chapter of *Discovering Statistics,* Third Edition.

Additional Resources Available with *Discovering Statistics,* Third Edition

Web site www.macmillanhighered.com/discostat3e. This open-access Web site includes data files and Chapter 14, "Nonparametric Statistics."

Special Software Packages. Student versions of JMP and Minitab are available for packaging with the printed text. JMP is also available inside LaunchPad at no additional cost. Contact your W. H. Freeman representative for information or visit www.whfreeman.com.

i-clicker is a two-way radio-frequency classroom response solution developed by educators for educators. Each step of i-clicker's development has been informed by teaching and learning. To learn more about packaging i-clicker with this textbook, please contact your local sales rep or visit www1.iclicker.com.

*SPSS was acquired by IBM in October 2009.

ACKNOWLEDGMENTS

I want to join W. H. Freeman and Company in thanking Ann Cannon at Cornell College for accuracy reviewing the book, Christina Morian at Lincoln University for authoring the back-of-book answers, and John Samons at Florida State College at Jacksonville for accuracy reviewing the back-of-book answers. I would also like to thank the instructors who authored the supplements for the book: Christina Morian (solutions), Becky Moening at Ivy Tech Community College (test bank), Tom Achatz at Washtenaw Community College (practice quizzes), Mark Gebert at University of Kentucky (lecture slides), and David French at Tidewater Community College (clicker questions). I'd also like to thank John Samons for accuracy reviewing the solutions and John Davis at Baylor University for accuracy reviewing the test bank, practice quizzes, and clicker questions.

I'd like to thank the reviewers who offered comments that assisted in the development and refinement of the Third Edition of *Discovering Statistics*:

Thomas Achatz, *Washtenaw Community College*
Victor Akatsa, *Chicago State University*
John Beyers, *University of Maryland University College*
Melissa Bingham, *University of Wisconsin–La Crosse*
Daniel Birmajer, *Nazareth College*
Mark Bollman, *Albion College*
Carolyn Chapel, *Western Technical College*
Seo-eun Choi, *Arkansas State University*
Carolyn Cuff, *Westminster College*
Dawn R. Dabney, *Northeast State Community College*
Kevin Dennis, *Saint Mary's University of Minnesota*
Geoffrey Dietz, *Gannon University*
Brandi Falley, *Texas Woman's University*
David French, *Tidewater Community College*
Petre Ghenciu, *University of Wisconsin–Stout*
Nicholas Gorgievski, *Nichols College*
Donna Gorton, *Butler Community College*
Lorraine Gregory, *Lake Superior State University*
Susan Herring, *Sonoma State University*

Martin Jones, *College of Charleston*
Joseph Karnowski, *Norwalk Community College*
Weiping Li, *Walsh University*
Catherine Matos, *Clayton State University*
Becky A. Moening, *Ivy Tech Community College*
Carla Monticelli, *Camden County College*
Shai Neumann, *Eastern Florida State College*
Lyn A. Noble, *Florida State College at Jacksonville*
William Radulovich, *Florida State College at Jacksonville*
Gina Reed, *University of Georgia*
Fary Sami, *Harford Community College*
Jason Samuels, *Borough of Manhattan Community College*
Connie Schrock, *Emporia State University*
Mohammed A. Shayib, *Prairie View A&M University*
Abdallah Shuaibi, *Harry S. Truman College*
James Stamey, *Baylor University*
Martha Tapia, *Berry College*
Wayne Tarrant, *Rose-Hulman Institute of Technology*
Karin Vorwerk, *Westfield State University*

I also want to thank the instructors who offered comments on the Second Edition:

Holly Ashton, *Pikes Peak Community College*
John Beyers, *University of Maryland University College*
Ferry Butar Butar, *Sam Houston State University*
Ann Cannon, *Cornell College*
Ayona Chatterjee, *University of West Georgia*
Zhao Chen, *Florida Gulf Coast University*
Geoffrey Dietz, *Gannon University*
Wanda Eanes, *Middle Georgia State College*
Elaine Fitt, *Bucks County Community College*
Elizabeth Flow-Delwiche, *Community College of Baltimore County*
Joe Gallegos, *Salt Lake Community College*
Dave Gilbert, *Santa Barbara City College*
David Gurney, *Southeastern Louisiana University*
Noal Harbertson, *California State University, Fresno and Sacramento*
Steve Hundert, *College of Southern Maryland*
Andreas Lazari, *Valdosta State University*

Ananda Manage, *Sam Houston State University*
Christina Morian, *Lincoln University*
John Nardo, *Oglethorpe University*
Michael Nasab, *Long Beach City College*
Fary Sami, *Harford Community College*
Jason Samuels, *Borough of Manhattan Community College*
Mohammed Shayib, *Prairie View A&M University*
Kim Sheppard, *Cecil College*
Marcia Siderow, *California State University, Northridge*
Karen Smith, *University of West Georgia*
Tabrina Smith, *Lake Erie College*
John Trimboli, *Middle Georgia State College*
Cameron Troxell, *Mt. San Antonio College*
Mahbobeh Vezvaei, *Kent State University*
Karin Vorwerk, *Westfield State University*
James Wan, *Long Beach City College*
Tanya Wojtulewicz, *Community College of Baltimore County*

The Third Edition of *Discovering Statistics* owes much to the untiring efforts of the team of professionals at W. H. Freeman and Company. I want to thank Terri Ward, Karen Carson, Jorge Amaral, Cara LeClair, Marie Dripchak, Laura Judge, Liam Ferguson, Victoria Garvey, Bailey James, Jodi Isman, Robin Fadool, Vicki Tomaselli, Diana Blume, and Paul W. Rohloff for contributing their talents to the creation of this book and its online resources.

I also want to thank Dr. Philip Halloran and Dr. Chun Jin, Chair and Assistant Chair of the Department of Mathematical Sciences at Central Connecticut State University, Dr. Dipak K. Dey, Distinguished Professor and Associate Dean, College of Liberal Arts and Sciences at the University of Connecticut, and Dr. John Judge, Professor of Statistics in the Department of Mathematics at Westfield State College. Thanks to my daughter Chantal Danielle (27) for carrying on the love of statistics to the next generation, and to my twin children Tristan Spring and Ravel Renaissance (16) for providing insight into what life is all about. Above all, I extend my deepest gratitude to my darling wife of 30 years, Debra J. Larose, for sharing with me "one love, one lifetime . . . ".

ABOUT THE AUTHOR

Since his days of collecting baseball cards as a youngster and scrutinizing the statistics of his favorite players, Dan Larose has loved statistics. He also loved language and writing, so when Dan went to college, he majored in French, then philosophy, and finally, in linguistics and computer science. This background in the liberal arts honed his writing ability. However, his love of statistics never left him, so he went on to earn an M.S. (1993) and a Ph.D. in statistics (1996) from the University of Connecticut. Today, Dan is Professor of Statistics in the Department of Mathematical Sciences at Central Connecticut State University (CCSU).

At CCSU, Dan developed and now directs the world's first online Master of Science degree and Graduate Certificate program in data mining. He has published four books on data mining and predictive analysis and one book on SAS programming. These books include:

- *Discovering Knowledge in Data: An Introduction to Data Mining,* Second Edition
- *Data Mining Methods and Models*
- *Data Mining the Web: Uncovering Patterns in Web Content, Structure, and Usage*
- *Data Mining and Predictive Analytics,* Second Edition

Discovering Knowledge in Data and *Data Mining and Predictive Analytics* are co-authored with his daughter Chantal Larose, who received her Ph.D. in Statistics from the University of Connecticut in 2015 and is an Assistant Professor of Decision Science and Statistics in the School of Business at the State University of New York at New Paltz.

Dan is the founder of DataMiningConsultant.com, and his consulting clients include *The Economist* magazine; Microsoft; *Forbes* magazine; the CIT Group; KPMG International; Computer Associates; Deloitte, Inc.; Sonalysts, Inc.; Booz Allen Hamilton; and the Hospital for Special Care. His consulting work includes a $750,000 Phase II grant from the Air Force Office of Research, *Storage-Efficient Data Mining of High-Speed Data Streams*. He is the series editor for the Wiley series on Methods and Applications in Data Mining.

However, his favorite work is imparting a love of statistics to a new generation, just as he did with his daughter Chantal, and he trusts that *Discovering Statistics* will help to do so for students around the country. Dan lives in Tolland, Connecticut, with his wife and children.

APPLICATIONS

Discovering Statistics, Third Edition, presents a wide variety of applications from diverse disciplines. The list below indicates examples, Your Turn features, and exercises related to different fields. Note that some items appear in more than one category.

Examples by Application

Biology and the Environment
Ch 1: 3, 21; **Ch 2:** 12–14, 17, 19, 25; **Ch 4:** 4 –11; **Ch 6:** 37; **Ch 7:** 1; **Ch 8:** 1; **Ch 9:** 4, 17, 31; **Ch 12:** 5–7; **Ch 14:** 1–3, 25

Business and Consumer Behavior
Ch 1: 5, 6, 12,13; **Ch 2:** 8, 20, 27; **Ch 3:** 1–5, 10–13, 21, 22, 24–28, 30–34, 37; **Ch 4:** 2, 3; **Ch 5:** 8, 11–13, 19, 26, 33, 36, 38, 46, 47; **Ch 6:** 23, 41, 42; **Ch 8:** 1, 11, 17, 18; **Ch 9:** 7, 9, 20, 24, 34–36; **Ch 10:** 1, 8–12; **Ch 12:** 13–15; **Ch 14:** 16, 17

College Life
Ch 1: 17; **Ch 2:** 8, 20, 27, 28; **Ch 3:** 8, 9, 17, 23; **Ch 5:** 17, 35, 41; **Ch 6:** 7–9, 11–13, 25; **Ch 12:** 1–4

Demographics and Characteristics of People
Ch 1: 2, 4, 11; **Ch 2:** 21; **Ch 3:** 7, 18–20, 35; **Ch 5:** 17, 22; **Ch 6:** 20; **Ch 7:** 2, 10; **Ch 9:** 19, 29; **Ch 10:** 16; **Ch 11:** 6–8; **Ch 14:** 7

Economics and Finance
Ch 1: 6; **Ch 3:** 3–5, 10–13, 21, 22, 24–28, 30, 31, 33, 34, 37; **Ch 4:** 2, 3; **Ch 5:** 11–13, 19, 26; **Ch 6:** 23, 24; **Ch 8;** 11; **Ch 9:** 17, 20, 24, 30; **Ch 10:** 8–12, 20; **Ch 14:** 6, 16, 17

Education and Child Development
Ch 1: 7, 8; **Ch 2:** 6, 7, 16, 28, 32; **Ch 3:** 15–17, 23, 36; **Ch 6:** 7–9, 11–13, 19, 38–40; **Ch 7:** 4; **Ch 8:** 4, 10; **Ch 9:** 16, 18, 30; **Ch 10:** 2, 3, 5, 6; **Ch 12:** 1–4, 8–12; **Ch 14:** 5, 6, 9, 10, 18, 23, 24

Health and Nutrition
Ch 1: 20, 21, 22; **Ch 3:** 29, 32, 38; **Ch 4:** 10; **Ch 5:** 28–32, 45; **Ch 6:** 16–18, 22, 27, 44, 45; **Ch 7:** 2, 3, 11, 12; **Ch 8:** 14–16; **Ch 9:** 5, 6, 8, 10, 11, 14, 15, 18, 23, 32, 33; **Ch 10:** 1, 4, 7, 21; **Ch 11:** 9; **Ch 13:** 11–15; **Ch 14:** 4, 7, 8, 13–15, 19–21

Humanities and Social Sciences
Ch 1: 4, 9, 10, 16, 19, 20; **Ch 2:** 6, 7; **Ch 4:** 12–17; **Ch 5:** 15, 16, 20; **Ch 9:** 19, 21, 22; **Ch 10:** 16; **Ch 13:** 5, 6; **Ch 14:** 22

International
Ch 2: 26, 32; **Ch 3:** 37; **Ch 14:** 5, 19–21

Law Enforcement and Crime
Ch 1: 1; **Ch 2:** 15, 18, 19, 24, 26, 30, 31; **Ch 8:** 20–22; **Ch 11:** 5; **Ch 14:** 11, 12

Politics and Voting
Ch 1: 18; **Ch 8:** 20–22

Sports, Leisure, and Entertainment
Ch 1: 5, 12, 14, 15; **Ch 2:** 1–4, 9–11, 22, 23, 29; **Ch 3:** 6, 8, 9; **Ch 5:** 3–7, 14, 18, 21, 23–25, 40, 48; **Ch 6:** 1, 3, 5, 6, 10, 15; **Ch 7:** 5, 6; **Ch 8:** 19; **Ch 9:** 1, 2, 28; **Ch 11:** 9; **Ch 13:** 16; **Ch 14:** 22

Technology and the Internet
Ch 1: 16; **Ch 2:** 1–4; **Ch 3:** 1, 2; **Ch 5:** 8, 22, 46, 47; **Ch 6:** 15, 20, 24; **Ch 7:** 5, 6, 10; **Ch 8:** 19; **Ch 9:** 25, 26, 28, 29; **Ch 10:** 13–15, 17, 18, 20; **Ch 11:** 1–4; **Ch 12:** 8–15; **Ch 13:** 1–4, 7–10; **Ch 14:** 18

Transportation and Fuel
Ch 2: 5; **Ch 3:** 14; **Ch 5:** 9; **Ch 6:** 41, 42; **Ch 7:** 7–9, 13–15; **Ch 8:** 5–9, 24; **Ch 9:** 13, 15, 27

Your Turn Features by Application

Biology and the Environment
Ch 3: 1–3; **Ch 8:** 1; **Ch 9:** 2

Business and Consumer Behavior
Ch 1: 1, 2, 4, 8; **Ch 3:** 2, 4, 11–14; **Ch 5:** 1, 12, 22, 24; **Ch 6:** 11, 12, 22; **Ch 8:** 1, 6; **Ch 9:** 3, 4, 9; **Ch 11:** 1–3

College Life
Ch 1: 10; **Ch 3:** 10; **Ch 5:** 11, 21, 26; **Ch 6:** 13

Demographics and Characteristics of People
Ch 5: 11; **Ch 7:** 1

Economics and Finance
Ch 1: 4, 8; **Ch 2:** 7–14; **Ch 3:** 4 –7; **Ch 5:** 12; **Ch 8:** 6; **Ch 9:** 9; **Ch 11:** 1–3

Education and Child Development
Ch 1: 7, 9; **Ch 2:** 5; **Ch 3:** 10; **Ch 4:** 1; **Ch 6:** 19–21; **Ch 8:** 4; **Ch 9:** 7; **Ch 10:** 1–4

Health and Nutrition
Ch 3: 5–7, 9; **Ch 4:** 2–10; **Ch 5:** 7, 18–20; **Ch 7:** 1, 7; **Ch 8:** 9, 10; **Ch 9:** 8, 10

Humanities and Social Sciences
Ch 1: 5, 6, 9; **Ch 2:** 5; **Ch 5:** 9, 10, 13

International
Ch 2: 7–14

Law Enforcement and Crime
Ch 2: 1–4, 6; **Ch 3:** 8; **Ch 8:** 13–15

Ch 3 Review: 18–21; **4.2:** 47; **5.1:** 70; **5.2:** 25–36; 74–93; **5.3:** 119–121; **5.4:** 52, 59; **Ch 5 Review:** 6; **7.1:** 106, 114–118; **8.1:** 72; **8.3:** 61, 62; **Ch 8 Review:** 3, 7; **9.6:** 27; **10.1:** 32, 36; **10.2:** 48–58; **11.2:** 39, 40; **14.6:** 21; **14.7:** 23

International

2.1 : 49–56, 61–70; **2.2:** 17–66, 102; **2.4:** 14–16; **Ch 2 Quiz:** 7–15; **3.1:** 15, 16, 21, 22, 27, 28, 33, 34, 39, 57–65; **3.2:** 13, 14, 19, 20, 75–78; **4.1:** 48, 72–82; **4.2:** 50, 69–75; **4.3:** 36; **5.1:** 84, 85; **5.3:** 9–22, 33–38, 49, 50; **10.1:** 39–44; **10.2:** 32; **12.1:** 48–50; **12.2:** 36; **12.3:** 15, 18; **13.3:** 35; **14.2:** 33; **14.4:** 16; **14.5:** 23, 24; **14.6:** 22, 24; **Ch 14 Quiz:** 11, 13

Law Enforcement and Crime

1.1: 1, 2; **2.1:** 99–101; 134–137; **2.2:** 33–40, 49–52, 85–90, 134–141; **2.3:** 11, 12, 15–18, 31, 34–36; **2.4:** 13; **Ch 2 Review:** 21, 22; **3.1:** 15, 17, 18, 21, 23, 24, 27, 29, 30, 33, 35, 36, 37, 74–76; **3.2:** 13, 15, 16, 19, 21, 22; **3.5:** 19–24; **5.3:** 54; **6.1:** 63, 64, 67, 68; **6.3:** 28; **6.5:** 32, 36; **Ch 6 Review:** 20; **Ch 6 Quiz:** 12; **9.1:** 26; **Ch 9 Review:** 5; **Ch 9 Quiz:** 11; **Ch 11 Quiz:** 14

Politics and Voting

1.3: 60; **Ch 1 Review:** 6; **2.1:** 11–22, 57; **3.1:** 46, 48; **6.2:** 45–48, 70; **8.3:** 63, 64; **Ch 8 Quiz:** 9; **Ch 14 Review:** 28

Sports, Leisure, and Entertainment

1.2: 11–22, 78–88; **1.3:** 7–22, 51–54, 58; **Ch 1 Quiz:** 10; **2.1:** 49–56, 71–76, 84–92, 105–119, 120, 121; **2.2:** 9–16, 67–70, 99; **2.3:** 32, 33; **2.4:** 8, 10, 12; **3.1:** 14, 20, 26, 32, 38, 45, 47, 66–68, 106–113; **3.2:** 12, 18, 41, 42, 49–52, 68–70, 79–81; **4.1:** 35, 40, 49, 57–59; **4.2:** 37, 42, 51, 52, 62–64; **4.3:** 23, 28, 37, 57–59; **5.1:** 66, 68, 69,

71–80, 84–98; **5.2:** 45–48, 53–57, 59, 65–81, 94–102; **5.3:** 67–74, 79–88, 93–100, 111; **5.4:** 57; **Ch 5 Quiz:** 9–20; **6.1:** 21, 29–32, 69–77; **6.2:** 60, 64, 69; **6.3:** 27, 29; **6.4:** 100, 117; **6.5:** 34, 38; **6.6:** 26; **Ch 6 Review:** 2; **Ch 6 Quiz:** 11; **7.2:** 40, 42, 44, 47; **8.1:** 80–85; **8.2:** 59; **8.3:** 48, 50, 52, 54; **8.4:** 39; **9.2:** 49, 52, 54; **9.4:** 43, 50; **9.5:** 31, 34; **11.2:** 32, 33; **Ch 11 Review:** 3, 7; **Ch 11 Quiz:** 15; **13.1:** 40, 42, 46, 50, 54, 57–60; **13.2:** 16, 18, 20, 22; **13.3:** 34, 40; **14.3:** 28; **14.4:** 15; **14.6:** 23, 25; **14.7:** 21, 22, 24; **Ch 14 Review:** 8

Technology and the Internet

1.2: 74, 75; **Ch 1 Review:** 5; **2.1:** 77–80; **2.4:** 11; **3.1:** 40, 49, 50; **3.4:** 7–14, 19–26, 33–40, 100–105; **3.5:** 7–12; **4.2:** 69–75; **5.1:** 63, 64, 70; **5.2:** 49–52, 58; **5.3:** 23–32, 39–48, 51, 52, 101–104, 109, 111, 122, 123; **5.4:** 53; **6.2:** 59, 63, 68; **6.3:** 24, 30, 31; **6.4:** 95, 96, 100; **6.5:** 34, 38; **7.1:** 101, 104; **7.2:** 39, 42, 43, 46, 47; **8.3:** 57, 60; **9.1:** 30; **9.2:** 49, 51, 54; **9.3:** 52, 60; **9.4:** 42, 43, 45, 46, 52; **9.5:** 31, 38; **Ch 9 Review:** 28; **10.2:** 29, 48, 49; **10.3:** 23, 27; **11.1:** 29, 34, 35; **11.2:** 27, 28, 32; **14.2:** 27, 33; **14.4:** 15

Transportation and Fuel

Ch 1 Review: 1–3; **2.1:** 83; **3.1:** 61–65, 77–79; **3.2:** 63–65, 82, 84, 89; **3.3:** 28–33; **3.4:** 45–50, 57–62, 69–72, 113–120; **3.5:** 13–18, 45–53; **4.1:** 42; **4.2:** 44, 57, 80–84; **4.3:** 30, 60–67; **5.2:** 61–64; **6.1:** 62, 66, 89–98; **6.2:** 80–84; **Ch 6 Review:** 9; **Ch 6 Quiz:** 10, 12; **8.1:** 73, 93–97; **8.2:** 61, 63, 64, 69–80; **8.3:** 76–85; **Ch 8 Quiz:** 8; **9.1:** 24–26; **9.2:** 53, 57, 58, 60–67; **9.3:** 51, 57, 59, 65; **9.4:** 52; **9.5:** 45–47; **Ch 9 Review:** 5; **Ch 9 Quiz:** 11; **10.1:** 31, 32, 35, 36; **10.2:** 30; **10.4:** 38; **Ch 11 Review:** 1; **12.1:** 48–54; **12.2:** 36, 40; **Ch 12 Quiz:** 10; **13.1:** 56, 75–78; **Ch 13 Review:** 3, 6; **14.2:** 27, 30

1 The Nature of Statistics

Introduction

The goal of Section 1.1 is to demonstrate to students that the field of statistics provides the tools to understand the experiences of the real people behind the numbers. Section 1.2 covers the building blocks of data analysis, including many basic but crucial concepts that the rest of the book depends on, such as *population*, *sample*, *variable*, *element*, and *observation*. Section 1.3 investigates ways of gathering data, including random sampling, questionnaires, and statistical studies.

From the Author

A main thrust of *Discovering Statistics*, Third Edition, is student motivation. For example, the new Chapter 1 Case Study, *Video Game Sales*, is intended to help you open the student's mind to statistics, and capture the student's interest in the subject.

Section 1.1 Data Stories: The People Behind the Numbers

- Section 1.1 tries to ease "stats anxiety" using a few illustrations of how statistics and graphs can be used to tell the story of real people going through real-life experiences. This may help some students relate to the subject more easily. For instance, the new Example 1, Declining Murder Rate in New York City, uses a time series plot to show some good news, for a change, about crime in America. Example 3, California Wildfires, is also new. Some instructors like to assign Section 1.1 as reading, and then discuss it in the next class.

Section 1.2 An Introduction to Statistics

- Using Example 4, The Four Phases of Statistics, you may wish to ask your students which best illustrates the difference in people's behavior on Friday the 13th, Table 2 or Figure 4. Hopefully, the response is Figure 4, where every green bar is longer than every yellow bar. Using graphs to help support hypotheses is *discovery-based* statistics, and that is why we titled the book *Discovering Statistics*.

- The students may enjoy perusing Table 3, looking for video games they have played.

- We have added a new feature, Now It's Your Turn. The first one occurs after Example 5. These are intended to check the student's understanding of the previous example.

- Examples 9 and 10 have been added to reinforce the difference between population and sample.

- There are a lot of definitions in this section. Most are very important and will be used throughout the course. Stress the importance of *statistical inference*, which is usually the goal of the intro stats course.

- Note that each exercise in the Clarifying the Concepts section provides the page number where the student can find the answer. This is intended to help the student complete the homework, as well as increase the probability that the student will actually do the homework.

- Note the new *Check It Out!* feature for the Practicing the Techniques exercises. This feature represents the "inverse" of the "Now You Can Do" feature. For a given example, the Now You

Can Do feature tells the student which exercises the example addresses. On the other hand, for a given set of exercises, the Check It Out! feature informs the students which example addresses these exercises. This will help students find which example they should study for a given set of exercises.

Section 1.3 Gathering Data

- You may wish to stress the very important topic of what constitutes a random sample. This phrase is used repeatedly throughout the book.

- New Examples 21 and 22 have been provided to help students differentiate between randomization and replication, as well as between experiments and observational studies.

Teaching Tips

Starting Examples: Use the students in the classroom to illustrate the different types of sampling. To illustrate random sampling, write each person's name on a slip of paper and place all the slips of paper in a box. Then randomly select a certain number of names from the box. To show stratified sampling, divide the students into males and females or freshmen, sophomores, juniors, and seniors. Then select a random sample from each group. For systematic sampling, have the class count off to some number, such as 1, 2, 3, 1, 2, 3, and so forth. Then all of the 3s can be sampled. Clustered sampling can be demonstrated by randomly selecting a sample of rows or a sample of tables in the classroom, and then sampling everyone in the selected rows or the selected tables.

In-Class Activities

1. Getting to know you. It is often helpful to get to know the students a little better at the beginning of the semester. In this way, *the instructor can tailor his or her pedagogy to the individual life experiences of the students.* For example, students who played basketball or baseball in high school may be interested in sports examples or analogies. Here are some questions that will help the instructor learn more about "where the students are coming from." The instructor should ask the students to write down their responses and pass them back to the instructor.

- What is your name?
- Where are you from (town)?
- What is (are) your major(s) (for example, communication, business administration)?
- Tell me a little about yourself, your interests, etc.
- What are your long-term goals? (What do you want to do after graduation?)
- What is something that you are really proud of about yourself?
- How do you feel about taking statistics?

2. **Sample statistics versus population parameters.**

   a. Break up into groups.

   b. What is the proportion of females in your group? Let us define our population to be all the students present in our class today. Is your group a sample or a population? Is the proportion of females in your group a statistic or a parameter?

   c. What is the population proportion of females in your class? In this activity, this proportion can be figured out by counting how many females are in the class and dividing by the number of students in the class. For interesting real-world problems, however, population parameters like this are usually unknown. Note how the value of this parameter is a fixed constant, in this case between 0% and 100%.

   d. Make a list of the sample proportions of females for all the different groups. Note how each statistic may be different, so that statistics are not fixed constants but may vary from sample to sample.

 3. **Generating Random Data.** Here is a fun way to generate random samples of students from your class by rolling a pair of dice! (For large lecture classes, it is probably easier to use technology or the *Simple Random Sample* applet to generate a random sample. See the Step-by-Step Technology Guide at the end of Section 1.3 for instructions.) Make a table constructed as follows:

	1	2	3	4	5	6
1						
2						
3						
4						
5						
6						

Print this table on a sheet of paper and pass the paper around the class, having each student choose one cell and write his or her name in the cell. The numbers across the top are for the results of tossing a red die, and the numbers down the left are for the results of tossing a black die. When a certain combination of red and black numbers is tossed, the student whose cell that combination refers to becomes part of the sample. In Chapter 5 we will learn that all the cells are equally likely to be picked if the dice are fair. If your class has fewer than 36 students, simply write *Roll Again* in the remaining empty cells after all students have chosen their cells. If your class has more than 36 students, just make more than one of these tables, and use a third colored die (for example, a green die) to choose which table is to be sampled. If a student is chosen more than once, simply roll again.

   Do you think this arrangement will produce random samples? Can you suggest any other arrangements that will produce random samples from the population of students in your class?

4. (This activity requires Activity 3 to have been performed first.) Now let's take samples of the class.

   a. Let the instructor call the first 10 names on an alphabetical list of those present for class today. Is this a random sample? Why or why not? Find the proportion of females in this sample.

   b. Let the instructor ask those students who watched ESPN this week to raise their hands. Is this a random sample? Why or why not? Find the proportion of females in this group. Is this proportion the same as in (**a**)? Why doesn't it have to be the same?

    c. Get the dice out and generate a sample using the table(s) you constructed. Is this a random sample? Why or why not? Find the sample proportion of females. Is this the same as in (**a**) or (**b**)?

    d. Choose another sample as in (**c**). Is this a random sample? Find the sample proportion of females. Is it the same as in (**a**) or (**b**) or (**c**)?

Supplements

The following resources are available for instructors and students at the book Web site: www.macmillanhighered.com/discostat3e: (a) statistical applets, (b) data sets in a variety of formats, (c) EESEE case studies (access code required), and (d) CrunchIt! (access code required).

    Further, the following additional resources are available in LaunchPad: (a) StatTutor tutorials; (b) Stats@Work simulations; (c) Student and Instructor's Solutions Manuals; (d) Video Technology Manuals for Excel, Minitab, TI-83/84, SPSS, JMP, R, Rcmdr, and CrunchIt!; and (e) Lecture Slides.

StatTutor: Note some of the terminology used in StatTutor differs from that used in *Discovering Statistics*. For example, the elements of a data set are called individuals in StatTutor.

Applets

The *Simple Random Sample* applet is referenced in Chapter 1 to generate random samples. Additionally, it is used for Exercise 51 in Section 1.3.

    A wealth of other activities, tools, and applets can be found at http://mathforum.org/mathtools.

Videos

Against All Odds: Inside Statistics is a telecourse consisting of 26 half-hour programs. It was prepared by COMAP for the Annenberg/Corporation for Public Broadcasting Project. For more information (including free streaming videos), go to www.learner.org/resources/series65.html. For this chapter, use the following video.

- Program 1: What Is Statistics?

Web Sites

- The Consortium for the Advancement of Undergraduate Statistics Education (CAUSE) provides a rich assortment of resources at its CAUSEweb site: https://www.causeweb .org/resources/.

- This site has several activities that generate random data: http://mathforum.org/mathtools/sitemap2/ps/.

- This site has simulations for random data: http://onlinestatbook.com/index.html.

1

The Nature of Statistics

OVERVIEW

© Erik Tham/Alamy

Video Game Sales

CASE STUDY   Do you love to play video games? Have you ever played *Minecraft*? If you have, you are not alone. According to VGChartz.com, for the week of May 17, 2014, in the United States alone, video game lovers purchased 36,732 copies of *Minecraft* for the PS3 platform and 33,887 copies of *Minecraft* for the Xbox 360 platform. Let's face it: video games are fun, with today's technology delivering ever more realistic thrills and spills. And the video game industry continues to pump out new platforms, such as Xbox One, Wii U, and the PS4, that promise to offer superior performance. In this chapter, we will examine some data concerning video game sales in the United States for the week of May 17, 2014, provided by VGChartz.com (www.vgchartz.com/weekly/41777/USA/).

- In Section 1.1, we examine an individual value plot of game sales, by platform, with the sales leaders highlighted (*Hint: Minecraft* is the overall leader).
- In Section 1.2, we use the data set to learn about the types of variables we use in this book.
- Finally, in Section 1.3, we use the data set to demonstrate random sampling. We hope you enjoy learning about statistics using this case study on video game sales.

THE BIG PICTURE

Where we are coming from and where we are headed . . .

- The objective of *Discovering Statistics* is to help you understand how to analyze and interpret data and, thereby, become a successful citizen in the Information Age.
- Chapter 1 introduces the basic ideas of the field of statistics and the methods for gathering data.
- In Chapter 2, we will learn to summarize the data we have gathered using graphs and tables.

1.1   Data Stories: The People Behind the Numbers

OBJECTIVES By the end of this section, I will be able to . . .

1   Realize that behind each data set lies a story about real people undergoing real-life experiences.

We begin *Discovering Statistics* by sharing some data stories. We start with some good news.

EXAMPLE 1   Declining murder rate in New York City

Our Chapter 2 Case Study, *Criminal Justice in New York City*, examines a wide range of criminal behavior throughout the police precincts of the five boroughs of New York City, from misdemeanors to murder. In this chapter, we briefly preview these data by looking at Figure 1. This figure is a time series plot of the murder rate (number of murders per year per 100,000 residents) for New York City for the years 1990–2014 (*Source:* New York City Police Department, www.nyc.gov). Note the steep decline from 1993 to 1998, followed by a flattening until 2010, when another slow descent began. Think about what this means: Thousands of men and women are living their lives who would not be alive today had the high murder rates of the early 1990s continued. And this heartening pattern is not restricted to New York City. Major cities across the country are seeing their crime rates drop over this same period (*Source:* FBI Uniform Crime Reports).

In the Chapter 2 Case Study, we examine other types of crime in New York City and look to see if further good news is available. We learn how to construct a time series plot similar to Figure 1 in Section 2.3.

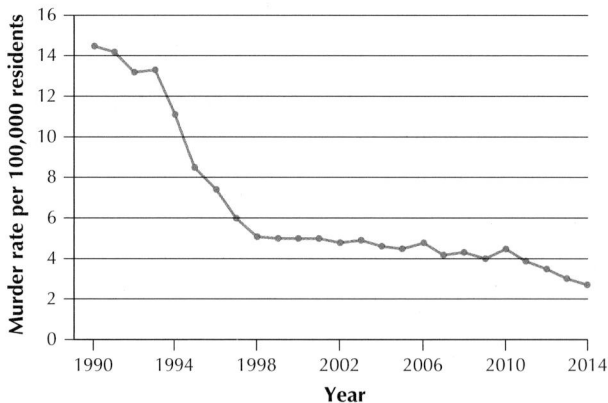

FIGURE 1 Time series plot. Murder rate in New York City, 1990–2014.

EXAMPLE 2   UFO sightings

Have you or any of your friends sighted any unidentified flying objects (UFOs)? Americans in each of the 50 states have reported seeing UFOs. Figure 2 represents a scatterplot of the number of UFO sightings versus state population, for each of the 50 states. Each dot represents a state. The straight line is a regression line that approximates the relationship between UFO sightings and state population. As the state population increases, the number of UFO sightings also tends to increase, which is not surprising.

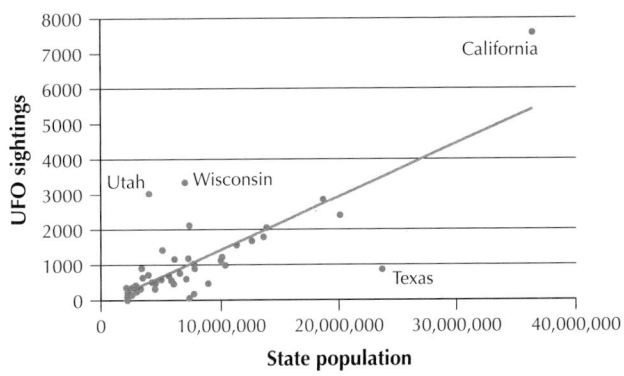

FIGURE 2
A scatterplot of the number of UFO sightings versus state population, showing that UFOs don't mess with Texas.

What may be surprising is that the UFOs seem to be attracted to certain states, while avoiding others. States considerably above the regression line have a larger than expected number of UFO sightings for their population size, whereas states below the line have a smaller than expected number of UFO sightings for their population size. So, there are more sightings than expected in California, Wisconsin, and Utah, given their population size, and fewer than expected in Texas. Why this might occur is open to discussion. Perhaps people in California are more likely to attribute unusual sightings to UFOs than most Americans; perhaps people in Texas are more pragmatic than most Americans. But if the sightings are valid (a big if!), it sure looks like the UFOs don't want to mess with Texas. We will learn how to construct and interpret scatterplots in Chapter 4, "Correlation and Regression," and we will learn how to quantify the relationship between two numerical variables in Chapter 4 and Chapter 13, "Inference in Regression."

EXAMPLE 3   California wildfires

California wildfires raged across the state in 2014.

Jonathan Alcorn/ Getty Images

In 2014, severe drought continued to batter the state of California and large areas of the western United States. The drought contributed to a series of wildfires that consumed hundreds of thousands of acres across the region. Table 1 contains a data set, which includes a listing of the uncontrolled wildfires raging around the state of California as of August 22, 2014. We will learn in Section 1.2 about how data sets are structured. Meanwhile, this is the first data set we look at, which gives us a chance to think about how these fires affected the lives of ordinary Californians: the lives lost, the homes destroyed, the forests and wildlife burned. As statisticians, we should always remember that behind the data lie the stories of real people. As statisticians, we will learn to apply the power of statistical analysis to improve the lives of the people behind the data. Let's get started.

Table 1 California wildfires in August 2014

Fire	Location	Size (acres)	Percent contained
Eiler	Lassen National Forest	32,416	97
Happy Camp Complex	Klamath National Forest	9,844	10
July Complex	Klamath National Forest	31,945	25
Junction	Merced-Mariposa Unit, Cal Fire	612	65
KNF Beaver	Klamath National Forest	32,307	93
Lodge Complex	Mendocino National Forest	12,535	95
Log	Klamath National Forest	3,629	95
Way	Central California District	3,858	48

Source: National Interagency Fire Center, www.nifc.gov.

Section 1.1 Exercises

Refer to Example 1 for Exercises 1 and 2.

1. Between which two years was there the largest drop in the murder rate?

2. Does the murder rate always go down year by year?

Refer to Example 2 for Exercises 3–6.

3. Estimate the following for the state of California.
   a. State population
   b. UFO sightings

4. Estimate the following for the state of Texas.
   a. State population
   b. UFO sightings

5. For a given population size, the expected number of UFO sightings falls on the regression line. For the state of California, what is the expected number of UFO sightings? (*Hint:* It's at the point on the line directly below the dot for California.)

6. For the state of Texas, what is the expected number of UFO sightings?

Refer to Example 3 for Exercises 7 and 8.

7. Which wildfire is the largest? The smallest?

8. Which wildfire is the most contained? The least contained?

1.2 An Introduction to Statistics

OBJECTIVES By the end of this section, I will be able to . . .

1 Describe the field of statistics, and state the meaning of the term *descriptive statistics*.

2 Explain what elements, variables, and observations are.

3 Describe the difference between qualitative and quantitative variables, and between discrete and continuous variables.

4 State the four levels of measurement.

5 Describe what is meant by a population, a sample, a parameter, and a statistic, and explain statistical inference.

1 What Is Statistics?

Is Facebook not cool anymore? Figure 3 shows that the percentage of Facebook users that are 13–24 years old has decreased from 40% to 29% in just three years: 2011 to 2014. This reflects a loss of 6 million Facebook users from the 13–24 age group (*Source: Facebook Social Ads Platform*, 2014). These numbers are examples of *statistics*—numbers that describe a group of people or things. Think about these numbers. Here are some questions we could ask about this survey:

• How did the researchers arrive at these figures?

• Are the figures accurate? Could they be inaccurate?

• The research found that fewer young people were on Facebook in 2014 than in 2011. But is this difference meaningful or just a product of random chance?

These are some of the types of questions we will be investigating throughout this book.

FIGURE 3
Pie charts comparing percentages of Facebook users, by age group.(*Source: http:// istrategylabs.com/2014/01/3 -million-teens-leave-facebook -in-3-years-the-2014-facebook -demographic-report/*).

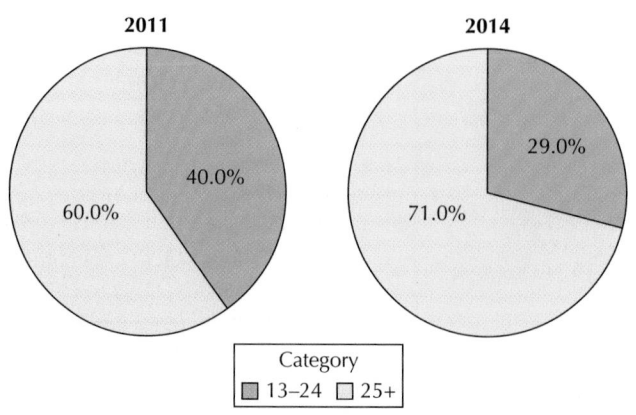

Examples of statistics include the following.

- Miguel Cabrera led Major League Baseball in 2013 with a batting average of .348.
- The Gallup polling organization reports that the average amount spent by Americans on a daily basis (not counting monthly bills or mortgage) in August 2014 is $98.
- The Centers for Disease Control report that 17% of American children and adolescents are obese.
- "Three out of four dentists surveyed recommend sugarless gum for their patients who chew gum."

You may have noticed that the section title, "What Is Statistics?" refers to statistics in the singular. Why? Because the *field of statistics* involves much more than just collecting and reporting numerical facts. The field of statistics may be defined as follows.

> The field of **statistics** is the *art* and *science* of
> - collecting data
> - analyzing data
> - presenting data
> - interpreting data

A statistician, then, is someone trained in the art and science of statistics. You may be surprised at the inclusion of the word *art* in the definition of statistics. But there is no question that judgment, experience, and even a little intuition are indispensable tools for any statistician's portfolio.

For today's college student, the field of statistics is especially relevant and useful. For example,

- A *nursing major* may want to learn whether or not homeopathic medicine really works.
- A *business major* may be interested in whether she should consider diversifying her portfolio to tech stocks, based on their price/earnings ratios.
- A *psychology major* may be interested in determining whether differences in therapeutic outcomes exist between traditional counseling methods and a new cognitive approach.
- An *education major* may be interested in whether listening to a Mozart sonata before taking an exam can significantly improve his grade.

The field of statistics can help solve each of these puzzles.

The following example is a classic illustration of how statistics represents the art and science of (1) collecting, (2) analyzing, (3) presenting, and (4) interpreting data.

EXAMPLE 4   The four phases of statistics: Does Friday the 13th change human behavior?

Superstitions affect most of us. Some people will never walk under a ladder, whereas others will alter their path to avoid a black cat. Do you think that people change their behavior on Friday the 13th? Perhaps, suspecting that it may be unlucky, some people might elect to stay home and watch television instead of venturing outdoors or driving on the highway. How would researchers go about studying whether superstitions change the way people behave? What kind of evidence would support the hypothesis that Friday the 13th causes a change in human behavior? T. J. Scanlon and his co-researchers

thought that if fewer vehicles were on the road on Friday the 13th than on the previous Friday, this would be evidence that some people were playing it safe on Friday the 13th and staying off the roads.[1] Note that the researchers didn't simply argue about the validity of the Friday the 13th superstition. Such discussions are interesting but largely subjective. What they deemed important is the effect of such a superstition on human behavior and how to measure such an effect as a change in behavior.

Phase 1 **Data collection.** The first phase of a statistical study, as in the definition of statistics, is to *collect* the data. The researchers obtained data kept by the British Department of Transport on the traffic flow through certain junctions of the M25 motorway in England.

Phase 2 **Data analysis.** Next is the analysis of the data. The authors compared the number of vehicles passing through certain junctions on the M25 motorway on Friday the 13th and the previous Friday.

fridaythe13th

Table 2 Traffic through M25 junctions

Friday the 6th	Friday the 13th	Difference
139,246	138,548	698
134,012	132,908	1104
137,055	136,018	1037
133,732	131,843	1889
123,552	121,641	1911
121,139	118,723	2416
128,293	125,532	2761
124,631	120,249	4382
124,609	122,770	1839
117,584	117,263	321

Table 2 shows that, in every instance, the number of vehicles passing through these junctions on Friday the 13th was less than on the preceding Friday. Now, let's examine the data graphically. The clustered bar graph in Figure 4 illustrates the difference in the number of vehicles traveling on the M25 motorway on Friday the 6th (in green) and the subsequent Friday the 13th (in yellow) for 10 pairs (clusters) of dates. Note that, *in every instance*, the green bar is longer than its partner yellow bar. This indicates that the number of vehicles on the motorway *decreased* on Friday the 13th when compared with the previous Friday in every instance.

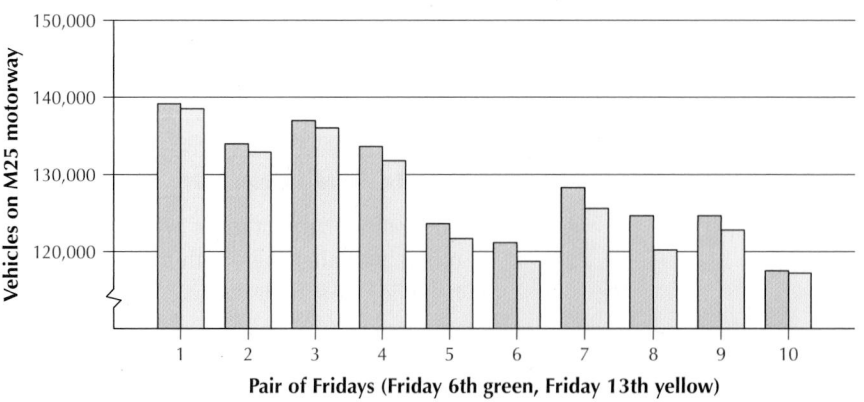

FIGURE 4 Clustered bar graph of motorway traffic.

Phase 3 **Data presentation.** The *presentation* of the results is important, and the researchers found a highly respectable journal, the *British Medical Journal*, in which to publish their findings. Other avenues for presentation are delivering a talk at a conference, writing up a report for one's supervisor, or presenting a class project.

Phase 4 **Data interpretation.** Finally, the last facet in our definition of statistics is *interpretation*. It is crucial for those who are performing a statistical study to make their results understandable to the general public. It is not sufficient for the statistician alone to understand the results. Instead, the statistician must communicate the results clearly, whether in writing or orally. In this case, the researchers chose the decrease in number of vehicles as the criterion on which to base support for their hypothesis that people changed their behavior on Friday the 13th. Their finding of an observable decrease in traffic on Friday the 13th is consistent with their hypothesis.

2 Elements, Variables, and Observations

Every data set holds within it a story waiting to be told, as we saw in Section 1.1, *Data Stories*. To provide us with the tools to uncover these stories we need to learn some simple concepts, *the building blocks of data analysis.*

> **Descriptive statistics** refers to methods for summarizing and organizing the information in a data set.

In **descriptive statistics** we use numbers (such as counts and percents), graphs, and tables to describe the data set, as a first step in data analysis. In Chapters 2 to 4, we will examine descriptive methods much more closely. But first we need to introduce a few terms. Suppose a data analyst for a health maintenance organization (HMO) is collecting data about the patients in a particular hospital, including the diagnosis, length of stay, gender, and total cost. The sources of the information (the patients) are called the **elements**. The patients' characteristics (for example, diagnosis, length of stay) are called the **variables**. Finally, the complete set of characteristics for a particular patient is called an **observation**.

> **Elements, Variables, and Observations**
>
> An **element** is a specific entity about which information is collected.
>
> A **variable** is a characteristic of an element, which can assume different values for different elements.
>
> An **observation** is the set of values of the variables for a given element.

When data are presented in tables and spreadsheets, it is typical practice to have the columns indicate the variables, and the rows to indicate the observations. So, for the hospital patients, the observation (specific values for the set of all the variables) for each element (patient) would appear as a row in the table.

EXAMPLE 5 Elements, variables, and observations

Video Game Sales

Table 3 contains the top 30 best-selling video games in the United States for the week of May 17, 2014, along with the game platform, publishing studio, type of game, sales that week, total sales, and how many weeks the game has been on the list. Use Table 3 to answer the following questions.

a. What are the variables?

b. State the first five elements.

c. List all the values that the variable *platform* takes.

d. Provide the observation for *Titanfall* for Xbox One.

Table 3 Top 30 best-selling video games in the United States for the week of May 17, 2014

Game	Platform	Studio	Type	Sales for week	Sales total	Weeks on list
Minecraft for PS3	PS3	Sony	Adventure	36,732	36,732	1
Minecraft for Xbox 360	Xbox 360	MS	Adventure	33,887	2,517,732	50
Kirby: Triple Deluxe for 3DS	3DS	Nintendo	Platform	28,184	116,658	3
MLB 14 The Show for PS4	PS4	Sony	Sports	27,088	161,770	2
Titanfall for Xbox One	Xbox One	Electronic Arts	Shooter	23,244	1,150,856	10
Call of Duty: Ghosts for Xbox 360	Xbox 360	Activision	Shooter	15,781	5,243,214	28
Bound by Flame for PS4	PS4	Focus	Action	15,346	15,346	2
Pokemon X/Y for 3DS	3DS	Nintendo	Role-Playing	14,543	3,442,714	32
Titanfall for Xbox 360	Xbox 360	Electronic Arts	Shooter	13,692	460,934	6
Grand Theft Auto V for Xbox 360	Xbox 360	Take-Two	Action	11,652	7,417,036	35
Grand Theft Auto V for PS3	PS3	Take-Two	Action	9,861	5,424,163	35
Call of Duty: Ghosts for PS4	PS4	Activision	Shooter	8,917	1,100,682	27
Super Luigi U for Wii U	Wii U	Nintendo	Platform	8,599	840,698	38
Super Mario Bros. U for Wii U	Wii U	Nintendo	Action	8,476	1,690,689	78
Call of Duty: Ghosts for PS3	PS3	Activision	Shooter	8,323	3,046,580	28
Borderlands 2 for PSV	PSV	Take-Two	Shooter	8,189	50,568	2
Battlefield 4 for Xbox 360	Xbox 360	Electronic Arts	Shooter	8,188	1,707,310	29
Forza Motorsport 5 for Xbox One	Xbox One	MS	Racing	7,910	736,743	26
Call of Duty: Ghosts for Xbox One	Xbox One	Activision	Shooter	7,777	1,139,310	26
inFamous: Second Son for PS4	PS4	Sony	Action	7,150	634,733	9
Battlefield 4 for PS3	PS3	Electronic Arts	Shooter	6,774	911,687	29
NBA 2K14 for Xbox 360	Xbox 360	Take-Two	Sports	6,593	1,597,734	33
Spiderman 2 for PS4	PS4	Activision	Action	6,510	49,292	3
Super Mario 3D World for Wii U	Wii U	Nintendo	Platform	6,064	835,941	26
Yoshi's New Island for 3DS	3DS	Nintendo	Action	6,006	172,680	10
Battlefield 4 for PS4	PS4	Electronic Arts	Shooter	5,875	786,607	27
Mario Golf for 3DS	3DS	Nintendo	Action	5,633	42,199	3
Nintendo Land for Wii U	Wii U	Nintendo	Action	5,428	1,550,278	78
Mario Kart 7 for 3DS	3DS	Nintendo	Racing	5,353	3,394,162	128
NBA 2K14 for PS4	PS4	Take-Two	Sports	5,290	608,899	27

 videogamesales

NOW YOU CAN DO
Exercises 11–13,
17–19, and 23–25.

Solution

a. The variables are the game platform, publishing studio, type of game, sales that week, total sales, and how many weeks the game has been on the list.

b. The first five elements are *Minecraft* for PS3, *Minecraft* for Xbox 360, *Kirby: Triple Deluxe* for 3DS, *MLB 14 The Show* for PS4, and *Titanfall* for Xbox One.

c. The variable *platform* takes the following values: PS3, Xbox 360, 3DS, PS4, Xbox One, Wii U, and PSV.

d. The observation for *Titanfall* for Xbox One is as follows:

Game	Platform	Studio	Type	Sales for week	Sales total	Weeks on list
Titanfall for Xbox One	Xbox One	Electronic Arts	Shooter	23,244	1,150,856	10

YOUR TURN #1

1. List the values that the variable *type* takes.
2. Provide the observation for *Spiderman 2* for PS4.

(The solutions are shown in Appendix A.)

3 Qualitative and Quantitative Variables; Discrete and Continuous Variables

Notice that we have variables that can take on various types of values, some of which are numbers and some of which are categories. For example, *Titanfall* for Xbox One had sales for the week of 23,244, and has been on the Top 30 list for 10 weeks. Each of these variables is numeric. On the other hand, the studio for *Titanfall* for Xbox One is Electronic Arts and the game type is shooter, which are characteristics that do not have numeric values but instead are categories. This leads us to define two types of variables: **qualitative** and **quantitative**.

CAUTION Not all numerical variables are quantitative. For example, a zip code such as 90210 is numerical. But we would not perform arithmetic on zip codes, such as adding two zip codes together. Therefore, zip codes are numerical but not quantitative.

A **qualitative variable** is a variable that may be classified into categories. A **quantitative variable** is a variable that takes numeric values and upon which arithmetical operations, such as addition or subtraction, may be meaningfully performed.

Qualitative variables are also called *categorical variables*, because they can be grouped into categories. For *Titanfall* for Xbox One, the qualitative variables are platform, studio, and type. The quantitative variables are sales for the week, sales total, and weeks on list.

EXAMPLE 6   Qualitative or quantitative?

Xinhua/eyevine/Redux

NASDAQ is an American stock exchange that includes many technology companies.

Some of the most widespread applications of statistical analysis occur in the business world. Managers examine patterns and trends in data, thereby hoping to increase profitability. Table 4 shows the five most active stocks on the New York Stock Exchange (NYSE) and NASDAQ (National Association of Securities Dealers Automated Quotations), as reported by *USA Today* for June 3, 2014. (**a**) What are the elements and the variables of this data set? (**b**) Which variables are qualitative? Which are quantitative? (**c**) Provide the observation for Bank of America.

Table 4 Most active stocks on NYSE and NASDAQ, June 3, 2014

Stock	Exchange	Last	Volume	Change
Quiksilver	NYSE	$3.41	59,328,858	−$2.38
Sirius XM	NASDAQ	$3.30	54,392,299	+$0.02
Bank of America	NYSE	$15.21	48,690,356	−$0.05
Newlead Holdings	NASDAQ	$0.76	46,703,983	−$0.19
Applied Materials	NASDAQ	$21.42	33,295,094	+$4.39

Solution

a. The *elements* are the five most active stocks traded on the NYSE and NASDAQ on this day in 2014. The *variables* are as follows:

- Exchange: the exchange where the stock was traded.
- Last: the most recent trading price for the stock.
- Volume: how many shares of the stock were traded that day.
- Change: the change in share price (in dollars) between the opening price and the closing price that day.

b. The exchange, because it can be categorized as either NYSE or NASDAQ, is qualitative. The other variables are quantitative.

c. The observation for Bank of America includes the exchange and the set of the day's stock data for that company. Bank of America is traded on the NYSE. Its last share price was $15.21 per share, 48,690,356 shares of its stock were traded, and the price decreased by $0.05 per share.

NOW YOU CAN DO
Exercises 14, 20, and 26.

Stock	Exchange	Last	Volume	Change
Bank of America	NYSE	$15.21	48,690,356	−$0.05

YOUR TURN
#2

1. Is Sirius XM an element or a variable?
2. What is another term for the variable *Exchange*?

(The solutions are shown in Appendix A.)

Hint: A quantitative variable that must be counted (not measured) is probably a discrete variable, whereas a quantitative variable that must be measured (not counted) is probably a continuous variable.

Quantitative variables can be classified as either **discrete** or **continuous**.

A **discrete variable** can take either a finite or a countable number of values. Each value can be graphed as a separate point on a number line, with space between each point. A **continuous variable** can take infinitely many values, forming an interval on the number line with no space between the points.

EXAMPLE 7 Discrete or continuous?

Suppose we collect data on a statistics student in your class, including (**a**) number of math courses taken, and (**b**) grade point average (GPA). Determine whether these variables are discrete or continuous.

Solution

a. Because the number of math courses is finite, the variable *number of math courses taken* is discrete.

NOW YOU CAN DO
Exercises 15, 21, and 27.

b. Because GPA can take an infinite number of possible values (for example, in the interval 0.0 to 4.0), the variable *GPA* is continuous.

YOUR TURN
#3

Consider a data set containing the competitors in the 100-meter dash in the Summer Olympics, where we keep track of the number of medals each competitor has won, along with their time in the 100-meter dash. State whether the following variables are discrete or continuous.

1. Number of medals won.
2. Racing time in the 100-meter dash.

(The solutions are shown in Appendix A.)

4 Levels of Measurement

Data may be classified according to the following four *levels of measurement*.

- *Nominal data* consist of names, labels, or categories. No natural or obvious ordering of nominal data (such as high to low) occurs. Arithmetic cannot be performed on nominal data.

- *Ordinal data* can be arranged in a particular order. However, no arithmetic can be performed on ordinal data.

- *Interval data* are similar to ordinal data, with the extra property that subtraction may be performed on interval data. No *natural zero* occurs for interval data.

- *Ratio data* are similar to interval data, with the extra property that division may be performed on ratio data. A natural zero does exist for ratio data.

EXAMPLE 8 Levels of measurement

Identify which level of measurement is represented by the following data.

a. Years covered in European History 101: 1066–1492

b. Annual income of students in Statistics 101 class: $0–$15,000

c. Course grades in English 101: A, B, C, D, F

d. Student gender: male, female

Solution

a. The years 1066 to 1492 represent interval data. No natural zero occurs (no "year zero"; the calendar goes from 1 B.C. to A.D. 1). Also, division (1492/1066) does not make sense in terms of years, so that the data are not ratio data. However, subtraction does make sense, in that the course covers 1492 − 1066 = 426 years.

b. Student income represents ratio data. Here, division does make sense. That is, someone who made $4000 last year made twice as much as someone who made $2000 last year. Also, some students probably had no income last year, so that $0, the natural zero, also makes sense.

c. Course grades represent ordinal data, because (a) they may be arranged in a particular order, and (b) arithmetic cannot be performed on them. The quantity A − B makes no sense.

d. Student gender represents nominal data, because the data cannot be ordered in a natural or obvious way. Also, no arithmetic can be performed on student gender.

NOW YOU CAN DO
Exercises 16, 22, and 28.

YOUR TURN #4

Using Table 4, identify which level of measurement is represented by the following variables.

1. Exchange

2. Last price

(The solutions are shown in Appendix A.)

5 Statistical Inference

Descriptive methods of data analysis are widespread and quite informative. However, the modern field of statistics involves much more than simply summarizing a data set. For example, suppose a medical researcher is investigating caffeine consumption

Do more than 75% of 19- to 22-year-old Americans consume caffeine?

among 19- to 22-year-old Americans, and claims that more than 75% of Americans who are 19 to 22 years old consume caffeine. How should the medical researcher go about collecting evidence to support her claim? One method would be to ask each and every person in the population of 19- to 22-year-old Americans whether he or she consumes caffeine. In general, a **population** is the collection of *all* elements (persons, items, or data) of interest in a particular study.

This proportion of caffeine consumers is one characteristic of the population of American 19- to 22-year-olds. A characteristic of a population is called a **parameter**. However, to ask every 19- to 22-year-old in America about his or her caffeine consumption would be a daunting task that is expensive, time-consuming, and, in the end, simply impossible. So, unfortunately, the population proportion of 19- to 22-year-olds who consume caffeine remains *unknown*. The actual value of a population parameter is often unknown.

A **sample** is a subset of the population from which information is collected. For example, from a sample of one hundred 19- to 22-year-olds, suppose that 76 of them consume caffeine. That is, the sample proportion of students who consume caffeine is 76/100 = 76%. This proportion is a characteristic of the sample and is called a **statistic**. The advantage here is that, because the sample is relatively small, the characteristics of the sample can be determined. On the other hand, if we take a different sample, we are likely to get a different value for the sample proportion.

> **Populations, Parameters, Samples, and Statistics**
>
> A **population** is the collection of *all* elements (persons, items, or data) of interest in a particular study. A **parameter** is a characteristic of a population.
>
> A **sample** is a subset of the population from which information is collected. A **statistic** is a characteristic of a sample.

A sample is a subset of a population.

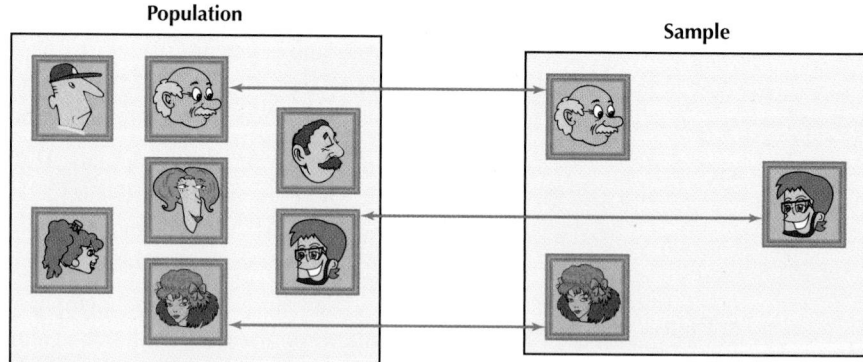

EXAMPLE 9   Populations and samples

For the following scenarios, state whether the data represent a population or a sample.

a. The seven continents: Asia, Africa, North America, South America, Europe, Antarctica, Australia

b. Europe and North America

c. All of Ludwig van Beethoven's nine symphonies

d. Beethoven's 5th, 6th, and 9th symphonies

Solution

a. Because *all* of the continents are listed, this group represents a population.

b. Europe and North America represent a *subset* of the population of all continents and, therefore, represent a sample.

NOW YOU CAN DO
Exercises 43–45
and 52–57.

c. Because *all* of Beethoven's symphonies are included, this represents a population.

d. Because the 5th, 6th, and 9th symphonies represent a subset of all of Beethoven's symphonies, this is a sample.

YOUR TURN #5

For the following scenarios, state whether the data represent a population or a sample.

1. The three largest counties in Florida.
2. All of the counties in Florida.

(The solutions are shown in Appendix A.)

EXAMPLE 10   Parameters and statistics

For the following scenarios, state whether the indicated measure is a statistic or a parameter.

a. The average income of the countries in all seven continents

b. The average income of the countries in Europe and North America

c. The shortest (in time) of all of Beethoven's nine symphonies

d. The shortest (in time) of Beethoven's 5th, 6th, and 9th symphonies

Solution

a. Because all seven continents represent a population, the average income is a parameter.

b. Only two of the seven continents are represented, so the average income of this sample is a statistic.

c. Since all of Beethoven's nine symphonies represent a population, the shortest time is a parameter.

d. Only three of the nine symphonies are represented, so the shortest time of this sample is a statistic.

NOW YOU CAN DO
Exercises 46–48.

YOUR TURN #6

For the following scenarios, state whether the indicated measure is a statistic or a parameter.

1. The most expensive hotel in the three largest counties in Florida
2. The most expensive hotel in all of Florida (that is, all of the counties in Florida)

(The solutions are shown in Appendix A.)

The U.S. Constitution requires that a census be conducted every 10 years. A **census** is the collection of data from every element in the population. As you can imagine, such a task is very difficult and very expensive. In fact, the Census Bureau estimates that the 2000 U.S. census "undercounted the actual U.S. population by over three million individuals."[2]

It is often best to gather data from a sample, a subset of that population, for the following reasons.

• The population you are interested in studying may be too large to allow you to elicit information from every element.

• Time and money often constrain the researcher to choosing a sample instead of studying the entire population.

- In some experiments, the resource is exhausted when testing is done (for example, in estimating the mean lifetime of light bulbs).

- Finally, it may be simply impossible to gather information from the entire population, such as when studying the quality of water in Lake Erie.

For instance, it would be impractical to contact every 19- to 22-year-old in the country for a survey about caffeine consumption. Instead, to estimate the proportion of all American 19- to 22-year-olds who consume caffeine, we can use statistical inference. **Statistical inference** refers to learning about the characteristics of a population by studying those characteristics in a subset of the population (that is, in a *sample*).

The journal *Pediatrics* reported[3] that a survey of 2600 Americans ages 19 to 22 found that 77% of them were consumers of caffeine. These 2600 teenagers and young adults represent a sample, and their characteristics can be known. Thus, at this point, the medical researcher can make the *inference* that the proportion of *all* American 19- to 22-year-olds who consume caffeine is 77%, because this is the proportion in the sample. In doing this, the medical researcher is performing *statistical inference*.

> **Statistical inference** consists of methods for estimating and drawing conclusions about population characteristics based on the information contained in a subset (sample) of that population.

"Now wait just a minute," you might object. "How can you say that the proportion of *all* 19- to 22-year-old Americans who consume caffeine is 77% just because your *sample* proportion is 77%?" Actually, you have a point. We *are* generalizing. We are taking what we know about a portion of the whole (a sample) and using it to draw a conclusion about the whole (the population). However, although the true proportion of 19- to 22-year-old Americans who consume caffeine is probably not exactly 77%, it is most likely not very far from 77%. The 77% is an *estimate*, an approximation based on sample data. In Chapter 8, we will learn how we can get the estimate as close as we wish to the actual value just by taking a large enough sample.

Finally, we need to point out one further attribute of parameters and statistics. The value of a parameter, although it is unknown, is a fixed constant. For example, the average age of all persons in your home state (population) at noon today is unknown, but it still exists, and it is a specific number. On the other hand, the value of a statistic depends on the sample. For example, a sample of 100 people in your hometown may produce an average age of 31. The average age of a sample of 100 people in a neighboring town may be 32. Later, we will learn that this is because a statistic is a *random variable*.

Of course, to deliver a valid estimate, the sample needs to be *representative* of the population. The sample should not differ systematically in any major characteristic from the population. We will learn more about this in Section 1.3, when we study sampling methods. Table 5 summarizes the attributes of a population and a sample.

Table 5 Summary of attributes of population and sample

	Population	**Sample**
Thumbnail definition	All elements	Subset of population
Characteristic	Parameter	Statistic
Value	Usually unknown	Usually known
Status	Constant	Depends on sample

Parameters are measures from a *population*, whereas *statistics* are measures from a *sample*. The characteristic associated with the population starts with the same letter, and the same is true for sample.

EXAMPLE 11   Descriptive statistics or statistical inference?

State whether the following situations illustrate the use of descriptive statistics or statistical inference.

a. In Baltimore County, Maryland, the average amount spent per week on gasoline consumption in a sample of 500 commuters was $75. The county government infers that the average amount spent weekly by all Baltimore County commuters is $75.

b. A sample of 100 residents of Broward County, Florida, yielded 27 residents who work for the government at the local, state, or federal level. Thus, 27% of these 100 residents work for the government.

c. The average age of a sample of 200 residents of Garden City, New York, was 34 years old.

d. In a survey of 1000 citizens in the Seattle, Washington, metropolitan area, 570 said they would pay higher prices in order to reduce greenhouse emissions. City planners conclude that 57% of all Seattle citizens would do so.

Solution

a. **Statistical inference.** A sample was taken, and a sample statistic ($75 per week) was calculated. Then the county government used this statistic to make the *statistical inference* that this was the average amount spent by all Baltimore County commuters.

b. **Descriptive statistics.** Though a sample was taken, there was no attempt to make an inference from this sample of 100 workers to the entire population of Broward County, Florida. So, no statistical inference is being made here.

c. **Descriptive statistics.** The average age of 34 years old is a descriptive statistic, because it describes the sample. However, no inference is made regarding a larger population.

d. **Statistical inference.** The survey found that 57% of the sample of 1000 citizens would pay higher prices in order to reduce greenhouse emissions. This 57% is a statistic. Then the city planners used this statistic in order to perform statistical inference about the population of all Seattle citizens.

NOW YOU CAN DO
Exercises 49–51 and 58–64.

YOUR TURN #7

State whether the following situations illustrate the use of descriptive statistics or statistical inference.

1. Your instructor states that the average grade on the first quiz for your class is 85.

2. In 10 games of ping pong, Jessica has lost to her friend Lu Li 8 times. Jessica sadly concludes that, going forward, she has only a 20% chance of winning a game of ping pong against her friend Lu Li.

(The solutions are shown in Appendix A.)

A Statistical Literacy Quiz

Regardless of major, every student in America (indeed, every citizen) needs to become *statistically literate* in order to survive in today's wired society. Why not take this quiz to find out if you are statistically literate? Answer each question true or false.

1. A fair coin is tossed five times and comes up heads each time. That means that tails is "due" and the chances of tails on the next toss is increased.

2. One politician says that the mean income is rising, whereas another politician says that the median income is falling. One of them has to be lying.

3. Jim is tested for HIV and the test comes back positive. Thus, Jim is HIV-positive.

The correct answer to each question is *false*. Question 1 deals with something called "the Gambler's Fallacy," and we will cover this, along with the explanation for Question 3, in Chapter 5, "Probability." We will deal with Question 2, the quirks of means and medians, in Chapter 3, "Describing Data Numerically."

Section 1.2 Summary

1. The field of statistics is the art and science of collecting, analyzing, presenting, and interpreting data.
2. Descriptive statistics refers to methods for summarizing and organizing the information in a data set. Data sets include information collected on elements. Variables are characteristics of an element and can take different values for different elements.
3. Variables may be either quantitative or qualitative. A qualitative variable may be classified into categories. A quantitative variable takes numeric values upon which arithmetical operations may meaningfully be performed. A discrete variable is a quantitative variable that can take

either a finite or a countable number of possible values. A continuous variable is a quantitative variable that can take an infinite number of possible values.
4. Data may be classified according to four levels of measurement: nominal, ordinal, interval, and ratio.
5. A population is a collection of all elements of interest, whereas a sample is a subset of the population. The characteristics for a population are called parameters, whereas the characteristics for a sample are called statistics. Inferential statistics consists of methods for estimating and drawing conclusions about population characteristics based on the information in the sample.

Section 1.2 Exercises

CLARIFYING THE CONCEPTS

1. Write a sentence describing in your own words the field of statistics. (p. 5)
2. What do we call the entities from which the data are collected? (p. 7)
3. Describe the difference between a qualitative and a quantitative variable. (p. 9)
4. What is another term for a qualitative variable? (p. 9)
5. True or false: The actual value of a population parameter is usually unknown. (p. 12)
6. What is the difference between a sample and a population? (p. 12)
7. Explain what a statistic is. (p. 12)
8. Describe one difference between a statistic and a parameter. (p. 12)
9. What is a census? (p. 13)
10. True or false: Statistical inference refers to methods for summarizing and organizing the information in a data set. (p. 14)

PRACTICING THE TECHNIQUES

✓ CHECK IT OUT!

To do	Check out	Topic
Exercises 11–13, 17–19, and 23–25	Example 5	Elements, variables, and observations
Exercises 14, 20, and 26	Example 6	Qualitative and quantitative variables
Exercises 15, 21, and 27	Example 7	Discrete and continuous variables
Exercises 16, 22, and 28	Example 8	Levels of measurement
Exercises 29–42	Examples 6, 7, and 8	Types of variables and levels of measurement
Exercises 43–45	Example 9	Population and sample
Exercises 46–48	Example 10	Parameter and statistic
Exercises 49–51	Example 11	Statistical inference
Exercises 52–57	Example 9	Population and sample
Exercises 58–64	Example 11	Statistical inference

For Exercises 11–16, answer the following questions about the data in Table 6.

TABLE 6 Information about four sports teams in a dormitory intramural league

Team	Captain's gender	Wins	Rank	Winning percentage
Dragonborn	Male	10	1	0.667
Sprites	Female	9	2	0.600
Enchanters	Female	7	3	0.467
Trolls	Male	4	4	0.267

11. What are the elements?
12. List the variables.
13. Do the following.
  a. List the values that the variable *Captain's gender* takes.
  b. Provide the observation for the Sprites.
14. List the quantitative variables and the qualitative variables.
15. Which variables are discrete and which variables are continuous?
16. For each variable, determine whether it represents nominal, ordinal, interval, or ratio data.

For Exercises 17–22, answer the following questions about the data in Table 7.

TABLE 7 Major League Baseball batting leaders, 2013

Player	Team	Batting average	Hits	Rank	Year of birth
Miguel Cabrera	Detroit Tigers	0.348	193	1	1983
Michael Cuddyer	Colorado Rockies	0.331	162	2	1979
Joe Mauer	Minnesota Twins	0.324	144	3	1991
Michael Trout	Los Angeles Angels	0.323	190	4	1983
Chris Johnson	Atlanta Braves	0.321	165	5	1984

Source: www.baseball-reference.com.

17. What are the elements?

18. List the variables.

19. Do the following.

  a. List the values that the variable *Team* takes.

  b. Provide the observation for Miguel Cabrera.

20. List the quantitative variables and the qualitative variables.

21. Which variables are discrete and which variables are continuous?

22. For each variable, determine whether it represents nominal, ordinal, interval, or ratio data.

For Exercises 23–28, answer the following questions regarding the data in Table 8, which contains the five universities with the most federal student loan recipients for the 2013–2014 academic year.

23. What are the elements?

24. List the variables.

25. Do the following.

  a. List the values that the variable *School type* takes.

  b. Provide the observation for Penn State University.

26. List the quantitative variables and the qualitative variables.

27. Which variables are discrete and which variables are continuous?

28. For each variable, determine whether it represents nominal, ordinal, interval, or ratio data.

TABLE 8 Federal student loan data[4]

School	State	School type	Recipients	Total loan amount ($ millions)
University of Phoenix	AZ	Proprietary	123,583	453
Devry University	IL	Proprietary	45,361	215
ITT Technical Institute	IN	Proprietary	43,671	155
Penn State University	PA	Public	42,011	151
Kaplan University	IA	Proprietary	36,001	140

For Exercises 29–42:

  a. State whether the variable is qualitative or quantitative. If the variable is quantitative, state whether it is discrete or continuous.

  b. Identify the level of measurement represented by the data.

29. The year you were born

30. Whether you own a cell phone or not

31. The price of tea in China

32. The SAT Math score of the person sitting next to you (scores range from 200 to 800)

33. The winning score in next year's Super Bowl

34. The winning team in next year's Super Bowl

35. The rank of the winning Super Bowl team in its division

36. The number of friends on a student's Facebook page

37. Your favorite television show

38. How many contacts you have on your cell phone

39. Your favorite ice cream

40. Your credit card balance

41. How old your car is

42. What model your car is

For Exercises 43–45, state whether the data in the indicated table represent a sample or a population.

43. Table 6: Note that the four teams represent all the teams in the intramural league.

44. Table 7

45. Table 8

For Exercises 46–48, state whether the indicated measure represents a statistic or a parameter.

46. Refer to Table 6 and Exercise 43. The most wins in the league is 10.

47. Refer to Table 7 and Exercise 44. The oldest player was born in 1979.

48. Refer to Table 8 and Exercise 45. Four out of five (80%) of the universities in Table 8 are proprietary.

For Exercises 49–51, state whether descriptive statistics or statistical inference is indicated.

49. Refer to Table 6. Half of the teams in the league have female team captains.

50. Refer to Table 7. Suppose we find the average number of hits of the players in Table 7, and infer this value to represent the average number of hits for all players in the league.

51. Refer to Table 8. Suppose we infer from Exercise 48 that 80% of all universities are proprietary.

For Exercises 52–57, identify the population and the sample.

52. A researcher is interested in the median home sales price in Tarrant County, Texas. He collects sales data on 100 home sales.

53. A psychologist is concerned about the health of veterans returning from war. She examines 20 veterans and assesses whether they show signs of post-traumatic stress disorder.

54. A sociologist wants to learn about the number of meetings per year of the 4-H clubs in Maricopa County, Arizona. He collects information from 10 different 4-H clubs in various parts of the county.

55. A physical therapist would like to determine whether a new exercise method can delay the onset of osteoporosis in older women. She chooses 10 of her patients to use the new method.

56. An educator asks a sample of students at Portland Community College whether they would be interested in taking a course online.

57. A financial adviser would like to assess the effect of mergers on price/earnings ratio. She collects data on 50 companies that recently underwent a merger.

For Exercises 58–64, state whether descriptive statistics or statistical inference was used, and explain why.

58. The average price in a sample of 15 homes sold in Jacksonville, Florida, for the week of April 21 was $253,200.

59. According to the Department of Transportation, 60% of all automobile passengers wear seat belts. This is based on a survey of 1000 automobile passengers, of whom 600 wore seat belts.

60. In a sample of 500 subjects, it was found that daily exercise lowered the average cholesterol level by 10%. A medical spokesperson then stated that daily exercise can lower everyone's cholesterol level by 10%.

61. In a sample of 140 traffic fatalities in New York, 75 involved alcohol.

62. The goals-against average for the Charlestown Chiefs hockey team in a sample of 20 games was 3.57 goals per game.

63. The Department of Health and Human Services conducted a survey, in which it was found that the percentage of 15- to 18-year-olds using illicit drugs has dropped in the last two years. The department concluded that illicit drug use has fallen among all 15- to 18-year-olds.

64. The average on the first statistics exam for a sample of 10 students in Ms. Reynolds' class was 70.

APPLYING THE CONCEPTS

For Exercises 65–71, do the following.
  a. List the elements and the variables.
  b. Identify the qualitative variables and the quantitative variables.
  c. For each quantitative variable, indicate whether it is discrete or continuous.
  d. For each variable, identify the level of measurement.
  e. Provide the observation for the indicated element.

65. Endangered Species. Refer to the following table, which lists four of the endangered animal species in the United States, as listed by www.earthsendangered.com. Do (**a**)–(**d**) and then provide the observation for the Florida panther.

Endangered species	Year listed as endangered	Estimated number remaining	Range
Pygmy rabbit	2001	20	Washington State
Florida panther	1973	50	Florida
Red wolf	1967	200	North Carolina
West Indian manatee	1967	2500	Florida

66. Top Five Employers in Santa Monica, CA. Refer to the following table. Do (**a**)–(**d**), and then provide the observation for the city of Santa Monica.

Company	Employees	Industry
City of Santa Monica	1892	Government
St. John's Health Center	1755	Health services
The Macerich Company	1605	Real estate
Fremont General Corp	1600	Insurance
Entravision Corp	1206	Media company

Source: Santa Monica Chamber of Commerce.

67. Genetically Engineered Crops. Genetically engineered (GE) crops are now planted on the majority of acreage in many states around the country. GE corn comes in three varieties: insect-resistant, herbicide-tolerant, and stacked genes. The following table contains the proportion of the corn grown in each of five states that is GE, along with the GE type most prevalent in each state, for 2013.[5] Do (**a**)–(**d**), and then provide the observation for the state of Texas.

State	Proportion of GE corn	Most prevalent type
Texas	89%	Herbicide-tolerant
Missouri	92%	Insect-resistant
Minnesota	91%	Herbicide-tolerant
Ohio	85%	Herbicide-tolerant
South Dakota	96%	Herbicide-tolerant

68. Hospitals Near Jackson, MS. Refer to the following table. Do (**a**)–(**d**). What is the observation for Rankin Medical Center?

Hospital	Beds	City	Zip
Hardy Wilson	49	Hazlehurst	39083
Humphreys County	34	Belzoni	39038
Jefferson County	30	Fayette	39069
Lackey Memorial	15	Forest	39074
Leake Memorial	25	Carthage	39051
Madison County	67	Canton	39046
Montfort Jones	72	Kosciusko	39090
Rankin Medical Center	134	Brandon	39042

69. Births and Maternal Age in Westchester County, NY. The following table represents the number of births and the average maternal age in 10 hospitals in northwest Westchester County, New York. Do (**a**)–(**d**). What is the observation for Sleepy Hollow?

Hospital	Births	Average maternal age
Briarcliff Manor	71	34.1
Buchanan	25	31.6
Cortlandt	348	32.2
Croton-on-Hudson	93	33.5
Mount Pleasant	277	32.8
Ossining 1	80	32.1
Ossining 2	371	29.2
Peekskill	365	29.0
Pleasantville	79	32.9
Sleepy Hollow	134	29.2

70. Commodity Prices. The financial company Bloomberg (www.bloomberg.com) reported that, on June 4, 2014, the prices in dollars for the following commodities were oil ($102.79, +0.13%), gold ($1243.62, −0.110%), and wheat ($616.25, +0.61%). Do (**a**)–(**d**). What is the observation for gold?

71. Worst Tornadoes. CNN.com reports[6] that the five worst tornadoes in American history in terms of death toll are as shown in the following table. Do (**a**)–(**d**). What is the observation for the St. Louis Tornado?

Tornado name	Deaths	Year
Tri-State	695	1925
Natchez	317	1840
St. Louis	255	1896
Tupelo	216	1936
Gainesville	203	1936

72. Top Five Employers in Santa Monica, CA. Refer to Exercise 66 to answer the following questions.
    a. Do these five employers represent a sample or a population?
    b. Could these five companies be considered a representative sample of the number of employees per company for all companies in Santa Monica? Explain.

73. Worst Tornadoes. Refer to Exercise 71 to answer the following questions.
    a. Do the data in the table represent a sample or a population?
    b. Could these data be considered a representative sample of the number of annual tornado deaths for all years? Explain.

Light Bulb Lifetime. Use the following information for Exercises 74 and 75. An electrical company has developed a new form of light bulb that it claims lasts longer than current models. The company has 1 million bulbs in its inventory.

74. How do you think the company found evidence for its claim?

75. Suppose you take a representative sample of 100 of the new light bulbs and find the average lifetime to be 2000 hours.
    a. Is this a statistic or a parameter?
    b. Write a sentence that estimates the average lifetime of all the new light bulbs.

Largest University Campuses. The National Center for Education Statistics reported that the colleges or university campuses with the largest enrollment in 2014 are as shown in the table. Use this information for Exercises 76 and 77.

Institution	State	Enrollment	Rank
Ashford University	Iowa	74,596	1
Arizona State University	Arizona	72,254	2
Liberty University	Virginia	64,096	3
Miami Dade College	Florida	63,736	4
Lone Star College System	Texas	63,029	5

76. Do the following.
    a. List the elements.
    b. List the variables.
    c. Identify the qualitative variables.
    d. Identify the quantitative variables.
    e. For each variable, identify the level of measurement.

77. Answer the following.
    a. Do these five campuses represent a sample or a population?
    b. Could these five campuses be considered a representative sample of the enrollment for all university campuses in the United States? Explain.
    c. Provide the observation for Arizona State University.

BRINGING IT ALL TOGETHER

Chapter 1 Case Study: Video Game Sales. Use Table 3 (page 8) to answer Exercises 78–88.

    videogamesales

78. Which of the variables are qualitative?

79. List the quantitative variables.

80. Is *Weeks on list* a discrete variable or a continuous variable?

81. Does the list in Table 3 represent a sample or a population? Explain.

82. The number for highest sales for the week is 36,732. Does this represent a parameter or a statistic?

83. State the nominal variables.

84. Are there any ordinal variables?

85. Which variables represent ratio data?

86. Is there a variable that can be viewed as interval data? Explain.

87. The Xbox 360 version of *Grand Theft Auto V* outsold the PS3 version of the game for the week of May 14, 2014. Is this considered descriptive statistics or statistical inference?

88. Refer to the previous question. Suppose we then predict that the Xbox 360 version of *Grand Theft Auto V* will outsell the PS3 version of the game for the following week. Does this represent descriptive statistics or statistical inference?

1.3 Gathering Data

OBJECTIVES By the end of this section, I will be able to . . .

1 Explain what a random sample is, and why we need one.
2 Identify systematic sampling, stratified sampling, cluster sampling, and convenience sampling.
3 Explain selection bias and good questionnaire design.
4 Understand the difference between an observational study and an experiment.

1 Random Sampling

We can use the information gathered from a sample to generalize about the population when it is impractical or impossible to take a census of the entire population. However, if we get a "bad" sample, the information gleaned from the sample will be misleading, with potentially catastrophic consequences. This section introduces a method of sampling that minimizes many potential biases, which could lead to incorrect generalizations about the population. This sampling method is called *random sampling*. Everyday examples of random sampling include:

- randomly selecting lottery numbers from a basket that continuously churns the number-balls,

- randomly choosing one card from a deck of playing cards that has been well shuffled, and

- randomly pulling a name out of a hat, after the names have been well stirred.

Because random samples are not always practical or desirable, this section also discusses some of the many alternative sampling methods available, including stratified sampling and cluster sampling.

What Is a Random Sample, and Why Do We Need It?

Survey sampling, or polling, has now become so widespread that hardly a day goes by without the results of some new poll or survey making the headlines. Polls are a good example of statistical sampling at work. The pollsters canvass about 1000 or so respondents, analyze the sample results, and then report their statistical inference that, for example, "32% of Americans have used a cell phone to access the Internet."

Today, many polls are conducted quite scientifically, and their results are usually very accurate. However, such was not always the case. In 1936, the *Literary Digest* had correctly predicted the past three presidential elections and went to work to predict the winner of the contest between Republican Alf Landon and Democrat Franklin Roosevelt. The magazine sent ballots to 10 million citizens. The results ran strongly in favor of Landon, leading the *Literary Digest* to predict Landon to win the election. About 25% of the ballots were returned, giving the newsweekly a sample size of 2.5 million. George Gallup, on the other hand, was working with a sample size that was about 1000 times *smaller* than the *Literary Digest*'s. However, Gallup predicted a victory for Roosevelt. Clearly, with more data, the *Literary Digest* should have been able to give a more accurate prediction, right? Not necessarily. Roosevelt won in a landslide, and the embarrassed *Literary Digest* later declared bankruptcy.

The problem stemmed from the way that the *Literary Digest* identified its sample. It used lists of people who owned cars and had telephones, which in the 1930s excluded millions of poor and underprivileged people, who overwhelmingly supported Roosevelt. Its sample of 2.5 million, therefore, was highly biased toward the richer folks, who were less likely to have any great fondness for Roosevelt and his New Deal policies. Gallup, on the other hand, chose his sample more scientifically, and even though his sample size was smaller, it was more representative of the population as a whole.

One inexpensive way of eliminating many types of bias is to make sure your sample is a **random sample**.

> A **random sample** (also known as a **simple random sample**) is a sample for which every element of the population has an equal chance of being selected.

Note: When we take a sample, we usually discard any repeated elements because we already have their information.

For example, today, the Gallup polling organization uses random digit dialing, a computer program that generates random four-digit numbers, which are then appended to the telephone exchanges and area codes. Thus, each household phone number in America has an equal chance of being included in the sample, regardless of whether it is listed or unlisted.

 Random samples may be generated using technology, using the *Simple Random Sample* applet, or using the random number table provided in Table A in the Appendix (page T-2). At the end of this section, we demonstrate how to generate random samples using the TI-83/84 graphing calculator, Excel, Minitab, SPSS, JMP, and CrunchIt!. The *Simple Random Sample* applet allows you to produce a random sample of up to 100 elements, in the form of a lotto.

EXAMPLE 12   Generating a random sample using technology

 Video Game Sales

Recall the top 30 best-selling video games in the United States for the week of May 17, 2014, shown in Table 3 on page 8; use the TI-83/84, Excel, Minitab, or SPSS to generate a random sample of 7 video games from this list.

Solution

We used the instructions provided in the Step-by-Step Technology Guide at the end of this section (page 30) to create four random samples, listed below. Note that each random sample is different, as yours will be.

Random sample 1 using the TI-83/84	Random sample 2 using Excel
28. Nintendo Land for Wii U	3. Kirby: Triple Deluxe for 3DS
7. Bound by Flame for PS4	17. Battlefield 4 for Xbox 360
25. Yoshi's New Island for 3DS	16. Borderlands 2 for PSV
6. Call of Duty: Ghosts for Xbox 360	5. Titanfall for Xbox One
4. MLB 14 The Show for PS4	25. Yoshi's New Island for 3DS
2. Minecraft for Xbox 360	24. Super Mario 3D World for Wii U
20. inFamous: Second Son for PS4	28. Nintendo Land for Wii U

Random sample 3 using Minitab	Random sample 4 using SPSS
23. Spiderman 2 for PS4	22. NBA 2K14 for Xbox 360
25. Yoshi's New Island for 3DS	28. Nintendo Land for Wii U
24. Super Mario 3D World for Wii U	18. Forza Motorsport 5 for Xbox One
30. NBA 2K14 for PS4	8. Pokemon X/Y for 3DS
15. Call of Duty: Ghosts for PS3	2. Minecraft for Xbox 360
9. Titanfall for Xbox 360	4. MLB 14 The Show for PS4
13. Super Luigi U for Wii U	7. Bound by Flame for PS4

NOW YOU CAN DO
Exercises 7–10.

2 More Sampling Methods

In certain circumstances, simple random sampling can have shortcomings. A simple random sample may not provide sufficient information about subgroups within the population. For example, suppose you are interested in knowing the proportion of 19- to 22-year-olds in Walnut, California, who consume caffeine. A random sample size of 100 of all the residents in Walnut may yield only a dozen 19- to 22-year-olds, which may be too small a sample to be useful for statistical inference. Therefore, the researcher needs other methods for obtaining samples, depending on the situation and the research question.

Systematic Sampling

Note: Most of the sampling methods mentioned here involve randomness. However, only the simple random sample is used throughout the text. Therefore, whenever you see the phrase *random sample,* it should be understood as *simple random sample.*

Perhaps the easiest method of sampling is systematic sampling, which is used when a random sample is unobtainable. In systematic sampling, each element of the population is numbered, and the sample is obtained by selecting every kth element, where k is some whole number. The first element selected corresponds to a random whole number between 1 and k. The ancient Romans understood well how to use systematic sampling. When a Roman legion mutinied or showed cowardice in battle, every 10th member was selected and summarily executed before his comrades. Literally, the legion was *decimated,* from the Latin *decem,* meaning "ten."

EXAMPLE 13   Systematic sampling

 20richest

Table 9 contains the top 20 richest people in the world for the year 2014, according to the annually published *Forbes 400* listing. Obtain a systematic sample from this list, using $k = 4$.

Table 9 Twenty richest people in the world

Rank	Name	Net worth ($ billions)	Rank	Name	Net worth ($ billions)
1	Bill Gates	72	11	Sheldon Adelson	28.5
2	Warren Buffett	58.5	12	Jeff Bezos	27.2
3	Larry Ellison	41	13	Larry Page	24.9
4	Charles Koch	36	14	Sergey Brin	24.4
5	David Koch	36	15	Forrest Mars	20.5
6	Christy Walton	35.4	16	Jacqueline Mars	20.5
7	Jim Walton	33.8	17	John Mars	20.5
8	Alice Walton	33.5	18	Carl Icahn	20.3
9	S. Robson Walton	33.3	19	George Soros	20
10	Michael Bloomberg	31	20	Mark Zuckerberg	19

Source: Forbes magazine, www.forbes.com/forbes-400/list/.

Solution

First, we randomly select a whole number between 1 and $k = 4$. Suppose we select 2. Thus, our systematic sample will consist of every 4th person in Table 9, starting with the 2nd person. That is, our systematic sample will consist of the 2nd, 6th, 10th, 14th, and 18th persons, shown here:

Systematic sample: Warren Buffett, Christy Walton, Michael Bloomberg, Sergey Brin, Carl Icahn.

NOW YOU CAN DO
Exercises 11–14.

YOUR TURN
#8

1. Generate a systematic sample with $k = 3$, selecting every 3rd person, starting from Bill Gates.

2. Obtain a systematic sample with $k = 5$, selecting every 5th person, starting with Larry Ellison.

(The solutions are shown in Appendix A.)

Stratified Sampling

Often, researchers are interested in investigating characteristics of a certain subgroup of a population, such as 19- to 22-year-olds in Walnut, California. In cases like this, the researcher divides the population into subgroups, or *strata*, according to some characteristic, such as race or gender. Then a random sample is taken from each stratum. In this way, the researcher knows that a sample will be obtained from each stratum and that it will be large enough to provide reliable statistical inference for each stratum.

EXAMPLE 14   Stratified sampling

Gregory Shamus/Getty Images

LeBron James, of the Cleveland Cavaliers.

A researcher is interested in analyzing whether differences in scoring exist among the basketball teams in the three divisions of the Eastern Conference of the National Basketball Association (Table 10). Obtain a stratified sample of two teams from each division.

Table 10 Teams in the three divisions of the Eastern Conference of the National Basketball Association

Atlantic Division	Central Division	Southeast Division
Boston Celtics	Chicago Bulls	Atlanta Hawks
Brooklyn Nets	Cleveland Cavaliers	Charlotte Bobcats
New York Knicks	Detroit Pistons	Miami Heat
Philadelphia 76ers	Indiana Pacers	Orlando Magic
Toronto Raptors	Milwaukee Bucks	Washington Wizards

Solution

A random sample of size two was drawn from the teams in each of the three divisions. These six teams are then combined to form our stratified sample of basketball teams. Note that each random sample is different, as yours will be.

Atlantic Division	Central Division	Southeast Division
Boston	Chicago	Atlanta
Brooklyn	Cleveland	Charlotte
New York	Detroit	Miami
Philadelphia	Indiana	Orlando
Toronto	Milwaukee	Washington

Stratified sample of six teams

Boston Celtics
Cleveland Cavaliers
Miami Heat
Milwaukee Bucks
New York Knicks
Orlando Magic

NOW YOU CAN DO
Exercises 15–18.

Cluster Sampling

Cluster sampling is used when the population is widely scattered geographically or poses other logistical difficulties. For example, if we were interested in estimating the mean income of Manhattan residents, it would be time-consuming and expensive to visit 1000 different locations in Manhattan to elicit sample information. In cluster sampling, the population is divided into *clusters,* such as precincts or city blocks. Then several clusters are chosen at random, and all of the elements within the chosen clusters are selected for the sample. One disadvantage of cluster sampling is that the respondents from within a certain cluster will tend to be more similar to each other than the elements of a random sample would be. For example, if one of the clusters in the Manhattan income survey was a Fifth Avenue block, the mean income of residents there would be at the higher end of the income scale.

EXAMPLE 15   Cluster sampling

Using Table 10, consider each division to be a cluster. Construct a cluster sample of the teams in the Eastern Conference by randomly selecting two of the three clusters (divisions).

Solution

Suppose that we randomly select our clusters to be the Atlantic Division and the Southeast Division. Our cluster sample then consists of *all* the teams in both of these divisions, as follows:

Cluster sample of 10 teams

Atlanta Hawks
Boston Celtics
Brooklyn Nets
Charlotte Bobcats
Miami Heat
New York Knicks
Orlando Magic
Philadelphia 76ers
Toronto Raptors
Washington Wizards

Atlantic Division	Central Division	Southeast Division
Boston	Chicago	Atlanta
Brooklyn	Cleveland	Charlotte
New York	Detroit	Miami
Philadelphia	Indiana	Orlando
Toronto	Milwaukee	Washington

NOW YOU CAN DO
Exercises 19 and 20.

Stratified Sampling versus Cluster Sampling

Stratified sampling and cluster sampling are sometimes confused. To obtain a stratified sample, we (a) divide the population into subgroups (strata, the divisions in Table 10), and (b) take a random sample from each subgroup, as shown by the shaded teams in Example 14. In cluster sampling, we (a) divide the population into subgroups (the divisions in Table 10, this time called clusters), (b) take a random sample of the clusters, as shown by the shaded divisions in Example 15, and (c) choose *all* the elements in the selected clusters for our cluster sample. In stratified sampling, we are randomly selecting elements from the subgroups; in cluster sampling, we are randomly selecting the clusters only, not the elements in the clusters.

Convenience Sampling

In convenience sampling, subjects are chosen based on what is convenient for the survey personnel. If you were to estimate the true proportion of females taking an introductory statistics course using only the people in your class, this would be considered a convenience sample. As we shall see in Example 16, convenience sampling usually does not result in a representative sample.

EXAMPLE 16   Convenience sampling using online polls

 Caution: Surveys, like online polls, that use convenience sampling should be treated with a healthy dose of skepticism. They are not statistically sound.

An online newspaper reports that, in an online poll of its readership, 60% say that they get most of their news from online sources. Does this number accurately reflect the proportion of all Americans who get most of their news from online sources?

Solution

No, the sample is not random. Only those Americans who are online already (and already using an online news source) can respond to this online poll. Therefore, the sample is not random, and it is biased. It overestimates the proportion of Americans who get their news from online sources. Further, no mechanism is available to guard against a single person responding repeatedly and getting his or her vote counted multiple times. Online polls are not scientific, and their results should not be considered a true reflection of the sentiments of all Americans.

NOW YOU CAN DO
Exercises 21 and 22.

YOUR TURN
#9

State whether the following scenarios represent convenience sampling.

1. You need to conduct a survey for sociology class. You generate a random sample of five students at your school.

2. You need to conduct a survey for sociology class. You obtain data from your five closest friends at school.

(The solutions are shown in Appendix A.)

EXAMPLE 17   Recognizing the sampling method

For each of the following, identify which type of sampling is represented.

a. Students in your class are divided into females and males. A random sample of size 5 is then drawn from each of the groups.

b. You are interested in estimating the average number of hours dormitory residents spend studying. In each dormitory, one floor is chosen at random and all the students on that floor are interviewed.

c.   You are researching the proportion of college students who prefer country music to other forms of music. You obtain a listing of all the students at your college and contact every 20th student on the list.

d.   Your campus statistical consulting center uses random digit dialing to locate potential subjects for a political survey.

e.   A student is investigating the prevalence of flu on campus this semester, and he asks 20 of his friends whether they have had the flu.

Solution

a.   Stratified sampling: (a) the population was divided into subgroups (females and males), and (b) a random sample was drawn from each of the groups.

b.   Cluster sampling: (a) the population was divided into clusters (dormitory floors), (b) a random sample of the clusters (floors) is taken, and (c) all students on that floor (cluster) were selected.

c.   Systematic sampling: where every kth member of the population is taken, with $k = 20$.

NOW YOU CAN DO
Exercises 23–26.

d.   An example of random sampling, as illustrated on page 21.

e.   Convenience sampling: the student is choosing a sample convenient for him.

YOUR TURN
#10

For each of the following, identify which type of sampling is represented.

1.   You are collecting data for a business project, and you ask five of your friends for their responses to the survey.

2.   Your campus meal services company asks every 10th student for his or her opinion on a new menu item.

3.   The school would like to determine student attitudes toward online learning. One class is chosen at random, and every student in the class is surveyed.

4.   Your professor wants to start a new nursing course. He divides students into two groups—nursing majors and all others—and then takes a random sample of 10 students from each group.

5.   Your professor wants help writing problems on the board, and he pulls two names out of a hat.

(The solutions are shown in Appendix A.)

3 Selection Bias and Questionnaire Design

Here, we learn about some common pitfalls in the design and implementation of a survey, including selection bias and the wording of a questionnaire.

> The **target population** is the complete collection of all elements that we are interested in studying.
>
> The **potential population** is the collection of elements from the target population that had a chance of being sampled.
>
> **Selection bias** occurs when the potential population from which the actual sample is drawn is not representative of the target population, due to an inappropriate sampling method.

EXAMPLE 18   Selection bias

Suppose Ashley would like to estimate the proportion of American voters who would favor abandoning the present system of Social Security in favor of a system where retirement funds would be invested in the stock market. Ashley goes to the mall with her clipboard and canvasses as many people as she can on Monday between 9 A.M. and 5 P.M. To each person, she asked the question "Do you favor or oppose abandoning the present Social Security system in favor of a system that invests retirement funds in the stock market?"

a. Identify Ashley's target population.

b. Identify Ashley's potential population.

c. Discuss any possible problems.

Solution

a. Ashley's **target population** is the population of all American voters.

b. The collection of all the American voters who visited the mall on Monday between 9 A.M. and 5 P.M. represent her **potential population**.

c. It appears that Ashley's survey may suffer from **selection bias**. The population of people who went to the mall on Monday between 9 A.M. and 5 P.M. is not representative of the target population of all American voters. Because many American voters work on Mondays between 9 A.M. and 5 P.M., they are not elements of the potential population. Further, the proportion of retirees at the mall during that time was likely larger than in the target population of all American voters. These retirees tend to oppose strongly any potential changes to the Social Security system and would probably tend to respond in the negative to the survey question.

NOW YOU CAN DO
Exercises 27–30.

Five Factors for Good Questionnaire Design

You may have heard of the aphorism "Be careful what you ask for; you may get it." This warning is certainly relevant to the issue of questionnaire design. The wording of questions can greatly affect the responses. Here are several factors to consider when designing a questionnaire.

1. **Remember: simplicity and clarity.** Do not use four-syllable words when one-syllable words will do. Respondents will be shy about asking you to clarify the question. The result will be confused responses and muddled data.

2. **When reporting results, include the actual question asked.** Be careful about drawing generalizations. The conclusions you draw may not have been what your respondents had in mind when they answered the questions.

3. **Avoid leading questions.** The respondent is often eager to please and will try to tell you what he or she thinks you want to hear. For example, a researcher is interested in determining the proportion of Americans who favor preserving the welfare system. A leading question would be "A child growing up poor in America faces more than his fair share of crime and negligence. Do you support preserving the welfare safety net to help ensure that children are given a fair chance?"

4. **Avoid asking two questions in one.** Avoid questions such as "Have you argued with your friends or family in the last month?" This is really two questions in one, and you will not know which question the respondents are answering.

5. **Avoid vague terminology.** Words mean different things to different people. Avoid using terminology like "often" or "sometimes." Instead, try to use specific terms such as "three times a week." If you use ambiguous terms, the data you collect will be ambiguous, and any conclusions you draw will probably not be valid.

EXAMPLE 19   Questionnaire design

For each of the following questionnaire items, identify which of the five factors for good questionnaire design is violated, if any.

a. Do you oppose the wasteful spending on foreign aid when so many problems confront us here at home?

b. Do you often feel lonely?

c. Do you espouse or disavow the conglomerative confluence of macroeconomic indicators?

d. Have you watched television or downloaded music in the past 24 hours?

e. Do you ever use a cell phone to access the Internet?

Solution

a. This is a leading question, which is clearly trying to influence the respondent's answer.

b. What is meant by "often"? Three times a week? Three times a day? This is vague terminology.

c. This question would only be understood by those who have studied economics, and is neither simple nor clear.

d. This is asking two questions in one. It is possible that respondents have done one or the other, or both.

e. This question is fine. In fact, it is an actual survey question from the Pew Research Center.

NOW YOU CAN DO
Exercises 31–34.

4 Experimental Studies and Observational Studies

Two major types of statistical studies are **experimental studies** and **observational studies**. We have seen that researchers can gather data by consulting existing sources, by distributing a questionnaire, or by taking a sample. However, you may not be able to obtain the information you require by using survey or sampling methods. In this case, you may prefer to conduct an experimental study.

Note: What is the difference between an element and a subject? *Subject* is a term usually reserved for statistical studies, whereas the term *element* can be used for any data set.

Note: In some experiments, especially in medicine, members of the control group receive a placebo, a nonfunctioning simulated treatment. Sometimes, the symptoms of the members of the control group improve simply by taking the placebo, a phenomenon known as the *placebo effect*.

> **Experimental Studies**
>
> In an **experimental study**, researchers investigate how varying the predictor variable affects the response variable.
>
> A **predictor variable** (also called an **explanatory variable**) is a characteristic intended to explain differences in the response variable.
>
> A predictor variable that takes the form of a purposeful intervention is called a **treatment**.
>
> A **response variable** is an outcome, a characteristic of the subjects of the experiment presumably brought about by differences in the predictor variable or treatment.
>
> The **subjects** in a statistical study represent the elements from which the data are drawn.
>
> A **control** is used as a standard of comparison for checking the results of the experiment. The treatment is calibrated against the control. Without the control, the treatment results cannot be compared.

We illustrate experimental studies using the following example.

EXAMPLE 20   Newborn babies and a heartbeat: An experimental study

SergiyN/istock

A psychologist wanted to test whether the sound of a human heartbeat would help newborn babies grow. A baby nursery at a hospital was set up so that the sound of a human heartbeat could be heard throughout the nursery. The heartbeat sound was played in the nursery for a large batch of newborn children, who were then weighed to determine their weight gain after four days in the nursery. Later, a second batch of children occupied the nursery, but no heartbeat sound was played. These children were also weighed after four days in the nursery. Babies were randomly placed into the two groups. Identify the following:

a. The subjects

b. The predictor variable

c. The treatment

d. The response variable

e. The control

Solution

a. The babies were the **subjects** of this experimental study.

b. The **predictor variable** is whether or not the heartbeat sound was played in the nursery.

c. The **treatment** is the sound of the human heartbeat.

d. The **response variable** is the baby's weight gain, which is the outcome of the study.

e. The **control** is the group of babies for whom the heartbeat sound was not played.

The results were consistent with the psychologist's conjecture; the babies who listened to the heartbeat sound had a greater average weight gain than the babies for whom no heartbeat sound was played.

NOW YOU CAN DO
Exercises 35–46.

Two additional factors should be considered when designing an experimental study: *randomization* and *replication*.

Randomization. Many biases can be introduced into an experiment. For example, a well-meaning doctor may want to place underweight high-risk babies in the group with the heartbeat, in the hope that such babies will flourish. To eliminate biases like these, the placement of the subjects into the treatment and control groups should be done randomly.

Replication. One major theme of statistical investigation is that larger samples are usually better, because they allow more precise inference. In a statistical study, the treatment and the control groups each must contain a large enough number of subjects to allow detection of meaningful differences between the treatment and control. For example, if a researcher examined only three babies with the heartbeat sound and three babies without the heartbeat sound, this would not be a sufficient number of replications. In Chapter 8, "Confidence Intervals," we will learn how large a sample size is sufficient for the needs of a particular study.

EXAMPLE 21   Randomization and replication

For each of the following scenarios, indicate (**a**) whether randomization is present, and (**b**) whether sufficient replication has been made.

a. An experiment is conducted to determine if a certain type of genetically engineered corn seed will outproduce traditional corn seed. Fifty rows of Farmer Brown's field are sown with the genetically engineered corn seed and 50 rows of Farmer Grey's field are sown with the traditional seed.

b. An experiment is conducted to see whether a new diabetes treatment will outperform a traditional treatment. Two mice each are randomly assigned to each of the treatment and control groups.

Solution

a. The 50-row sample is probably enough replication. However, randomization is lacking. Perhaps Farmer Brown is a better farmer than Farmer Grey, in which case, this *lurking variable* or *confounding variable* will confound the results.

NOW YOU CAN DO
Exercises 47 and 48.

b. Here, the randomization is fine, but two mice each is probably insufficient replication to uncover any strong statistical results.

Observational Studies

Circumstances exist where it could be impossible, impractical, or unethical for the researcher to place subjects into treatment and control groups. For example, suppose we are interested in whether women who work outside the home suffer less depression than women who remain at home with the children. The explanatory variable here is whether or not a woman works outside the home. However, it is not possible for the researcher to take women and randomly separate them into groups that either work outside the home or do not work outside the home.

Sometimes an experimental study is not possible for ethical reasons. Suppose you are interested in whether babies born to chemically dependent mothers display differences in cognitive skills from babies born to mothers who are not chemically dependent. It is clearly not ethical to randomly assign half of the mothers in the study to become chemically dependent during their pregnancy. Therefore, researchers need another type of statistical study: the observational study. In an observational study, the researcher observes whether the subjects' differences in the predictor variable are associated with differences in the response variable. No attempt is made to create differences in the predictor variable.

A sample survey is an example of an observational study. Data about a response variable may be obtained through the survey, along with information about possible predictor variables. No attempt is made to manipulate the variables. The researcher analyzes the information to determine whether differences in the predictor variable are associated with differences in the response variable.

EXAMPLE 22   Experiment or observational study

Examine the following scenarios, and state whether each represents an experiment or an observational study.

a. A total of 50 female tobacco smokers and 50 female non-tobacco smokers are chosen. Lung capacity of the women is compared.

b. A total of 100 asthma sufferers are randomly assigned to receive either a new treatment or a control. Asthma relief is measured.

Solution

a. It would be unethical to assign a woman to smoke tobacco as part of an experiment. Therefore, scientists need to use observational studies, as this one is.

b. The subjects were randomized into the treatment and control groups, making this an experiment.

NOW YOU CAN DO
Exercises 49 and 50.

STEP-BY-STEP TECHNOLOGY GUIDE: Generating a Random Sample

We illustrate using Example 12 (page 21).

TI-84

Set up the TI-84 Menus (For calculators without STATWIZARD, you may skip these steps.)

Step 1   Press 2nd then Catalog.
Step 2   Scroll down to **STATWIZARD OFF**. Highlight it and press **ENTER** (see Figure 5). Press **ENTER** again.

FIGURE 5

TI-83/84

Generate a random sample

Step 1   Enter a "seed," which can be any nonzero number.

Step 2   Press **STO** ⇒.

Step 3   Press **MATH**, highlight **PRB**, select 1: **rand**, and press **ENTER** (see Figure 6, which uses 1776 for the seed). Your seed number is now in the calculator's memory.

Step 4   Press **MATH**, highlight **PRB**, and select 5: **randInt(**.

Step 5   Enter 1, N, *two times* n, where **N** = population size and n = sample size. We enter twice the sample size in case repeats occur. For Example 12, because n = 7, we enter **randInt(1, 30, 14)** and press **ENTER** (Figure 7).

Step 6   Store the random sample in list **L1** as follows: press **STO** ⇒, then **2nd**, then **L1** (Figure 7). Then press Enter.

Step 7   View the random sample by pressing **STAT**, highlighting **EDIT**, and pressing **ENTER** (Figure 8). The random sample for Example 12 is therefore **28, 7, 25, 6, 4, 2, 20**.

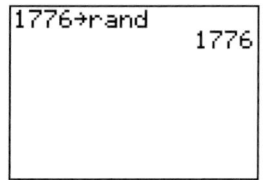

FIGURE 6

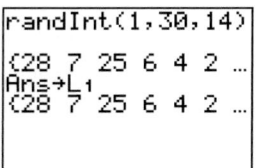

FIGURE 7

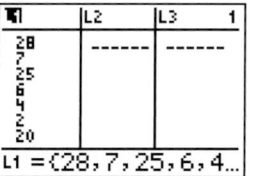

FIGURE 8

EXCEL

Step 1   Select cell **A1**. Click the **Insert Function** icon f_x.

Step 2   For "Search for a function," enter **randbetween**. Click **Go**, then **OK**.

Step 3   For **Bottom**, enter 1. For **Top**, enter population size **N**. For Example 12, **N** = 30. Click **OK**.

Step 4   Cell **A1** now contains a random integer between 1 and **N**. Copy and paste cell **A1** into *twice* as many cells as needed for the sample size n, just in case repeats occur. For Example 12, copy and paste into cells **A2** to **A14**. The results are shown in Figure 9. Note that **17, 16,** and **5** are repeated, so that our random sample is **3, 17, 16, 5, 25, 24, 28**.

	A	B	C	D	E	F
1	3					
2	17					
3	17					
4	16					
5	16					
6	5					
7	25					
8	24					
9	5					
10	28					
11	12					
12	29					
13	9					
14	25					

A1 · : × ✓ f_x =RANDBETWEEN(1,30)

FIGURE 9 Excel random sample.

MINITAB

Step 1   Click on **Calc > Random Data > Integer …**

Step 2   In the **Number of rows of data to generate** section, enter *twice* your desired sample size, just in case repeats occur. For example, if your desired sample size is **7**, enter **14**.

Step 3   In the **Store in column(s)** section, enter whichever column is convenient for you, such as **C1**.

Step 4   For **Minimum value**, enter 1. For **Maximum value**, enter your population size, **N**. Click **OK**.

Step 5   The random integers appear in column **C1**. Start from the top and go down the list, omitting any repeats, until you have your sample of size **n**. Our random sample (Figure 10) is therefore **23, 25, 24, 30, 15, 9, 13**.

Excel and Minitab base the seed on the current time, so that you need not set it yourself.

↓	C1	C2
1	23	
2	25	
3	24	
4	30	
5	30	
6	15	
7	9	
8	13	
9	13	
10	9	
11	23	
12	2	
13	1	
14	12	
15		

Worksheet 1 ***

FIGURE 10 Minitab random sample.

SPSS

Step 1   Select the top-right cell. Enter **0**, which appears as **.00**. Click the **Variable View** tab, select the **Name** of the variable (currently **VAR00001**) and type **Sample** to rename the variable. Press **ENTER**.

Step 2   Select the **Data View** tab, and fill in rows 1 through 14 of **Sample** with the value **0**.

Step 3   Click **Transform > Compute Variable …**

Step 4   In **Target Variable**, enter **Sample**. In the **Function Group** menu, click **Random Numbers**. In **Functions and Special Variables**, double-click **Rv.Uniform**.

Step 5   In **Numeric Expression**, replace the first question mark with **0** and the second question mark with **30**. Click **OK**, then **OK** again.

Step 6   Minimize the output window, and go back to your dataset. The results are shown in Figure 11. Round the values up to the nearest integer, skipping any repeats. Our random sample is then **22, 28, 18, 8, 2, 4, 7**.

	Sample
1	21.46
2	27.85
3	17.34
4	7.86
5	21.74
6	1.11
7	3.02
8	21.98
9	6.76
10	6.64
11	18.30
12	28.62
13	7.11
14	10.16

FIGURE 11 SPSS Random Sample.

JMP

Step 1   Click **File**, then **New**, then **Data Table**.

Step 2   Click **Rows**, then **Add Rows …**; in **How many rows to add**, enter *twice* your desired sample size. Click OK.

Step 3   Click **Column 1**, above the first row in the first column. Type **Sample** to rename the variable, and press **ENTER**.

Step 4   Right-click **Sample** and select **Formula …**

Step 5   In the list under **Functions (grouped)**, click **Random**, then **Random Uniform**. Double-click **Random Uniform()** inside the red box to edit the formula. Type ***30**, so that the entire formula is **Random Uniform() * 30**. Click **OK**.

Step 6   The results are in Figure 12. Round the values up to the nearest integer to obtain the random sample, omitting repeats. Our random sample is **17, 26, 30, 11, 14, 6, 25**.

	Sample
1	16.26257088
2	25.957574691
3	29.054666595
4	10.13052671
5	13.269112052
6	5.0429479033
7	25.635380584
8	24.840021729
9	16.229105741
10	5.5036720843
11	6.073313097
12	25.835461321
13	4.5117695653
14	11.223771488

FIGURE 12 JMP random sample.

CRUNCHIT!

Step 1   Select **Insert**, then **Random Numbers**, then **Uniform**.

Step 2   For **Minimum**, enter **1**. For **Maximum**, enter **30**. For **Samples**, enter **14**. Click **Sample**. The results are in Figure 13.

Step 3   Round the numbers *up* to the nearest integer, and skip repeated numbers. Our random sample is then **14, 30, 11, 12, 7, 15**, and **10**.

	Var1	Var2	Var3	Var4	Var5
1					13.2962303643…
2					29.0921119432…
3					10.1763872026…
4					11.6178985420…
5					12.0401815779…
6					6.58471923950…
7					14.4112223873…
8					9.90451646596…
9					12.8794438699…
10					13.5899610286…
11					25.5066422421…
12					13.0487911421…
13					1.09878805372…
14					4.28687658091…

FIGURE 13 CrunchIt! random sample.

Section 1.3 Summary

1. A random sample is a sample for which every element has an equal chance of being included. A random sample can minimize many potential biases, which could lead to incorrect generalizations about the population.

2. Other sampling methods include stratified sampling, systematic sampling, cluster sampling, and convenience sampling.

3. When constructing a survey, avoid selection bias and follow the five factors for good questionnaire design.

4. Two types of statistical studies are used: experimental studies and observational studies. In an experimental study, researchers investigate how varying the predictor variable affects the response variable. It is not always possible to conduct an experimental study, however, and sometimes an observational study is used instead.

Section 1.3 Exercises

CLARIFYING THE CONCEPTS

1. Explain why convenience sampling usually does not result in a representative sample. (p. 25)

2. What type of bias did the *Literary Digest* poll (pp. 20–21) exhibit? How did it affect the results? (p. 26)

3. How could the *Literary Digest* have decreased the bias in its poll? (p. 21)

4. Was the *Literary Digest* poll a random sample? (p. 21)

5. Describe what a random sample is. (p. 21)

6. Describe the difference between an observational study and an experimental study. (p. 28)

PRACTICING THE TECHNIQUES

 CHECK IT OUT!

To do	Check out	Topic
Exercises 7–10	Example 12	Random sampling
Exercises 11–14	Example 13	Systematic sampling
Exercises 15–18	Example 14	Stratified sampling
Exercises 19–20	Example 15	Cluster sampling
Exercises 21–22	Example 16	Convenience sampling
Exercises 23–26	Example 17	Recognize the sampling method
Exercises 27–30	Example 18	Selection bias
Exercises 31–34	Example 19	Questionnaire design
Exercises 35–46	Example 20	Experimental study
Exercises 47–48	Example 21	Randomization and replication
Exercises 49–50	Example 22	Observational study

Refer to Table 11 to obtain random samples in Exercises 7–10.

TABLE 11   College football teams in four major conferences

Big Ten	Southeastern	Atlantic Coast	Pacific 12
Illinois	Alabama	Boston College	Arizona
Indiana	Arkansas	Clemson	Ariz. State
Iowa	Auburn	Duke	California
Michigan	Florida	Florida State	Colorado
Mich. State	Georgia	Georgia Tech	Oregon
Minnesota	Kentucky	Maryland	Oregon State
Nebraska	Louisiana State	Miami	Stanford
Northwestern	Mississippi	North Carolina	UCLA
Ohio State	Miss. State	NC State	USC
Penn State	Missouri	Pittsburgh	Utah
Purdue	South Carolina	Syracuse	Washington
Wisconsin	Tennessee	Virginia	Wash. State
	Texas A&M	Virginia Tech	
	Vanderbilt	Wake Forest	

7. Obtain a random sample of size 4 teams from the Big Ten Conference.

8. Obtain a random sample of size 5 teams from the Southeastern Conference.

9. Obtain a random sample of size 6 teams from the Atlantic Coast Conference.

10. Obtain a random sample of size 7 teams from the Pacific 12 Conference.

Use Table 11 to obtain systematic samples in Exercises 11–14.

11. Obtain a systematic sample of teams from the Big Ten Conference. Use $k = 2$ and start with the 1st team.

12. Obtain a systematic sample of every 3rd team from the Southeastern Conference. Start with the 2nd team.

13. Obtain a systematic sample of every 4th team from the Southeastern Conference. Start with the 1st team.

14. Obtain a systematic sample of teams from the Pacific 12 Conference. Use $k = 3$ and start with the 3rd team.

Use Table 11 to obtain stratified samples in Exercises 15–18.

15. Obtain a stratified sample of two teams from each of the Big Ten Conference and the Southeastern Conference.

16. Obtain a stratified sample of three teams from each of the Atlantic Coast Conference and the Pacific 12 Conference.

17. Obtain a stratified sample of four teams from each of the Big Ten Conference and the Atlantic Coast Conference.

18. Obtain a stratified sample of two teams from each of the four conferences.

Use Table 11 to obtain cluster samples in Exercises 19–20.

19. Obtain a cluster sample of all the teams from two randomly selected conferences.

20. Obtain a different cluster sample of all the teams from two randomly selected conferences.

Use Table 11 to obtain convenience samples in Exercises 21–22.

21. It is convenient for the researcher to choose the first five teams from the Southeastern Conference for the sample, using alphabetical order. Is this likely to result in a representative sample?

22. It is convenient for the researcher to choose only those teams in the state of Washington for the sample from the Pacific 12 conference. Is this likely to result in a representative sample?

For Exercises 23–26, state which type of sampling is represented.

23. Students in your class are divided into freshmen, sophomores, juniors, and seniors. One of the groups is selected at random and all the students in that group are selected.

24. An instructor in a large lecture course of 300 students wants to get a student sample, and he selects every 10th name from the class roster.

25. You are researching the proportion of college students who prefer country music to other forms of music. You survey all the students in all the classes you are taking this semester.

26. An instructor in a large lecture course of 300 students (two lectures, one lab per week) wants to get a student sample. He takes a random sample of three of the 15 lab sections and selects all of the students from those three sections.

Use the following information for Exercises 27 and 28. Brandon is trying to estimate the proportion of all college students who are physically fit. He obtains a sample of students working out at the gymnasium on Monday night.

27. Identify the target population and the potential population.

28. Does selection bias exist? Explain why or why not.

Use the following information for Exercises 29 and 30. Michelle would like to determine the proportion of small businesses that employ at least one college student part-time. She obtains a sample of businesses near the state university.

29. Identify the target population and the potential population.

30. Does selection bias exist? Explain why or why not.

For Exercises 31–34, identify which of the five factors for good questionnaire design is violated, if any.

31. Do you sometimes feel anxiety about your health?

32. Do you support the valiant efforts of our mayor to dispel the lies spread by the corrupt opposition?

33. Do you espouse the diminution of the graduated income tax?

34. Do you support laws restricting invasion of privacy and locking up those responsible for doing so?

For Exercises 35–38, do the following: **(a)** State which type of study is involved: experimental or observational. **(b)** Identify the response variable and the predictor variable.

35. A sociologist is interested in whether large families (at least four children) attend religious services more often than smaller families do.

36. A financial researcher is interested in whether companies that give large bonuses to their chief executive officers (at least $1 million per year) have a higher stock price.

37. A manufacturer is interested in whether a new computer processor will improve the performance of its electronics equipment.

38. A pharmaceutical company wants to see if its new drug will lower high blood pressure.

Use the following information for Exercises 39–42. Agricultural researchers are investigating whether a new form of pesticide will lead to lower levels of insect damage to crops than the traditional pesticide.

39. Identify the response variable.

40. Identify the predictor variable.

41. What is the treatment?

42. What is the control?

Use the following information for Exercises 43–46. Cholesterol researchers are investigating whether any difference exists between a new medication and a placebo (inactive pill) in lowering LDL cholesterol levels in the bloodstream.

43. Identify the response variable.

44. Identify the predictor variable.

45. What is the treatment?

46. What is the control?

For each of the following scenarios in Exercises 47 and 48, indicate **(a)** whether randomization is present, and **(b)** whether sufficient replication has been made.

47. A total of 100 subjects with high LDL cholesterol levels are randomly assigned to receive a new medication or the traditional medication.

48. Four subjects, two each, are randomly assigned to receive either a new Alzheimer's-resistant drug or a control.

Examine the following scenarios in Exercises 49 and 50, and state whether each represents an experiment or an observational study.

49. Patients with heart conditions were randomly assigned to receive either a dose of aspirin or a placebo. Occurrences of myocardial infarctions were measured.

50. Groups of Democrats, Independents, and Republicans were surveyed, and their opinions regarding Second Amendment rights were compared.

APPLYING THE CONCEPTS

CASE STUDY

Case Study: Video Game Sales. Use the 30 top-selling video games in Table 3 (page 8) for Exercises 51–54.

Use the *Simple Random Sample* applet for Exercise 51.

51. Random Sampling.
   a. Generate a random sample of size 5 video games.
   b. Before you generate another random sample of 5 video games, is it likely that it will be the same as in **(a)**? Why or why not?
   c. Think of your favorite video game from Table 3. Before we generate another random sample, is there any way of foretelling whether your game will be in the random sample? Explain.
   d. Go ahead and generate another random sample of 5 video games. Was it the same as the first sample? Was your game in the sample?

52. Obtain a systematic sample of every 4th video game. Start with the first video game.

53. Obtain a stratified sample of two video games from each platform (omit the PSV platform).

54. Obtain a cluster sample of all the video games from two randomly selected platforms.

55. Contradicting Ann Landers. "If you had to do it over again, would you have children?" This is the question that advice columnist Ann Landers once asked her readers. It turns out that nearly 70% of the 10,000 responses she received were "No." A professional poll by *Newsday* found that 91% of respondents would have children again. Explain the apparent contradiction between these two surveys using what you have learned in this section.

56. High School Dropouts. For the following survey, describe the target population and the potential population, and discuss the potential for selection bias. Researchers are interested in the proportion of high school students in New England who drop out (leave school before graduating). A survey is made of 15 high schools in Greater Boston.

57. Living Below the Poverty Level. For the following survey, describe the target population and the potential population, and discuss the potential for selection bias. A sociologist is interested in the proportion of people living below the poverty level in Chicago. He takes a random sample of phone numbers from the Chicago phone directory and asks each respondent his or her annual household income.

58. Rap or Hip-Hop? Describe what is wrong, if anything, with the following survey question. "Do you enjoy listening to rap or hip-hop music?"

59. Financial Ruin. Describe what is wrong, if anything, with the following survey question: "Do you think that we should tax and spend our way into financial ruin?"

60. Abortion. Suppose 67% of female respondents respond affirmatively to the question "Do you support the right of a woman to terminate a pregnancy when her life is in danger?" Would the researcher be justified in reporting, "Two-thirds of women support abortion"?

61. Most Active Stocks. Here is a list of the five most active stocks on the NYSE on June 5, 2014.

Stock	Price
Rite Aid	$7.87
Bank of America	$15.43
Twitter	$33.89
Sprint	$9.02
Ford Motors	$16.68

  a. We are about to select a random sample and determine the lowest price in the sample. Do we know what this price will be before we select the sample? Why or why not?
  b. Select a random sample of size 2 from the table.
  c. If you take another sample of size 2, is it likely to comprise the same two companies? Why or why not?
  d. Which stock in your sample has the lowest price? What is that price?

62. Most Active Stocks. Refer to the list of stocks from the previous exercise.

  a. We are about to select another random sample and determine the lowest price in the sample. Do we know what this price will be before we select the sample? Do we know whether it will be the same as in the previous exercise? Why or why not?
  b. Select another random sample of size 2.
  c. Which stock in your new sample has the lowest price? What is that price?

63. Challenge Exercise. Compare your answers in Exercise 61(d) with those in Exercise 62(c). What can we say about a quantity like "the lowest price in a random sample of stocks"?

64. Mediterranean Diet. The American Heart Association reported the following results of an experimental study.[7] Patients who ate a Mediterranean diet had a significantly lower risk of having a second heart attack than patients who ate a Western diet. Identify the response variable and the predictor variable in this experimental study.

65. Secondhand Smoking and Illness in Children. A Surgeon General's report found that "the evidence is sufficient to infer a causal relationship" between secondhand tobacco smoke exposure from parental smoking and respiratory illnesses in infants and children.[8]

  a. Given the health risks associated with tobacco use, discuss the ethics of forcing the parents of a treatment group to smoke tobacco.
  b. State whether this report was based on an experimental study or an observational study.

66. Ethics, Experiments, and Observational Studies. According to the British medical journal *The Lancet*, experimental studies performed on animals (nonhuman primates, squirrel monkeys, and rodents) have revealed that large doses of the drug Ecstasy (methylene-dioxy-methamphetamine, or MDMA) produce "large and possibly permanent damage" to neural axons in the brain.[9] Explain why the researchers did not perform their experiment on humans.

BRINGING IT ALL TOGETHER

Evidence for an Alternative Therapy? Use the following information for Exercises 67–69. A company called QT, Inc. sells "ionized bracelets," called Q-Ray Bracelets, that it claims help to ease pain by balancing the body's flow of "electromagnetic energy." QT, Inc. claims that Q-Ray Bracelets can ease pain caused by cancer, restore well-being, and provide many other health benefits. The Mayo Clinic decided to conduct a statistical study to determine whether the extravagant claims for Q-Ray Bracelets were justified.[10] In the study, 305 subjects wore the Q-Ray "ionized" bracelet and 305 wore a placebo bracelet (identical to the ionized bracelet except for the ionization) for four weeks, at the end of which certain measures of pain were evaluated and compared between the treatments. The subjects, upon entry to the study, were randomly assigned to receive either the ionized bracelet or the placebo bracelet.

67. Identify the following aspects of this study.
  a. The control
  b. The randomization
  c. The replication
68. Identify the following aspects of this study.
  a. The predictor variable

b. The treatment
c. The response variable
69. Does this statistical study represent an experimental study or an observational study? Write a sentence explaining why.

Chapter 1 Vocabulary

SECTION 1.2
- **Census** (p. 13)
- **Continuous variable** (p. 10)
- **Descriptive statistics** (p. 7)
- **Discrete variable** (p. 10)
- **Element** (p. 7)
- **Observation** (p. 7)
- **Parameter** (p. 12)
- **Population** (p. 12)
- **Qualitative variable** (p. 9)
- **Quantitative variable** (p. 9)
- **Sample** (p. 12)
- **Statistic** (p. 12)
- **Statistical inference** (p. 14)

- **Statistics** (p. 5)
- **Variable** (p. 7)

SECTION 1.3
- **Experimental study** (p. 28)
- **Observational study** (p. 28)
- **Potential population** (p. 26)
- **Predictor variable (explanatory variable)** (p. 28)
- **Random sample** (p. 21)
- **Response variable** (p. 28)
- **Selection bias** (p. 26)
- **Subjects** (p. 28)
- **Target population** (p. 26)
- **Treatment** (p. 28)

Chapter 1 Review Exercises

SECTION 1.2
Refer to the following table for Exercises 1–3. The table contains information on some sports cars, as reported by the Environmental Protection Agency for model year 2014.

Make/Model	Cylinders	Transmission	Combined mileage
Chevrolet Corvette	8	Manual	21
Ferrari 458 Italia	8	Automatic	14
Honda CR-Z	4	Manual	34
Jaguar F Convertible	6	Automatic	23
Porsche Boxster S	6	Automatic	24

1. Use the table of sports cars to find each of the following.
  a. List the elements.
  b. Identify the variables.
2. Use the table of sports cars to answer the following.
  a. Identify the qualitative variables.
  b. Identify the quantitative variables.
  c. For each variable, state the level of measurement.
3. Provide the observation for the Chevrolet Corvette.

4. The following table contains population figures for the five most populous states.

State	Pop. (1960, in 1000s)	Pop. (2013, in 1000s)	Increase
California	15,717	38,333	22,616
Texas	9,580	26,448	16,868
New York	16,782	19,651	2,869
Florida	4,952	19,553	14,601
Illinois	10,081	12,882	2,801

  a. Identify the elements and the variables.
  b. Are the variables qualitative or quantitative?
  c. Provide the observation for the state of Florida.
  d. Which three states had the largest population increases? Which two states had the smallest population increases?
5. An electrical company has developed a new form of light bulb that it claims lasts longer than current models. The company has 1 million bulbs in its inventory. Consider the population average lifetime.
  a. What is the only way to find out the population average lifetime of the 1 million bulbs in the inventory?
  b. Suppose someone who worked for you wrote you a memo suggesting that it was crucial to know the exact value of the population average lifetime of all

1 million new light bulbs. How would you respond? What might you suggest instead?

SECTION 1.3

6. Refer to the *Literary Digest* poll discussed in Section 1.3.
   a. What was the target population?
   b. What was the potential population?
   c. What was the sample?
   d. Discuss whether the sample was similar to the target population in all important characteristics.
7. Suppose you are interested in finding out how the statistics grades for your class compare with those of the college as a whole.
   a. Would you use an experimental study or an observational study?

   b. Discuss how this study situation would preclude effective randomization.
8. A long-running television advertisement claimed that "Three out of four dentists surveyed recommend sugarless gum for their patients who chew gum."
   a. If in fact only four dentists were surveyed, which of the study factors were violated?
   b. Use this situation to discuss why replication is important.
9. Suppose we are interested in determining whether differences exist in the cognitive levels of children from single-parent families and those from two-parent families. Would we use an observational study or an experimental study? Clearly describe why.

Chapter 1 QUIZ

TRUE OR FALSE

1. True or false: Statistical inference consists of methods for estimating and drawing conclusions about sample characteristics based on the information contained in the population.
2. True or false: A parameter is a characteristic of a sample.

FILL IN THE BLANK

3. Statistics is the art and science of _____, analyzing, presenting, and interpreting data.
4. An _____ is the set of values of all variables for a given element.
5. A statistic is a characteristic of a _____.

SHORT ANSWER

6. Is a sample survey examining the effects of secondhand smoke an example of an experimental study or an observational study?
7. State which type of statistical study is involved in the following. A large pharmaceutical company is interested in

whether a new drug will reduce Alzheimer's disease symptoms in elderly patients.
8. For the study in the previous exercise, identify the predictor variable and the response variable.

CALCULATIONS AND INTERPRETATIONS

9. Suppose we are interested in the proportion of left-handed statistics students, and we take a sample to estimate the percent of students in our class who are left-handed.
   a. What is the population?
   b. What is the sample?
   c. What is the variable? Is it quantitative or qualitative?
   d. Is the sample proportion likely to be exactly the same as the population proportion? Is it likely to be very far away from the population proportion? Explain.
10. Describe what is wrong, if anything, with the following survey question. "How often would you say that you attend the movie theater: often, occasionally, sometimes, seldom, or never?"

2

Describing Data Using Graphs and Tables

Introduction

In Chapter 2, we apply the adage "A picture is worth a thousand words." Before a researcher can draw conclusions from a data set, he or she must first organize the data, usually into some sort of a table or graph.

Section 2.1 discusses graphs and tables for categorical data, including frequency distributions, relative frequency distributions, bar graphs, Pareto charts, and pie charts.

Sections 2.2 and 2.3 discuss graphs and tables for quantitative data, including frequency distributions, relative frequency distributions, histograms, frequency polygons, stem-and-leaf displays, dotplots, cumulative frequency distributions, cumulative relative frequency distributions, frequency ogives, relative frequency ogives, and time series graphs.

Section 2.4 covers graphical misrepresentations of data.

From the Author

The new Chapter 2 Case Study, *Criminal Justice in New York City*, provides a theme for the chapter. Throughout Sections 2.1–2.3, we examine a series of data sets provided by the New York Police Department, analyzing with graphs and tables the prevalence of various misdemeanors and felonies across the various precincts and boroughs.

Section 2.1 Graphs and Tables for Categorical Data

- Table 1 represents the top 20 free IOS apps, as reported by Apple.com. Again, the hope is that this will help you capture the student's interest.

- Pie charts are challenging to construct manually. Instructors may wish to leave their construction to technology.

- Note the new topic, Working with Tabular Data.

- The topic of Clustered Bar Graphs may be omitted if desired.

Section 2.2 Graphs and Tables for Quantitative Data

- The music celebrity data in Table 17 is intended to enhance student interest.

- Frequency polygons. You may wish to mention that these graphs are not technically polygons in a geometric sense, but that this terminology is now widespread.

- Note the new topic, Obtaining Information from Graphs and Tables.

- Example 21 draws Section 2.2 together. Students should realize that the use of a particular graph depends, in part, on the objective of the data analyst.

Section 2.3 Further Graphs and Tables for Quantitative Data

- Table 36 is helpful for learning the meaning of *cumulative frequency*.

- Figure 49 shows two types of behavior: yearly seasonal cycle and overall increase in carbon dioxide over time.

Section 2.4 Graphical Misrepresentations of Data

- Section 2.4 is important for *statistical literacy*, which is the ability to understand the use of graphs, tables, and statistics in the everyday media. One resource in this area is www.statlit.org.

Teaching Tips

Starting Example: It will be helpful for students to see all the graphs and tables for categorical data using one small data set to illustrate construction of the plots by hand. Examples can be collected from students by asking for volunteer information, such as year in school or major, or small data sets may be easily obtained from the Web.

It is also useful for students to see all the graphs and displays for quantitative data using one data set. Once again, the data set should be small and can be collected from students or obtained online. Quiz or exam scores or how long it took them to find a parking space are possible examples.

Using the same data set for both the tables and the graphs gives students an opportunity to comment on the different information that can be obtained from each type of display.

In-Class Activities

1. Ask students to collect data from the class about the number of children in their families and in the family of five other married relatives (uncles, aunts, and so on) or people they know (neighbors, parents of friends, and so on) with children. They can use technology (such as Minitab or TI-83/84) to construct a histogram with five to seven classes. From the information displayed in the histogram, ask students to construct a table showing frequency, cumulative frequency, relative frequency, and cumulative relative frequency distribution. Then have them use the cumulative frequencies and cumulative relative frequencies to construct ogives. They can compare to see if differences can be found.

2. A possible activity for constructing crosstabulations and/or clustered bar graphs would be to determine whether gender and major are related by collecting information on those variables from the class.

3. Construct a frequency distribution and a relative frequency distribution of the genders of the students in your classroom.

4. Where do you think your favorite sport would fall in the distribution of favorite sports? Construct a frequency distribution and a relative frequency distribution of the favorite sports of the students in your classroom.

5. Where do you think your height would fall in a stem-and-leaf display of the heights of students in your class? Construct such a stem-and-leaf-display, using 1s as leaf units. Did you fall about where you thought you would in this display? How would you describe the shape of the distribution?

↗ For the following activities, you may randomize using (a) technology, (b) the *Simple Random Sample* applet, or (c) two dice and a 6 × 6 table of students' names.

6. Generate a random sample of 10 students, recording the gender for each student.

  a. Is the distribution of genders the same as for the entire class (viewed as a population)?

  b. If we generated another random sample of 10 students, would the distribution of genders be the same? Why or why not? Would you expect it to be roughly the same or wildly different? How could we make the sample distribution more similar to the population distribution?

7. Generate a random sample of 10 students, recording the height for each student. Construct a dotplot or histogram of the heights.

  a. Is the shape similar to the distribution for the entire class?

  b. If we generated another sample of 10 students, would the shape of the dotplot or histogram be the same as previously? Why or why not? Would you expect it to be roughly the same or wildly different? How could we make the sample distribution more similar to the population distribution?

Supplements

See the descriptions in the Preface for more information about these resources.

- StatTutor:
  - Categorical variables: pie charts
  - Quantitative variables: histograms
  - Interpreting histograms
  - Quantitative variables: stemplots
  - Marginal distributions
  - Time series plots

- Stats@Work Simulation
  - Picture This, Which Graph Is Most Appropriate?; Jan Pepperoni

- EESEE case studies for displaying distributions with graphs
  - Acorn Size and Oak Tree Page (Question 7 on boxplots; to be revisited in Chapter 3)
  - Billionaires in 1992 (covers variables, histograms, skewness)
  - Brain Size and Intelligence (Question 2 on histograms)
  - Influenza Outbreak of 1918 (Question 1 on time series)
  - The State of SAT (Question 5 on histograms)
  - Historical Farm Data (questions on pie charts and time series)
  - How Many Poets Wrote Cleanness? (questions cover frequency distributions, pie charts, and time series)
  - Surviving the *Titanic* (Question 3 on the bar graph)
  - Nutrition and Breakfast Cereals (Question 1 on histograms)
  - Habitat of Spotted Owls (Question 1 on the dotplot)

Applets

The *One Variable Statistical Calculator* and *Simple Random Sample* applets are referenced in Chapter 2 to display histograms and stem-and-leaf displays.

The *One Variable Statistical Calculator* applet is used for Exercises 126–131 in Section 2.2 and Exercises 14–16 in Section 2.4.

Activities and applets can be found at http://mathforum.org/mathtools.

Videos

- *Against All Odds: Inside Statistics:* www.learner.org/resources/series65.html
    - Program 1: What Is Statistics?
    - Program 2: Picturing Distributions

Web Sites

- The Consortium for the Advancement of Undergraduate Statistics Education (CAUSE) provides a rich assortment of resources at their CAUSEweb site: https://www.causeweb.org/resources/.

- The following Web site has a collection of class projects, at least one of which deals with the topics of this chapter: www.amstat.org/publications/jse/v6n3/smith.html.

- This Texas Instruments Web site has a host of TI-83/84 statistics activities: http://education.ti.com/educationportal/sites/US/nonProductSingle/activitybook_83_statistics.html.

2 Describing Data Using Graphs and Tables

OVERVIEW

Jean-nicolas Nault/iStock/Getty Images

Criminal Justice in New York City

CASE STUDY Is it safer to walk the streets in New York City than it was, say, in the 1990s? In the Chapter 2 Case Study, *Criminal Justice in New York City*, we examine whether rates of crime have fallen in the past couple of decades, along with many other questions.

- In Section 2.1, we examine a bar chart of the various misdemeanor offenses that occurred throughout all police precincts of the city, and ask which category of misdemeanors is the most common. We also explore, in the *Your Turn* examples, tables and graphics regarding a random sample of traffic violations in Manhattan and Brooklyn.

- In the Section 2.1 exercises, we compare bar graphs for the number of petit ("petty") larcenies that took place in 2000 and 2013, and ask whether the results represent good news.

- In Section 2.2, we construct a stem-and-leaf display for the number of misdemeanor dangerous weapons cases in 20 Manhattan precincts. We also examine a comparison dotplot of third-degree assault and criminal trespass cases across all the police precincts of New York City. Then, we note that a histogram of the criminal trespass data represents a right-skewed distribution.

- In the Section 2.2 exercises, we compare the frauds taking place in Brooklyn in 2000 and 2013, using distributions, histograms, and a comparison dotplot, and ask whether the difference in the graphs represents welcome news. We also construct a frequency polygon, a stem-and-leaf display, and a dotplot of the 2013 Brooklyn fraud data. Then, we compare the number of petit larceny cases city-wide for 2000 and 2013, using the previously named graphs. At the end of the section we ask which graph is preferable, depending on the objective of the analysis.

- In Section 2.3, we construct a time series plot of the murder rate in New York City from 1990 to 2014, and we find some truly good news in the results.

- Finally, in the Section 2.3 exercises, we construct time series plots of the number of third-degree assaults city-wide and describe the patterns we see.

THE BIG PICTURE

Where we are coming from and where we are headed . . .

- In Chapter 1, we learned the basic concepts of statistics, such as population, sample, and types of variables, along with methods of collecting data.

- Here, in Chapter 2, we learn about graphs and tables for summarizing qualitative data and quantitative data, and we examine how to prevent our graphics from becoming misleading.

- Later, in Chapter 3, we will learn how to describe a data set using numerical measures instead of graphs and tables.

2.1 Graphs and Tables for Categorical Data

OBJECTIVES By the end of this section, I will be able to . . .

1 Construct and interpret a frequency distribution and a relative frequency distribution for qualitative data.
2 Build and interpret bar graphs and Pareto charts.
3 Construct and interpret pie charts.
4 Build crosstabulations to describe the relationship between two variables.
5 Work with tabular data to construct graphs and distributions.
6 Construct a clustered bar graph to describe the relationship between two variables.

In Chapter 2, we apply the adage "A picture is worth a thousand words." The human mind can assess information presented in a graph or table better than it can through words and numbers alone. Psychologists sometimes call this innate ability *pattern recognition*. Statistical graphs and tables take advantage of this ability to quickly summarize data.

1 Frequency Distributions and Relative Frequency Distributions

Recall from Chapter 1 that categorical (qualitative) data take values that are nonnumeric and are usually classified into categories. In this section, we learn graphical and tabular methods for handling categorical data. Let us begin with an example.

Table 1 shows the 20 most downloaded free apps for the IOS platform, as reported by Apple.com, along with the app type, for June 2014. We will analyze the variable *app type*, which is a qualitative, not quantitative, variable.

Table 1 Top 20 free IOS apps, June 2014, as reported by Apple.com

Rank	App	App type	Rank	App	App type
1	Two Dots	Games	11	Facebook	Social networking
2	The Line	Games	12	NBC Sports Live	Sports
3	Traffic Racer	Games	13	Twitter	Social networking
4	Rival Knights	Games	14	FIFA Official App	Sports
5	Piano Tiles	Games	15	Pandora	Music
6	Snap Chat	Photo and video	16	Spotify	Music
7	Instagram	Photo and video	17	Pinterest	Social networking
8	The Test	Games	18	Emoji Keyboard 2	Social networking
9	Republique	Games	19	WhatsApp	Social networking
10	YouTube	Photo and video	20	SoundCloud	Music

From this data set, it is not immediately clear which app type is the most popular choice among the 20 apps in the sample. That is why we need ways to summarize the values in a data set. One popular method used to summarize the values in a data set is the **frequency distribution** (or *frequency table*).

> The **frequency**, or **count**, of a category refers to the number of observations in each category. A **frequency distribution** for a qualitative variable is a listing of all the values (for example, categories) that the variable can take, together with the frequencies for each value.

EXAMPLE 1   Frequency distributions

Create a frequency distribution for the variable *app type* from Table 1.

Solution

For each app type, we compute the **frequency**; that is, we **count** (or tally) how many apps were of that particular app type. Table 2 shows the frequency distribution for the variable *app type*. For example, five of the apps were social networking apps. The frequency distribution summarizes the data set so that quick observations can be made, such as "The most popular app type in the Apple.com top 20 list of the most downloaded free apps is the *Games* app type."

Roberto Westbrook/Blend Images/Getty Images

Note: Check that the sum of the frequencies equals the sample size, *n*.

Table 2 Frequency distribution of app type

App type	Tally	Frequency
Games	ⵑⵑ II	7
Social networking	ⵑⵑ	5
Music	III	3
Photo and video	III	3
Sports	II	2

YOUR TURN
#1

The New York City Police Department tracks the number and type of traffic violations. Table 3 contains a random sample of 12 traffic violations and the borough in which they occurred (Manhattan or Brooklyn).

1. Build a frequency distribution of *Borough*.
2. Construct a frequency distribution of *Violation type*.

Table 3 Violation type and borough of 12 traffic violations

Violation type	Borough	Violation type	Borough
Cell phone	Brooklyn	Disobey sign	Manhattan
Safety belt	Manhattan	Speeding	Brooklyn
Cell phone	Brooklyn	Safety belt	Manhattan
Cell phone	Manhattan	Disobey sign	Manhattan
Speeding	Brooklyn	Disobey sign	Brooklyn
Safety belt	Manhattan	Cell phone	Manhattan

(The solutions are shown in Appendix A.)

As the data set gets larger, the need for summarization gets more and more acute. (Imagine if the Apple.com listing consisted of 1000 apps instead of 20.) Take a moment to add up the frequencies in Table 2. What do they add up to? This number is the sample size: *n* = 20. Now, is this just a coincidence, or does this happen every time?

Actually, this happens every time: the sum of the frequencies equals the sample size, *n*. One way to check if you made a mistake in forming your frequency distribution table is to add up the frequencies and see if the sum equals the sample size.

Relative Frequency Distributions

Next, suppose you didn't know the size of the sample in the survey. Suppose you were told only that seven apps were games. The logical question is "Is that a lot?" If our sample size was only 10 apps, then 7 of those apps being games is certainly a lot. However, if our sample size was 1000 apps, then only 7 of those apps being games is *not* a lot. So, the number's significance depends on what you compare the seven apps to—that is, "relative to what?" or "compared to what?" In statistics, we compare the frequency of a category with the total sample size to get the **relative frequency**.

> The **relative frequency** of a particular category of a qualitative variable is its frequency divided by the sample size. A **relative frequency distribution** for a qualitative variable is a listing of all values that the variable can take, together with the relative frequencies for each value.

EXAMPLE 2 Relative frequency distributions

Create a **relative frequency distribution** for the variable *app type* using Table 2.

Solution

The relative frequency of the games app type is the frequency 7 divided by the sample size 20:

$$\text{relative frequency of games} = \frac{\text{frequency}}{\text{sample size}} = \frac{7}{20} = 0.35$$

The relative frequency of games apps is 0.35, or 35%. So, if someone told you that 35% of the apps were games, without telling you the sample size, you would have a better idea of the relative popularity of that app type. To construct the relative frequency distribution in Table 4, divide each frequency in the frequency distribution in Table 2 by the sample size 20.

Note: The relative frequencies always add up to 1.00, which represents 100%.

Table 4 Relative frequency distribution of app type

App type	Relative frequency
Games	7/20 = 0.35
Social networking	5/20 = 0.25
Music	3/20 = 0.15
Photo and video	3/20 = 0.15
Sports	2/20 = 0.10

NOW YOU CAN DO
Exercises 11–14 and 23–28.

YOUR TURN
#2

Use Table 3 to construct relative frequency distributions for the following categorical variables.

1. *Borough*

2. *Violation type*

(The solutions are shown in Appendix A.)

2 Bar Graphs and Pareto Charts

Frequency distributions and relative frequency distributions are tabular and thus useful for summarizing data sets. The graphical equivalent of a frequency distribution or a relative frequency distribution is called a **bar graph** (or **bar chart**).

> A **bar graph** is used to represent the frequencies or relative frequencies for categorical data. It is constructed as follows:
>
> **1.** On the horizontal axis, provide a label for each category.
> **2.** Draw rectangles (bars) of equal width for each category. The height of each rectangle represents the frequency or relative frequency for that category. Ensure that the bars are not touching each other.

EXAMPLE 3 Constructing bar graphs

Construct a frequency bar graph and a relative frequency bar graph for the distributions of app type in Tables 2 and 4.

Solution

The bar graphs are provided in Figures 1a and 1b. Across the horizontal axis are the five app type categories. Next, draw rectangles, the heights of which represent either the frequency or the relative frequency for that category represented on the vertical axis. For example, in Figure 1a, the first rectangle (Games) reaches a height of 7, while the second rectangle reaches only to 5. Note that the rectangles are of equal width, and none of them touch each other. Also notice that the two bar graphs are exactly alike, except for the scale indicated on the vertical axis. This is because we divide each frequency by the same number, the sample size, to get the relative frequency.

NOW YOU CAN DO
Exercises 15–18 and 29–34.

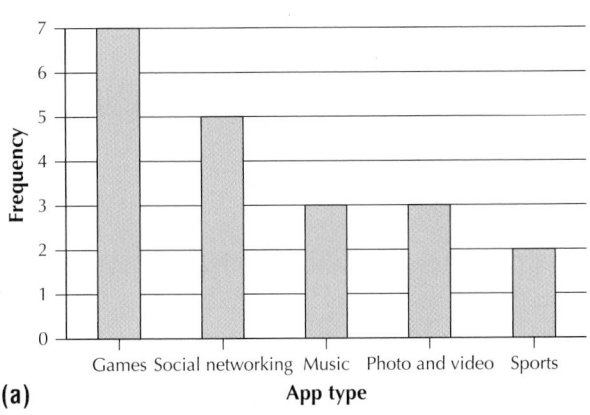
(a)

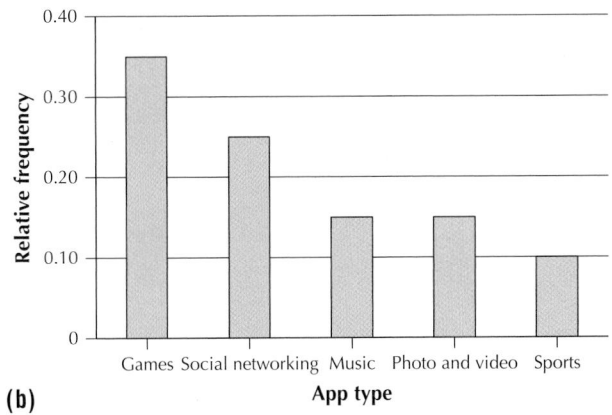
(b)

FIGURE 1 (a) Frequency bar graph; (b) relative frequency bar graph.

YOUR TURN #3

Use Table 3 to construct the following graphs.

1. Frequency bar graph for *Borough*
2. Relative frequency bar graph for *Borough*
3. Frequency bar graph for *Violation type*
4. Relative frequency bar graph for *Violation type*

(The solutions are shown in Appendix A.)

The bars in a bar graph may be presented horizontally, especially when the category names are long. Figure 2 contains a horizontal bar chart of the top celebrities with the most Twitter followers, as of June 8, 2014.

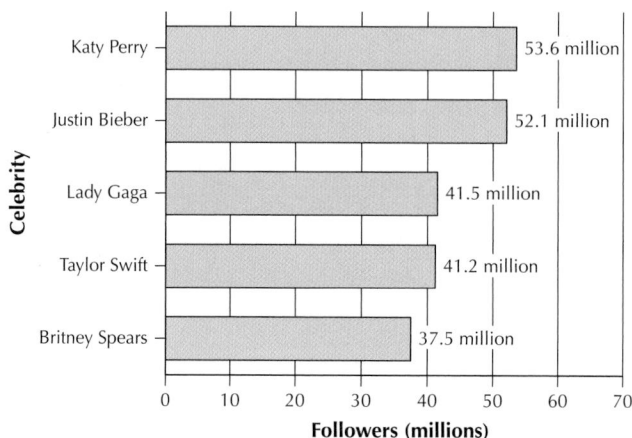

FIGURE 2 Horizontal bar chart of top five celebrities with the most Twitter followers, June 8, 2014. (*Source:* http://friendorfollow.com/twitter/most-followers/)

> A **Pareto chart** is a bar graph in which the rectangles are presented in decreasing order from left to right.

Figures 1a and 1b are both examples of Pareto charts. Had the bars for *Games* and *Social networking* switched places, then those figures would no longer have been Pareto charts because they would no longer have been in decreasing order.

3 Pie Charts

Pie charts are a common graphical device for displaying the relative frequencies of a categorical variable.

> A **pie chart** is a circle divided into sections (that is, slices or wedges), with each section representing a particular category. The size of the section is proportional to the relative frequency of the category.

Pie charts are typically made using technology. However, one can construct a pie chart using a protractor and a compass. A circle contains 360 degrees; therefore, we need to multiply the relative frequency for each category by 360°. This will tell us how large a slice to make for each category, in terms of degrees.

EXAMPLE 4 Constructing a pie chart

Construct a pie chart for the app type data from Example 2.

Solution

The relative frequencies from Example 2 are shown in Table 5. We multiply each relative frequency by 360° to get the number of degrees for that section (slice) of the pie chart.

Table 5 Finding the number of degrees for each slice of the pie chart

Variable: app type	Relative frequency	Multiply by 360°	Degrees for that section
Games	7/20 = 0.35	0.35 × 360° =	126°
Social networking	5/20 = 0.25	0.25 × 360° =	90°
Music	3/20 = 0.15	0.15 × 360° =	54°
Photo and video	3/20 = 0.15	0.15 × 360° =	54°
Sports	2/20 = 0.10	0.10 × 360° =	36°
Total	**20/20 = 1.00**		**360°**

Our pie chart will have five slices—one for each app type category. Use the compass to draw a circle. Then, use the protractor to construct the appropriate angles for each section. From the center of the circle, draw a line to the top of the circle. Measure your first angle using this line. For the Games app type, we need an angle of 126°. This angle is shown in Figure 3. Then, from there, measure your second angle—in this case, the 90° right angle for Social networking apps. Continue until your circle is complete.

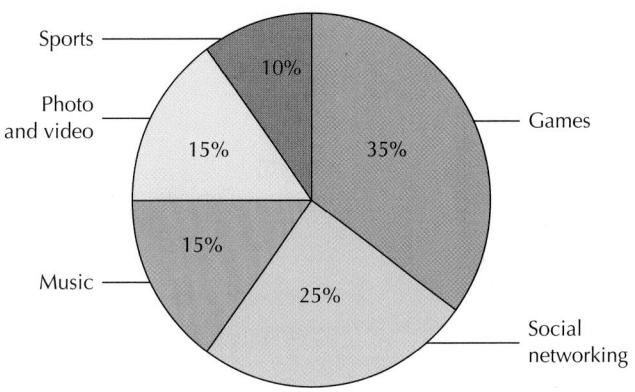

FIGURE 3 Pie chart of app type data.

NOW YOU CAN DO
Exercises 19, 20, 35, and 36.

4 Crosstabulations

So far, we have analyzed only one variable at a time. **Crosstabulation** is a tabular method for simultaneously summarizing the data for two categorical (qualitative) variables.

Steps for Constructing a Crosstabulation

Step 1 Put the categories of one variable at the top of each column and the categories of the other variable at the beginning of each row.

Step 2 For each row and column combination, enter the number of observations that fall in the two categories.

Step 3 The bottom of the table gives the column totals, and the right-hand column gives the row totals.

Crosstabulations are also known as **two-way tables** or **contingency tables**. We will introduce crosstabulations using an example.

EXAMPLE 5 Constructing a crosstabulation

Table 6 contains information about the size (compact, midsize, or large) and the recommended gasoline (regular or premium) for a sample of ten 2014 automobiles.

a. Construct a crosstabulation of the variables *size* and *gasoline*.

b. Identify any patterns.

Table 6 Size and recommended gasoline for ten 2014 automobiles

Car	Car size	Recommended gasoline
BMW 328i	Compact	Premium
Chevrolet Camaro	Compact	Regular
Honda Accord	Compact	Regular
Cadillac CTS	Midsize	Premium
Nissan Sentra	Midsize	Regular
Subaru Legacy AWD	Midsize	Premium
Toyota Camry	Midsize	Regular
Ford Taurus	Large	Regular
Hyundai Genesis	Large	Premium
Rolls-Royce	Large	Premium

Source: www.fueleconomy.gov.

Solution

a. Step 1 We use the values of the two variables to create the crosstabulation given in Table 7. Note that the categories for the variable *gasoline* are shown at the top, whereas the categories for the variable *size* are shown on the left. Each car in the sample is associated with a certain *cell* in the crosstabulation, in the appropriate row and column. For example, the Chevrolet Camaro is one of the two cars that appear in the "Compact" car size row and the "Regular" gasoline column.

Step 2 For each row and column combination in the crosstabulation, enter the number of observations that fall in the two categories.

Step 3 The "Total" column contains the sum of the counts of the cells in each row (category) of the *size* variable and represents the frequency distribution for this variable. Similarly, the "Total" row along the bottom sums the counts of the cells in each column (category) of the *gasoline* variable and represents the frequency distribution for this variable. In the lower right-hand corner we have the grand total, which should equal the sample size.

 carsizegas

Table 7 Crosstabulation of car size and recommended gasoline

Car size	Recommended gasoline		Total
	Regular	Premium	
Compact	2	1	3
Midsize	2	2	4
Large	1	2	3
Total	5	5	10

b. We can use the crosstabulation to look for patterns in the data set. One possible pattern is the following: Compact cars tend to use regular gasoline, whereas large cars tend to use premium gasoline. Of course, this sample size is too small to form any conclusions about such a relationship.

NOW YOU CAN DO
Exercises 21, 37, 41, and 45.

 Use Table 3 to construct a crosstabulation for *Borough* and *Violation type*.

(The solution is shown in Appendix A.)

5 Working with Tabular Data

In earlier examples, we worked with raw data, such as the automobiles in Table 6, and developed graphs and distributions using the raw data. However, data often comes to us already summarized in a table. Here we show how to use tabular data to construct graphs and distributions.

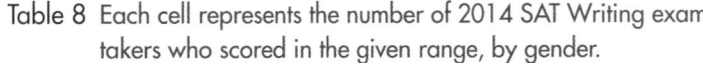
EXAMPLE 6 Working with tabular data

Commercial Eye/Iconica/Getty Images

Table 8 is a crosstabulation showing the SAT Writing exam score for females and males in 2014. Note that the data are not presented in raw form, such as the automobiles in Table 6. Instead, the *frequencies* for each cell have been presented. Use Table 8 to construct the following:

a. Frequency distribution of the Gender of SAT Writing exam takers, for females and males

b. Frequency bar graph of the Gender of SAT Writing exam takers, for females and males

Table 8 Each cell represents the number of 2014 SAT Writing exam takers who scored in the given range, by gender.

| | SAT Writing score | | | |
Gender	200–390	400–590	600–800	Total
Female	169,314	550,172	164,469	883,955
Male	178,606	464,949	132,537	776,092
Total	347,920	1,015,121	297,006	1,660,047

Source: The College Board.

Solution

a. The total column on the right-hand side of Table 8 provides us with the frequency distribution, which is shown in Table 9.

Table 9 Frequency distribution of SAT Writing exam takers across the genders

Gender	Frequency
Female	883,955
Male	776,092
Total	1,660,047

b. We use the frequency distribution in Table 9 to produce the bar graph shown in Figure 4.

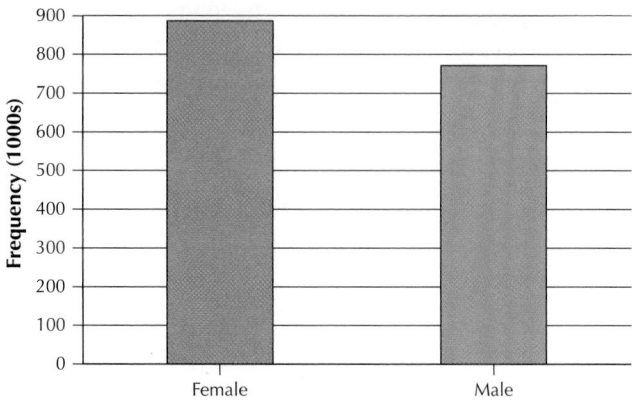

FIGURE 4 Bar graph of Gender of SAT Writing exam takers.

NOW YOU CAN DO
Exercises 22, 38–40,
42–44, and 46–56.

We work with the tabular data in Table 8 to construct further graphs and distributions in the exercises.

**YOUR TURN
#5**

Use Table 8 to construct the following:

a. Frequency distribution of the Score of SAT Writing exam takers, for the three score categories

b. Frequency bar graph of the Score of SAT Writing exam takers, for the three score categories

(The solutions are shown in Appendix A.)

6 Clustered Bar Graphs

Clustered bar graphs are useful for comparing two categorical variables and are often used in conjunction with crosstabulations. Each set of bars in a clustered bar graph represents a single category of one variable across all the categories of the other categorical variable (see Figure 5). This allows the analyst to make comparisons easily. One can construct clustered bar graphs using either frequencies or relative frequencies. To construct a clustered bar graph, identify which of the two categorical variables will define the cluster of bars. Then, for each category of the other variable, draw bars for each category of the clustering variable.

EXAMPLE 7   Clustered bar graphs

Use the tabular data in Table 8 to construct a clustered bar chart of SAT Writing exam scores clustered by gender.

Solution

Gender is given as the clustering variable. Thus, for each category of the variable *SAT Writing exam score*, we will draw two bars: one representing females and the other representing males. For the first category, Low = SAT score between 200 and 390, we draw the female rectangle going up to 169.314 (because Figure 5 is given in 1000s), and we draw the male rectangle going up to 178.606. These two rectangles should touch each other but should not touch any other rectangles. Continue to draw two rectangles for each SAT score category: one for each of the females' and males' frequencies. The resulting clustered bar graph is shown here as Figure 5. We say that the Writing exam scores are *clustered* by gender. (*Note:* Medium = SAT score between 400 and 590; High = SAT score between 600 and 800.)

FIGURE 5
Clustered bar graph of SAT Writing score, clustered by gender.

NOW YOU CAN DO
Exercises 57–60.

YOUR TURN #6

Use Table 3 (page 41) to construct the following graphs:

1. Clustered bar graph of *Violation type* clustered by *Borough*
2. Clustered bar graph of *Borough* clustered by *Violation type*

(The solutions are shown in Appendix A.)

EXAMPLE 8   Comparison bar charts

Student-Run Café Business

Have you ever thought what it must be like to run a business? Business students at a Midwestern public university found out because they volunteered to manage a student-run café in the business school, which replaced a for-profit vendor that closed. In this chapter, we examine sales data from the student-run café, such as the number of coffees sold and the number of sodas sold, to help us learn about how to describe data using graphs and tables. The data set is called **Café**, a brief excerpt of which is shown in Table 10.

Table 10 Excerpt of café data

Month	Time period	Day of week	...	Sodas	Coffees	Sales	Max. daily temperature (°F)
Jan	1	2 - Tue	...	20	41	199.95	36
Jan	1	3 - Wed	...	13	33	195.74	34
Jan	1	4 - Thu	...	23	34	102.68	39
Jan	1	5 - Fri	...	13	27	162.88	40
Jan	1	1 - Mon	...	13	20	101.76	36
Jan	1	2 - Tue	...	33	23	186.94	26
Jan	1	3 - Wed	...	15	32	120.18	34
Jan	1	4 - Thu	...	27	31	228.78	33
Jan	1	5 - Fri	...	12	30	88.02	20
Feb	1	1 - Mon	...	19	27	119.57	37
⋮	⋮	⋮	⋮	⋮	⋮	⋮	⋮

Source: Journal of Statistics Education.[1]

There are 47 days of sales data. The variable *time period* is a categorical variable that equals 1 for the first 16 days (late January to early February), 2 for the second 16 days (late February to early March), and 3 for the last 15 days (late March to early April). This variable allows us to examine changes in sales behavior over time. Construct a *comparison bar chart* of the number of sodas sold and the number of coffees sold, over the three time periods, and comment on the change in customer purchasing behavior as the winter months turned to spring.

Solution

Figure 6 shows the comparison bar chart of the number of sodas sold and the number of coffees sold, over the three time periods. As winter turned to spring, soda sales increased from 332 to 576, while coffee sales decreased from 445 to 131.

FIGURE 6
Comparison bar chart shows soda sales increased, while coffee sales decreased, as winter turned to spring.

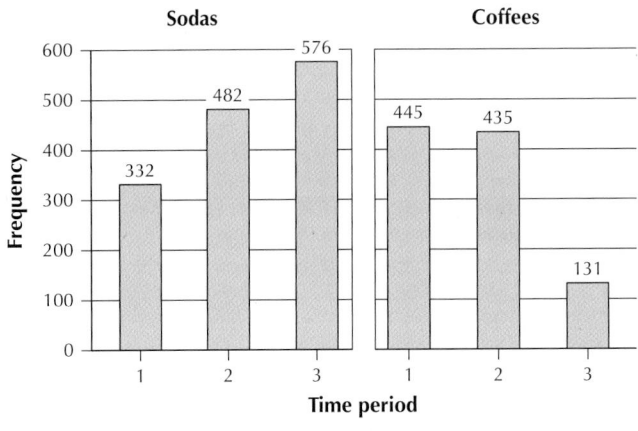

STEP-BY-STEP TECHNOLOGY GUIDE: Frequency Distributions, Bar Graphs, and Pie Charts

If you enjoy statistics, you may want to become a statistician! Demand is high and you get to work with lots of interesting data. A job search conducted on the job site Monster.com on August 21, 2014, using the keyword *statistics*, and restricting the search to *bachelor's degree* and *posted in the past week*, resulted in 20 hits. Table 11 contains the section of the country where the available jobs were located. We use the data set in Table 11 to demonstrate how to use technology to construct a frequency distribution, relative frequency distribution, bar graph, and pie chart.

countrysections   **TABLE 11   Section of the country for statistics jobs**

Northeast	South	Midwest	Midwest	West
Midwest	Northeast	Northeast	West	Northeast
Midwest	West	West	West	South
South	Northeast	Midwest	Northeast	Northeast

EXCEL

Frequency Distributions

Step 1   Enter the data in column A, with the topmost cell indicating the variable name, *Section*.

Step 2   Select cells A1–A21, click **Insert > PivotTable**, and click OK.

Step 3   Under **Choose fields to add to report**, select *Section*.

Step 4   Click on *Section* and drag to the **Values** box at the lower right of the screen. The resulting frequency distribution is shown in Figure 7. In Excel, this takes the form of a *pivot table*, which is an interactive tabular format.

Bar Graphs and Pie Charts

Note: Excel can make bar graphs or pie charts using frequency distributions but not from the raw data.

Step 1   Enter the frequency distribution as shown in Figure 8.

Step 2   Select cells A1 to B5. For a bar graph, click **Insert > Insert ColumnChart > Clustered Column**. For a pie chart, click **Insert > Insert Pie or Doughnut Chart > Pie**.

Step 3   The resulting frequency bar graph and pie chart are shown in Figures 9 and 10.

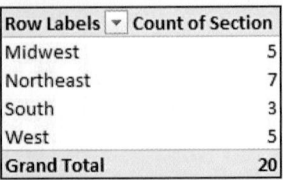

FIGURE 7 Excel pivot table

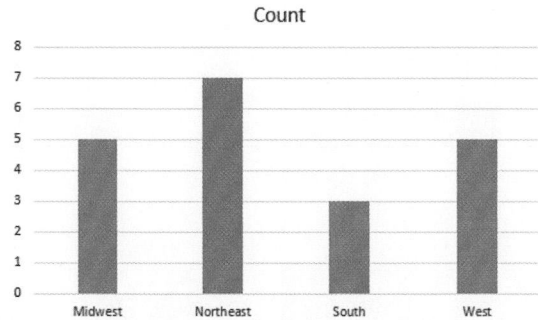

FIGURE 8 Excel frequency distribution.

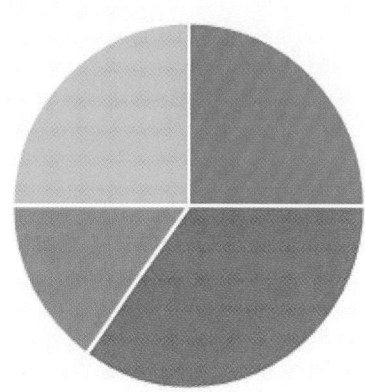

FIGURE 10 Excel pie chart.

Count

FIGURE 9 Excel frequency bar graph

Crosstabulation of Car Size Data

Step 1   Enter the data from Table 6 (page 46) into two columns, with the topmost cell indicating the variable names *Size* and *Gasoline*.

Step 2   Highlight cells **A1** to **B11**, click **Insert > Pivot Table**, and click **OK**.

Step 3   Under **Pivot Table Fields**, click and drag *Size* to **Rows**, *Gasoline* to **Columns**, and *Gasoline* to **Values**.

Clustered Bar Graph for Car Size Data

Step 1   Enter the data from Table 6 (page 46) into two columns, with the topmost cell indicating the variable names *Size* and *Gasoline*.

Step 2   Obtain the crosstabulation of the data. Select the three rows and two columns in the crosstabulation.

Step 3   Click **Insert > Insert Column Chart** (in the **Chart** section). Click **Clustered Column**, which is the first option.

MINITAB

Frequency Distributions

Step 1   Name your variable *Section* and enter the data into the C1 column.

Step 2   Click **Stat > Tables > Tally Individual Variables**.

Step 3   Under **Display**, select **Counts** and **Percents**.

Step 4   Click inside the **Variables** box. Select the variable C1 *Section*, and click **Select**. Then click **OK**.

Bar Graphs

Step 1   Name your variable *Section* and enter the data into the C1 column.

Step 2   Click **Graph > Bar Chart**. For raw data select **Bars Represent: Counts of unique values**, select **Simple**, and click **OK**. (For summarized data such as a frequency distribution, select **Bars Represent: Values from a table**, and select **Simple**. Then click **OK**.)

Step 3   Click inside the **Categorical variables** box, click on the *Section* variable, and click **Select**. Then click **OK**.

Pie Charts

Step 1   Name your variable *Section* and enter the data into the C1 column.

Step 2   Click **Graph > Pie Chart**. For raw data, select **Chart counts of unique values**. Then click in the **Categorical variables** box, select the variable *Section*, click **Select**, and click **OK**. (For summarized data such as a frequency distribution, select **Chart** values from a table. Then select the category variable for **Categorical variable**, and select the variable with the frequencies or relative frequencies for the **Summary variable**. Then click **OK**.)

Crosstabulation of Car Size Data

Step 1   Enter the data from Table 6 (page 46) into two columns, named *Size* and *Gasoline*.

Step 2   Click **Stat > Tables > Crosstabulation and Chi-Square…**

Step 3   Select **Raw data (categorical variables)**. For **Rows**, select *Size*; for **Columns**, select *Gasoline*. Select **Counts** under **Display**. Then click **OK**.

Step 4   The resulting crosstabulation is shown in Figure 11. The rows and columns are in alphabetical order.

```
Rows: Size     Columns: Gasoline

            Premium   Regular   Missing   All

Compact        1         2         0        3
Large          2         1         0        3
Midsize        2         2         0        4
Missing        0         0        10        *
All            5         5         *       10
```

FIGURE 11 Minitab crosstabulation.

Clustered Bar Graphs

If you have the original data set:
Step 1   Click **Graph > Bar Chart**.
Step 2   Select **Bars Represent: Counts of unique values**, and select **Cluster**. Then click **OK**.
Step 3   Select your two categorical variables, and click **OK**.

If you have only the crosstabulation and not the original data:
Step 1   Click **Graph > Bar Chart**.
Step 2   Select **Bars Represent: Values from a Table**, and select **Cluster** under **Two-way table**. Then click **OK**.
Step 3   For **Graph variables**, select the variables that contain the frequencies or relative frequencies. For **Row labels**, select the variable that holds the row names of your crosstabulation. Then click **OK**.

SPSS

Frequency Distributions

Step 1   Enter the data from Table 11 into the first column. Rename the column *Section*.
Step 2   Click **Analyze > Descriptive Statistics > Frequencies...**
Step 3   Select *Section*, click the arrow to move it to **Variable(s):**, then click **OK**.

Bar Graphs and Pie Charts

Step 1   Enter the data from Table 11 into the first column. Rename the column *Section*.
Step 2   Click **Analyze > Descriptive Statistics > Frequencies...**
Step 3   Select *Section*, click the arrow to move it to **Variable(s):**, then click **OK**.
Step 4   Click **Charts...**, and select **Bar charts** (or **Pie Charts**) under **Chart Type**. Choose **Frequencies** (or **Percentages**) under **Chart Values**. Click **Continue** and then **OK**.

Crosstabulation and Clustered Bar Graphs of Car Size data

Step 1   Enter the data from Table 6 (page 46) into two columns. Rename the columns *Size* and *Gasoline*.
Step 2   Click **Analyze > Descriptive Statistics > Crosstabs...**
Step 3   Select the variable for the rows (Size), and click the first arrow to move it to the **Row(s)** box. Select the variable for the columns (Gasoline), and click the second arrow to move it to the **Column(s)** box.

Step 4   Select **Display clustered bar charts**, and click **OK**. The resulting graph is shown in Figure 12.

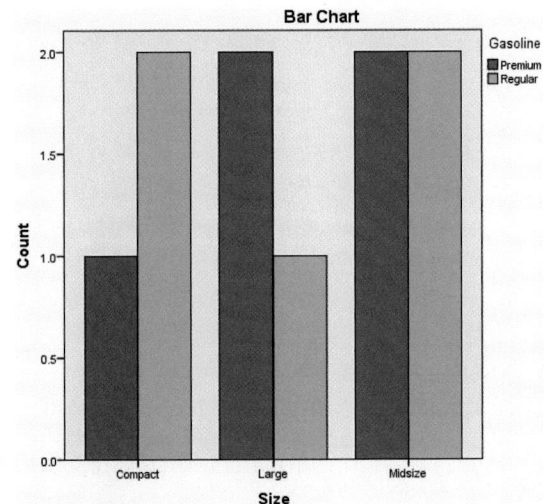

FIGURE 12 SPSS clustered bar chart.

JMP

Frequency Distributions

Step 1   Select **File > New > Data Table**. Enter the data from Table 11 into **Column 1**. Rename **Column 1** *Section*.
Step 2   Click **Analyze > Distribution**.
Step 3   Select *Section* from **Select Columns** and click **Y, Columns**. Click **OK**. The result is shown in Figure 13.

Bar Graphs and Pie Charts

Step 1   Select **File > New > Data Table**. Enter the data from Table 11 into **Column 1**. Rename **Column 1** *Section*.

⊿ **Frequencies**		
Level	**Count**	**Prob**
Midwest	5	0.25000
Northeast	7	0.35000
South	3	0.15000
West	5	0.25000
Total	20	1.00000
N Missing	0	
4 Levels		

FIGURE 13 JMP frequency distribution.

Step 2   Click **Graph > Graph Builder**. Drag *Section*, under **Variables**, to where **Drag variables into drop zones** is shown.
Step 3   For a bar graph, select **Bar** from the series of options above the graph. For a pie chart, select **Pie**. Click **Done**.

Crosstabulation

Step 1   Select **File > New > Data Table**. Enter the data from Table 6 (page 46) into two columns. Name the columns *Size* and *Gasoline*.
Step 2   Click **Analyze > Tabulate**.
Step 3   Drag *Gasoline* from the list on the left to **Drop zone for columns**. A frequency table appears. Drag *Size* from the list to the first column of the third row of the table. The result is the crosstabulation.

Clustered Bar Graphs

Step 1   Select **File > New > Data Table**. Enter the data from Table 6 (page 46) into two columns. Name the columns *Size* and *Gasoline*.
Step 2   Click **Graph > Graph Builder**. Drag both variables from Variables to the *x* axis.
Step 3   Select **Bar** from the graph options shown above the graph. Click **Done**.

CRUNCHIT!

Frequency Distributions
Step 1 Click **File**, highlight **Load from Larose, Discostat3e > Chapter 2**, and click on **Table 01_11**.
Step 2 Click **Statistics**, and select **Frequency Table**. For **Sample**, select *Section*. Then click **Calculate**.

Bar Graphs and Pie Charts
Step 1 Click **File**, highlight **Load from Larose, Discostat3e > Chapter 2**, and click on **Table 01_11**.
Step 2 Click **Graphics** and select **Bar Chart with Raw Data**. For a pie chart, select **Pie Chart with Raw Data**.

Step 3 For **Sample**, select *Section*. Then click **Calculate**.

Crosstabulation of Car Size Data
Step 1 Click **File**, then highlight **Load from Larose, Discostat3e > Chapter 2**, and click on **Example 01_05**.
Step 2 Click **Statistics**, highlight **Contingency Table**, and select get frequencies.
Step 3 For **Row Variable**, select *Car size*, and for **Column variable**, select *Recommended gasoline*. Then click **Calculate**.

Section 2.1 Summary

In this section, we learned about tabular and graphical methods for summarizing qualitative (categorical) data.

1. Frequency distributions and relative frequency distributions list all the values that a qualitative variable can take, along with the frequencies (counts) or relative frequencies (percents) for each value.

2. A bar graph is the graphical equivalent of a frequency distribution or a relative frequency distribution. When the rectangles are presented in decreasing order from left to right, the result is a Pareto chart.

3. Pie charts are a common graphical device for displaying the relative frequencies of a categorical variable. A pie chart is a circle divided into sections (that is, slices or wedges),

with each section representing a particular category. The size of the section is proportional to the relative frequency of the category.

4. Crosstabulation summarizes the relationship between two categorical variables. A crosstabulation is a table that gives the counts for each row-column combination, with totals for the rows and columns.

5. Data often comes to us already summarized in a table. We can use this tabular data to construct graphs and distributions.

6. Clustered bar graphs are useful for comparing two categorical variables and are often used in conjunction with crosstabulations.

Section 2.1 Exercises

CLARIFYING THE CONCEPTS

1. Why do we use graphical and tabular methods to summarize data? What's wrong with simply reporting the raw data? (p. 40)

2. What's the difference between a frequency distribution and a relative frequency distribution? (p. 42)

3. True or false: For a given data set, a frequency bar graph and a relative frequency bar graph look alike, except for the scale on the vertical axis. (p. 43)

4. True or false: A pie chart is used to represent quantitative data. (p. 44)

5. What should be the sum of the frequencies in a frequency distribution? (p. 41)

6. What should be the sum of the relative frequencies in a relative frequency distribution? (p. 42)

7. In a crosstabulation, the "Total" column represents what? How about the "Total" row? (p. 46)

8. What does the number in the lower right corner of the crosstabulation represent? What should this number be equal to? (p. 48)

9. True or false: Each set of bars in a clustered bar graph represents a single category of one variable across all the categories of the other categorical variable. (p. 48)

10. True or false: A clustered bar graph cannot be constructed for relative frequency data. (p. 48)

PRACTICING THE TECHNIQUES

 CHECK IT OUT!

To do	Check out	Topic
Exercises 11–14 and 23–28	Examples 1 and 2	Frequency distributions and relative frequency distributions.
Exercises 15–18 and 29–34	Example 3	Bar graphs
Exercises 19–20 and 35–36	Example 4	Pie charts
Exercises 21, 37, 41, and 45	Example 5	Constructing crosstabulations
Exercises 22, 38–40, 42–44, and 46–56	Example 6	Working with tabular data
Exercises 57–60	Example 7	Clustered bar graphs

Table 12 contains data regarding how all the counties in Nevada voted in the 2012 presidential election (Dem = Democrat,

Rep = Republican), as well as the size of the county in terms of number of voters (Small = at most 10,000; Medium = 10,001 to 100,000; Large = over 100,000). Use Table 12 to construct the indicated distribution or graph for Exercises 11–22.

TABLE 12   2012 Presidential election results for all Nevada counties ᴵᴵᴵ nevadacounties

County	Vote	Size	County	Vote	Size
Carson City	Rep	Medium	Lincoln	Rep	Small
Churchill	Rep	Medium	Lyon	Rep	Medium
Clark	Dem	Large	Mineral	Rep	Small
Douglas	Rep	Medium	Nye	Rep	Medium
Elko	Rep	Medium	Pershing	Rep	Small
Esmerelda	Rep	Small	Storey	Rep	Small
Eureka	Rep	Small	Washoe	Dem	Large
Humboldt	Rep	Small	White Pine	Rep	Small
Lander	Rep	Small			

11. Frequency distribution of the variable *vote*.
12. Relative frequency distribution of the variable *vote*.
13. Frequency distribution of the variable *size*.
14. Relative frequency distribution of the variable *size*.
15. Frequency bar graph of *vote*.
16. Relative frequency bar graph of *vote*.
17. Frequency bar graph of *size*.
18. Relative frequency bar graph of *size*.
19. Pie chart for *vote*.
20. Pie chart for *size*.
21. Crosstabulation of the variables *vote* and *size*.
22. Nevada's electoral votes actually went to the Democratic candidate in 2012. Use the crosstabulation from the previous exercise to explain how this might have happened, even though the Republican candidate carried all but two of the counties.

The sinking of the *Titanic* was one of the greatest disasters in maritime history. The data set **Titanic** contains information on 2202 passengers and crew, including their class (1st class, 2nd class, 3rd class, or crew), their gender, and whether they survived. Table 13 contains a random sample of size 20 from this data set. Use Table 13 for Exercises 23–48. For Exercises 23–37, construct the indicated distribution or graph.

TABLE 13   Class, gender, and survival status of 20 passengers and crew of the *Titanic* ᴵᴵᴵ titanicsample

Class	Gender	Survived	Class	Gender	Survived
3rd	Female	No	1st	Female	Yes
Crew	Male	No	1st	Male	Yes
3rd	Female	Yes	Crew	Male	No
Crew	Male	No	3rd	Male	Yes
Crew	Male	Yes	Crew	Male	No
3rd	Male	Yes	2nd	Male	Yes
Crew	Male	No	3rd	Male	No
Crew	Male	No	Crew	Male	No
2nd	Male	No	2nd	Female	Yes
Crew	Male	Yes	2nd	Male	No

23. Frequency distribution of the variable *gender*.
24. Relative frequency distribution of the variable *gender*.
25. Frequency distribution of the variable *survived*.
26. Relative frequency distribution of the variable *survived*.
27. Frequency distribution of the variable *class*.
28. Relative frequency distribution of the variable *class*.
29. Frequency bar graph of *gender*.
30. Relative frequency bar graph of *gender*.
31. Frequency bar graph of *survived*.
32. Relative frequency bar graph of *survived*.
33. Frequency bar graph of *class*.
34. Relative frequency bar graph of *class*.
35. Pie chart for *survived*.
36. Pie chart for *class*.
37. Crosstabulation of the variables *gender* and *class*.

Use the crosstabulation in Exercise 37 for Exercises 38–41.
38. For which categories of *class* did females outnumber males?
39. For which category of *class* did males most outnumber females? Explain how this might make sense.
40. For which category of *class* did the number of females come closest to the number of males?
41. Build a crosstabulation of the variables *gender* and *survived*.

Use the crosstabulation in Exercise 41 for Exercises 42–45.
42. What is the proportion of females who survived?
43. Find the proportion of males who survived.
44. It turns out that a greater proportion of females survived than males. Would we have been able to uncover this fact using only frequency distributions of the variables *gender* and *survived*, and not using a crosstabulation? Explain.
45. Construct a crosstabulation of the variables *class* and *survived*.

Use the crosstabulation in Exercise 45 for Exercises 46–48.
46. What is the proportion of 1st class passengers who survived?
47. Find the proportion of 2nd class passengers, 3rd class passengers, and crew who survived.
48. Compare the proportions you found in the previous two exercises, and comment on the results.

Table 14 shows the number of injuries in various winter sports to occur to females and males at the 2010 Winter Olympics.[2] Use Table 14 to construct the graph or distribution indicated in Exercises 49–52, for the number of alpine skiing injuries, for female and male athletes.
49. Frequency distribution
50. Relative frequency distribution
51. Pie chart
52. Relative frequency bar graph

Use Table 14 to construct the graph or distribution indicated in Exercises 53–56, for the number of Olympic injuries, for male athletes only.
53. Frequency distribution
54. Relative frequency distribution
55. Frequency bar graph
56. Relative frequency bar graph

TABLE 14   Number of injuries suffered, by sport, at the 2010 Winter Olympics 🏔 winterolympics

	Sport			
Gender	Alpine skiing	Figure skating	Ice hockey	Total
Female	20	12	38	70
Male	21	9	44	74
Total	41	21	82	144

57. Use Table 12 to produce a clustered bar graph of *size*, clustered by *vote*.

Use Table 13 to produce clustered bar graphs for the variables indicated in Exercises 58–60.

58. Clustered bar graph of *gender*, clustered by *survived*
59. Clustered bar graph of *class*, clustered by *survived*
60. Clustered bar graph of *survived*, clustered by *gender*

APPLYING THE CONCEPTS

World Water Usage. See Table 15 for Exercises 61–64. For the indicated variable, construct the following: 🏔 worldwater
   a. Frequency distribution
   b. Relative frequency distribution
   c. Frequency bar graph
   d. Relative frequency bar graph
   e. Pareto chart, using relative frequencies
   f. Pie chart

TABLE 15   World water usage

Country	Continent	Climate	Main use
Iraq	Asia	Arid	Irrigation
United States	North America	Temperate	Industry
Pakistan	Asia	Arid	Irrigation
Canada	North America	Temperate	Industry
Madagascar	Africa	Tropical	Irrigation
North Korea	Asia	Temperate	Not reported
Chile	South America	Arid	Irrigation
Bulgaria	Europe	Temperate	Not reported
Afghanistan	Asia	Arid	Irrigation
Iran	Asia	Arid	Irrigation

61. The variable *continent*
62. The variable *climate*
63. The variable *main use*
64. Explain why it is not appropriate to construct a frequency distribution for *country*.

Use Table 15 for Exercises 65–70.
65. Construct a crosstabulation of the variables *continent* and *climate*.
66. Construct a crosstabulation of the variables *continent* and *main use*.
67. Construct a crosstabulation of the variables *climate* and *main use*.
68. Construct a clustered bar graph of the variable *continent* clustered by *climate*.
69. Construct a clustered bar graph of the variable *main use* clustered by *continent*.

70. Construct a clustered bar graph of the variable *main use* clustered by *climate*.

Video Game Sales. Open the **VideoGameSales** data set that we used for the Chapter 1 Case Study. See Table 3 on page 8 in Chapter 1. Recall that this data set contains the top 30 best-selling video games in the United States for the week of May 17, 2014, along with the game platform, publishing studio, type of game, sales that week, total sales, and how many weeks the game has been on the list. Use the data set to answer Exercises 71–76. 🏔 videogamesales
71. Construct a frequency distribution and a relative frequency distribution for the variable *type*. Which is the most common game type? The least common?
72. Build a frequency bar graph of the variable *platform*. Which is the most popular platform?
73. Make a relative frequency bar graph of the variable *platform*. What percentage of platforms is the *PS4?*
74. Construct a pie chart for the variable *type*.
75. Build a crosstabulation of *type* and *platform*.
76. Refer to the crosstabulation. Which platform has the most *action* games? The most *shooter* games?

Cell Phone Ownership. Figure 14 shows the percentage of cell phone ownership categorized by level of education.[3] Use Figure 14 to answer Exercises 77 and 78.

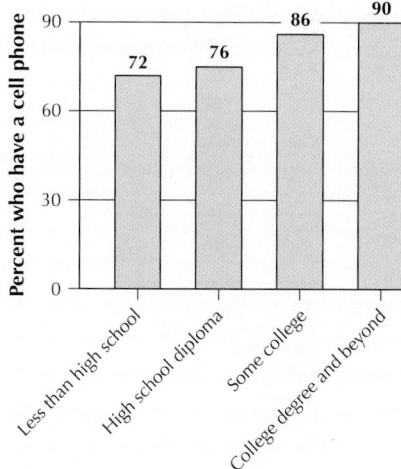

FIGURE 14 Cell phone ownership.

77. Can we use the information in Figure 14 to construct a pie chart? Explain why or why not.
78. Is Figure 14 a Pareto chart? Explain why or why not.

Cell Phones and the Internet. Figure 15 is a pie chart representing the percentage of Americans who access the Internet or email using their cell phones. Use Figure 15 to answer Exercises 79 and 80.
79. According to this survey:
   a. What is the most common response? What percentage does this represent?
   b. What is the least common response? What percentage does this represent?

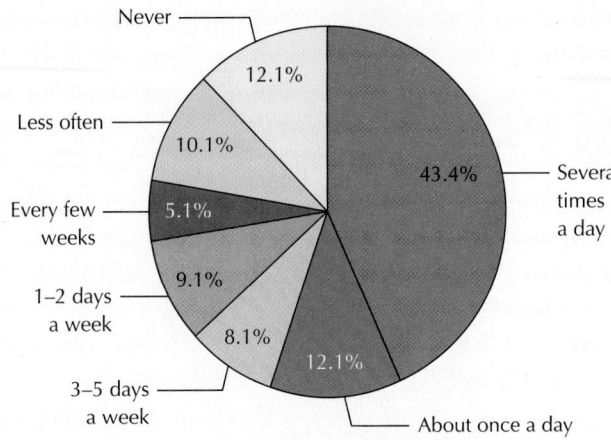

FIGURE 15 Percentage using cell phones for Internet or email.[4]

80. According to this survey:
  a. What percentage uses the cell phone to access the Internet or email about once a day?
  b. What percentage never uses the cell phone to access the Internet or email?
  c. If 1000 adults were surveyed, how many never use their cell phones for email or Internet?

Sledding Injuries. Every year, about 20,000 children and teenagers visit the emergency room with injuries sustained from snow sledding.[5] Use the horizontal bar graph in Figure 16 to answer Exercises 81 and 82.

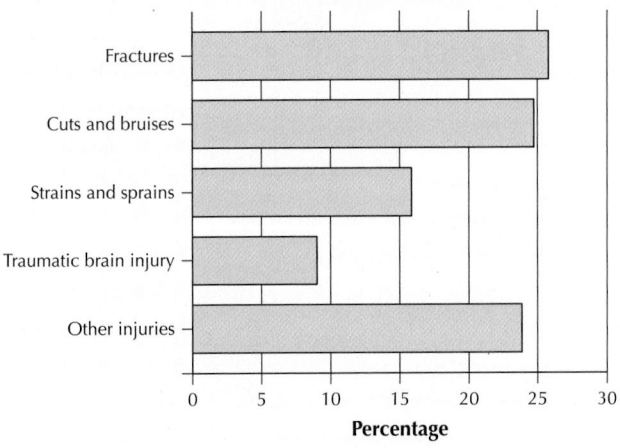

FIGURE 16 Most common injuries from sledding.

81. According to this study:
  a. What is the most common category of injury? Estimate the percentage.
  b. Of the specific injuries shown, what is the least common category of injury? What is the percentage?
  c. Is it possible for there to be an injury type that has a lower percentage than traumatic brain injury? Explain.

82. According to this study:
  a. What is the percentage for cuts and bruises?
  b. What is the percentage for strains and sprains?

83. Table 16 shows the numbers of vehicle models, which are categorized by vehicle type, examined each year by the U.S. Department of Energy to determine vehicle gas mileage. Use Table 16 to construct the following: **🖪 cartypemodel**
  a. Relative frequency distribution
  b. Frequency bar graph
  c. Relative frequency bar graph
  d. Pareto chart, using relative frequencies
  e. Pie chart of the relative frequencies

TABLE 16   Frequency distribution of vehicle type

Vehicle type	Number of models
SUVs	370
Compact cars	128
Midsize cars	120
Subcompact cars	110
Standard pickup trucks	106
Large cars	76
Station wagons	62
Small pickup trucks	59
Other types	151
Total	1182

Musical Activities. Use the following information for Exercises 84 and 85. *USA Weekend* conducted a survey of the nation's teenagers, asking them various questions about lifestyle and music. Nearly 60,000 teenagers responded to the poll, which was conducted in part through *USA Weekend's* Web site. One question was, "Do you listen to music while you are . . . ?" and listed several options. Respondents were asked to select all that apply. The most common responses are shown in the table below.

Doing chores	79%
On the computer	73%
Doing homework	72%
Eating meals at home	33%
In the classroom	18%

84. Do you think that the sample is representative of all U.S. teenagers? How might the sample systematically introduce bias?

85. Can you construct a pie chart of the five activities listed in the table? Explain precisely why or why not.

Music and Violence. Another poll question asked by *USA Weekend* was "Do you think shock rock and gangsta rap are partly to blame for violence such as school shootings or physical abuse?" The results are shown in the table below. Use this information to answer Exercises 86–89.

Yes	31%
No	45%
I've never thought about it	24%

86. Assuming the sample size is 6000, construct a frequency distribution of the responses.

87. Make a relative frequency distribution of the responses.

88. Construct a frequency bar graph of the responses.

89. Construct a relative frequency pie chart of the responses.

Astrological Signs. Use the following information for Exercises 90–92. The General Social Survey collects data on social aspects of life in America. Here, 1464 respondents reported their astrological signs. A pie chart of the results is shown below.

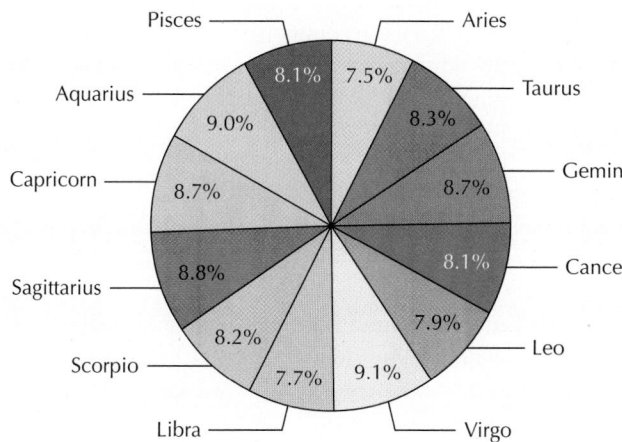

Pie chart of astrological signs.

90. Answer the following:
   a. What is the most common astrological sign?
   b. What is the least common astrological sign?

91. Use the percentages in the pie chart to do the following:
   a. Construct a relative frequency bar graph of the astrological signs.
   b. Construct a relative frequency bar graph, but, this time, have the *y* axis begin at 7% instead of zero. Describe the difference between the two bar graphs. When would this one be used as opposed to the earlier bar graph?

92. Construct a frequency distribution of the astrological signs. Which sign occurs the least? The most?

Satisfaction with Family Life. Use the following information for Exercises 93–95. The General Social Survey also asked respondents about their amount of satisfaction with family life. Here, 1002 respondents reported their satisfaction levels with family life, as shown in the frequency distribution below. **satisfaction**

Satisfaction	Frequency
Very great deal	415
Great deal	329
Quite a bit	99
A fair amount	90
Some	27
A little	25
None	17
Total	1002

93. Construct a Pareto graph of the categories. Do you think that, overall, people are happy with their family lives?

94. Construct a frequency or a relative frequency bar graph of the categories. Do you think that this bar graph reinforces the positive message in the data?

95. What is your opinion of the categories? Do you think that everyone interprets these phrases in the same way? For example, what is the difference between "quite a bit" and "great deal"?

Litter. Don't Mess with Texas (http://dontmesswithtexas .org) is a Texas statewide anti-littering organization that identified paper, plastic, metals, and glass as the top four categories of litter by composition. The report also identified tobacco, household, food, and beverages as the top four categories of litter by use. A sample of 12 items of litter had the following characteristics. Use the table to answer Exercises 96–98. **litter**

Litter item	Composition	Use
1	Paper	Tobacco
2	Plastic	Household
3	Glass	Beverages
4	Paper	Tobacco
5	Metal	Household
6	Plastic	Food
7	Glass	Beverages
8	Paper	Household
9	Metal	Household
10	Plastic	Beverages
11	Paper	Tobacco
12	Plastic	Food

96. Construct a crosstabulation of litter composition by litter use.

97. Construct a clustered bar graph of litter composition and litter use. Cluster by use.

98. Identify any patterns you discover.

Misdemeanors in New York City. For Exercises 99–101, refer to Figure 17, a bar chart of the total number of misdemeanor crimes committed in New York City in 2013 categorized by misdemeanor type.

99. Which category leads all categories of misdemeanor crime in New York City?

100. About how many petit larceny cases were there in 2013?

101. Is Figure 17 in the form of a Pareto graph? Explain why or why not.

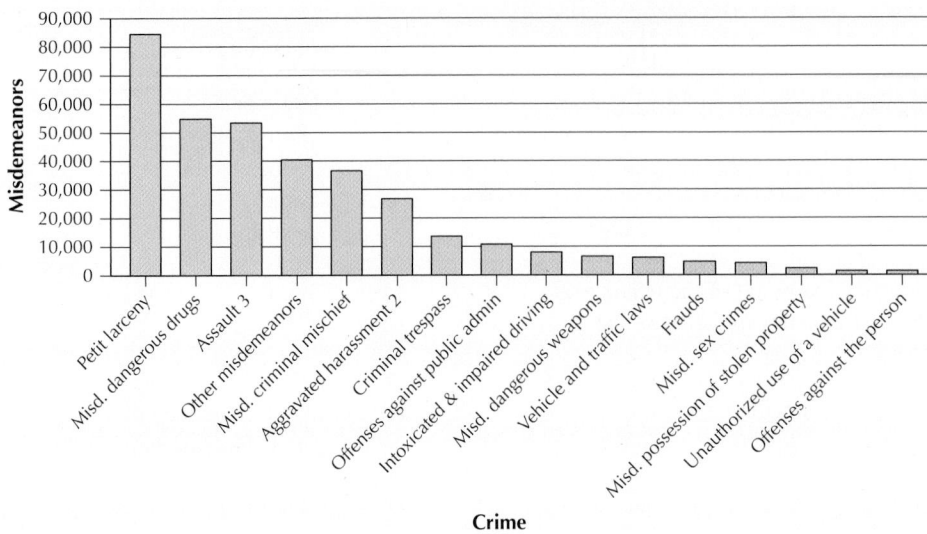

FIGURE 17 Bar chart of misdemeanors in New York City.

Comparison Pie Chart. The Centers for Disease Control and Prevention (CDC) recommends routine vaccination with Gardasil for boys and girls ages 11 and 12, in order to combat the HPV virus. For example, it has been shown that 99.7% of cervical cancer patients have been infected with the HPV virus.[6] The large data set **Gardasil** (*Source: Journal of Statistics Education*)[7] contains information regarding 1414 young women who underwent vaccination treatment. To complete the vaccination, a series of three shots must be undertaken. Researchers were interested in whether the location of the clinic was related to the proportion of patients who completed treatment. Figure 18 shows a *comparison pie chart*—a pair of pie charts of the proportion of patients completing the vaccine treatment, with one pie chart for suburban locations and one pie chart for urban locations.

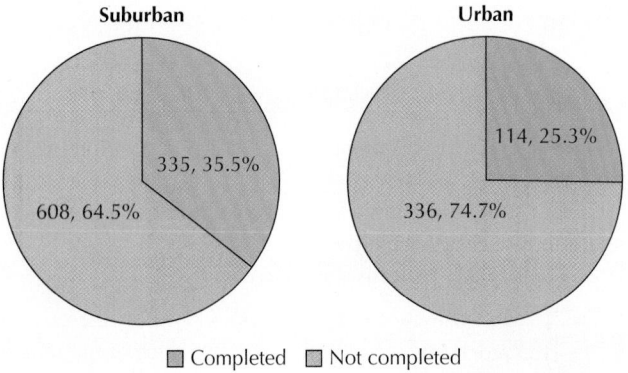

FIGURE 18 Comparison pie chart of treatment completion proportions.

102. What proportion of females from suburban locations completed the treatment? From urban locations?

103. Use the data from the comparison pie chart to make the following relative frequency bar charts.
    a. Treatment status for suburban females
    b. Treatment status for urban females

104. Do you think that the 35.5% to 25.3% difference in vaccine completion is just due to chance, or do you think that it represents a real disparity between the two groups?

BRINGING IT ALL TOGETHER

Shopping Enjoyment and Gender. Use the information in the crosstabulation for Exercises 105–119. The Pew Internet and American Life Project surveyed 4514 American men and women and asked them, "How much, if at all, do you enjoy shopping?" The results shown in the crosstabulation are missing some entries.

Crosstabulation of shopping enjoyment by gender

	Gender		Total
	Male	**Female**	
A lot	—	950	1338
Some	582	—	1255
Only a little	662	497	—
Not at all	497	—	717
Don't know/refused	—	25	45
Total	2149		4514

105. Fill in the missing entries.
106. Convert the table to a relative frequency crosstabulation. Make it so that the "Male" and "Female" proportions in each row add up to 1.0.
107. Did men or women have the higher proportion of respondents who enjoy shopping
    a. a lot?
    b. some?
    c. only a little?
    d. not at all?

108. Construct a frequency distribution of *gender.*

109. Build a frequency distribution of *response.*

110. Make a relative frequency distribution of *gender.*

111. Construct a relative frequency distribution of *response.*

112. Build a bar graph of *gender.*

113. Make a bar graph of *response.*

114. Construct a pie chart of *gender.*

115. Build a pie chart of *response.*

116. Construct a clustered bar graph of *gender* clustered by *response.*

117. Build a clustered bar graph of *response* clustered by *gender.*

118. What proportion of the respondents is female?

119. *What if* we doubled each cell count? How would that affect the following?

    a. Frequency distribution of *gender*

    b. Relative frequency of *gender*

    c. Pie chart of *gender*

WORKING WITH LARGE DATA SETS

Educational Goals in Sports. Use your knowledge of technology to solve Exercises 120 and 121. Open the **Goals** data set. The subjects are students in grades four, five, and six from three school districts in Michigan. The students were asked which of the following was most important to them: good grades, sports, or popularity. Information about the students' ages, genders, races, and grades was also gathered, as well as whether their schools were in an urban, suburban, or rural setting.[8] **goals**

120. Generate bar graphs for the following variables.

    a. *Gender.* Estimate the relative frequency of girls, and boys, in the sample.

    b. *Goals.* About what percentage of the students chose "grades" as most important? About what percentage chose "popular"? About what percentage chose "sports"?

121. Generate relative frequency distributions for the following variables.

    a. *Gender.* How close were your estimates in the previous exercise?

    b. *Goals.* How close were your estimates in the previous exercise?

WORKING WITH LARGE DATA SETS

Analysis of Households. For Exercises 122–124, use your knowledge of technology. Open the data set **Household**. **household**

122. How many observations are in this data set? How many variables?

123. Which of the variables are qualitative? Which of the variables are quantitative?

124. What would a relative frequency distribution of the variable *state* look like? A pie chart? A bar graph?

WORKING WITH LARGE DATA SETS

Open the data set **Titanic**. Use technology for Exercises 125–130. **titanic**

125. Construct (i) a frequency distribution and (ii) a relative frequency distribution of each of the following variables:

    a. *Class*

    b. *Age*

    c. *Sex*

    d. *Survived*

126. Build (i) a frequency bar graph and (ii) a relative frequency bar graph of each of the following variables:

    a. *Class*

    b. *Age*

    c. *Sex*

    d. *Survived*

127. Construct pie charts of each variable in the previous exercise.

128. Construct a clustered bar graph of the variable *survived* clustered by gender. What does this graph tell us about the relative proportions of females and males who survived?

129. Construct a clustered bar graph of the variable *survived* clustered by age. Did you stand a better chance of surviving as an adult or a child?

130. Construct a clustered bar graph of the variable *survived* clustered by class. Which class had the best chance of surviving? (*Hint:* For each class, compare the heights of the "Yes" rectangles to the heights of the "No" rectangles.)

CONSTRUCT YOUR OWN DATA SETS

Environmental Club. Use the following information for Exercises 131–133. You are the president of the College Environmental Club, which has members among all four classes: freshmen, sophomores, juniors, and seniors. The total number of members in the club is 20.

131. Set the frequency of each class so that each class has an equal number of members.

    a. Construct a frequency distribution of the variable *class.*

    b. Construct a relative frequency distribution of the variable *class.*

132. Set the frequency of each class so that there are more sophomores than freshmen, more juniors than sophomores, and more seniors than juniors.

    a. Construct a Pareto chart of the variable *class.*

    b. Construct a pie chart of the variable *class.*

133. Set the frequency of each class so that there are more seniors than any other class, while the other three classes have equal numbers.

    a. Construct a frequency bar graph of the variable *class.*

    b. Construct a relative frequency bar graph of the variable *class.*

WORKING WITH LARGE DATA SETS

CASE STUDY

Open the data set **Petit Larceny**, which contains the number of petit larceny misdemeanors, per precinct, for the years 2000–2013. For the years 2000 and 2013, these frequencies have been categorized into four categories: 1 – Low: 800 or less, 2 – Medium: 800 to less than 1050, 3 – High: 1050 to less than 1500, and 4 – Very High: 1500 or more. Use technology to answer the following: 🏛 **petitlarceny**

134. Construct bar graphs of "2000 categorized" and "2013 categorized."

135. Discuss the differences between the bar graphs. Which categories are larger or smaller? On the whole, do you think this represents good news or bad news?

136. Construct pie charts of "2000 categorized" and "2013 categorized."

137. Discuss whether the bar graphs or the pie charts express the results more clearly in this case.

2.2 Graphs and Tables for Quantitative Data

OBJECTIVES By the end of this section, I will be able to . . .

1 Construct and interpret a frequency distribution and a relative frequency distribution for discrete and continuous data.
2 Use histograms and frequency polygons to summarize quantitative data.
3 Construct and interpret stem-and-leaf displays and dotplots.
4 Use graphs and tables to obtain useful information.
5 Recognize distribution shape, symmetry, and skewness.

1 Frequency Distributions and Relative Frequency Distributions for Discrete and Continuous Data

In Section 2.1, we introduced tables and graphs for summarizing qualitative data. However, most of the data sets that we will encounter in this book are quantitative instead of qualitative. Recall from Chapter 1 that quantitative data take on numerical values on which arithmetic can be meaningfully performed. We can apply frequency and relative frequency distributions to quantitative data, just as we did for the qualitative data in Section 2.1

EXAMPLE 9   Frequency distribution and relative frequency distribution for discrete data

The Recording Industry Association of America (RIAA) awards multi-platinum status for any musical recording that sells more than 2 million copies. Table 17 contains a random sample of 20 of the musical artists with the most multi-platinum singles.

Table 17 Number of multi-platinum singles

Artist	Multi-platinums	Artist	Multi-platinums
Beyoncé	4	Linkin Park	2
Bruno Mars	4	Madonna	2
Chris Brown	2	Michael Jackson	1
Elton John	1	Nicki Minaj	2
Fergie	3	Red Hot Chili Peppers	2
Jay-Z	4	Shakira	1
Justin Timberlake	1	Sugarland	1
Kanye West	7	Taylor Swift	8
Katy Perry	8	The Beatles	4
Lady Gaga	6	Tim McGraw	2

Source: RIAA.

Use this raw data to construct a **frequency distribution** and a **relative frequency distribution** of the number of multi-platinum singles.

Solution

We begin by making a tally of how many artists had one multi-platinum, how many had two, and so on. We then construct the frequency distribution for the variable *Multi-platinums*. Finally, we construct the relative frequency distribution by dividing the frequency by the total number of observations, 20. See Table 18.

Table 18 Frequency distribution and relative frequency distribution of *Multi-platinums*

Multi-platinums	Tally	Frequency	Relative frequency
1	⦀⦀	5	5/20 = 0.25
2	⦀⦀ ⏐	6	6/20 = 0.30
3	⏐	1	1/20 = 0.05
4	⏐⏐⏐⏐	4	4/20 = 0.20
5		0	0/20 = 0.00
6	⏐	1	1/20 = 0.05
7	⏐	1	1/20 = 0.05
8	⏐⏐	2	2/20 = 0.10
Total		**20**	**20/20 = 1.00**

NOW YOU CAN DO
Exercises 9–12.

We can combine several numbers of multi-platinums together into "classes," in order to produce a more concise distribution. **Classes** represent a range of data values and are used to group the elements in a data set.

EXAMPLE 10   Frequency and relative frequency distributions using classes

Combine the data from Table 18 into three classes, and construct frequency and relative frequency distributions.

Solution

Let us define the following classes for the numbers of multi-platinum records:

- 1–3 Multi-platinums
- 4–6 Multi-platinums
- 7–9 Multi-platinums

For each class, we group together all the musical artists in the class. Table 19 provides the frequency distribution and relative frequency distribution for these three classes.

Table 19 Frequency distribution and relative frequency distribution of *Multi-platinums*, after combining into classes

Class	Frequency	Relative frequency
1–3	12	12/20 = 0.60
4–6	5	5/20 = 0.25
7–9	3	3/20 = 0.15
Total	20	20/20 = 1.00

NOW YOU CAN DO
Exercises 13–16.

The class with the highest frequency (and therefore highest relative frequency) is the class with 1–3 multi-platinum singles: 12, or 60% of the total. Therefore, we can say that most of the artists in our sample had between one and three multi-platinum singles.

Developing Your Statistical Sense

Choosing Which Distribution to Use

So which frequency distribution is the "right" one—Table 18 or Table 19? There is no absolute answer. It depends on the goals of the analysis, as well as other factors. For example, from Table 19, we can see that most of the artists in our sample had between one and three multi-platinum singles, a finding that was not immediately apparent from Table 18. Therefore, combining data values into classes can lead to interesting overall findings. However, whenever data values are combined into classes, some information is lost. For example, it is not possible, using Table 19 alone, to determine that the number of multi-platinums occurring with the greatest frequency in our sample is two.

We use the following definitions to construct frequency distributions and histograms (for a discussion of histograms, see page 65).

> The **lower class limit** of a class equals the smallest value within that class.
>
> The **upper class limit** of a class equals the largest value within that class.
>
> The **class width** equals the difference between the lower class limits of two successive classes.
>
> The **class boundary** of two successive classes is found by taking the sum of the upper class limit of a class and the lower class limit of the class to its right, and dividing this sum by two. The lower class boundary of the leftmost class equals its upper class boundary minus the class width. The upper class boundary of the rightmost class equals its lower class boundary plus the class width.

EXAMPLE 11 Class limits, class widths, and class boundaries

For the classes in Example 10, find the following:

a. The lower class limits and the upper class limits

b. The class width

c. The class boundaries

Solution

a. The following table shows the lower class limits and the upper class limits for the classes in Example 10.

Class	Lower class limit (smallest value)	Upper class limit (largest value)
1–3	1	3
4–6	4	6
7–9	7	9

b. Our lower class limits are 1, 4, and 7; therefore, the class width of each class is 3 because the lower class limits differ by 3. For example, $4 - 1 = 3$.

c. To find the class boundary of the first and second class, we find the sum of the upper class limit of the first class and the lower class limit of the second class, and divide this sum by 2, giving us $(3 + 4)/2 = 3.5$. Similarly, the class boundary of the second class with the third class is $(6 + 7)/2 = 6.5$. The lower class boundary of the leftmost class equals its upper class boundary minus the class width (that is, $3.5 - 3 = 0.5$). The upper class boundary of the rightmost class equals its lower class boundary plus the class width (that is, $6.5 + 3 = 9.5$).

Next, we show how to construct frequency distributions for continuous data.

> To construct a frequency distribution for continuous data:
> **1.** Choose the number of classes.
> **2.** Determine the class width. It is best (though not required) to use the same width for all classes.
> **3.** Find the upper and lower class limits. Make sure the classes are nonoverlapping.
> **4.** Calculate the class boundaries.
> **5.** Find the frequencies of each class.

EXAMPLE 12   Constructing a frequency distribution for continuous data

 carbon

The U.S. Department of Energy reported the total 2011 carbon emissions emitted for a sample of 20 states, in millions of metric tons of carbon dioxide (Table 20).

Table 20 Total carbon emissions (millions of metric tons of carbon dioxide) for a sample of states, 2011

State	Carbon emissions	State	Carbon emissions
Arizona	93.28	New Jersey	117.56
Arkansas	67.56	New Mexico	56.60
Colorado	91.98	Oklahoma	107.92
Iowa	87.42	South Carolina	80.21
Kansas	72.36	Tennessee	105.73
Maryland	65.80	Virginia	99.86
Massachusetts	68.89	Washington	70.81
Minnesota	92.69	West Virginia	95.97
Mississippi	61.21	Wisconsin	98.05
Nebraska	52.26	Wyoming	63.89

Construct a frequency distribution of the carbon emissions data.

Solution

Step 1   **Choose the number of classes.**
It is generally recommended that between 5 and 20 classes be used, with the number of classes increasing with the sample size; a small data set such as this will do just

fine with 7 classes. In general, choose the number of classes to be large enough to show the variability in the data set, but not so large that many classes are nearly empty.

Step 2 **Determine the class widths.**

First, find the range of the data, that is, the difference between the largest and smallest data points. Then, divide this range by the number of classes you chose in Step 1. This gives an estimate of the class width. Here, our largest data value is 117.56 and our smallest is 52.26, giving us a *range* of 117.56 − 52.26 = 65.3. In Step 1, we chose 7 classes, so that our estimated class width is 65.3/6 ≈ 10.9. For convenience, we will round this to a class width of 10. It is recommended that each class have the same width.

Step 3 **Find the upper and lower class limits.**

Choose limits so that each data point belongs to only one class. For example, suppose we chose one class to be 50–60 and the next class to be 60–70. Then, to which class would an emissions value of exactly 60 belong? The classes should not overlap. Therefore, we define the following classes.

> The notation "50 to < 60" indicates that this class contains values from 50 (inclusive) up to but not including 60.

50 to < 60, 60 to < 70, 70 to < 80, 80 to < 90, 90 to < 100, 100 to < 110, 110 to < 120

Note that the lower class limit of the first class, 50, is slightly below that of the smallest value in the data set, 52. Also note that the class width equals 60 − 50 = 10, as desired.

Step 4 **Calculate the class boundaries.**

The class boundary for the first two classes is (60 + 60)/2 = 60. Similarly, the other class boundaries are 70, 80, 90, 100, and 110. The lower class boundary of the leftmost class is 60 − 10 = 50. The upper class boundary of the rightmost class is 110 + 10 = 120.

Step 5 **Find the frequencies for each class.**

Using these seven classes, we now proceed to construct the frequency and relative frequency distributions (see Table 21) for the carbon emissions data. We count the number of data values that fall into each class, and we divide each frequency by the sample size (20) to obtain the relative frequency.

Table 21 Distributions for the carbon emissions data

Class: $x =$	Tally	Frequency	Relative frequency
50 to < 60	||	2	2/20 = 0.10
60 to < 70	⊞	5	5/20 = 0.25
70 to < 80	||	2	2/20 = 0.10
80 to < 90	||	2	2/20 = 0.10
90 to < 100	⊞ |	6	6/20 = 0.30
100 to < 110	||	2	2/20 = 0.10
110 to < 120	|	1	1/20 = 0.05
Total		**20**	**20/20 = 1.00**

NOW YOU CAN DO
Exercises 17–40.

YOUR TURN
#7

A country's unemployment rate is an indicator of its general economic health. Table 22 contains the unemployment rates in August 2014 for a sample of 20 countries from around the world.

Table 22 Unemployment rates for 20 countries

Country	Unemployment rate	Country	Unemployment rate
Britain	6.4	Japan	3.7
Canada	7.0	Mexico	4.8
China	4.1	Pakistan	6.2
France	10.2	Philippines	7.0
Germany	6.7	Poland	12.0
India	8.8	Russia	4.9
Indonesia	5.7	Singapore	2.0
Ireland	11.5	South Korea	3.4
Israel	6.3	Turkey	8.8
Italy	12.3	United States	6.2

Source: The Economist, www.economist.com/node/21604509.

Use five classes of class width 2.5 each, and define the leftmost class as follows: 1.0 to < 3.5. Calculate the class boundaries. Find the frequencies and relative frequencies for each class, and construct a frequency distribution and a relative frequency distribution of the unemployment rates.

(The solution is shown in Appendix A.)

2 Histograms and Frequency Polygons

Histograms

Many different methods are available for graphically summarizing numeric data. One example of a graphical summary for quantitative data is a **histogram**.

> A **histogram** is constructed using rectangles for each class of data. The heights of the rectangles represent the frequencies or relative frequencies of the class. The widths of the rectangles represent the class widths of the corresponding frequency distribution. The class boundaries are placed on the horizontal axis, so that the rectangles are touching each other.

EXAMPLE 13   Constructing a histogram

Construct a histogram of the frequency of the carbon emissions data from Example 12.

Solution

Step 1   **Find the class limits, and draw the horizontal axis.**
Note that the class boundaries for these data were found in Example 12: 60, 70, 80, 90, 100, 110, and 120. Draw the horizontal axis, with the numbers 60, 70, 80, 90, 100, 110, and 120 equally spaced along it. The numbers indicate where the rectangles will touch each other.

Step 2   **Determine the frequencies, and draw the vertical axis.**
Use the frequencies given in Table 21. These will indicate the heights of the five rectangles along the vertical axis. Find the largest frequency, which is 6. It is a good idea to provide a little bit of extra vertical space above the tallest rectangle, so make 7 your highest label along the vertical axis. Then provide equally spaced labels along the vertical axis between 0 and 7.

Step 3 **Draw the rectangles.**

Draw your first rectangle from 50 to 60, with height 2, the first frequency, and your second rectangle from 60 to 70, with height 5. Proceed to draw the remaining rectangles similarly.

The resulting frequency histogram is shown in Figure 19a. The relative frequency histogram is shown in Figure 19b. Note that the two histograms have identical shapes and differ only in the labeling along the vertical axis.

NOW YOU CAN DO
Exercises 41–52.

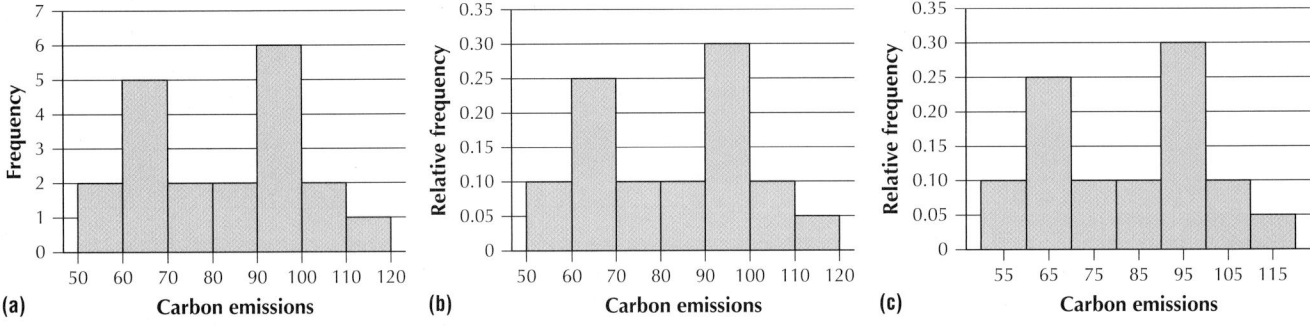

(a)   Carbon emissions   (b)   Carbon emissions   (c)   Carbon emissions

FIGURE 19 (a) Frequency histogram; (b) relative frequency histogram; (c) histogram using midpoints.

YOUR TURN #8

Using the unemployment rate data from Table 22, construct a histogram of unemployment rates, using the same classes you used previously.

(The solution is shown in Appendix A.)

Note: Histograms are often presented using class midpoints instead of class boundaries. The class boundaries can then be inferred by splitting the difference between the class midpoints. In Figure 19c, the upper class boundary for the leftmost class is halfway between 55 and 65, that is, 60. Otherwise, Figure 19c is equivalent to Figure 19b.

Note that the histogram, unlike the frequency distribution, provides us with a graphical impression of the data distribution. This characteristic will be crucial later on, when we evaluate the fitness of data sets to undergo certain data analysis methods. Also, notice that the rectangles are contiguous (touching), unlike the rectangles of the bar graphs in Section 2.1. Because the data are quantitative, the horizontal axis in a histogram should be considered as the number line.

A **class midpoint** is the average of two consecutive lower class limits. For example, the class midpoint for the leftmost class in Figure 19 is $(50 + 60)/2 = 55$.

WHAT IF ?

What If Scenario: Give the Calculator a Rest!

What if scenarios offer you a chance to reflect on how changes in the initial conditions will percolate through the various aspects of a problem. The only requirement is to put your calculator down and think through the problem. You are asked to find the answers by using your knowledge of what the statistics *represent.*

Shifting the Histogram to the Left

What if we subtracted 10 million metrics tons from each state's carbon emissions? How would that affect the frequency histogram in Figure 19a? Assume that the number of classes and the class width would stay the same.

Solution

The new class limits and class boundaries would each be 10 points lower than the corresponding class limits and class boundaries from Example 12. However, the frequencies for each corresponding class would be the same as those from Example 12. Thus, the rectangles would look the same, with the only difference being that they are "shifted left" 10 points along the number line (Figure 20). We discuss more about the shapes of histograms later in this section.

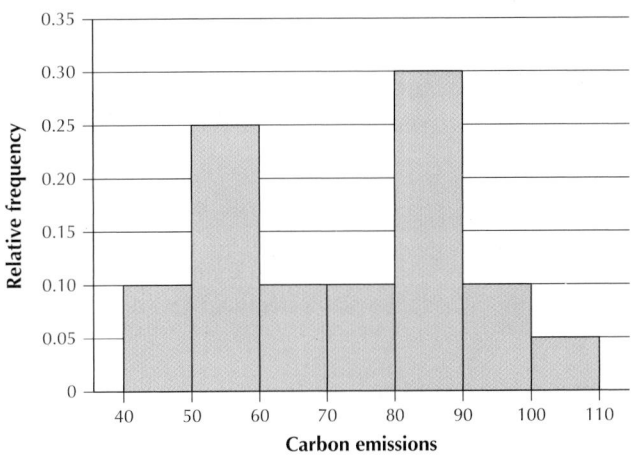

FIGURE 20   *"Shape" of histogram is unchanged if shifted left.*

The *One Variable Statistics and Graphs* applet can display histograms for a selection of data sets in this textbook, including the carbon emissions data. The applet allows you to experiment with different class widths.

Frequency Polygons

Frequency polygons provide the same information as histograms, but in a slightly different format.

> A **frequency polygon** is constructed as follows. For each class, plot a point at the class midpoint, at a height equal to the frequency for that class. Then join each consecutive pair of points with a line segment.

EXAMPLE 14   Constructing a frequency polygon

 carbon

Construct a frequency polygon for the carbon emissions data in Example 12.

Solution

The midpoints for the classes are shown in Figure 19c. Plot a point for each frequency above each midpoint, and join consecutive points. The result is the frequency polygon in Figure 21.

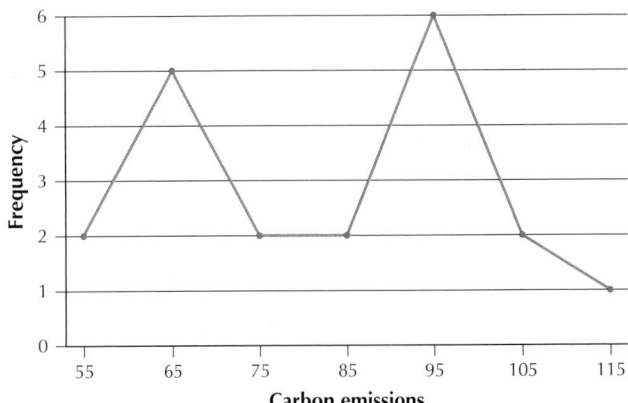

NOW YOU CAN DO
Exercises 53–56.

FIGURE 21   Frequency polygon.

Alfred Eisenstaedt/Time Life Pictures/Getty Images

Professor John Tukey.

YOUR TURN
#9

Using the unemployment rate data from Table 22, construct a frequency polygon of unemployment rates, using the same classes as last time.

(The solution is shown in Appendix A.)

3 Stem-and-Leaf Displays and Dotplots

Stem-and-Leaf Displays

Stem-and-leaf displays were developed by Professor John Tukey of Princeton University in the late 1960s. This type of display generally contains more information than either a frequency distribution or a histogram. We will demonstrate how to construct a stem-and-leaf display in Example 15.

EXAMPLE 15   Constructing a stem-and-leaf display

 dangerousweapons

CASE STUDY

We will construct a stem-and-leaf display for the number of misdemeanors: dangerous weapons cases in 20 Manhattan precincts, shown here in Table 23.

Table 23 Dangerous weapons cases in Manhattan police precincts

Precinct	Dangerous weapons cases	Precinct	Dangerous weapons cases
1	19	20	24
5	24	22	9
6	42	23	91
7	79	24	22
9	52	25	109
10	21	26	67
13	93	28	40
17	12	32	90
18	45	33	82
19	23	34	88

```
 0
 1
 2
 3
 4
 5
 6
 7
 8
 9
10
```

FIGURE 22 The stems for the dangerous weapons data.

```
      0
      1  9
Stem  2       Leaf
      3
      4
      5
      6
      7
      8
      9
     10
```

FIGURE 23 The 1 "tens" leaf takes a 9 "ones" leaf, making the number 19.

Solution

Each of these data values may be broken down into a certain number of "tens" and a certain number of "ones." For example, take the number of dangerous weapons cases for Precinct 1: 19. Notice that $19 = (1 \cdot 10) + (9 \cdot 1)$, where we have 1 ten and 9 ones. Also note for Precinct 25: $109 = (10 \cdot 10) + (9 \cdot 1)$, where we have 10 tens and 9 ones. For 19, the tens value is 1, whereas for 109, the tens value is 10. So we begin constructing a stem-and-leaf display by first collecting all the tens values and putting them in order in a column (Figure 22). These are called the **stems**.

Next, consider the ones place of each data value. For example, the first data value, 19, has 1 in the tens place (the stem) and 9 in the ones place. Place this number, called the **leaf**, next to its stem (Figure 23).

The second data value, 24, has 2 in the tens place and 4 in the ones place, and the third data value, 42, has 4 in the tens place and 2 in the ones place. Place the leaf 4 on the same line as its stem, 2, to make 24. Place the leaf 2 on the same line as its stem, 4, to make 42 (Figure 24). Continue this process with the remaining data, placing each ones value next to its stem. Then, for each stem, order the leaves from left to right in increasing order. This produces the stem-and-leaf display in Figure 25.

```
 0 |
 1 | 9
 2 | 4
 3 |
 4 | 2
 5 |
 6 |
 7 |
 8 |
 9 |
10 |
```
Key: Leaf units = ones

FIGURE 24 The numbers 19, 24, and 42 in the stem-and-leaf display.

```
 0 | 9
 1 | 29
 2 | 12344
 3 |
 4 | 025
 5 | 2
 6 | 7
 7 | 9
 8 | 28
 9 | 013
10 | 9
```
Key: Leaf units = ones

FIGURE 25 Stem-and-leaf display for the dangerous weapons data.

```
 0 |
 0 | 9
 1 | 2
 1 | 9
 2 | 12344
 2 |
 3 |
 3 |
 4 | 02
 4 | 5
 5 | 2
 5 |
 6 |
 6 | 7
 7 |
 7 | 9
 8 | 2
 8 | 8
 9 | 013
 9 |
10 |
10 | 9
```

FIGURE 26 Split-stem version of stem-and-leaf display.

In general, the leaf units represent the smallest decimal place represented in the data values. Then, the stem unit consists of the remainder of the number. For example, suppose we have a data value of 127. Then, the 7 is the leaf unit, and the 12 is the stem. Or else, suppose our data value is 0.146. Then, our leaf unit is the 6, and the stem is the 14. Note that the stem-and-leaf display contains all the information that a histogram turned on its side contains, but it also contains more information than a histogram because the stem-and-leaf display shows the original values.

Split stems may sometimes be used in a stem-and-leaf display to provide a clearer idea of the data distribution when too many data points fall on just a few stems. When using split stems, each stem appears twice, with the leaves 0 to 4 on the upper stem and the leaves 5 to 9 on the lower stem. Figure 26 shows how the stem-and-leaf display of dangerous weapons data would appear when using split stems.

NOW YOU CAN DO
Exercises 57–60.

YOUR TURN
#10

Using the unemployment rate data from Table 22, do the following:

1. Construct a stem-and leaf display of unemployment rates.

2. Build a stem-and-leaf display of unemployment rates, using split stems.

(The solution is shown in Appendix A.)

EXAMPLE 16   Stem-and-leaf displays of decimal-valued data

Here follows a sample of the grade point averages of the author's advisees. Construct a stem-and-leaf display of the data.

2.9	3.5	3.7	4.0	2.8	3.9	3.2	4.0	3.5	3.8

Solution

Here, we define the "ones" decimal place to represent the stem, and the "tenths" decimal place to represent the leaves. The resulting stem-and-leaf display is shown in Figure 27. Note that the two "0" leaves next to the "4" stem represent two different students whose GPA is 4.0.

```
2 | 89
3 | 255789
4 | 00
```
Key: Leaf unit = tenths

FIGURE 27 Stem-and-leaf display of decimal-valued data.

The *One Variable Statistics and Graphs* applet can display stem-and-leaf displays for a selection of data sets in this textbook. The applet allows you to experiment with split stems if you want to.

Dotplots

A simple but effective graphical display is a **dotplot**. In a dotplot, each data point is represented by a dot above the number line. When the sample size is large, each dot may represent more than one data point.

EXAMPLE 17   Dotplots

Construct a dotplot of the carbon emissions data.

Solution

Figure 28 is a dotplot of the carbon emissions data.

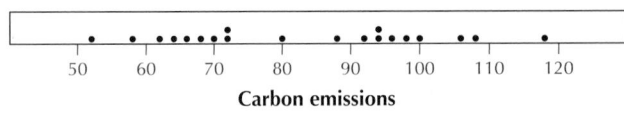

Carbon emissions

NOW YOU CAN DO
Exercises 61–64.

FIGURE 28 Dotplot of the carbon emissions data.

 Using the unemployment rate data from Table 22, construct a dotplot of the unemployment rates.

(The solution is shown in Appendix A.)

Dotplots are useful for comparing two variables, as shown in the following example.

EXAMPLE 18   Comparison dotplots

 Suppose a criminal justice researcher wants to compare the occurrences of third-degree assault and criminal trespass across the police precincts of New York City. The researcher could construct a comparison dotplot, as shown in

Note: All of the graphs and tables learned in this section may be used for quantitative data. But histograms, frequency polygons, stem-and-leaf displays, and dotplots cannot be used for qualitative data.

Figure 29. This graph shows that the number of criminal trespass misdemeanors per precinct lie mostly below 300, whereas the number of third-degree assaults per-precinct frequencies lie mostly above 300.

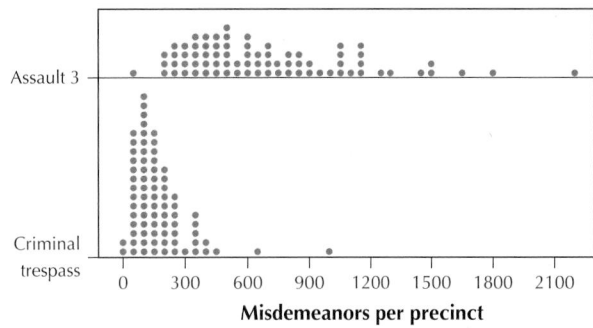

FIGURE 29 Comparison dotplot of third-degree assault and criminal trespass.

Note that the two groups are graphed using the same number line, which makes comparison easier.

NOW YOU CAN DO
Exercises 65–66.

4 Obtaining Information from Graphs and Tables

Graphs and tables make it easy to acquire useful information about the data that we could not get just by looking at the raw data. In fact, that is the whole point of constructing the graphs and tables in the first place: to allow us to more easily observe patterns and trends in the data. Example 19 illustrates this.

EXAMPLE 19   Obtaining information from graphs and tables

a. Using the relative frequency distribution in Table 21, what proportion of states in the sample has carbon emissions of 100 million or more metric tons?

b. Using the histogram in Figure 19a, how many states have carbon emissions between 70 (inclusive) and 90 (not inclusive) million metric tons?

c. Using the stem-and-leaf display in Figure 25 on page 69, how many precincts had between 22 and 24 (inclusive) dangerous weapons cases?

d. Now, using the histogram in Figure 19a, can you determine how many states had between 94 and 104 million metric tons of carbon emissions?

Solution

a. From Table 21, 100 (million) or more metric tons is represented by two classes: 100–109 and 110–119. Adding their relative frequencies gives us the proportion of states with 100 million or more metric tons of carbon emissions: $0.10 + 0.05 = 0.15$.

b. From Figure 19a, there are two classes whose values together contain the values 70 to < 90, each of which has a frequency of two. Therefore, there are $2 + 2 = 4$ states with carbon emissions between 70 (inclusive) and 90 (not inclusive) million metric tons.

c. The stem-and-leaf display shows us that there are precincts with 22, 23, 24, and 24 dangerous weapons cases. This makes four precincts.

d. It is not possible to answer this question using the histogram because the histogram does not tell us how many states are above or below 94—just how many are in the 90s.

NOW YOU CAN DO
Exercises 67–78.

YOUR TURN

#12

Using the graphs you constructed for the unemployment rate data from Table 22, answer the following:

1. Using the histogram you constructed earlier, how many countries had unemployment rates below 6.0?

2. Using the histogram alone, can you determine how many countries had unemployment rates between 6.5 and 7.5 (inclusive)? Explain.

3. Now, using the stem-and-leaf display, how many countries had unemployment rates between 6.5 and 7.5 (inclusive)?

(The solutions are shown in Appendix A.)

5 Distribution Shape, Symmetry, and Skewness

Frequency distributions are tabular summaries of the set of values that a variable takes. We now generalize the concept of **distribution**.

> The **distribution** of a variable is a table, graph, or formula that identifies the variable values and frequencies for all elements in the data set.

For example, a frequency distribution is a distribution because it is a table that specifies each of the values that a variable can take, along with the frequencies. However, our definition of "distribution" also includes histograms, stem-and-leaf displays, dotplots, and other graphical summaries. (In Chapter 6, we will introduce distributions defined by formulas.) These graphical distributions invite us to consider the shape of a distribution. The *shape* of a distribution is the overall form of a graphical summary approximated by a smooth curve.

The Bell-Shaped Curve

Figure 30 contains the relative frequency histogram of the heights of 1000 college women. Note that relatively fewer women are in both the left-hand tail (shorter women) and the right-hand tail (taller women). Instead, as height increases from left to right, the relative frequency gradually increases until it reaches a peak near 65 inches tall and then gradually decreases. Thus, the distribution of heights is said to be *bell-shaped*.

The rectangles represent the actual data. However, the smoothed curve represents an approximation of the overall form of the distribution, and thus the smoothed curve represents the shape of the distribution, which is bell-shaped. The formal name of this bell-shaped distribution is the *normal* distribution. In Chapter 6, we will learn much more about this important distribution, which occurs often in nature and the real world. For example, student heights (within a given gender) follow a bell-shaped distribution. In Chapter 7, we will learn how to assess whether or not a particular distribution is

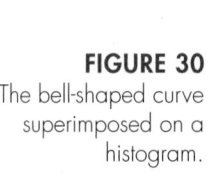

FIGURE 30
The bell-shaped curve superimposed on a histogram.

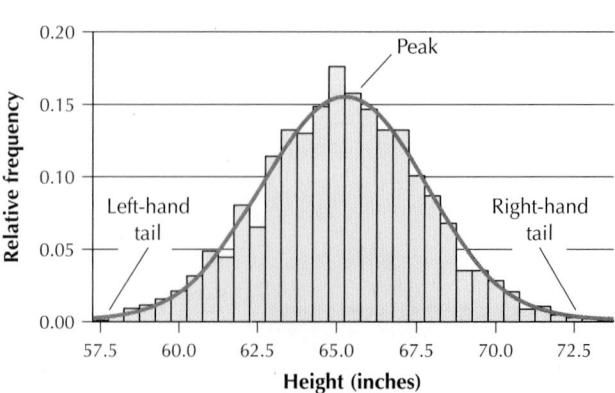

FIGURE 31 This butterfly is symmetric.

Note: Only quantitative data, not qualitative data, may be described as symmetric or skewed.

 In Figure 32, do not say that the distribution is right-skewed because most of the data are on the right. This is a common mistake. Instead, skewness follows the longer tail, making Figure 32 left-skewed.

normal (bell-shaped). Starting in Chapter 8, many of the methods for statistical inference that we will learn will depend on this distribution.

Analyzing the Shape of a Distribution

Next, we learn some tools for analyzing the shape of a distribution. An image has *symmetry* (or is **symmetric**) if a line (axis of symmetry) splits the image in half, so that one side is the mirror image of the other. For example, the butterfly in Figure 31 has symmetry because a line drawn down the middle of the butterfly would create two mirror images of each other. It is important to develop the talent for recognizing which distribution shapes are symmetric.

For example, the smoothed curve in Figure 30 is perfectly symmetric. However, the histogram rectangles reflecting the actual data are only nearly symmetric, because a vertical line drawn down the middle of the distribution would not result in two perfect mirror images. *Due to random variation, data from the real world rarely exhibit perfect symmetry.* With this in mind, the data analyst is usually content with the approximate symmetry exhibited by the data (the rectangles) in Figure 30.

However, not all distributions are symmetric. Instead, a distribution may be **skewed**. Skewed distributions may be either **left-skewed** or **right-skewed**. Left-skewed distributions have a longer "tail" on the left than on the right. For example, the distribution of quiz grades from the author's Summer 2014 Business Statistics course has a longer tail on the left than on the right (Figure 32). This can happen with quizzes because several students bump up against the 100% boundary on the right, most students are somewhere in the middle, and only a few students are in the left tail.

FIGURE 32
The author's quiz grades are left-skewed.

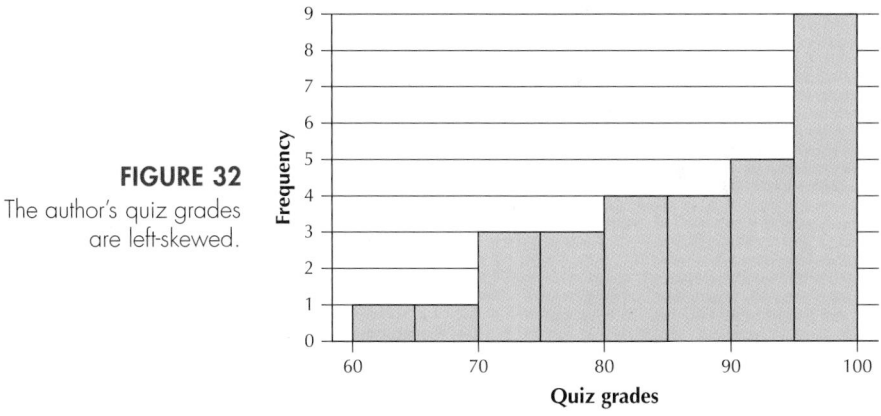

Alternatively, a data distribution may be right-skewed, where the distribution has a longer tail on the right than on the left. For example, in the Chapter 2 case study data, the number of criminal trespass misdemeanors per precinct is right-skewed because the right tail is longer than the left tail (Figure 33).

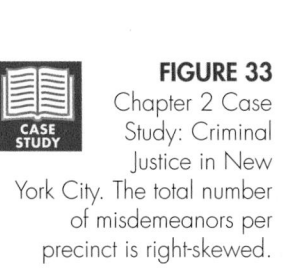

FIGURE 33
Chapter 2 Case Study: Criminal Justice in New York City. The total number of misdemeanors per precinct is right-skewed.

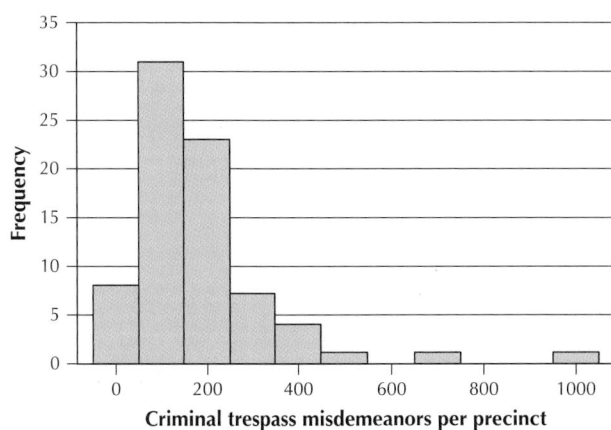

EXAMPLE 20   Identifying the shape of a distribution

Student-Run Café Business

We return to the student-run café business data set first introduced in Example 8 on page 49 of Section 2.1. Figure 34 contains a histogram of the daily sales (in dollars) for the student-run café business. Identify whether the distribution is symmetric, right-skewed, or left-skewed.

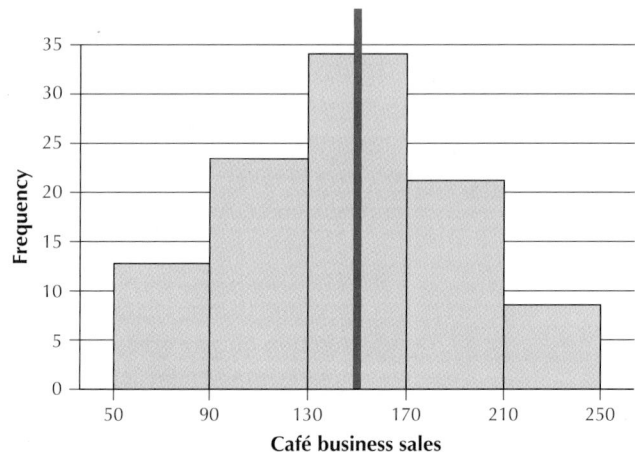

FIGURE 34 Histogram of café business sales.

Solution

NOW YOU CAN DO

Exercises 79–82.

The histogram in Figure 34 has a vertical axis of symmetry (red vertical line). Although the two halves are not perfect mirror images of each other, they are close. In the real world, data are seldom perfectly symmetric, due to random variation. Thus, we can accept that the café business sales data are symmetric.

Different graphical summaries may be appropriate for different situations. Example 21 illustrates this.

EXAMPLE 21   Choosing the appropriate graphical summary

Statistically literate citizens recognize that one may select different graphical summaries, depending on the intention of the presenter. Figures 35a, 35b, and 35c contain a dotplot, a histogram, and a stem-and-leaf display, respectively, of the average size of households in the 50 states and the District of Columbia. Which graphical summary—the dotplot, the histogram, or the stem-and-leaf display—is most useful if our primary objective is to

a.   assess symmetry and skewness?

b.   be able to construct it quickly using paper and pencil?

c.   retain complete knowledge of the original data set?

d.   give a presentation to people who have never taken a stats course?

(a)

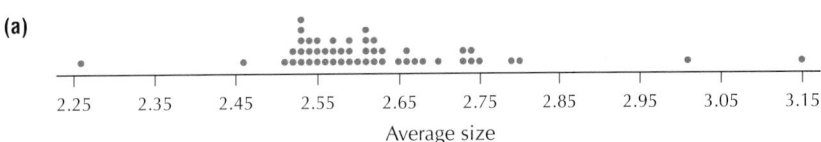

Average size

(b)

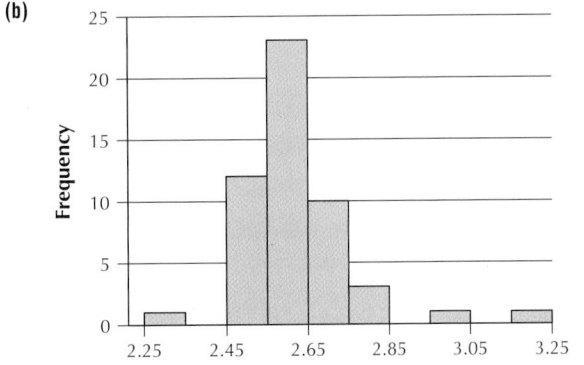

(c)

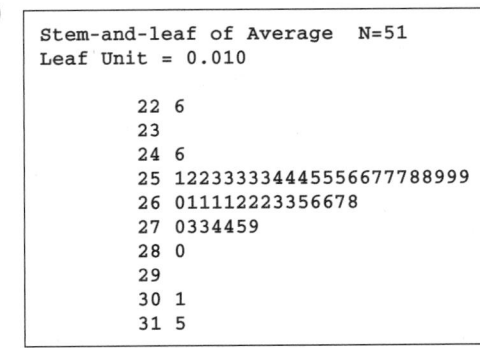

FIGURE 35 (a) Dotplot; (b) histogram; (c) stem-and-leaf display. Which is most useful?

Solution

a. All three graphics are good at assessing symmetry and skewness.

b. The dotplot's great asset is its simplicity. It can be quickly drawn, with minimal preparation, in contrast to the other two summaries, which require some organization or calculation.

c. The stem-and-leaf display was invented in order to retain complete knowledge of the data set. Histograms are the least effective in this regard.

d. The histogram is widely used in the real world and is probably the best choice for a presentation to those who have never taken a stats course.

STEP-BY-STEP TECHNOLOGY GUIDE: Quantitative Data

Suppose we want to produce a histogram of the carbon emissions data from Example 12 (page 63).   **carbon**

TI-83/84

Entering a Data Set
Step 1   Press **STAT**, then press **ENTER**. Highlight the **L1** list.
Step 2   Clear out any old data in **L1**. Press the **up arrow** key, then **CLEAR**, then **ENTER**.
Step 3   Enter the first data value **93.28**, and press **ENTER**.
Step 4   Continue entering data until the entire data set is in **L1** (Figure 36).

Constructing a Histogram
Step 1   Press **2nd**, then **Y =**. In the STAT PLOTS menu, highlight 1:, and press **ENTER**.
Step 2   Highlight **ON**, and press **ENTER**. Select the histogram icon (Figure 37), and press **ENTER**.
Step 3   Press **ZOOM**, then select **9:ZOOMSTAT**.
Step 4   Press **TRACE**. Selecting each class, in turn, provides class limits and class frequency. The histogram is given in Figure 38.

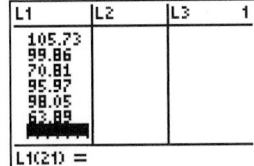

FIGURE 36 All data entered.

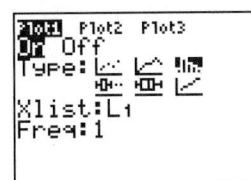

FIGURE 37 Selecting the histogram icon.

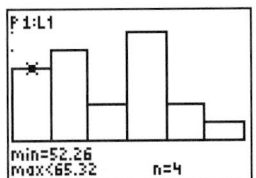

FIGURE 38 Histogram with leftmost class selected.

EXCEL

Constructing a Histogram
Make sure the Data Analysis package has been installed on your version of Excel.
Step 1 Input the carbon emissions data into column **A**. Click **Data > Data Analysis**.

Step 2 Select **Histogram** and click **OK**.
Step 3 For the *input range*, select the cells in which the data set resides. If the topmost cell has the variable name, select **Labels**. Select **Chart Output**, then click **OK**.

MINITAB

Constructing a Histogram
Step 1 Enter the carbon emissions data into column **C1**. Name the column *Carbon*.
Step 2 Click **Graph > Histogram**. Choose **Simple**, and click **OK**.
Step 3 Select **C1** *Carbon*, and click **Select**. Then click **OK**.
Step 4 The histogram is shown in Figure 39. Note that by default, Minitab uses midpoints instead of class limits to define the classes. Double-clicking anywhere on the midpoint values (50, 60, . . .) brings up a dialog box providing a wide range of options for changing the number of classes, class limits, etc.

Constructing a Stem-and-Leaf Display
Step 1 Enter the carbon emissions data into column **C1**. Name the column *Carbon*.

Step 2 Click **Graph > Stem-and-Leaf**.
Step 3 Click inside the space indicated **Graph variables**, select **C1** *Carbon*, and click **Select**. Then click **OK**.
Step 4 The output shown in Figure 40 tells us that the leaf unit is defined to be ones (1.0). Therefore, the stem unit is tens. (Ignore the leftmost column, which simply provides a cumulative count of the data points from the minimum and maximum.) The first row shows 526, indicating two data points, 52 and 56.

Dotplots
Step 1 Enter the carbon emissions data into column **C1**. Name the column *Carbon*.
Step 2 Click **Graph > Dotplot**. Select **Simple**. Click **OK**.
Step 3 Click in the **Graph Variables** section, click **C1** *Carbon*, and **Select**. Click **OK**.

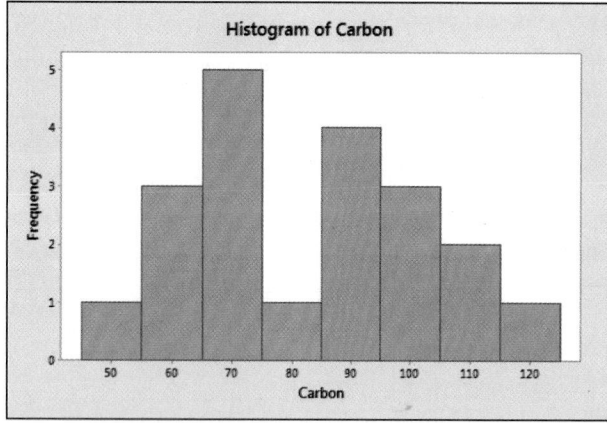

FIGURE 39 Minitab histogram.

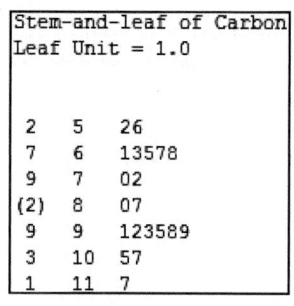

FIGURE 40 Minitab stem-and-leaf display.

SPSS

Constructing a Histogram and a Stem-and-Leaf Display
Step 1 Input the carbon emissions data into the first column, and name the column *Carbon*.
Step 2 Click **Analyze > Descriptive Statistics > Explore…**
Step 3 Select *Carbon* on the left, and click the first arrow to move it to the **Dependent List**.
Step 4 Click **Plots…**, select **Stem-and-leaf** and **Histogram**. Click **Continue**.

Step 5 Select **Plots** from the **Display** options. Click **OK**.

Dotplots
Step 1 Input the carbon emissions data into the first column, and name the column *Carbon*.
Step 2 Click **Graphs > Legacy Dialogs > Scatter/Dot…**
Step 3 Select **Simple Dot**, and click Define. Click *Carbon*, and move it to the **X Axis Variable**. Click **OK**.

JMP

Constructing a Histogram and a Dotplot
Step 1 Click **File > New > Data Table**. Input the carbon emissions data into **Column 1**. Name Column 1 *Carbon*.
Step 2 Click **Graph > Graph Builder**. Click and drag *Carbon* from the **Variables** box to the **X** axis.
Step 3
a. For a histogram, select **Histogram** from the options above the plot. The plot and graph options are shown in Figure 41. Click **Done**.

b. For a dotplot, select **Points** from the options above the plot. Make sure **Jitter**, under **Points**, is checked. Click **Done**.

Constructing a Stem-and-Leaf Display
Step 1 Click **File > New > Data Table**. Input the carbon emissions data into Column 1. Name Column 1 *Carbon*.
Step 2 Click **Analyze > Distribution**. Click *Carbon* in the **Select Columns** box, then click **Y, Columns**. Click **OK**.
Step 3 Click the red triangle next to *Carbon*. Click **Stem and leaf**. The output is shown in Figure 42.

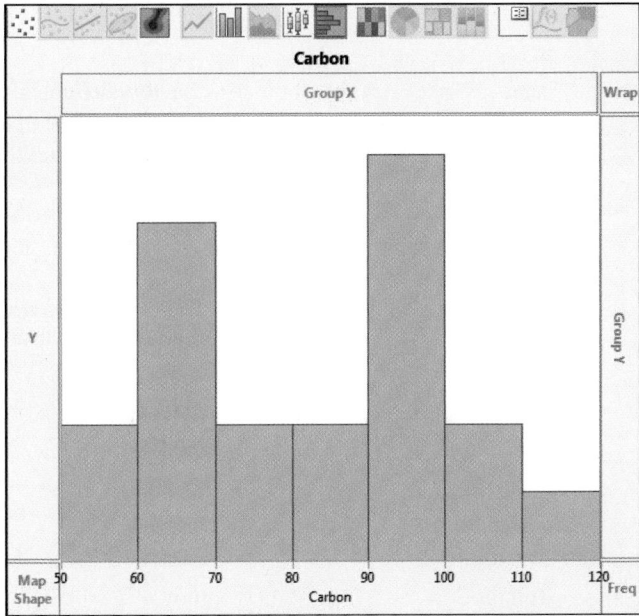

FIGURE 41 JMP histogram.

Stem and Leaf

Stem	Leaf	Count
11	8	1
11		
10	68	2
10	0	1
9	68	2
9	233	3
8	7	1
8	0	1
7		
7	12	2
6	689	3
6	14	2
5	7	1
5	2	1

5|2 represents 52

FIGURE 42 JMP stem-and-leaf.

CRUNCHIT!

Constructing a Histogram
Step 1   Click **File**, highlight **Load from Larose, Discostat3e > Chapter 2**, and click on **Example 02_12**.
Step 2   Click **Graphics**, and select **Histogram**. For **Sample**, select **Carbon**. (You may optionally select the number of bins, the bin width, and the location for the leftmost lower class limit.) Then click **Calculate**.

Constructing a Dotplot
Step 1   Click **File**, highlight **Load from Larose, Discostat3e > Chapter 2**, and click on **Example 02_12**.
Step 2   Click **Graphics**, and select **Dot Plot**. For **Sample**, select **Carbon**. Then click **Calculate**.

Section 2.2 Summary

In this section, we learned about using graphs and tables for summarizing quantitative (numerical) data.

1. Quantitative variables can be summarized using frequency and relative frequency distributions.

2. Histograms are a graphical display of a frequency or a relative frequency distribution with class intervals on the horizontal axis and the frequencies or relative frequencies on the vertical axis. A frequency polygon is constructed as follows: for each class, plot a point at the class midpoint, at a height equal to the frequency for that class; then join each consecutive pair of points with a line segment.

3. Stem-and-leaf displays contain more information than either a frequency distribution or a histogram because they retain the original data values in the display. In a dotplot, each data point is represented by a dot above the number line.

4. The point of making graphs and tables is so that we may obtain information from them. We illustrate how to obtain useful information from the graphs and tables we have made in this section.

5. An image or distribution has symmetry (or is symmetric) if a line (axis of symmetry) splits the image in half, so that one side is the mirror image of the other. Nonsymmetric distributions with a long right-hand tail are called right-skewed, whereas those with a long left-hand tail are called left-skewed.

Section 2.2 Exercises

CLARIFYING THE CONCEPTS

1. Which of the methods for displaying data introduced in this section (frequency and relative frequency distributions, histograms, frequency polygons, stem-and-leaf displays, and dotplots) can be used with both quantitative and qualitative data? Which can be used for quantitative data only? (p. 71)

2. Describe at least one potential benefit of combining classes when constructing a frequency distribution. Describe at least one potential benefit from retaining a larger number of classes. (p. 62)

3. In general, how many classes should be used when constructing a frequency distribution? (p. 63)

4. Describe at least one drawback of choosing class limits that overlap. (p. 64)

5. Describe at least one way that a dotplot may be useful. (p. 70)

6. In your own words, describe what is meant by "symmetry." Provide an example of a shape that is symmetric and an example of a shape that is not symmetric. (p. 73)

7. What are some examples of data sets that are often right-skewed? Left-skewed? (p. 73)

8. True or false: When the objective is to retain complete knowledge of the data set, the best graphical summary to use is the histogram. (p. 75)

PRACTICING THE TECHNIQUES

✅ **CHECK IT OUT!**

To do	Check out	Topic
Exercises 9–12	Example 9	Frequency and relative frequency distributions
Exercises 13–16	Example 10	Frequency distributions using classes
Exercises 17–40	Examples 11 and 12	Class limits, widths, boundaries, and frequency distributions for continuous data
Exercises 41–52	Example 13	Constructing histograms
Exercises 53–56	Example 14	Frequency polygons
Exercises 57–60	Example 15	Stem-and-leaf displays
Exercises 61–64	Example 17	Dotplots
Exercises 65–66	Example 18	Comparison dotplots
Exercises 67–78	Example 19	Acquiring information from graphs and tables
Exercises 79–82	Example 20	Identifying the shape of the distribution

Business Insider reported the list of actors in Table 24 who have received more than one Oscar nomination but have never won an Oscar. Use the data to construct the table or graph indicated in Exercises 9 and 10.

TABLE 24 Oscar nominations for actors who have never won an Oscar 🎰 nooscars

Actor	Nominations	Actor	Nominations
Peter O'Toole	8	Tom Cruise	3
Richard Burton	7	Will Smith	2
Glenn Close	6	John Travolta	2
Leonardo DiCaprio	5	Edward Norton	2
Julianne Moore	4	Judy Garland	2
Sigourney Weaver	3	James Dean	2
Johnny Depp	3		

Source: www.businessinsider.com/actors-who-have-never-won-oscar-2014-3.

9. Frequency distribution
10. Relative frequency distribution

The data in Table 25 represent the top 18 players in baseball history for the number of career grand slams (home run with the bases loaded). Use the data to construct the table or graph indicated in Exercises 11 and 12.

TABLE 25 Most career grand slams 🎰 grandslams

Player	Grand slams	Player	Grand slams
Alex Rodriguez	24	Hank Aaron	16
Lou Gehrig	23	Dave Kingman	16
Manny Ramírez	21	Babe Ruth	16
Eddie Murray	19	Ken Griffey, Jr.	15
Willie McCovey	18	Richie Sexson	15
Robin Ventura	18	Jason Giambi	14
Carlos Lee	17	Gil Hodges	14
Jimmie Foxx	17	Mark McGwire	14
Ted Williams	17	Mike Piazza	14

11. Frequency distribution
12. Relative frequency distribution

For Exercises 13 and 14, use the *Oscar nomination* data from Table 24.

13. Define the following classes: 0–3, 4–6, and 7–9. Use these classes to construct a frequency distribution.

14. Using the classes in the previous exercise, construct a relative frequency distribution.

For Exercises 15 and 16, use the *grand slams* data from Table 25.
15. Define the following classes: 11–15, 16–20, and 21–25. Use these classes to construct a frequency distribution.
16. Using the classes in the previous exercise, construct a relative frequency distribution.

The United States currently maintains a negative trade balance with many countries around the world, meaning that we import more from those countries than we export to them. This tends to increase unemployment here in the United States. The data in Table 26 represent the exports, imports, and trade balance of the United States with a sample of 11 countries, for the month of June 2014.

TABLE 26   Exports, imports, and trade balance (in $ billions) 🔣 tradebalance

Country	Exports to	Imports from	Trade balance
Brazil	3.5	2.5	1
France	2.8	4	−1.2
Germany	4.5	10	−5.6
India	1.9	3.2	−1.3
Italy	1.2	3.7	−2.4
Japan	5.6	11.3	−5.6
South Korea	3.8	5.6	−1.8
Saudi Arabia	1.7	3.5	−1.8
United Kingdom	4.4	4.4	0

Source: Foreign Trade Division, U.S. Census Bureau.

Use the *exports to* data for Exercises 17–20. Use five classes with class widths equal to 1. Define the leftmost class as: 1 to < 2.
17. Determine the class limits.
18. Determine the class boundaries.
19. Construct a frequency distribution.
20. Build a relative frequency distribution.

For Exercises 21–24, use the *imports from* data from Table 26. Use five classes with class widths equal to 2. Let the leftmost class be: 2 to < 4.
21. Determine the class limits.
22. Determine the class boundaries.
23. Construct a frequency distribution.
24. Build a relative frequency distribution.

For Exercises 25–28, again use the *imports from* data from Table 26. But this time, use seven classes with class widths equal to 1.5. Let the leftmost class be: 2.0 to < 3.5.
25. Determine the class limits.

26. Determine the class boundaries.
27. Construct a frequency distribution.
28. Build a relative frequency distribution.

For Exercises 29–32, use the *trade balance* data from Table 26. Use eight classes with class widths equal to 1. Let the leftmost class be: −6 to < −5.
29. Determine the class limits.
30. Determine the class boundaries.
31. Construct a frequency distribution.
32. Build a relative frequency distribution.

Table 27 contains the motor vehicle theft rate for the top 20 countries in the world for motor vehicle theft, for 2012. The theft rate equals the number of motor vehicles stolen in 2012 per 100,000 residents. 🔣 theftrate20

TABLE 27   Motor vehicle theft rate for top 20 countries

Country	Motor vehicle theft rate	Country	Motor vehicle theft rate
Italy	208.0	Trinidad	61.7
France	174.1	Jordan	58.3
USA	167.8	Hungary	56.5
Sweden	117.2	Lithuania	45.7
Belgium	106.0	Slovakia	45.2
Greece	100.2	Latvia	37.8
Norway	94.1	Switzerland	30.9
Netherlands	75.2	Serbia	28.9
Spain	75.1	Austria	27.2
Cyprus	66.0	Barbados	24.0

Source: United Nations Office on Drugs and Crime.

Use the *motor vehicle theft rate* data for Exercises 33–36. Use nine classes with class widths equal to 25. Define the leftmost class as: 0 to < 25.
33. Determine the class limits.
34. Determine the class boundaries.
35. Construct a frequency distribution.
36. Build a relative frequency distribution.

For Exercises 37–40, again use the *motor vehicle theft rate* data from Table 27. But this time, use eight classes with class widths equal to 25. Define the leftmost class as: 20 to < 45.
37. Determine the class limits.
38. Determine the class boundaries.
39. Construct a frequency distribution.
40. Build a relative frequency distribution.

For Exercises 41 and 42, use the *exports to* data from Table 26. Construct the indicated histogram using five classes with class widths equal to 1. Your work from Exercises 17–20 should be helpful.

41. Frequency histogram
42. Relative frequency histogram

For Exercises 43 and 44, use the *imports from* data from Table 26. Construct the indicated histogram, using five classes with class widths equal to 2. Your work from Exercises 21–24 should be helpful.
43. Frequency histogram
44. Relative frequency histogram

For Exercises 45 and 46, again use the *imports from* data from Table 26. This time, construct the indicated histogram using seven classes with class widths equal to 1.5. Use your work from Exercises 25–28.
45. Frequency histogram
46. Relative frequency histogram

For Exercises 47 and 48, use the *trade balance* data from Table 26. Construct the indicated histogram, this time using eight classes with class widths equal to 1. Use your work from Exercises 29–32.
47. Frequency histogram
48. Relative frequency histogram

For Exercises 49 and 50, use the *motor vehicle theft rate* data from Table 27. Construct the indicated histogram, using nine classes with class widths equal to 25. Use your work from Exercises 33–36.
49. Frequency histogram
50. Relative frequency histogram

For Exercises 51 and 52, again use the *motor vehicle theft rate* data from Table 27. But this time, construct the indicated histogram, using eight classes with class widths equal to 25. Use your work from Exercises 37–40.
51. Frequency histogram
52. Relative frequency histogram

For Exercises 53–56, construct a frequency polygon with the indicated data.
53. The *exports to* data from Table 26. Calculate the midpoints using five classes with class width equal to 1.
54. The *imports from* data from Table 26. Get the midpoints using five classes with class width equal to 2.
55. The *imports from* data from Table 26. But this time compute the midpoints using seven classes with class width equal to 1.5.
56. The *motor vehicle theft rate* data from Table 27. Calculate the midpoints using nine classes with class widths equal to 25.

For Exercises 57–60, construct a stem-and-leaf display of the indicated data.
57. *Exports to* data from Table 26.
58. *Imports from* data from Table 26.
59. *Trade balance* data from Table 26.
60. *Motor vehicle theft rate* data from Table 27.

For Exercises 61–64, construct a dotplot of the indicated data.
61. *Exports to* data from Table 26.
62. *Imports from* data from Table 26.

63. *Trade balance* data from Table 26.
64. *Motor vehicle theft rate* data from Table 27.

For Exercises 65 and 66, construct a comparison dotplot of the indicated data.
65. *Exports to* data and *imports from* data from Table 26.
66. *Exports to* data and *trade balance* data from Table 26.

The cost of the last music download (in dollars) for 100 college students is summarized in Figure 43. Use this information for Exercises 67–70.

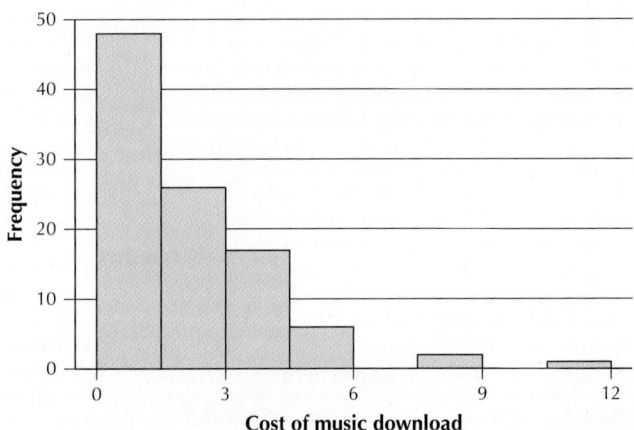

FIGURE 43 Cost of music download.

67. What are the class midpoints?
68. About how many students paid less than $1.50 for their last music download? What is the relative frequency?
69. About how many students paid $10.50 or more on their last music download?
70. Use your answers to the last two questions to estimate about how many students paid between $1.50 and $10.50 for their last music download.

Figure 44 shows a histogram of a set of statistics quiz scores. Use this information for Exercises 71–74.

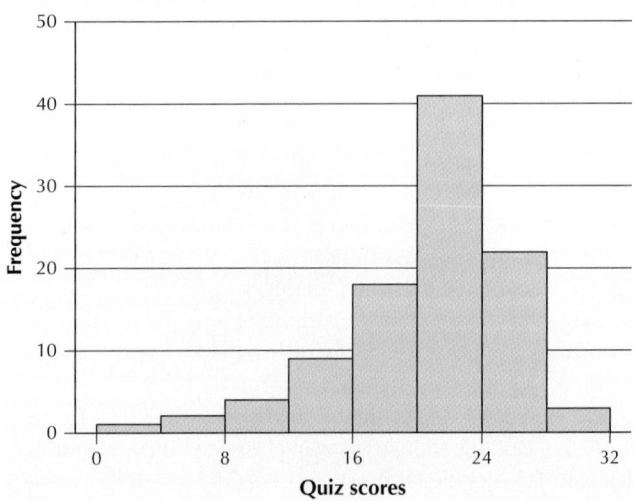

FIGURE 44 Quiz scores.

71. What are the class midpoints?

72. Between which two scores did the most quiz scores occur?

73. Can we tell what the highest grade on the quiz was? Why or why not? Would a stem-and-leaf display of this data be able to tell us what the highest grade was?

74. Estimate the relative frequency of quiz scores below 8.

Figure 45 shows a histogram of a set of women's heights. Use this information for Exercises 75–78.

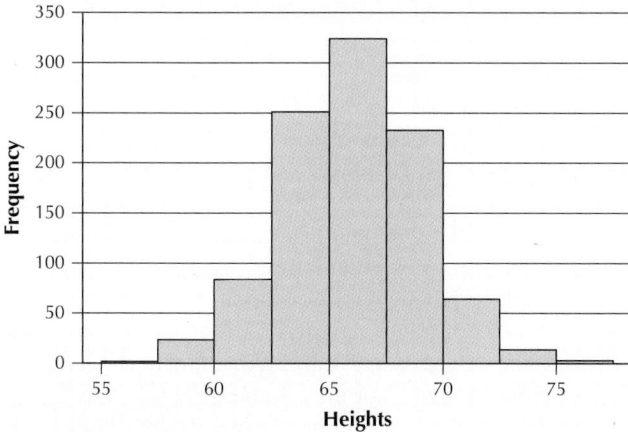

FIGURE 45 Women's heights.

75. Calculate the class midpoints.

76. Between which two values did the most heights occur?

77. If we added one inch to every woman's height, how would Figure 45 change?

78. If we added one inch to every woman's height, which aspects of Figure 45 would stay the same?

For Exercises 79–82, identify the shape of the distribution as either symmetric, right-skewed, or left-skewed.

79. The data represented in Figure 43.

80. The distribution of the data in Figure 44.

81. The data represented in Figure 45.

82. The data from Table 19 on page 61.

APPLYING THE CONCEPTS

83. Fruit Cup Sales for the Student-Run Café. Table 28 contains the number of fruit cups sold per day for the student-run café business. **fruitcups**

TABLE 28 Fruit cups sold per day

1	1	1	1	2	2
0	2	2	4	2	2
0	1	2	1	1	1
3	2	2	4	0	0
2	3	3	0	4	4
2	0	1	2	1	3
2	1	3	2	2	1
0	2	0	3	2	

a. Construct a frequency distribution and a relative frequency distribution similar to Table 18.

b. Use the following classes to construct a frequency distribution and a relative frequency distribution using classes, similar to Table 19. Use the following classes: 0–1, 2–3, 4–5.

c. What is the relative frequency of days that three or more fruit cups were sold? Did you use your work from (**a**) or (**b**) to answer this? Explain how we could not have used the distributions in (**b**) to answer this.

84. Sandwich Sales for the Student-Run Café. Table 29 shows the number of sandwiches sold per day for the student-run café business.

TABLE 29 Number of sandwiches sold sandwiches

5	6	8	4	3	7
6	0	3	2	3	4
9	1	3	8	7	8
2	3	8	6	4	4
6	7	6	5	2	3
8	4	4	6	7	3
5	2	4	8	4	6
1	2	6	4	4	

a. Construct a frequency distribution and a relative frequency distribution similar to Table 18.

b. Use the following classes to construct a frequency distribution and a relative frequency distribution using classes, similar to Table 19. Use the following classes: 0–2, 3–5, 6–8, 9–11.

c. Would you say that the distributions in (**b**) are symmetric? Explain.

d. What is the proportion (relative frequency) of days that fewer than six sandwiches were sold?

85. Frauds in Brooklyn in 2000. Table 30 provides the number of misdemeanor fraud cases in each of Brooklyn's 23 police precincts in 2000. **brooklynfrauds2000**

TABLE 30 Fraud cases in Brooklyn precincts, 2000

Precinct	Frauds	Precinct	Frauds
40	60	75	198
41	198	76	45
42	92	77	83
43	109	78	33
44	79	79	156
45	240	81	69
46	130	83	78
47	89	84	54
48	210	88	107
49	89	90	45
50	103	94	46
52	95		

a. Use 7 classes of width 40 each, starting at the leftmost class: 0 to < 40. Find the frequencies for each class.

b. Find the relative frequencies for each class.

c. Construct a frequency histogram of the number of fraud cases in Brooklyn in 2000.

d. Construct a relative frequency histogram of the number of fraud cases in Brooklyn in 2000.

 Frauds in Brooklyn in 2013. Table 31 provides the number of misdemeanor fraud cases in each of Brooklyn's 23 police precincts in 2013. Use this data for Exercises 86–90. ▐▌ **brooklynfrauds2013**

TABLE 31   Fraud cases in Brooklyn precincts, 2013

Precinct	Frauds	Precinct	Frauds
60	33	75	133
61	53	76	19
62	76	77	41
63	52	78	15
66	23	79	63
67	57	81	36
68	44	83	68
69	16	84	41
70	42	88	48
71	73	90	51
72	27	94	21
73	90		

86. Constructing and comparing histograms. Do the following:

a. Use 7 classes of width 40 each, starting at the leftmost class: 0 to < 40. Find the class boundaries.

b. Find the frequencies for each class.

c. Find the relative frequencies for each class.

d. Construct a frequency histogram of the number of fraud cases in Brooklyn in 2013.

e. Construct a relative frequency histogram of the number of fraud cases in Brooklyn in 2013.

f. Compare your frequency histograms from Exercises 85c and 86d. Describe any differences between the two graphs. Do you think this represents good news, bad news, or no news?

87. Getting Information from Histograms. Refer to your work from Exercise 86 to answer the following questions.

a. Find the relative frequency of precincts where 80 or more frauds occurred.

b. Compute the proportion (relative frequency) of precincts where 40 or more frauds occurred.

c. Compute the relative frequency of precincts that had fewer than 40 frauds. Explain how you could have used your answer from (b) to calculate this.

88. Constructing Frequency Polygons. Use Table 31 to answer the following:

a. Using the same class boundaries you calculated in Exercise 86, find the class midpoints.

b. Construct a frequency polygon of the number of misdemeanor fraud cases in Brooklyn in 2013.

c. Use (b) to construct a relative frequency polygon.

d. Use the relative frequency polygon to find the relative frequency of precincts where more than 100 frauds occurred.

89. Stem-and-Leaf Displays. Use Table 31 to answer the following:

a. Construct a stem-and-leaf display of the number of frauds in Brooklyn in 2013.

b. Build a split-stem stem-and-leaf display of the number of frauds.

c. Using the stem-and-leaf display, find how many precincts had 52 or more frauds. Explain whether this could have been found using the histogram from Exercise 86.

90. Dotplots. Use Tables 30 and 31 for the following:

a. Make a dotplot of the number of frauds in Brooklyn in 2013.

b. Construct a comparison dotplot of the number of frauds in Brooklyn in 2000 and in 2013.

c. Describe any differences between the two distributions in (b).

91. Coffee Sales for the Student-Run Café. In Table 32, we see the number of coffees sold per day for the student-run café business. Consider this to be continuous data.

TABLE 32   Coffees sold per day ▐▌ **coffees**

41	30	21	25	21	4
33	27	28	35	8	13
34	30	23	33	8	4
27	27	31	35	4	16
20	26	29	16	4	14
23	24	48	24	3	10
32	18	25	20	5	11
31	22	31	11	6	

a. Use 7 classes of width 8 each, starting with the following leftmost class: 0 to < 8. Calculate the class boundaries.

b. Find the frequencies for each class.

c. Find the relative frequencies for each class.

d. Construct a frequency histogram of coffees sold.

e. Construct a relative frequency histogram of coffees sold.

92. Soda Sales for the Student-Run Café. In Table 33, we see the number of sodas sold per day for the student-run café business. Consider this to be continuous data.

TABLE 33   Sodas sold per day 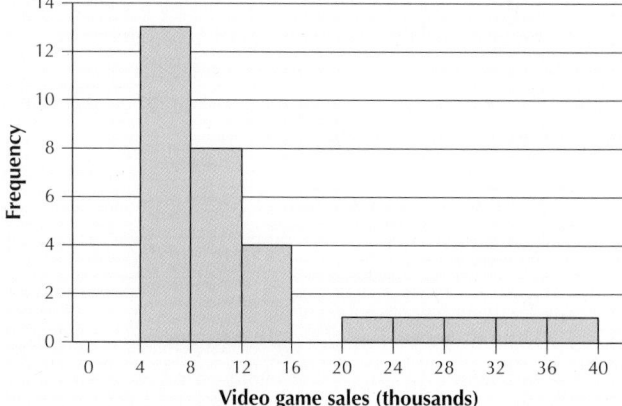 sodas

20	12	24	25	43	45
13	19	31	36	24	50
23	33	15	33	48	26
13	20	26	37	35	26
13	29	39	22	33	55
33	14	24	36	24	42
15	17	35	54	30	45
27	31	11	34	50	

a. Use 10 classes of width 5 each, starting with the following leftmost class: 10 to < 15. Calculate the class boundaries.
b. Find the frequencies for each class.
c. Find the relative frequencies for each class.
d. Construct a frequency histogram of sodas sold.
e. Construct a relative frequency histogram of sodas sold.

93. Coffee Sales. Refer to your work from Exercise 91 to answer the following questions.
a. Find the relative frequency of days that more than 39 coffees were sold.
b. Compute the proportion (relative frequency) of days that more than 31 coffees were sold.
c. Compute the relative frequency of days that 31 or fewer coffees were sold. Explain how you could have used your answer from (**b**) to calculate this.
d. Find the proportion of days that between 16 and 31 coffees (inclusive) were sold.

94. Soda Sales. Refer to your work from Exercise 92 to answer the following questions.
a. Find the relative frequency of days that more than 34 sodas were sold.
b. Find the relative frequency of days that fewer than 20 sodas were sold.
c. Compute the proportion (relative frequency) of days that 20 or more sodas were sold. Explain how you could have used your answer from (**b**) to calculate this.
d. Find the proportion of days that between 20 and 39 sodas (inclusive) were sold.

95. Coffee Sales. Use Table 32 to answer the following:
a. Use 7 classes of width 8 each, starting with the following leftmost class: 0 to < 8. Use the class boundaries you calculated in Exercise 91. Find the class midpoints.
b. Construct a frequency polygon of the number of coffees sold.
c. Use (**b**) to construct a relative frequency polygon of the number of coffees sold.
d. Use the relative frequency polygon to find the relative frequency of days that fewer than 17 coffees were sold.

96. Soda Sales. Use Table 33 to answer the following:
a. Use 10 classes of width 5 each, starting with the following leftmost class: 10 to < 15. Use the class boundaries you calculated in Exercise 92. Find the class midpoints.
b. Construct a frequency polygon of the number of sodas sold.
c. Use (**b**) to construct a relative frequency polygon of the number of sodas sold.
d. Use the relative frequency polygon to find the relative frequency of days that more than 24 sodas were sold.

97. Coffee Sales. Use Table 32 to answer the following:
a. Construct a stem-and-leaf display of the number of coffees sold.
b. Build a split-stem stem-and-leaf display of the number of coffees sold.
c. Make a dotplot of the number of coffees sold.
d. Using the stem-and-leaf display, find the number of days that 28 or more coffees were sold. Explain whether this could have been found using the histogram from Exercise 91.

98. Soda Sales. Use Table 33 to answer the following:
a. Construct a stem-and-leaf display of the number of sodas sold.
b. Build a split-stem stem-and-leaf display of the number of sodas sold.
c. Make a dotplot of the number of sodas sold.
d. Using the stem-and-leaf display, find the number of days that 32 or more sodas were sold. Explain whether this could have been found using the histogram from Exercise 92.

99. Best-selling video games. Figure 46 contains a histogram of the top 30 best-selling video game sales for the week of May 17, 2014. Use Figure 46 to answer the following:

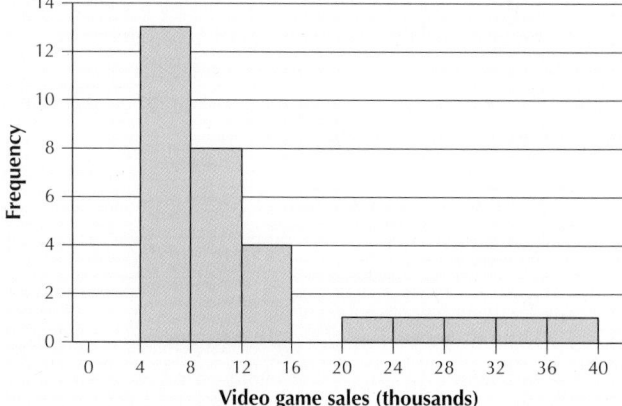

FIGURE 46 Histogram of video game sales.

a. Identify whether the distribution is symmetric, left-skewed, or right-skewed. Explain your answer.
b. Use the histogram to construct a relative frequency histogram.

c. Explain whether we could use the information in Figure 46 to construct a stem-and-leaf display of video game sales.

100. Small Businesses. The U.S. Census Bureau tracks the number of small businesses per city. The accompanying frequency polygon represents the numbers of small businesses per city (in thousands) for 266 cities nationwide.
  a. What is the class width?
  b. What is the lower class limit of the leftmost class? (*Hint:* Don't forget about the units.)
  c. Which class has the highest frequency?
  d. Which class has the lowest frequency?

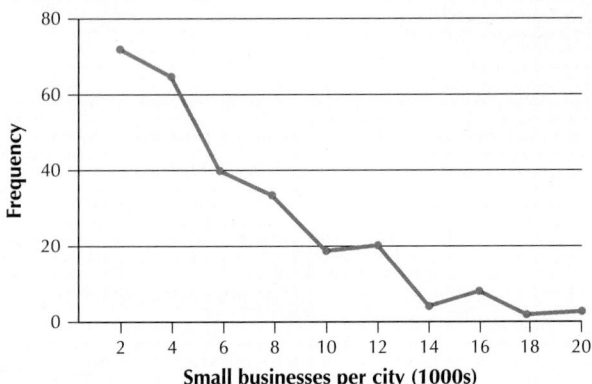

Small businesses per city (1000s)

101. Refer to the frequency polygon of small businesses per city.
  a. About how many cities have between 1000 and 3000 small businesses?
  b. About how many cities have more than 19,000 small businesses?
  c. About how many cities have between 9000 and 11,000 small businesses?

102. Countries and Continents. Suppose we are interested in analyzing the variable *continent* for the 10 countries in Table 34. Construct each of the following tabular or graphical summaries. If not appropriate, explain clearly why we can't use that method. **countrycont**

TABLE 34 Countries and continents

Country	Continent
Iraq	Asia
United States	North America
Pakistan	Asia
Canada	North America
Madagascar	Africa
North Korea	Asia
Chile	South America
Bulgaria	Europe
Afghanistan	Asia
Iran	Asia

a. Frequency distribution
b. Relative frequency distribution
c. Frequency histogram
d. Dotplot
e. Stem-and-leaf display

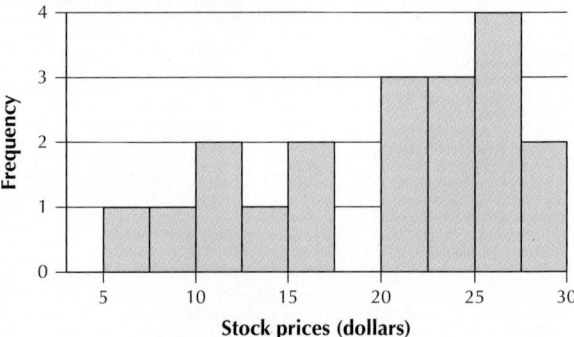

Stock prices (dollars)

103. Stock Prices. Refer to the histogram of stock prices of 19 technology firms.
  a. How could we turn this into a relative frequency histogram? Would the classes or the rectangles be affected?
  b. Suppose we were given a relative frequency histogram instead. How could we turn it into a frequency histogram?
  c. What is the sample size?

104. Refer to the histogram of stock prices.
  a. How many stocks were priced above $27.50?
  b. What is the relative frequency of stocks priced above $27.50?
  c. How many stocks had a price below $15?
  d. What is the relative frequency of stocks with a price below $15?

105. Refer to the histogram of stock prices.
  a. How many stocks are priced between $17.50 and $20?
  b. What is the relative frequency of stocks priced below $5?
  c. Which class has the largest relative frequency? Calculate this relative frequency.
  d. What is the frequency of stocks priced between $10 and $14.99?
  e. How many stocks had a price of $40?

106. Would you characterize the shape of the stock prices distribution as (a) tending to be symmetric, (b) tending to be right-skewed, or (c) tending to be left-skewed?

107. Stem-and-Leaf Display. Refer to the accompanying stem-and-leaf display. Reconstruct the data set.

```
Stem-and-leaf of Data   N  = 20
Leaf Unit = 1.0

   2   3
   2   45
   2   67
   2   889
   3   011
   3   2223
   3   5
   3   67
   3   9
   4   0
```

108. Refer to the stem-and-leaf display in Exercise 107. Construct a relative frequency distribution, using appropriate values for the class width and the lower class limit of the leftmost class.

109. Refer to the stem-and-leaf display in Exercise 107. Construct a frequency histogram.

110. Refer to the stem-and-leaf display in Exercise 107. Construct a dotplot.

111. Frequency Polygon. The following frequency polygon represents the quiz scores for a course in introductory statistics.

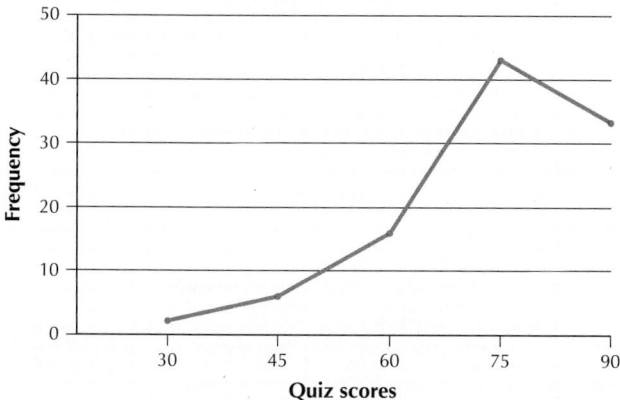

a. What is the class width?
b. What is the lower class limit of the class that has 45 as its midpoint?
c. What is the upper class limit of the class that has 45 as its midpoint?
d. Which class has the highest frequency?
e. Which class has the lowest frequency?

112. Refer to the frequency polygon of quiz scores.
a. About how many students scored higher than 82.5?
b. About how many students scored lower than 52.5?
c. Can we say how many students scored in the 90s? Why or why not?

WORKING WITH LARGE DATA SETS

New York Townspeople. Use the following information for Exercises 113–116. For towns in New York State, the following histogram provides information on the percentage of the townspeople who are between 18 and 65 years old.

newyork

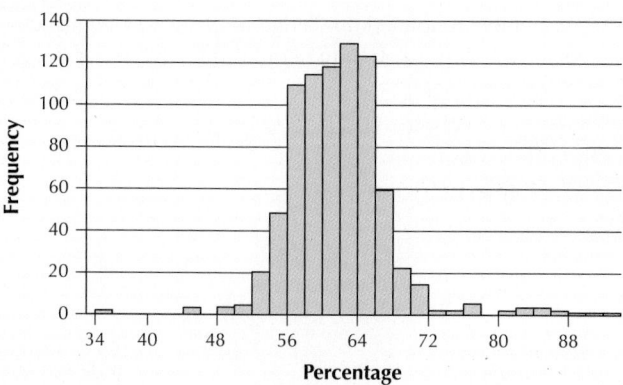

113. Would you characterize the distribution as left-skewed, right-skewed, or fairly symmetrical?

114. Provide an estimate of the "typical" percentage of townspeople who are between 18 and 65 years old. Is this typical value near the middle or near one of the "tails" of the distribution?

115. Provide a rough estimate of the sample size.

116. Would it be possible to construct a stem-and-leaf display using the information from the histogram? Explain.

Number of Businesses. Use the following data for Exercises 117–120. The data represent the number of business establishments in a sample of states. **statebusinesses**

State	Businesses (1000s)	State	Businesses (1000s)
Alabama	3.8	Michigan	7.5
Arizona	7.9	Minnesota	6.1
Colorado	8.9	Missouri	5.9
Connecticut	3.1	Ohio	9.5
Georgia	10.3	Oklahoma	3.8
Illinois	11.9	Oregon	5.4
Indiana	5.6	South Carolina	4.6
Iowa	2.7	Tennessee	5.4
Maryland	5.7	Virginia	8.6
Massachusetts	6.3	Washington	9.3

Source: U.S. Census Bureau.

117. Construct the following:
a. A frequency distribution
b. A relative frequency distribution
c. A relative frequency histogram

118. Construct the following:
a. A dotplot
b. A frequency polygon
c. A stem-and-leaf display

119. Compare and contrast the relative usefulness of each of four graphical presentation methods—dotplot, histogram, stem-and-leaf display, and frequency polygon—if our primary objective is to
a. assess symmetry and skewness.
b. be able to construct it quickly using paper and pencil.
c. retain complete knowledge of the data set.
d. give a presentation to people who have never taken a stats course.

WHAT IF ? **120.** *What if* we subtract the same amount (say, 1000) from each state's number of businesses. Explain how this would affect the following: What would change? What would stay the same?
a. Relative frequency histogram
b. Dotplot
c. Stem-and-leaf display
d. Frequency polygon

WORKING WITH LARGE DATA SETS

Fats and Cholesterol. For Exercises 121–125, use your knowledge of technology. Open the **Nutrition** data set. ▦ nutrition

121. How many observations are there in the data set? How many variables?

122. The variable *fat* contains the fat content in grams for each food. Construct a histogram of *fat*. Comment on the symmetry or the skewness of the histogram.

123. Is there a particular type of food with a fat content that is particularly large? Which type of food item (or set of similar food items) is this?

124. The variable *cholesterol* contains the cholesterol content in milligrams for each food. Construct a histogram of *cholesterol*. Comment on the symmetry or the skewness of the histogram.

125. Which food item (or set of similar food items) is highest in cholesterol?

Earthquakes. Use the *One Variable Statistics and Graphs* applet for Exercises 126–131. Work with the **Earthquakes** data set, which shows the magnitude on the Richter scale of 57 earthquakes that occurred during the week of October 15–22, 2007. ▦ earthquakes

126. Click on the **Histogram** tab.
  a. How many classes are included in the histogram?
  b. What is the class width?

127. Click on the leftmost rectangle in the histogram.
  a. What is the frequency for this class?
  b. What are the lower and upper class limits?

128. Click on the number line and drag slowly all the way to the left.
  a. What happens to the number of classes as you drag to the left?
  b. What happens to the class widths as you drag to the left?

129. Click on the number line and drag slowly all the way to the right.
  a. What happens to the number of classes as you drag to the right?
  b. What happens to the class widths as you drag to the right?

130. Click on the **Stem-and-Leaf** tab.
  a. How many stems are there?
  b. Without counting, how many leaves are there? How do we know this?

131. Select **Split Stems**.
  a. Now how many stems are there?
  b. How many leaves are there?
  c. Which stem-and-leaf display is preferable for the **Earthquakes** data—regular or split stems? Why?

CONSTRUCT YOUR OWN DATA SETS

132. Construct your own right-skewed data set of about 20 values. Just make up the data points, but be sure you know what the data represent (income, housing costs, etc.).
  a. Construct a stem-and-leaf display of your data set.
  b. Construct a dotplot of your data set.

133. Construct your own symmetric data set of about 20 values. Just make up the data points, but be sure you know what the data represent (for example, runs in a baseball game, number of right answers on a quiz).
  a. Construct a stem-and-leaf display of your data set.
  b. Construct a dotplot of your data set.

WORKING WITH LARGE DATA SETS

  **Petit Larceny.** Use the **Petit Larceny** data set for Exercises 134–141. ▦ petitlarceny

134. Build a frequency histogram of the number of petit larceny cases, per precinct, for 2000.

135. Build a relative frequency histogram of the number of petit larceny cases, per precinct, for 2000.

136. Build a histogram of the number of petit larceny cases, per precinct, for 2013. Make sure you use the same scale for the *x* axis and the *y* axis in each case, so that you may compare the two histograms more easily.

137. Compare the two histograms of petit larceny cases in 2000 and 2013. Describe any differences between the histograms. Would you say this reflects good news or bad news?

138. Identify the precinct with the unusual number of petit larceny cases in each case. Using the Internet, research where this precinct lies in New York City.

139. Construct a comparison dotplot of the number of petit larcenies for the years 2000 and 2013. Describe any differences between the two groups.

140. Build stem-and-leaf displays of the number of petit larcenies for the years 2000 and 2013. Describe any difference between the two groups.

141. Which graph—the histogram, the dotplot, or the stem-and-leaf display—is preferable if your objective is to:
  a. assess symmetry and skewness?
  b. be able to construct it quickly using paper and pencil?
  c. retain complete knowledge of the original data set?
  d. give a presentation to people who have never taken a stats course?

2.3 Further Graphs and Tables for Quantitative Data

OBJECTIVES By the end of this section, I will be able to . . .

1 Build cumulative frequency distributions and cumulative relative frequency distributions.
2 Create frequency ogives and relative frequency ogives.
3 Construct and interpret time series graphs.

1 Cumulative Frequency Distributions and Cumulative Relative Frequency Distributions

Quantitative data can be put in ascending order, so we can keep track of the accumulated counts at or below a certain value using a **cumulative frequency distribution** or **cumulative relative frequency distribution**. For example, if we list the prices of homes for sale in a neighborhood, a cumulative frequency distribution tells us how many homes are priced at $300,000 or less.

> For a discrete variable, a **cumulative frequency distribution** shows the total number of observations *less than or equal to* the category value. For a continuous variable, a **cumulative frequency distribution** shows the total number of observations *less than or equal to* the upper class limit.
>
> A **cumulative relative frequency distribution** shows the proportion of observations less than or equal to the category value (for a discrete variable) or the proportion of observations less than or equal to the upper class limit (for a continuous variable).

EXAMPLE 22 **Constructing cumulative frequency and cumulative relative frequency distributions**

Table 35 contains the total 2013 attendance for 25 Major League Baseball teams.

Table 35 Total 2013 attendance for 25 Major League Baseball teams (in millions)

1.5	1.6	1.6	1.7	1.8
1.8	1.8	1.8	2.1	2.1
2.3	2.4	2.5	2.5	2.5
2.6	2.6	2.8	2.8	3.0
3.0	3.1	3.2	3.3	3.7

Source: http://mlb.mlb.com.

The first three columns in Table 36 below contain the frequency distribution and relative frequency distribution for the attendance data. Construct a cumulative frequency distribution and a cumulative relative frequency distribution for the attendance figures.

Solution

To find the cumulative frequency for a class, add the frequencies of the classes equal to or below the upper class limit of that class. For example, the cumulative frequency for the class $2.5 \leq x < 3$ is the sum of the frequency for this class and for the class $1.5 \leq x < 2$. The procedure for the cumulative relative frequencies is similar. The results are shown in the last two columns of Table 36, where we can see that more than three-quarters (0.76) of these teams had attendance of less than 3 million.

Table 36 Cumulative frequency distribution and cumulative relative frequency distribution

Attendance	Frequency	Relative frequency	Cumulative frequency	Cumulative relative frequency
$1.5 \leq x < 2$	8	0.32	8	0.32
$2 \leq x < 2.5$	4	0.16	$8 + 4 = 12$	$0.32 + 0.16 = 0.48$
$2.5 \leq x < 3$	7	0.28	$12 + 7 = 19$	$0.48 + 0.28 = 0.76$
$3 \leq x < 3.5$	5	0.20	$19 + 5 = 24$	$0.76 + 0.20 = 0.96$
$3.5 \leq x < 4$	1	0.04	$24 + 1 = 25$	$0.96 + 0.04 = 1.00$
Total	25	1.00		

NOW YOU CAN DO
Exercises 9–12.

YOUR TURN #13

Using the unemployment rate data from Table 22 on page 65, construct a cumulative frequency distribution and cumulative relative frequency distribution of the unemployment rates.

(The solution is shown in Appendix A.)

2 Ogives

Just as histograms and frequency polygons are the graphical equivalent of frequency distributions, we have the following graphical equivalent of a cumulative frequency distribution.

> An **ogive** (pronounced "oh jive") is the graphical equivalent of a cumulative frequency distribution or a cumulative relative frequency distribution. Similar to a frequency polygon, an ogive consists of a set of plotted points connected by line segments. The x coordinates of these points are the upper class limits; the y coordinates are the cumulative frequencies or cumulative relative frequencies.

EXAMPLE 23   Constructing an ogive

 ballattend

Construct a relative frequency **ogive** for the attendance data in Table 36.

Solution

For the x coordinates, we use the upper class limits for attendance, and for the y coordinates, we use the cumulative relative frequencies. The result is shown in Figure 47.

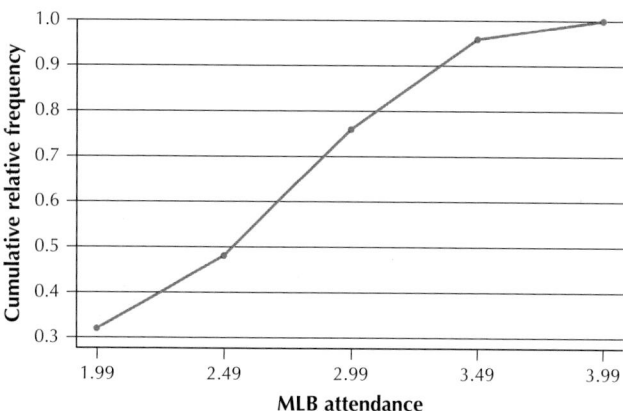

FIGURE 47 Ogive for baseball attendance.

NOW YOU CAN DO
Exercises 13–16.

YOUR TURN
#14

Using the unemployment data in Table 22 on page 65, construct a frequency ogive of the unemployment rates.

(The solution is shown in Appendix A.)

What Does This Graph Mean?

The ogive is a graphical representation of a cumulative relative frequency distribution. Thus, the first point (1.99, 0.32) indicates that 32% of the teams had total attendance at or below 1.99 million. The cumulative nature of the graph means that it can never decrease from left to right. The cumulative attendance increases until the rightmost point (3.99, 1.0) indicates that 100% (all) of the teams had total attendance at or below 3.99 million.

---

3 Time Series Graphs

Data analysts are often interested in how the value of a variable changes over time. Data that are analyzed with respect to time are called *time series data*.

A graph of time series data is called a **time series plot**. The horizontal axis of a time series plot represents time (for example, hours, days, months, years). The values of the time series data are plotted on the vertical axis, and line segments are drawn to connect the points.

EXAMPLE 24   Constructing a time series plot

 murderrate

Table 37 contains the murder rate per 100,000 residents for New York City, from 1990 to 2014.

a. Construct a time series plot of the data.

b. Describe any patterns you see.

Solution

a. We indicate the years 1990–2014 on the horizontal axis of the time series plot (Figure 48). Then, for each year, we plot the murder rate per 100,000 residents. Finally, we join the points using line segments.

Table 37 Murder rate in New York City, 1990–2014

Year	Rate	Year	Rate	Year	Rate
1990	14.5	1999	5	2008	4.3
1991	14.2	2000	5	2009	4
1992	13.2	2001	5	2010	4.5
1993	13.3	2002	4.8	2011	3.9
1994	11.1	2003	4.9	2012	3.5
1995	8.5	2004	4.6	2013	3
1996	7.4	2005	4.5	2014	2.7
1997	6	2006	4.8		
1998	5.1	2007	4.2		

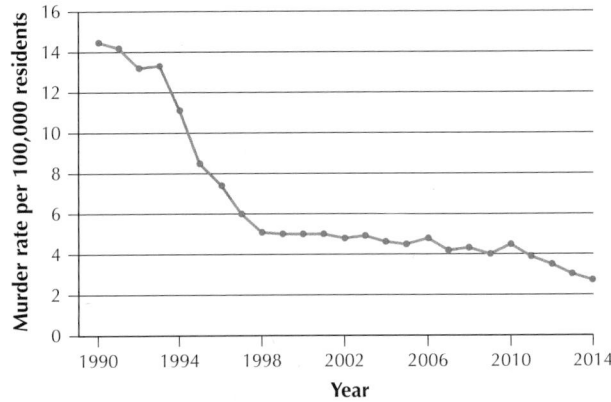

FIGURE 48 Time series plot. Murder rate in New York City, 1990–2014.

NOW YOU CAN DO
Exercises 17 and 18.

b. Note that the murder rate fell quickly from 1993 to 1998 and then tended to flatten out until 2010, when it began a slow descent until 2014. In the Step-by-Step Technology Guide, we illustrate how to construct this time series graph using technology.

EXAMPLE 25 Constructing a time series plot using technology

 maunaloa1

The data set **Mauna Loa 1** contains the carbon dioxide levels at Mauna Loa from May 2000 to May 2014. Use technology to construct a time series plot of the data.

Solution

We use the instructions provided in the Step-by-Step Technology Guide at the end of this section. The resulting Minitab time series plot is shown in Figure 49. (The year on the horizontal axis indicates the month of May of each year. For example "2014" refers to May 2014.)

In Figure 49, we observe both a seasonal pattern and a long-term trend. Every autumn and winter, the carbon dioxide level increases, and every summer it decreases. In autumn and winter, leaves and other deciduous vegetation decay, releasing their store of carbon back into the atmosphere. In the spring and summer, the new year's leaves require carbon to grow and extract it from the atmosphere, thereby reducing the

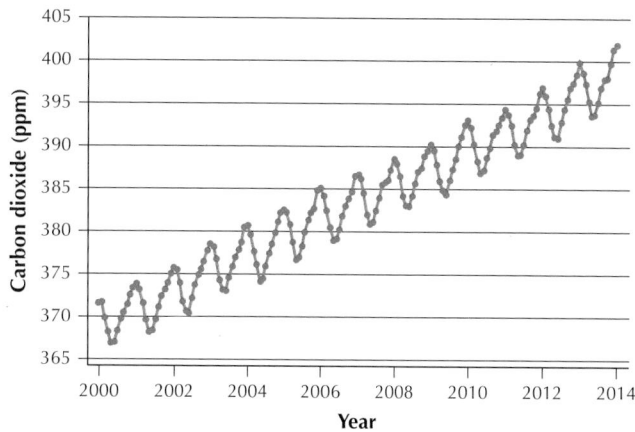

FIGURE 49 Minitab plot of carbon dioxide levels at Mauna Loa, Hawaii, 2000–2014

atmosphere's carbon dioxide level. Thus, the biosphere "inhales" carbon each summer and "exhales" it each winter. However, the low point of each successive cycle does not quite reach the level of the previous cycle before heading up again. This leads to an overall increasing trend in the amount of carbon dioxide in the atmosphere as we move from 2000 to 2014.

STEP-BY-STEP TECHNOLOGY GUIDE: Time Series Plots

We illustrate how to construct a time series plot using Example 24 (page 89).

TI-83/84

Step 1   Enter your time index (integers 1, 2, . . .) into list **L1**.
Step 2   Enter the values of your time series variable into list **L2**.
Step 3   Press **2nd**, then **Y =**. In the STAT PLOTS menu, highlight **1:** and press **ENTER**.
Step 4   Highlight **ON**, and press **ENTER**. Select the time series icon (Figure 50), and press **ENTER**.
Step 5   Press **ZOOM > 9:ZOOMSTAT**, and press **ENTER**. The time series plot is shown in Figure 51.

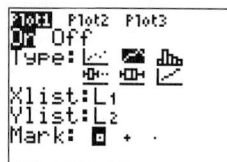

FIGURE 50 Selecting the time series icon.

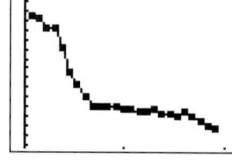

FIGURE 51 TI-83/84 time series plot.

EXCEL

Step 1   Enter the values of your time series variable into column **A**. Put the variable name at the top of the column.
Step 2   Select cells A1–A26. Click **Insert > Insert Line Chart** (in the **Chart** section).

Step 3   Choose **Line with Markers**.

MINITAB

Step 1   Enter the values of your time series variable into column C1.
Step 2   Click **Graph > Time Series Plot...**
Step 3   Select **Simple** and click **OK**.
Step 4   For **Series**, double-click on **C1**.

Step 5   Click **Time/Scale**. Select **Calendar > Year**.
Step 6   For **Start value**, enter **1990**. For **Data Increment**, enter 1.
Step 7   Click **OK** and ,**OK**.

SPSS

Step 1 Enter the year data into the first column and the values of the time series into the second column. In the **Variable View** tab, name the columns *Year* and *Rate* and assign *Year* a **Decimals** value of zero.

Step 2 Click **Analyze > Forecasting > Sequence Charts...**

Step 3 Click the name of the column that contains the time series values (*Rate*), then click the first arrow to move it to the **Variables** box.

Step 4 Click the name of the column that contains the months (*Year*), then click the second arrow to move it to the **Time Axis Labels** box.

Step 5 Click **OK**. The time series plot is shown in Figure 52.

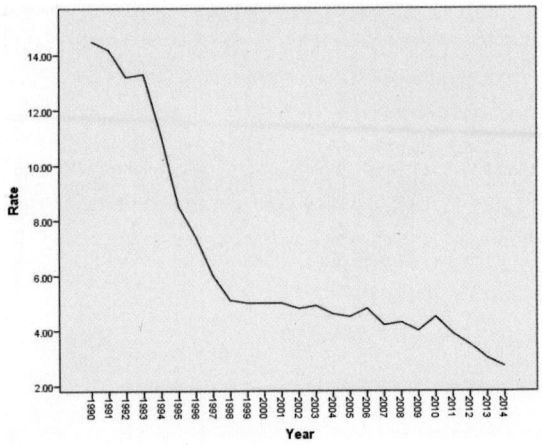

FIGURE 52 SPSS time series plot.

JMP

Step 1 Click **File > New > Data Table**.

Step 2 Enter the year data into **Column 1**, and the time series values into **Column 2**. Name the columns *Year* and *Rate*.

Step 3 Click **Graph > Graph Builder**.

Step 4 Drag *Year* (in the **Variables** section) to the **X** axis, and *Rate* to the **Y** axis.

Step 5 Select **Line** from the graph options shown above the plot. Click **Done**.

CRUNCHIT!

Step 1 Click **File**, highlight **Load from Larose, Discostat3e > Chapter 2**, and click **Table 03_37**.

Step 2 Click **Graphics**, then **Scatterplot**.

Step 3 For **X**, choose **Month**. For **Y**, choose **Rate**. Under **Parameters**, for **Show** choose **Both**. Click **Calculate**.

Section 2.3 Summary

1. A cumulative frequency distribution shows the total number of observations less than or equal to the category value (for a discrete variable) or the upper class limit (for a continuous variable). A cumulative relative frequency distribution shows the proportion of observations less than or equal to the category value (for a discrete variable) or the upper class limit (for a continuous variable).

2. An ogive is the graphical equivalent of a cumulative frequency distribution or a cumulative relative frequency distribution. The *x* coordinates of the points are the upper class limits; the *y* coordinates are the cumulative frequencies or cumulative relative frequencies.

3. Data that are analyzed with respect to time are called time series data. A graph of time series data is called a time series plot. The horizontal axis of a time series plot represents time (for example, hours, days, months, years). The values of the time series data are plotted on the vertical axis, and line segments are drawn to connect the points.

Section 2.3 Exercises

CLARIFYING THE CONCEPTS

1. Explain the difference between a frequency distribution and a cumulative frequency distribution. (p. 87)

2. Explain the difference between a cumulative frequency distribution and a cumulative relative frequency distribution. (p. 87)

3. What is the graphical equivalent of a cumulative frequency distribution? (p. 88)

4. Explain how to construct an ogive. (p. 88)

5. What do we call data that are analyzed with respect to time? (p. 89)

6. Explain how to construct a time series plot. (p. 89)

PRACTICING THE TECHNIQUES

 CHECK IT OUT!

To do	Check out	Topic
Exercises 9–12	Example 22	Cumulative frequency and cumulative relative frequency distributions
Exercises 13–16	Example 23	Ogives
Exercises 17–18	Examples 24 and 25	Time series plots

The U.S. Census Bureau reported in 2014 that the relative frequency of the age of the head of the household is as shown in Table 38 (for those with heads of households younger than age 65). Use Table 38 to construct the following graphical summaries of the variable *age*.

TABLE 38   Relative frequency distribution of *age*

Age	Frequency (millions)	Relative frequency
15 to < 35	22.7	0.24
35 to < 45	22.2	0.24
45 to < 55	25.8	0.27
55 to < 65	23.2	0.25

7. Cumulative frequency distribution
8. Cumulative relative frequency distribution

For Exercises 9–12, do the following:
   a. Construct a cumulative frequency distribution for the indicated data.
   b. Build a cumulative relative frequency distribution for the indicated data.
9. Carbon emissions data from Table 20 on page 63.
10. Unemployment data from Table 22 on page 65.
11. Dangerous weapons data from Table 23 on page 68.
12. Brooklyn frauds 2013 data from Table 31 on page 82.

For Exercises 13–16, do the following:
   a. Construct a frequency ogive for the indicated data.
   b. Build a relative frequency ogive for the indicated data.
13. Carbon emissions data from Table 20 on page 63.
14. Unemployment data from Table 22 on page 65.
15. Dangerous weapons data from Table 23 on page 68.
16. Brooklyn frauds 2013 data from Table 31 on page 82.

17. The following time series data represent the number of aggravated harassment cases handled by New York City Police Precinct 1 from 2000 to 2013.
   a. Construct the time series graph of the data.
   b. Describe any patterns you see.   **harassment**

Year	2000	2001	2002	2003	2004	2005	2006
Cases	547	568	476	475	450	445	379
Year	2007	2008	2009	2010	2011	2012	2013
Cases	424	404	425	429	343	400	400

18. The following time series data represent the number of petit larceny cases handled by New York City Police Precinct 5 from 2000 to 2013.
   a. Construct the time series graph of the data.
   b. Describe any patterns you see.   **petitlarceny5**

Year	2000	2001	2002	2003	2004	2005	2006
Cases	909	846	834	793	798	871	808
Year	2007	2008	2009	2010	2011	2012	2013
Cases	859	1020	1014	1263	1197	1240	1288

APPLYING THE CONCEPTS

19. Unemployment Rate. The frequency ogive below represents the unemployment rate (in percentages) for 367 cities nationwide.[9]

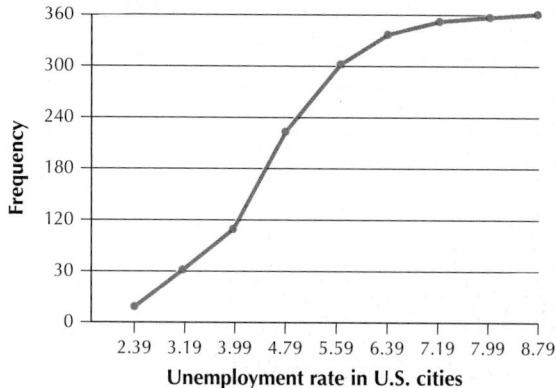

   a. What is the class width?
   b. What is the upper class limit of the leftmost class?
   c. What is the class midpoint of the leftmost class?
20. Refer to the frequency ogive of unemployment rates.
   a. About how many cities have unemployment rates 3.99 and below?
   b. About how many cities have unemployment rates 5.59 and below?
   c. About how many cities have unemployment rates 5.6 and above?
21. Atmospheric Carbon Dioxide. Table 39 contains the amount of carbon dioxide in parts per million (ppm) found in the atmosphere above Mauna Loa, Hawaii, measured monthly from October 2012 to September 2013.
   a. Construct a **time series plot** of these data.
   b. Describe the pattern you see.

maunaloa2

TABLE 39   Atmospheric carbon dioxide at Mauna Loa, October 2012 to September 2013

Month	Carbon dioxide (ppm)	Month	Carbon dioxide (ppm)
Oct.	391.01	Apr.	398.35
Nov.	392.81	May	399.76
Dec.	394.28	June	398.58
Jan.	395.54	July	397.20
Feb.	396.80	Aug.	395.15
Mar.	397.31	Sept.	393.51

Source: Dr. Pieter Tans, Earth System Research Laboratory, National Oceanic and Atmospheric Administration, www.esrl.noaa.gov/gmd/ccgg/trends.

22. Medicare. Table 40 contains a time series of the number of enrollees (in millions) in Medicare from 1987 to 2012. Construct a time series plot.

TABLE 40 Medicare enrollees (in millions)

Year	Enrollees	Year	Enrollees	Year	Enrollees
1987	30	1996	35	2005	40
1988	31	1997	36	2006	40
1989	31	1998	36	2007	41
1990	32	1999	37	2008	43
1991	33	2000	38	2009	43
1992	33	2001	38	2010	45
1993	33	2002	39	2011	47
1994	34	2003	40	2012	49
1995	35	2004	40		

Source: U.S. Census Bureau.

23. Refer to your time series plot from the preceding exercise. The increase in the number of enrollees is fairly constant and then becomes steeper. In what year does this change occur?

Agricultural Exports. For Exercises 24–26, refer to Table 41. The table gives the value of agricultural exports (in billions of dollars) from the top 20 U.S. states in 2009.

TABLE 41 Agricultural exports (in billions of dollars)

State	Exports	State	Exports
California	12.5	Arkansas	2.6
Iowa	6.5	North Dakota	5.2
Texas	4.7	Ohio	2.7
Illinois	5.5	Florida	2.1
Nebraska	4.8	Wisconsin	2.2
Kansas	4.7	Missouri	2.7
Minnesota	4.3	Georgia	1.8
Washington	3.0	Pennsylvania	1.7
North Carolina	2.9	Michigan	1.6
Indiana	3.1	South Dakota	2.3

Source: U.S. Department of Agriculture. **agriexports**

24. Construct a cumulative frequency distribution of agricultural exports. Start at $0 and use class widths of $2 billion.
 a. How many states have exports of less than $4 billion?
 b. How many states have exports of less than $6 billion?
 c. How many states have exports of at least $6 billion?
25. Construct a cumulative relative frequency distribution of agricultural exports. Start at $0 and use class widths of $2 billion.
 a. What proportion of states has exports of less than $4 billion?
 b. What proportion of states has exports of less than $6 billion?
 c. What proportion of states has exports of at least $6 billion?
26. Use your cumulative relative frequency distribution to construct a relative frequency ogive of agricultural exports.

27. Per Capita Income. The following data represent the per capita income in the United States from 1967 to 2012, in thousands of constant (2012) dollars. **percapitaincome**

Year	Per capita income $1000s	Year	Per capita income $1000s	Year	Per capita income $1000s
1967	15	1983	21	1998	28
1968	16	1984	22	1999	29
1969	17	1985	22	2000	30
1970	17	1986	23	2001	30
1971	17	1987	24	2002	29
1972	18	1988	24	2003	29
1973	19	1989	25	2004	29
1974	19	1990	25	2005	29
1975	19	1991	24	2006	30
1976	19	1992	24	2007	30
1977	20	1993	25	2008	29
1978	21	1994	25	2009	28
1979	21	1995	26	2010	28
1980	21	1996	26	2011	28
1981	21	1997	27	2012	28
1982	21				

Source: U.S. Census Bureau.

 a. Construct a time series plot of the per capita income.
 b. A fairly constant increasing trend occurs. In what year does this trend appear to end?
28. Rainfall in Fort Lauderdale. The following data represent the total monthly rainfall (in inches) in 2013 in Fort Lauderdale, Florida, as reported by the U.S. Historical Climatology Network. **flrainfall**

Jan.	0.55	July	15.54
Feb.	2.39	Aug.	3.33
Mar.	0.15	Sept.	6.78
Apr.	3.99	Oct.	5.8
May	13.63	Nov.	11.61
June	13.63	Dec.	1.11

 a. Construct a time series plot of the data.
 b. Is it wetter in summer or winter in Fort Lauderdale?
WHAT IF ? **29.** In Exercise 28, *what if* we add 3 inches to each month's rainfall amount. Describe how this would affect the time series plot. What would change? What would stay the same?
30. Cigarette Use Among 12th-Graders. Table 42 presents the percentages of 12th-graders who smoke cigarettes, for the years 1980–2009.

 12thsmokers

 a. Construct a time series plot of the data.
 b. Describe any trends that you see.

TABLE 42   12th-graders who smoke

Year	Percent	Year	Percent
1980	30.5	1995	33.5
1981	29.4	1996	34.0
1982	30.0	1997	36.5
1983	30.3	1998	35.1
1984	29.3	1999	34.6
1985	30.1	2000	31.4
1986	29.6	2001	29.5
1987	29.4	2002	26.7
1988	28.7	2003	24.4
1989	28.6	2004	25.0
1990	29.4	2005	23.2
1991	28.3	2006	21.6
1992	27.8	2007	21.6
1993	29.9	2008	20.4
1994	31.2	2009	20.1

Source: Monitoring the Future study, University of Michigan.

31. Miami Arrests. The Miami-Dade Department of Corrections and Rehabilitation publishes its monthly average daily population of inmates in its Annual Report. Table 43 shows the average daily number of inmates from October 2010 through September 2012. Construct a time series graph of the data.

TABLE 43   Average monthly inmate population, Miami-Dade Department of Corrections miamideptcorrections

Month	Inmates	Month	Inmates	Month	Inmates
Oct 2010	5753	Jun 2011	5500	Feb 2012	5138
Nov 2010	5600	Jul 2011	5486	Mar 2012	5111
Dec 2010	5387	Aug 2011	5515	Apr 2012	5097
Jan 2011	5388	Sep 2011	5406	May 2012	5117
Feb 2011	5471	Oct 2011	5304	Jun 2012	5175
Mar 2011	5504	Nov 2011	5201	Jul 2012	5185
Apr 2011	5538	Dec 2011	5141	Aug 2012	5214
May 2011	5567	Jan 2012	5129	Sept 2012	5229

Source: www.miamidade.gov/corrections/library/Annual-Report-2011-2012.pdf.

Individual Value Plot. Have you played *Minecraft*? On which platform did you play it? Figure 53 represents an *individual value plot* of video game sales for the week of May 17, 2014, separated by platform. An individual value plot is similar to a dotplot for different categories, which is rendered vertically. Each dot represents the sales for a particular video game. The top five sellers are labeled. *Minecraft* was the biggest seller that week, with the PS3 version slightly outselling the Xbox 360 version. Use this information for Exercises 32 and 33.

32. Which platform is indicated by MLB 14?

33. Are sales higher for the top-performing title for Xbox One ("Xone") or 3DS?

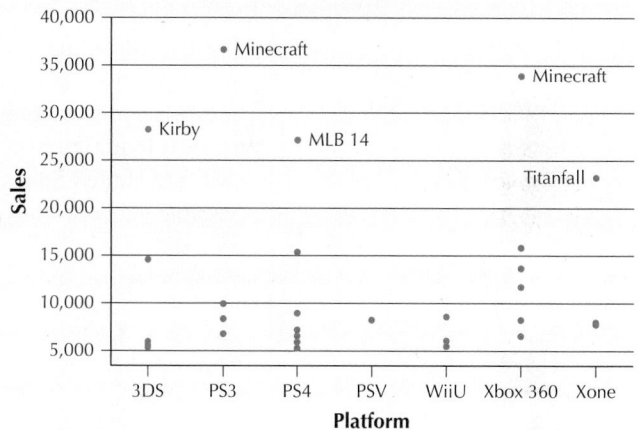

FIGURE 53 Individual value plot of video games sales, by platform.

WORKING WITH LARGE DATA SETS

Assault. Open the **Assault** data set, which contains the number of third-degree assaults per precinct for the years 2000–2013. Use technology to do the following:

34. Construct a time series plot of the number of third-degree assaults in Precinct 1 from 2000 to 2013. Describe any patterns you see.

35. For each year, calculate the sum of the number of third-degree assaults, across all precincts.

36. Build a time series plot of the total number of third-degree assaults, across all precincts. Describe any patterns you see.

2.4   Graphical Misrepresentations of Data

OBJECTIVE By the end of this section, I will be able to . . .

1   Avoid eight common practices that can make a graph misleading, confusing, or deceptive.

In the Information Age, when our world is awash in data, it is important for citizens to understand how graphics may be made misleading, confusing, or deceptive. Such an understanding enhances our statistical literacy and makes us less prone to being deceived by misleading graphics.

Eight Common Methods for Making a Graph Misleading

1. Graphing/selecting an inappropriate statistic.
2. Omitting the zero on the relevant scale.
3. Manipulating the scale.
4. Using two dimensions (area) to emphasize a one-dimensional difference.
5. Careless combination of categories in a bar graph.
6. Inaccuracy in relative lengths of bars in a bar graph.
7. Biased distortion or embellishment.
8. Unclear labeling.

EXAMPLE 26   Inappropriate choice of statistic

The United Nations Office on Drugs and Crime reports the statistics, given in Table 44, on the top five nations in the world ranked by numbers of cars stolen in 2012. The car thieves seem to be preying on cars in the United States, which has endured more than the next four highest countries put together. (See also the bar graph in Figure 54.) However, the United States has a much greater population than these other countries. Is it possible that, *per capita* (per person), the car theft rate in the United States is not so bad?

Table 44 Top five nations for total number of cars stolen in 2012

Country	Cars stolen
United States	532,900
Italy	126,627
France	111,305
Spain	35,131
Netherlands	12,575

Solution

In this case, the total number of cars stolen is an *inappropriate statistic* because the population of the United States is greater than the populations of the other countries.

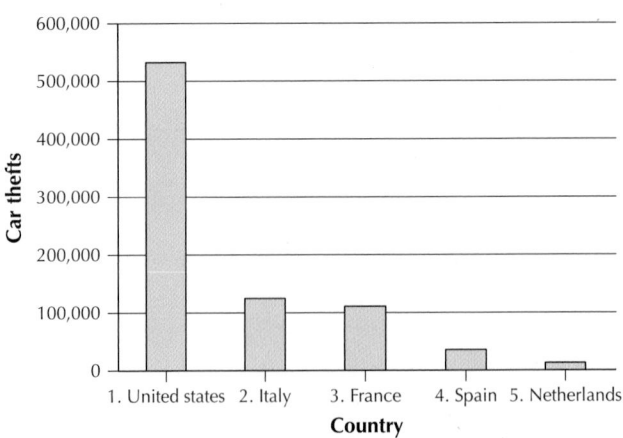

FIGURE 54
Bar graph of the top five nations for number of cars stolen in 2012.

To find the per capita car theft rate, divide the number of cars stolen in a country by that country's population. The resulting list in Table 45 of the top five countries for per capita car theft contains a few surprises. Note that the United States has dropped to third on the revised list.

Table 45 Top five nations for total number of cars stolen per capita in 2012

Country	Cars stolen per capita
Italy	0.00208
France	0.00174
United States of America	0.00168
Sweden	0.00117
Belgium	0.00106

Developing Your Statistical Sense

Choose the Appropriate Statistic

The bottom line is that *we need to be careful how we use statistics*. Put in an extreme form, "Figures don't lie, but liars figure." One table of statistics tells us the car theft epidemic is striking the United States with special vehemence. The other table asserts the contrary. An American insurance company looking to increase car insurance rates could point to the first table to support its rate request. A citizens group opposing the request could cite the second table. Which table of statistics is true? They both are! We need to be careful how we phrase our research questions and how we choose the types of statistical evidence we use to investigate research questions.

NOW YOU CAN DO
Exercises 3–5.

EXAMPLE 27   Omitting the zero

Student-Run Café Business

Suppose someone wanted to make the point that the students at the university with the student-run café business are drinking too much soda, and he or she produced Figure 55 to support this argument. Figure 55 is a bar graph of the total number of sodas sold over the 47 days compared with the total number of coffees sold. However, Figure 55 is misleading because it exaggerates the difference. Explain how Figure 55 is misleading, and produce the proper bar graph.

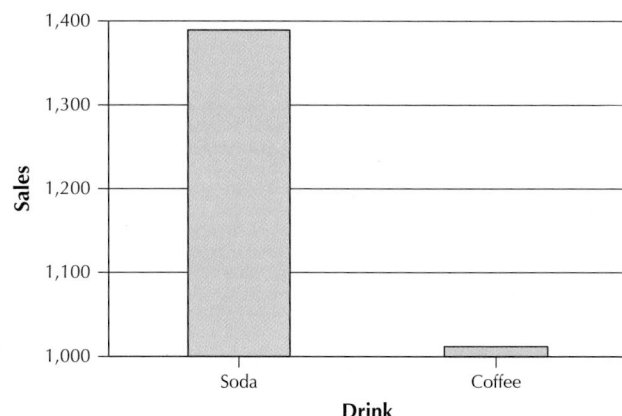

FIGURE 55 Omitting the zero is inappropriate because it exaggerates the difference.

Solution

Figure 55 is misleading because the vertical scale does not begin at zero. Instead, as we see in Figure 56, when zero is included on the vertical scale, the difference between the numbers of soda and coffee sold is not so dramatic.

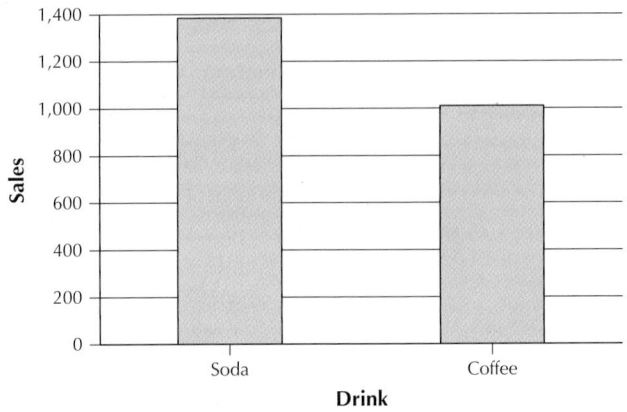

FIGURE 56 Appropriate graph.

EXAMPLE 28   Manipulating the scale

Figure 57 shows a relative frequency bar graph of the majors chosen by 25 business school students. Explain how we could manipulate the scale to de-emphasize the differences.

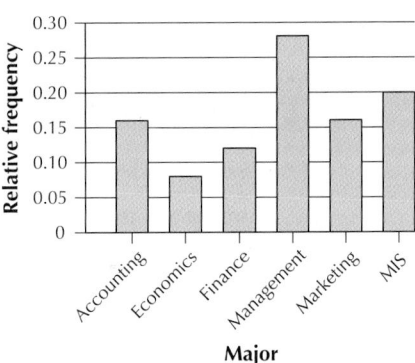

FIGURE 57 Well-constructed bar graph.

Solution

If we wanted to de-emphasize the differences, we could extend the vertical scale up to its maximum, 1.0 = 100%, to produce the graph in Figure 58.

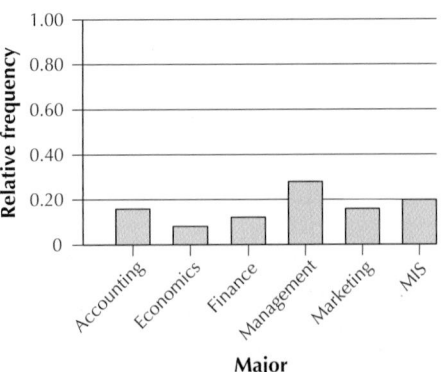

FIGURE 58 Inappropriate overextension of vertical scale.

EXAMPLE 29   Using two dimensions for a one-dimensional difference and unclear labeling

Figure 59 compares the leaders in career playoff points scored in the NBA playoffs, as of June 2014. Explain how this graph may be misleading.

| Kobe Bryant | Tim Duncan | LeBron James | Tony Parker |
| 5,640 points | 4,988 points | 4,419 points | 3,705 points |

FIGURE 59 This graph uses two dimensions (height and width) to overemphasize a one-dimensional (points) difference.

Solution

The height of the balls is supposed to represent the total points, but this is not clearly labeled. Points should be indicated using a vertical axis, but the vertical axis is not labeled at all. Further, note that the ball for Kobe Bryant is larger both in height and in width. This is misleading because it overemphasizes the difference in points scored between Kobe Bryant and Tim Duncan. In a bar graph, the bars for all four players should have the same width.

EXAMPLE 30   Careless combination of categories in a bar graph and biased embellishment

Figure 60 shows a bar graph of how often people have observed drivers running red lights. Explain how this bar graph may be considered both confusing and biased.

FIGURE 60
Careless combination of categories.

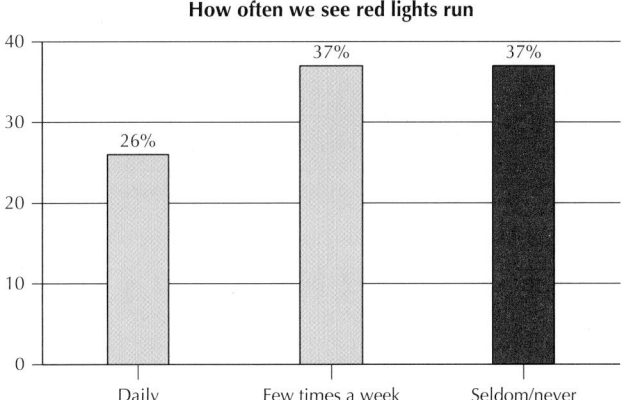

Solution

One problem with this bar graph is that the categories of *seldom* and *never* have been combined, which may not be appropriate. Also, as we learned in Chapter 1, what is "seldom" to one person may not be "seldom" to someone else. A third problem is that the bar of the *Seldom/never* category is highlighted in a different color, which may be evidence of bias on the part of the designer of the bar graph.

EXAMPLE 31 Inaccuracy in relative lengths of bars in a bar graph and unclear labeling

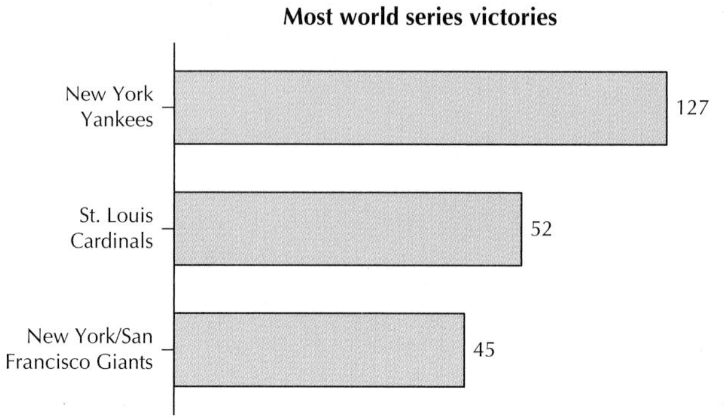

FIGURE 61
Inaccuracy in bar length.

Figure 61 is a horizontal bar graph of the three teams with the most World Series victories in baseball history. Explain what is unclear or misleading about this graph.

Most world series victories

Solution

Note that 127 is more than twice as many as 52, and so the Yankees' bar should be more than twice as long as the Cardinals' bar, which it is not. Finally, note the absence of a horizontal axis.

When constructing a histogram, changing the number of classes or the width of the interval can sometimes lead to a completely different-looking distribution. Thus, we need to exercise care when someone shows us a histogram because it presents, not the data themselves, but one of many ways of classifying the data.

EXAMPLE 32 Presenting the same data set as both symmetric and left-skewed

The National Center for Education Statistics sponsors the Trends in International Mathematics and Science Study (TIMSS). Science tests were administered to eighth-grade students in countries around the world (see Table 46). Construct two different histograms: one that shows the data as almost symmetric and one that shows the data as left-skewed.

Table 46 Science test scores

Country	Score	Country	Score	Country	Score
Singapore	578	New Zealand	520	Bulgaria	479
Taiwan	571	Lithuania	519	Jordan	475
South Korea	558	Slovak Republic	517	Moldova	472
Hong Kong	556	Belgium	516	Romania	470
Japan	552	Russian Federation	514	Iran	453
Hungary	543	Latvia	513	Macedonia	449
Netherlands	536	Scotland	512	Cyprus	441
United States	527	Malaysia	510	Indonesia	420
Australia	527	Norway	494	Chile	413
Sweden	524	Italy	491	Tunisia	404
Slovenia	520	Israel	488	Philippines	377

Solution

Figure 62 is nearly symmetric, but Figure 63 is clearly left-skewed. It is important to realize that *both figures are histograms of the very same data set.* Clever choices for the number of classes and the class limits can affect how a histogram presents the data. The reader must therefore beware! The histogram represents a summarization of the data set, not the data set itself. Analysts may wish to supplement the histogram with other graphical methods, such as dotplots and stem-and-leaf displays, in order to gain a better understanding of the distribution of the data.

The *One Variable Statistics and Graphs* applet allows you to experiment with the class width and number of classes when constructing a histogram.

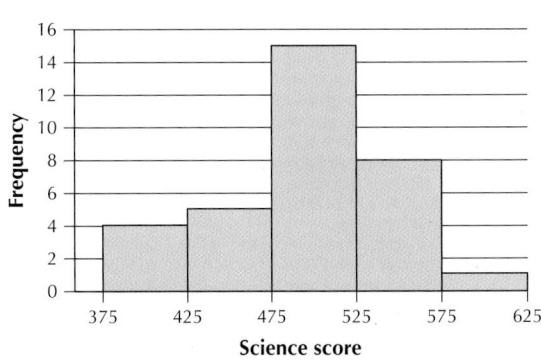

FIGURE 62 Nearly symmetric histogram of science test scores.

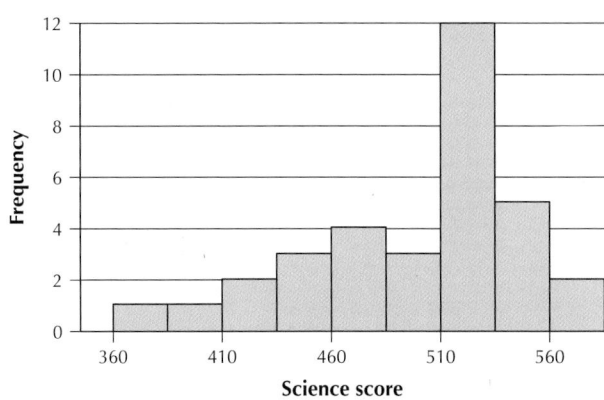

FIGURE 63 Left-skewed histogram of the same science test scores.

Section 2.4 Summary

1. Understanding how graphics are constructed will help you avoid being deceived by misleading graphics. Some common methods for making a graph misleading include manipulating the scale, omitting the zero on the relevant scale, and biased distortion or embellishment.

Section 2.4 Exercises

CLARIFYING THE CONCEPTS

1. Explain in your own words why it is important to be aware of the methods that can be used to make graphics misleading. (p. 95)

2. True or false: What we have learned in this chapter proves that all statistics are misleading.

PRACTICING THE TECHNIQUES

Refer to Example 26 for the following exercises.

3. Which do you think is more effective at convincing the American public that a problem exists, Table 44 or Figure 54?

4. How would factoring in the *number of cars per country* affect the rankings, in your view?

5. If you were an insurance claims adjuster arguing for higher car insurance rates in the United States, would you prefer Table 44 or Table 45? Why?

APPLYING THE CONCEPTS

6. Eating Bread. Consider the accompanying graphic of the types of bread people eat.

a. What type of graph is it supposed to represent, among the graphs that we have learned in this chapter?
b. Consider how the *wheat* category dominates the graph. Which of the eight common methods for misrepresenting data is present here?
c. Construct a graphic that is not misleading in this way.

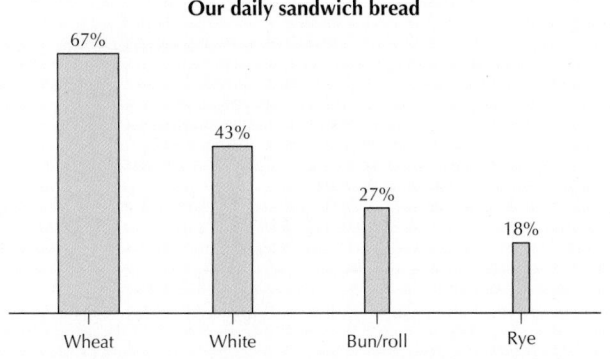

7. Child-Rearing Costs. Consider the accompanying graphic of child-rearing costs by type of cost.
   a. Identify one problem with the graphic that makes it misleading.
   b. Construct a graphic that is not misleading in this way.

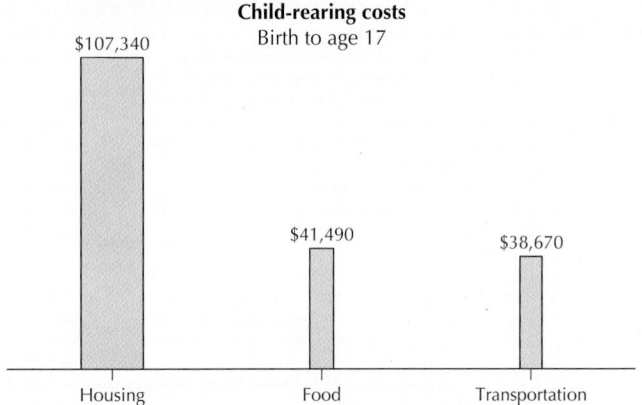

Child-rearing costs
Birth to age 17

$107,340 — Housing
$41,490 — Food
$38,670 — Transportation

8. Going to the Game. Consider the accompanying graphic of the proportions of people who go to see professional sports events.
   a. Identify two problems with the graphic that make it misleading.
   b. Construct a graphic that is not misleading in these ways.

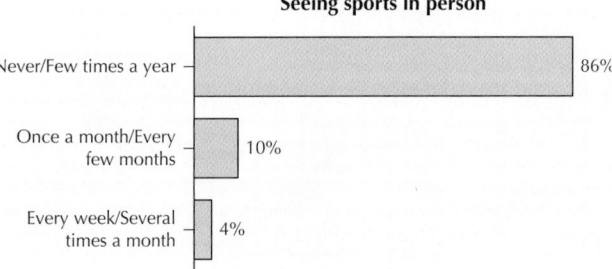

Seeing sports in person

Never/Few times a year — 86%
Once a month/Every few months — 10%
Every week/Several times a month — 4%

9. Living with AIDS. Consider the accompanying graphic.
   a. What point is the graphic trying to make?
   b. Which of the eight common problems is most obviously present here?
   c. Construct a graphic that is not misleading in this way.

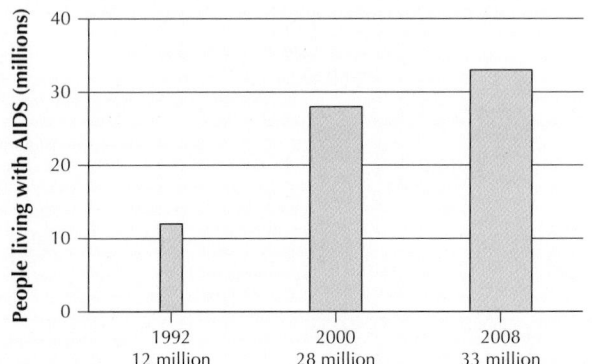

People living with AIDS (millions)

1992 — 12 million
2000 — 28 million
2008 — 33 million

10. What's Your Sign? The General Social Survey collects data on social aspects of life in America. Consider the accompanying bar graph of the results of asking 1464 people what their astrological sign is.
   a. Which of the eight common problems is most obviously present here?
   b. Construct a graphic that is not misleading in this way.

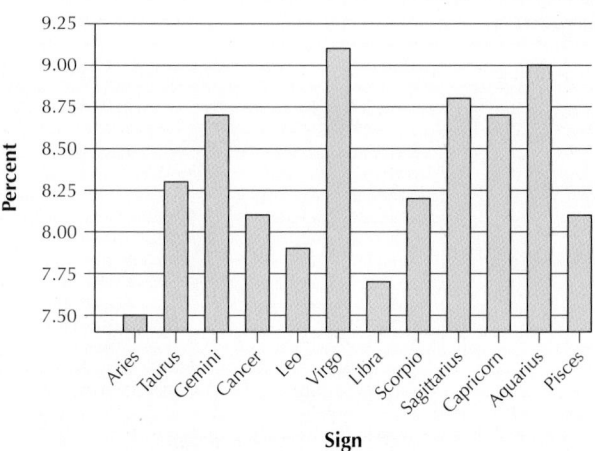

(vertical axis: Percent; horizontal axis: Sign — Aries, Taurus, Gemini, Cancer, Leo, Virgo, Libra, Scorpio, Sagittarius, Capricorn, Aquarius, Pisces)

11. Apps. Refer to the app type data in Table 2 on page 41.
   a. Construct a bar graph that overemphasizes the difference among the app types.
   b. Which of the common methods for making graphics misleading are you using in (**a**)?
   c. Construct a bar graph that underemphasizes the difference among the app types.
   d. Which of the common methods for making graphics misleading are you using in (**c**)?

12. Music and Violence. *USA Weekend* conducted a poll that asked, "Do you think shock rock and gangsta rap are partly to blame for violence such as school shootings or physical abuse?" The results are shown in the following table.

Yes	31%
No	45%
I've never thought about it	24%

   a. Construct a bar graph that overemphasizes the difference among the responses.
   b. Construct a bar graph that underemphasizes the difference among the responses.
   c. Construct a bar graph that fairly represents the data.

13. Inmate Population. Refer to Table 43 on page 95, which contains the average monthly inmate population for Miami-Dade Department of Corrections. There seems to have been a moderate drop in the monthly average inmate population over the two years.
   a. Alter the vertical axis so as to overemphasize this drop.
   b. Change the vertical axis so as to de-emphasize this drop.

Use the *One Variable Statistics and Graphs* applet for Exercises 14–16. Work with the TIMSS scores from Example 32.

14. Click on the **Histogram** tab. Experiment with the class widths by clicking and dragging on the number line. Produce a histogram that is nearly symmetric, like Figure 62.

15. Produce a histogram that is somewhat left-skewed, like Figure 63.

16. Click on the **Stem-and-Leaf** tab. The previous two exercises left us with two different ideas as to the shape of the distribution.

a. Now produce a stem-and-leaf display of the TIMSS scores.

b. Compare the regular stem-and-leaf display with the split-stem stem-and-leaf display. Which is preferable for this data set?

c. Use your preferred stem-and-leaf display from (**b**) to describe the shape of the distribution.

d. Which of the two histograms does your description in (**c**) support?

Chapter 2 Vocabulary

SECTION 2.1
- **Bar graph (bar chart)** (p. 43)
- **Clustered bar graph** (p. 48)
- **Crosstabulation (two-way table, contingency table)** (p. 45)
- **Frequency (count)** (p. 41)
- **Frequency distribution** (for qualitative data) (p. 41)
- **Pareto chart** (p. 44)
- **Pie chart** (p. 44)
- **Relative frequency** (for a qualitative variable) (p. 42)
- **Relative frequency distribution** (for qualitative data) (p. 42)

SECTION 2.2
- **Bell-shaped curve** (p. 72)
- **Class** (p. 61)
- **Class boundary** (p. 62)
- **Class limit (lower)** (p. 62)
- **Class limit (upper)** (p. 62)

- **Class midpoint** (p. 66)
- **Class width** (p. 62)
- **Distribution of a variable** (p. 72)
- **Dotplot** (p. 70)
- **Frequency distribution** (for quantitative data) (p. 61)
- **Frequency polygon** (p. 67)
- **Histogram** (p. 65)
- **Relative frequency distribution** (for quantitative data) (p. 61)
- **Skewed distribution** (p. 73)
- **Stem-and-leaf display** (p. 68)
- **Symmetric** (p. 73)

SECTION 2.3
- **Cumulative frequency distribution** (p. 87)
- **Cumulative relative frequency distribution** (p. 87)
- **Ogive** (p. 88)
- **Time series plot (time series graph)** (p. 89)

Chapter 2 Review Exercises

SECTION 2.1

1. Parts of Speech. The accompanying bar graph summarizes the frequencies for the various parts of speech in a sample of English words. Should we be interested in determining whether this graph is symmetric or skewed? Clearly explain why or why not.

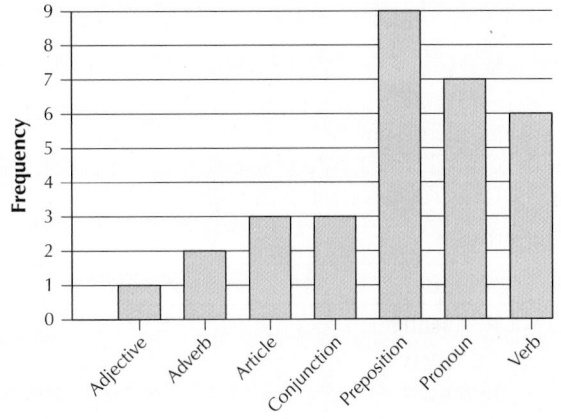

Parts of Speech

For Exercises 2–6, refer to the bar graph from Exercise 1 to construct the following for the variable *parts of speech*.

2. Relative frequency bar graph

3. Frequency distribution

4. Relative frequency distribution

5. Frequency pie chart

6. Relative frequency pie chart

Happiness in Marriage. The General Social Survey tracks trends in American society through annual surveys. Use the following contingency table for Exercises 7–11.

	Happiness of marriage			
Respondents' gender	Very happy	Pretty happy	Not too happy	Total
Male	242	115	9	366
Female	257	149	17	423
Total	499	264	26	789

7. What proportion of the males responded that they were very happy in their marriage?

8. What proportion of the females responded that they were very happy in their marriage?

9. What proportion of the males responded that they were not too happy in their marriage?

10. What proportion of the females responded that they were not too happy in their marriage?

11. Construct a clustered bar graph of the data.

SECTION 2.2

Student-Run Café Business. Use the Minitab relative frequency distribution of potato chip sales for Exercises 12–14. The distribution shows the number of days each number of chips was sold. For example, four bags of chips were sold on three different days.

Chips	Count	Percent
0	3	6.38
1	1	2.13
2	2	4.26
3	2	4.26
4	3	6.38
6	1	2.13
7	4	8.51
8	4	8.51
9	6	12.77
10	3	6.38
11	7	14.89
12	2	4.26
13	2	4.26
14	2	4.26
16	1	2.13
20	2	4.26
21	1	2.13
25	1	2.13

Minitab relative frequency distribution of potato chip sales

12. Construct a relative frequency distribution of 4 classes, each of class width 7. Let the leftmost class be: 0 to < 7.

13. Use the relative frequency distribution to construct a histogram of potato chip sales.

14. For what percentage of days were more than 20 bags of chips sold?

Student-Run Café Business. Use the data in the following table for Exercises 15–20. The data represent the total sales (in dollars) for the first 20 days of sales for the student-run café. You will construct a histogram of the sales data, with 8 classes, each with class width 25.

Total sales for the first 20 days ▥ **totalsales20**

199.95	186.94	172.31	181.43
195.74	120.18	137.65	125.57
102.68	228.78	197.56	180.63
162.88	88.02	70.00	75.87
101.76	119.57	97.00	150.51

15. Find the class limits, and draw the horizontal axis. Let the leftmost lower class limit equal 50.

16. Determine the frequencies, and draw the vertical axis.

17. Draw the rectangles.

18. What percentage of days had sales over $200?

19. What proportion of days had sales of at most $149.99?

20. Find the percentage of days with sales between $150 and $199.99 inclusive.

SECTION 2.3

21. Fraud Cases in Brooklyn. Use the data from Table 31 on page 82 of Section 2.2 to

    a. construct a cumulative frequency distribution.

    b. construct a cumulative relative frequency distribution.

22. Fraud Cases in Brooklyn. Use the data from Table 31 on page 82 of Section 2.2 to

    a. construct a frequency ogive.

    b. construct a relative frequency ogive.

23. Trade Deficit. Table 47 presents the annual trade deficits (imports minus exports) for the United States from 1991 to 2013, in billions of dollars. ▥ **tradedeficits**

    a. Construct a time series plot of the data.

    b. Describe any trends that you see.

TABLE 47 U.S. trade deficits

Year	Trade deficit in $ billions	Year	Trade deficit in $ billions
1991	31	2003	494
1992	39	2004	609
1993	70	2005	714
1994	98	2006	759
1995	96	2007	702
1996	104	2008	698
1997	108	2009	374
1998	166	2010	495
1999	264	2011	548
2000	379	2012	538
2001	364	2013	476
2002	421		

Source: U.S. Census Bureau.

SECTION 2.4

24. Carbon Dioxide Emissions. The following table contains the energy-related carbon dioxide emissions (in millions of metric tons), by end-use sector, as reported by the U.S. Energy Information Administration.

▥ **co2emissions**

Sector	Emissions
Residential	1213.9
Commercial	1034.1
Industrial	1736.0
Transportation	1939.2

a. Construct a bar graph that overemphasizes the differences among the sectors.

b. Which of the common methods for making misleading graphics are you using in (**a**)?

c. Construct a bar graph that underemphasizes the differences among the sectors.

d. Which of the common methods for making graphics misleading are you using in (**c**)?

e. Construct a bar graph that fairly represents the data.

Chapter 2 **QUIZ**

TRUE OR FALSE

1. True or false: Histograms are superior to stem-and-leaf displays because histograms retain the information contained in the data set.

2. True or false: A histogram always provides a realistic summary of the symmetry or skewness of a data set.

FILL IN THE BLANK

3. The frequencies in a frequency distribution must add up to the _____ _____ [two words].

4. A _____ _____ [two words] for a qualitative variable is a listing of all values that the variable can take, together with the frequencies for each value.

SHORT ANSWER

5. If there is a line that splits an image in half, so that one side is the mirror image of the other, we say that the image is what?

6. If the right tail of a distribution is longer than the left tail, we say that the distribution is what?

CALCULATIONS AND INTERPRETATIONS

Life Expectancy. For Exercises 7–15, refer to the following table, which shows the life expectancy at birth in 2010, as reported by the World Health Organization.[10] **lifeexpect**

Country	Life expectancy
Afghanistan	42
Canada	81
China	74
Ghana	62
India	64
Israel	81
Mexico	76
Russia	68
United Kingdom	80
United States	78

Construct the following:

7. Frequency distribution

8. Relative frequency distribution

9. Cumulative frequency distribution

10. Cumulative relative frequency distribution

11. Frequency bar graph

12. Relative frequency bar graph

13. Pie chart of the relative frequencies

14. Ogive of the frequencies

15. Relative frequency ogive of the frequencies

3 Describing Data Numerically

Introduction

In Chapter 3, students develop numerical summaries to help them discover important characteristics about a data set. They also become acquainted with some powerful and widespread methodologies for applying the tools of descriptive statistics.

Section 3.1 introduces measures of center—the mean, the median, and the mode. Section 3.2 introduces measures of variability—the range, the variance, and the standard deviation, as well as their applications: the Empirical Rule and Chebyshev's Rule. Section 3.3 discusses how to work with grouped data. Section 3.4 introduces us to measures of position, including z-scores, percentiles, percentile ranks, and quartiles, and how to use z-scores to detect outliers. Section 3.5 discusses the five-number summary, boxplots, and how to use the IQR method to detect outliers.

From the Author

The Chapter 3 Case Study (*Can the Financial Experts Beat the Darts*?) has been extended throughout the chapter.

Section 3.1 Measures of Center

- Stress the notion of the mean representing the "balance point" of the data, so that students may check their calculations throughout the remainder of the course.
- Early in Section 3.1, you may wish to review the definitions of *population* and *sample*.
- The *What if* scenario, page 115. Usually, this feature is structured in such a way that a calculator will not help. Instead, students need to think about how a change in one aspect of the problem will affect other aspects of the situation.
- Construct Your Own Data Sets, page 125 and page 148. This is a good way for students to apply their understanding of the concepts, by making up their own list of numbers that satisfies a particular set of conditions.

Section 3.2 Measures of Variability

- While many (most?) students now learn mean, median, and mode (Section 3.1) in elementary school, not so many learn about the standard deviation or the variance (Section 3.2). So, for most students, most of the material in this section (and subsequent sections) will be new.
- *Discovering Statistics* stresses what the statistics *mean*. This can be helpful when checking calculations, such as the standard deviation. If the student understands what a deviation means, and understands that the standard deviation represents a typical deviation, then the student may catch a calculation error.

Section 3.3 Working with Grouped Data

- Some instructors find that they do not have time to cover Section 3.3. If you choose to omit this section, you may wish to cover Objective 1, The Weighted Mean, using the grading policy in your syllabus as an example.

Section 3.4 Measures of Relative Position and Outliers

- Example 21 has been provided to underscore the fact that z-scores do not have to follow a bell-shaped distribution.
- The dance score data set, which was not real, has been replaced by an exports data set, which represents real data. This data is used for several examples in Sections 3.4 and 3.5.

Section 3.5 Five-Number Summary and Boxplots

- We have moved the section on five-number summary and boxplots ahead of the section on the Empirical Rule and Chebyshev's Rule. This is because the boxplot uses the quartiles and the IQR, which were learned in the previous section.

Teaching Tips

Students may experience a steeper learning curve beginning at Section 3.2. The material in Section 3.1—mean, median, and mode—is often covered in high school or earlier. The material in Section 3.2 on variability is not usually covered in high school and is very important to an understanding of statistics. Stress the concept of *spread*—how spread out a data set is. The more spread out the data, the larger the measure of spread will be, whether it is the range, variance, standard deviation, or interquartile range.

In-Class Activities

1. Access the data set for Old Faithful at the following Web site: www.stat.cmu.edu/~larry/all-of -statistics/=data/faithful.dat.

The data set consists of 225 values in an Excel file for the variables duration and interruption time. The Yellowstone National Park Web site states that "Old Faithful erupts every 35–120 minutes for 1.5–5 minutes" (www.yellowstone.net/geysers/old-faithful). Use the data to construct appropriate graphs for the duration times and interruption times for Old Faithful. Ask students, "What can you say about the distribution of these variables?" Ask them to compute numerical summaries for these two variables. Ask which data set has more variability.

Measures of Center

2. What is your guess of the typical height of all students in your class?

3. Make a dotplot of the heights of the students in your class.

4. Discuss where to place the center of this distribution of student heights. Without crunching any numbers, form a consensus on the location of the center.

5. Calculate the mean, median, and mode of the student heights.

6. Which measure (mean, median, or mode) comes closest to the consensus of where the center is located in (**4**)?

7. What is the relation between these measures and your guess of the typical height in (**2**)?

8. Which measure (mean, median, mode, class consensus, your guess) do you think is the best measure of the center of student heights?

Measures of Spread

9. Do you think that the distribution of the heights of all students in your class is more spread out or less spread out than the distribution of the heights of only the females in your class?

10. Would the values of our measures of spread (range, standard deviation) be larger for the entire class or for only the females?

11. Make a dotplot of the heights of only the females in the class. Make sure it uses the same scale as the dotplot for the heights of all the students in the class.

12. Use the two dotplots to assess which group has greater variability.

13. Back up your intuition by calculating and comparing our measures of spread (range, standard deviation) for the two groups.

Supplements

- StatTutor 2.1–2.10
- EESEE case studies for describing data numerically
    - Weighing Trucks in Motion (Question 2 on mean, median, and standard deviation)
    - Acorn Size and Oak Tree Range (Question 7 on boxplots, Question 2 on mean and standard deviation, Question 3 on range and standard deviation)
    - Faculty Salary Comparison (Question 1 on boxplots, Question 3 on weighted averages, Question 4 on ranking and means)

Applets

The *Mean and Median* applet is referenced in Chapter 3 to compute values for the mean and median and for Exercises 104 and 105 in Section 3.1.

Activities and applets that relate to measures of center, spread, and boxplots can be found at http://mathforum.org/mathtools/tool/12489/.

The site Online Statistics: An Interactive Multimedia Course of Study has numerous applets and activities: http://onlinestatbook.com/index.html.

Videos

- *Against All Odds: Inside Statistics:* www.learner.org/resources/series65.html
  - Program 3: Histograms

Web Sites

- CAUSEweb provides resources for statistics education: https://www.causeweb.org/resources/.
- The following Web site has a collection of 20 class projects: www.amstat.org/publications /jse/v6n3/smith.html.
- This Texas Instruments Web site has a host of TI-83/84 statistics activities: http://education .ti.com/educationportal/sites/US/nonProductSingle/activitybook_83_statistics.html.
- This Web site has a host of activities, simulations, and so on, which relate to elementary statistics: http://davidmlane.com/hyperstat/ch2_contents.html.
- This Web site lists other sites that do statistical calculations: http://statpages.org/.

3 Describing Data Numerically

...RQP 25.11 ↑ 0.3

42 ↑ 1.93 ...BJG 18

7 ...DRZ 29.33 ↑ 2.

Mark Hooper/Getty Images

Can the Financial Experts Beat the Darts?

CASE STUDY

Have you ever wondered whether a bunch of monkeys throwing darts to choose stocks could select a portfolio that performed as well as the stocks carefully chosen by Wall Street experts? The *Wall Street Journal* (www.wsj.com) apparently believed that the comparison was worth a look. The *Journal* ran a contest between stocks chosen randomly by *Journal* staff members (instead of monkeys) throwing darts at the *Journal* stock pages (mounted on a board) and stocks chosen by a team of four professional financial experts. At the end of six months, the *Journal* compared the percentage change in the price of the experts' stocks and the dartboard's stocks and compared both to the Dow Jones Industrial Average as well. So, who do you think did better? Did the six-figure-salary financial experts put the random dart selections to shame?

- In Section 3.1, we do some graphical exploration with the data, comparing the balance points (means) of each group using comparative dotplots. We then determine whether the student's intuition of the location of the means is confirmed by the statistics.
- In Section 3.2, we compare the variability of the three groups and find that different measures of spread can disagree about which data set has more variability.
- In the Section 3.2 exercises, we calculate the coefficient of skewness for each group.
- In the Section 3.3 exercises, we examine how close the estimated mean, variance, and standard deviation for grouped data are to their true values.
- In the Section 3.4 exercises, we use the case study data to examine measures of relative position such as *z*-scores and percentiles.
- Finally, in the Section 3.5 exercises, we construct boxplots and identify outliers for each group in the case study data set.

THE BIG PICTURE

Where we are coming from and where we are headed . . .

- Chapter 2 showed us graphical and tabular summaries of data.
- Here, in Chapter 3, we "crunch the numbers," that is, we develop numerical summaries of data. We examine measures of center, measures of variability, and measures of relative position.
- In Chapter 4, we will learn how to summarize the relationship between two quantitative variables.

3.1 Measures of Center

OBJECTIVES By the end of this section, I will be able to . . .

1 Calculate the mean for a given data set.
2 Find the median, and describe why the median is sometimes preferable to the mean.
3 Find the mode of a data set.
4 Describe how skewness and symmetry affect these measures of center.

Do you like to make money? Then you might want to stay in school and finish your Bachelor's degree. The Pew Research Center reports that the median annual earnings among young people ages 25–32 with a Bachelor's degree was $45,500, compared with $30,000 for those who did not finish their college degree (*Source:* Pew Research Center: *The Rising Cost of Not Going to College*[1]). The $45,500 is a *sample median*, which was calculated from the sample taken by the researchers. As such, it *summarizes* the earnings of over 1000 different young people from all over the country. In Chapter 3, we learn how to do this: to summarize an entire dataset with just a few numbers. In Section 3.1, we will learn about three numerical measures that tell us where the center of the data lies: the mean, the median, and the mode.

1 The Mean

The most well-known and widely used **measure of center** is the **mean**. In everyday usage, the word *average* is often used to denote the **mean**.

The mean is often called the arithmetic mean.

> To find the **mean** of the values in a data set, simply add up all the numbers and divide by how many numbers you have.

EXAMPLE 1 Calculating the population mean

The Web site CNET.com provides reviews and prices for gadgets and electronics, including cell phones. In Table 1, you will find all eight of the cell phones in CNET's "Editors' Picks" for June 27, 2014. Recall from Chapter 1 that a population is the collection of all elements of interest in a particular study. Thus, the data in Table 1 represents a population. Find the mean price of all the cell phones.

Table 1 Prices for a population of cell phones

Samsung Galaxy S5 Standard	$200
Samsung Galaxy S5 Active	$200
Sony Xperia Z2	$600
Nokia Lumia Icon	$200
LG G3	$800
Apple iPhone 5s	$250
HTC One M8	$200
Samsung Galaxy Note 3	$300

Source: www.cnet.com/topics/phones/best-phones.

Solution

To find the mean, we add up the prices of all eight cell phones and divide by the number of phones:

NOW YOU CAN DO
Exercises 13–18.

$$\text{Mean cell phone price} = \frac{200+200+600+200+800+250+200+300}{8} = \$343.75$$

The population mean price for all eight cell phones is $343.75.

YOUR TURN
#1

Table 2 contains the number of tropical storms reported by the National Oceanic and Atmospheric Administration for 2006–2013. All years in this period are represented, so this can be considered a population. Find the population mean number of tropical storms.

Table 2 Number of tropical storms

Year	2006	2007	2008	2009	2010	2011	2012	2013
Tropical storms	10	15	16	9	19	19	19	14

(The solution is shown in Appendix A.)

Before we proceed, we need to learn some notation.

Notation

Statisticians like to use specialized notation. It is worth learning because it saves a lot of writing, and certain concepts can best be understood by using this special notation.

- The **population size**, the number of observations in your population, is always denoted as N. We have a population with eight observations in Example 1, so $N = 8$.

- The **sample size**, which refers to how many observations you have in your sample data set, is always denoted as n.

- The shorthand notation for "the sum of all the data" is Σx, where x refers to the data, and Σ (capital sigma), which is the Greek letter for "S," stands for "Summation." Note in Example 1 that we added up the prices of all the cell phones. This summing is denoted as Σx.

- The **population mean** is denoted as μ (pronounced "mew"), which is the Greek letter for m. As we saw in Example 1, to calculate the population mean, we add up all the data and divide by the population size, N. Thus, the formula for the population mean is:

$$\mu = \frac{\sum x}{N}$$

- For Example 1, we therefore have:

$$\mu = \frac{\sum x}{N} = \frac{200 + 200 + 600 + 200 + 800 + 250 + 200 + 300}{8} = \$343.75$$

- The **sample mean** is denoted as $\bar{x}$ (pronounced "x-bar"). You should try to commit this to long-term memory because $\bar{x}$ may be the most important symbol used in this book and will return again and again in nearly every chapter. The sample mean is calculated just like the population mean, except that we divide by the sample size n instead of the population size N. Thus, the formula for the sample mean is:

$$\bar{x} = \frac{\sum x}{n}$$

EXAMPLE 2 Calculating the sample mean

Suppose the cell phones in Table 3 represent a random sample of size four from the population in Table 1. Calculate the sample mean price of this sample of cell phones.

Table 3 Prices for a sample of cell phones

Samsung Galaxy S5 Active	$200
Sony Xperia Z2	$600
Apple iPhone 5s	$250
Samsung Galaxy Note 3	$300

Solution

The sample mean price of this sample of four cell phones is calculated like this:

$$\bar{x} = \frac{\sum x}{n} = \frac{200 + 600 + 250 + 300}{4} = \$337.50$$

NOW YOU CAN DO
Exercises 19–24.

The sample mean cell phone price for this particular sample is $337.50. Of course, a different sample would have yielded a different value for $\bar{x}$.

YOUR TURN
#2

Suppose we took a sample of size three instead and obtained the same sample as in Table 3, except that the Sony Xperia Z2 was not included.

a. Would you expect that the sample mean price would be higher or lower than $337.50? Explain.

b. Calculate the sample mean price for the sample of three cell phones. Was your intuition in (**a**) confirmed?

(The solutions are shown in Appendix A.)

What Does This Number Mean?

The Mean as the Balance Point of the Data

Let's explore our sample cell phone price data a bit further. Consider the dotplot of the cell phone prices in Figure 1. To find out where the mean price lies on this number line, imagine that the dots are little blocks on a ruler or a seesaw and that you must decide where to place the support (like the triangle in Figure 1) so that the ruler balances perfectly. *The place where the data set balances perfectly is the location of the mean.* Placing the fulcrum too far to the right or left would create an imbalance. This data set balances precisely at the sample mean, $\bar{x} = \$337.50$

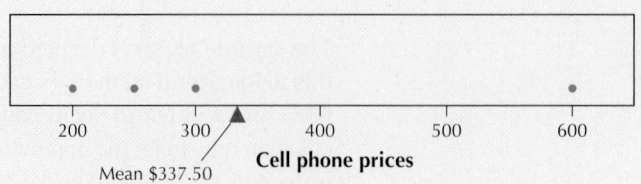

FIGURE 1 The price data balance at the mean.

Checking Your Results Against Experience and Common Sense

When you have found the balance point, you have found the mean. When you calculate the mean, or have a computer or calculator do it for you, don't just accept whatever value pops out. Make sure the result makes sense. Because the mean always indicates the place where the data values are in balance, the mean is often near the center of the data. If the value you have calculated lies nowhere near the center of the data, then you may want to check your calculations.

For example, suppose we were finding the mean of the cell phone data, and we accidentally entered 6000 instead of 600 for the price of the Sony Xperia Z2. Then, our value for the mean resulting from this *incorrect* calculation would be

$$\bar{x} = \frac{\sum x}{n} = \frac{200 + 6000 + 250 + 300}{4} = \$1687.50$$

The mean price cannot equal \$1687.50 because all the values in the data set are less than \$1687.50. The mean can never be larger or smaller than all the values in the data set.

Don't automatically accept the result you get from a computer or calculator. Remember GIGO: Garbage In Garbage Out. If you enter the wrong data, the calculator or computer will not bail you out. Human error is one reason for the explosion of faulty statistical analyses in the newspapers and on the Internet. Now more than ever, data analysts must use good judgment. When you calculate a mean, always have an idea of what you *expect* the sample mean to be, that is, at least a ballpark figure.

For calculating the mean, we will adopt the convention of rounding our final calculation, if necessary, to one more decimal place than that in the original data.

The Mean Is Sensitive to Extreme Values

One drawback of using the mean to measure the center of the data is that the mean is sensitive to the presence of extreme values in the data set. We illustrate this phenomenon with the following example.

EXAMPLE 3 Sensitivity of the mean to extreme values

Table 4 contains a sample of six home sales prices for Broward County, Florida, for June 27, 2014. We want to get an idea of the typical home sales price in Broward County.

a. Find the mean sales price of the homes in Table 4.

b. Suppose we add a seventh home in Hillsborough Beach, selling for \$6 million. Calculate the mean sales price of all seven homes. Comment on how the extreme value affected the mean sales price.

Table 4 Home sales prices in Broward County, Florida  homesales

Location	Price
Pembroke Pines	\$300,000
Weston	\$350,000
Hallandale	\$360,000
Miramar	\$425,000
Davie	\$500,000
Fort Lauderdale	\$600,000

Source: www.homes.com (prices rounded to nearest \$1000).

Solution

a. The mean sales price of the homes in Table 4 is:

$$\bar{x} = \frac{\sum x}{n} = \frac{300{,}000 + 350{,}000 + 360{,}000 + 425{,}000 + 500{,}000 + 600{,}000}{6}$$

$$= \$422{,}500$$

b. Now, suppose that we append a seventh home to our sample: a home in Hillsborough Beach listed for $6 million, which is much more expensive than any of the other homes in the sample. Recalculating the mean, we get

$$\bar{x} = \frac{\sum x}{n} = \frac{300{,}000 + 350{,}000 + 360{,}000 + 425{,}000 + 500{,}000 + 600{,}000 + 6{,}000{,}000}{7}$$

$$= \$1{,}220{,}000$$

Note that the mean sales price nearly tripled from $422,500 to $1,220,000 when we added this extreme value. Also, this new mean is much higher than every price in the original sample. Thus, it is highly unlikely that this new mean of about $1.2 million is representative of the *typical* sales price of homes in Broward County. This example shows how the mean is sensitive to the presence of extreme values. For situations like this, we prefer a measure of center that is not so sensitive to extreme values. Fortunately, the *median* is just such a measure.

NOW YOU CAN DO
Exercises 25–30.

2 The Median

Recall that the median strip on a highway is the slice of land in the *middle* of the two lanes of the highway. In statistics, the **median** of a data set is the *middle data value* when the data are put into ascending order. There are two cases, depending on whether the sample size is odd or even.

The Median

The **median** of a data set is the *middle data value* when the data are put into ascending order. Half of the data values lie below the median, and half lie above.

- If the sample size n is odd, then the median is the middle value and lies at the $\left(\frac{n+1}{2}\right)^{\text{th}}$ position when the data are put in ascending order.
- If the sample size n is even, then the median is the mean of the two middle data values that lie on either side of the $\left(\frac{n+1}{2}\right)^{\text{th}}$ position.

The case when the sample size is even is clear if you hold up four fingers on one hand. Notice that there is no unique finger in the middle. No middle value exists when the sample size is even, so we take the two data values in the middle and split the difference.

The Median Is Not Sensitive to Extreme Values

Unlike the mean, the median is not sensitive to extreme values. If the expensive home is included in the sample, the median price should not change much, even though, as we saw in Example 3, the mean sales price nearly tripled. Let's look at an example of how this would occur.

EXAMPLE 4   Median is not sensitive to extreme values

Show that the median is not sensitive to extreme values by doing the following:

a. Find the median sales price of the homes in Table 4.

b. Add the seventh home in Hillsborough Beach, selling for $6 million. Calculate the median sales price of all seven homes.

Solution

a. Fortunately, the data are already presented in ascending order in the table. Because $n = 6$ is even, the median is the mean of the two data values that lie on either side of the $\left(\frac{n+1}{2}\right)^{th} = \left(\frac{6+1}{2}\right)^{th} = 3.5$th position. That is, the median is the mean of the 3rd and 4th data values, $360,000 and $425,000. Splitting the difference between these two, we get

$$\text{median price} = \frac{\$360,000 + \$425,000}{2} = \$392,500$$

We note that, in Table 4, there are exactly as many homes with prices lower than $392,500 as homes with prices higher than $392,500.

b. Now, what happens to the median when we add in the $6 million home from Hillsborough Beach? Because $n = 7$ is odd, the median is the unique $\left(\frac{n+1}{2}\right)^{th} = \left(\frac{7+1}{2}\right)^{th} = 4$th observation, given by the home in Miramar for $425,000. The extreme value increased the median only from $392,500 to $425,000. In Example 3, we showed that the value of the mean price nearly tripled when the expensive home was added. Thus, the median home sales price is a better measure of center because it more accurately reflects the typical sales prices of homes in Broward County.

Because the median is not sensitive to extreme values, we say that it is a *robust, or resistant,* measure of center. The mean is neither robust nor resistant.

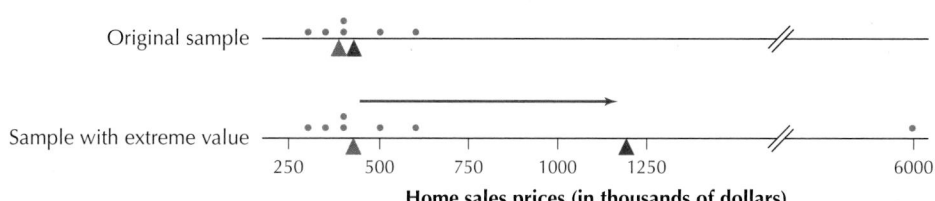

FIGURE 2 The mean (red triangles) is sensitive to extreme values, but the median (green triangles) is not.

NOW YOU CAN DO
Exercises 31–36.

 The *Mean and Median* applet allows you to insert your own data values and see how changes in these values affect both the mean and the median.

EXAMPLE 5   Using technology to find the mean and median

Find the mean and median of the home sales prices in Table 4, using **(a)** the TI-83/84, **(b)** Excel, **(c)** Minitab, and **(d)** JMP.

Solution

Using the instructions in the Step-by-Step Technology Guide on page 117, we get the following output:

a. The first TI-83/84 screen shows $\bar{x} = 422,500$ and $n = 6$. The second screen shows the median, Med $= 392,500$.

 CAUTION Note that the formula $\frac{n+1}{2}$ gives the *position*, not the *value*, of the median. For example, the median home sales price for Table 4 is *not* $\frac{n+1}{2} = \frac{6+1}{2} = 3.5$.

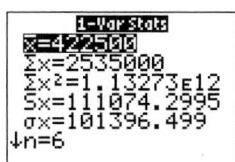 

b. The mean and median are shown in the Excel output.

Home Sales Price	
Mean	422500
Standard Error	45345.89
Median	392500
Mode	#N/A

c. The mean and median are shown in the Minitab output.

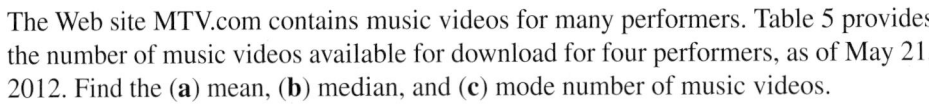

Descriptive Statistics: Home Price

```
Variable         Mean   Median
Home Price     422500   392500
```

d. The mean and median are shown in the JMP output.

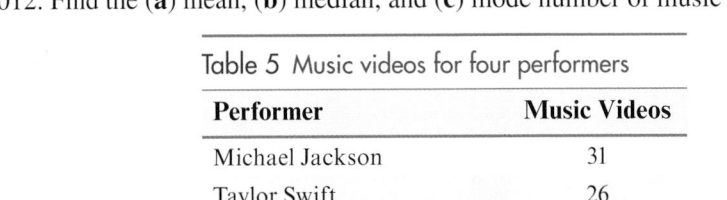

	N Rows	Mean(Home Price)	Median(Home Price)
1	6	422500	392500

3 The Mode

Sometimes the mode does not indicate the center of a data set. For example, suppose we have the following set of biology lab scores: 60, 80, 100, 100. The mode is 100, but it is not near the center of the data.

A third measure of center is called the **mode**. French speakers will recognize that the term *mode* in French refers to *fashion*. The popularity of clothing, cosmetics, music, and even basketball shoes often depends on just which style is in fashion. In a data set, the value that is most "in fashion" is the value that occurs the most.

> The **mode** of a data set is the data value that occurs with the greatest frequency.

EXAMPLE 6 Finding the mean, median, and mode: Music videos

Taylor Swift.

The Web site MTV.com contains music videos for many performers. Table 5 provides the number of music videos available for download for four performers, as of May 21, 2012. Find the (**a**) mean, (**b**) median, and (**c**) mode number of music videos.

Table 5 Music videos for four performers

Performer	Music Videos
Michael Jackson	31
Taylor Swift	26
Usher	26
Katy Perry	15

Solution

a. The sample mean number of music videos is

$$\bar{x} = \frac{\sum x}{n} = \frac{31 + 26 + 26 + 15}{4} = 24.5$$

The mean number of music videos is 24.5.

b. Because $n = 4$ is even, the median is the mean of the two middle data values:

$$\text{Median} = \frac{26 + 26}{2} = 26 \text{ music videos.}$$

c. The mode is the data value that occurs with the greatest frequency. Two performers have 26 music videos: Taylor Swift and Usher. No other data value occurs more than once. Therefore, the mode is 26 music videos, as shown in Figure 3.

NOW YOU CAN DO
Exercises 37–40.

FIGURE 3 Dotplot of music videos, showing 26 as the mode.

YOUR TURN
#3

Take a sample from Table 2 that consists of the number of tropical storms from the even-numbered years. Find the mean, median, and mode number of tropical storms.

(The solutions are shown in Appendix A.)

One of the strengths of the mode is that it can also be used with categorical, or qualitative, data. Suppose you asked your friends to name their favorite flower. Six of them answered "rose," three answered "lily," and one answered "daffodil." Note that these data are categorical, not numerical. The most frequently occurring flower is "rose"; therefore, the rose represents the mode of the variable *favorite flower.* Unfortunately, we cannot use arithmetic with categorical variables, and thus the mean or median for this variable cannot be found.

It may happen that no value occurs more than once, in which case we say there is *no mode*. On the other hand, more than one data value could occur with the greatest frequency, in which case we would say there is more than one mode. Data sets with one mode are *unimodal;* data sets with more than one mode are *multimodal*.

WHAT IF
?

What If Scenario

Consider Example 6 once again. Now imagine: *what if* there was an incorrect data entry, such as a typo, and the number of Michael Jackson's videos was greater than 31 by some unspecified amount?

Describe how and why this change would have affected the following, if at all:

a. The mean number of music videos

b. The median number of music videos

c. The mode number of music videos

Solution

a. Consider Figure 4, a dotplot of the number of music videos, with the triangle indicating the mean, or balance point, at 24.5. Recall that this represents the balance point of the data. As the number of Michael Jackson's videos increases (arrow), the point at which the data balance (the mean) also moves somewhat to the right. Thus, the mean number of followers will increase.

b. Recall from Example 6 that the median is the mean of the middle two data values. In other words, the mean *ignores* most of the data values, including the largest value, which is the only one that has increased. Therefore, the median will remain unchanged.

c. The mode also remains unchanged, because the only data value that occurs more than once is the original mode—26 music videos—and this remains unchanged.

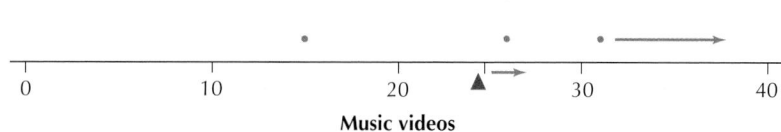

FIGURE 4 As the number of Michael Jackson's videos increases, so does the mean, but not the median or mode.

4 Skewness and Measures of Center

The skewness of a distribution can often tell us something about the relative values of the mean, median, and mode (see Figure 5).

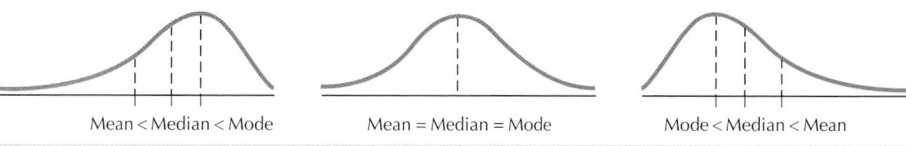

FIGURE 5 How skewness affects the mean and median.

> **How Skewness Affects the Mean and Median**
>
> - For a right-skewed distribution, the mean is larger than the median.
> - For a left-skewed distribution, the median is larger than the mean.
> - For a symmetric unimodal distribution, the mean, median, and mode are fairly close to each other.

EXAMPLE 7   Mean, median, and skewness

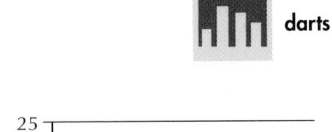

The histogram of the average size of households in the 50 states and the District of Columbia from Example 21 of Chapter 2 (page 74) is reproduced here as Figure 6.

a. Based on the skewness of the distribution, state the relative values of the mean, median, and mode.

b. Use Minitab to verify your claim in (**a**).

Solution

a. The distribution of average household size is somewhat right-skewed. Thus, from Figure 6, we would expect the mean to be greater than the median, which is greater than the mode.

b. The Minitab descriptive statistics are shown here. Note that the mean is greater than the median, which is greater than the mode.

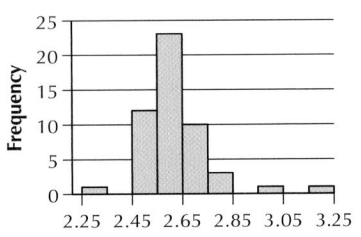

FIGURE 6 Household size is somewhat right-skewed.

```
Descriptive Statistics: Size

Variable   Mean    Median   Mode
Size       2.619   2.590    2.530
```

NOW YOU CAN DO
Exercises 41–44.

Mark Hooper/Getty Images

Remember: It is often helpful to have a "ballpark" estimate of the mean or other statistics as a reality check of your calculations.

Can the Financial Experts Beat the Darts?

CASE STUDY

Recall the contest held by the *Wall Street Journal* to compare the performance of stock portfolios chosen by financial experts and stocks chosen at random by throwing darts at the *Journal* stock pages. We will examine the results of 100 such contests in various ways, using the methods we have learned thus far, and will return to examine them further as we acquire more analysis tools. Let's start by reporting the raw result data. The percentage increase or decrease in stock prices was calculated for the portfolios chosen by the professional financial advisers and by the randomly thrown darts, and was compared with the percentage net change in the Dow Jones Industrial Average (DJIA).

Exploratory Data Analysis

Figure 7 shows comparative dotplots of the percentage net change in price for the professionally selected portfolio, the randomly selected darts portfolio, and the DJIA, over the course of the 100 contests. First, estimate the mean of each distribution by choosing the balance point of the data. This balance spot is the *mean*. For fun, write down your guess for the mean for the professionals so you can see how close you were when we provide the descriptive statistics later. Now compare this with where you would find the balance spot (mean) for the darts dotplot. Which numerical value is larger: the balance spot for the pros or the darts? Just think: you are comparing the

Note: In exploratory data analysis, we use graphical methods to compare numerical statistics.

mean portfolio performances for the professionals and the darts without using a formula or a calculator. This is *exploratory data analysis*. You are using *graphical* methods to compare *numerical* statistics.

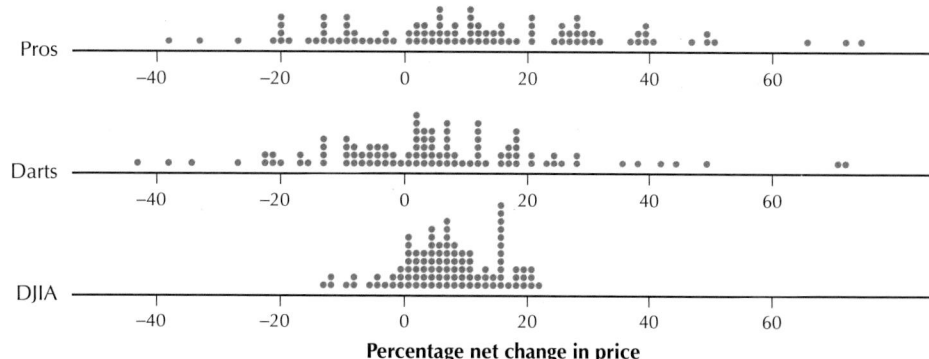

FIGURE 7 Dotplot of the percentage net price change for the professionally selected portfolio, the randomly selected darts portfolio, and the DJIA.

Hopefully, you discovered that the estimated mean for the pros is greater than the estimated mean for the darts. This is not particularly surprising, is it? Next, find the balance point for the DJIA dotplot. Compare the numerical value for the DJIA balance spot with the mean you found for the dotplot for the pros. Write down your estimate of the means for the DJIA and darts dotplots, so you can see how close you were later. Again, hopefully, you found that the estimated professionals' mean was higher than that of the DJIA. Now, a tougher comparison is to compare the estimated DJIA mean with that of the darts. Which of these two do you think is higher?

Finally, Minitab provides us with the mean percentage net price changes, as shown in Figure 8. Over the course of 100 contests, the mean price for the portfolios chosen by the professional financial advisers increased by 10.95%, by 6.793% for the DJIA, and by 4.52% for the random darts portfolio.

This is evidence in support of the view that financial experts can consistently outperform the market.

Variable	N	Mean
Pros	100	10.95
Darts	100	4.52
DJIA	100	6.79

FIGURE 8 Mean percentage net price change for the professionals, darts, and DJIA.

STEP-BY-STEP TECHNOLOGY GUIDE: Descriptive Statistics

TI-83/84

Step 1   Press STAT > 1: Edit. Enter the data in L1 using the instructions found in the Step-by-Step Technology Guide in Section 2.2.
Step 2   Press STAT. Use the right **arrow** button to move the cursor so that **CALC** is highlighted.

Step 3   Select 1-Var Stats, and press ENTER.
Step 4   On the home screen, the command 1-Var Statistics is shown. Press **2nd**, then **L1** (above the 1 key), and press ENTER.

EXCEL

Step 1   Enter the data in column **A**.
Step 2   Select **Data > Data Analysis**.
Step 3   Select **Descriptive Statistics**, and click **OK**.

Step 4   For the **Input Range**, click and drag to select the data in column A. If the variable name is at the top of the column, click **Labels in the First Row**.
Step 5   Check **Summary Statistics**, and click **OK**.

MINITAB

Step 1   Enter the data in column **C1**.
Step 2   Select **Stat > Basic Statistics > Display Descriptive Statistics...**
Step 3   The variable selection dialog box appears. Select the variable you want to summarize by double-clicking on it until it appears in the **Variables** box.

Step 4   Click **Statistics...**
Step 5   Select the desired statistics, and click **OK**. Then click **OK**.

SPSS

Step 1   Enter the data in the first column.
Step 2   Click **Analyze > Descriptive Statistics > Frequencies…**
Step 3   Click the variable name, then click the arrow to move it to the **Variable(s)** box.

Step 4   Click **Statistics…** and choose the desired statistics. Click **Continue**, and then **OK**.

JMP

Step 1   Click **File > New > DataTable**. Enter the data in **Column 1**.
Step 2   Click **Tables > Summary**.

Step 3   Select the column, and then select the desired statistics from the **Statistics** drop-down menu one by one. Click **OK**.

CRUNCHIT!

We will use the data from Example 3 (page 111).
Step 1   Click **File**, highlight **Load from Larose, Discostat3e > Chapter 3**, and click on **Example 01_03**.

Step 2   Click **Statistics** and select **Descriptive Statistics**. For Data, select *Price*, and then click **Calculate**.

Section 3.1 Summary

1. Measures of center are introduced in Section 3.1. The sample mean ($\bar{x}$) represents the sum of the data values in the sample divided by the sample size (n). The population mean (μ) represents the sum of the data values in the population divided by the population size (N). The mean is sensitive to the presence of extreme values.

2. The *median* occupies the middle position when the data are put in ascending order and is not sensitive to extreme values.

3. The *mode* is the data value that occurs with the greatest frequency. Modes can be applied to categorical data as well as numerical data but are not always reliable as measures of center.

4. The skewness of a distribution can often tell us something about the relative values of the mean and the median.

Section 3.1 Exercises

CLARIFYING THE CONCEPTS

1. Explain what a measure of center is. (p. 108)
2. Which measure may be used as the balance point of the data set? Explain how this works. (p. 110)
3. Explain what we mean when we say that the mean is sensitive to the presence of extreme values. Explain whether the median is sensitive to extreme values. (pp. 111–112)
4. What are the three measures of center that we learned about in this section? (p. 108)

For Exercises 5–12, either state what is being described or provide the notation.
5. The number of observations in your sample data set (p. 109)
6. The number of observations in your population data set (p. 109)
7. Notation denoting "sum all the data" (p. 109)
8. Notation for what we get when we add up all the data values in the population, and divide by how many observations there are in the population (p. 109)
9. Notation for what we get when we add up all the data values in the sample, and divide by how many observations there are in the sample (p. 109)
10. The middle data value when the data are put in ascending order (p. 112)

11. The data value that occurs with the greatest frequency (p. 114)
12. The sample mean (p. 109)

PRACTICING THE TECHNIQUES

 CHECK IT OUT!

To do	Check out	Topic
Exercises 13–18	Example 1	Population mean
Exercises 19–24	Example 2	Sample mean
Exercises 25–30	Example 3	Sensitivity of mean
Exercises 31–36	Example 4	Median
Exercises 37–40	Example 6	Mode
Exercises 41–44	Example 7	Mean, median, and skewness

For the data in Exercises 13–18:
    a. Find the population size N.
    b. Calculate the population mean μ.
13. State exports to other countries are shown in the table for the population of all New England states, for the month of June 2014, expressed in billions of dollars.

State	Exports	State	Exports
Connecticut	1.4	New Hampshire	0.4
Maine	0.3	Rhode Island	0.2
Massachusetts	2.4	Vermont	0.3

Source: U.S. Census Bureau.

14. The number of wins for each baseball team in the population of the American League West division for 2013 is shown in the table.

Team	Wins	Team	Wins
Oakland Athletics	96	Seattle Mariners	71
Texas Rangers	91	Houston Astros	51
Los Angeles Angels	78		

Source: MLB.mlb.com.

15. The table provides the motor vehicle theft rate for the population of the top 10 countries in the world for motor vehicle theft, for 2012. The theft rate equals the number of motor vehicles stolen in 2012 per 100,000 residents.

Country	Theft rate	Country	Theft rate
Italy	208.0	Greece	100.2
France	174.1	Norway	94.1
USA	167.8	Netherlands	75.2
Sweden	117.2	Spain	75.1
Belgium	106.0	Cyprus	66.0

Source: United Nations Office on Drugs and Crime.

16. The National Center for Education Statistics sponsors the Trends in International Mathematics and Science Study (TIMSS). The table contains the mean science scores for the eighth-grade science test for the populations of all Asian-Pacific countries that took the exam.

Country	Science score	Country	Science score
Singapore	578	Australia	527
Taiwan	571	New Zealand	520
South Korea	558	Malaysia	510
Hong Kong	556	Indonesia	420
Japan	552	Philippines	377

17. The table contains the number of petit larceny cases for the population of all police precincts in South Manhattan in 2013.

Precinct	Petit larcenies	Precinct	Petit larcenies
1	2014	10	995
5	1288	13	2094
6	1555	14	4551
7	584	17	823
9	1607	18	2071

Source: New York City Police Department.

18. The table contains the number of criminal trespass cases for the population of all police precincts in South Manhattan in 2013.

Precinct	Criminal trespasses	Precinct	Criminal trespasses
1	108	10	207
5	105	13	135
6	113	14	340
7	233	17	74
9	219	18	120

Source: New York City Police Department.

For the data in Exercises 19–24:
  a. Find the sample size n.
  b. Calculate the sample mean $\bar{x}$.

19. A sample of the state export data from Exercise 13 is provided in the table.

State	Exports
Connecticut	1.4
Massachusetts	2.4
Rhode Island	0.2

20. A sample from the baseball data in Exercise 14 is shown here.

Team	Wins
Texas Rangers	91
Los Angeles Angels	78
Seattle Mariners	71

21. A sample from the motor vehicle theft data in Exercise 15 is as follows.

Country	Theft rate
Italy	208.0
USA	167.8
Greece	100.2

22. A sample from the science score data in Exercise 16 is given here.

Country	Science score
South Korea	558
Hong Kong	556
Japan	552
Australia	527

23. The following sample is taken from the petit larceny data in Exercise 17.

Precinct	Petit larcenies
1	2014
6	1555
9	1607
14	4551
17	823

24. A sample taken from the criminal trespass data in Exercise 18 is as follows.

Precinct	Criminal trespasses
1	108
7	233
14	340
18	120

For Exercises 25–30, use the data from the indicated exercise, along with the indicated extreme, to show that the mean is more sensitive to extreme values. For each exercise, find the sample mean including the extreme value. Compare your answer to the mean calculated without the extreme value from the earlier exercise.

25. Data from Exercise 19. Extreme value = 10
26. Data from Exercise 20. Extreme value = 20
27. Data from Exercise 21. Extreme value = 1000
28. Data from Exercise 22. Extreme value = 0
29. Data from Exercise 23. Extreme value = 20,000
30. Data from Exercise 24. Extreme value = 1500

For Exercises 31–36, use the data from the indicated exercise, along with the indicated extreme, to show that the mean is more sensitive to extreme values than the median is. Do the following:

    a. Calculate the median of the data without the extreme value.
    b. Find the median of the data including the extreme value. Compare your answers from (**a**) and (**b**). Note that the median did not change as much as the mean did in Exercises 25–30.

31. Data from Exercise 19. Extreme value = 10
32. Data from Exercise 20. Extreme value = 20
33. Data from Exercise 21. Extreme value = 1000
34. Data from Exercise 22. Extreme value = 0
35. Data from Exercise 23. Extreme value = 20,000
36. Data from Exercise 24. Extreme value = 1500

For the data in Exercises 37–40, find the mode.
37. The table contains the number of dangerous weapons cases for four police precincts in Manhattan.

Precinct	Dangerous weapons cases
1	19
5	24
20	24
22	9

38. The Recording Industry Association of America (RIAA) awards multi-platinum status for any musical recording that sells more than 2 million copies. The table contains a random sample of 10 of the musical artists with the most multi-platinum singles.

Artist	Multi-platinums	Artist	Multi-platinums
Beyoncé	4	Linkin Park	2
Bruno Mars	4	The Beatles	4
Jay-Z	4	Michael Jackson	1
Katy Perry	8	Taylor Swift	8
Lady Gaga	6	Tim McGraw	2

Source: RIAA.

39. The table contains the unemployment rates in August 2014 for 10 countries.

Country	Unemployment rate	Country	Unemployment rate
Britain	6.4	Japan	3.7
Canada	7.0	Mexico	4.8
China	4.1	Pakistan	6.2
India	8.8	South Korea	3.4
Italy	12.3	United States	6.2

Source: The Economist, www.economist.com/node/21604509.

40. The table contains the top 10 most downloaded free apps for the IOS platform, as reported by Apple.com, along with the app type, for June 2014. Find the mode of *App Type*.

Rank	App	App type	Rank	App	App type
1	Two Dots	Games	6	Snap Chat	Photo and video
2	The Line	Games	7	Instagram	Photo and video
3	Traffic Racer	Games	8	The Test	Games
4	Rival Knights	Games	9	Republique	Games
5	Piano Tiles	Games	10	YouTube	Photo and video

For Exercises 41–44, consider the accompanying distributions. What can we say about the values of the mean, median, and mode in relation to one another for the given histograms?

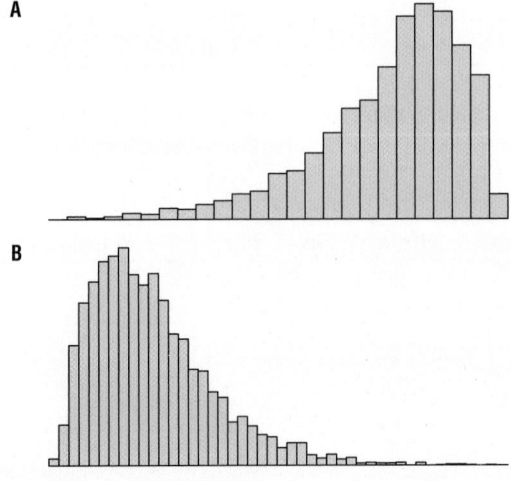

A

B

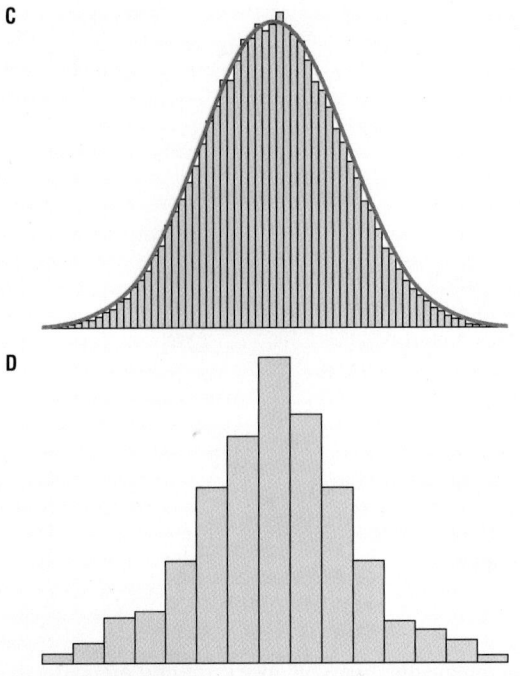

C

D

41. The distribution in A
42. The distribution in B
43. The distribution in C
44. The distribution in D

APPLYING THE CONCEPTS

45. NFL Football, Southern Style. The table contains the population of all the teams in the National Football Conference South Division, along with the number of wins in the 2013 season.
  a. What is the population size, N, where the population is the NFC South Division?
  b. What is the population mean number of wins, μ?

NFC South team	Wins
Carolina Panthers	12
New Orleans Saints	11
Atlanta Falcons	4
Tampa Bay Buccaneers	4

46. New England Electoral Votes. The table contains the population of all the New England states, along with their electoral votes.
  a. What is the population size, N?
  b. Calculate the population mean number of electoral votes, μ.

	Electoral votes
Connecticut	7
Maine	4
Massachusetts	11
New Hampshire	4
Rhode Island	4
Vermont	3

47. NFL Football, Southern Style. Refer to the population data in Exercise 45. Suppose we take a sample from the population, and we get the Carolina Panthers and the Atlanta Falcons.
  a. What is the sample size n?
  b. Calculate the sample mean number of wins, $\bar{x}$.

48. New England Electoral Votes. Refer to the population data in Exercise 46. Suppose we take a sample from the population, and get Massachusetts, Rhode Island, and Vermont.
  a. What is the sample size n?
  b. Calculate the sample mean number of electoral votes, $\bar{x}$.

Video Game Sales. The Chapter 1 Case Study looked at video game sales for the top 30 video games. The following table contains the total sales (in game units) and weeks on the top 30 list for a sample of five randomly selected video games. Use this information for Exercises 49 and 50.

Video game	Total sales in millions of units	Weeks
Super Mario Bros. U for WiiU	1.7	78
NBA 2K14 for PS4	0.6	27
Battlefield 4 for PS3	0.9	29
Titanfall for XBoxOne	1.2	10
Yoshi's New Island for 3DS	0.2	10

Source: www.vgchartz.com.

 videogamereg

49. Find the following measures of center for total sales.
  a. Mean
  b. Median
50. Calculate the following measures of center for weeks.
  a. Mean
  b. Median

Darts and the Dow Jones. The following table contains a random sample of eight days from the Chapter 3 Case Study data set, indicating the stock market gain or loss for the portfolio chosen by the random darts, as well as the Dow Jones Industrial Average gain or loss for that day. Use this information for Exercises 51 and 52.

51. Find the following measures of center for the darts stock returns.
  a. Mean
  b. Median
52. Find the following measures of center for the DJIA.
  a. Mean
  b. Median

Darts	DJIA
−27.4	−12.8
18.7	9.3
42.2	8
−16.3	−8.5
11.2	15.8
28.5	10.6
1.8	11.5
16.9	−5.3

Source: Wall Street Journal.

 **dartsdjia**

Age and Height. The following table provides a random sample from the Chapter 4 Case Study data set **body_females**, showing the age and height of the eight women. Use this information for Exercises 53 and 54.

Age	Height
40	63.5
28	63
25	64.4
34	63
26	63.8
21	68
19	61.8
24	69

Source: Journal of Statistics Education.

 **ageheight**

53. Find the following measures of center for the women's ages.
   a. Mean
   b. Median
54. Find the following measures of center for the women's heights.
   a. Mean
   b. Median

Saturated Fat and Calories. The table contains the calories and saturated fat in a sample of ten food items. Use this information for Exercises 55 and 56.
55. Find the following measures of center for calories.
   a. Mean
   b. Median
56. Find the following measures of center for the grams of saturated fat.
   a. Mean
   b. Median

 **satfatcorr**

Food item	Calories	Grams of saturated fat
Chocolate bar (1.45 ounces)	216	7.0
Meat & veggie pizza (large slice)	364	5.6
New England clam chowder (1 cup)	149	1.9
Baked chicken drumstick (no skin, medium size)	75	0.6
Curly fries, deep-fried (4 ounces)	276	3.2
Wheat bagel (large)	375	0.3
Chicken curry (1 cup)	146	1.6
Cake doughnut hole (one)	59	0.5
Rye bread (1 slice)	67	0.2
Raisin bran cereal (1 cup)	195	0.3

Source: Food-a-Pedia.

Table 6 contains the trade balance currently maintained by the United States with a sample of 9 countries, for the month of June 2014. Use this data for Exercises 57–60.

TABLE 6   Trade balance

Country	Trade balance ($ billions)
Brazil	1
France	−1.2
Germany	−5.6
India	−1.3
Italy	−2.4
Japan	−5.6
South Korea	−1.8
Saudi Arabia	−1.8
United Kingdom	0

Source: Foreign Trade Division, U.S. Census Bureau.

57. Find the sample size, n.
58. Calculate the sample mean trade balance, $\bar{x}$.
59. Find the median.
60. Find the modes.

Table 7 contains the number of cylinders, the engine size (in liters), the fuel economy (miles per gallon [mpg], city driving), and the country of manufacture for six 2011 automobiles. Use this information for Exercises 61–65.

 **cylinderengine**

TABLE 7   Cylinders, engine size, and fuel economy for six cars

Vehicle	Cylinders	Engine size	City mpg	Country of manufacture
Cadillac CTS	6	3.0	18	USA
Ford Fusion Hybrid	4	2.5	41	USA
Ford Taurus	6	3.5	18	USA
Honda Civic	4	1.8	25	Japan
Rolls Royce	12	6.7	11	UK
Toyota Camry Hybrid	4	2.4	31	Japan

Source: www.fueleconomy.gov.

61. Find the following for the number of cylinders:
   a. Mean   **b.** Median   **c.** Mode
62. Refer to your work in Exercise 61. Which measure of center do you think is most representative of the typical number of cylinders? Explain.
63. Find the following for the engine size:
   a. Mean   **b.** Median   **c.** Mode
64. Find the following for the city mpg:
   a. Mean   **b.** Median   **c.** Mode
65. Find the mode for country of manufacture.

Use the information in Table 8 to answer Exercises 66–68, which gives the number of wins for the top 10 NASCAR racing drivers in various categories.

nascar

TABLE 8   Top 10 NASCAR winners in the modern era

Rank	Driver	Total	Super speedways	Short tracks
1	Darrell Waltrip	84	18	47
2	Dale Earnhardt	76	29	27
3	Jeff Gordon	75	15	15
4	Cale Yarborough	69	15	29
5	Richard Petty	60	19	23
6	Bobby Allison	55	24	12
7	Rusty Wallace	55	5	25
8	David Pearson	45	20	1
9	Bill Elliott	44	16	2
10	Mark Martin	35	5	7

Source: www.nascar.com.

66. Refer to the super speedways data. Find the following:
   a. Mean   **b.** Median   **c.** Mode
67. Refer to the short tracks data. Find the following:
   a. Mean   **b.** Median   **c.** Mode
68. Refer to the totals data. Find the following:
   a. Mean   **b.** Median   **c.** Mode

For Exercises 69–73, refer to Table 9, which lists the top five mass market paperback fiction books for the week of July 1, 2014, as reported by the *New York Times*.

TABLE 9   Top five best-sellers in paperback trade fiction

Rank	Title	Author	Price
1	A Game of Thrones	George R. R. Martin	$7.83
2	Takedown Twenty	Janet Evanovich	$7.64
3	Inferno	Dan Brown	$8.48
4	A Dance with Dragons	George R. R. Martin	$6.71
5	The 9th Girl	Tami Hoag	$8.47

69. Find the mean, median, and mode for the price of these five books on the best-seller list. Suppose a salesperson claimed that the price of a typical book on the best-seller list is less than $14. How would you use these statistics to respond to this claim?
70. Linear Transformations. Add $10 to the price of each book.
   a. Now find the mean of these new prices.
   b. How does this new mean relate to the original mean?
   c. Construct a rule to describe this situation in general.
71. Linear Transformations. Multiply the price of each book by 5.
   a. Now find the mean of these new prices.
   b. How does this new mean relate to the original mean?
   c. Construct a rule to describe this situation in general.
72. Find the mode for the following variables:
   a. Price
   b. Author
73. Explain whether it makes sense to find the mean or median of the variable *author*.

Mode of Categorical Data. The New York City Police Department tracks the number and type of traffic violations. The table contains a random sample of 12 traffic violations and the borough in which they occurred (Manhattan or Brooklyn). Use the data for Exercises 74–76.

Violation type	Borough	Violation type	Borough
Cell phone	Brooklyn	Disobey sign	Manhattan
Safety belt	Manhattan	Speeding	Brooklyn
Cell phone	Brooklyn	Safety belt	Manhattan
Cell phone	Manhattan	Disobey sign	Manhattan
Speeding	Brooklyn	Disobey sign	Brooklyn
Safety belt	Manhattan	Cell phone	Manhattan

74. Find the mode for *violation type*. Does this mean that most violations are of this type?
75. Calculate the mode for *borough*.
76. Does the idea of the mean or median of these two variables make any sense? Explain clearly why not.

Car Model Years. The dotplot in Figure 9 represents the model year for a sample of cars in a used car lot. Refer to the dotplot for Exercises 77–79.

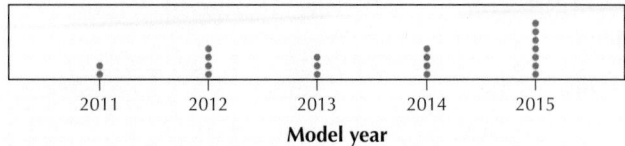

FIGURE 9 Dotplot of model year.

77. What are the mean, median, and mode of the model year?
78. Calculate a new statistic "age of the car in 2015" as follows: take the model year and subtract it from 2015.
    a. Find the mode of the car ages.
    b. Find the mean and median of the car ages.
79. What will be the mean, median, and mode of the car ages in 2025?
80. Five friends have just had dinner at the local pizza joint. The total bill came to $30.60. What is the mean cost of each person's meal?
81. Lindsay just bought four shirts at the boutique in the mall, costing a total of $84.28. What was the mean cost of each shirt?

Dealing with Missing Data. Exercises 82–85 ask you to calculate measures of center when one of the values is missing.
82. The mean cost of a sample of five items is $20. The cost of four of the items is as follows: $25, $15, $15, $20. What is the cost of the 5th item?
83. The mean size of four downloaded music files is 3 Mb (megabytes). The size of three of the files is as follows: 5 Mb, 2 Mb, 3 Mb. What is the length of the 4th music file?
84. The median number of students in a sample of seven statistics classes is 25. The ordered values are: 20, 22, 24, __, 27, 27, 28. What is the missing value?
85. The median number of academic credits taken in a sample of six students is 15. The ordered values are: 12, 12, 14, __, 17, 17. What is the missing value?

Nutrition Ratings of Breakfast Cereals. Refer to the following information for Exercises 86–89. (Note that Minitab denotes both the sample size and the population size as *N*.) The data represent the nutrition rate of 59 cereals based on sugar content, vitamin content, and so on.

```
Descriptive Statistics: Rating

Variable        N        Mean     Median    TrMean     StDev    SE Mean
Rating          59       45.46    42.00     44.76      14.39    1.87

Variable     Minimum   Maximum      Q1        Q3
Rating        22.40     93.70      35.25     53.37
```

86. Find the following sample statistics.
    a. The sample size
    b. The sample mean
    c. The sample median
    d. The highest and lowest ratings in the sample

87. What do these statistics tell us about the skewness of the distribution?
88. Linear Transformations. If we take each cereal rating and subtract 5 from it, how would that affect the mean, median, and mode? Would it affect each of the measures equally?
89. Linear Transformations. If we cut each of the cereals' ratings in half, how would that affect the mean, median, and mode? Would it affect each of the measures equally?

BRINGING IT ALL TOGETHER

Pulse Rates for Men and Women. To answer Exercises 90–93, refer to Figure 10, which includes comparative dotplots of the pulse rates for males and females.[2]

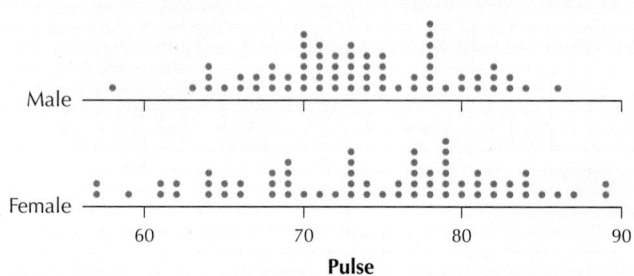

FIGURE 10 Comparative dotplots of pulse rates, by gender.

90. Examine Figure 10.
    a. Without doing any calculations, what is your impression of which gender, if any, has the higher overall pulse rate?
    b. Find the mean pulse rate for the males by estimating the location of the balance point.
    c. Find the mean pulse rate for the females by estimating the location of the balance point.
    d. Based on (b) and (c), which gender has the higher mean pulse rate? Does this agree with your earlier impression?
91. Find the following medians:
    a. The median pulse rate for the males
    b. The median pulse rate for the females
    c. Which gender has the higher median pulse rate? Does this agree with your findings for the mean earlier?
92. Find the following modes:
    a. The mode pulse rate for the males
    b. The mode pulse rate for the females
    c. Which gender has the higher mode pulse rate? Does this agree with your findings for the mean earlier?

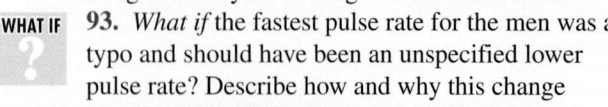 **93.** *What if* the fastest pulse rate for the men was a typo and should have been an unspecified lower pulse rate? Describe how and why this change would have affected the following, if at all. Would they increase, decrease, or remain unchanged? Or is there insufficient information to tell what would happen? Explain your answers.
    a. The mean men's pulse rate
    b. The median men's pulse rate
    c. The mode men's pulse rate

94. Trimmed Mean. Because the mean is sensitive to extreme values, the *trimmed mean* was developed as another measure of center. To find the 10% trimmed mean for a data set, omit the largest 10% of the data values and the smallest 10% of the data values, and calculate the mean of the remaining values. Because the most extreme values are omitted, the trimmed mean is *less sensitive, or more robust (resistant),* than the mean as a measure of center. For the data in the table, calculate the following:

    a. The mean
    b. The 10% trimmed mean
    c. The 20% trimmed mean

The data represent the number of business establishments in a sample of states.

statebusinesses

State	Businesses (1000s)	State	Businesses (1000s)
Alabama	3.8	Michigan	7.5
Arizona	7.9	Minnesota	6.1
Colorado	8.9	Missouri	5.9
Connecticut	3.1	Ohio	9.5
Georgia	10.3	Oklahoma	3.8
Illinois	11.9	Oregon	5.4
Indiana	5.6	South Carolina	4.6
Iowa	2.7	Tennessee	5.4
Maryland	5.7	Virginia	8.6
Massachusetts	6.3	Washington	9.3

Source: U.S. Census Bureau.

95. Challenge Exercise. In general, would you expect the trimmed mean to be larger, smaller, or about the same as the mean for data sets with the following shapes?

    a. Right-skewed data
    b. Left-skewed data
    c. Symmetric data

96. Midrange. Another measure of center is the *midrange*.

$$midrange = \frac{\text{largest data value} + \text{smallest data value}}{2}$$

Because the midrange is based on the maximum and minimum values in the data set, it is not a robust statistic, but it is sensitive to extreme values. Calculate the midrange for the following data:

    a. The price data from Table 9 on page 123.
    b. The car model year data from Figure 9 on page 124.

97. Harmonic Mean. The *harmonic mean* is a measure of center most appropriately used when dealing with rates, such as miles per hour (mph). The harmonic mean is calculated as

$$\frac{n}{\sum \frac{1}{x}}$$

where *n* is the sample size and the *x*'s represent rates, such as the speeds in mph. Emily walked five miles today, but her walking speed slowed as she walked farther. Her walking speed was 5 mph for the first mile, 4 mph for the second mile, 3 mph for the third mile, 2 mph for the fourth mile, and 1 mph for the fifth mile. Calculate her harmonic mean walking speed over the entire five miles.

98. Challenge Exercise. The (arithmetic) mean for Emily's five-mile walk in Exercise 97 is 3 mph. Explain clearly why the value you calculated for the harmonic mean in Exercise 97 makes more sense than this arithmetic mean of 3 mph. (*Hint:* Consider time.)

99. Geometric Mean. The *geometric mean* is a measure of center used to calculate growth rates. Suppose that we have *n* positive values; then the geometric mean is the *n*th root of the product of the *n* values. Jamal has been saving money in an account that has had 4% growth, 6% growth, and 10% growth over the last three years. Calculate the *average growth rate* over these three years. (*Hint:* Find the geometric mean of 1.04, 1.06, and 1.10 and subtract 1.)

CONSTRUCT YOUR OWN DATA SETS

100. Construct your own data set with $n = 10$, where the mean, the median, and the mode are all the same. Yes, just make up your own list of numbers, as long as the mean, median, and mode are all the same. Draw a dotplot. Comment on the skewness of the distribution.

101. Construct your own data set with $n = 10$, where the mean is greater than the median, which is greater than the mode. Draw a dotplot. Comment on the skewness of the distribution.

102. Construct your own data set with $n = 10$, where the mode is greater than the median, which is greater than the mean. Draw a dotplot. Comment on the skewness of the distribution.

103. Construct your own data set with $n = 3$. Let the mean and median be equal. Now, alter the three data values so that the mean of the altered data set has increased, while the median of the altered data set has decreased.

Use the *Mean and Median* applet for Exercises 104 and 105.

104. Insert three points on the line by clicking just below it: two near the left side and one near the middle.

    a. Click and drag the rightmost point to the right.
    b. Describe what happens to the mean when you do this.
    c. Describe what happens to the median when you do this.

105. Explain why each of the measures behaves the way it does in the previous exercise.

WORKING WITH LARGE DATA SETS

Open the **VideoGameSales** data set from the Chapter 1 Case Study. The data set represents a sample. Use technology to do the following. **videogamesales**

106. Find the mean and median weekly sales.

107. Suppose we remove the biggest seller for the week, *Minecraft for PS3*, from the data. Given what you have learned about the sensitivity of the mean to the presence of extreme values, which measure do you expect will change the most, the mean or the median?

108. Recalculate the mean and the median of weekly sales, this time omitting *Minecraft for PS3*. Was your intuition in Exercise 107 confirmed?

109. Compute the mean and median total sales for the 30 games.

110. Identify the video game with the largest total sales. Omit this video game, and recompute the mean and median total sales. Which measure of center was more sensitive to the removal of the extreme value?

111. Find the mode for each of the following variables:
  a. Platform
  b. Studio
  c. Game type

112. Compute the mean, median, and mode for the variable weeks on list.

WHAT IF
113. *What if* we add a certain unknown amount x to each value in the variable weeks on list? Describe what will happen to the following measures of center.
  a. Mean
  b. Median
  c. Mode

3.2   Measures of Variability

OBJECTIVES By the end of this section, I will be able to . . .

1 Find the range of a data set.
2 Calculate the variance and the standard deviation for a population.
3 Compute the variance and the standard deviation for a sample.
4 Use the Empirical Rule to find approximate percentages for a bell-shaped distribution.
5 Apply Chebyshev's Rule to find minimum percentages.

1 The Range

In Section 3.1, we learned how to find the center of a data set. Is that all there is to know about a data set? Definitely not! Two data sets can have exactly the same mean, median, and mode and yet be quite different. We need measures that summarize the data set in a different way, namely, the variation or variability of the data. In Section 3.2, we will learn *measures of variability* that will help us answer the question: "How spread out is the data set?"

EXAMPLE 8   Different data sets with the same measures of center

Martin Meissner/AP Photo

Table 10 contains the heights (in inches) of the players on two volleyball teams.

Table 10 Women's volleyball team heights (in inches)

Western Massachusetts University	Northern Connecticut University
60	66
70	67
70	70
70	70
75	72

 volleyball

a. Describe in words and graphs the variability of the heights of the two teams.

b. Verify that the means, medians, and modes for the two teams are equal.

Solution

a. There are some distinct differences between the teams. The Western Massachusetts (WMU) team has a player who is relatively short (60 inches: 5 feet tall) and a player who is very tall (75 inches: 6 feet, 3 inches tall). The Northern Connecticut (NCU) team has players whose heights are all within 6 inches of each other.

b. But despite the differences in (**a**), the mean, median, and mode of the heights for the two teams are precisely the same. As illustrated in Figure 11, the mean height (red triangle) for each team is 69 inches, the median height (green triangle) for each team is 70 inches, and the mode height (yellow triangle) for each team is 70 inches.

$$\bar{x}_{\text{WMU}} = \frac{60 + 70 + 70 + 70 + 75}{5} = \frac{345}{5} = 69$$

$$\bar{x}_{\text{NCU}} = \frac{66 + 67 + 70 + 70 + 72}{5} = \frac{345}{5} = 69$$

Clearly, these measures of location do not give us the whole picture. We need **measures of variability** (or **measures of spread** or **measures of dispersion**) that will describe how spread out the data values are. Figure 11 illustrates that the heights of the WMU team are *more spread out* than the heights of the NCU team.

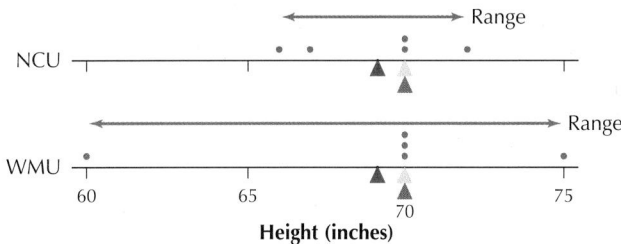

FIGURE 11 Comparative dotplots of the heights of two volleyball teams.

Just as there were several measures of the center of a data set, there are also a variety of ways to measure how spread out a data set is. The simplest measure of variability is the **range**.

> The **range** of a data set is the difference between the largest value and the smallest value in the data set:
>
> range = largest value − smallest value = maximum − minimum

A larger range is an indication of greater variability, or greater spread, in the data set.

EXAMPLE 9   Range of the volleyball teams' heights

Calculate the range of player heights for each of the WMU and NCU teams.

Solution

What Results Might We Expect?

From Figure 11, it is intuitively clear that the heights of the WMU team are more spread out than the heights of the NCU team. Therefore, we would expect the range of the WMU team to be larger than the range of the NCU team, reflecting its greater variability.

$$\text{range}_{\text{WMU}} = \text{largest value} - \text{smallest value} = 75 - 60 = 15 \text{ inches}$$

$$\text{range}_{\text{NCU}} = \text{largest value} - \text{smallest value} = 72 - 66 = 6 \text{ inches}$$

As we expected, the range for WMU players is indeed larger than the range for NCU players, reflecting WMU's players' greater variability in height.

Table 11 contains a sample from the data set for the Chapter 3 Case Study. The percent increase or decrease in stock portfolio is recorded for the set of stocks chosen by throwing darts at the stock pages, along with the Dow Jones Industrial Average (DJIA) for the same day.

Table 11 Sample set of stock market returns

Darts	DJIA
11.2	15.8
72.9	16.2
16.6	17.3
28.7	17.7

1. Construct a comparison dotplot of the darts returns and the DJIA returns.
2. Using the dotplot, which group would you say has the larger range?
3. Calculate the range for each group. Is your intuition from (2) confirmed?

(The solutions are shown in Appendix A.)

The range is quite simple to calculate; however, it does have its drawbacks. For example, the range is quite sensitive to extreme values, because it is calculated from the difference of the two most extreme values in the data set. *It completely ignores all the other data values in the data set.* We would prefer our measure of variability to quantify spread with respect to the center, as well as to actually use all the available data values. Two such measures are the **variance** and the **standard deviation**.

2 Population Variance and Population Standard Deviation

Before we learn about the variance and the standard deviation, we need to get a firm understanding of what a **deviation** means, in the statistical sense.

Deviation

A **deviation** for a given data value x is the difference between the data value and the mean of the data set. For a sample, the deviation equals $x - \bar{x}$. For a population, the deviation equals $x - \mu$.

- If the data value is larger than the mean, the deviation will be positive.
- If the data value is smaller than the mean, the deviation will be negative.
- If the data value equals the mean, the deviation will be zero.

The deviation can roughly be thought of as the distance between a data value and the mean, except that the deviation can be negative, whereas distance is always positive.

EXAMPLE 10   Calculating deviations

Catherine Yeulet/Getty Images

Ashley and Brandon, certified public accountants.

Ashley and Brandon are certified public accountants who work for a large accounting firm, preparing tax returns for small business clients. Because tax returns are often filed close to the deadline, it is important that the returns be prepared in a timely fashion, with not a lot of variability in the length of time it takes to prepare a return. The chief accountant kept careful track of the amount of time (in hours, Table 12) for all the tax returns prepared by Ashley and Brandon during the last week of March.

a. Find the mean preparation time for each accountant.

b. Use comparative dotplots to compare the variability of Ashley and Brandon's tax preparation times.

c. Calculate the deviations for each of Ashley and Brandon's tax preparation times.

Table 12 Preparation times (in hours) for Ashley and Brandon

Ashley	5	7	8	9	11
Brandon	3	5	7	11	14

Solution

Because the data represent *all* the tax returns for the indicated period, they may be considered a population.

a. For Ashley:

$$\mu = \frac{\sum x}{N} = \frac{5 + 7 + 8 + 9 + 11}{5} = 8 \text{ hours}$$

For Brandon:

$$\mu = \frac{\sum x}{N} = \frac{3 + 5 + 7 + 11 + 14}{5} = 8 \text{ hours}$$

So the two accountants spent the same mean amount of time in tax preparation.

b. Figure 12 contains comparative dotplots of Ashley and Brandon's tax preparation times. Note that Brandon's preparation times vary more than Ashley's. Compared to Ashley, we can say that Brandon's tax preparation times

- are *more spread out*,
- show *greater variability*,
- have *more variation*, and
- are *more dispersed*.

The chief accountant probably prefers a more consistent tax preparation time, with less variability.

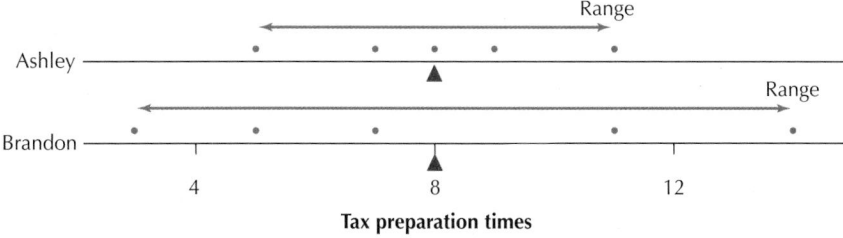

FIGURE 12 Brandon's tax preparation times are more spread out.

c. Here we find the deviations, $x - \mu$.

- Ashley's mean preparation time is $\mu = 8$ hours. Her first tax return took $x = 5$ hours, so the deviation for this first tax return is $x - \mu = 5 - 8 = -3$. Note that, when $x < \mu$, the deviation is negative.

- Ashley's last tax return took 11 hours, so the deviation for this last return is $x - \mu = 11 - 8 = 3$. Note that, when $x > \mu$, the deviation is positive.

- Continuing in this way, we find the deviations for all of Ashley's and Brandon's tax preparation times, as recorded in Table 13.

Table 13 Tax preparation times and their deviations

Ashley's times	5	7	8	9	11
Ashley's deviations	$5 - 8 = -3$	$7 - 8 = -1$	$8 - 8 = 0$	$9 - 8 = 1$	$11 - 8 = 3$
Brandon's times	3	5	7	11	14
Brandon's deviations	$3 - 8 = -5$	$5 - 8 = -3$	$7 - 8 = -1$	$11 - 8 = 3$	$14 - 8 = 6$

NOW YOU CAN DO
Exercises 11c–16c.

These deviations are used for the most widespread measures of spread: the variance and the standard deviation. However, we cannot use the mean deviation, because the mean deviation always equals zero. For example,

- Ashley's mean deviation: $\dfrac{(-3) + (-1) + 0 + 1 + 3}{5} = 0$

- Brandon's mean deviation: $\dfrac{(-5) + (-3) + (-1) + 3 + 6}{5} = 0$

The mean deviation always equals zero for any data set because the positive and negative deviations cancel each other out. Thus, the mean deviation is not a useful measure of spread. To avoid this problem, we will work with the squared deviations.

Table 14 shows the squared deviations for Ashley and Brandon. Note that Brandon's squared deviations are, on average, larger than Ashley's, reflecting the greater spread in Brandon's preparation times. It is therefore logical to build our measure of spread using the *mean squared deviation*.

Table 14 Squared deviations of tax preparation times

Ashley's deviations	−3	−1	0	1	3
Ashley's squared deviations	9	1	0	1	9
Brandon's deviations	−5	−3	−1	3	6
Brandon's squared deviations	25	9	1	9	36

The Population Variance, σ^2

For populations, the mean squared deviation is called the **population variance** and is symbolized by σ^2. This is the lowercase Greek letter *sigma*, not to be confused with the uppercase *sigma* (Σ) used for summation.

> The **population variance**, σ^2, is the mean of the squared deviations in the population and is given by the formula
>
> $$\sigma^2 = \frac{\sum (x - \mu)^2}{N}$$

Notice that the numerator in σ^2 is a sum of squares. Squared numbers can never be negative, so a sum of squares also can never be negative. The denominator, N, which is the population size, also can never be negative. Thus, σ^2 can never be negative. The only time $\sigma^2 = 0$ is when all the population data values are equal.

EXAMPLE 11 Calculating the population variances for Ashley and Brandon

Calculate the population variances of the tax preparation times for Ashley and Brandon.

Solution

Using the squared deviations from Table 14, we have

$$\sigma^2 = \frac{\sum(x-\mu)^2}{N} = \frac{9+1+0+1+9}{5} = \frac{20}{5} = 4$$

for Ashley, and

$$\sigma^2 = \frac{\sum(x-\mu)^2}{N} = \frac{25+9+1+9+36}{5} = \frac{80}{5} = 16$$

for Brandon. The population variance of the tax preparation times for Brandon is greater than the variance for Ashley, thus indicating that Brandon's tax preparation times are more variable than Ashley's.

NOW YOU CAN DO
Exercises 11d–16d.

YOUR TURN #5

Table 15 contains the funding provided by the Centers for Disease Control (CDC) to all the states in New England, in order to fight HIV/AIDS.[3] This includes all the states in New England, so we may consider this a population.

1. Find the population mean funding, μ.
2. Calculate the population variance of the funding, σ^2.

(The solution is shown in Appendix A.)

Table 15 CDC funding to fight HIV/AIDS for New England states

State	Funding (in millions)
Connecticut	7.8
Maine	1.9
Massachusetts	14.9
New Hampshire	1.5
Rhode Island	2.7
Vermont	1.6

However, what is the *meaning* of the values we obtained for σ^2, 4, and 16, apart from their comparative value? The problem is that the units of these values represent *hours squared*, which is not a useful measure. Unfortunately, the intuitive meaning of the population variance is not self-evident.

The Population Standard Deviation, σ

In practice, the *standard deviation* is easier to interpret than the variance. The standard deviation is simply the square root of the variance, and by taking the square root, we return the units of measure back to the original data unit (for example, "hours" instead of "hours squared"). The symbol for the **population standard deviation** is σ. Conveniently, $\sqrt{\sigma^2} = \sigma$.

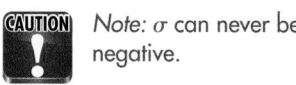

Note: σ can never be negative.

> The **population standard deviation**, σ, is the positive square root of the population variance and is found by
>
> $$\sigma = \sqrt{\frac{\sum (x - \mu)^2}{N}}$$

EXAMPLE 12   Calculating the population standard deviations for Ashley and Brandon

Calculate the population standard deviations of the tax preparation times for Ashley and Brandon.

Solution

Brandon's population variance of 16 is larger than Ashley's population variance of 4, so Brandon's population standard deviation will also be larger because we are simply taking the square root. We have

$$\sigma = \sqrt{\sigma^2} = \sqrt{4} = 2$$

for Ashley and

$$\sigma = \sqrt{\sigma^2} = \sqrt{16} = 4$$

for Brandon.

The population standard deviation of Brandon's tax preparation times is 4 hours, which is larger than Ashley's 2 hours. As expected, the greater variability in Brandon's preparation times leads to a larger value for his population standard deviation, σ.

NOW YOU CAN DO
Exercises 11e–16e.

YOUR TURN #6

Calculate the population standard deviation of the CDC from Table 15.

(The solution is shown in Appendix A.)

What Do These Numbers Mean?

The Standard Deviation

So how do we interpret these values for σ? One quick thumbnail interpretation of the standard deviation is that it represents a "typical" deviation. That is, *the value of σ represents a distance from the mean that is representative for that data set.* For example, the typical distance from the mean for Ashley's and Brandon's tax preparation times is 2 hours and 4 hours, respectively.

Developing Your Statistical Sense

Communicating the Results

As you study statistics, keep in mind that during your career you will likely need to explain your results to others who have never taken a statistics course. Therefore, you should always keep in mind *how to interpret your results to the general public.* Communication and interpretation of your results can be as important as the results themselves.

3 Compute the Sample Variance and Sample Standard Deviation

The Sample Variance, s^2, and the Sample Standard Deviation s

In the real world, we usually cannot determine the exact value of the population mean or the population standard deviation. Instead, we use the sample mean and **sample standard deviation** to estimate the population parameters. The **sample variance** also depends on the concept of the mean squared deviation. If the sample mean is $\bar{x}$, and the sample size is n, then we would expect the formula for the sample variance to resemble the formula for the population variance, namely

$$\frac{\sum (x - \bar{x})^2}{n}$$

Note: In this book, we will work with sample statistics unless the data set is identified as a population.

However, this formula has been found to underestimate the population variance, so that we need to replace the n in the denominator with $n - 1$. We therefore have the following.

The **sample variance, s^2**, is approximately the mean of the squared deviations in the sample and is found by

$$s^2 = \frac{\sum (x - \bar{x})^2}{n - 1}$$

The sample standard deviation is perhaps the second most important statistic you will encounter in this book (after the sample mean, $\bar{x}$). It is the most commonly used measure of spread. The sample standard deviation is simply the square root of the sample variance and takes as its symbol the letter s, which is the Roman letter for the Greek σ. Again, $s = \sqrt{s^2}$.

Neither s^2 nor s can ever be negative. Both the variance and standard deviation are equal to zero only when all the data values in the data set are the same.

The **sample standard deviation, s**, is the positive square root of the sample variance s^2:

$$s = \sqrt{s^2} = \sqrt{\frac{\sum (x - \bar{x})^2}{n - 1}}$$

The value of s may be interpreted as the typical distance between a data value and the sample mean, for a given data set.

EXAMPLE 13   Calculating the sample variance and the sample standard deviation

Suppose we obtain a sample of size $n = 3$ from Ashley's population of tax preparation times, as follows: 5 hours, 8 hours, 11 hours, as shown.

Ashley's Population	5	7	8	9	11
Ashley's Sample	5		8		11

a. Calculate the sample variance of the tax preparation times.

b. Compute the sample standard deviation of the tax preparation times.

c. Interpret the sample standard deviation.

Solution

a. We first find the sample mean, $\bar{x} = \frac{\sum x}{n} = \frac{5 + 8 + 11}{3} = 8$. It so happens that the value for this sample mean equals the population mean $\mu = 8$, but this is only a coincidence.

Then the sample variance is

$$s^2 = \frac{\sum (x - \bar{x})^2}{n - 1} = \frac{(5 - 8)^2 + (8 - 8)^2 + (11 - 8)^2}{2} = \frac{9 + 0 + 9}{2} = 9$$

The sample variance is $s^2 = 9$ hours squared.

b. Then the sample standard deviation is

$$s = \sqrt{s^2} = \sqrt{9} = 3 \text{ hours}$$

NOW YOU CAN DO
Exercises 17–22.

c. For this sample of Ashley's tax returns, the typical difference between a tax preparation time and the mean preparation time is 3 hours.

YOUR TURN
#7

Suppose we take as our sample from the CDC funding data set in Table 15 the three northernmost (and least populated) New England states: Maine, New Hampshire, and Vermont.

1. Look at the funding values for the sample states. Would you expect our measures of spread to be larger or smaller than those of all the New England states? Why?

2. Find the variance of this sample. Express it in dollars squared.

3. Use your answer from (2) to calculate the standard deviation. Express it in dollars.

4. Interpret the value of the standard deviation.

(The solutions are shown in Appendix A.)

Developing Your Statistical Sense

Less Variation Is Better

In most real-world applications, consistency is a great advantage. In statistical data analysis, less variation is often better, even though variability is natural and cannot be eliminated. Throughout the text, you will find that smaller variability will lead to

• more precise estimates and

• higher confidence in conclusions.

In the exercises, you will find alternative **computational formulas** for the variance and standard deviation.

EXAMPLE 14 **Using technology to find the sample variance and sample standard deviation**

Find the sample standard deviation and the sample variance of the city gas mileage for the 2015 cars shown in the following table. Use (**a**) the TI-83/84, (**b**) Excel, (**c**) Minitab, (**d**) JMP, and (**e**) SPSS.

gasmileage

Vehicle	City mpg
Subaru Forester	22
Lexus RX 350	18
Ford Taurus	19
Mini Cooper	25
Cadillac Escalade	14
Mazda MX-5	21

Source: www.fueleconomy.gov.

Solution

Using the instructions in the Step-by-Step Technology Guide on page 117, we obtain the following output:

a. The TI-83/84 output is shown in Figure 13. The sample standard deviation, s, is given as $Sx = 3.763863264$. The sample variance is $s^2 = (3.763863264)^2 = 14.16667$.

b. The Excel output is provided in Figure 14. The sample standard deviation and sample variance are highlighted.

c. The Minitab output is provided in Figure 15. Note that Minitab rounds s to two decimal places.

d. The JMP output is shown in Figure 16.

e. The SPSS results are provided in Figure 17.

CAUTION For the TI-83/84, do not confuse Sx, the TI's notation for the sample standard deviation, with σx, which the TI-83/84 uses to label the population standard deviation.

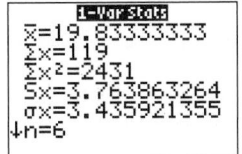

```
1-Var Stats
x̄=19.83333333
Σx=119
Σx²=2431
Sx=3.763863264
σx=3.435921355
↓n=6
```

FIGURE 13 TI-83/84 output.

City mpg	
Mean	19.83333
Standard Error	1.536591
Median	20
Mode	#N/A
Standard Deviation	3.763863
Sample Variance	14.16667

FIGURE 14 Excel output.

Descriptive Statistics: City mpg

Variable	Mean	StDev	Variance	Range
City mpg	19.83	3.76	14.17	11.00

FIGURE 15 Minitab output.

N Rows	Mean(City mpg)	Std Dev(City mpg)	Variance(City mpg)
6	19.833333333	3.7638632635	14.166666667

FIGURE 16 JMP output.

CityMPG		
N	Valid	6
	Missing	0
Mean		19.8333
Std. Deviation		3.76386
Variance		14.167

FIGURE 17 SPSS output.

Next, we turn to methods for applying the standard deviation.

4 The Empirical Rule

If the data distribution is bell-shaped, we may apply the Empirical Rule to find the approximate percentage of data that lies within k standard deviations of the mean, for $k = 1$, 2, or 3.

The Empirical Rule

If the data distribution is bell-shaped:

- About 68% of the data values will fall within 1 standard deviation of the mean.
    - For a population, about 68% of the data will lie between $\mu - 1\sigma$ and $\mu + 1\sigma$.
    - For a sample, about 68% of the data will lie between $\bar{x} - 1s$ and $\bar{x} + 1s$.
- About 95% of the data values will fall within 2 standard deviations of the mean.
    - For a population, about 95% of the data will lie between $\mu - 2\sigma$ and $\mu + 2\sigma$.
    - For a sample, about 95% of the data will lie between $\bar{x} - 2s$ and $\bar{x} + 2s$.
- About 99.7% of the data values will fall within 3 standard deviations of the mean.
    - For a population, about 99.7% of the data will lie between $\mu - 3\sigma$ and $\mu + 3\sigma$.
    - For a sample, about 99.7% of the data will lie between $\bar{x} - 3s$ and $\bar{x} + 3s$.

Figure 18 illustrates these approximate percentages.

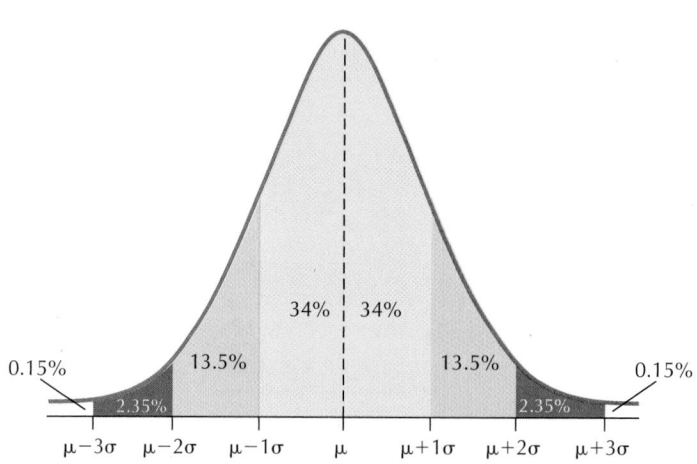

FIGURE 18 Empirical Rule, with approximate percentages.

CAUTION *Remember:* The Empirical Rule may be applied only if the data distribution is bell-shaped.

EXAMPLE 15   Using the Empirical Rule to find percentages

The College Board reports that the population mean Math SAT score for 2014 is $\mu = 514$, with a population standard deviation of $\sigma = 118$. Assume the distribution of Math SAT scores is bell-shaped.

a. Find the percentage of Math SAT scores between 396 and 632.

b. Compute the percentage of Math SAT scores that are above 750.

Solution

a. We see that a Math SAT score of 396 represents 1 standard deviation below the mean, because

$$\mu - 1\sigma = 514 - 1(118) = 396.$$

Similarly, a Math SAT score of 632 represents 1 standard deviation above the mean, because

$$\mu + 1\sigma = 514 + 1(118) = 632.$$

Thus, "Math SAT scores between 396 and 632" represents between $\mu - 1\sigma$ and $\mu + 1\sigma$, that is, within 1 standard deviation of the mean. The data distribution is bell-shaped, so we may use the Empirical Rule. Therefore, about 68% of the Math SAT scores lie between 396 and 632, as shown in Figure 19.

b. We note that a Math SAT score of 750 represents 2 standard deviations above the mean, because

$$\mu + 2\sigma = 514 + 2(118) = 750.$$

We know from the Empirical Rule that about 95% of the Math SAT scores lie within 2 standard deviations of the mean, so that about 95% of the Math SAT scores lie between 278 and 750. The left-over area of about 5% in the two tails in Figure 19 is the percentage of Math SAT scores above 750 or below 278. Because the bell-shaped curve is symmetric, the two tail areas are equal in area, which means that about 2.5% of the Math SAT scores lie above 750 (Figure 19).

Remember: The English word "about" is not optional; it is required. The Empirical Rule is an approximation of normal distribution probabilities that we will examine more closely in Chapter 6.

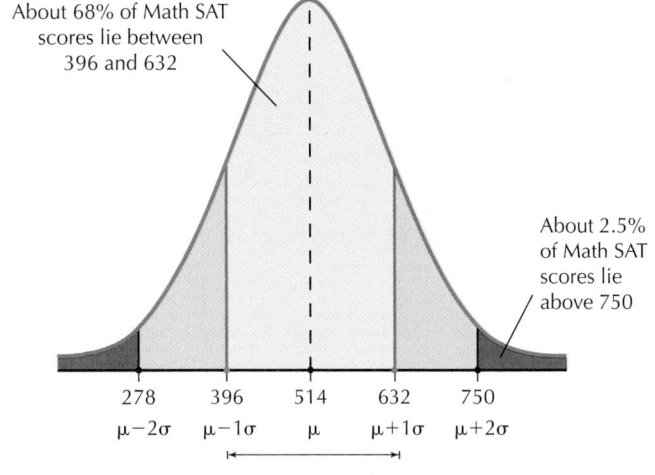

NOW YOU CAN DO
Exercises 23–30.

FIGURE 19 Example of Empirical Rule applied to Math SAT scores.

YOUR TURN
#8

Suppose vehicle speeds on the local interstate highway are bell-shaped, with a mean of $\mu = 70$ mph and a standard deviation of $\sigma = 5$ mph.

1.   Find the percentage of vehicle speeds between 65 mph and 75 mph.

2.   Compute the percentage of vehicles that are obeying the speed limit of at most 65 mph.

(The solutions are shown in Appendix A.)

5 Chebyshev's Rule

P. L. Chebyshev (1821–1894, Russia) derived a result, called **Chebyshev's Rule**, that can be applied to any continuous data set.

Portrait of Pafnuty Chebyshev–
Yaroslav Sergeyevich (1821–1894).

> **Chebyshev's Rule**
>
> The proportion of values from a data set that will fall within k standard deviations of the mean will be *at least*
>
> $$\left(1 - \frac{1}{k^2}\right)100\%$$
>
> where $k > 1$. Chebyshev's Rule may be applied to either samples or populations. For example:
>
> - When $k = 2$, at least 3/4 (or 75%) of the data values will fall within 2 standard deviations of the mean.
> - When $k = 3$, at least 8/9 (or 88.89%) of the data values will fall within 3 standard deviations of the mean.

Because of the phrase "at least," we say that Chebyshev's Rule provides minimum percentages, instead of the approximate percentages provided by the Empirical Rule. The actual percentage may be much greater than the minimum percentage provided by Chebyshev's Rule.

EXAMPLE 16   Using Chebyshev's Rule to find minimum percentages

The College Board reports that the population mean SAT Writing exam score for 2014 is $\mu = 488$, with a population standard deviation of $\sigma = 114$. However, assume we do not know the data distribution. Find the minimum percentage of exam scores that is

a. between 260 and 716.

b. between 317 and 659.

c. between 374 and 602.

Solution

The data distribution is unknown, so we cannot apply the Empirical Rule.

a. Because 260 lies 2 standard deviations below the mean

$$\mu - 2\sigma = 488 - 2(114) = 260$$

and 716 lies 2 standard deviations above the mean

$$\mu + 2\sigma = 488 + 2(114) = 716,$$

this question is really asking what is the minimum percentage within $k = 2$ standard deviations of the mean. From Chebyshev's Rule, the minimum percentage is

$$\left(1 - \frac{1}{k^2}\right)100\% = \left(1 - \frac{1}{2^2}\right)100\% = \left(\frac{3}{4}\right)100\% = 75\%$$

Thus, *at least* 75% of the SAT Writing exam scores will lie between 260 and 716.

b. The exam scores 317 and 659 lie $k = 1.5$ standard deviations below and above the mean, respectively. Therefore, at least

$$\left(1 - \frac{1}{1.5^2}\right)100\% = \left(1 - \frac{1}{2.25}\right)100\% = 55.6\%$$

of the SAT Writing exam scores will lie between 317 and 659.

NOW YOU CAN DO
Exercises 31–38.

c. The scores 374 and 602 lie $k = 1$ standard deviation below and above the mean, respectively. Unfortunately, Chebyshev's Rule is restricted to situations where $k > 1$. Thus, we cannot answer this question.

Developing Your
Statistical Sense

Strengths and Weaknesses of the Empirical Rule and Chebyshev's Rule

Example 16 shows that the lack of knowledge of a bell-shaped distribution can have a cost.

a. For part (**a**), using the Empirical Rule with $k = 2$ would have given us an answer of "about 95%," which is more precise than "at least 75%." However, this extra precision comes only if we know the distribution is bell-shaped.

b. For part (**b**), however, the Empirical Rule does not apply to any values other than 1, 2, or 3, so would have been no help here.

c. Finally, had we been able to apply the Empirical Rule in part (**c**), then we could have gotten an answer of "about 68%" for $k = 1$.

YOUR TURN
#9

Suppose systolic blood pressure in a population of senior citizens has a mean of $\mu = 130$ and a standard deviation of $\sigma = 10$. Find the minimum percentage of systolic blood pressure readings between 110 and 150.

(The solution is shown in Appendix A.)

If a given data set is bell-shaped, either the Empirical Rule or Chebyshev's Rule may be applied to it.

CASE STUDY Can the Financial Experts Beat the Darts?

Recall from the Case Study at the beginning of this chapter, the *Wall Street Journal* competition between stocks chosen randomly by *Journal* staff members throwing darts and stocks chosen by a team of four financial experts. Note from Figure 20 that the DJIA exhibits less variability than the other two portfolios. This smaller variability is due to the fact that the DJIA is made up of 30 component stocks, whereas each portfolio is made up of only four stocks. Smaller sample sizes can be associated with increased variability, because an unusual result in one value has a relatively strong effect on the mean when it is not offset by a large sample.

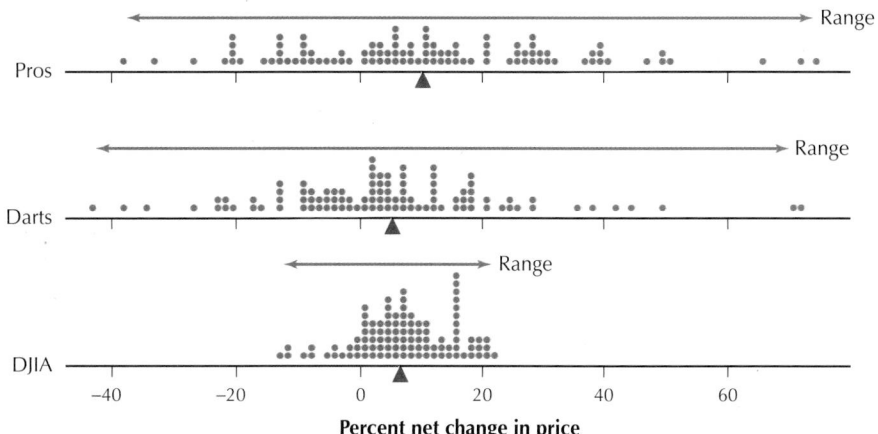

FIGURE 20 Comparative dotplots of the net change in prices.

Which of the portfolios, pros or darts, shows greater variability? It is difficult to determine which has the greater standard deviation, just by examining Figure 20. We therefore turn to the Minitab descriptive statistics in Figure 21. The range for the darts, 115.90, is greater than the range for the pros, 112.80. But the standard deviation for the darts (19.39) is less than that of the pros (22.25).

Descriptive Statistics: Pros, Darts, DJIA

Variable	Mean	StDev	Variance	Range
Pros	10.95	22.25	494.91	112.80
Darts	4.52	19.39	375.91	115.90
DJIA	6.793	8.031	64.505	35.600

FIGURE 21 Descriptive statistics for the portfolios.

Measures of spread may disagree about which data set has more variability. However, the range takes into account only the two most extreme data values; therefore, the standard deviation is the preferred measure of spread because it uses all the data values. Our conclusion, therefore, is that the returns for the professionals exhibit a greater variability.

Why did the pros have more variability than the darts? After all, in finance, high variability is not necessarily advantageous because it is associated with greater *risk*. The professionals evidently chose higher-risk stocks with greater potential for high returns—but also greater potential for losing money.

Section 3.2 Summary

1. The simplest measure of variability, or measure of spread, is the range. The range is simply the difference between the maximum and minimum values in a data set, but the range has drawbacks because it relies on the two most extreme data values.

2. The variance and standard deviation are measures of spread that utilize all available data values. The population variance can be thought of as the mean squared deviation. The standard deviation is the square root of the variance. We interpret the value of the standard deviation as the typical deviation, that is, the typical distance between a data value and the mean.

3. The variance and standard deviation may also be calculated for a sample. Again, we interpret the value of the standard deviation as the typical deviation, that is, the typical distance between a data value and the mean.

4. For bell-shaped distributions, the Empirical Rule may be applied. The Empirical Rule states that, for bell-shaped distributions, about 68%, 95%, and 99.7% of the data values will fall within 1, 2, and 3 standard deviations of the mean, respectively.

5. Chebyshev's Rule allows us to find the minimum percentage of data values that lie within a certain interval. Chebyshev's Rule states that the proportion of values from a data set that will fall within k standard deviations of the mean will be at least $[1 - 1/(k)^2]100\%$, where $k > 1$.

Section 3.2 Exercises

Unless a data set is identified as a population, you can assume that it is a sample.

CLARIFYING THE CONCEPTS

1. Explain what a deviation is. (p. 128)
2. What is the interpretation of the value of the standard deviation? (p. 132)
3. State one benefit and one drawback of using the range as a measure of spread. (p. 128)
4. True or false: If two data sets have the same mean, median, and mode, then they are identical. (p. 127)
5. What is one benefit of using the standard deviation instead of the range as a measure of spread? What is one drawback? (p. 128)
6. Which measure of spread represents the mean squared deviation for the population? (p. 130)
7. True or false: Chebyshev's Rule provides exact percentages. (p. 138)

8. When can the sample standard deviation, *s*, be negative? (p. 133)

9. When does the sample standard deviation, *s*, equal zero? (p. 133)

10. When may the Empirical Rule be used? (p. 135)

PRACTICING THE TECHNIQUES

 CHECK IT OUT!

To do	Check out	Topic
Exercises 11a–16a	Example 9	Range
Exercises 11c–16c	Example 10	Calculating deviations
Exercises 11d–16d	Example 11	Population variance
Exercises 11e–16e	Example 12	Population standard deviation
Exercises 17–22	Example 13	Sample variance and sample standard deviation
Exercises 23–30	Example 15	Empirical Rule
Exercises 31–38	Example 16	Chebyshev's Rule

For the population data in Exercises 11–16, do the following:

a. Compute the range.
b. Find the population mean, μ.
c. Calculate the deviations, $x - \mu$.
d. Compute the population variance, σ^2.
e. Find the population standard deviation, σ.

11. State exports to other countries are shown in the table for the population of all New England states, for the month of June 2014, expressed in billions of dollars.

State	Exports	State	Exports
Connecticut	1.4	New Hampshire	0.4
Maine	0.3	Rhode Island	0.2
Massachusetts	2.4	Vermont	0.3

Source: U.S. Census Bureau.

12. The number of wins for each baseball team in the population of the American League West division for 2013 is shown in the table.

Team	Wins	Team	Wins
Oakland Athletics	96	Seattle Mariners	71
Texas Rangers	91	Houston Astros	51
Los Angeles Angels	78		

Source: MLB.mlb.com.

13. The table provides the motor vehicle theft rate for the population of the top 10 countries in the world for motor vehicle theft, for 2012. The theft rate equals the number of motor vehicles stolen in 2012 per 100,000 residents.

Country	Theft rate	Country	Theft rate
Italy	208.0	Greece	100.2
France	174.1	Norway	94.1
USA	167.8	Netherlands	75.2
Sweden	117.2	Spain	75.1
Belgium	106.0	Cyprus	66.0

Source: United Nations Office on Drugs and Crime.

14. The National Center for Education Statistics sponsors the Trends in International Mathematics and Science Study (TIMSS). The table contains the mean science scores for the eighth-grade science test for the population of all Asian-Pacific countries that took the exam.

Country	Science Score	Country	Science score
Singapore	578	Australia	527
Taiwan	571	New Zealand	520
South Korea	558	Malaysia	510
Hong Kong	556	Indonesia	420
Japan	552	Philippines	377

15. The table contains the number of petit larceny cases for the population of all police precincts in South Manhattan in 2013.

Precinct	Petit larcenies	Precinct	Petit larcenies
1	2014	10	995
5	1288	13	2094
6	1555	14	4551
7	584	17	823
9	1607	18	2071

Source: New York City Police Department.

16. The table contains the number of criminal trespass cases for the population of all police precincts in South Manhattan in 2013.

Precinct	Criminal trespass	Precinct	Criminal trespass
1	108	10	207
5	105	13	135
6	113	14	340
7	233	17	74
9	219	18	120

Source: New York City Police Department.

For the sample data in Exercises 17–22, do the following:

a. Calculate the sample variance.
b. Compute the sample standard deviation.
c. Interpret the sample standard deviation.

17. A sample of the state export data from Exercise 11 is provided in the table.

State	Exports
Connecticut	1.4
Massachusetts	2.4
Rhode Island	0.2

18. A sample from the baseball data in Exercise 12 is shown here.

Team	Wins
Texas Rangers	91
Los Angeles Angels	78
Seattle Mariners	71

19. A sample from the motor vehicle theft data in Exercise 13 is as follows.

Country	Theft rate
Italy	208.0
USA	167.8
Greece	100.2

20. A sample from the science score data in Exercise 14 is given here.

Country	Science score
South Korea	558
Hong Kong	556
Japan	552
Australia	527

21. The following sample is taken from the petit larceny data in Exercise 15.

Precinct	Petit larcenies
1	2014
6	1555
9	1607
14	4551
17	823

22. A sample taken from the criminal trespass data in Exercise 16 is as follows.

Precinct	Criminal trespass
1	108
7	233
14	340
18	120

For Exercises 23–26, use the following information. A data distribution is bell-shaped, with a mean of 50 and a standard deviation of 5. Use the Empirical Rule to approximate the percentage of data.

23. Between 45 and 55
24. Between 40 and 60
25. Between 35 and 65
26. Less than 45

For Exercises 27–30, use the following information. A data distribution is bell-shaped, with a mean of 0 and a standard deviation of 1. Use the Empirical Rule to approximate the percentage of data.

27. Between –1 and 1
28. Greater than 2
29. Less than –2
30. Between –2 and 2

For Exercises 31–34, use the following information. A data set has an unknown distribution, with a mean of 20 and a standard deviation of 2. Use Chebyshev's Rule to estimate the minimum possible percentage of data.

31. Between 16 and 24
32. Between 14 and 26
33. Between 12 and 28
34. Between 13 and 27

For Exercises 35–38, use the following information. A data set has an unknown distribution, with a mean of 20 and a standard deviation of 5. If possible, use Chebyshev's Rule to estimate the minimum possible percentage of data.

35. Between 0 and 40
36. Between 5 and 35
37. Between 12.5 and 27.5
38. Between 15 and 25

APPLYING THE CONCEPTS

39. Match the histograms in (**a**)–(**d**) to the statistics in (**i**)–(**iv**).

    i. Mean = 75, standard deviation = 20
    ii. Mean = 75, standard deviation = 10
    iii. Mean = 50, standard deviation = 20
    iv. Mean = 50, standard deviation = 10

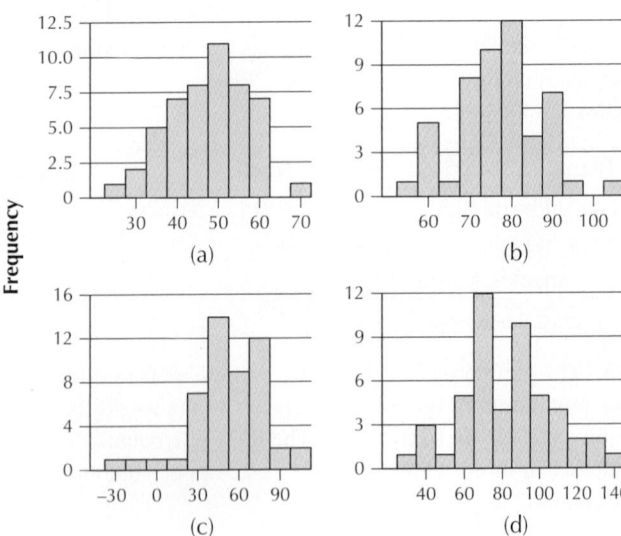

40. Match the histograms in (**a**)–(**d**) to the statistics in (**i**)–(**iv**).

    i. Mean = 1, standard deviation = 1
    ii. Mean = 1, standard deviation = 0.1
    iii. Mean = 0, standard deviation = 1
    iv. Mean = 0, standard deviation = 0.1

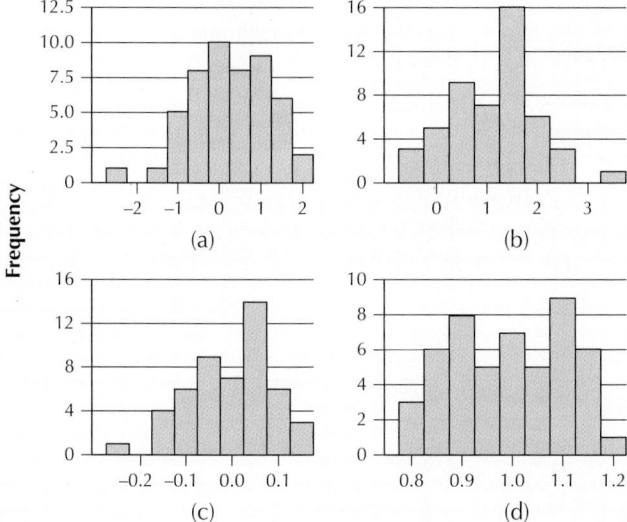

For the following exercises, make sure to state your answers in the proper units, such as "years" or "years squared."

Video Game Sales. The Chapter 1 Case Study looked at video game sales for the top 30 video games. The following table contains the total sales (in game units) and weeks on the top 30 list for a sample of five randomly selected video games. Use this information for Exercises 41 and 42.

41. Find the following measures of spread for total sales:
    a. Range
    b. Sample variance
    c. Sample standard deviation

42. Calculate the following measures of spread for the number of weeks on the top 30 list:
    a. Range
    b. Sample variance
    c. Sample standard deviation

Video Game	Total sales in millions of units	Weeks on list
Super Mario Bros. U for WiiU	1.7	78
NBA 2K14 for PS4	0.6	27
Battlefield 4 for PS3	0.9	29
Titanfall for XboxOne	1.2	10
Yoshi's New Island for 3DS	0.2	10

Source: www.vgchartz.com.

 **videogamereg**

 Darts and the DJIA. The following table contains a random sample of eight days from the Chapter 3 Case Study data set, indicating the stock market gain or loss for the portfolio chosen by the random darts, as well as the DJIA gain or loss for that day. Use this information for Exercises 43 and 44.

43. Find the following measures of spread for the darts:
    a. Range
    b. Sample variance
    c. Sample standard deviation

44. Calculate the following measures of spread for the DJIA:
    a. Range
    b. Sample variance
    c. Sample standard deviation

dartsdjia

Darts	DJIA
−27.4	−12.8
18.7	9.3
42.2	8
−16.3	−8.5
11.2	15.8
28.5	10.6
1.8	11.5
16.9	−5.3

Source: Wall Street Journal.

Age and Height. The following table provides a random sample from the Chapter 4 Case Study data set **body_females**, showing the age and height of the eight women. Use this information for Exercises 45 and 46.

45. Find the following measures of spread for age:
    a. Range
    b. Sample variance
    c. Sample standard deviation

46. Calculate the following measures of spread for height:
    a. Range
    b. Sample variance
    c. Sample standard deviation

ageheight

Age	Height
40	63.5
28	63.0
25	64.4
34	63.0
26	63.8
21	68.0
19	61.8
24	69.0

Source: Journal of Statistics Education.

Saturated Fat and Calories. The table contains the calories and saturated fat in a sample of 10 food items. Use this information for Exercises 47 and 48.

47. Find the following measures of spread for calories:
   a. Range
   b. Sample variance
   c. Sample standard deviation

48. Calculate the following measures of spread for saturated fat:
   a. Range
   b. Sample variance
   c. Sample standard deviation satfatcorr

Food item	Calories	Grams of saturated fat
Chocolate bar (1.45 ounces)	216	7.0
Meat & veggie pizza (large slice)	364	5.6
New England clam chowder (1 cup)	149	1.9
Baked chicken drumstick (no skin, medium size)	75	0.6
Curly fries, deep-fried (4 ounces)	276	3.2
Wheat bagel (large)	375	0.3
Chicken curry (1 cup)	146	1.6
Cake doughnut hole (one)	59	0.5
Rye bread (1 slice)	67	0.2
Raisin bran cereal (1 cup)	195	0.3

Source: Food-a-Pedia.

Video Game Sales. Refer to the video game sales data in Exercises 41 and 42 for Exercises 49–52.

49. The sample variance of sales was expressed in "game units squared." Do you find this concept easy to understand? Which measure do you find to be more easily understood and interpreted for these data, the variance or the standard deviation?

50. Consider the histogram of total units sold for all the top 30 video games.
   a. Is the distribution bell-shaped?
   b. Can we apply the Empirical Rule?
   c. Can we apply Chebyshev's Rule?

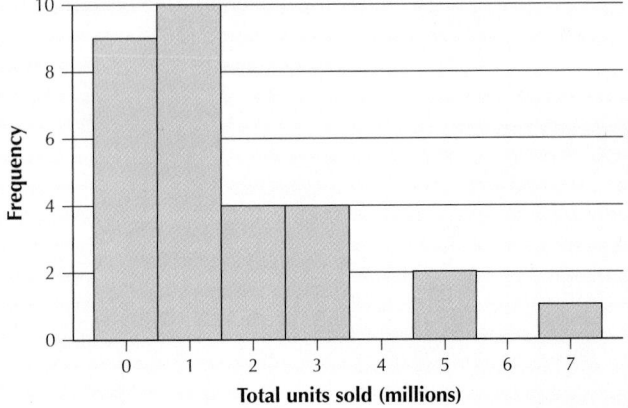

51. Use the sample of size five and Chebyshev's Rule to find the minimum percentage of total sales that are between 0.0048 million and 1.8352 million.

52. Refer to Table 3 of Chapter 1 on page 8. Calculate the actual proportion of total sales that are between 0.0048 million and 1.8352 million. Does this fit the answer you got using Chebyshev's Rule?

Darts and the DJIA. Refer to the darts and DJIA data in Exercises 43 and 44 for Exercises 53–56.

53. Based on your measures of spread in Exercises 43 and 44, which stock market return reflects greater variability, the darts or the DJIA?

54. The histogram shows the population distribution of the stock market changes for the darts. Can we live with the assumption that the distribution is bell-shaped?

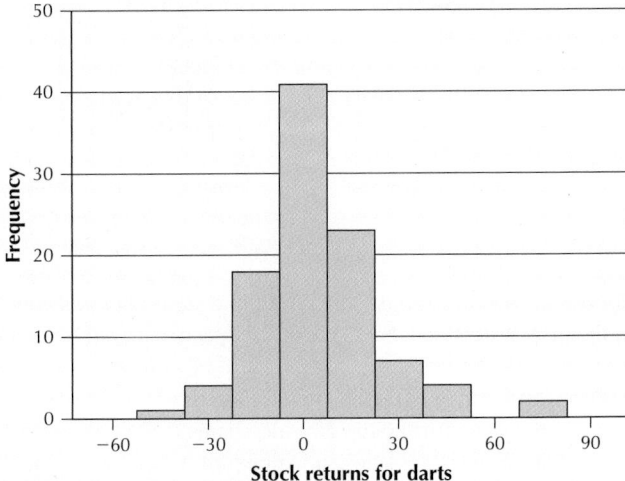

55. Based on the sample of size 8, use the Empirical Rule to approximate the percentage of darts stock returns that lie between −13.41 and 32.31.

56. Can the Empirical Rule tell us what approximate percentage of the darts stock returns lie between −1.98 and 20.88? Explain.

Age and Height. Refer to the age and height data in Exercises 45 and 46 for Exercises 57–60.

57. The histogram shows the population distribution of the women's ages.

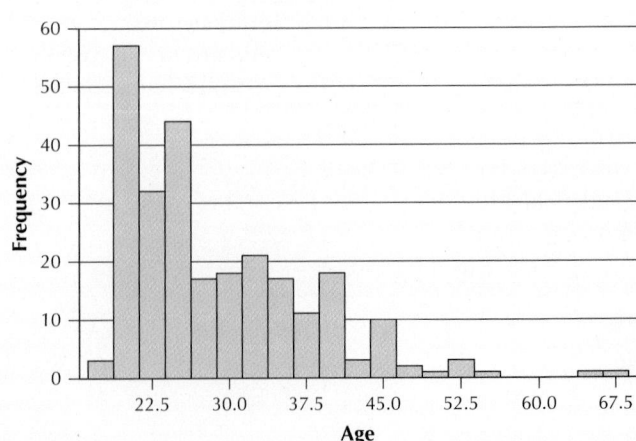

a. Is the distribution bell-shaped?
b. Can we apply the Empirical Rule?
c. Can we apply Chebyshev's Rule?

58. Based on the sample of size 8, use Chebyshev's Rule to find the minimum percentage of the women's ages that lie between 16.78 and 37.48.

59. The histogram shows the population distribution of the women's heights.

    a. Though it's not perfect, can we live with the assumption that the distribution is bell-shaped?
    b. Can we apply the Empirical Rule?
    c. Can we apply Chebyshev's Rule?

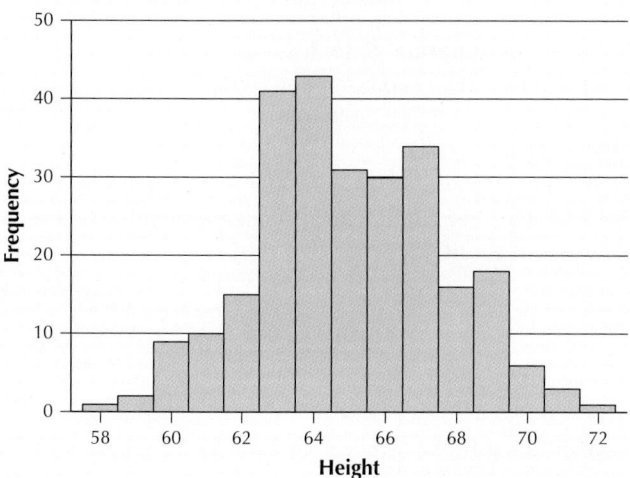

60. Based on the sample of size 8, use the Empirical Rule to approximate the percentage of the women's heights that lie between 59.449 inches and 69.677 inches.

Saturated Fat and Calories. Refer to the food data in Exercises 47 and 48 for Exercises 61 and 62.

61. The histogram contains the grams of saturated fat for the 10 foods in the sample.

    a. Is the distribution bell-shaped?
    b. Can we apply the Empirical Rule?
    c. Can we apply Chebyshev's Rule?

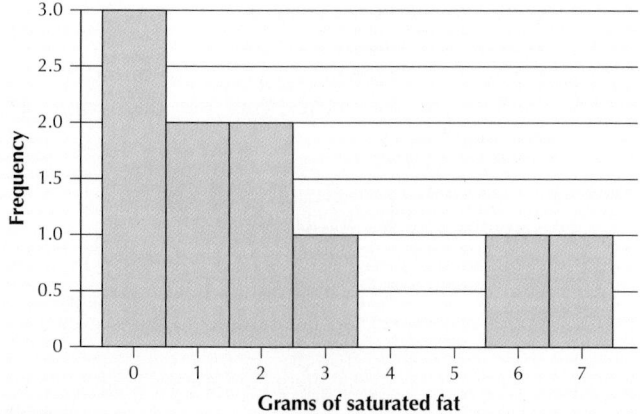

62. Use Chebyshev's Rule to find the minimum percentage of food items with saturated fat between −1.51 and 5.75. (Note that, because grams of saturated fat cannot be negative, this is the same as between 0 and 5.75.)

Fuel Economy. Refer to Table 7 on page 123 to answer Exercises 63–65. The data represent a sample.

63. Find the following measures of spread for the number of cylinders:
    a. Range
    b. Variance
    c. Standard deviation

64. Find the following measures of spread for the engine size:
    a. Range
    b. Variance
    c. Standard deviation

65. Find the following measures of spread for the fuel economy:
    a. Range
    b. Variance
    c. Standard deviation

Ant Size. Use the following information for Exercises 66 and 67. A study compared the size of ants from different colonies. The masses (in milligrams) of samples of ants from two different colonies are shown in the accompanying table.[4]

🐜 antcolony

Colony A		Colony B	
109	134	148	115
120	94	110	101
94	113	110	158
61	111	97	67
72	106	136	114

66. Calculate the range for each ant colony.
    a. Which has the greater range?
    b. Which colony has the greater variability according to the range?

67. Calculate the standard deviation for each colony.
    a. Which has the greater standard deviation?
    b. Which colony has the greater variability according to the standard deviation? Does this concur with your answer from the previous exercise?
    c. Without calculating the variances, say which colony has the greater variance. How do you know this?

68. Computational Formula for the Population Variance and Standard Deviation: Wins in Baseball. The following table provides the number of wins for all the teams in the American League East Division for the 2013 season, which we can consider to be a population.

Team	Wins
Boston Red Sox	97
Tampa Bay Rays	92
Baltimore Orioles	85
New York Yankees	85
Toronto Blue Jays	74

Source: MLB.mlb.com.

An alternative computational formula for the population variance is as follows:

$$\sigma^2 = \frac{\sum x^2 - \left(\sum x\right)^2/N}{N}$$

a. Use the computational formula to find the population variance for the number of wins.
b. Use your result from (a) to find the population standard deviation for the number of wins.

Note: $\sum x^2$ means that you square each data value and then add up the squared data values, and $(\sum x)^2$ means that you add up all the data values and then square the sum.

69. Computational Formula for the Sample Variance and Standard Deviation. Refer to the previous exercise. Suppose a random sample of size $n = 3$ from these teams yields the New York Yankees, the Tampa Bay Rays, and the Baltimore Orioles.

An alternative computational formula for the sample variance is as follows:

$$s^2 = \frac{\sum x^2 - \left(\sum x\right)^2/n}{n - 1}$$

a. Use the computational formula to find the sample variance for the number of wins.
b. Use your result from (a) to find the sample standard deviation for the number of wins.
c. Interpret your result from (b).

70. Challenge Exercise. Refer to the table in Exercise 68. Suppose we are taking a sample of size $n = 2$.
a. Which sample of two teams will yield the largest sample standard deviation? Explain your reasoning.
b. Which sample of two teams will yield the smallest sample standard deviation? Explain your reasoning.

71. Empirical Rule: October in Santa Monica. The National Climate Data Center reports that the mean October temperature in Santa Monica, California, is 63 degrees Fahrenheit, with a standard deviation of 3 degrees. Suppose the data distribution is bell-shaped. If possible, estimate the percentage of October days with temperatures within the following ranges. If not possible, explain why.
a. Between 60 and 66 degrees
b. Between 57 and 69 degrees
c. Between 55 and 71 degrees

72. Empirical Rule: Energy Consumption. The U.S. Department of Energy reports that the mean annual energy consumption per person in the United States is 1400 watts.

Assume that the standard deviation is 200 watts and the data distribution is bell-shaped. Estimate the percentage of Americans with energy consumption within the following ranges.
a. Between 1200 and 1600 watts
b. Between 1000 and 1800 watts
c. Above 1000 watts

73. Chebyshev's Rule. Refer to Exercise 71. Suppose that we did not know that the October temperature in Santa Monica is bell-shaped. If possible, find minimums for (**a**)–(**c**) in Exercise 71.

74. Chebyshev's Rule. Refer to Exercise 72. Suppose that we did not know that the annual energy consumption is bell-shaped. If possible, find minimums for (**a**)–(**c**) in Exercise 72.

Energy Consumption. Refer to Table 16, which shows the per capita energy consumption (watts per person) for samples of countries on three continents for Exercises 75–78.

ɪɪɪ **energyconsumption**

TABLE 16   Per capita energy consumption for three samples of countries

Asia	Europe	North America
China 447	Germany 861	USA 1402
Japan 774	France 804	Canada 1871
South Korea 1038	United Kingdom 622	Mexico 131

Source: The World Factbook.

75. Construct dotplots of the energy consumption for each continent. Which continent would you say has the greatest spread (variability)? Why?

76. Find the range and variance of the per capita energy consumption for each of the continents. Do your findings agree with your judgment from the previous exercise?

77. *Without performing any calculations,* use your results from the previous exercise to state which continent has (a) the largest standard deviation, and (b) the smallest standard deviation.

WHAT IF ? **78.** Now suppose we omit Mexico from the data.
a. *Without recalculating them,* describe how this would affect the values of the measures of spread you found for the North American countries.
b. Now recalculate the three measures of spread for the North American countries. Was your judgment in (**a**) supported?

Women's Volleyball Team Heights. Refer to Table 10 on page 126 for Exercises 79–81.

79. Suppose a new player joins the NCU team. She is 7 feet tall (84 inches) and replaces the 72-inch-tall player.
a. Would you expect the standard deviation to go up or down, and why?
b. Now find the standard deviation for the team including the new player. Was your intuition correct?

80. Linear Transformations. Add 4 inches to the height of each player on the WMU team.
  a. Recalculate the range and standard deviation.
  b. Formulate a rule for the behavior of these measures of variability when a constant (such as 4) is added to each member of the data set.
81. Linear Transformations. Starting with the original data, double the height of each player on the NCU team.
  a. Recalculate the range and standard deviation.
  b. Formulate a rule for the range and standard deviation when the data values are doubled.

Coefficient of Variation. The *coefficient of variation* enables analysts to compare the variability of two data sets that are measured on different scales. The coefficient of variation (CV) itself does not have a unit of measure. Larger values of CV indicate greater variability or spread. The coefficient of variation is given as

$$CV = \frac{\text{standard deviation}}{\text{mean}} \cdot 100\%$$

Use this measure of variability for Exercises 82 and 83.
82. Coefficient of Variation for Fuel Economy Data. Refer to Table 7 on page 123.
  a. Calculate the coefficient of variation for the following variables: *cylinders*, *engine size*, and *city mpg*.
  b. According to the coefficient of variation, which variable has the greatest spread? The least variability?
83. Coefficient of Variation for Energy Consumption. Refer to Table 16 on page 146.
  a. Calculate the coefficient of variation for the per capita energy consumption for each continent.
  b. According to the coefficient of variation, which continent has the greatest spread? Does this agree with your measures of spread from Exercise 76?

Mean Absolute Deviation. Recall that the variance and standard deviation use squared deviations because the mean deviation for any data set is zero. Another way to avoid negative deviations offsetting positive ones is to use the absolute value of the deviations. The *mean absolute deviation (MAD)* is a measure of spread that looks at the average of the absolute values of the deviations:

$$MAD = \frac{\sum |x_i - \bar{x}|}{n}$$

Use this measure of variability for Exercises 84 and 85.
84. Mean Absolute Deviation for the Fuel Economy Data. Refer to Table 7 on page 123.
  a. Find the mean absolute deviation for *cylinders*, *engine size*, and *city mpg*.
  b. According to the mean absolute deviation, which variable has the greatest variability? The least variability?

85. Mean Absolute Deviation for Energy Consumption. Refer to Table 16 on page 146.
  a. Calculate the mean absolute deviation for each continent.
  b. According to the mean absolute deviation, which continent has the greatest spread? Does this agree with your measures of spread from Exercise 76?

Coefficient of Skewness. The coefficient of skewness quantifies the skewness of a distribution. It is defined as

$$\text{skewness} = \frac{3(\text{mean} - \text{median})}{\text{standard deviation}}$$

Most skewness values lie between -3 and 3. Negative values of skewness are associated with left-skewed distributions, whereas positive values are associated with right-skewed distributions. Values close to zero indicate distributions that are nearly symmetric. Use this information for Exercises 86–88.
86. Coefficient of Skewness. For the following distributions, compute the coefficient of skewness and comment on the skewness of the distribution.
  a. Mean = 0, Median = 0, Standard deviation = 1
  b. Mean = 1, Median = 0, Standard deviation = 1
  c. Mean = 0, Median = 1, Standard deviation = 1
  d. Mean = 75, Median = 80, Standard deviation = 10
  e. Mean = 100, Median = 100, Standard deviation = 15
  f. Mean = 3.2, Median =3.0, Standard deviation = 1.0
87. What is the coefficient of skewness for any distribution where the mean equals the median, regardless of the nonzero value of the standard deviation?

88. Coefficient of Skewness for the Case Study Data. The median price change for the professional analysts is 9.60, the median for the dart throwers is 3.25, and the median for the DJIA is 7.00. Use this information, along with the information in Figure 21 on page 140 to answer the following.
  a. Calculate the coefficient of skewness for each of the Pros, the Darts, and the DJIA.
  b. Comment on the skewness of each distribution.

BRINGING IT ALL TOGETHER

In Exercises 89 and 90, we bring together all the measures of spread we have learned in the chapter and the new ones we learned in the exercises.
89. Fuel Economy Data. You calculated the range, variance, and standard deviation for this data in Exercises 63–65. You calculated the coefficient of variation in Exercise 82 and the mean absolute deviation in Exercise 84. Use this information to do the following.
  a. Construct a table of the five measures of dispersion (range, sample variance, sample standard deviation, coefficient of variation, and mean absolute deviation) for the number of cylinders, the engine size, and the city mpg.

b. Which measures of dispersion suggest that the *city mpg* is the most dispersed variable? *Engine size*? Number of *cylinders*?

90. Energy Consumption Data. You calculated the range and variance for this data in Exercise 76. You calculated the coefficient of variation in Exercise 83 and the mean absolute deviation in Exercise 85. Use this information to do the following:

a. Using the variance, calculate the standard deviation energy consumption for each continent.

b. Construct a table of the five measures of spread (range, sample variance, sample standard deviation, coefficient of variation, and mean absolute deviation) for each continent.

c. Do the measures of spread agree on which distribution has the greatest variability?

d. Bringing together all your statistics about measures of spread, what is your conclusion about the variability in Europe, compared with the other two continents?

CONSTRUCT YOUR OWN DATA SETS

91. Construct two data sets, A and B, that you make up on your own, so that the range of A is greater than the range of B. Verify this.

92. Construct two data sets, A and B, that you make up on your own, so that the standard deviation of A is greater than the range of B. Verify this.

93. Construct two data sets, A and B, that you make up on your own, so that the mean of A is greater than the mean of B, but the standard deviation of B is greater than that of A. Verify this.

94. Construct two data sets, A and B, that you make up on your own, so that the mean of A is greater than the mean of B,

and the standard deviation of A is greater than that of B. Verify this.

95. Construct two data sets, A and B, that you make up on your own, so that the range of A is greater than the range of B, but the standard deviation of B is greater than that of A. Verify this. (*Hint:* Remember the sensitivity of the standard deviation to extreme values.)

WORKING WITH LARGE DATA SETS

 The Professionals versus the Darts. We will assess how well the Empirical Rule performs, using the Chapter 3 Case Study data set. Open the **Darts** data set. Use technology to do the following.

 darts

96. Find the mean and standard deviation for each of the *Pros*, the *Darts*, and the *DJIA*.

97. Construct histograms of each of the *Pros*, the *Darts*, and the *DJIA*. Conclude that we can live with the assumption of a bell-shaped distribution for all three groups.

98. For the Pros, do the following:

a. Calculate the following quantities: $\mu - 1\sigma$, $\mu + 1\sigma$, $\mu - 2\sigma$, $\mu + 2\sigma$, $\mu - 3\sigma$, and $\mu + 3\sigma$.

b. State what approximate percentages lie within those intervals, according to the Empirical Rule.

c. Count how many stock returns actually lie within each of those intervals. Divide these counts by the population size 100 to obtain the actual percentages.

d. Compare the approximate percentages estimated by the Empirical Rule with the actual percentages from the population data.

99. Repeat the same comparison (**a**)–(**d**) from Exercise 98, but this time for the Darts.

100. Repeat the same comparison (**a**)–(**d**) from Exercise 98, but this time for the DJIA.

3.3 Working with Grouped Data

OBJECTIVES By the end of this section, I will be able to . . .

1 Calculate the weighted mean.
2 Estimate the mean for grouped data.
3 Estimate the variance and standard deviation for grouped data.

1 The Weighted Mean

Note: Before tackling this section, you may wish to review Section 2.2, "Graphs and Tables for Quantitative Data" (page 60).

Sometimes, not all the data values in a data set are of equal importance. Certain data values may be assigned greater importance or weight than others when calculating the mean. For example, have you ever figured out what your final grade for a course was based on the percentages listed in the syllabus? What you actually found was the **weighted mean** of your grades.

Note: The weights, *w*, do not have to be percentages, nor do they have to add up to 1.

In the special case when all the weights equal 1, the weighted mean equals the sample mean $\bar{x}$ from Section 3.1.

Weighted Mean

To find the **weighted mean**:

1. Multiply each weight, *w*, by its corresponding data value, *x*.
2. Add up the products to get $\Sigma(w \cdot x)$.
3. Divide the result by the sum of the weights, Σw.

$$\bar{x} = \frac{\sum (w \cdot x)}{\sum w}$$

EXAMPLE 17   Weighted mean of course grades

The syllabus for the Introduction to Management course at a local college specifies that the midterm exam is worth 30%, the term paper is worth 20%, and the final exam is worth 50% of your course grade. Now, say you did not get serious about the course until after Halloween, so that you got a 40 on the midterm. You then began working harder, and got a 70 on the term paper. Finally, you remembered that you had to pay for the course again if you did not pass and had to retake it, so you worked really hard for the last month of the course and got a 90 on the final exam. Calculate your course average, that is, the weighted mean of your grades.

Solution

The data values are 40, 70, and 90. The weights are 0.30, 0.20, and 0.50. Your course weighted mean is then calculated as follows:

$$\bar{x} = \frac{\sum (w \cdot x)}{\sum w} = \frac{(0.30)(40) + (0.20)(70) + (0.50)(90)}{0.30 + 0.20 + 0.50} = \frac{71}{1.0} = 71$$

Because the final exam had the most weight, you were able to raise your course weighted mean to 71, and you passed the course.

NOW YOU CAN DO
Exercises 4–8.

YOUR TURN #10

The author's syllabus for his Business Statistics I course during Summer 2014 stated that the quiz average was worth 50% of the course grade, with the midterm worth 20% and the final exam worth 30%. One of the students had a 90 quiz average, a 70 midterm grade, and an 85 final exam grade. Calculate the student's course grade.

(The solution is shown in Appendix A.)

2 Estimating the Mean for Grouped Data

Thus far in Chapter 3, we have computed measures of center and spread from a raw data set. However, data are often reported using grouped frequency distributions. Without the original data, we cannot calculate the exact values of the measures of center and spread. The remainder of this section examines methods for approximating the mean, variance, and standard deviation of *grouped data*—that is, population data summarized using frequency distributions.

For each class in the frequency distribution, we estimate the class mean using the *class midpoint*. The class midpoint, denoted *x*, is defined as the mean of two adjoining lower class limits.

The product of the class frequency, *f,* and class midpoint, *x,* is used as an estimate of the sum of the data values within that class. Summing these products across all classes and dividing by the size of the data set thus provides us with an **estimated mean for data grouped into a frequency distribution.**

Note: Even though we are working with population data, we will notate these values using $\bar{x}$ and s because we are estimating the values of the mean and standard deviation.

> **Estimated Mean for Data Grouped into a Frequency Distribution**
>
> Given a population frequency distribution, the **estimated mean** for the variable is given by
>
> $$\bar{x} = \frac{\sum(f \cdot x)}{\sum f}$$
>
> where x and f represent the class midpoints and class frequencies, respectively.

EXAMPLE 18   Calculating the estimated mean for grouped data

The first two columns of Table 17 contain the frequency distribution of the number of Americans younger than 85 years old who were living in the United States in 2013, by age group, as reported by the U.S. Census Bureau.

a. Find the class midpoints.

b. Calculate the product of each class frequency with its midpoint.

c. Find the sum of the frequencies, $\sum f$, and the sum of the products, $\sum(f \cdot x)$.

d. Divide $\sum(f \cdot x)$ by $\sum f$ to find the estimated mean age of all Americans under the age of 85.

Solution

a. The midpoint for the first class (ages 0–20) is the mean of the lower class limits for this class (0) and the adjoining class (20). That is, the midpoint is $(0 + 20)/2 = 10$. Similarly, the midpoint for the second class (ages 20–40) is $(20 + 40)/2 = 30$. The remainder of the class midpoints are calculated in the same way and are shown in Table 17.

b. We multiply the frequency for the first age group by its midpoint to get $83.3 \cdot 10 = 833$. We do the same for the other age groups, as shown in Table 17.

c. We add up all the frequencies to get $\sum f = 303.3$. Also, we add up all the products from (**b**) to obtain $\sum(f \cdot x) = 11{,}338$.

d. Finally, obtain the estimated mean, as follows:

$$\bar{x} = \frac{\sum(f \cdot x)}{\sum f} = \frac{11{,}338}{303.3} = 37.4$$

The estimated mean age of all Americans under 85 is 37.4 years.

Table 17 Frequency distribution of Americans, by age group, in millions

Class: age	Frequency f	Midpoint x	Product $f \cdot x$
$0 \le$ age < 20	83.3	10	$83.3 \cdot 10 = 833$
$20 \le$ age < 40	82.8	30	$82.8 \cdot 30 = 2{,}484$
$40 \le$ age < 60	85.6	50	$85.6 \cdot 50 = 4{,}280$
$60 \le$ age < 85	51.6	72.5	$51.6 \cdot 72.5 = 3{,}741$
Total	$\sum f = 303.3$		$\sum(f \cdot x) = 11{,}338$

NOW YOU CAN DO
Exercises 9–14.

3 Estimating the Variance and Standard Deviation for Grouped Data

We also use class midpoints and class frequencies to calculate the **estimated variance for data grouped into a frequency distribution** and the **estimated standard deviation for data grouped into a frequency distribution**.

> **Estimated Variance and Standard Deviation for Population Data Grouped into a Frequency Distribution**
>
> The **estimated variance** for data grouped into a frequency distribution is given by
>
> $$s^2 = \frac{\sum (x - \bar{x})^2 \cdot f}{\sum f}$$
>
> and the **estimated standard deviation** is given by
>
> $$s = \sqrt{s^2} = \sqrt{\frac{\sum (x - \bar{x})^2 \cdot f}{\sum f}}$$
>
> where x represents the class midpoints, f represents the class frequencies, and $\bar{x}$ is the estimated mean.

You should carry as many decimal places as you can for the value of $\bar{x}$ when calculating s^2, and for s^2 when calculating s.

EXAMPLE 19 Calculating the estimated variance and standard deviation for grouped data

Calculate the estimated variance and standard deviation of the ages of Americans under age 85 from Table 17.

Solution

Table 18 contains the calculations required for finding $\sum (x - \bar{x})^2 \cdot f = 144{,}234$. The variance is therefore estimated as

$$s^2 = \frac{\sum (x - \bar{x})^2 \cdot f}{\sum f} = \frac{144{,}234}{303.3} \approx 475.5487$$

and the standard deviation is estimated as

$$s = \sqrt{s^2} = \sqrt{475.5487} \approx 21.8$$

Table 18 Calculating $\Sigma (x - \bar{x})^2 \cdot f$

Class: age	Midpoint x	Frequency f	$\bar{x}$	$x - \bar{x}$	$(x - \bar{x})^2 \cdot f$
$0 \le \text{age} < 20$	10.0	83.3	37.4	−27.4	62,538.31
$20 \le \text{age} < 40$	30.0	82.8	37.4	−7.4	4,534.13
$40 \le \text{age} < 60$	50.0	85.6	37.4	12.6	13,589.86
$60 \le \text{age} < 85$	72.5	51.6	37.4	35.1	63,571.72
					$\Sigma (x - \bar{x})^2 \cdot f = 144{,}234$

NOW YOU CAN DO
Exercises 15–20.

In other words, the age of Americans under 85 typically differs from the mean age of 37.4 years by about 21.8 years.

EXAMPLE 20 Using technology to find the estimated mean, variance, and standard deviation for grouped data

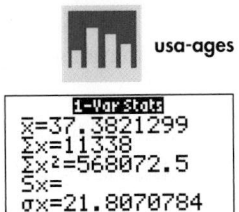

usa-ages

Use the TI-83/84 calculator to find the estimated mean, variance, and standard deviation for the frequency distribution in Table 17.

Solution

Following the instructions in the Step-by-Step Technology Guide, we get the estimated mean, $\bar{x} = 37.3821299$ (which we round to 37.4), the estimated standard deviation, s (shown in the output as σ_x) = 21.8070784, and the estimated variance as $(21.8070784)^2 = 475.5487$.

STEP-BY-STEP TECHNOLOGY GUIDE: Estimating the Mean, Variance, and Standard Deviation for Grouped Data

TI-83/84

Step 1   Press **STAT** and select **1: Edit**. Enter the class midpoints in L1 and the frequencies or relative frequencies in L2.
Step 2   Press **STAT**, select the **CALC** menu, and choose **1: 1-Var Stats**.

Step 3   Press **2nd 1 Comma 2nd 2**, so that the following appears on the home screen: **1-Var Stats L1, L2**.
Step 4   Press **ENTER**.

Section 3.3 Summary

1. The weighted mean is the sum of the products of the data points with their respective weights, divided by the sum of the weights.
2. We do not have access to the original raw data, so it is not possible to find exact values for the mean, variance, and standard deviation of data that have been grouped into a

frequency distribution. The estimated mean, $\bar{x}$, in this case is the sum of the products of the class frequencies, f, and class midpoints, x, divided by the sum of the frequencies, Σf.
3. Class midpoints and class frequencies are also used to find the estimated variance, s^2, and estimated standard deviations of grouped data.

Section 3.3 Exercises

CLARIFYING THE CONCEPTS

1. Explain why the formula for the mean of grouped data will provide an estimate only and not the exact value of the mean if the data were not grouped. (p. 149)
2. Describe how the weighted mean is calculated. (p. 149)
3. Suppose we calculate the weighted mean of the following data: 2, 7, 4. Let each of the weights equal 1. To what measure of center from Section 3.1 does this weighted mean simplify when all the weights equal 1? (p. 149)

PRACTICING THE TECHNIQUES

 CHECK IT OUT!

To do	Check out	Topic
Exercises 4–8	Example 17	Weighted mean
Exercises 9–14	Example 18	Estimated mean for grouped data
Exercises 15–20	Example 19	Estimated variance and standard deviation for grouped data

For Exercises 4–8, the data values and weights are provided. Find the weighted mean.

4. $x_1 = 60, x_2 = 70; x_3 = 80; w_1 = 0.25, w_2 = 0.50, w_3 = 0.25$.
5. $x_1 = 100, x_2 = 60, x_3 = 90; w_1 = 0.25, w_2 = 0.40, w_3 = 0.35$.
6. $x_1 = 10, x_2 = 10, x_3 = 100; w_1 = 10, w_2 = 20, w_3 = 5$.
7. $x_1 = 2.0, x_2 = 3.5, x_3 = 2.5, x_4 = 3.0, x_5 = 2.0; w_1 = w_2 = w_3 = w_4 = 3, w_5 = 8$.
8. $x_1 = 70, x_2 = 80, x_3 = 85, x_4 = 95; w_1 = 0.25, w_2 = 0.25, w_3 = 0.25, w_4 = 0.25$.

For Exercises 9–14, the frequency distribution is provided for a particular variable. Do the following:
   a. Find the class midpoints.
   b. Calculate the product of each class frequency with its midpoint.
   c. Find the sum of the frequencies, Σf, and the sum of the products $\Sigma(f \cdot x)$.
   d. Divide $\Sigma(f \cdot x)$ by Σf to find the estimated mean of the variable, $\bar{x}$.

9.

Class	Frequency f
$0 \leq$ GPA < 1.0	2
$1.0 \leq$ GPA < 2.0	10
$2.0 \leq$ GPA < 3.0	13
$3.0 \leq$ GPA < 4.0	5

10.

Class	Frequency f
$-10 \leq$ golf score < -5	3
$-5 \leq$ golf score < 0	7
$0 \leq$ golf score < 5	7
$5 \leq$ golf score < 10	3

11.

Class	Frequency f
$0 \leq$ score < 2	10
$2 \leq$ score < 4	20
$4 \leq$ score < 6	30
$6 \leq$ score < 8	20
$8 \leq$ score < 10	10

12.

Class	Frequency f
$0 \leq$ grade < 50	5
$50 \leq$ grade < 70	10
$70 \leq$ grade < 80	15
$80 \leq$ grade < 90	20
$90 \leq$ grade < 100	20

13.

Class	Frequency f
$0 \leq$ cost < 5	100
$5 \leq$ cost < 10	150
$10 \leq$ cost < 15	200
$15 \leq$ cost < 20	250
$20 \leq$ cost < 30	300
$30 \leq$ cost < 50	350
$50 \leq$ cost < 100	400
$100 \leq$ cost < 200	450

14.

Class	Frequency f
$0 \leq$ cash < 10	15
$10 \leq$ cash < 20	10
$20 \leq$ cash < 30	5
$30 \leq$ cash < 40	4
$40 \leq$ cash < 50	4
$50 \leq$ cash < 75	2
$75 \leq$ cash < 100	1
$100 \leq$ cash < 200	1

For Exercises 15–20, find the estimated variance and standard deviation for the frequency distribution given in the indicated Exercise.

15. Exercise 9.
16. Exercise 10.
17. Exercise 11.
18. Exercise 12.
19. Exercise 13.
20. Exercise 14.

APPLYING THE CONCEPTS

21. Dupage County Age Groups. The Census Bureau reports the following frequency distribution of population by age group for Dupage County, Illinois, for residents who are less than 65 years old. **dupageage**

Class	Residents
$0 \leq$ age < 5	63,422
$5 \leq$ age < 18	240,629
$18 \leq$ age < 65	540,949

a. Find the class midpoints.
b. Find the estimated mean age of residents of Dupage County.
c. Find the estimated variance and standard deviation of ages.

22. Broward County House Values. Table 19 gives the frequency distribution of the dollar value of the owner-occupied housing units in Broward County, Florida. **browardhouse**

TABLE 19   Broward County house values

Class (1000s)	Housing units
$0 \leq$ value < 50	5,430
$50 \leq$ value < 100	90,605
$100 \leq$ value < 150	90,620
$150 \leq$ value < 200	54,295
$200 \leq$ value < 300	34,835
$300 \leq$ value < 500	15,770
$500 \leq$ value < 1000	5,595

a. Find the class midpoints.
b. Find the estimated mean dollar value for housing units in Broward County.
c. Find the estimated variance and standard deviation of the dollar value.

23. Lightning Deaths. Table 20 gives the frequency distribution of the number of deaths due to lightning nationwide over a 67-year period. Find the estimated mean and standard deviation of the number of lightning deaths per year. **lightningdeath**

TABLE 20   Lightning deaths

Class	Years
$20 \leq$ deaths < 60	13
$60 \leq$ deaths < 100	21
$100 \leq$ deaths < 140	10
$140 \leq$ deaths < 180	6
$180 \leq$ deaths < 260	10
$260 \leq$ deaths < 460	7

Source: National Oceanic and Atmospheric Administration.

24. Calculating a Course Grade. An introductory statistics syllabus has the following grading system. The weekly quizzes are worth a total of 25% toward the final course grade. The midterm exam is worth 32%; the final exam is worth 33%; and attendance/participation is worth 10% toward the final course grade. Anthony's weekly quiz average is 70. He got an 80 on the midterm and a 90 on the final exam. He got a 100 for attendance/participation. Calculate Anthony's final course grade.

25. Wages for Computer Managers. The U.S. Bureau of Labor Statistics (BLS) publishes wage information for various occupations. For the occupation "computer and information systems management," Table 21 gives the wages reported by the BLS for the top-paying states. Find the weighted mean wage across all five states, using the employment figures as weights. **compwage**

TABLE 21   Wages for computer managers

State	Employment	Hourly mean wage
New Jersey	12,380	$60.32
New York	18,580	$60.25
Virginia	9,540	$59.39
California	35,550	$57.98
Massachusetts	10,130	$55.95

26. Salaries of Scientists and Engineers. The National Science Foundation compiles statistics on the annual salaries of full-time employed doctoral scientists and engineers in universities and four-year colleges. The mean annual salary for the fields of science, engineering, and health are $67,000, $82,200, and $70,000, respectively. Suppose we have a sample of 10 professors, 5 of whom are in science, 2 in engineering, and 3 in health, and each of whom is making the mean salary for his or her field. Find the weighted mean salary of these 10 professors.

27. Challenge Exercise. Assign the weights, w, to show that the formula for the sample mean from Section 3.1, $\bar{x} = \Sigma x/n$, is a special case of the formula for the weighted mean, $\bar{x} = \Sigma(wx)/\Sigma w$.

BRINGING IT ALL TOGETHER

Wait Times at Los Angeles Airport. Use the following table for Exercises 28–33. The data represent the number of passengers whose flights were delayed at the Tom Bradley Terminal of Los Angeles Airport (LAX), on July 2, 2014, between 4 P.M. and 5 P.M. Counts are given based on how long their flights were delayed.

Delay (minutes)	Passengers
0 to < 16	665
16 to < 31	551
31 to < 46	497
46 to < 61	399
61 to < 91	355
91 to < 120	27

Source: U.S. Customs and Border Protection: awt.cbp.gov.

28. Find the delay midpoints.

29. Construct a table similar to Table 17, showing the frequencies, f, the midpoints, x, the products, $f \cdot x$, the sum of the frequencies, Σf, and the sum of the products, $\Sigma(f \cdot x)$.

30. Use the quantities from Exercise 29 to calculate the estimated mean delay time.

31. Extend your table from Exercise 29 so that it is similar to Table 18, including columns for $\bar{x}$, $x - \bar{x}$, and $(x - \bar{x})^2 \cdot f$. Calculate $\Sigma(x - \bar{x})^2 \cdot f$.

32. Use the statistics from Exercise 31 to compute the estimated variance.

33. Calculate the estimated standard deviation of delay times.

WORKING WITH LARGE DATA SETS

 Financial Experts versus the Darts. This set of exercises examines how close the estimated mean, variance, and standard deviation are to their true values. Use the **Darts** data set from the Chapter 3 Case Study for Exercises 34–37. **darts**

34. Use the following classes to construct a frequency distribution for the **Professionals**, **Darts**, and the **DJIA** data sets.

Class
$-50 \le$ price change < -25
$-25 \le$ price change < 0
$0 \le$ price change < 25
$25 \le$ price change < 50
$50 \le$ price change < 75
$75 \le$ price change < 100

35. Use the frequency distribution from Exercise 34 to calculate the estimated mean stock price change for the **Professionals**, **Darts**, and the **DJIA** data sets.

36. Use the information from the two previous exercises to compute the estimated variance and standard deviation for the stock price changes for the **Professionals**, **Darts**, and the **DJIA** data sets.

37. Using technology, find the mean, variance, and standard deviation for the **Professionals**, **Darts**, and the **DJIA** data sets. Calculate the difference between the estimated values and the actual values.

WORKING WITH LARGE DATA SETS

Year-by-year age distribution. Open the **Age Distribution 100** data set, and use it for Exercises 38–42. This data set shows the year-by-year age distribution of Americans under age 100, as reported by the U.S. Census Bureau, for 2011. Use technology to answer the following: **agedistribution100**

38. How many tiny tots have yet to reach their first birthday?

39. Find the mean age of Americans under 100.

40. Calculate the estimated standard deviation of Americans under 100.

41. Use the Empirical Rule (see Section 3.2) to find two age values between which lie about 68% of the ages of all Americans under 100.

42. Compute the actual proportion between the age values found in the previous exercise. Compare the actual number to the estimate in the previous exercise.

3.4   Measures of Relative Position and Outliers

OBJECTIVES   By the end of this section, I will be able to . . .

1   Calculate z-scores, and explain why we use them.
2   Detect outliers using the z-score method.
3   Find percentiles and percentile ranks for both small and large data sets.
4   Compute quartiles and the interquartile range.

In this section, we learn about *measures of relative position*, which tell us the position that a particular data value has relative to the rest of the data set. For example, a prestigious nursing school may grant admission to only the top 10% of applicants. How high a score would you need to enter? This is one type of question we will answer in this section.

1 z-Scores

Recall that the standard deviation is a common measure of the variability, or spread, of a data set, and its value is interpreted as a typical deviation from the mean.

Our first measure of relative position is the z-score. Recall that the standard deviation is a common measure of the variability, or spread, of a data set. The **z-score** indicates how many standard deviations a particular data value is from the mean. If the z-score is positive, then the data value is above the mean. If the z-score is negative, then the data value is below the mean.

> **z-Score**
>
> The **z-score** for a particular data value from a *sample* is
>
> $$z\text{-score} = \frac{\text{data value} - \text{mean}}{\text{standard deviation}} = \frac{x - \bar{x}}{s}$$
>
> where $\bar{x}$ is the sample mean, and s is the sample standard deviation.
> The z-score for a particular data value from a *population* is
>
> $$z\text{-score} = \frac{\text{data value} - \text{mean}}{\text{standard deviation}} = \frac{x - \mu}{\sigma}$$
>
> where μ is the population mean, and σ is the population standard deviation.
> z-scores can be positive or negative.
>
> • A positive z-score indicates that the data value, x, lies above the mean.
> • A negative z-score implies that x lies below the mean.
> • A z-score equal to zero indicates that x equals the mean.

In this section, we will use the sample z-score unless otherwise indicated.

EXAMPLE 21   Calculating z-scores, given data values

People thinking about applying for a loan should take care that they maintain a healthy credit score, which comes from paying monthly bills on time and paying off previous loans without any problems. Figure 22 shows a histogram of the credit scores of over 150,000 loan applicants (*Source: Data Mining and Predictive Analytics*, by Daniel Larose and Chantal Larose, Wiley, 2015). The mean of this population of credit scores is $\mu = 670$, with a standard deviation of $\sigma = 70$. Calculate and interpret the z-scores for the following loan applicants:

a. Jasmine has been taking care to pay all her bills on time, so she has a healthy credit score of 740.

b. Jeremy was laid off, defaulted on a previous loan, and so has a credit rating of 439.

c. May-Chang always pays her bills on time and has already paid off several loans. Her credit score is 817.

Solution

Note that here we have population values, with $\mu = 670$ and $\sigma = 70$.

a. Jasmine's credit score is $x = 740$. Her z-score is

$$z\text{-score} = \frac{\text{data value} - \text{mean}}{\text{standard deviation}} = \frac{x - \mu}{\sigma} = \frac{740 - 670}{70} = 1$$

We interpret Jasmine's z-score of 1 to mean that her credit score of 740 lies 1 standard deviation above the mean $\mu = 670$. See Figure 22.

b. The z-score for Jeremy's credit score of 439 is

$$z\text{-score} = \frac{\text{data value} - \text{mean}}{\text{standard deviation}} = \frac{x - \mu}{\sigma} = \frac{439 - 670}{70} = -3.3$$

Jeremy's credit score lies 3.3 standard deviations below the mean.

c. The z-score for May-Chang's credit score of 817 is

$$z\text{-score} = \frac{\text{data value} - \text{mean}}{\text{standard deviation}} = \frac{x - \mu}{\sigma} = \frac{817 - 670}{70} = 2.1$$

May-Chang's credit score lies 2.1 standard deviations above the mean.

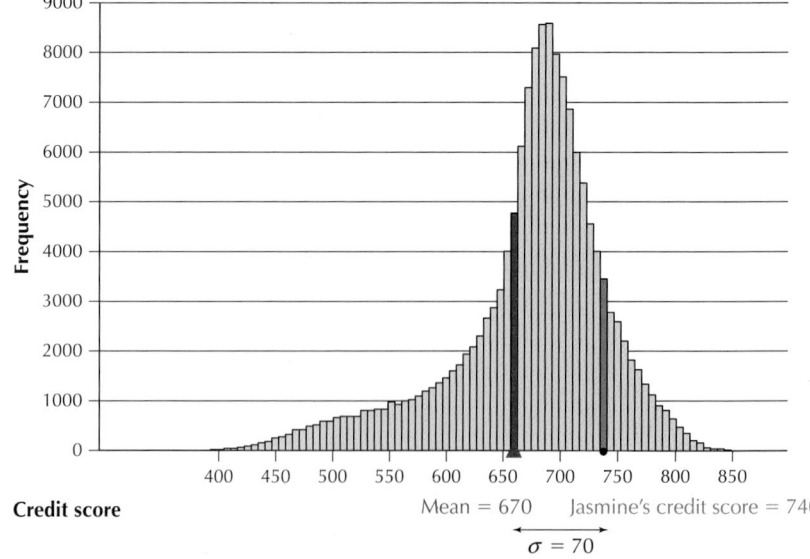

FIGURE 22
Jasmine's *z*-score of 1 places her 1 standard deviation above the mean.

NOW YOU CAN DO
Exercises 7–18.

YOUR TURN
#11

The IBM Digital Analytics Benchmark reports that tablet users (for tablets such as the *iPad*) spent a mean of $96 per order for their 2013 online holiday shopping. Assume that the standard deviation is $40. Find the z-scores for the following tablet-using holiday shoppers:

1. Austin spent $136 on video games.

2. Brian spent $16 on music downloads.

3. Courtney spent $256 on gifts for her friends.

(The solutions are shown in Appendix A.)

Alternatively, we may be given a z-score and asked to find its associated data value, x. To do so, use the following formulas.

Given a z-score, to find its associated data value x:

For a sample: $x = z\text{-score} \cdot s + \bar{x}$

For a population: $x = z\text{-score} \cdot \sigma + \mu$

where μ is the population mean, $\bar{x}$ is the sample mean, σ is the population standard deviation, and s is the sample standard deviation.

Note: We arrive at these formulas simply by taking the z-score formula and using algebra to solve for x.

EXAMPLE 22   Finding data values given z-scores

Continuing with the credit score data from Example 21, find the credit scores (the x-values) associated with the following z-scores:

a. -1 **b.** 0 **c.** 0.5

Solution

We have population data, with $\mu = 670$, $\sigma = 70$.

a. For a z-score of -1, we have

$$x = z\text{-score} \cdot \sigma + \mu = (-1) \cdot (70) + (670) = 600.$$

A credit score of 600 is associated with a z-score of -1, and therefore lies 1 standard deviation below the mean.

b. For a z-score of 0, we have

$$x = z\text{-score} \cdot \sigma + \mu = (0) \cdot (70) + 670 = 670.$$

As noted earlier, a z-score of zero exactly equals the mean $\mu = 670$.

c. For a z-score of 0.5, we have

$$x = z\text{-score} \cdot \sigma + \mu = (0.5) \cdot (70) + 670 = 705.$$

A z-score of 0.5 is associated with a credit score of 705.

NOW YOU CAN DO
Exercises 19–30.

YOUR TURN
#12

Continuing the online holiday shopping example from Your Turn #11 on page 156, find the spending amounts associated with the following z-scores.

1. David's z-score was -1.5. How much did he spend?

2. Emily had a z-score of 2.5. What was her spending amount?

3. Frances had a z-score of zero. What did she spend?

(The solutions are shown in Appendix A.)

EXAMPLE 23   Using the z-score to compare data from different data sets

Andrew is bragging to his friend Brittany that he did better than she did on the last statistics test. Andrew got a 90, while Brittany got an 80. Andrew's class mean was 80, with a standard deviation of 10. Brittany's class mean was 60, with a standard deviation of 10. The professors in both classes grade "on a curve" using z-scores. Who did better relative to his or her class?

Solution

Brittany can use *z*-scores to show that she did better *relative to her class*. Figure 23 shows comparative dotplots of the scores in the two classes. The red dots represent Brittany's and Andrew's scores. Brittany found her *z*-score by subtracting her class mean from her score of 80 and then dividing by the standard deviation $s = 10$:

$$z\text{-score}_{\text{Brittany}} = \frac{x - \bar{x}}{s} = \frac{80 - 60}{10} = 2$$

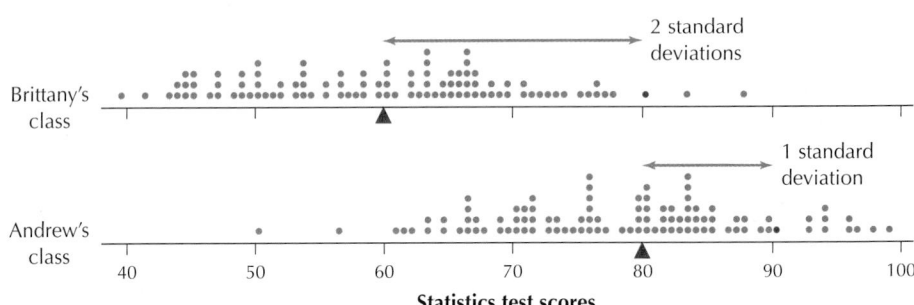

FIGURE 23
Brittany actually did better relative to her class.

Brittany's *z*-score is 2. What does that mean? It means that *Brittany scored 2 standard deviations above the mean* of 60. Brittany then found the *z*-score for Andrew:

$$z\text{-score}_{\text{Andrew}} = \frac{x - \bar{x}}{s} = \frac{90 - 80}{10} = 1$$

z-Scores enable the data analyst to compare data values from two different distributions.

Andrew's *z*-score was 1, which means that Andrew scored 1 standard deviation above the mean. From Figure 23, we can observe that Andrew's exam score of 90 lies closer to the mean exam score of 80 for his class. That is, the arrow is shorter for Andrew than for Brittany. Finally, note that 10 of the 100 students who took the exam in his class did better than he did, whereas only two did better than Brittany in her class. So, relative to her class, Brittany did better than Andrew, even though Andrew got a higher score. The *z*-scores allowed her to compare their grades, even though they were in different classes.

NOW YOU CAN DO
Exercises 31 and 32.

YOUR TURN
#13

Continuing the online holiday shopping example from Your Turn #11, the IBM Digital Analytics Benchmark also reported that cell phone users spent a mean of an average of $85 per order for their 2013 online holiday shopping. Assume the standard deviation is $40. Gisele is a tablet user, whereas Hong is a cell phone user. They both spent the same amount for an online holiday shopping order: $120. Who spent more, *relative to his or her group*?

(The solution is shown in Appendix A.)

2 Detecting Outliers Using the *z*-Score Method

Note: If an outlier is detected, it does not automatically follow that it should be discarded. Outliers often indicate the presence of something interesting going on in the data that would call for further investigation. On the other hand, it could simply be a typo. The analyst should check with the data source.

An **outlier** is a data value that is very much greater than or less than the mean. It may represent a data entry error, or it may be genuine data. One way of identifying an outlier is to determine whether it is farther than 3 standard deviations from the mean, that is, its *z*-score is less than -3 or greater than 3.

Guidelines for Identifying Outliers

1. A data value whose *z*-score lies in the following range is considered *not unusual*:

$$-2 < z\text{-score} < 2$$

2. A data value whose z-score lies in either of the following ranges may be considered *moderately unusual*:

$$-3 < z\text{-score} \le -2 \quad \text{or} \quad 2 \le z\text{-score} < 3$$

3. A data value whose z-score lies in either of the following ranges may be considered an *outlier*:

$$z\text{-score} \le -3 \quad \text{or} \quad z\text{-score} \ge 3$$

EXAMPLE 24   Detecting outliers using the z-score method

For the three loan applicants in Example 21 on page 155, determine whether each of their credit scores represents an outlier.

Solution

- Jasmine's z-score is 1, which lies in the range, $-2 < z\text{-score} < 2$. Therefore, Jasmine's credit rating is not considered unusual.

- Jeremy's z-score is -3.3, which is ≤ -3. Thus, Jeremy's credit score may be considered an outlier.

NOW YOU CAN DO
Exercises 33–44.

- May-Chang's z-score is 2.1, which lies in the range, $2 \le z\text{-score} < 3$. Thus, May-Chang's credit score may be considered moderately unusual.

YOUR TURN
#14

Refer to the z-scores you calculated for Austin, Brian, and Courtney in Your Turn #11 on page 156. Determine whether each of their spending amounts represents an outlier.

(The solutions are shown in Appendix A.)

In Section 3.5, we will learn about the IQR method of detecting outliers.

3 Percentiles and Percentile Ranks

The next measure of relative position we consider is the **percentile**, which shows the location of a data value relative to the other values in the data set.

Some analysts prefer to define the *p*th percentile to be a data value at which at least *p* percent of the values in the data set are less than or equal to this value, *and* at least (1 − *p*) percent of the values are greater than or equal to this value.

> **Percentile**
>
> Let *p* be any integer between 0 and 100. The *p*th **percentile** of a data set is the data value at which *p* percent of the values in the data set are less than or equal to this value.

EXAMPLE 25   Meaning of a percentile

Jasmine's credit score of 740 represents the 88th percentile of the 150,000 credit scores. What does "88th percentile" mean?

Solution

To say that 740 is the 88th percentile means that 88% of all credit scores fell at or below Jasmine's credit score of 740. We call the percentile a *measure of relative position* because it indicates the position of Jasmine's credit score relative to all other credit scores. Figure 24 indicates the position of Jasmine's credit score relative to the rest of the loan applicants.

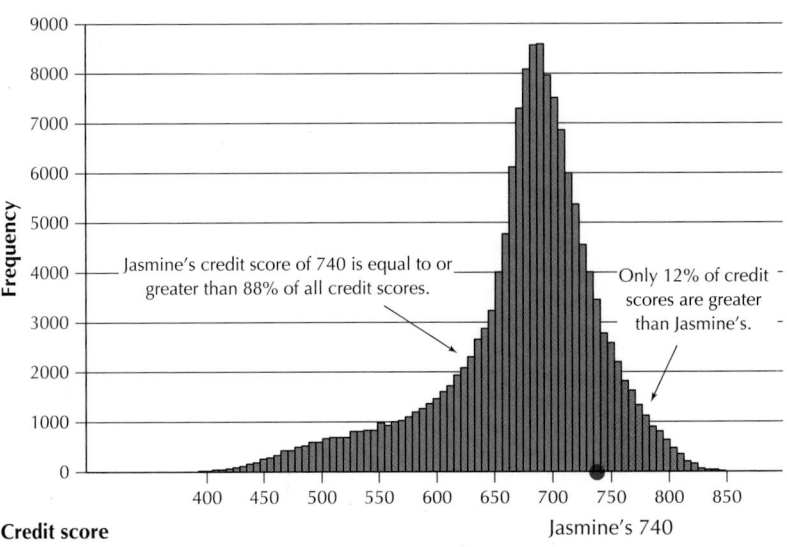

FIGURE 24 Jasmine's credit score of 740 represents the 88th percentile. The 88th percentile is the score with 88% of the data values at or below its value.

For large data sets, calculation of the percentiles is best left to computers. However, for small data sets, we can use the following step-by-step method to calculate the related position of any percentile.

> **Step 1** Sort the data into ascending order (from smallest to largest).
>
> **Step 2** Calculate
>
> $$i = \left(\frac{p}{100}\right)n$$
>
> where p is the particular percentile you wish to calculate, and n is the sample size.
>
> **Step 3**
>
> **a.** If i is an integer (a whole number with no decimal part), the pth percentile is the mean of the data values in positions i and $i + 1$.
>
> **b.** If i is not an integer, round up to the next integer and use the value in this position.

CAUTION These steps do not give the value of the pth percentile itself, but rather the *position* of the pth percentile in the data set when the data set is in ascending order.

EXAMPLE 26   Finding percentiles

Table 22 contains the value of international exports for a sample of 12 states for the month of June 2014, expressed in millions of dollars. Find the 75th percentile of the exports.

Table 22 Exports for 12 states

State	VA	NC	NJ	GA	PA	OH	MI	FL	LA	IL	WA	NY
Exports ($ millions)	1.6	2.7	3.3	3.5	3.5	4.6	4.7	4.8	5.0	5.8	7.5	7.7

Solution

stateexports

Step 1 Sort the data into ascending order. Fortunately, Table 22 is already presented in ascending order of exports.

Step 2 The particular percentile we wish to calculate is the 75th percentile, so $p = 75$. Our data set includes 12 values, so $n = 12$. Calculate

$$i = \left(\frac{p}{100}\right)n = \left(\frac{75}{100}\right)12 = 9$$

So, $i = 9$

Step 3 Here, i is an integer, so the 75th percentile is the mean of the data values in positions 9 and 10.

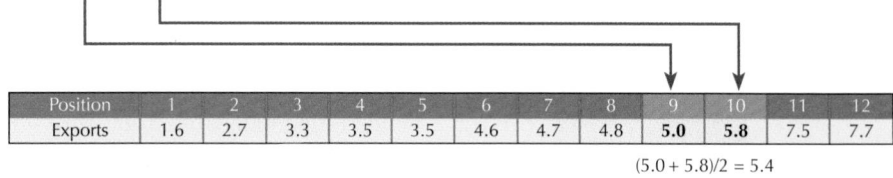

Position	1	2	3	4	5	6	7	8	9	10	11	12
Exports	1.6	2.7	3.3	3.5	3.5	4.6	4.7	4.8	**5.0**	**5.8**	7.5	7.7

$(5.0 + 5.8)/2 = 5.4$

NOW YOU CAN DO
Exercises 45–56.

Counting from left to right, the data value in the 9th position is Louisiana's 5.0, and the data value in the 10th position is Illinois' 5.8. The mean of these two values is 5.4. Thus, the 75th percentile is 5.4, representing $5.4 million in exports.

YOUR TURN
#15

Jason is doing a class project on some of the lowest-rated movies on the movie database IMDB. He will use movies whose ratings are in the 20th percentile or lower. A sample of movie ratings follows. Calculate the 20th percentile rating.

8.7 5.4 7.1 3.6 1.9 5.7 4.2 9.3 2.5

(The solution is shown in Appendix A.)

Remember: A percentile is a data value, whereas a percentile rank is a percentage.

The **percentile rank** of a data value, x, equals the percentage of values in the data set that are less than or equal to x. In other words:

$$\text{percentile rank of data value } x = \frac{\text{number of values in data set} \leq x}{\text{total number of values in data set}} \cdot 100$$

EXAMPLE 27   Finding percentile ranks

For the state export data in the previous example, calculate the percentile ranks for the following export values:

a. $5.4 million

b. $3.4 million

Solution

a. Here, $x = 5.4$. Nine states have x-values at or below 5.4, so the percentile rank of a state with $5.4 million in exports is

$$\text{percentile rank of data value } (x = 5.4) = \frac{\text{number of values in data set} \le 5.4}{\text{total number of values in data set}} \cdot 100$$

$$= \frac{9}{12} \cdot 100 = 75\%$$

Note that, therefore, $5.4 million represents the 75th percentile of state exports.

b. Here $x = 3.4$. Three states have x-values at or below 3.4, so the percentile rank of a state with $3.4 million in exports is

$$\text{percentile rank of data value } (x = 3.4) = \frac{\text{number of values in data set} \le 3.4}{\text{total number of values in data set}} \cdot 100$$

$$= \frac{3}{12} \cdot 100 = 25\%$$

NOW YOU CAN DO
Exercises 57–68.

Thus, $3.4 million represents the 25th percentile of state exports.

**YOUR TURN
#16**

For the movie rating data from Your Turn #15 on page 161, calculate the percentile rank for a movie with a rating of 9.0.

(The solution is shown in Appendix A.)

4 Quartiles and the Interquartile Range

Just as the median divides the data set into halves, the **quartiles** are the percentiles that divide the data set into quarters (Figure 25).

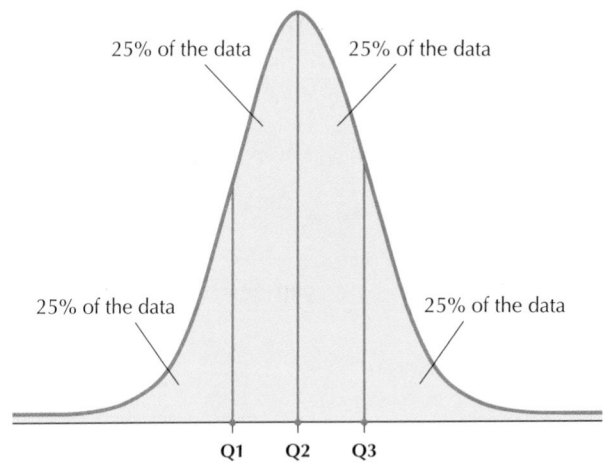

25% of the data   25% of the data

25% of the data   25% of the data

Q1   Q2   Q3

FIGURE 25 The quartiles Q1, Q2, and Q3 divide the data set into four quarters.

The Quartiles

The **quartiles** of a data set divide the data set into four parts, each containing 25% of the data.

* The *first quartile* (Q1) is the 25th percentile.
* The *second quartile* (Q2) is the 50th percentile, that is, the median.
* The *third quartile* (Q3) is the 75th percentile.

For small data sets, the division may be into four parts of only approximately equal size.

EXAMPLE 28   Finding the quartiles for a small data set

Note: It may be helpful to note that the phrase *third quartile* is akin to the phrase *three quarters*, which is 75%, representing the 75th percentile. Also, the phrase *first quartile* is akin to the phrase *one quarter*, which is 25%, representing the 25th percentile.

In Example 26 (page 160), we found the 75th percentile of the export data to be $5.4 million. By definition, the 75th percentile is the third quartile Q3. Therefore, this export value of $5.4 million is also the third quartile (Q3) of the export values. Now calculate the first quartile and the median (second quartile) of export values.

Solution

To find the quartiles, we use the steps for finding percentiles (page 160). First, arrange the data set in ascending order, which they already are in Table 22.

Here, $n = 12$. To find Q1, plug $p = 25$ into the equation $i = \left(\dfrac{p}{100}\right)n$, where $n = 12$. We get $i = \left(\dfrac{p}{100}\right)n = \left(\dfrac{25}{100}\right)12 = 3$. Since 3 is an integer, we know that the 25th percentile is the mean of the export values in the 3rd and 4th positions. New Jersey's export value of 3.3 is in the 3rd position, while Georgia's export value of 3.5 is in the 4th position. Since $(3.3 + 3.5)/2 = 3.4$, we get the 25th percentile of the export data to be 3.4, representing $3.4 million in exports (Figure 26).

Position	1	2	3	4	5	6	7	8	9	10	11	12
Exports	1.6	2.7	3.3	3.5	3.5	4.6	4.7	4.8	5.0	5.8	7.5	7.7

Q1 = 3.4

FIGURE 26 The 25th percentile splits the difference between 3.3 and 3.5.

To find the median (the second quartile, Q2), plug $p = 50$ into your steps for finding the percentiles: $i = \left(\dfrac{p}{100}\right)n = \left(\dfrac{50}{100}\right)12 = 6$. Since 6 is an integer, we know that the 50th percentile is the mean of the export data in the 6th and 7th positions, that is, 4.6 and 4.7. Since $(4.6 + 4.7)/2 = 4.65$, the 50th percentile of the export data is 4.65, representing $4.65 million in exports (Figure 27). This agrees with the method we learned for finding the median, on page 112.

Position	1	2	3	4	5	6	7	8	9	10	11	12
Exports	1.6	2.7	3.3	3.5	3.5	4.6	4.7	4.8	5.0	5.8	7.5	7.7

Median = Q2 = 4.65

FIGURE 27 The 50th percentile splits the difference between 4.6 and 4.7.

The quartiles may be found on the TI-83/84 by using the instructions for descriptive statistics shown on page 117.

In Example 26, we determined that the 75th percentile was 5.4. Therefore, the quartiles for the export data are Q1 = 3.4, median = Q2 = 4.65, and Q3 = 5.4. Note that these quartiles divide the data set into four equal sections, with three observations each (Figure 28).

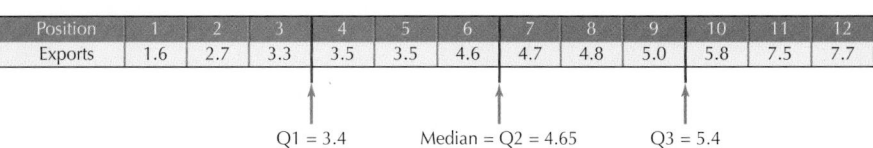

Position	1	2	3	4	5	6	7	8	9	10	11	12
Exports	1.6	2.7	3.3	3.5	3.5	4.6	4.7	4.8	5.0	5.8	7.5	7.7

Q1 = 3.4   Median = Q2 = 4.65   Q3 = 5.4

NOW YOU CAN DO
Exercises 69–76.

FIGURE 28 The quartiles for the export data.

> **YOUR TURN**
> **#17**
>
> As a follow-up to his project, Jason is dividing movie ratings into Great (at or above Q3), Good (from Q2 to Q3), Mediocre (from Q1 to Q2), and Awful (lower than Q1). Find Q1, Q2, and Q3 from the following sample of movie ratings.
>
> 8.7   5.4   7.1   3.6   1.9   5.7   4.2   9.3   2.5
>
> (The solution is shown in Appendix A.)

Of course, for small data sets, the division into quarters is not always exact. For example, what if our data set consisted of 11 states instead of 12? Eleven data values cannot be divided equally into four quarters. In this case, therefore, the quartiles would divide the data set into four sections of approximately equal size. However, for large data sets, which the data analyst most often encounters, this becomes less of an issue.

EXAMPLE 29   Finding quartiles of a large data set: Cholesterol levels in food

 nutrition

The U.S. Department of Agriculture recommends a diet low in cholesterol to reduce the risk of heart disease. The data set **Nutrition** contains information on the cholesterol content (in milligrams) of 961 different foods. Find the mean, standard deviation, and quartiles.

Solution

The Minitab descriptive statistics for the cholesterol data are shown in Figure 29. Note that the mean cholesterol content is 32.55 mg and the standard deviation is about 120 mg. A standard deviation that is much larger than the mean may be associated with strongly skewed distributions. Compare the value for the mean with the values for the quartiles as follows:

- Q1, the first quartile, or 25th percentile, is 0 mg of cholesterol.

- The median, or Q2, the second quartile (50th percentile), is also 0 mg of cholesterol.

- Q3, the third quartile, or 75th percentile, is 20 mg of cholesterol.

Note: Minitab uses a different way to calculate the quartiles than the way we have learned, which results in different values than our hand-calculation methods. However, for large data sets, the difference is minimal.

Variable	N	Mean	StDev	Min	Q1	Median	Q3	Max
Cholesterol	961	32.55	119.96	0	0	0	20	2053

FIGURE 29 Descriptive statistics for the cholesterol data.

Figure 30 shows that the data distribution is extremely right-skewed. Only a few foods have over 1000 mg cholesterol, and another handful have over 500 (see data on disk). Therefore, it appears that we have outliers in this data set. What is the effect of

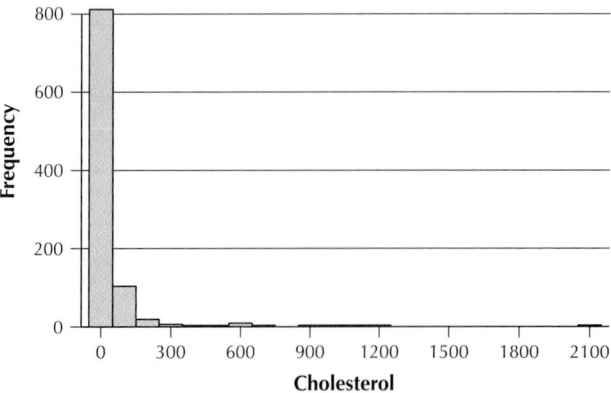

FIGURE 30 Cholesterol content (mg) of 961 foods.

these outliers on the mean and standard deviation? Does the mean represent a truly typical cholesterol content level for the data set, or is its value unduly increased by the outliers? Let's find out.

Developing Your Statistical Sense

The Mean Is Not Always Representative

Note that the median is 0 mg of cholesterol, meaning that at least half of the food items tested by the U.S. Department of Agriculture in this data set had no cholesterol at all. We are intrigued by this result and ask Minitab to provide us with a frequency distribution for the cholesterol content, along with the cumulative percentages ("CumPct"). Figure 31 provides a portion of this frequency distribution, with the following results:

- 61.91% of the food items have no cholesterol at all, which explains why Q1 and the median are both zero.

- The 75th percentile, Q3, is verified as 20 mg cholesterol.

- The 81st percentile of the data set is 32 mg cholesterol.

	Chol	Count	CumPct		Chol	Count	CumPct	
61.91% of food items had zero cholesterol. Thus, Q1 = 0 and median = 0.	0	595	61.91		20	5	75.23	75th percentile (Q3) = 20 mg cholesterol
	1	12	63.16		21	5	75.75	
	2	8	64.00		22	6	76.38	
	3	7	64.72		23	3	76.69	
	4	11	65.87		24	4	77.11	
	5	10	66.91		25	3	77.42	
	6	6	67.53		26	4	77.84	
	7	5	68.05		27	9	78.77	
	8	7	68.78		28	3	79.08	
	9	4	69.20		29	4	79.50	
	10	8	70.03		30	3	79.81	
	11	3	70.34		31	6	80.44	
	12	4	70.76		32	5	80.96	81st percentile is 32 mg. The mean is 32.55 mg.
	13	4	71.18		33	3	81.27	
	14	5	71.70		34	4	81.69	
	15	7	72.42		35	2	81.89	
	16	5	72.94		36	1	82.00	
	17	6	73.57		37	4	82.41	
	18	7	74.30		38	1	82.52	
	19	4	74.71		40	1	82.62	
	⋮	⋮	⋮		⋮	⋮	⋮	

FIGURE 31 Partial frequency distribution of cholesterol content.

Think about these results for a moment. We found that the 81st percentile is 32 mg cholesterol. In other words, 81% of the food items have a cholesterol content of 32 mg or less. And yet, this 32 mg is still *less than the mean* cholesterol content, reported by Minitab to be 32.55 mg. In other words, the mean of this data set is larger than 81% of the data values in the data set.

It seems clear, therefore, that *the mean 32.55 mg cannot be considered as typical or representative* of the data set. Its value has been exaggerated by the presence of the outliers, to such an extent that it is now larger than 81% of the data. We need another, more *robust* measure of center—one that is *resistant* to the undue influence of outliers, such as the median. Here, the value of the median is 0 mg cholesterol. An argument may certainly be made that this is indeed typical and representative of the data set, because 61.91% of the food items have no cholesterol content at all.

Recall from Section 3.2 that the variance and standard deviation are measures of spread that are sensitive to the presence of extreme values. A more robust (less sensitive) measure of variability is the **interquartile range**, or **IQR**.

> **Interquartile Range**
> The **interquartile range (IQR)** is a robust measure of variability. It is calculated as
> $$IQR = Q3 - Q1.$$
> The interquartile range is interpreted to be the spread of the middle 50% of the data.

The Latin word *inter* means "between," so the *inter*quartile range is the *difference between* the quartiles Q3 and Q1. The IQR represents how spread out the "middle half" of the data set is. A larger IQR implies a greater degree of variability, or spread, in the data set. The IQR ignores both the highest 25% and the lowest 25% of the data set, so it is completely unaffected by outliers and is thus quite robust.

EXAMPLE 30   Finding the interquartile range

In Example 28, we found that, for the export data, Q1 = 3.4 and Q3 = 5.4. Find the IQR for the export data, and explain what it means.

Solution

Because Q1 = 3.4 and Q3 = 5.4, the IQR = Q3 − Q1 = 5.4 − 3.4 = 2.0, which represents $2 million. We would say that the middle 50%, or middle half, of the export data ranged over $2 million (Figure 32).

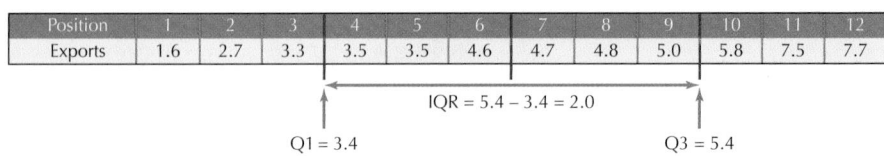

Position	1	2	3	4	5	6	7	8	9	10	11	12
Exports	1.6	2.7	3.3	3.5	3.5	4.6	4.7	4.8	5.0	5.8	7.5	7.7

IQR = 5.4 − 3.4 = 2.0

Q1 = 3.4   Q3 = 5.4

NOW YOU CAN DO
Exercises 77 and 78.

FIGURE 32 The interquartile range for the exports data.

YOUR TURN #18

Find the interquartile range for Jason's follow-up movie ratings project.

    8.7   5.4   7.1   3.6   1.9   5.7   4.2   9.3   2.5

(The solution is shown in Appendix A.)

WHAT IF
?

What If Scenario

For the state export data, consider the following two scenarios, and explain how the change would affect the quartiles and the IQR.

a. New York's imports are increased by an unknown amount.

b. Illinois' imports are increased by an unknown amount.

Solution

The IQR pays attention only to the middle half of the data, and it ignores what goes on in the upper 25% and the lower 25%.

a. New York is the maximum value for the data set, so any change in New York's exports would leave the quartiles unaffected; therefore, the IQR would also be unaffected.

b. Recall that Illinois' $5.8 million was used in the calculation of Q3, which is (5.8 − 5.0)/2 = 5.4. Increasing Illinois' exports would therefore increase the value of Q3; therefore, the IQR would also increase. However, Q1 and the median would remain unaffected.

STEP-BY-STEP TECHNOLOGY GUIDE: Percentiles and Quartiles

TI-83/84
The quartiles are provided using the instructions for descriptive statistics shown on page 117.

EXCEL
Step 1   Enter the data into column **A**.
Step 2   Select **Data > Data Analysis**.
Step 3   Select **Rank and Percentile** and click **OK**.

Step 4   Click in the **Input Range** cell. Then highlight the data in column **A**. If the variable name is in the column, select **Labels in First Row**. Click **OK**.

MINITAB
The quartiles are provided using the instructions for descriptive statistics shown on page 117.

SPSS
Step 1   Enter the data in the first column.
Step 2   Click **Analyze > Descriptive Statistics > Frequencies**....
Step 3   Click the column name and the arrow to move it to the **Variable(s)** box.

Step 4   Click **Statistics**.... For quartiles, select **Quartiles**. For other percentiles, select **Percentile(s)**, type in the desired percentile in the small box, and click **Add**. Click **Continue**, then **OK**.

JMP
Step 1   Click **File > New > DataTable**. Enter the data in **Column 1**.
Step 2   Click **Tables > Summary**.
Step 3   Click the variable name under **Select Columns**. For each desired percentile, enter its value in the **For quantile**

statistics, enter value (%) box. Then select **Quantiles** from the **Statistics** drop-down menu. After all have been entered, click **OK**.

CRUNCHIT!
We will use the data from Example 26 (page 160).
Step 1   Click **File**, then highlight **Load from Larose, Discostat3e > Chapter 3**, and click **Example 03_26**.
Step 2   Click **Statistics** and select **Descriptive statistics**. For **Columns**, select *Exports*.

Step 3   In the **Percentiles (comma-separated)** cell, enter the percentiles that you would like to find. For example, to find the 5th and 95th percentiles, enter 5, 95.
Step 4   Click **Calculate**.

Section 3.4 Summary

1. In this section, we learned about measures of relative position, which tell us the position that a particular data value holds relative to the rest of the data set. The z-score indicates how many standard deviations a particular data value is from the mean. The z-score equals the data value minus the mean, divided by the standard deviation. We may also calculate a data value, given its z-score.

2. An outlier is a value that is very much greater than or less than the mean. An outlier can be identified when its z-score is less than −3 or greater than 3.

3. The pth percentile of a data set is the value at which p percent of the values in the data set are less than or equal to this value. The percentile rank of a data value equals the percentage of values in the data set that are less than or equal to that value.

4. Quartiles divide the data set into approximately equal quarters. The interquartile range (IQR) is a measure of spread found by subtracting the first quartile from the third quartile.

Section 3.4 Exercises

CLARIFYING THE CONCEPTS

1. What does it mean for a z-score to be positive? Negative? Zero? (p. 155)

2. Explain in your own words what the 95th percentile of a data set means. (p. 160)

3. Why doesn't it make sense for there to be a 120th percentile of a data set? (p. 159)

4. Is it possible for the 1st percentile of a data set to equal the 99th percentile? Explain when this would happen. (p. 159)

5. Explain the difference between a percentile and a percentile rank. (p. 161)

6. True or false: The IQR is sensitive to the presence of outliers. (p. 165)

PRACTICING THE TECHNIQUES

CHECK IT OUT!

To do	Check out	Topic
Exercises 7–18	Example 21	Calculate z-score, given data
Exercises 19–30	Example 22	Find data value, given z-score
Exercises 31–32	Example 23	Use z-scores to compare different data sets
Exercises 33–44	Example 24	Identify outliers using z-scores
Exercises 45–56	Example 26	Percentiles
Exercises 57–68	Example 27	Percentile rank
Exercises 69–76	Example 28	Quartiles
Exercises 77–78	Example 30	Interquartile range

Use the following information for Exercises 7–10. Facebook reports that the average number of friends per Facebook user is 130. Assume the standard deviation is 30. Calculate the z-score for the indicated number of Facebook friends.

7. 190 Facebook friends

8. 145 Facebook friends

9. 100 Facebook friends

10. Zero Facebook friends

For Exercises 11–14, use the following information. Social Strand Media reports that the mean amount of video uploaded to YouTube every minute by users around the world is 100 hours. Assume the standard deviation is 25 hours. Calculate the z-score for the indicated number of hours of video uploaded to YouTube.

11. 125 hours

12. 50 hours

13. 200 hours

14. 87.5 hours

Use the following information for Exercises 15–18. Suppose the mean blood sugar level is 100 mg/dl (milligrams per deciliter), with a standard deviation of 10 mg/dl.

15. Alyssa has a blood sugar level of 90 mg/dl. How many standard deviations is Alyssa's blood sugar level below the mean?

16. Benjamin has a blood sugar level of 135 mg/dl. How many standard deviations is Benjamin's blood sugar level above the mean?

17. Chelsea has a blood sugar level of 125 mg/dl.
 a. If we calculate Chelsea's z-score, what is the scale?
 b. Calculate Chelsea's z-score.
 c. Interpret her z-score.

18. David has a blood sugar level of 85 mg/dl.
 a. Calculate David's z-score.
 b. Interpret his z-score.

For Exercises 19–22, use the following information. Facebook reports that the average number of friends per Facebook user is 130. Assume the standard deviation is 30. Find the number of Facebook friends represented by the following z-scores.

19. z-score = -1.0

20. z-score = 1.5

21. z-score = 0.0

22. z-score = -3.5

Use the following information for Exercises 23–26. Social Strand Media reports that the mean amount of video uploaded to YouTube every minute by users around the world is 100 hours. Assume the standard deviation is 25 hours. Find the number of hours of YouTube video uploaded per minute for the following z-scores.

23. z-score = 2.0

24. z-score = -2.0

25. z-score = -0.5

26. z-score = 0.0

Use the following information for Exercises 27–30. Suppose the mean blood sugar level is 100 mg/dl (milligrams per deciliter), with a standard deviation of 10 mg/dl. Find the blood sugar levels associated with the following z-scores.

27. z-score = 1.96

28. z-score = -2.576

29. z-score = -1.96

30. z-score = 2.576

31. Elizabeth's statistics class had a mean quiz score of 70 with a standard deviation of 15. Fiona's statistics class had a mean quiz score of 75 with a standard deviation of 5. Both Elizabeth and Fiona got an 85 on the quiz. Who did better relative to her class?

32. Juan's business class had a mean quiz score of 60 with a standard deviation of 15. Luis's business class had a mean quiz score of 70 with a standard deviation of 5. Both Juan and Luis got a 75 on the quiz. Who did better relative to his class?

For Exercises 33–44, determine whether the data value represents an outlier, using the z-score method.

33. The 190 Facebook friends from Exercise 7

34. The 145 Facebook friends from Exercise 8

35. The 100 Facebook friends from Exercise 9.
36. The zero Facebook friends from Exercise 10.
37. The 125 hours of YouTube video from Exercise 11.
38. The 50 hours of YouTube video from Exercise 12.
39. The 200 hours of YouTube video from Exercise 13.
40. The 87.5 hours of YouTube video from Exercise 14.
41. Alyssa's blood sugar level from Exercise 15.
42. Benjamin's blood sugar level from Exercise 16.
43. Chelsea's blood sugar level from Exercise 17.
44. David's blood sugar level from Exercise 18.

Use the following data for Exercises 45–50. The variable is *Highway MPG*, which is the number of miles a vehicle can travel on a highway on one gallon of gas. The sample is taken from the Chapter 8 Case Study, *Motor Vehicle Fuel Efficiency*. Find the highway MPG represented by the indicated percentiles.

Vehicle	Highway MPG	Vehicle	Highway MPG
Honda CR-V	30	Subaru Impreza	25
Nissan Pathfinder	26	Ford Mustang	26
Chevrolet Chevy SS	21	Cadillac ATS	31
Dodge Charger	27	Chevrolet Camaro	24
Jeep Compass	23	Ford Taurus	29
Lincoln MKT	25	Ford Expedition	20

Source: www.fueleconomy.gov.

45. 75th
46. 5th
47. 95th
48. 90th
49. 10th
50. 99th

Use the following data for Exercises 51–56. Research has shown that the amount of sodium consumed in food has been associated with hypertension (high blood pressure). The table provides a list of 16 breakfast cereals, along with their sodium content, in milligrams per serving. Find the amount of sodium represented by the indicated percentiles. **cereals**

Cereal	Sodium	Cereal	Sodium
Apple Jacks	125	Grape Nuts Flakes	140
Cap'n Crunch	220	Kix	260
Cinnamon Toast Crunch	210	Life	150
Corn Flakes	290	Lucky Charms	180
Count Chocula	180	Raisin Bran	210
Cream of Wheat	80	Rice Chex	240
Fruit Loops	125	Special K	230
Fruity Pebbles	135	Total Whole Grain	200

51. 75th
52. 10th
53. 90th
54. 30th

55. 5th
56. 95th

Using the highway MPG data above, calculate the percentile rank for the indicated highway MPG in Exercises 57–62.
57. 30
58. 31
59. 20
60. 25
61. 27
62. 29

Use the cereal sodium data above to calculate the percentile rank for the indicated amount of sodium (in mg) in Exercises 63–68.
63. 80
64. 290
65. 260
66. 125
67. 230
68. 220

Use the highway MPG data above for Exercises 69–72.
69. Find Q1, the first quartile.
70. Calculate Q2, the second quartile.
71. Compute Q3, the third quartile.
72. Find the median, and compare it to Q2.

For Exercises 73–76, use the cereal sodium data above.
73. Find Q1, the first quartile.
74. Calculate Q2, the second quartile.
75. Compute Q3, the third quartile.
76. Find the median, and compare it to Q2.
77. Use your work in Exercises 69 and 71 to compute the IQR for the highway MPG data. What does this number mean?
78. Use your work in Exercises 73 and 75 to compute the IQR for the cereal sodium data. What does this number mean?

APPLYING THE CONCEPTS
Breakfast Calories. Refer to Table 23 for Exercises 79–86.
breakfastcal

TABLE 23 Calories in 12 breakfast cereals

Cereal	Calories
Apple Jacks	110
Basic 4	130
Bran Chex	90
Bran Flakes	90
Cap'n Crunch	120
Cheerios	110
Cinnamon Toast Crunch	120
Cocoa Puffs	110
Corn Chex	110
Corn Flakes	100
Corn Pops	110
Count Chocula	110

79. Find the *z*-scores for the calories for the following cereals:
  a. Corn Flakes   **c.** Bran Flakes
  b. Basic 4   **d.** Cap'n Crunch

80. Find the number of calories associated with the following *z*-scores:
  a. 0   **b.** 1   **c.** −1   **d.** 0.5

81. Determine whether any of the cereals is an outlier.

82. Find the following percentiles:
  a. 25th   **b.** 50th   **c.** 75th   **d.** 95th

83. Find the percentile rank for each of the following:
  a. 90 calories   **c.** 110 calories
  b. 120 calories   **d.** 100 calories

84. Find the following:
  a. Q1   **b.** Q2   **c.** Q3   **d.** IQR

85. Explain what the IQR value from Exercise 84(**d**) means.

86. Suppose that a weight-control organization recommended eating breakfast cereals with the lowest 10% of calories.
  a. How many calories does this cutoff represent?
  b. Which cereals are recommended?

Dietary Supplements. Refer to Table 24 for Exercises 87–94. The table gives the number of American adults who have used the indicated "nonvitamin, nonmineral, natural products."

 dietarysupp

TABLE 24   Use of dietary supplements

Product	Usage (in millions)	Product	Usage (in millions)
Echinacea	14.7	Ginger	3.8
Ginseng	8.8	Soy	3.5
Ginkgo biloba	7.7	Chamomile	3.1
Garlic	7.1	Bee pollen	2.8
Glucosamine	5.2	Kava kava	2.4
St. John's wort	4.4	Valerian	2.1
Peppermint	4.3	Saw palmetto	2.0
Fish oil	4.2		

Source: Centers for Disease Control and Prevention, Vital and Health Statistics.

87. Find the *z*-scores for usage of the following products:
  a. Echinacea   **c.** Valerian
  b. Saw palmetto **d.** Ginseng

88. Find the usage associated with each of the following *z*-scores.
  a. 0   **b.** 3   **c.** −3   **d.** 1

89. Identify any outliers in the data set.

90. Find the following percentiles:
  a. 10th   **b.** 90th   **c.** 5th   **d.** 95th

91. Find the percentile rank for each of the following usages:
  a. 14.7 million   **c.** 8.8 million
  b. 2.0 million   **d.** 2.1 million

92. Find the following:
  a. Q1   **b.** Q2   **c.** Q3   **d.** IQR

93. Interpret the IQR value from Exercise 92(**d**) so that a nonspecialist could understand it.

94. Suppose an advertising agency is interested in the top 15% of supplements.
  a. What usage does this represent?
  b. Which supplements would be of interest?

95. **Expenditure per Pupil.** The 5th percentile expenditure per pupil nationwide in 2005 was $6,381, the 50th percentile was $8,998, and the 95th percentile was $17,188.[5]
  a. Determine whether the distribution of expenditures is symmetric, left-skewed, or right-skewed.
  b. Would we expect the mean expenditure per pupil to be less than, equal to, or greater than $8,998? Explain.
  c. Draw a distribution curve that matches this information.

For Exercises 96–99, consider whether the scenarios are possible. If it is possible, then clearly describe what the data set would look like. If it is not possible, explain why.

96. A scenario where the first and second quartiles of a data set are equal

97. A scenario where the mean of a data set is larger than Q3

98. A scenario where the median of a data set is smaller than Q1

99. A scenario where the IQR is negative

Twitter Followers. Are you on Twitter? How many Twitter followers do you have? Jon Bruner from O'Reilly Media reported[6] the information in Table 25. For selected percentiles, Table 25 shows the number of Twitter followers that each percentile represents. For example, the 50th percentile is 61 Twitter followers. Use Table 25 for Exercises 100–105. Twitter reports that there are 400 million active Twitter users worldwide who actually tweet (post messages).

100. What percent of Twitter accounts have three or fewer followers?

101. What percent of Twitter accounts have between three and 19 followers?

102. How many active Twitter users have between 2,991 and 24,964 followers?

103. How many active Twitter users have more than 24,964 followers?

104. Is it possible using Table 25 to find what percent of Twitter accounts have 100 or fewer followers? How might we estimate it?

105. What is the percentile rank of 819 Twitter followers?

TABLE 25   Table of percentiles of Twitter followers

Percentile	Number of Twitter followers
10	3
20	9
30	19
40	36
50	**61**
60	98
70	154
80	246
90	458
95	819
99	2,991
99.9	24,964

 twitterpercentile

WORKING WITH LARGE DATA SETS

 Financial Experts versus the Darts. This set of exercises examines measures of relative position using the **Darts** data set from the Chapter 3 Case Study. Open the **Darts** data set. Use technology to do Exercises 106–112. darts

106. Find the median for each of the **Professionals**, the **Darts**, and the **DJIA**. To those who would say that using darts is better, what do the relative values of the medians say?

107. Calculate the z-score for the median for each of the three groups. What does the sign of the z-score for each group indicate about the relationship between the median and the mean?

108. For each group, compute the stock price change represented by the following z-scores.
   a. 2
   b. −2

109. For each group, what percentage of the data lies between the values you found in the previous exercise?

110. For each group, calculate the first quartile and the third quartile.

111. Calculate and interpret the IQR for each group.

112. For each group, compare the IQR with the range and standard deviation. Do all these measures of spread agree regarding which group has the least variability? The most variability?

BRINGING IT ALL TOGETHER

Pedestrian Fatalities. The Department of Transportation releases statistics on the number of pedestrians killed by vehicles in the United States. The following table contains the pedestrian fatality rate (number of fatalities per 100,000 population) for 2013 for six states. Use this information for Exercises 113–120.

pedestrians

State	Pedestrian fatality rate
Nebraska	0.38
Ohio	0.90
Tennessee	1.25
Texas	1.64
California	1.66
Florida	2.57

Source: U.S. Department of Transportation: www-fars.nhtsa .dot.gov/Main/index.aspx.

113. Find the z-scores for the pedestrian fatality rate for the following states:
   a. Ohio   **b.** Texas   **c.** Florida

114. Find the pedestrian fatality rates indicated by the following z-scores:
   a. −2   **b.** 1   **c.** 3

115. Determine whether the pedestrian fatality rates for any of the states represents an outlier.

116. If the pedestrian fatality rate for Nebraska and Florida do not represent outliers, explain why we need not check whether the pedestrian fatality rates for the other states are outliers.

117. Find the following percentiles:
   a. 50th   **b.** 75th   **c.** 25th

118. Calculate the percentile rank for the following pedestrian fatality rates:
   a. 0.38   **b.** 1.25   **c.** 2.57

119. Find the following:
   a. Q1   **b.** Q2   **c.** Q3   **d.** IQR

120. Interpret the IQR value from Exercise 119(**d**).

3.5 Five-Number Summary and Boxplots

OBJECTIVES By the end of this section, I will be able to . . .

1 Calculate the five-number summary of a data set.
2 Construct and interpret a boxplot for a given data set.
3 Detect outliers using the IQR method.

1 The Five-Number Summary

Because the mean and the standard deviation are sensitive to the presence of outliers, data analysts sometimes prefer a less sensitive set of statistics to summarize a data set. The **five-number summary** is an alternative method of summarizing a data set. It includes the median and the quartiles, which are less sensitive to the presence of outliers than are the mean and standard deviation. On the other hand, it also includes the minimum and maximum data values, which are very sensitive to outliers. The five-number summary consists of five measures we have already seen.

> The **five-number summary** consists of the following set of statistics:
> **1.** Minimum; the smallest value in the data set
> **2.** First quartile, Q1
> **3.** Median, Q2
> **4.** Third quartile, Q3
> **5.** Maximum; the largest value in the data set

EXAMPLE 31 The five-number summary for a small data set

Find the five-number summary for the state export data from Table 22, which is repeated here for convenience as Table 26.

Table 26 State export data, in millions of dollars, June 2014

State	VA	NC	NJ	GA	PA	OH	MI	FL	LA	IL	WA	NY
Exports ($ millions)	1.6	2.7	3.3	3.5	3.5	4.6	4.7	4.8	5.0	5.8	7.5	7.7

Solution

From Example 28, we have the quartiles of the export data: Q1 = 3.4, median = Q2 = 4.65, and Q3 = 5.4. From Table 26, the minimum is Virginia's 1.6 and the maximum is New York's 7.7, which are all in millions of dollars. Thus, the five-number summary is:

1. Minimum = 1.6
2. First quartile, Q1 = 3.4
3. Median = Q2 = 4.65
4. Third quartile, Q3 = 5.4
5. Maximum = 7.7

NOW YOU CAN DO
Exercises 7–8, 13–14, and 19–20.

YOUR TURN
#19

Jason is analyzing the movie ratings in the accompanying sample. Find the five-number summary of movie ratings.

| 8.7 | 5.4 | 7.1 | 3.6 | 1.9 | 5.7 | 4.2 | 9.3 | 2.5 |

(The solution is shown in Appendix A.)

EXAMPLE 32   The five-number summary for a large data set: Cholesterol levels in food

nutrition

Find the five-number summary for the cholesterol data from Example 29 on page 164.

Solution

Minitab's reporting of the descriptive statistics makes it particularly straightforward to report the five-number summary, as shown here in Figure 33 (repeated from page 164) for the cholesterol data.

Variable	N	Mean	StDev	Min	Q1	Median	Q3	Max
Cholesterol	961	32.55	119.96	0	0	0	20	2053

FIGURE 33 Descriptive statistics for the cholesterol data.

The five-number summary for the cholesterol data set is:

1. Smallest value in the data set = Min = 0
2. First quartile, Q1 = 0
3. Median = 0
4. Third quartile, Q3 = 20
5. Largest value in the data set = Max = 2053

Or, simply, Min = 0, Q1 = 0, Med = 0, Q3 = 20, Max = 2053.

The five-number summary is associated with a certain type of graphical summary of data, called a *boxplot,* which we examine next.

2 The Boxplot

The **boxplot** (sometimes called a box-and-whisker plot) is a convenient graphical display of the five-number summary of a data set. The boxplot allows the data analyst to evaluate the symmetry or skewness of a data set.

EXAMPLE 33   The characteristics of a boxplot

Interpret the boxplot for the export data in Figure 34.

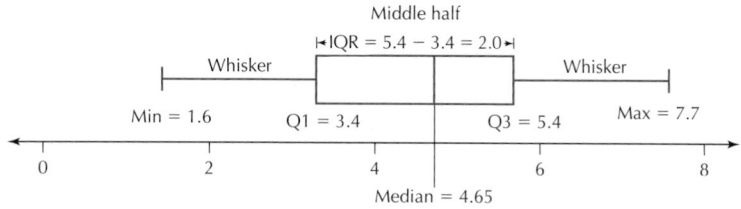

FIGURE 34 Boxplot of the state export data.

Solution

Let's examine this boxplot carefully. The horizontal axis represents the export values. The red box itself represents the middle half of the data set. The left-hand side of the box, called the *lower hinge,* is located at Q1, which is 3.4. The right-hand side of the box, called the *upper hinge,* is located at Q3, which is 5.4. The solid vertical line inside the box is located at the median, which is 4.65. The horizontal lines emanating from the left and right of the box are called the *whiskers.* If no outliers exist, the whiskers extend as far as the maximum and minimum values of the data set, which are represented by the vertical lines at Min = 1.6 and Max = 7.7.

Constructing a Boxplot by Hand

1. The lower and upper fences (represented by brackets in Figure 35b below) represent limits, beyond which data values are considered outliers. Determine the lower and upper fences as follows:
   a. Lower fence = Q1 − 1.5(IQR)
   b. Upper fence = Q3 + 1.5(IQR), where IQR = Q3 − Q1

2. Draw a horizontal number line that encompasses the range of your data, including the fences. Above the number line, draw vertical lines at Q1, the median, and Q3. Connect the lines for Q1 and Q3 to each other so as to form a box.

3. Temporarily indicate the fences as brackets ([and]) above the number line.

4. Draw a horizontal line from Q1 to the smallest data value greater than the lower fence. This is the lower whisker. Draw a horizontal line from Q3 to the largest data value smaller than the upper fence. This is the upper whisker.

5. Indicate any data values smaller than the lower fence or larger than the upper fence using an asterisk (*). These data values are outliers. Remove the temporary brackets.

EXAMPLE 34   Constructing a boxplot by hand

Construct a boxplot by hand for the export data.

Solution

From Example 31, the five-number summary for the state export data is Min =1.6, Q1 = 3.4, Med = 4.65, Q3 = 5.4, Max = 7.7. The interquartile range for the state export data is IQR = Q3 − Q1 = 5.4 − 3.4 = 2.0.

Step 1   Determine the lower and upper fences:

a.   Lower fence = Q1 − 1.5(IQR) = 3.4 − 1.5(2) = 0.4

b.   Upper fence = Q3 + 1.5(IQR) = 5.4 + 1.5(2) = 8.4

Step 2   Draw a horizontal number line that encompasses the range of your data, including the fences. Above the number line, draw vertical lines at Q1 = 3.4, median = 4.65, and Q3 = 5.4. Connect the lines for Q1 and Q3 to each other so as to form a box, as shown in Figure 35a.

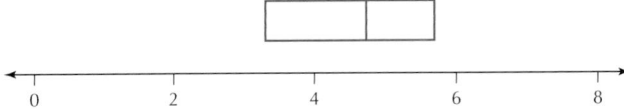

FIGURE 35A Constructing a boxplot by hand: Steps 1 and 2.

Step 3   Temporarily indicate the fences (lower fence = 0.4 and upper fence = 8.4) as brackets above the number line. (See Figure 35b.)

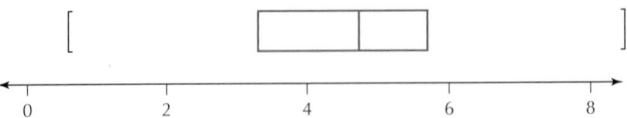

FIGURE 35B Constructing a boxplot by hand: Step 3.

Step 4   Draw a horizontal line from Q1 = 3.4 to the smallest data value greater than the lower fence. The lowest data value is Min = 1.6. This is greater than the lower fence = 0.4, so draw the line from 3.4 to 1.6. Draw a horizontal line from Q3 = 5.4 to the largest data value smaller than the upper fence. The largest data value is Max = 7.7, which is smaller than the upper fence, so draw the line from 5.4 to 7.7. (See Figure 35c.)

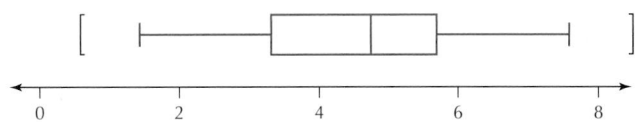

FIGURE 35C Constructing a boxplot by hand: Step 4.

Step 5 No data values are lower than the lower fence or greater than the upper fence. Thus, no outliers exist in this data set. Therefore, simply remove the temporary brackets, and the boxplot is complete, as shown in Figure 35d.

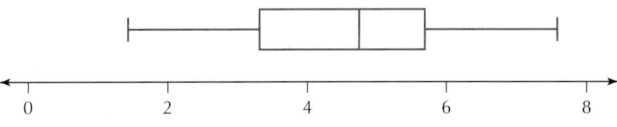

FIGURE 35D The completed boxplot.

NOW YOU CAN DO
Exercises 9–10, 15–16, and 21–22.

YOUR TURN #20

Previously, you found the five-number summary for Jason's movie ratings. Use the five-number summary to construct a boxplot of the data.

| 8.7 | 5.4 | 7.1 | 3.6 | 1.9 | 5.7 | 4.2 | 9.3 | 2.5 |

(The solution is shown in Appendix A.)

The next examples show how to recognize when boxplots indicate that a data set is right-skewed, left-skewed, or symmetric.

EXAMPLE 35 Boxplot for right-skewed data

The population of the 50 U.S. states in 2013 (*Source:* U.S. Census Bureau) is a right-skewed distribution, as shown in the histogram of the data in Figure 36, where the results are shown in millions of people living in the state. The five-number summary is Min = 0.6, Q1 = 1.8, Med = 4.5, Q3 = 7.1, and Max = 37.7. Note that, in the right-skewed boxplot (Figure 37), the upper whisker is much longer than the lower whisker. Also, it is often the case that the median is closer to Q1 than to Q3 in right-skewed data, but that didn't happen with this data.

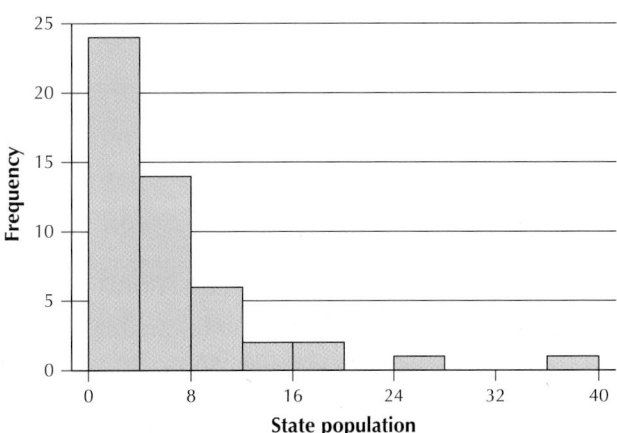

FIGURE 36 State population is right-skewed.

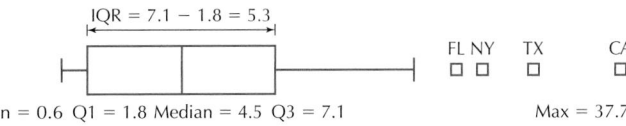

FIGURE 37 TI-83/84 boxplot of state population: right-skewed.

The four little boxes at the right represent outliers. (The TI-83/84 uses little boxes instead of asterisks.) These states are California, Texas, New York, and Florida. When no outliers exist, the whiskers extend as far as the minimum and maximum values. However, when outliers exist, the whiskers extend only as far as the most extreme data value that is not an outlier.

EXAMPLE 36   Boxplot for left-skewed data

Figure 38 is a histogram of 650 exam scores. Clearly, the data are left-skewed, with many students getting scores in the 90s and fewer getting grades in the 70s or 80s.

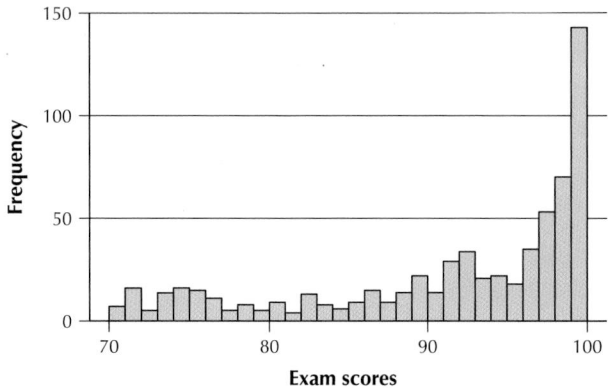

FIGURE 38 Histogram of exam scores.

Solution

The five-number summary is Min = 70, Q1 = 86, Med = 94, Q3 = 98, and Max = 100. So, this time, with left-skewed data, the median is closer to Q3 than to Q1. Bet you guessed it!

In the boxplot (Figure 39), notice that the median (94) is closer to the upper hinge (Q3, 98) than to the lower hinge (Q1, 86), and the lower whisker is much longer than the upper whisker. This combination of characteristics indicates a left-skewed data set.

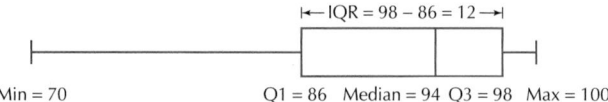

FIGURE 39 TI-83/84 boxplot of the exam scores.

NOW YOU CAN DO
Exercises 25 and 26.

**What Results
Might We Expect?**

Symmetric Data and Boxplots

So, can you now predict how a boxplot of *symmetric* data will look? The median will be about the same distance from Q1 (lower hinge) and Q3 (upper hinge). And the upper and lower whiskers will be about the same length. An example of a boxplot of symmetric data is shown in Figure 40.

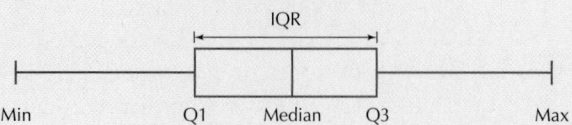

FIGURE 40 Boxplot of symmetric data.

3 Detecting Outliers Using the IQR Method

When using the mean and standard deviation as your summary measures, in most cases, outliers occur more than 3 standard deviations from the mean. However, due to the sensitivity of these measures to the outliers themselves, we often use a more robust method of detecting outliers. Earlier, we mentioned that, when constructing a boxplot, data values lower than the lower fence and higher than the upper fence are considered outliers. We can use this method to detect outliers without constructing a boxplot.

> **IQR Method to Detect Outliers**
>
> A data value is an outlier if
> **a.** it is located 1.5(IQR) or more below Q1, or
> **b.** it is located 1.5(IQR) or more above Q3.

EXAMPLE 37 IQR method for detecting outliers

Table 27 contains the value of exports by the United States to a sample of 12 countries around the world. Determine if there are any outliers in the country export data.

Table 27 U.S. Exports

Country	U.S. exports ($ millions)
Italy	1.2
Saudi Arabia	1.7
India	1.9
France	2.8
Brazil	3.5
South Korea	3.8
United Kingdom	4.4
Germany	4.5
Japan	5.6
China	9.7
Mexico	20.3
Canada	26.3

Solution

The TI *1-Var Stats* analysis provides the five-number summary shown in Figure 41.

Using these statistics, we calculate the IQR to be Q3 − Q1 = 7.65 − 2.35 = 5.3. The quantity 1.5(IQR) = 1.5(5.3) = 7.95. We next find the two quantities Q1 − 1.5(IQR) and Q3 + 1.5(IQR):

$$Q1 - 1.5(IQR) = 2.35 - 7.95 = -5.6$$

$$Q3 + 1.5(IQR) = 7.65 + 7.95 = 15.6$$

Thus, for this data set, a data value would be an outlier if it were −5.6 or less or 15.6 or more. No data values are −5.6 or less. However, both Mexico (20.3) and Canada (26.3) have values greater than 15.6. Therefore, both the $20.3 million in exports to Mexico and the $26.3 in exports to Canada may be considered outliers, using the IQR method.

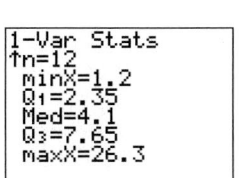

```
1-Var Stats
↑n=12
 minX=1.2
 Q₁=2.35
 Med=4.1
 Q₃=7.65
 maxX=26.3
```

FIGURE 41 Five-number summary.

NOW YOU CAN DO
Exercises 11–12,
17–18, and 23–24.

YOUR TURN
#21

Use the IQR method to determine whether any outliers exist in the movie review data.

| 8.7 | 5.4 | 7.1 | 3.6 | 1.9 | 5.7 | 4.2 | 9.3 | 2.5 |

(The solution is shown in Appendix A.)

The next example shows how comparison boxplots may be used to compare two data sets side-by-side.

EXAMPLE 38 **Comparison boxplots: Comparing body temperatures for women and men**

Determine whether the body temperatures of women or men exhibit greater variability.

Solution

Consider the comparison boxplots in Figure 42. The box for females (on top) lies slightly to the right of that for the males, meaning that the first quartile, the median, and the third quartile are each higher for the women than the men. Therefore, the middle 50% of the body temperatures is higher for women than for men.

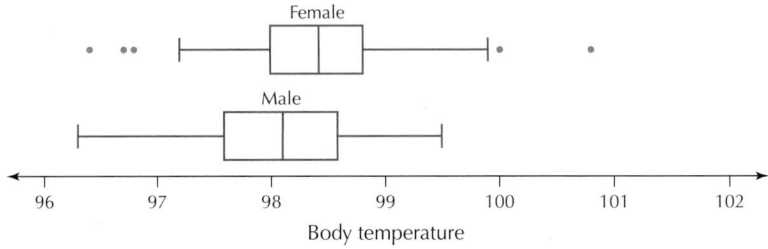

FIGURE 42 Comparison of boxplots of female and male body temperatures.

This figure seems to offer some evidence that the mean body temperature for women may be higher than that for men. The location of the box is an indication of the center of the data, but where would we look for a difference in the variability of body temperatures between women and men? From Figure 43, for the females we have

$$IQR = Q3 - Q1 = 98.8 - 98.0 = 0.8.$$

For the males, we have

$$IQR = Q3 - Q1 = 98.6 - 97.6 = 1.0.$$

Therefore, the IQR for males is greater.

We will formally test whether a difference exists in the true mean body temperature between women and men in Chapter 10.

Let's determine which data set has greater variability based on the three different measures of spread that we have learned: the range, the standard deviation, and the IQR.

Gender	N	Mean	Median	StDev	Min	Max	Q1	Q3
female	65	98.394	98.4	0.743	96.4	100.8	98.0	98.8
male	65	98.105	98.1	0.699	96.3	99.5	97.6	98.6

FIGURE 43 Descriptive statistics for body temperature, by gender.

NOW YOU CAN DO
Exercises 27–30.

Range for women = 100.8 − 96.4 = 4.4   Range for men = 99.5 − 96.3 = 3.2
Standard deviation for women = 0.743   Standard deviation for men = 0.699
IQR for women = 0.8   IQR for men = 1.0

Developing Your
Statistical Sense

When Measures of Spread Disagree

Two measures of spread that are sensitive to the presence of extreme values—range and standard deviation—find that the female body temperatures are more variable. The measure of spread that is resistant to the effects of extreme values—IQR—finds that the male body temperatures are more variable. How do we resolve this apparent inconsistency? What appears to be happening is that, for the middle 50% of each data set, the men are more variable, but as we move toward the tails, the women are more spread out.

Note that outliers exist for the women but not for the men. In part, this may be because the IQR for the women is smaller, and thus the distance 1.5(IQR) is also smaller. For example, the woman whose body temperature is 100 degrees is identified as an outlier because 100 is the same as the outlier threshold Q3 + 1.5(IQR) = 98.8 + 1.5(0.8) = 100. *The same temperature in a man would not be classified as an outlier, even though the male temperatures are lower overall* (and Q3, specifically, is lower). This is because the temperature of 100 is not higher than Q3 + 1.5(IQR) = 98.6 + 1.5(1.0) = 100.1, which is the male outlier threshold. Thus, the measures of spread that are sensitive to outliers indicate that women have greater variability, whereas the measure of spread that is not sensitive to outliers indicates that men have greater variability.

STEP-BY-STEP TECHNOLOGY GUIDE: Boxplots

We will make boxplots for the exports data from Section 3.4, Example 28 on page 163.

TI-83/84
Step 1   Enter the data in list **L1**.
Step 2   Press **2nd Y =**, and choose **1: Plot 1**.
Step 3   Highlight **On** and press **ENTER**. Highlight the boxplot icon, as shown in Figure 44. Press **ENTER**.
Step 4   Press **ZOOM**, and choose **9: ZoomStat**.

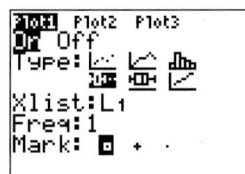

FIGURE 44

MINITAB
Step 1   Enter the data in column **C1**, and name your data *Exports*.
Step 2   Click **Graph > Boxplot....** Select **Simple**, and click **OK**.
Step 3   Double-click on C1 *Exports*, and click **OK**, as shown in Figure 45.

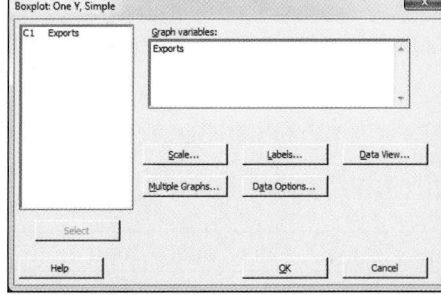

FIGURE 45

SPSS
Step 1   Input the scores into the first column. Name the column *Exports*.
Step 2   Click **Graphs > Chart Builder....** Click **OK**, then **Scan Data**.
Step 3   In the **Gallery** tab, find the **Choose from** menu and select **Boxplot**.

Step 4   Click and drag the **Simple boxplot** to where it says "Drag a Gallery chart here..." Close the Element Properties box.
Step 5   Click and drag *Exports* to where it says "Y-Axis?" in the chart preview.
Step 6   Click **OK**.

JMP

Step 1 Click **File > New > Data Table**. Enter the data into Column 1, and rename the column *Exports*.
Step 2 Click **Graph > Graph Builder**.

Step 3 Drag *Exports* from the **Variables** box to the **Y** axis. Select **Box Plot** from the graph options above the plot. Click **Done**.

CRUNCHIT!

Step 1 Click **File**, highlight **Load from Larose, Discostat3e > Chapter 3**, and click on **Example 05_26**.

Step 2 Click **Graphics**, and select **Box Plot**. For **Data** select *Exports*. Click **Calculate**.

Section 3.5 Summary

1. The five-number summary is an alternative to the usual mean-and-standard-deviation method of summarizing a data set. It consists of simply reporting the minimum, first quartile, median, third quartile, and maximum of the data set.

2. A boxplot is a graphical representation of the five-number summary, and is useful for investigating skewness and the presence of outliers.

3. The IQR method of detecting outliers is to consider a data value an outlier if it is located 1.5(IQR) or more below Q1, or it is located 1.5(IQR) or more above Q3.

Section 3.5 Exercises

CLARIFYING THE CONCEPTS

1. True or false: The five-number summary consists of the minimum, Q1, Mean, Q3, Maximum. (p. 172)
2. Explain what we mean when we say that the five-number summary is associated with the boxplot. (p. 173)
3. Explain how we can use a boxplot to recognize the following:
    a. Symmetric distribution (p. 176)
    b. Right-skewed distribution (p. 175)
    c. Left-skewed distribution (p. 176)
4. When is it possible for outliers to be found inside the box of a boxplot? (p. 177)
5. Explain the IQR method for detecting outliers. (p. 177)
6. Why do we need the IQR method for detecting outliers when we already have the *z*-score method? (p. 177)

PRACTICING THE TECHNIQUES

 **CHECK IT OUT!**

To do	Check out	Topic
Exercises 7–8, 13–14, and 19–20.	Example 31	Five-number summary
Exercises 9–10, 15–16, and 21–22.	Example 34	Boxplots
Exercises 11–12, 17–18, and 23–24.	Example 37	IQR method for identifying outliers
Exercises 25 and 26	Examples 35 and 36	Boxplots and skewness
Exercises 27–30	Example 38	Comparison boxplots

Use the following cell phone price data for Exercises 7–12.

Samsung Galaxy S5 Standard	$200
Samsung Galaxy S5 Active	$200
Sony Xperia Z2	$600
Nokia Lumia Icon	$200
LG G3	$800
Apple iPhone 5s	$250
HTC One M8	$200
Samsung Galaxy Note 3	$300

Source: www.cnet.com/topics/phones/best-phones.

7. Find the quartiles.
8. Compute the five-number summary.
9. Calculate the interquartile range for cell phone price.
10. Construct a boxplot for cell phone price.
11. Use the IQR method to determine whether $200 is an outlier.
12. Use the IQR method to determine whether $600 is an outlier.

The Environmental Protection Agency calculates the estimated annual fuel cost for motor vehicles, with the resulting data provided in the variable *annual fuel cost* of the Chapter 8 Case Study data set **FuelEfficiency**. A sample of the annual fuel cost (in dollars) is provided for 12 vehicles. Use this data to answer Exercises 13–18.

Annual fuel cost (dollars)			
1750	2500	2400	2350
2150	3100	2950	2500
2550	2750	2300	2800

13. Find the quartiles.
14. Compute the five-number summary.

15. Calculate the interquartile range for annual fuel cost.

16. Construct a boxplot for annual fuel cost.

17. Use the IQR method to determine whether $1750 is an outlier.

18. Use the IQR method to determine whether $3100 is an outlier.

Here are the numbers of criminal trespass cases for the police precincts in Brooklyn in 2013. Use this data set to answer Exercises 19–24.

Criminal trespass cases	
150	451
98	111
55	166
41	67
68	258
101	190
32	145
101	49
88	131
55	223
111	48
363	

19. Find the quartiles.

20. Compute the five-number summary.

21. Calculate the interquartile range.

22. Construct a boxplot for the number of criminal trespass cases.

23. Use the IQR method to determine whether 32 criminal trespass cases is an outlier.

24. Use the IQR method to determine whether 451 criminal trespass cases is an outlier.

For Exercises 25 and 26, do the following:
  a. Identify the shape of the distribution.
  b. Use the boxplot to find the five-number summary.

25.

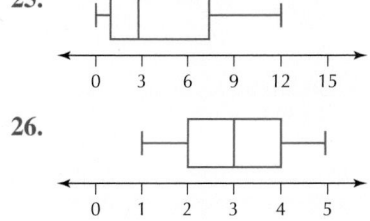

26.

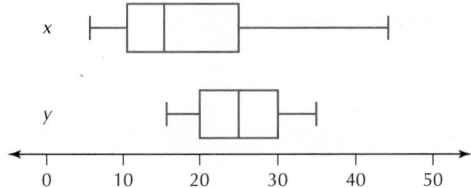

Use the comparison boxplots shown to answer Exercises 27–30.

27. For the variable *x*:
  a. Identify the shape of the distribution.
  b. Use the boxplot to find the five-number summary.

28. For the variable *y*:
  a. Identify the shape of the distribution.
  b. Use the boxplot to find the five-number summary.

29. Which variable has greater variability, according to the IQR?

30. Which variable has greater variability, according to the range?

APPLYING THE CONCEPTS

Most Active Stocks. Use Table 28 for Exercises 31–38. These companies represent the 10 most actively traded stocks on the NASDAQ stock exchange as of 10:00 A.M. on July 11, 2014. The variables are the stock price and the net change in stock price, with both variables in dollars.

nasdaqstock

TABLE 28 The most active stocks on NASDAQ

Company	Price	Change
Facebook	65.28	+0.41
Apple	95.18	+0.15
Cisco Systems	25.28	+0.14
Intel	31.25	−0.01
Fifth Street Finance	9.66	−0.36
QQQQ Trust	94.75	+0.09
Microsoft	41.54	−0.15
Sirius XM	3.38	−0.01
eBay	51.43	+1.09
Yahoo	35.02	+0.09

Source: www.nasdaq.com.

31. Find the five-number summary for *price*.

32. Find the interquartile range for *price*. Interpret what this value means.

33. Use the IQR method to investigate the presence of outliers in *price*.

34. Construct a boxplot for *price*.

35. Find the five-number summary for *change*.

36. Find the interquartile range for *change*. Interpret what this value means.

37. Use the IQR method to investigate the presence of outliers in *change*.

38. Construct a boxplot for *change*.

Dietary Supplements. Refer to Table 24 (page 170) for Exercises 39–44. **dietarysupp**

39. Find the five-number summary for *usage*.

40. Find the interquartile range for *usage*. Interpret what this value actually means, so that a nonspecialist could understand it.

41. Use the IQR method to investigate the presence of outliers in *usage*.

42. Construct a boxplot for *usage*.

43. Calculate the mean and standard deviation of *usage*.

44. Find the *z*-score for echinacea, and use it to determine whether the product is an outlier. Compare the result with that from the IQR method.

BRINGING IT ALL TOGETHER

Honda or Lexus? The following data represent the combined (city and highway) fuel efficiency in miles per gallon for independent random samples of models manufactured by Honda and Lexus. Use this data for Exercises 45–53. ᴴᵀᴴ **hondalexus**

Honda car	mpg	Lexus car	mpg
Accord	24	GX 470	15
Odyssey	18	LS 460	18
Civic Hybrid	42	RX 350	19
Fit	31	IS 350	20
CR-V	23	GS 450	23
Ridgeline	17	IS 250	24
S2000	21		

45. Compute the five-number summary for each of the Honda cars and the Lexus cars.

46. Construct comparison boxplots for the Honda cars and the Lexus cars.

47. Describe the shapes of the distribution for the Honda cars and the Lexus cars.

48. Based on your descriptions in the previous exercise, would you expect the mean to be larger or smaller or about the same as the median for the Honda cars? The Lexus cars?

49. Calculate the mean for the Honda cars and the Lexus cars. Do they concur with your expectations from the previous exercise?

50. Describe the difference between the Honda cars and the Lexus cars, in terms of the location of the box. Which make of vehicle seems to have the greater overall combined mpg? Does this agree with what a comparison of the means from the previous exercise is telling you?

51. Describe the difference of the combined mpg between the Honda cars and the Lexus cars, in terms of the IQR measure of spread.

52. Based on your answer to the previous exercise, which make of car has greater variability?

53. Identify any outliers for the Honda cars and the Lexus cars, using the IQR method.

WORKING WITH LARGE DATA SETS

Nutrition. Use the data set Nutrition for Exercises 54–57.

54. Open the data set **Nutrition**. ᴴᵀᴴ **nutrition**
  a. How many observations are in the data set?
  b. How many variables?

55. Use a statistical computing package (like Minitab) to explore the variable *iron*.
  a. Find the mean and standard deviation for the amount of iron in the food.
  b. Find the five-number summary, the range, and the interquartile range.

56. Which food item has the maximum amount of iron? Does this surprise you?

57. Use the computer to generate a boxplot. Also, comment on the symmetry or the skewness of the boxplot.

WORKING WITH LARGE DATA SETS

Financial Experts versus the Darts. This set of exercises uses the **Darts** data set from the Chapter 3 Case Study to examine the methods and techniques we have learned in this section. Open the **Darts** data set. Use technology to do the following in Exercises 58–63. ᴴᵀᴴ **darts**

58. Find the five-number summary for each group.

59. Construct a comparison boxplot of all three groups. From the boxplot, which group has the greatest variability? The smallest variability?

60. Calculate the range and standard deviation for each group. Does the relative variability of the groups agree with your answer from Exercise 59?

61. For which groups are there no outliers?

62. How many outliers are there for the **Darts**? Verify using the IQR method that these data values are indeed outliers.

63. Check whether the outliers you found in Exercise 62 are also identified as outliers using the *z*-score method.

Chapter 3 Formulas and Vocabulary

SECTION 3.1
- **Mean** (p. 108)
- **Measure of center** (p. 108)
- **Median** (p. 112)
- **Mode** (p. 114)
- **Population mean** (p. 109). $\mu = \Sigma x / N$.
- **Population size** (p. 109). Denoted by N.
- **Sample mean** (p. 109). $\bar{x} = \Sigma x / n$.
- **Sample size** (p. 109). Denoted by n.

SECTION 3.2
- **Chebyshev's Rule** (p. 138). The proportion of values from a data set that will fall within k standard deviations of the mean will be *at least* $\left(1 - \frac{1}{k^2}\right)100\%$, where $k > 1$.
- **Deviation** (p. 128). $x - \bar{x}$.
- **Empirical Rule** (p. 136). If the data distribution is bell-shaped:
    About 68% of the data values will fall within 1 standard deviation of the mean.

About 95% of the data values will fall within 2 standard deviations of the mean.

About 99.7% of the data values will fall within 3 standard deviations of the mean.

- **Measure of variability (measure of spread, measure of dispersion)** (p. 127)
- **Population standard deviation** (p. 132).

$$\sigma = \sqrt{\frac{\Sigma (x - \mu)^2}{N}}$$

- **Population variance** (p. 130).

$$\sigma^2 = \frac{\Sigma (x - \mu)^2}{N}$$

- **Range** (p. 127)
- **Sample standard deviation** (p. 133).

$$s = \sqrt{\frac{\Sigma (x - \bar{x})^2}{n - 1}}$$

- **Sample variance** (p. 133).

$$s^2 = \frac{\sum (x - \bar{x})^2}{n - 1}$$

- **Standard deviation** (p. 128)

SECTION 3.3
- **Estimated mean for data grouped into a frequency distribution** (p. 150).

$$x = \frac{\sum (f \cdot x)}{\sum f}$$

- **Estimated standard deviation for data grouped into a frequency distribution** (p. 151).

$$s = \sqrt{s^2} = \sqrt{\frac{\sum (x - x)^2 \cdot f}{\sum f}}$$

- **Estimated variance for data grouped into a frequency distribution** (p. 151).

$$s^2 = \frac{\sum (x - \bar{x})^2 \cdot f}{\sum f}$$

- **Weighted mean** (p. 149).

$$x = \frac{\sum (w \cdot x)}{\sum w}$$

SECTION 3.4
- **Finding a data value x given its z-score** (p. 157)

Sample: $x = z\text{-score} \cdot s + \bar{x}$

Population: $x = z\text{-score} \cdot \sigma + \mu$

- **Interquartile range (IQR)** (p. 166).

$$IQR = Q3 - Q1$$

- **Outlier** (p. 158)
- **Percentile** (p. 159)
- **Percentile rank** (p. 161)
- **Quartiles** (p. 162)
- **z-Score** (p. 155)
  a. Sample:

$$z\text{-score} = \frac{\text{data value} - \text{mean}}{\text{standard deviation}} = \frac{x - \bar{x}}{s}$$

  b. Population:

$$z\text{-score} = \frac{\text{data value} - \text{mean}}{\text{standard deviation}} = \frac{x - \mu}{\sigma}$$

SECTION 3.5
- **Boxplot** (p. 173)
- **Five-number summary** (p. 172)
- **IQR method of detecting outliers** (p. 177)

Chapter 3 Review Exercises

SECTION 3.1
CDC Funding. The following table contains the funding provided by the Centers for Disease Control (CDC) to all the states in New England, in order to fight HIV/AIDS. Use the data for Exercises 1–3.

CDC funding to fight HIV/AIDS for New England states

State	Funding ($ millions)
Connecticut	7.8
Maine	1.9
Massachusetts	14.9
New Hampshire	1.5
Rhode Island	2.7
Vermont	1.6

Source: Centers for Disease Control and Prevention: www.cdc.gov/nchhstp/stateprofiles/usmap.htm.

1. Find the mean.
2. Calculate the median.
3. Suppose we added California, with $62.1 million in funding, to the data set. Recompute the mean and the median. Which is more affected by the presence of California? What can we say about each of the mean and the median, with respect to extreme values?

Calories in Cereal. For Exercises 4–8, refer to the calories in breakfast cereals given in Table 23 (page 169).
4. Compute the mean.
5. Calculate the median.
6. Find the mode
7. If we eliminated the cereals with 90 or less calories from the sample, which measure would not be affected at all? Why?
8. If we added 10 calories to each cereal, how would that affect the mean, median, and mode? Would it affect each of the measures equally?

SECTION 3.2
CDC Funding to Fight HIV/AIDS. Refer to the CDC funding data above for Exercises 9–14. Omit California.
9. Find the range of the data set.
10. For each state, find its deviation from the population mean.
11. Calculate the average deviation. Would the average deviation be a good measure of spread? Why or why not?
12. Compute the sum of squared deviations. Then divide by the number of states. The result is the population variance, σ^2.
13. Take the square root of the population variance to find the population standard deviation, σ.

14. Interpret the value for the standard deviation.

Calories in Cereal. For Exercises 14–17, refer to the calories in breakfast cereals given in Table 23 (page 169).

15. Calculate the standard deviation of the sample.

16. Suppose we consider the cereals in Table 23 to be representative of all breakfast cereals. Use the mean from Exercise 4 and the standard deviation from Exercise 15, along with Chebyshev's Rule, to find two values between which at least 75% of cereal calories will fall.

17. Refer to the previous exercise. Now further assume the data distribution is bell-shaped. Find two values between which about 95% of cereal calories will fall.

Common Syllables in English. Refer to the table shown here of some common syllables in English for Exercises 18–21. syllables

Syllable	Frequency
an	462
bi	621
sit	104
ed	907
its	293
est	186
wil	470
tiv	136
en	675
biz	114

18. Find the mean and the range of the syllable frequencies.

19. Would you say that a typical distance from the mean for the frequencies is about 900, about 500, about 300, or about 100?

20. What is your best guesstimate of the value of a typical distance from the mean for the syllable frequencies?

21. Find the sample variance and the sample standard deviation of syllable frequencies.

   a. How far is each from your estimate of the typical deviation earlier?

   b. Interpret the meaning of this value for the standard deviation so that someone who has never studied statistics would understand it.

SECTION 3.3
Age Distribution of Twenty-Somethings. The following table shows the number of Americans (in millions) between 20 and 29 years old in 2011. Use this data for Exercises 22–25.

22. Find the estimated mean age of twenty-somethings.

23. Calculate the estimated standard deviation of Americans in their 20s.

24. Use the Empirical Rule to find two age values between which fall about 68% of all American twenty-somethings.

25. Compare your answer in the previous exercise to the actual proportion of twenty-somethings whose ages lie between the values found in the previous exercise. What does this discrepancy mean, regarding the distribution of ages in the table?

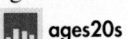 ages20s

Age	Number (millions)
20	4.5
21	4.4
22	4.3
23	4.2
24	4.2
25	4.3
26	4.2
27	4.2
28	4.2
29	4.2

Source: U.S. Census Bureau.

SECTION 3.4
Ragweed Pollen. Use the table of ragweed pollen index in New York localities for Exercises 26–41. Are you allergic to ragweed pollen? You are not alone. The American Academy of Allergy maintains the ragweed pollen index, which details the severity of the pollen problem for hundreds of communities across the nation. The following table contains the ragweed pollen index on a particular day for 10 localities in New York State. ragweed

Locality	Ragweed pollen index
Albany	48
Binghamton	31
Buffalo	59
Elmira	43
Manhattan	25
Rochester	60
Syracuse	25
Tupper Lake	8
Utica	26
Yonkers	38

Find the following percentiles of total ragweed pollen index.

26. 10th percentile

27. 50th percentile

28. 90th percentile

For Exercises 29–31, find the *z*-scores for the following localities for the ragweed pollen index.

29. Albany

30. Rochester

31. Tupper Lake

32. Identify any outliers or moderately unusual observations in the ragweed pollen index.

For Exercises 33–35, find the percentile rank for the given ragweed pollen index.

33. 25

34. 59

35. 48

36. Find the first, second, and third quartiles of the ragweed pollen index.

37. Find the interquartile range. Interpret what this value means.

38. Detect any outliers using the IQR method.

SECTION 3.5

39. Let's draw a boxplot of the ragweed pollen index.
   a. What is the five-number summary?
   b. By hand, draw a boxplot.
   c. Is the data set left-skewed, right-skewed, or symmetric?

   d. What should the symmetry or skewness mean in terms of the relative values of the mean and median?
   e. Find the mean and standard deviation. Is your prediction in (**d**) supported?

40. Detect any outliers using the IQR method. Compare with Exercise 32. Do the two methods concur or disagree?

41. Suppose the ragweed pollen index in Rochester were 600 instead of 60. How would this outlier affect the quartiles and the IQR? What property of these measures is this behavior an example of?

Chapter 3 **QUIZ**

TRUE OR FALSE

1. True or false: If two data sets have the same mean, median, and mode, then the two data sets are identical.

2. True or false: The variance is the square root of the standard deviation.

3. True or false: The Empirical Rule applies for any data set.

FILL IN THE BLANK

4. A(n) _____ is an extremely large or extremely small data value relative to the rest of the data set.

5. The mean can be viewed as the _____ point of the data.

6. The measure of center that is sensitive to the presence of extreme values is the _____.

SHORT ANSWER

7. What do we call summary descriptive measures that are not sensitive to the presence of outliers?

8. Which of the mean, median, and mode may be used for categorical data?

9. For any data set, what is the average of the deviations?

10. What do we use to estimate the mean for each class in a frequency distribution?

CALCULATIONS AND INTERPRETATIONS

11. Calculating a Grade Point Average. At a certain college in Texas, student grade point averages are calculated as follows. For each credit hour, an A is worth 4.0 quality points, an A− is worth 3.7 quality points, a B+ is worth 3.3 quality points, a B is worth 3.0, a B− is worth 2.7, a C+ is worth 2.3, and so on. To find the grade point average, the number of credits for each course is multiplied by the quality points earned for that course; the results are added together; and the sum is divided by the number of credits. This semester, Angelita's grades are as follows. She got an A in her four-credit honors biology course, an A− in her three-credit calculus course, a B+ in her three-credit English course, a B− in her three-credit anthropology course, and a

C+ in her two-credit physical education course. Calculate Angelita's grade point average for this semester.

12. A sample of 30 Americans yielded a sample mean consumption of carbonated beverages this year of 60 gallons, with a sample standard deviation of 40 gallons. Find the z-scores for the following amounts of carbonated beverage consumption.
   a. 120 gallons
   b. 20 gallons
   c. 100 gallons
   d. 0 gallons
   e. 60 gallons

13. Refer to the information in Exercise 12. Assume the distribution is bell-shaped. (*Hint:* Use your knowledge about the Empirical Rule to give a range for the proportions in parts (**b**) and (**d**)).
   a. Find the 50th percentile.
   b. Estimate the proportion of Americans who drink between 20 and 100 gallons per year.
   c. Discuss whether we could find the estimate in (**b**) without assuming that the distribution is bell-shaped.
   d. Estimate the proportion of Americans who drink more than 100 gallons per year.

Use the following SAT 1 Math score for Exercises 14–18.

   510,   515,   523,   514,   521,   501,   502,   499

🔲 satmath

14. Find the following quartiles for SAT 1 Math score:
   a. Q1
   b. Q2
   c. Q3

15. Find the interquartile range of SAT 1 Math score.

16. Find the five-number summary for SAT 1 Math score.

17. Use robust methods to investigate the presence of outliers.

18. Construct a boxplot for SAT 1 Math score.

4 Correlation and Regression

Introduction

In Chapter 4, students learn to examine the relationship between two quantitative variables. Section 4.1 discusses the naturally related topics of scatterplots and correlation. Section 4.2 introduces us to linear regression, while Section 4.3 provides further topics in regression analysis. Both correlation and regression quantify the linear relationship between two quantitative variables. This edition features a new Chapter 4 Case Study, *Measuring the Human Body*, which is used throughout the chapter.

From the Author

We are aware that not every professor prefers to cover correlation and regression at this point in the course. Therefore, we have written Chapter 4 Correlation and Regression in such a way that instructors may omit the chapter without loss of continuity. No topics covered in this chapter are required for any other chapter in the book except Chapter 13, Inference in Regression.

Section 4.1 Scatterplots and Correlation

- Note that we avoid the terminology "independent variable" and "dependent variable" for the x and y variables. This avoids confusion with the notions of independence that we will learn in Chapter 5, "Probability," and Chapter 11, "Further Topics in Inference."

- The temperature data set has been reduced to five observations, for ease of calculation.

- The properties of the correlation coefficient r are now paired with illustrative graphics, for each case.

Section 4.2 Introduction to Regression

- Note that the temperature example provides nice round numbers for the values of the slope and y intercept of the regression line. We thought it more important for students to concentrate on the new concepts instead of worry about extra decimal places. The bonus is that this percolates throughout the remainder of the chapter, as we calculate the estimated value $\hat{y}$ and the prediction error $y - \hat{y}$.

- Ease of interpretation using graphics has been emphasized in this revision throughout Chapter 4.

Section 4.3 Further Topics in Regression Analysis

- Instructors who prefer not to spend too much time on regression early in the course may decide to delay or omit Section 4.3, "Further Topics in Regression Analysis."

- Note that our examples in Section 4.3 also provide nice round numbers for several statistics, including SSE, SST, and SSR.

- We now provide two ways to calculate SST: (a) the usual method summing the $(y - \bar{y})^2$, and (b) a helpful new method, using the fact that the SST is proportional to the variance of the y data. This underscores the method that SST measures the variability in y.

Extra Case Study Example

 body_males

For this example, we investigate the relationship between shoulder girth and chest girth in the Chapter 4 Case Study data set **body_males**. Use the scatterplot of shoulder girth (y) against chest girth (x), and the Minitab output of the regression analysis of shoulder girth versus chest girth, to answer the following questions.

a. In the following scatterplot characterize the relationship as positive, negative, or not apparent.

b. Write a sentence that describes what tends to happen to shoulder girth as chest girth increases.

c. Will the correlation coefficient r be positive, negative, or zero?

d. State the regression equation in words that a friend who didn't take this course would understand. Make sure to use the word "estimated."

e. Interpret the value of the slope of the regression equation so that your friend would understand.

f. Interpret the value of the y-intercept.

g. Use the regression equation to predict the shoulder girth for a male with chest girth of 100 centimeters.

h. State whether the prediction in (**g**) represents extrapolation.

i. Interpret the value of s, the standard error of the estimate.

j. Interpret the value of r^2, the coefficient of determination.

k. Use the value of r^2 to calculate the correlation coefficient r between shoulder girth and chest girth. Does this confirm your answer in (**c**)?

l. Interpret the value of the correlation coefficient r.

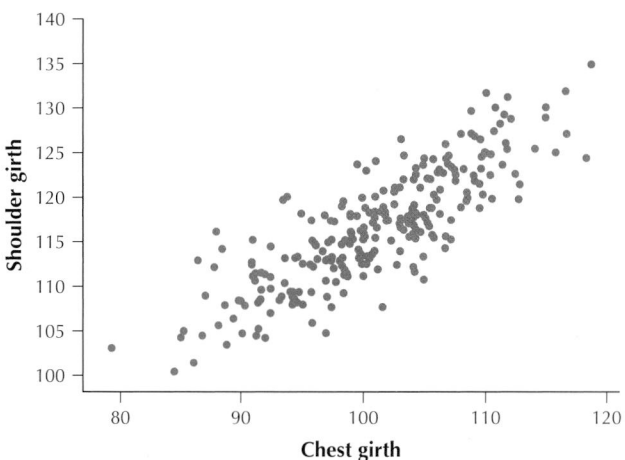

Scatterplot of shoulder girth (y) versus chest girth (x)

Regression Analysis: Shoulder versus Chest girth

Model Summary

S	R-sq	R-sq(adj)	R-sq(pred)
3.58376	69.71%	69.58%	69.19%

Coefficients

Term	Coef	SE Coef	T-Value	P-Value	VIF
Constant	40.50	3.21	12.62	0.000	
Chest girth	0.7526	0.0317	23.74	0.000	1.00

Regression Equation

Shoulder = 40.50 + 0.7526 Chest girth

Output from regression analysis of shoulder girth versus chest girth

Solution

    a. Positive.

    b. As chest girth increases, shoulder girth tends to increase.

    c. Positive.

    d. The estimated shoulder girth equals 40.5 centimeters plus 0.7526 times chest girth.

    e. For each increase of one centimeter in chest girth, the estimated increase in shoulder girth is 0.7526 centimeters.

    f. When chest girth is zero, estimated shoulder girth is 40.5 centimeters.

    g. $\hat{y} = 40.50 + 0.7526(100) = 115.76$. The estimated shoulder girth for a male with a chest girth of 100 centimeters is 115.76 centimeters.

    h. Since $x = 100$ does not lie outside the range of x, the prediction estimate in (**g**) does not represent extrapolation.

    i. The typical error in predicting shoulder girth from chest girth using this regression equation is $s = 3.58376$ centimeters.

    j. $r^2 = 0.6971$. In other words, 69.71% of the variability in shoulder girth is accounted for by the linear relationship between shoulder girth and chest girth.

    k. Since the slope is positive, $r = +\sqrt{r^2} = \sqrt{0.6971} = 0.8349$. Yes, r is positive.

    l. Shoulder girth and chest girth are positively correlated.

Teaching Tips

Stress the idea that, in this chapter, every element has two variable values associated with it. For example, for each student, there is a height and a weight associated with that student. As a result, there must be the same number of x-values as y-values.

In-Class Activities

1. For Sections 4.2 and 4.3, construct a scatterplot and calculate the correlation coefficient and the regression equation for the same data set. One possible example is to use high school GPA (x) and college GPA (y).

 2. For each student in the class, obtain the following information: the distance (in miles) his or her (off-campus) home is from the statistics classroom and the time (in minutes) it takes to commute that distance. If the class is large, take a random sample of the students in the class using either (**a**) technology or (**b**) the *Simple Random Sample* applet.

    a. Discuss which variable is the predictor (x) variable and which variable is the response (y) variable.

    b. Construct a scatterplot of time versus distance.

    c. Is there a positive relationship? Negative relationship? No apparent relationship?

    d. Estimate the value of the correlation coefficient. Will it be positive or negative?

    e. Calculate the correlation coefficient.

    f. Interpret the correlation coefficient.

    g. Next, consider the regression of time on distance. Will the slope be positive or negative?

    h. Using the scatterplot, visually estimate the slope and y intercept of the regression line.

    i. Calculate the actual values of the slope and y intercept.

j. Carefully interpret the meaning of the value for the slope, using words that someone who has not taken statistics would understand.

k. Interpret the meaning of the y intercept.

l. For the person who lives farthest from campus, use the regression equation to estimate the time it takes him or her to get to class.

m. Calculate the prediction error for this student (the difference between the actual commuting time and the estimated commuting time).

n. Can you estimate commuting time for a new student who lives farther away than any other student in the class? If not, clearly explain why not.

Supplements

- StatTutor 4.0–4.5, 5.1–5.4, 5.7, 5.8
- Stats@Work Simulation
  - Picture This, Scatterplots and Correlation; Jan Pepperoni
- EESEE case studies for describing data numerically
  - Visibility of Highway Signs (Question 6(a) on regression)
  - The State of SAT (Questions 6, 7, and 8 on scatterplots, correlation, and regression)
  - Meddling with Olympic Medals (Questions 2, 4, and 6 on scatterplots)
  - Blood Alcohol Content (Questions 1–10 on scatterplots and regression)
  - Is Old Faithful Faithful? (Questions 3 and 4 on scatterplots and regression)

Applets

The *Correlation and Regression* applet is referenced in Chapter 4 to compute correlations and to perform regression analysis. It is used for Exercises 91–93 in Section 4.1, Exercises 87–88 in Section 4.2, and Exercises 76–78 in Section 4.3. Activities and applets that relate to correlation and regression can be found at http://mathforum.org/mathtools/sitemap2/ps/. The site Online Statistics: An Interactive Multimedia Course of Study has the *Prediction* applet that deals with correlation and regression: http://onlinestatbook.com/index.html.

Videos

- *Against All Odds: Inside Statistics:* www.learner.org/resources/series65.html
  - Program 10: Scatterplots
  - Program 11: Fitting Lines to Data
  - Program 12: Correlation

Web Sites

- CAUSEweb provides resources for teaching statistics: https://www.causeweb.org/.
- This site has a collection of 20 class projects, at least one of which deals with the topics of this chapter: www.amstat.org/publications/jse/v6n3/smith.html.
- This Texas Instruments Web site has a host of TI-83/84 statistics activities: http://education.ti.com/educationportal/sites/US/nonProductSingle/activitybook_83_statistics.html.
- This site lists other sites that do statistical calculations: http://statpages.org/.

4 Correlation and Regression

OVERVIEW

Kris Hanke/Vetta/Getty Images

Measuring the Human Body

CASE STUDY

Many more measurements can be made of the human body than height and weight alone. The Chapter 4 Case Study examines 18 measurements of the bodies of 260 females and 247 males who are physically active, with data obtained from a study reported in the *Journal of Statistics Education*.[1] Two large data sets are used: **body_females** and **body_males**.

- In Section 4.1, we visit the case study data in a series of **Your Turn** exercises, where we do the following for the heights and weights of the women in the data set **body_females**:
  - Construct a scatterplot.
  - Characterize the relationship between height and weight.
  - Calculate the correlation coefficient *r* for the height and weight of the women.
- We work with the large data set **body_males** in the Section 4.1 exercises, by repeating the analysis done for the women earlier and comparing the results.
- In Section 4.2, we continue to apply regression techniques to the case study data, in a series of **Your Turn** exercises, where we:
  - Calculate the regression coefficients, and express the regression equation in English.
  - Interpret the values of the regression coefficients.
  - Use the regression equation to make predictions.
  - Calculate the prediction error for our predictions.
  - Determine whether our predictions represent extrapolation.
- In the Section 4.2 exercises, we work with the large data sets **body_females** and **body_males**, performing regression analysis using the techniques learned in the section.
- Finally, in the Section 4.3 exercises, we again work with the large data sets **body_females** and **body_males**, and apply the further regression techniques learned in the section.

THE BIG PICTURE

Where we are coming from and where we are headed . . .

- Chapter 3 showed us methods for summarizing data using descriptive statistics, but only one variable at a time.
- In Chapter 4, we learn how to analyze the relationship between two quantitative variables using scatterplots, correlation, and regression.
- In Chapter 5, we will learn about probability, which we will need to know about in order to perform statistical inference.

4.1 Scatterplots and Correlation

OBJECTIVES By the end of this section, I will be able to . . .

1   Construct and interpret scatterplots for two quantitative variables.
2   Calculate and interpret the correlation coefficient.

So far, most of our work has looked at ways to describe only one quantitative variable at a time. But there may exist a relationship between two quantitative variables (for example, *height* and *weight*) that we want to graph or quantify. We may also want to use the value of one variable, say, *height,* to predict the value of the other variable, *weight.* In Section 4.1, we explore scatterplots, which are graphs of the relationship between two quantitative variables, and we learn about correlation, which quantifies this relationship.

1 Scatterplots

Whenever you are examining the relationship between two quantitative variables, your best bet is to start with a scatterplot. A **scatterplot** is used to summarize the relationship between two quantitative variables that have been measured on the same element. An example of a scatterplot is given in Figure 1.

Note: The predictor variable and response variable are sometimes referred to as the independent variable and dependent variable, respectively. This textbook avoids this terminology because it may be confused with the definition of independent and dependent events and variables in probability (Chapter 5) and categorical data analysis (Chapter 11).

> A **scatterplot** is a graph of points (x, y), each of which represents one observation from the data set. One of the variables is measured along the horizontal axis and is called the *x variable.* The other variable is measured along the vertical axis and is called the *y variable.*

Often, the value of the *x* variable can be used to predict or estimate the value of the *y* variable. For this reason, the *x* variable is referred to as the *predictor* variable, and the *y* variable is called the *response* variable. We also say that the value of the response variable depends on the value of the predictor variable.

EXAMPLE 1   Predictor variables and response variables

For the following pairs of variables, identify which is the predictor (*x*) variable and which is the response (*y*) variable:

a.   The cost of an engagement ring, and the size of the diamond (in carats)

b.   The heights of primary school children, and their ages

Solution

a.   The cost of an engagement ring depends in part on the size of the diamond. We can use the size of the diamond to predict the cost of the ring. Thus, the diamond size is the predictor (*x*) variable, and the cost is the response (*y*) variable.

b.   Because the response variable depends on the predictor variable, and because a child's age depends on nothing but the calendar, then age cannot be the response variable. Age must therefore be the predictor (*x*) variable, with height as the response (*y*) variable.

NOW YOU CAN DO
Exercises 9–12.

YOUR TURN
#1

For the following variables, identify which is the predictor (x) variable and which is the response (y) variable: the number of hours spent studying for an exam, and the grade on the exam.

(The solution is shown in Appendix A.)

EXAMPLE 2 Constructing a scatterplot

 sqrfootsale

Suppose you are interested in moving to Glen Ellyn, Illinois, and want to purchase a lot upon which to build a new house. Table 1 contains a random sample of eight lots for sale in Glen Ellyn, with their square footage and prices.

a. Identify the predictor variable and the response variable.

b. Construct a scatterplot.

Note: The square footage is expressed in 100s of square feet, so that "90" represents 90 × 100 = 9000 square feet. Similarly, the sales price is expressed in $1000s, so that "200"= 200 × 1000 = $200,000.

Table 1 Lot square footage and sales price

Lot	x = square footage (100s of sq. ft.)	y = sales price ($1000s)
Harding St.	75	155
Newton Ave.	125	210
Stacy Ct.	125	290
Eastern Ave.	175	360
Second St.	175	250
Sunnybrook Rd.	225	450
Ahlstrand Rd.	225	530
Eastern Ave.	275	635

Solution

a. It is reasonable to expect that the price of a new lot depends in part on the size of the lot. Thus, we define our predictor variable x to be x = *square footage* and our response variable y to be y = *sales price.*

b. Next, we construct the scatterplot using the data from Table 1. Draw the horizontal axis so that it can contain all the values of the predictor (x) variable, and similarly for the vertical axis. Then, at each data point (x, y), draw a dot. For example, for the Harding Street lot, move along the x axis to 75, then go up until you reach a spot level with y = 155, at which point you draw a dot. Proceed similarly for the other seven properties. The result should look similar to the scatterplot in Figure 1.

FIGURE 1
Scatterplot of sales price versus square footage.

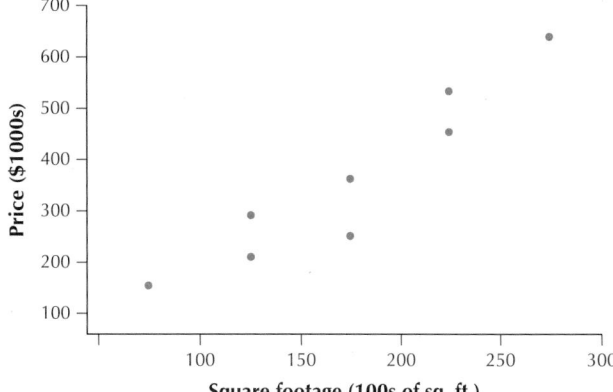

NOW YOU CAN DO
Exercises 13a–20a.

From this scatterplot, we can see that larger lots tend to have higher prices. This is not the case for each observation. For example, the Second Street property is larger than the Stacy Court property, but it has a lower price. Nevertheless, the overall tendency remains.

YOUR TURN
#2

Measuring the Human Body

Table 2 contains the heights in inches and weights in pounds of the first eight women in the **body_females** Case Study data set. Do the following:

a. Identify the predictor variable and the response variable.

b. Construct a scatterplot.

Table 2 Heights and weights of eight women

x = Height (inches)	y = Weight (pounds)
63.5	113.8
65.9	130.1
62.8	108.5
61.8	138.9
61.3	118.2
66.9	130.1
62.6	104.9
65.4	153.9

(The solutions are shown in Appendix A.)

Developing Your
Statistical Sense

Scatterplot Terminology

Note the terminology in the caption to Figure 1. When describing a scatterplot, always indicate the y variable first, and then use the term *versus* (*vs.*) or *against* the x variable. This terminology reinforces the notion that the y variable depends on the x variable.

The relationship between two quantitative variables can take many different forms. We illustrate four of the most common relationships.

Note the phrase, "as x increases in value . . .". When interpreting scatterplots, we always move from left to right.

Positive linear relationship. Figure 2 shows a positive linear relationship between x = height and y = weight of 25 middle-school children.

- Smaller values of height (x) are associated with smaller values of weight (y).
- Larger values of height (x) are associated with larger values of weight (y).
- As height (x) increases, weight (y) also tends to increase.

Negative linear relationship. Figure 3 illustrates a negative linear relationship between x = age and y = cost of 25 used cars.

- Smaller values of age (x) are associated with larger values of cost (y).
- Larger values of age (x) are associated with smaller values of cost (y).
- As x increases, y tends to decrease.

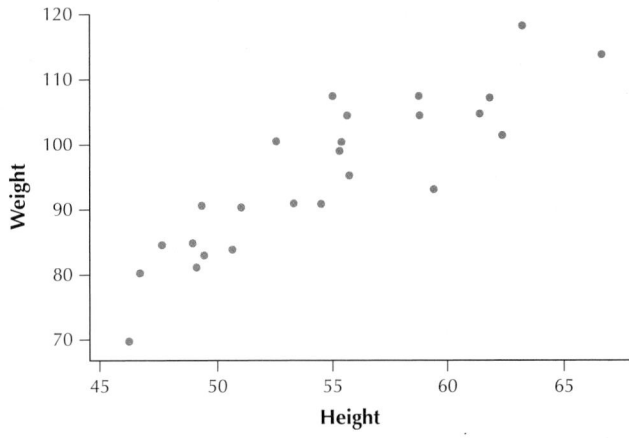

FIGURE 2 Height and weight have a positive linear relationship.

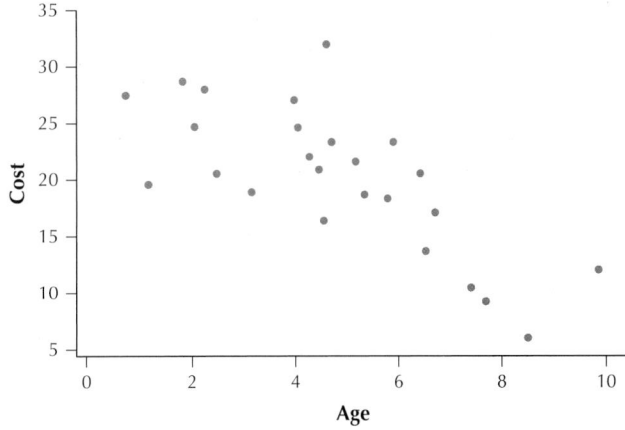

FIGURE 3 Age of used cars and cost have a negative linear relationship.

No apparent relationship. Figure 4 shows that no apparent relationship exists between the height of people who purchase used cars (x) and the cost of the used car (y).

- Smaller values and larger values of height (x) are associated with essentially similar values for vehicle cost (y).
- As x increases, y tends to remain unchanged.

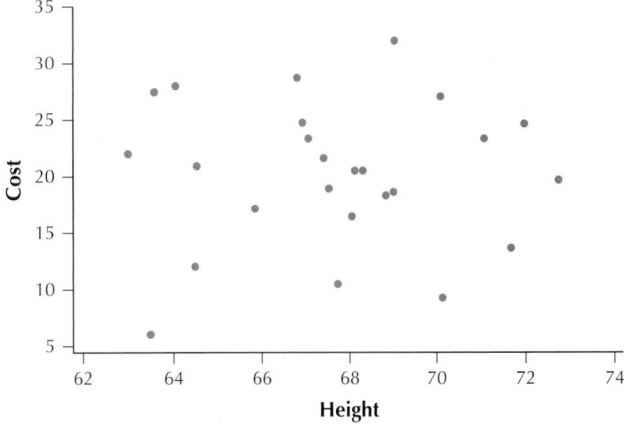

FIGURE 4 Height of car purchasers and car cost have no apparent relationship.

Nonlinear relationship. Figure 5 illustrates an example of a nonlinear relationship. When there is either too little salad dressing (x), or too much salad dressing, the tastiness (y) of a salad can be lower than when a moderate amount of salad dressing is used. Thus, in this case, as the salad dressing increases, at first the tastiness also tends to increase, but then it tends to decrease as too much salad dressing is applied. This is only one example of many different types of nonlinear relationships.

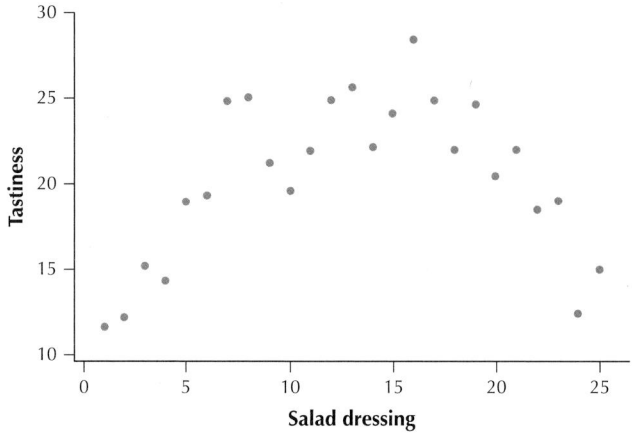

FIGURE 5 The amount of salad dressing and the tastiness of salad have a nonlinear relationship.

EXAMPLE 3 Characterize the relationship between two variables using a scatterplot

Using Figure 1 on page 189, characterize the relationship between lot square footage and lot price.

Solution

NOW YOU CAN DO
Exercises 13b–20b
and 21–26.

The scatterplot in Figure 1 most resembles Figure 2 on page 191, where a positive linear relationship exists between the variables. Thus, smaller lot sizes tend to be associated with lower prices, and larger lot sizes tend to be associated with higher prices. Put another way, as the lot size increases, the lot price also tends to increase.

YOUR TURN
#3

Measuring the Human Body
Characterize the relationship between height and weight, using the scatterplot you constructed from Table 2.

(The solution is shown in Appendix A.)

2 Correlation Coefficient *r*

Scatterplots provide a visual description of the relationship between two quantitative variables. The *correlation coefficient* is a numerical measure for quantifying the linear relationship between two quantitative variables. Table 3 contains the low and high temperatures in degrees Fahrenheit (°F) for five American cities on a particular day. The variables are $x = low\ temperature$ and $y = high\ temperature$. Applying what we have just learned, we construct a scatterplot of the data set, which is presented in Figure 6.

Figure 6 shows us that a positive relationship exists between the high temperature and the low temperature of a city. That is, colder low temperatures are associated with colder high temperatures. Warmer low temperatures are associated with warmer high temperatures. In this section, we seek to *quantify this relationship* between two numerical variables, using the **correlation coefficient *r***. The correlation coefficient *r* (sometimes

Table 3 Low and high temperatures, in degrees Fahrenheit, of five American cities

City	x = low temperature	y = high temperature
Boston	30	50
Chicago	35	55
Philadelphia	40	70
Washington, DC	45	65
Dallas	50	80

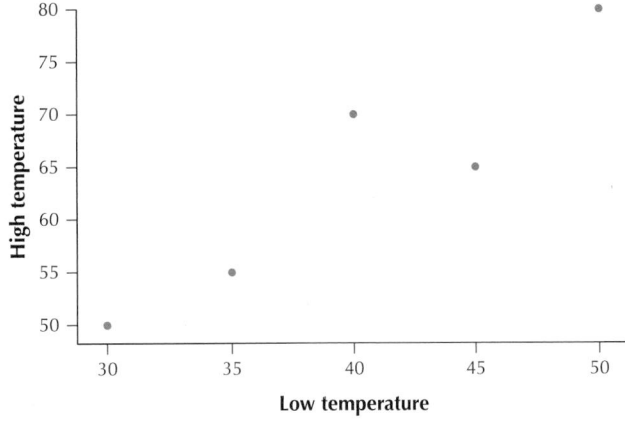

FIGURE 6 Scatterplot of high versus low temperatures for five American cities.

known as the *Pearson product moment correlation coefficient*) measures the strength and direction of the linear relationship between two variables. By *linear,* we mean *straight line.* The correlation coefficient does not measure the strength of a curved relationship between two variables.

> The **correlation coefficient r** measures the strength and direction of the linear relationship between two variables. The correlation coefficient r is
>
> $$ r = \frac{\sum (x - \bar{x})(y - \bar{y})}{(n - 1)\, s_x s_y} $$
>
> where s_x is the sample standard deviation of the x data values, and s_y is the sample standard deviation of the y data values.

EXAMPLE 4 Calculating the correlation coefficient *r*

highlowtemp

Find the value of the correlation coefficient *r* for the temperature data in Table 3.

Solution

We will outline the steps used in calculating the value of *r* using the temperature data.

Step 1 Calculate the respective sample means, $\bar{x}$ and $\bar{y}$.

$$ \bar{x} = \frac{\sum x}{n} = 40, \qquad \bar{y} = \frac{\sum y}{n} = 64 $$

Step 2 Construct a table, as shown here in Table 4.

Table 4 Calculation table for the correlation coefficient r

City	x	y	$(x - \bar{x})$	$(x - \bar{x})^2$	$(y - \bar{y})$	$(y - \bar{y})^2$	$(x - \bar{x})(y - \bar{y})$
Boston	30	50	−10	100	−14	196	140
Chicago	35	55	−5	25	−9	81	45
Philadelphia	40	70	0	0	6	36	0
Washington, DC	45	65	5	25	1	1	5
Dallas	50	80	10	100	16	256	160
				$\sum(x - \bar{x})^2 = 250$		$\sum(y - \bar{y})^2 = 570$	$\sum(x - \bar{x})(y - \bar{y}) = 350$

Note on Rounding: Whenever you calculate a quantity that will be needed for later calculations, do not round. Round only when you arrive at the final answer. Here, because the quantities s_x and s_y are used to calculate the correlation coefficient r, neither of them is rounded until the end of the calculation.

Step 3 Calculate the respective sample standard deviations s_x and s_y. Using the sums calculated from Table 4, we have

$$s_x = \sqrt{\frac{\sum(x - \bar{x})^2}{n - 1}} = \sqrt{\frac{250}{5 - 1}} \approx 7.90569415 \quad \text{and}$$

$$s_y = \sqrt{\frac{\sum(y - \bar{y})^2}{n - 1}} = \sqrt{\frac{570}{5 - 1}} \approx 11.93733639$$

Step 4 Put these values all together in the formula for the correlation coefficient r:

$$r = \frac{\sum(x - \bar{x})(y - \bar{y})}{(n - 1)s_x s_y} = \frac{350}{(4)(7.90569415)(11.93733639)} \approx 0.92717265 \approx 0.9272$$

NOW YOU CAN DO
Exercises 13c–20c.

The correlation coefficient r for the high and low temperatures is 0.9272.

YOUR TURN #4

Measuring the Human Body
Use Steps 1–4 to calculate the correlation coefficient r between height and weight for the data in Table 2.

(The solution is shown in Appendix A.)

What Does This Formula Mean?

The Correlation Coefficient r

Let's analyze the definition formula for the correlation coefficient r. When would r be positive, and when would it be negative? We see that the formula

$$r = \frac{\sum(x - \bar{x})(y - \bar{y})}{(n - 1)s_x s_y}$$

consists of a ratio.

- Note that the denominator can never be negative because it is the product of three non-negative values (standard deviations can never be negative). Therefore, the numerator determines whether r will be positive or negative.

- We know that $x - \bar{x}$ is positive whenever the data value x is greater than $\bar{x}$, and it is negative when x is less than $\bar{x}$. This relationship is similar for $y - \bar{y}$.

Four cases (or regions, which are illustrated in Figure 7) describe when the product $(x - \bar{x})(y - \bar{y})$ will be positive or negative, as shown in Table 5.

Table 5 When the product $(x - \bar{x})(y - \bar{y})$ will be positive or negative

Region	$(x - \bar{x})$	$(y - \bar{y})$	$(x - \bar{x})(y - \bar{y})$
1	Positive	Positive	Positive
2	Negative	Positive	Negative
3	Negative	Negative	Positive
4	Positive	Negative	Negative

- If most of the data values fall in Regions 1 and 3, then r will tend to be positive.

- If most of the data values fall in Regions 2 and 4, then r will tend to be negative.

Let's explore how our high and low temperature data fit into the above framework. The mean low temperature is $\bar{x} = 40°F$, whereas the mean high temperature is $\bar{y} = 64°F$. We find the point $(\bar{x}, \bar{y}) = (40, 64)$ in our scatterplot of the high and low temperatures, draw the lines $x = \bar{x} = 40$ and $y = \bar{y} = 64$, and mark out our four regions, as shown in Figure 7. Note that four of the five data points fall in Regions 1 and 3, with the fifth falling exactly on a boundary line. Therefore, we expect the value of r for this data set to be positive, which is indeed the case, because we observed $r = 0.9272$ in Example 4.

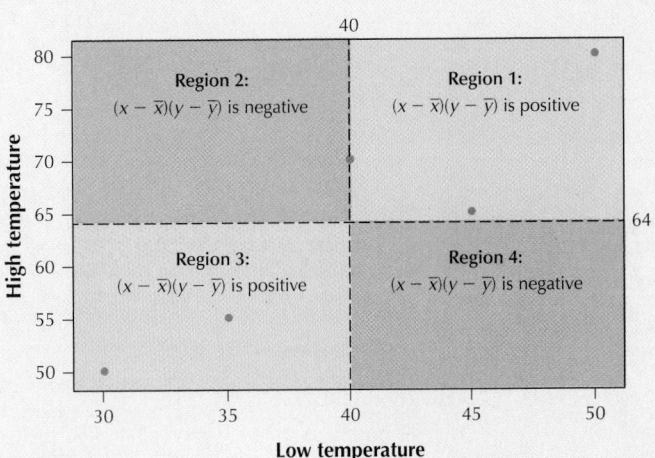

FIGURE 7 Nearly all of the temperature data points lie in Regions 1 and 3, making r positive.

Next, we outline the properties of the correlation coefficient r.

Properties of the correlation coefficient r

1. The correlation coefficient r always takes on values between -1 and 1, inclusive.

   That is, $-1 \leq r \leq 1$.

2. When $r = +1$, a perfect positive relationship exists between x and y. Figure 8 illustrates the perfect positive relationship between $x =$ number of hours worked at a part-time job, and $y =$ the income from that job at $15 per hour.

CAUTION If your calculations give you a value of r outside this range, try it again.

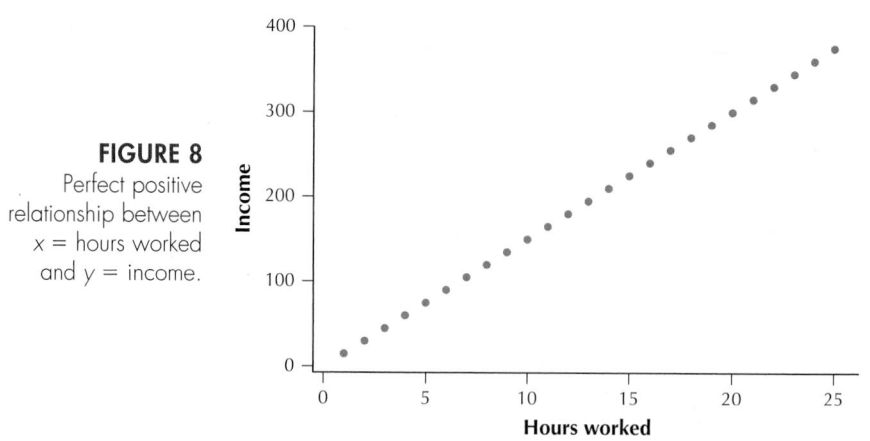

FIGURE 8
Perfect positive relationship between x = hours worked and y = income.

3. Positive values of r indicate a positive relationship between x and y (Figures 9 and 10):

   - The closer r gets to $+1$, the stronger the evidence for a positive relationship.
   - The variables are said to be **positively correlated**.
   - As x increases, y tends to increase.

Figure 9 repeats Figure 1, the scatterplot of sales price versus square footage, for which $r = 0.943$. Figure 10 repeats Figure 2, the scatterplot of height and weight of middle school children, for which $r = 0.597$.

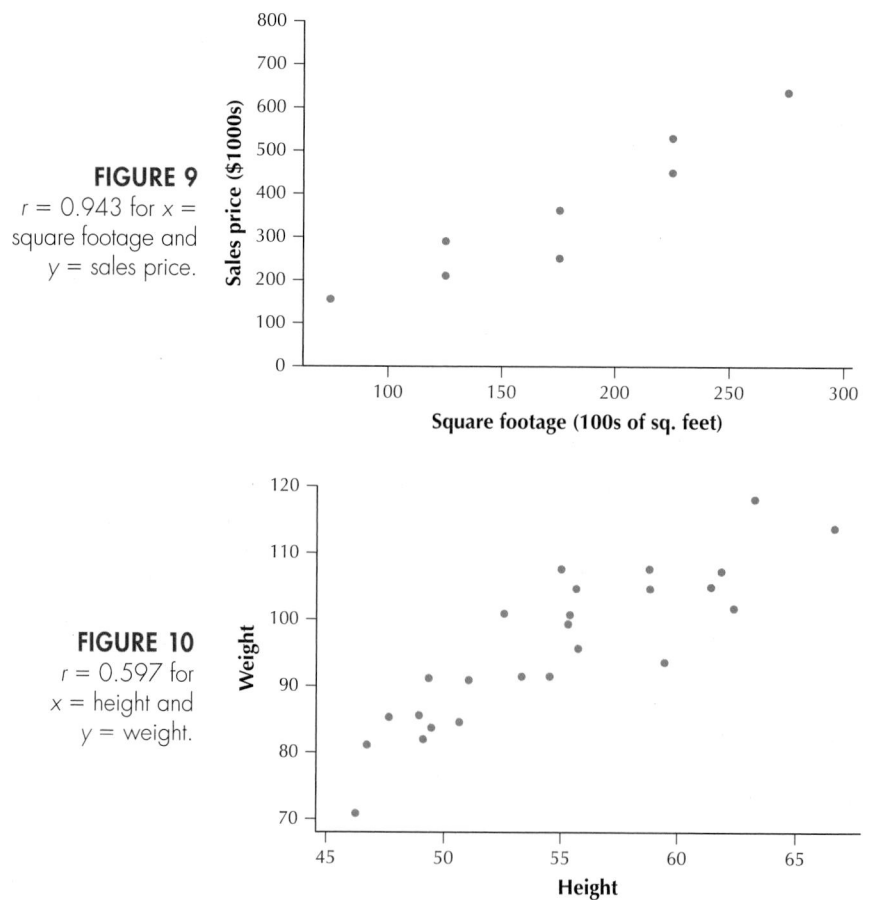

FIGURE 9
$r = 0.943$ for x = square footage and y = sales price.

FIGURE 10
$r = 0.597$ for x = height and y = weight.

4. When $r = -1$, a perfect negative relationship exists between x and y. Figure 11 illustrates the perfect negative relationship between x = the number of $100 ATM withdrawals from a bank account, and y = the account balance.

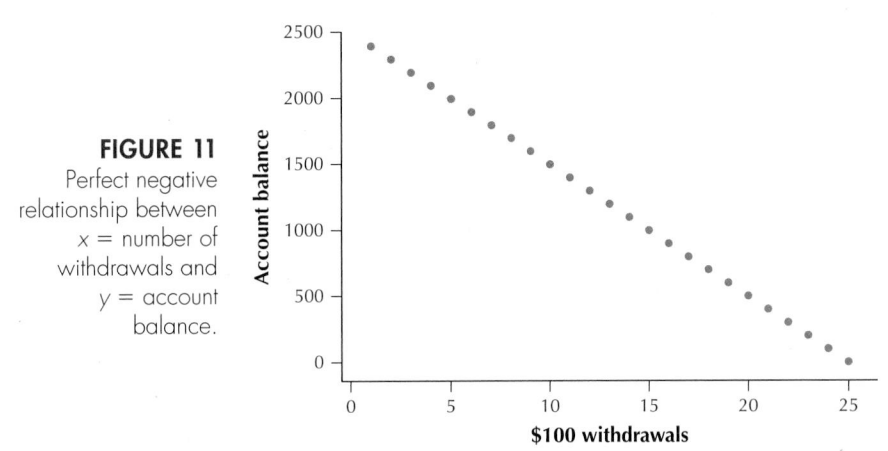

FIGURE 11
Perfect negative relationship between x = number of withdrawals and y = account balance.

5. Negative values of r indicate a negative relationship between x and y (Figures 12 and 13):

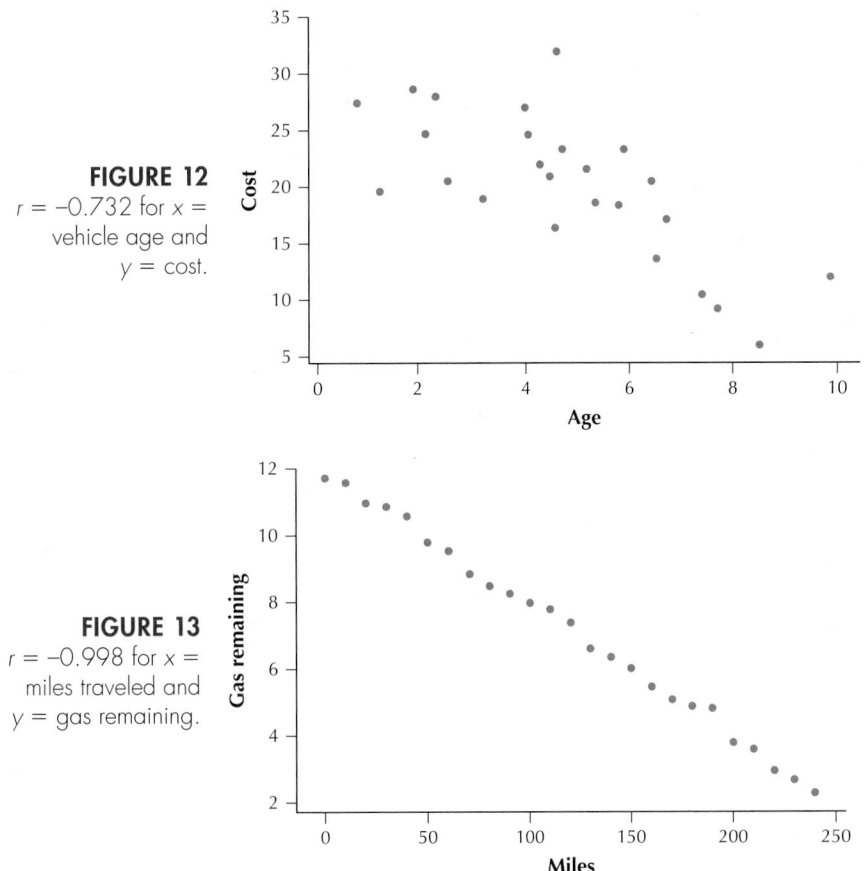

FIGURE 12
$r = -0.732$ for x = vehicle age and y = cost.

FIGURE 13
$r = -0.998$ for x = miles traveled and y = gas remaining.

- The closer r gets to -1, the stronger the evidence for a negative relationship.
- The variables are said to be **negatively correlated**.
- As x increases, y tends to decrease.

Figure 12 repeats Figure 3, the scatterplot of the cost of used cars versus their age, for which $r = -0.732$. Figure 13 shows a scatterplot of x = number of miles traveled on a tank of gas, and y = number of gallons of gas remaining, for a trip combining city and highway travel. The correlation is $r = -0.998$.

6. Values of r near 0 indicate that no linear relationship exists between x and y (Figure 14):
- The closer r gets to 0, the weaker the evidence for a linear relationship.
- The variables are **not linearly correlated**.
- A *nonlinear* relationship may exist between x and y.

Figure 14 repeats Figure 4, the scatterplot of x = the heights of car purchasers and y = the vehicle price, for which $r = 0.023$.

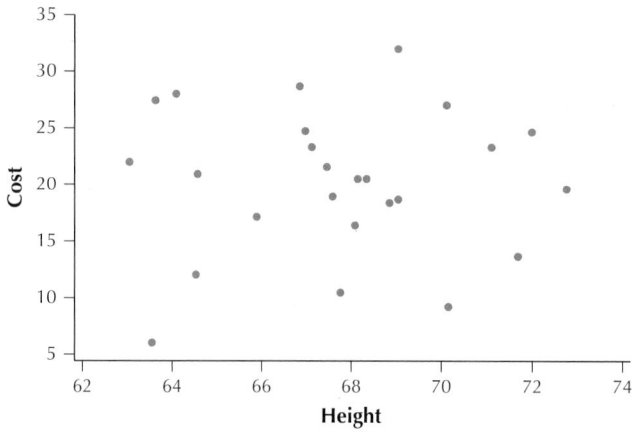

FIGURE 14 $r = 0.023$ for x = height of car purchaser, and y = cost of vehicle.

EXAMPLE 5   Interpreting the correlation coefficient

Interpret the value of the correlation coefficient found in Example 4.

Solution

In Example 4, we found the correlation coefficient for the relationship between high and low temperatures to be $r = 0.9272$. This value of r is very close to the maximum value $r = 1$. We would therefore say that high and low temperatures for these five American cities are strongly positively correlated. As low temperature increases, high temperature also tends to increase.

NOW YOU CAN DO
Exercises 13d–20d.

YOUR TURN
#5

Measuring the Human Body

Interpret the value of the correlation coefficient you found for the data in Table 2.

(The solution is shown in Appendix A.)

Developing Your
Statistical Sense

Note: The Correlation and Regression applet allows you to insert your own data values and see how the regression line changes.

Correlation Is Not Causation

If we conclude that two variables are correlated, it does not necessarily follow that one variable *causes* the other to occur. For example, in the late 1940s, before the development of a vaccine for the disease polio, analysts noticed a strong correlation between the amounts of ice cream consumed nationwide and higher levels of the onset of polio. Some doctors went on to recommend eliminating ice cream as a way to fight polio. But did ice cream really *cause* polio? No. Ice cream consumption and polio outbreaks both peaked in the hot summer months, and so were *correlated* seasonally. Ice cream did not cause polio. After the development of the polio vaccine by Jonas Salk in the 1950s, the disease disappeared from most countries in the world.

STEP-BY-STEP TECHNOLOGY GUIDE: Scatterplots and Correlation  sqrfootsale

TI-83/84

Constructing a Scatterplot for Data in Table 1 (p. 189)

Step 1   Enter the *x* variable (square footage) into **L1** and the *y* variable (sales price) into **L2**.

Step 2   Press **2nd**, then *Y* = for the **STAT PLOTS** menu.

Step 3   Select **1:** and press **ENTER**. Select **ON**, and press **ENTER**.

Step 4   Select the scatterplots icon (see Figure 15), and press **ENTER**.

Step 5   Select **L1** for **Xlist**, and **L2** for **Ylist**.

Step 6   Press **ZOOM**, choose **9:ZoomStat**, and press **ENTER**. The scatterplot is shown in **Figure 16**.

Correlation Coefficient *r*

Step 1   Turn on the diagnostics as follows:. Press **2nd 0** (catalog). Then scroll down and select **DiagnosticOn**. Press **ENTER**

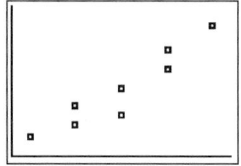

FIGURE 15   **FIGURE 16**

twice to turn the diagnostics on. This will give you more output results for regression and correlation.

Step 2   Enter your *x* data into **L1** and your *y* data into **L2**.

Step 3   Press **STAT**, select **CALC**, and select **LinReg** (ax+b). Press **ENTER** twice to get the results. The correlation coefficient *r* is given, among other statistics used in regression.

EXCEL

Scatterplots

Step 1   Enter your *x* variable and your *y* variable in two neighboring columns, with the *x* variable on the left. Make sure the first entry in each column is the variable name. Select the two columns.

Step 2   Click **Insert > Insert Scatter (X,Y) or Bubble Chart** (in **Chart** section), then click **Scatter**. See Figure 17.

Correlation Coefficient *r*

Step 1   Make sure the Data Analysis add-in is activated. Click **Data > Data Analysis**, then select **Correlation** and click **OK**.

Step 2   Click on the box next to **Input Range**, then highlight the data, select **Labels in First Row**, and click **OK**.

	A	B	C
1	Lot Location	Square Footage	Sales Price
2	Harding St.	75	155
3	Newton Ave.	125	210
4	Stacy Ct.	125	290
5	Eastern Ave.	175	360
6	Second St.	175	250
7	Sunnybrook Rd.	225	450
8	Ahlstrand Rd.	225	530
9	Eastern Ave.	275	635

FIGURE 17 Excel scatterplot.

MINITAB

Scatterplots

Step 1   Enter the data into two columns.

Step 2   Click **Graph > Scatterplot….** Select **Simple** and click **OK**.

Step 3   Click on the cell under **Y variables**, and double-click on your *y* variable; then click on the cell under **X variables**, and double-click on your *x* variable. Then click **OK**.

Correlation Coefficient *r*

Step 1   Enter your *x* data into column **C1** and your *y* data into column **C2**.

Step 2   Click on **Stat**, highlight **Basic Statistics**, and select **Correlation…**

Step 3   Choose C1 and C2 and click **OK**.

SPSS

Scatterplots

Step 1   Enter the data into two columns.

Step 2   **Graphs > Chart Builder**. Click **OK**, then **Scan Data**.

Step 3   In the *Gallery* tab, find the **Choose from** menu and select **Scatter/Dot**. Drag **Simple scatter** to the chart preview area, and close the **Element Properties** box.

Step 4   Drag the *x* variable to "X-Axis?" and the *y* variable to "Y-Axis?" Click **OK**.

Correlation Coefficient *r*

Step 1   Enter the data into two columns.

Step 2   Click on **Analyze > Correlate > Bivariate….**

Step 3   Move the variables to the Variables box, and click **OK**.

JMP

Scatterplots

Step 1 Click **File > New > Data Table**. Enter the data into two columns.
Step 2 **Graph > Chart Builder**. Drag the *x* variable to the X box, and the *y* variable to the Y box. De-select the **Smoother** option above the graph. Click **Done**.

Correlation Coefficient *r*

Step 1 Click **File > New > Data Table**. Enter the data into two columns.
Step 2 Click **Analyze > Multivariate Methods > Multivariate**. Move both variables into the **Y, Columns** area. Click **OK**. The results are displayed in the Correlations section.

CRUNCHIT!

We will use the data from Example 2 (p. 189).

Scatterplots

Step 1 Click **File**, highlight **Load from Larose, Discostat3e > Chapter 4**, and click on **Example 01_02**.
Step 2 Click **Graphics** and select **Scatterplot**. For **X**, select the predictor (*x*) variable *Square footage (100s of sq.ft.)*. For **Y**, select the response (*y*) variable *Sales price ($1000s)*. In the **Show** menu under **Parameters**, choose **Points**. Click **Calculate**.

Correlation Coefficient *r*

Step 1 Click **File**, highlight **Load from Larose, Discostat3e > Chapter 4**, and click on **Example 01_02**.
Step 2 Click **Statistics** and select **Correlation**.
Step 3 Click the boxes next to *Square footage (100s of sq.ft.)* and *Sales price ($1000s)*. Then click **Calculate**.

Section 4.1 Summary

1. For two quantitative variables, scatterplots summarize the relationship by plotting all the (*x*, *y*) points.
2. The correlation coefficient *r* is a measure of the strength of linear association between two numeric variables. Values of *r* close to 1 indicate that the variables are positively correlated. Values of *r* close to −1 indicate that the variables are negatively correlated. Values of *r* close to 0 indicate that the variables are not linearly correlated.

Section 4.1 Exercises

CLARIFYING THE CONCEPTS

1. When investigating the relationship between two quantitative variables, what graph should you use first? (p. 188)
2. In your own words, explain what the correlation coefficient measures. What is the symbol that we use for the correlation coefficient? (p. 192)
3. What is the range of values the correlation coefficient can take? (p. 195)
4. What do the following values of *r* indicate about the relationship between two variables? What can we say about the variables?
    a. A value of *r* close to 1 (p. 196)
    b. A value of *r* close to −1 (p. 197)
    c. A value of *r* close to 0 (p. 197)
5. Why do we call *x* the predictor variable? (p. 188)
6. Suppose two quantitative variables have a positive relationship. What can we say about the values of the *y* variable as the *x* variable increases? (p. 190)
7. Suppose two quantitative variables have a negative relationship. What can we say about the values of the *y* variable as the *x* variable increases? (p. 190)

8. Suppose that the correlation coefficient *r* equals 0. Does this mean that *x* and *y* have no relationship? Explain. (pp. 192, 197)

PRACTICING THE TECHNIQUES

 **CHECK IT OUT!**

To do	Check out	Topic
Exercises 9–12	Example 1	Predictor variables and response variables
Exercises 13a–20a	Example 2	Constructing a scatterplot
Exercises 13b–20b and 21–26	Example 3	Describing the relationship between *x* and *y*
Exercises 13c–20c and 31–34	Example 4	Calculating the correlation coefficient *r*
Exercises 13d–20d and 27–30	Example 5	Interpreting the correlation coefficient *r*

For Exercises 9–12, two variables are given. Determine which variable is the predictor variable (x) and which is the response variable (y).

9. The heights (in inches) and weights (in pounds) of a sample of five women are recorded.

10. The number of days absent from a class, and the course grade.

11. The cost of a repair job and the number of hours spent on the repair.

12. Attendance at a baseball game (in thousands), and the amount of rainfall at a baseball stadium that day (in inches).

For Exercises 13–20, do the following for the indicated data set:
  a. Construct a scatterplot of the relationship between x and y.
  b. Interpret the scatterplot.
  c. Calculate the correlation coefficient r.
  d. Interpret the value of the correlation coefficient r.

13. The values of x and y are as follows:

x	y
10	2
20	2
30	3
40	4
50	4

14. The values of x and y are given as follows:

x	y
1	10
2	9
3	8
4	8
5	7

15. The predictor variable (x) and the response variable (y) take the following values:

x	y
–5	–5
–5	–15
0	–20
5	–25
5	–30

16. The predictor variable (x) and the response variable (y) take the following values:

x	y
0	10
20	10
40	15
60	20
80	20

17. The heights (in inches) and weights (in pounds) of a sample of five women are recorded:

Height	Weight
66	122
67	133
69	153
68	138
65	125

18. The number of days absent from a class, and the course grade:

Days absent	Course grade
0	95
2	90
4	85
6	70
8	60

19. The cost of a repair job, and the number of hours spent on the repair:

Cost	Hours
120	2
180	3
230	5
350	8
380	10

20. Attendance at a baseball game (in thousands), and the amount of rainfall at a baseball stadium that day (in inches):

Rain	Attendance
0.0	40
0.0	42
0.1	38
0.5	30
1.0	20

For Exercises 21–26, do the following:
  a. Characterize the relationship between x and y.
  b. State what happens to the values of the y variable as the x-values increase.

21.

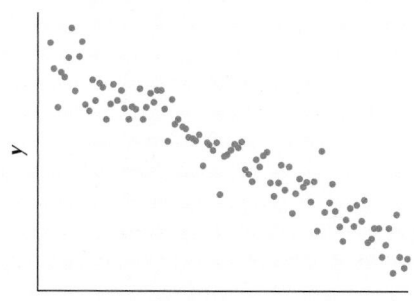

22.

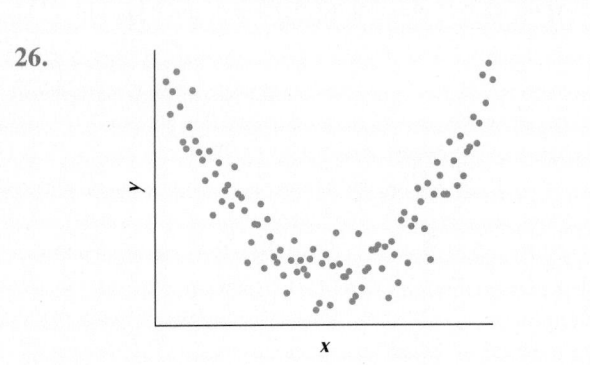

23.

24.

25.

26.

For Exercises 27–30, identify which of the scatterplots in **i–iv** represents the data set with the following correlation coefficients:

i.

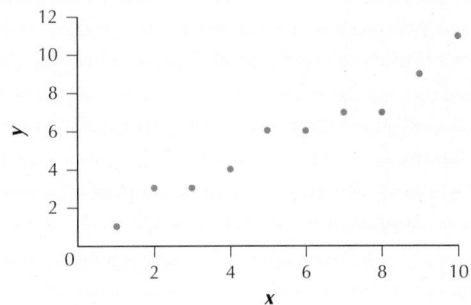

ii.

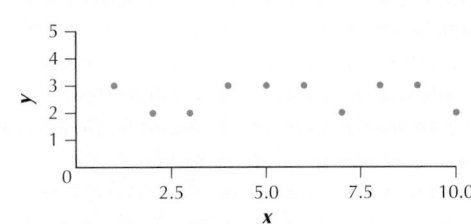

iii.

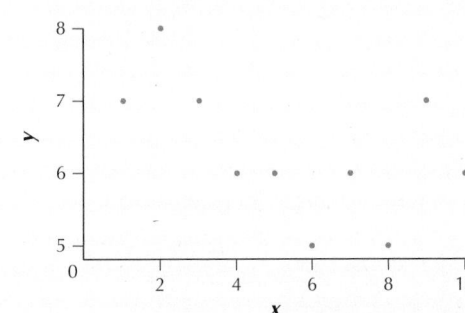

iv.

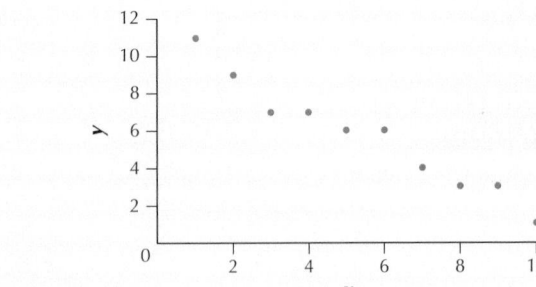

27. Near 1
28. Near zero
29. Near -0.5
30. Near -1

In Exercises 31–34, the values for x and y in each scatterplot are integer-valued. For each scatterplot, (**a**) reconstruct the original data set, and (**b**) calculate the correlation coefficient for the data.

31. The data in scatterplot **i**
32. The data in scatterplot **ii**
33. The data in scatterplot **iii**
34. The data in scatterplot **iv**

APPLYING THE CONCEPTS

For Exercises 35–49, do the following:

    a. Construct a scatterplot of the relationship between *x* and *y*.

    b. Interpret the scatterplot.

    c. Calculate the correlation coefficient *r*.

    d. Interpret the value of the correlation coefficient *r*.

35. Video Game Sales. The Chapter 1 Case Study looked at video game sales for the top 30 video games. The following table contains the weeks on the top 30 list (*x*) and the total sales (*y*, in millions of game units).

 **videogamereg**

Video game	Weeks (*x*)	Total sales in millions of units (*y*)
Super Mario Bros. U for WiiU	78	1.7
NBA 2K14 for PS4	27	0.6
Battlefield 4 for PS3	29	0.9
Titanfall for Xbox One	10	1.2
Yoshi's New Island for 3DS	10	0.2

Source: www.vgchartz.com.

36. Does It Pay to Stay in School? The U.S. Census Bureau reported the following unemployment rates (*y*) associated with the given years of education (*x*).

 **edunemploy**

x = years of education	*y* = unemployment rate
5.0	16.8
7.5	17.1
8.0	15.3
10.0	20.6
12.0	11.7
14.0	8.1
16.0	3.8

37. Darts and the Dow Jones. The following table contains a random sample of eight days from the Chapter 3 Case Study data set, indicating the stock market gain or loss for the portfolio chosen by the random darts (*y*), as well as the Dow Jones Industrial Average (DJIA) gain or loss for that day (*x*).

 **dartsdjia**

Darts (*y*)	DJIA (*x*)
−27.4	−12.8
18.7	9.3
42.2	8.0
−16.3	−8.5
11.2	15.8
28.5	10.6
1.8	11.5
16.9	−5.3

Source: The Wall Street Journal.

 38. Age and Height. The following table provides a random sample from the Chapter 4 Case Study data set **body_females**, showing the age (*x*) and height (*y*) of eight women.

 **ageheight**

Age (*x*)	Height (*y*)
40	63.5
28	63.0
25	64.4
34	63.0
26	63.8
21	68.0
19	61.8
24	69.0

Source: Journal of Statistics Education.

39. Gardasil Shots and Age. The accompanying table shows a random sample of 10 patients from the Chapter 5 Case Study data set, Gardasil, including the age of the patient (*x*) and the number of shots taken by the patient (*y*).

 **gardasilreg**

Age (*x*)	Shots (*y*)
13	3
21	3
16	3
17	2
17	3
18	1
25	2
15	3
12	1
16	1

Source: Journal of Statistics Education.

40. NCAA Power Ratings. The accompanying table shows the top 10 teams' winning percentages (*x*) and power ratings (*y*) for the 2013–2014 NCAA basketball season, according to www.teamrankings.com.  **ncaa2014**

Team	Winning proportion (x)	Power rating (y)
Florida	0.923	121.2
Wichita State	0.971	119.1
Arizona	0.868	118.8
Louisville	0.838	117.9
Connecticut	0.800	117.2
Virginia	0.811	116.8
Wisconsin	0.789	116.6
Villanova	0.853	116.4
Michigan State	0.763	115.9
Michigan	0.757	115.9

41. Saturated Fat and Calories. The table contains the calories and saturated fat in a sample of 10 food items.

satfatcorr

Food item	Calories	Grams of saturated fat
Chocolate bar (1.45 ounces)	216	7.0
Meat & veggie pizza (large slice)	364	5.6
New England clam chowder (1 cup)	149	1.9
Baked chicken drumstick (no skin, medium size)	75	0.6
Curly fries, deep-fried (4 ounces)	276	3.2
Wheat bagel (large)	375	0.3
Chicken curry (1 cup)	146	1.6
Cake doughnut hole (one)	59	0.5
Rye bread (1 slice)	67	0.2
Raisin bran cereal (1 cup)	195	0.3

Source: Food-a-Pedia.

42. Engine Displacement and Gas Mileage. The table provides the engine displacement (size, in liters) and the city mpg (miles per gallon) gas mileage of a random sample of 12 vehicles taken from the Chapter 6 Case Study data set, **FuelEfficiency**.

displacement

Vehicle	Engine displacement	City mpg
GMC Yukon Denali	6.2	13
Ford E350 Wagon	5.4	11
BMW435i Coupe	3.0	20
Land Rover Range Rover	5.0	13
Infiniti Q50a	3.7	19
Dodge Journey	3.6	17
Jaguar XF	5.0	15
Dodge Challenger	6.4	14
Toyota Highlander Hybrid	3.5	28
Mercedes-Benz S550	4.7	17
Ford Fiesta	1.6	29
Hyundai Elantra	2.0	24

43. Completing College. The twenty-first century economy not only needs students to attend college; it needs students to complete their college degrees, in order to compete in the information age. The table contains a sample of 10 states, with data on the percentage of residents who have attended college (x) and the percentage of college attendees who have completed their college degrees (y).

collegecompleters

State	x = college attenders	y = college completers
California	30.9	38.8
Florida	26.6	35.5
Georgia	30.1	34.1
Illinois	35.7	39.1
Massachusetts	45.2	45.9
New York	38.7	42.8
North Carolina	30.1	35.5
Ohio	28.4	37.1
Pennsylvania	32.5	40.2
Texas	26.2	32.2

Source: American Community Survey.

44. Walking or Biking to Work. In these days of high gas prices, it is worth considering alternative methods of commuting. The table contains, for a sample of 10 American cities, the percentage of people who walk to work (x) and the percentage of people who bike to work (y).

walkbike

City	x = walk to work	y = bike to work
Anaheim	1.8	0.9
Baltimore	6.5	0.8
Buffalo	6.2	0.9
Cincinnati	5.4	0.5
Detroit	3.1	0.3
Jacksonville	1.4	0.4
Las Vegas	1.9	0.4
New Orleans	5.1	2.1
Orlando	1.9	0.4
Sacramento	3.2	2.5

Source: U.S. Census Bureau.

45. Teenage Birth Rate. The National Center for Health Statistics publishes data on state birth rates. The table contains the overall birth rate and the teenage birth rate for eight randomly chosen states. The overall birth rate is defined by the NCHS as "live births per 1000 women," and the teenage birth rate is defined as "live births per 1000 women aged 15–19."

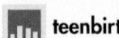 teenbirth

State	x = overall birth rate	y = teen birth rate
California	62.0	23.6
Florida	59.3	24.6
Georgia	61.6	30.5
New York	58.8	17.7
Ohio	62.7	27.2
Pennsylvania	58.4	20.9
Texas	69.9	41.0
Virginia	60.9	20.1

46. Brain and Body Weight. A study compared the body weight (in kilograms) and brain weight (in grams) for a sample of mammals, with the results shown in the following table.[2]  **brainbody**

x = body weight (kg)	y = brain weight (g)
52.16	440.0
60.00	81.0
27.66	115.0
85.00	325.0
36.33	119.5
100.00	157.0
35.00	56.0
62.00	132.0
83.00	98.2
55.50	175.0

47. Consumer Sentiment. Would you expect the consumer sentiment (a measure of how upbeat a consumer feels about his or her personal economic condition) of those with lower incomes to be correlated with that of those with higher incomes, over time? The University of Michigan's Survey of Consumers published the data in the following table, showing the consumer sentiment in 2013 month by month for the two groups. **consumersentiment**

Month	x = consumer sentiment for incomes under $75,000	y = consumer sentiment for incomes $75,000 or higher
Jan	71.6	80.2
Feb	75.7	82.4
Mar	78.3	83.7
Apr	74.5	79.8
May	80.3	94.1
Jun	76.1	98.9
Jul	82.4	90.0
Aug	78.0	89.6
Sep	72.3	86.2
Oct	71.4	77.0
Nov	67.9	88.7
Dec	78.9	88.0

48. SAT Scores, by Foreign Language. The table contains the mean 2014 SAT Critical Reading and Math scores, which are categorized by the foreign language taken in high school or spoken at home.

satlanguages

Language	SAT Critical Reading score	SAT Mathematics score
Chinese	535	606
French	519	525
German	530	540
Greek	526	543
Hebrew	526	541
Italian	497	509
Japanese	521	552
Korean	490	576
Latin	556	556
Russian	483	535
Spanish	498	508

49. Batting Average and Runs Scored. The table shows the top 10 hitters in the American League of Major League Baseball for 2014. We are interested in estimating the number of runs scored (y) using the player's batting average (x).

batters2014

Batter	Team	Runs scored	Batting average
Jose Altuve	Houston Astros	85	0.341
Victor Martinez	Detroit Tigers	87	0.335
Michael Brantley	Cleveland Indians	94	0.327
Adrian Beltre	Texas Rangers	79	0.324
Jose Abreu	Chicago White Sox	80	0.317
Robinson Cano	Seattle Mariners	77	0.314
Miguel Cabrera	Detroit Tigers	101	0.313
Melky Cabrera	Toronto Blue Jays	81	0.301
Adam Eaton	Chicago White Sox	76	0.300
Howie Kendrick	Los Angeles Angels	85	0.293

Correlation in Accounting. A company's current ratio measures its ability to pay its short-term obligations. Use the data in the table, which contains a random sample of large technology companies in 2010, for Exercises 50–56. Total assets and total liabilities are in billions of dollars.

accountingcorr

Company	Current ratio	Price-earnings ratio	Assets	Liabilities
Microsoft	1.82	12.51	77.9	38.3
Intel	2.79	18.44	53.1	11.4
Dell	1.28	10.95	33.7	28.0
Apple	1.88	24.57	53.9	26.0
Google	10.62	18.87	40.5	4.5

Source: Lexis Nexis.

50. Provide and interpret a scatterplot of liabilities versus assets.
51. Calculate and interpret the correlation coefficient between liabilities and assets.
52. Provide and interpret a scatterplot of current ratio versus price-earnings ratio.
53. Calculate and interpret the correlation coefficient between current ratio and price-earnings ratio.
54. Compute a new variable, called *net worth*, which equals *assets – liabilities*.
55. Provide and interpret a scatterplot of net worth versus current ratio.
56. Calculate and interpret the correlation coefficient between net worth and current ratio.

Best Places for Dating. Sperling's Best Places published the list of best places for dating in America. The table shows the top 10 places, along with the overall dating score (y) and a set of predictor variables. Use this information for Exercises 57–59. **bestdating**

City	$y =$ overall dating score	Percentage 18–24 years old	Percentage 18–24 who are single	Online dating score
Austin	100.0	13.4%	81.2%	77.8
Colorado Springs	88.7	10.5%	74.2%	88.9
San Diego	84.0	11.3%	79.4%	77.4
Raleigh	80.7	11.6%	82.9%	79.2
Seattle	78.7	9.0%	83.9%	100.0
Charleston	78.7	11.2%	82.7%	66.9
Norfolk	77.0	11.2%	75.6%	82.9
Ann Arbor	75.5	12.9%	90.3%	51.1
Springfield	75.2	11.7%	89.8%	63.5
Honolulu	75.2	10.1%	82.3%	50.2

57. Construct and interpret a scatterplot for each of the following predictor variables versus the overall dating score:
   a. Percentage 18–24 years old
   b. Percentage 18–24 years old who are single
   c. Online dating score
58. Calculate and interpret the correlation coefficient r between each of the following predictor variables and the overall dating score:

   a. Percentage 18–24 years old
   b. Percentage 18–24 years old who are single
   c. Online dating score
59. Based on your work in Exercises 57 and 58, which predictor variable is the best indicator of the overall dating score, and thus an indicator of the best places for dating?

Virginia Weather. The table contains data on weather in a sample of cities in the state of Virginia. Use this information for Exercises 60–65. **vaweather**

Data on the weather in Virginia

City	Average January temperature	Heating degree-days	Average July temperature	Cooling degree-days
Alexandria	34.9	4055	79.2	1531
Arlington	34.9	4055	79.2	1531
Blacksburg	30.9	5559	71.1	533
Charlottesville	35.5	4103	76.9	1212
Chesapeake	40.1	3368	79.1	1612
Danville	36.6	3970	78.8	1418
Hampton	39.4	3535	78.5	1432
Harrisonburg	30.5	5333	73.5	758
Leesburg	31.5	5031	75.2	911
Lynchburg	34.5	4354	75.1	1075
Manassas	31.7	4925	75.7	1075
Newport News	41.2	3179	80.3	1682
Norfolk	40.1	3368	79.1	1612
Petersburg	39.7	3334	79.6	1619
Portsmouth	40.1	3368	79.1	1612
Richmond	36.4	3919	77.9	1435
Roanoke	35.8	4284	76.2	1134
Suffolk	39.6	3467	78.5	1427
Virginia Beach	40.7	3336	78.8	1482

Source: National Oceanic and Atmospheric Administration.

60. Construct and interpret a scatterplot for the average January temperature versus the cooling degree-days.
61. Based on your scatterplot, what would you expect the sign of the correlation coefficient to be? Why?
62. Calculate the correlation coefficient r between the average January temperature and the cooling degree-days.
63. Build and interpret a scatterplot for the average July temperature versus the heating degree-days.
64. Calculate the correlation coefficient r between the average July temperature and the heating degree-days.
65. Which relationship would you say is stronger: the relationship between the average January temperatures with the cooling degree-days, or the relationship between the average July temperatures and the heating degree-days?

What's Your Major? The table contains the percentages of students majoring in (i) Mathematics/Statistics/Computer Science, (ii) Biological Sciences, and (iii) Psychology, for a sample of 10 states. Use this information for Exercises 66–71. ⊞ **statemajors**

State	Math/Stat/ Comp Sci	Biological Sci	Psychology
Alaska	2.0	11.0	5.5
Connecticut	4.1	5.3	5.7
Hawaii	2.9	6.9	5.2
Idaho	3.7	9.6	3.2
Maryland	5.7	7.1	4.8
Montana	2.7	11.1	4.1
New Jersey	5.3	5.3	4.7
Oregon	3.2	8.4	5.4
Virginia	5.5	5.4	4.7
Wyoming	2.2	12.2	3.6

Source: American Community Survey.

66. Construct a scatterplot of the Math/Stat/Comp Sci majors versus the Psychology majors.

67. Based on your scatterplot, would you say that a relationship exists between the two sets of majors?

68. Calculate the correlation coefficient r between the Math/Stat/Comp Sci majors and the Psychology majors. Was your intuition in Exercise 67 confirmed?

69. Construct a scatterplot of the Biological Science majors versus the Psychology majors.

70. Based on your scatterplot in Exercise 69, would you say that a relationship exists between the two sets of majors?

71. Calculate the correlation coefficient r between the Biological Science majors and the Psychology majors. Was your intuition in Exercise 70 confirmed?

Worldwide Indicators of Well-Being. The Statistics Online Computational Resource (SOCR) provides resources for statistics students and educators. The following table, sampled from data provided by SOCR, includes 10 countries, along with 3 indicators of their well-being: an indicator of economic dynamism, a measure of literacy, and a health indicator. Use this information for Exercises 72–82.

⊞ **wellbeing**

Country	Economy	Literacy	Health
Australia	71.5	91.5	95.2
Canada	68.5	96.7	92.8
Germany	61.9	91.1	92.8
India	50.0	66.2	51.7
Japan	69.0	94.0	100.0
Mexico	42.0	74.7	78.3
Pakistan	41.5	66.9	49.3
South Korea	73.0	96.7	87.9
United Kingdom	72.9	92.8	90.3
United States	77.8	89.4	85.5

72. Some analysts claim that literacy depends on the economy of a country. Provide a scatterplot of literacy (y) versus economy (x). Describe any relationship between the variables.

73. Calculate and interpret the correlation coefficient r for the linear relationship between $x = $ economy and $y = $ literacy.

74. Other analysts state that a country's economy depends on the country's literacy. Build a scatterplot of economy (y) versus literacy (x). Describe any relationship between the variables. Compare this scatterplot with the one in Exercise 72. Discuss similarities and differences.

75. Do you expect that the correlation coefficient between $x = $ economy and $y = $ literacy will be the same as that between $x = $ literacy and $y = $ economy? Calculate and interpret the correlation coefficient r for the linear relationship between $x = $ literacy and $y = $ economy. Was your intuition confirmed?

76. Provide a scatterplot of literacy (y) versus health (x). Describe any relationship between the variables.

77. Calculate and interpret the correlation coefficient r for the linear relationship between $x = $ health and $y = $ literacy.

78. Without calculating it, state what the correlation coefficient r would be for the linear relationship between $x = $ literacy and $y = $ health.

79. Provide a scatterplot of health (y) versus economy (x). Describe any relationship between the variables.

80. Calculate and interpret the correlation coefficient r for the linear relationship between $x = $ economy and $y = $ health.

81. Without calculating it, state what the correlation coefficient r would be for the linear relationship between $x = $ health and $y = $ economy.

82. Based on your work in the previous exercises, state a general rule about the correlation between x and y, and the correlation between y and x.

83. Computational Formula for r. The following computational formula may be used as an equivalent of the definition formula for the correlation coefficient r:

$$r = \frac{\sum xy - \left(\sum x \sum y\right)/n}{(n-1)s_x s_y}$$

Use the computational formula and the TI-83/84 to calculate the correlation coefficient r for the relationship between square footage and sales price of the eight home lots for sale in Glen Ellyn from Example 2 (page 189).

⊞ **sqrfootsale**

BRINGING IT ALL TOGETHER

Lead and Zinc Concentrations in River Fish. Like to go fishing? In some areas, it may not be healthy to eat the fish you catch, due to the pollutants in the river that the fish ingest. Use the information in the table for Exercises 84–88. The table contains the lead and zinc concentrations in river fish from the Spokane River in Washington State, in parts per million.[3]

leadzincfish

Fish	x: lead (ppm)	y: zinc (ppm)
1	0.73	45.3
2	1.14	50.8
3	0.60	40.2
4	1.59	64.0
5	4.34	150.0
6	1.98	106.0
7	3.12	90.8
8	1.80	58.8
9	0.65	35.4
10	0.56	28.4

84. Investigate the relationship.
   a. Construct a scatterplot of the variables. Make sure the y variable goes on the y axis.
   b. What type of relationship do these variables have: positive, negative, or no apparent linear relationship?
   c. Will the correlation coefficient be positive, negative, or near zero?

85. Calculate and interpret the correlation coefficient.
   a. Compute the value of the correlation coefficient.
   b. Does this value for r concur with your judgment in part (**c**) of the previous exercise?
   c. Interpret the meaning of this value of the correlation coefficient.

86. Determine whether we can conclude that x and y are correlated.

87. Transformation. Add 5 to each value for y.
   a. Redraw the scatterplot. Comment on the similarity or difference from the scatterplot in Exercise 84(**a**).
   b. Recalculate the correlation coefficient.
   c. Compare your answers from Exercises 85(**a**) and 87(**b**).
   d. Compose a rule that states the behavior of the correlation coefficient r when a constant is added to each y data value.

WHAT IF 88. Transformation. *What if*, starting with the original data in the table, we added a certain unknown constant amount to each value for x?
   a. *Without redrawing the scatterplot*, describe how this change would affect the scatterplot you drew in Exercise 84(**a**).
   b. *Without recalculating the correlation coefficient*, state what you think the effect of this change would be on the correlation coefficient. Why do you think that?
   c. Compose a rule that states the behavior of the correlation coefficient r when a constant is added to each x data value.

CONSTRUCT YOUR OWN DATA SETS

89. Describe two variables from real life that would have a value of r close to 1. Explain why they are positively correlated.

90. Create a sample of five observations from each of your variables in the previous exercise, and put them into a table similar to Table 4 (page 194). Next, construct a scatterplot of the variables. Finally, draw a single straight line through the data points in the plot in a manner that you think best approximates the relationship between the variables.

Use the *Correlation and Regression* applet for Exercises 91–93.

91. Create a set of $n = 10$ points such that the correlation coefficient r takes approximately the following values. Note that you can drag points up or down to adjust your value of r.
   a. $r = 0.90$
   b. $r = -0.90$
   c. $r = 0.00$

92. Describe the relationship between the variables for each of the sets of points in the previous exercise.

93. Select "Show mean X and mean Y lines." Create a set of $n = 4$ points such that the correlation coefficient r takes approximately the following values. Note that you can drag points up or down to adjust your value of r.
   a. $r = 0.70$
   b. $r = -0.70$
   c. $r = 0.00$

WORKING WITH LARGE DATA SETS

Chapter 4 Case Study: Measuring the Human Body. Open the data sets **body_females** and **body_males.** We shall explore the relationships between height and weight for the women and men in these data sets using the tools and techniques we have learned in this section. Use technology to do the following: **body_females** **body_males**

94. Construct a scatterplot of weight versus height for the men.

95. Characterize the relationship between male height and weight.

96. Find the correlation coefficient between height and weight for the males.

97. Interpret the correlation coefficient for men's height and weight.

98. Discuss similarities and differences among your results for the females (from the **Your Turn** exercises in this section) and the males, as follows:
   a. The scatterplots
   b. The relationships between height and weight
   c. The correlation coefficients

4.2   Introduction to Regression

OBJECTIVES By the end of this section, I will be able to . . .

1   Calculate the value and understand the meaning of the slope and the y intercept of the regression line.

2   Predict values of y for given values of x, and calculate the prediction error for a given prediction.

1 The Regression Line

In Section 4.1 we learned about the *correlation coefficient*. Here, in Section 4.2, we will learn how to approximate the linear relationship between two numerical variables using the regression line and the regression equation. For convenience, we repeat Table 3 here.

Table 3 Low and high temperatures, in degrees Fahrenheit, of five American cities

City	$x = \text{low}$ temperature	$y = \text{high}$ temperature
Boston	30	50
Chicago	35	55
Philadelphia	40	70
Washington, DC	45	65
Dallas	50	80

Consider again Figure 6 (page 193), the scatterplot of the high and low temperatures for five American cities, from Table 3. The data points generally seem to follow a roughly linear path. We may in fact draw a straight line from the lower left to the upper right to approximate this relatively linear path. Such a straight line, called a **regression line**, is shown in Figure 18.

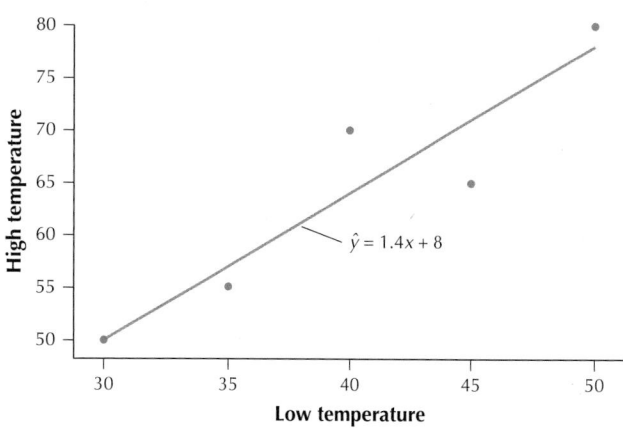

FIGURE 18 Scatterplot of high versus low temperatures, with regression line.

As you may recall from high school algebra, the equation of a straight line may be written as $y = mx + b$. We will write the **equation of the regression line** similarly as $\hat{y} = b_1 x + b_0$.

Note: The "hat" over the *y* (pronounced "*y*-hat") indicates that this is an estimate of *y* and not necessarily an actual value of *y*.

Equation of the Regression Line

The **equation of the regression line** that approximates the relationship between *x* and *y* is

$$\hat{y} = b_1 x + b_0$$

where the *regression coefficients* are the **slope**, b_1, and the *y* **intercept**, b_0. Do not let $\hat{y}$ and $\bar{y}$ be confused. $\hat{y}$ is the predicted value of *y* from the regression equation. $\bar{y}$ represents the mean of the *y*-values in the data set. The equations of these coefficients are

$$b_1 = r \cdot \frac{s_y}{s_x} \qquad b_0 = \bar{y} - (b_1 \cdot \bar{x})$$

where s_x and s_y represent the sample standard deviation for the *x* and *y* data, respectively.

An infinite number of different straight lines could approximate the relationship between high and low temperatures. Why did we choose this one? Because this is the *least-squares* regression line, which is the most widely used linear approximation for bivariate relationships. We will learn more about least squares in Section 4.3.

EXAMPLE 6 Calculating the regression coefficients b_0 and b_1

a. Find the value of the regression coefficients b_0 and b_1 for the temperature data in Table 3.

b. Write out the equation of the regression line for the temperature data.

c. Clearly explain the meaning of the regression equation.

Solution

a. We will outline the steps used in calculating the value of b_1 using the temperature data.

Step 1 Calculate the respective sample means $\bar{x}$ and $\bar{y}$. We have already done this in Example 4 (page 193): $\bar{x} = 40$ and $\bar{y} = 64$.

Step 2 Calculate the respective sample standard deviations s_x and s_y. We have already done this in Example 4: $s_x \approx 7.90569415$ and $s_y \approx 11.93733639$.

Step 3 Find the correlation coefficient *r*. This was computed in Example 4: $r \approx 0.92717265$.

Step 4 Combine the statistics from Steps 2 and 3 to calculate b_1:

$$b_1 = r \cdot \frac{s_y}{s_x} = 0.92717265 \cdot \frac{11.93733639}{7.90569415} = 1.4$$

Step 5 Use the statistics from Steps 1–4 to calculate b_0:

$$b_0 = \bar{y} - (b_1 \cdot \bar{x}) = 64 - (1.4)(40) = 8$$

b. Thus, the equation of the regression line for the temperature data is

$$\hat{y} = 1.4x + 8$$

NOW YOU CAN DO
Exercises 7–12 and
13a–24a.

c. Because y and x represent high and low temperatures, respectively, this regression equation is read as follows: "The estimated high temperature for an American city is 1.4 times the low temperature for that city plus 8 degrees Fahrenheit."

**YOUR TURN
#6**

Measuring the Human Body

a. Find the value of the regression coefficients b_0 and b_1 for the height and weight data from Table 2 (page 190).

b. Write out the equation of the regression line for the height and weight data.

c. Clearly explain the meaning of the regression equation.

(The solutions are shown in Appendix A.)

EXAMPLE 7   Interpreting the slope and the y intercept

Interpret the following values of the regression line we obtained in Example 6:

a. The y intercept $b_0 = 8$

b. The slope $b_1 = 1.4$

Solution

a. In statistics, we interpret the slope of the regression line as the *estimated change in* y *per unit increase in* x. In our temperature example, the units are degrees Fahrenheit, so we interpret our value $b_1 = 1.4$ as follows:

"For each increase of 1°F in low temperature, the estimated high temperature increases by 1.4°F."

b. The y intercept is interpreted as the *estimated value of* y *when* x *equals zero*. Here, we interpret our value $b_0 = 8$ as follows:

"When the low temperature is 0°F, the estimated high temperature is 8°F."

NOW YOU CAN DO
Exercises 13b–24b.

**YOUR TURN
#7**

Measuring the Human Body

Interpret the following values of the regression line for the height and weight data from Table 2 (page 190):

a. The y intercept b_0

b. The slope b_1

(The solutions are shown in Appendix A.)

Recall from Section 4.1 that the correlation coefficient for the temperature data is $r = 0.9272$. Is it a coincidence that both the slope and the correlation coefficient are positive? Not at all.

This relationship holds because
$b_1 = r \cdot \dfrac{s_y}{s_x}$ and neither s_y nor s_x can be negative.

Relationship Between Slope and Correlation Coefficient

The slope b_1 of the regression line and the correlation coefficient r always have the same sign.

- b_1 is positive if and only if r is positive.
- b_1 is negative if and only if r is negative.

Thus, when we found in Section 4.1 that the correlation coefficient between high and low temperatures was positive, we could have immediately concluded that the slope of the regression line was also positive

Other ways to describe regression include:

- "Perform a regression of the y variable versus the x variable."
- "Regress the y variable on the x variable."

Note that the first variable is always the y variable and the second variable is always the x variable. For example, in Example 7 we could write, "Perform a regression of high temperature against low temperature."

EXAMPLE 8   Correlation and regression using technology

Use technology to find the correlation coefficient r and the regression coefficients b_1 and b_0 for the temperature data in Table 3 (page 193).

Solution

The instructions for using technology for correlation and regression are provided in the Step-by-Step Technology Guide at the end of this section (page 217). The TI-83/84 scatterplot is shown in Figure 19, and the TI-83/84 results are shown in Figure 20. (Note that the TI-83/84 indicates the slope b_1 as a, and the y intercept b_0 as b.) Figures 21 and 22 show the Excel results, with the y intercept ("Intercept") and the slope ("Low") highlighted. Figures 23 and 24 show excerpts from the Minitab results, with the y intercept ("Constant") and the slope ("Low") highlighted. Figures 25 and 26 show excerpts from the SPSS results.

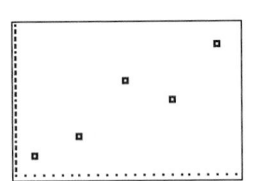

FIGURE 19 TI-83/84 scatterplot.

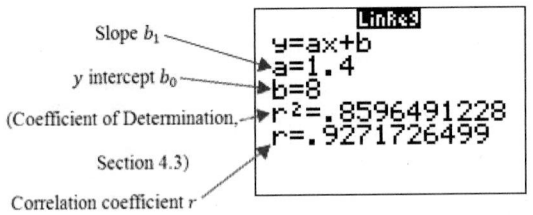

FIGURE 20 TI-83/84 correlation and regression results.

	Low	High
Low	1	
High	0.927173	1

FIGURE 21 Excel correlation results.

	Coefficients	Standard Error	t Stat	P-value
Intercept	8	13.26649916	0.603022689	0.589031761
Low	1.4	0.326598632	4.28660705	0.023333162

FIGURE 22 Excel regression results.

Correlation: Low, High

Pearson correlation of Low and High = 0.927
P-Value = 0.023

FIGURE 23 Minitab correlation results.

Coefficients

Term	Coef	SE Coef	T-Value	P-Value	VIF
Constant	8.0	13.3	0.60	0.589	
Low	1.400	0.327	4.29	0.023	1.00

FIGURE 24 Minitab regression results.

Correlations

		High	Low
Pearson Correlation	High	1.000	.927
	Low	.927	1.000
Sig. (1-tailed)	High	.	.012
	Low	.012	.
N	High	5	5
	Low	5	5

FIGURE 25 SPSS correlation results.

Coefficients^a

Model		Unstandardized Coefficients		Standardized Coefficients	t	Sig.
		B	Std. Error	Beta		
1	(Constant)	8.000	13.266		.603	.589
	Low	1.400	.327	.927	4.287	.023
a. Dependent Variable: High						

FIGURE 26 SPSS regression results.

2 Predictions and Prediction Error

We can use the regression equation to make estimates or predictions. For any particular value of x, the predicted value for y lies on the regression line.

EXAMPLE 9   Using the regression equation to make a prediction

Suppose we are moving to a city that has a low temperature of 40°F on this particular day. Use the regression equation in Example 6 to find the predicted high temperature for this city.

Solution

To generate an estimate of the high temperature, we plug in the value of 40°F for the x variable *low*:

$$\hat{y} = 1.4(low) + 8 = 1.4(40) + 8 = 64$$

NOW YOU CAN DO
Exercises 25a–36a.

We would say, "The estimated high temperature for an American city with a low temperature of 40°F is 64°F."

YOUR TURN
#8

CASE STUDY

Measuring the Human Body
Use the regression equation you generated for the height and weight data in Table 2 to make a prediction of the weight for a female who is 63.5 inches tall.

(The solution is shown in Appendix A.)

Developing Your
Statistical Sense

Actual Data versus Predicted (Estimated) Data

The city of Philadelphia in Table 3 has a low temperature of 40°F. The actual high temperature for Philadelphia is $y = 70°F$, but our predicted high temperature is $\hat{y} = 64°F$. Note the important difference in concept between y and $\hat{y}$. The actual high temperature in Philadelphia, $y = 70°F$, is an established fact: real, observed data. On the other hand, our prediction, $\hat{y} = 64°F$, is just an estimate based on a formula, the regression equation.

Prediction Error

Our prediction for Philadelphia's high temperature was too low by

$$y - \hat{y} = 70 - 64 = 6°F$$

The difference $y - \hat{y}$ is the vertical difference from the Philadelphia data point (x, y) to the regression line. This difference is called the *prediction error*.

Prediction errors are also called estimation errors or residuals.

The **prediction error** or **residual** $(y - \hat{y})$ measures how far the predicted value of $\hat{y}$ is from the actual value of y observed in the data set. The prediction error may be positive or negative.

- Positive prediction error: The data value lies *above* the regression line, so the observed value of y is greater than predicted for the given value of x.
- Negative prediction error: The data value lies *below* the regression line, so the observed value of y is lower than predicted for the given value of x.
- Prediction error equal to zero: The data value lies *directly* on the regression line, so the observed value of y is exactly equal to what is predicted for the given value of x.

All values of $\hat{y}$ (the predicted values of y) lie on the regression line.

EXAMPLE 10 Calculating and interpreting prediction errors (residuals)

Use the regression equation from Example 9 to calculate and interpret the prediction error (residual) for the following cities.

a. Philadelphia: low = 40, high = 70
b. Washington, DC: low = 45, high = 65

Solution

a. The actual data point for Philadelphia is shown in the scatterplot in Figure 27 (denoted as "Phil"). In Example 9, we calculated the predicted high temperature for Philadelphia to be $\hat{y} = 64°F$. In Figure 27, $\hat{y} = 64°F$ represents the y-value of the point on the regression line where it intersects $x = 40°F$. That is, the actual high temperature $y = 70°F$ lies directly above predicted temperature $\hat{y} = 64°F$ for low temperature $x = 40°F$.

b. The actual high temperature in Washington that day was $y = 65$. Using the regression equation, the predicted high temperature is $\hat{y} = 1.4(45) + 8 = 71$. So the prediction error is $y - \hat{y} = 65 - 71 = -6°F$. The data point lies below the regression line, so that its actual high temperature of 65°F is lower than predicted given its low temperature of 45°F.

Travelif/The Image Bank/Getty Images

Philadelphia, Pennsylvania

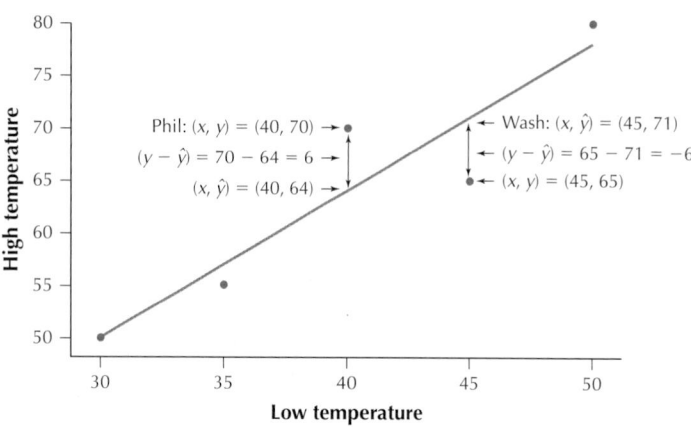

NOW YOU CAN DO
Exercises 25b–36b.

FIGURE 27 Prediction error for Philadelphia and Washington, DC.

 YOUR TURN #9

 Measuring the Human Body

Recall the prediction you made in the previous Your Turn for the weight for a female who is 63.5 inches tall. Calculate the prediction error (residual) for this woman. Does her actual weight lie above or below the regression line?

(The solution is shown in Appendix A.)

Of course, we need not restrict our predictions to values of x (low temperature) that are in our data set (but please see the warning on extrapolation below). For example, the estimated high temperature for a city in which $low = 32°F$ is

$$\hat{y} = 1.4(low) + 8 = 1.4(32) + 8 = 52.8°F$$

Note that we cannot calculate the prediction error for this estimate because we do not have a city with a low temperature of 32°F to compare it to.

Extrapolation

The y intercept b_0 is the estimated value for y when x equals zero. However, in many regression problems, a value of zero for the x variable would not make sense. For example, a lot for sale of $x = 0$ square feet does not make sense, so the y intercept would not be meaningful. On the other hand, a value of zero for the low temperature does make sense. Therefore, we would be tempted to predict $\hat{y} = 1.4(0) + 8 = 8°F$ as the high temperature for a city with a low of zero degrees. However, $low = 0°F$ is not within the range of the data set. Making predictions based on x-values that are beyond the range of the x-values in our data set is called *extrapolation*. It may be misleading and should be avoided.

> **Extrapolation** consists of using the regression equation to make estimates or predictions based on x-values that are outside the range of the x-values in the data set.

Extrapolation should be avoided, if possible, because the relationship between the variables may no longer be linear outside the range of x. Consider Figure 28, which is the scatterplot of the tastiness versus amount of salad dressing we encountered in Section 4.1. Suppose we only had the data values shown in green, which show the tastiness tending to increase as the amount of salad dressing increases. These green data have x-values ranging from $x = 1$ to $x = 16$. The regression equation for these 16 points is $\hat{y} = 0.911x + 12.87$. Now, suppose we made a prediction for $x = 25$, which lies outside the range of the x-values. The predicted tastiness is a tastiness score of $\hat{y} = 0.911(25) + 12.87 = 35.645$. However, the actual tastiness score for that much salad dressing is only 15, giving us a large prediction error of $(y - \hat{y}) = 15 - 35.645 = -20.645$. This large prediction error is due to our *extrapolation*, where we made a prediction for y for a value of x for which we had no data, and which lay outside the range of the x-values.

The further outside the range of x, the more unreliable the prediction becomes.

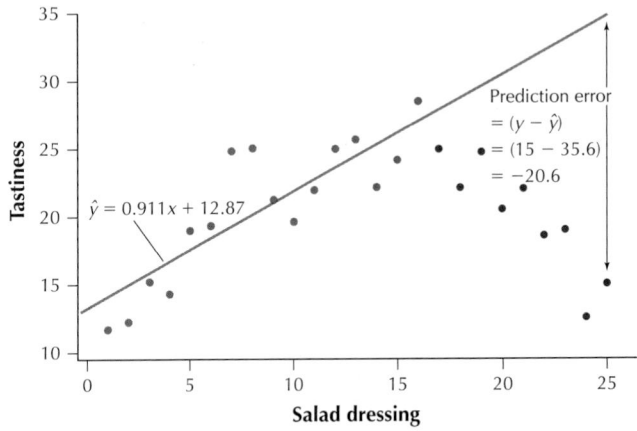

FIGURE 28 Dangers of extrapolation.

EXAMPLE 11   Identifying when extrapolation occurs

Using the regression equation from Example 10, estimate the high temperature for the following low temperatures. If the estimate represents extrapolation, indicate so.

a. 50°F

b. 60°F

Solution

From Table 3, the smallest value of x is 30°F and the largest is 50°F, so estimates for any value of x between 30°F and 50°F, inclusive, would not represent extrapolation.

a. $\hat{y} = 1.4(50) + 8 = 78°F$. Because $x = 50°F$ lies between 30°F and 50°F, inclusive, this estimate does not represent extrapolation.

b. $\hat{y} = 1.4(60) + 8 = 92°F$. Because $x = 60°F$ does not lie between 30°F and 50°F, this estimate represents extrapolation.

NOW YOU CAN DO
Exercises 25c–36c.

YOUR TURN
#10

Measuring the Human Body

Using the regression equation you constructed for the height and weight data from Table 2, estimate the weight for females with the following heights. If the estimate represents extrapolation, indicate so.

a. 65 inches

b. 70 inches

(The solutions are shown in Appendix A.)

 The *Correlation and Regression* applet allows you to insert your own data values and see how the regression line changes.

STEP-BY-STEP TECHNOLOGY GUIDE: Regression

Using Table 3, we illustrate the temperature data (page 193).

TI-83/84
Step 1   Turn diagnostics on as follows: Press **2nd 0**. Scroll down and select **DiagnosticOn** (Figure 29). Press **ENTER** twice to turn diagnostics on.
Step 2   **Enter** the x *(low temp)* data in **L1**, and the y *(high temp)* data in **L2**.
Step 3   Press **STAT** and highlight **CALC**.
Step 4   Select **LinReg(ax+b)**.

Step 5   On the home screen, **LinReg(ax+b)** appears. Press **ENTER**.

FIGURE 29

EXCEL
Step 1   Enter the x variable in column **A** and the y variable in column **B**, with the topmost cells indicating the variable names, *Low* and *High*.

Step 2   Click on **Data > Data Analysis > Regression** and click **OK**.
Step 3   For **Input Y Range**, select cells **B1–B6**. For **Input X Range**, select cells **A1–A6**. Check the **Labels** option, and click **OK**.

MINITAB
Step 1   Enter the x *(low temp)* in **C1** and the y *(high temp)* in **C2**.
Step 2   Click on **Stat > Regression > Regression > Fit Regression Model…**.

Step 3   Select the y variable for the **Responses** and the x variable for the **Continuous predictors**. Click **OK**.

SPSS
Step 1   Enter the data into the first two columns. Name the columns *Low* and *High*.
Step 2   Click **Analyze > Regression > Linear…**.

Step 3   Move *High* to the **Dependent** box, and *Low* to the **Independent(s)** box using the arrows. Click **OK**.
Step 4   The **Coefficients** table has the regression coefficients.

JMP
Step 1   Click **File > New > Data Table**. Input the x data into **Column 1** and the y data **Column 2**. Name Column 1 *Low Temp*, and Column 2 *High Temp*.
Step 2   Click **Analyze > Fit Y by X**.

Step 3   Click *Low Temp* under **Select Columns**, then click **X, Factor**. Click *High Temp* under **Select Columns**, then click **Y, Response**. Click **OK**. A scatterplot appears.
Step 4   Click the red triangle next to "Bivariate Fit of High Temp by Low Temp," and select **Fit Line**. Regression output is in the Parameter Estimates table.

CRUNCHIT!
Step 1   Click **File**, highlight **Load from Larose, Discostat3e > Chapter 4**, and click on **Example 02_03**.
Step 2   Click **Statistics**, highlight **Regression**, and select **Simple linear**.

Step 3   For **Dependent Variable**, select *High Temp*. For **Independent Variable**, select *Low Temp*.
Step 4   For **Display**, make sure **Numeric Results** is selected. Then click **Calculate**.

Section 4.2 Summary

1. Section 4.2 introduces regression, where the linear relationship between two numerical variables is approximated using a straight line, called the regression line. The equation of the regression line is written as $\hat{y} = b_1 x + b_0$,

where the regression coefficients are the y intercept, b_0, and the slope, b_1.

2. The regression equation can be used to make predictions about values of y for particular values of x.

Section 4.2 Exercises

CLARIFYING THE CONCEPTS

1. What is the objective of regression analysis? (p. 209)
2. What is the regression equation? (p. 210)
3. Describe how we use the regression equation to make predictions. (p. 213)

4. Explain the difference between y and $\hat{y}$. (p. 214)
5. Describe what is meant by extrapolation. (p. 215)
6. What is the relationship between the slope of the regression line and the correlation coefficient? (p. 210)

PRACTICING THE TECHNIQUES

✅ **CHECK IT OUT!**

To do	Check out	Topic
Exercises 7–12, and 13a–24a	Example 6	Calculating the regression coefficients.
Exercises 13b–24b	Example 7	Interpreting slope and y intercept
Exercises 25a–36a	Example 9	Making predictions
Exercises 25b–36b	Example 10	Calculating and interpreting prediction errors
Exercises 25c–36c	Example 11	Extrapolation

Exercises 7–12 refer to scatterplots in the Section 4.1 exercises. For each indicated scatterplot, state whether the slope b_1 of the regression line would be positive, negative, or near zero.

7. Exercise 21
8. Exercise 22
9. Exercise 23
10. Exercise 24
11. Exercise 25
12. Exercise 26

For Exercises 13–24, do the following:
   a. Calculate the slope b_1 and the y intercept b_0 of the regression line. Write the regression equation.
   b. Interpret the values for b_1 and b_0.

13.

x	10	20	30	40
y	2	5	9	12

14.

x	0	2	4	6
y	50	60	50	40

15.

x	−5	−4	−3	−2	−1
y	10	18	18	26	26

16.

x	−3	−1	1	3	5
y	−25	−35	−40	−45	−50

17.

x	5	10	15	20	25	30
y	7	8	8	8	7	8

18.

x	6	7	8	9	11	13
y	9	9	9	9	9	9

19.

x	−70	−60	−50	−40	−30	−20	−10	0
y	5	10	15	20	20	15	10	5

20.

x	−30	−23	−15	−12	−1	5	14	29
y	103	88	76	62	54	47	30	20

21. The heights (in inches) and weights (in pounds) of a sample of five women are recorded.

x = height	y = weight
66	122
67	133
69	153
68	138
65	125

22. The number of days absent from a class, and the course grade.

x = days absent	y = course grade
0	95
2	90
4	85
6	70
8	60

23. The cost of a repair job and the number of hours spent on the repair.

x = hours	y = cost
2	120
3	180
5	230
8	350
10	380

24. Attendance at a baseball game (in thousands), and the amount of rainfall at a baseball stadium that day (in inches).

x = rain	y = attendance
0.0	40
0.0	42
0.1	38
0.5	30
1.0	20

For Exercises 25–36, do the following for the indicated data:
   a. Predict the value of y for the given value of x.
   b. Calculate and interpret the prediction error.
   c. State whether or not the prediction represents extrapolation.

25. Data from Exercise 13; $x = 30$
26. Data from Exercise 14; $x = 2$
27. Data from Exercise 15; $x = −5$
28. Data from Exercise 16; $x = 3$
29. Data from Exercise 17; $x = 0$

30. Data from Exercise 18; $x = 5$
31. Data from Exercise 19; $x = 0$
32. Data from Exercise 20; $x = 5$
33. Data from Exercise 21; $x = 68$
34. Data from Exercise 22; $x = 10$
35. Data from Exercise 23; $x = 400$
36. Data from Exercise 24; $x = 0.25$

APPLYING THE CONCEPTS

For Exercises 37–51, do the following for the indicated data sets from the Section 4.1 exercises:

   a. Calculate the slope b_1 and the y intercept b_0 of the regression line.
   b. State the regression equation in words.
   c. Interpret the value for the slope b_1 of the regression line, in terms of the variables from the particular exercise.
   d. Interpret the value for the y intercept b_0 of the regression line, in terms of the variables from the particular exercise.

37. **Video Game Sales.** The Chapter 1 Case Study looked at video game sales for the top 30 video games. The following table contains the total sales (x, in game units) and weeks on the top 30 list (y) of 5 randomly chosen video games.

videogamereg

Video game	Total sales in millions of units (y)	Weeks (x)
Super Mario Bros. U for WiiU	1.7	78
NBA 2K14 for PS4	0.6	27
Battlefield 4 for PS3	0.9	29
Titanfall for Xbox One	1.2	10
Yoshi's New Island for 3DS	0.2	10

Source: www.vgchartz.com.

38. **Does It Pay to Stay in School?** The U.S. Census Bureau reported the following unemployment rates (y) associated with the given years of education (x).

edunemploy

x = years of education	y = unemployment rate
5.0	16.8
7.5	17.1
8.0	15.3
10.0	20.6
12.0	11.7
14.0	8.1
16.0	3.8

39. **Darts and the Dow Jones.** The following table contains a random sample of eight days from the Chapter 3 Case Study data set, indicating the stock market gain or loss for the portfolio chosen by the random darts (y), as well as the Dow Jones Industrial Average (DJIA) gain or loss for that day (x).

dartsdjia

Darts (y)	DJIA (x)
−27.4	−12.8
18.7	9.3
42.2	8.0
−16.3	−8.5
11.2	15.8
28.5	10.6
1.8	11.5
16.9	−5.3

Source: The Wall Street Journal.

40. **Age and Height.** The following table provides a random sample from the Chapter 4 Case Study data set *body_females,* showing the age (x) and height (y) of eight women.

ageheight

Age (x)	Height (y)
40	63.5
28	63.0
25	64.4
34	63.0
26	63.8
21	68.0
19	61.8
24	69.0

Source: Journal of Statistics Education.

41. **Gardasil Shots and Age.** The accompanying table shows a random sample of 10 patients from the Chapter 5 Case Study data set, Gardasil, including the age of the patient (x) and the number of shots taken by the patient (y).

gardasilreg

Age (x)	Shots (y)
13	3
21	3
16	3
17	2
17	3
18	1
25	2
15	3
12	1
16	1

Source: Journal of Statistics Education.

42. **NCAA Power Ratings.** The accompanying table shows the top 10 teams' winning percentages (x) and power ratings (y) for the 2013–2014 NCAA basketball season, according to www.teamrankings.com. **ncaa2014**

Team	Winning proportion (x)	Power rating (y)
Florida	0.923	121.2
Wichita State	0.971	119.1
Arizona	0.868	118.8
Louisville	0.838	117.9
Connecticut	0.800	117.2
Virginia	0.811	116.8
Wisconsin	0.789	116.6
Villanova	0.853	116.4
Michigan State	0.763	115.9
Michigan	0.757	115.9

43. Saturated Fat and Calories. The table contains the calories and saturated fat in a sample of 10 food items.

🏨 satfatcorr

Food item	Calories	Grams of saturated fat
Chocolate bar (1.45 ounces)	216	7.0
Meat & veggie pizza (large slice)	364	5.6
New England clam chowder (1 cup)	149	1.9
Baked chicken drumstick (no skin, medium size)	75	0.6
Curly fries, deep-fried (4 ounces)	276	3.2
Wheat bagel (large)	375	0.3
Chicken curry (1 cup)	146	1.6
Cake doughnut hole (one)	59	0.5
Rye bread (1 slice)	67	0.2
Raisin bran cereal (1 cup)	195	0.3

Source: Food-a-Pedia.

44. Engine Displacement and Gas Mileage. The table provides the engine displacement (size, in liters) and the city mpg (miles per gallon) gas mileage of a random sample of 12 vehicles taken from the Chapter 6 Case Study data set, **FuelEfficiency**. 🏨 displacement

Vehicle	Engine displacement	City mpg
GMC Yukon Denali	6.2	13
Ford E350 Wagon	5.4	11
BMW435i Coupe	3.0	20
Land Rover Range Rover	5.0	13
Infiniti Q50a	3.7	19
Dodge Journey	3.6	17
Jaguar XF	5.0	15
Dodge Challenger	6.4	14
Toyota Highlander Hybrid	3.5	28
Mercedes-Benz S550	4.7	17
Ford Fiesta	1.6	29
Hyundai Elantra	2.0	24

45. Completing College. The twenty-first century economy not only needs students to attend college; it needs students to complete their college degrees, in order to compete in the information age. The table contains a sample of 10 states, with data on the percentage of residents who have attended college (x) and the percentage of college attendees who have completed their college degrees (y). 🏨 collegecompleters

State	x = college attenders	y = college completers
California	30.9	38.8
Florida	26.6	35.5
Georgia	30.1	34.1
Illinois	35.7	39.1
Massachusetts	45.2	45.9
New York	38.7	42.8
North Carolina	30.1	35.5
Ohio	28.4	37.1
Pennsylvania	32.5	40.2
Texas	26.2	32.2

Source: American Community Survey.

46. Walking or Biking to Work. The table contains, for a sample of 10 American cities, the percentage of people who walk to work (x) and the percentage of people who bike to work (y). 🏨 walkbike

City	x = walk to work	y = bike to work
Anaheim	1.8	0.9
Baltimore	6.5	0.8
Buffalo	6.2	0.9
Cincinnati	5.4	0.5
Detroit	3.1	0.3
Jacksonville	1.4	0.4
Las Vegas	1.9	0.4
New Orleans	5.1	2.1
Orlando	1.9	0.4
Sacramento	3.2	2.5

Source: U.S. Census Bureau.

47. Teenage Birth Rate.

a. The National Center for Health Statistics publishes data on state birth rates. The table contains the overall birth rate and the teenage birth rate for eight randomly chosen states. The overall birth rate is defined by the NCHS as "live births per 1000 women," and the teenage birth rate is defined as "live births per 1000 women aged 15–19." 🏨 teenbirth

State	x = overall birth rate	y = teen birth rate
California	62.0	23.6
Florida	59.3	24.6
Georgia	61.6	30.5
New York	58.8	17.7
Ohio	62.7	27.2
Pennsylvania	58.4	20.9
Texas	69.9	41.0
Virginia	60.9	20.1

48. Brain and Body Weight. A study compared the body weight (in kilograms) and brain weight (in grams) for a sample of mammals, with the results shown in the following table.[4] **brainbody**

x = body weight (kg)	y = brain weight (g)
52.16	440.0
60.00	81.0
27.66	115.0
85.00	325.0
36.33	119.5
100.00	157.0
35.00	56.0
62.00	132.0
83.00	98.2
55.50	175.0

49. Consumer Sentiment. The University of Michigan's Survey of Consumers published the data in the following table, showing the consumer sentiment in 2013 month by month for the two groups. **consumersentiment**

Month	x = consumer sentiment for incomes under $75,000	y = consumer sentiment for incomes $75,000 or higher
Jan	71.6	80.2
Feb	75.7	82.4
Mar	78.3	83.7
Apr	74.5	79.8
May	80.3	94.1
Jun	76.1	98.9
Jul	82.4	90.0
Aug	78.0	89.6
Sep	72.3	86.2
Oct	71.4	77.0
Nov	67.9	88.7
Dec	78.9	88.0

50. SAT Scores, by Foreign Language. The table contains the mean 2014 SAT Critical Reading and Math scores, which are categorized by the foreign language taken in high school or spoken at home. **satlanguages**

Language	SAT Critical Reading score	SAT Mathematics score
Chinese	535	606
French	519	525
German	530	540
Greek	526	543
Hebrew	526	541
Italian	497	509
Japanese	521	552
Korean	490	576
Latin	556	556
Russian	483	535
Spanish	498	508

51. Batting Average and Runs Scored. The table shows the top 10 hitters in the American League of Major League Baseball for 2014. We are interested in estimating the number of runs scored (y) using the player's batting average (x).

batters2014

Batter	Team	Runs scored	Batting average
Jose Altuve	Houston Astros	85	0.341
Victor Martinez	Detroit Tigers	87	0.335
Michael Brantley	Cleveland Indians	94	0.327
Adrian Beltre	Texas Rangers	79	0.324
Jose Abreu	Chicago White Sox	80	0.317
Robinson Cano	Seattle Mariners	77	0.314
Miguel Cabrera	Detroit Tigers	101	0.313
Melky Cabrera	Toronto Blue Jays	81	0.301
Adam Eaton	Chicago White Sox	76	0.300
Howie Kendrick	Los Angeles Angels	85	0.293

52. Video Game Sales. Refer to your work in Exercise 37.
   a. Compute the predicted total sales for a video game that has been on the list for 27 weeks.
   b. Find the predicted total sales for a video game that has been on the list for 10 weeks.
   c. Calculate the prediction error for NBA 2K14. Does NBA 2K14 lie above or below the regression line? How can we tell?
   d. Calculate the prediction errors for Titanfall for Xbox One and Yoshi's New Island for 3DS. Where do each of these games lie in the scatterplot, with respect to the regression line? Where do these games fall with respect to each other?

53. Does it Pay to Stay in School? Refer to your work from Exercise 38. For parts (**a**)–(**c**), if appropriate, use your regression equation to estimate the unemployment for individuals with the following years of education. If it is not appropriate, clearly state why not.
   a. 10 years
   b. 15 years
   c. 20 years
   d. Calculate the prediction error for your prediction in part (**a**). Does this data point lie above or below the regression line, and what does that mean?

54. Darts and the Dow Jones. Refer to your work from Exercise 39. For parts (**a**)–(**c**), if appropriate, use your regression equation to estimate the Darts stock returns based on the following values of the Dow Jones Industrial Average. If it is not appropriate, clearly state why not.
   a. 8
   b. 20

c. −10

d. Calculate the prediction error for your prediction in part (**a**). Does this data point lie above or below the regression line, and what does that mean?

55. Age and Height. Refer to your work from Exercise 40.

a. Estimate the height of a 40-year-old person.

b. Does the interpretation of the y intercept from Exercise 40 make sense? Explain.

c. Is it OK, or is it misleading to use the regression equation to predict the height of a 50-year-old person? Explain.

d. What is the distinction between your result from part (**a**) and the height of the first person in the data set?

e. Calculate and interpret the prediction error for your prediction in part (**a**).

56. Brain and Body Weight. Refer to your work from Exercise 48.

a. Estimate the brain weight for a mammal with a body weight of 100 kilograms.

b. Is the interpretation of the y intercept from Exercise 48 useful? Explain.

c. Is it OK, or is it misleading to use the regression equation to predict the brain weight for a mammal with body weight of 10 kg? Explain.

d. Explain the distinction between your result from part (**a**) and the actual brain weight of 157 grams for the mammal from the data table.

e. Calculate and interpret the prediction error for your prediction in part (**a**).

WHAT IF
? **57.** Consider again your work on the engine displacement and gas mileage data set in Exercise 44. *What if* there was a typo, and all of the engine displacements in the data set needed to be adjusted downward by the same amount? Explain how this change would affect the following, and why. Increase, decrease, or no change?

a. $\bar{x}$

b. $\bar{y}$

c. y intercept b_0

d. Slope b_1

e. Correlation coefficient r

58. Computational Formula for the Slope. The following computational formula is equivalent to the definition formula for the slope b_1:

$$b_1 = \frac{\sum xy - \left(\sum x \sum y\right)/n}{\sum x^2 - \left(\sum x\right)^2/n}$$

Use the computational formula to calculate the slope b_1 for the relationship between square footage and sales price of the eight home lots for sale in Glen Ellyn from Example 2 (page 189). Then find the y intercept b_0 and the regression equation.

WORKING WITH LARGE DATA SETS

DC Households. Use the following information for Exercises 59–61. The data set **Households**, located on the

text Web site, contains information on the number and type of households in the 50 states and the District of Columbia. For each state, there are seven variables. Two of these variables are the percentage of households headed by women ($y = HHLD\_WOMEN$) and the total number of households in the state ($x = TOT\_HHLD$). Minitab provides the following regression equation:

households

> **Regression Analysis**
> The regression equation is
> HHLD_Women = 10.5 + 2.82E-07 TOT_HHLD

Note: Minitab shows its regression equations as $y = b_0 + b_1x$ instead of $\hat{y} = b_1x + b_0$. Also, the notation 2.82E-07 refers to the scientific notation method of writing numbers. Often, software and calculators will present you with this type of notation, so you need to know how to read it. The number 2.82E-07 represents 2.82 times 10^{-7}, or 0.000000282.

59. In this exercise, we explore the regression coefficients and the regression equation.

a. Find and interpret the meaning of the value for the y intercept. Does it make sense?

b. Would the estimate in (**a**) be considered extrapolation? Why or why not?

c. Find and interpret the meaning of the slope coefficient as the total number of households in the state increases.

d. Write the regression equation. Now state in words what the regression equation means.

e. Is the correlation coefficient positive or negative? How do you know?

60. Estimate the increase or decrease in the percentage of households headed by women, using a sentence, for the following situations:

a. Suppose State A has 1 million more households than State B.

b. Suppose State C has 5 million fewer households than State D.

61. The number of households per state ranges from about 170,000 to about 10 million.

a. Estimate the percentage of households headed by women for a state with 7 million households, if appropriate.

b. Estimate the percentage of households headed by women for a state with 100,000 households, if appropriate.

Invalid Application of Linear Regression. Use the information in the following scatterplot for Exercises 62–64. Scrabble (hasbro.com/scrabble/en_US/) is one of the most popular games in the world. We are interested in approximating the relationship between the frequency (x) of the letter tiles in the game and their point value (y). The scatterplot shows the point value versus the letter frequency, along with the regression line, as plotted by Minitab.

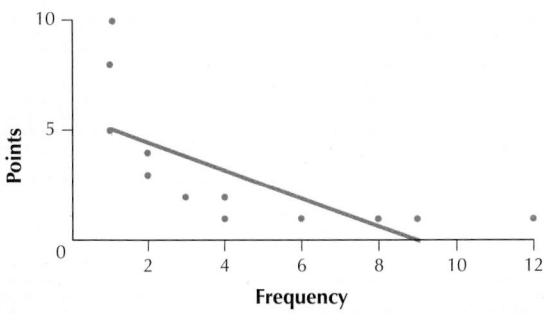

Scatterplot of Scrabble point value versus letter frequency, with regression line.

62. Does the relationship between frequency and points seem to be positive or negative?

63. Does the relationship between frequency and points seem to be a straight-line relationship or a curved relationship?

64. Do you think that we should use a regression line to approximate the relationship between frequency and points? If not, explain clearly what you think is wrong.

Chapter 3 Case Study (Continued). Use the following information for Exercises 65–68. Shown here is the regression equation for the linear relationship between the randomly selected Darts portfolio and the Dow Jones Industrial Average (DJIA), from the Chapter 3 Case Study. (*Note:* This is for the entire data set, not the small sample taken in the Exercise 39 data set.)

Regression Analysis

```
The regression equation is
Darts = -2.49 + 1.032 DJIA
```

---

65. In this exercise, we examine the y intercept.

   a. Which variable is the x variable and which is the y variable? Work by analogy with the previous exercises.

   b. What is the value of the y intercept?

   c. Interpret the meaning of this value for the y intercept. Does it make sense?

   d. Would the estimate in (**c**) be considered extrapolation? Why or why not?

66. In this exercise, we look at the slope and the regression equation.

   a. Find and interpret the meaning of the value of the slope coefficient, as the DJIA increases.

   b. Write the regression equation. Now state in words what the regression equation means.

   c. Is the correlation coefficient positive or negative? How do we know?

67. Estimate the increase or decrease in the net price of the Darts portfolio, using a sentence, for the following situations:

   a. Suppose that for Contest A the DJIA increased by 10% more than for Contest B.

   b. Suppose that for Contest C the DJIA decreased by 5% more than for Contest D.

68. The net change in the DJIA ranged from –13.1% to 22.5%.

   a. Estimate the net price change for the Darts portfolio when the DJIA is up by 22%, if appropriate.

   b. Estimate the net price change for the Darts portfolio when the DJIA is down by 10%, if appropriate.

   c. Estimate the net price change for the Darts portfolio when the DJIA is down by 22%, if appropriate.

Cell Phone Use for Internet Access, Worldwide. Would you expect that residents of richer countries tend to use their cell phones to browse the Internet more often than residents of poorer countries? The Pew Global Attitudes Project conducted a study[5] of cell phone usage in countries around the world. The table shows $x =$ the per capita gross domestic product (GDP, a measure of the wealth of the country), and $y =$ the percentage of cell phone owners who use their cell phones to browse the Internet for a random sample of 10 countries. Use this information for Exercises 69–75.

cellregression

Nation	$x =$ per capita GDP($)	$y =$ percentage who use cell phone to browse Internet
USA	48,147	43
Britain	35,974	38
France	35,048	28
Russia	16,687	27
Poland	20,136	30
Israel	31,004	47
China	8,394	37
Japan	34,362	47
India	3,703	10
Mexico	15,121	18

69. Construct and interpret a scatterplot of the data in the table.

70. Based on your interpretation in Exercise 69, would the value for the correlation coefficient r be positive or negative?

71. Calculate the correlation coefficient r.

72. Find the slope and y intercept of the regression line. Write the regression equation in a sentence.

73. Interpret the values of the slope and the y intercept. Determine whether the interpretation of the y intercept represents extrapolation in this case.

74. Calculate the estimated percentage using their cell phones to browse the Internet for a nation with a per capita GDP of $48,147.

75. Identify the country with a per capita GDP of $48,147. Calculate and interpret the prediction error for this country.

Unidentified Flying Objects. Have you or any of your friends sighted any unidentified flying objects (UFOs)? Americans in each of the 50 states have reported seeing UFOs. Figure 30 represents a scatterplot of the number of UFO sightings versus state population, for each of the

50 states. Each dot represents a state. The straight line is a regression line, which approximates the relationship between UFO sightings and state population. As the state population increases, the number of UFO sightings also tends to increase, which is not surprising.

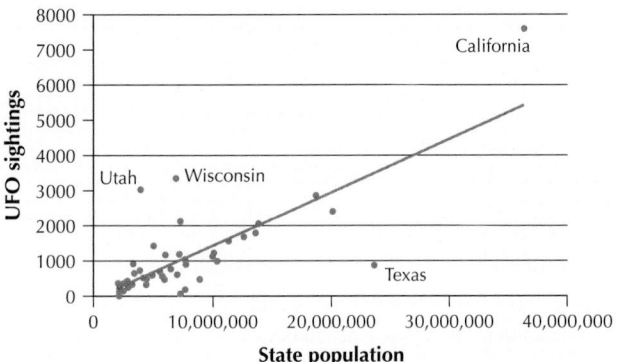

FIGURE 30 A scatterplot of the number of UFO sightings versus state population, showing that UFOs don't mess with Texas.

What may be surprising is that the UFOs seem to be attracted to certain states, yet avoid others. States considerably above the regression line have a larger than expected number of UFO sightings for their population size, whereas states below the line have a smaller than expected number of UFO sightings for their population size. So, there are more sightings than expected in California, Wisconsin, and Utah, given their population size, and fewer than expected in Texas. Why this might occur is open to discussion. Perhaps people in California are more likely to attribute unusual sightings to UFOs than most Americans; perhaps people in Texas are more pragmatic than most Americans. But if the sightings are valid (a big if!), it sure looks like the UFOs don't want to mess with Texas. Refer to Figure 30 for Exercises 76–79.

76. Provide a rough estimate of the following for the state of California.
   a. State population
   b. UFO sightings

77. Provide a rough estimate of the following for the state of Texas.
   a. State population
   b. UFO sightings

78. For the state of California, what is the estimated number of UFO sightings? (*Hint:* It's at the point on the line directly below the dot for California.)

79. For the state of Texas, what is the estimated number of UFO sightings?

BRINGING IT ALL TOGETHER

Fuel Economy. Refer to the following table of fuel economy data for a sample of 10 vehicles for Exercises 80–84. The predictor variable is x = engine size, expressed in liters; the response variable is y = combined (city/highway) gas mileage, expressed in miles per gallon (mpg). **ᴀᴛ⋀ enginempg**

Vehicle	x = engine size (liters)	y = combined mpg
Mini Cooper	1.6	31
Ford Focus	2.0	28
Toyota Camry	2.5	26
Honda Accord	2.4	26
Subaru Forester	2.5	23
Toyota Highlander	2.7	22
Ford Taurus	3.5	20
Chevrolet Equinox	3.0	19
Dodge Nitro	4.0	17
Cadillac limousine	4.6	14

80. Exploring the Data.
   a. Look at the data table. As the engine size values increase, what seems to be happening to the combined mpg?
   b. Construct a scatterplot of the data.
   c. Interpret the scatterplot. Is your insight from part (**a**) supported?

81. What Results Do You Expect? Based on your scatterplot in Exercise 80, answer the following:
   a. Will the correlation coefficient be positive or negative?
   b. Do you expect that the correlation will be closer to -0.9 or -0.5? Why?
   c. Do you think that the slope b_1 will be positive or negative? Why?

82. Correlation. Do the following:
   a. Calculate the correlation coefficient r. Does this concur with your predictions from Exercises 81(**a**) and 81(**b**)?
   b. Interpret the correlation between engine size and combined mpg.

83. Regression. Answer the following:
   a. Calculate the slope b_1 of the regression equation. Does the sign of b_1 agree with your prediction from Exercise 81(**c**)?
   b. Calculate the y intercept b_0.
   c. Interpret the values you calculated in parts (**a**) and (**b**), so that a nonstatistician would understand them.

84. Making Predictions. Answer the following:
   a. Note that the Chevrolet Equinox has an engine size of 3 liters. Predict the combined mpg for a vehicle with an engine size of 3 liters.
   b. Is your prediction error positive or negative? Thus, does the data value lie above or below the regression line? What does this mean?

CONSTRUCT YOUR OWN DATA SETS

85. Describe two variables from real life whose regression line would have a positive slope b_1.
   a. Explain why the y variable depends on the x variable.
   b. Explain why the slope is positive.

86. Create a sample of five observations from each of your variables from Exercise 85, and put them into a table similar to Table 1 in Section 4.1.

    a. Construct a scatterplot of the variables.

    b. Draw a single straight line through the data points in the plot in a manner that you think best approximates the relationship between the variables.

    c. Using your regression line from (**b**), estimate the slope b_1 and the y intercept b_0.

    d. Write your results from (**c**) in the form of a regression equation.

 Use the *Correlation and Regression* applet for Exercises 87 and 88.

87. Create a set of $n = 10$ points, such that the slope of the regression line has the following characteristics. Note that you can drag points up or down to adjust your regression line.

    a. The slope is positive.

    b. The slope is negative.

    c. The slope is neither positive nor negative.

88. Describe the relationship between the variables for each of the sets of points in the previous exercise.

WORKING WITH LARGE DATA SETS

 **Chapter 4 Case Study: Measuring the Human Body.** Open the data sets *body_females* and *body_males*. We will apply what we have learned in this section regarding regression analysis to some of the measurements in these data sets. Use technology for the Exercises 89–103. 📊 **body_females** 📊 **body_males**

89. Perform a regression of weight on height for the females.

90. Interpret the slope and y intercept values of the regression in the previous exercise.

91. Use the regression equation from Exercise 89 to estimate the weight of a female who is 63.5 inches tall.

92. Find the prediction error for the first woman in the data set, with height 63.5 inches and weight 113.8 pounds.

93. Estimate the weight for females with the following heights. If the estimate represents extrapolation, indicate so.

    a. 65 inches

    b. 70 inches

94. Perform a regression of weight on height for the men.

95. Interpret the slope and y intercept values of the regression in the previous exercise.

96. Use the regression equation from Exercise 94 to estimate the weight of a man who is 68.5 inches tall.

97. Find the prediction error for the first male in the data set, with height 68.5 inches and weight 144.6 pounds.

98. Estimate the weight for males with the following heights. If the estimate represents extrapolation, indicate so.

    a. 60 inches

    b. 65 inches

99. Perform a regression of hip girth on waist girth for the women. State the regression equation in words. Note that both variables are expressed in centimeters (cm).

100. Interpret the slope and y intercept values of the regression of hip girth on waist girth for the women.

101. Perform a regression of hip girth on waist girth, this time for the men. State the regression equation in words. Note that both variables are expressed in cm.

102. Interpret the slope and y intercept values of the regression of hip girth on waist girth for the males.

103. Compare the similarities and differences in the regression coefficients between the women and men for the regression of hip girth on waist girth.

4.3 Further Topics in Regression Analysis

OBJECTIVES By the end of this section, I will be able to . . .

1 Calculate the sum of squares error (SSE), and use the standard error of the estimate s as a measure of a typical prediction error.

2 Describe how total variability, prediction error, and improvement are related to the total sum of squares (SST), the sum of squares error (SSE), and the sum of squares regression (SSR).

3 Explain the meaning of the coefficient of determination r^2 as a measure of the usefulness of the regression.

In Section 4.2, we were introduced to regression analysis, which uses an equation to approximate the linear relationship between two quantitative variables. Here in Section 4.3, we learn some further topics that will enable us to better apply the tools of regression analysis for a deeper understanding of our data.

1 Sum of Squares Error (SSE) and Standard Error of the Estimate *s*

Table 6 shows the results for 10 student subjects who were given a set of short-term memory tasks to perform within a certain amount of time. These tasks included memorizing nonsense words and random patterns. Later, the students were asked to repeat the words and patterns, and the students were scored according to the number of words and patterns memorized and the quality of their memories. Partially remembered words and patterns were given partial credit, so the score was a continuous variable. Figure 31 displays the scatterplot of y = score versus x = time, together with the regression line $\hat{y} = 2x + 7$, as calculated by Minitab.

shortmemory

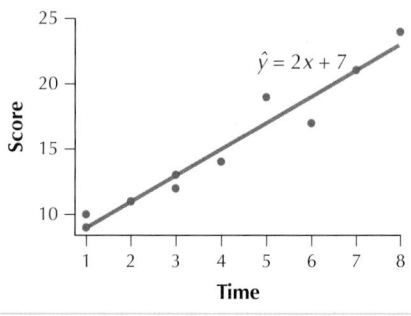

FIGURE 31 Scatterplot with regression line.

```
The regression equation is
Score = 7.00 + 2.00 Time
```

Minitab regression results (excerpt).

Table 6 Results of short-term memory test

Student	Time to memorize (in minutes) (x)	Short-term memory score (y)
1	1	9
2	1	10
3	2	11
4	3	12
5	3	13
6	4	14
7	5	19
8	6	17
9	7	21
10	8	24

In Section 4.2, we learned that the difference $y - \hat{y}$ represented the prediction error or residual between the actual data value y and the predicted value $\hat{y}$. For example, for a student who is given $x = 5$ minutes to study, the predicted score is $\hat{y} = 2\,(time) + 7 = 17$.

For Student 7, who was given 5 minutes to study and got a score of 19, the prediction error is $y - \hat{y} = 19 - 17 = 2$.

We can calculate the prediction errors for every student who was tested. If we wish to use the regression to make useful predictions, we want to keep all our prediction errors small. To measure the prediction errors, we calculate the sum of squared prediction errors, or more simply, the **sum of squares error (SSE)**:

Sum of Squares Error (SSE)

$$SSE = \sum (y - \hat{y})^2 = \sum (\text{prediction error})^2 = \sum (\text{residual})^2$$

We want our prediction errors to be small, therefore, it follows that we want SSE to be as small as possible.

Least-Squares Criterion

The least-squares criterion states that the regression line will be the *line for which the SSE is minimized*. That is, out of all possible straight lines, the least-squares criterion chooses the line with the smallest SSE to be the regression line.

EXAMPLE 12   Calculating SSE, the sum of squares error

a. Construct a scatterplot of the memory score data, indicating each residual.

b. Calculate SSE for the memory score data.

Solution

a. The brackets (}) in the scatterplot in Figure 32 indicate the residual for each student's score. The quantities represented by these brackets are the residuals $y - \hat{y}$.

b. Table 7 shows the $\hat{y}$-values and residuals for the data in Table 6. The SSE is then found by squaring each residual and taking the sum. Thus,

$$\text{SSE} = \sum (y - \hat{y})^2 = 12$$

We know that $\hat{y} = 2x + 7$ is the regression line, according to the least-squares criterion, so no other possible straight line would result in a smaller SSE.

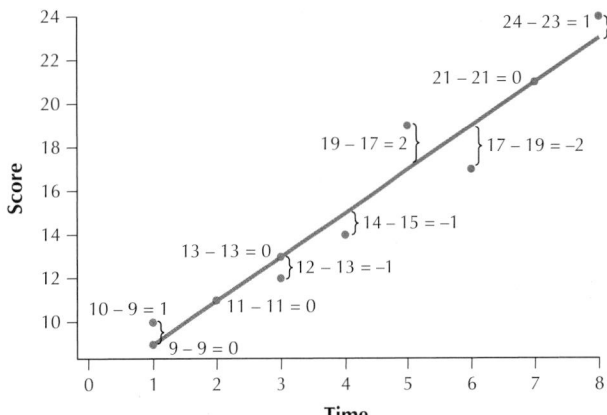

FIGURE 32 Scatterplot showing the prediction errors or residuals $y - \hat{y}$.

Table 7 Calculation of the SSE for the short-term memory test example

Student	Time (x)	Actual score (y)	Predicted score ($\hat{y} = 2x + 7$)	Residual ($y - \hat{y}$)	(Residual)² ($y - \hat{y}$)²
1	1	9	9	0	0
2	1	10	9	1	1
3	2	11	11	0	0
4	3	12	13	−1	1
5	3	13	13	0	0
6	4	14	15	−1	1
7	5	19	17	2	4
8	6	17	19	−2	4
9	7	21	21	0	0
10	8	24	23	1	1

$$\text{SSE} = \sum (y - \hat{y})^2 = 12$$

NOW YOU CAN DO
Exercises 11a–22a.

A useful interpretive statistic is s, the **standard error of the estimate**. The formula for s follows.

Don't confuse this use of the s notation for the standard error of the estimate with the use of the s notation for the sample standard deviation.

Standard Error of the Estimate s

$$s = \sqrt{\frac{SSE}{n - 2}}$$

The standard error of the estimate gives a measure of the typical residual. That is, s is a measure of the size of the *typical prediction error*, which is the typical difference between the predicted value of y and the actual observed value of y. If the typical prediction error is large, then the regression line may not be useful.

EXAMPLE 13   Calculating and interpreting *s*, the standard error of the estimate

Note: Here, we are rounding $s = 1.2247$ for reporting purposes. However, when we use s for calculating other quantities later, we will not round until the last calculation.

Calculate and interpret the standard error of the estimate s for the memory score data.

Solution
$SSE = 12$ and $n = 10$, so

$$s = \sqrt{\frac{SSE}{n - 2}} = \sqrt{\frac{12}{8}} \approx 1.2247$$

Thus, the typical error in prediction is 1.2247 points. In other words, if we know the amount of time (x) a given student spent memorizing, then our estimate of the student's score on the short-term memory test will typically differ from the student's actual score by only 1.2247 points.

NOW YOU CAN DO
Exercises 11b–22b.

2 SST, SSR, and SSE

The least-squares criterion guarantees that the value of $SSE = 12$ that we found in Example 12 is the smallest possible value for SSE, given the data in Table 6. However, this guarantee in itself does not tell us that the regression is useful. For the regression to be useful, the prediction error (and therefore SSE) must be small. But, we cannot yet tell whether the value of $SSE = 12$ is indeed small because we can't compare it to anything.

Suppose for a moment that we want to estimate short-term memory scores, but we have no knowledge of the amount of time (x) for memorization. Then the best estimate for y is simply $\bar{y} = 15$, the mean of the sample of short-term memory test scores. The graph of $\bar{y} = 15$ is the horizontal line in Figure 33.

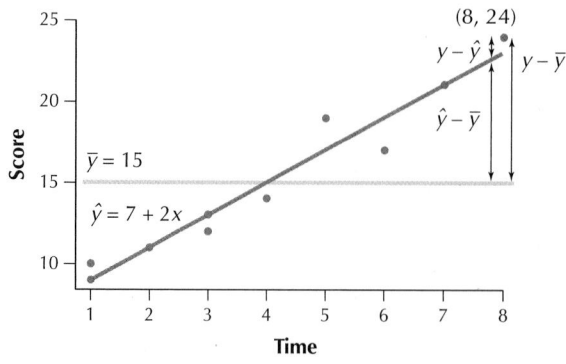

FIGURE 33 Comparing $(y - \hat{y})$ and $(y - \bar{y})$.

In general, the data points are closer to the regression line than they are to the horizontal line $\bar{y} = 15$, indicating that the errors in prediction are smaller when using the regression equation. Consider Student 10, who had a short-term memory score of $y = 24$ after memorizing for $x = 8$ minutes. Using $\bar{y} = 15$ as the estimate, the error for Student 10 is

$$(y - \bar{y}) = 24 - 15 = 9$$

This error is shown in Figure 33 as the vertical distance $(y - \bar{y})$.

Suppose we found this value $(y - \bar{y})$ for every student in the data set and summed the squared $(y - \bar{y})$, just as we did for the $(y - \hat{y})$ when finding SSE. The resulting statistic is called the **total sum of squares (SST)** and is a measure of the total variability in the values of the y variable:

$$SST = \sum (y - \bar{y})^2$$

Developing Your Statistical Sense	**Relationship Between SST and the Variance of the y's**

Relationship Between SST and the Variance of the y's

Note that SST ignores the presence of the x information; it is simply a measure of the variability in y. Recall (see page 133) that the *variance* of a sample of y-values is given by $s^2 = \Sigma(y - \bar{y})^2/(n - 1)$. Thus,

$$SST = (n - 1)\, s^2$$

Thus, SST is proportional to the variance of the y's and, as such, is a measure of the variability in the y data.

EXAMPLE 14   Calculating SST, the total sum of squares, in two ways

Calculate SST, the total sum of squares, for the memory score data in two ways:

a. By using Table 8

b. By using the fact that the sample variance of the score data (the y-values) equals $25\frac{1}{3}$

Solution

a. Table 8 shows the values for $(y - \bar{y}) = (y - 15)$ for the data in Table 7. Thus,
$SST = \Sigma(y - \bar{y})^2 = 228$.

Table 8 Calculation of SST

Student	Score (y)	($y - \bar{y}$)	($y - \bar{y}$)^2
1	9	−6	36
2	10	−5	25
3	11	−4	16
4	12	−3	9
5	13	−2	4
6	14	−1	1
7	19	4	16
8	17	2	4
9	21	6	36
10	24	9	81

$$SST = \sum (y - \bar{y})^2 = 228$$

NOW YOU CAN DO
Exercises 11c–22c.

b. When we are given the variance of y, we may calculate SST as follows:

$$SST = (n - 1)s^2 = (10 - 1)\left(25\frac{1}{3}\right) = 228$$

Consider Figure 33 once again. For Student 10, note that the error in prediction when ignoring the x data is $(y - \bar{y}) = 9$, while the error in prediction when using the regression equation is $(y - \hat{y}) = 1$. (Recall that $\hat{y} = 2(8) + 7 = 23$ because Student 10's time is $x = 8$.) The amount of improvement (that is, the amount by which the prediction error is diminished) is the difference between $\hat{y}$ and $\bar{y}$:

$$(\hat{y} - \bar{y}) = 23 - 15 = 8$$

Once again, we can find $(\hat{y} - \bar{y})$ for each observation in the data set, square them, and sum the squared results to obtain $\Sigma(\hat{y} - \bar{y})^2$. The resulting statistic is **SSR**, the **sum of squares regression**.

$$SSR = \sum(\hat{y} - \bar{y})^2$$

SSR measures the amount of *improvement* in the accuracy of our estimates when using the regression equation compared with relying only on the y-values and ignoring the x information. Note in Figure 33 that the distance $(y - \bar{y})$ is the same as the sum of the distances $(\hat{y} - \bar{y})$ and $(y - \hat{y})$. It can be shown, by using algebra, that the following also holds true.

Note: None of these sums of squares can ever be negative.

> **Relationship Among SST, SSR, and SSE**
>
> SST = SSR + SSE

If any two of these sums of squares are known, the third can also be calculated, as shown in the following example.

EXAMPLE 15 Using SST and SSE to find SSR

Use SST and SSE to find the value of SSR for the data from Examples 12–14.

Solution

From Example 12, we have SSE = 12, and from Example 14 we have SST = 228. That leaves us with just one unknown in the equation SST = SSR + SSE, so we can solve for the unknown SSR:

$$SSR = SST - SSE = 228 - 12 = 216$$

NOW YOU CAN DO
Exercises 11d–22d.

3 Coefficient of Determination r^2

SSR represents the amount of variability in the response variable that is accounted for by the regression equation, that is, by the linear relationship between y and x. SSE represents the amount of variability in the y that is left unexplained after accounting for the relationship between x and y (including random error). We know that SST represents the sum of SSR and SSE; therefore, it makes sense to consider the *ratio* of SSR and SST, which is called the **coefficient of determination r^2**.

> The **coefficient of determination r^2** = SSR/SST measures the goodness of fit of the regression equation to the data. We interpret r^2 as the proportion of the variability in y that is accounted for by the linear relationship between y and x. The values that r^2 can take are $0 \leq r^2 \leq 1$. Note that the coefficient of determination r^2 is the square of the correlation coefficient r. Thus, $\pm\sqrt{r^2} = r$, the correlation coefficient.

EXAMPLE 16   Calculating and interpreting the coefficient of determination r^2

Calculate and interpret the value of the coefficient of determination r^2 for the memory score data.

Solution

From Example 14, we have SST = 228, and from Example 15 we have SSR = 216. Thus,

$$r^2 = \frac{\text{SSR}}{\text{SST}} = \frac{216}{228} \approx 0.9474$$

NOW YOU CAN DO
Exercises 11e–22e.

Therefore, 94.74% of the variability in the memory test score (y) is accounted for by the linear relationship between score (y) and the time given for study (x).

What Does This Number Mean?

What does the value of $r^2 \approx 0.9474$ mean? Consider that the memory test scores have a certain amount of variability: some scores are higher than others. In addition to the amount of time (x) given for memorizing, there may be several other factors that might account for variability in the scores, such as the memorizing ability of the students, how much sleep the students had, and so on. However, $r^2 \approx 0.9474$ indicates that 94.74% of this variability in memory scores (y) is explained by the single factor "amount of time given for study" (x). All other factors, including factors such as amount of sleep, account for only 100% − 94.74% = 5.26% of the variability in the memory test scores.

Suppose that the regression equation was a perfect fit to the data, so that every observation lies exactly on the regression line. No errors in prediction would occur; therefore, SSE would equal 0, which would imply that

$$\text{SST} = \text{SSR} + 0 = \text{SSR}$$

In this case, SST = SSR, then

$$r^2 = \frac{\text{SSR}}{\text{SST}} = \frac{\text{SST}}{\text{SST}} = 1$$

Conversely, if SSR = 0, then *no improvement at all* is gained by using the regression equation. That is, the regression equation accounts for no variability at all, and $r^2 = 0/\text{SST} = 0$.

The closer the value of r^2 is to 1, the better the fit of the regression equation to the data set. A value near 1 indicates that the regression equation fits the data extremely well. A value near 0 indicates that the regression equation fits the data extremely poorly.

Recall from Section 4.1 that the correlation coefficient r is given by

$$r = \frac{\sum (x - \bar{x})(y - \bar{y})}{(n - 1)s_x s_y}$$

where s_x and s_y represent the sample standard deviation of the x data and the y data, respectively. We can express the correlation coefficient r as

$$r = \pm \sqrt{r^2}$$

where r^2 is the coefficient of determination. The correlation coefficient r takes the same sign as the slope b_1. If the slope b_1 of the regression equation is positive, then $r = \sqrt{r^2}$; if the slope b_1 of the regression equation is negative, then $r = -\sqrt{r^2}$.

EXAMPLE 17 Calculate the correlation coefficient using r^2

Use r^2 to calculate the value of the correlation coefficient r for the memory score data.

Solution

The slope $b_1 = 2$, which is positive, tells us that the sign of the correlation coefficient r is positive. Thus,

$$r = \sqrt{r^2} = \sqrt{0.9474} \approx 0.9733$$

NOW YOU CAN DO
Exercises 11f–22f.

Therefore, student scores on the short-term memory test are strongly positively correlated with the amount of time allowed for memorization.

Section 4.3 Summary

1. The sum of squared prediction errors is referred to as the sum of squares error, $\text{SSE} = \Sigma(y - \hat{y})^2$. The standard error of the estimate, $s = \sqrt{\frac{\text{SSE}}{n-2}}$, is an indicator of the precision of the estimates derived from the regression equation because it provides a measure of the typical residual or prediction error.

2. The total variability in the y variable is measured by the total sum of squares, $\text{SST} = \Sigma(y - \bar{y})^2$, and may be divided into the sum of squares regression, $\text{SSR} = \Sigma(\hat{y} - \bar{y})^2$, and the sum of squares error, $\text{SSE} = \Sigma(y - \hat{y})^2$. SSR measures the

amount of improvement in the accuracy of estimates when using the regression equation compared with ignoring the x information.

3. The coefficient of determination, $r^2 = \text{SSR}/\text{SST}$, measures the goodness of fit of the regression equation as an approximation of the relationship between x and y. Finally, the correlation coefficient r may be expressed as $r = \pm\sqrt{r^2}$, taking the positive or negative sign of the slope b_1.

Section 4.3 Exercises

CLARIFYING THE CONCEPTS

1. What does s measure? Would we want s to be large or small? Why? (p. 228)

2. How does the least-squares criterion choose the "best" line to approximate the relationship between x and y? (p. 226)

3. What does SSE measure? Would we want SSE to be large or small? Why? (p. 226)

4. What does SSR measure? Would we want SSR to be large or small? Why? (p. 230)

5. What does SST measure? What statistic is it proportional to? (p. 229)

6. What does it mean when r^2 is close to 1? How about when it is close to 0? (p. 231)

7. Do the values of x affect SST at all? (p. 229)

8. Suppose we performed a regression analysis that resulted in $r^2 = 0.64$. Without further information, would it be possible to calculate the correlation coefficient r? Explain. (p. 231)

9. Suppose we performed a regression analysis on a data set that resulted in $r^2 = 0.64$. Interpret this statistic in terms of the amount of variance in y explained by the linear relationship between x and y. (p. 231)

10. True or false: When the prediction errors are too small, the sum of squared error SSE can be negative. (p. 230)

PRACTICING THE TECHNIQUES

 CHECK IT OUT!

To do	Check out	Topic
Exercises 11a–22a	Example 12	SSE, the sum of squares error
Exercises 11b–22b	Example 13	Standard error of the estimate, s
Exercises 11c–22c	Example 14	SST, the total sum of squares
Exercises 11d–22d	Example 15	SSR, the regression sum of squares
Exercises 11e–22e	Example 16	r^2, the coefficient of determination
Exercises 11f–22f	Example 17	Calculating r using r^2

For Exercises 11–22, use the regression equations you calculated in Exercises 13–24 in Section 4.2. Do the following:

a. Calculate the sum of squares error, SSE.

b. Compute and interpret the standard error of the estimate, s.

c. Calculate the total sum of squares, SST.

d. Find the sum of squares regression, SSR.

e. Calculate and interpret the coefficient of determination, r^2.

f. Use r^2 to calculate the correlation coefficient, r.

11.

x	10	20	30	40
y	2	5	9	12

12.

x	0	2	4	6
y	50	60	50	40

13.

x	−5	−4	−3	−2	−1
y	10	18	18	26	26

14.

x	−3	−1	1	3	5
y	−25	−35	−40	−45	−50

15.

x	5	10	15	20	25	30
y	7	8	8	8	7	8

16.

x	6	7	8	9	11	13
y	9	9	9	9	9	9

17.

x	−70	−60	−50	−40	−30	−20	−10	0
y	5	10	15	20	20	15	10	5

18.

x	−30	−23	−15	−12	−1	5	14	29
y	103	88	76	62	54	47	30	20

19. The heights (in inches) and weights (in pounds) of a sample of five women are recorded.

x = height	y = weight
66	122
67	133
69	153
68	138
65	125

20. The number of days absent from a class, and the course grade.

x = days absent	y = course grade
0	95
2	90
4	85
6	70
8	60

21. The number of hours spent on a repair job, and the cost of the repair.

x = hours	y = cost
2	120
3	180
5	230
8	350
10	380

22. The amount of rainfall at a baseball stadium (in inches), and attendance at the baseball game that day (in thousands).

x = rain	y = attendance
0.0	40
0.0	42
0.1	38
0.5	30
1.0	20

APPLYING THE CONCEPTS

For Exercises 23–37, follow these steps. You have already calculated the regression equation in Exercises 37–51 in Section 4.2. Do the following.

a. Calculate the sum of squares error, SSE.

b. Compute and interpret the standard error of the estimate, s.

c. Calculate the total sum of squares, SST.

d. Find the sum of squares regression, SSR.

e. Calculate and interpret the coefficient of determination, r^2.

f. Use r^2 to calculate and interpret the correlation coefficient, r.

23. Video Game Sales. See Exercise 37 in Section 4.2.

24. Does It Pay to Stay in School? See Exercise 38 in Section 4.2.

25. Darts and the Dow Jones. See Exercise 39 in Section 4.2.

26. Age and Height. See Exercise 40 in Section 4.2.

27. Gardasil Shots and Age. See Exercise 41 in Section 4.2.

28. NCAA Power Ratings. See Exercise 42 in Section 4.2.

29. Saturated Fat and Calories. See Exercise 43 in Section 4.2.

30. Engine Displacement and Gas Mileage. See Exercise 44 in Section 4.2.

31. Completing College. See Exercise 45 in Section 4.2.

32. Walking or Biking to Work. See Exercise 46 in Section 4.2.

33. Teenage Birth Rate. See Exercise 47 in Section 4.2.

34. Brain and Body Weight. See Exercise 48 in Section 4.2.

35. Consumer Sentiment. See Exercise 49 in Section 4.2.

36. SAT Scores, by Foreign Language. See Exercise 50 in Section 4.2.

37. Batting Average and Runs Scored. See Exercise 51 in Section 4.2.

Regression in Accounting. Use the accounting data from Exercises 50–56 in Section 4.1 to answer Exercises 38–44.

38. Compute a new variable, called *net worth*, which equals *assets – liabilities*.

39. Perform a regression of current ratio on net worth. Interpret the coefficients.

40. Regress current ratio on price-earnings ratio. Interpret the coefficients.

41. What proportion of the variability in current ratio is accounted for by the following?
  a. Net worth
  b. Price-earnings ratio

42. How large is the typical residual when predicting current ratio using the following predictors?
  a. Net worth
  b. Price-earnings ratio

43. Based on your answers to Exercises 41 and 42, which predictor, net worth or price-earnings ratio, is more useful for predicting current ratio?

44. Use the value of SST from Exercise 39 to calculate the variance of the current ratio data.

Best Places for Dating. Use the dating data from Exercises 57–59 in Section 4.1 to answer Exercises 45–48.

45. Perform the following regressions:
  a. Overall dating score on percentage 18–24 years old
  b. Overall dating score on percentage 18–24 years old who are single
  c. Overall dating score versus online dating score

46. What proportion of the variability in overall dating score is accounted for by the following?
  a. Percentage 18–24 years old
  b. Percentage 18–24 years old who are single
  c. Online data score

47. How large is the typical residual when predicting overall dating score using the following predictors?
  a. Percentage 18–24 years old
  b. Percentage 18–24 years old who are single
  c. Online data score

48. Based on your answers to Exercises 46 and 47, which predictor is most helpful for predicting overall dating score?

Virginia Weather. Use the data from Exercises 60–65 in Section 4.1 to answer Exercises 49–56.

49. Perform a regression of heating degree-days on average January temperature.

50. Interpret the slope of the regression line from Exercise 49.

51. What is the size of the typical error in estimating heating degree-days using average January temperature?

52. What proportion of the variability in heating degree-days is accounted for by average January temperature?

53. Regress cooling degree-days on average July temperature.

54. How do we interpret the value of the slope of the regression equation from Exercise 53?

55. Calculate the size of the typical residual when using average July temperature to predict cooling degree-days.

56. Find the proportion of the variability in cooling degree-days that is accounted for by the average July temperature.

Does It Pay to Stay in School? Refer to your work in Exercise 24 for Exercises 57 and 58.

57. Answer the following:
  a. Which data value has the largest residual? Describe what is unusual about this observation.
  b. Suppose a public figure stated that 50% of the variability in the unemployment rate was due to competition from abroad. How would you use the regression results to respond to this claim?
  c. Suppose a politician claimed that using the years of education alone could allow us to predict the unemployment rate to within 1%. How would you use the regression results to respond to this claim?
  d. Suppose a newspaper claimed that each additional year of education brought down the unemployment rate by "more than 1%." How would you use the regression results to either support or refute this claim?

WHAT IF
?
58. *What if* the unemployment rate for individuals with 5 years of education was not 16.8% but a much higher percentage? Describe how this would affect the slope and *y* intercept of the regression line. Explain your reasoning.

59. Computational Formula for SST and SSR. The alternate computational formulas for finding SST and SSR are as follows:

$$\text{SST} = \sum y^2 - \left(\sum y\right)^2 / n \qquad \text{SSR} = \frac{\left[\sum xy - \left(\sum x\right)\left(\sum y\right)/n\right]^2}{\sum x^2 - \left(\sum x\right)^2 / n}$$

Use the computational formulas to find SSR and SST for the memory score data on page 226. Assume we have the following summary statistics: $\Sigma x = 40$, $\Sigma y = 150$, $\Sigma xy = 708$, $\Sigma x^2 = 214$, $\Sigma y^2 = 2478$.

BRINGING IT ALL TOGETHER

Fuel Economy. For Exercises 60–67, refer to the table of fuel economy data from Exercises 80–84 in Section 4.2. The predictor variable is x = engine size, expressed in liters; the response variable is y = combined (city/highway) gas mileage, expressed in miles per gallon (mpg).

60. Calculating and interpreting the residuals and SSE and s.
  a. Compute the residual for each data value. Form a table similar to Table 7 of the residuals and squared residuals. Sum the squared residuals to get SSE.
  b. What does SSE measure? At this point, do we know whether SSE is large or small? Why or why not?
  c. Which vehicle has the largest absolute residual? Clearly explain why this vehicle is unusual.

61. Calculating and Interpreting s.
  a. Calculate the value of s, the standard error of the estimate.
  b. Interpret the value of s, so that a nonstatistician could understand it.

62. Computing and Interpreting SST, SSR, and r^2.
  a. Calculate the sample variance of the y data, s^2. Then use s^2 to calculate SST.

b. Use SSE and SST to find SSR. Explain clearly what it is that SSR is measuring.

c. Calculate and interpret the coefficient of determination, r^2.

63. Correlation. Do the following:

a. Use r^2 and b_1 to find the correlation coefficient, r.

b. Interpret the correlation between engine size and combined mpg.

WHAT IF ? **64.** *What if* we added one new vehicle to the data set, and its value was exactly $(\bar{x}, \bar{y})$? How would this affect the slope and the y intercept?

WHAT IF ? **65.** Refer to the previous exercise. *What if* we added an unknown amount to the engine size of the new vehicle? Describe how this change would affect the slope and the y intercept.

66. Challenge Exercise. Suppose we increased the combined mpg for the Cadillac limousine, so that the slope of the regression line would be exactly zero. What would the combined mpg for the Cadillac limousine have to be to accomplish this?

67. Challenge Exercise. Refer to the previous exercise. Describe how this change to the fuel economy of the Cadillac limousine would affect each of the following, and why: SSE, SSR, SST, s, r^2, r.

WORKING WITH LARGE DATA SETS

For Exercises 68–70, use technology and follow steps **(a)**–**(e)**.

a. Construct the scatterplot.

b. Compute and interpret the regression equation.

c. Calculate and interpret the coefficient of determination, r^2.

d. Compute and interpret s, the standard error of the estimate.

e. Find r, using r^2.

68. Open the *darts* data set, which we used for the Chapter 3 Case Study. Let $x =$ the Dow Jones Industrial Average, and let $y =$ the pros' performance. **▮▮▮ darts**

69. Open the *nutrition* data set. Let $x =$ the amount of fat per gram, and let $y =$ the number of calories per gram. **▮▮▮ nutrition**

70. Open the *pulse and temp* data set. Let $x =$ heart rate, and let $y =$ body temperature. **▮▮▮ pulseandtemp**

CONSTRUCT YOUR OWN DATA SETS

Suppose we have a tiny data set with the following (x, y) pairs:

x	y
1	?
2	?
3	?

For Exercises 71–75, create a set of y-values that would fulfill each specification.

71. The slope of the line is positive.

72. The slope of the line is negative.

73. The slope of the line is 0.

74. The slope of the line is equal to 2.

75. The slope of the line is equal to -3.

Use the *Correlation and Regression* applet for Exercises 76–78.

76. In these applet exercises, use the "thermometer" above the graph (where it says "Sum of squares $=$") to help find the least-squares regression line interactively.

a. Select five points so that the correlation coefficient is about 0.8. Then select "Draw line."

b. Make your best guess about where the least-squares regression line should be, and draw the line there.

77. The blue section of the thermometer is a measure of the sum of squares error, which is the total squared vertical distance from the data points to the actual regression line. Recall that the least-squares regression line minimizes this distance. The green section of the thermometer tells you how much "extra" squared error you get from using the line you constructed in Exercise 76(**a**).

a. Adjust the line you drew in Exercise 76(**a**) by clicking and dragging on the points until the green section of the thermometer has disappeared.

b. What does the disappearance of the green part tell you about the adjusted line you constructed?

c. Will the line now coincide with the least-squares regression line?

78. Verify that your adjusted line from Exercise 77 coincides with the least-squares regression line by selecting "Show least-squares line."

WORKING WITH LARGE DATA SETS

Chapter 4 Case Study: Measuring the Human Body. Open the data sets **body_females** and **body_males**. We shall apply what we have learned in this section regarding regression analysis to some of the measurements in these data sets. Use technology for Exercises 79–92. **▮▮▮ body_females**
▮▮▮ body_males

79. For the regression of weight on height for females, find and interpret the standard error of the estimate, s.

80. For the regression of weight on height for females, calculate and interpret the coefficient of determination, r^2

81. Use r^2 and the sign for the slope to compute the correlation coefficient for women's heights and weights.

82. For the regression of weight on height for men, find and interpret the standard error of the estimate, s.

83. For the regression of weight on height for males, calculate and interpret the coefficient of determination, r^2.

84. Use r^2 and the sign for the slope to compute the correlation coefficient for men's heights and weights.

85. Discuss similarities and differences among your results for the females and the males, with respect to the regression of weight on height, as follows:
    a. The standard error of the estimate, s
    b. The coefficient of determination, r^2.

86. Perform a regression of bicep girth on thigh girth for females. State the regression equation in words. Interpret the slope and y intercept values.

87. Find and interpret the standard error of the estimate for the regression of bicep girth on thigh girth for females.

88. For the regression of bicep girth on thigh girth for females, calculate and interpret the coefficient of determination, r^2.

89. Now run a regression of bicep girth on thigh girth for the men. State the regression equation in words. Interpret the slope and y intercept values.

90. Find and interpret the standard error of the estimate for the regression of bicep girth on thigh girth for males.

91. For the regression of bicep girth on thigh girth for males, calculate and interpret the coefficient of determination, r^2.

92. Discuss similarities and differences among your results for the females and the males, with respect to the regression of bicep girth on thigh girth, as follows:
    a. The slope
    b. The y intercept
    c. The standard error of the estimate, s
    d. The coefficient of determination, r^2.

Chapter 4 Formulas and Vocabulary

SECTION 4.1
- **Correlation coefficient, r** (p. 192).

Definition formula:

$$r = \frac{\sum (x - \bar{x})(y - \bar{y})}{(n-1)s_x s_y}$$

Computational formula:

$$r = \frac{\sum xy - \left(\sum x \sum y\right)/n}{\sqrt{\left[\sum x^2 - \left(\sum x\right)^2/n\right]\left[\sum y^2 - \left(\sum y\right)^2/n\right]}}$$

- **Positive and negative correlation** (p. 196).
- **Scatterplot** (p. 188)

SECTION 4.2
- **Extrapolation** (p. 215)
- **Prediction error, or residual** (p. 214).

$$(y - \hat{y})$$

- **Regression equation (regression line)** (p. 210).

$$\hat{y} = b_1 x + b_0$$

- **Slope of the regression line** (p. 210).

Definition formula:

$$b_1 = \frac{\sum (x - \bar{x})(y - \bar{y})}{\sum (x - \bar{x})^2}$$

Computational formula:

$$b_1 = \frac{\sum xy - \left(\sum x \sum y\right)/n}{\sum x^2 - \left(\sum x\right)^2/n}$$

- **y Intercept of the regression line** (p. 210).

$$b_0 = \bar{y} - (b_1 \cdot \bar{x})$$

SECTION 4.3
- **Coefficient of determination, r^2** (p. 230).

$$r^2 = SSR/SST$$

- **Least-squares criterion** (p. 226)
- **SSE, sum of squares error** (p. 226).

$$SSE = \sum (y - \hat{y})^2$$

- **Standard error of the estimate, s** (p. 228).

$$s = \sqrt{\frac{SSE}{n-2}}$$

- **SSR, sum of squares regression** (p. 230).

Definition formula:

$$SSR = \sum (\hat{y} - \bar{y})^2$$

Computational formula:

$$SSR = \frac{\left[\sum xy - \left(\sum x\right)\left(\sum y\right)/n\right]^2}{\sum x^2 - \left(\sum x\right)^2/n}$$

- **SST, total sum of squares** (p. 229).

Definition formula:

$$SST = \sum (y - \bar{y})^2$$

Computational formula:

$$SST = \sum y^2 - \left(\sum y\right)^2/n$$

Chapter 4 Review Exercises

SECTION 4.1

Square footage and home price in Broward County, Florida. The following table contains the square footage (x) and price (y) of a sample of homes in Broward County, Florida. Use this data for Exercises 1–17. 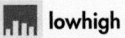 **browardprices**

Home	Square feet (x)	Price (y)
1	1500	$156,250
2	2000	$225,000
3	2000	$325,000
4	2500	$412,500
5	2500	$275,000
6	3000	$525,000
7	3000	$625,000
8	3500	$756,250

1. Construct a scatterplot of price versus square feet.
2. Refer to your scatterplot from Exercise 1. Characterize the relationship as positive, negative, or not apparent.
3. Write a sentence that describes the behavior of the price as the square footage increases.
4. Calculate the value of the correlation coefficient r between price and square feet.
5. Interpret the value for r.

SECTION 4.2

6. Calculate the regression coefficients b_0 and b_1, and write the regression equation.
7. State the regression equation in words.
8. Interpret the value of the slope b_1.
9. Interpret the value of the y intercept b_0.
10. Use the regression equation to predict the price for a home with 3500 square feet.
11. Calculate and interpret the prediction error in Exercise 10.
12. Use the regression equation to predict the price for a home with 4500 square feet. State whether the prediction represents extrapolation.

SECTION 4.3

13. Calculate SSE.
14. Calculate s, the standard error of the estimate. What does this number mean?
15. Calculate SST, and then use SSE and SST to find SSR.
16. Calculate r^2, the coefficient of determination. Comment on how useful square footage is for predicting price.
17. Use r^2 to calculate the correlation coefficient. Comment on the relationship between square footage and home price.

Chapter 4 QUIZ

TRUE OR FALSE

1. True or false: Scatterplots are constructed with the y variable on the horizontal axis and the x variable on the vertical axis.
2. True or false: The y intercept measures the strength of the linear relationship between two numerical variables.

FILL IN THE BLANK

3. The "hat" over the y in $\hat{y}$ indicates that it is a(n) _____ of y.
4. We interpret the slope of the regression line as the estimated change in y per _____ increase in x.

SHORT ANSWER

5. Making predictions based on x-values that are beyond the range of the x-values in our data set is called what?
6. Values of r close to -1 indicate what type of relationship between the two variables?

CALCULATIONS AND INTERPRETATIONS

High and Low Temperature. Use the following information for Exercises 7–14. The low (x) and high (y) temperatures (in degrees Fahrenheit) for a sample of eight cities are shown in the following table.

 lowhigh

Low temperature	High temperature
0	24
10	32
10	38
20	36
20	44
30	45
30	53
40	59
40	65

7. Construct a scatterplot of high temperature versus low temperature.
8. Based on your scatterplot, would you characterize the linear relationship, if any, as positive or negative?
9. Compute the regression equation of high temperature versus low temperature.
10. Calculate the three sums of squares: SSR, SST, and SSE.
11. Calculate s, the standard error of the estimate. What does this number mean?
12. Calculate r^2. Comment on how useful the low temperatures are in predicting the high temperatures.
13. Use r^2 to calculate and interpret the correlation coefficient.
14. Find the prediction error for when the low temperature is zero.

5 Probability

Introduction

Chapter 5 explains the tools of probability, which enable data analysts to quantify the level of uncertainty in statistical inference. Section 5.1 introduces experiments, outcomes, events, and sample spaces. The classical method, the relative frequency method, and the subjective method of assigning probabilities are introduced, as well as the Law of Large Numbers. In Section 5.2, students learn about combining events using the concepts of union, intersection, and complement. Section 5.3 examines conditional probability, independent events, and sampling with and without replacement. In Section 5.4, counting methods are used to find the number of combinations and permutations of a set of objects.

From the Author

Section 5.1 Introducing Probability

- You may want to stress the "difference in perspective" of this chapter from previous chapters. See the Developing Your Statistical Sense feature on page 241.

- Two new examples are provided to ease the student's learning curve into probability: Example 1 Probability Models, and Example 2 Determining Unusual or Likely Outcomes.

- You may note the increased use of contingency tables (crosstabulations) in Chapter 5 to help explain various aspects of probability, such as Example 10 Relative Frequency Method, which uses the Chapter 5 Case Study: The Gardasil Vaccine.

- The probability chapter offers many opportunities to increase student motivation by talking about the application of probability to games of chance. For example, many students enjoy working with the two-dice sample space or the deck of cards sample space.

Section 5.2 Combining Events

- Students who have trouble with the concept of intersection may benefit from a look at Table 6 on page 261. Consider the row to be a "street" and the column to be a different street, then female survivors of the *Titanic* lie in the *intersection* of these two streets. Contingency tables work well for this.

Section 5.3 Conditional Probability

- In conditional probability, students often confuse "*B* given *A*" with "*B* and *A*." Stress the Developing Your Statistical Sense feature along with Figure 17 on page 272, to help dispel this confusion.

- It may save your gambling students a lot of money if you discuss the Gambler's Fallacy in Example 21 on page 275.

- Coverage of Bayes' Rule has been added to this section. See page 282 and Example 31.

Section 5.4 Counting Methods

- Many business and engineering students may need to learn quality control methods. *Acceptance sampling* on page 300 is a widespread method for businesses to track quality across their enterprises.

Teaching Tips

In Chapters 1–4, the students used statistics, tables, and graphs to describe a data set. This is descriptive statistics. However, here in Chapter 5, we need to shift our point of view. No longer are we given a data set to describe. Instead, we are presented with an experiment, with certain outcomes and events defined. We need to find probabilities of these outcomes and events. This is the realm of probability, which has a different perspective from descriptive statistics.

In-Class Activities

1. To explain the difference between the classical method and the relative frequency method of assigning probabilities, first ask students, "What is the probability of rolling a 72 with a 100-sided die? This is an example of an event where we use the classical method of probability." Then ask, "What is the probability of an event that uses the relative frequency method, such as selecting a student at random from a college in your state?" and "What is the probability that the student is enrolled in an elementary statistics class that semester? These are examples of events where we use the relative frequency method of probability." What is the difference in how we find the probabilities in these cases? Why is the first a use of the classical method and the last the relative frequency method. Students will be able to find the first probability even if they have never rolled a 100-sided die before, but they won't be able to find the second probability without knowing the number of students enrolled at a college in their state and the number of students enrolled in an elementary statistics class.

2. To explain the counting rules, ask students to think about how they would perform the task. Suppose you wanted to line up six people. Ask them how many ways can be used to select a first person in line. Have them select a first person. Now ask them how many ways can be used to select a second person in line. Have them select a second person. Continue this process until all six people are lined up. Then show that there are $6 \cdot 5 \cdot 4 \cdot 3 \cdot 2 \cdot 1 = 720$ ways to line up six people.

3. To illustrate the Law of Large Numbers, have students do an experiment in class, such as flipping a coin or rolling a die. Divide them into groups of 2 or 3, and perform the experiment 50 times. Ask students to keep track of the number of heads and tails if they flip coins or the number of 1s, 2s, . . . 6s if they roll a die. Pool all the results, and have students calculate the relative frequencies. Ask them to compare the relative frequencies to the theoretical probabilities.

4. Have you ever taken a multiple-choice test where you couldn't decide between the alternatives and wished you could flip a coin or roll a die to decide? Suppose you are facing a four-question multiple-choice exam, with three alternatives (A, B, or C) for each question, and you have no clue as to which answers are correct. In this activity, you will calculate the probability of getting one or more answers right.

  a. Why does it make sense to use the classical method?

  b. Draw a tree diagram for this four-question test—the experiment.

  c. What is the probability that you will answer a particular question correctly?

  d. What is the probability that you will answer two questions correctly?

  e. What is the probability that you will answer three questions correctly?

  f. What is the probability that you will answer all four questions correctly?

  g. Suppose that you need to answer at least three out of the four questions correctly to pass the test. What is the probability that you will pass?

5. We use simulation to estimate the probabilities we calculated in Activity 4. We use a single fair die to determine our responses.

  a. How should we "code" the die results, so that there is an even chance of choosing A, B, or C?

  b. Your answer to each exam question will be either right or wrong. Why not use a coin toss instead of a die roll for this simulation?

  c. We need to code the probability of getting a correct answer. For each question, A, B, or C is correct. We want to simulate the probability of guessing correctly, which is 1/3. How do you code the die roll so that the probability of guessing correctly is 1/3?

  d. Perform 20 trials of the experiment. Rolling the die one time is considered as answering one test question. Therefore, rolling the die four times is considered as one "trial" of the experiment (taking the test). For each trial, keep track of the number of "right" and "wrong" guesses. Answer questions (**e**)–(**h**) using your simulation results.

  e. Estimate the probability that you will answer the first and only the first question correctly.

  f. Estimate the probability that you will answer exactly two questions correctly.

  g. Estimate the probability that you will answer exactly three questions correctly.

  h. Estimate the probability that you will answer all four questions correctly.

  i. Estimate the probability that you will pass the test.

Supplements

- StatTutor 6.1, 6.2, 10.1–10.3, 12.1–12.6

Applets

The *Law of Large Numbers* applet is referenced in Chapter 5 to compute relative frequencies for experiments and for Exercises 99 and 100 in Section 5.1. Activities on probability simulation can be found at http://mathforum.org/mathtools/sitemap2/ps/. The site Online Statistics:

An Interactive Multimedia Course of Study has a *Probability* applet that demonstrates several simulations: http://onlinestatbook.com/index.html.

Videos

- *Against All Odds: Inside Statistics:* www.learner.org/resources/series65.html
  - Program 18: Introduction to Probability
  - Program 19: Probability Models

Web Sites

- CAUSEweb provides resources for teaching statistics: https://www.causeweb.org/.
- This Web site has a collection of 20 class projects: www.amstat.org/publications/jse/v6n3/smith.html.
- This Texas Instruments Web site has a host of TI-83/84 statistics activities: http://education.ti.com/educationportal/sites/US/nonProductSingle/activitybook_83_statistics.html.
- This Web site lists other sites that do statistical calculations: http://statpages.org/.
- This Web site lists numerous activities on all topics: http://mathforum.org/mathtools/sitemap2/ps/.

5 Probability

OVERVIEW

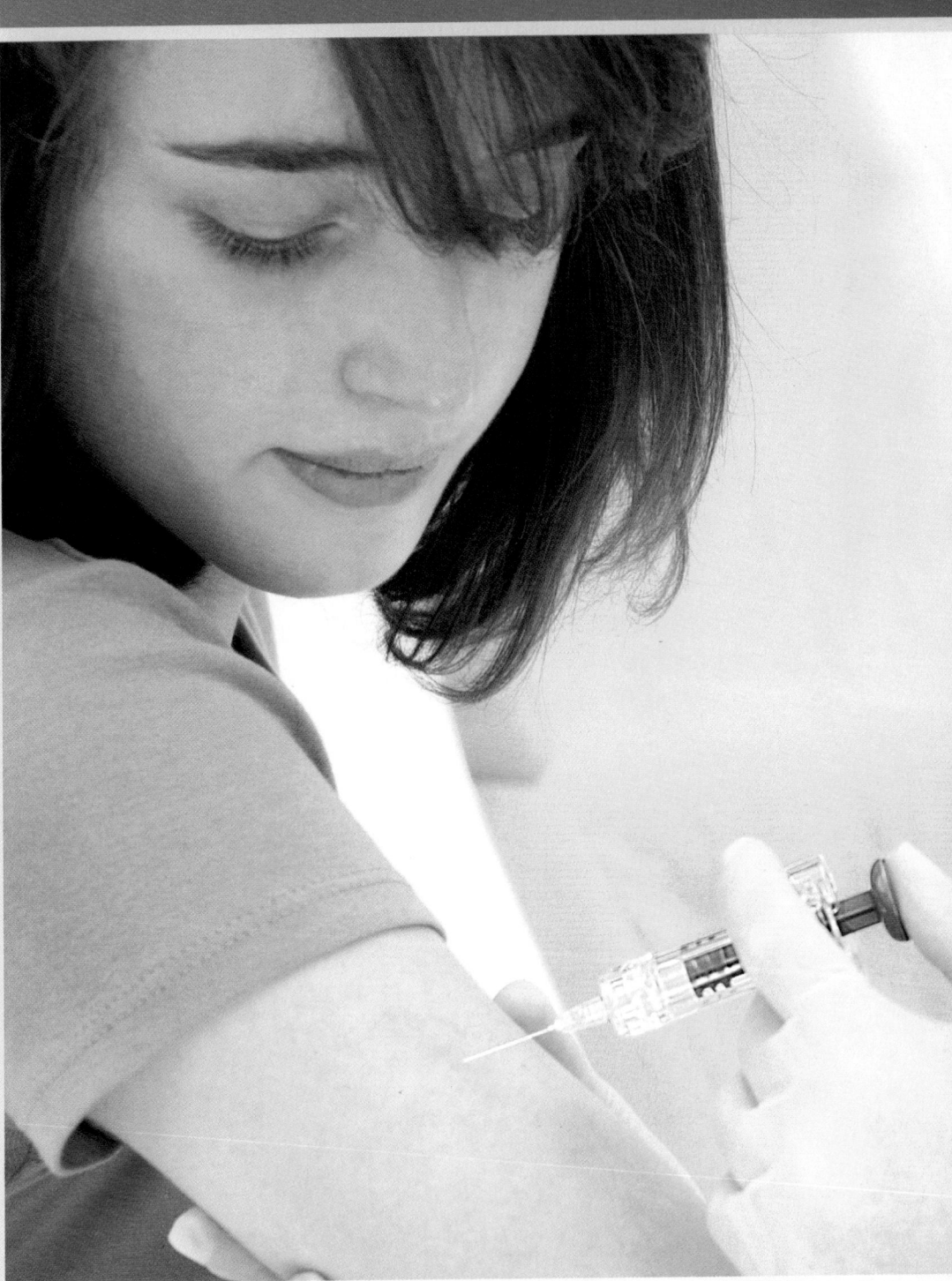

MAY/BSIP/Superstock

The Gardasil Vaccine

The Centers for Disease Control and Prevention (CDC) recommends routine vaccination with Gardasil for boys and girls ages 11 and 12, in order to combat the HPV virus. One reason for this is that it has been shown that 100% of cervical cancer patients have been infected with the HPV virus.[1] The large data set **Gardasil** (*Source: Journal of Statistics Education*[2]) contains information regarding 1413 young women who underwent vaccination treatment. To complete the vaccination, a series of three shots must be undertaken. In the Chapter 5 Case Study, The Gardasil Vaccine, we explore whether patient location, insurance type, gender, and other factors affect a patient's probability of completing the vaccine treatment.

- In Section 5.1, we use the relative frequency method to assign probabilities to whether or not patients completed their Gardasil treatment.

- In the Section 5.1 exercises, working with the large data set **Gardasil**, we explore some probabilities about clinic location and patient insurance type, using the tools and techniques learned in the section.

- Again working with the large data set **Gardasil**, in the Section 5.2 exercises we examine probabilities regarding patient completion of the treatment and whether the patient received medical assistance, using contingency tables.

- In an example in Section 5.3, we apply the 1% Guideline to help find the probability that two randomly selected patients completed the treatment.

- Later in Section 5.3, we apply Bayes' Rule using a crosstabulation of the variables *completed* and *insurance type*.

- Then, in the Section 5.3 exercises, we use a contingency table of *completed* and *practice type* (pediatric, family, or OB/GYN) to calculate some conditional probabilities. Later in the Section 5.3 exercises, we work with the large data set **Gardasil** to explore some conditional probabilities about patient completion and the *age group* of the patient.

THE BIG PICTURE

Where we are coming from and where we are headed . . .

- Chapters 1–4 dealt with descriptive statistics that summarize data. In later chapters, we will learn inferential statistics, which generalize from a sample to a population. But generalizing involves *uncertainty*.

- Chapter 5 teaches us the language of uncertainty: **probability**. We will learn how to quantify uncertainty, using experiments, events, outcomes, rules for combining events, conditional probability, and counting methods.

- In Chapter 6, "Probability Distributions," we learn about the two most important probability distributions, the normal and the binomial, which will be our companions for the remainder of the text.

5.1 Introducing Probability

OBJECTIVES By the end of this section, I will be able to . . .

1 Understand the meaning of an experiment, an outcome, an event, and a sample space.
2 Describe the classical method of assigning probability.
3 Explain the Law of Large Numbers and the relative frequency method of assigning probability.

Imagine you are striding down the midway of your local town fair, when a particular game of chance catches your eye. The object of this game is to roll a six on a single roll of a single fair die. If you do so, you win $5. It costs $1 to play the game. What is the likelihood of winning?

To show how to solve this problem, we must first introduce the building blocks of probability.

1 Building Blocks of Probability

Our daily lives are filled with *uncertainty,* seemingly governed by *chance.* We try to cope with uncertainty by estimating the *chances* that a particular event will occur. We are expected, on a daily basis, to make intelligent decisions about probabilities. Consider the following scenarios, and think about how the italicized words all refer to uncertainty:

- What is the *chance* that there will be a speed trap on this stretch of I-95 on a particular day?
- What is the *likelihood* that this lottery ticket will make me rich?
- What is the *probability* that this throw of the dice will come up a seven?

Sometimes, the amount of uncertainty in our daily lives is so great that there appears to be no order to the world whatsoever. However, if you look closely, there are *patterns in randomness.* In this chapter, we learn to become better decision makers by becoming acquainted with the tools of probability in order to quantify many of the uncertainties of everyday life.

> The **probability** of an outcome represents the chance or likelihood that the outcome will occur.

Let us acquaint ourselves with the building blocks of probability, starting with the concept of an experiment. In probability, an **experiment** is any activity for which the outcome is uncertain. Consider the stock market, for example. Suppose you own 100 shares of Consolidated Widgets and are interested in what the share price will be at the end of trading tomorrow. Will the share price increase or decrease? The actual result is uncertain, so this is an example of an experiment. Each of the possible results of the experiment is called an **outcome**. Another example of an experiment is when you toss a coin. In the coin-toss experiment, the result may be heads or it may be tails. The collection of all possible outcomes is called the **sample space**. The sample space for the coin-toss experiment is {heads, tails} or {H, T}. Following are some common experiments, together with their sample spaces.

We use brackets like these { } to enclose a set of outcomes.

Experiment	Sample space
Roll a single six-sided die	{1, 2, 3, 4, 5, 6}
Toss two coins	{HH, HT, TH, TT}
Play a video game	{win, lose}

We use the building blocks of probability to investigate the likelihood of an outcome or **event**.

Building Blocks of Probability

An **experiment** is any activity for which the outcome is uncertain.

An **outcome** is the result of a single performance of an experiment.

The collection of all possible outcomes is called the **sample space**. We denote the sample space S.

An **event** is a collection of outcomes from the sample space. To find the probability of an event, add up the probabilities of all the outcomes in the event.

The following table shows some typical events for the experiments in the table on page 240.

Experiment	Sample space	Typical events
Roll a single die	{1, 2, 3, 4, 5, 6}	E: roll an even number = {2, 4, 6} L: roll a 4 or larger = {4, 5, 6}
Toss two coins	{HH, HT, TH, TT}	H: exactly one head = {HT, TH} T: at most one tail = {HH, HT, TH}
Play a video game	{win, lose}	W: win = {win} L: lose = {lose}

When we talk about the probability of some outcome, we are referring to a number that indicates how likely the particular outcome is. The notation $P(A)$ stands for "the probability that outcome A occurred." Say we define outcome W to be "you win the video game." Then "the probability that you win the video game" can be denoted as $P(W)$. Probabilities abide by the following rules.

Rules of Probability

1. The probability $P(E)$ for any event E is always between 0 and 1, inclusive. That is, $0 \le P(E) \le 1$.

2. **Law of Total Probability:** For any experiment, the sum of all the outcome probabilities in the sample space must equal 1.

Developing Your Statistical Sense

A Different Perspective

As you read this chapter, notice that the perspective differs from that in previous chapters. Earlier, we were looking at a data set and trying to describe it graphically and numerically. Now, instead of trying to describe a data set, we are faced with an experimental situation, and our task is to calculate probabilities associated with various outcomes in the experiment.

CAUTION If the probability that you calculated is negative or greater than 1, then you should try again.

From the definition, the probability of an event is a proportion. Therefore, probability cannot be negative because proportions cannot be negative, and probability cannot be greater than 1 (100%) because an event cannot occur more than 100% of the time. A **probability model** is a table or listing of all the possible outcomes of an experiment, together with the probability of each outcome. A probability model must follow the Rules of Probability.

EXAMPLE 1 Probability models

For the following three scenarios, determine whether or not a valid probability model exists:

a. You ask your friend what his favorite ice cream flavor is.

Outcome	Probability
Mint chocolate chip	0.25
Pistachio	−0.25
Vanilla	0.50
Other	0.50

b. You ask another friend to go to a concert.

Outcome	Probability
Yes	0.5
No	0.5
Maybe	0.5

c. You play a single game of roulette and bet on red.

Outcome	Probability
Win	0.47
Lose	0.53

Solution

To be a valid probability model, each scenario must meet both rules of probability.

a. The probability that your friend likes pistachio ice cream may not be very high, but it cannot be negative. This violates the first rule of probability that every event must have probability between zero and one. Note that, even though the second rule of probability is fulfilled (the probabilities sum to one), we nevertheless do not have a valid probability model here.

b. Here, the first rule of probability is met, but the second one is violated. The sum of the three probabilities is $0.5 + 0.5 + 0.5 = 1.5$. This is not equal to one, so that we do not have a valid probability model here either.

c. Here, both probabilities lie between zero and one, so that the first rule of probability is met. Also, the sum of the two probabilities is $0.47 + 0.53 = 1.00$, which fulfills the second rule of probability. Therefore, (**c**) represents a valid probability model.

NOW YOU CAN DO
Exercises 11–16.

Throughout the remainder of this book, you will often be asked to calculate the probability of various events. Table 1 contains the meanings of some values of probability.

Table 1 Meaning of some probability values

Probability value	Meaning
Near 0	Outcome or event is very unlikely.
Equal to 0	Outcome or event cannot occur.
Near 1	Outcome or event is nearly certain to occur.
Equal to 1	Outcome or event is certain to occur. It's "a sure thing."
Low	Outcome or event is unusual, say, with probability less than 0.05.
High	Outcome or event is likely to happen.

The threshold of an unusual event depends on the specific experiment; the 0.05 probability for an unusual event is not set in stone.

Higher probability values are associated with higher likelihood of occurrence. An outcome with probability 0.5 will happen about half of the time. An outcome with probability 0.95 is very likely. We say that an outcome or event is *unusual* if its probability is below a certain threshold, say, 0.05.

EXAMPLE 2   Determining unusual or likely outcomes

For the following probabilities, determine the meaning of the indicated probability:

a. The weatherman said that the chance of rain tonight was near 100%.

b. The coach said that after that last loss, the team had no chance to make the playoffs.

c. The doctor said that the chance of a negative reaction to the planned procedure was 3%.

Solution

a. A value of near 100% is near 1. Therefore, it is nearly certain to occur that we will have rain tonight.

b. No chance means a probability equal to zero. Therefore, it cannot occur that the team will make the playoffs.

c. The chance of a negative reaction is 3%, or 0.03. This probability is low (for example, below a certain threshold, such as 0.05), so we conclude that it would be unusual for a negative reaction to the planned procedure to occur.

NOW YOU CAN DO
Exercises 17–20.

YOUR TURN
#1

For the following probabilities, determine the meaning of the indicated probability:

a. The stockbroker said that the chance for a recession was near zero.

b. The salesperson said that your chance of satisfaction with the purchase of a new car was 100%.

(The solutions are shown in Appendix A.)

When we perform an experiment, it is a "sure thing" that one of the outcomes in the sample space will occur. For example, when you toss a coin, you know that it will be either heads or tails. Put into probability terms, the sum of the probabilities of all the individual outcomes must equal 1, which is the Law of Total Probability we mentioned above.

2 Classical Method of Assigning Probability

Many people have a certain degree of intuition when it comes to assigning probabilities. For example, when asked what the chances are of rolling a 6 on a single toss of a fair die, many people would quite correctly answer 1/6. However, intuition can often let us down. For example, when asked what the chances are of observing two heads when you toss a fair coin twice, many people would incorrectly respond 1/3. ("Well, it's either both heads or both tails or one of each." The correct answer is in fact 1/4.) In this section, we learn how to quantify our methods of assigning probabilities, so that we don't have to depend on intuition alone.

Did you know? People have been tossing dice for a long time. Archaeologists have dug up dice from Roman ruins looking just the same as ours. These three dice were uncovered from the ruins of Pompeii buried by the eruption of Mount Vesuvius in the first century A.D.

Three methods for assigning probabilities are available:

- Classical method
- Relative frequency method
- Subjective method

We first take a close look at the classical method. Later in this section, we will examine the relative frequency method and the subjective method.

Many experiments are structured so that each experimental outcome is equally likely. *Equally likely outcomes* are outcomes that have the same probability of occurring. For example, if you toss a fair coin, the probability of observing either of the outcomes—heads or tails—is the same. The **classical method of assigning probabilities** is used when an experiment has equally likely outcomes.

Classical Method of Assigning Probabilities

Let $N(E)$ and $N(S)$ denote the number of outcomes in event E and the sample space S, respectively. If the experiment has equally likely outcomes, then the probability of event E is

$$P(E) = \frac{\text{number of outcomes in } E}{\text{number of outcomes in sample space}} = \frac{N(E)}{N(S)}$$

EXAMPLE 3   Classical method: Single draw from a deck of cards

Find the probability of drawing an ace when drawing a single card at random from a deck of cards.

Solution

The sample space for the experiment in which a subject chooses a single card at random from a deck of cards is given in Figure 1. If the card is chosen truly at random, then it is reasonable to assume that each card has the same chance of being drawn. Each card is equally likely to be drawn, so we can use the classical method to assign probabilities.

FIGURE 1

Sample space for drawing a card at random from a deck of cards.

A total of 52 outcomes are included in this sample space, so $N(S) = 52$. Let E be the event that an ace is drawn. Event E consists of the four aces $\{A\heartsuit, A\diamondsuit, A\clubsuit, A\spadesuit\}$, so $N(E) = 4$. Therefore, the probability of drawing an ace is

$$P(E) = \frac{N(E)}{N(S)} = \frac{4}{52} = \frac{1}{13}$$

NOW YOU CAN DO
Exercises 21–24.

**YOUR TURN
#2**

Find the probability of the following events when drawing a single card at random from a deck of cards:

a. a heart

b. a black card

(The solutions are shown in Appendix A.)

EXAMPLE 4   Classical method: Single fair die

Recall the town fair example (page 240). In the game, you win if you roll a 6 on a single roll of a single fair die. Find the probability of winning the game.

Solution

The sample space for a single die toss consists of six outcomes, $\{1, 2, 3, 4, 5, 6\}$. When the six outcomes are equally likely, we say that the die is *fair*. If the outcomes are not equally likely, then the die is *loaded* or defective. If we assume the die is fair, then, because the sum of the probabilities of the $n = 6$ outcomes must equal 1, the probability of any particular outcome must equal 1/6, using the classical method. We write

NOW YOU CAN DO
Exercises 25–30.

$$\text{probability of winning} = P(W) = 1/6$$

**YOUR TURN
#3**

For the experiment in Example 4, find the following probabilities:

a. Not winning the game

b. Rolling a 5

(The solutions are shown in Appendix A.)

Tree Diagrams

A **tree diagram** is a graphical display that allows us to list all the outcomes in the sample space of a multistage experiment. The next example shows how to construct a tree diagram.

EXAMPLE 5   List all outcomes in a sample space using a tree diagram

Suppose our experiment is to toss a fair coin twice.

a. Construct a tree diagram.

b. Use the tree diagram to list all the outcomes in the sample space.

Solution

a. Think of this experiment as a two-stage process:

- Stage 1: Toss the coin the first time.

- Stage 2: Toss the coin the second time.

Figure 2 shows the tree diagram for the experiment of tossing a fair coin twice. Note the branches for Stage 1: the first time the coin is tossed, it can come up heads or tails. At Stage 2, the tree diagram again has branches for either heads or tails.

b. The sample space for the experiment of tossing a coin twice is {HH, HT, TH, TT}. There are $N(S) = 4$ outcomes in the sample space.

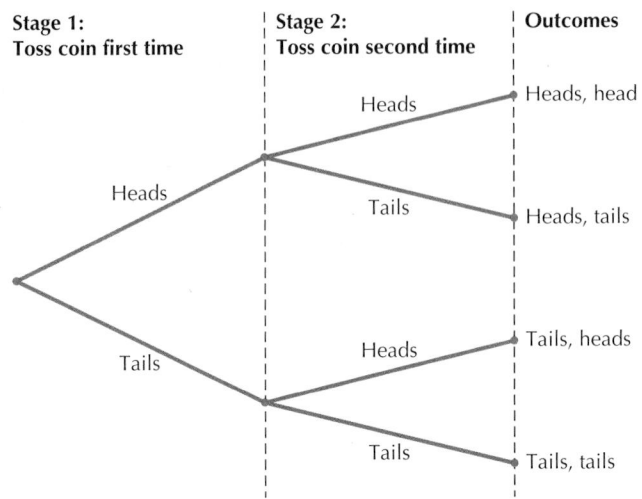

NOW YOU CAN DO
Exercises 31–38.

FIGURE 2 Tree diagram for the experiment of tossing a fair coin twice.

YOUR TURN
#4

Suppose our experiment is to toss a fair coin three times.

a. Construct a tree diagram.

b. Use the tree diagram to list all the outcomes in the sample space.

(The solutions are shown in Appendix A.)

Note that two possible outcomes can occur at Stage 1 of this two-stage experiment and two possible outcomes when flipping the coin at Stage 2. To determine the number of outcomes in the entire experiment, the *counting rule* is simply to multiply the number of possible outcomes at each stage. In this two-stage experiment, $2 \times 2 = 4$ possible outcomes, which is the number of outcomes we see in the sample space.

EXAMPLE 6   Classical method: Tossing a fair coin twice

Find the probability of obtaining one heads and one tails when a fair coin is tossed twice.

Solution

It is reasonable to assume that the $N(S) = 4$ outcomes in the sample space {HH, HT, TH, TT} are equally likely. The coin doesn't remember what occurred at Stage 1, so the probabilities at Stage 2 are precisely the same as at Stage 1. Also, recall from the Law of Total Probability that the sum of the probabilities of all the outcomes in the sample space must equal 1. Thus, each of the four outcomes must have probability 1/4. Let E be the event that one heads and one tails is obtained. Then $E = $ {HT, TH}, so $N(E) = 2$. Thus,

NOW YOU CAN DO
Exercises 39–42.

$$P(E) = \frac{\text{number of outcomes in } E}{\text{number of outcomes in sample space}} = \frac{N(E)}{N(S)} = \frac{2}{4} = \frac{1}{2}$$

YOUR TURN

#5

For the experiment in Example 6, find the following probabilities:

a. Two heads

b. Two tails

(The solutions are shown in Appendix A.)

EXAMPLE 7 Finding probabilities for the experiment of tossing two fair dice

Punchstock/CutandDeal

Imagine that you are playing Monopoly with your dormitory roommate, and the loser has to do the laundry for both of you for the rest of the semester. You have a hotel on Boardwalk, and if your roommate lands on it, you will surely win. Right now your roommate's piece is on Short Line: if he or she rolls a 4, you will win and get your laundry done free for the remainder of the semester. Put into statistical terms, the experiment is to toss two fair dice and observe the sum of the two dice. Find the probability of rolling a sum of 4 when tossing two fair dice.

Solution

It is reasonable to assume that each of these $N(S) = 36$ outcomes in the sample space (Figure 3) is equally likely. The experiment of tossing two dice can be viewed as a two-stage experiment, where we add the result from the first die to the result from the second die. If a 5 appears on the first (say, dark green) die, and a 3 appears on the second (light green) die, the overall outcome is (5,3), with the resulting sum equal to 8. Note that the outcome (5,3) is not the same as the outcome (3,5), where the dark green die comes up 3 and the light green die comes up 5.

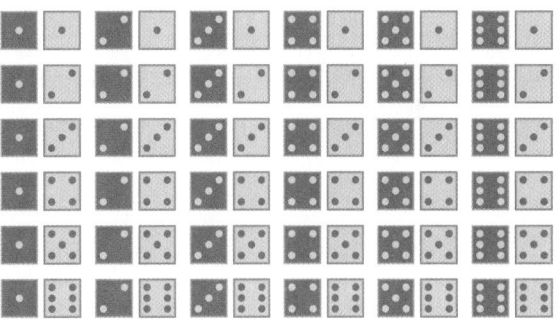

FIGURE 3 Sample space for tossing two fair dice.

Let E denote the event that your roommate rolls a sum equal to 4. Then the outcomes that belong in this event are E: {(3,1) (2,2) (1,3)}, so $N(E) = 3$. The outcomes are equally likely, so we can use the classical method for finding probabilities of events.

$$P(E) = \frac{\text{number of outcomes in } E}{\text{number of outcomes in sample space}} = \frac{N(E)}{N(S)} = \frac{3}{36} = \frac{1}{12}$$

NOW YOU CAN DO

Exercises 43–50.

The probability that your roommate will land on Boardwalk on this throw of the dice is 1/12.

YOUR TURN

#6

For the experiment in Example 7, find the following probabilities:

a. Rolling a sum of 7 with the two dice

b. Rolling boxcars (sum of 12 with the two dice)

c. Rolling snake eyes (sum of 2 with the two dice)

(The solutions are shown in Appendix A.)

| EXAMPLE 8 | Inappropriate use of the classical method |

Watch a lot of YouTube? Hate the in-stream ads? According to Social Strand Media, 75% of in-stream ads are skippable (www.slideshare.net/SocialStrand/social-media -stats-2014). Suppose we choose one in-stream ad at random. Define the following events:

> *C:* The randomly chosen in-stream ad is skippable.
>
> *D:* The randomly chosen in-stream ad is not skippable.

Determine whether the classical method can be used to assign probability to events *C* and *D*.

Solution

Because more than half of in-stream ads are skippable, if we choose one at random, we are more likely to find a skippable one than a non-skippable one. Therefore, the events *C* and *D* are not equally likely. It would be inappropriate to use the classical method of assigning probabilities for this experiment because the classical method can be used only when all the outcomes of an experiment are equally likely.

The proper method for solving this problem is the relative frequency method, which we discuss next.

3 Relative Frequency Method

In Example 4, we need the classical method to find that the probability of rolling a 6 with a fair die is 1/6. What does this probability mean? Remember that the definition of *probability* included the phrase "long-term proportion." The next example demonstrates what we mean by "long-term."

| EXAMPLE 9 | Simulating the long-term proportion of 6s in a fair die roll |

Suppose we want to investigate the proportion of 6s we observe if we roll a fair die 100 times. We can use technology, such as the TI-83/84 used here, to help us simulate rolling a fair die a large number of times. A **simulation** uses methods such as rolling dice or computer generation of random numbers to generate results from an experiment. The actual die rolls from our simulation are shown here, in order, with the 6s in boldface.

$$1\,4\,4\,\mathbf{6}\,2\,4\,3\,2\,1\,3\,4\,3\,3\,4\,3\,3\,\mathbf{6}\,3\,5\,5\,1\,5\,3\,5\,5\,2\,1\,3\,1\,1\,1\,5\,5\,\mathbf{6}\,3\,\mathbf{6}\,2\,1\,\mathbf{6}\,5\,5\,4\,4\,\mathbf{6}\,5\,4\,1\,1\,4\,\mathbf{6}$$
$$4\,2\,2\,2\,\mathbf{6}\,3\,2\,5\,5\,\mathbf{6}\,1\,1\,3\,1\,\mathbf{6}\,5\,4\,\mathbf{6}\,\mathbf{6}\,5\,5\,5\,2\,5\,5\,3\,4\,2\,4\,\mathbf{6}\,4\,5\,5\,1\,\mathbf{6}\,3\,1\,1\,1\,3\,5\,4\,2\,3\,3\,\mathbf{6}\,2\,5\,3$$

Thus, the first die roll was a 1, so the proportion of 6s was 0/1. The second and third die rolls were 4s, so the proportion of 6s after 3 rolls was 0/3. On the fourth roll, a 6 appeared, so the proportion of 6s after the fourth roll was 1/4. Figure 4 provides a

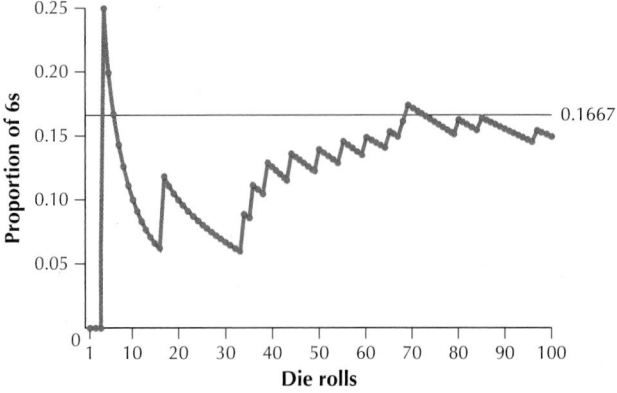

FIGURE 4 Proportion of 6s, 100 die rolls.

graph of the proportion of 6s in this simulation as the number of die rolls increased. Note that as the number of die rolls increases, the proportion of 6s tends to get closer to the horizontal line: 0.1667 ≈ 1/6.

The simulation was rerun—this time with 1000 die rolls. The resulting graph of the proportion of 6s is provided in Figure 5. Note that as the number of die rolls increases, the proportion of 6s approaches the line 0.1667 ≈ 1/6, and the fit is tighter with 1000 die rolls than with 100. This is what we mean by "long-term proportion."

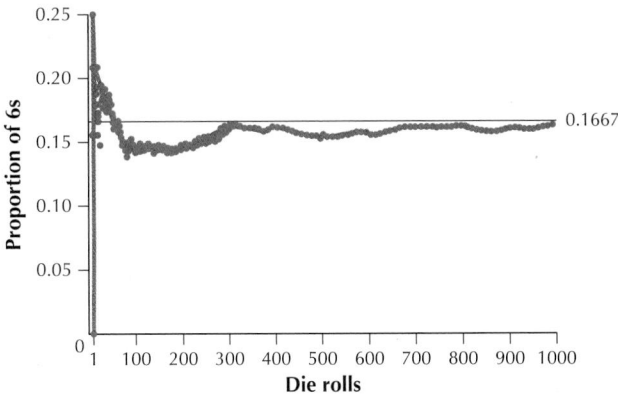

FIGURE 5 Proportion of 6s, 1000 die rolls.

This example leads directly to the following law.

Law of Large Numbers

As the number of times that an experiment is repeated increases, the relative frequency (proportion) of a particular outcome tends to approach the *probability* of the outcome.

- For quantitative data, as the number of times that an experiment is repeated increases, the mean of the outcomes tends to approach the population mean.

- For categorical (qualitative) data, as the number of times that an experiment is repeated increases, the proportion of times a particular outcome occurs tends to approach the population proportion.

 The *Law of Large Numbers for Proportions* applet allows you to simulate coin tossing and observe the proportion of heads as the number of tosses increases.

Note: The *relative frequency method* may be used when previous information is available about the relative frequency of an event.

Relative Frequency Method

If we can't use the classical method for assigning probabilities, then the **Law of Large Numbers** gives us a hint about how we can estimate the probability of an event. It often happens that previous information is available about the relative frequency of an event. Relative frequency information can be used to estimate the probability of the event.

Note: Tree diagrams can be used for the relative frequency method as well as the classical method of assigning probability.

Relative Frequency Method of Assigning Probabilities

The probability of event E is approximately equal to the relative frequency of event E. That is,

$$P(E) \approx \text{relative frequency of } E = \frac{\text{frequency of } E}{\text{number of trials of experiment}}$$

The relative frequency method is also known as the **empirical method**.

EXAMPLE 10 Relative frequency method

The Gardasil Vaccine

The Case Study data set **Gardasil** provides the following relative frequency distribution for the variable *Completed,* indicating whether or not the patient completed the course of three Gardasil vaccinations. Use the relative frequency method to find the probability that a randomly chosen patient completed the treatment.

Completed	Frequency	Relative frequency
Yes	469	0.3319
No	944	0.6681
Total	1413	1.0000

Solution

Define the event.

C: Patient completed treatment.

We use the relative frequency method to find the probability of event *C:*

NOW YOU CAN DO
Exercises 51–60.

$$P(C) \approx \text{relative frequency of } C = \frac{\text{frequency of } C}{\text{number of trials in experiment}} = \frac{469}{1413} \approx 0.3319$$

YOUR TURN #7

Using the experiment from Example 10, do the following:

a. Use a capital letter to define the event that a patient did not complete the treatment.

b. Find the relative frequency of your event from (**a**).

c. Assign your relative frequency from (**b**) as the probability that a patient did not complete the treatment.

(The solutions are shown in Appendix A.)

We can also use the relative frequency method to build a probability model with data that have been summarized in a table.

EXAMPLE 11 Probability models based on frequency tables

Table 2 contains the employment type for a sample of 1000 employed citizens of Fairfax County, Virginia.[3] Use the data to construct the probability model by generating the relative frequencies and using the relative frequencies to estimate the probabilities for each employment type.

Table 2 Employment types

Employment type	Count
Private company	597
Federal government	141
Self-employed	97
Private nonprofit	92
Local government	59
State government	12
Other	2

Solution

We calculate the relative frequencies of each employment group by dividing the count (frequency) for each group by the sample size 1000. For example, the relative frequency for "Private Company" is $\frac{597}{1000} = 0.597$. The relative frequency is then used to estimate the probability of selecting citizens who work at private companies in Fairfax County, Virginia. Filling in the remaining calculations produces the *probability model* in Table 3. Note that the table follows the Rules of Probability in that (a) each outcome has probability between 0 and 1, and (b) the sum of the probabilities of all the outcomes equals 1.0.

 fairfaxemploy

Table 3 Probability model

Employment type	Probability
Private company	0.597
Federal government	0.141
Self-employed	0.097
Private nonprofit	0.092
Local government	0.059
State government	0.012
Other	0.002

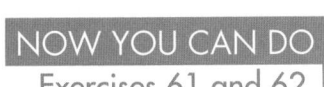

NOW YOU CAN DO
Exercises 61 and 62.

EXAMPLE 12   Random draws using a probability model

Suppose we consider the probabilities in Table 3 as population values. Use technology to simulate random draws using the probability model in Table 3.

Solution

Using the Step-by-Step Technology Guide on page 252, we drew samples of sizes 10, 100, 1,000, and 10,000 from the probability model in Table 3. The results are shown in Table 4.

Table 4 Relative frequencies from random draws of different sizes

Employment type	Rel freq $n = 10$	Rel freq $n = 100$	Rel freq $n = 1,000$	Rel freq $n = 10,000$
Private company	0.60	0.62	0.566	0.596
Federal government	0.20	0.15	0.15	0.143
Self-employed	0.10	0.11	0.109	0.991
Private nonprofit	0.10	0.07	0.106	0.914
Local government	0.00	0.04	0.055	0.056
State government	0.00	0.01	0.012	0.012
Other	0.00	0.00	0.002	0.002

Note that each relative frequency tends to approach its respective probability as the sample sizes grow larger.

Subjective method for assigning probability

Some cases have outcomes that are not equally likely (so the classical method does not apply) and there has been no previous research (so the relative frequency approach does not apply). For example, what is the probability that the Dow Jones Industrial Average will decrease today? In cases like this, there is no absolutely correct probability. Reasonable people can disagree reasonably over these probabilities. The idea is to consider all available information, tempered by our experience and intuition, and then assign a probability value that expresses our estimate of the likelihood that the outcome will occur. Finally, it should be noted that the subjective method should be used when the event is not (even theoretically) repeatable.

> **Subjective probability** refers to the assignment of a probability value to an outcome based on personal judgment.

EXAMPLE 13　Subjective method for assigning probability

A financial analyst is interested in the probability that the Dow Jones Industrial Average will go down today. Suppose the Chairman of the Federal Reserve warned against inflation in a major speech yesterday. How would the financial analyst use this information to subjectively assign the probability that the stock market will go down today?

Solution

Through long professional experience, the financial analyst has seen the Dow go down—often, the day after speeches by the Federal Reserve Chairman warning against inflation. She therefore assigns a subjective probability that the probability the Dow Jones Industrial Average will go down today is 75%. Although the exact subjective probability assigned by other financial analysts may differ somewhat, most will probably share the opinion that the Dow will decrease today.

STEP-BY-STEP TECHNOLOGY GUIDE: Probability Simulations Using Technology

TI-83/84
Simulating 100 Die Rolls
Step 1 Set the random number seed as follows. (The random number seed is a number that the calculator uses to generate random numbers.) Enter any number on the home screen. Press **STO →**, then **MATH**, highlight **PRB**, select **1: rand**, and press **ENTER**. On the home screen press **ENTER**.
Step 2 Press **MATH**, highlight **PRB**, select **5: randInt(**, and press **ENTER**.
Step 3 Enter 1, comma, 6, comma, 100, close parenthesis (Figure 6).
Step 4 Store the data in list **L1** as follows. Press **STO →**, then **2nd**, then **1**, then press **ENTER**.
Step 5 To examine the die rolls, press **STAT**, select **1: EDIT**, and press **ENTER** (Figure 7).

Simulating Coin Flips
You can simulate coin flips instead of die rolls by coding "heads" as 1 and "tails" as 0. Use the instructions for simulating 100 die rolls with the following changes: Enter 0, comma, 1, comma, 100, close parenthesis, so that the home screen shows **randInt(0, 1, 100)**.

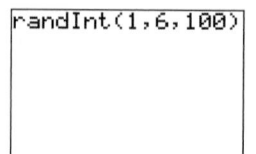

FIGURE 6　　　　**FIGURE 7**

EXCEL
Simulating 100 Die Rolls
Step 1 Select cell **A1**. Click the **Insert Function** icon f_x.
Step 2 For **Search for a Function**, type **randbetween**, click **Go**, then click **OK**.

Step 3 For **Bottom**, enter 1. For **Top**, enter 6 (Figure 8). Click **OK**. Cell **A1** now contains a simulated random die roll.
Step 4 Select cell **A1**, copy it, and paste the contents into cells **A2** through **A100**.

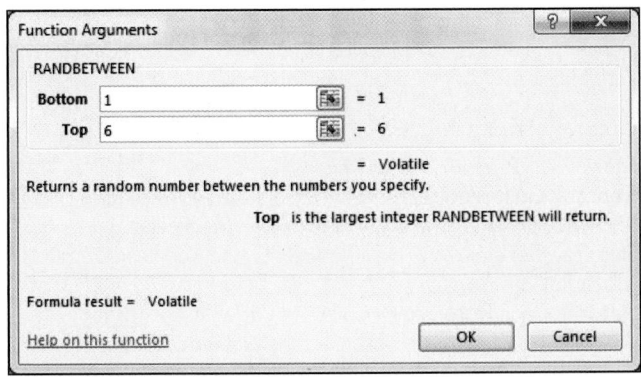

FIGURE 8 Random die rolls in Excel.

Simulating the Sum of Two Dice
Step 1   Generate 100 die rolls in column A and another 100 die rolls in column B.
Step 2   Select cell C1. Enter =(A1+B1), and press **ENTER**.
Step 3   Select cell C1, copy it, and paste the contents into cells C2 through C100. Column C then represents 100 randomly generated sums of two dice.

Simulating Random Draws from a Probability Table
We illustrate using Example 11 (page 250). Excel requires that the categories in the probability model be coded as numeric. We therefore code "Private company" as 1, "Federal government" as 2, and so on.
Step 1   Type the model categories (for example, "Employment type") in column A, their numeric codes in column B, and the respective probabilities in column C.
Step 2   Click **Data > Data Analysis > Random Number Generation**, then **OK**.
Step 3   For **Number of Variables**, enter **1**.
Step 4   For **Number of Random Numbers**, enter the desired sample size.
Step 5   For **Distribution**, select **Discrete**.
Step 6   For **Value & Probability Input Range**, click and drag to select the coded categories and their probabilities, for example, **B1:C7**. Click **OK**.

Repeat Steps 1–6 for increasing sample sizes.

Simulating Coin Flips
You can simulate coin flips instead of die rolls by coding "heads" as 1 and "tails" as 0. Use the die roll instructions with the following changes: For **Bottom**, enter **0**. For **Top**, enter **1**.

MINITAB
Simulating 100 Die Rolls
Step 1   Click on **Calc > Random Data > Integer…**
Step 2   For **Number of rows of data to generate**, enter **100**.
Step 3   For **Store in column(s)**, type **C1**.
Step 4   For **Minimum value**, enter **1**. For **Maximum value**, enter **6**.
Step 5   Click **OK**.

Simulating the Sum of Two Dice
Step 1   Generate 100 die rolls in C1 and another 100 die rolls in C2.
Step 2   Click **Calc > Calculator**. For **Store result in variable**, enter **C3**. For **Expression**, enter **C1 + C2**. Click **OK**. Column **C3** then represents 100 randomly generated sums of two dice.

Simulating Random Draws from a Probability Table
We illustrate using Example 11 (page 250). Minitab requires that the categories in the probability model be coded as numeric. We therefore code "Private company" as 1, "Federal government" as 2, and so on.
Step 1   Type the model categories in C1, their numeric codes in C2, and the respective probabilities in C3 (Figure 9).
Step 2   Click on **Calc > Random Data > Discrete…**
Step 3   For **Number of rows of data to generate**, enter the desired sample size.
Step 4   For **Store in column(s)**, select the next available column, such as **C4**.
Step 5   For **Values in**, enter the column with the numerically coded categories, such as **C2**.

Step 6   For **Probabilities in**, enter the column with the probabilities, such as **C3**.
Step 7   Click **OK**.

Repeat Steps 1–7 for increasing sample sizes, as shown in Figure 9.

Simulating Coin Flips
You can simulate coin flips instead of die rolls by coding "heads" as 1 and "tails" as 0. Use the die roll instructions with the following changes: For **Minimum value**, enter **0**. For **Maximum value**, enter **1**.

↓	C1-T	C2	C3	C4	C5	C6	C7
	Type	Type-N	Prob	10	100	1000	10000
1	Private company	1	0.597	4	2	1	1
2	Federal government	2	0.141	2	7	4	1
3	Self-employed	3	0.097	6	1	1	1
4	Private nonprofit	4	0.092	5	1	3	1
5	Local government	5	0.059	2	3	1	1
6	State government	6	0.012	2	4	1	1
7	Other	7	0.002	1	2	1	1
8				1	2	1	2
9				2	1	1	1
10				1	5	1	1
11					4	1	1
12					1	1	1
13					4	1	1

FIGURE 9 Random draws in Minitab.

SPSS
Simulating 100 Die Rolls
Step 1   Select the header of the first column, and click **Edit > Insert Variable**. Highlight between 1 and 30 rows and click **Edit > Insert Cases**. More cases can be added by selecting more rows and choosing **Edit > Insert Cases** again.
Step 2   Go to **Variable View**. Name the first column **Roll1**.
Step 3   Click **Transform > Compute Variable…**. For **Target Variable**, put **Roll1**. Click in the **Numeric Expression** box, then

under **Function Group** click **Arithmetic**, and under **Functions and Special Variables** double-click **Rnd(1)**.
Step 4   Under **Function Group**, click **Random Numbers**, and under **Functions and Special Variables** double-click **Rv.Uniform**. Replace the two question marks in **Numeric Expression** with 1 and 6, as shown in Figure 10. Click **OK** and **OK**. Minimize the output window. The die rolls can be seen in the **Data View** tab.

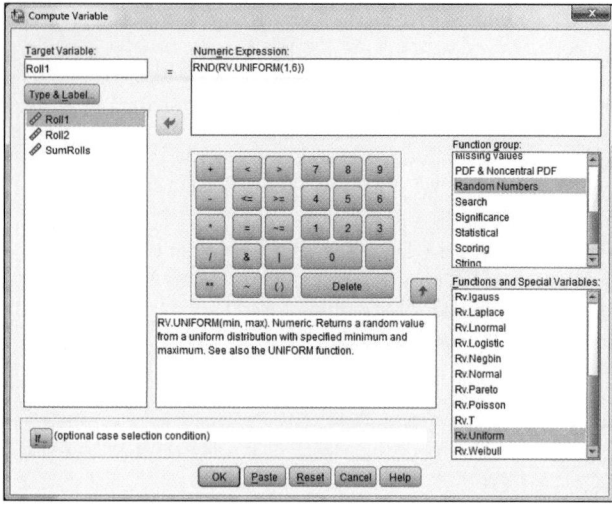

FIGURE 10 Random die rolls in SPSS

Simulating the Sum of Two Dice

Step 1 Select the header of the first column, and click **Edit > Insert Variable**. Repeat for the second and third column. Highlight between 1 and 30 rows, and click **Edit > Insert Cases**. More cases can be added by selecting more rows and choosing **Edit > Insert Cases** again.

Step 2 Go to **Variable View**. Name the first column **Roll1**, the second column **Roll2**, and the third column **SumRolls**.

Step 3 Generate 100 die rolls in **Roll1** and another 100 rolls in **Roll2**.

Step 4 Click **Transform > Compute Variable**. For **Target Variable**, put **SumRolls**. Click inside the **Numeric Expression** box, then double-click **Roll1**, click the **plus sign**, and double-click **Roll2**. Click **OK** and **OK**, and minimize the output window.

Simulating Coin Flips

You can simulate coin flips instead of die rolls by coding "heads" as 1 and "tails" as 0. Use the die roll instructions with the following changes: replace the two question marks with 0 and 1.

JMP

Simulating 100 Die Rolls

Step 1 Click **File > New > Data Table**. Click **Rows > Add Rows...**, and enter **100** rows.

Step 2 Select **Column 1**. Click **Cols > Formula...**. In the list under **Functions(grouped)**, select **Random > Random Integer**.

Step 3 In the formula, replace **n1** with **1,6**, so the formula is **Random Integer(1,6)**. Click **OK**.

Simulating the Sum of Two Dice

Step 1 Generate 100 die rolls in **Column 1** and another 100 rolls in **Column 2**.

Step 2 Doubleclick the header of the third column, and select **Column 3**. Click **Cols > Formula...**. Select **Column 1**, the **plus sign**, then **Column 2**, as in Figure 11. Click **OK**.

Simulating Coin Flips

You can simulate coin flips instead of die rolls by coding "heads" as 1 and "tails" as 0. Use the die roll instructions, but change the formula to **Random Integer(0,1)**.

FIGURE 11 Sum of two variables in JMP.

Section 5.1 Summary

1. Section 5.1 introduces the building blocks of probability, including the concepts of probability, outcome, experiment, and sample space. Probabilities always take values between 0 and 1, where 0 means that the outcome cannot occur and 1 means that the outcome is certain.

2. The classical method of assigning probability is used if all outcomes are equally likely. The classical method states that the probability of an event *A* equals the number of outcomes in *A* divided by the number of outcomes in the sample space.

3. The Law of Large Numbers states that, as an experiment is repeated many times, the relative frequency (proportion) of a particular outcome tends to approach the probability of the outcome. The relative frequency method of assigning probability uses prior knowledge about the relative frequency of an outcome. The subjective method of assigning probability is used when the other methods are not applicable.

Section 5.1 Exercises

CLARIFYING THE CONCEPTS

1. Describe in your own words how chance and uncertainty affect you in your life. List some synonyms that we use in everyday life for the word *probability*. (p. 240)

2. Why do you think we use numerical values for probability instead of only qualitative terms such as "likely" or "impossible"? (p. 242)

3. Give three examples from your own life of experiments, as the term is used in this chapter. (p. 241)

a. For each experiment, what are some of the outcomes?
b. Write out the sample space of one of these experiments.
c. Describe how the Law of Total Probability applies to the sample.

4. List the three methods for assigning probabilities. (pp. 243, 248, 251)

5. What assumption do we need to make to use the classical method? (p. 244)

6. When can we use the relative frequency method? (p. 249)

7. If we can't use either the classical method or the relative frequency method, explain how we go about using the subjective method. (p. 252)

8. The experiment is to toss 10 fair coins 25 times each. Which methods can we use to assign probabilities? (pp. 243, 248)

9. How would you find the probability that a randomly chosen student at your college likes hip-hop music? What method would you use? (p. 248)

10. Describe the meaning of the following probabilities: (p. 242)
    a. Near 0
    b. 0
    c. Near 1
    d. 1

PRACTICING THE TECHNIQUES

 CHECK IT OUT!

To do	Check out	Topic
Exercises 11–16	Example 1	Probability models
Exercises 17–20	Example 2	Determining unusual or likely outcomes
Exercises 21–24	Example 3	Single draw from deck of cards
Exercises 25–30	Example 4	Single fair die
Exercises 31–38	Example 5	Tree diagrams and sample space
Exercises 39–42	Example 6	Tossing a fair coin twice
Exercises 43–50	Example 7	Tossing two fair dice
Exercises 51–60	Example 10	Relative frequency method.
Exercises 61–62	Example 11	Probability models based on frequency tables

Determine whether each table in Exercises 11–16 is a probability model. If not, clearly explain why it is not a probability model.

11. Customers at a clothing store at the mall

Gender	Probability
Females	1.1
Males	−0.2

12. Singers in the church choir

Voice	Probability
Soprano	0.50
Alto	0.25
Tenor	−0.25
Bass	0.25

13. Voters at a town meeting

Party	Probability
Democrat	0.25
Republican	0.25
Independent	0.3
Green	0.1
Libertarian	0.1
Other	0.1

14. Majors of students taking introductory statistics

Major	Probability
Business	0.50
Nursing	0.20
Social sciences	0.20
Science	0.20
Math	−0.10

15. Students taking undergraduate introductory statistics

Class	Probability
Freshmen	0.15
Sophomores	0.25
Juniors	0.40
Seniors	0.20

16. Reasons why Hurricane Katrina survivors did not evacuate

Reason	Probability
I did not have a car or a way to leave.	0.36
I thought the storm and its aftermath would not be as bad as they were.	0.29
I just didn't want to leave.	0.10
I had to care for someone who was physically unable to leave.	0.07
All other reasons	0.18

For the probabilities in Exercises 17–20, determine the meaning of the indicated probability.

17. The doctor said the chances of recovery from the surgery were near 100%.

18. The parent said that, if the student's grades did not improve, there was no chance of getting the new video game system for a birthday present.

19. The stockbroker said the chances were high that the blue chip stock would gain in value this year.

20. The political analyst gloomily told the candidate that his chances of winning the election were about 2%.

For Exercises 21–24, the experiment is to draw a card at random from a shuffled deck of 52 cards. Find the following probabilities:

21. Drawing a jack
22. Drawing a club
23. Drawing the jack of clubs
24. Drawing a red card

For Exercises 25–30, the experiment is to roll a fair die once. Find the following probabilities:

25. Observing a 2
26. Observing an odd number
27. Observing a number greater than 2
28. Observing a number less than 2
29. Observing a 2 or a 3
30. Observing a 2 and a 3

For Exercises 31 and 32, consider the experiment of tossing a fair die two times, with the outcomes being the observation of either an even number or an odd number.

31. Construct a tree diagram for the experiment.
32. Construct the sample space for the experiment.

For Exercises 33 and 34, let the experiment be tossing a fair die two times, with the outcomes being observing either a number less than 4 or a number greater than or equal to 4.

33. Construct a tree diagram for the experiment.
34. Construct the sample space for the experiment.

For Exercises 35–38, consider the experiment of tossing a fair coin three times and observing either heads or tails.

35. Construct a tree diagram for the experiment.
36. Construct the sample space for the experiment.
37. How does the tree diagram help to construct the sample space?
38. How do we find each outcome using the tree diagram?

For Exercises 39–42, consider the experiment of tossing a fair coin twice.

39. Find the probability of observing zero heads.
40. Find the probability of observing exactly one head.
41. Find the probability of observing two heads.
42. Use your results from Exercises 39–41 to construct the probability model for the number of heads observed when tossing a fair coin twice.

For Exercises 43–50, consider the experiment of tossing two fair dice and observing the sum of the two dice. (*Hint:* Use the sample space in Figure 3 on page 247.)

43. What is the probability that the sum of the dice equals 9?
44. Find the probability that the dark green die equals 9.
45. Calculate the probability that the sum of the dice equals 11.
46. Find the probability that the light green die equals 5.
47. What is the probability that the sum of the dice equals 1?
48. Construct the probability model for the sum of the dice.
49. Use the probability model to find which event has the greatest probability.
50. Which events have the lowest probability?

For Exercises 51–56, suppose that, in a sample of 100 students who drink hot caffeinated beverages, 35 preferred regular coffee, 20 preferred latte, 20 preferred cappuccino, and 25 preferred tea. Find the probability that a randomly selected student prefers the following:

51. Regular coffee
52. Latte
53. Cappuccino
54. Tea
55. For Exercises 51–54, which method of assigning probability are you using?

56. Construct the probability model for hot caffeinated beverages.

For Exercises 57–60, suppose that, in a sample of 200 college students, 80 live on campus, 80 live with family off campus, and 40 live in an apartment off campus. Find the probability that a randomly selected student lives in the following places:

57. On campus
58. With family off campus
59. In an apartment off campus
60. Construct the probability model for where these students live.
61. Use the following frequency table to estimate the probabilities for each color and construct the probability model. A sample of 100 students was asked to name their favorite color.

Favorite color	Frequency
Red	25
Blue	25
Green	20
Black	10
Violet	10
Yellow	10

62. Use the following frequency table to estimate the probabilities for each season and construct the probability model. A sample of 200 students was asked to name their favorite season.

Favorite season	Frequency
Summer	70
Spring	70
Autumn	50
Winter	10

APPLYING THE CONCEPTS

63. **Technology Adopters.** The Gallup organization published a study[4] in which they categorized American technology adopters as either Super Tech Adopters (31%), Smart Phone Reliants (19%), Mature Technophiles (22%), or Tech-Averse Olders (28%). Consider the experiment of two American technology adopters at random.

    a. Construct the tree diagram for the experiment.
    b. What is the sample space?

64. **Facebook Females and Males.** The Facebook Social Ads Platform reported the following number of users in the United States in 2014, by gender: females: 96 million, males: 82 million, unknown: 2 million.

    a. Construct a relative frequency distribution.
    b. Use the relative frequency method to assign probabilities to each outcome.
    c. Consider the experiment of choosing three Facebook users at random. Construct the tree diagram for the experiment.
    d. What is the sample space for the experiment in (**c**)?

65. **Owning a Time Machine.** The Pew Research Internet Project reported that 14 of the 144 people ages 18–29 that it surveyed would someday like to own a time machine.[5]

a. What is the probability that a randomly chosen 18- to 29-year-old would someday like to own a time machine?

b. What is the probability that a randomly chosen 18- to 29-year-old would *not* someday like to own a time machine?

c. Which method of assigning probability did you use?

66. Basketball. Your college's basketball team is playing a game next week.

a. What is the probability that the team will win the game?

b. Which method did you use?

67. Brisbane Babies. The table shows the births of babies at a Brisbane, Australia, hospital on a particular day.[6]

Girl	Girl	Boy	Boy	Boy	Girl	Girl	Boy	Boy
Boy	Boy	Boy	Girl	Girl	Boy	Girl	Girl	Boy
Boy	Boy	Boy	Girl	Girl	Girl	Girl	Boy	Boy
Boy	Girl	Boy	Girl	Boy	Boy	Boy	Boy	Boy
Girl	Boy	Boy	Boy	Boy	Girl	Girl	Girl	

a. Construct a relative frequency distribution of the numbers of girls and boys born.

b. Use the relative frequencies to construct a probability model.

c. Confirm that your probability model follows the Rules of Probability.

68. Draw an Ace. If you draw the ace of spades from a deck of cards, you win $200.

a. What is the probability of winning this game?

b. What would be a fair price for playing this game? (*Hint:* A fair price might be determined by balancing out the winnings and the price in the long run.)

69. A Bazaar Game. Lenny has gone to the church bazaar with his family. In one of the games at the bazaar, if Lenny rolls two dice and gets a sum of at least 10, he wins $10; otherwise, he wins nothing.

a. Find the probability of winning $10.

b. Find the probability of winning nothing.

c. What would you suggest would be a fair (break-even) price for playing this game?

70. Sharing Social Media Profiles. The Pew Research Center reported that 98 of the 889 social media users in committed relationships that it surveyed shared a social media profile with their partners.[7]

a. Explain whether or not you can use the classical method of assigning probability to find the probability of sharing a social media profile with a partner, and why.

b. Find the probability that a social media user in a committed relationship shares a social media profile with his or her partner. Which method did you use?

For Exercises 71–75, consider the experiment of tossing a fair coin three times and observing either heads or tails.

71. Find the probability of zero heads.

72. What is the probability of exactly one head?

73. Calculate the probability of exactly two heads.

74. Find the probability of exactly three heads.

75. Use your results from Exercises 71–74 to construct a probability model for the number of heads observed.

76. For Exercises 71–74, which method of assigning probability are you using?

For Exercises 77–80, consider the experiment of tossing a fair coin three times. Find the indicated probabilities. (*Hint:* Use a tree diagram similar to the one on page 246, but add one more stage.)

77. Observing 3 heads

78. Not observing 3 heads

79. Observing 2 tails

80. Not observing 2 tails

81. Fairfax County Income. The following table contains a probability model for the distribution of income in Fairfax County, Virginia. 🎹 **fairfaxincome**

a. Use technology to draw random samples of sizes 10, 100, 1,000, and 10,000 from this probability model.

b. What can you conclude about the relative frequencies as the sample size increases?

Annual income	Probability
Under $25,000	0.083
$25,000 to $49,999	0.166
$50,000 to $74,999	0.169
$75,000 to $99,999	0.160
$100,000 to $149,999	0.200
$150,000 or more	0.222

Teenagers' Music and Lifestyle. Every year, *USA Weekend* conducts a survey of the nation's teenagers, asking them various questions about lifestyle and music. Nearly 60,000 teenagers responded to the survey, conducted in part through a Web site. Use this information for Exercises 82 and 83.

82. One *USA Weekend* survey question asked teenagers, "Do you listen to music while you are . . . ?" and listed several options. The most common responses are shown in the table. Respondents could choose more than one response. Explain why you cannot construct a probability model with these percentages.

Doing chores	79%
On the computer	73%
Doing homework	72%
Eating meals at home	33%
In the classroom	18%

83. Another survey question asked by *USA Weekend* was "If you had to choose just one type of music to listen to exclusively, which would it be?" The results are shown in the table. 🎹 **teenmusic**

Hip-hop/rap	27%
Pop	23%
Rock/punk	17%
Alternative	7%
Christian/gospel	6%
R&B	6%
Country	5%
Techno/house	4%
Jazz	1%
Other	4%

a. Construct a probability model.

b. Is it unusual for a respondent to prefer jazz?

c. Use technology to draw random samples of sizes 10, 100, 1,000, and 10,000 from your probability model.

d. What can you conclude about the relative frequencies as the sample size increases?

84. Paul the Predicting Octopus. During the 2010 World Cup, Paul the Octopus picked the correct outcome of each soccer match in which Germany was involved, and then he continued his winning streak right to the end, picking champion Spain in the final match. Paul indicated his choice by swimming over and choosing one of two containers of mussels (his favorite food). Each container had the flag of a country in the day's match. Paul predicted the outcome correctly eight times in a row.

a. Find the probability of Paul predicting all eight matches correctly, assuming a success probability of 50% for each match. Express this answer in the form $(1/2)^k$, where k is the number of matches.

b. Let's say that Paul had predicted nine matches in a row correctly. Find the probability of doing so, assuming 50% probability of success for each match. Express this answer in the form $(1/2)^k$.

c. What is the ratio of your results from **(b)** to your results from **(a)**?

d. Use your results from **(a)**–**(c)** to form a general rule for the probability of correctly predicting k matches in a row when the probability of success for each match is 50%.

WHAT IF ? **85.** Refer to the previous exercise. *What if* the probability of Paul predicting correctly was larger than 50%? Would your answers to the following be greater or less than what you calculated in the previous exercise?

a. The probability of Paul predicting all eight matches in a row correctly

b. The probability of Paul predicting nine matches in a row correctly

Use the following information for Exercises 86–91. Consider the experiment where a fair die is rolled twice. Define the following events for each roll: low = {1, 2}, medium = {3, 4}, high = {5, 6}, odd = {1, 3, 5}, even = {2, 4, 6}.

86. Construct a tree diagram for this experiment. Make sure you use the outcomes and not the events.

87. Use the tree diagram to construct the sample space. To which sample space discussed in Section 5.1 is the sample space for this experiment similar? Explain why this is so.

88. The sample space is the collection of all possible outcomes of an experiment. Explain why the sample space is not defined as the collection of all possible events.

89. Find the probability of observing a 1, followed by another 1. What method of assigning probability are you using? Why?

90. Find the probability of observing two high die rolls. What method of assigning probability are you using? Why?

91. Find the following probabilities:

a. Two high die results

b. Exactly one medium die result

c. No low die results

d. At least one high die result

e. At most one medium die result

BRINGING IT ALL TOGETHER

Best-Selling Video Games. Table 5 contains the top 20 best-selling video games in the United States for the week of May 17, 2014, along with the game platform, publishing studio, and the type of game. Use this information for Exercises 92–98.

TABLE 5 Top 20 best-selling video games in the United States for the week of May 17, 2014

Game/Platform	Studio	Type
Minecraft for PS3	Sony	Adventure
Minecraft for Xbox 360	MS	Adventure
Kirby: Triple Deluxe for 3DS	Nintendo	Platform
MLB 14 The Show for PS4	Sony	Sports
Titanfall for Xbox One	Electronic Arts	Shooter
Call of Duty: Ghosts for Xbox 360	Activision	Shooter
Bound by Flame for PS4	Focus	Action
Pokemon X/Y for 3DS	Nintendo	Role-playing
Titanfall for Xbox 360	Electronic Arts	Shooter
Grand Theft Auto V for Xbox 360	Take-Two	Action
Grand Theft Auto V for PS3	Take-Two	Action
Call of Duty: Ghosts for PS4	Activision	Shooter
Super Luigi U for WiiU	Nintendo	Platform
Super Mario Brothers U for WiiU	Nintendo	Action
Call of Duty: Ghosts for PS3	Activision	Shooter
Borderlands 2 for PSV	Take-Two	Shooter
Battlefield 4 for Xbox 360	Electronic Arts	Shooter
Forza Motorsport 5 for Xbox One	MS	Racing
Call of Duty: Ghosts for Xbox One	Activision	Shooter
inFamous: Second Son for PS4	Sony	Action

92. Suppose our experiment is to select one video game at random from Table 5 and observe its studio. What are the possible outcomes of this experiment?

93. Find the probability that the video game studio is *Sony*.

94. Continue to find the probability of each outcome in Exercise 93. Combine these to construct the probability model for this experiment. Which method of assigning probability did you use?

95. Verify that your probability model in Exercise 94 meets the Rules of Probability.

96. Next, suppose our experiment is to select, at random, two video games in succession and to observe the type of game.

   a. Construct a tree diagram for this experiment.

   b. Use the tree diagram to generate the sample space.

97. Construct a relative frequency distribution of the type of game.

98. Use your relative frequency distribution from Exercise 97 to construct a probability model of *type*. Verify that your probability model meets the Rules of Probability.

 Use the *Law of Large Numbers for Proportions* applet for Exercises 99 and 100.

99. Set the probability of heads to 0.5 and the number of tosses to 40. Click **Toss**.

   a. Record the proportion of heads observed.

   b. Without pressing **Reset**, continue to click **Toss** until the total number of tosses is 120. Again, record the proportion of heads.

   c. Without pressing **Reset**, continue to click **Toss** until the total number of tosses is 240. Again, record the proportion of heads.

   d. Without pressing **Reset**, continue to click **Toss** until the total number of tosses is 480. Again, record the proportion of heads.

100. The proportions you recorded in Exercise 99 are relative frequencies of heads. What can you conclude about the relative frequencies as the sample size increases?

WORKING WITH LARGE DATA SETS

 **Chapter 5 Case Study: The Gardasil Vaccine.** Open the data set **Gardasil**. We shall explore some probabilities about clinic location and patient insurance type, using the tools and techniques we have learned in this section. Use technology to do Exercises 101–106. **gardasil**

101. Construct a frequency distribution of the variable *Clinic Location*.

102. Use the frequency distribution from Exercise 101 to construct a probability model for the location of the clinic.

103. What is the probability that a randomly chosen clinic will be urban? What method are you using to assign this probability?

104. Construct a frequency distribution of the variable *Insurance Type*.

105. Use the frequency distribution from Exercise 104 to construct a probability model for type of insurance.

106. What is the probability that a randomly chosen patient will have a private payer insurance type? What method are you using to assign this probability?

5.2   Combining Events

OBJECTIVES By the end of this section, I will be able to . . .

1   Combine events using complement, union, and intersection.

2   Apply the Addition Rule to events in general and to mutually exclusive events in particular.

1 Complement, Union, and Intersection

In Example 7, if your roommate rolled a 4, then your roommate was to do your laundry for the rest of the semester. Your roommate is keenly interested in *not* rolling a 4. If *A* is an event, then the collection of outcomes not in event *A* is called the **complement of** *A*, denoted A^C. The term *complement* comes from the word "to complete," meaning that any event and its complement together make up the complete sample space.

EXAMPLE 14   **Finding the probability of the complement of an event**

If *A* is the event "observing a sum of 4 when the two fair dice are rolled," then your roommate is interested in the probability of A^C, the event that a sum of 4 is not rolled. Find the probability that your roommate does not roll a sum of 4.

Solution

Which outcomes belong to A^C? By the definition, A^C is all the outcomes in the sample space that do not belong in *A*. The following outcomes are included in *A*: {(3,1), (2,2), (1,3)}.

Figure 12 shows all the outcomes, except the outcomes from A in the two-dice sample space. There are 33 outcomes in A^C and 36 outcomes in the sample space. The classical probability method then gives the probability of not rolling a sum of 4 to be

$$P(A^C) = \frac{N(A^C)}{N(S)} = \frac{33}{36} = \frac{11}{12}$$

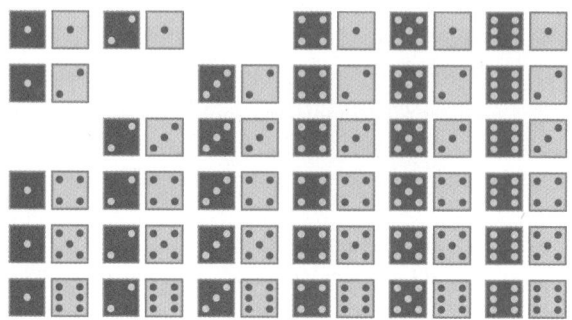

NOW YOU CAN DO
Exercises 7–12.

FIGURE 12 Outcomes in A^C.

YOUR TURN
#8

Refer to the two-dice sample space on page 247. Define event B as rolling a sum of 8.

a. Find $P(B)$.

b. Calculate $P(B^C)$.

(The solutions are shown in Appendix A.)

The probability is high that, on this roll at least, your roommate will not land on Boardwalk.

For event A in Example 14, note that

$$P(A) + P(A^C) = \frac{1}{12} + \frac{11}{12} = 1$$

Is this a coincidence, or does the sum of the probabilities of an event and its complement always add to 1? Recall the Law of Total Probability (Section 5.1), which states that the sum of all the outcome probabilities in the sample space must be equal to 1. Because any event A and its complement A^C together make up the entire sample space, then it always happens that $P(A) + P(A^C) = 1$.

Probabilities for Complements

For any event A and its complement A^C, $P(A) + P(A^C) = 1$. Applying a touch of algebra gives the following:

- $P(A) = 1 - P(A^C)$
- $P(A^C) = 1 - P(A)$

Sometimes we need to find the probability of a combination of events. For example, consider the casino game of craps in which you roll two dice. One way of winning is by rolling the sum 7 or 11. We can find the probability of the following two events: the sum is 7 or the sum is 11. First, we need some tools for finding the probability of a combination of events.

> **Union and Intersection of Events**
>
> The **union** of two events A and B is the event representing all the outcomes that belong to A or B or both. The union of A and B is denoted as A ∪ B and is associated with "or." In other words, (A ∪ B) = (A or B).
>
> The **intersection** of two events A and B is the event representing all the outcomes that belong to both A and B. The intersection of A and B is denoted as A ∩ B and is associated with "and." In other words, (A ∩ B) = (A and B).

If you are asked to find the probability of "*A* or *B*," you should find the probability of *A* ∪ *B*. Figure 13 shows the union of two events, with the red dots indicating the outcomes. Note from Figure 13 that the union of the events *A* and *B* refers to all outcomes in *A* or *B* or both. Figure 14 shows that the intersection of the two events is the part where *A* and *B* overlap. Both union and intersection are commutative. That is, (*A* ∪ *B*) = (*B* ∪ *A*) and (*A* ∩ *B*) = (*B* ∩ *A*). In other words, (*A* or *B*) = (*B* or *A*) and (*A* and *B*) = (*B* and *A*).

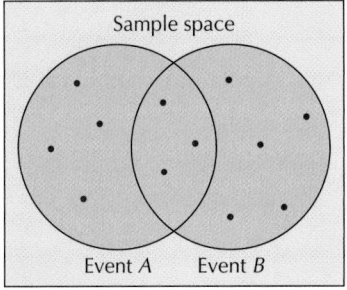

FIGURE 13 Union of events *A* and *B*, (*A* ∪ *B*) = (*A* or *B*).

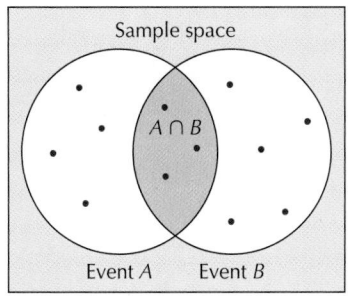

FIGURE 14 Intersection of events *A* and *B*, (*A* ∩ *B*) = (*A* and *B*).

EXAMPLE 15 Union and intersection

Recall from Section 2.1 that a *contingency table* (also known as a *crosstabulation*) is a tabular summary of the relationship between two categorical variables. Table 6 contains a contingency table summarizing the gender and survival status of the passengers and crew of RMS *Titanic*.

Table 6 Contingency table of gender and survival status of passengers and crew of RMS *Titanic*. The intersection of *Female* and *Survived* is highlighted.

	Female	Male	Total
Did not survive	126	1364	1490
Survived	344	367	711
Total	470	1731	2201

Intersection of *Female* and *Survived*

Source: Report on the Loss of the "Titanic" (S.S.). British Board of Trade Inquiry Report (reprint), Allan Sutton Publishing, Gloucester, United Kingdom, 1990.

Suppose our experiment is to select one person at random from the passengers and crew. Define the following events:

    F: Person is a female

    S: Person survived

a. Find the intersection of these events, (*F* and *S*).

b. Find the union of these events, (*F* or *S*).

Solution

a. The intersection of *F* and *S* is the event containing the outcomes that are common to both *F* and *S*. Note in Table 6 that the *female* outcomes form a column and the *survived* outcomes form a row. The intersection of *F* and *S* lies at the "intersection" of this column and this row, as illustrated in Table 6. The green cell belongs both to the *Female* column and the *Survived* row, and thus belongs to both events. The green cell therefore represents the intersection, (*F* and *S*), which includes the 344 passengers and crew who were both female *and* survived.

b. The union of *F* and *S* is the event containing all the people who were either female or who survived, or both. That is, the union (*F* or *S*) contains the following groups:

- *F:* Those who were female (126 who did not survive and 344 who did survive)

- *S:* Those who survived (344 females and 367 males).

Thus, in Table 6, the union (*F* or *S*) is represented by the cells containing 126, 344, and 367.

NOW YOU CAN DO
Exercises 13–18.

YOUR TURN
#9

Refer to the experiment in Example 15. Define the following events:

M: Person is a male

N: Person did not survive

a. Find (M and N).

b. Find (M or N).

(The solutions are shown in Appendix A.)

2 Addition Rule

We are often interested in finding the probability that either one event *or* another event may occur. The formula for finding these kinds of probabilities is called the **Addition Rule**.

Addition Rule

$$P(A \cup B) = P(A) + P(B) - P(A \cap B)$$

In other words,

$$P(A \text{ or } B) = P(A) + P(B) - P(A \text{ and } B)$$

What Does the Addition Rule Mean?

We can use Figure 15 to understand the Addition Rule. We are trying to find *P(A or B)*, the probability of all the outcomes in *A* or *B* or both. The first part of the formula says to add the probabilities of the outcomes in *A* to those of the outcomes in *B*. But what about the overlap between *A* and *B*, which includes outcomes that belong to both events? To avoid counting the outcomes in the overlap (intersection) twice, we have to subtract the probability of the intersection, *P(A and B)*.

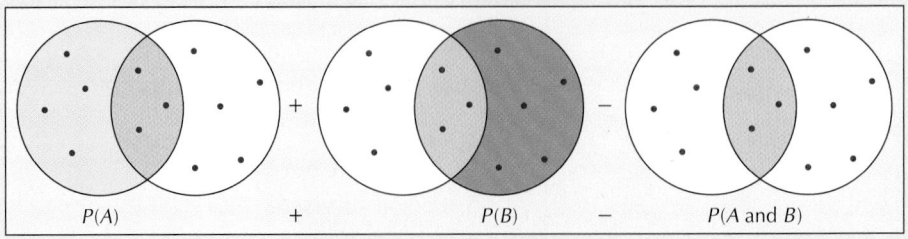

P(A)   +   *P(B)*   −   *P(A and B)*

FIGURE 15 How the Addition Rule works.

EXAMPLE 16 Addition Rule

Continuing with the RMS *Titanic* data from Example 15, Table 6 is reproduced here. Find the probability that a randomly chosen passenger or crew member had the following characteristics:

a. Was female

b. Survived

c. Was female and survived

d. Was female or survived

Table 6 Contingency table of gender and survival status of passengers and crew of RMS *Titanic*. The intersection of *Female* and *Survived* is highlighted.

	Female	Male	Total
Did not survive	126	1364	1490
Survived	344	367	711
Total	470	1731	2201

Intersection of *Female* and *Survived*

Solution

a. Here, we seek $P(F)$. There were 470 female passengers and crew among the 2201 people on board the *Titanic*. Therefore, $P(F) = 470/2201 = 0.2135$.

b. We are looking for $P(S)$. Of the 2201 people on board the *Titanic*, 711 survived. Thus, $P(S) = 711/2201 = 0.3230$.

c. Those who were female *and* survived represent the intersection (F and S). In Example 15, we found that these were represented by the green cell with 344 people in it, lying in the intersection of the *Female* column and the *Survived* row. Therefore, $P(F$ and $S) = 344/2201 = 0.1563$.

NOW YOU CAN DO
Exercises 19–44.

d. Here, we seek $P(F$ or $S)$. By the Addition Rule,

$$P(F \text{ or } S) = P(F) + P(S) - P(F \text{ and } S) = 0.2135 + 0.3230 - 0.1563 = 0.3802$$

YOUR TURN
#10

Refer to the experiment in Example 15. Define the following events:

M: Person was a male

N: Person did not survive

Find the probability that a randomly chosen passenger or crew member has the following characteristics:

a. Was male

b. Did not survive

c. Was a male who did not survive

d. Was a male or did not survive

(The solutions are shown in Appendix A.)

Mutually Exclusive Events

When drawing a card at random from a deck of 52 cards, the events "a heart is drawn" and "a diamond is drawn" have no outcomes in common. That is, no card is both a heart and a diamond. We say that these two events are **mutually exclusive**.

Two events are said to be **mutually exclusive**, or **disjoint**, if they have no outcomes in common.

Note that any event and its complement are always mutually exclusive. Other examples of mutually exclusive events are given in Table 7.

Table 7 Examples of mutually exclusive events

Experiment	Mutually exclusive events
Draw a single card from a deck of 52 cards	Card is red; card is a spade.
Buy a stock	Stock will increase in value; stock will not change in value.
Select a student at random	Student is 30 years old or older; student is under 18 years old.
Select a college course at random	Course has 3 credits; course has 4 credits.

Figure 16 shows how mutually exclusive events are represented graphically. It shows the events

$$A = \{1, 3, 5, 7, 9\} \quad \text{and} \quad B = \{0, 2, 4, 6, 8\}$$

with sample space $S = \{0, 1, 2, 3, 4, 5, 6, 7, 8, 9\}$.

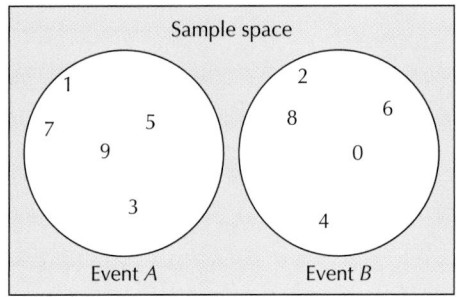

FIGURE 16 Even and odd digits are mutually exclusive. Note that the overlap (A and B) is empty, so that P(A and B) = 0.

Note that no overlap exists between the two events. When two events are mutually exclusive, they share no outcomes, and therefore the intersection of mutually exclusive events is empty. Because the intersection (A and B) is empty, then for mutually exclusive events, $P(A \text{ and } B) = 0$. Therefore, we can formulate a special case of the **Addition Rule for Mutually Exclusive Events** A and B:

$$P(A \text{ or } B) = P(A) + P(B) - P(A \text{ and } B) = P(A) + P(B) - 0 = P(A) + P(B)$$

> **Addition Rule for Mutually Exclusive Events**
> If A and B are mutually exclusive events, P(A or B) = P(A) + P(B).

EXAMPLE 17 Addition Rule for mutually exclusive events

The National Center for Education Statistics conducted a survey regarding the living arrangements of 19,735 college students. The results are shown in Table 8.

Table 8 Contingency table of college student living arrangements

	On campus	Off campus	With parents	Total
Females	1,368	5,741	4,103	11,212
Males	1,240	4,170	3,113	8,523
Total	2,608	9,911	7,216	19,735

Calculate the following probabilities for a randomly chosen college student:

a. Lives on campus, $P(C)$.

b. Lives with parents, $P(W)$.

c. Lives on campus or with parents, $P(C \text{ or } W)$.

Solution

a. A total of 2,608 of the 19,735 students live on campus, so $P(C) = 2{,}608/19{,}735 = 0.1322$.

b. Of the 19,735 students, 7,216 live with parents, so $P(W) = 7{,}216/19{,}735 = 0.3656$.

c. Living on campus and living with parents are mutually exclusive. So, by the Addition Rule for Mutually Exclusive Events, $P(C \text{ or } W) = P(C) + P(W) = 0.1322 + 0.3656 = 0.4978$.

NOW YOU CAN DO
Exercises 45–48.

YOUR TURN
#11

Refer to Example 17. Find the probability that a randomly chosen student lives on campus or lives off campus.

(The solution is shown in Appendix A.)

Section 5.2 Summary

1. We may combine events using the concepts of complement, union, and intersection.

2. The Addition Rule provides the probability of event A or event B to be the sum of their two probabilities minus the probability of their intersection. Mutually exclusive events have no outcomes in common.

Section 5.2 Exercises

CLARIFYING THE CONCEPTS

1. Describe in your own words what it means for two events to be mutually exclusive. (p. 263)

2. Describe the intersection of two mutually exclusive events. (p. 264)

3. Describe the union of two mutually exclusive events. (p. 264)

4. Is it true that the union of two events always contains at least as many outcomes as the intersection of two events? Use Figures 13 and 14 to help you visualize this problem. (p. 261)

5. If we choose a student at random from your college or university, is it more likely that we choose a male or a male football player? Why? (p. 261)

6. What is your personal estimate of the probability that it will rain on any given day? How about the probability that it won't rain? Why do these numbers have to add up to 1 (or 100%)? (p. 260)

PRACTICING THE TECHNIQUES

 **CHECK IT OUT!**

To do	Check out	Topic
Exercises 7–12	Example 14	Finding the probability of the complement of an event
Exercises 13–18	Example 15	Union and intersection
Exercises 19–44	Example 16	Addition Rule
Exercises 45–48	Example 17	Addition Rule for mutually exclusive events

For Exercises 7–12, consider the experiment of rolling a fair die once. Find the indicated probabilities:

7. Observing a number that is not 6

8. Observing some other number than 1

9. The complement of the event E, where E: $\{1, 3, 5\}$

10. L^C, where L: $\{5, 6\}$

11. E^C, where E: $\{1, 3, 5\}$

12. Not rolling an even number

For Exercises 13–18, consider the experiment of drawing a single card at random from a deck of cards. Define the following events. Find the indicated unions and intersections:

J: The card is a jack.
B: The card is a black suit.
S: The card is a spade.

13. $J \cap B$

14. $J \cap S$

15. $B \cap S$

16. $J \cup B$

17. $J \cup S$

18. $B \cup S$

For Exercises 19–24, consider the experiment of drawing a single card at random from a deck of cards. Define the following events. Find the indicated probabilities:

J: The card is a jack.
B: The card is a black suit.
S: The card is a spade.

60. Use Table 13 to complete the contingency table shown in Table 14.

TABLE 14 Contingency table of car size and recommended gasoline

		Recommended gasoline		
		Regular	Premium	Total
Car Size	Compact			
	Midsize			
	Large			
	Total			

For Exercises 61–64, the experiment is to select a car at random.

61. Use the completed Table 14 to find the probability of the following:
  a. Choosing a compact car
  b. Choosing a car that is not a compact car
  c. Selecting a midsize car
  d. Selecting a car that is not a midsize car

62. Use the completed Table 14 to find the probability of the following:
  a. Choosing a car that uses regular gasoline
  b. Choosing a car that does not use regular gasoline
  c. Selecting a compact car that uses premium gasoline
  d. Selecting a compact car that uses regular gasoline

63. Use the completed Table 14 to find the following probabilities:
  a. Choosing a midsize car that uses premium gasoline
  b. Choosing a midsize car that uses regular gasoline
  c. Selecting a large car that uses premium gasoline
  d. Selecting a large car that uses regular gasoline

64. Use the completed Table 14 to find the following probabilities:
  a. Choosing a compact car or a car that uses premium gasoline
  b. Choosing a compact car or a car that uses regular gasoline
  c. Selecting a car that uses regular gasoline and is not a compact car
  d. Selecting a car that uses regular gasoline or is not a compact car
  e. Selecting a car that uses premium gasoline or a large car

Top 10 Movies of All Time. Table 15 contains the top 10 domestic movies of all time, according to Box Office Mojo (http://boxofficemojo.com), along with the studio, whether the movie grossed over or under $500 million, and the century in which the movie was produced.

TABLE 15 Top 10 domestic movies of all time

Rank	Movie	Studio	Gross over or under $500 million	Century produced
1	Avatar	Fox	Over	21st
2	Titanic	Other	Over	20th
3	Marvel's The Avengers	Other	Over	21st
4	The Dark Knight	Warner	Over	21st
5	Star Wars: Episode I - The Phantom Menace	Fox	Under	20th
6	Star Wars	Fox	Under	20th
7	The Dark Knight Rises	Warner	Under	21st
8	Shrek 2	Other	Under	21st
9	E.T.: The Extra-Terrestrial	Other	Under	20th
10	The Hunger Games: Catching Fire	Other	Under	21st

For Exercises 65–69, find the following probabilities regarding the studio and what the movie grossed:
65. Studio was Fox and the movie grossed over $500 million.
66. Studio was not Fox and the movie grossed over $500 million.
67. Studio was Fox or the movie grossed over $500 million.
68. Studio was not Fox or the movie grossed over $500 million.
69. Studio was Warner and the movie grossed over $500 million.

For Exercises 70–73, find the following probabilities about the studio and the century produced:
70. Studio was Warner and the movie was produced in the 20th century.
71. Studio was Warner and the movie was produced in the 21st century.
72. Studio was not Warner and the movie was produced in the 20th century.
73. Studio was not Warner and the movie was produced in the 21st century.

Wheel of Fortune! Use Table 16 for Exercises 74–81. Imagine yourself on the television game show *Wheel of Fortune,* where contestants guess the letters contained in a hidden phrase. You want to ask for the letters that have the greatest chance of occurring, so you want to know the various probabilities of the letters in the English alphabet.

The experiment is to choose one letter at random from a sample of 1000 letters. The sample space is the 26 letters of the alphabet. The total sample size is 1000, so you can find the relative frequencies of the letters in English simply by dividing each frequency by the total sample size (remember this from Chapter 2?).

TABLE 16   Frequency distribution of English letters

A	B	C	D	E	F	G
73	9	30	44	130	28	16
H	I	J	K	L	M	N
35	74	2	3	35	25	78
O	P	Q	R	S	T	U
74	27	3	77	63	93	27
V	W	X	Y	Z		
13	16	5	19	1		

74. Construct a probability model of the frequency distribution in Table 16.

75. Find the probability of selecting an E.

76. Find the probability of selecting a vowel, that is, either A, E, I, O, or U.

77. Note that selecting a consonant may be considered the complement of selecting a vowel. Find the probability of selecting a consonant.

78. Choosing one letter at random, which is it more likely to be, a consonant or a vowel?

79. Of the eight letters with the highest frequencies, how many are vowels?

80. Compare your answers to Exercises 76 and 77, and explain why *Wheel of Fortune* contestants are not allowed to guess vowels (they have to buy them).

81. If you were a contestant on *Wheel of Fortune* and had no money to buy a vowel, what would be your first five letter choices?

Gender and GPA. The National Center for Education Statistics reported the contingency table results in Table 17 for the gender and the grade point average of the 21,504 undergraduate students it surveyed. The experiment is to select one random selected student. Use Table 17 to calculate the probabilities indicated in Exercises 82–93.

TABLE 17   Contingency table of gender and GPA

	GPA < 2.5	GPA 2.5 to < 3	GPA 3 to < 3.5	GPA 3.5 or higher	Total
Females	2,698	2,439	3,314	3,868	12,319
Males	2,535	1,975	2,324	2,351	9,185
Total	5,233	4,414	5,638	6,219	21,504

82. Student is female.

83. Student is male.

84. Student has GPA < 2.5.

85. Student is female and has a GPA < 2.5.

86. Student is male and has a GPA < 2.5.

87. Student is female or has a GPA < 2.5.

88. Student is male or has GPA < 2.5.

89. Student does not have a GPA < 2.5.

90. Student is female and does not have a GPA < 2.5.

91. Student is male and does not have a GPA < 2.5.

92. Student is female or does not have a GPA < 2.5.

93. Student is male or does not have a GPA < 2.5.

BRINGING IT ALL TOGETHER

Best-Selling Video Games. Table 18 contains the top 20 best-selling video games in the United States for the week of May 17, 2014, along with the game platform, publishing studio, and the type of game. Use this information for Exercises 94–102.

TABLE 18   Top 20 best-selling video games in the United States for the week of May 17, 2014

Game/Platform	Studio	Type
Minecraft for PS3	Sony	Adventure
Minecraft for Xbox 360	MS	Adventure
Kirby: Triple Deluxe for 3DS	Nintendo	Platform
MLB 14 The Show for PS4	Sony	Sports
Titanfall for Xbox One	Electronic Arts	Shooter
Call of Duty: Ghosts for Xbox 360	Activision	Shooter
Bound by Flame for PS4	Focus	Action
Pokemon X/Y for 3DS	Nintendo	Role-Playing
Titanfall for Xbox 360	Electronic Arts	Shooter
Grand Theft Auto V for Xbox 360	Take-Two	Action
Grand Theft Auto V for PS3	Take-Two	Action
Call of Duty: Ghosts for PS4	Activision	Shooter
Super Luigi U for WiiU	Nintendo	Platform
Super Mario Brothers U for WiiU	Nintendo	Action
Call of Duty: Ghosts for PS3	Activision	Shooter
Borderlands 2 for PSV	Take-Two	Shooter
Battlefield 4 for Xbox 360	Electronic Arts	Shooter
Forza Motorsport 5 for Xbox One	MS	Racing
Call of Duty: Ghosts for Xbox One	Activision	Shooter
inFamous: Second Son for PS4	Sony	Action

94. Suppose our experiment is to select one video game at random from Table 18, and observe its studio. List the video games belonging to the following events:
   a. S: Sony
   b. N: Nintendo.

95. For each studio in Exercise 94, find the probability of a video game being made by that studio.

96. Describe in words the complement of the following events (save space by not listing them all):

    a. S: Sony

    b. N: Nintendo.

97. For each complement in Exercise 96, find its probability.

98. What is the sum of the following two probabilities: (i) Video game made by Sony, (ii) Video game made by some other studio.

99. Next, suppose our experiment is to select a video game at random. Define the following events: S: Studio is Sony, N: Studio is Nintendo, A: Type = Action, P: Type = Platform. List the video games in the following events:

    a. (S and A)

    b. (S or A)

    c. (N and P)

    d. (N or P)

    e. (N and A)

    f. (N or A)

100. Find the following probabilities:

    a. P (S and A)

    b. P (S or A)

    c. P (N and P)

    d. P (N or P)

    e. P (N and A)

    f. P (N or A)

101. List the video games in the following events:

    a. (S and N)

    b. (S or N)

    c. (A and P)

    d. (A or P)

102. Find the following probabilities:

    a. P (S and N)

    b. P (S or N)

    c. P (A and P)

    d. P (A or P)

WORKING WITH LARGE DATA SETS

Chapter 5 Case Study: The Gardasil Vaccine. Open the data set **Gardasil**. We will explore some probabilities about the patient completion of the vaccination treatment and whether the patient received medical assistance, using the tools and techniques we have learned in this section. Use technology to complete Exercises 103–107. **gardasil**

103. Construct a relative frequency distribution of the variables *medical assistance* and *completed*. Use these to find the probability that a randomly selected patient:

    a. had medical assistance.

    b. completed the vaccine treatment.

104. Construct a contingency table (crosstabulation) of *medical assistance* with *completed*. Include the row totals, column totals, and grand total.

105. Use the contingency table from Exercise 104 to find the probability that a randomly chosen patient:

    a. had medical assistance and completed the treatment.

    b. had medical assistance and did not complete the treatment.

    c. did not have medical assistance and did complete the treatment.

    d. did not have medical assistance and did not complete the treatment.

106. Think about the four probabilities you found in Exercise 105. Do you think that patients who had medical assistance were more likely to complete the treatment? Provide support for your statement.

107. Find the probability that a randomly chosen patient:

    a. had medical assistance or completed the treatment.

    b. had medical assistance or did not have medical assistance.

5.3   Conditional Probability

OBJECTIVES   By the end of this section, I will be able to . . .

1   Calculate conditional probabilities.

2   Explain independent and dependent events.

3   Solve problems using the Multiplication Rule and recognize the difference between sampling with replacement and sampling without replacement.

4   Approximate probabilities for dependent events.

5   Apply Bayes' Rule to solve probability problems.

1   Introduction to Conditional Probability

As we progress through this book, you will notice a recurring theme: *the more informa-tion available, the better*. Very often, when we are investigating the probability of a certain event *A*, we learn that another event *B* has occurred. If events *A* and *B* are related, then the occurrence of event *B* often influences the probability that event *A* will occur.

EXAMPLE 18 . **Having more information often affects the probability of an event**

Table 19 contains the Academy Award winners for Best Actress and Best Actor for 2009–2014, along with their ages at the time of the award.

Table 19 Oscar-winning actresses and actors, and their ages

Year	Best actress	Film	Age	Best actor	Film	Age
2009	Kate Winslet	The Reader	33	Sean Penn	Milk	48
2010	Sandra Bullock	The Blind Side	45	Jeff Bridges	Crazy Heart	60
2011	Natalie Portman	Black Swan	29	Colin Firth	The King's Speech	50
2012	Meryl Streep	Iron Lady	62	Jean Dujardin	The Artist	39
2013	Jennifer Lawrence	Silver Linings Playbook	22	Daniel Day-Lewis	Lincoln	55

Source: www.oscars.org.

Table 20 is a contingency table, summarizing the information in Table 19, providing the counts of the performers' genders and whether the performer was under age 40.

Table 20 Contingency table of Oscar-winning performers

	Female	Male	Total
Under age 40	3	1	4
Age 40 or older	2	4	6
Total	5	5	10

Now, if we choose a performer at random from Table 20, the probability of choosing a female is $P(F) = \frac{5}{10} = \mathbf{0.5}$. But what if we were *given the extra information* that the performer is age 40 or older? How does this extra information affect the probability of selecting a female?

Solution

Notice that when we are given that the person is age 40 or older, we may restrict our attention to the performers who are age 40 or older in Table 20 (highlighted). In other words, this extra information reduces the number of possible outcomes in the sample space from the 10 performers to the 6 performers who are age 40 or older. Of these six performers, two of them are female. Thus, the probability of selecting a female, *given that* the performer is age 40 or older, is 2/6 = 1/3 ≈ 0.33. The *extra information we were given* changed the probability of selecting a female, from 1/2 to 1/3.

The extra information about a related event changed the probability of the event of interest. This type of probability is an example of what is called **conditional probability**.

Note: The vertical line | is read as the word "given." Whatever follows the word "given" is assumed to have occurred already.

> For two related events A and B, the probability of B given A is called a **conditional probability** and is denoted as $P(B|A)$, or $P(B$ given $A)$.

Thus, if we let F represent the event that a female is selected and A represent the event that the performer is age 40 or older, then

$$P(F) = \frac{5}{10} = \frac{1}{2} \quad \text{but} \quad P(F \text{ given } A) = P(F \mid A) = \frac{2}{6} = \frac{1}{3}$$

Figure 17 can help us visualize how conditional probability works. The idea is that, *once event A has occurred*, the only chance for event F to occur is in the overlap—the intersection (F and A). Therefore, the conditional probability that F will occur, given that event A has already taken place, is found by taking the ratio $P(F \text{ and } A)/P(A)$.

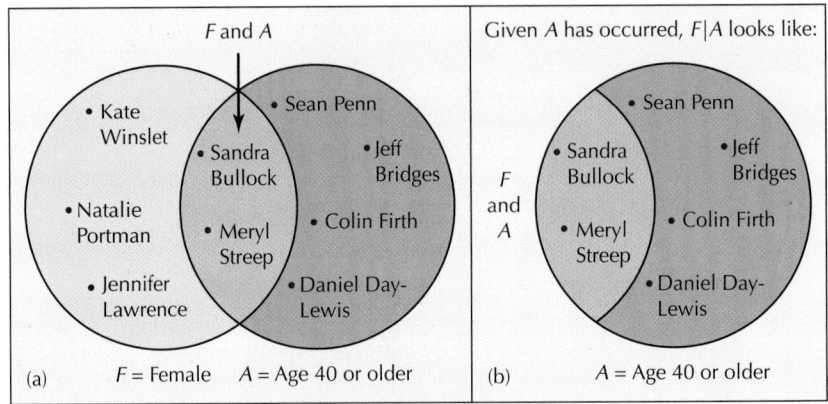

FIGURE 17 How conditional probability works.

Developing Your Statistical Sense

The difference between $P(F \text{ given } A)$ and $P(F \text{ and } A)$

Students sometimes confuse the meanings of $P(F \text{ given } A)$ with $P(F \text{ and } A)$. The two are very different.

- For $P(F \text{ and } A)$, neither A nor F has occurred. We don't know if either will or will not occur. Instead, we need to determine the probability that events F and A both occurred. Look at Figure 17a. For finding $P(F \text{ and } A)$, we don't know if our randomly chosen performer is female or is age 40 or older. Two of our 10 performers are both female and age 40 or older: Sandra Bullock and Meryl Streep. Thus, $P(F \text{ and } A) = 2/10 = 1/5$.

- For $P(F \text{ given } A) = P(F \mid A)$, *we assume that the event A has already occurred* and now need to find the probability of event F, given event A. There is no probability involved with the event A, because it has already occurred. Look at Figure 17b. For $P(F \text{ given } A) = P(F \mid A)$, we are given that the performer is age 40 or older; we know this event occurred. Therefore, we can restrict our attention solely to performers age 40 or older (event A). Of the six performers age 40 or older, two are female. Thus, $P(F \text{ given } A) = P(F \mid A) = 2/6 = 1/3$.

In general, the formula for computing conditional probability is given as follows.

Calculating Conditional Probability

The conditional probability that B will occur, given that event A has already taken place, equals

$$P(B \text{ given } A) = P(B \mid A) = \frac{P(A \text{ and } B)}{P(A)} = \frac{N(A \text{ and } B)}{N(A)}$$

EXAMPLE 19   Calculating conditional probability

Science Photo library/Alamy

Marketing companies are looking to statistical analysis and data mining in an effort to increase the customer response to their promotions. Therefore, these marketing companies are looking to hire more statistics-savvy students. Table 21 is adapted from a study on direct mail marketing. It contains the numbers of customers who either responded or did not respond to a direct mail marketing campaign, along with whether they had a credit card on file with the company. The two events are:

R: Responded to direct mail marketing campaign

C: Has a credit card on file

Table 21 Credit card status and marketing response

Response	Credit card on file?		
	No	Yes	Total
Did not respond	161	79	240
Did respond	17	31	48
Total	178	110	288

Source: Data Mining and Predictive Analytics, by Daniel Larose and Chantal Larose, Wiley Interscience, 2015.

a. Find the probability that a randomly chosen customer responded to the marketing campaign.

b. Calculate the probability that a randomly chosen customer both responded to the marketing campaign *and* has a credit card on file.

c. Find the conditional probability that a randomly selected customer responded, *given that* the customer has a credit card on file.

Solution

a. $P(R) = \frac{N(R)}{N(S)}$. There are $N(R) = 48$ customers who did respond, and there are $N(S) = 288$ customers in this experiment. Thus,

$$P(R) = \frac{N(R)}{N(S)} = \frac{48}{288} \approx 0.1667$$

b. Here, we are looking for $P(R$ and $C)$, which is the intersection between R and C. Earlier, we learned that this is represented by the cell at the intersection of the "Did respond" row and the "Credit card yes" column. In Table 21, this cell contains $N(R$ and $C) = 31$ such customers. Thus, the probability that a randomly chosen customer both responded to the marketing campaign *and* has a credit card on file = $P(R$ and $C) = 31/288 = 0.1076$.

c. We will use $P(R$ given $C) = P(R \mid C) = N(R$ and $C)/N(C)$ because, in this example, it is easier to work directly with the numbers of outcomes instead of the probabilities. From (**b**), $N(R$ and $C) = 31$. Also, there are $N(C) = 110$ customers in total who had a credit card on file. Therefore,

$$P(R \text{ given } C) = P(R \mid C) = \frac{N(R \text{ and } C)}{N(C)} = \frac{31}{110} \approx 0.2818$$

That is, the probability that a randomly chosen customer responded to the direct mail marketing campaign, given that the customer had a credit card on file, is 0.2818

NOW YOU CAN DO

Exercises 9–32.

YOUR TURN
#12

Using Table 21, find the probability that a randomly selected customer

a. Did not respond to the marketing campaign.

b. Did not respond to the marketing campaign, given that the customer did not have a credit card on file.

(The solutions are shown in Appendix A.)

What Do These
Numbers Mean?

Conditional Probability

Conditional probabilities can often be interpreted as percentages of some *subset* of a population. For example, the conditional probability that a customer responded, given that the customer has a credit card on file, may be interpreted as the percentage of customers with credit cards who responded.

2 Independent Events

Having a credit card on file increased the probability of a customer responding from 0.1667 to 0.2818, so we can say that the probability of responding *depends* in part on whether the customer has a credit card on file. In other words, the events R and C are *dependent events*.

On the other hand, if the probability of responding had been unaffected by whether the customer had a credit card on file, then we would have said that R and C were **independent events**. That is, R and C would have been independent events had $P(R \mid C)$ equaled $P(R)$. In general, if the occurrence of an event does not affect the probability of a second event, then the two events are independent.

> Events A and B are **independent** if
>
> $$P(A \text{ given } B) = P(A) \qquad \text{or if} \qquad P(B \text{ given } A) = P(B)$$
>
> Equivalently, using symbols, events A and B are **independent** if
>
> $$P(A \mid B) = P(A) \qquad \text{or if} \qquad P(B \mid A) = P(B)$$
>
> Otherwise, the events are said to be **dependent**.

Developing Your
Statistical Sense

Dependent and Independent Events

It may help to think of it this way: Andrew Luck is a superb quarterback for the Indianapolis Colts. Let B = the event the Colts win, and let A be the event that Andrew Luck gets injured. The Colts are certainly better with their great quarterback than without him. So $P(B)$ is greater than $P(B \mid A)$, because their chances of winning decrease if Luck gets injured. Because $P(B) \neq P(B \text{ given } A)$, the two events A and B are dependent. This makes sense, because the probability of the Colts winning is dependent on whether Andrew Luck gets injured. Now, do the same calculations—this time letting event C be the event that the water boy for the Colts gets injured. It is not likely that the water boy getting injured will affect the Colts' chances of winning, so that $P(B) = P(B \text{ given } C)$. Thus, the events B = Colts winning and C = water boy gets injured are independent.

Michael Hickey/Getty Images

Andrew Luck, quarterback for the Indianapolis Colts.

Alternatively, in Step 1 you can find $P(A)$ and in Step 2 you can find $P(A$ given $B)$. Then compare these two quantities for Step 3.

Strategy for Determining Whether Two Events Are Independent
1. Find $P(B)$.
2. Find $P(B$ given $A)$.
3. Compare the two probabilities. If they are equal, then A and B are independent events. Otherwise, A and B are dependent events.

EXAMPLE 20 Determining whether two events are independent

Table 22 contains a contingency table summarizing the gender and survival status of the passengers and crew of RMS *Titanic*.

Table 22 Contingency table of gender and survival status of passengers and crew of RMS *Titanic*. Notice that when we are given that the person is a female, we may restrict our attention to the females in the table.

	Female	Male	Total
Did not survive	126	1364	1490
Survived	344	367	711
Total	470	1731	2201

Source: Report on the Loss of the "Titanic" (S.S.). British Board of Trade Inquiry Report (reprint), Allan Sutton Publishing, Gloucester, United Kingdom, 1990.

Suppose our experiment is to select one person at random from the passengers and crew. Define the following events:

  F: Person is a female

  S: Person survived

Determine whether events *F* and *S* are independent.

Solution

We use the strategy for determining whether two events are independent.

Step 1 Find $P(F) = P$(Person is a female). There are 470 females out of a total of 2201 passengers and crew. Thus, $P(F) = 470/2201 \approx 0.21$.

Step 2 We need to find $P(F$ given $S)$, which is the probability that the person survived, given the person was female. We are told that the person was female, so we may restrict our attention to the females in the table. In other words, our sample space is reduced when we know that the person was a female. Of the 470 females, 344 survived, so that $P(F$ given $S) = 344/470 \approx 0.73$.

NOW YOU CAN DO
Exercise 33–52.

Step 3 Because, $P(F) \neq P(F$ given $S)$, we conclude that *F* and *S* are dependent events.

**YOUR TURN
#13**

In Example 20, determine whether being a male and not surviving are independent.

(The solution is shown in Appendix A.)

Developing Your
Statistical Sense

Don't Confuse Independent Events and Mutually Exclusive Events

It is important to stress the difference between independent events and mutually exclusive events. Mutually exclusive events have no outcomes in common. For two events to be independent means that the occurrence of one does not affect the probability of the other. The concepts are different.

EXAMPLE 21 Gambler's Fallacy

Suppose we have tossed a fair coin 10 times and have observed heads come up every time. Find the probability of tails on the next toss.

Solution

We have observed an unusual number of heads, so we might think that the probability of tails on the next toss is increased. However, the short answer is "Not so." Successive tosses of a fair coin are independent because the coin has no memory of its previous tosses. Thus, what happened on the first 10 tosses has no effect on the next toss. Probability theory tells us that, in the long run, the proportion of heads and tails will eventually even out if the coin is fair. Therefore, the probability of tails on the next toss is 0.5. This is an example of the Gambler's Fallacy.

3 Multiplication Rule

Just as the Addition Rule is used to find probabilities of unions of events, the **Multiplication Rule** is used to find probabilities of intersections of events. Recall the formula for the conditional probability of event B given event A:

$$P(B \text{ given } A) = \frac{P(A \text{ and } B)}{P(A)} \text{ where } P(A) \neq 0$$

We solve for $P(A \text{ and } B)$ by multiplying each side by $P(A)$:

$$P(A \text{ and } B) = P(A) \cdot P(B \text{ given } A)$$

Similarly, consider the conditional probability of event A given event B:

$$P(A \text{ given } B) = \frac{P(A \text{ and } B)}{P(B)} \text{ where } P(B) \neq 0$$

Solving for $P(A \text{ and } B)$ gives us a second equation for $P(A \text{ and } B)$:

$$P(A \text{ and } B) = P(B) \cdot P(A \text{ given } B)$$

The two equations for $P(A \text{ and } B)$ lead directly to the Multiplication Rule.

Multiplication Rule

$P(A \text{ and } B) = P(A) \cdot P(B \text{ given } A)$ or equivalently $P(A \text{ and } B) = P(B) \cdot P(A \text{ given } B)$

EXAMPLE 22   Multiplication Rule

©Alex Segre/Alamy

According to the Pew Internet and American Life Project,[11] 35% of American adults have cell phones with apps, but only 68% of those who have apps on their cell phones actually use the apps. Define the following events:

  A: American adult has a cell phone with apps.

  U: American adult uses the apps on his or her cell phone.

a. Find $P(A)$.

b. Find $P(U \text{ given } A)$, the probability that an American adult uses the apps, given that he or she has a cell phone with apps.

c. Use the multiplication rule to calculate $P(A \text{ and } U)$, the probability that an American adult has a cell phone with apps *and* uses the apps on his or her cell phone.

Solution

a. According to the study, 35% of American adults have a cell phone with apps. So $P(A) = 0.35$.

b. The research says that 68% of those who have apps actually use them, so $P(U \text{ given } A) = 0.68$.

c. Using the Multiplication Rule, we have

$$P(A \text{ and } U) = P(A) \cdot P(U \text{ given } A) = 0.35(0.68) = 0.238$$

The probability that an American adult has a cell phone with apps *and* uses them is 0.238.

NOW YOU CAN DO
Exercises 53–58.

YOUR TURN
#14

Suppose our experiment is to toss a fair die twice. Find the probability of rolling two sixes in a row.

(The solution is shown in Appendix A.)

By the definition of independent events (page 274), when events A and B are independent, $P(A$ given $B) = P(A)$ or $P(B$ given $A) = P(B)$. Using these identities, we can formulate a special case of the Multiplication Rule. Using $P(A$ given $B) = P(A)$, we can write the Multiplication Rule as

$$P(A \text{ and } B) = P(B) \cdot P(A|B) = P(B)\,P(A) = P(A)\,P(B)$$

Equivalently, the Multiplication Rule also states that $P(A \text{ and } B) = P(A) \cdot P(B \text{ given } A)$, but if A and B are independent, $P(B \text{ given } A) = P(B)$, so, again, $P(A \text{ and } B) = P(A)\,P(B)$.

Multiplication Rule for Two Independent Events

If A and B are any two independent events, $P(A$ and $B) = P(A)\,P(B)$.

EXAMPLE 23 Multiplication Rule for two independent events

Successive spins of the roulette wheel are independent because the wheel does not remember its result from the previous spin. Suppose the experiment is to spin the roulette wheel two times and to bet on red each time. There are 18 red numbers out of a total of 38 numbers on the wheel. What is the probability that red will occur twice in succession?

Solution

Define the following events:

    A: Roulette wheel comes up red on the first spin.

    B: Roulette wheel comes up red on the second spin.

We have the probability of each event given as:

$$P(A) = P(B) = \frac{18}{38}$$

Then, because the successive spins are independent, the **Multiplication Rule for Two Independent Events** tells us:

NOW YOU CAN DO
Exercises 59–66.

$$P(\text{Winning}) = P(A \text{ and } B) = P(A) \cdot P(B) = \left(\frac{18}{38}\right)\left(\frac{18}{38}\right) \approx 0.2244$$

YOUR TURN
#15

For the roulette experiment in Example 23, what is the probability of not getting red on the first spin and not getting red on the second spin?

(The solution is shown in Appendix A.)

Sampling With and Without Replacement

The relationship between two events can be determined by the way the samples are chosen. Two methods of choosing samples are *sampling with replacement* and *sampling without replacement*.

In **sampling with replacement**, the randomly selected unit is returned to the population after being selected. When sampling with replacement, it is possible for the same unit to be sampled more than once.

In **sampling without replacement**, the randomly selected unit is not returned to the population after being selected. When sampling without replacement, it is not possible for the same unit to be sampled more than once.

EXAMPLE 24   Sampling with replacement

We draw a card at random from a shuffled deck, observe the card, and return it to the deck. The deck is then reshuffled, and we draw another card at random. What is the probability that both cards we select will be aces?

Solution

Define the following events:

    A: Observe an ace on the first draw.

    B: Observe an ace on the second draw.

We want to find $P(A \text{ and } B)$, the probability of observing an ace on the first draw *and* an ace on the second draw. From the Multiplication Rule, $P(A \text{ and } B) = P(A) \cdot P(B|A)$. To find $P(A)$, recall that there are 4 aces in the deck of 52 cards. It is reasonable to assume that all cards are equally likely to be selected, so using the classical method, $P(A) = 4/52$. Similarly, $P(B) = 4/52$.

Next, we need to find $P(B \text{ given } A)$, the probability of observing an ace on the second draw, given that we observe an ace on the first draw. Because *the deck of 52 cards has not changed* (except for shuffling), there are still 52 cards—4 of which are aces. Therefore, $P(B \text{ given } A) = 4/52$. Thus, the probability that both cards we select will be aces is $P(A \text{ and } B) = P(A) \cdot P(B \text{ given } A) = (4/52)(4/52) \approx 0.0059$.

Note that $P(B \text{ given } A) = P(B) = 4/52$. Thus, by the alternative method for determining independence, *A* and *B* are independent events when sampling with replacement.

NOW YOU CAN DO
Exercises 67–70.

YOUR TURN
#16

For the experiment in Example 24, what is the probability of observing hearts on the first draw and observing hearts on the second draw, if we sample with replacement?

(The solution is shown in Appendix A.)

We can generalize this result as follows.

> When sampling with replacement, successive draws can be considered **independent**.

EXAMPLE 25   Sampling without replacement

Suppose we alter the experiment in Example 24 as follows: We draw a card at random from a shuffled deck, hold onto the card (do not replace it) while the deck is reshuffled, and then select another card at random. What is the probability that both cards we select will be aces?

Solution

Define events *A* and *B* as in Example 24. Again we use the Multiplication Rule to find $P(A \text{ and } B)$. The difference in this experiment comes when finding $P(B|A)$, the probability of observing an ace on the second draw given an ace on the first draw. Once we select the first ace, we do not replace it in the deck. Therefore, when the deck is reshuffled, it has only 51 cards left, only 3 of which are aces. The classical method then gives the probability of observing an ace on the second draw:

$$P(B \text{ given } A) = \frac{\text{number of aces in the deck}}{\text{number of cards in the deck}} = \frac{3}{51}$$

Thus, the probability that both cards we select will be aces is

$$P(A \text{ and } B) = P(A) \cdot P(B \text{ given } A) = \frac{4}{52} \cdot \frac{3}{51} = \frac{12}{2652} \approx 0.0045$$

NOW YOU CAN DO
Exercises 71–74.

This probability is somewhat less than the probability that both cards will be aces when sampling with replacement. Note that here we found that $P(B \text{ given } A)$ was not equal to $P(B)$. Thus, by the alternative method for determining independence, A and B are not independent events; they are dependent events.

**YOUR TURN
#17**

For the experiment in Example 25, what is the probability of observing hearts on the first draw and observing hearts on the second draw, if we sample without replacement?

(The solution is shown in Appendix A.)

We can generalize this result as follows.

> When sampling without replacement, successive draws should be considered **dependent**.

Note that the Multiplication Rule for Independent Events provides us with an alternative method for determining whether two events are indeed independent.

Alternative Method for Determining Independence
If $P(A) \cdot P(B) = P(A \text{ and } B)$, then events A and B are **independent**.
If $P(A) \cdot P(B) \neq P(A \text{ and } B)$, then events A and B are **dependent**.

EXAMPLE 26 Determining independence using the alternative method

We return to the direct mail marketing data from Example 19, reproduced here in Table 23. Use the alternative method for determining independence to determine whether the following two events are independent:

$R:$ Responded to direct mail marketing campaign.

$C:$ Has a credit card on file.

Table 23 Credit card status and marketing response

Response	Credit card on file?		Total
	No	Yes	
Did not respond	161	79	240
Did respond	17	31	48
Total	178	110	288

Source: Daniel Larose and Chantal Larose, *Data Mining and Predictive Analytics* (Wiley Interscience, 2015).

Solution

Using Table 23, we may find the following probabilities:

$$P(R) = \frac{48}{288} \qquad P(C) = \frac{110}{288} \qquad P(R \text{ and } C) = \frac{31}{288} \approx 0.1076$$

$$P(R) \cdot P(C) = \frac{48}{288} \cdot \frac{110}{288} \approx 0.0637$$

NOW YOU CAN DO
Exercises 79–84.

Because $0.0637 \neq 0.1076$, we have $P(R) \cdot P(C) \neq P(R \text{ and } C)$; therefore, R and C are dependent.

The next example illustrates the relationship between mutually exclusive events and independence.

EXAMPLE 27   **Conditional probability for mutually exclusive events**

Suppose two events A and B are mutually exclusive, with $P(A) > 0$ and $P(B) > 0$.

a. Find $P(B$ given $A)$.

b. Are events A and B independent or dependent?

Solution

a. Because A and B are mutually exclusive, $P(A$ and $B) = 0$. Then

$$P(B \text{ given } A) = \frac{P(A \text{ and } B)}{P(A)} = 0$$

That is, *if event A has occurred, then event B cannot occur.* This is a natural consequence of events A and B being mutually exclusive.

> ### What Results Might We Expect?
>
> Two events are independent if the occurrence of one does not affect the probability that the other will occur. However, as we saw in (**a**), if event A occurs, then the probability that event B will occur is 0. Thus, we would expect events A and B to be dependent.

In other words, if two events are mutually exclusive, then they are dependent.

b. We are given that $P(A) > 0$ and $P(B) > 0$. Thus, the product $P(A) \cdot P(B)$ is also greater than 0. However, from (**a**), $P(A$ and $B) = 0$. Thus, $P(A) \cdot P(B) \neq P(A$ and $B)$, and from the alternative method for determining independence, we conclude that events A and B are dependent.

NOW YOU CAN DO
Exercises 85–92.

We can extend the Multiplication Rule to cover n independent events.

> **Multiplication Rule for *n* Independent Events**
> If A, B, C, ... are independent events, then $P(A$ and B and C and ...$) = P(A) \cdot P(B) \cdot P(C)$ ...

EXAMPLE 28   **Multiplication rule for *n* independent events**

According to the National Health Interview Survey, 24% of Americans ages 18–44 smoke tobacco. In a random sample of $n = 3$ Americans ages 18–44, find the probability that all three smoke.

Solution

The U.S. Census Bureau estimates that over 100 million Americans are ages 18–44. We will shortly see that, because the sample of 3 is so small compared to the population of over 100 million, it is reasonable to assume that the successive draws are independent. Let S_i denote the event that the ith American in the 18–44 age group smokes.

$$P(S_1) = P(S_2) = P(S_3) = 0.24$$

Then, using the Multiplication Rule for *n* Independent Events,

NOW YOU CAN DO
Exercises 93–100.

$$P(S_1 \text{ and } S_2 \text{ and } S_3) = P(S_1) \cdot P(S_2) \cdot P(S_3) = (0.24)(0.24)(0.24) = (0.24)^3 = 0.013824$$

YOUR TURN
#18

Refer to Example 28. In a random sample of $n = 10$ Americans ages 18–44, find the probability that all 10 smoke.

(The solution is shown in Appendix A.)

Next, we turn to solving an "at least" problem. This involves a technique that uses complements in order to give us a nice shortcut for calculating a probability.

EXAMPLE 29   Solving an "at least" problem

Using information in Example 28, find the probability that, in a random sample of 5 Americans ages 18–44, at least 1 of them smokes.

Solution

The phrase "at least" means that one or more of the five Americans smoke, so the probability we are looking for is:

$$P(1 \text{ or } 2 \text{ or } 3 \text{ or } 4 \text{ or } 5 \text{ Americans smoke})$$

Now, each of these events is mutually exclusive, meaning that our sample can't yield exactly 1 American who smokes and exactly 2 Americans who smoke. Thus, by the Addition Rule for Mutually Exclusive Events, the above probability equals:

$$P(1 \text{ American smokes}) + P(2 \text{ Americans smoke}) + \cdots + P(5 \text{ Americans smoke})$$

Calculating all these probabilities would take a while. So, we can use the probability of the complement we learned earlier to get us a shortcut. Note that "at least 1 American smokes" is the *complement* of "no Americans smoke." Then, because the complement rule for probability is: $P(A^C) = 1 - P(A)$, we get:

$$P(\text{At least 1 of the 5 Americans smokes})$$

$$= 1 - P(\text{None of the 5 Americans smokes because the events are independent})$$

$$= 1 - P(\text{1st doesn't smoke and 2nd doesn't smoke and } \cdots \text{ and 5th doesn't smoke})$$

$$= 1 - P(\text{1st doesn't smoke}) \cdot P(\text{2nd doesn't smoke}) \cdot \cdots \cdot P(\text{5th doesn't smoke})$$

NOW YOU CAN DO
Exercises 101–104.

$$= 1 - (0.76)^5 \text{ because } P(\text{not smoking}) = 0.76$$

$$= 0.7464$$

YOUR TURN
#19

For the experiment in Example 29, find the probability that, in a random sample of four Americans ages 18–44, at least one of them smokes.

(The solution is shown in Appendix A.)

4 Approximating Probabilities for Dependent Events

When the sample size is small compared to the population size, we can estimate the probability of a dependent event as if it were independent. The question is: How small is a small sample? We will use the following **1% Guideline**.

> **The 1% Guideline**
>
> Suppose successive draws, such as those for a random sample, are being made from a population. If the sample size is no larger than 1% of the size of the population, then the probability of dependent successive draws from the population may be approximated using the assumption that the draws are independent.

EXAMPLE 30   Applying the 1% Guideline

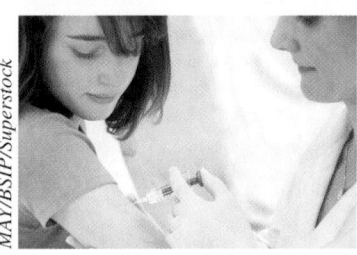

MAY/BSIP/Superstock

The Gardasil Vaccine

CASE STUDY

In the Section 5.2 exercises, we found the probability that a randomly chosen patient completed the treatment to be $(469/1413) \approx 0.3319$. Suppose we take a random sample of size $n = 2$ patients.

a. Calculate the probability that both patients complete the treatment, using sampling without replacement.

b. Confirm that the 1% Guideline applies.

c. Approximate the probability that both patients complete the treatment using the 1% Guideline.

d. Find the approximation error by comparing your answers from (**a**) and (**c**).

Solution

Define the following events:

    A: First patient completes the treatment

    B: Second patient completes the treatment

a. We have $P(A) = (469/1413)$. Then, *given* that the first patient completed the treatment, we have, by sampling without replacement, $P(B \mid A) = (468/1412)$. Then the Multiplication Rule gives us $P(A \text{ and } B) = P(A) \cdot P(B \text{ given } A) = (469/1413) \cdot (468/1412) \approx 0.110012$.

b. The sample of size 2 represents $\frac{2}{1413} = 0.0014 = 0.14\%$ of the population. Thus, the 1% Guideline applies, and we may treat the successive draws as independent.

c. Thus, we can use the Multiplication Rule for Independent Events to solve this problem.

$$P(A \text{ and } B) = P(A) \cdot P(B) = (469/1413)(469/1413) = 0.110169$$

d. The approximation error, the difference between (**a**) and (**c**), is $0.110169 - 0.110012 = 0.000157$. For most applications, this small approximation error is acceptable.

> The 1% Guideline is also helpful when we do not know the size of the population, but may presume that the population is very large compared to the sample size.

NOW YOU CAN DO
Exercises 75–78.

5 Bayes' Rule

When new information becomes available, our estimates of probability need to be revised to reflect the presence of the new information. An English Presbyterian Minister, Thomas Bayes (1702–1761) developed a rule, called **Bayes' Rule**, for updating previous probabilities using new information to arrive at revised probabilities.

Bayes' Rule

For any two events A and B,

$$P(A \text{ given } B) = \frac{P(A) \cdot P(B \text{ given } A)}{P(A) \cdot P(B \text{ given } A) + P(A^C) \cdot P(B \text{ given } A^C)}$$

We apply Bayes' Rule in the following example.

EXAMPLE 31   Bayes' Rule

The Gardasil Vaccine

Table 24 contains a contingency table (crosstabulation) of the variables *completed* and *insurance type* for the data set **Gardasil**.[12]

Table 24 Contingency table of the variables *completed* and *insurance type*

		Hospital-based	Medical assistance	Military	Private payer	Total
		Insurance type				
Completed	No	45	220	209	470	944
	Yes	39	55	122	253	469
	Total	84	275	331	723	1413

Define the following events:

  M: Insurance type = Military

  C: Completed the vaccination treatment

Use Bayes' Rule to find the probability that a randomly selected patient used military insurance, given that he or she completed the treatment.

Solution

We are interested in the probability $P(M \text{ given } C)$. Substituting M for A and C for B in the formula for Bayes' Rule, we obtain:

$$P(M \text{ given } C) = \frac{P(M) \cdot P(C \text{ given } M)}{P(M) \cdot P(C \text{ given } M) + P(M^C) \cdot P(C \text{ given } M^C)}$$

From Table 24, we have $P(M) = \frac{331}{1413}$ and $P(C \text{ given } M) = \frac{122}{331}$. The event M^C consists of all other insurance types besides military. So we have $P(M^C) = \frac{1082}{1413}$, and $P(C \text{ given } M^C) = \frac{347}{1082}$. Thus,

$$P(M \text{ given } C) = \frac{\dfrac{331}{1413} \cdot \dfrac{122}{331}}{\dfrac{331}{1413} \cdot \dfrac{122}{331} + \dfrac{1082}{1413} \cdot \dfrac{347}{1082}} = \frac{\dfrac{122}{1413}}{\dfrac{122}{1413} + \dfrac{347}{1413}} = \frac{122}{469} \approx 0.26$$

NOW YOU CAN DO
Exercises 105–108.

Note that this result is confirmed by the direct calculation of $P(M \mid C)$ using Table 24, which is the single cell of 122 military completers divided by the total number of completers.

YOUR TURN
#20

Using Table 24 and Bayes' Rule, find the probability that a randomly selected patient used hospital-based insurance, given that he or she completed the treatment.

(The solution is shown in Appendix A.)

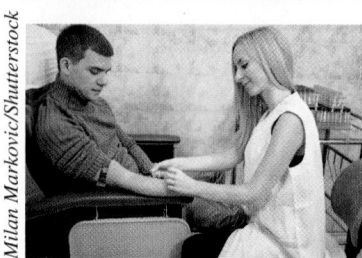

Milan Markovic/Shutterstock

EXAMPLE 32   The ELISA test for the presence of HIV

The ELISA test is used to screen blood for the presence of HIV. Like most diagnostic procedures, the test is not foolproof.

- When a blood sample contains HIV, the ELISA test will give a positive result 99.6% of the time. That is, the *false-negative rate,* the percentage of tests returning a negative result when the HIV virus is actually present, is $1 - 0.996 = 0.004$.

- When the blood does not contain HIV, the ELISA test will give a negative result 98% of the time. That is, the *false-positive rate,* the percentage of tests returning a positive result when the HIV virus is not actually present, is $1 - 0.98 = 0.02$.

A positive result means that the test says that the person has the HIV infection. A negative result means that the test says that the person does not have the virus. The *prevalence rate* for HIV in the general population is 0.5%. That is, 5 of 1000 persons in the general population have HIV.

Suppose we have samples of blood from 100,000 randomly chosen people.

Problem 1.   How many people in the sample of 100,000 have HIV? How many do not?

Solution

The prevalence rate of 0.5% means that 0.005 (100,000) = 500 people in the sample have HIV. The remainder—99,500—do not.

Problem 2.   A positive result is given 99.6% of the time for blood containing HIV. For the 500 people with HIV, how many positive results will the ELISA test return? How many of the 500 people with HIV will receive a negative result?

Solution

The ELISA test will return a positive result for 0.996 (500) = 498 of the 500 people. Thus, two people who actually have HIV will receive a test result indicating that they do not have the virus.

Problem 3.   A negative result is given 98% of the time for blood without HIV. For the 99,500 people without HIV, how many negative results will the ELISA test return? Positive results?

Solution

The ELISA test will return a negative result for 0.98 (99,500) = 97,510 of the 99,500 people without HIV. The remaining 2%, or 1990 people, will receive positive ELISA test results, even though they do not have the virus.

We can use the counts we found to fill in Table 25.

Table 25 ELISA test contingency table

ELISA test results	In reality		Total
	Person has HIV	**Person does not have HIV**	
Positive	498	1,990	2,488
Negative	2	97,510	97,512
Total	500	99,500	100,000

We will use the information in the ELISA test contingency table to solve Problems 4 and 5. If a person is chosen at random from the sample of 100,000, define the following events:

A: Person has HIV.

A^C: Person does not have HIV.

Pos: ELISA test returned positive results.

Neg: ELISA test returned negative results.

Problem 4.   What is the probability that a randomly chosen person actually does have HIV, given that the ELISA results are negative? In other words, find $P(A\mid\text{Neg})$.

Solution

$$P(A \text{ given Neg}) = \frac{N(A \text{ and Neg})}{N(\text{Neg})} = \frac{2}{97{,}512} \approx 0.0000205$$

Problem 5.   What is the probability that a randomly chosen person actually does not have HIV, given that the ELISA test results are positive? In other words, find $P(A^C \text{ given Pos})$.

Solution

$$P(A^C \text{ given Pos}) = \frac{N(A^C \text{ given Pos})}{N(\text{Pos})} = \frac{1990}{2488} \approx 0.7998 \approx 0.80$$

Developing Your Statistical Sense

Which Error Is More Dangerous?

In Problems 4 and 5, we examined the probabilities of the two ways that the ELISA test can be wrong. Which error do you think is more dangerous? $P(A \text{ given Neg})$ represents the probability that HIV is present, even though the ELISA test says otherwise. $P(A^C \text{ given Pos})$ represents the probability that HIV is not present, even though the ELISA test says it is present. The designers of the ELISA test worked hard to reduce the false-negative rate $P(A \text{ given Neg})$ to as low a level as possible. They rightly considered that it is the more dangerous type of error because of the epidemic nature of the illness. A person who receives a false-negative ELISA result could spread the infection further. Therefore, the designers tried to keep this probability as low as they could.

There is a price to be paid, however, which is the high false-positive rate, $P(A^C \text{ given Pos})$, a very high 80%. Thus, if a random person receives a positive ELISA test result, the probability that the person does *not* have HIV is 80%. When the ELISA test comes back positive, a second batch of tests that have a more reasonable false-positive rate is usually administered.

Section 5.3 Summary

1. Section 5.3 discusses conditional probability $P(B$ given $A)$, which is the probability of an event B given that an event A has occurred.

2. We can compare $P(B$ given $A)$ to $P(B)$ to determine whether the events A and B are independent. Events are independent if the occurrence of one event does not affect the probability that the other event will occur.

3. The Multiplication Rule for Independent Events is the product of the individual probabilities. Sampling with replacement is associated with independence, whereas sampling without replacement means that the events are not independent.

4. We can use the 1% Guideline for approximating probabilities of dependent events.

5. Bayes' Rule allows us to revise probabilities of events in light of new data.

EXAMPLE 33 Design your own T-shirt

Bruce Laurance/The Image Bank/Getty Images

A store at the local mall allows customers to design their own T-shirts. The store offers the following options to its customers:

- **Sleeve type:** Long-sleeve or short-sleeve
- **Color:** White, black, or red
- **Image:** Stock picture or uploaded photo

List the possible T-shirt options.

Solution

Figure 18 is a tree diagram that shows all the different T-shirts that can be designed. Two choices are available for the type of sleeve. For each sleeve type, there are three choices for color. For each color, there are two choices of image: stock picture or uploaded photo. Altogether, customers have a choice from among

$$2 \cdot 3 \cdot 2 = 12$$

different T-shirt options.

NOW YOU CAN DO
Exercises 7–10.

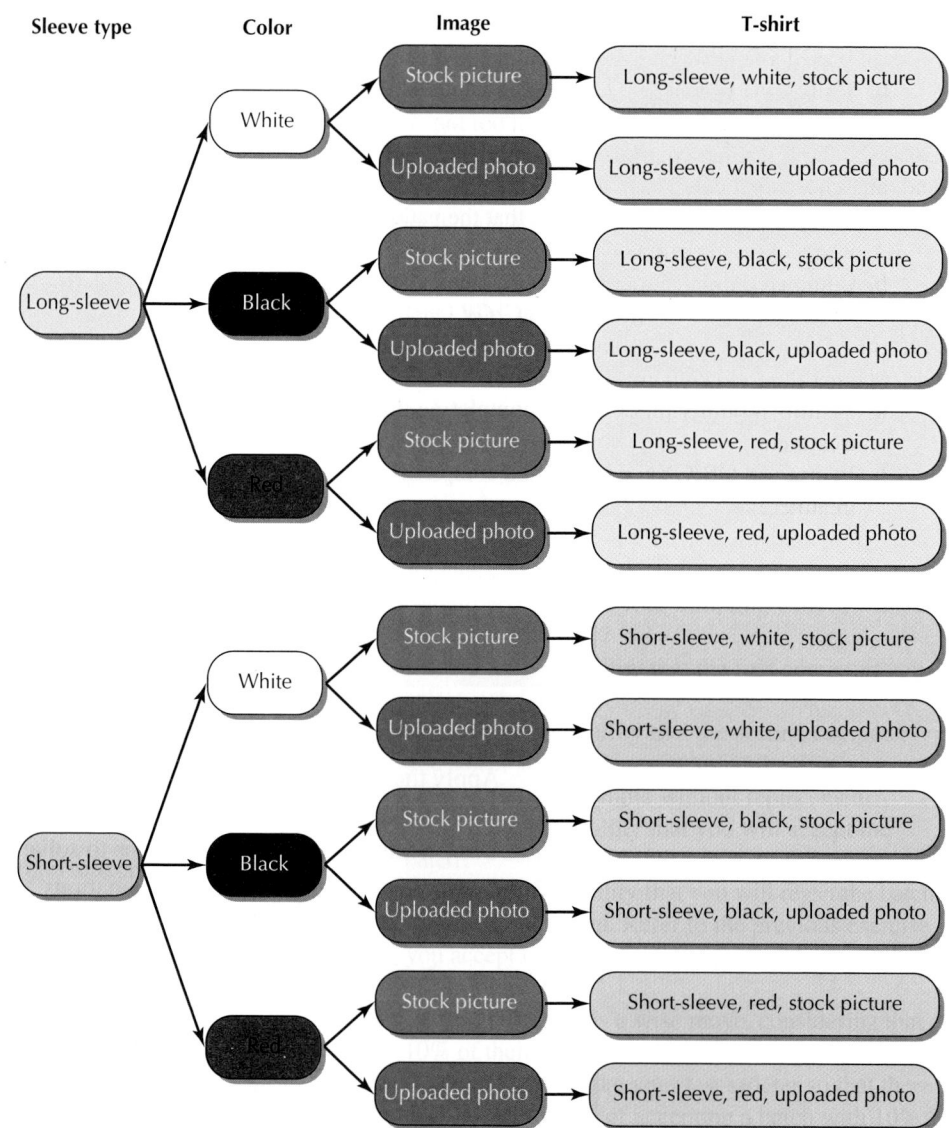

FIGURE 18
Tree diagram for the different T-shirt options.

We can generalize the result from Example 33 as the **Multiplication Rule for Counting**.

> **Multiplication Rule for Counting**
>
> Suppose an activity consists of a series of events in which there are a possible outcomes for the first event, b possible outcomes for the second event, c possible outcomes for the third event, and so on. Then the total number of different possible outcomes for the series of events is
>
> $$a \cdot b \cdot c \dots$$

EXAMPLE 34   Counting with repetition: Famous initials

Some Americans in history are uniquely identified by their initials. For example, "JFK" stands for John Fitzgerald Kennedy, and "FDR" stands for Franklin Delano Roosevelt. How many different possible sets of initials are there for people with a first, middle, and last name?

Solution

Let us consider the three initials as an activity consisting of three events. Note that a particular letter may be repeated, as in "AAM" for A. A. Milne, author of *Winnie the Pooh*. Then there are $a = 26$ ways to choose the first initial, $b = 26$ ways to choose the second initial, and $c = 26$ ways to choose the third initial. Thus, by the Multiplication Rule for Counting, the total number of different sets of initials is

$$26 \cdot 26 \cdot 26 = 17{,}576$$

NOW YOU CAN DO
Exercises 11 and 12.

EXAMPLE 35   Counting without repetition: Intramural singles tennis

Note: To summarize the key difference between Examples 34 and 35: If repetitions are allowed, then $a = b = c$. If repetitions are not allowed, then the numbers being multiplied decrease by one from left to right.

A local college has an intramural singles tennis league with five players: Ryan, Megan, Nicole, Justin, and Kyle. The college presents a trophy to the top three players in the league. How many different possible sets of three trophy winners are there?

Solution

The major difference between Example 34 and this example is that, in this example there can be no repetition. Ryan cannot finish in first place *and* second place. So we proceed as follows: Five possible players could finish in first place, so $a = 5$. Now there are only four players left, one of whom will finish in second place, so $b = 4$. That leaves only three players, one of whom will finish in third place, giving $c = 3$. Thus, by the Multiplication Rule for Counting, the number of different possible sets of trophy winners is

$$5 \cdot 4 \cdot 3 = 60$$

NOW YOU CAN DO
Exercises 13 and 14.

YOUR TURN
#21

For the situation in Example 35, suppose there are six players, and the college presents a trophy to the top four players in the league. How many different possible sets of four trophy winners are there?

(The solution is shown in Appendix A.)

EXAMPLE 36 **Traveling salesman problem**

A Southeast regional salesman has eight destinations that he must travel to this month: Atlanta, Raleigh, Charleston, Nashville, Jacksonville, Richmond, Mobile, and Jackson. How many different possible routes could he take?

Solution

The salesman has $a = 8$ different choices for where to go first. Once the first destination has been chosen, there are only $b = 7$ choices for where to go second. And once the first two destinations have been chosen, there are only $c = 6$ choices for where to go third, and so on. Thus, by the Multiplication Law for Counting, the number of different possible routes for the salesman is

NOW YOU CAN DO
Exercises 15 and 16.

$$a \cdot b \cdot c \cdot d \cdot e \cdot f \cdot g \cdot h = 8 \cdot 7 \cdot 6 \cdot 5 \cdot 4 \cdot 3 \cdot 2 \cdot 1 = 40,320$$

YOUR TURN
#22

For the traveling salesman in Example 36, suppose there are ten destinations he must travel to. How many different possible routes could he take?

(The solution is shown in Appendix A.)

The calculation in Example 36 leads us to introduce the **factorial symbol**, which is used for the counting rules we will learn in the remainder of this section.

> For any integer $n \geq 0$, the **factorial symbol $n!$** is defined as follows:
> - $0! = 1$
> - $1! = 1$
> - $n! = n(n-1)(n-2) \ldots 3 \cdot 2 \cdot 1$

EXAMPLE 37 **Factorials**

Calculate the following factorials:

a. 2! **b.** 3! **c.** 4! **d.** 5! **e.** 6! **f.** 7! **g.** 8!

Solution

a. $2! = 2 \cdot 1 = 2$

b. $3! = 3 \cdot 2 \cdot 1 = 6$

c. $4! = 4 \cdot 3 \cdot 2 \cdot 1 = 24$

d. $5! = 5 \cdot 4 \cdot 3 \cdot 2 \cdot 1 = 120$

e. $6! = 6 \cdot 5 \cdot 4 \cdot 3 \cdot 2 \cdot 1 = 720$

NOW YOU CAN DO
Exercises 17–22.

f. $7! = 7 \cdot 6 \cdot 5 \cdot 4 \cdot 3 \cdot 2 \cdot 1 = 5040$

g. $8! = 8 \cdot 7 \cdot 6 \cdot 5 \cdot 4 \cdot 3 \cdot 2 \cdot 1 = 40,320$, as in Example 36

YOUR TURN
#23

Find the following factorials:

a. 9!

b. 10!

(The solutions are shown in Appendix A.)

2 Permutations and Combinations

EXAMPLE 38   Traveling to some but not all of the cities

Example 37 calculated the number of possible routes for traveling to $n = 8$ cities. However, suppose we are interested in traveling to *some but not all* of the cities? For example, suppose that the salesman is traveling to three of the eight cities. Find the number of possible routes.

Solution

Eight choices are available for the first city, seven choices for the second city, and six choices for the third city. The salesman is traveling to three cities only, so the number of possible routes is thus

$$8 \cdot 7 \cdot 6 = 336$$

This result may be rewritten using factorial notation, as follows:

$$8 \cdot 7 \cdot 6 = \frac{8 \cdot 7 \cdot 6 \cdot (5 \cdot 4 \cdot 3 \cdot 2 \cdot 1)}{(5 \cdot 4 \cdot 3 \cdot 2 \cdot 1)} = \frac{8!}{5!} = \frac{8!}{(8-3)!}$$

NOW YOU CAN DO
Exercises 23 and 24.

YOUR TURN #24

For the traveling salesman in Example 38, suppose there are 10 destinations, but he is traveling only to 5 of the 10 destinations. Find the number of possible routes.

(The solution is shown in Appendix A.)

Example 38 leads us to the following definition.

Permutations

A **permutation** is an arrangement of items, such that

- r items are chosen at a time from n distinct items.
- repetition of items is not allowed.
- the order of the items is important.

The number of permutations of n items chosen r at a time is denoted as ${}_nP_r$ and given by the formula

$${}_nP_r = \frac{n!}{(n-r)!}$$

In Example 38, we are looking for the number of permutations of 8 cities taken 3 at a time. We have $n = 8$, $r = 3$:

$${}_nP_r = {}_8P_3 = \frac{n!}{(n-r)!} = \frac{8!}{(8-3)!} = \frac{8!}{5!} = 8 \cdot 7 \cdot 6 = 336$$

EXAMPLE 39   Calculating numbers of permutations

Find the following numbers of permutations:
a. ${}_5P_2$ **b.** ${}_6P_2$ **c.** ${}_6P_6$

Solution

a. ${}_5P_2 = \frac{5!}{(5-2)!} = \frac{5 \cdot 4 \cdot 3!}{3!} = 20$

NOW YOU CAN DO
Exercises 25–32.

b. $_6P_2 = \dfrac{6!}{(6-2)!} = \dfrac{6 \cdot 5 \cdot 4!}{4!} = 30$

c. $_6P_6 = \dfrac{6!}{(6-6)!} = \dfrac{6 \cdot 5 \cdot 4 \cdot 3 \cdot 2 \cdot 1}{0!} = 720$

YOUR TURN
#25

Find the following numbers of permutations:

a. $_5P_3$ **b.** $_6P_4$ **c.** $_2P_2$

(The solution is shown in Appendix A.)

EXAMPLE 40   Counting permutations: Secret Santas

"Secret Santa" refers to a method whereby each member of a group anonymously buys a holiday gift for another member of the group. Each person is secretly assigned to buy a gift for another randomly chosen person in the group. Suppose Jessica, Laverne, Samantha, and Luisa share a dorm suite and want to do Secret Santa this holiday season.

a. Verify that in this instance one woman purchasing a gift for another woman represents a permutation.

b. Calculate how many possible different permutations of gift buying there are for the four women.

Solution

a. • There are $n = 4$ women, and $r = 2$ people are associated with each gift (the giver and the receiver).

• Each person can buy only one gift, so repetition is not allowed.

• Finally, there is a difference between Jessica buying for Laverne and Laverne buying for Jessica. Thus, order is important, and thus, buying a gift represents a permutation.

b. The number of permutations is calculated as follows:

$$_nP_r = {}_4P_2 = \dfrac{4!}{(4-2)!} = \dfrac{4 \cdot 3 \cdot 2!}{2!} = 12$$

In a permutation, order is important. For example, in Example 40, there was a difference between Jessica buying a gift for Laverne and Laverne buying one for Jessica. However, what if we consider shaking hands instead? Then Jessica shaking hands with Laverne is considered the same as Laverne shaking hands with Jessica. Thus, sometimes order is not important. What is important here is the **combination** of Jessica and Laverne.

Combinations

A **combination** is an arrangement of items in which

• r items are chosen from n distinct items.

• repetition of items is not allowed.

• the order of the items is not important.

The number of combinations of r items chosen from n different items is denoted as

$$_nC_r$$

EXAMPLE 41   How many combinations in the intramural tennis league?

We return to the intramural singles tennis league at the local college. There are five players: Ryan, Megan, Nicole, Justin, and Kyle. Each player must play each other once.

a. Confirm that a match between two players represents a combination.

b. How many matches will be held?

Solution

a. Let {Ryan, Megan} denote a tennis match between Ryan and Megan. *Note:*

- There are $r = 2$ players chosen from $n = 5$ players.

- Each player plays each other player once, so repetition is not allowed.

- There is no difference between {Ryan, Megan} and {Megan, Ryan}, so order is not important.

Thus, a tennis match between two players represents a combination.

b. The list of all matches is as follows.

{Ryan, Megan}	{Megan, Nicole}	{Nicole, Justin}
{Ryan, Nicole}	{Megan, Justin}	{Nicole, Kyle}
{Ryan, Justin}	{Megan, Kyle}	{Justin, Kyle}
{Ryan, Kyle}		

Thus, there are $_5C_2 = 10$ possible matches of $r = 2$ players chosen from $n = 5$ players.

YOUR TURN #26

For the intramural league in Example 41, suppose there are 10 players. How many matches will be held?

(The solution is shown in Appendix A.)

We saw in Example 39 that $_5P_2 = 20$ and in Example 41 that $_5C_2 = 10$. Permutations and combinations differ only in that ordering is ignored for combinations. To calculate the number of combinations $_nC_r$, we have to take into consideration how many different rearrangements there are of the same items. For example, in Example 41, there are $r! = 2! = 2$ rearrangements of the same players, such as {Ryan, Megan} and {Megan, Ryan}. Thus,

$$_5C_2 = \frac{_5P_2}{2!} = \frac{20}{2} = 10$$

Note: Following are some special combinations you may find useful. For any integer n:

$$_nC_n = 1$$
$$_nC_0 = 1$$
$$_nC_1 = n$$
$$_nC_{n-1} = n$$

In general, the *number of combinations* can be computed as the number of permutations divided by the factorial of the number of items chosen.

Formula for the Number of Combinations

The number of combinations of r items chosen from n different items is given by

$$_nC_r = \frac{n!}{r!(n-r)!}$$

For instance, in Example 41, the formula for the number of combinations is

$$_5C_2 = \frac{5!}{2!(5-2)!} = \frac{5!}{2!3!} = \frac{5 \cdot 4 \cdot 3!}{2 \cdot 1 \cdot 3!} = \frac{20}{2} = 10$$

Thus, the relation: $_5C_2 = {_5P_2}/2!$ is verified.

EXAMPLE 42 Calculating numbers of combinations

Find the following numbers of combinations:

a. $_6C_2$ **b.** $_6C_3$ **c.** $_6C_4$

Solution

a. $_6C_2 = \dfrac{6!}{2!(6-2)!} = \dfrac{6 \cdot 5 \cdot 4!}{2 \cdot 1 \cdot 4!} = \dfrac{30}{2} = 15$

b. $_6C_3 = \dfrac{6!}{3!(6-3)!} = \dfrac{6 \cdot 5 \cdot 4 \cdot 3!}{3 \cdot 2 \cdot 1 \cdot 3!} = \dfrac{120}{6} = 20$

NOW YOU CAN DO
Exercises 33–40.

c. $_6C_4 = \dfrac{6!}{4!(6-4)!} = \dfrac{6!}{(6-4)!4!} = \dfrac{6 \cdot 5 \cdot 4!}{2 \cdot 1 \cdot 4!} = \dfrac{30}{2} = 15$

YOUR TURN
#27

Find the following numbers of combinations:

a. $_5C_4$ **b.** $_5C_3$ **c.** $_5C_2$

(The solutions are shown in Appendix A.)

Note that in (**c**), we used the commutative property of multiplication ($a \cdot b = b \cdot a$) and found that $_6C_4 = {}_6C_2 = 15$. In general, $_nC_r = {}_nC_{n-r}$ for this reason.

EXAMPLE 43 Calculating the number of permutations and combinations using technology

Use the TI-83/84 and Excel to calculate the following:

a. $_9P_6$ **b.** $_{10}C_7$

Solution

We use the instructions provided in the Step-by-Step Technology Guide at the end of this section (page 302).

a. From Figures 19a and 19b, we find that $_9P_6 = 60{,}480$.

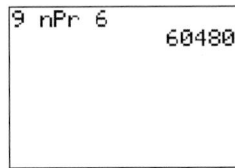

FIGURE 19a TI-83/84 permutation results.

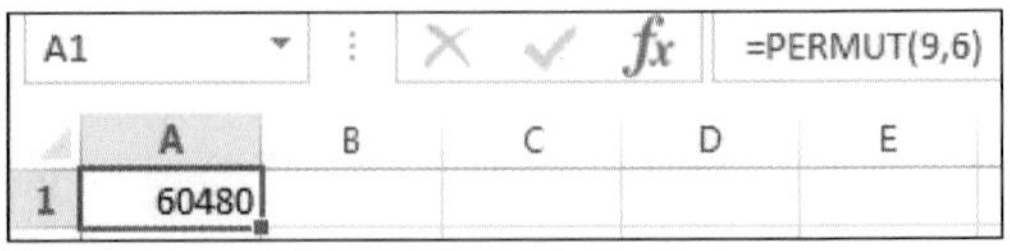

FIGURE 19b Excel permutation results.

b. From Figures 19c and 19d, we find that $_{10}C_7 = 120$.

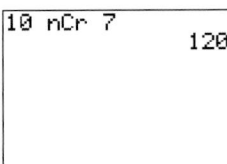

FIGURE 19c TI-83/84 combination results.

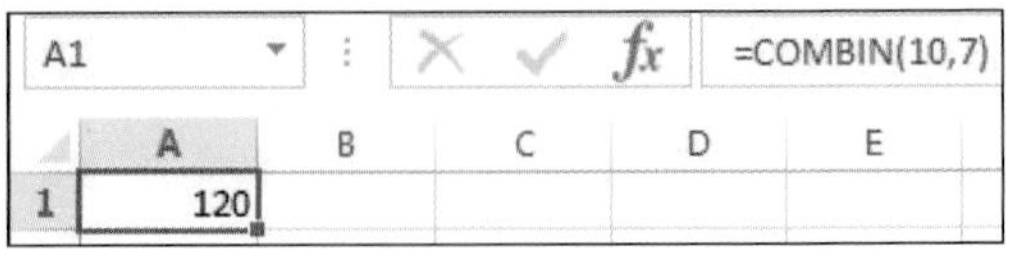

FIGURE 19d Excel combination results.

Sometimes we wish to find the number of permutations of items where some of the items are not distinct.

EXAMPLE 44   **Permutations with nondistinct items**

How many distinct strings of letters can we make by using all the letters in the word STATISTICS?

Solution

Each string will be 10 letters long and include 3 S's, 3 T's, 2 I's, 1 A, and 1 C. The 10 positions shown here need to be filled.

$$\underline{}\ \underline{}\ \underline{}\ \underline{}\ \underline{}\ \underline{}\ \underline{}\ \underline{}\ \underline{}\ \underline{}$$
$$1\quad 2\quad 3\quad 4\quad 5\quad 6\quad 7\quad 8\quad 9\quad 10$$

The string-forming process is as follows:

Step 1   Choose the positions for the three S's.

Step 2   Choose the positions for the three T's.

Step 3   Choose the positions for the two I's.

Step 4   Choose the position for the one A.

Step 5   Choose the position for the one C.

There are $_{10}C_3$ ways to place the three S's in Step 1. Once Step 1 is done, seven slots are left, leaving $_7C_3$ positions for the three T's. Once Step 2 is done, four slots are left, so there are $_4C_2$ ways to place the two I's. Once Step 3 is done, only two slots are left, so there are only $_2C_1$ ways to position the A. Finally, there is only $_1C_1$ way to place the C.

Putting Steps 1–5 together, we calculate the number of distinct letter strings as

$$_{10}C_3 \cdot {}_7C_3 \cdot {}_4C_2 \cdot {}_2C_1 \cdot {}_1C_1 = \frac{10!}{3!\,7!} \cdot \frac{7!}{3!\,4!} \cdot \frac{4!}{2!\,2!} \cdot \frac{2!}{1!\,1!} \cdot \frac{1!}{1!\,0!}$$

$$= \frac{10!}{3!\,3!\,2!\,1!\,1!} = \frac{3{,}628{,}800}{72}$$

$$= 50{,}400$$

There are 50,400 distinct strings of letters that can be made using the letters in the word STATISTICS.

This example can be generalized in the following result.

Permutations of Nondistinct Items

The number of permutations of n items of which n_1 are of the first kind, n_2 are of the second kind, ..., and n_k are of the kth kind is calculated as

$$\frac{n!}{n_1! \cdot n_2! \cdot\ \cdots\ \cdot n_k!}$$

where $n = n_1 + n_2 + \cdots + n_k$.

EXAMPLE 45 Number of permutations of nondistinct items

Brandon brings a healthy snack to school each day, consisting of 5 carrot sticks, 4 celery sticks, and 2 cherry tomatoes. If Brandon eats one item at a time, in how many different ways can he eat his snack?

Solution

We are seeking the number of permutations of $n = 11$ items, of which $n_1 = 5$ are carrot sticks, $n_2 = 4$ are celery sticks, and $n_3 = 2$ are cherry tomatoes. Using the formula for the number of permutations of nondistinct items,

$$\frac{n!}{n_1! \cdot n_2! \cdot n_3!} = \frac{11!}{5! \cdot 4! \cdot 2!} = \frac{39,916,800}{120 \cdot 24 \cdot 2} = 6930$$

There are 6930 distinct ways in which Brandon can eat his snack.

NOW YOU CAN DO
Exercises 41 and 42.

Acceptance sampling refers to the process of (1) selecting a random sample from a batch of items, (2) evaluating the sample for defectives, and (3) either accepting or rejecting the entire batch based on the evaluation of the sample.

EXAMPLE 46 Acceptance sampling uses combinations

Suppose we have a batch of 20 cell phones, of which, unknown to us, 3 are defective and 17 are nondefective; we will take a random sample of size 2 and evaluate both items once.

a. Are the arrangements in acceptance sampling permutations or combinations?

b. Find the number of ways that both sampled cell phones are defective.

Solution

a. Both permutations and combinations require the following:

- r items are chosen from n distinct items. Here, we are selecting $r = 2$ phones from a batch of $n = 20$.

- Repetition of the items is not allowed. Each item is evaluated only once. The difference between permutations and combinations is that, for permutations, order is important, whereas for combinations, order is not important. In acceptance sampling, the order of the items is not important. Thus, acceptance sampling uses combinations.

b. The number of ways of choosing two of the three defectives is

$$_3C_2 = \frac{3!}{2!(3-2)!} = \frac{3 \cdot 2!}{2! \cdot 1!} = 3$$

Selecting 2 defectives means that we are choosing 0 of the 17 nondefectives. The number of ways this can happen is

$$_{17}C_0 = \frac{17!}{0!(17-0)!} = \frac{17!}{1 \cdot 17!} = 1$$

By the Multiplication Rule for Counting, the number of ways that both sampled cell phones are defective is

$$_3C_2 \cdot {}_{17}C_0 = 3 \cdot 1 = 3$$

3 Computing Probabilities Using Combinations

The counting methods we have learned in this section may be used to compute probabilities. We assume that each possible outcome in a random sample is *equally likely,* and thus we use the classical method for assigning the probability of an event *E:*

$$P(E) = \frac{\text{number of outcomes in } E}{\text{number of outcomes in sample space}} = \frac{N(E)}{N(S)}$$

EXAMPLE 47 Probability using combinations: Acceptance sampling

Continuing with Example 46, if both cell phones in the sample of size 2 are defective, we will reject the batch and cancel our contract with the supplier.

a. What is the number of ways that both cell phones will be defective?

b. What is the number of outcomes in this sample space?

c. What is the probability that both cell phones will be defective?

Solution

a. From Example 46, the number of ways that both cell phones will be defective is

$$_3C_2 \cdot {}_{17}C_0 = 3 \cdot 1 = 3$$

b. The number of outcomes in the sample space is given by the number of ways of selecting 2 cell phones out of a batch of 20; that is,

$$N(S) = {}_{20}C_2 = \frac{20!}{2!(20 - 2)!} = \frac{20 \cdot 19 \cdot 18!}{2! \cdot 18!} = \frac{380}{2} = 190$$

c. Therefore, the probability that both cell phones will be defective is given by

$$P(\text{Both defective}) = \frac{\text{number of ways both defective}}{\text{number of outcomes in sample space}} = \frac{3}{190} \approx 0.01579$$

EXAMPLE 48 Florida Lotto

You can win the jackpot in the Florida Lotto by correctly choosing all 6 winning numbers out of the numbers 1–53.

a. What is the number of ways of winning the jackpot by choosing all 6 winning numbers?

b. What is the number of outcomes in this sample space?

c. If you buy a single ticket for $1, what is your probability of winning the jackpot?

d. If you mortgage your house and buy 500,000 tickets, what is your probability of winning the jackpot (assuming that all the tickets are different)?

Solution

a. The number of ways of winning the jackpot by correctly choosing all 6 of the winning numbers and none of the losing numbers is

$$N(\text{Jackpot}) = {}_6C_6 \cdot {}_{47}C_0 = 1 \cdot 1 = 1$$

b. The size of the sample space is

$$N(S) = {}_{53}C_6 = \frac{53!}{6!(53-6)!} = \frac{53 \cdot 52 \cdot 51 \cdot 50 \cdot 49 \cdot 48 \cdot 47!}{6! \cdot 47!}$$

$$= \frac{16{,}529{,}385{,}600}{720} = 22{,}957{,}480$$

c. Therefore, if you buy a single ticket for \$1, your probability of winning the jackpot is given by

$$P(\text{Jackpot}) = \frac{1}{22{,}957{,}480} \approx 0.00000004356$$

d. If you buy 500,000 tickets and they are all unique, then your probability of winning becomes

$$P(\text{Jackpot}) = \frac{500{,}000}{22{,}957{,}480} \approx 0.02178$$

This is because the unique tickets are mutually exclusive, and the Addition Rule for Mutually Exclusive Events allows us to add the probabilities of the 500,000 tickets. After mortgaging your \$500,000 house and buying lottery tickets with the proceeds, there is a better than 97% probability that you will *not* win the lottery.

STEP-BY-STEP TECHNOLOGY GUIDE: Factorials, Permutations, and Combinations

TI-83/84

Factorials $n!$
Step 1 On the home screen, enter the value of n.
Step 2 Press **MATH**, highlight **PRB**, and select **4: !** (Figure 20).
Step 3 Press **ENTER** two times.
Permutations ${}_nP_r$ **and Combinations** ${}_nC_r$
Step 1 On the home screen, enter the value of n.
Step 2 **a.** For permutations, press **MATH**, highlight **PRB**, and select **2:** ${}_nP_r$.
b. For combinations, press **MATH**, highlight **PRB**, and select **3:** ${}_nC_r$.

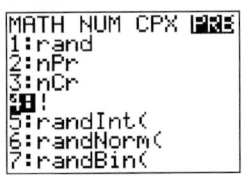

FIGURE 20

Step 3 On the home screen, enter the value of r.
Step 4 Press **ENTER** (see Figure 19a and Figure 19c in Example 43).

EXCEL

Factorials $n!$
Calculate 9!
Step 1 Select an empty cell, and type = FACT(9).
Step 2 Press **ENTER**.

Permutations ${}_nP_r$
We illustrate Example 43a (page 298): ${}_9P_6$.
Step 1 Select an empty cell and type = PERMUT(9,6).

Step 2 Press **ENTER**. See Figure 19b in Example 43 for the result.

Combinations ${}_nC_r$
We illustrate Example 43b (page 298): ${}_{10}C_7$.
Step 1 Select an empty cell and type = COMBIN(10,7).
Step 2 Press **ENTER**. See Figure 19d in Example 43 for the result.

MINITAB

Factorials $n!$
Step 1 Click **Calc > Calculator...**
Step 2 In **Store result in variable**, type **C1**. From **Functions**, choose **Arithmetic**, then double-click **Factorial**. Under **Expression**, replace **number of items** with n. Click **OK**.

Permutations ${}_nP_r$
We illustrate Example 43a (page 298): ${}_9P_6$.
Step 1 Click **Calc > Calculator...**
Step 2 In **Store result in variable**, type **C1**. From **Functions**, choose **Arithmetic**, then double-click **Permutations**. Under

Expression, replace **number of items** with 9 and **number to choose** with 6. Click **OK**.

Combinations ${}_nC_r$
We illustrate Example 43b (page 298): ${}_{10}C_7$.
Step 1 Click **Calc > Calculator...**
Step 2 In **Store result in variable**, type **C1**. From **Functions**, choose **Arithmetic**, then double-click **Combinations**. Under **Expression**, replace **number of items** with 10 and **number to choose** with 7. Click **OK**.

Section 5.4 Summary

1. The Multiplication Rule for Counting provides the total number of different possible outcomes for a series of events.
2. A permutation $_nP_r$ is an arrangement in which
- r items are chosen from n distinct items.
- repetition of items is not allowed.
- the order of the items is important.

In a permutation, order is important. In a combination, order does not matter. A combination $_nC_r$ is an arrangement in which
- r items are chosen from n distinct items.
- repetition of items is not allowed.
- the order of the items is not important.

3. Combinations may be used to calculate certain probabilities. For such problems, use the following steps:
Step 1   Confirm that the desired probability involves a combination.
Step 2   Find $N(E)$, the number of outcomes in event E.
Step 3   Find $N(S)$, the number of outcomes in the sample space.
Step 4   Assuming that each possible combination is equally likely, find the probability of event E as follows:

$$P(E) = \frac{N(E)}{N(S)}$$

Section 5.4 Exercises

CLARIFYING THE CONCEPTS

1. What type of diagram is helpful in itemizing the possible outcomes of a series of events? (p. 292)
2. Explain in words how 5! is calculated. (p. 294)
3. What is the difference between a permutation and a combination? (pp. 295–296)
4. Does $_8P_9$ make sense? Explain why or why not. (p. 295)
5. Describe in your own words what is meant by acceptance sampling. (p. 300)
6. The counting methods that we have learned in this section may be used to compute probabilities. (p. 301)
   a. For assigning probability, which method is used: classical, relative frequency, or subjective?
   b. Referring to part (**a**), what assumption must be made to apply the method?

PRACTICING THE TECHNIQUES

 CHECK IT OUT!

To do	Check out	Topic
Exercises 7–10	Example 33	Tree diagrams
Exercises 11–12	Example 34	Counting with repetition
Exercises 13–14	Example 35	Counting without repetition
Exercises 15–16	Example 36	Traveling salesman problem, Part 1
Exercises 17–22	Example 37	Factorials
Exercises 23–24	Example 38	Traveling salesman problem, Part 2
Exercises 25–32	Example 39	Calculating numbers of permutations
Exercises 33–40	Example 42	Calculating numbers of combinations
Exercises 41–42	Example 45	Number of permutations of nondistinct items

7. A pizza store offers the following options to its customers. Use a tree diagram to list all the possible options from which a customer may choose.
- Cheese: no cheese, regular cheese, double cheese
- Pepperoni: no pepperoni, regular pepperoni, double pepperoni

8. An ice cream shop offers the following options to its customers. Use a tree diagram to list all the possible options from which a customer may choose.
- Ice cream: vanilla, chocolate, mint chocolate chip
- Toppings: hot fudge, butterscotch, sprinkles

9. A particular baseball pitcher has to choose from the following options on each pitch. Use a tree diagram to list all the possible options.
- Type of pitch: fastball, curve, slider
- Horizontal position: inside corner, over the plate, outside corner
- Vertical position: high, low

10. A women's clothing store tracks its sales transactions according to the following options. Use a tree diagram to list all the possible options.
- Payment method: credit card, debit card, check, cash
- Size category: Juniors, Misses, Women's
- Type of clothing: tops, pants

11. Our 41st president, George Herbert Walker Bush, had four names, with initials GHWB. How many different possible sets of initials are there for people with four names?

12. NCAA ice hockey games can have the following outcomes: win (W), lose (L), or tie (T). In a tournament of five games, how many different possible sets of outcomes are there for a particular team? (*Hint:* LLTWW is one possible set.)

13. A college dining service conducted a survey in which it asked students to select their first and second favorite flavors of ice cream from a list of five flavors: vanilla, chocolate, mint chocolate chip, strawberry, and maple walnut. How many different possible sets of two favorites are there?

14. A town library is considering loaning video games, and surveyed its membership to ask their four favorite PlayStation 3 games from among the following six games: Gran Turismo, Call of Duty 4, Metal Gear Solid 4, Little Big Planet, Grand Theft Auto IV, and Final Fantasy XIII. How many different possible sets of four favorites are there?

15. A woman is considering four sororities to rush this year. How many possible orderings are there?

16. Students working for the college newspaper have six drop locations around campus at which they must drop off newspapers. How many different possible routes are there for the students to do so?

For Exercises 17–22, find the value of each factorial.

17. 6!
18. 9!
19. 0!
20. 11!
21. 1!
22. 15!

23. A woman is considering four sororities to rush this year, but only has time to rush two. How many possible orderings are there?

24. Students working for the college newspaper have six drop locations around campus at which they must drop off newspapers, but they only have enough time to get to four locations. How many different possible routes are there for the students to do so?

For Exercises 25–32, find the value of each permutation $_nP_r$.

25. $_7P_3$
26. $_7P_4$
27. $_8P_5$
28. $_8P_3$
29. $_{100}P_1$
30. $_{100}P_0$
31. $_{100}P_{100}$
32. $_{100}P_{99}$

For Exercises 33–40, find the value of each combination $_nC_r$. Then answer Exercises 43 and 44.

33. $_7C_3$
34. $_7C_4$
35. $_{11}C_8$
36. $_{11}C_9$
37. $_{11}C_{10}$
38. $_{11}C_{11}$
39. $_{100}C_0$
40. $_{100}C_1$

41. How many distinct strings of letters can we make by using all the letters in the word PIZZA?

42. How many distinct strings of letters can we make by using all the letters in the word PEPPERONI?

43. Explain why the answers to Exercises 33 and 34 are equal. Use the commutative property of multiplication (for example, $2 \cdot 7 = 7 \cdot 2$) in your answer.

44. Use the idea behind your answer to Exercise 43 to find a combination that is equal to $_{11}C_8$. Verify your answer.

45. List all the permutations of the following people taken three at a time: Amy, Bob, Chris, Danielle. What is $_4P_3$?

46. List all the combinations of the following people taken three at a time: Amy, Bob, Chris, Danielle. What is $_4C_3$?

47. Explain in your own words why $_4P_3$ is larger than $_4C_3$.

48. What quantity do we divide $_4P_3$ by to get $_4C_3$? Express this quantity as a factorial. (*Hint:* For example, if the quantity were 120, we would express it as 5!)

49. In general, what do we divide $_nP_r$ by to get $_nC_r$?

APPLYING THE CONCEPTS

50. Fast Food. A fast-food restaurant has three types of sandwiches: chicken sandwich, fish sandwich, and beef burger. The restaurant has two types of side dishes: French fries and salad.

    a. Draw a tree diagram to find all the different meals a customer can order at this restaurant.

    b. How many different meals can a customer order at this restaurant?

51. What to Eat? A sit-down restaurant has two types of appetizers: garden salad and Buffalo wings. It has three entrees: spaghetti, steak, and chicken. And it offers three kinds of desserts: ice cream, cake, and pie.

    a. Draw a tree diagram to find all the different meals a customer can order at this restaurant.

    b. How many different meals can a customer order at this restaurant?

52. Greek Alphabet. The ancient Greek alphabet had 24 letters. How many different possible initials are there for people with a first and last name?

53. Facebook Friends. A student has 10 friends on her Facebook page. How many ways can she arrange her 10 friends top to bottom?

54. Document Delivery. A document delivery person must deliver documents to five different destinations within a particular city. How many different routes are possible?

55. Traveler Fellow. A corporate sales executive must travel to the following countries this quarter: China, Russia, Germany, Brazil, India, and Nigeria. How many different routes are possible?

56. Sales Traveler. A corporate sales executive has the choice of traveling to four of the following six countries this quarter: China, Russia, Germany, Brazil, India, and Nigeria. How many different routes are possible?

57. Playing Catch. Five children are playing catch with a ball. How many different ways can one child throw a ball to another child once?

58. Chimp Grooming. Six chimpanzees are grooming each other at the city zoo. In how many different ways can one chimp groom another?

59. Shake Hands. In an ice-breaker exercise, each of 25 students is asked to shake hands with each of the other students. How many handshakes will there be in all?

60. Statistics Competition. Three students from the Honors Statistics class of 15 students will be chosen to represent the school at the state statistics competition. How many different possible groupings of 3 students are there?

61. How many random samples of size 1 can be chosen from a population of size 20?

62. How many random samples of size 20 can be chosen from a population of size 20?

63. How many random samples of size 10 can be chosen from a population of size 20?

64. How many distinct strings of letters can be made using all the letters in the word MATHEMATICS?

65. How many distinct strings of letters can be made using all the letters in the word BUSINESS?

66. Acceptance Sampling. A shipment of 25 personal digital assistants (PDAs) contains 3 that are defective.

A quality control specialist inspects 2 of the 25 PDAs. If both are defective, then the shipment is rejected.

    a. Explain whether a permutation or a combination is being used.

    b. Find the number of ways that both PDAs will be defective.

    c. Find the probability of rejecting the shipment.

Chapter 5 Formulas and Vocabulary

SECTION 5.1

- **Classical method of assigning probabilities** (p. 244).

$$P(E) = \frac{\text{number of outcomes in } E}{\text{number of outcomes in sample space}} = \frac{N(E)}{N(S)}$$

- **Event** (p. 241)
- **Experiment** (p. 241)
- **Law of Large Numbers** (p. 249)
- **Law of Total Probability** (p. 241)
- **Outcome** (p. 241)
- **Probability** (p. 240)
- **Probability model** (p. 241)
- **Relative frequency method of assigning probabilities** (also known as the empirical method) (p. 249).

$$P(E) \approx \frac{\text{frequency of } E}{\text{number of trials of experiment}}$$

- **Sample space** (p. 241)
- **Simulation** (p. 248)
- **Subjective probability** (p. 252)
- **Tree diagram** (p. 245)

SECTION 5.2

- **Addition Rule** (p. 262).
  $P(A \text{ or } B) = P(A \cup B) = P(A) + P(B) - P(A \cap B)$
- **Addition Rule for Mutually Exclusive Events** (p. 264). If A and B are mutually exclusive, then $P(A \cup B) = P(A) + P(B)$.
- **Complement of an event A** (p. 259). Denoted as A^C.
- **Intersection of two events A and B** (p. 261). Denoted as $A \cap B$ or as "A and B."
- **Mutually exclusive (disjoint) events** (p. 263)
- **Probabilities for complements** (p. 260).
  $P(A) + P(A^C) = 1$, $P(A) = 1 - P(A^C)$, and $P(A^C) = 1 - P(A)$
- **Union of two events A and B** (p. 261). Denoted as $A \cup B$ or as "A or B."

SECTION 5.3

- **Conditional probability** (p. 271).

$$P(B|A) = \frac{P(A \cap B)}{P(A)} = \frac{N(A \cap B)}{N(A)}$$

- **Independent events** (p. 274). Events A and B are *independent* if $P(A|B) = P(A)$ or if $P(B|A) = P(B)$.
- **Multiplication Rule** (p. 276). $P(A \cap B) = P(B) P(A|B)$ or, equivalently, $P(A \cap B) = P(A) P(B|A)$.
- **Multiplication Rule for Independent Events** (p. 277). If events A and B are independent, then $P(A \cap B) = P(A) P(B)$.
- **Multiplication Rule for *n* Independent Events** (p. 280). If $A, B, C, \ldots$ are independent events, then $P(A \cap B \cap C \cap \ldots) = P(A) P(B) P(C) \ldots$.
- **Sampling with replacement** (p. 277)
- **Sampling without replacement** (p. 277)

SECTION 5.4

- **Acceptance sampling** (p. 300)
- **Combination** (p. 296).

$$_nC_r = \frac{n!}{r!(n-r)!}$$

- **Factorial symbol *n*!** (p. 294). $0! = 1$; $1! = 1$; $n! = n(n-1)(n-2) \ldots 3 \cdot 2 \cdot 1$
- **Multiplication Rule for Counting** (p. 293)
- **Permutation** (p. 295).

$$_nP_r = \frac{n!}{(n-r)!}$$

- **Permutations of nondistinct items** (p. 299).

$$_nP_r = \frac{n!}{n_1! \cdot n_2! \cdot \cdots \cdot n_k!}$$

Chapter 5 Review Exercises

SECTION 5.1

For Exercises 1–5, consider the experiment of tossing a fair coin three times and find the probabilities of the following events.

  1. 2 heads

  2. At least 2 heads

  3. 4 heads

  4. 2 tails

  5. At most 1 tail

  6. A New Sonnet. Literature researchers have unearthed a sonnet that they know to be by either William Shakespeare or Christopher Marlowe. The probability that the sonnet is by Marlowe is 25%.

a. What is the probability that the sonnet is by Shakespeare?

b. What method of assigning probability do you think was used here? Why was this method used, and not the others?

SECTION 5.2

7. Farmworkers' Educational Level. The U.S. Department of Agriculture reports on the demographics of hired farmworkers.[15] An excerpt of the results is provided in the table, showing the percentage of noncitizen and citizen farmworkers who attained various educational levels. The educational levels are mutually exclusive. Find the following probabilities:

a. The probability that a noncitizen farmworker is a high school graduate or has some college

b. The probability that a citizen farmworker is a high school graduate or has some college

c. The probability that a noncitizen farmworker has less than a ninth-grade education and has some college

d. The probability that a farmworker is not a citizen

	Noncitizens	Citizens
Less than 9th grade	238,008	61,776
9th–12th grade (no diploma)	57,904	152,880
High school graduate	59,784	222,144
Some college	20,304	187,200

SECTION 5.3

8. Drug Research Studies. The *Annals of Internal Medicine* reported that 39 of the 40 research studies sponsored by a drug company had outcomes favoring a certain drug. Find the following probabilities, assuming independence:

a. Three randomly selected research studies all favor this drug.

b. None of the three randomly selected research studies favors this drug.

c. At least one of three randomly selected research studies favors this drug.

9. Drug Research Studies. Use the information in Exercise 8. Suppose we sample two research studies without replacement. Find the probability that the second study does not favor this drug, given that the first study does not favor this drug.

Gender and Pet Preference. Do you think your gender affects what type of pet you own? For Exercises 10–13, use the following table, showing preferences for various pets by owner gender.

Gender of owner	Cats	Dogs	Other pets	Total
Female	100	50	30	180
Male	50	50	20	120
Total	150	100	50	300

10. Find the probability that a randomly chosen person has the following characteristics:

a. Owns a cat, $P(C)$ **b.** Owns a dog, $P(D)$

11. Find the probability that a randomly chosen person has the following characteristics:

a. Is female and owns a dog, $P(F \cap D)$

b. Is male and owns a dog, $P(M \cap D)$

12. Find the following conditional probabilities for a randomly chosen person:

a. Owns a dog, given that the person is female, $P(D|F)$

b. Owns a dog, given that the person is male, $P(D|M)$

13. If you were a dog-food manufacturer, would you advertise more on a men's TV channel or a women's TV channel? Why? Cite your evidence.

SECTION 5.4

14. How many distinguishable strings of letters can be made using all the letters in the word MISSISSIPPI?

15. Statistics Quiz. On a statistics quiz, there are five true/false questions, four fill-in-the-blank questions, and three short-answer questions. How many different ways are there of taking this quiz?

16. Inspection Time. A U.S. Army drill instructor will perform inspection on 2 soldiers in a squad of 18 soldiers. If both soldiers fail the inspection because their rifles are not clean, the entire squad will have to run a five-mile course in full gear. Three of the 18 soldiers have rifles that are not clean.

a. Explain whether the drill instructor is using a permutation or a combination.

b. Find the number of ways that both soldiers will fail the inspection.

c. Find the probability that the entire squad will have to run a five-mile course in full gear.

Chapter 5 **QUIZ**

TRUE OR FALSE

1. True or false: An outcome is a collection of a series of events from the sample space of an experiment.

2. True or false: For any event A (even events like A: the moon is made of green cheese), the probability of A plus the probability of A^C always add up to 1.

FILL IN THE BLANK

3. The minimum value that a probability can take is _____ and the maximum value is _____.

4. The union of two events is associated with the English word _____, and the intersection of two events is associated with the English word _____.

5. Someone has told you that there is a 50-50 chance of rain tomorrow. This means that the probability of rain tomorrow equals _____.

SHORT ANSWER

6. For any experiment, what is the sum of all the outcome probabilities in the sample space?

7. For which type of sampling are consecutive draws independent?

8. For two events A and B, what do we call the event containing only those outcomes that belong to both A and B?

CALCULATIONS AND INTERPRETATIONS

9. Consider the experiment of rolling a fair die twice. Find the following probabilities:
   a. Sum of the two dice equals 5.
   b. Sum of the two dice does not equal 5.
   c. One of the dice shows 2.
   d. Sum of the two dice equals 5 and one of the dice shows 2.
   e. Sum of the two dice equals 5 or one of the dice shows 2.

10. Suppose that A and B are any two events, with $P(B) = 0.75$ and $P(A \cap B) = 0.15$. Find $P(A|B)$.

11. Suppose that A and B are any two events, with $P(B) = 0.85$ and $P(A|B) = 0.25$. Find $P(A \cap B)$.

12. Pick a Card. Consider the experiment of drawing a single card from a deck of 52 cards. Find the probability of observing the following events:
   a. Heart
   b. Face card (king, queen, or jack)
   c. Seven
   d. Red card
   e. Seven of hearts
   f. Red queen

For Exercises 13–18, let the experiment be to toss two fair dice. Use the sample space in Figure 3 on page 247. Define the following events:
   X: Roll a sum equal to 9.
   Y: Roll a sum equal to 10.
   Z: Roll doubles, where the dark green die equals the light green die.
   W: Light green die equals 6.

Use the strategy for determining whether two events are independent (page 274) to determine whether the following pairs of events are independent:

13. X and Z

14. Y and Z

15. X and W

16. Y and W

17. X and Y

18. Z and W

19. Football Teams. The four teams in the AFC South division of the National Football League are Indianapolis Colts, Jacksonville Jaguars, Tennessee Titans, and Houston Texans. Suppose the top three teams in the division this year will make the playoffs. How many different sets of teams are making the playoffs?

20. State Lottery. In a state lottery, balls numbered 1 to 20 are placed in an urn. To win, you must choose numbers that match the three balls chosen in the order that they're chosen.
   a. Explain whether a permutation or a combination is being used.
   b. How many possible outcomes are there?
   c. Find the probability of winning this lottery if your ticket contains a single ordering of three numbers.

6 Probability Distributions

Introduction

In Chapter 5, students were introduced to probability. In Chapter 6, students encounter *random variables* and *probability distributions*. With these new tools, they can increase the efficiency of their decision making. Sections 6.1–6.3 introduce discrete random variables, including the binomial random variable, and the Poisson probability distribution. In Section 6.4, students learn about continuous random variables, including the normal random variable. Section 6.4 also covers the special case of the standard normal distribution. Section 6.5 provides some applications of the normal distribution to everyday problems. Finally, Section 6.6 covers the normal approximation to the binomial distribution.

From the Author

Section 6.1 Discrete Random Variables

- Sometimes students have a hard time moving from Chapter 5 to Chapter 6. Example 1 will help them express the concepts they learned in Chapter 5 using the language of Chapter 6.

- Your more eager students may enjoy learning that the formula for the mean of a discrete random variable is a special case of a weighted mean (page 148 in Chapter 3).

- Note that the mean as the balance point applies to probability distributions just as it did for the graphs in Chapter 3.

- The examples for calculating the variance and standard deviation of X provide a whole-number answer for both. This topic is a computational challenge to many students, and they don't need lots of decimals getting in the way of the concepts.

Section 6.2 Binomial Probability Distribution

- Instructors may wish to stress that the key to a binomial distribution is knowing the sample size n and the probability of success p. When faced with a word problem, students should find the values of these quantities, and then find the values of X about which the problem is asking.

- Example 20 shows how to find the *mode* of a binomial distribution.

- A new Applying the Concepts exercise explores when the mean, median, and mode of a binomial random variable are equal.

- The following topics are covered in the Section 6.2 exercises: (a) *geometric probability distribution*, (b) *negative binomial probability distribution*, (c) *hypergeometric probability distribution*, and (d) *multinomial distribution*.

Section 6.3 Poisson Probability Distribution

- Some instructors may find that they do not have time to cover Section 6.3.

Section 6.4 Continuous Random Variables and the Normal Probability Distribution

- Note that we have merged the Standard Normal Distribution section (formerly with its own section) into this section. This helps streamline the coverage.

- Instructors who will be covering the *p*-value method in later chapters should make sure that students understand how to calculate the areas in Table 8 on page 355.

Section 6.5 Applications of the Normal Distribution

- Many of the examples in this section were rewritten to take advantage of the new Chapter 6 Case Study: *SAT Scores and AP Exam Scores.*

- The topic of assessing normality using normal probability plots was moved here from Chapter 7.

Section 6.6 Normal Approximation to the Binomial Probability Distribution

- Some instructors may find that they do not have the time to cover this topic.

Teaching Tips

Describing the number of students in class today is a way to convey the meaning of a discrete variable. For example, there are either 20 or 21 students in class but nothing in between (there are not 20½ students or 20.23 students in the class, for instance). A continuous variable can be explained by using an example of distance. How far can someone throw an eraser in the room? It is possible to throw an eraser anywhere from 0 feet to the back of the room. For instance, it is possible to throw the eraser 10½ feet or 8.645 feet. Another example of a continuous variable is time. If students have ever competed in a timed event or watched the timed events at the Olympics, they can see that once the clock starts, it can stop at any time.

In this chapter, we are working with probability distributions. This means that, as in Chapter 5, we are not given a data set to describe; instead, we are provided with a probability distribution, with a known population mean and known population standard deviation, and asked to find certain probabilities.

We use the cumulative normal distribution table, which may be new to some instructors. The author finds that it is easier to use because the students don't have to remember to add or subtract 0.5.

You may want to make sure that your students are expert at finding the tail area probability under a standard normal curve. This skill is needed for finding *p*-values for *Z* tests in later chapters.

In-Class Activities

1. Elicit the heights (in inches) of all the students in the class.

    a. Would you expect the distribution of heights to be normal? Why or why not?

    b. Construct a graph (dotplot, histogram, or stem-and-leaf display) of student heights. Is your expectation in (a) supported?

    c. Next, consider the heights of only the males in the class. Might this distribution be more normally distributed than the heights of all the students? Why or why not?

    d. Construct a graph (dotplot, histogram, or stem-and-leaf display) of male student heights. Is your expectation in (c) supported?

    e. If desired, repeat (c) and (d) for the females in the class.

2. Go to http://lib.stat.cmu.edu/DASL/Stories/ChestsizesofMilitiamen.html and download the Scottish Militiamen chest size data set. Ask students to construct histograms with different numbers of classes and discuss their observations. In addition, they can compute descriptive statistics for the mean, median, and mode and compare their values. Discussions may be focused on the shape of the distribution and the measures of central tendencies.

Probability Distributions and a Deck of Cards

3. Bring in a deck of cards and perform the following experiment:

    a. Take out the 12 face cards (kings, queens, and jacks).

    b. Construct the probability distribution of the face value of the remaining cards (ace → $X = 1$, deuce → $X = 2$, and so on).

    c. Find the population mean μ and standard deviation s of this probability distribution.

    d. Designate one student as shuffler and another student as selector. Have the selector draw one card at random from the abridged deck, observe it, record the value of X, and replace it.

    e. The shuffler then shuffles the cards. The selector then selects another card at random, until 20 cards have been sampled with replacement.

    f. Using the relative frequencies from (e), construct a relative frequency distribution of your observed face values. Is it very different from the population distribution from (b)?

    g. Calculate the mean and standard deviation of your relative frequency distribution. Are they very different from the population values calculated in (c)? How would you define "very different"?

A Binomial Probability Activity

4. There is a famous problem in binomial probability. Which parents would wind up with the greater proportion of boy babies: the parents who simply stopped having babies after a certain number of babies or the parents who had babies until they had a boy? It might seem that the answer is that the parents who stopped after the first boy would have more boys, but the answer is that both sets of parents could expect to have the same proportion of boys. You will use simulation to try to confirm this. Assume that the probability that a randomly chosen newborn is a boy is 50%. Then, clearly, the parents who simply stopped having babies after a certain number of babies could expect to have the same proportion (long-run probability!) of boys and

girls. Now, take a single die and define the results of your rolls as follows: an odd-number result will represent a boy, and an even-number result will represent a girl.

   a. Have your group roll the die. If you get a girl the first time, record it and roll again. Continue rolling and recording until you get a boy. That ends that particular trial.

   b. Repeat until you have a large number (say, 30) of trials.

   c. Then find the proportions of boys and girls in your 30 trials (representing 30 reproducing families). Your results shouldn't be too far away from a 50–50 split between boys and girls.

Supplements

- StatTutor 3.1–3.8, 10.2, 10.4–10.6, and 13.1–13.6
- EESEE case studies
  - Is Caffeine Dependence Real? (Questions 1, 2, 3, 4, and 5(**a**) on probability)

Applets

The *Normal Density Curve* applet is referenced in Chapter 6 to compute normal probabilities for Exercise 125 in Section 6.4 and Exercise 43 in Section 6.5. The *Normal Approximation to the Binomial Distribution* applet is used for Exercise 29 in Section 6.6.

Videos

- *Against All Odds: Inside Statistics:* www.learner.org/resources/series65.html
  - Program 7: Normal Curves
  - Program 8: Normal Calculations
  - Program 9: Checking Assumption of Normality
  - Program 20: Random Variables
  - Program 21: Binomial Distributions

Web Sites

- CAUSEweb provides resources for teaching statistics: https://www.causeweb.org/
- This Texas Instruments Web site has a host of TI-83/84 statistics activities: http://education.ti.com/en/us/activities/explorations-series-books/activitybook_83_statistics.html.
- This Web site lists other sites that do statistical calculations: http://statpages.org/.
- This Web site lists numerous activities on all topics: http://mathforum.org/mathtools/sitemap2/ps.
- This site has several activities using Excel and Minitab: www.mathspace.com/NSF_ProbStat/Teaching_Materials/Primarily_Statistics.htm.
- The site Online Statistics: An Interactive Multimedia Course of Study has a *Normal Distribution* applet that demonstrates simulations and computes normal probabilities, and it has a link to the Rice Virtual Lab in Statistics with numerous simulation demonstrations, along with other topics related to statistics: http://onlinestatbook.com/index.html.

6.1 Discrete Random Variables

OBJECTIVES By the end of this section, I will be able to . . .

1 Identify random variables.
2 Explain what a discrete probability distribution is and construct probability distribution tables and graphs.
3 Calculate the mean, variance, and standard deviation of a discrete random variable.

1 Random Variables

In Chapter 5, we calculated the probabilities of outcomes from experiments. If the experiment is tossing a fair coin twice, the outcomes are HH, HT, TH, and TT. The probability of observing exactly one head in two tosses is the probability of the event $A = \{HT, TH\}$. Because the outcomes were equally likely, we used the classical method of assigning probability. The probability of $\{HT, TH\}$ is $N(A)/N(S) = 2/4 = 0.5$, where S is the sample space.

In this chapter, we develop a different approach that analyzes probability problems more efficiently. Recall from Chapter 1 that a *variable* is a characteristic that can assume different values. Suppose we define a variable $X =$ number of heads observed when two fair coins are tossed. In this experiment we may observe zero heads, one head, or two heads, so that the possible values of X are 0, 1, and 2. Clearly, before we conduct our experiment, we do not know how many heads we will observe. Thus, randomness plays a role in the value of the variable X, and so we call X a *random variable*.

> A **random variable** is a variable that takes on quantitative values representing the results of a probability experiment, and thus its values are determined by chance. We denote random variables using capital letters such as X, Y, or Z.

In Chapter 5 (page 246), we found that the probability of observing exactly $X =$ one head was 0.5. We denote this probability using the notation

$$P(X = 1) = 0.5$$

Similarly, the probability of observing zero heads is $P(X = 0) = 0.25$, and the probability of two heads is $P(X = 2) = 0.25$.

Developing Your Statistical Sense

Random Variables Must Be Random!

The role of chance in the definition of a random variable is crucial. For example, is your age a random variable? If we are just talking about you and no one else, and we know your age, then there is no chance involved. In that case, your age is not a random variable. On the other hand, what if we select students *at random* by picking names from a hat? Then the age of the person drawn is a random variable because its value depends at least partly on chance (on which name is drawn at random).

Let's start with an example aimed at helping you move from the language of probability (experiments and outcomes) to the language of random variables.

EXAMPLE 1 Notation for random variables

Comstock/Stockbyte/Jupiterimages

Suppose our experiment is to toss a single fair die, and we are interested in the number rolled. We define our random variable X to be the outcome of a single die roll.

a. Why is the variable X a random variable?

b. What are the possible values that the random variable X can take?

c. What is the notation used for rolling a 5?

d. Use random variable notation to express the probability of rolling a 5.

Solution

a. We don't know the value of X before we toss the die, which introduces an element of chance into the experiment, thereby making X a random variable.

b. The possible values for X are 1, 2, 3, 4, 5, and 6.

c. When a 5 is rolled, then X equals the outcome 5, and we write $X = 5$.

d. Recall from Section 5.1 that the probability of rolling a 5 for a fair die is 1/6. In random variable notation, we denote this as $P(X = 5) = 1/6$.

There are two main types of random variables: **discrete random variables** and **continuous random variables**. The difference between the two types relates to the possible values that each type of random variable can assume.

Discrete and Continuous Random Variables

- A **discrete random variable** can take either a finite or a countable number of values. These values may be written as a list of numbers, so each value can be graphed as a separate point on a number line, with space between each point. (See Figure 1a.)

FIGURE 1a Discrete random variable.

- A **continuous random variable** can take uncountably infinite different values. Because of this, the values of a continuous random variable form an interval on the number line. (See Figure 1b.)

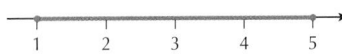

FIGURE 1b Continuous random variable.

Discrete random variables usually need to be counted, such as 1, 2, 3, and so forth. Continuous random variables usually need to be measured, not counted, such as measuring the amount of gasoline purchased.

Examples of discrete random variables include the number of children a randomly selected person has and the number of times a randomly chosen student has been pulled over for speeding on the interstate. Continuous random variables often need to be measured, not counted. For example, the temperature in Atlanta, Georgia, at noon today may be reported as 77 degrees, but this value represents actual temperatures that may lie anywhere between 76.5 degrees and 77.5 degrees.

EXAMPLE 2 Identifying discrete and continuous random variables

For the following random variables, (i) determine whether they are discrete or continuous, and (ii) indicate the possible values they can take:

a. The number of automobiles owned by a family

b. The width of your desk in this classroom

c. The number of games played in the next World Series

d. The weight of model year 2015 SUVs

Solution

a. The possible number of automobiles owned by a family is finite and may be written as a list of numbers, so it represents a discrete random variable. The possible values are $\{0, 1, 2, 3, 4, \ldots\}$.

b. Width is something that must be measured, not counted. Width can take infinitely many different possible values, with these values forming an interval on the number line. Thus, the width of your desk is a continuous random variable. The possible values might be $1 \text{ ft} \leq W \leq 10 \text{ ft}$.

c. The number of games played in the next World Series can be counted and thus represents a discrete random variable. The possible values are finite and may be written as a list of numbers: $\{4, 5, 6, 7\}$.

d. The weight of model year 2015 SUVs must be measured, not counted, and thus represents a continuous random variable. Weight can take infinitely many different possible values, with these values forming an interval on the number line: $2500 \text{ lbs} \leq Y \leq 7000 \text{ lbs}$.

NOW YOU CAN DO
Exercises 7–16.

YOUR TURN #1

For the following random variables, (i) determine whether they are discrete or continuous, and (ii) indicate the possible values they can take:

a. Your best friend's height

b. The number of cats you own

(The solutions are shown in Appendix A.)

We will return to continuous random variables in Section 6.4. Sections 6.1, 6.2, and 6.3 concentrate on discrete random variables.

2 Discrete Probability Distributions

For every random variable, there is a probability distribution that allows us to view all possible values of the random variable at a glance. Discrete probability distributions show the probabilities associated with the various values that the discrete random variable can take.

> A **probability distribution of a discrete random variable** provides all the possible values that the random variable can assume, together with the probability associated with each value. The probability distribution can take the form of a table, graph, or formula. Probability distributions describe populations, not samples.

When constructing the tabular form of a **probability distribution of a discrete random variable**, create a table with two rows:

- The top row will contain all the possible values of *X*.
- The bottom row will contain the probability associated with each value of *X*.

EXAMPLE 3   Probability distribution table

Construct the probability distribution table of the number of heads observed when tossing a fair coin twice.

Solution

NOW YOU CAN DO
Exercises 17a–24a.

The probability distribution table given in Table 1 uses the probabilities we found on page 246.

The probabilities in Table 1 were assigned using the classical method, because we assumed that tossing a fair coin would result in equally likely outcomes.

Table 1 Probability distribution table of the number of heads on two fair coin tosses

X = **number of heads observed**	0	1	2
P(X) = **probability of observing that many heads**	1/4	1/2	1/4

Note that the probabilities in the bottom row of Table 1 add up to 1. Also, note that because each value in the bottom row is a probability, each value must be between 0 and 1, inclusive, that is, $0 \leq P(X) \leq 1$. We can generalize this as follows.

Rules for a Discrete Probability Distribution

- The sum of the probabilities of all the possible values of a discrete random variable must equal 1. That is, $\Sigma P(X) = 1$.
- The probability of each value of *X* must be between 0 and 1, inclusive. That is, $0 \leq P(X) \leq 1$.

This first rule derives from the Law of Total Probability from Section 5.1 (page 241).

EXAMPLE 4 Recognizing valid discrete probability distributions

Identify which of the following is a valid discrete probability distribution.

a.

X	1	10	100	1000
P(X)	0.2	0.4	0.3	0.2

b.

X	−10	0	10	20
P(X)	0.5	0.3	0.4	−0.2

c.

X	Red	Green	Blue	Yellow
P(X)	0.1	0.3	0.4	0.2

d.

X	−5	0	5	10
P(X)	0.1	0.3	0.4	0.2

Solution

a. This is not a valid probability distribution because the probabilities add up to 1.1, which is greater than 1.

b. This is not a valid probability distribution because $P(X = 20)$ is negative.

c. This is not a valid probability distribution for a discrete random variable because the values of *X* are not quantitative.

d. This is a valid probability distribution because the probabilities sum to 1, and each probability $P(X)$ takes a value between 0 and 1.

NOW YOU CAN DO
Exercises 25–28.

YOUR TURN
#2

Identify whether the following is a valid discrete probability distribution.

X	−4	−2	0	2
$P(X)$	0.25	0.30	0.30	0.20

(The solution is shown in Appendix A.)

Probability distributions can also take the form of a probability distribution graph.

EXAMPLE 5   Discrete probability distribution as a graph

The number of points a soccer team gets for a game is a random variable because it is not certain, before the game, how many points the team will get.

Jamie Sabau/Getty Images

The probabilities in Table 2 were assigned to the random variable *X* using the relative frequency (empirical) method.

Given a graph of a probability distribution, you should know how to construct the probability distribution table, and vice versa.

In Major League Soccer (MLS), teams are awarded 3 points in the standings for a win, 1 point for a tie, and 0 points for a loss. In the 34-game 2013 MLS season, the New York Red Bulls had 17 wins, 9 losses, and 8 ties.

a. Construct a probability distribution table of the number of points per game, based on the team's performance during the 2013 MLS season.

b. Construct a probability distribution graph of the number of points per game.

Solution

a. Let X = points awarded. Then the probability distribution table is given in Table 2.

Table 2 Probability distribution table of points awarded for New York Red Bulls

X = points	0	1	3
$P(X)$	9/34 = 0.26	8/34 = 0.24	17/34 = 0.5

b. The probability distribution graph is given in Figure 2.

- The horizontal axis is the usual *x* axis (the number line), and it shows all the possible values that the random variable *X* can take, such as $X = 0$, 1, or 3. The horizontal axis gives the same information as the top row of the table.

- The vertical axis represents probability, and is the information in the bottom row in the table. A vertical bar is drawn at each value of *X*, with the height representing the probability of that value of *X*. For example, the bar of probability at $X = 0$ goes up to 0.26 and represents the probability that the New York Red Bulls will lose a game.

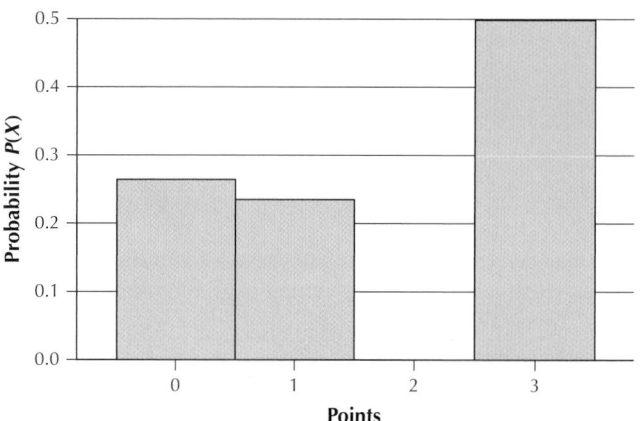

NOW YOU CAN DO
Exercises 17b–24b.

FIGURE 2 Probability distribution graph of points awarded for New York Red Bulls.

YOUR TURN

#3

Construct a probability distribution graph of the number of heads observed in Table 1 on page 313.

(The solution is shown in Appendix A.)

We may use probability distributions to calculate probabilities for multiple values of X. In discrete probability distributions, the outcomes are always mutually exclusive. For example, it is not possible to observe both zero heads ($X = 0$) and two heads ($X = 2$) when tossing two fair coins. Thus, we always use the Addition Rule for Mutually Exclusive Events to find the probability of two or more outcomes for a discrete random variable. For example, $P(X = 0 \text{ or } 2) = P(X = 0) + P(X = 2)$.

EXAMPLE 6 Calculating probabilities for multiple values of X

Use the probability distribution from Example 5 to find the following probabilities:

a. Probability that the New York Red Bulls are awarded either 0 or 3 points in a game

b. Probability that the New York Red Bulls are awarded both 0 and 3 points in a game

c. Probability that the New York Red Bulls are awarded at least 1 point in a game

d. Probability that the New York Red Bulls are awarded at most 1 point in a game

Solution

a. $P(X = 0 \text{ points } or \text{ 3 points}) = P(X = 0) + P(X = 3) = 0.26 + 0.5 = 0.76$. For a randomly selected game, the probability that the Red Bulls either lose the game or win the game is 0.76.

b. The outcomes $X = 0$ and $X = 3$ are mutually exclusive. Therefore, $P(X = 0 \text{ points } and \text{ 3 points}) = 0$.

c. The phrase *at least* means "that many or more." Thus, we need to find: $P(X \geq 1) = P(X = 1 \text{ point } or \text{ 3 points}) = P(X = 1) + P(X = 3) = 0.24 + 0.5 = 0.74$.

NOW YOU CAN DO
Exercises 29–44.

d. The phrase *at most* means "that many or fewer." Thus, $P(X \leq 1) = P(X = 1 \text{ point } or \text{ 0 points}) = P(X = 1) + P(X = 0) = 0.24 + 0.26 = 0.5$.

YOUR TURN

#4

For the situation in Example 6, what is the probability that the New York Red Bulls are awarded the following number of points in a game?

a. Either 1 point or 3 points

b. Both 1 point and 3 points

c. At most 3 points

d. At least 3 points

(The solutions are shown in Appendix A.)

3 Mean and Variability of a Discrete Random Variable

Just as we can compute the mean and standard deviation of quantitative data, we can calculate the mean and standard deviation of a random variable X.

The **mean μ of a discrete random variable X** represents the mean result when the experiment is repeated an indefinitely large number of times.

> **Finding the Mean of a Discrete Random Variable _X_**
>
> The mean μ of a discrete random variable X is found as follows:
>
> 1. Multiply each possible value of X by its probability.
> 2. Add the resulting products.
>
> This procedure is denoted as
>
> $$\mu = \sum [X \cdot P(X)]$$

EXAMPLE 7 Calculating the mean of a discrete probability distribution

Note: These 10 friends constitute a population, not a sample, so the mean is μ, not $\bar{x}$

Carla has 10 friends in school. She took a census of all 10 friends, asking each how many credits they had registered for that semester. Five of her friends were taking 15 credits, with one each taking 12, 13, 14, 16, and 20 credits. The relative frequency distribution is shown in Table 3.

Table 3 Relative frequency distribution for the number of credits

Credits	Frequency	Relative frequency
12	1	0.1
13	1	0.1
14	1	0.1
15	5	0.5
16	1	0.1
20	1	0.1

a. Construct the probability distribution table for X = number of credits taken.

b. Calculate the mean μ number of credits taken.

Solution

a. Our random variable is X = number of credits taken. We use the relative frequencies from Table 3 to assign probabilities to the various values of X. The resulting probability distribution table is shown in the first two columns of Table 4.

b. To find the mean μ, we first need to multiply each possible outcome (value of X) by its probability $P(X)$. This is shown in the right-hand column in Table 4. We

Table 4 Probability distribution table of X = number of credits

X = number of credits	$P(X)$	$X \cdot P(X)$
12	0.1	$12 \cdot (0.1) = 1.2$
13	0.1	$13 \cdot (0.1) = 1.3$
14	0.1	$14 \cdot (0.1) = 1.4$
15	0.5	$15 \cdot (0.5) = 7.5$
16	0.1	$16 \cdot (0.1) = 1.6$
20	0.1	$20 \cdot (0.1) = 2.0$
Total	1.0	$\mu = \sum X \cdot P(X) = 15$

multiply the value $X = 12$ by its probability $P(X) = 0.1$, the value $X = 13$ by its probability $P(X) = 0.1$, and so on. Then we add these five products to find the mean:

$$\mu = \sum X \cdot P(X) = 15$$

The mean number of credits taken by Carla's friends is 15.

NOW YOU CAN DO
Exercises 45a–52a.

YOUR TURN
#5

Refer to Table 1 on page 313. Calculate the mean μ number of heads.

(The solution is shown in Appendix A.)

What Does This Number Mean?

What does it mean to say that $\mu = 15$ is the mean of the random variable X = number of credits? First of all, the mean of the random variable X is definitely not the same as the mean of a sample of Carla's friends, which is a sample mean. For example, suppose that a sample of 4 of Carla's 10 friends were taking the following number of credits: 15, 14, 13, 12. The mean of this sample of four friends is $\bar{x} = 13.5$. However, if we were to consider an *infinite number* of friends, then the mean of this very large sample would converge to $\mu = 15$. So the mean μ of a discrete random variable is interpreted as the mean of the results from the *population* of all possible repetitions of the experiment, which is why we denote the mean of a random variable as μ.

Note: The population mean μ does not need to equal any values of X, nor does it need to be an integer.

Developing Your Statistical Sense

Why Does This Formula Work?

The formula for the mean of a discrete random variable works because it is a special case of the weighted mean (page 149 of Chapter 3). Of the population of 10 friends, 1 was taking 12 credits. Thus, the first weight is $w_1 = 1$. Similarly, $w_2 = 1$, $w_3 = 1$, $w_4 = 5$, $w_5 = 1$, and $w_6 = 1$. Thus, the population weighted mean is

$$\mu = \frac{\sum w_i x_i}{\sum w_i} = \frac{(1)(12) + (1)(13) + (1)(14) + (5)(15) + (1)(16) + (1)(20)}{10} = 15$$

Dividing through and rearranging terms give us

$$(12)(0.1) + (13)(0.1) + (14)(0.1) + (15)(0.5) + (16)(0.1) + (20)(0.1)$$
$$= \sum X \cdot P(X) = 15$$

We may also interpret the mean μ as the *balance point* of the distribution.

EXAMPLE 8 Mean μ as balance point of the distribution

Graph the probability distribution of the random variable X = credits taken, and insert a pivot (a balance point) at the value of the mean, $\mu = 15$.

Solution

The probability distribution graph of X = age is given in Figure 3. Note that the distribution is balanced at the point $\mu = 15$.

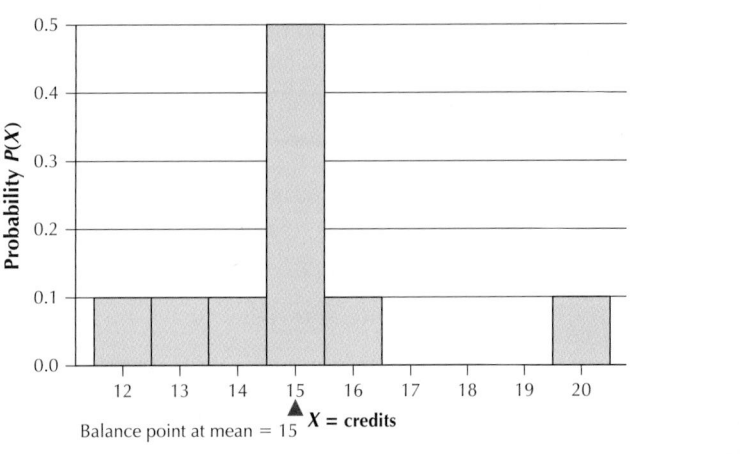

FIGURE 3
Probability distribution graph balances at $\mu = 15$.

YOUR TURN
#6

Refer to Table 1 on page 313. Graph the probability distribution of the random variable X = number of heads, and insert a pivot (a balance point) at the value of the mean, $\mu = 1$.

(The solution is shown in Appendix A.)

In certain situations, we may need to identify the most likely value of the random variable X.

EXAMPLE 9   Identifying the most likely value of a discrete random variable

If one of the friends represented in the table in Example 7 is chosen at random, what is *the most likely* number of credits taken by that friend?

Solution

NOW YOU CAN DO
Exercises 45b–52b.

The largest probability in the probability table is $P(X = 15)$, and the tallest bar in the probability graph (Figure 3) is for $X = 15$, so 15 is the most likely number of credits.

YOUR TURN
#7

Refer to Table 1 on page 313. Identify the most likely number of heads when a fair coin is tossed twice.

(The solution is shown in Appendix A.)

Note: In Example 9, the most likely number of credits equals the mean, but this is not typical. Very often, the most likely value of a random variable is not equal to the mean.

The mean μ of a random variable is also called the **expected value** or the **expectation of the random variable** X. It does not necessarily follow that the expected value of X is the most likely value of X. However, the expected value of X (that is, the mean μ) is often a good indication of the center of the distribution of the random variable.

The **expected value, or expectation, of a random variable X** is the mean μ of X. It is denoted as $E(X)$. This definition holds for both discrete and continuous random variables.

EXAMPLE 10   Expected value of a discrete random variable X

Find the expected value $E(X)$ of the following discrete random variables:

a. X = number of heads in Example 3

b. X = number of points awarded in Example 5

c. X = number of credits in Example 7

Solution

a. Using the probabilities in Table 1, we have

$$E(X) = \mu = \sum [X \cdot P(X)] = 0(0.25) + 1(0.5) + 2(0.25) = 1$$

The expected number of heads is 1.

b. Using Table 2, we have

$$E(X) = \mu = \sum [X \cdot P(X)] = 0(0.26) + 1(0.24) + 3(0.5) = 1.74$$

The expected number of points is 1.74.

NOW YOU CAN DO

Exercises 45c–52c.

c. From Example 7, $E(X) = \mu = 15$. The expected number of credits is 15.

Note from Example 10(**b**) that the mean or expected value of a random variable need not be a particular value of X. Instead, it represents the mean of a very large number of repetitions of the experiment.

Variability of a Discrete Random Variable

Because a discrete random variable takes on quantitative values, we use the **variance** or **standard deviation of a random variable** X to help us determine whether a particular value of that random variable is unusual. Just as a random variable X has a mean (μ), which is a measure of center, so a random variable X also has a standard deviation (σ) and variance (σ^2), which are measures of spread.

Variance and Standard Deviation of a Discrete Random Variable X

Variance of X: $\sigma^2 = \sum [(X - \mu)^2 \cdot P(X)]$

Standard deviation of X: $\sigma = \sqrt{\sum [(X - \mu)^2 \cdot P(X)]}$

Notice that these formulas include μ as one of the terms, so that you must first find the mean μ of the discrete random variable before you find the variance (or standard deviation). Recall from Chapter 3 that the standard deviation is simply the square root of the variance.

EXAMPLE 11 Calculating the variance and standard deviation of a discrete random variable

The probability distribution for the number of credits is repeated here as Table 5. In Example 7, we calculated the mean number of credits as $\mu = 15$. Calculate the variance and standard deviation.

Table 5 Probability distribution of the number of credits

X = number of credits taken	$P(X)$
12	0.1
13	0.1
14	0.1
15	0.5
16	0.1
20	0.1

Solution

Refer to Table 6. The first two columns correspond to the probability distribution of X = number of credits taken. The third column represents the calculations needed to find $(X - \mu)^2 \cdot P(X)$. Summing the values in the rightmost column provides the variance $\sigma^2 = 4$. Taking the square root of the variance gives us the standard deviation $\sigma = \sqrt{\sigma^2} = \sqrt{4} = 2$ credits.

 credits

Table 6 Calculating σ^2 and σ

X	P(X)	$(X - \mu)^2 \cdot P(X)$
12	0.1	$(12 - 15)^2 \cdot 0.1 = 0.9$
13	0.1	$(13 - 15)^2 \cdot 0.1 = 0.4$
14	0.1	$(14 - 15)^2 \cdot 0.1 = 0.1$
15	0.5	$(15 - 15)^2 \cdot 0.5 = 0.0$
16	0.1	$(16 - 15)^2 \cdot 0.1 = 0.1$
20	0.1	$(20 - 15)^2 \cdot 0.1 = 2.5$
		$\sigma^2 = \sum (X - \mu)^2 \cdot P(X) = 4$

NOW YOU CAN DO
Exercises 53a,b–60a,b.

Now that we have calculated the standard deviation σ, we may use it along with the mean to determine whether values of X are outliers or moderately unusual, using the Z-score method.

EXAMPLE 12 Z-score method for determining an unusual value

a. Using the information from Example 11, determine whether $X = 20$ is an unusual number of credits to take this semester.

b. Construct a probability distribution graph of X, illustrating how $X = 20$ credits is moderately unusual.

Solution

a. Recall from Section 3.4 (page 159) that a data value with a Z-score between 2 and 3 may be considered moderately unusual. The Z-score for $X = 20$ credits is

$$Z = \frac{X - \mu}{\sigma} = \frac{20 - 15}{2} = 2.5$$

Thus, among Carla's friends, it would be considered moderately unusual to take 20 credits this semester.

b. Figure 4 shows the probability distribution graph of $X =$ number of credits. The mean $\mu = 15$ is indicated, along with the distances $\mu \pm 1\sigma$, $\mu \pm 2\sigma$, and $\mu \pm 3\sigma$.

FIGURE 4
$X = 20$ credits is moderately unusual because it lies $Z = 2.5$ standard deviations above the mean.

NOW YOU CAN DO
Exercises 53c–60c.

[Figure 4: Probability distribution graph with Probability on the y-axis (0.0 to 0.5) and Credits on the x-axis (9 to 21). Bars at 12, 13, 14 (height 0.1), 15 (height 0.5), 16 (height 0.1), and 20 (height 0.1). Below axis: $\mu - 3\sigma$, $\mu - 2\sigma$, $\mu - 1\sigma$, $\mu = 15$, $\mu + 1\sigma$, $\mu + 2\sigma$, $\mu + 3\sigma$.]

EXAMPLE 13 **Compute the mean and standard deviation of a discrete random variable using technology**

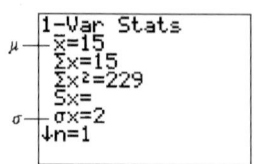

FIGURE 5 TI-83/84 results for mean and standard deviation of a discrete random variable.

Compute the mean and standard deviation of the probability distribution given in Example 11 using the TI-83/84 graphing calculator.

Solution

We use the instructions provided in the following Step-by-Step Technology Guide. The results are shown in Figure 5. Be careful! The calculator indicates that the mean is $\bar{x}$. It is not $\bar{x}$ but μ.

STEP-BY-STEP TECHNOLOGY GUIDE: Mean and Standard Deviation of a Discrete Random Variable

We illustrate using Example 13.

TI-83/84

Step 1 Enter the X values in list **L1** and the corresponding P(X) values in list **L2**. See Figure 6a.
Step 2 Press **STAT**, highlight **CALC**, and select **1-Var Stats**.
Step 3 Type L1 followed by a comma, followed by L2, as shown in Figure 6b. Press **ENTER**. The results are shown in Figure 5 above.

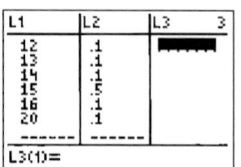

FIGURE 6a

FIGURE 6b

EXCEL

Mean

Step 1 Enter the X values in Column **A** and the corresponding P(X) values in Column **B**.
Step 2 Click **C1**, type =A1*B1, and press **Enter**.
Step 3 Copy **C1** and paste it in cells **C2** through **C6**.
Step 4 Click on the first empty element of Column C (here, that is C7). Select the **Formula** icon, type **Sum**, click **GO**, then click **OK**. Highlight **C1** through **C6**, and click **OK**. The number in **C7** is the mean of the random variable.

Standard Deviation

Step 1 Enter the X values in Column **A** and the corresponding P(X) values in Column **B**. Calculate the mean of the random variable. In our case, the mean is 15.
Step 2 Select **D1**, and type =(A1-15)^2*B1. We use 15 because it is the mean of the random variable.
Step 3 Copy **D1**, and paste it in cells **D2** through **D6**.
Step 4 Click on the first empty element of Column D (here, that is D7). Type =SQRT(SUM(D1:D6)), then hit **Enter**. The number in D7 is the standard deviation of the random variable.

MINITAB

Mean

Step 1 Enter the X values in **C1** and the corresponding P(X) values in **C2**. Label your columns **X** and **P(X)**.
Step 2 Click **Calc > Calculator....** In **Store result in variable**, type **C3**. In **Expression**, doubleclick *C1 X*, click the multiply sign *, then click *C2 P(X)*. Click **OK**.
Step 3 Click **Calc > Calculator....** In **Store result in variable**, type **C4**. Delete the previous formula from the Expression box. Under **Functions**, select **Statistics**, then double-click **Sum**, double-click *C3*, and click **OK**.

Standard Deviation

Step 1 Enter the X values in **C1** and the corresponding P(X) values in **C2**. Label your columns **X** and **P(X)**. Calculate the mean.
Step 2 Click **Calc > Calculator....** In **Store result in variable**, type **C5**. In **Expression**, type ('X'-15)^2*'P(X)'. Click **OK**.
Step 3 Click **Calc > Calculator....** In **Store result in variable**, type **C6**. Delete the previous formula from the Expression box. In the **Functions** menu, select **Arithmetic**, then double-click **Square Root**. From the **Functions** menu, select **Statistics**, and double-click **Sum**. Double-click **C5**, and click **OK**.

78. Explain why the Statistics AP exam score is a random variable.

79. Explain why the Statistics AP exam score is a discrete, not a continuous, variable.

80. Construct the probability distribution table for $X =$ Statistics AP exam score.

81. Confirm that your probability distribution in Exercise 80 is valid.

82. Draw a probability distribution graph of X.

83. Calculate the following probabilities:
   a. $P(X \geq 2)$
   b. $P(X > 1)$
   c. $P(X \leq 2)$
   d. $P(X < 3)$
   e. Compare your answers for (**a**) and (**b**). Compare your answers for (**c**) and (**d**). Explain why this is happening.

84. Calculate the mean Statistics AP exam score, μ.

85. Find the most likely value of X.

86. What is the expected Statistics AP exam score?

87. Calculate the variance and standard deviation of X.

88. Use the Z-score method to identify any unusual values.

WORKING WITH LARGE DATA SETS

Motor Vehicle Fuel Efficiency. Open the data set **FuelEfficiency**. We will explore some characteristics of this data set, using the tools and techniques we have learned in this section. Use technology to do Exercises 89–98.

fuelefficiency

89. Obtain a relative frequency distribution of the variable $X = cylinders$.

90. Use the relative frequency distribution in Exercise 89 to construct a probability distribution table of $X = cylinders$.

91. Construct a probability distribution graph of the number of cylinders.

92. Calculate the mean number of cylinders. Insert this value as the balance point in your probability distribution graph.

93. What is the most likely value of X?

94. Construct a probability distribution table of $X = gears$.

95. Draw a probability distribution graph of $X = gears$.

96. Find the mean number of gears. Insert this value as the balance point in your probability distribution graph.

97. Use a spreadsheet to help you calculate the variance and standard deviation of X.

98. Identify any unusual or somewhat unusual number of gears.

6.2 Binomial Probability Distribution

OBJECTIVES By the end of this section, I will be able to . . .

1 Explain what constitutes a binomial experiment.

2 Compute probabilities using the binomial probability formula, binomial tables, and technology.

3 Calculate the mean, variance, and standard deviation of the binomial random variable and find the mode of the distribution.

1 Binomial Experiment

Many different types of discrete probability distributions are used. Perhaps the most important is the *binomial* distribution, which we will learn about in this section. Life is full of situations where there are *only two possible outcomes* to a process.

• A baby is about to be born. Will it be a boy or a girl?

• A basketball player is about to attempt a free throw. Will she make it or miss?

• A friend of yours is also taking statistics. Will he pass or fail?

Because situations for which there are only two possible outcomes are so widespread, methods have been developed to make it more convenient to analyze them. These methods begin with the definition of a **binomial experiment**.

> **Binomial Experiment**
>
> A probability experiment that satisfies the following four requirements is said to be a **binomial experiment**:
>
> 1. Each trial of the experiment has only *two possible mutually exclusive outcomes* (or is defined in such a way that the number of outcomes is reduced to two). One outcome is denoted a *success* and the other a *failure*.
> 2. A *fixed number of trials exists*, which is known in advance of the experiment.
> 3. The experimental outcomes are *independent* of each other.
> 4. The *probability* of observing a success remains the same from trial to trial.

Let's take a moment to discuss what these requirements really mean.

Many experiments having more than two outcomes can often be defined so that only two outcomes are possible. For example, the answer to a multiple-choice question that has five answer choices may be recorded as either correct or incorrect.

1. A *success* denotes simply the outcome in which we are interested, without necessarily implying that the outcome is desirable. For example, for a researcher investigating college dropout rates, a dropout would be considered a success in the context of a binomial experiment.

2. Tossing a coin 10 times is a binomial experiment because we know the *fixed number of trials*. A salesman contacting customers one-by-one until he makes a sale is not a binomial experiment because he doesn't know how many customers he will have to contact.

3. Sampling without replacement would technically violate the *independence* requirement. However, recall that we may apply the 1% Guideline from Section 5.3, so that when the sample is small compared to the population, successive trials can be considered independent.

4. Suppose four friends are wondering how many of them will get an A in statistics. This is not a binomial experiment because the four friends presumably do not all have the same probability of success.

The outcomes of a binomial experiment, together with their probabilities, generate a special discrete probability distribution called the **binomial probability distribution**. For binomial probability distributions, only two outcomes are always possible, and each outcome has a probability associated with it. The *binomial random variable*, denoted by X, represents the number of successes observed in the n trials. Note that $0 \leq X \leq n$.

EXAMPLE 14   Recognizing binomial experiments

Determine whether each of the following experiments fulfills the conditions for a binomial experiment. If the experiment is binomial, identify the random variable X, the number of trials, the probability of success, and the probability of failure. If the experiment is not binomial, explain why not.

a. A fisherman is going fishing and will continue to fish until he catches a rainbow trout.

b. We flip a fair coin three times and observe the number of heads.

c. A market researcher at a shopping mall is asking consumers whether they use Fib detergent. She asks a sample of four men, one of whom is clearly the employer of the other three.

d. The National Burglar and Fire Alarm Association reports that 34% of burglars get in through the front door. A random sample of 36 burglaries is taken, and the number of entries through the front door is noted.

Solution

a. This is not a binomial experiment because you don't know how many fish he will catch before the rainbow trout shows up, so a fixed number of trials isn't known in advance.

b. This is a binomial experiment because it fulfills the requirements:

    i. Only two possible outcomes are possible on each trial, with heads defined as success and tails as failure.

    ii. We know in advance that we are tossing the coin three times.

    iii. The coin doesn't remember its result from toss to toss, and so the trials are independent.

    iv. The coin is fair on each toss, and so the probability of observing heads is the same on each toss.

The binomial random variable X is the number of heads observed on the three trials; because the coin is fair, the probability of success is 0.5 and the probability of failure is 0.5. The possible values for X are 0, 1, 2, or 3.

c. This is not a binomial experiment because the responses are not independent. The response given by the employer is likely to affect the employees' responses.

d. This is a binomial experiment because it fulfills the requirements:

    i. Only two possible outcomes are possible on each trial: entering through the front door or not entering through the front door.

    ii. We know in advance that the size of the random sample is 36 burglaries.

    iii. The sample is random, so the trials are independent.

    iv. The sample is quite small compared to the size of the population, so the probability of entering through the front door remains the same from burglary to burglary.

The binomial random variable X is the number of front-door-entry burglaries noted for the 36 break-ins; the probability of success is 0.34 and the probability of failure is $1 - 0.34 = 0.66$.

NOW YOU CAN DO
Exercises 5–14.

Table 7 gives some notation regarding binomial experiments and the binomial distribution.

Table 7 Notation for binomial experiments and the binomial distribution

Symbol	Meaning
S	The outcome denoted as a success
F	The outcome denoted as a failure
$P(\text{success}) = P(S) = p$	The probability of observing a success
$P(\text{failure}) = P(F) = 1 - p = q$	The probability of observing a failure
n	The number of trials

Using this notation in the experiment in Example 14(d), we have $S =$ burglary through front door ($P(S) = p = 0.34$), and $F =$ burglary not through front door ($P(F) = 1 - p = 1 - 0.34 = 0.66 = q$).

Note: In Section 5.4, we used $_nC_r$ to indicate the number of combinations. Now that we have learned about random variables, which can be denoted X, we use $_nC_x$ to represent the number of combinations.

2 Computing Binomial Probabilities

We demonstrate three ways of computing binomial probabilities: (a) the binomial probability formula, (b) binomial tables, and (c) technology. Before we examine the binomial probability distribution formula, let us recall from Section 5.4 (page 296) the formula for the **number of combinations**.

Note: You may find the following special combinations useful. For any integer *n*:

$$_nC_n = 1 \qquad _nC_0 = 1$$
$$_nC_1 = n \qquad _nC_{n-1} = n$$

The **number of combinations** of *X* items chosen from *n* different items is given by

$$_nC_X = \frac{n!}{X!(n-X)!}$$

where *n*! represents *n* **factorial**, which equals $n(n-1)(n-2) \ldots (2)(1)$, and 0! is defined to be 1.

We are often interested in finding probabilities associated with a binomial experiment.

EXAMPLE 15   Constructing a binomial probability distribution

Lori Lee Miller/Alamy

A recent study reported that about 40% of online dating survey respondents are "hoping to start a long-term relationship" (LTR).[2] Consider the experiment of choosing three online daters at random, and let

$$X = \text{the number of "LTRers"}$$

so that a success is defined as choosing someone hoping to start a long-term relationship.

a. Construct a tree diagram for this experiment.

b. Suppose that we are interested in finding the probability that exactly two of the three online daters would be LTRers, $P(X = 2)$. In the tree diagram, highlight in blue the outcomes where exactly two of the three online daters are LTRers. Find the probability for each outcome, and use these to find $P(X = 2)$.

c. Suppose that we are interested in finding $P(X = 1)$. In the tree diagram, highlight in red the outcomes where exactly one of the three online daters is an LTRer. Find the probability for each outcome, and use these to find $P(X = 1)$.

Solution

a. Figure 7 shows the tree diagram for this experiment.

b. As we can see from Figure 7, there are $(_nC_X) = (_3C_2) = 3$ different ways that exactly two of the three online daters could be LTRers (highlighted in blue).

FIGURE 7
Tree diagram and binomial probabilities.

1st Trial	2nd Trial	3rd Trial	Outcome	Number of successes, *X*	Probability of outcome
		S	S, S, S	3	$(0.4) \cdot (0.4) \cdot (0.4) = 0.064$
	S	F	S, S, F	2	$(0.4) \cdot (0.4) \cdot (0.6) = 0.096$
S		S	S, F, S	2	$(0.4) \cdot (0.6) \cdot (0.4) = 0.096$
	F	F	S, F, F	1	$(0.4) \cdot (0.6) \cdot (0.6) = 0.144$
		S	F, S, S	2	$(0.6) \cdot (0.4) \cdot (0.4) = 0.096$
	S	F	F, S, F	1	$(0.6) \cdot (0.4) \cdot (0.6) = 0.144$
F		S	F, F, S	1	$(0.6) \cdot (0.6) \cdot (0.4) = 0.144$
	F	F	F, F, F	0	$(0.6) \cdot (0.6) \cdot (0.6) = 0.216$

For each of these three outcomes, the probability that $X = 2$ is $(0.4)^2(0.6) = 0.096$.

- The outcome S, S, F (second row in Figure 7) has probability $(p)(p)(q) = (0.4)(0.4)(0.6) = 0.096$.

- The outcome S, F, S has probability $(p)(q)(p) = (0.4)(0.6)(0.4) = 0.096$.

- The outcome F, S, S has probability $(q)(p)(p) = (0.6)(0.4)(0.4) = 0.096$.

Note that each of these products equals $(p)^2 \cdot q$, with p having exponent $X = 2$, and (q) having exponent $n - X = 3 - 2 = 1$. Thus,

$$P(X = 2) = (_3C_2)(0.4)^2(0.6)$$
$$= 3(0.096) = 0.288$$

c. Similarly, suppose that we are interested in whether exactly one ($X = 1$) of the three online daters is an LTRer. Then, Figure 7 shows us, highlighted in red, that there are $(_nC_X) = (_3C_1) = 3$ different ways this could happen. Each of these outcomes has probability $(p) \cdot (q)^2 = (0.4)(0.6)^2 = 0.144$, where p has exponent $X = 1$, and q has exponent $n - X = 3 - 1 = 2$. Thus,

$$P(X = 1) = (_3C_1)(0.4)(0.6)^2$$
$$= 3(0.144) = 0.432$$

We can generalize these procedures and use the **binomial probability distribution formula** to find probabilities for the number of successes for any binomial experiment.

Remember: $P(S) = p$ and $P(F) = q$.

The Binomial Probability Distribution Formula

The probability of observing exactly X successes in n trials of a binomial experiment is

$$P(X) = (_nC_X)\, p^X\, (q)^{n-X}$$

That is,

$$P(X) = (_nC_X)\, [P(success)^{number\ of\ successes} \cdot P(failure)^{number\ of\ failures}].$$

We often call this the binomial probability formula.

Steps for Solving Binomial Probability Problems

To solve a binomial probability distribution problem, follow these steps:

Step 1 Find the number of trials n, and the probability of success on a given trial p.

Step 2 Find the number of successes X about which the question is asking.

Step 3 Using the values from Steps 1 and 2, find the required probabilities using either the binomial probability formula, the binomial tables (which we learn below), or technology.

EXAMPLE 16 Applying the binomial probability distribution formula

A report from SleepFoundation.org reported that 20% of Americans are sleep-deprived (defined as getting less than six hours sleep per night, on average). This has serious consequences for our nation's highways and productivity. Suppose we take a random sample of four Americans. Find the probability that the following numbers of people are sleep-deprived:

a. None

b. At least one

c. Between one and three, inclusive

d. Five

Solution

We apply the steps for solving binomial probability problems.

Step 1   We have a random sample of four Americans, so the number of trials is $n = 4$. "Success" is denoted as a particular American being sleep-deprived. The report states that 20% of Americans are sleep-deprived, so $p = 0.2$ and $q = 1 - 0.2 = 0.8$.

Step 2   For **(a)**, $X = 0$. For **(b)**, $X \geq 1$; that is, $X = 1, 2, 3$, or 4. For **(c)**, $1 \leq X \leq 3$; that is, $X = 1, 2$, or 3. For **(d)**, $X = 5$.

Step 3   We apply Step 3 for each of **(a)**–**(d)** as follows:

a.   *Step 3*   To find the probability that none ($X = 0$) of the Americans are sleep-deprived, we use the binomial probability formula:

$$P(X = 0) = ({}_4C_0)(0.2)^0\,(0.8)^{4-0} = (1)(1)(0.4096) = 0.4096$$

Therefore, the probability that none of the Americans in the sample are sleep-deprived is 0.4096.

b.   *Step 3*   Note that "at least one" includes all possible values of X except $X = 0$. In other words, the two events ($X = 0$) and ($X \geq 1$) are complements of each other. Therefore, from the formula for the probability for complements in Section 5.2 (page 260), we have

$$P(X \geq 1) = 1 - P(X = 0) = 1 - 0.4096 = 0.5904$$

The probability that at least one of the Americans is sleep-deprived is 0.5904.

c.   *Step 3*   We need to find the probability that either $X = 1$ or $X = 2$ or $X = 3$ of the Americans are sleep-deprived. Because these three values of X are mutually exclusive, we find the required probability by using the Addition Rule for Mutually Exclusive Events.

$$P(1 \leq X \leq 3) = P(X = 1 \text{ or } X = 2 \text{ or } X = 3)$$
$$= P(X = 1) + P(X = 2) + P(X = 3)$$

So we calculate the following:

$$P(X = 1) = ({}_4C_1)(0.2)^1\,(0.8)^{4-1} = (4)(0.2)(0.512) = 0.4096$$
$$P(X = 2) = ({}_4C_2)(0.2)^2\,(0.8)^{4-2} = (6)(0.04)(0.64) = 0.1536$$
$$P(X = 3) = ({}_4C_3)(0.2)^3\,(0.8)^{4-3} = (4)(0.008)(0.8) = 0.0256$$

Thus, $P(1 \leq X \leq 3) = 0.4096 + 0.1536 + 0.0256 = 0.5888$. The probability is 0.5888 that between one and three, inclusive, of the Americans in the sample of four are sleep-deprived.

d.   *Step 3*   In a binomial experiment, the number of successes X can never exceed the number of trials n. In other words, $X \leq n$, always. So, if our sample has only $n = 4$ Americans, then $P(X = 5) = 0$. It is not possible for there to be five Americans who are sleep-deprived.

NOW YOU CAN DO
Exercises 15–28.

YOUR TURN
#8

For a binomial experiment with $n = 3$ and $p = 0.5$, find the probability that X equals the following:

a.   0

b.   1

c.   At most 1

(The solutions are shown in Appendix A.)

As you can imagine, calculations involving binomial probabilities can sometimes get tedious. For example, to find the probability of observing at least 60 heads on 100 tosses of a fair coin, we would have to use the binomial formula for $X = 60$, $X = 61$, $X = 62$, and so on, right up to $X = 100$. For this type of problem, you can use Table B, Binomial Distribution, in the Appendix. If you are trying to answer a question involving unusual values of n, such as 103, or unusual values of p, such as 0.47, then you can use technology instead.

EXAMPLE 17 Finding probabilities using the binomial table

Use the binomial table and the binomial distribution from Example 16 to find the following probabilities:

a. No Americans are sleep-deprived.

b. At least one American is sleep-deprived.

Solution

a. From Example 16, we have a binomial distribution with $n = 4$ and $p = 0.2$. We next find n and p in the binomial table. In Figure 8:

- Look under the n column until you find $n = 4$. That is the portion of the table you will use.

- Then go across the top of the table until you get to $p = 0.20$.

- For part (**a**), $X = 0$, so go down the X column until you see 0 under the X column on the left (and in the subgroup with $n = 4$).

- The number in the p column is 0.4096 (see Figure 8), which is the same answer we calculated in Example 16(**a**).

					p	
n	X	0.10	0.15	0.20	0.25	0.30
2	0	0.8100	0.7225	0.6400	0.5625	0.4900
	1	0.1800	0.2550	0.3200	0.3750	0.4200
	2	0.0100	0.0225	0.0400	0.0625	0.0900
3	0	0.7290	0.6141	0.5120	0.4219	0.3430
	1	0.2430	0.3251	0.3840	0.4219	0.4410
	2	0.0270	0.0574	0.0960	0.1406	0.1890
	3	0.0010	0.0034	0.0080	0.0156	0.0270
4	0	0.6561	0.5220	0.4096	0.3164	0.2401
	1	0.2916	0.3685	0.4096	0.4219	0.4116
	2	0.0486	0.0975	0.1536	0.2109	0.2646
	3	0.0036	0.0115	0.0256	0.0469	0.0756
	4	0.0001	0.0005	0.0016	0.0039	0.0081

FIGURE 8
Excerpt from the binomial tables.

$X = 1$ →
$X = 2$ →
$X = 3$ →
$X = 4$ →

b. In this case, "at least 1" means 1 or 2 or 3 or 4. So, by the Addition Rule for Mutually Exclusive Events, find the probabilities for $X = 1$, $X = 2$, $X = 3$, and $X = 4$, and add them up. Using the same column with column head 0.20 in the table as in part (**a**), we add up the four probabilities.

$$P(X \geq 1) = P(X = 1) + P(X = 2) + P(X = 3) + P(X = 4)$$

$$= 0.4096 + 0.1536 + 0.0256 + 0.0016 = 0.5904$$

This is the same answer we calculated in Example 16(**b**), but it is arrived at in a different way.

Next, a word about *cumulative probability*. Cumulative probability refers to the probability of, *at most,* a particular value of X. For example, what is the probability that, at most, X = 2 Americans are sleep-deprived? This is the cumulative probability that X = 0, X = 1, or X = 2. Statistical software and the TI-83/84 graphing calculator each have a function that will find cumulative binomial probabilities for you.

EXAMPLE 18   Using technology to find binomial probabilities

Using the binomial distribution from Example 16, use the TI-83/84 and CrunchIt! to find the following probabilities:

a. P(X = 4), the probability that all four Americans are sleep-deprived.

b. P(X ≤ 2), the (cumulative) probability that, *at most,* two Americans are sleep-deprived.

Solution

We use the instructions in the Step-by-Step Technology Guide at the end of this section (page 336).

a. Figure 9 shows that we use the TI-83/84 function **binompdf** with n = 4, p = 0.2, and X = 4. Figure 10 shows the result: P(X = 4) = 0.0016. Figure 11 shows the same input and final answer using CrunchIt!.

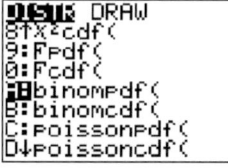

FIGURE 9 TI-83/84 menu.

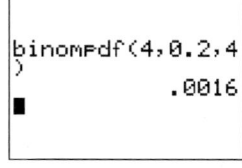

FIGURE 10 TI-83/84 result.

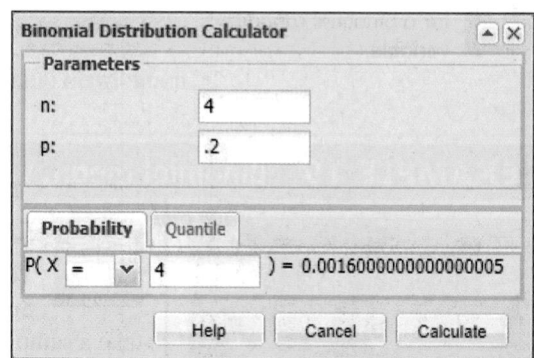

FIGURE 11 CrunchIt!

b. With the TI-83/84, we use the function **binomcdf** with n = 4, p = 0.2, and X = 2. Figure 12 shows the result: P(X ≤ 2) = 0.9728. Figure 13 shows the input and final answer using CrunchIt!.

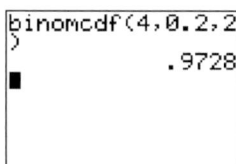

FIGURE 12 TI-83/84 result.

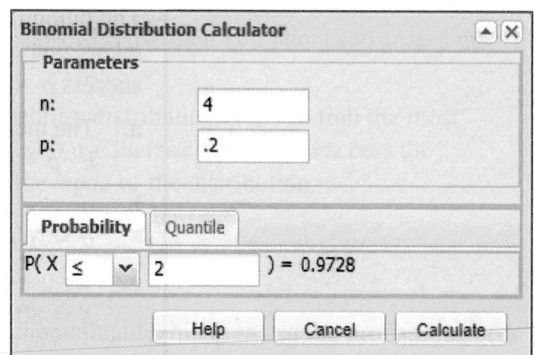

FIGURE 13 CrunchIt!

NOW YOU CAN DO
Exercises 29–48.

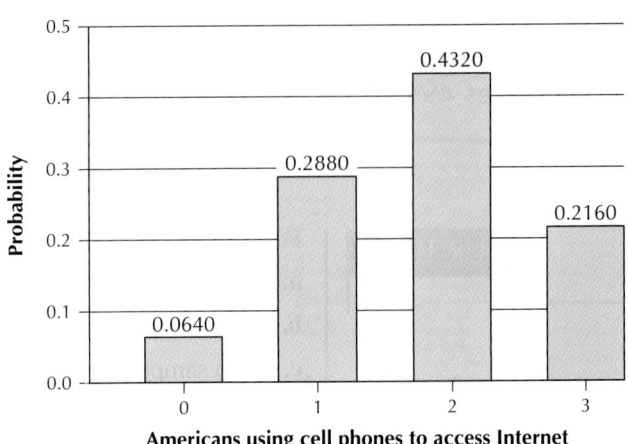

n	x	0.55	(0.60)
2	0	0.2025	0.1600
	1	0.4950	0.4800
	2	0.3025	0.3600
(3)	0	0.0911	0.0640
	1	0.3341	0.2880
	2	0.4084	0.4320
	3	0.1664	0.2160

FIGURE 14 Probabilities for $X = 0, 1, 2, 3$.

FIGURE 15 Probability distribution graph of X.

c. The most likely number of Americans using their cell phones to access the Internet is associated with the largest probability in the boxed section of Figure 14, 0.4320, which is $P(X = 2)$. Note from Figure 15 that $X = 2$ has the tallest bar of probability. Thus, $X = 2$ is the most likely number of American adults using their cell phones to access the Internet. We say that $X = 2$ is the *mode* of the distribution of X.

NOW YOU CAN DO
Exercises 53–56.

STEP-BY-STEP TECHNOLOGY GUIDE: Finding Binomial Probabilities

For Example 18 (page 333):

TI-83/84
Step 1 Press **2nd > DISTR** (the **VARS** key).
Step 2 Do one of **(a)** or **(b)**:
a. For individual binomial probabilities, highlight **binompdf(**, and press **ENTER**.

b. For cumulative binomial probabilities, highlight **binomcdf(**, and press **ENTER**.
Step 3 Enter the values for n, p, and K, separated by commas.
Step 4 Press **ENTER**. (See Figures 9 and 10 on page 333.)

EXCEL
Step 1 Select cell **A1**. Click the **Insert Function** icon f_x.
Step 2 In the **Search for a function** area, type BINOMDIST, click **Go** then **OK**.
Step 3 For **Number_s**, enter the number of successes, **K**. For **Trials**, enter the sample size, **n**. For **Probability_s**, enter the probability of success, **p**.
Step 4 Do one of **(a)** or **(b)**:

a. For individual binomial probabilities, next to **Cumulative**, enter **false**.
b. For cumulative binomial probabilities, next to **Cumulative**, enter **true**.
Step 5 Click **OK**. See Figures 16 and 17 for illustrations using Example 18.

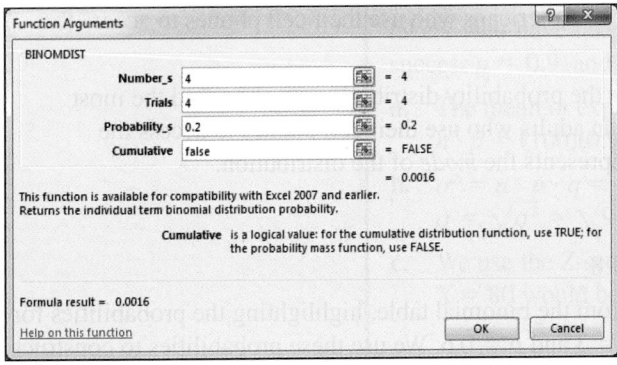

FIGURE 16 Example 18(a) using Excel.

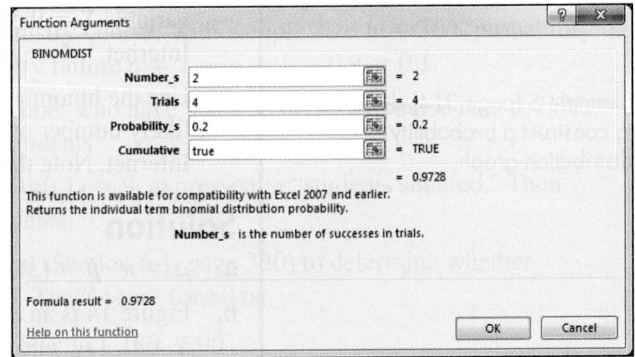

FIGURE 17 Example 18(b) using Excel.

MINITAB

Step 1   Click **Calc > Probability Distributions > Binomial...**
Step 2   Do one of (a) or (b):
a. For individual binomial probabilities, select **Probability** and enter the number of trials *n* and probability of success *p*.

b. For cumulative binomial probabilities, select **Cumulative Probability** and enter the number of trials *n* and probability of success *p*.
Step 3   Select **Input Constant**, enter *K* and click **OK**.

SPSS

Step 1   Input any one number into the first element of the spreadsheet. Rename the variable **Prob**.
Step 2   Select **Transform > Compute Variable....** In Target Variable, type **Prob**.
Step 3   Do one of (a) or (b):
a. For individual binomial probabilities, under **Function group** select **PDF & Noncentral PDF**. Under **Functions and Special Variables**, double-click **Pdf.Binom**. Replace the three question

marks with the number of successes **K**, the number of trials **n**, and the probability of success **p**, in that order.
b. For cumulative binomial probabilities, under **Function group** select **CDF & Noncentral CDF**. Under **Functions and Special Variables**, double-click **Cdf.Binom**. Replace the three question marks with the number of successes **K**, the number of trials **n**, and the probability of success **p**, in that order.
Step 3   Click **OK** and **OK**. Minimize the output window.

JMP

Step 1   **File > New > Data Table**. Double-click the first element of **Column 1**. Right-click the header of Column 1. Select **Column Info...** and under **Column Properties**, choose **Formula**. Click **Edit Formula**.
Step 2   Click **Discrete Probability** under **Functions (grouped)**.
Step 3   Do one of (a) or (b):

a. For individual binomial probabilities, select **Binomial Probability**. Input probability of success **p**, number of trials **n**, and number of successes **K**, in that order.
b. For cumulative binomial probabilities, select **Binomial Distribution**. Input probability of success **p**, number of trials **n**, and number of successes **K**, in that order.
Step 3   Click **OK** twice.

CRUNCHIT!

Step 1   Click **Distribution calculator > Binomial**.
Step 2   For **n**, enter **4**. For **p**, enter **0.2**.
Step 3   Do one of (a) or (b):
a. For individual binomial probabilities, select **=** from the drop-down menu next to **P(X**, and enter **4** in the box beside it. Click **Calculate**. See Figure 12 for the result.

b. For cumulative binomial probabilities, select **≤** from the drop-down menu, and enter **2**. Click **Calculate**. See Figure 13 for the result.

Section 6.2 Summary

1. The most important discrete distribution is the binomial distribution, where two outcomes are possible, each with probability of success *p* and *n* independent trials.

2. The probability of observing a particular number of successes can be calculated using the binomial

probability distribution formula, binomial tables, or technology.

3. There are formulas for finding the mean, variance, and standard deviation of a binomial random variable, *X*. The mode is the value of *X* with the largest probability.

Section 6.2 Exercises

CLARIFYING THE CONCEPTS

1. State the four requirements for a binomial experiment. (p. 327)

2. What is meant by a "success" in a binomial experiment? Is a success always a good thing? (p. 328)

3. In a binomial experiment, explain why it is not possible for *X* to exceed *n*. (p. 327)

4. Restate the binomial probability distribution formula using the following terms: $_nC_x$, the probability of success, the number of successes, the probability of failure, and the number of failures. (p. 330)

PRACTICING THE TECHNIQUES

 CHECK IT OUT!

To do	Check out	Topic
Exercises 5–14	Example 14	Recognizing binomial experiments
Exercises 15–28	Example 16	Applying the binomial probability distribution formula
Exercises 29–48	Examples 17 and 18	Finding binomial probabilities using table or technology
Exercises 49–52	Example 19	Binomial mean, variance, and standard deviation
Exercises 53–56	Example 20	The binomial mode: the most likely outcome of a binomial experiment

For Exercises 5–14, determine whether the experiment is binomial or not. If the experiment is binomial, identify the random variable X, the number of trials n, the probability of success p and the probability of failure q. If the experiment is not binomial, explain why not.

5. Ask 10 of your friends to come to your party (remember the independence assumption on page 327).

6. Toss a fair die three times, and note the total number of spots.

7. Answer a random sample of eight multiple-choice questions either correctly or incorrectly by random guessing. There are four choices, (a)–(d), for each question.

8. Toss a fair die three times, and note the number of 6s.

9. Select a student at random in the class until you come across a left-handed student.

10. Four cards are selected at random with replacement from a deck of cards, and the number of queens is observed.

11. Four cards are selected at random without replacement from a deck of cards, and the number of queens is observed.

12. Four cards are selected at random with replacement from a deck of cards, and the total number of blackjack-style points (number cards = number of points; face cards = 10 points; aces = either 1 or 11) is calculated.

13. Bob has paid to play two games at a carnival. The probability that he wins a particular game is 0.25.

14. Bob is playing a game at a carnival where he gets to play until he loses. The probability that he wins a particular game is 0.25.

For Exercises 15–28, calculate the probability of X successes for the binomial experiments with the following characteristics:

15. $n = 5, p = 0.25, X = 1$
16. $n = 5, p = 0.25, X = 0$
17. $n = 10, p = 0.5, X = 7$
18. $n = 10, p = 0.5, X = 8$
19. $n = 12, p = 0.9, X = 10$
20. $n = 12, p = 0.9, X = 11$
21. $n = 5, p = 0.25, X \leq 1$
22. $n = 5, p = 0.25, X \geq 1$
23. $n = 10, p = 0.5, X = 7$ or $X = 8$
24. $n = 10, p = 0.5, X = 7$ and $X = 8$
25. $n = 12, p = 0.9, X \geq 10$
26. $n = 12, p = 0.9, X < 10$ (*Hint:* Use the result from Exercise 25.)
27. $n = 12, p = 0.9, 9 \leq X \leq 12$
28. $n = 12, p = 0.9, 8 \leq X \leq 12$

According to the National Center for Education Statistics, business majors accounted for 25% of the proportion of all Master's degrees granted in 2012. For Exercises 29–34, the binomial experiment is to select three Master's degrees at random and to observe X = number of business majors. Calculate the indicated probabilities.

29. Observe no business majors
30. Observe one business major
31. Observe two business majors

32. Observe at most two business majors
33. Observe at least one business major
34. Observe between zero and two business majors, inclusive

A study by the Centers for Disease Control and Prevention (Use of Medication Prescribed for Emotional or Behavioral Difficulties Among Children Aged 6–17 Years in the United States, 2011–2012, by LaHeana Howie et al., NCHS Data Brief Number 148, April 2014) showed that 7.5% of children ages 6–17 used prescribed medication during the past six months for emotional or behavioral difficulties. For Exercises 35–40, the binomial experiment is to select four children ages 6–17 at random, and to find the probability that X takes the following values, where X = the number of children using prescription medication for emotional or behavioral difficulties:

35. $X = 3$
36. $X \geq 3$
37. $X = 0$
38. $X \leq 1$
39. $1 \leq X \leq 4$
40. $1 < X < 4$

According to the Current Population Survey, 10% of Americans ages 25–29 live alone. For Exercises 41–44, the binomial experiment is to take a random sample of five Americans ages 25–29 and observe X = the number living alone. Find the indicated probabilities.

41. None are living alone.
42. At least one is living alone.
43. At most two are living alone.
44. Six are living alone.

A 2014 study by the Harvard University Institute of Politics found that 40% of 18- to 29-year olds had a Twitter account. For Exercises 45–48, suppose we take a random sample of six 18- to 29-year-olds and find X = the number who have a Twitter account. Find the following probabilities:

45. $X = 6$
46. $X \leq 6$
47. $3 \leq X \leq 5$
48. $3 < X < 5$

For each of the following binomial experiments, do the following:
  a. Find and interpret the mean μ of X.
  b. Calculate the variance σ^2 of X.
  c. Compute the standard deviation σ of X.

49. Business majors accounted for 25% of the proportion of all Master's degrees granted in 2012. Select three Master's degrees at random. Let X = the number of business majors.

50. The CDC found that 7.5% of children ages 6–17 used prescribed medication during the past six months for emotional or behavioral difficulties. Select four children ages 6–17 at random, and let X = the number of children using prescription medication for emotional or behavioral difficulties.

51. Ten percent of Americans ages 25–29 live alone. Take a random sample of five Americans ages 25–29, and observe X = the number living alone.

52. Forty percent of 18- to 29-year-olds have a Twitter account. Take a random sample of six 18- to 29-year-olds, and find X = the number who have a Twitter account.

For each of the following binomial experiments, do the following:
   a. Construct the probability distribution graph of X.
   b. Identify the mode of X.
53. The binomial experiment in Exercise 49
54. The binomial experiment in Exercise 50
55. The binomial experiment in Exercise 51
56. The binomial experiment in Exercise 52

APPLYING THE CONCEPTS

57. Random Guessing on a Quiz. Suppose that you are taking a quiz of five multiple-choice questions (the instructor chose the questions randomly), with each question having four possible responses. You did not study at all for the quiz and will randomly guess the correct response for each question. The random variable X is the number of correct responses.
   a. If each question has four possible responses, why is this a valid binomial experiment?
   b. State the values of n and p.
   c. Calculate the probability that you will pass this quiz by correctly responding to at least three of the five questions. Is this good news for you?
   d. Use your answer to **(c)** to find the probability that you will not pass the quiz.

58. Women in Management. According to the U.S. Government Accountability Office, women hold 40% of the management positions in the United States.[3] Suppose we take a random sample of 20 people in management positions.
   a. Find the probability that the sample contains exactly 10 women.
   b. Find the probability that the sample contains, at most, one woman.
   c. Find the probability that the sample contains between 8 and 10 women, inclusive.

59. Abandoning Landlines. The Centers for Disease Control reported in 2014 that 41% of U.S. households use only cell phones (no landline) (*Source*: http://www.pewresearch.org/fact-tank/2014/07/08/two-of-every-five-u-s-households-have-only-wireless-phones/). Suppose we take a random sample of 12 telephone users.
   a. Find the probability that the sample contains exactly four users who use cell phones only.
   b. Find the probability that the sample contains, at most, four users who use cell phones only.
   c. Find the probability that the sample contains between four and six users inclusive who use cell phones only.

60. Online Dating. The Pew Research Center reported in 2014 that 22% of 25- to 34-year-olds have used online dating. Suppose we select fifteen 25- to 34-year-olds at random.

   a. Explain why we cannot use the binomial table to solve probability problems for this binomial experiment.
   b. Find the probability that the sample contains exactly three 25- to 34-year-olds who have used online dating.
   c. Find the probability that the sample contains at most three 25- to 34-year-olds who have used online dating.

61. Random Guessing on a Quiz. Refer to Exercise 57.
   a. Compute the mean, variance, and standard deviation of X. Interpret the mean.
   b. Use the Z-score method to determine which numbers of correct responses should be considered outliers.
   c. Use technology or the binomial table to construct a probability distribution graph of X. Then state the mode of X, that is, the most likely number of correct responses.
   d. Find the probability that X = the mode.

62. Women in Management. Refer to Exercise 58.
   a. Find the mean, variance, and standard deviation of the number of women in management positions.
   b. Suppose that the sample contains six women in management positions. Use the Z-score method to determine whether this outcome is unusual or not.
   c. Use technology or the binomial table to determine the most likely number of women in management positions.
   d. Compute the probability that the sample contains the mode number of women in management positions.

63. Abandoning Landlines. Refer to Exercise 59.
   a. Calculate the mean, variance, and standard deviation of the number of users in the sample who have abandoned their landlines. Interpret the mean.
   b. Suppose the sample contains no users who have abandoned their landlines. Is this outcome unusual or an outlier? Use the Z-score method to find out.

64. Online Dating. Refer to Exercise 60.
   a. Find the mean, variance, and standard deviation of the number of 25- to 34-year-olds who have used online dating.
   b. Suppose that the sample contains eight 25- to 34-year-olds who have used online dating. Use the Z-score method to determine whether this outcome is unusual or not.

65. Women and Depression. According to the National Institute of Mental Health, nearly twice the proportion of women (12%) as men (6.6%) are affected by a depressive disorder each year. Suppose that random samples of five women and five men are taken. Let X represent the number of women affected by a depressive disorder.
   a. Find and interpret the mean of X.
   b. If possible, find the probability that X equals the mean. If not possible, explain why it is not possible to do so.
   c. Construct the probability distribution graph of X, and identify the mode of X.
   d. Find the probability that X equals the mode of X.

66. Men and Depression. Refer to Exercise 65. Let Y represent the number of men affected by a depressive disorder in a random sample of size 5.
   a. Find and interpret the mean of Y.
   b. If possible, find the probability that Y equals the mean. If not possible, explain why it is not possible to do so.
   c. Construct the probability distribution graph of Y, and identify the mode of Y.
   d. Find the probability that Y equals the mode of Y.

67. Mean, Median, Mode. For a binomial distribution, if the mean $\mu = n \cdot p$ is a whole number, then mean of $X =$ median of $X =$ mode of X. Use this equation to answer the following questions:
   a. Find the median of X for the binomial distribution in Example 19.
   b. Find the mode of X for the binomial distribution in Example 19.
   c. What is the most likely value of X for the binomial distribution in Example 19?

68. Geometric Probability Distribution. Refer to Example 14(**a**), where a fisherman is going fishing and will continue to fish until he catches a rainbow trout. This is an example of the *geometric probability distribution,* which has the same requirements as the binomial distribution, except that there is not a fixed number of trials n. Instead, the geometric random variable X represents the number of trials until a success is observed. The geometric probability distribution formula is

$$P(X) = p(1 - p)^{X-1}$$

where p represents the probability of success. The possible values of X are $X = 1, 2, 3, \ldots$. The U.S. Census Bureau reported in 2010 that 30% of U.S. households have no access at all to the Internet. A random sample is taken of U.S. households. Let the random variable X represent the number of trials until a household is found that has access to the Internet.
   a. Find the probability that $X = 1$, that is, the first household sampled has access to the Internet.
   b. Find the probability that $X = 2$, that is, the first household sampled does not have access, but the second household sampled does have access to the Internet.
   c. Find the probability that $X = 3$, that is, the first two households sampled do not have access, but the third household sampled does have access to the Internet.

69. Hypergeometric Probability Distribution. If samples are drawn from a relatively small finite population, and the sample size is larger than 1% of the population, so that the 1% Guideline (page 282) does not apply, we should not use the binomial distribution because the samples are not independent. Instead, if we are sampling without replacement, and there are two mutually exclusive categories, then you should use the *hypergeometric probability distribution.* Suppose that N_1 objects belong to the first category ("successes"), and N_2 objects belong to the second

category ("failures"). Then the probability of getting X successes and $n - X$ failures is given by the hypergeometric probability distribution formula:

$$P(X) = \frac{\left(_{N_1}C_X\right)\left(_{N_2}C_{n-X}\right)}{\left(_NC_n\right)}$$

where $N_1 + N_2 = N$, N is the population size, and n is the sample size. You are dealt 5 cards at random from a deck of 52 cards.
   a. Find the probability that all 5 cards are spades.
   b. Find the probability that exactly 4 cards are spades.
   c. Find the probability that at least 4 cards are spades.
   d. Find the probability that exactly 3 cards are spades.
   e. Find the probability that, at most, 2 cards are spades.

70. Multinomial Distribution. The *multinomial probability distribution* is similar to the binomial distribution, except that the binomial involves only two categories, whereas the multinomial involves more than two categories. Suppose we have three mutually exclusive outcomes, A, B, and C, where $p_A = P(A)$, $p_B = P(B)$, and $p_C = P(C)$. If we have a sample of n independent trials, then the probability that we get X_A outcomes of category A, X_B outcomes of category B, and X_C outcomes of category C is given by the following formula:

$$P(X_A, X_B, X_C) = \frac{n!}{X_A! \, X_B! \, X_C!} \cdot P_A^{X_A} \cdot P_B^{X_B} \cdot P_C^{X_C}$$

Suppose that 30% of students on a particular college campus are Democrats, 30% are Republicans, and 40% are Independents. Suppose we take a random sample of 10 students.
   a. Find the probability that 3 are Democrat, 3 are Republican, and 4 are Independent.
   b. Find the probability that 3 are Democrat, 4 are Republican, and 3 are Independent.
   c. Find the probability that 4 are Democrat, 3 are Republican, and 3 are Independent.

BRINGING IT ALL TOGETHER

Small Business Jobs. According to the U.S. Small Business Administration, small businesses provide 75% of the net new jobs added to the economy. Consider a random sample of 10 new jobs. Let X represent the number of the new jobs added to the economy that are provided by small businesses. Use this information for Exercises 71–79.

71. Confirm that this situation represents a binomial experiment.

72. Use the binomial distribution formula to find the following probabilities:
   a. That 7 new jobs are provided by small businesses
   b. That 8 new jobs are provided by small businesses

73. Use the binomial tables or technology to find the following probabilities. Then explain why the probabilities in (**a**) and (**b**) are equal, as are the probabilities in (**c**) and (**d**).
   a. $P(X > 8)$
   b. $P(X \geq 9)$
   c. $P(X \leq 4)$
   d. $P(X < 5)$

74. Find the following parameters of the binomial distribution for this experiment. Interpret the mean and standard deviation.
  a. Mean μ
  b. Variance
  c. Standard deviation

75. If possible, find the probability that X equals μ. If not possible, explain why it is not possible to do so.

76. Construct the probability distribution graph of X.

77. Identify the mode of X.

78. Find the probability that X equals the mode of X.

79. Following Example 19(c), identify all values of this binomial distribution that are unusual or somewhat unusual.

WORKING WITH LARGE DATA SETS

Motor Vehicle Fuel Efficiency. Open the data set **FuelEfficiency**. We will explore some characteristics of this data set, using the tools and techniques we have learned in this section. Use technology to do Exercises 80–84.

 fuelefficiency

80. Examine the variable *class*. Make this into a binomial random variable as follows: Consider all vehicles that are compact cars to be a success, and all other vehicles to be a failure. If we use all 1141 vehicles, what is the probability of success?

81. If we take a random sample of 100 vehicles, what is the expected number of compact cars?

82. If we take a random sample of 100 vehicles, what is the standard deviation of the number of compact cars?

83. Use technology to obtain a random sample of 100 vehicles. How many compact cars are in the sample?

84. Use the population mean and standard deviation from Exercises 81 and 82 to determine whether the observed number of compact cars in your sample in Exercise 83 is unusual, using the Z-score method.

6.3 Poisson Probability Distribution

OBJECTIVES By the end of this section, I will be able to . . .

1 Explain the requirements for the Poisson probability distribution.

2 Compute probabilities for a Poisson random variable.

3 Calculate the mean, variance, and standard deviation of a Poisson random variable.

4 Use the Poisson distribution to approximate the binomial distribution.

1 Requirements for the Poisson Distribution

The Poisson distribution was developed in 1838 by Siméon Denis Poisson (1781–1840), a French mathematician and physicist, who published more than 300 works in mathematical physics.

The Poisson distribution, like the binomial distribution, is a discrete probability distribution. The Poisson probability distribution is used when we wish to find the probability of observing a certain number of occurrences (X) of a particular event within a fixed interval of space or time. For example, the number of calls X per hour to a 911 emergency center follows a Poisson distribution, as does the number of typographical errors X per chapter in a book.

> **Poisson Probability Distribution**
>
> The **Poisson probability distribution** is a discrete probability distribution that is used when observing the number of occurrences of an event within a fixed interval of space or time. The random variable X represents the number of occurrences of the event in the interval.
>
> **Requirements for the Poisson Probability Distribution**
>
> **1.** The occurrences must be random.
> **2.** Each occurrence must be independent.
> **3.** The occurrences must be uniformly distributed over the given interval.

For a binomial random variable, the maximum number of successes X is the number of trials n. But for the Poisson random variable, there is no upper limit to the number of occurrences X. For example, there is no upper limit to the number of calls X per hour to a 911 emergency center.

EXAMPLE 21 Recognizing when to use the Poisson distribution

For each of the following situations, state whether or not the random variable X follows a Poisson probability distribution. If not, state why not.

a. X is the number of telephone poles along a particular one-mile stretch of highway.

b. X is the number of rabbits living in a particular five-acre plot of land scheduled for development.

c. X is the number of calls to a radio station phone line in the hour following the announcement of a two-free-ticket giveaway to a sold-out concert.

d. X is the number of customers who purchase gasoline at a particular filling station from 2 P.M. to 2:15 P.M.

Solution

a. X does not follow a Poisson distribution. Telephone poles are spaced equidistantly from each other. Knowing the distance between the first two poles gives us the distance to each of the succeeding poles. This violates the requirement that the occurrences be random.

b. X does not follow a Poisson distribution. Related rabbits live together in warrens or dens. Thus, if there is one rabbit living within the specified area, there is probably more than one. This violates the independence requirement.

c. X does not follow a Poisson distribution. Presumably, the radio station would be inundated with calls within the first few minutes of the announcement. Later, there would be fewer calls to the station. This violates the requirement that the occurrences be uniformly distributed throughout the interval.

d. X does follow a Poisson distribution. The customers are random, independent, and occur uniformly over the 15-minute period.

NOW YOU CAN DO
Exercises 5–8.

2 Computing Probabilities for a Poisson Random Variable

In Section 6.2, we used the binomial probability distribution formula to compute binomial probabilities. Similarly, we may use the following formula for calculating probabilities for a Poisson random variable X.

Poisson Probability Distribution Formula

If the requirements are met, the probability that a particular event occurs X times within a given interval is

$$P(X) = \frac{\mu^X \cdot e^{-\mu}}{X!}$$

where

μ = the mean of the Poisson probability distribution

e = a constant approximately equal to 2.718281828

X = the number of occurrences of the event within the interval

---

EXAMPLE 22 Finding probabilities using the Poisson distribution

A study was done of the number of cardiac arrests to occur per week in a particular hospital of 850 beds over a period of 5 years.[4] The number of cardiac arrests fulfills the requirements for the Poisson probability distribution and has a mean of $\mu = 1.09$ cardiac arrests per week. Calculate the following probabilities:

a. The probability of two cardiac arrests in a given week

b. The probability of fewer than two cardiac arrests in a week

c. The probability of at most two cardiac arrests in a week

d. The probability of more than two cardiac arrests in a week

e. The probability of at least three cardiac arrests in a week

Solution

The number of cardiac arrests fulfills the requirements for the Poisson probability distribution, so we may use the Poisson probability distribution formula to calculate the desired probabilities. To use this formula, we must determine the values of μ and X. For each of (a)–(e), we have $\mu = 1.09$. The Poisson distribution with $\mu = 1.09$ for $x = 1 \ldots 6$ is shown in Figure 18.

FIGURE 18
The Poisson distribution for $\mu = 1.09$. (Note that there is no upper limit to the value that X may take, but that the probabilities for $X \geq 6$ are very small.)

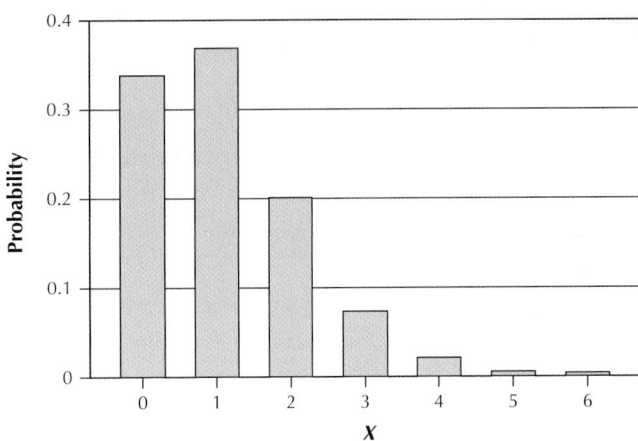

a. Here, $X = 2$, so the probability that $X = 2$ is

$$P(2) = \frac{1.09^2 \cdot e^{-1.09}}{2!} \approx 0.1997$$

b. The Poisson distribution is a discrete distribution, so that "fewer than 2" means $X = 0$ or $X = 1$. We thus find $P(0)$ and $P(1)$, and add the resulting probabilities to arrive at the answer.

$$P(0) = \frac{1.09^0 \cdot e^{-1.09}}{0!} \approx 0.3362 \qquad P(1) = \frac{1.09^1 \cdot e^{-1.09}}{1!} \approx 0.3665$$

So the probability of fewer than two cardiac arrests in a week equals $0.3362 + 0.3665 = 0.7027$.

c. The phrase "at most 2" means $X = 0$ or $X = 1$ or $X = 2$. We thus find $P(0)$, $P(1)$, and $P(2)$ and add the resulting probabilities to arrive at the answer. We have already found $P(0)$ and $P(1)$ from part (b) and have found $P(2)$ from part (a). Thus, the probability of, at most, two cardiac arrests in a week equals

$$P(X \leq 2) = P(0) + P(1) + P(2) = 0.3362 + 0.3665 + 0.1997 = 0.9024$$

d. The phrase "more than 2" means $X = 3$, $X = 4$, and so on to infinity. In Figure 18, these probabilities are represented by the rectangles for $X = 3$, $X = 4$, and so on. Now the Law of Total Probability says that the sum of the probabilities of all values of X equals 1. Thus, the probabilities represented by the rectangles for $X = 3$, $X = 4$, and so on equals 1 minus the sum of the rectangles for $X = 0$, $X = 1$, and $X = 2$. In other words, $P(X > 2) = 1 - P(X \leq 2)$. We found $P(X \leq 2) = 0.9024$ in part **(c)**. Thus, $P(X > 2) = 1 - 0.9024 = 0.0976$.

e. The phrase "at least 3" means $X = 3$, $X = 4$, and so on to infinity. These are the same X-values represented in part **(d)** by the phrase "more than 2." Thus, $P(X \geq 3) = P(X > 2) = 0.0976$.

NOW YOU CAN DO
Exercises 9–14.

YOUR TURN #11

Suppose the number of customers to a boutique crafts shop follows a Poisson distribution with mean $\mu = 10$ per hour. Find the following probabilities:

a. Getting 10 customers in an hour

b. Getting 12 customers in an hour

(The solutions are shown in Appendix A.)

3 The Mean, Variance, and Standard Deviation for a Poisson Distribution

Just as the binomial distribution has a mean, variance, and standard deviation, we can calculate these quantities for a Poisson probability distribution. The mean, variance, and standard deviation for a Poisson distribution are as follows.

Parameters of the Poisson Distribution

$$\text{Mean} = \mu$$
$$\text{Variance} = \sigma^2 = \mu$$
$$\text{Standard deviation} = \sigma = \sqrt{\mu}$$

EXAMPLE 23 Applying the mean and standard deviation

Zillow.com indicated that the number of homes for sale in Storrs, Connecticut, in July 2014 was 25. Suppose that the number of homes for sale in Storrs meets the requirements for the Poisson probability distribution, with mean $\mu = 25$.

a. Calculate the mean, variance, and standard deviation of the number of homes for sale in Storrs.

b. Determine which values of $X =$ the number of homes for sale in Storrs would be considered moderately unusual.

Solution

a. Mean $= \mu = 25$. Variance $= \sigma^2 = \mu = 25$. Standard deviation $= \sigma = \sqrt{\mu} = \sqrt{25} = 5$.

b. Recall from Section 3.4 (page 159) that a data value farther than 2 standard deviations from the mean is considered moderately unusual. The numbers of homes that lie 2 standard deviations above and below the mean are calculated as follows:

$$\mu + 2\sigma \approx 25 + (2)5 = 35$$

$$\mu - 2\sigma \approx 25 - (2)5 = 15$$

NOW YOU CAN DO
Exercises 15–18.

Thus, if 15 or fewer homes were sold in Storrs in 1 month, this would be considered moderately unusual. Similarly, it would be moderately unusual if 35 or more homes were sold.

YOUR TURN

#12

Suppose the number of customers to a boutique crafts shop follows a Poisson distribution with mean $\mu = 10$ per hour.

a. Calculate the mean, variance, and standard deviation of the number of customers per hour.

b. Determine which values of $X =$ the number of customers per hour would be considered unusual.

(The solutions are shown in Appendix A.)

4 Using the Poisson Distribution to Approximate the Binomial Distribution

We can use the Poisson distribution to approximate the binomial distribution when the number of trials n is large and the probability of success p is small, as measured by the following requirements:

> **Requirements for Using the Poisson Distribution to Approximate the Binomial Distribution**
>
> $$n \geq 100 \quad \text{and} \quad np \leq 10$$
>
> where n is the number of trials and p is the probability of success for the binomial distribution.

If the requirements are met, then the mean of the Poisson distribution used to approximate the binomial distribution is given as

$$\mu = np$$

EXAMPLE 24 Poisson approximation to the binomial distribution

Two percent of online e-commerce transactions are fraudulent.[5] **(a)** In a sample of 100 online e-commerce transactions, approximate the probability that three fraudulent transactions occur. **(b)** Measure the accuracy of the approximation.

Solution

a. We first verify that the requirements are met. The number of trials is $n = 100 \geq 100$. The probability of "success" (that is, fraud) on any particular transaction is $p = 0.02$, so that $np = (100)(0.02) = 2 \leq 10$. The requirements are met. Next, we find the mean of the Poisson distribution used to approximate the binomial distribution: $\mu = np = (100)(0.02) = 2$. Then the probability that $X = 3$ fraudulent transactions occur is

d. Approximate the probability that chromosome 1 will have at least one mutation.

e. Approximate the probability that chromosome 1 will have at most two mutations.

26. **Due-Date Babies.** Only 4% of babies are born on their due dates.[9] Massachusetts General Hospital in Boston assists in the delivery of 142 babies per 2-week period.[10] Do the following:

   a. Verify that the requirements are met for using the Poisson distribution to approximate the binomial distribution.

   b. Find the mean of the Poisson distribution used to approximate the binomial distribution.

   c. Approximate the probability that $X = 2$ babies will be born on their due dates during a two-week period.

   d. Approximate the probability that at least one baby will be born on its due date during a two-week period.

27. **Red Sox Runs.** In 2013, the Boston Red Sox led all Major League Baseball teams with a mean number of runs per game of $\mu = 5.19$ (*Source:* www.teamrankings.com). Let X refer to the number of runs per game for the Boston Red Sox. Find the following probabilities:

   a. Probability that $X = 5$

   b. $P(X = 4)$

   c. Probability that the Red Sox get shut out (score zero runs)

   d. Probability that $X = \mu$

28. **Broward Burglaries.** The mean number of burglaries taking place at the Central Campus of Broward Community College in Davie, Florida, is $\mu = 23.3$ per year.[11] Let X refer to the number of burglaries taking place on that campus in a year. Find the following probabilities:

   a. Probability that $X = 23$

   b. Probability that X is either 23 or 24

   c. Probability that no burglaries occur

   d. Probability that $X = \mu$

29. **Football Fatalities.** The mean number of fatalities from playing high school football in the United States is 3.8.[12] Let X refer to the number of fatalities from playing high school football in one year. Find the following probabilities:

   a. Probability that $X = 3$

   b. $P(X = 2)$

   c. Probability that no fatalities occur in a year

   d. Probability that $X > 3$

30. **Social Media–Driven Holiday Sales.** IBM reported that 1% of Black Friday purchases on ecommerce Web sites were driven directly by social media. Suppose that we take a sample of 200 such purchases. Do the following:

   a. Verify that the requirements are met for using the Poisson distribution to approximate the binomial distribution.

   b. Find the mean of the Poisson distribution used to approximate the binomial distribution.

   c. Approximate the probability that $X = 1$ purchase was driven by social media.

   d. Approximate the probability that at least one purchase was driven by social media.

31. **Teens in Virtual Worlds.** According to a Pew Research study, 10% of teenagers participate in online virtual worlds such as Second Life, Gaia, or Habbo Hotel.[13] Suppose that we take a sample of 100 teenagers. Do the following:

   a. Verify that the requirements are met for using the Poisson distribution to approximate the binomial distribution.

   b. Find the mean of the Poisson distribution used to approximate the binomial distribution.

   c. Approximate the probability that $X = 10$ teenagers participate in such online virtual worlds.

   d. Approximate the probability that either 10, or 11, or 12 teenagers participate in such online virtual worlds.

32. **Challenge Exercise.** For the given exercise, identify the values of X that would be considered (i) moderately unusual, and (ii) outliers.

   a. Exercise 27

   b. Exercise 28

   c. Exercise 29

   d. Exercise 30

6.4 Continuous Random Variables and the Normal Probability Distribution

OBJECTIVES By the end of this section, I will be able to . . .

1   Identify a continuous probability distribution and state the requirements.

2   Calculate probabilities for the uniform probability distribution.

3   Explain the properties of the normal probability distribution.

4   Find areas under the standard normal curve, given a Z-value.

5   Compute the standard normal Z-value, given an area.

Sections 6.1–6.3 dealt with discrete random variables, such as the binomial random variable. Next, we turn to continuous random variables.

1 Continuous Probability Distributions

Continuous random variables assume infinitely many possible values, with no gap between the values. For example, the height of a randomly chosen classmate of yours is a continuous random variable because it can take an infinite number of possible values.

For a given continuous random variable X, we are not interested in whether X equals any particular value. Instead, we are interested in whether X is

- greater than a particular value, or

- less than a particular value, or

- between two particular values.

That is, we are interested in whether X is located in an *interval*.

We are not interested in the probability that X equals some particular value because this probability *always equals zero*. If this sounds crazy, then consider the following example: How much soda does a "12-ounce can" of soda actually contain? Are you sure it's 12 ounces and not 11.99999999 ounces? Or could it contain 12.00000001 ounces? In fact, the can could contain any of the infinite number of possible amounts of soda, say between 11.9 and 12.1 ounces (see Figure 19). Thus, any given amount of soda in the can is so unlikely that the probability that you will get exactly 12.00000000 ounces of soda in your 12-ounce can is essentially zero.

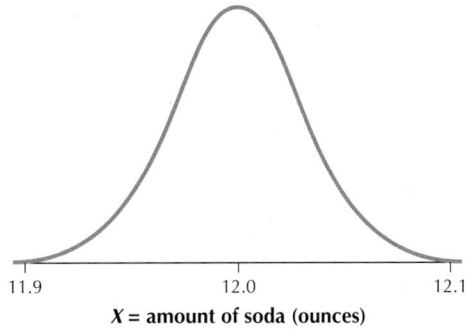

In contrast to the graph for a discrete distribution, the graph for a continuous probability distribution is "smooth" because it represents probability at infinitely many points along an interval.

FIGURE 19 X = amount of soda is a continuous random variable with a continuous probability distribution.

The graph in Figure 19 is called a **continuous probability distribution**, defined as follows.

Continuous Probability Distribution

A **continuous probability distribution** is represented by a graph that indicates on the horizontal axis the range of values that the continuous random variable X can take, and above which is drawn a curve, called the **density curve**. A continuous probability distribution must meet the following requirements.

Requirements for a Continuous Probability Distribution

1. The total area under the density curve must equal 1 (this is the **Law of Total Probability for Continuous Random Variables**).

2. The vertical height of the density curve can never be negative. That is, the density curve never goes below the horizontal axis.

---

2 Calculating Probabilities for the Uniform Probability Distribution

To learn how to calculate probabilities for continuous random variables, we turn to the **uniform probability distribution**.

> The **uniform probability distribution** is a continuous distribution that has constant probability from left endpoint *a* to right endpoint *b*. Its curve is a flat, straight line, so that the shape of the uniform distribution is a rectangle.

For example, suppose the waiting time X for the campus shuttle bus follows a uniform distribution, with waiting times ranging from $a = 0$ minutes to $b = 10$ minutes. Then the uniform probability distribution is given in Figure 20.

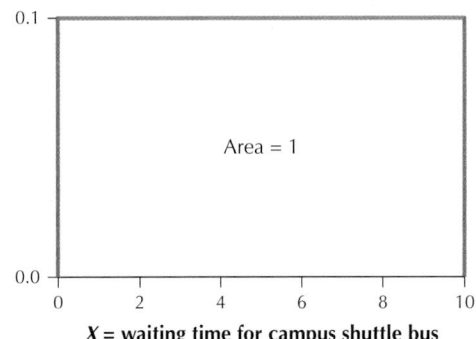

X = waiting time for campus shuttle bus

FIGURE 20 Waiting time *X* has a rectangular shape.

Note that the width of the rectangle in Figure 20 is $b - a = 10 - 0 = 10$. The total area under the density curve must equal 1 by the Law of Total Probability for Continuous Distributions; therefore, the height of the rectangle must equal $1/10 = 0.1$.

So how do we represent probability for the uniform distribution, or for continuous distributions in general?

> **Probability for Continuous Distributions**
>
> The probability that a continuous random variable *X* takes a value in an interval is equal to the *area under the density curve above that interval.*

EXAMPLE 25 Uniform probability distribution

Using the uniform probability distribution in Figure 20, calculate the probability that you will wait the following amount of time for the campus shuttle bus:

a. Between 2 and 4 minutes

b. More than 6 minutes

c. Exactly 8 minutes

Solution

a. We are interested in the interval between $X = 2$ and $X = 4$ minutes. The area above this interval forms a rectangle, shown in Figure 21. The area of this green rectangle represents the probability that X is between 2 and 4 minutes. The base of the rectangle equals $b - a = 4 - 2 = 2$. The height of the rectangle equals 0.1, so we find that the area of this rectangle is

$$\text{area} = \text{base} \times \text{height} = 2 \times 0.1 = 0.2$$

Because *area represents probability,* we conclude that the probability is 0.2 that you will wait between 2 and 4 minutes for the campus shuttle bus.

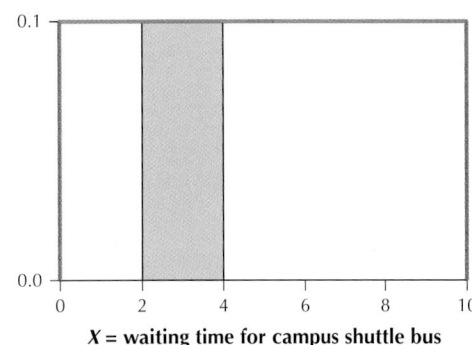

FIGURE 21
Probability that X is between 2 and 4 equals the area of the green rectangle.

b. The assumption that the distribution follows a uniform distribution, with waiting times ranging from $a = 0$ minutes to $b = 10$ minutes, means that the maximum waiting time is 10 minutes. Thus, we are interested in the interval between $X = 6$ and $X = 10$. The base of this rectangle equals $b - a = 10 - 6 = 4$. Multiplied by the height of the rectangle, 0.1, the resulting area equals $= 4 \times 0.1 = 0.4$. Because area represents probability, this means that the probability we will wait between 6 and 10 minutes equals 0.4.

c. Here, we are not given an interval, only a single point, exactly 8 minutes. We can express this as the "interval" from 8 to 8, so that both $a = 8$ and $b = 8$. Thus, the width of this "interval" is $b - a = 8 - 8 = 0$. Multiplied by the height of the rectangle, 0.1, gives us an area of $0 \times 0.1 = 0$. Probability equals area, so the probability that we will wait *exactly* 8 minutes (and not 7.99999 minutes or 8.000001 minutes) is zero. This is an example of our earlier discussion where we learned that, for continuous distributions, *the probability that X exactly equals some particular value is always zero.*

NOW YOU CAN DO
Exercises 11–20.

YOUR TURN
#13

For the scenario in Example 25(**a**), find the probability that you will wait between 4 and 8 minutes for the campus shuttle.

(The solution is shown in Appendix A.)

Notice from Example 25 that the probability 0.2 equals $\frac{4 - 2}{10 - 0}$. We generalize this as follows.

> The probability that a uniform random variable with left endpoint a and right endpoint b takes a value in the interval $[c, d]$ is given by
>
> $$P(c \leq X \leq d) = \frac{d - c}{b - a}$$

For example, the probability that you would wait between $c = 0$ and $d = 5$ minutes for the campus shuttle bus is

$$P(0 \leq X \leq 5) = \frac{5 - 0}{10 - 0} = 0.5$$

Now, because X is a continuous random variable, $P(X = 0) = 0$ and $P(X = 5) = 0$. Thus, $P(0 \leq X \leq 5) = P(0 < X < 5)$. In fact, for any continuous random variable, the inequalities $\leq$ and $<$ are interchangeable, as are $\geq$ and $>$.

3 Introduction to Normal Probability Distribution

We now turn to what is considered to be the most important probability distribution in the world: the **normal probability distribution**. Sometimes referred to as the bell-shaped curve (Chapter 3), the normal distribution is a continuous distribution that has been found to model accurately such phenomena as

- the amount of rainfall in Imperial Valley, California;

- the heights and weights of high-risk infants in New York City; and

- the errors in manufacturing machine bolts in a Pennsylvania factory.

Remember that, as with all probability distributions, we are dealing with a *population* of data values.

Similar to a discrete random variable, a continuous random variable has a mean and a standard deviation. The parameters of the normal distribution are the mean μ, which determines the center of the distribution on the number line, and the standard deviation σ, which determines the spread or shape of the distribution curve. The mean μ can be positive, negative, or zero; the standard deviation σ can never be negative.

From Figure 22, we can see that the normal distribution curve is symmetric about μ. If you slice the curve neatly in half at the mean μ, the result will be two pieces that are perfect mirror images of each other, as in Figure 22.

FIGURE 22
The normal distribution is symmetric about its mean μ.

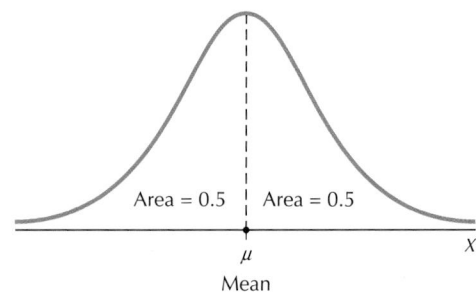

Area = 0.5   Area = 0.5

μ
Mean

X

Properties of the Normal Density Curve (Normal Curve)

1. It is symmetric about, and centered at, the mean μ.

2. The highest point occurs at $X = \mu$ because symmetry implies that the mean equals the median, which equals the mode of the distribution.

3. The total area under the curve equals 1.

4. Symmetry also implies that the area under the curve to the left of μ and the area under the curve to the right of μ are both equal to 0.5 (Figure 22).

5. The normal distribution is defined for values of X extending indefinitely in both the positive and negative directions. As X moves farther from the mean, the curve approaches but never quite touches the horizontal axis.

6. Values of X are always found on the horizontal axis. Probabilities are represented by areas under the curve.

EXAMPLE 26   Normal distribution mean and standard deviation

a. Figure 23 shows two normal distributions, with different means but the same standard deviation. Which distribution has mean $\mu = 6$ and which distribution has mean $\mu = 2$?

b. Figure 24 shows two normal distributions, with the same mean but different standard deviations. Which distribution has $\sigma = 1$ and which distribution has $\sigma = 2$?

Solution

a. Note that the two distributions have precisely the same spread or shape because each distribution has the same standard deviation, $\sigma = 2$. However, the yellow distribution is symmetric about an axis drawn at 2, and it is centered at 2. Therefore, it has mean $\mu = 2$. The green distribution is symmetrical about, and centered at $\mu = 6$. The two curves are essentially identical, with the green one shifted four units to the right.

b. Because σ is a measure of spread, the larger the value of σ, the more spread out the distribution of X will be. This is illustrated in Figure 24. The normal distribution with the smaller standard deviation ($\sigma = 1$) has a curve with a higher peak in the center and thinner "tails" than the distribution with a larger standard deviation ($\sigma = 2$). Thus, the green distribution has $\sigma = 1$ and the yellow distribution has $\sigma = 2$.

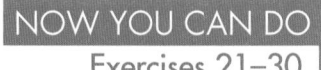

NOW YOU CAN DO
Exercises 21–30.

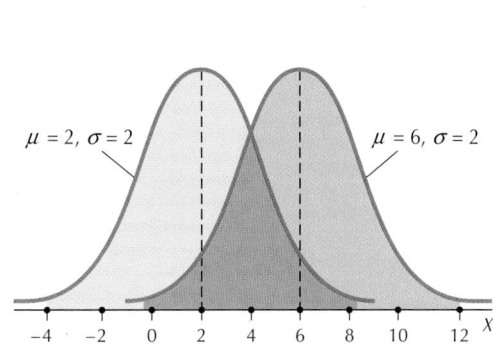

FIGURE 23 Different μ, same σ.

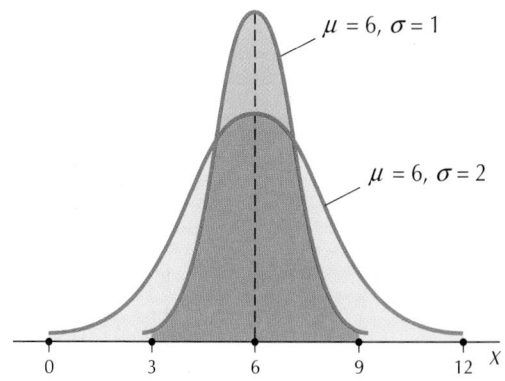

FIGURE 24 Same μ, different σ.

EXAMPLE 27   Properties of the normal curve

A statistical study found that when nurses made home visits to pregnant teenagers to provide support services, discourage smoking, and otherwise provide care, the mean birth weight of the babies was higher for this treatment group (3285 grams) than for a control group of teenagers who were not visited (2922 grams), when the visits began before midgestation.[14] The birth weights of babies are known to follow a normal distribution.[15]

Suppose the birth weights for the babies whose mothers were visited by the nurses (treatment group) also follow a normal distribution. Then our random variable is

$$X = \text{birth weight of babies in the treatment group}$$

The mean is $\mu = 3285$ grams. Assume that the standard deviation is $\sigma = 500$ grams. Graph the normal curve of $X =$ birth weights and describe some properties of this distribution.

Solution

Hint: Draw a bell-shaped curve with center at $\mu = 3285$. Label the horizontal axis in increments equal to the standard deviation $\sigma = 500$. Make sure the areas to the left and right of μ are equal.

Figure 25 shows the probability graph of $X =$ birth weights. Note that the curve has the following properties:

1. It is symmetric about the mean $\mu = 3285$ grams.

2. The highest point occurs at $\mu = 3285$ grams, which is also the median and the mode.

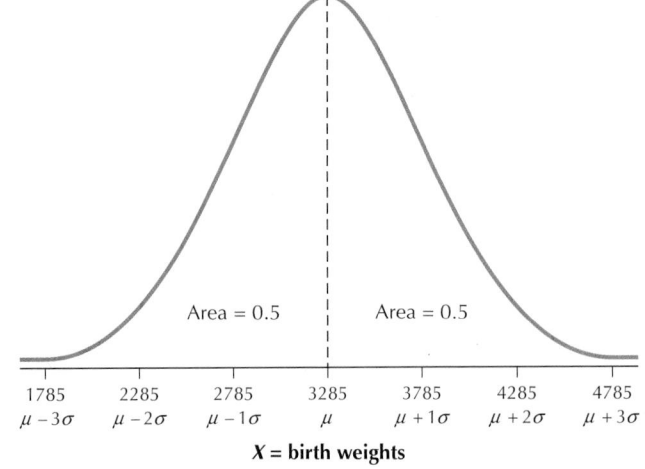

FIGURE 25
The normal curve of X = birth weights is symmetric about its mean $\mu = 3285$.

3. The total area under the curve equals 1.
4. The area under the curve to the left of $\mu = 3285$ equals 0.5, as does the area under the curve to the right of $\mu = 3285$.

NOW YOU CAN DO
Exercises 31–36.

4 Finding Areas Under the Standard Normal Curve for a Given Z-Value

Note: Understanding the techniques explained in this section will allow you to analyze a whole world of data sets, even those that are not normally distributed (see the Central Limit Theorem in the next chapter). Beyond this chapter, these techniques help you to calculate and understand *p*-values in Chapters 9–13.

Many populations in the world are normally distributed, from test scores to student heights with different means and standard deviations. But there is one very special normal distribution called the **standard normal distribution**. The mean and standard deviation of the standard normal distribution make it unique.

The **standard normal (Z) distribution** is a normal distribution with

• mean $\mu = 0$ and
• standard deviation $\sigma = 1$.

Because of its importance, the standard normal random variable is always denoted as a capital Z. The graph of the standard normal random variable Z is given in Figure 26. The standard normal curve is symmetric about its mean $\mu = 0$.

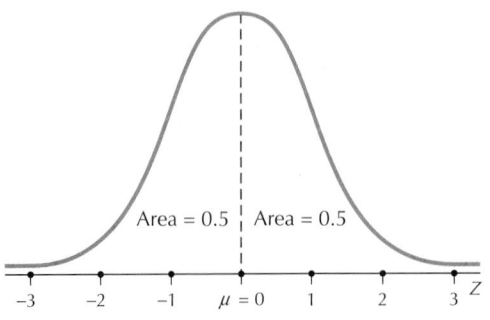

FIGURE 26 Z is symmetric about its mean $\mu = 0$.

Note: Although your Z table contains only values between $Z = -3.49$ and $Z = 3.49$, there is no upper or lower limit to the values that Z may take. The curve essentially goes on forever in both the positive and the negative directions, always getting closer and closer to the horizontal axis but never quite touching it (there's a great plot for a love story in there somewhere).

We will discuss two methods for finding probabilities associated with Z, using (a) the table for finding standard normal probabilities, called the **Z table**, and (b) technology. For the Z table, see Table C in the Appendix. *The Z table provides areas under the standard normal curve to the left of a specified value of Z, denoted as Z_1* (see Figure 27).

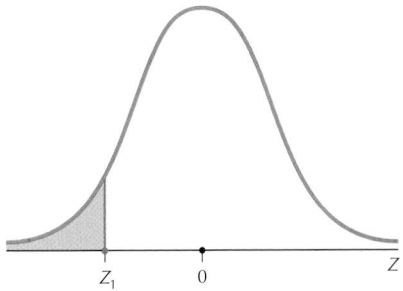

FIGURE 27 The Z table provides areas under the curve to the left of a specified value Z_1.

Let's get acquainted with the Z table (see excerpt in Figure 29 on page 356). Along the left side and across the top of the Z table are possible values of Z. These numbers, which in the table run from -3.49 to 3.49, are the values of Z found on the number line when you draw a graph. Down the left are the ones and tenths digits of the Z-value, and across the top, is the hundredths digit. The body of the Z table contains areas (probabilities). These numbers, which run from 0.0002 to 0.9998, are areas under the standard normal curve that represent probabilities *to the left* of the specified value of Z. Table 8 shows the steps for finding areas under the standard normal curve, that is, for finding probabilities for specified values of Z.

Table 8 Steps for finding areas under the standard normal curve

Case 1 **Find the area to the left of Z_1.** ***Step 1*** Draw the standard normal curve. Label the Z-value Z_1. ***Step 2*** Shade in the area to the left of Z_1.	**Case 2** **Find the area to the right of Z_1.** ***Step 1*** Draw the standard normal curve. Label the Z-value Z_1. ***Step 2*** Shade in the area to the right of Z_1.	**Case 3** **Find the area between Z_1 and Z_2.** ***Step 1*** Draw the standard normal curve. Label the Z-values Z_1 and Z_2. ***Step 2*** Shade in the area between Z_1 and Z_2.
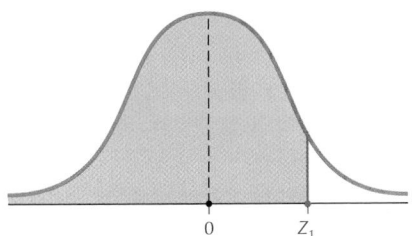	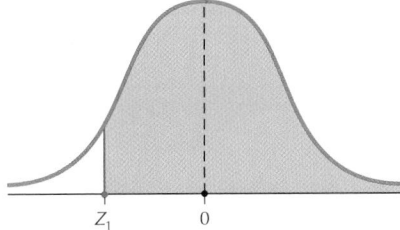	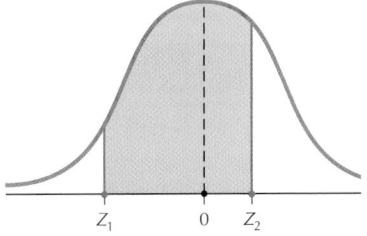
Step 3 Use the Z table to find the area to the left of Z_1.	***Step 3*** Use the Z table to find the area to the left of Z_1. The area to the right of Z_1 is then equal to $1 -$ (area to the left of Z_1).	***Step 3*** Use the Z table to find the area to the left of Z_1 and the area to the left of Z_2. The area between Z_1 and Z_2 is then equal to (area to the left of Z_2) $-$ (area to the left of Z_1).

EXAMPLE 28   Case 1: Find the area to the left of a value of Z

Find the area to the left of $Z = 0.57$.

Solution

Step 1 First draw the standard normal curve and label $Z = 0.57$.

Step 2 Shade the area to the left of 0.57, as shown in Figure 28.

Step 3 In the Z table, excerpted as Figure 29, go down the left-hand column to 0.5 and select that row. Then go across the top row (representing the hundredth's digit) to 0.07 and select that column. The quantity at the intersection of this row and column represents the area to the left of $Z = 0.57$. That is, the area to the left of $Z = 0.57$ is 0.7157.

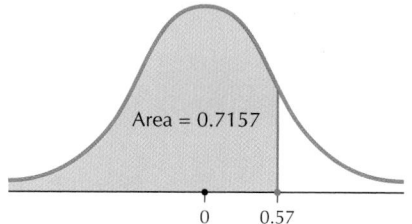

NOW YOU CAN DO
Exercises 37–44.

FIGURE 28 Finding the area to the left of Z.

Standard Normal Distribution

Z	0.00	0.01	0.02	0.03	0.04	0.05	0.06	0.07	0.08	0.09
0.0	0.5000	0.5040	0.5080	0.5120	0.5160	0.5199	0.5239	0.5279	0.5319	0.5359
0.1	0.5398	0.5438	0.5478	0.5517	0.5557	0.5596	0.5636	0.5675	0.5714	0.5753
0.2	0.5793	0.5832	0.5871	0.5910	0.5948	0.5987	0.6026	0.6064	0.6103	0.6141
0.3	0.6179	0.6217	0.6255	0.6293	0.6331	0.6368	0.6406	0.6443	0.6480	0.6517
0.4	0.6554	0.6591	0.6628	0.6664	0.6700	0.6736	0.6772	0.6808	0.6844	0.6879
0.5	0.6915	0.6950	0.6985	0.7019	0.7054	0.7088	0.7123	0.7157	0.7190	0.7224
0.6	0.7257	0.7291	0.7324	0.7357	0.7389	0.7422	0.7454	0.7486	0.7517	0.7549

FIGURE 29 Using the Z table to find the area to the left of Z.

YOUR TURN #14

Find the area to the left of $Z = 1.32$.

(The solution is shown in Appendix A.)

EXAMPLE 29   Case 2: Find the area to the right of a value of Z

Find the area to the right of $Z = -1.25$.

Solution

Step 1 First draw the standard normal curve and label $Z = -1.25$.

Step 2 Shade the area to the right of -1.25, as shown in Figure 30.

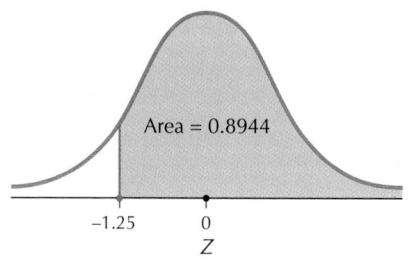

FIGURE 30 Finding the area to the right of Z.

NOW YOU CAN DO
Exercises 45–48.

CAUTION Remember that, although values of Z can be negative, *probabilities (or areas) can never be negative.*

Step 3 In the Z table, excerpted in Figure 31, go down the left-hand column to −1.2 and select that row. Then go across the top row to 0.05 and select that column. The area to the left of Z = −1.25 is therefore 0.1056. From Case 2 in Table 8, the area to the right of −1.25 is then

$$1 - (\text{area to the left of } -1.25) = 1 - 0.1056 = 0.8944$$

YOUR TURN
#15

Find the area to the right of Z = 1.28.

(The solution is shown in Appendix A.)

Developing Your Statistical Sense

Checking That Your Answer Makes Sense

As you are finding probabilities for values of Z, you should always be checking to see that your answer makes sense. For instance, in Example 29, what if we had added the table area to 1 instead of subtracted the table area from 1? We would know that this answer is incorrect because the resulting probability would then have exceeded 1, and no probability can ever exceed 1.

Standard Normal Distribution

Z	0.00	0.01	0.02	0.03	0.04	0.05	0.06	0.07	0.08	0.09
−3.4	0.0003	0.0003	0.0003	0.0003	0.0003	0.0003	0.0003	0.0003	0.0003	0.0002
−3.3	0.0005	0.0005	0.0005	0.0004	0.0004	0.0004	0.0004	0.0004	0.0004	0.0003
−3.2	0.0007	0.0007	0.0006	0.0006	0.0006	0.0006	0.0006	0.0005	0.0005	0.0005
−3.1	0.0010	0.0009	0.0009	0.0009	0.0008	0.0008	0.0008	0.0008	0.0007	0.0007
−3.0	0.0013	0.0013	0.0013	0.0012	0.0012	0.0011	0.0011	0.0011	0.0010	0.0010
−1.4	0.0808	0.0793	0.0778	0.0764	0.0749	0.0735	0.0721	0.0708	0.0694	0.0681
−1.3	0.0968	0.0951	0.0934	0.0918	0.0901	0.0885	0.0869	0.0853	0.0838	0.0823
−1.2	0.1151	0.1131	0.1112	0.1093	0.1075	0.1056	0.1038	0.1020	0.1003	0.0985
−1.1	0.1357	0.1335	0.1314	0.1292	0.1271	0.1251	0.1230	0.1210	0.1190	0.1170
−1.0	0.1587	0.1562	0.1539	0.1515	0.1492	0.1469	0.1446	0.1423	0.1401	0.1379

FIGURE 31 Using the Z table to find the area to the right of Z.

EXAMPLE 30 **Case 3: Find the area between two Z-values (checking the accuracy of the Empirical Rule)**

Recall that the Empirical Rule (page 135 of Chapter 3) states that about 68% of the area under the curve lies within 1 standard deviation of the mean, that is, between $\mu - \sigma$ and $\mu + \sigma$. Check this result for the standard normal distribution by using the Z table.

Solution

For the standard normal random variable Z, $\mu = 0$ and $\sigma = 1$, so that $\mu - \sigma = 0 - 1 = -1$ and $\mu + \sigma = 0 + 1 = 1$. Thus, using Case 3, we have $Z_1 = -1$ and $Z_2 = 1$.

Step 1　Draw the standard normal curve. Label the Z-values $Z_1 = -1$ and $Z_2 = 1$.

Step 2　Shade the area between -1 and 1, as shown in Figure 32a.

NOW YOU CAN DO
Exercises 49–58.

Step 3　Find the area to the left of $Z_1 = -1$ and the area to the left of $Z_2 = 1$. The Z table gives these areas as follows: area to the left of $Z_1 = -1$ is 0.1587, and area to the left of $Z_2 = 1$ is 0.8413. We subtract the smaller area from the larger to give us the area between -1 and 1, as shown in Figures 32a–32c.

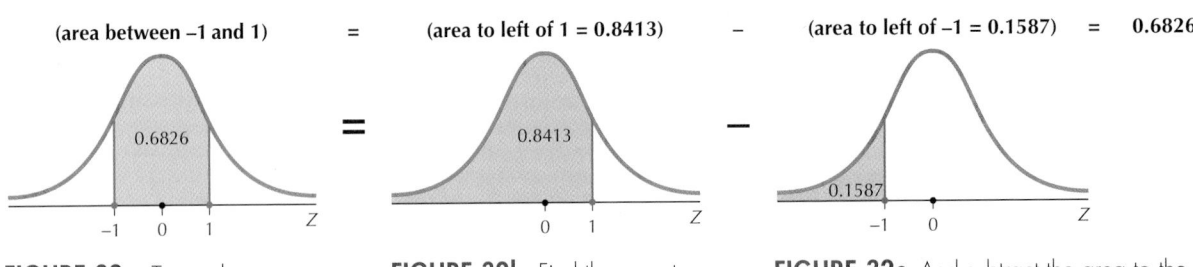

FIGURE 32a To get the area we are looking for…

FIGURE 32b Find the area to the left of 1…

FIGURE 32c And subtract the area to the left of -1.

Thus, the area under the Z curve within 1 standard deviation of the mean equals 0.6826. The Empirical Rule does very well for an approximation, missing the actual area by only 0.0026. Checking the accuracy of the Empirical Rule for other values of Z is left as an exercise.

EXAMPLE 31 **Using technology to find the area under a standard normal curve**

In Example 28, we found the area under the standard normal curve to the left of $Z = 0.57$ to be 0.7157. Confirm this result using technology.

Solution

The *Normal Density Curve* applet allows you to find areas associated with various values of Z.

We follow the instructions in the Step-by-Step Technology Guide at the end of Section 6.5 (page 380). Figures 33–36 show the results from TI-83/84, Excel, Minitab, and CrunchIt!, respectively.

　　The word "cumulative" in the Minitab output means "less than or equal to." Each of these results provides the area under the standard normal curve for values of Z that are less than or equal to 0.57. Each technology rounds to a different number of decimal places.

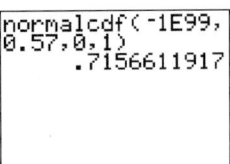

```
normalcdf(-1E99,
0.57,0,1)
          .7156611917
```

	=NORMDIST(0.57,0,1,TRUE)	
D	**E**	**F**
0.7156612		

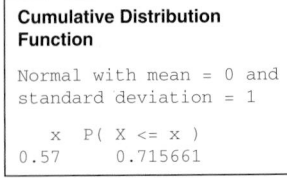

Cumulative Distribution Function

Normal with mean = 0 and
standard deviation = 1

```
   x   P( X <= x )
0.57     0.715661
```

FIGURE 33 TI-83/84 results.

FIGURE 34 Excel results.

FIGURE 35 Minitab results.

FIGURE 36
CrunchIt! results.

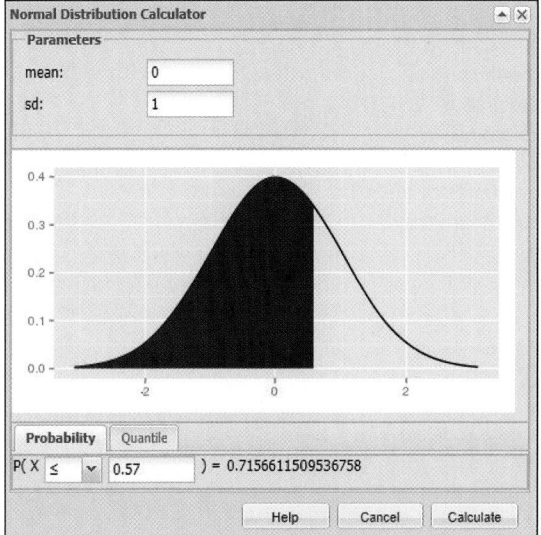

Note that the areas we have been finding in this section may also be expressed as probabilities. For continuous distributions, probabilities are represented by areas under the curve above an interval. Specifically, for the standard normal distribution, probability is represented as the area above an interval under the standard normal curve. For instance, in Example 28, we found that the area under the standard normal curve to the left of $Z = 0.57$ is 0.7157. This may be re-expressed as follows:

"The probability that Z is less than 0.57 is 0.7157"

or

$$P(Z < 0.57) = 0.7157$$

EXAMPLE 32   Expressing areas under the standard normal curve as probabilities

Re-express the following areas as probabilities:

a. In Example 29, we found the area under the standard normal curve to the right of $Z = -1.25$ to be 0.8944.

b. In Example 30, we found the area under the standard normal curve between $Z = -1$ and $Z = 1$ to be 0.6826.

Solution

a. The probability that Z is greater than -1.25 is 0.8944. That is, $P(Z > -1.25) = 0.8944$.

b. The probability that Z is between -1 and 1 is 0.6826. That is, $P(-1 < Z < 1) = 0.6826$.

NOW YOU CAN DO
Exercises 59–70.

5 Finding Standard Normal Z-Values for a Given Area

In previous examples, we were given a Z-value and asked to find an area or probability. What if we turned this around, so that we are given an area, and asked to find its associated Z-value? We may call these "backward" problems because we would need to use the Z table in reverse (unless we are using technology to solve the problem). Let's check out an example.

EXAMPLE 33 Finding the *Z*-value with given area to its left

Recall that the *r*th *percentile* is the value in the data set such that *r* percent of the data values fall at or below that value. Thus, $Z = 1.28$ represents the 90th percentile of the *Z* distribution because it is greater than 90% of *Z*-values.

Find the *Z*-value with area 0.90 to its left.

Solution

Step 1   Draw the standard normal curve. Label the *Z*-value Z_1.

Step 2   Shade the area to the left of Z_1. Remember that *we are given an area and are looking for a value of Z*. Label the area to the left of Z_1 with the given area (0.90), as shown in Figure 37.

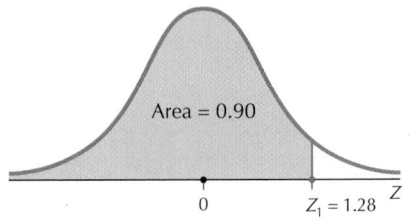

FIGURE 37
$Z_1 = 1.28$ is the value of *Z* with area 0.90 to the left of it.

Step 3   Look for 0.90 on the inside of the *Z* table (that is, in the body of the table), because the values *inside the table* represent areas. Because there is no 0.90 inside the table, by convention we take the area that is closest to 0.90, which is 0.8997. Next is the trick of the backward problems and the reason for that name. Move from 0.8997 to the left until you reach 1.2 in the first column, and then move up from 0.8997 until you get to 0.08 (see Figure 38). Putting these values together, we get $Z = 1.2 + 0.08 = 1.28$.

NOW YOU CAN DO
Exercises 71–78.

Standard Normal Distribution

Z	0.00	0.01	0.02	0.03	0.04	0.05	0.06	0.07	0.08	0.09
0.0	0.5000	0.5040	0.5080	0.5120	0.5160	0.5199	0.5239	0.5279	0.5319	0.5359
0.1	0.5398	0.5438	0.5478	0.5517	0.5557	0.5596	0.5636	0.5675	0.5714	0.5753
0.2	0.5793	0.5832	0.5871	0.5910	0.5948	0.5987	0.6026	0.6064	0.6103	0.6141
0.3	0.6179	0.6217	0.6255	0.6293	0.6331	0.6368	0.6406	0.6443	0.6480	0.6517
0.4	0.6554	0.6591	0.6628	0.6664	0.6700	0.6736	0.6772	0.6808	0.6844	0.6879
0.5	0.6915	0.6950	0.6985	0.7019	0.7054	0.7088	0.7123	0.7157	0.7190	0.7224
0.6	0.7257	0.7291	0.7324	0.7357	0.7389	0.7422	0.7454	0.7486	0.7517	0.7549
0.7	0.7580	0.7611	0.7642	0.7673	0.7704	0.7734	0.7764	0.7794	0.7823	0.7852
0.8	0.7881	0.7910	0.7939	0.7967	0.7995	0.8023	0.8051	0.8078	0.8106	0.8133
0.9	0.8159	0.8186	0.8212	0.8238	0.8264	0.8289	0.8315	0.8340	0.8365	0.8389
1.0	0.8413	0.8438	0.8461	0.8485	0.8508	0.8531	0.8554	0.8577	0.8599	0.8621
1.1	0.8643	0.8665	0.8686	0.8708	0.8729	0.8749	0.8770	0.8790	0.8810	0.8830
1.2	0.8849	0.8869	0.8888	0.8907	0.8925	0.8944	0.8962	0.8980	0.8997	0.9015
1.3	0.9032	0.9049	0.9066	0.9082	0.9099	0.9115	0.9131	0.9147	0.9162	0.9177

FIGURE 38 Using the *Z* table to find a value of *Z* for a given area.

YOUR TURN
#16

Find the *Z*-value with area 0.975 to its left.

(The solution is shown in Appendix A.)

EXAMPLE 34   Find the *Z*-value with given area to its right

Find the standard normal Z-value that has area 0.03 to the right of it.

Solution

Step 1   Draw the standard normal curve. Label the Z-value Z_1. Shade the area to the right of it with the given area, as shown in Figure 39.

Step 2   The Z table contains areas to the left of values of Z, so we must find the area to the left of the specific value Z_1, as follows:

$$\text{area to left of } Z_1 = 1 - \text{area to right of } Z_1$$

So the area to the left of Z_1 is $1 - 0.03 = 0.97$.

FIGURE 39
$Z_1 = 1.88$ has an area 0.03 to the right of it.

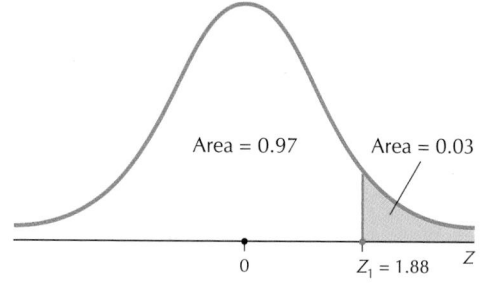

Step 3   Look up 0.97 on the inside of the Z table. The closest area is 0.9699. Move from 0.9699 to the left until you reach 1.8, and then move up from 0.9699 until you get to 0.08 (see Figure 40). Putting these values together, we get $Z = 1.8 + 0.08 = 1.88$. In other words, the Z-value with area 0.03 to its right is $Z = 1.88$.

NOW YOU CAN DO
Exercises 79–86.

Standard Normal Distribution

Z	0.00	0.01	0.02	0.03	0.04	0.05	0.06	0.07	0.08	0.09
0.0	0.5000	0.5040	0.5080	0.5120	0.5160	0.5199	0.5239	0.5279	0.5319	0.5359
0.1	0.5398	0.5438	0.5478	0.5517	0.5557	0.5596	0.5636	0.5675	0.5714	0.5753
0.2	0.5793	0.5832	0.5871	0.5910	0.5948	0.5987	0.6026	0.6064	0.6103	0.6141
1.6	0.9452	0.9463	0.9474	0.9484	0.9495	0.9505	0.9515	0.9525	0.9535	0.9545
1.7	0.9554	0.9564	0.9573	0.9582	0.9591	0.9599	0.9608	0.9616	0.9625	0.9633
1.8	0.9641	0.9649	0.9656	0.9664	0.9671	0.9678	0.9686	0.9693	0.9699	0.9706
1.9	0.9713	0.9719	0.9726	0.9732	0.9738	0.9744	0.9750	0.9756	0.9761	0.9767

FIGURE 40 Using the Z table to find a value of Z for a given area.

YOUR TURN
#17

Find the Z-value with area 0.975 to its right.

(The solution is shown in Appendix A.)

When we learn statistical inference in later chapters, we will need to identify which Z-values divide the middle 90%, 95%, or 99% of the area under the standard normal curve from the tail area.

EXAMPLE 35 Find the values of *Z* that mark the boundaries of the middle 95% of the area

Find the two values of *Z* that mark the boundaries of the middle 95% of the area under the standard normal curve.

Solution

Step 1 Draw the standard normal curve, showing the desired middle area (95%) with boundaries labeled as Z_1 and Z_2, as shown in Figure 41. By symmetry, there is area = $(1 - 0.95)/2 = 0.025$ in each tail.

Step 2 Look up 0.025 on the inside of the *Z* table. Find Z_1 by moving to the left and up from 0.025 in the *Z* table, giving us $Z_1 = -1.96$.

Step 3 The area in the right tail is also 0.025, so the area to the left of Z_2 is $1 - 0.025 = 0.975$. Looking up 0.975 in the *Z* table gives us $Z_2 = 1.96$.

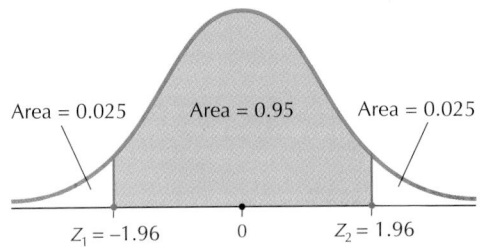

Note: Is it a coincidence that the two values of *Z* that determine the middle 95% of the area under the standard normal curve are 1.96 and −1.96? Not at all. The standard normal curve is symmetric about the mean 0, so the values −1.96 and 1.96 that form the boundaries of the middle 95% must be *equidistant from zero.*

FIGURE 41
Z_1 and Z_2 mark the middle 95% of the *Z* distribution.

NOW YOU CAN DO
Exercises 87–90.

Thus, the two *Z*-values that mark the boundaries of the middle 95% of the area under the standard normal curve are −1.96 and 1.96. This is a more precise result, which states that about 95% lies between −2 and 2.

YOUR TURN
#18

Find the two values of *Z* that mark the boundaries of the middle 90% of the area under the standard normal curve.

(The solution is shown in Appendix A.)

EXAMPLE 36 Using technology to find values of *Z*, given an area

In Example 33, we found that the value of *Z* with area 0.90 to its left is $Z = 1.28$. Confirm this result with technology.

Solution

We follow the instructions in the Step-by-Step Technology Guide at the end of Section 6.5 (page 380). Figures 42–45 show the results from TI-83/84, Excel, Minitab, and SPSS, respectively. Note that technology usually provides a more precise solution than the *Z* table.

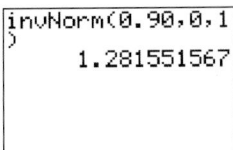

FIGURE 42 TI-83/84 results.

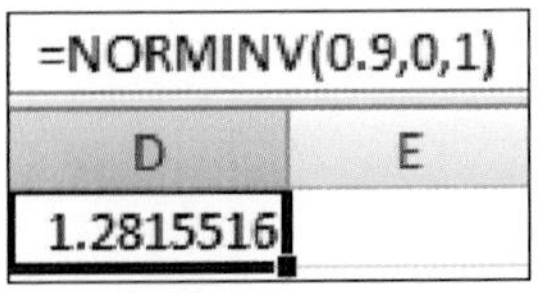

FIGURE 43 Excel results.

Inverse Cumulative Distribution Function
Normal with mean = 0 and standard deviation = 1
P(X <= x)   x   0.9 1.28155

FIGURE 44 Minitab results.

1 : VAR00001		1.28155515655446
	VAR00001	var
1	1.28	
2		

FIGURE 45 SPSS results.

Section 6.4 Summary

1. Continuous random variables assume infinitely many possible values, with no gap between the values. Probability for continuous random variables consists of the area above an interval on the number line and under the distribution curve.

2. The uniform probability distribution has constant probability from its left to its right endpoints and is therefore shaped like a rectangle.

3. The normal distribution is the most important continuous probability distribution. It is symmetric about its mean μ and has standard deviation σ.

4. The standard normal distribution has mean $\mu = 0$ and standard deviation $\sigma = 1$. This distribution is often called the Z distribution. The Z table and technology can be used to find areas under the standard normal curve. In the Z table, the numbers on the outside are values of Z, and the numbers inside are areas to the left of values of Z.

5. The Z table and technology can also be used to find a value of Z, given a probability or an area under the curve.

Section 6.4 Exercises

CLARIFYING THE CONCEPTS

1. For a continuous random variable X, why are we not interested in whether X equals some particular value? (p. 349)

2. In the graph of a probability distribution, what is represented on the number line? (p. 349)

3. How is probability represented in the graph of a continuous probability distribution? (p. 350)

4. What are the possible values for the mean of a normal distribution? For the standard deviation? (p. 352)

5. True or false: The graph of the uniform distribution is always shaped like a square. (p. 350)

6. For continuous probability distributions, what is the difference between $P(X > 1)$ and $P(X \geq 1)$? (p. 351)

7. What is the value for the mean of the standard normal distribution? (p. 354)

8. What is the value for the standard deviation of the standard normal distribution? (p. 354)

9. True or false: The area under the Z curve to the right of $Z = 0$ is 0.5. (p. 354)

10. True or false: $P(Z = 0) = 0$. (p. 349)

PRACTICING THE TECHNIQUES

 **CHECK IT OUT!**

To do	Check out	Topic
Exercises 11–20	Example 25	Uniform probability distribution
Exercises 21–30	Example 26	Normal distribution mean and standard deviation

To do	Check out	Topic
Exercises 31–36	Example 27	Properties of the normal curve
Exercises 37–44	Example 28	Find the area to left of the Z-value
Exercises 45–48	Example 29	Find the area to right of the Z-value
Exercises 49–58	Example 30	Find the area between two Z-values
Exercises 59–70	Example 32	Expressing areas under the standard normal curve as probabilities
Exercises 71–78	Example 33	Finding the Z-value, given area to its left
Exercises 79–86	Example 34	Finding the Z-value, given area to its right
Exercises 87–90	Example 35	Find two Z-values that mark the boundaries of the middle 95% of the area

For Exercises 11–16, assume that X is a uniform random variable, with left endpoint 0 and right endpoint 100. Find the following probabilities:

11. $P(50 < X < 100)$
12. $P(50 \leq X \leq 100)$
13. $P(25 < X < 90)$
14. $P(15 \leq X \leq 35)$
15. $P(24 < X < 25)$
16. $P(25 < X < 25)$

For Exercises 17–20, assume that X is a uniform random variable, with left endpoint -5 and right endpoint 5. Compute the following probabilities:

17. $P(0 \le X \le 5)$
18. $P(-5 \le X \le 5)$
19. $P(-5 \le X \le -4)$
20. $P(-1 \le X \le 5)$

21. The two normal distributions in the accompanying figure have the same standard deviation of 5 but different means. Which normal distribution has mean 10 and which has mean 25? Explain how you know this.

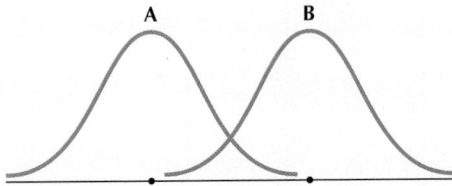

22. The two normal distributions in the figure below have the same mean of 100 but different standard deviations. Which normal distribution has standard deviation 3 and which has standard deviation 6? Explain how you know this.

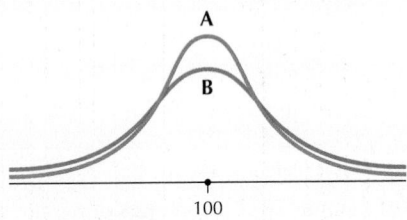

For Exercises 23–30, use the graph of the normal distribution to determine the mean and standard deviation. (*Hint:* The distance between dotted lines in the figures represents 1 standard deviation.)

23.

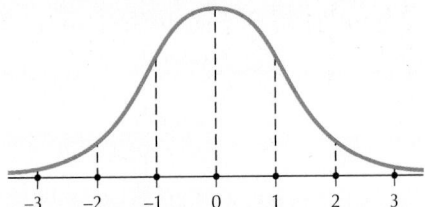

24.

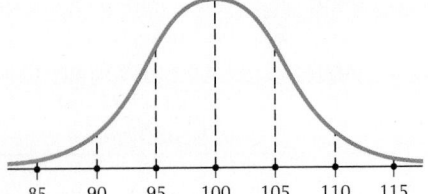

25.

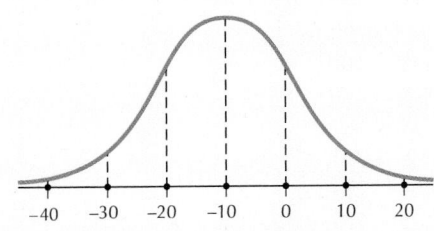

26.

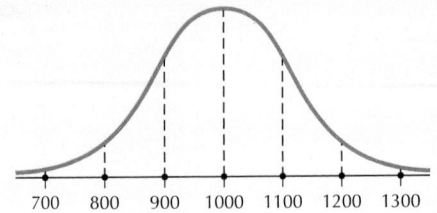

27.

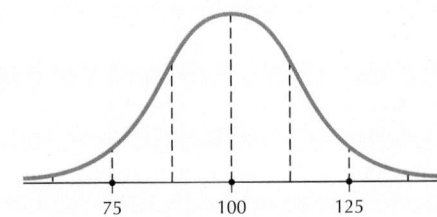

28.

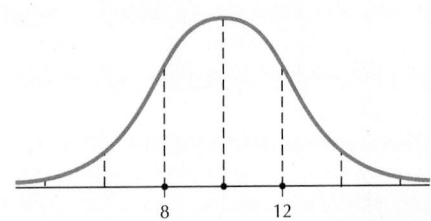

29.

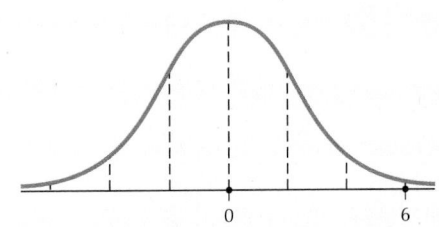

30.

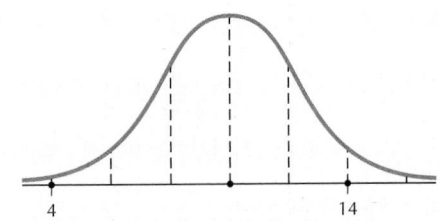

Use the normal distribution from Example 27 for Exercises 31–36. Birth weights are normally distributed with a mean weight of $\mu = 3285$ grams and a standard deviation of $\sigma = 500$ grams.

31. What is the probability of a birth weight equal to 3285 grams?

32. What is the probability of a birth weight more than 3285 grams?

33. What is the probability of a birth weight of at least 3285 grams?

34. What can we say about the area to the left of 3285 grams and the area to the right of 3285 grams?

35. Is the area to the right of $X = 4285$ grams greater than or less than 0.5? How do you know this?

36. Is the area to the left of $X = 4285$ grams greater than or less than 0.5? How do you know this?

For Exercises 37–58,
   a. draw the graph.
   b. find the area using the Z table or technology.

Find the area under the standard normal curve that lies to the left of the following:
37. $Z = 1$
38. $Z = 2$
39. $Z = 3$
40. $Z = 0.5$
41. $Z = -2.7$
42. $Z = -0.9$
43. $Z = -0.2$
44. $Z = -1.2$

Find the area under the standard normal curve that lies to the right of the following:
45. $Z = 1.27$
46. $Z = 2.12$
47. $Z = -3.01$
48. $Z = -0.69$

Find the area under the standard normal curve that lies between the following:
49. $Z = 0$ and $Z = 1$
50. $Z = 1$ and $Z = 2$
51. $Z = 2$ and $Z = 3$
52. $Z = 1.28$ and $Z = 1.96$
53. $Z = -1$ and $Z = 0$
54. $Z = -2$ and $Z = -1$
55. $Z = -3$ and $Z = -2$
56. $Z = -1.96$ and $Z = -1.28$
57. $Z = -1.28$ and $Z = 1.28$
58. $Z = -2.01$ and $Z = 2.37$

For Exercises 59–70, find the indicated probability for the standard normal Z.
   a. Draw the graph.
   b. Find the area using the Z table or technology.
59. $P(Z = 0)$
60. $P(Z < 0)$
61. $P(Z < 10)$
62. $P(Z > 1.29)$
63. $P(Z < -2.17)$
64. $P(Z < 0.57)$
65. $P(-1.96 < Z < 1.96)$
66. $P(-2.07 < Z < 0.46)$
67. $P(-3.05 < Z < -0.94)$
68. $P(1.54 < Z < 2.20)$
69. $P(-100 < Z < 0)$
70. $P(-1.72 < Z < -1.57)$

For Exercises 71–78, find the Z-value with the following areas under the standard normal curve to its left. Draw the graph, and then find the Z-value.
71. 0.3336
72. 0.4602
73. 0.3264
74. 0.4247

75. 0.95
76. 0.975
77. 0.98
78. 0.99

For Exercises 79–86, find the Z-value with the following areas under the standard normal curve to its right. Draw the graph, and then find the Z-value.
79. 0.8078
80. 0.3085
81. 0.9788
82. 0.5120
83. 0.90
84. 0.975
85. 0.9988
86. 0.9998

For Exercises 87–94, find the values of Z that mark the boundaries of the indicated areas.
87. The middle 80%
88. The middle 95%
89. The middle 98%
90. The middle 85%
91. Find the 50th percentile of the Z distribution. (*Hint*: See margin note on page 360.)
92. Find the 75th percentile of the Z distribution.
93. Find the value of Z that is larger than 99.5% of all values of Z.
94. Find the value of Z that is smaller than 99.5% of all values of Z.

APPLYING THE CONCEPTS

95. **Uniform Distribution: Web Page Loading Time.** Suppose that the Web page loading time for a particular home network is uniform, with left endpoint 1 second and right endpoint 5 seconds.
   a. What is the probability that a randomly selected Web page will take between 3 seconds and 4 seconds to load?
   b. Find the probability that a randomly selected Web page will take between 1 second and 2 seconds to load.
   c. How often does it take less than 1 second for a Web page to load?
   d. What is the probability that it takes exactly 2 seconds for a page to load? Explain.

96. **Uniform Distribution: Random Number Generation.** Computers and calculators use the uniform distribution to generate random numbers. Suppose we have a calculator that randomly generates numbers between 0 and 1, so that they form a uniform distribution.
   a. What is the probability that the random number generated is less than 0.3?
   b. Find the probability that a random number is generated that is between 0.27 and 0.92.
   c. What is the probability that a random number greater than 1 is generated?

6.5 Applications of the Normal Distribution

OBJECTIVES By the end of this section, I will be able to . . .

1 Compute probabilities for a given value of any normal random variable.
2 Find the appropriate value of any normal random variable, given an area or probability.
3 Use normal probability plots to assess normality.

1 Finding Probabilities for Any Normal Distribution

To *standardize* things means to make them all the same. For example, college applicants take standardized tests so that the admissions officers can compare students according to a consistent assessment tool. Here, we standardize many different normal random variables X into the same standard normal Z.

The data in problems that we face in the real world do not usually follow the standard normal distribution, Z. Instead, a problem may be stated in terms of some normal random variable X that has a mean other than 0 or a standard deviation other than 1. In cases like these, X needs to be standardized to Z so that we can use the Section 6.4 techniques.

Standardizing X to Z

To standardize a normal random variable X, we *transform* that normal random variable X into the standard normal random variable Z.

Suppose that X is a normal random variable with population mean μ and population standard deviation σ. We standardize X by subtracting the mean μ and dividing by the standard deviation σ. The result of this transformation is the familiar **standard normal random variable Z**.

Standardizing a Normal Random Variable

Any normal random variable X can be transformed into the standard normal random variable Z by *standardizing* X with the formula

$$Z = \frac{X - \mu}{\sigma}$$

The key here is the following: *for a given area of interest for a normal random variable X, the corresponding area after the transformation to Z is exactly the same.* For any normal random variable X

the area between a and b

is exactly the same as

the area between $Z_a = \dfrac{(a - \mu)}{\sigma}$ and $Z_b = \dfrac{(b - \mu)}{\sigma}$ (see Figure 46)

So we can solve problems about areas under the nonstandard normal X curve by using the corresponding area under the Z curve.

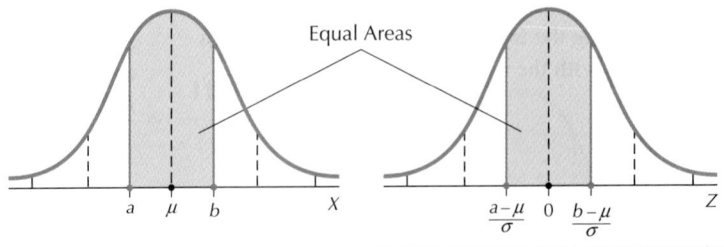

FIGURE 46 Corresponding areas are equal.

EXAMPLE 37   April in Georgia

April in Georgia.

The state of Georgia reports that the mean temperature statewide for the month of April is $\mu = 61.5°F$. Assume that the standard deviation is $\sigma = 8°F$ and that temperature in Georgia in April is normally distributed. Draw the normal curve for temperatures between 45.5°F and 77.5°F, and the corresponding Z curve. Find the probability that the temperature is between 45.5°F and 77.5°F in April in Georgia.

Solution

Here, we have $a = 45.5$ and $b = 77.5$, giving us

$$Z_a = \frac{a - \mu}{\sigma} = \frac{45.5 - 61.5}{8} = -2 \quad \text{and} \quad Z_b = \frac{b - \mu}{\sigma} = \frac{77.5 - 61.5}{8} = 2$$

In Figure 47, the area between $X = 45.5°F$ and $X = 77.5°F$ is the same as between $Z = -2$ and $Z = 2$. In other words,

$$P(45.5 < X < 77.5) = P(-2 < Z < 2)$$

This is a Case 3 problem from Table 8 (page 355). The Z table tells us that the area to the left of $Z_1 = -2$ is 0.0228, and the area to the left of $Z_2 = 2$ is 0.9772. The area between -2 and 2 is then equal to $0.9772 - 0.0228 = 0.9544$. The probability that the temperature is between 45.5°F and 77.5°F in April in Georgia is 0.9544.

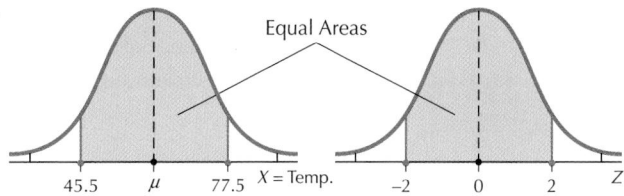

FIGURE 47 Find the area under the Z curve and we have found the area under the X curve.

Finding Probabilities for Any Normal Distribution

Step 1 Determine the random variable X, the mean μ, and the standard deviation σ. Draw the normal curve for X, and shade the desired area.

Step 2 Standardize by using the formula $Z = (X - \mu)/\sigma$ to find the values of Z corresponding to the X-values.

Step 3 Draw the standard normal curve and shade the area corresponding to the shaded area in the graph of X.

Step 4 Find the area under the standard normal curve using either the Z table or technology. This area is equal to the area under the normal curve for X drawn in Step 1.

CAUTION **Check Your Answer!** According to the Empirical Rule, almost all Z-values lie between -3 and 3, so it is unlikely that a randomly selected value of Z lies outside this range. You should remember this when you are doing your calculations. If you are standardizing a normal random variable X and get a very large Z-value (such as $Z = 50$), you should recheck your calculations because the probability that Z takes such a large value is very small.

*Chris Ryan/OJO Images/
Getty Images*

EXAMPLE 38 Finding probability for a normal random variable X

SAT Scores and AP Exam Scores

The College Board reports that the population mean Math SAT score in 2013 was = 514, with a population standard deviation of $\sigma = 118$, and that the scores follow a normal distribution. Suppose that a local college wants to identify at-risk math students, which it considers to be students scoring below 396 on the Math SAT. Find the proportion of students who score below 396 on the Math SAT.

Solution

Step 1 **Determine X, μ, and σ.**

We are given that the normal random variable X = Math SAT score has mean $\mu = 514$ and standard deviation $\sigma = 118$. In the center of the number line, mark the mean $\mu = 514$. Also mark on the number line the value of X about which the problem is asking. Figure 48 shows the graph of X (the Math SAT scores) with the mean of 514 and the score of 396 marked.

Remember that you may solve problems asking for proportions or percentages by finding the appropriate probability.

You need to know the proportion of scores below 396, so shade the area under the curve to the left of 396. We can express this proportion as a probability, the probability that a randomly chosen student will score less than 396, or $P(X < 396)$. Just by looking at Figure 48, you should be able to get a rough idea of what the proportion of these scores will be. Certainly, this proportion will be less than 50%. If you get an answer such as "60%" for your proportion, you should recognize that it is wrong.

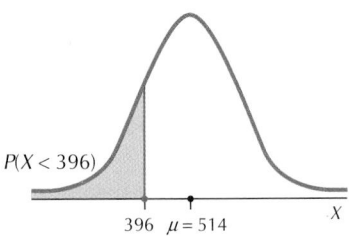

FIGURE 48 Graph of proportion of Math SAT scores lower than 396.

Step 2 **Standardize.**

Now standardize the random variable X to the standard normal Z:

$$Z = \frac{X - \mu}{\sigma} = \frac{X - 514}{118}$$

Find the Z-value corresponding to the Math SAT score of 396:

$$\frac{396 - \mu}{\sigma} = \frac{396 - 514}{118} = -1$$

So the Z-value associated with a score of 396 is -1, which indicates that the score of 396 is 1 standard deviation below the mean of 514.

Step 3 **Draw the standard normal curve.**

Scores less than 396 are more than 1 standard deviation below the mean, so shade the area to the left of -1 in Figure 49. Now find the area to the left of $Z = -1$ using the methods of Section 6.4.

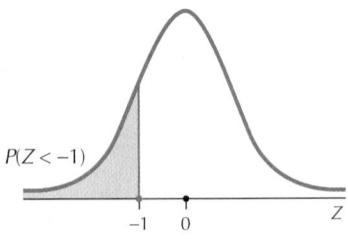

FIGURE 49 Graph of $P(Z < -1)$.

Step 4 **Find the area under the standard normal curve.**

The Z table tells us that the area to the left of $Z = -1$ is 0.1587. The proportion of scores below 396 is 0.1587, or 15.87%. Note that this value for $P(X < 396)$ agrees with our earlier intuition that the proportion was less than 50%.

NOW YOU CAN DO
Exercises 3–9.

YOUR TURN
#19

For the scenario in Example 38, find the proportion of Math SAT scores greater than 600.

(The solution is shown in Appendix A.)

EXAMPLE 39   Finding the probability that *X* lies between two given values

SAT Scores and AP Exam Scores

Continuing the Math SAT score problem, what percentage of students score between 215 and 595?

Solution

Step 1   Determine *X*, *μ*, and *σ*.

We have already seen that *X* = Math SAT score has mean $\mu = 514$ and standard deviation $\sigma = 118$. Once again, draw a graph of the distribution of scores *X*, with the mean 514 in the middle, the score 215 to the left of the mean, and the score 595 to the right of the mean, as in Figure 50.

Step 2   Standardize.

This is a "between" example, where two values of *X* are given, and we are asked to find the area between them. In this case, just standardize both of these values of *X* to get a *Z*-value for each:

$$Z = \frac{215 - \mu}{\sigma} = \frac{215 - 514}{118} \approx -2.53 \quad \text{and} \quad Z = \frac{595 - \mu}{\sigma} = \frac{595 - 514}{118} \approx 0.69$$

> The *Normal Density Curve* applet allows you to find areas associated with various values of any normal random variable.

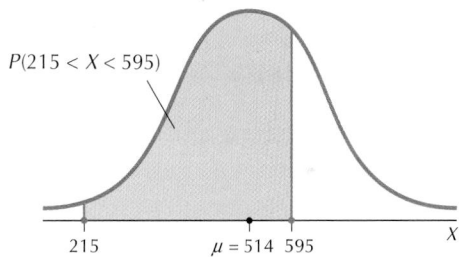

FIGURE 50
Graph of percentage of students scoring between 215 and 595 on the Math SAT.

Step 3   Draw the standard normal curve.

Draw a graph of *Z*, shading the area between $Z = -2.53$ and $Z = 0.69$, as shown in Figure 51. Again, the key is that the area between $Z = -2.53$ and $Z = 0.69$ is exactly the same as the area between $X = 215$ and $X = 595$.

Step 4   Find area under the standard normal curve.

Figure 51 is a Case 3 problem from Table 8 (page 355). Find the area to the left of 0.69, which is 0.7549, and the area to the left of -2.53, which is 0.0057. Subtracting the smaller from the larger gives us

$$P(-2.53 < Z < 0.69) = 0.7549 - 0.0057 = 0.7492$$

Thus, the percentage of Math SAT scores that are between 215 and 595 is 74.92%.

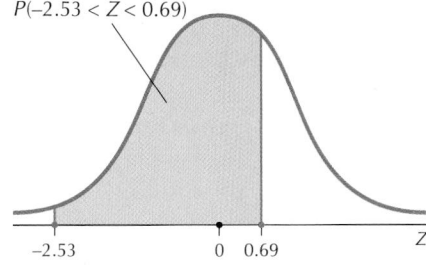

NOW YOU CAN DO
Exercises 10–14.

FIGURE 51   Graph of percentage of *Z*-values between -2.53 and 0.69.

YOUR TURN #20

For the scenario in Example 39, find the proportion of Math SAT scores between 305 and 605.

(The solution is shown in Appendix A.)

2 Finding a Normal Data Value for a Given Area or Probability

Sometimes we are given a probability (or proportion or area), and we are asked to find the associated value of X. Questions like these are similar to the "backwards" problems of Section 6.4, which are so called because we must use the Z table backward or inside out. The formula for standardizing X gives the value for Z, so we need to use our algebra skills to find the equation for X: Start with the standard normal formula $Z = (X - \mu)/\sigma$. Multiply both sides by σ to get $Z\sigma = X - \mu$. Then add μ to both sides, giving us $X = Z\sigma + \mu$.

Finding Normal Data Values for a Given Area or Probability

Step 1 **Determine X, μ, and σ, and draw the normal curve for X.** Shade the desired area. Mark the position of X_1, the unknown value of X.

Step 2 **Find the Z-value corresponding to the desired area.** Look up the area you identified in Step 1 on the *inside* of the Z table. If you do not find the exact value of your area, by convention choose the area that is closest.

Step 3 **Transform this value of Z into a value of X, which is the solution.** Use the formula $X_1 = Z\sigma + \mu$.

EXAMPLE 40   Finding a normal data value for a given area

SAT Scores and AP Exam Scores

Suppose the students in the top 1% of Math SAT scores won a fellowship to an Ivy League university. What is the score that students will have to obtain to win this fellowship?

Solution

Notice that we are not asked to find a probability (or proportion or area). Instead, we are given a percentage (1%) and asked to find the value of X (the Math SAT score) that is associated with this 1%.

Step 1   **Determine X, μ, and σ, and draw the normal curve for X.**
We already know that X = Math SAT score, $\mu = 514$, and $\sigma = 118$. The value of X in which we are interested refers to high scores, so that X_1 will be at the far right of the distribution of X. Only 1% of scores will be greater than this score, so the area to the right of X_1 is 0.01, as shown in Figure 52.

FIGURE 52
X_1 is the cutoff value (or critical value) of X, at which students will win a fellowship to an Ivy League university.

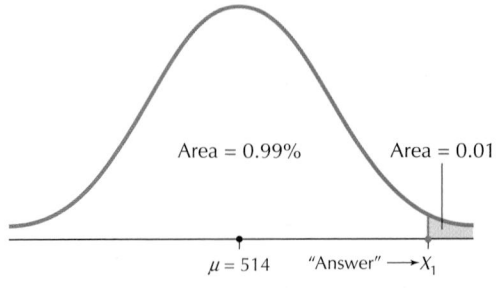

Step 2   **Find the Z-value corresponding to the desired area.**
The area to the right of X_1 equals 0.01, so that the area to the left of X_1 equals $1 - 0.01 = 0.99$. Looking up 0.99 on the inside of the Z table gives us $Z = 2.33$.

Step 3   **Transform using the formula $X_1 = Z\sigma + \mu$.**
We calculate

$$X_1 = Z\sigma + \mu = (2.33)(118) + 514 = 788.94$$

NOW YOU CAN DO
Exercises 15–22.

The cutoff value for the top 1% of Math SAT scores for winning a fellowship to an Ivy League university is 788.94. It won't be easy getting that fellowship.

YOUR TURN
#21

For the situation in Example 40, what is the Math SAT score that separates the lowest 2.5% of the scores from the others?

(The solution is shown in Appendix A.)

EXAMPLE 41   Finding the X-values that mark the boundaries of the middle 95% of X-values

Edmunds.com reported that the average amount that people were paying for a 2015 Toyota Camry XLE was $28,720. Let X = price, and assume that price follows a normal distribution with $\mu = \$28{,}720$ and $\sigma = \$1000$. Find the prices that separate the middle 95% of 2015 Toyota Camry XLE prices from the bottom 2.5% and the top 2.5%.

Solution

Step 1   **Determine X, μ, and σ, and draw the normal curve for X.**
Let X = price, $\mu = \$28{,}720$, and $\sigma = \$1000$. The middle 95% of prices are between X_1 and X_2, as shown in Figure 53.

Step 2   **Find the Z-values corresponding to the desired area.**
The area to the left of X_1 equals 0.025, and the area to the left of X_2 equals 0.975. Looking up area 0.025 on the inside of the Z table gives us $Z_1 = -1.96$. Looking up area 0.975 on the inside of the Z table gives us $Z_2 = 1.96$.

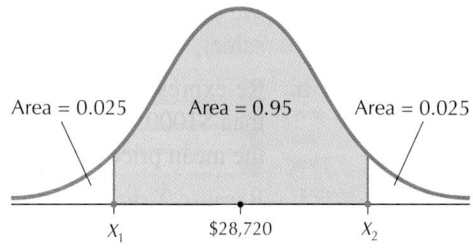

Area = 0.025   Area = 0.95   Area = 0.025

X_1 \$28,720 X_2

FIGURE 53 X_1 and X_2 mark the middle 95% of Camry XLE prices.

Step 3   **Transform using the formula $X_1 = Z\sigma + \mu$.**
We calculate

$$X_1 = Z_1\sigma + \mu = (-1.96)(1000) + 28{,}720 = 26{,}760$$
$$X_2 = Z_2\sigma + \mu = (1.96)(1000) + 28{,}720 = 30{,}680$$

NOW YOU CAN DO
Exercises 23–26.

The prices that separate the middle 95% of 2015 Toyota Camry XLE prices from the bottom 2.5% of prices and the top 2.5% of prices are $26,760 and $30,680.

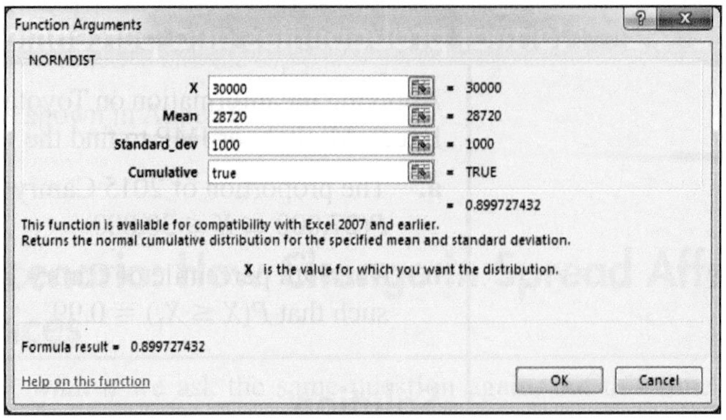

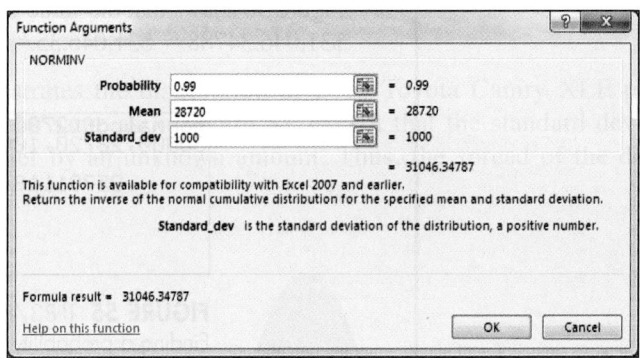

FIGURE 58 Excel: $P(X \le 30,000)$.

b. Excel provides the result shown in Figure 59: $X_1 = \$31,046.34788 \approx \$31,046.35$

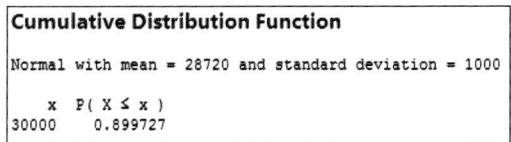

FIGURE 59 Excel: Finding a value of X.

Minitab

a. Similar to Excel, Minitab asks you to take the difference of two cumulative probabilities: $P(X \le 27,000)$ in Figure 60 and $P(X \le 30,000)$ in Figure 61:

$$P(27,000 \le X \le 30,000) = 0.899727 - 0.0427162 = 0.8570108 \approx 0.8570$$

Cumulative Distribution Function

Normal with mean = 28720 and standard deviation = 1000

```
     x   P( X ≤ x )
27000    0.0427162
```

FIGURE 60 Minitab: $P(X \le 27,000)$.

Cumulative Distribution Function

Normal with mean = 28720 and standard deviation = 1000

```
     x   P( X ≤ x )
30000    0.899727
```

FIGURE 61 Minitab: $P(X \le 30,000)$.

b. The results are given in Figure 62: $X_1 = \$31,046.30$

Inverse Cumulative Distribution Function

Normal with mean = 28720 and standard deviation = 1000

```
P( X ≤ x )        x
     0.99   31046.3
```

FIGURE 62 Minitab: Finding a value of X.

JMP

a. JMP also asks you to take the difference of two cumulative probabilities: $P(X \le 27{,}000)$ in Figure 63 and $P(X \le 30{,}000)$ in Figure 64:

$$P(27{,}000 \le X \le 30{,}000) = 0.899727432 - 0.0427162208 = 0.8570112112 \approx 0.8570$$

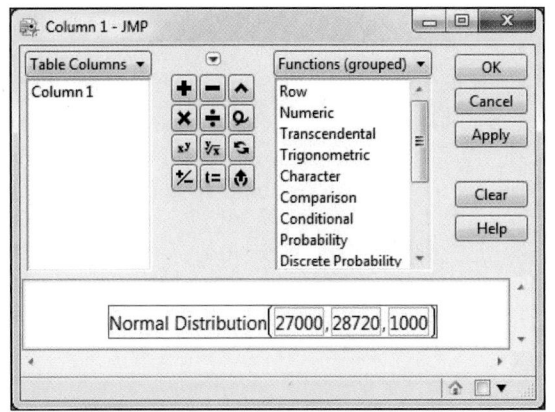
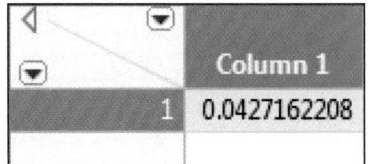

FIGURE 63 JMP: $P(X \le 27{,}000)$.

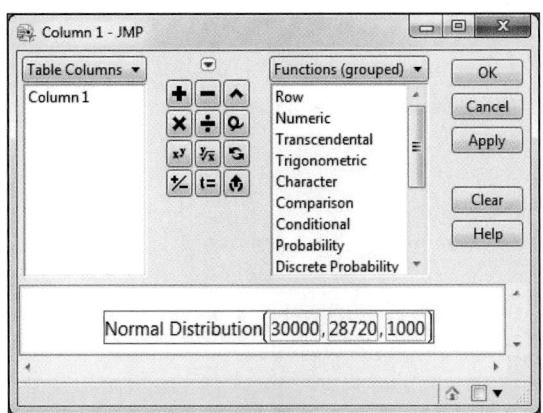

FIGURE 64 JMP: $P(X \le 30{,}000)$.

b. The results are given in Figure 65: $X_1 = \$31{,}046.30$

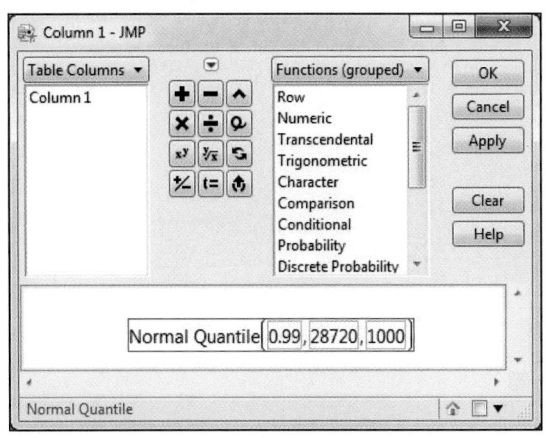
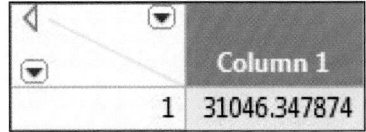

FIGURE 65 JMP: Finding a value of X.

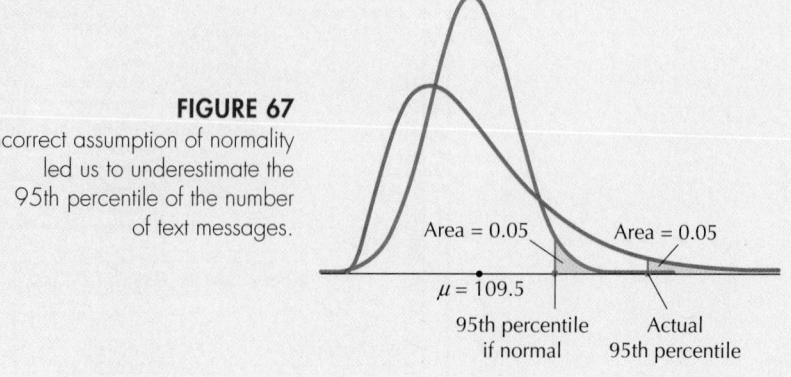

Developing Your Statistical Sense

Text Messaging: Be Careful What You Assume

The Pew Internet and American Life Project reported in 2011 that the mean number of text messages sent per day by 18- to 24-year-old Americans is 109.5. Assume that the distribution of the number of text messages is normal, with $\mu = 109.5$ and standard deviation $\sigma = 35$.

Problem 1. Suppose that cell phone customers get a special rate if the number of text messages they send per day is at or above the 95th percentile. Find the number of text messages represented by the 95th percentile.

Solution to Problem 1. On the assumption that the number of text messages is normally distributed, and working similarly to Example 42b, we find the 95th percentile of text messages to be about 167, as shown in Figure 66.

FIGURE 66
95th percentile of text messages.

```
invNorm(0.95,109
.5,35)
        167.0698769
```

Problem 2. Pew reports further that the median number of text messages sent per day by 18- to 24-year-old Americans is 50.

a. What does this say about our assumption of normality for the distribution of text messages?

b. What shape does the distribution of the number of text messages actually take?

c. Is the actual 95th percentile of text messages greater or less than 167, and why?

Solution to Problem 2.

a. In Chapter 3, we learned that, for symmetric distributions (such as the normal distribution), the mean and the median were about equal (see Figure 5 on page 115). The mean number of text message 109.5 is much larger than the median of 50 text messages, so the distribution of text messages is not symmetric and thus cannot be normal.

b. The number of text messages takes a shape like Figure 33 on page 73. Thus, the distribution of the number of text messages is actually right-skewed.

c. Figure 67 shows the (wrongly) assumed normal distribution in green and the actual right-skewed distribution in orange. Both distributions have the same mean: $\mu = 109.5$. The 95th percentile for each distribution is shown. Because the right tail of the right-skewed distribution is extended, the 95th percentile of the right-skewed distribution is greater than the 95th percentile of the normal distribution. Thus, the actual 95th percentile of the number of text messages sent per day by 18- to 24-year-old Americans is greater than 167.

FIGURE 67
Incorrect assumption of normality led us to underestimate the 95th percentile of the number of text messages.

Area = 0.05   Area = 0.05

$\mu = 109.5$

95th percentile if normal   Actual 95th percentile

3 Assessing Normality Using Normal Probability Plots

Much of the analysis we conduct in this text requires that the sample data come from a population that is normally distributed. But how do we assess whether a data set is normally distributed? Histograms, dotplots, and stem-and-leaf displays may be used. But a more precise graphical tool for assessing normality is the **normal probability plot**. A normal probability plot is a scatterplot of the estimated cumulative normal probabilities (expressed as percents) against the corresponding data values in the data set.

Analyzing Normal Probability Plots

If the points in the normal probability plot either cluster around a straight line or nearly all fall within the curved bounds, then it is likely that the data set is normal. Systematic deviations off the straight line are evidence against the claim that the data set is normal.

Professional statistical analysts always use technology to construct normal probability plots. We show how this is done in the Step-by-Step Technology Guide at the end of this section.

EXAMPLE 43 Normal probability plots

Figures 68 and 69 show normal probability plots for two different data sets. Analyze these plots for evidence for or against the normality of each data set.

Solution

In Figure 68, the points are arrayed nicely along the straight line, and all the points lie within the curved bounds. We therefore conclude that the data represented in Figure 68 are normally distributed. (In fact, the underlying data are drawn from a normal distribution.) In Figure 69, the points do not line up in a straight line, and many points lie outside the curved bounds, indicating that the data set is not normal. We therefore conclude that the data represented in Figure 69 are not normally distributed. (In reality, the underlying data set is right-skewed.)

NOW YOU CAN DO
Exercises 27–30.

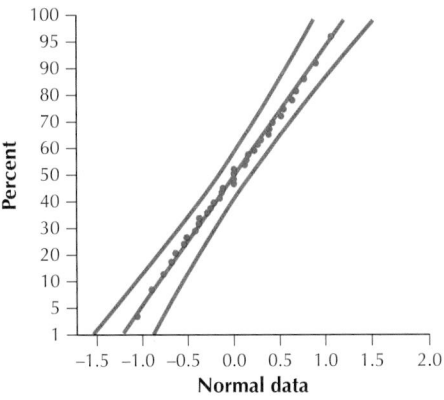

FIGURE 68 Normal probability plot of normal data.

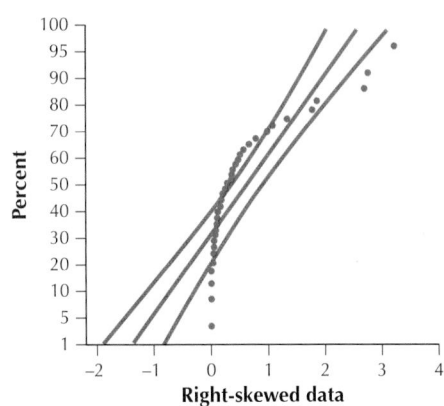

FIGURE 69 Normal probability plot of right-skewed data.

Solution

Once again, we have a binomial experiment with $n = 64$ and $p = 0.2$.

a. "At most" 12 children means 12 or fewer children. That is, $X = 12$ and $X = 11$ and $X = 10$, and so on; that is, $P(X_{binomial} \leq 12)$. In this case, we see that $X = 12$ is included in the probability we seek, as shown in Figure 74. From Table 9, we see that $P(X_{binomial} \leq 12)$ is of the form $P(X_{binomial} \leq a)$. Thus, our continuity correction takes the form $P(Y_{normal} \leq a + 0.5)$, where we add 0.5 to 12, so that

$$P(X_{binomial} \leq 12) \approx P(Y_{normal} \leq 12.5)$$

Recall that $\mu_X = 12.8$ and $\sigma_X = 3.2$. We use the TI-83/84, as shown in Figures 75 and 76, and find that the probability that, at most, 12 children lack immunizations is $0.462653813 \approx 0.4627$.

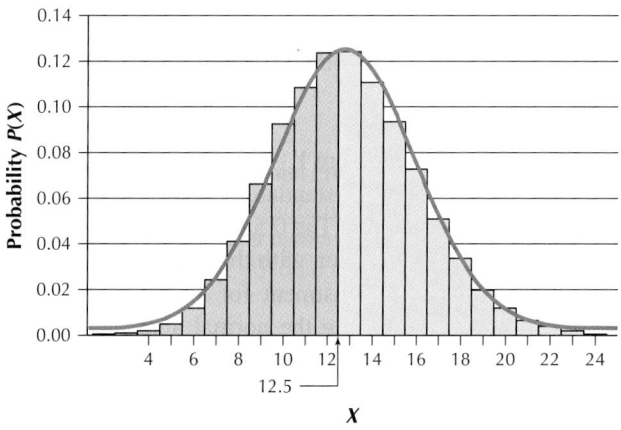

FIGURE 74 Approximates a binomial probability with a normal probability.

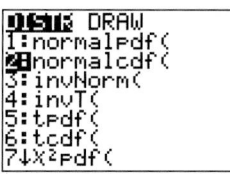

FIGURE 75 TI-83/84.

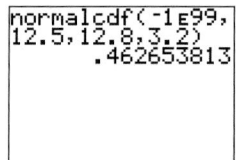

FIGURE 76 TI-83/84 results.

The *Normal Approximation to the Binomial Distributions* applet allows you to choose your own values of n and p and see how changes in these values affect the normal approximation to the binomial distribution.

b. "More than" 12 children means $X = 13$ and $X = 14$, and so on. In other words, $X = 12$ is not included. That is, we want $P(X_{binomial} > 12)$. From Table 9, we see that $P(X_{binomial} > 12)$ is of the form $P(X_{binomial} > a)$. Thus, our continuity correction takes the form $P(Y_{normal} > a + 0.5)$, where we add 0.5 to 12, so that

$$P(X_{binomial} > 12) \approx P(Y_{normal} > 12.5)$$

The desired area is the complement of the green area in Figure 74, so we can find the answer like this:

$$P(X_{binomial} > 12) \approx P(Y_{normal} > 12.5) = 1 - P(Y_{normal} \leq 12.5) = 1 - 0.4627 = 0.5373$$

The probability that more than 12 preschool children will not have the required immunizations is 0.5373.

NOW YOU CAN DO
Exercises 9–24.

For the scenario in Example 45, approximate the probability that there are less than 12 children without immunizations.

(The solution is shown in Appendix A.)

Section 6.6 Summary

1. For certain values of n, p, and X, it may be inconvenient to calculate probabilities for the binomial distribution. The normal distribution can be used to approximate binomial probabilities when $n \cdot p \geq 5$ and $n \cdot q \geq 5$.

Section 6.6 Exercises

CLARIFYING THE CONCEPTS

1. Provide an example of why we would need to use the normal approximation to the binomial distribution. (p. 385)
2. What are the requirements for using the normal approximation to the binomial distribution? (p. 386)

PRACTICING THE TECHNIQUES

 **CHECK IT OUT!**

To do	Check out	Topic
Exercises 3–8	Example 44	Requirements for a normal approximation to the binomial distribution
Exercises 9–24	Example 45	The normal approximation to the binomial distribution

For Exercises 3–8, determine whether the requirements are met for using the normal approximation to the binomial probability distribution.

3. X is a binomial random variable with $n = 10$ and $p = 0.5$.
4. X is a binomial random variable with $n = 8$ and $p = 0.5$.
5. X is a binomial random variable with $n = 10$ and $p = 0.4$.
6. X is a binomial random variable with $n = 13$ and $p = 0.4$.
7. X is a binomial random variable with $n = 45$ and $p = 0.1$.
8. X is a binomial random variable with $n = 50$ and $p = 0.1$.

For Exercises 9–16, let X be a binomial random variable with $n = 40$ and $p = 0.5$. Use the normal approximation to find the following probabilities:

9. $P(X = 20)$
10. $P(X \geq 20)$
11. $P(X > 20)$
12. $P(X \leq 20)$
13. $P(X < 20)$
14. $P(18 \leq X \leq 22)$
15. $P(18 < X < 22)$
16. $P(18 \leq X < 22)$

For Exercises 17–24, let X be a binomial random variable with $n = 120$ and $p = 0.1$. Use the normal approximation to find the following probabilities:

17. $P(X \geq 10)$
18. $P(X \geq 9)$
19. $P(X > 10)$
20. $P(X \leq 8)$
21. $P(X < 8)$
22. $P(9 \leq X \leq 11)$
23. $P(9 < X < 11)$
24. $P(9 < X \leq 11)$

APPLYING THE CONCEPTS

25. **e-Cigarettes.** A 2014 study[17] found that 6.5% of adolescents have tried e-cigarettes. (Though e-cigarettes have been marketed as cessation helpers, the study found that use of e-cigarettes does not discourage, and may encourage, conventional cigarette use among adolescents.) For a sample of 1000 U.S. adolescents, approximate the following probabilities:
    a. More than 65 have tried e-cigarettes.
    b. At least 65 have tried e-cigarettes.
    c. Less than 65 have tried e-cigarettes.
    d. Between 60 and 70 (inclusive) have tried e-cigarettes.

26. **Make Mine Medium.** fivethirtyeight.com reported that 31% of those surveyed like their steaks cooked medium.[18] For a sample of 200 people, approximate the following probabilities:
    a. At least 70 people like their steaks cooked medium.
    b. More than 69 people like their steaks cooked medium.
    c. At most 69 people like their steaks cooked medium.
    d. Between 70 and 75 (inclusive) like their steaks cooked medium.

27. **Hurricane Response.** A survey found that 19% of respondents in New Orleans rated the overall response by government and volunteer agencies to major hurricanes in the past three years as good or excellent, whereas 57% of those living in other areas did so.[19] Suppose that we have a sample

7 Sampling Distributions

Introduction

In Chapter 7, students are introduced to point estimation, sampling distributions, and one of the most important results in statistical inference, the Central Limit Theorem. Section 7.1 introduces sampling distributions and develops the Central Limit Theorem for Means. Section 7.2 develops the Central Limit Theorem for Proportions. Note that we have merged the former Section 7.1 Introduction to Sampling Distributions into the new Section 7.1 Central Limit Theorem for Means. This will help to streamline the coverage.

From the Author

Section 7.1 Central Limit Theorem for Means

- The topic *Using normal probability plots to assess normality* has been moved to Section 6.5, Applications of the Normal Distribution.

- The Olympic medals demonstration at the beginning of the section provides a simple illustration of a sampling distribution.

- Students should examine Figure 2, which summarizes the difference between a distribution of X and a sampling distribution of $\bar{x}$.

- The inference for the population mean in subsequent chapters all comes down to the Three Possible Cases for the Sampling Distribution of the Sample Mean $\bar{x}$, in the box on page 401. Either the population is normal (Case 1), or the sample size is large (Case 2), or neither condition is fulfilled (Case 3).

- Example 8, Finding Two Symmetric Sample Mean Values Using the Central Limit Theorem for Means, is intended as a warm-up for learning about confidence intervals for the population mean.

Section 7.2 Central Limit Theorem for Proportions

- If you cover confidence intervals for proportions, you may want to check out Figure 17 in Example 15 on page 420.

Teaching Tips

Chapter 7 is a pivotal chapter. The preceding two chapters showed us the rules of probability and the characteristics of the normal distribution. The reason we learned these things is so that we can perform statistical inference in Chapters 8–11. That statistical inference depends on the sampling distributions that we learn about here in Chapter 7.

It is important for students to understand that sample statistics, such as the sample mean, have patterns of behavior when the sample size is large. If we repeatedly take samples from a population, then the mean of all the sample means will equal the mean of the population.

We can also find what the standard deviation of all these sample means will be. Armed with this knowledge of the mean and standard deviation of the sample mean, we can determine what is an *unusual* or *extreme* value of the sample mean. This is the basis for statistical inference.

The finite population correction factor. For the data in Table 1 on page 396, the population variance as calculated by software equals $\sigma = 43.3667$. Fact 2 on page 398 tells us that the standard deviation of the sampling distribution for a sample of size 2 is $\sigma_{\bar{x}} = \frac{\sigma}{\sqrt{n}} = \frac{43.3667}{\sqrt{2}} = 30.6649$. However, direct calculation of the standard deviation of the sample means in Table 1 yields a value of 21.6833. An enterprising student may notice this discrepancy. The response is that, where the population is not much larger than the sample, the finite population correction factor is applied, so that the relationship between the two standard deviations becomes $\sigma_{\bar{x}} = \sqrt{\frac{N-n}{N-1}} \cdot \frac{\sigma}{\sqrt{n}}$. Here, $N = 3$ and $n = 2$, so that $\sigma_{\bar{x}} = \sqrt{\frac{3-2}{3-1}} \cdot \frac{43.3667}{\sqrt{2}} = 21.6833$, which concurs with the direct calculation. This finite population correction factor does not apply when sampling with replacement, and its value tends to zero as the sample size approaches the population size. However, for most real-world problems, and for the remainder of this book, we dispense with this coefficient and assume that the population size is very large compared to the sample size.

Derivation of Fact 2: For those familiar with mathematical statistics, Fact 2 may be formally derived as follows:

$$Variance\ (\bar{x}) = Var\left(\frac{\sum x}{n}\right) = Var\left[\frac{1}{n}(x_1) + \cdots + \frac{1}{n}(x_n)\right]$$

$$= \frac{1}{n^2}(\sigma^2) + \cdots + \frac{1}{n^2}(\sigma^2) = \frac{1}{n^2}(n\sigma^2) = \frac{\sigma^2}{n}$$

$$\text{Then, } SD(\bar{x}) = \sqrt{Var\ (\bar{x})} = \frac{\sigma}{\sqrt{n}}.$$

In-Class Activities

1. Use class test scores or homework scores as a population, and then determine all the samples of size 2 and their corresponding sample means. Show that the population mean of the samples of test scores is equal to the population mean of all the individual test scores, but that the standard deviation of the sample means is less than the standard deviation of the individual test scores.

2. Simulate the rolling of a six-sided die, in Minitab, for instance, using the integer distribution. Simulate several columns of data. Row means for different sample sizes can then be computed and histograms for these means can be displayed. The shapes of these histograms should generate discussion. Also, the descriptive statistics for these sample means can be computed. In-class discussions about the means and standard deviations for the sample means can be related to the required formulas for the Central Limit Theorem for Means.

✈ Sampling Distributions

3. Perform the following activity as a class. Use the *Simple Random Sample* applet to generate the random samples requested.

a. Each student in the class should write down an estimate of the length of the following line segment. Use your eyes only! No rulers.

_____

b. Consider your class to be a population. Find the mean μ and the standard deviation σ of all student estimates.

c. Choose one student at random from the class. Ask for this student's estimate.

d. Take a random sample of two students from the class, and find the mean $\bar{x}$ of their estimates. What is the mean $\mu_{\bar{x}}$ and standard deviation $\sigma_{\bar{x}}$ of the sampling distribution of $\bar{x}$? Given these values for $\mu_{\bar{x}}$ and $\sigma_{\bar{x}}$, which values for the length of the line segment would be considered unusual? (*Hint:* Which would be outliers?)

e. Take a random sample of 10 students from the class, and find the mean $\bar{x}$ of their estimates. What are $\mu_{\bar{x}}$ and $\sigma_{\bar{x}}$ for $n = 10$? Now, which values for the length of the line segment would be considered unusual?

f. Take a random sample of 20 students, and find their estimates. What are $\mu_{\bar{x}}$ and $\sigma_{\bar{x}}$? Now, which values for the length of the line segment would be considered unusual?

g. As the sample size increases, what is happening to $\sigma_{\bar{x}}$, and what is happening to the estimates that are considered unusual? How can you state this phenomenon in terms of sample size and precision of the estimate?

Supplements

- StatTutor 11.1–11.5(a) and (b), 20.1, and 20.2(a) and (b)

Applets

The *Law of Large Numbers Central Limit Theorem* and *Simple Random Sample* applets are referenced in Chapter 7.

The *Central Limit Theorem* applet is used for Exercises 119 and 120 in Section 7.1.

This Web site has an applet for the Central Limit Theorem: http://wise.cgu.edu/.

Videos

- *Against All Odds: Inside Statistics:* www.learner.org/resources/series65.html
  - Program 22: Sampling Distributions
  - Program 23: Control Charts

Web Sites

- This Web site has a collection of 20 class projects, one of which deals with sampling distribution: www.amstat.org/publications/jse/v6n3/smith.html.

- The site Online Statistics: An Interactive Multimedia Course of Study has a *Sampling Distribution* applet that demonstrates several simulations, and it has a link to the Rice Virtual Lab in Statistics with numerous simulation demonstrations, along with other topics related to statistics: http://onlinestatbook.com/index.html.

- This site lists other sites that do statistical calculations: http://statpages.org/.

- This site lists numerous activities on all topics: http://mathforum.org/mathtools/sitemap2/ps/.

- This site has several activities using Excel and Minitab: www.mathspace.com/NSF_ProbStat/Teaching_Materials/Primarily_Statistics.htm.

7 Sampling Distributions

OVERVIEW

© Garry Gay/Alamy

 CASE STUDY

Trial of the Pyx: How Much Gold Is in Your Gold Coins?

The kings of bygone England had a problem: How much gold should they put into their gold coins? After all, the very commerce of the kingdom depended on the purity of the currency. How did the lords of the realm ensure that the coins floating around the kingdom contained reliable amounts of gold?

From the year 1282, the Trial of the Pyx has been held annually in London to ensure that newly minted coins adhere to the standards of the realm. It is the responsibility of the presiding judge to ensure that the trial proceeds lawfully and to inform Her Majesty's Treasury of the verdict. Six members of the Company of Goldsmiths compose the jury, who are given two months to test the coins. It works like this: A ceremonial boxwood chest, called the Pyx, is brought forth, and a sample of 100 of the coins cast that year at the mint is put into it. The Pyx is then weighed. In times past, each gold coin, called a guinea, had an expected weight of 128 grams, so the total weight of the guineas in the Pyx was expected to be 12,800 grams.

If the weight of the coins in the Pyx was much less than 12,800 grams, the jury concluded that the Master of the Mint was cheating the crown by pocketing the excess gold, and he was severely punished. On the other hand, if the coins in the Pyx weighed much more than 12,800 grams, that wasn't good either, because it cut down on the profits produced by the kings' coin-minting monopoly.

By how much could the Master of the Mint debase the coinage before getting caught? We shall see in this chapter's Case Study, *Trial of the Pyx*, which unfolds in Section 7.1.

THE BIG PICTURE

Where we are coming from and where we are headed . . .

- In Chapters 1–4, we learned ways to describe data sets using numbers, tables, and graphs. Then, in Chapters 5 and 6, we learned the tools of probability and probability distributions that allow us to quantify uncertainty.
- Here, in Chapter 7, "Sampling Distributions," we will discover that seemingly random statistics, such as the sample mean $\bar{x}$, have *predictable behaviors*. The special type of distribution we use to describe these behaviors is called the *sampling distribution*. This leads us to perhaps the most important result in statistical inference: the Central Limit Theorem.
- The sampling distributions we learn about in this chapter form the basis for most of the statistical inference we perform in the remainder of the book. For example, in Chapter 8, "Confidence Intervals," we will learn how to estimate an unknown parameter with a certain level of confidence.

OBJECTIVES By the end of this section, I will be able to . . .

1 Describe the sampling distribution of the sample mean $\bar{x}$ when the population is normal.
2 Describe the sampling distribution of $\bar{x}$ for skewed and symmetric populations as the sample size increases.
3 Solve probability questions about the sample mean.
4 Find percentiles for the sample mean.

In Chapter 6, we dealt with probability distributions, which describe populations. Here, in Chapter 7, we return to the use of sample data, in order to show how populations and their samples are connected.

1 Sampling Distribution of $\bar{x}$ for a Normal Population

In this chapter, we will develop methods that will allow us to quantify the *behavior* of statistics like $\bar{x}$. For the sample mean $\bar{x}$, this behavior is expressed in the **sampling distribution of the sample mean**.

> The **sampling distribution of the sample mean** $\bar{x}$ for a given sample size n consists of the collection of the means of all possible random samples of size n from the population.

For example, consider the population of all three North American countries, the USA, Canada, and Mexico. In the 2012 Olympics, the USA won 104 medals, Canada won 18, and Mexico won 7. So, the population mean number of medals is

$$\mu = \frac{104 + 18 + 7}{3} = 43$$

Table 1 shows all possible random samples of size $n = 2$ from this population.

Table 1 Sampling distribution of $\bar{x}$ for samples of size $n = 2$

Sample 1	Sample 2	Sample 3
USA and Canada	**USA and Mexico**	**Canada and Mexico**
$\bar{x} = \dfrac{104 + 18}{2} = 61$	$\bar{x} = \dfrac{104 + 7}{2} = 55.5$	$\bar{x} = \dfrac{18 + 7}{2} = 12.5$

So, for this tiny population, the sampling distribution of $\bar{x}$ for $n = 2$ is the collection of sample means, {61, 55.5, 12.5}. Note from Table 1 that the value for the sample mean $\bar{x}$ varies from sample to sample. Thus, $\bar{x}$ is a *random variable*, because its value depends on chance; that is, its value depends on which countries are drawn in the random sample. We say that $\bar{x}$ exhibits *sampling variability* because its value changes from sample to sample. Fortunately, there are patterns (predictable behaviors) in how the sample mean $\bar{x}$ varies. These patterns underlie all the statistical inference about μ that we will perform in Chapters 8–10. And learning those patterns is what this chapter is all about.

First, like any distribution, the sampling distribution of the sample mean has a balance point and, therefore, a mean. Let's find the mean of the sample means in Table 1, which is denoted $\mu_{\bar{x}}$:

$$\mu_{\bar{x}} = \frac{61 + 55.5 + 12.5}{3} = 43$$

Figure 1 provides a dotplot of the sample means in Table 1, along with the mean of these sample means, indicated at the balance point $\mu = 43$.

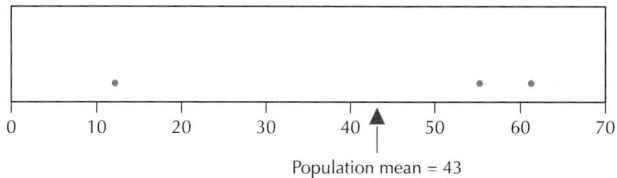

FIGURE 1 The mean of the sample means $\mu_{\bar{x}}$ equals the population mean $\mu = 43$.

Note that the mean of the sample means $\mu_{\bar{x}}$ equals the population mean $\mu = 43$. This is always true. So, we generalize it as follows.

Note: It is convenient to number a set of important *facts* as we build toward the Central Limit Theorem for Means and the Central Limit Theorem for Proportions.

Fact 1: Mean of the Sampling Distribution of $\bar{x}$

The mean of the sampling distribution of the sample mean $\bar{x}$ is the value of the population mean μ. It can be denoted as $\mu_{\bar{x}} = \mu$ and read as "the mean of the sampling distribution of $\bar{x}$ is μ."

Now, the populations and samples you will encounter, both here and in the real world, are too large to list out all the possible sample means, as we did in Table 1. Instead, we will examine population distributions of data, with a mean μ and standard deviation σ, just as we did in Chapter 6. However, in Chapter 7, the population may or may not be normally distributed.

Next let's explore some more behavior of the sampling distribution of $\bar{x}$, using a graphical example. In Example 1 we compare the population distribution of X with the sampling distribution of $\bar{x}$.

EXAMPLE 1   Sampling distribution of $\bar{x}$ for a normal population

In Example 37 in Chapter 6 (page 369), we saw that $X =$ temperature in Georgia in the month of April was *normally distributed* with a mean of $\mu = 61.5°F$ and a standard deviation of $\sigma = 8°F$. We want to compare this population distribution of X with the sampling distribution of $\bar{x}$. Using Minitab, 1000 samples of size $n = 4$ were generated from this normal distribution, and the sample means $\bar{x}$ were calculated for each sample. The histogram of green rectangles in Figure 2 was constructed, representing the sampling distribution of $\bar{x}$ for $n = 4$.

a. Identify the shape of the sampling distribution of $\bar{x}$.

b. Find the mean of the sampling distribution of $\bar{x}$.

c. Use the Empirical Rule to estimate the standard deviation of the sampling distribution of $\bar{x}$.

d. Compare the population distribution of X with the sampling distribution of $\bar{x}$.

Solution

Figure 2 shows the histogram (rectangles) of the means from the 1000 samples of size $n = 4$.

a. As you may have expected, the histogram of rectangles representing the sampling distribution of $\bar{x}$ is normal.

b. The balance point of the sampling distribution is located at 61.5. Thus, the mean $\mu_{\bar{x}}$ of the sampling distribution of $\bar{x}$ equals the population mean $\mu = 61.5$.

c. Figure 2 shows that almost all the sample means lie between 49.5 and 73.5. Recall that the Empirical Rule states that almost all the data from a normal distribution lie within 3 standard deviations of the mean. Thus, the distance $61.5 - 49.5 = 12$ represents 3 standard deviations. Thus, we estimate that the standard deviation of the sampling distribution of $\bar{x}$ is: $\sigma_{\bar{x}} = 12/3 = 4$.

d. Note the orange curve in Figure 2. This represents the distribution of the original normal temperature data, with $\mu = 61.5$ and $\sigma = 8$. Compare this to the green curve, which represents the sampling distribution of $\bar{x}$, with mean $\mu_{\bar{x}} = 61.5$ and $\sigma_{\bar{x}} = 4$. Note that the spread (variability) of the sampling distribution is less than the spread of the original distribution of X.

FIGURE 2

The sampling distribution of $\bar{x}$ (in green) has less spread (variability) than the original distribution X (in orange).

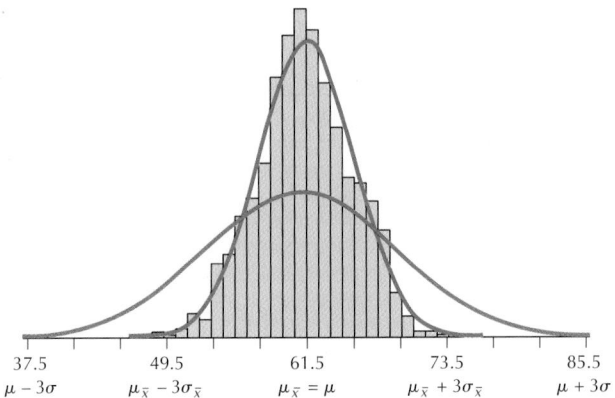

37.5	49.5	61.5	73.5	85.5
$\mu - 3\sigma$	$\mu_{\bar{x}} - 3\sigma_{\bar{x}}$	$\mu_{\bar{x}} = \mu$	$\mu_{\bar{x}} + 3\sigma_{\bar{x}}$	$\mu + 3\sigma$

In Example 1, we estimated the standard deviation of the sampling distribution of $\bar{x}$ to be $\sigma_{\bar{x}} = 4$, compared with the original standard deviation $\sigma = 8$. Fact 2 shows the relationship between these two quantities in general.

Note: Fact 2 is derived using methods found in texts on advanced statistics.

> **Fact 2: Standard Deviation of the Sampling Distribution of $\bar{x}$**
>
> The standard deviation of the sampling distribution of the sample mean $\bar{x}$ is
>
> $$\sigma_{\bar{x}} = \sigma/\sqrt{n}$$
>
> where σ is the population standard deviation and n is the sample size. $\sigma_{\bar{x}}$ is called the **standard error of the mean**.

Thus, for Example 1, $\sigma_{\bar{x}} = \sigma/\sqrt{n} = 8/\sqrt{4} = 8/2 = 4$, so our estimate was correct. Note the $\sqrt{n}$ in the denominator of the formula. Because of this factor, the larger the sample size, the tighter the resulting sampling distribution. Larger sample sizes lead to smaller variability, which results in more precise estimation.

EXAMPLE 2   Finding the mean and standard deviation of the sampling distribution of $\bar{x}$

According to the American Time Use Survey, the mean amount of sleep that 20- to 24-year-old women get per night is 9.3 hours (*Source:* Bureau of Labor Statistics[1]). Assume that the population standard deviation is $\sigma = 1$ hour. Find the mean and standard deviation for the sampling distribution of $\bar{x}$ for the following sample sizes: **(a)** 4, **(b)** 100, **(c)** 400.

Solution

We have $\mu_{\bar{x}} = \mu = 9.3$. Note that this value for $\mu_{\bar{x}}$ does not depend on the sample size, so the value is true for any sample size. We also have $\sigma = 1$.

a. $n = 4$. Then, $\sigma_{\bar{x}} = \dfrac{\sigma}{\sqrt{n}} = \dfrac{1}{\sqrt{4}} = 0.5$. The standard error of the mean for $n = 4$ is $\sigma_{\bar{x}} = 0.5$ hours.

b. $n = 100$. Then, standard error $\sigma_{\bar{x}} = \dfrac{\sigma}{\sqrt{n}} = \dfrac{1}{\sqrt{100}} = 0.1$.

c. $n = 400$. Then, standard error $\sigma_{\bar{x}} = \dfrac{\sigma}{\sqrt{n}} = \dfrac{1}{\sqrt{400}} = 0.05$.

NOW YOU CAN DO

Exercises 9–16.

YOUR TURN #1

For the situation in Example 2, find the mean and standard deviation for the sampling distribution of $\bar{x}$ for $n = 900$.

(The solution is shown in Appendix A.)

What Does This Number Mean?

Consider $\sigma_{\bar{x}} = 0.1$ for $n = 100$ in Example 2(**b**). This is a measure of the variability of the sampling distribution of $\bar{x}$ for this sample size. That is, if we take samples of size 100, our estimation of the population mean amount of sleep μ of all 20- to 24-year-old women will be within 0.1 hour (6 minutes) of the true population mean most of the time.

Recall in Example 1 that when the original population distribution is normal, the sampling distribution of $\bar{x}$ is also normal. This is true in general, as stated in Fact 3, which also includes what we have learned so far in Fact 1 and Fact 2.

Note: Let the notation

$$\text{normal } (\mu, \sigma/\sqrt{n})$$

denote a normal distribution with mean of μ and standard deviation of $\sigma/\sqrt{n}$.

Fact 3: Sampling Distribution of the Sample Mean for a Normal Population

- When random samples of size n are taken from a data distribution X that is normally distributed, the sampling distribution of $\bar{x}$ is also normally distributed.

- Thus, when random samples of size n are taken from a normal population with mean μ and standard deviation σ, the sampling distribution of the sample mean $\bar{x}$ is distributed as normal with mean $\mu_{\bar{x}} = \mu$ and standard error $\sigma_{\bar{x}} = \sigma/\sqrt{n}$.

- In other words, the sampling distribution of $\bar{x}$ is distributed as normal $(\mu, \sigma/\sqrt{n})$.

The good news is that, once we know that the sampling distribution is normal, we can use the methods we learned in Section 6.5 to solve probability problems. Specifically, we can standardize and produce Z, just as we would for any normal random variable, using Fact 4.

Fact 4: Standardizing a Normal Sampling Distribution for Means

When the sampling distribution of $\bar{x}$ is normal, we may standardize to produce the standard normal random variable Z as follows:

$$Z = \frac{\bar{x} - \mu_{\bar{x}}}{\sigma_{\bar{x}}} = \frac{x - \mu}{\sigma/\sqrt{n}}$$

where μ is the population mean, σ is the population standard deviation, and n is the sample size.

---

2 Central Limit Theorem for Means

The discussion so far has been for the case when the original population is normally distributed. However, what if the population (distribution of X) is not normal? In this section, we use a simulation study to learn how the sampling distribution of the sample mean $\bar{x}$ for non-normal populations becomes approximately normal as the sample size increases.

EXAMPLE 3   Simulation study: Sample means from a strongly skewed population

nutrition

The data set **Nutrition** contains nutrition information on a population of 961 foods.

a. Construct a histogram of the potassium content of these 961 foods, and describe the shape of the population distribution.

b. Using Minitab, take 500 random samples of sizes $n = 10$, 20, and 30 from the population. Assess the normality of the resulting sampling distributions of $\bar{x}$ using histograms and normal probability plots.

Solution

a. A histogram of the potassium content of these foods is shown in Figure 3, revealing a strongly right-skewed, non-normal data set.

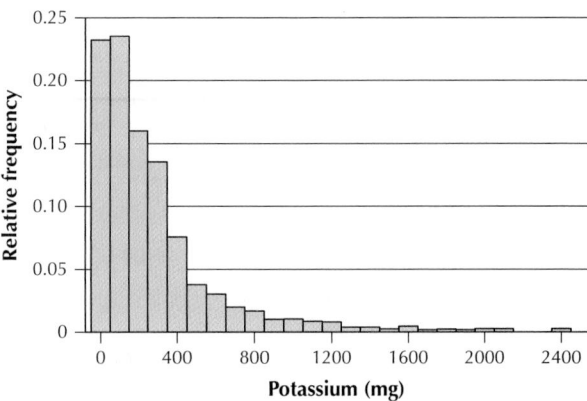

FIGURE 3
Potassium content is strongly
right-skewed, not normal.

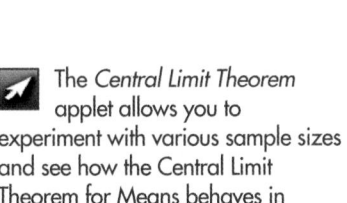

The *Central Limit Theorem*
applet allows you to
experiment with various sample sizes
and see how the Central Limit
Theorem for Means behaves in
action.

b. Using Minitab, we take 500 random samples of size $n = 10$ from the population. The histograms of each set of 500 means are shown in Figure 4.

- $n = 10$: The sampling distribution $\bar{x}$ is skewed (Figure 4a).
- $n = 20$: The sampling distribution of $\bar{x}$ is still somewhat skewed (Figure 4b).
- $n = 30$: Despite a few outliers, the sampling distribution of $\bar{x}$ is approximately normal (Figure 4c).

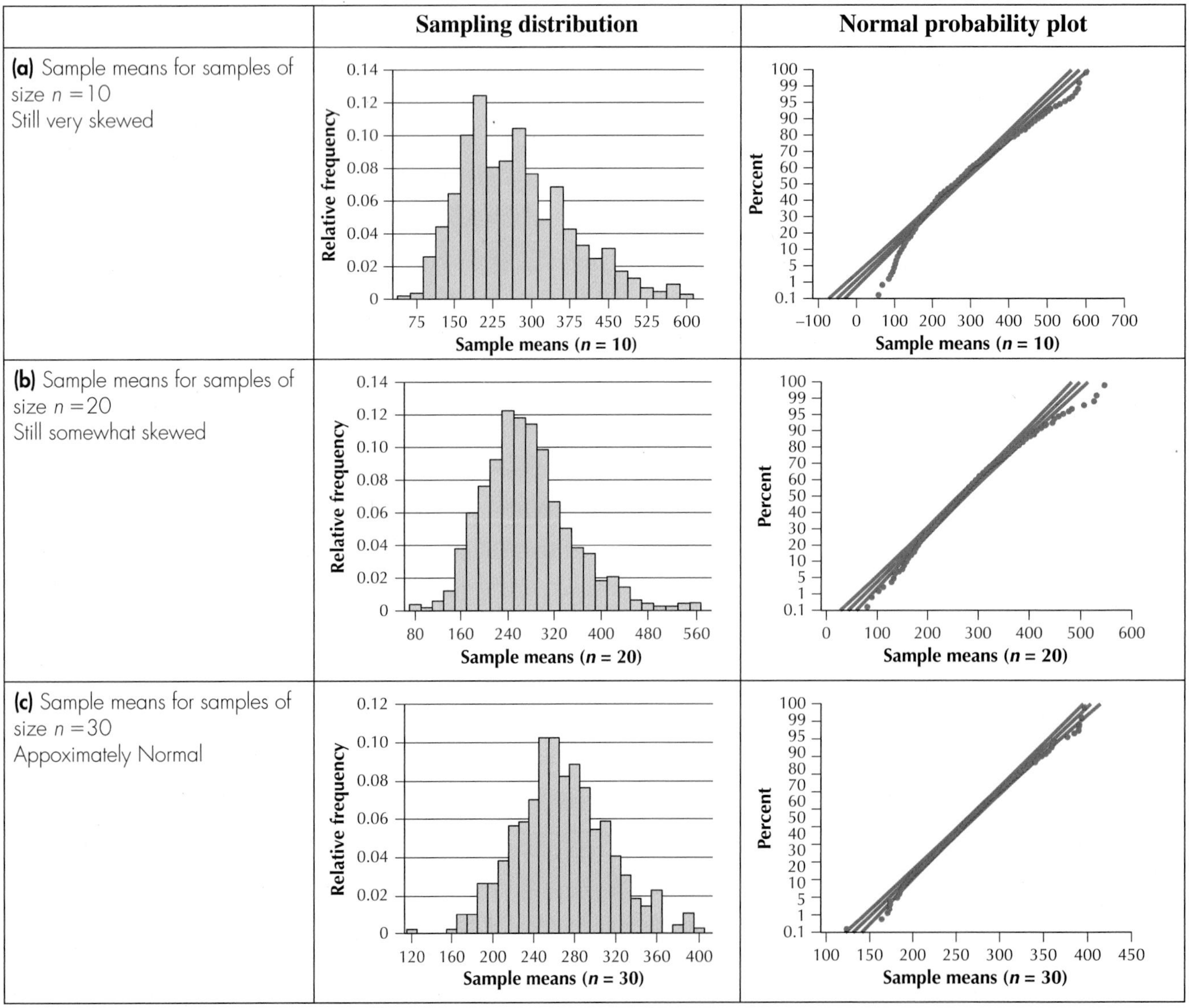

	Sampling distribution	**Normal probability plot**
(a) Sample means for samples of size $n = 10$ Still very skewed		
(b) Sample means for samples of size $n = 20$ Still somewhat skewed		
(c) Sample means for samples of size $n = 30$ Approximately Normal		

FIGURE 4 Sampling distribution of $\bar{x}$ and normal probability plots for $n = 10$, 20, and 30.

For a skewed population, we have seen that the sampling distribution of the sample mean becomes approximately normal as the sample size reaches 30. For a less skewed population, we can expect that the sampling distribution of $\bar{x}$ approximates a normal distribution for smaller sample sizes. Our simulation study has shown us that *regardless of the population, the sampling distribution of the sample mean becomes approximately normal as the sample size gets larger*. We can then combine this statement with Fact 3 (page 399) to form the **Central Limit Theorem for Means**.

Central Limit Theorem for Means

When random samples of size n are taken from a population with mean μ and standard deviation σ, the sampling distribution of the sample mean $\bar{x}$ becomes approximately normal $(\mu, \sigma/\sqrt{n})$ as the sample size gets larger, regardless of the shape of the population.

How large does the sample size have to be before the Central Limit Theorem for Means takes effect? In general, it depends on the degree of symmetry, or skewness, of the population. In the simulation study (Figure 4), we saw that the sampling distribution of $\bar{x}$ was approximately normal, even for a skewed population when $n = 30$. Thus, we shall abide by the following rule of thumb.

Rule of Thumb for When to Use the Central Limit Theorem for Means

We consider $n \geq 30$ as large enough to apply the Central Limit Theorem for Means for any population.

Developing Your Statistical Sense

The Central Limit Theorem

The Central Limit Theorem (CLT) is one of the most important results in statistics. Worldwide, much statistical inference is based on the CLT. It actually makes fairly intuitive sense, doesn't it? If we find the mean of a sample of data values, in many cases the extreme values will tend to balance out. However, remember that the mean is very sensitive to outliers. In a small sample, there may not be enough nonextreme values to balance the influence of the outliers. This is what was happening early in the potassium simulation (for example, Figure 4a). However, as the sample sizes increase, the influence of extreme values diminishes and the resulting sample means start to migrate toward the center.

Combining Fact 3 and the Central Limit Theorem for Means, we can identify three possible cases for the sampling distribution of $\bar{x}$.

Three Possible Cases for the Sampling Distribution of the Sample Mean $\bar{x}$

Case 1.   The population is normal. Therefore, the sampling distribution of $\bar{x}$ is *normal* (Fact 3, page 399).

Case 2.   The population is either non-normal or of unknown distribution *and* the sample size is at least 30. Therefore, the sampling distribution of $\bar{x}$ is *approximately normal* (Central Limit Theorem for Means).

Case 3.   The population is either non-normal or of unknown distribution *and* the sample size is less than 30. Therefore, we have *insufficient information* to conclude that the sampling distribution of the sample mean $\bar{x}$ is either normal or approximately normal.

Of course, in the real world, no one will tell you which of the three cases applies. You need to investigate the assumptions of each of the cases to determine for yourself which one applies.

What Does This Probability Mean?

There is essentially no chance that the sample mean number of page likes $\bar{x}$ will be greater than 80. Compare this to the nearly 16% chance that a *particular* Facebook user's number of page likes would be above 80. Figure 5 shows the graphs of the distributions of (a) the individual Facebook users, and (b) the means of sets of 25 users. Both distributions are centered at $\mu_{\bar{x}} = \mu = 70$, but the standard deviations differ. The arrow in Figure 5a represents the standard deviation of X, $\sigma = 10$, and it shows that $x = 80$ is only 1 standard deviation above the mean $\mu = 70$. The arrows in Figure 5b represent the standard error of the mean, $\sigma_{\bar{x}} = 2$, and they illustrate that $\bar{x} = 80$ lies 5 standard errors above the mean $\mu_{\bar{x}} = 70$. Thus, group means are less variable than individual data values.

In Chapter 6, we found the percentiles of normally distributed random variables. The sampling distribution of $\bar{x}$ is normal, so we are able to find the percentiles of the $\bar{x}$'s, too. Once the appropriate Z-value is found, we use the following equation to transform the Z-value into an $\bar{x}$-value:

$$\bar{x} = Z \cdot \sigma_{\bar{x}} + \mu_{\bar{x}} = Z \cdot \frac{\sigma}{\sqrt{n}} + \mu$$

EXAMPLE 6 Finding a value of $\bar{x}$, given a probability or area

Use the information in Example 5 to do the following:

a. Find the 95th percentile of the sample mean number of Facebook page likes from samples of size $n = 25$.

b. Find the 5th percentile of the sample means.

c. What two symmetric values for the sample mean contain the middle 90% of all sample means between them?

d. Verify that $P(66.71 \leq \bar{x} \leq 73.29) = 0.90$.

Solution

The 95th percentile of the sample mean number of Facebook likes is the value of $\bar{x}$ with area 0.95 to the left of it.

a. We want the 95th percentile, so we seek 0.95 on the *inside* of the Z table. Because 0.95 is not in the Z table, we take the closest value. The two closest values, 0.9495 and 0.9505, are equally close, so we split the difference. Working backward from 0.9495, we find $Z = 1.64$, and for 0.9505 we find $Z = 1.65$. Splitting the difference, we get $Z = 1.645$. This value of $Z = 1.645$ is the 95th percentile of the standard normal distribution.

Because we are looking for a sample mean number of page likes, 1.645 is probably not the answer. We need to "unstandardize" by transforming this value of Z to an $\bar{x}$-value:

$$\bar{x} = Z \cdot \sigma_{\bar{x}} + \mu = 1.645(2) + 70 = 73.29$$

Thus, the 95th percentile of the sample means for the number of Facebook page likes is 73.29.

b. The sampling distribution is normal, so it is also symmetric. Thus, the 95th percentile and the 5th percentile are the same distance away from the mean. Because the 95th percentile is $(73.29 - 70) = 3.29$ above the mean, the 5th percentile must be 3.29 below the mean, or $(70 - 3.29) = 66.71$.

c. This is just another way of asking for the 5th and 95th percentiles, which we found in parts (**a**) and (**b**). (See Figure 6.) The answer is 66.71 and 73.29.

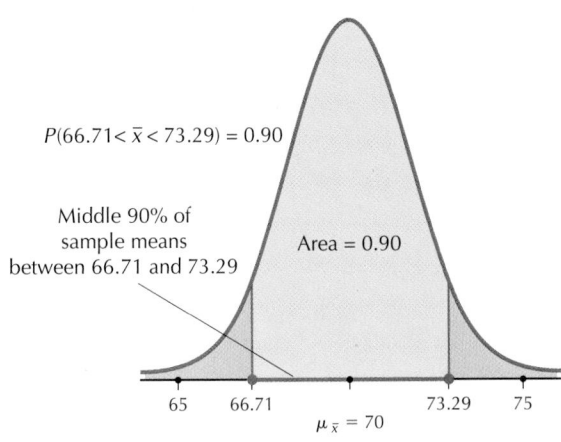

NOW YOU CAN DO
Exercises 33–58.

FIGURE 6
Middle 90% of the sample means.

$P(66.71 < \bar{x} < 73.29) = 0.90$

Middle 90% of
sample means
between 66.71 and 73.29

Area = 0.90

d.   We seek $P(66.71 < \bar{x} < 73.29)$, as shown in Figure 6. Proceeding with the calculations, we have, as expected,

$$P(66.71 < \bar{x} < 73.29) = P\left(\frac{66.71 - 70}{2} < \frac{\bar{x} - 70}{2} < \frac{73.29 - 70}{2}\right)$$

$$= P(-1.645 < Z < 1.645) = 0.95 - 0.05 = 0.90$$

YOUR TURN
#3

Using the information in Example 5, find the 2.5th percentile of the mean number of Facebook page likes.

(The solution is shown in Appendix A.)

EXAMPLE 7   Finding probabilities using the Central Limit Theorem for Means

fuelefficiency

The data set **FuelEfficiency** contains information about the fuel efficiency of more than one thousand 2014 model year vehicles (*Source:* fueleconomy.gov). Figure 7 shows a histogram of the *city miles per gallon* (city mpg) of these vehicles, illustrating that the distribution is not normal, but somewhat right-skewed. The mean city mpg for all vehicles is $\mu = 20$, and the standard deviation is $\sigma = 6$. Find the probability that a random sample of size $n = 36$ vehicles will have a mean city mpg greater than 21.

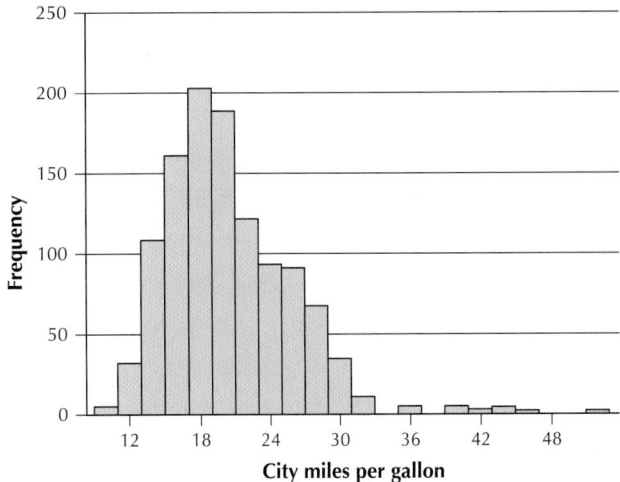

FIGURE 7
Population is skewed, so a large sample is needed to apply the Central Limit Theorem.

Solution

Clearly, the population is not normal, but the sample size $n = 36$ is large enough, so the Central Limit Theorem applies. The sampling distribution of the sample mean $\bar{x}$ is approximately normal. This is Case 2 from the Three Cases (page 401). Next, we need to find $\mu_{\bar{x}}$ and $\sigma_{\bar{x}}$. Facts 1 and 2 tell us that

$$\mu_{\bar{x}} = \mu = 20 \qquad \text{and} \qquad \sigma_{\bar{x}} = \frac{\sigma}{\sqrt{n}} = \frac{6}{\sqrt{36}} = 1$$

Therefore, as the CLT indicates, the sampling distribution of $\bar{x}$ is approximately normal ($\mu_{\bar{x}} = 20$, $\sigma_{\bar{x}} = 1$). We are then left to solve a normal probability problem using the methods of Section 6.5. Figure 8 shows the sampling distribution of $\bar{x}$ and the probability we are interested in, $P(\bar{x} > 21)$. Using Fact 4, we standardize:

$$Z = \frac{21 - \mu_{\bar{x}}}{\sigma_{\bar{x}}} = \frac{21 - 20}{1} = 1$$

Thus, $P(\bar{x} > 21) = P(Z > 1)$, as shown in Figure 9. We therefore look up $Z = 1$ in the Z table and subtract this table area (0.8413) from 1 to get the desired tail area:

$$P(Z > 1) = 1 - 0.8413 = 0.1587$$

NOW YOU CAN DO
Exercises 59–70.

The probability is 0.1587 that a random sample of 36 vehicles will have a mean city mpg greater than 21.

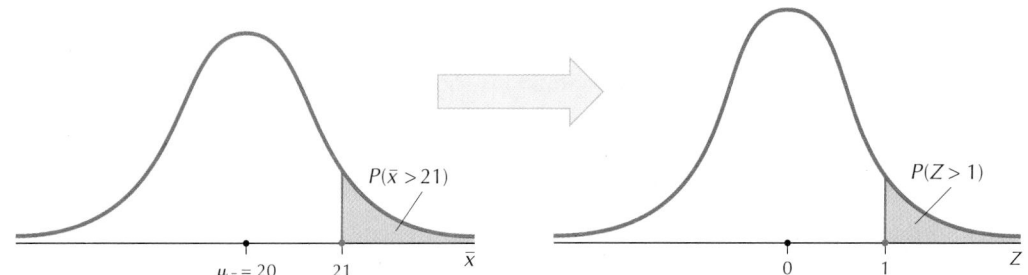

FIGURE 8 Area to the right of $\bar{x} = 21$ equals . . .   **FIGURE 9** Area to the right of $Z = 1$.

YOUR TURN #4

For the situation in Example 7, find the probability that a random sample of size $n = 36$ vehicles will have a mean city mpg that is less than 18.5.

(The solution is shown in Appendix A.)

EXAMPLE 8 Finding two symmetric sample mean values using the Central Limit Theorem for Means

FIGURE 10

NOW YOU CAN DO
Exercises 71–80.

Refer to Example 7. If we take samples of size $n = 36$, find the two symmetric values of $\bar{x}$ that contain the middle 95% of sample mean city mpg values.

Solution

Here, we use the instructions from the **Finding Percentiles for Any Normal Distribution** portion of the Section 6.5 Step-by-Step Technology Guide on page 380. We have $\mu_{\bar{x}} = 20$ and $\sigma_{\bar{x}} = 1$. We want the middle 95% of sample means, so we ask the TI-83/84 calculator to give us the sample mean with 2.5% to the left of it and the sample mean with 97.5% to the left of it, as shown in Figure 10. We get $\bar{x}_1 \approx 18.04$ as the 2.5th percentile of sample means and $\bar{x}_2 \approx 21.96$ as the 97.5th percentile of sample means. These are the two symmetric values of the sample mean that contain the middle 95% of all sample mean city mpg values.

YOUR TURN
#5

For the situation in Example 7, if we take samples of size $n = 36$, find the two symmetric values of $\bar{x}$ that contain the middle 90% of sample mean city mpg values.

(The solution is shown in Appendix A.)

EXAMPLE 9   Sometimes there is insufficient information to solve the problem

Using the same data set as in Example 8, suppose the sample size is only $n = 10$. Now try again to find the probability that a random sample of size $n = 10$ vehicles will have a mean city mpg greater than 21.

Solution

The population is skewed (not normal) and the sample size $n = 10$ is less than the minimum $n = 30$ required to apply the Central Limit Theorem. Therefore, we have insufficient information to conclude that the sampling distribution of the sample mean $\bar{x}$ is either normal or approximately normal. This is Case 3 from the Three Cases (page 401). Unfortunately, we cannot find the probability that a random sample of $n = 10$ vehicles will have a mean city mpg greater than 21.

© Garry Gay/Alamy

Trial of the Pyx: How Much Gold Is in Your Gold Coins?

CASE STUDY

Medieval English kings devised a procedure to ensure that the coins of the realm contained the proper amount of gold. A sample of 100 of the gold coins that were cast each year was placed in a ceremonial box called the Pyx. At the chosen time, the Company of Goldsmiths jury weighed the gold coins. The mean weight of the coins was supposed to be 128 grams. If the mean weight was much less than 128 grams, the jury concluded that the Master of the Mint was cheating the crown by pocketing the excess gold, and he was severely punished. If the mean weight of the coins was within 0.32 gram of the expected 128 grams, the jury accepted the year's gold as pure. Thus, the mean weight had to lie between 127.68 grams and 128.32 grams.

Problem 1.   Can we estimate what the jury used for a standard deviation?

Solution to Problem 1.   Let's assume that "much less than" indicated a measurement that is 2 or more standard deviations below average. For the sampling distribution of $\bar{x}$, then, this would indicate a range of $0.32 = 2\sigma_{\bar{x}}$ between 127.68 and the mean 128. Therefore, $\sigma_{\bar{x}} = 0.16$. And therefore, by the Empirical Rule, for instance, approximately 95% of the sample mean observations for the Trial of the Pyx would have been between 127.68 and 128.32. Because $\sigma_{\bar{x}} = \sigma/\sqrt{n}$, it follows that $\sigma = \sqrt{100} \cdot 0.16 = 1.6$ grams.

Problem 2.   What were the chances that the Master of the Mint would have been caught and punished if he were in fact cheating the throne?

Solution to Problem 2.   What if the Master of the Mint set the mean amount of gold per coin in the population of all coins to be $\mu = 127.9$ grams instead of the required 128, shortchanging the crown by a tenth of a gram of gold per coin? The jury would never have noticed this, would they?

Let's calculate the probability that the Master of the Mint would have passed the Trial of the Pyx if the mean amount of gold per coin had been only 127.9 grams. We've seen that the Master of the Mint would have passed the Trial of the Pyx if $127.68 < \bar{x} < 128.32$. Now, because 100 is a large sample size, the Central Limit Theorem tells us that the sampling distribution of $\bar{x}$ is approximately normal, with

$$\mu_{\bar{x}} = \mu = 127.9 \quad \text{and} \quad \sigma_{\bar{x}} = \frac{\sigma}{\sqrt{n}} = \frac{1.6}{\sqrt{100}} = 0.16$$

Standardizing using Fact 5:

$$Z = \frac{127.68 - \mu_{\bar{x}}}{\sigma_{\bar{x}}} = \frac{127.68 - 127.9}{0.16} \approx -1.38 \quad \text{and}$$

$$Z = \frac{128.32 - \mu_{\bar{x}}}{\sigma_{\bar{x}}} = \frac{128.32 - 127.9}{0.16} \approx 2.63$$

Solving using Table 8 in Chapter 6 (page 355):

$$P(-1.38 < Z < 2.63) = 0.9957 - 0.0838 = 0.9119$$

That is, the chances of the crown accepting the coins as pure, even if the Master of the Mint had been shortchanging by a tenth of a gram per coin, were over 91% (Figure 11).

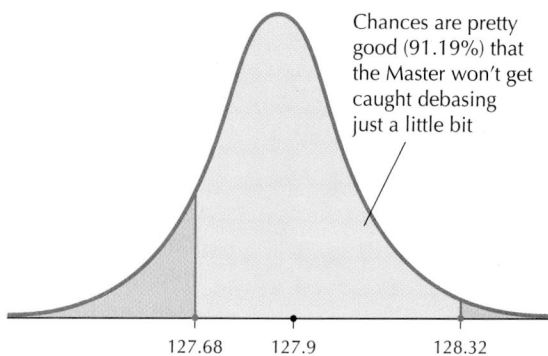

Chances are pretty good (91.19%) that the Master won't get caught debasing just a little bit

127.68   127.9   128.32

FIGURE 11 Sampling distribution if population mean gold weight is reduced to 127.9 grams.

Note: Sir William Sharington, 1493–1553, Master of the Mint during the turbulent Tudor era in England. He debased the currency, issued worthless coinage, and diverted the real gold to fund Thomas Seymour's conspiracy to topple the government and seize young King Edward VI. Sharington was arrested in 1548 or 1549, but he later received pardon and became Sheriff of Wiltshire for a short time before he died.

Problem 3. Would the Master of the Mint have been satisfied with this small amount of debasement? Would he have quit while he was ahead?

Solution to Problem 3. No way! The following year, the Master of the Mint decided to debase the currency even further, setting the mean amount of gold in the coins to be $\mu = 127.3$ grams per coin.

We need to find the probability of the Master passing the Trial of the Pyx if the mean amount of gold in a coin was 127.3 grams instead of the required 128 grams per coin. We use the same calculations, with $\mu_{\bar{x}} = 127.3$ grams. Standardizing:

$$Z = \frac{127.68 - \mu_{\bar{x}}}{\sigma_{\bar{x}}} = \frac{127.68 - 127.3}{0.16} \approx 2.38 \text{ and}$$

$$Z = \frac{128.32 - \mu_{\bar{x}}}{\sigma_{\bar{x}}} = \frac{128.32 - 127.3}{0.16} \approx 6.38$$

Then, $P(2.38 < Z < 6.38) \approx 1 - 0.9913 = 0.0087$.

In other words, the Master of the Mint actually would have stood very little chance—less than 1% probability—of passing the Trial of the Pyx if he cheated by this much (Figure 12).

England is a great country for retaining fine old traditions. Today, England's Company of Goldsmiths still operates the London Assay Office where the purity of the kingdom's coin is tested at the annual Trial of the Pyx.

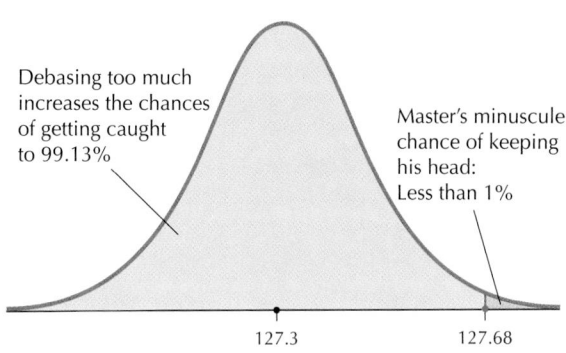

FIGURE 12 Sampling distribution if population mean gold weight is reduced to 127.3 grams.

Section 7.1 Summary

1. The sampling distribution of the sample mean $\bar{x}$ for a given sample size n consists of the collection of the means of all possible samples of size n from the population. The mean of the sampling distribution of $\bar{x}$ is the value of the population mean μ (Fact 1). The standard error is $\sigma_{\bar{x}} = \sigma/\sqrt{n}$, where σ is the population standard deviation (Fact 2). For a normal population, the sampling distribution of $\bar{x}$ is distributed as normal $(\mu, \sigma/\sqrt{n})$, where μ is the population mean and σ is the population standard deviation (Fact 3).

2. A simulation study showed that the sampling distribution of $\bar{x}$ for a skewed population achieved approximate normality when n reached 30. The Central Limit Theorem is one of the most important results in statistics and is stated as follows: given a population with mean μ and standard deviation σ, the sampling distribution of the sample mean $\bar{x}$ becomes approximately normal $(\mu, \sigma/\sqrt{n})$ as the sample size gets larger, regardless of the shape of the population.

3. We can use Facts 3 and 4 to find probabilities for problems involving sample means.

4. Similarly, we can find percentiles for the sample means.

Section 7.1 Exercises

CLARIFYING THE CONCEPTS

1. Explain what a sampling distribution is. Why are sampling distributions so important? (p. 396)

2. For a normal population, what can we say about the sampling distribution of the sample mean? (p. 399)

3. True or false: $\mu_{\bar{x}} = \mu$ and $\sigma_{\bar{x}} = \sigma/\sqrt{n}$ regardless of whether or not the sampling distribution of $\bar{x}$ is normal. (pp. 397, 398)

4. Use the Central Limit Theorem to explain what happens to the sampling distribution of $\bar{x}$ as the sample size gets larger. (p. 401)

5. According to our rule of thumb, what is the minimum sample size for approximate normality of the sampling distribution of $\bar{x}$? (p. 401)

6. State the three possible cases for the sampling distribution of $\bar{x}$. (p. 401)

7. Suppose we want to decrease the size of the standard error to half its original size. How much do we have to increase the sample size? (p. 398)

8. State the conditions when the sampling distribution of $\bar{x}$ is neither normal nor approximately normal. (p. 401)

PRACTICING THE TECHNIQUES

 CHECK IT OUT!

To do	Check out	Topic
Exercises 9–16	Examples 1 and 2	Mean, standard deviation, and shape of the sampling distribution of $\bar{x}$
Exercises 17–26	Example 4	Is the sampling distribution of $\bar{x}$ normal?
Exercises 27–32	Example 5	Finding probabilities for the sample mean
Exercises 33–58	Example 6	Finding a value of $\bar{x}$, given a probability or area
Exercises 59–70	Example 7	Finding probabilities using the Central Limit Theorem for Means
Exercises 71–80	Example 8	Finding two symmetric sample mean values using the Central Limit Theorem for Means
Exercises 81–90	Examples 4, 6, and 8	When appropriate, calculating two symmetric sample mean values

> The **sampling distribution of the sample proportion** $\hat{p}$ for a given sample size n consists of the collection of the sample proportions of all possible samples of size n from the population.
>   In general, the **sampling distribution of any particular statistic** for a given sample size n consists of the collection of the values of that sample statistic across all possible samples of size n.

Recall that in Section 7.1, we found that the mean of the sampling distribution of the sample mean $\bar{x}$ is $\mu_{\bar{x}} = \mu$ and the standard error of the mean is $\sigma_{\bar{x}} = \sigma/\sqrt{n}$. We now learn the mean and standard error of the sampling distribution of the sample proportion $\hat{p}$.

> **Fact 5: Mean of the Sampling Distribution of the Sample Proportion $\hat{p}$**
>
> The mean of the sampling distribution of the sample proportion $\hat{p}$ is the value of the population proportion p. This may be denoted as $\mu_{\hat{p}} = p$ and read as "the mean of the sampling distribution of $\hat{p}$ is p."

Fact 5 provides a measure of center for the sampling distribution of the sample proportion $\hat{p}$, and Fact 6 provides a measure of spread.

> **Fact 6: Standard Deviation of the Sampling Distribution of the Sample Proportion $\hat{p}$**
>
> The standard deviation of the sampling distribution of the sample proportion $\hat{p}$ is $\sigma_{\hat{p}} = \sqrt{\frac{p \cdot q}{n}}$, where p is the population proportion, $q = 1 - p$, and n is the sample size. $\sigma_{\hat{p}}$ is called the **standard error of the proportion**.

EXAMPLE 11   Mean and standard error of $\hat{p}$

The National Institutes of Health reported that color blindness linked to the X chromosome afflicts 8% of men. Suppose we take a random sample of 100 men and let p denote the proportion of men in the population who have color blindness linked to the X chromosome. Find $\mu_{\hat{p}}$ and $\sigma_{\hat{p}}$.

Solution

First, we note that this is a binomial experiment with $p = 0.08$ and $n = 100$. Fact 5 tells us that $\mu_{\hat{p}} = p$; that is, the sampling distribution of the sample proportion $\hat{p}$ has a mean of $p = 0.08$. Fact 6 states that the standard error is

$$\sigma_{\hat{p}} = \sqrt{\frac{p \cdot q}{n}} = \sqrt{\frac{0.08 \cdot (1 - 0.08)}{100}} = \sqrt{0.000736} \approx 0.02713$$

YOUR TURN #7

Refer to Example 11. Suppose we take a random sample of 400. Find $\mu_{\hat{p}}$ and $\sigma_{\hat{p}}$.

(The solution is shown in Appendix A.)

Imagine that we repeatedly draw random samples of 100 men and observe the proportion of men $\hat{p}$ in each sample who have color blindness linked to the X chromosome. Each sample provides us with a value for $\hat{p}$. Eventually, the values for $\hat{p}$, when graphed, form the sampling distribution shown in Figure 13.

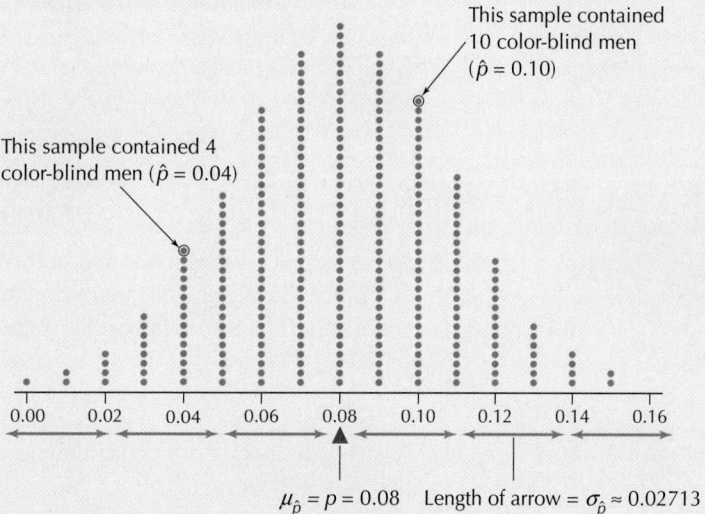

This sample contained 10 color-blind men ($\hat{p} = 0.10$)

This sample contained 4 color-blind men ($\hat{p} = 0.04$)

$\mu_{\hat{p}} = p = 0.08$ Length of arrow $= \sigma_{\hat{p}} \approx 0.02713$

FIGURE 13 Sampling distribution of sample proportion $\hat{p}$.

Note that $\mu_{\hat{p}} = p = 0.08$ is located at the balance point of this distribution, which we should expect because the mean proportion of these samples is $\mu_{\hat{p}} = p = 0.08$. Each arrow represents 1 standard error $\sigma_{\hat{p}} = 0.02713$. Note that nearly all the sample proportions lie within 3 standard errors of the mean.

Unfortunately, the sampling distribution of $\hat{p}$ is not always normal. Recall from Section 7.1 that the approximate normality provided by the Central Limit Theorem for Means was a useful tool for solving probability problems for the sample mean $\bar{x}$. Similarly, in order to solve probability problems for the sample proportion $\hat{p}$, we need a way to achieve approximate normality for the sampling distribution of $\hat{p}$. Conditions for the approximate normality of the sampling distribution of $\hat{p}$ are as follows.

Fact 7: Conditions for Approximate Normality of the Sampling Distribution of the Sample Proportion $\hat{p}$

The sampling distribution of the sample proportion $\hat{p}$ may be considered approximately normal only if both the following conditions hold:

$$n \cdot p \geq 5 \quad \text{and} \quad n \cdot q \geq 5$$

Alternatively, the conditions may be expressed as follows: $x \geq 5$ and $(n - x) \geq 5$.

The **minimum sample size** required to produce approximate normality in the sampling distribution of $\hat{p}$ is the *larger* of either

$$n_1 = \frac{5}{p} \quad \text{or} \quad n_2 = \frac{5}{q}$$

(rounded up to the next integer).

2 Applying the Central Limit Theorem for Proportions

Using information from Facts 5, 6, and 7, we express the Central Limit Theorem for Proportions.

Alternatively, the conditions may be expressed as follows: $x \geq 5$ and $(n - x) \geq 5$.

> **Central Limit Theorem for Proportions**
> The sampling distribution of the sample proportion $\hat{p}$ follows an approximately normal distribution with mean $\mu_{\hat{p}} = p$ and standard deviation $\sigma_{\hat{p}} = \sqrt{\frac{p \cdot q}{n}}$ when both the following conditions are satisfied: $n \cdot p \geq 5$ and $n \cdot q \geq 5$.

EXAMPLE 12   Determining whether the Central Limit Theorem for Proportions applies

In Example 11, we learned that color blindness linked to the X chromosome afflicts 8% of men. Determine the approximate normality of the sampling distribution of $\hat{p}$, the proportion of men who have color blindness linked to the X chromosome, for samples of size (a) 50 and (b) 100.

Solution

We need to check both conditions to find whether the sampling distribution of $\hat{p}$ is approximately normal.

a. We are given that $p = 0.08$ and $n = 50$.

$$n \cdot p = 50 \cdot 0.08 = 4 \quad \text{and} \quad n \cdot q = 50 \cdot (0.92) = 46$$

Because 4 is not ≥ 5, the first condition is not satisfied. The Central Limit Theorem for Proportions cannot be used. We cannot conclude that the sampling distribution of $\hat{p}$ is approximately normal.

b. Here, $p = 0.08$ and $n = 100$.

$$n \cdot p = 100 \cdot 0.08 = 8 \quad \text{and} \quad n \cdot q = 100 \cdot (0.92) = 92$$

Because both 8 and 92 are ≥ 5, both conditions are satisfied. The Central Limit Theorem for Proportions applies, and we can conclude that the sampling distribution of $\hat{p}$ is approximately normal. From Example 11, we have $\mu_{\hat{p}} = 0.08$ and $\sigma_{\hat{p}} = 0.02713$. Thus, the sampling distribution of $\hat{p}$ is approximately normal with $\mu_{\hat{p}} = 0.08$ and $\sigma_{\hat{p}} = 0.02713$.

NOW YOU CAN DO
Exercises 7–18.

EXAMPLE 13   Minimum sample size for approximate normality

According to George Washington University, 4.3% of all vehicles on the road are large trucks. Let $p = 0.043$ represent the population proportion.

a. Find the minimum size of the samples that produces a sampling distribution of $\hat{p}$ that is approximately normal.

b. Describe the sampling distribution of $\hat{p}$ if we use this minimum sample size.

Solution

a. Using Fact 7, the minimum sample size required is the *larger* of either

$$n_1 = \frac{5}{p} \quad \text{or} \quad n_2 = \frac{5}{q}$$

Here,

$$n_1 = \frac{5}{p} = \frac{5}{0.043} \approx 116.3 \quad \text{and} \quad n_2 = \frac{5}{q} = \frac{5}{0.957} \approx 5.2$$

The larger of n_1 and n_2 is $n_1 = 116.3$. However, it is unclear what "0.3" of a vehicle means. So we round up to the next integer: $n = 117$. Therefore, the minimum sample size required to produce a sampling distribution of $\hat{p}$ that is approximately normal is $n = 117$ vehicles. We confirm that this satisfies our conditions:

$$n \cdot p = (117)(0.043) = 5.031 \geq 5 \quad \text{and} \quad n \cdot q = (117)(0.957) = 111.969 \geq 5$$

b. We have $\mu_{\hat{p}} = 0.043$ and

$$\sigma_{\hat{p}} = \sqrt{\frac{pq}{n}} = \sqrt{\frac{0.043(0.957)}{117}} \approx \sqrt{0.00035172} \approx 0.01875$$

NOW YOU CAN DO
Exercises 19–24.

Because the conditions are met, the Central Limit Theorem for Proportions applies. The sampling distribution of $\hat{p}$ is approximately normal ($\mu_{\hat{p}} = 0.043$, $\sigma_{\hat{p}} = 0.01875$).

In those cases where we determine that the sampling distribution of $\hat{p}$ is approximately normal, we can standardize using Fact 8 to obtain the standard normal Z. Then we may proceed to apply the normal distribution methods we learned in Chapter 6. Fact 8 for proportions is similar to Fact 4 for means.

Fact 8: Standardizing a Normal Sampling Distribution for Proportions

When the sampling distribution of $\hat{p}$ is approximately normal, we can standardize to produce the standard normal Z:

$$Z = \frac{\hat{p} - \mu_{\hat{p}}}{\sigma_{\hat{p}}} = \frac{\hat{p} - p}{\sqrt{\frac{pq}{n}}}$$

where p is the population proportion of successes, $q = 1 - p$, and n is the sample size.

EXAMPLE 14   Finding probabilities using the Central Limit Theorem for Proportions

Using the information in Example 13, find the probability that a sample of vehicles will have a proportion of large trucks greater than 9% for samples of size (**a**) 30 vehicles and (**b**) 117 vehicles.

Solution

a. We found in Example 13(**a**) that this sample size of $n = 30$ does not meet the minimum sample size required for the sampling distribution of $\hat{p}$ to be approximately normal, so we cannot conclude that the sampling distribution of $\hat{p}$ is approximately normal. Thus, we cannot solve this problem.

b. From Example 13(**b**), the sampling distribution of $\hat{p}$ is approximately normal with mean $\mu_{\hat{p}} = 0.043$ and standard deviation $\sigma_{\hat{p}} = 0.01875$. We are then faced with a normal probability problem similar to those in Section 6.5. Figure 14 shows the sampling distribution of $\hat{p}$ and the probability we are interested in, $P(\hat{p} > 0.09)$. Using Fact 8, we standardize as follows:

Again, we can use our normal distribution methods from Section 6.5 because the Central Limit Theorem for Proportions gives us approximate normality.

$$Z = \frac{0.09 - \mu_{\hat{p}}}{\sigma_{\hat{p}}} = \frac{0.09 - 0.043}{0.01875} \approx 2.51$$

Thus, $P(\hat{p} > 0.09) = P(Z > 2.51)$, as shown in Figure 15.

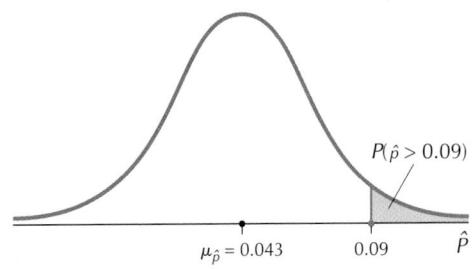

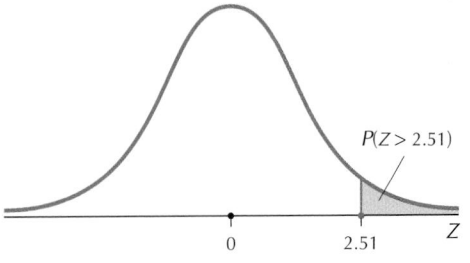

FIGURE 14 Area to the right of $\hat{p} = 0.09$ equals . . .

FIGURE 15 Area to the right of $Z = 2.51$.

NOW YOU CAN DO
Exercises 25–32.

Following Table 8 in Chapter 6 (page 355), we look up $Z = 2.51$ in the Z table and subtract this table area (0.9940) from 1 to get the desired tail area. That is,

$$P(Z > 2.51) = 1 - 0.9940 = 0.006$$

So the probability that the sample proportion of large trucks will exceed 0.09 is 0.006.

YOUR TURN
#8

Using the information in Example 13, find the probability that a sample of vehicles will have a proportion of large trucks smaller than 4% for a sample of size $n = 225$ vehicles.

(The solution is shown in Appendix A.)

EXAMPLE 15 Finding percentiles using the Central Limit Theorem for Proportions

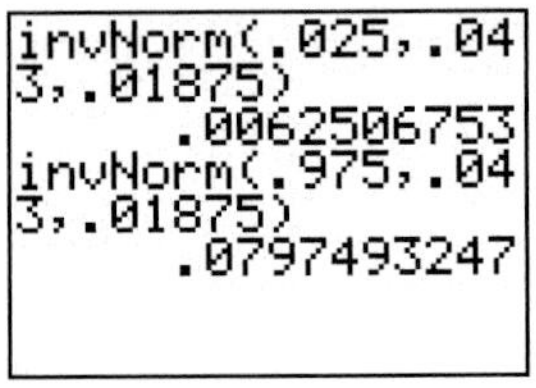

```
invNorm(.025,.04
3,.01875)
        .0062506753
invNorm(.975,.04
3,.01875)
        .0797493247
```

FIGURE 16 Finding the percentiles using the TI-83/84.

Using the information from Example 13, find the 2.5th and the 97.5th percentiles of sample proportions for $n = 117$.

Solution

We use the inverse normal function on the TI-83/84, just as we did in Example 8 (page 406). The results, using mean $\mu_{\hat{p}} = 0.043$ and standard deviation $\sigma_{\hat{p}} = 0.01875$, are shown in Figure 16. We have the 2.5th percentile ≈ 0.006251 and the 97.5th percentile ≈ 0.079475.

The 2.5th and the 97.5th percentiles contain the middle 95% of sample proportions, as shown in Figure 17.

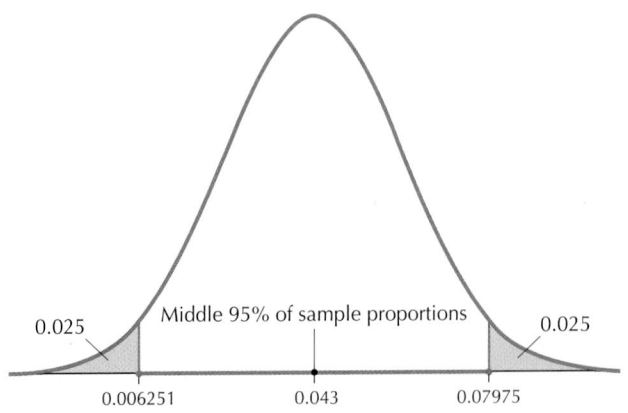

Sampling distribution of the sample proportion

NOW YOU CAN DO
Exercises 33–38.

FIGURE 17 The 2.5th percentile and the 97.5th percentile contains the middle 95% of sample proportions.

Using the information from Example 13, find the two percentiles that contain the middle 90% of sample proportions.

(The solution is shown in Appendix A.)

Developing Your
Statistical Sense

Note: What can we do to estimate the 1st percentile? One way is to use simulation. Generate samples of size $n = 117$ from the population of the original survey respondents, record the sample proportion from each, and simply choose the 1st percentile. Proceeding in this manner, we estimate the 1st percentile as 0.0128.

Pitfalls of Using an Approximation

Let's use symmetry and the results from Example 15 to find the 1st percentile of the sampling distribution of $\hat{p}$ for $n = 117$. By symmetry, the 1st percentile will be the same distance below the mean that the 99th percentile is above the mean. The 99th percentile, 0.0867, lies $(0.0867 - 0.043) = 0.0437$ above the mean. Therefore, the 1st percentile lies 0.0437 below the mean:

$$\hat{p} = (0.043 - 0.0437) = -0.0007$$

However, this value of −0.0007 is negative and cannot represent a sample proportion. This negative result is obtained because the normality of the sampling distribution of $\hat{p}$ is only approximate and not exact.

Section 7.2 Summary

1. The sampling distribution of the sample proportion $\hat{p}$ for a given sample size n consists of the collection of the sample proportions of all possible samples of size n from the population.

2. According to the Central Limit Theorem for Proportions, the sampling distribution of the sample proportion $\hat{p}$ follows an approximately normal distribution with mean $\mu_{\hat{p}} = p$ and standard deviation $\sigma_{\hat{p}} = \sqrt{pq/n}$ when both the following conditions are satisfied: $n \cdot p \geq 5$ and $n \cdot q \geq 5$.

Section 7.2 Exercises

CLARIFYING THE CONCEPTS

1. Explain what a sample proportion is, using as an example the courses for which you got an A last semester. (p. 415)
2. What is the mean of the sampling distribution of $\hat{p}$? (p. 416)
3. Give the formula for the standard error of the proportion. (p. 416)
4. What are the requirements for the sampling distribution of $\hat{p}$ to be approximately normal? (p. 417)
5. Suppose you double the sample size. What happens to the standard error of the proportion? (p. 416)
6. For the following values of X and n, calculate the sample proportion $\hat{p}$: (p. 415)
   a. $X = 10, n = 40$
   b. $X = 25, n = 75$
   c. Number of successes = 27, number of trials = 54
   d. Number of successes = 1000, number of trials = 1 million

PRACTICING THE TECHNIQUES

 **CHECK IT OUT!**

To do	Check out	Topic
Exercises 7–18	Example 12	Determining whether the CLT for Proportions applies
Exercises 19–24	Example 13	Minimum sample size for approximating normality
Exercises 25–32	Example 14	Finding probabilities using the CLT for Proportions
Exercises 33–38	Example 15	Finding percentiles using the CLT for Proportions

In Exercises 7–18, samples are taken. Find (**a**) $\mu_{\hat{p}}$ and (**b**) $\sigma_{\hat{p}}$, and (**c**) determine whether the sampling distribution of $\hat{p}$ is approximately normal or unknown.

Chapter 7 Review Exercises

SECTION 7.1

For Exercises 1–5, find $\mu_{\bar{x}}$ and $\sigma_{\bar{x}}$, the mean and standard deviation of the sampling distribution of $\bar{x}$.

1. $\mu = 100, \sigma = 15, n = 25$
2. $\mu = 100, \sigma = 15, n = 36$
3. $\mu = 100, \sigma = 15, n = 49$
4. $\mu = 0, \sigma = 1, n = 4$
5. $\mu = 0, \sigma = 1, n = 16$

For Exercises 6 and 7, if possible find the indicated probability. If it is not possible, explain why not.

6. Scores on a psychological test are not normally distributed, with $\mu = 100$ and $\sigma = 15$. A sample of size 25 is taken. Find $P(100 < \bar{x} < 105)$.

7. Scores on a psychological test are normally distributed, with $\mu = 100$ and $\sigma = 15$. A sample of size 25 is taken. Find $P(100 < \bar{x} < 105)$.

8. **Cocaine and Heart Attacks.** The American Medical Association reported: "During the first hour after using cocaine, the user's risk of heart attack increases nearly 24 times. The average age of people in the study who suffered heart attacks soon after using cocaine was only 44. That's about 17 years younger than the average heart attack patient. Of the 38 cocaine users who had heart attacks, 29 had no prior symptoms of heart disease."[9] Assume that the standard deviation of the age of people who suffered heart attacks soon after using cocaine was 10 years and we take a sample of size 38.

   a. Find the 99.5th percentile of the mean age at heart attack after using cocaine.
   b. Find the 0.5th percentile of the mean age at heart attack after using cocaine.

   c. Between which two sample mean ages that are symmetric about the population mean lie 99% of mean ages of all people who suffered heart attacks soon after using cocaine?
   d. By hand, sketch a plot of how this sampling distribution would look.

SECTION 7.2

For Exercises 9 and 10, if possible find the indicated probability. If it is not possible, explain why not.

9. $p = 0.9, n = 40, P(\hat{p} < 0.88)$
10. $p = 0.9, n = 50, P(\hat{p} < 0.88)$

For Exercises 11 and 12, find the indicated value of $\hat{p}$. If it is not possible, explain why not.

11. $p = 0.98, n = 400$, the value of $\hat{p}$ smaller than 75% of all p values
12. $p = 0.98, n = 625$, the value of $\hat{p}$ smaller than 75% of all p values

13. **Women and Depression.** According to the National Institute for Mental Health, 12% of women are affected by a depressive disorder each year. Suppose we take samples of 100 women. Answer the following:

   a. Find $P(\hat{p} > 0.16)$, where $\hat{p}$ represents the sample proportion of women who are affected by a depressive disorder each year.
   b. Calculate $P(0.12 < \hat{p} < 0.16)$.
   c. Use your answer to (a) to calculate $P(\hat{p} < 0.16)$.
   d. Find the 5th and 95th percentiles of the sample proportion.

Chapter 7 QUIZ

TRUE OR FALSE

1. True or false: For a normal population, the sampling distribution of the sample mean is always normal.
2. True or false: The Central Limit Theorem takes effect at $n = 30$, so it doesn't make sense to get larger samples.

FILL IN THE BLANK

3. The distance between the point estimate and its target parameter is called the _____ _____ [two words].
4. If the population is either non-normal or of unknown distribution and the sample size is large, then the sampling distribution of $\bar{x}$ is _____ _____ (two words).

SHORT ANSWER

5. If the population is either non-normal or of unknown distribution and the sample size is small, then do we know the sampling distribution of $\bar{x}$?
6. The sampling distribution of the sample proportion $\hat{p}$ may be considered approximately normal only if *both* the following conditions hold: (1) _____ and (2) _____.

CALCULATIONS AND INTERPRETATIONS

Soybean Crop. Protein content in a particular farmer's soybean crop is normally distributed, with a mean of 40 grams and a standard deviation of 20 grams. Suppose we take samples of size 100 soy plants. Use this information for Exercises 7 and 8.

7. a. Find the probability that the sample mean protein content will be less than 38 grams.
   b. Find the probability that the sample mean protein content will be between 36.08 and 43.92 grams.
   c. Find the probability that the sample mean protein content will be greater than 42.5 grams.

8. Refer to Exercise 7.
   a. Find the sample mean protein content higher than 95% of all such sample means.
   b. Find the sample mean protein content lower than 95% of all such sample means.
   c. Between which two values does the middle 90% of sample mean protein content lie?

Comparing Per Capita Incomes. Use this information for Exercises 9 and 10. According to the U.S. Census Bureau, the mean per capita income in Texas is $25,800, whereas the mean per capita income in Florida is $26,500. Suppose both distributions are normal, each with a standard deviation of $8000. Suppose we take samples of 100 residents of each state.

9. **a.** Find the probability that the sample mean per capita income in Texas will exceed $27,000.

   b. Find the probability that the sample mean per capita income in Florida will exceed $27,000.

10. **a.** Find the 95th percentile of sample mean per capita income in Texas.

   b. Find the 95th percentile of sample mean per capita income in Florida.

11. **Men and Depression.** According to the National Institute for Mental Health, 6.6% of men are affected by a depressive disorder each year.

   a. If we take samples of 400 men, find $P(\hat{p} < 0.066)$.

   b. If we take samples of 400 men, find $P(0.05 < \hat{p} < 0.066)$.

   c. If we take samples of 400 men, find the 2.5th and 97.5th percentiles of the sample proportion.

8 Confidence Intervals

Introduction

In Chapter 8, students learn about confidence interval estimations, using their knowledge of sampling distributions from Chapter 7. By studying the patterns implicit in the sampling distribution of a statistic, such as the sample mean or sample proportion, students can infer with a certain degree of confidence that the associated population parameters lie within a certain interval.

This chapter builds confidence intervals for a population mean, a population proportion, and the population variance and standard deviation. The relationship between sample size and the precision of the interval estimate is also discussed.

A new Case Study, *Fuel Efficiency*, explores a large data set containing 13 different measurements on a population of 1141 vehicle models for model year 2014. We return to this case study again and again throughout Sections 8.1, 8.2, and 8.3.

From the Author

Section 8.1 Z Interval for the Population Mean

- Instructors may wish to stress that the $\pm$ notation refers to two different quantities, and that $\bar{x} \pm E$ refers to an interval on the number line.

- There is a new Example 2: Determining Whether the Z Interval for μ May Be Used.

- Instructors may wish to refer to previous examples in earlier chapters where intervals on the number line were discussed, such as Example 30 in Chapter 6 (page 358), Example 35 in Chapter 6 (page 362), and Example 6(**c**) in Chapter 7 (page 404).

- Another new example is Example 10: Finding the Margin of Error, Given the Lower and Upper Bounds.

Section 8.2 t Interval for the Population Mean

- Instructors may point out that, if the population standard deviation σ is unknown, it is wrong to use the Z interval. Use the t interval instead.

- Note that the t table used in *Discovering Statistics* provides the confidence levels for finding the values of $t_{\alpha/2}$, so students need not calculate $\alpha/2$.

- Newer models of the TI-84 can calculate $t_{\alpha/2}$.

- There is a new Example 13: Checking Whether the Conditions Are Met for the t Interval for μ.

Section 8.3 Z Interval for the Population Proportion

- Instructors may wish to highlight Example 20 because many students may have heard of the phrase "plus or minus 3 percentage points."

Section 8.4 Confidence Intervals for the Population Variance and Standard Deviation

- This is where the chi-square distribution is introduced. Instructors who plan on covering Sections 11.1 and 11.2 on chi-square procedures, but are not covering confidence intervals for the population standard deviation, may want to return to this section for a quick review of the chi-square distribution before starting Chapter 11.

Teaching Tips

- See Example 20 on page 466.

- You may want to make sure that students understand the $\pm$ notation. That is, $a \pm b$ refers to two quantities: $a + b$ and $a - b$.

- For t intervals, note that the t table has the confidence levels provided at the top of the table, so that it is not necessary to find $1 - \alpha$ or $1 - \frac{\alpha}{2}$.

In-Class Activities

1. **Estimate the proportion of females at your university.**

    a. Guess the population proportion of females at your school.

    b. Find the sample proportion of females in your class.

    c. What would you say is the probability that the value for the sample proportion from (**b**) is exactly equal to the actual population proportion of females at your school?

    d. If appropriate, construct a 90% confidence interval for the population proportion of females at your school, using the sample size and sample proportion from your class.

    e. Does your guess from (**a**) fall within the confidence interval from (**d**)?

2. To illustrate what "confidence" means as it relates to a confidence interval, you can simulate 100 confidence intervals (or more) by sampling from a normal distribution with a known mean and standard deviation. If the appropriate technology (for example, Minitab) is available, students also can simulate these confidence intervals for different means and standard deviations. The percentages of the number of intervals that contain the population mean can then be computed and related to the level of confidence. If students participate, the class can pool the percentages to find the average percent, which is a better estimate for the level of confidence.

3. **Estimating the Mean Shoe Size for All Female and Male Students at Your School**

    a. Make a guess at what the population mean shoe size might be for all female students at your school. Do the same for the males.

b. Collect data on the shoe sizes of the female students in your class. Find the mean of this sample.

c. What would you say is the probability that the value for the sample mean from (**b**) is exactly equal to the mean foot size of all female students at your school?

d. Use technology to determine if the female shoe sizes are normally distributed. If not, and if the number of females is not at least 30, discuss whether it is appropriate to construct a t confidence interval for the population mean shoe size of all females at your school, and go to (**g**).

e. Construct a 95% confidence interval for the population mean shoe size for all female students at your school.

f. Does your guess for the population mean shoe size from (**a**) fall within the confidence interval from (**e**)?

g. Repeat (**b**)–(**f**) for the males in your class.

Supplements

- StatTutor 14.1–14.4, 18.1–18.3, and 20.3–20.5
- Stats@Work Simulations
  - What Does It Really Mean? Confidence Level; Mike Mobile
  - What Does It Really Mean? Confidence Level; Vicki Vitamin
  - What Does It Really Mean? Confidence Level; Ruby Sweet
  - What Does It Really Mean? Confidence Level; Justine Red
- EESEE case studies
  - Columbus's 1993 Election Poll (Questions 1, 2, 3, 4, and 5 on estimation and confidence interval for a population proportion)
  - Radar Detectors and Speeding (Questions 1, 2, and 3 on confidence interval for a population proportion)

Applets

The *Normal Density Curve* and *Confidence Intervals* applets are referenced in Chapter 8 for demonstrations.

The *Confidence Intervals* applet is used for Exercise 91 in Section 8.1 and the *Normal Density Curve* applet is used for Exercise 92 in Section 8.1.

This Web site has an applet for confidence interval: http://wise.cgu.edu/.

Videos

- *Against All Odds: Inside Statistics:* www.learner.org/resources/series65.html
  - Program 24: Confidence Intervals

Web Sites

- The site Online Statistics: An Interactive Multimedia Course of Study has information on confidence intervals and a link to the Rice Virtual Lab in Statistics, with numerous simulation demonstrations and other topics related to statistics: http://onlinestatbook.com/index.html.

- This Web site lists other sites that do statistical calculations: http://statpages.org/.

- This Web site lists activities on all topics: http://mathforum.org/mathtools/sitemap2/ps/.

- This site has several activities using Excel and Minitab: www.mathspace.com/NSF_ProbStat/Teaching_Materials/Primarily_Statistics.htm.

| **8.1** | Z Interval for the Population Mean |

OBJECTIVES By the end of this section, I will be able to . . .

1 Calculate a point estimate of the population mean.
2 Calculate and interpret a Z interval for the population mean when the population is normal and when the sample size is large.
3 Find ways to reduce the margin of error.
4 Calculate the sample size needed to estimate the population mean.

1 Calculate a Point Estimate of the Population Mean

Recall from Section 1.2 that characteristics of a sample, such as the sample mean $\bar{x}$, are called **statistics**, whereas characteristics of a population, such as the population mean μ, are called **parameters**. **Statistical inference** consists of methods for estimating and drawing conclusions about parameters, based on the corresponding statistic. For example, we use the known value of $\bar{x}$ to estimate the unknown value of μ.

Suppose a random sample of 30 male students at your school produced a sample mean height of $\bar{x} = 70$ inches. We could then use this statistic $\bar{x} = 70$ to *infer* that the population mean height μ of all male students at your school was close to 70 inches. This value of $\bar{x} = 70$ is called a **point estimate** of the population mean μ.

> **Point estimation** is the process of estimating unknown population parameters by known sample statistics. The value of each sample statistic used as an estimate is called a **point estimate**.

| **EXAMPLE 1** | Calculating a point estimate |

Daniel Acker/Bloomberg via Getty Images

Farmers work hard to increase the yield of their acreage. Yield represents the number of bushels of a crop produced per acre. Suppose we are interested in estimating the population mean yield for winter wheat across all 50 states. Shown here is the mean July 2014 yield for a sample of five states, in bushels, as published by the U.S. Department of Agriculture (USDA).

a. Find the sample mean yield $\bar{x}$.

b. Express $\bar{x}$ as the point estimate of μ, the unknown population mean winter wheat yield for all 50 states.

State	Yield (bushels)
California	85
Georgia	55
Illinois	67
Ohio	68
Texas	25

Solution

a. The sample mean yield is calculated as

$$\bar{x} = \frac{\sum x}{n} = \frac{85 + 55 + 67 + 68 + 25}{5} = 60$$

b. The point estimate of μ, the unknown nationwide mean winter wheat yield for all 50 states, is 60 bushels per acre.

NOW YOU CAN DO
Exercises 11–14.

YOUR TURN
#1

See Example 1. The USDA reports the yields for Colorado, Indiana, Maryland, Michigan, and Pennsylvania to be 36, 68, 65, 70, and 63 bushels, respectively.

a. Find the sample mean yield $\bar{x}$.

b. Express $\bar{x}$ as the point estimate of μ, the unknown population mean winter wheat yield for all 50 states.

(The solutions are shown in Appendix A.)

However, because a sample is only a small subset of the population, generalizing from a sample to the population carries the risk that the point estimate may not be very accurate. For example, do you think that the population mean yield of winter wheat μ exactly equals our point estimate of 60 bushels per acre? It's not likely, because we learned in Example 1 of Chapter 7 (page 397) that different samples will produce different sample means, and thus different point estimates of μ. Our point estimate $\bar{x} = 60$ may be close to μ or it may be far from μ. In other words, *we have no measure of confidence* that our particular point estimate is close to μ. There has to be a better way, and there is: *confidence intervals*, the subject of this chapter.

2 The *Z* Interval for the Population Mean

Although we cannot measure how confident we are of $\bar{x}$ as a point estimate for μ, we can use the point estimate $\bar{x}$ to find an interval that is likely to contain μ. Suppose we are interested in estimating the mean height of the students at your school. The students in your class are a sample of the population of students at your school, so we can calculate the sample mean height of the students in your class to be $\bar{x} = 67.5$ inches (5 feet 7.5 inches tall).

We may then use $\bar{x} = 67.5$ inches as a point estimate of the unknown population mean height of all students at your school. However, this estimate is not likely to be exactly correct. To address this uncertainty in our estimate, we can use a range of heights instead, such as 67.5 inches, give or take an inch, which we write

67.5 inches $\pm$ 1 inch

and would equal the interval

(66.5 inches, 68.5 inches).

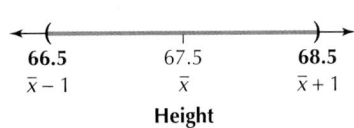

66.5   67.5   68.5
$\bar{x} - 1$ $\bar{x}$ $\bar{x} + 1$

Height

We are 90% confident that μ lies between 66.5 inches and 68.5 inches.

The "1 inch" is called the *margin of error*. We might then say that we are 90% confident that the mean height of all students at our school lies in the

interval 67.5 inches $\pm$ 1 inch (see the figure in the margin)

To increase the confidence in our estimate, we increase the margin of error, so that we might say we are 95% confident that the mean height of all students at our school lies in the interval 67.5 inches $\pm$ 2 inches or the interval (65.5 inches, 69.5 inches). These two intervals are examples of what are called **confidence intervals**.

Two important results from Chapter 7 form the conditions that allow us to construct the Z interval for μ:

- The first condition comes from Fact 3 in Section 7.1: if the population is normal, then the sampling distribution of $\bar{x}$ is also normal.

- The second condition is a result of the Central Limit Theorem for Means (from Section 7.1): if the sample size is large, then the sampling distribution of $\bar{x}$ is approximately normal.

Table 1 provides a listing of $Z_{\alpha/2}$ values for the most common confidence levels.

Table 1 $Z_{\alpha/2}$ values for common confidence levels

Confidence level $(1 - \alpha)100\%$	α	$\alpha/2$	$Z_{\alpha/2}$
$100(1 - 0.20)\% = 80\%$	0.20	0.10	1.28
$100(1 - 0.10)\% = 90\%$	0.10	0.05	1.645
$100(1 - 0.05)\% = 95\%$	0.05	0.025	1.96
$100(1 - 0.01)\% = 99\%$	0.01	0.005	2.576

The *Normal Density Curve* applet may be used to find $Z_{\alpha/2}$ critical values for confidence levels not listed in Table 1.

EXAMPLE 3   Finding the value of $Z_{\alpha/2}$

For the following situations, find the value of $Z_{\alpha/2}$:

a. Confidence level = 95%

b. $\alpha = 0.01$

Solution

a. From Table 1, we have $Z_{\alpha/2} = 1.96$. We mentioned this case earlier (page 430), and in Example 35 of Chapter 6 (page 362).

b. Table 1 gives us $Z_{\alpha/2} = 2.576$.

NOW YOU CAN DO
Exercises 21–26.

YOUR TURN
#3

For the following situations, find the value of $Z_{\alpha/2}$:

a. Confidence level = 99%

b. $\alpha/2 = 0.05$

(The solutions are shown in Appendix A.)

EXAMPLE 4   Constructing a confidence interval for the mean of a normal population

The College Board reports that the scores on the 2014 SAT Math test were normally distributed. A sample of 25 SAT scores had a mean of $\bar{x} = 510$. Assume that the population standard deviation of such scores is $\sigma = 118$. Construct a 90% confidence interval for the population mean score on the 2014 SAT Math test.

Be careful! In order to use the *Z* interval for μ, the *population standard deviation* σ must be known, not just the sample standard deviation. If the word problem provides the sample standard deviation *s* but not the population standard deviation σ, then you cannot use the *Z* interval. You might be able to use the *t* confidence interval for μ (Section 8.2).

Solution

Because the population is normal and the population standard deviation σ is known, the requirements for the *Z* interval are met:

$$\text{lower bound} = \bar{x} - Z_{\alpha/2}\left(\sigma/\sqrt{n}\right) \qquad \text{upper bound} = \bar{x} + Z_{\alpha/2}\left(\sigma/\sqrt{n}\right)$$

We are given $\bar{x} = 510$, $\sigma = 118$, and $n = 25$. From Table 1, we have $Z_{\alpha/2} = 1.645$. Thus,

$$\text{lower bound} = 510 - 1.645\left(118/\sqrt{25}\right) = 471.2$$
$$\text{upper bound} = 510 + 1.645\left(118/\sqrt{25}\right) = 548.8$$

We are 90% confident that the population mean score on the 2014 Mathematics SAT test lies between 471.2 and 548.8.

 **NOW YOU CAN DO**
Exercises 27–30.

YOUR TURN
#4

For the scenario in Example 4, construct a 95% confidence interval for the population mean score on the 2014 SAT Math test.

(The solution is shown in Appendix A.)

What Does This Confidence Interval Mean?

What does the 90% in the phrase *90% confidence interval* mean? If we take sample after sample for a very long time, then in the long run, the proportion of intervals that will contain the population mean μ will equal 90%.

Interpreting Confidence Intervals

You may use the following generic interpretation for the confidence intervals that you construct: "We are 90% (or 95% or 99% and so on) confident that the population mean _____ (for example, SAT Math score) lies between _____ (lower bound) and _____ (upper bound)."

The *Z* interval for the population mean μ takes the form

$$\text{point estimate} \pm \text{margin of error } E$$

where the point estimate equals the sample mean $\bar{x}$ and the margin of error *E* equals $Z_{\alpha/2}\left(\sigma/\sqrt{n}\right)$.

The **margin of error** *E* is a measure of the *precision* of the confidence interval estimate. For the *Z* interval, the margin of error takes the form $E = Z_{\alpha/2}\left(\sigma/\sqrt{n}\right)$. Smaller values of *E* indicate smaller margin of error, and therefore, greater precision.

For example, the confidence interval from Example 4 has the form

$$\text{point estimate} \pm \text{margin of error } E$$
$$= \bar{x} \pm E$$
$$= \bar{x} \pm Z_{\alpha/2}\left(\sigma/\sqrt{n}\right)$$
$$= 510 \pm 38.8$$

Later in this section (page 437) we learn ways to reduce the margin of error.

Motor Vehicle Fuel Efficiency

One of the variables in our case study is *City MPG*, which is the number of miles a vehicle can travel in city conditions on one gallon of gas. Because we have information on the entire population of 1141 vehicles, we know the population standard deviation $\sigma = 5.637$ mpg. We obtained a sample of 100 vehicles and observed a sample mean city gas mileage of $\bar{x} = 20.71$ mpg.

a. Determine whether the requirements are met for constructing the *Z* interval for μ.

b. Construct a 90% confidence interval for μ, the population mean City MPG for all vehicles.

c. Interpret the confidence interval.

Solution

a. We are not given any information about the distribution of the population, so we don't know if the population is normally distributed. However, the sample size $n = 100$ is greater than 30, and the value of $\sigma = 5.637$ is known; therefore, we can proceed to construct the confidence interval.

b. The formula for the confidence interval is given by

$$\text{lower bound} = \bar{x} - Z_{\alpha/2}\left(\sigma/\sqrt{n}\right)$$

$$\text{upper bound} = \bar{x} + Z_{\alpha/2}\left(\sigma/\sqrt{n}\right)$$

Note: As a check on your arithmetic, make sure that
$$\frac{(\text{lower bound} + \text{upper bound})}{2} = \bar{x}.$$
In other words, the sample mean should lie exactly midway between the lower bound and the upper bound.

We are given $n = 100$, $\bar{x} = 20.71$, and $\sigma = 5.637$. For a confidence level of 90%, Table 1 provides the value of $Z_{\alpha/2} = Z_{0.025} = 1.645$. Plugging into the formula:

$$\text{lower bound} = 20.71 - 1.645\left(5.637/\sqrt{100}\right) \approx 20.71 - 0.93 = 19.78$$

$$\text{upper bound} = 20.71 + 1.645\left(5.637/\sqrt{100}\right) \approx 20.71 + 0.93 = 21.64$$

c. We are 90% confident that μ, the population mean City MPG for all motor vehicles, lies between 19.78 mpg and 21.64 mpg. (See Figure 2.)

NOW YOU CAN DO
Exercises 31–34.

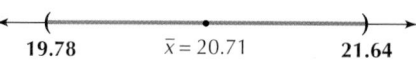

FIGURE 2 90% confidence interval for the population mean City MPG.

YOUR TURN #5

For the scenario in Example 5, construct a 99% confidence interval for μ, the population mean City MPG for all vehicles.

(The solution is shown in Appendix A.)

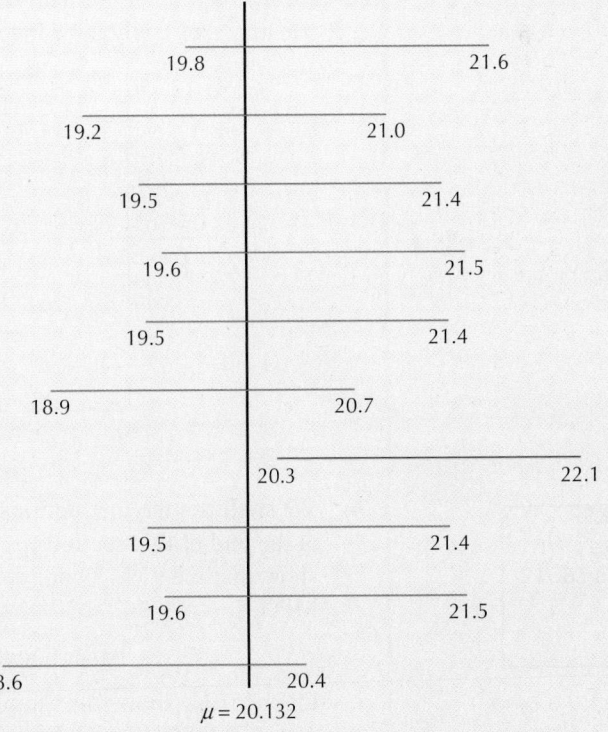 The *Confidence Interval* applet
allows you to see for yourself
how individual samples generate
intervals that either do or do not
contain the population mean.

What Is Random Here?

It is important to understand that *it is the interval that is random, not the population mean* μ. The interval is formed by sample statistics such as $\bar{x}$, and for each different sample we get different values for the statistics. So the interval is random because it is constructed using $\bar{x}$, which is also random. The population mean μ, though usually unknown, is nevertheless *constant*.

We generated 10 samples of size 100 vehicles from the **Fuel Efficiency** data set, and observed the City MPG of each vehicle. For each sample, a 90% Z confidence interval for the population mean City MPG was constructed. The results are shown in Figure 3. Note that, because we have the entire population of 1141 vehicles, we know the population mean City MPG is $\mu = 20.132$ mpg, which is also shown in Figure 3. Note that the confidence intervals are random, whereas μ is constant. The confidence intervals are random because they are based on the different values that the sample mean $\bar{x}$ takes with each sample. The randomness involved in the sampling leads to the randomness of the values of $\bar{x}$. (This relates to what we learned in Chapter 7: the sample mean is a random variable that has its own distribution, the sampling distribution.)

Now, the confidence interval from our sample in Example 5 is shown as the first confidence interval, and is rounded to (19.8, 21.6). Note that this confidence interval happened to "capture" the population mean $\mu = 20.132$. However, one of the confidence intervals did not capture the population mean (the red one). It turns out that 9 out of 10 of the samples (90%) produced confidence intervals that contained μ. But it did not have to turn out this way. The 90% refers to the proportion of intervals that will contain μ after a great many samples are taken.

FIGURE 3 The confidence intervals are random; μ is constant.

EXAMPLE 6   Z intervals for μ using technology

highwaympg16

Motor Vehicle Fuel Efficiency

Another of the variables in our case study is *Highway MPG*, which is the number of miles a vehicle can travel on a highway on one gallon of gas. We know the population standard deviation $\sigma = 6.326$ mpg. The sample of 16 vehicles, shown here, has a sample mean highway gas mileage of $\bar{x} = 26.94$ mpg.

Vehicle	Highway MPG	Vehicle	Highway MPG
Honda CR-V	30	Subaru Impreza	25
Nissan Pathfinder	26	Ford Mustang	26
Acura MDX	28	Cadillac ATS	31
Porsche Cayenne	29	Chevrolet Camaro	24
Mercedes-Benz GLK 250	33	Ford Taurus	29
Chevrolet Chevy SS	21	Ford Expedition	20
Dodge Charger	27	Lincoln MKT	25
Jeep Compass	23	BMW X1	34

a. Determine whether the Z interval for μ may be applied.

b. Use the TI-83/84, Minitab, and JMP to construct a 95% Z confidence interval for the population mean Highway MPG.

Solution

a. The sample size $n = 16$ is not large (≥ 30), so we need to check if the data follow a normal distribution. The normal probability plot of the data in Figure 4 supports the assumption of normality. Further, the population standard deviation is known. We may thus apply the n interval for μ.

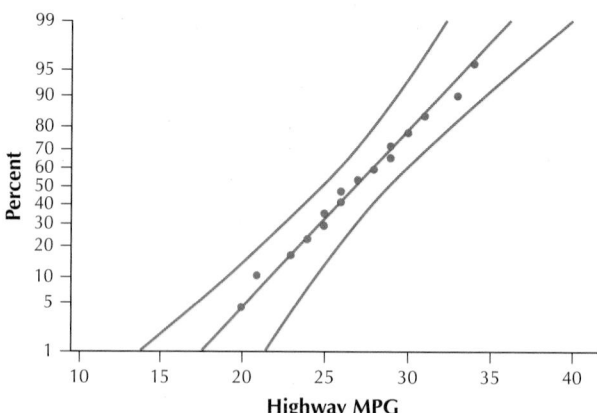

FIGURE 4
Normal probability plot of the Highway MPG data.

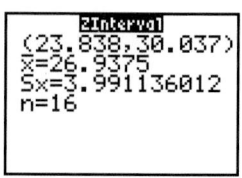

FIGURE 5 TI-83/84 results.

b. We shall use the instructions provided in the Step-by-Step Technology Guide at the end of this section (page 441). The results for the TI-83/84 in Figure 5 show that the 95% Z confidence interval for the population mean Highway MPG is

lower bound = 23.838, upper bound = 30.037

Figure 5 also shows the sample mean $\bar{x} = 26.9375$, the sample standard deviation $\sigma = 3.991136012$, and the sample size $n = 16$.

The Minitab results are provided in Figure 6. The "assumed standard deviation" is indicated to be $\sigma = 6.326$. Then the sample size $n = 16$, the sample mean $\bar{x} = 26.94$, and the sample standard deviation $s = 3.99$ are displayed. "SE Mean" refers to the standard error of the mean, but we don't need it here. Finally, the 95% confidence interval is given as (lower bound = 23.84, upper bound = 30.04).

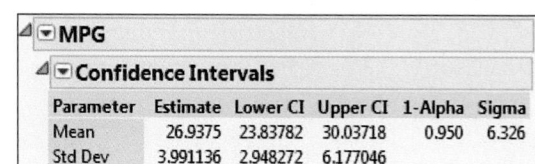

FIGURE 6
Minitab results.

```
One-Sample Z: MPG

The assumed standard deviation = 6.326


Variable   N   Mean  StDev  SE Mean      95% CI
MPG        16  26.94  3.99     1.58   (23.84, 30.04)
```

The JMP results are shown in Figure 7. The sample mean $\bar{x} = 26.9375$ is shown in the first column, with the sample standard deviation $s = 3.991136$ below it. The 95% confidence interval is given in the row labeled Mean, with lower bound = 23.84 (rounded) and upper bound = 30.04 (rounded).

FIGURE 7
JMP results.

MPG					
Confidence Intervals					
Parameter	Estimate	Lower CI	Upper CI	1-Alpha	Sigma
Mean	26.9375	23.83782	30.03718	0.950	6.326
Std Dev	3.991136	2.948272	6.177046		

3 Ways to Reduce the Margin of Error

Recall that the *Z* interval for μ takes the form

$$\text{point estimate} \pm \text{margin of error} = \bar{x} \pm E$$

where $E = Z_{\alpha/2}\left(\sigma/\sqrt{n}\right)$. We interpret the margin of error E for a $(1 - \alpha)100\%$ confidence interval for μ as follows:

CAUTION Remember that the "$\pm$" notation *always* represents a pair of numbers.

"We can estimate μ to within E units with $(1 - \alpha)100\%$ confidence."

EXAMPLE 7   Finding and interpreting the margin of error

In Example 5, the *Z* interval for the population mean city gas mileage for all motor vehicles is:

$$\text{lower bound} = 20.71 - 1.645\left(5.637/\sqrt{100}\right) \approx 20.71 - 0.93 = 19.78$$

$$\text{upper bound} = 20.71 + 1.645\left(5.637/\sqrt{100}\right) \approx 20.71 + 0.93 = 21.64$$

a. Find the margin of error E.

b. Express the confidence interval in the form "point estimate $\pm$ margin of error."

c. Interpret the margin of error E.

Solution

a. We find the margin of error as follows:

$$E = Z_{\alpha/2}\left(\sigma/\sqrt{n}\right) = 1.645\left(5.637/\sqrt{100}\right) \approx 0.93$$

b. The point estimate is $\bar{x} = 20.71$. Thus, the 95% confidence interval for the population mean city gas mileage for all motor vehicles takes the following form:

$$\text{point estimate} \pm \text{margin of error}$$

$$= \bar{x} \pm Z_{\alpha/2}\left(\sigma/\sqrt{n}\right)$$

$$= 20.71 \pm 0.93$$

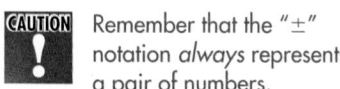

NOW YOU CAN DO
Exercises 35–42.

c. We interpret the margin of error E by saying that we can estimate the population mean city gas mileage for all vehicles to *within 0.93 mpg with 90% confidence.*

Note: When it comes to the margin of error *E*, smaller is better!

Of course, we want our confidence interval estimates to be as precise as possible. Therefore, we want the margin of error to be as small as possible, which would in turn result in a tighter confidence interval. Tighter confidence intervals are better, because the likely maximum difference between the sample mean and the population mean is reduced.

So how do we reduce the size of the margin of error? Let's look at the margin of error for the *Z* interval:

$$E = Z_{\alpha/2}\left(\sigma/\sqrt{n}\right)$$

The population standard deviation σ is fixed, so only $Z_{\alpha/2}$ and *n* can vary. There are therefore two strategies for decreasing the margin of error:

- *Decrease the confidence level,* which would decrease the value of $Z_{\alpha/2}$ (see Table 1), and

- *Increase the sample size n,* because dividing by a larger $\sqrt{n}$ will reduce *E*.

EXAMPLE 8 Decreasing the margin of error by decreasing the confidence level

For the confidence interval for the population mean city gas mileage in Example 5, suppose we reduce the confidence level from 90% to 80% and leave everything else unchanged. Find the new margin of error. Describe how the margin of error has changed.

Solution

From Example 5, we have the margin of error for the 90% confidence interval for μ as follows:

$$E = Z_{\alpha/2}\left(\sigma/\sqrt{n}\right) = 1.645\left(5.637/\sqrt{100}\right) \approx 0.93$$

Decreasing the confidence level from 90% to 80% decreases $Z_{\alpha/2}$ from 1.645 to 1.28. This gives us the margin of error for the 90% confidence interval as:

$$E = Z_{\alpha/2}\left(\sigma/\sqrt{n}\right) = 1.28\left(5.637/\sqrt{100}\right) \approx 0.72$$

Decreasing the confidence level from 90% to 80% decreases the margin of error from 0.93 mpg to 0.72 mpg.

Developing Your
Statistical Sense

There's No Free Lunch

The margin of error in Example 8 is smaller than the one in Example 5, which is good because it gives a more precise estimate of μ. However, this smaller margin of error is due entirely to the decrease in the confidence level, which is not good. In statistical data analysis, there is rarely a free lunch. The trade-off here is that, while the margin of error went down, so did the confidence level, from 90% to 80%. On the other hand, confidence intervals that are too wide can be useless. For example, we can be 99.9999% confident that the population mean age of college students in Florida lies between 15 and 75 years old. But, so what? The interval is too wide to be of practical use. More useful would be a 95% confidence interval that the population mean age of college students in Florida lies between 20 and 27.

This leads us to Strategy 2 for reducing the margin of error: increase the sample size. *The only way to have both high confidence and a tight interval is to boost the sample size.*

EXAMPLE 9 Decreasing the margin of error by increasing the sample size

For the confidence interval for the population mean city gas mileage in Example 5, suppose the results were based on a sample of size $n = 400$ instead of $n = 100$. Leaving everything else unchanged, find the new margin of error, and describe how the margin of error has changed.

Solution

For $n = 400$, the margin of error is

$$E = Z_{\alpha/2}\left(\sigma/\sqrt{n}\right) = 1.645\left(5.637/\sqrt{400}\right) \approx 0.46$$

Increasing the sample size from $n = 100$ to $n = 400$ has decreased the margin of error from 0.93 mpg to 0.46 mpg.

"More data" is a familiar refrain in statistical analysis. Of course, increasing the sample size often raises pocketbook issues, because large samples can get very expensive ("We want a large-sample estimate of the amount of damage sustained by Corvettes hitting a wall at 90 mph"). Sometimes obtaining large samples is simply impossible. Suppose an astronomer has developed a new technique for predicting corona effects during solar eclipses; she will have to wait a while (say, a few hundred years) to build up a large sample. So, take samples as large as realistically possible to keep the width of the confidence interval as narrow as possible.

Increasingly, technology is being used to perform statistical analysis, including confidence intervals. Therefore, it is important to know how to read and interpret confidence intervals provided by software output. For instance, Example 10 shows how to calculate the margin of error E, when the software gives you only the lower bound and upper bound of the confidence interval.

EXAMPLE 10 Finding the margin of error, given the lower and upper bounds

Figure 8 shows the results for a 95% Z confidence interval for μ, where μ represents the population mean score on the SAT Math test. Do the following:

a. Report the confidence interval in the form "(lower bound, upper bound)."

b. Interpret the confidence interval.

c. Calculate the margin of error E for the confidence interval.

d. Interpret the margin of error.

Solution

a. The TI-83/84 output gives us the following confidence interval:

(lower bound, upper bound) = (482.28, 537.72)

b. We interpret this confidence interval as follows: We are 95% confident that the population mean score on the SAT Math test lies between 482.28 and 537.72.

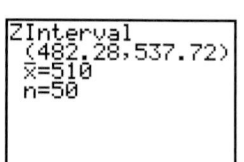

FIGURE 8 TI-83/84 output for a Z-interval for μ.

c. Here, we show how to calculate the margin of error, given the lower bound and upper bound of the confidence interval. The confidence interval from **(a)** is illustrated in Figure 9.

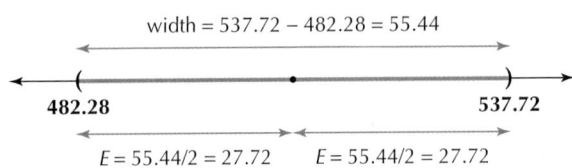

Now, the *width* of the margin of error is:

$$\text{width} = \text{upper bound} - \text{lower bound}$$

In Figure 9, the width of our confidence interval is:

$$\text{width} = 537.72 - 482.28 = 55.44$$

Then, the margin of error is half this width, as shown in Figure 9. This gives us a margin of error of

$$E = \text{margin of error} = \text{width}/2 = 55.44/2 = 27.72$$

NOW YOU CAN DO
Exercises 43–46.

d. We interpret the margin of error E by saying that we can estimate the population mean Math SAT score to *within 27.72 points with 95% confidence.*

In general, when the lower bound and upper bound of the confidence interval for μ have already been found, then the margin of error may be calculated as follows.

$$E = \text{margin of error} = (\text{upper bound} - \text{lower bound})/2$$

4 Sample Size for Estimating the Population Mean

In general, more data implies more precise results. In fact, when samples are plentiful and cheap, *arbitrarily precise confidence intervals with arbitrarily high confidence are possible simply by taking sufficiently large samples.*

Therefore, the question arises: *How large a sample size do I need* to get a tight confidence interval with a high confidence level?

Note: We solve for *n* as follows:

$$E = (Z_{\alpha/2})(\sigma/\sqrt{n})$$

Multiply both sides by $\sqrt{n}$:

$$\sqrt{n}\,(E) = (Z_{\alpha/2})\sigma$$

Divide both sides by E:

$$\sqrt{n} = \left(\frac{(Z_{\alpha/2})\sigma}{E}\right)$$

Square both sides to get the formula for *n*:

$$n = \left(\frac{(Z_{\alpha/2})\sigma}{E}\right)^2$$

Sample Size for Estimating the Population Mean

The sample size for a Z interval that estimates the population mean μ to within a margin of error E with confidence $100(1 - \alpha)\%$ is given by

$$n = \left(\frac{(Z_{\alpha/2})\sigma}{E}\right)^2$$

where $Z_{\alpha/2}$ is the value associated with the desired confidence level (Table 1), E is the desired margin of error, and σ is the population standard deviation. By convention, whenever this formula yields a sample size with a decimal, *always round up to the next whole number.*

EXAMPLE 11 Sample size for estimating the population mean

Suppose we want to estimate to within $1000 the mean salary μ of all college graduates who were business majors. Assume $\sigma = \$5000$. How many business majors would we sample to estimate the mean salary to within $1000 with 95% confidence?

Solution

"Within $1000" means that the desired margin of error E is $1000, and 1.96 is the $Z_{\alpha/2}$ value associated with 95% confidence. Substituting into the formula for the sample size, we get:

$$n = \left(\frac{1.96 \cdot 5000}{1000} \right)^2 = 96.04$$

Now, when finding the required sample size, if the formula results in a decimal, we always round up to the next whole number. Thus, we need a sample size of $n = 97$ for a confidence level of 95%.

We round up because (**a**) the sample size *n* must be a whole number and (**b**) rounding down will lead to a value of *n* with less than the desired confidence level.

NOW YOU CAN DO
Exercises 47–54.

YOUR TURN #6

For the situation in Example 11, suppose we now needed our estimate to be within only $100 the mean salary μ. How many business majors would we sample to estimate the mean salary to within $100 with 95% confidence?

(The solution is shown in Appendix A.)

STEP-BY-STEP TECHNOLOGY GUIDE: *Z* Confidence Intervals

We illustrate how to construct the confidence interval for Example 6 (page 436).

TI-83/84

If you have the data values:
Step 1 Enter the data into list **L1** (Figure 10).
Step 2 Press **STAT**, highlight **TESTS**.
Step 3 Press **7** (for ZInterval).
Step 4 For input (**Inpt**), highlight **Data** and press **ENTER** (Figure 11).
a. For σ, enter the assumed value of **6.326**.
b. For **List**, press **2nd** then **L1**.
c. For **Freq**, enter **1**.
d. For **C-Level** (confidence level), enter the appropriate confidence level (e.g., **0.95**), and press **ENTER**.
e. Highlight **Calculate** and press **ENTER**. The results are shown in Figure 5 in Example 6.

If you have the summary statistics:
Step 1 Press **STAT**, and highlight **TESTS**.
Step 2 Press **7** (for ZInterval).
Step 3 For input (**Inpt**), highlight **Stats** and press **ENTER** (Figure 12).
a. For σ, enter the assumed value of **6.326**.
b. For $\bar{x}$, enter the sample mean **26.9375**.
c. For *n*, enter the sample size **16**.
d. For **C-Level** (confidence level), enter the appropriate confidence level (e.g., **0.95**), and press **ENTER**.
e. Highlight **Calculate** and press **ENTER**. The results are shown in Figure 5 in Example 6.

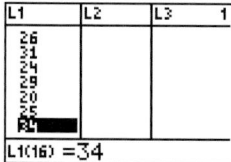

FIGURE 10

FIGURE 11

FIGURE 12

determines the shape of the t distribution, just as the mean and variance uniquely determine the shape of the normal distribution. All t curves have several characteristics in common.

Characteristics of the t Distribution

- Centered at zero. The mean of t is zero, just as with Z.
- Symmetric about its mean zero, just as with Z.
- As the degrees of freedom decreases, the t curve gets flatter, and the area under the t curve decreases in the center and increases in the tails. That is, the t curve has heavier tails than the Z curve.
- As degrees of freedom increases toward infinity, the t curve approaches the Z curve, and the area under the t curve increases in the center and decreases in the tails.

Similar to the definition of $Z_{\alpha/2}$ in Section 8.1, we can define $t_{\alpha/2}$ to be the value of the t distribution with area $\alpha/2$ to the right of it, as seen in Figure 15. Table 1 in Section 8.1 provides the $Z_{\alpha/2}$ values for certain common confidence levels. Unfortunately, because there is a different t curve for each sample size, there are many possible $t_{\alpha/2}$ values. You will need to use the t table (Table D in the Appendix) to find the value of $t_{\alpha/2}$, as follows.

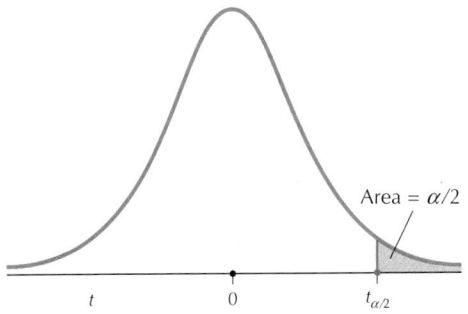

FIGURE 15 $t_{\alpha/2}$ has area to the right of it.

Procedure for Finding $t_{\alpha/2}$

Step 1 Go across the row marked "Confidence level" in the t table (Table D in the Appendix) until you find the column with the desired confidence level at the top. The $t_{\alpha/2}$ value is in this column somewhere.

Step 2 Go down the column until you see the correct number of degrees of freedom on the left. The number in that row and column is the desired value of $t_{\alpha/2}$.

EXAMPLE 12 Finding $t_{\alpha/2}$

Find the value of $t_{\alpha/2}$ that will produce a 95% confidence interval for μ if the sample size is $n = 20$.

Solution

Step 1 We go across the row labeled "Confidence level" in the t table (Figure 16) until we see the 95% confidence level. Our $t_{\alpha/2}$ is somewhere in this column.

Step 2 The degrees of freedom are df $= n - 1 = 20 - 1 = 19$. We go down the column until we see 19 on the left. The number in that row is our $t_{\alpha/2}$, 2.093.

Note: For the newer TI-84s

1. Press 2nd DISTR and select **4:invT**.
2. Enter the area to the *left* of the t value, then **comma**, then df $= n - 1$.
3. Press **ENTER**.

For example, **invT(0.975,19)** gives 2.093024022. The TI-83 does not have this function.

t Distribution

df	80%	90%	95%	98%	99%
	0.10	0.05	0.025	0.01	0.005
	0.20	0.10	0.05	0.02	0.01
1	3.078	6.314	12.706	31.821	63.657
2	1.886	2.920	4.303	6.965	9.925
3	1.638	2.353	3.182	4.541	5.841
14	1.345	1.761	2.145	2.624	2.977
15	1.341	1.753	2.131	2.602	2.947
16	1.337	1.746	2.120	2.583	2.921
17	1.333	1.740	2.110	2.567	2.898
18	1.330	1.734	2.101	2.552	2.878
19	1.328	1.729	2.093	2.539	2.861
20	1.325	1.725	2.086	2.528	2.845
21	1.323	1.721	2.080	2.518	2.831

Confidence level / Area in one tail / Area in two tails

FIGURE 16
Use the confidence level and the degrees of freedom to find $t_{\alpha/2}$.

NOW YOU CAN DO
Exercises 5–8.

YOUR TURN #7
Find the value of $t_{\alpha/2}$ that will produce a 90% confidence interval for μ if the sample size is $n = 20$.

(The solution is shown in Appendix A.)

2 *t* Interval for the Population Mean

The *t* distribution provides the following confidence interval for the unknown population mean μ, called the *t* interval.

Note: Suppose that σ is unknown, and the population is either non-normal or of unknown distribution, and the sample size is not large. Then we should not use the *t* interval. Instead, we need to turn to nonparametric methods, for example, the *sign interval* or the *Wilcoxon interval*. (See Chapter 14: *Nonparametric Statistics*, available online.)

t Interval for μ

The *t* interval for μ may be constructed whenever *either* of the following conditions is met:

- The population is normal.
- The sample size is large ($n \geq 30$).

Suppose a random sample of size n is taken from a population with unknown mean μ and unknown standard deviation σ. A $100(1 - \alpha)\%$ confidence interval for μ is given by the interval

$$\text{lower bound} = \bar{x} - t_{\alpha/2}(s/\sqrt{n}), \text{ upper bound} = \bar{x} + t_{\alpha/2}(s/\sqrt{n})$$

where $\bar{x}$ is the sample mean, $t_{\alpha/2}$ is associated with the confidence level and $n - 1$ degrees of freedom, and s is the sample standard deviation. The *t* interval may also be written as

$$\bar{x} \pm t_{\alpha/2}(s/\sqrt{n})$$

and is denoted

(lower bound, upper bound)

EXAMPLE 13 Checking whether the conditions are met for the *t* interval for μ

For each of the following, we are taking a random sample from a population with σ unknown. Determine whether the conditions are met for constructing the indicated *t* interval for μ. If not, explain why not.

a. Confidence level 99%, $n = 16$, $\bar{x} = 35$, $s = 8$
b. Confidence level 95%, $n = 25$, $\bar{x} = 42$, $s = 10$, normal population

Solution

a. The sample size is not large (n is not ≥ 30), and we are not told that the population is normal. Therefore, the conditions are not met for the *t* interval for μ. It is not okay to construct the *t* interval.

b. Again the sample size is not large, but this time we are told that the population is normal. Thus, the conditions are met for the *t* interval for μ. It is okay to construct the *t* interval.

Never assume normality unless it is indicated or evidence for it exists.

NOW YOU CAN DO
Exercises 9–12.

YOUR TURN
#8

For each of the following, we are taking a random sample from a population with σ unknown. Determine whether the conditions are met for constructing the indicated *t* interval for μ. If not, explain why not.

a. Confidence level 95%, $n = 25$, $\bar{x} = 100$, $s = 10$
b. Confidence level 95%, $n = 36$, $\bar{x} = 50$, $s = 6$

(The solutions are shown in Appendix A.)

EXAMPLE 14 Constructing a *t* confidence interval for μ

Research has shown that the amount of sodium consumed in food has been associated with hypertension (high blood pressure). The table provides a list of 16 breakfast cereals, along with their sodium contents, in milligrams per serving.

a. Determine whether the conditions are met for constructing a *t* interval for the population mean sodium content per serving for all breakfast cereals.

b. Find the value of $t_{\alpha/2}$ for 99% confidence and degrees of freedom $n - 1 = 15$.

c. Construct a 99% confidence interval for the population mean sodium content.

d. Interpret the meaning of this confidence interval.

cerealsodium

Cereal	Sodium (grams)	Cereal	Sodium (grams)
Apple Jacks	125	Grape Nuts Flakes	140
Cap'n Crunch	220	Kix	260
Cinnamon Toast Crunch	210	Life	150
Corn Flakes	290	Lucky Charms	180
Count Chocula	180	Raisin Bran	210
Cream of Wheat	80	Rice Chex	240
Fruit Loops	125	Special K	230
Fruity Pebbles	135	Total Whole Grain	200

Solution

a. Figure 17 contains the normal probability plot for the data set. Though not perfect, all points lie within the bounds, indicating acceptable normality. Thus, we proceed to construct the 99% confidence interval.

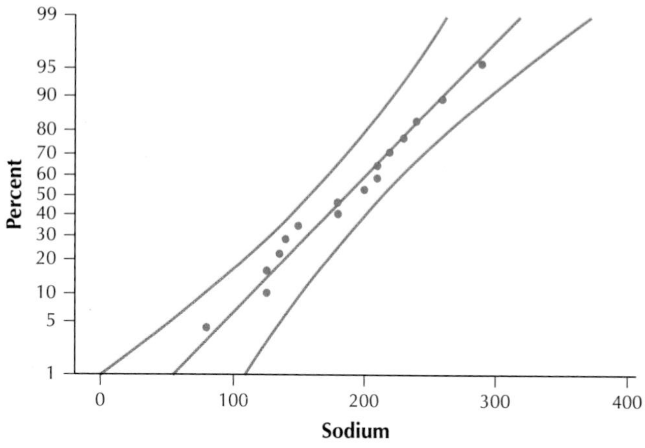

FIGURE 17
Normal probability plot for sodium in cereal.

b. The value of $t_{\alpha/2}$ for 99% confidence and 15 degrees of freedom is 2.947.

c. A 99% confidence interval for μ is given by the interval

$$\text{lower bound} = \bar{x} - t_{\alpha/2}\,(s/\sqrt{n}), \quad \text{upper bound} = \bar{x} + t_{\alpha/2}\,(s/\sqrt{n})$$

From the Minitab output in Figure 18, we have $n = 16$, $\bar{x} = 185.9$, and $s = 56.8$. Substituting, we get:

$$\text{lower bound} = 185.9 - (2.947)(56.8/\sqrt{16}) = 185.9 - 41.8 = 144.1,$$

$$\text{upper bound} = 185.9 + (2.947)(56.8/\sqrt{16}) = 185.9 + 41.8 = 227.7$$

Descriptive Statistics: Sodium

Variable	Total Count	Mean	StDev
Sodium	16	185.9	56.8

FIGURE 18 Minitab ouput.

NOW YOU CAN DO
Exercises 13–28.

d. We are 99% confident that μ, the population mean sodium content per serving of all breakfast cereals, lies between 144.1 grams and 227.7 grams.

YOUR TURN
#9

Find and interpret a 95% confidence interval for μ, which is the population mean sodium content per serving of all breakfast cereals.

(The solution is shown in Appendix A.)

Developing Your
Statistical Sense

t Intervals May Offer More Peace of Mind Than *Z* Intervals

In Example 14, if we had assumed that the population standard deviation σ was known ($\sigma = 56.8$), then the 99% *Z* interval for the population mean amount of sodium would have been

$$\text{lower bound} = 185.9 - (2.576)(56.8/\sqrt{16}) = 185.9 - 36.6 = 149.3,$$

$$\text{upper bound} = 185.9 + (2.576)(56.8/\sqrt{16}) = 185.9 + 36.6 = 222.5$$

Note that this *Z* interval (149.3, 222.5) is only slightly more precise than the *t* interval (144.1, 227.7). However, the *Z* interval depends on prior knowledge of the value of σ. If the value of σ is inaccurate, then the *Z* interval will be misleading and overly optimistic. With even moderate sample sizes, reporting the *t* interval instead of the *Z* interval may offer peace of mind to the data analyst.

If the degrees of freedom needed to find $t_{\alpha/2}$ do not appear in the df column of the t table, a conservative solution is to take the next row with smaller df in the t table. For example, if we have a data set such that df $= n - 1 = 47$, we find that df $= 47$ is not in the table. Instead, we assign df $= 40$. (Even though 50 is closer, it will lead to an interval that overstates the precision in the data.) For df > 1000, use the associated Z critical values, because the t distribution approaches the Z distribution as n gets very large.

Margin of Error

Recall that the margin of error for the Z interval equals $Z_{\alpha/2} \cdot (\sigma/\sqrt{n})$. For the t interval, because σ is unknown, the margin of error is given as follows.

> **Margin of Error for the *t* Interval**
>
> $$E = t_{\alpha/2} \cdot \left(\frac{s}{\sqrt{n}}\right)$$
>
> The margin of error E for a $(1 - \alpha)100\%$ t interval for μ can be interpreted as follows:
> "We can estimate μ to within E units with $(1 - \alpha)100\%$ confidence."

EXAMPLE 15   Margin of error

Use the statistics observed in Example 14.

a. Find the margin of error for the 99% confidence interval for mean sodium content per serving of all breakfast cereals.

b. Interpret the margin of error.

Solution

a. From Example 14c, we have:

$$E = (2.947)(56.8/\sqrt{16}) = 41.8$$

The margin of error for mean sodium content is 41.8 grams.

NOW YOU CAN DO
Exercises 29–40.

b. We can estimate the population mean sodium content per serving of all breakfast cereals to within 41.8 grams with 99% confidence.

YOUR TURN
#10

Find and interpret the margin of error for the 95% confidence interval for mean sodium content found in the Your Turn #9 after Example 14.

(The solution is shown in Appendix A.)

What Does the Margin of Error Mean?

The margin of error $E = 41.8$ grams provides an indication of the accuracy of the confidence interval estimate for confidence level $= 99\%$. That is, if we repeatedly take many samples of size 16 breakfast cereals, our sample mean $\bar{x}$ will be within $E = 41.8$ grams of the unknown population mean μ in 99% of those samples.

EXAMPLE 16   *t* intervals for μ using technology

cerealsodium

For the breakfast cereal data in Example 14, construct a 99% confidence interval for the population mean sodium content, using the TI-83/84, Minitab, and SPSS.

Solution

We use the instructions provided in the Step-by-Step Technology Guide below. The sample size $n = 16$ is not large (≤ 30), so it is necessary to check for normality. Figure 17 indicates acceptable normality.

The results for the TI-83/84 in Figure 19 display the 95% *t* confidence interval for the population mean sodium content to be

$$(\text{lower bound} = 155.67, \text{upper bound} = 216.21)$$

They also show the sample mean $\bar{x} = 185.9375$, the sample standard deviation $s = 56.81017368$, and the sample size $n = 16$.

The Minitab results are shown in Figure 20, providing the sample size $n = 16$, the sample mean $\bar{x} = 185.9$, the sample standard deviation $s = 56.8$, the standard error (SE mean) 14.2, and the 95% *t* confidence interval (155.7, 216.2).

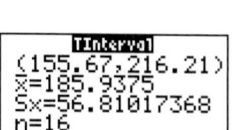

FIGURE 19 TI-83/84 results.

One-Sample T: Sodium

Variable	N	Mean	StDev	SE Mean	95% CI
Sodium	16	185.9	56.8	14.2	(155.7, 216.2)

FIGURE 20 Minitab results.

The SPSS results are shown in Figure 21, providing the sample mean $\bar{x} = 185.9375$, the standard error 14.20254, and the 95% *t* confidence interval (155.6655, 216.2095).

Descriptives

			Statistic	Std. Error
Sodium	Mean		185.9375	14.20254
	95% Confidence Interval for Mean	Lower Bound	155.6655	
		Upper Bound	216.2095	

FIGURE 21 SPSS results.

STEP-BY-STEP TECHNOLOGY GUIDE: *t* Confidence Intervals

We illustrate how to construct the *t* confidence interval for Example 16 (page 454).

TI-83/84

If you have the data values:
Step 1   Enter the data into list **L1**.
Step 2   Press **STAT**, highlight **TESTS**.
Step 3   Press **8** (for TInterval, see Figure 22).
Step 4   For input (Inpt), highlight **Data** and press **ENTER** (Figure 23).
a. For **List**, press **2nd** then **L1**.
b. For **Freq**, enter 1.

c. For **C-Level** (confidence level), enter the appropriate confidence level (for example, **0.95**), and press **ENTER**.
d. Highlight **Calculate** and press **ENTER**. The results are shown in Figure 19 in Example 16.

If you have the summary statistics:
Step 1   Press **STAT**, highlight **TESTS**.
Step 2   Press **8** (for TInterval, see Figure 22).

FIGURE 22

FIGURE 23

FIGURE 24

Step 3   For input (**Inpt**), highlight **Stats** and press **ENTER** (Figure 24).
a. For $\bar{x}$, enter the sample mean **185.9375**.
b. For **Sx**, enter the sample standard deviation **56.81017368**.
c. For *n*, enter the sample size **16**.

d. For **C-Level** (confidence level), enter the appropriate confidence level (for example, **0.95**), and press **ENTER**.
e. Highlight **Calculate** and press **ENTER**. The results are shown in Figure 19 in Example 16.

EXCEL

If you have the data values:
Step 1   Enter the data into Column **A**.
Step 2   Label **C1** through **F1** X-bar, Std Err, X-bar − StdErr, and X-bar + Std Err.
Step 3   Click **Data > Data Analysis > Descriptive Statistics**. For Input Range, highlight **A1** through **A16**. Click **Summary Statistics**, click **OK**. Note the mean, **185.9375**, and standard deviation, **56.81017**.
Step 4   Click cell **C2** and enter the mean, **185.9375**.
Step 5   Click cell **D2**, and click the **Insert Function** icon f_x. Under **Search for a function**, type **confidence.t**. Click **Go**, and **OK**.
Step 6   Enter the **Alpha** value, **0.01**, the **Standard_dev**, **56.81017**, and the sample **Size 16**. Click **OK**.
Step 7   In cell **E2**, type **=C2-D2**, and press **ENTER**. In cell **F2**, type **=C2+D2**, and press **ENTER**. Cells E2 and F2 are your confidence interval. Results are shown in Figure 25.

If you have the summary statistics:
Step 1   Label **C1** through **F1** X-bar, Std Err, X-bar − StdErr, and X-bar + Std Err.

$\times$ $\checkmark$ f_x	=CONFIDENCE.T(0.01,56.81017,16)		
C	D	E	F
X-bar	**Std Err**	**X-bar - StdErr**	**X-bar + StdErr**
185.9375	41.85081	144.086685	227.788315

FIGURE 25   Excel results.

Step 2   Click cell **C2** and enter the mean, **185.9375**.
Step 3   Click cell **D2**, and click the **Insert Function** icon f_x. Under **Search for a function**, type **confidence.t**. Click **Go**, and **OK**.
Step 4   Enter the **Alpha** value, **0.01**, the **Standard_dev**, **56.81017**, and the sample **Size 16**. Click **OK**.
Step 5   In cell **E2**, type **=C2-D2**, and press **ENTER**. In cell **F2**, type **=C2+D2**, and press **ENTER**. Cells E2 and F2 are your confidence interval.

MINITAB

If you have the data values:
Step 1   Enter the data into column **C1**.
Step 2   Click **Stat > Basic Statistics > 1-Sample t…**
Step 3   Select **One or more samples, each in a column** from the drop-down menu. Click the box below the menu, click **C1**, and click **Select**.
Step 4   Click **Options…**, enter **95** as the **Confidence level**, and click **OK** twice.

The results are shown in Figure 20 in Example 16.

If you have the summary statistics:
Step 1   Click **Stat > Basic Statistics > 1-Sample t…**
Step 2   Click **Summarized Data** from the drop-down menu.
Step 3   For **Sample Size** enter **16**, for **Sample Mean** enter **185.9**, and for **Standard Deviation** enter **56.8**.
Step 4   Click **Options…**, enter **95** as the **Confidence Level**, click **OK**, and click **OK** again.

SPSS

If you have the data values:
Step 1   Enter the data into the first column. Click **Analyze > Descriptive Statistics > Explore…**
Step 2   Move the variable to the **Dependent List**. Under **Display**, select **Statistics**.

Step 3   Click **Statistics…**, select **Descriptives**, and in the box for **Confidence Interval for Mean**, enter **95**. Click **Continue**, then click **OK**. The results are shown in Figure 21 in Example 16.

JMP

If you have the data values:
Step 1   Click **File > New > Data Table**. Enter the data into Column 1.
Step 2   Click **Analyze > Distribution**. Click **Column 1** under **Select Columns**, then click **Y, Columns**. Click **OK**.
Step 3   Click the red triangle beside Column 1, and click **Confidence Interval > 0.95**. The output is shown in Figure 26.

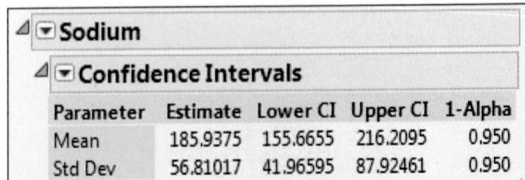

◢ ▼ **Sodium**				
◢ ▼ **Confidence Intervals**				
Parameter	**Estimate**	**Lower CI**	**Upper CI**	**1-Alpha**
Mean	185.9375	155.6655	216.2095	0.950
Std Dev	56.81017	41.96595	87.92461	0.950

FIGURE 26   JMP results.

CRUNCHIT!

If you have the data values:
Step 1   Click **File**, then highlight **Load from Larose, Discostat3e >** Chapter 8, and click on **Example 02_17**.
Step 2   Click **Statistics**, highlight **t** and select **1-sample**.
Step 3   With the **Columns** tab chosen, for **Sample** select Sodium.
Step 4   Select the **Confidence Interval** tab, and enter **95** for the Confidence Interval Level (%). Then click **Calculate**.

If you have the summary statistics:
Step 1   Click **Statistics**, highlight **t**, and select **1-sample**.
Step 2   Choose the **Summarized** tab. For **n** enter the sample size **16**; for **Sample Mean**, enter **185.9375**. For **Standard Deviation**, enter **56.8**.
Step 3   Select the **Confidence Interval** tab, and enter **95** for the Confidence Interval Level (%). Then click **Calculate**.

Section 8.2 Summary

1. For a normal population, the distribution of

$$t = \frac{\bar{x} - \mu}{s/\sqrt{n}}$$

follows a *t* distribution, with $n - 1$ degrees of freedom, where $\bar{x}$ is the sample mean, μ is the unknown population mean, *s* is the sample standard deviation, and *n* is the sample size. The *t* distribution is symmetric about its mean 0, just like the *Z* distribution. However, the *t* distribution is flatter.

2. A $100(1 - \alpha)\%$ confidence interval for μ is given by the interval

$$\bar{x} \pm t_{\alpha/2} \left(s/\sqrt{n} \right)$$

where $\bar{x}$ is the sample mean, $t_{\alpha/2}$ is associated with the confidence level and $n - 1$ degrees of freedom, *s* is the sample standard deviation, and *n* is the sample size. We can construct a *t* interval whenever *either* of the following conditions is met: the population is normal, or the sample size is large ($n \geq 30$).

Section 8.2 Exercises

CLARIFYING THE CONCEPTS

1. Why do we need the *t* interval? Why can't we always use *Z* intervals? (pp. 448–449)
2. Suppose that σ is known. Should we still use a *t* interval? (p. 453)
3. As the sample size gets larger and larger, what happens to the *t* curve? (p. 450)
4. State the formula for the margin of error for the *t* interval. (p. 454)

PRACTICING THE TECHNIQUES

 CHECK IT OUT!

To do	Check out	Topic
Exercises 5–8	Example 12	Finding $t_{\alpha/2}$
Exercises 9–12	Example 13	Checking whether the conditions are met for the *t* interval for μ
Exercises 13–28	Example 14	Constructing a *t* confidence interval for μ
Exercises 29–40	Example 15	Margin of error

5. For the following scenarios, we are taking a random sample from a normal population with σ unknown. Find $t_{\alpha/2}$.

a. Confidence level 90%, sample size 21
b. Confidence level 95%, sample size 21
c. Confidence level 99%, sample size 21

6. For the following scenarios we are taking a random sample from a normal population with σ unknown. Find $t_{\alpha/2}$.
a. Confidence level 95%, sample size 11
b. Confidence level 95%, sample size 21
c. Confidence level 95%, sample size 31

7. Refer to Exercise 5.
a. Describe what happens to the value of $t_{\alpha/2}$, as the confidence level increases, for a given sample size.
b. Draw a sketch of the *t* curve for sample size $n = 21$, and explain why the value of $t_{\alpha/2}$ changes as it does.

8. Refer to Exercise 6.
a. Describe what happens to the value of $t_{\alpha/2}$, as the sample size increases, for a given confidence level.
b. Draw a sketch of the *t* curve for a confidence level of 95%, and explain why the value of $t_{\alpha/2}$ changes as it does.

For each of Exercises 9–12, we are taking a random sample from a population with σ unknown. Check whether the conditions are met for constructing the indicated *t* interval for μ. If not, explain why not.

9. Confidence level 95%, $n = 16$, $\bar{x} = 250$, $s = 20$
10. Confidence level 99%, $n = 225$, $\bar{x} = 10$, $s = 5$, normal population

57. Carbon Emissions. The Excel output shows the carbon emissions (in millions of tons) from consumption of fossil fuels for a random sample of five nations.[12] The sample mean (using the function "average") and the sample standard deviation (using the function "stdev") are also given. The highlighted cell is the margin of error given by the function CONFIDENCE.T, which needs the following values: $1 - \alpha$, s, and n. Below, the normal probability plot is given. **carbon**

fx	=CONFIDENCE.T(0.05,194.812,5)	
	D	E
Country		Carbon Emissions
Brazil		361
Germany		844
Mexico		398
Great Britain		577
Canada		631
Xbar		562.2
s		194.812
E		241.89
Xbar - E		320.31
Xbar + E		804.09

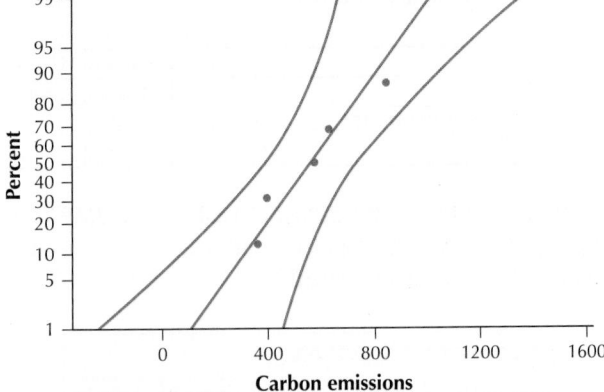

a. Check the normality assumption.
b. Use the Excel output to report and interpret a 95% t confidence interval for the population mean carbon emissions.
c. Use the Excel output to report and interpret the margin of error for the confidence interval in part (**b**).
d. Explain two ways we could decrease the margin of error. Which method is preferable, and why?

58. *Deepwater Horizon* Cleanup Costs. The Excel output shows the amount of money disbursed by BP to a random sample of six Florida counties for cleanup of the *Deepwater Horizon* oil spill, in millions of dollars.[13] The functions used to provide the confidence interval are similar to those in Exercise 57. The normal probability plot is also given. **deepwaterclean**

fx	=CONFIDENCE.T(0.05,0.348,6)	
D	E	F
County	Cleanup costs ($ millions)	
Broward	0.85	
Escambia	0.7	
Franklin	0.5	
Pinellas	1.15	
Santa Rosa	0.5	
Walton	1.35	
Xbar	0.842	
s	0.348	
E	0.37	
Xbar - E	0.48	
Xbar + E	1.21	

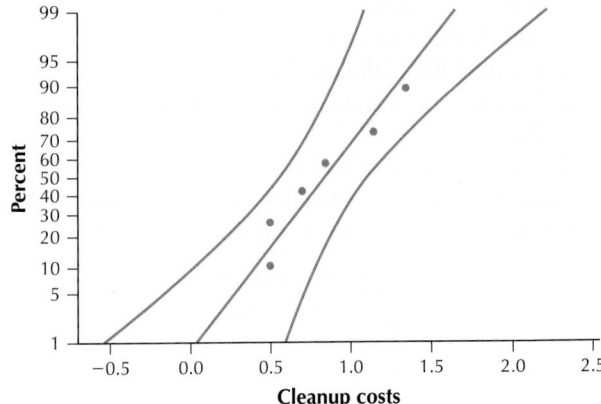

a. Check the normality assumption.
b. Use the Excel output to report and interpret a 95% t confidence interval for the population mean cleanup cost.
c. Use the Excel output to report and interpret the margin of error for the confidence interval in part (**b**).
d. Explain two ways we could decrease the margin of error. Which method is preferable, and why?

59. Wii Game Sales. The following table represents the number of units sold (in 1000s) in the United States for the week ending March 26, 2011, for a random sample of 8 Wii games.[14] The normal probability plot is given. **wiisales**

Game	Units (1000s)	Game	Units (1000s)
Wii Sports Resort	65	Zumba Fitness	56
Super Mario All Stars	40	Wii Fit Plus	36
Just Dance 2	74	Michael Jackson	42
New Super Mario Brothers	16	Lego Star Wars	110

Source: www.vgchartz.com, April 1, 2011.

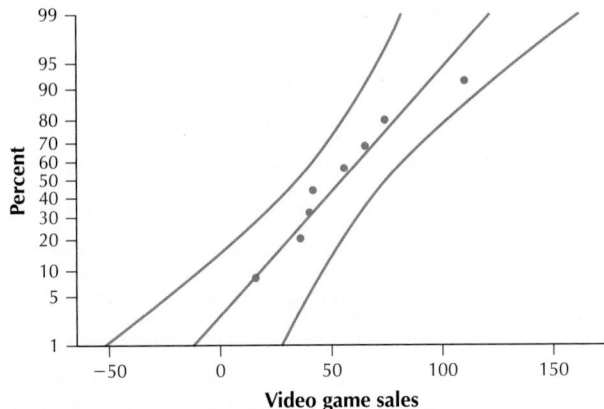

a. Confirm that the normality condition is met.
b. Construct and interpret a 95% confidence interval for the population mean number of units sold.
c. Calculate and interpret the margin of error.
d. How could we decrease the margin of error of our confidence interval without decreasing the confidence level?

60. A Rainy Month in Georgia? The following table represents the total rainfall (in inches) for the month of February 2011 for a random sample of 10 locations in Georgia.[15] **georgiarain**

Location	Rainfall (inches)	Location	Rainfall (inches)
Athens	4.72	Atlanta	4.25
Augusta	4.31	Cartersville	3.03
Dekalb	2.96	Fulton	4.36
Gainesville	4.06	Lafayette	3.75
Marietta	3.20	Rome	3.26

Source: National Weather Service.

a. Use technology to construct a normal probability plot to confirm that the data exhibit acceptable normality.
b. Construct and interpret a 90% confidence interval for the population mean rainfall in inches.
c. Calculate and interpret the margin of error.
d. How could we decrease the margin of error of our confidence interval without decreasing the confidence level?

61. Electric Cars. The accompanying table shows the miles-per-gallon equivalent (MPGe) for five electric cars, as reported by www.hybridcars.com in 2014. **electricmiles**

Electric Vehicle	MPGe
Tesla Model S	89
Nissan Leaf	99
Ford Focus	105
Mitsubishi i-MiEV	112
Chevrolet Spark	119

a. Use technology to construct a normal probability plot of MPGe. Confirm that the distribution is normal.
b. Find $t_{\alpha/2}$ for a confidence interval with 90% confidence.
c. Compute and interpret the margin of error E for a confidence interval with 90% confidence.
d. Construct and interpret a 90% confidence interval (*t* interval) for the population mean mileage.

62. Calories in Breakfast Cereals. What is the mean number of calories in a bowl of breakfast cereal? A random sample of six well-known breakfast cereals yielded the following calorie data: **cerealcalories**

Cereal	Calories
Apple Jacks	110
Cocoa Puffs	110
Mueslix	160
Cheerios	110
Corn Flakes	100
Shredded Wheat	80

a. Use technology to construct a normal probability plot of the number of calories.
b. Is there evidence that the distribution is not normal?
c. Can we proceed to construct a *t* interval? Why or why not?

63. Commuting Distances. A university is trying to attract more commuting students from the local community. As part of the research into the modes of transportation students use to commute to the university, a survey was conducted asking how far commuting students commuted from home to school each day. A random sample of 30 students provided the distances (in miles) shown. **commutedist**

14	10	14	12	12	11	5	6	9	14	9	9	4	7	15
9	7	7	12	10	15	10	6	11	9	11	10	11	7	12

a. Find $t_{\alpha/2}$ for a confidence interval with 90% confidence.
b. Compute and interpret the margin of error for a confidence interval with 90% confidence.
c. Construct and interpret a 90% *t* confidence interval for the population mean commuting distance.

WHAT IF ? **64.** Refer to the previous exercise. *What if* we increased the sample size to some unspecified value but everything else stayed the same. Describe what, if anything, would happen to each of the following measures and why:
a. $t_{\alpha/2}$
b. Margin of error E
c. Width of the confidence interval

Cigarette Consumption. Use the following information for Exercises 65–67. Health officials are interested in estimating the population mean number of cigarettes smoked annually per capita in order to evaluate the efficacy of their antismoking campaign. A random sample of eight U.S. counties yielded the following numbers of cigarettes smoked per capita: 2206, 2391, 2540, 2116, 2010, 2791, 2392, 2692.

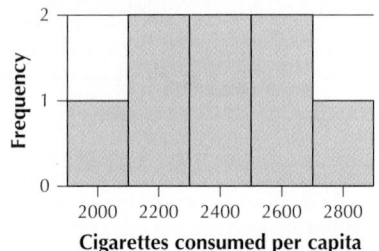

65. Evaluate the normality assumption using the accompanying histogram. Is it appropriate to construct a *t* interval using this data set? Why or why not? What is it about the histogram that tells you one way or the other?

66. Compute and interpret the margin of error *E* for a confidence interval with 90% confidence. What is the meaning of this number?

67. Construct and interpret a 90% confidence interval for the population mean number of cigarettes smoked per capita.

68. Baby Weights. A random sample of 20 babies born in Brisbane (Australia) Hospital had the histogram of the babies' weights shown here. The sample mean is 3227 grams, and the sample standard deviation is 560 grams. Discuss the normality of the data. Do the data appear acceptably normal? Is it appropriate to apply the *t* interval or not? Explain why or why not.

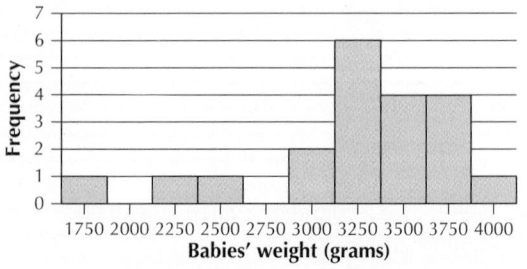

BRINGING IT ALL TOGETHER

Chapter 8 Case Study: Motor Vehicle Fuel Efficiency. Use the following information for Exercises 69–74. The Environmental Protection Agency calculates the estimated annual fuel cost for motor vehicles, with the resulting data provided in the variable *annual fuel cost* of the Chapter 8 Case Study data set **Fuel Efficiency**. A sample of the annual fuel cost is provided for 12 vehicles.

Annual fuel cost			
1750	2500	2400	2350
2150	3100	2950	2500
2550	2750	2300	2800

69. Construct a normal probability plot of the data (see pages 381–384). Evaluate the normality assumption using the accompanying histogram. Is it appropriate to construct a *t* interval using this data set? Why or why not?

70. Find the point estimate of μ, the population mean annual fuel cost.

71. Compute the sample standard deviation *s*.

72. Find $t_{\alpha/2}$ for a confidence interval with 90% confidence.

73. Construct and interpret a 90% confidence interval for the population mean annual fuel cost.

74. Compute and interpret the margin of error *E* for a confidence interval with 90% confidence. What is the meaning of this number?

WORKING WITH LARGE DATA SETS

Chapter 8 Case Study: Motor Vehicle Fuel Efficiency. Open the Chapter 8 Case Study data set **Fuel Efficiency**. Here, we will examine confidence intervals for the population mean amount of carbon dioxide generated by motor vehicles in city driving. We will then see whether these confidence intervals succeeded in capturing the population mean amount of carbon dioxide. Use technology to do the following: **⊞ fuelefficiency**

75. Obtain a random sample of size 100 from the data set.

76. Suppose we are interested in constructing a 95% *t* interval for the population mean amount of carbon dioxide generated by these vehicles in city driving (the variable *city CO2*), using our sample from the previous exercise. Do we need to check for normality?

77. Construct and interpret a 90% *t* interval for the population mean amount of city CO2.

78. Did your interval in Exercise 77 capture the population mean? Check by finding the mean city CO2 of all the vehicles.

79. Generate a second sample of size 100 from the data set. Construct a second 90% *t* interval for the population mean amount of city CO2. Did this confidence interval capture the population mean?

80. If we keep on obtaining new samples all day long, about what proportion of the 90% confidence intervals will capture the population mean?

8.3   Z Interval for the Population Proportion

OBJECTIVES By the end of this section, I will be able to . . .

1   Calculate the point estimate $\hat{p}$ of the population proportion p.
2   Construct and interpret a Z interval for the population proportion p.
3   Compute and interpret the margin of error for the Z interval for p.
4   Determine the sample size needed to estimate the population proportion.

1 Point Estimate $\hat{p}$ of the Population Proportion p

So far, we have dealt with interval estimates of the population mean μ only. However, we may also be interested in an interval estimate for the population proportion of successes, p. Recall from Section 7.2 that the sample proportion of successes

$$\hat{p} = \frac{x}{n} = \frac{\text{number of successes}}{\text{sample size}}$$

is a point estimate of the population proportion p.

EXAMPLE 17   Point estimate $\hat{p}$ of the population proportion p

Suppose that a random sample of 100 Starbucks' sales transactions is taken, and that 10 of these transactions were made using a cell phone. Calculate the sample proportion $\hat{p}$, and use it as a point estimate of the population proportion p.

Solution

We have $n = 100$ transactions and $x = 10$. Thus,

$$\hat{p} = \frac{x}{n} = \frac{10}{100} = 0.1$$

The point estimate of the population proportion p of Starbucks' transactions made using a cell phone is 0.1. (This sample proportion of 0.1 reflects the results from a survey made by the *Wall Street Journal* in 2013.[16])

NOW YOU CAN DO
Exercises 3–6.

YOUR TURN #11

For the following values of n and x, calculate the sample proportion $\hat{p}$, and use it as a point estimate of the population proportion p.

a. $n = 100, x = 50$

b. $n = 160, x = 90$

(The solutions are shown in Appendix A.)

Of course, different samples of Starbucks' customers may turn up different sample proportions $\hat{p}$. These are point estimates, and thus they carry no measure of confidence in their accuracy. The point estimates are probably close to the true values, but it's possible that they are not. They may be far from the true values. Only by using confidence intervals can we make probability statements about the accuracy of the estimates.

2 Z Interval for the Population Proportion p

Recall the Central Limit Theorem for Proportions in Section 7.2.

> ### Central Limit Theorem for Proportions
>
> The sampling distribution of the sample proportion $\hat{p}$ follows an approximately normal distribution with mean $\mu_{\hat{p}} = p$ and standard deviation $\sigma_{\hat{p}} = \sqrt{\frac{p \cdot q}{n}}$ when *both* the following conditions are satisfied: (1) $n \cdot p \geq 5$ and (2) $n \cdot q \geq 5$, where $q = 1 - p$.

Alternatively, the conditions may be expressed as follows: $x \geq 5$ and $(n - x) \geq 5$, that is, the number of successes ≥ 5 and the number of failures ≥ 5. Feel free to use these alternative conditions when the calculations are easier.

We can use the Central Limit Theorem for Proportions to construct confidence intervals for the population proportion p. Because the confidence interval for p is based on the standard normal Z distribution, it is called the **Z interval for the population proportion p**. Because p is unknown, the conditions and the formula for $\sigma_{\hat{p}}$ substitute $\hat{p}$ for p.

> ### Z Interval for p
>
> The Z interval for p may be performed only if *both* the following conditions are met: $n \cdot \hat{p} \geq 5$ and $n \cdot \hat{q} \geq 5$ (alternatively, $x \geq 5$ and $(n - x) \geq 5$) where $\hat{q} = 1 - \hat{p}$. When a random sample of size n is taken from a binomial population with unknown population proportion p, the $100(1 - \alpha)\%$ confidence interval for p is given by
>
> $$\text{lower bound} = \hat{p} - Z_{\alpha/2} \sqrt{\frac{\hat{p} \cdot \hat{q}}{n}}$$
>
> $$\text{upper bound} = \hat{p} + Z_{\alpha/2} \sqrt{\frac{\hat{p} \cdot \hat{q}}{n}}$$
>
> Alternatively,
>
> $$\hat{p} \pm Z_{\alpha/2} \sqrt{\frac{\hat{p} \cdot \hat{q}}{n}}$$
>
> where $\hat{p}$ is the sample proportion of successes, $\hat{q} = 1 - \hat{p}$, n is the sample size, and $Z_{\alpha/2}$ depends on the confidence level.

For convenience, we repeat Table 1 here, showing the $Z_{\alpha/2}$ values for the most common confidence levels.

Table 1 $Z_{\alpha/2}$ values for common confidence levels

Confidence level	α	$\alpha/2$	$Z_{\alpha/2}$
80%	0.20	0.10	1.28
90%	0.10	0.05	1.645
95%	0.05	0.025	1.96
99%	0.01	0.005	2.576

EXAMPLE 18 Z interval for the population proportion p

Using the Starbucks' data from Example 17, (**a**) verify that the conditions for constructing the Z interval for p have been met, and (**b**) construct a 95% confidence interval for the population proportion of all Starbucks' transactions that are made using a cell phone.

Solution

a. We have $n = 100$ transactions and $x = 10$. We check the conditions for the confidence interval. There are $x = 10$ successes, which is ≥ 5, and there are $n - x = 90$ failures, which is also ≥ 5. The conditions for constructing the *Z* interval for *p* have been met.

b. From Table 1, the confidence level of 95% gives $Z_{\alpha/2} = 1.96$. Thus, the confidence interval is

$$\text{lower bound} = \hat{p} - Z_{\alpha/2}\sqrt{\frac{\hat{p} \cdot \hat{q}}{n}} = 0.1 - 1.96\sqrt{\frac{0.1(0.9)}{100}}$$

$$= 0.1 - 1.96(0.03) = 0.1 - 0.0588 = 0.0412$$

$$\text{upper bound} = \hat{p} + Z_{\alpha/2}\sqrt{\frac{\hat{p} \cdot \hat{q}}{n}} = 0.1 + 1.96\sqrt{\frac{0.1(0.9)}{100}}$$

$$= 0.1 + 1.96(0.03) = 0.1 + 0.0588 = 0.1588$$

We are 95% confident that the population proportion of Starbucks' sales transactions made using a cell phone lies between 0.0412 and 0.1588. (See Figure 27.)

0.0412	$\hat{p}$ 0.1	0.1588

NOW YOU CAN DO
Exercises 7–20.

FIGURE 27 95% Confidence interval for the population proportion of Starbucks' sales transactions made using a cell phone.

YOUR TURN
#12

For the following values of *n* and *x*, (i) confirm that the conditions have been met, and (ii) construct a 95% confidence interval for the population proportion *p*.

a. $n = 100, x = 50$

b. $n = 160, x = 90$

(The solutions are shown in Appendix A.)

EXAMPLE 19 *Z* intervals for *p* using technology

A Pew Research Center survey of 1895 Internet users found 1118 who agree that "online dating is a good way to meet people." Use technology to find a 95% confidence interval for the population proportion of all Internet users who agree that online dating is a good way to meet people.

Solution

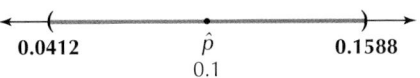

FIGURE 28 TI-83/84 results.

We use the instructions provided in the Step-by-Step Technology Guide at the end of this section (page 469). The results for the TI-83/84 in Figure 28 display the 95% confidence interval for the population proportion of Americans who agree that online dating is a good way to meet people to be

(lower bound = 0.56783, upper bound = 0.61212)

They also show the sample proportion $\hat{p} = 0.5899736148$ and the sample size $n = 1895$.

4 Sample Size for Estimating the Population Proportion

Next, we consider the question: How large a sample size do I need to estimate the population proportion p to within margin of error E with $100(1 - \alpha)\%$ confidence? The margin of error of the confidence interval for proportions equals

$$E = Z_{\alpha/2} \cdot \sqrt{\frac{\hat{p} \cdot \hat{q}}{n}}$$

Solving for n gives us

$$n = \hat{p} \cdot \hat{q} \left(\frac{Z_{\alpha/2}}{E}\right)^2 \qquad \text{(Equation 1)}$$

Unfortunately, Equation 1 depends on prior knowledge of $\hat{p}$. So, if we have such information about $\hat{p}$ available from some earlier sample, then we use Equation 1 to determine the required sample. However, what if we do not know the value of $\hat{p}$?

Figure 31 plots the sample size requirements for a 95% confidence interval for p, with a desired margin of error of 0.03, for values of $\hat{p}$ ranging from 0.01 to 0.99, representing all sample proportions from 1% to 99%. Note that the plot is symmetric, and therefore the largest required sample size occurs at the midpoint $\hat{p} = 0.5$. Thus, $\hat{p} = 0.5$ is the most conservative value for $\hat{p}$. When the actual value of $\hat{p}$ is not known, we use the following formula:

$$n = \left(\frac{0.5 \cdot Z_{\alpha/2}}{E}\right)^2$$

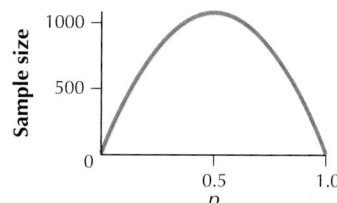

FIGURE 31 Sample size required with a margin of error of 0.03 for the range of values p.

Sample Size for Estimating a Population Proportion

When $\hat{p}$ is known, the sample size needed to estimate the population proportion p to within a margin of error E with confidence $100(1 - \alpha)\%$ is given by

$$n = \hat{p} \cdot \hat{q} \left(\frac{Z_{\alpha/2}}{E}\right)^2$$

where $Z_{\alpha/2}$ is the value associated with the desired confidence level, E is the desired margin of error, and $\hat{p}$ is the sample proportion of successes available from some earlier sample and $\hat{q} = 1 - \hat{p}$. Round up to the next integer.

When $\hat{p}$ is unknown, we use

$$n = \left(\frac{0.5 \cdot Z_{\alpha/2}}{E}\right)^2$$

These formulas are illustrated using the following two examples.

EXAMPLE 21 Sample size for estimating p when $\hat{p}$ is known

Refer to Example 20. Suppose that the Gallup Organization now wanted to estimate the population proportion of those who think there should be a law that would ban the possession of handguns to within a margin of error of $E = 0.01$ with 95% confidence. How large a sample size is needed?

Solution

From Example 20, we have the sample proportion $\hat{p} = 0.37$. The confidence level of 95% implies that our $Z_{\alpha/2} = 1.96$, and the desired margin of error is $E = 0.01$. Thus, the required sample size is

$$n = \hat{p} \cdot \hat{q} \left(\frac{Z_{\alpha/2}}{E} \right)^2 = 0.37(0.63) \left(\frac{1.96}{0.01} \right)^2 \approx 8954.77$$

NOW YOU CAN DO
Exercises 33–38.

Rounding up, this gives us a minimum required sample size of 8955. The smaller margin of error requires a larger sample size.

YOUR TURN
#14

For the situation in Example 21, suppose Gallup wants the estimate to be within a margin of error of 0.03 with 99% confidence. How large a sample size is needed?

(The solution is shown in Appendix A.)

EXAMPLE 22   Sample size for estimating *p* when $\hat{p}$ is unknown

Suppose your state wants to take a poll on the proportion of its citizens who support a single statewide primary instead of primaries for each party. No poll on this subject has been taken before, so no prior information is available on the value of the sample proportion, $\hat{p}$. How large a sample size does the state need to estimate the proportion to within plus or minus 3 percentage points ($E = 0.03$) with 95% confidence?

Solution

The 95% confidence implies that the value for $Z_{\alpha/2}$ is 1.96. Because no information is available about the value of the population proportion of all state citizens who support a single statewide primary, we use 0.5 as our most conservative value of *p*:

$$n = \left[\frac{0.5 \cdot Z_{\alpha/2}}{E} \right]^2 = \left[\frac{(0.5)(1.96)}{0.03} \right]^2 \approx 1067.11$$

NOW YOU CAN DO
Exercises 39–46.

So if the pollsters want to estimate the population proportion of all state citizens who support a single statewide primary to within 3% with 95% confidence, they will need a sample of 1068 voters (don't forget to round up!).

YOUR TURN
#15

For the scenario in Example 22, suppose the state does not have the funds to contact 1068 voters, and it wants the estimate to be within a margin of error of 0.05 with 95% confidence. How large a sample size is needed?

(The solution is shown in Appendix A.)

STEP-BY-STEP TECHNOLOGY GUIDE: *Z* Confidence Intervals for *p*

We illustrate how to construct the *Z* confidence interval for *p* from Example 19 (page 465).

TI-83/84

Step 1   Press **STAT** and highlight **TESTS**.
Step 2   Scroll down to **1-PropZInt** (see Figure 32), and press **ENTER**.
Step 3   For **x**, enter the number of success, 1118.
Step 4   For **n**, enter the sample size 1895.
Step 5   For **C-Level** (confidence level), enter the appropriate confidence level (e.g., **0.95**), and press **ENTER** (Figure 33).
Step 6   Highlight **Calculate** and press **ENTER**. The results are shown in Figure 28 in Example 20.

FIGURE 32   **FIGURE 33**

tool manufacturer relies on a quality control technician (who has a strong background in statistics) to make sure that the tools the company is making do not vary appreciably from the required specifications. Otherwise, the tools may be too large or too small. Data analysts therefore construct confidence intervals to estimate the unknown value of the population parameters that measure variability: the population variance σ^2 and the population standard deviation σ.

We first need to become acquainted with the χ^2 **(chi-square) distribution**, which is used to construct these confidence intervals.

1 Properties of the χ^2 (Chi-Square) Distribution

The χ^2 (pronounced *ky-square*, to rhyme with "my square") distribution was discovered in 1875 by the German physicist Friedrich Helmert and further developed in 1900 by the English statistician Karl Pearson. It is a continuous distribution, so the χ^2 random variable is continuous.

Just as we did with the normal and t distributions, we can find probabilities associated with values of χ^2, and vice versa. Similar to any continuous distribution, probability is represented by area below the curve above an interval. We examine the properties of the χ^2 distribution and then learn how to use the χ^2 table to find the critical values of the χ^2 distribution.

Properties of the χ^2 Distribution

- Just as for any continuous random variable, the total area under the χ^2 curve equals 1.
- The value of the χ^2 random variable is never negative, so the χ^2 curve starts at 0. However, it extends indefinitely to the right, with no upper bound.
- Because of the characteristics just described, the χ^2 curve is right-skewed.
- There is a different curve for every different degrees of freedom, $n-1$. As the number of degrees of freedom increases, the χ^2 curve begins to look more symmetric (Figure 34).

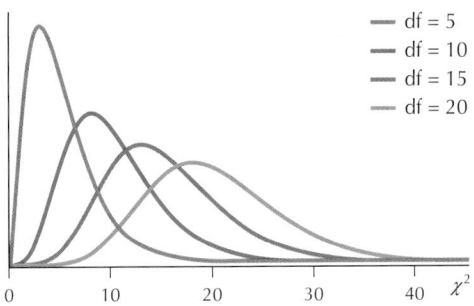

FIGURE 34 Shape of the χ^2 distribution for different degrees of freedom.

To construct the confidence intervals in this section, we will need to find the critical values of a χ^2 distribution for the given confidence level $100(1-\alpha)\%$, using either the χ^2 table (Table E in the Appendix) or technology. The χ^2 table is somewhat similar to the t table (Table D in the Appendix); both tables show the degrees of freedom in the left column. The area to the right of the χ^2 critical value is given across the top of the table.

The χ^2 distribution is not symmetric, so we cannot construct the confidence interval for σ^2 using the "point estimate ± margin of error" method. Instead, the lower bound and upper bound for the confidence interval are determined using two χ^2 critical values:

$\chi^2_{1-\alpha/2}$ = the value of the χ^2 distribution with area $1-\alpha/2$ to its right (Figure 35)

$\chi^2_{\alpha/2}$ = the value of the χ^2 distribution with area $\alpha/2$ to its right (Figure 35)

For instance, for a 95% confidence interval $(1 - \alpha) = 0.95$, $\alpha/2 = 0.025$ and $1 - \alpha/2 = 0.975$. Thus, $\chi^2_{0.975}$ represents the value of the χ^2 distribution with area $1 - \alpha/2 = 0.975$ to the right of the χ^2 critical value. The second critical value $\chi^2_{0.025}$ represents the value of the χ^2 distribution with area $\alpha/2 = 0.025$ to the right of the χ^2 critical value.

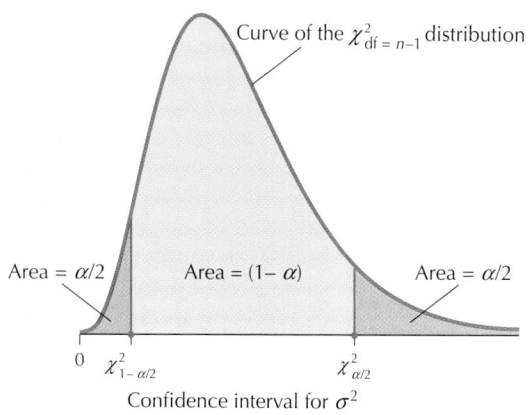

Area = $\alpha/2$ Area = $(1 - \alpha)$ Area = $\alpha/2$

$0 \quad \chi^2_{1 - \alpha/2} \qquad\qquad \chi^2_{\alpha/2}$

Confidence interval for σ^2

FIGURE 35 χ^2 critical values.

EXAMPLE 23 Finding the χ^2 critical values

Find χ^2 critical values for a 90% confidence interval, where we have a sample size of size $n = 10$.

Solution

For a 90% confidence interval,

$$(1 - \alpha) = 0.90 \qquad \frac{\alpha}{2} = \frac{0.10}{2} = 0.05 \qquad 1 - \frac{\alpha}{2} = 1 - 0.05 = 0.95$$

Note: If the appropriate degrees of freedom are not given in the χ^2 table, the conservative solution is to take the next row with the smaller df.

So we are seeking (1) $\chi^2_{0.95}$, the critical value with area $1 - \alpha/2 = 0.95$ to the right of it, and (2) $\chi^2_{0.05}$, the critical value with area $\alpha/2 = 0.05$ to the right of it.

Because $n = 10$, the degrees of freedom is df $= n - 1 = 10 - 1 = 9$. To find $\chi^2_{0.95}$ for df $= 9$, go across the top of the χ^2 table (Table E in the Appendix) until you see 0.95 (Figure 36). $\chi^2_{0.95}$ is somewhere in that column. Now go down that column until you see your number of degrees of freedom df $= 9$. Thus, for df $= 9$, $\chi^2_{0.95} = 3.325$. For a χ^2 distribution with 9 degrees of freedom, there is area $= 0.95$ to the right of 3.325.

Degrees of Freedom	Chi-Square (χ^2) Distribution Area to the Right of Critical Value									
	0.995	0.99	0.975	0.95	0.90	0.10	0.05	0.025	0.01	0.005
1	—	—	0.001	0.004	0.016	2.706	3.841	5.024	6.635	7.879
2	0.010	0.020	0.051	0.103	0.211	4.605	5.991	7.378	9.210	10.597
3	0.072	0.115	0.216	0.352	0.584	6.251	7.815	9.348	11.345	12.838
4	0.207	0.297	0.484	0.711	1.064	7.779	9.488	11.143	13.277	14.860
5	0.412	0.554	0.831	1.145	1.610	9.236	11.071	12.833	15.086	16.750
6	0.676	0.872	1.237	1.635	2.204	10.645	12.592	14.449	16.812	18.548
7	0.989	1.239	1.690	2.167	2.833	12.017	14.067	16.013	18.475	20.278
8	1.344	1.646	2.180	2.733	3.490	13.362	15.507	17.535	20.090	21.955
9	1.735	2.088	2.700	3.325	4.168	14.684	16.919	19.023	21.666	23.589
10	2.156	2.558	3.247	3.940	4.865	15.987	18.307	20.483	23.209	25.188

FIGURE 36 Finding $\chi^2_{0.95}$ and $\chi^2_{0.05}$ using the χ^2 table.

Similarly, $\chi^2_{0.05}$ is found in the column labeled "0.05" and the row corresponding to df = 9. We find that $\chi^2_{0.05} = 16.919$, as shown in Figure 37.

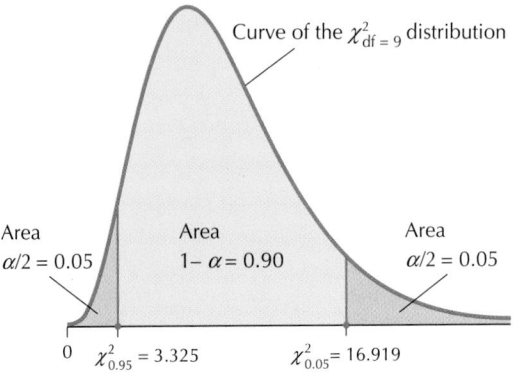

NOW YOU CAN DO
Exercises 9–16.

FIGURE 37 χ^2 critical values for the χ^2 distribution with df = 9.

Find χ^2 critical values for a 95% confidence interval, where we have a sample size of size $n = 20$.

(The solutions are shown in Appendix A.)

2 Constructing Confidence Intervals for the Population Variance and Standard Deviation

We derive the formula for a $100(1 - \alpha)\%$ confidence interval for the population variance σ^2. Suppose we take a random sample of size n from a normal population with mean μ and standard deviation σ. Then the statistic

$$\chi^2 = \frac{(n - 1)s^2}{\sigma^2}$$

follows a χ^2 distribution with $n - 1$ degrees of freedom, where s^2 represents the sample variance. From Figure 35, we see that $100(1 - \alpha)\%$ of the values of χ^2 lie between $\chi^2_{1-\alpha/2}$ and $\chi^2_{\alpha/2}$. These values are described as

$$\chi^2_{1-\alpha/2} < \frac{(n - 1)s^2}{\sigma^2} < \chi^2_{\alpha/2}$$

Rearranging this inequality so that σ^2 is in the numerator gives us the formula for the $100(1 - \alpha)\%$ confidence interval for σ^2:

$$\frac{(n - 1)s^2}{\chi^2_{\alpha/2}} < \sigma^2 < \frac{(n - 1)s^2}{\chi^2_{1-\alpha/2}}$$

Thus, the lower bound of the confidence interval for σ^2 is $\frac{(n - 1)s^2}{\chi^2_{\alpha/2}}$, and the upper bound is $\frac{(n - 1)s^2}{\chi^2_{1-\alpha/2}}$. Taking the square root of each gives us the lower and upper bounds for the confidence interval for σ.

Confidence Interval for the Population Variance σ^2

Suppose we take a sample of size n from a normal population with mean μ and standard deviation σ. Then a $100(1 - \alpha)\%$ confidence interval for the population variance σ^2 is given by

$$\text{lower bound} = \frac{(n - 1)s^2}{\chi^2_{\alpha/2}}, \quad \text{upper bound} = \frac{(n - 1)s^2}{\chi^2_{1-\alpha/2}}$$

where s^2 represents the sample variance and $\chi^2_{1-\alpha/2}$ and $\chi^2_{\alpha/2}$ are the critical values for a χ^2 distribution with $n - 1$ degrees of freedom.

Confidence Interval for the Population Standard Deviation σ

A $100(1 - \alpha)\%$ confidence interval for the population standard deviation σ is then given by

$$\text{lower bound} = \sqrt{\frac{(n-1)s^2}{\chi^2_{\alpha/2}}}, \qquad \text{upper bound} = \sqrt{\frac{(n-1)s^2}{\chi^2_{1-\alpha/2}}}$$

EXAMPLE 24   Constructing confidence intervals for the population variance σ^2 and population standard deviation σ

The accompanying table shows the miles-per-gallon equivalent (MPGe) for five electric cars, as reported by www.hybridcars.com in 2014. The normal probability plot in Figure 38 indicates that the data are normally distributed.

 electricmiles

Electric Vehicle	Mileage (MPGe)
Tesla Model S	89
Nissan Leaf	99
Ford Focus	105
Mitsubishi i-MiEV	112
Chevrolet Spark	119

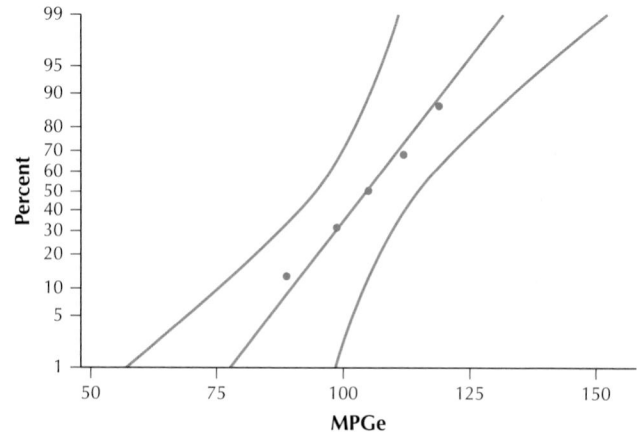

FIGURE 38   Normal probability plot of miles-per-gallon equivalent for five electric cars.

a. Find the critical values $\chi^2_{1-\alpha/2}$ and $\chi^2_{\alpha/2}$ for a confidence interval with a 95% confidence level.

b. Construct and interpret a 95% confidence interval for the population variance of electric car MPG.

c. Construct and interpret a 95% confidence interval for the population standard deviation of electric car MPG.

Solution

 electricmiles

a. There are $n = 5$ electric cars in our sample, so the degrees of freedom equal $n - 1 = 4$.

   For a 95% confidence interval,

   $$(1 - \alpha) = 0.95 \qquad \alpha/2 = 0.025 \qquad 1 - \alpha/2 = 0.975$$

   From the χ^2 table (Table E in the Appendix), therefore,

   $$\chi^2_{1-\alpha/2} = \chi^2_{0.975} = 0.484 \qquad \chi^2_{\alpha/2} = \chi^2_{0.025} = 11.143$$

   Figures 39 through 41 show these results using Excel, Minitab, and JMP.

b. Figure 42 shows the descriptive statistics for MPGe, as obtained by the TI-83/84. The sample standard deviation is $s = 11.58447237$.

For Exercises 15–17, we are estimating p and we know the value of $\hat{p}$. Find the required sample size.

15. Confidence level 99%, margin of error 0.05, $\hat{p} = 0.1$
16. Confidence level 95%, margin of error 0.05, $\hat{p} = 0.01$
17. Confidence level 95%, margin of error 0.05, $\hat{p} = 0.001$

For Exercises 18–20, we are estimating p and we do not know the value of $\hat{p}$. Find the required sample size.
18. Confidence level 95%, margin of error 0.06
19. Confidence level 95%, margin of error 0.05
20. Confidence level 95%, margin of error 0.04

SECTION 8.4

For Exercises 21–24, a random sample is drawn from a normal population. The sample of size $n = 100$ has a sample variance of $s^2 = 256$. Construct the specified confidence interval.
21. 90% confidence interval for the population variance σ^2
22. 95% confidence interval for the population variance σ^2
23. 90% confidence interval for the population standard deviation σ

24. 95% confidence interval for the population standard deviation σ

25. Union Membership. The table contains the total union membership for seven randomly selected states. Construct and interpret a 95% confidence interval for σ. Assume the data are normally distributed. **unionmember**

State	Union membership (1000s)
Florida	397
Indiana	334
Maryland	342
Massachusetts	414
Minnesota	395
Texas	476
Wisconsin	386

Source: U.S. Bureau of Labor Statistics, 2010.

Chapter 8 QUIZ

TRUE OR FALSE

1. True or false: In Figure 3 (page 435), the confidence level is 90%, so 90% of the intervals must contain μ. Explain your answer.
2. True or false: The t curve is symmetric about 0, just like the Z curve is. Therefore, we can use all our symmetry techniques with the t curve, too.

FILL IN THE BLANK

3. Suppose we cut a margin of error in half. The sample size requirement then becomes _____ times larger.
4. Our estimate of μ is _____ precise using the t curve instead of the Z curve.

SHORT ANSWER

5. α is used to find the value of $Z_{\alpha/2}$. Is α a probability or a value of x or a value of Z?
6. What are the conditions for constructing a t interval?

CALCULATIONS AND INTERPRETATIONS

7. College Education Costs. A random sample of 225 colleges yielded a mean cost of college education of $30,500 per year. Assume that the population standard deviation is $5000.
   a. Compute and interpret the margin of error for a confidence interval with 95% confidence.
   b. Construct and interpret a 95% confidence interval for the population mean cost of college education.
8. Crash Test Data. The National Highway Traffic Safety Administration collects data on crash tests for new motor vehicles. They reported that the mean femur

load (force applied to the femur) in a frontal crash for the passenger in a Chevrolet Equinox SUV was 1003 pounds. Assume the data are normally distributed, the population standard deviation was 300 pounds, and the sample size was 9.
   a. Compute and interpret the margin of error for a confidence interval with 90% confidence.
   b. Construct and interpret a 90% confidence interval for the population mean femur load in a frontal crash for the passenger in a Chevrolet Equinox SUV.
9. Independence for Quebec? A poll conducted by the newspaper *La Presse* reported that 340 of 1000 randomly chosen Quebec adults surveyed would vote "Yes" in a referendum for independence from Canada.
   a. If appropriate, find the margin of error for confidence level 99%. What does this number mean?
   b. If appropriate, find a 99% confidence interval for the population proportion of all Quebec residents who favor independence for the province of Quebec.
10. Tax Returns. In Example 10 in Section 3.2, Ashley and Brandon worked at an accounting firm preparing tax returns. Their Chief Accountant kept careful track of the amount of time (in hours) for all the tax returns that they prepared in the last week of March, shown in the accompanying table. Assume both data sets are normally distributed. **taxreturn**

Ashley	5	7	8	9	11
Brandon	3	5	7	11	14

a. Construct and interpret a 95% confidence interval for the population standard deviation of Ashley's preparation time.

b. Construct and interpret a 95% confidence interval for the population standard deviation of Brandon's preparation time.

11. **Quality of Education in America.** The National Assessment of Educational Progress (NAEP) administers exams to a nationwide sampling of students to assess the quality of education in America. Suppose NAEP wants to estimate the population proportion of American schoolchildren who would answer a given question correctly. Find a sample size that would give a margin of error of 0.03 with 90% confidence.

Hypothesis Testing

Introduction

Chapter 9 introduces *statistical hypothesis testing*. In Section 9.1, students learn how to make decisions about the value of a population parameter, and they examine different types of errors that can be made. The Z test for the population mean is covered in Section 9.2 (critical-value method) and Section 9.3 (p-value method). The t test for the population mean is covered in Section 9.4, followed by the Z test for the population proportion in Section 9.5 and the chi-square test for the population standard deviation in Section 9.6. Finally, Section 9.7 covers the probability of Type II error and the power of a hypothesis test.

We have a new Chapter 9 Case Study, *Clothing Store Sales*, which is excerpted from a case study in *Data Mining and Predictive Analytics*, by Daniel Larose and Chantal Larose (2015). In this case study, we examine the purchasing behavior of 5000 customers of a large clothing retail store in the Northeast, over a period of six months. We return again and again to this case study throughout Sections 9.1–9.5.

From the Author

Some instructors like to cover the critical-value method but not the p-value method. Some like to cover the p-value method but not the critical-value method. Others like to cover both. We offer the following suggestions:

- For those who like to cover the critical-value method but not the p-value method:
  - Simply cover Section 9.2, "Critical-Value Method," but not Section 9.3, "p-Value Method." However, you may wish to cover Objective 4 of Section 9.3, "Use the Z Confidence Interval for the Mean to Perform the Two-Tailed Z Test for the Mean."

- For those who like to cover the p-value method but not the critical-value method:
  - Cover only Objectives 1 and 2 in Section 9.2.
  - Then cover Section 9.3, "p-Value Method."

- For those who like to cover both methods: Simply cover both Sections 9.2 and 9.3.
  - Note that we have moved coverage of the critical-value method ahead of the p-value method. This aligns our coverage with that of most other textbooks, which makes it easier for instructors to use *Discovering Statistics*, Third Edition, even if they have used a different textbook previously.

Section 9.1 Introduction to Hypothesis Testing

- The introduction to Section 9.1 has been heavily reworked for clarity. A new Example 1, Are these Dice Loaded?, introduces students to the concept of looking to the data to determine the truth of an unknown claim. Another new example, Example 2, An Example of Hypotheses, was inserted to ease the student into the topic of hypothesis testing. A further new example in the introduction is Example 3, Identifying Valid and Invalid Hypotheses.

- Students usually relate well to the analogy of the criminal trial when learning about hypothesis testing. It is a good way to learn the difference between the null and alternative hypotheses, the two types of error, and the two ways of making a right decision.

- I have found that the "key words" strategy for constructing hypotheses works well for students who need to construct hypotheses from a word problem.

- Students are sometimes overwhelmed by all the notation in the hypotheses. You may want to stress that these are the only possible three forms for the hypotheses, that equal signs always appear in H_0, and that the three types of inequalities appear once each in H_a.

- There is a new example on stating what a Type I error and a Type II error means.

Section 9.2 Z Test for the Population Mean: Critical-Value Method

- There are new Objectives in Section 9.2: calculate the test statistic, Z_{data}, and find the critical region(s) and critical value(s) for a hypothesis test.

- We use Z_{data} to denote the test statistic, primarily to distinguish it from the critical value Z_{crit}. We call it Z_{data} because its value depends on the data, whereas Z_{crit} does not depend on the sample data.

- We provide a template for interpreting the conclusion. Note that the interpretation for not rejecting H_0 differs from that for rejecting H_0 by the inclusion of the single word "insufficient" before the word "evidence."

- The value of the test statistic for each example in this section is a whole number. This lets the students concentrate on learning hypothesis testing instead of dealing with decimals.

Section 9.3 Z Test for the Population Mean: p-Value Method

- Instructors who want to omit the critical-value method should cover Objective 1 in Section 9.2 before moving on to Section 9.3.

- Instructors who cover both methods may want to mention that the critical-value method works by comparing one Z-value with another Z-value, whereas the p-value method works by comparing one probability with another probability.

- Students may question why they have to learn the p-value method if it gives the same conclusions as the critical-value method. Instructors may want to respond by teaching Objective 2 in this section, Assessing the Strength of Evidence Against the Null Hypothesis. Here, we may dispense with the level of significance α and assess the strength of evidence against the null hypothesis using the p-value alone.

- Instructors who teach both methods may find Objective 3 of this section useful. Figures 13 and 14 illustrate the intimate relationships among Z_{data}, Z_{crit}, the p-value, and α.

- My doctoral adviser used to say that a single confidence interval was worth 1000 hypothesis tests. Instructors may want to show students how this may be true by teaching Objective 4 of this section, "Use the Z Confidence Interval for the Mean to Perform the Two-Tailed Z Test for the Mean."

- There is a new example on Interpreting Software Output, Example 17.

Section 9.4 *t* Test for the Population Mean

- Students may want to review the characteristics of the *t* distribution, from Section 8.2.

- Instructors should stress that, if the population standard deviation σ is unknown, it is wrong to use the *Z* test.

- The value of the test statistic t_{data} represents the number of standard errors that the sample mean lies above or below the hypothesized mean.

- From this point on, you will find that the method to be used to solve the Applying the Concepts exercises will no longer be specified. However, I still specify either the critical-value method or *p*-value method for the Practicing the Techniques exercises.

- There is a new example on Interpreting Software Output, Example 24.

Section 9.5 *Z* test for the Population Proportion

- Instructors may want to point out that, although the test statistic in Section 9.5 has the same name, Z_{data}, as the test statistic in Sections 9.2 and 9.3, the formula is different. On the other hand, once Z_{data} is calculated, we may compare it with Z_{crit} just as we did in those earlier sections.

- There is a new example on Interpreting Software Output, Example 30.

Section 9.6 Chi-Square Test for the Population Standard Deviation

- Instructors may want to review the chi-square distribution in Section 8.4.

Section 9.7 Probability of a Type II Error and the Power of a Hypothesis Test

- We formed this material into its own section.

Teaching Tips

Discuss how hypothesis testing is similar to a court case. Just as a person is innocent until proven guilty, the null hypothesis is assumed true unless the sample evidence indicates that the alternate hypothesis is true instead. Sometimes, an innocent person is found guilty, which is analogous to making a Type I error. Sometimes, the court fails to convict a guilty person, which is analogous to making a Type II error.

Hypothesis testing can be a bit scary for some students. You may want to stress the essential idea: If the observed value of $\bar{x}$ is extreme under the hypothesis that H_0 is correct, then the null hypothesis is rejected. All the rest is just mechanics. Emphasize that a hypothesis is just an idea or a claim; and that when there is a conflict between a hypothesis and the data, the data win.

Unlike what is taught in some textbooks, it is not correct to use the *Z* test for μ unless the population standard deviation σ is known. If σ is not known, then one may use the *t* test *if* the sample size is large or the population is normal.

If you are pressed for time, then the material on calculating a Type II error and the power of a test may be omitted without loss of continuity.

In-Class Activity

Ask students to consider this scenario: Your friend has made a claim that the average price for sneakers for men is no more than $75. You feel the average price is higher. To illustrate how your friend's claim can be tested, direct students to collect data from males for the price they paid for their most recent pair of sneakers. A large-enough sample ($n > 30$) should be taken. Use this example to help motivate the idea of hypotheses and hypothesis testing. Students can find the average for their collected data and compare this value to $75.

Center the discussion on the question of whether any difference is "significant." The idea of large sample tests for a population mean can be addressed and the Z test can be introduced. Also, the t test can be introduced for the case when the population standard deviation is unknown. If each student participates, then the students can pool their data to help find an estimate for the population standard deviation to be used in the Z test. The data should be collected before the topic is presented in class.

Supplements

- StatTutor 15.1–15.6(a) and (b), 16.1–16.7(c), 18.4, 18.5, and 20.6
- Stats@Work Simulations
  - Picture This, p-Value; Gary Pop
  - Picture This, p-Value; Bodhi Behav
  - Picture This, p-Value; Ruby Sweet
  - Picture This, p-Value; Mike Mobile

Note: Not all parts of the following simulations deal with a single population parameter.

  - Which Method to Use? Hypothesis Testing: Means and Proportions; Gary Pop
  - Which Method to Use? Hypothesis Testing: Means and Proportions; Sam Sport
  - Which Method to Use? Hypothesis Testing: Means and Proportions; Justine Red
  - Which Method to Use? Hypothesis Testing: Means and Proportions; Mindy Admin
- EESEE case studies
  - Psychic Probability (Questions 1 and 2 on estimation and hypothesis tests for a population proportion)

Applets

The *p-Value* applet is referenced in Chapter 9 for demonstrations and for Exercise 72 in Section 9.3.

Videos

- *Against All Odds: Inside Statistics:* www.learner.org/resources/series65.html
  - Program 25: Tests of Significance
  - Program 26: Small Sample Inference for One Mean
  - Program 28: Inference for Proportions

Web Sites

- The site Online Statistics: An Interactive Multimedia Course of Study addresses hypothesis testing and has a link to the Rice Virtual Lab in Statistics, with numerous simulation demonstrations and other topics related to statistics: http://onlinestatbook.com/index.html.
- This Web site lists other sites that do statistical calculations: http://statpages.org/.
- This Web site lists activities on topics relating to hypothesis testing: http://mathforum.org/mathtools/sitemap2/ps/.
- This site has several activities using Excel and Minitab: www.mathspace.com/NSF_ProbStat/Teaching_Materials/Primarily_Statistics.htm.

9.1   Introduction to Hypothesis Testing

OBJECTIVES By the end of this section, I will be able to . . .

1   Construct the null hypothesis and the alternative hypothesis from the statement of the problem.

2   State the two types of errors made in hypothesis tests: the Type I error, made with probability α, and the Type II error, made with probability β.

Researchers are interested in investigating many different types of questions, such as the following:

- An accountant may want to examine whether evidence exists for corporate tax fraud.

- A Department of Homeland Security executive may want to test whether a new surveillance method will uncover terrorist activity.

- A sociologist may want to examine whether the mayor's economic policy is increasing poverty in the city.

Questions such as these can be tackled using statistical **hypothesis testing**, which is a statistical inference process for using sample data to render a decision about claims regarding the unknown value of a population parameter. In this section, we will learn how to make decisions about the values of a population mean.

1 Constructing the Hypotheses

Let's start with an example.

EXAMPLE 1   Are these dice loaded?

Suppose you are playing a dice game, where you roll a pair of dice and win the sum of the two dice in dollars. A fair price to pay to play this game is $7 a throw, because the long-run mean when tossing two fair dice is 7. Now suppose you have played this game 10 times (paying a total of $70), with the following 10 results from throwing the two dice:

4	6	2	7	8	3	5	4	9	2

These 10 dice rolls add up to 50, meaning that, for your outlay of $70, you have only received $50 in return. You wonder:

- Are these dice fair but you have just had a streak of bad luck, or

- Are these dice not fair, that is, loaded (weighted) to provide low outcomes?

This is a basic example of hypothesis testing, where we have two competing ideas, and we turn to observed data (the dice rolls) to provide evidence in favor of one idea or the other.

We examine this question in more detail in the exercises and again in Section 9.2.

So, what is a hypothesis?

A **hypothesis** is a statement made about the value of a parameter. (A parameter is a characteristic of a population, such as the population mean μ.)

Examples of hypotheses might be the following:

a. The population mean μ of the dice tossed in Example 1 equals 7, meaning that you just had a run of bad luck.

b. The population mean μ of the dice tossed in Example 1 is less than 7, meaning that the dice were loaded.

c. The population proportion p of adults owning a tablet computer in 2014 was 42%. A media technology researcher states that this proportion is still the same today.

d. A different researcher states that the population proportion p of adults owning a tablet computer has increased since 2014.

Note: A hypothesis is not necessarily true. It is simply a statement. We need to look to the data for evidence either for it or against it.

Note that the statements in (**a**) and (**b**) are competing ideas, which can't both be right. Similarly, the statements in (**c**) and (**d**) are competing ideas.

The problem is that the value of the parameter is *unknown*, because it is a characteristic of a population, and we do not have access to the entire population. For example, we do not know the proportion of all people in the world today who own tablets, because new people are buying them all the time. If the true value of the parameter was known, there would be no need to perform a hypothesis test about it. This is why two reasonable people can have different ideas about the value of a population parameter. We must leave it up to the observed (sample) data to provide evidence in favor of a particular hypothesis.

To summarize, we have the following definition of hypothesis testing.

Hypothesis testing is a procedure for:

1. stating two competing hypotheses about the unknown value of a population parameter, such as the population mean μ,

2. analyzing the evidence collected from sample data, and

3. rendering a decision about which hypothesis the sample data support.

The two competing statements about the parameter are called the **null hypothesis** and **alternative hypothesis**, and they are described below.

The Hypotheses

- The **null hypothesis** represents what has been tentatively assumed about the value of the parameter. Thus, it represents no change, no effect, or no difference. The null hypothesis is denoted as H_0 (pronounced "H-naught"), and it is assumed true unless the sample data provide evidence against it.

- The **alternative hypothesis**, or **research hypothesis**, denoted as H_a, represents an alternative claim about the value of the parameter. If the alternative hypothesis is to be chosen over the null hypothesis, it requires sample evidence in its favor.

Hypothesis testing is like conducting a criminal trial. In a trial in the United States, the defendant is innocent until proven guilty, and the jury must evaluate the truth of two competing hypotheses:

$$H_0 : \text{defendant is not guilty} \quad \text{versus} \quad H_a : \text{defendant is guilty}$$

The not-guilty hypothesis is considered the **null hypothesis H_0** because the jurors must assume it is true until proven otherwise. The **alternative hypothesis H_a**, that the defendant is guilty, *must be demonstrated* to be true, beyond a reasonable doubt. How does a court of law determine whether the defendant is convicted or acquitted? This judgment is based upon the *evidence*, the hard facts heard in court. Similarly, in hypothesis testing, the researcher draws a conclusion based on the evidence provided by the sample data.

In Sections 9.1–9.4, we will examine hypotheses for the unknown mean μ. The null hypothesis will be a claim about a certain specified value for μ denoted μ_0, and the

Step 2 **Determine the form of the hypotheses.**
From Table 1, we see that the symbol $<$ means that we use a left-tailed test:

$$H_0 : \mu = \mu_0 \quad \text{versus} \quad H_a : \mu < \mu_0$$

Step 3 **Find the value for μ_0 and write your hypotheses.**
The alternative hypothesis H_a states that the mean annual rainfall in Arizona is less than some value μ_0. Less than what? Eight inches per year. Write the two hypotheses with $\mu_0 = 8$.

$$H_0 : \mu = 8 \quad \text{versus} \quad H_a : \mu < 8$$

NOW YOU CAN DO
Exercises 13–18.

YOUR TURN #1

Use Steps 1–3 in the **Strategy for Constructing Hypotheses About μ** to construct the hypotheses for the following scenario: Nielsen reports that iPhone and Android users spent 30 hours a month using apps on their devices in 2013. A media technology analyst states that the mean amount of time has increased since 2013. Write a null hypothesis and an alternative hypothesis for this situation.

(The solution is shown in Appendix A.)

Now that we know how to construct hypotheses, we next consider when sufficient evidence exists to reject the null hypothesis.

> **Statistical Significance**
> A result is said to be **statistically significant** if it is unlikely to have occurred due to chance.

EXAMPLE 5 Statistical significance

Suppose that you are a researcher for a pharmaceutical research company. You are investigating the side effects of a new cholesterol-lowering medication and want to determine whether the medication will *decrease* the population mean systolic blood pressure level from the current population mean of $\mu = 110$. If so, then a warning will have to be given not to prescribe the new medication to patients whose blood pressure is already low.

To determine which of these hypotheses is correct, we take a sample of randomly selected patients who are taking the medication. We record their systolic blood pressure levels and calculate the sample mean $\bar{x}$ and sample standard deviation s. Most likely, the mean of this sample of patients' systolic blood pressure levels will not be exactly equal to 110, even if the null hypothesis is true. Now, suppose that the sample mean blood pressure $\bar{x}$ is less than the hypothesized population mean of 110. *Is the difference due simply to chance variation, or is it evidence of a real side effect of the cholesterol medication?*

a. Construct the appropriate hypotheses.

b. For $\bar{x} = 109$ and $\bar{x} = 90$, discuss whether each result would be statistically significant or due to chance.

Solution

a. The key word "decrease" means we have a left-tailed test. "Less than what?" The current population mean systolic blood pressure of $\mu = 110$. Thus, our hypotheses are:

$$H_0 : \mu = 110 \quad \text{versus} \quad H_a : \mu < 110$$

where μ represents the population mean systolic blood pressure and $\mu_0 = 110$.

b. For $\bar{x} = 109$, the difference between $\bar{x}$ and μ_0 is only 1. Depending on the variability present in the sample, the researcher would likely not reject the null hypothesis because this small difference is probably due to chance variation. The result is probably *not statistically significant*. But, for $\bar{x} = 90$, the difference between $\bar{x}$ and μ_0 is 20. Depending on the variability present in the sample, the researcher would probably conclude that this difference is so large that it is unlikely that it is due to chance variation. Thus, the researcher would probably reject the null hypothesis H_0 in favor of the alternative hypothesis H_a. The result is *statistically significant*.

Note: When we reject H_0, we say that the results are statistically significant. If we do not reject H_0, the results are not statistically significant.

To summarize:

- In a hypothesis test, we compare the sample mean $\bar{x}$ with the value μ_0 of the population mean used in the H_0 hypothesis.
- If the difference between $\bar{x}$ and μ_0 is large, then the null hypothesis H_0 is rejected.
- If the difference between $\bar{x}$ and μ_0 is not large, then H_0 is not rejected.

The question is, "Where do you draw the line?" Just how large a difference between $\bar{x}$ and μ_0 is large enough to reject the null hypothesis? We answer this question starting in Section 9.2.

Note that there are only two possible hypothesis-testing conclusions:

- Reject H_0, or
- Do not reject H_0.

Developing Your Statistical Sense

A Decision Is Not Proof

It is important to understand that the decision to reject or not reject H_0 does not prove anything. The decision represents whether or not there is sufficient evidence against the null hypothesis. This is our best judgment, given the available data, similar to the best judgment of a jury, given the available evidence. You cannot claim to have *proven* anything about the value of a population parameter unless you elicit information from the entire population, which is usually not possible.

We can make decisions about population parameters using the limited information available in a sample because we base our decisions on *probability*. When the difference between the sample mean $\bar{x}$ and the hypothesized population mean μ_0 is large, then the null hypothesis is *probably* not correct. When the difference is small, then the data are *probably* consistent with the null hypothesis. But we don't know for sure.

2 Type I and Type II Errors

Next, we take a closer look at some of the thorny issues involved in performing a hypothesis test. Let's return to the example of a criminal trial. The jury will convict the defendant if they find evidence compelling enough to reject the null hypothesis of "not guilty" *beyond a reasonable doubt*. However, jurors are only human; sometimes their decisions are correct and sometimes they are not. Thus, the jury's verdict will represent one of the following outcomes:

1. An innocent defendant is wrongfully convicted.
2. A guilty defendant is convicted.
3. A guilty defendant is wrongfully acquitted.
4. An innocent defendant is acquitted.

Recall that we can write the two hypotheses for a criminal trial as

$$H_0 : \text{defendant is not guilty} \quad \text{versus} \quad H_a : \text{defendant is guilty}$$

Table 3 shows the possible verdicts on the left and the two hypotheses across the top.

Table 3 Four possible outcomes of a criminal trial

		Reality	
		H_0 true: Defendant did not commit the crime	H_0 false: Defendant did commit the crime
Jury's decision	Reject H_0: Find defendant guilty	Type I error	Correct decision
	Do not reject H_0: Find defendant not guilty	Correct decision	Type II error

A jury's decision can be correct or incorrect. The same is true for the conclusion of a hypothesis test.

Let's look at the two possible decisions the jury can make. It can find the defendant guilty: the jury *rejects the claim* in the null hypothesis H_0. Alternatively, the jury can find the defendant not guilty: the jury *does not reject* the null hypothesis H_0. The jury can render the *correct decision* in two ways.

Two Ways of Making the Correct Decision

- To not reject H_0 when H_0 is true.
  Example: To find the defendant not guilty when, in reality, he did not commit the crime.
- To reject H_0 when H_0 is false.
  Example: To find the defendant guilty when, in reality, he did commit the crime.

Unfortunately, the jury can also render an incorrect decision in two ways. In statistics, the two incorrect decisions are called **Type I** and **Type II** errors.

Two Types of Errors

- **Type I error:** To reject H_0 when H_0 is true.
  Example: To find the defendant guilty when, in reality, he did not commit the crime.
- **Type II error:** To not reject H_0 when H_0 is false.
  Example: To find the defendant not guilty when, in reality, he did commit the crime.

EXAMPLE 6   Type I and Type II errors

For the medication hypothesis test in Example 5, explain what it would mean if the following errors were made:

a.   Type I error

b.   Type II error

Solution

The hypotheses in Example 5 were the following:

$$H_0 : \mu = 110 \quad \text{versus} \quad H_a : \mu < 110$$

where μ represents the population mean systolic blood pressure.

a. A Type I error occurs when we reject H_0 when H_0 is true. This would be to conclude that μ had decreased when, in reality, it had stayed the same. In other words, a Type I error would be to conclude that the population mean systolic blood pressure had decreased when, in reality, it had not decreased. The pharmaceutical company, afraid of this possible side effect, might not continue production of the drug when, in reality, there is no side effect.

b. A Type II error occurs when we do not reject H_0 when H_0 is false. This would be to conclude that μ had stayed the same when, in reality, it had decreased. In this case, this is a very dangerous error to make, because the pharmaceutical company might then conclude that the side effect does not exist when, in reality, it does exist, and it could lead to dangerous lowering of blood pressure. This is why the Food and Drug Administration requires that strict protocols are followed regarding Type I and Type II errors when approving new medications for the market.

NOW YOU CAN DO
Exercises 19–22.

YOUR TURN
#2

Explain what it would mean to make a Type I error and a Type II error for the hypothesis test in the following examples:

a. Example 2

b. Example 4

(The solutions are shown in Appendix A.)

> The probability of a Type I error is denoted as α **(alpha)**. We set the value of α to be some small constant, such as 0.01, 0.05, or 0.10, so that only a small probability of rejecting a true null hypothesis exists.

To say that $\alpha = 0.05$ means that, if this hypothesis test were repeated over and over again, the long-term probability of rejecting a true null hypothesis would be 5%. The **level of significance** of a hypothesis test is another name for α, the probability of rejecting H_0 when H_0 is true. A smaller α makes it harder to wrongfully reject H_0 just by chance. If the consequences of making a Type I error are serious, then the level of significance should be small, such as $\alpha = 0.01$. If the consequences of making a Type I error are not so serious, then one may choose a larger value for the level of significance, such as $\alpha = 0.05$ or $\alpha = 0.10$.

The probability of a Type II error is denoted as β **(beta)**. This is the probability of not rejecting H_0 when H_0 is false, such as acquitting someone who is really guilty. Making α smaller inevitably makes β larger (for a fixed sample size). Of course, our goal is to simultaneously minimize both α and β. Unfortunately, the only way to do this is to increase the sample size.

Z_{data} standardizes the distance between the sample mean $\bar{x}$ and the hypothesized population mean μ, so that this distance is now on the standard normal scale. Thus, we can *sometimes* tell with a glance at Z_{data}, using our knowledge of the standard normal Z distribution (Section 6.4), whether $\bar{x}$ is extreme or not, and therefore whether to reject. Specifically, we recall that almost all values of Z lie between -3 and 3. Here are two examples:

- Say our data provides us with a value of $Z_{data} = 12$, which is far into the tail of the Z distribution. This represents a very extreme value of $\bar{x}$, and so we will reject H_0.
- Suppose the data set gives us a value of $Z_{data} = 0.27$, which is near the center of the Z distribution. This represents a value of $\bar{x}$ fairly close to μ_0, and so we will not reject H_0.

Of course, not all cases are as obvious as these, thus the need for the hypothesis testing procedure.

3 Critical Regions and Critical Values

In the critical-value method for the Z test, we compare Z_{data} with a *threshold value*, or **critical value** of Z, called $\mathbf{Z_{crit}}$. The value of Z_{crit} separates Z into two regions (see Table 4):

- **Critical region:** the values of Z_{data} for which we reject H_0
- **Noncritical region:** the values of Z_{data} for which we do not reject H_0

> - The **critical region** consists of the range of values of the test statistic Z_{data} for which we reject the null hypothesis.
> - The **noncritical region** consists of the range of values of the test statistic Z_{data} for which we do not reject the null hypothesis.
> - The value of Z that separates the critical region from the noncritical region is called the **critical value Z_{crit}**.

Z_{crit} represents the *boundary* between values of Z_{data} that are statistically significant and those that are not statistically significant. The value of Z_{crit} depends on the value of α, the probability of wrongly rejecting H_0. A smaller value of α will make it harder to reject H_0, that is, harder to find statistical significance. Thus, α is called the **level of significance** of the hypothesis test.

The value of Z_{crit} depends on (**a**) the form of the hypothesis test, and (**b**) the level of significance α. Table 4 shows values of Z_{crit} for the most commonly used levels of significance α. It also shows the location of the critical region.

Table 4 Table of critical values Z_{crit} for common values of the level of significance α

	Form of hypothesis test		
Level of significance α	Right-tailed $H_0: \mu = \mu_0$ $H_a: \mu > \mu_0$	Left-tailed $H_0: \mu = \mu_0$ $H_a: \mu < \mu_0$	Two-tailed $H_0: \mu = \mu_0$ $H_a: \mu \neq \mu_0$
0.10	$Z_{crit} = 1.28$	$Z_{crit} = -1.28$	$Z_{crit} = 1.645$
0.05	$Z_{crit} = 1.645$	$Z_{crit} = -1.645$	$Z_{crit} = 1.96$
0.01	$Z_{crit} = 2.33$	$Z_{crit} = -2.33$	$Z_{crit} = 2.58$
Critical region			
Rejection rule:	Reject H_0 if $Z_{data} \geq Z_{crit}$	Reject H_0 if $Z_{data} \leq Z_{crit}$	Reject H_0 if $Z_{data} \leq -Z_{crit}$ or $Z_{data} \geq Z_{crit}$

EXAMPLE 8   Finding Z_{crit} and the critical region

For the hypotheses,

$$H_0 : \mu = 110 \quad \text{versus} \quad H_a : \mu < 110$$

where μ represents the population mean systolic blood pressure, let the level of significance $\alpha = 0.05$.

a.   Find the critical value Z_{crit}.

b.   Graph the distribution of Z, showing the critical region.

Solution

We have a left-tailed test and level of significance $\alpha = 0.05$, so Table 4 tells us that the critical value is $Z_{crit} = -1.645$. The graph showing the critical region is provided in Figure 2. We would reject H_0 for values of Z_{data} that are $\leq Z_{crit} = -1.645$.

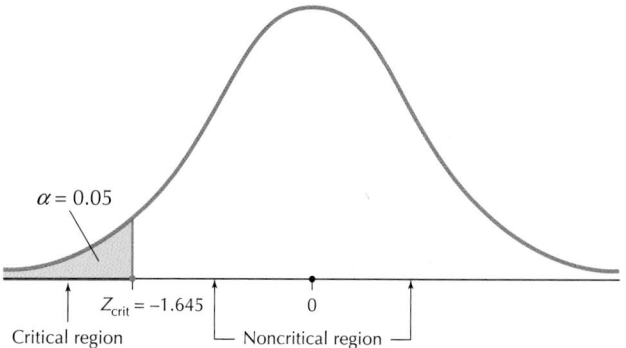

$\alpha = 0.05$

$Z_{crit} = -1.645$

0

Critical region

Noncritical region

FIGURE 2 Critical region for a left-tailed test lies in the left (lower) tail.

NOW YOU CAN DO
Exercises 23–26.

YOUR TURN
#4

In Example 7, we had the hypothesis test:

$$H_0 : \mu = 413 \quad \text{versus} \quad H_a : \mu > 413$$

where μ represents the population mean total sales per customer. Let the level of significance $\alpha = 0.10$.

a.   Find the critical value Z_{crit}.

b.   Graph the distribution of Z, showing the critical region.

(The solutions are shown in Appendix A.)

Developing Your
Statistical Sense

Why Is It Called a Left-Tailed Test Mean? Right-Tailed Test? Two-Tailed Test?

A hypothesis test of the form

$$H_0 : \mu = \mu_0 \quad \text{versus} \quad H_a : \mu < \mu_0$$

is called a *left-tailed* test because the critical region lies in the left (lower) tail. Similarly, a hypothesis test of the form

$$H_0 : \mu = \mu_0 \quad \text{versus} \quad H_a : \mu > \mu_0$$

is called a *right-tailed test* because its critical region lies in the right (upper) tail. Finally, a hypothesis test of the form

$$H_0 : \mu = \mu_0 \quad \text{versus} \quad H_a : \mu \neq \mu_0$$

is called a *two-tailed test* because its critical region occupies both the lower and upper tails.

4 Performing the *Z* Test for the Mean Using the Critical-Value Method

We are now ready to learn the steps for performing the *Z* test for the population mean using the critical-value method.

Z Test for the Population Mean μ: Critical-Value Method

When a random sample of size *n* is taken from a population where the population standard deviation σ is known, you can use the *Z* test if (a) the population is normal, or (b) the sample size is large ($n \geq 30$).

Step 1   State the hypotheses.
Use one of the forms from Table 4. State the meaning of μ.

Step 2   Find Z_{crit} and state the rejection rule.
Use Table 4 and the given level of significance α.

Step 3   Calculate Z_{data}.

$$Z_{data} = \frac{\bar{x} - \mu_0}{\sigma/\sqrt{n}}$$

Step 4   State the conclusion and the interpretation.
If Z_{data} falls in the critical region, then reject H_0; otherwise, do not reject H_0. Interpret your conclusion.

What Does This Conclusion Mean?

Interpreting Your Conclusion

Recall that a data analyst needs to interpret the results so that the general public can understand them. You can use the following generic interpretation for the two possible conclusions. Just remember that generic interpretations are no substitute for thinking clearly about the problem and the implications of the conclusion.

Interpreting the Conclusion

- If you reject H_0, the interpretation is: *There is evidence at level of significance α that [whatever H_a says].*

- If you do not reject H_0, the interpretation is: *There is insufficient evidence at level of significance α that [whatever H_a says].*

For example, suppose our conclusion for the hypotheses in Example 8

$$H_0 : \mu = 110 \quad \text{versus} \quad H_a : \mu < 110$$

was to reject H_0. Then the interpretation of this conclusion would be: *There is evidence at level of significance $\alpha = 0.05$ that the population mean systolic blood pressure reading is less than 110.*

Next, we illustrate the critical-value method of performing a right-tailed *Z* test, a left-tailed *Z* test, and a two-tailed *Z* test for μ.

EXAMPLE 9   Z Test for μ, critical-value method, right-tailed test

CASE STUDY

Clothing Store Sales

For the situation in Example 7, test at level of significance $\alpha = 0.01$ whether the population mean total sales per customer is more than $413.

Solution

We may apply the *Z* test because the sample is large ($n \geq 30$), and the population standard deviation σ is known.

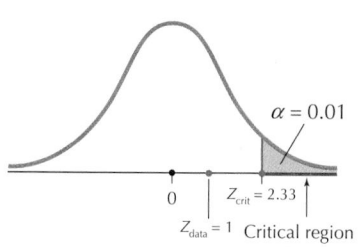

FIGURE 3 Critical region for a right-tailed test.

Step 1   State the hypotheses.
From Example 7, our hypotheses are

$$H_0 : \mu = 413 \quad \text{versus} \quad H_a : \mu > 413$$

where μ represents the population mean total sales per customer.

Step 2   Find Z_{crit} and state the rejection rule.
We have a right-tailed test and level of significance $\alpha = 0.01$, which, from Table 4, tell us that $Z_{\text{crit}} = 2.33$. Because we have a right-tailed test, the rejection rule will be "Reject H_0 if $Z_{\text{data}} \geq Z_{\text{crit}}$," that is, "Reject H_0 if $Z_{\text{data}} \geq 2.33$" (see Figure 3).

Step 3   Find Z_{data}.
From Example 7, we have $Z_{\text{data}} = 1$.

Step 4   State the conclusion and interpretation.
Our rejection rule states that we will reject H_0 if $Z_{\text{data}} \geq 2.33$. Because $Z_{\text{data}} = 1$, which is *not* ≥ 2.33, the conclusion is to *not* reject H_0 (Figure 4). Even though the sample mean of $\bar{x} = 480$ exceeds $\mu_0 = 413$, it does not do so by a wide enough margin to overcome the reasonable doubt that the difference between $\bar{x}$ and μ_0 may have been due to chance. We interpret our conclusion as follows: "There is insufficient evidence at the 0.01 level of significance that the population mean total sales is greater than \$413 per customer over the six-month period."

NOW YOU CAN DO
Exercises 37–40.

EXAMPLE 10   Z Test for μ, critical-value method, left-tailed test

For the hypotheses in Example 8, perform the Z test for the population mean, using level of significance $\alpha = 0.05$. Assume systolic blood pressure is normally distributed.

Solution

We may use the Z test, because the population of systolic blood pressure readings is normally distributed, and the population standard deviation σ is known.

Step 1   State the hypotheses.
From Example 8, we have

$$H_0 : \mu = 110 \quad \text{versus} \quad H_a : \mu < 110$$

where μ represents the population mean systolic blood pressure reading.

Step 2   Find Z_{crit} and state the rejection rule.
Example 8 gives us the critical value $Z_{\text{crit}} = -1.645$, and Table 4 tells us that, for level of significance $\alpha = 0.05$, we will reject H_0 if $Z_{\text{data}} \leq Z_{\text{crit}}$, that is, if $Z_{\text{data}} \leq -1.645$ (Figure 4).

Step 3   Calculate Z_{data}.
From page 499, we know that

$$Z_{\text{data}} = \frac{\bar{x} - \mu_0}{\sigma / \sqrt{n}} = \frac{104 - 110}{10 / \sqrt{25}} = -3$$

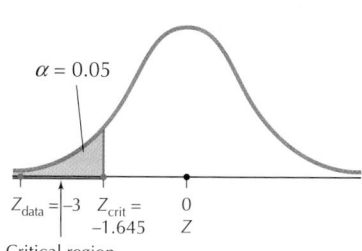

FIGURE 4 Critical region for a left-tailed test.

Step 4   State the conclusion and the interpretation.
In Step 2, we stated that we would reject H_0 if $Z_{\text{data}} \leq -1.645$. Our Z_{data} of $-3 \leq -1.645$, therefore, we reject H_0. Our interpretation is: "There is evidence at level of significance $\alpha = 0.05$ that the population mean systolic blood pressure reading is less than 110."

NOW YOU CAN DO
Exercises 41–44.

EXAMPLE 11 *Z* Test for μ, critical-value method, two-tailed test

When the level of hemoglobin in the blood is too low, a person is anemic. Unusually high levels of hemoglobin are also undesirable and can be associated with dehydration. The optimal hemoglobin level is 13.8 grams per deciliter (g/dl). Suppose a random sample of $n = 25$ women at a certain college showed a sample mean hemoglobin of $\bar{x} = 11.8$ g/dl, the population standard deviation of hemoglobin level is $\sigma = 5$ g/dl, and hemoglobin level is normally distributed. We are interested in testing whether the population mean hemoglobin level differs from 13.8 g/dl. Perform the appropriate hypothesis test, using level of significance $\alpha = 0.10$.

Solution

We may use the *Z* test, because the population of hemoglobin levels is normally distributed, and the population standard deviation σ is known.

Step 1 State the hypotheses.
The key words "differs from" indicate a two-tailed test, with $\mu_0 = 13.8$. Thus, our hypotheses are

$$H_0 : \mu = 13.8 \quad \text{versus} \quad H_a : \mu \neq 13.8$$

where μ represents the population mean hemoglobin level.

Step 2 Find Z_{crit} and state the rejection rule.
We have a two-tailed test and level of significance $\alpha = 0.10$. Using this information, Table 4 tells us that the critical value $Z_{crit} = 1.645$ and that we will reject H_0 if $Z_{data} \leq -1.645$ or if $Z_{data} \geq 1.645$ (Figure 5).

Step 3 Calculate Z_{data}.
We have $\bar{x} = 11.8$, $n = 25$, $\sigma = 5$, and $\mu_0 = 13.8$. Substituting:

$$Z_{data} = \frac{\bar{x} - \mu_0}{\sigma/\sqrt{n}} = \frac{11.8 - 13.8}{5/\sqrt{25}} = -2$$

Step 4 State the conclusion and the interpretation.
$Z_{data} = -2$, which is ≤ -1.645. Therefore we reject H_0. There is evidence at level of significance $\alpha = 0.10$ that the population mean hemoglobin level differs from 13.8 g/dl.

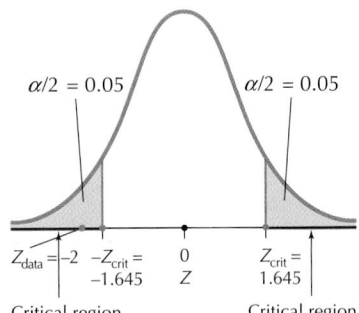

FIGURE 5 Critical region for a two-tailed test.

NOW YOU CAN DO
Exercises 45–48.

STEP-BY-STEP TECHNOLOGY GUIDE: *Z* Test for μ

To learn how to use technology to perform the *Z* test for the mean, see the Step-by-Step Technology Guide on page 519.

Section 9.2 Summary

1. The essential idea about hypothesis testing for the mean is as follows: When the observed value of $\bar{x}$ is unusual or extreme in the sampling distribution of $\bar{x}$ that assumes H_0 is true, we should reject H_0. Otherwise, we should not reject H_0.

2. The test statistic Z_{data} summarizes the information in the data set regarding the hypothesis test.

3. The critical region consists of the range of values of the test statistic Z_{data} for which we reject the null hypothesis. The value of Z that separates the critical region from the noncritical region is called the critical value Z_{crit}.

4. In the critical-value method for the *Z* test for the mean, we compare Z_{data} with Z_{crit}.

Section 9.2 Exercises

CLARIFYING THE CONCEPTS

1. What is the essential idea about hypothesis testing for the mean? (p. 498)

2. What does Z_{data} represent? (p. 499)

3. Explain what a test statistic is. (p. 499)

4. Describe the difference between the critical region and the noncritical region. (p. 500)

5. Clearly describe what Z_{crit} is. (p. 500)

6. Suppose we reject H_0 for the hypothesis test $H_0 : \mu = 5$ versus $H_a : \mu < 5$. Provide the generic interpretation. (p. 502)

7. How did the right-tailed test get its name? (p. 501)

8. True or false: The value of Z_{crit} does not depend at all on the sample data. (p. 500)

PRACTICING THE TECHNIQUES

CHECK IT OUT!

To do	Check out	Topic
Exercises 9–22	Example 7	Calculating Z_{data}
Exercises 23–36	Example 8	Finding Z_{crit} and the critical region
Exercises 37–40	Example 9	Right-tailed Z test for μ
Exercises 41–44	Example 10	Left-tailed Z test for μ
Exercises 45–48	Example 11	Two-tailed Z test for μ

For Exercises 9–48, assume that the conditions for performing the Z test are met.

For Exercises 9–20, calculate Z_{data}:

9. $H_0 : \mu = 75$ vs. $H_a : \mu > 75$, $\bar{x} = 79$, $\sigma = 20$, $n = 25$

10. $H_0 : \mu = 75$ vs. $H_a : \mu > 75$, $\bar{x} = 81$, $\sigma = 20$, $n = 25$

11. Right-tailed test with $\mu_0 = 50$ and $\sigma = 10$. A sample of size 100 has a mean of 51.

12. Right-tailed test with $\mu_0 = 50$ and $\sigma = 10$. A sample of size 100 has a mean of 54.

13. $H_0 : \mu = 98.6$ vs. $H_a : \mu < 98.6$, $\bar{x} = 98.35$, $\sigma = 1$, $n = 16$

14. $H_0 : \mu = 98.6$ vs. $H_a : \mu < 98.6$, $\bar{x} = 97.85$, $\sigma = 1$, $n = 16$

15. Left-tailed test with $\mu_0 = 20$ and $\sigma = 5$. A sample of size 100 has a mean of 19.

16. Left-tailed test with $\mu_0 = 20$ and $\sigma = 5$. A sample of size 100 has a mean of 18.5.

17. $H_0 : \mu = 1000$ vs. $H_a : \mu \neq 1000$, $\bar{x} = 1005$, $\sigma = 100$, $n = 400$

18. $H_0 : \mu = 1000$ vs. $H_a : \mu \neq 1000$, $\bar{x} = 995$, $\sigma = 100$, $n = 400$

19. Two-tailed test with $\mu_0 = 2.5$ and $\sigma = 0.5$. A sample of size 100 has a mean of 2.45.

20. Two-tailed test with $\mu_0 = 2.5$ and $\sigma = 0.5$. A sample of size 100 has a mean of 2.55.

21. Consider your results from Exercises 9–12. Describe what happens to Z_{data} for a right-tailed test as $\bar{x}$ increases its distance above μ_0, and everything else stays the same.

22. Consider your results from Exercises 13–16. Describe what happens to Z_{data} for a left-tailed test as $\bar{x}$ increases its distance below μ_0, and everything else stays the same.

For Exercises 23–34, do the following:
   a. Find the critical value Z_{crit}.
   b. Sketch the critical region, using the figures in Table 4 as a guide.
   c. State the rejection rule.

23. $H_0 : \mu = 75$ vs. $H_a : \mu > 75$, level of significance $\alpha = 0.10$

24. $H_0 : \mu = 75$ vs. $H_a : \mu > 75$, level of significance $\alpha = 0.05$

25. Right-tailed test with $\mu_0 = 50$, level of significance $\alpha = 0.05$

26. Right-tailed test with $\mu_0 = 50$, level of significance $\alpha = 0.01$

27. $H_0 : \mu = 98.6$ vs. $H_a : \mu < 98.6$, level of significance $\alpha = 0.05$

28. $H_0 : \mu = 98.6$ vs. $H_a : \mu < 98.6$, level of significance $\alpha = 0.01$

29. Left-tailed test with $\mu_0 = 20$, level of significance $\alpha = 0.10$

30. Left-tailed test with $\mu_0 = 20$, level of significance $\alpha = 0.05$

31. $H_0 : \mu = 1000$ vs. $H_a : \mu \neq 1000$, level of significance $\alpha = 0.05$

32. $H_0 : \mu = 1000$ vs. $H_a : \mu \neq 1000$, level of significance $\alpha = 0.01$

33. Two-tailed test with $\mu_0 = 2.5$, level of significance $\alpha = 0.10$

34. Two-tailed test with $\mu_0 = 2.5$, level of significance $\alpha = 0.05$

35. Consider your results from Exercises 23–26. Describe what happens to (**a**) Z_{crit} and (**b**) the critical region, for a right-tailed test when the only change is the decrease in the level of significance α.

36. Consider your results from Exercises 27–30. Explain what happens to (**a**) Z_{crit} and (**b**) the critical region, for a left-tailed test as the level of significance α decreases but everything else stays the same.

For Exercises 37–48, use the hypotheses and data from the indicated exercises to perform the Z test for μ by doing the following steps:
   a. State the hypotheses.
   b. Find Z_{crit} and state the rejection rule.
   c. State the value of Z_{data} from the indicated exercise.
   d. State the conclusion and the interpretation.

37. Use Z_{data} from Exercise 9 and Z_{crit} from Exercise 23.

38. Use Z_{data} from Exercise 10 and Z_{crit} from Exercise 24.

39. Use Z_{data} from Exercise 11 and Z_{crit} from Exercise 25.

40. Use Z_{data} from Exercise 12 and Z_{crit} from Exercise 26.

41. Use Z_{data} from Exercise 13 and Z_{crit} from Exercise 27.

42. Use Z_{data} from Exercise 14 and Z_{crit} from Exercise 28.

43. Use Z_{data} from Exercise 15 and Z_{crit} from Exercise 29.

44. Use Z_{data} from Exercise 16 and Z_{crit} from Exercise 30.

45. Use Z_{data} from Exercise 17 and Z_{crit} from Exercise 31.
46. Use Z_{data} from Exercise 18 and Z_{crit} from Exercise 32.
47. Use Z_{data} from Exercise 19 and Z_{crit} from Exercise 33
48. Use Z_{data} from Exercise 20 and Z_{crit} from Exercise 34.

APPLYING THE CONCEPTS

For Exercises 49–56, do the following:
   a. State the hypotheses.
   b. Find Z_{crit} and the critical region.
   c. Find Z_{data}. Also, draw a standard normal Z curve showing Z_{crit}, the critical region, and Z_{data}.
   d. State the conclusion and the interpretation.

49. **Facebook Connections.** According to Facebook.com, the mean number of community pages, groups, and events that users are connected to is 80. A random sample of 64 Facebook users showed a mean of 86 connections to community pages, groups, and events. Assume $\sigma = 48$. Test, using level of significance $\alpha = 0.05$, whether the population mean number of connections to community pages, groups, and events is greater than 80.

50. **Nurses' Study Time.** The National Survey of Student Engagement reported that the mean number of hours spent studying per week for nursing majors is 18. Suppose a medical researcher obtained a random sample of 100 nursing majors, which yielded a sample mean study time of 19 hours. Assume $\sigma = 10$ hours. Test, using level of significance $\alpha = 0.05$, whether the population mean study time for nursing majors is greater than 18 hours per week.

51. **Text Messages.** The Pew Internet and American Life Project reports that young people ages 12–17 send a mean of 60 text messages per day. A random sample of 100 young people showed a mean of 69 text messages per day. Assume $\sigma = 30$. Test, using level of significance $\alpha = 0.01$, whether the population mean number of text messages per day differs from 60.

52. **Video Gamers.** Can't pry the PlayStation away from your dad? In 2015, the Entertainment Software Association reported that the mean age of video gamers was 35 years old. A recent random sample of 36 video gamers had a mean age of 34. Assume $\sigma = 6$. Test, using level of significance $\alpha = 0.05$, whether the population mean age of video gamers is less than 35.

53. **Gas Prices.** The American Automobile Association reported in June 2011 that the mean price for a gallon of regular gasoline was $3.70. A recent random sample of 25 gas stations had a mean price of $3.90. Assume normality and $\sigma = \$0.50$. Test, using level of significance $\alpha = 0.05$, whether the population mean price for a gallon of regular gasoline has risen since June 2011.

54. **Online Shopping.** The Nielsen organization reports that smartphone owners spent a mean of 93 minutes using shopping apps on their smartphones in the fourth quarter (last three months) of 2013. A random sample of 225 smartphone owners showed a mean of 99 minutes using shopping apps on their smartphones in the fourth quarter of last year. Assume $\sigma = 45$ minutes. Conduct a hypothesis test, using level of significance $\alpha = 0.10$, to determine whether the population mean amount of time has changed.

55. **Americans' Height.** A random sample of 400 American adults yields a mean height of 176 centimeters. Assume $\sigma = 2.5$. Conduct a hypothesis test to investigate whether the population mean height of American adults has changed from 175 centimeters, using level of significance $\alpha = 0.10$.

56. **Price of Milk.** The U.S. Bureau of Labor Statistics reported that the mean price for a gallon of milk in 2011 was $3.34. A random sample of 100 retail establishments this year provides a mean price of $3.39. Assume $\sigma = \$0.25$. Perform a hypothesis test, using level of significance $\alpha = 0.05$, to investigate whether the population mean price of milk this year has increased from the 2011 value.

57. **Automobile Operation Cost.** The Bureau of Transportation Statistics reports that the mean cost of operating an automobile in the United States, including gas and oil, maintenance and tires, is 5.9 cents per mile. Suppose that a sample taken this year of 100 automobiles shows a mean operating cost of 6.2 cents per mile, and assume that the population standard deviation is 1.5 cents per mile. Test whether the population mean cost is greater than 5.9 cents per mile, using level of significance $\alpha = 0.05$.
   a. Is it appropriate to apply the Z test? Why or why not?
   b. We have a sample mean that is greater than the mean in the null hypothesis of 5.9 cents. Isn't this enough by itself to reject the null hypothesis? Explain why or why not.
   c. How many standard deviations above the mean is the 6.2 cents per mile? Do you think this is extreme?

58. **Automobile Operation Cost.** Refer to Exercise 57.
   a. Construct the hypotheses.
   b. Find the Z critical value and state the rejection rule.
   c. Calculate the value of the test statistic Z_{data}.
   d. State the conclusion and the interpretation.

59. **Accountants' Salaries.** According to accountingWEB.com, the mean starting salary for a new accountant right out of college was $52,900 in 2014. A random sample of 16 new accountants has a mean salary of $54,000. We assume that the population standard deviation equals $4000. The histogram of the salary (in $1000s) is shown here. If it is appropriate to apply the Z test, then do so, using the critical-value method and level of significance $\alpha = 0.05$. If not, then explain clearly why not.

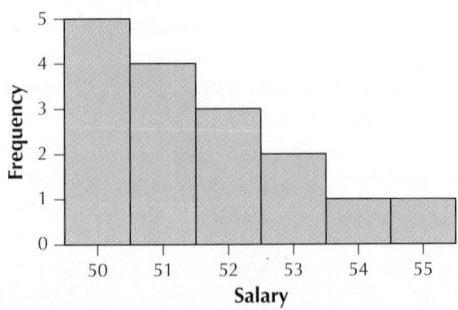

BRINGING IT ALL TOGETHER

Toyota Prius Gas Mileage. Use the following information for Exercises 60–67. Cars.com reported in 2014 that the mean city/highway combined gas mileage for the Toyota Prius hybrid car was 50 mpg. This year, a random sample of 16 Toyota Prius cars had the gas mileages shown here (in miles per gallon (mpg)). Assume $\sigma = 2$ mpg. The research question is to test whether mean gas mileage has increased since 2014.

48.1	49.9	50.6	51.8
48.8	50.0	50.9	52.2
49.3	50.2	51.3	52.9
49.5	50.4	51.6	53.9

60. Are the conditions met for performing the Z test for μ? Use technology to obtain a normal probability plot of the data.

61. Construct the appropriate hypotheses. State the meaning of μ.

62. Find the Z critical value and state the rejection rule. Use level of significance $\alpha = 0.10$.

63. Calculate the value of Z_{data}.

64. State and interpret your conclusion.

WHAT IF
?
65. Try to answer the following questions by thinking about the relationship between the statistics instead of by redoing all the calculations. Note the mpg of the worst-performing Prius: 48.1. *What if the 48.1 mpg is a typo?* We are not sure what the actual mpg is, but it is greater than 48.1 mpg. How would this change affect the following?

    a. $\bar{x}$
    b. σ (*Hint:* This is not the sample standard deviation.)
    c. n
    d. Z_{data}
    e. Z_{crit}
    f. The conclusion

66. Now, instead of 0.10, use level of significance $\alpha = 0.05$. Find the new Z critical value and state the rejection rule.

There is no need to recalculate the value of Z_{data}. State and interpret your conclusion with this new level of significance $\alpha = 0.05$.

67. Note the contradiction in your conclusions from Exercises 64 and 66. Can you suggest any way to resolve this?

WORKING WITH LARGE DATA SETS

68. Sodium. Work with the **Nutrition** data set.   **nutrition**
    a. Use technology to explore the variable *sodium*.
    b. Use technology to test at level of significance $\alpha = 0.05$ whether the population mean amount of sodium is greater than 280 mg. Let $\sigma = 625$ mg.
    c. Use technology to test at level of significance $\alpha = 0.05$ whether the population mean amount of sodium is greater than 290 mg. Let $\sigma = 625$ mg.

WORKING WITH LARGE DATA SETS

Chapter 9 Case Study: Clothing Store Sales. Open the Chapter 9 Case Study data set, **Clothing Store**. Here, we will perform a hypothesis test for the population mean number of coupons used per customer. We will then see whether this hypothesis test made the correct decision. Use technology to do the following. **clothingstore**

69. Obtain a random sample of size 100 from the data set.

70. Using your sample, test whether the population mean number of coupons used per customer differs from 0.75, using level of significance $\alpha = 0.05$. Assume $\sigma = 1.74$ coupons.

71. Using the 5000 customers in the entire data set as a population, find the actual value of the population mean number of coupons used per customer. Did your hypothesis test in Exercise 71 make the right decision? Explain.

9.3   Z Test for the Population Mean: *p*-Value Method

OBJECTIVES By the end of this section, I will be able to . . .

1   Perform the Z test for the mean, using the *p*-value method.
2   Assess the strength of evidence against the null hypothesis.
3   Describe the relationship between the *p*-value method and the critical-value method.
4   Use the Z confidence interval for the mean to perform the two-tailed Z test for the mean.

1   The *p*-Value Method of Performing the Z Test for the Mean

In Section 9.2, we considered the critical-value method for performing the Z test, which works by comparing one Z-value (Z_{data}) with another Z-value (Z_{crit}). In this section, we introduce the *p*-value method, which works by comparing one probability (the *p*-value) to another probability (α). The two methods are equivalent for the same level of significance α, giving you the same conclusion.

The **p-value** is a measure of how well (or how poorly) the data fit the null hypothesis.

> **p-Value**
>
> The **p-value** is the probability of observing a sample statistic (such as $\bar{x}$ or Z_{data}) at least as extreme as the statistic actually observed if we assume that the null hypothesis is true.
>
> Roughly speaking, the *p*-value represents the probability of observing the sample statistic if the null hypothesis is true. The term *p-value* means "probability value," so its value must always lie between 0 and 1.

A *p*-value is a probability associated with Z_{data} and tells us whether or not Z_{data} is an extreme value. The method for calculating *p*-values depends on the form of the hypothesis test (Table 5).

- For a right-tailed test, the *p*-value is in the right (or upper) tail area.
- For a left-tailed test, the *p*-value is in the left (or lower) tail area.
- For a two-tailed test, the *p*-value lies in both tails.

Remember that probability is represented by the area under the curve.

Table 5 Finding the *p*-value depends on the form of the hypothesis test

Type of hypothesis test	Right-tailed test	Left-tailed test	Two-tailed test
Hypotheses	$H_0 : \mu = \mu_0$ $H_a : \mu > \mu_0$	$H_0 : \mu = \mu_0$ $H_a : \mu < \mu_0$	$H_0 : \mu = \mu_0$ $H_a : \mu \neq \mu_0$
p-Value is tail area associated with Z_{data}	$p\text{-value} = P(Z > Z_{data})$ Area to right of Z_{data}	$p\text{-value} = P(Z < Z_{data})$ Area to left of Z_{data}	$p\text{-value} = P(Z > \lvert Z_{data}\rvert)$ $+ P(Z < -\lvert Z_{data}\rvert)$ $= 2 \cdot P(Z > \lvert Z_{data}\rvert)$ Sum of the two tail areas

---

EXAMPLE 12 Finding the *p*-value

For each of the following hypothesis tests, calculate and graph the *p*-value.

a. $H_0 : \mu = 3.0$ versus $H_a : \mu > 3.0, Z_{data} = 1$

b. $H_0 : \mu = 10$ versus $H_a : \mu < 10, Z_{data} = -1.45$

c. $H_0 : \mu = 100$ versus $H_a : \mu \neq 100, Z_{data} = -2$

Solution

a. We have a right-tailed test, so that the *p*-value equals the area in the right tail:

$$p\text{-value} = P(Z > Z_{data}) = P(Z > 1)$$

The *Z* table gives the probability for $P(Z < 1)$. Thus,

$$p\text{-value} = P(Z > 1) = 1 - P(Z < 1) = 1 - 0.8413 = 0.1587 \text{ (Figure 6a)}.$$

To review how to calculate these probabilities, see Table 8 in Chapter 6 on page 355.

b. We have a left-tailed test, so that the *p*-value equals the area in the left tail:

$$p\text{-value} = P(Z < Z_{data}) = P(Z < -1.45) = 0.0735 \text{ (Figure 6b)}$$

c.   Here, we have a two-tailed test, so that the *p*-value equals the sum of the areas in the two tails:

$$p\text{-value} = P(Z > |Z_{\text{data}}|) + (Z < -|Z_{\text{data}}|)$$
$$= P(Z > |-2|) + (Z < -|-2|)$$
$$= P(Z > 2) + (Z < -2)$$
$$= 0.0228 + 0.0228 = 0.0456 \text{ (Figure 6c)}$$

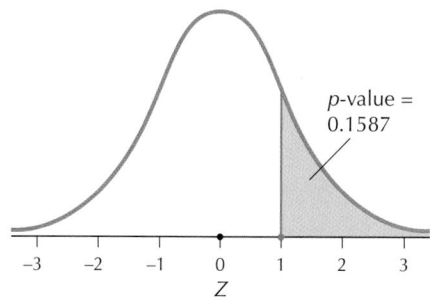

FIGURE 6a   *p*-Value for a right-tailed test.

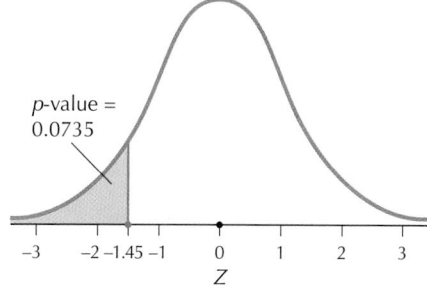

FIGURE 6b   *p*-Value for a left-tailed test.

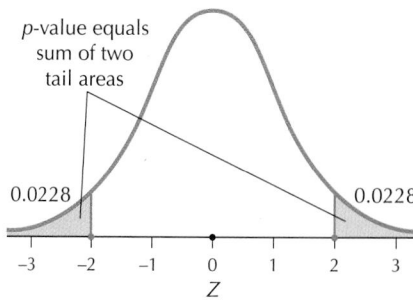

FIGURE 6c   *p*-Value for a two-tailed test.

NOW YOU CAN DO
Exercises 7–20.

YOUR TURN #5

The *p-Value* applet allows you to experiment with various hypotheses, means, standard deviations, and sample sizes in order to see how changes in these values affect the *p*-value.

For each of the following hypothesis tests, calculate and graph the *p*-value.

a. $H_0 : \mu = 75$ versus $H_a : \mu > 75, Z_{\text{data}} = 0.5$

b. $H_0 : \mu = 50$ versus $H_a : \mu < 50, Z_{\text{data}} = -1.2$

c. $H_0 : \mu = 1$ versus $H_a : \mu \neq 1, Z_{\text{data}} = -0.1$

(The solutions are shown in Appendix A.)

A *p*-value is based on the value of Z_{data}, so the *p*-value tells us whether or not Z_{data} is an extreme value. Unusual and extreme values of $\bar{x}$, and therefore of Z_{data}, will have a small *p*-value, whereas values of $\bar{x}$ and Z_{data} nearer to the center of the distribution will have a large *p*-value.

Assuming H_0 is true:

Unusual and extreme values of $\bar{x}$ and Z_{data} ⟷   Small *p*-value (close to 0; see Figure 6c)

Values of $\bar{x}$ and Z_{data} near center   ⟷   Large *p*-value (greater than, say, 0.15; see Figure 6a)

A small *p*-value indicates a conflict between your sample data and the null hypothesis, and will thus *lead us to reject H_0*. However, how small is small? We learned in Section 9.1 that the probability of Type I error α is chosen by the researcher to be small, usually 0.01, 0.05, or 0.10. Thus, a *p*-value is small if it is $\leq \alpha$. This leads us to the **rejection rule** that tells us when we may reject the null hypothesis.

This rejection rule can be applied to any type of hypothesis test we perform in Chapters 9–14 using the *p*-value method.

Rejection Rule When Using *p*-Value Method

The rejection rule for performing a hypothesis test using the *p*-value method is:

Reject H_0 when the *p*-value $\leq \alpha$. Otherwise, do not reject H_0.

The value of α represents the boundary between results that are statistically significant (where we reject H_0) and results that are not statistically significant (where we do not reject H_0). Thus, α is called the *level of significance* of the hypothesis test.

Here are the steps for performing the *Z* test for μ using the *p*-value method.

Z Test for the Population Mean μ: p-Value Method

When a random sample of size n is taken from a population where the standard deviation σ is known, you can use the Z test if either (a) the population is normal, or (b) the sample size is large ($n \geq 30$).

Step 1 **State the hypotheses and the rejection rule.**
Use one of the forms from Table 5 to write the hypotheses. State the meaning of μ. The rejection rule is "Reject H_0 if the p-value $\leq \alpha$."

Step 2 **Calculate Z_{data}.**

$$Z_{data} = \frac{\bar{x} - \mu_0}{\sigma/\sqrt{n}}$$

where the sample mean $\bar{x}$ and the sample size n represent the sample data, and the population standard deviation σ represents the population data.

Step 3 **Find the p-value.**
Either use technology to find the p-value, or calculate it using the form in Table 5 that corresponds to your hypotheses.

Step 4 **State the conclusion and interpretation.**
If the p-value $\leq \alpha$, then reject H_0. Otherwise do not reject H_0. Interpret your conclusion so that a nonspecialist (someone who has not had a course in statistics) can understand, as follows:

- Interpretation when you reject H_0: *There is evidence at level of significance α that [whatever H_a says].*
- Interpretation when you do not reject H_0: *There is insufficient evidence at level of significance α that [whatever H_a says].*

EXAMPLE 13   The Z test for the mean using the p-value method: One-tailed test

FlightStats.com compiles user ratings for airports worldwide. The mean rating for JFK International Airport in New York for July 2014 was 3.0 (out of 5). Assume that the population standard deviation of user ratings is known to be $\sigma = 1$. A random sample taken this year of $n = 36$ user ratings for JFK Airport showed a mean of $\bar{x} = 2.75$. Using level of significance $\alpha = 0.05$, test whether the population mean user rating for JFK Airport has fallen since 2014.

Solution

The sample size $n = 36$ is large, and the population standard deviation σ is known. We may therefore perform the Z test for the mean.

Step 1   **State the hypotheses and the rejection rule.**
The key words here are "has fallen," which means "is less than." The answer to the question "Less than what?" gives us $\mu_0 = 3.0$. Thus, our hypotheses are

$$H_0 : \mu = 3.0 \quad \text{versus} \quad H_a : \mu < 3.0$$

where μ refers to the population mean user rating for JFK Airport. We will reject H_0 if the p-value $\leq \alpha = 0.05$.

Step 2   **Calculate Z_{data}.**
We have $\bar{x} = 2.75$, $\mu_0 = 3.0$, $n = 36$, and $\sigma = 1$. Thus, our test statistic is

$$Z_{data} = \frac{\bar{x} - \mu_0}{\sigma/\sqrt{n}} = \frac{2.75 - 3.0}{1/\sqrt{36}} = -1.5$$

Step 3   **Find the *p*-value.**
Our hypotheses represent a left-tailed test from Table 5. Thus,

$$p\text{-value} = P(Z < Z_{data}) = P(Z < -1.5)$$

This is a Case 1 problem from Table 8 in Chapter 6 (page 355). The *Z* table (Appendix Table C) provides us with the area to the left of $Z = -1.5$ (Figure 7):

$$P(Z < -1.5) = 0.0668$$

Thus, the *p*-value is 0.0668.

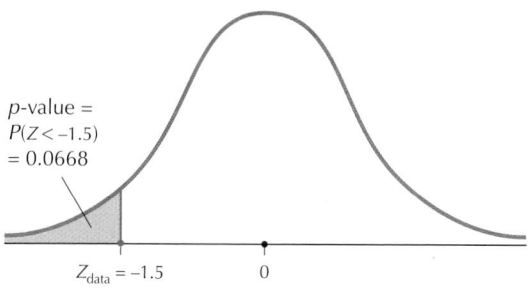

p-value =
$P(Z < -1.5)$
$= 0.0668$

$Z_{data} = -1.5$ 0

FIGURE 7 The *p*-value 0.0668 is not $\leq$ 0.05, so do not reject H_0.

Step 4   **State the conclusion and interpretation.**
Our level of significance is $\alpha = 0.05$ (from Step 1). The *p*-value $= 0.0668$ is *not* ≤ 0.05, therefore, we do *not* reject H_0. There is insufficient evidence at the level of significance $\alpha = 0.05$ that the population mean user rating for JFK Airport is less than 3.0.

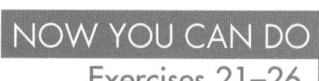

NOW YOU CAN DO
Exercises 21–26.

WHAT IF
?

What If Scenario

What if the sample mean in Example 13 was not $\bar{x} = 2.75$ but was instead some *unknown* value smaller than $\bar{x} = 2.75$? All other statistics and parameters remain the same. Suppose we wanted to perform the same hypothesis test as in Example 13. How would this decrease in the value of $\bar{x}$ affect the following, if at all?

a. Z_{data}

b. *p*-value

c. The conclusion

Solution

a. In Example 13, $\bar{x} = 2.75$ is smaller than $\mu_0 = 3.0$, which is why Z_{data} is negative. If we decrease $\bar{x}$ to an even smaller value, this will move Z_{data} further into negative territory (leftward on the number line).

b. For a left-tailed test, the *p*-value is the area to the left of Z_{data}. So, if Z_{data} is further to the left, there is less area to the left of it. Thus, the new *p*-value will be smaller.

c. We know from (**b**) that the *p*-value is decreasing, but not by how much, because we don't know how much smaller $\bar{x}$ and Z_{data} are. If the *p*-value decreases just a little bit, it will still be greater than $\alpha = 0.05$, and so we will still not reject H_0. However, if the *p*-value decreases by a lot, it will then be less than $\alpha = 0.05$, and so then we will reject H_0. Without further information, we just don't know.

EXAMPLE 14 The *p*-value method using technology: Two-tailed test

brisbane

The birth weights, in grams (1000 grams = 1 kilogram ≈ 2.2 pounds), of a random sample of 44 babies from Brisbane, Australia, have a sample mean weight $\bar{x} = 3276$ grams. Formerly, the mean birth weight of babies in Brisbane was 3200 grams. Assume that the population standard deviation $\sigma = 528$ grams. Is there evidence that the population mean birth weight of Brisbane babies now differs from 3200 grams? Use technology to perform the appropriate hypothesis test, with level of significance $\alpha = 0.10$.

What Results Might We Expect?

Note from Figure 8 that the sample mean birth weight $\bar{x} = 3276$ grams is close to the hypothesized mean birth weight of $\mu_0 = 3200$ grams. This value of $\bar{x}$ is not extreme and thus does not seem to offer strong evidence that the hypothesized mean birth weight is wrong. Therefore, we might expect to *not reject* the hypothesis that $\mu_0 = 3200$ grams.

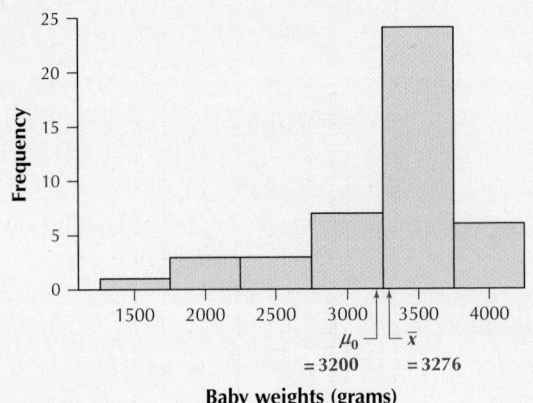

FIGURE 8 Sample mean, $\bar{x} = 3276$, is close to hypothesized mean, $\mu_0 = 3200$, so we expect to not reject the null hypothesis.

Solution

The sample size $n = 44$ is large and $\sigma = 528$ is known, so we may proceed with the Z test for μ.

Step 1 **State the hypotheses and the rejection rule.**
The key words "differs from" mean that we have a two-tailed test:

$$H_0: \mu = 3200 \quad \text{versus} \quad H_a: \mu \neq 3200$$

where μ refers to the population mean birth weight of Brisbane babies. We will reject H_0 if the *p*-value $\leq \alpha = 0.10$.

Step 2 **Calculate Z_{data}.**
We will use the instructions provided in the Step-by-Step Technology Guide at the end of this section (page 519). Figure 9 shows the TI-83/84 results from the Z test for μ:

FIGURE 9
TI-83/84 results.

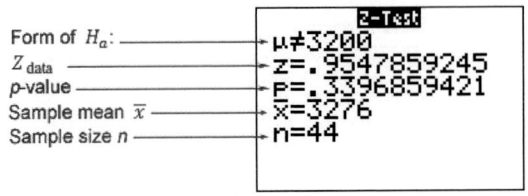

Form of H_a:
Z_{data}
p-value
Sample mean $\bar{x}$
Sample size n

$$Z_{\text{data}} = \frac{\bar{x} - \mu_0}{\sigma/\sqrt{n}} = \frac{3276 - 3200}{528/\sqrt{44}} = 0.9547859245 \approx 0.9548$$

Figure 10 shows the Minitab results, where

- "Test of $\mu = 3200$ versus $\neq 3200$" refers to the hypotheses being tested, $H_0 : \mu = 3200$ versus $H_a : \mu \neq 3200$.

- "The assumed standard deviation = 528" refers to our assumption that $\sigma = 528$.

- SE Mean refers to the standard error of the mean, that is, $\sigma/\sqrt{n}$. You can see that $528/\sqrt{44} \approx 79.6$.

- 90% CI represents a 90% *Z* confidence interval for μ.

- *Z* refers to our test statistic:

$$Z_{\text{data}} = \frac{\bar{x} - \mu_0}{\sigma/\sqrt{n}} = (3276 - 3200)/(528/\sqrt{44}) = 0.9547859245 \approx 0.95$$

- *P* represents our *p*-value of 0.340.

```
Test of µ = 3200 vs ≠ 3200
The assumed standard deviation = 528


 N    Mean   SE Mean       90% CI          Z      P
44   3276.0    79.6   (3145.1, 3406.9)   0.95   0.340
```

Different software rounds the results to different numbers of decimal places.

FIGURE 10
Minitab results.

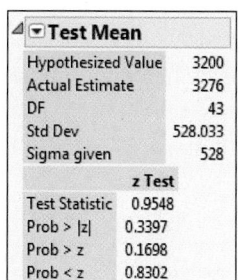

FIGURE 11 JMP results.

Figure 11 shows the JMP results, where

- "Hypothesized Value" refers to the hypotheses being tested: $H_0 : \mu = 3200$ versus $H_a : \mu \neq 3200$.

- "Actual Estimate" refers to the sample mean, $\bar{x} = 3276$.

- "Sigma given" refers to our assumption that $\sigma = 528$.

- "Test statistic" refers to Z_{data}, our test statistic.

- "Prob > |z|," "Prob > z," and "Prob < z" refers to the *p*-value of a two-sided, right-tailed, and left-tailed test, respectively. We want the two-sided *p*-value, which is "Prob > |z|" = 0.3397.

Step 3   **Find the *p*-value.**
We have a two-tailed test from Step 1, so that from Table 5 our *p*-value is (Figure 12)

$$p\text{-Value} = 2 \cdot P(Z > |Z_{\text{data}}|) = 2 \cdot P(Z > 0.9548) \approx 2 \cdot (0.1698)$$

$$= 0.3396$$

FIGURE 12
p-Value is the sum of two tail areas: 0.1698 + 0.1698 = 0.3396.

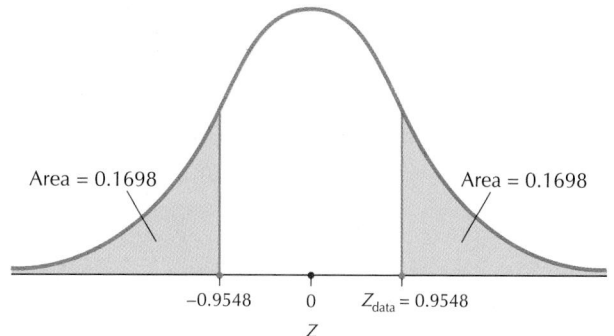

NOW YOU CAN DO

Exercises 27–30.

Step 4 **State the conclusion and interpretation.**
Because 0.3396 is not ≤ 0.10, we do not reject H_0. There is insufficient evidence that the population mean birth weight differs from 3200 grams. This conclusion is just as we expected.

2 Assessing the Strength of Evidence Against the Null Hypothesis

The hypothesis-testing methods we have shown so far deliver a simple "yes-or-no" conclusion: either "Reject H_0," or "Do not reject H_0." There is no indication of how strong the evidence is for rejecting the null hypothesis. Was the decision close? Was it a no-brainer? On the other hand, the p-value itself represents the *strength of evidence against the null hypothesis*. There is extra information here, which we should not ignore.

For instance, we can directly compare the results of hypothesis tests. Suppose that we have two hypothesis tests that both result in not rejecting the null hypothesis, with level of significance $\alpha = 0.05$. However, Test A has a p-value of 0.06, whereas Test B has a p-value of 0.57. Clearly, Test A came very close to rejecting the null hypothesis and shows a fair amount of evidence against the null hypothesis, whereas Test B shows no evidence at all against the null hypothesis. A simple statement of the "yes-or-no" conclusion misses the clear distinction between these two situations.

The p-value provides us with the smallest level of significance at which the null hypothesis would be rejected, that is, the smallest value of α at which the results would be considered significant.

Of course, we are free to determine whether the results are significant using whatever α level we want. For example, Test A would have rejected H_0 for any α value 0.06 or higher. Some data analysts in fact do not think in terms of rejecting or not rejecting the null hypothesis. Rather, they think completely in terms of *assessing the strength of evidence against the null hypothesis*.

For many (though not all) data domains, Table 6 provides a thumbnail impression of the strength of evidence against the null hypothesis for various p-values. For certain domains (such as the physical sciences), however, alternative interpretations are appropriate.

Table 6 Strength of evidence against the null hypothesis for various levels of p-value

p-**Value**	**Strength of evidence against H_0**
p-value ≤ 0.001	Extremely strong evidence
$0.001 < p$-value ≤ 0.01	Very strong evidence
$0.01 < p$-value ≤ 0.05	Solid evidence
$0.05 < p$-value ≤ 0.10	Moderate evidence
$0.10 < p$-value ≤ 0.15	Slight evidence
$0.15 < p$-value	No evidence

Note: Use Table 6 for all exercises that ask for an assessment of the strength of evidence against the null hypothesis.

EXAMPLE 15 Assessing the strength of evidence against H_0

Assess the strength of evidence against H_0 shown by the p-values in (**a**) Example 13 and (**b**) Example 14.

Solution

a. In Example 13, we tested $H_0 : \mu = 3.0$ versus $H_a : \mu < 3.0$, where μ refers to the population mean user rating for JFK International Airport. Our p-value of 0.0668 implies that there is *moderate* evidence against the null hypothesis that the population mean user rating for JFK Airport equals 3.0.

b. In Example 14, we tested $H_0 : \mu = 3200$ versus $H_a : \mu \neq 3200$, where μ refers to the population mean birth weight of Brisbane babies (in grams). Our *p*-value of 0.3397 implies that there is *no* evidence against the null hypothesis that the population mean birth weight of Brisbane babies equals 3200 grams.

NOW YOU CAN DO
Exercises 31–40.

YOUR TURN #6

Each of the following *p*-values was calculated in Example 12. For each, assess the strength of evidence against the null hypothesis.

a. $H_0 : \mu = 3.0$ versus $H_a : \mu > 3.0$, *p*-value = 0.1587

b. $H_0 : \mu = 10$ versus $H_a : \mu < 10$, *p*-value = 0.0735

c. $H_0 : \mu = 100$ versus $H_a : \mu \neq 100$, *p*-value = 0.0456

(The solutions are shown in Appendix A.)

Developing Your Statistical Sense

The Role of the Level of Significance α

Suppose that in Example 13, our level of significance α was 0.10 instead of 0.05. Would this have changed anything? Certainly. Our *p*-value of 0.0668 is less than the new $\alpha = 0.10$, so we would reject H_0. Think about that for a moment. *The data haven't changed at all, but our conclusion is reversed simply by changing α.* What is a data analyst to make of a situation like this? Two alternatives are available.

1. We don't want the choice of α to dictate our conclusion, so perhaps we should turn to a direct assessment of the strength of evidence against the null hypothesis, as provided in Table 6. In this case, the *p*-value of about 0.0668 would offer moderate evidence against the null hypothesis, *regardless of the value of α*.

2. Obtain more data, perhaps through a call for further research.

3 The Relationship Between the *p*-Value Method and the Critical-Value Method

Figure 13 shows the relationships between the *p*-value method and the critical-value method. The top half represents values of *Z* and the critical-value method that we studied in Section 9.2. The bottom half represents probabilities and the *p*-value method that we studied in this section. The left half represents statistics associated with the observed sample data. The right half represents critical-value thresholds for significance to which these statistics are compared.

Because Z_{data} helps us to determine the *p*-value, these two values are related. Similarly, because the level of significance α helps to determine the value of Z_{crit}, these two values are related. Moreover, just as we compare Z_{data} with the threshold Z_{crit}, we compare the *p*-value statistic with the α threshold to determine significance. Thus, the two methods for conducting hypothesis tests are equivalent and, in fact, are quite thoroughly interwoven.

Figures 14a and 14b illustrate this equivalence for a right-tailed test. The rejection rule for the *p*-value method is to reject H_0 when the *p*-value $\leq \alpha$. The rejection rule for the critical-value method is to reject H_0 when $Z_{\text{data}} \geq Z_{\text{crit}}$. Note in Figures 14a and 14b

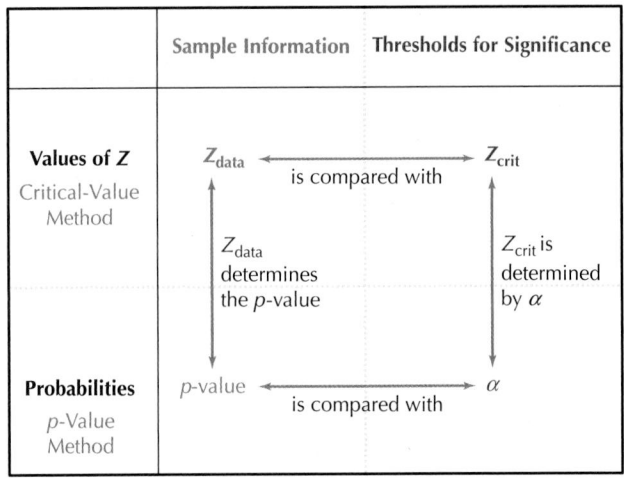

FIGURE 13 Critical-value method and *p*-value method are equivalent.

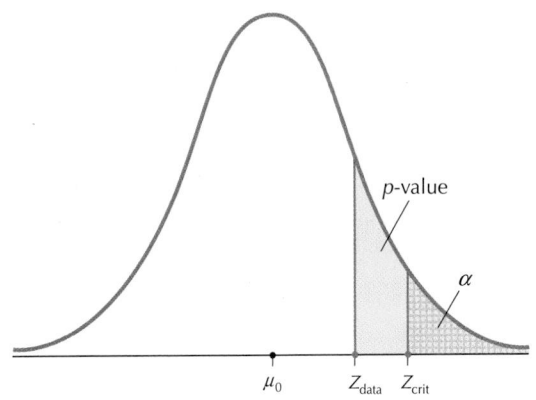

FIGURE 14a For a right-tailed test, $Z_{data} < Z_{crit}$ only when the *p*-value $> \alpha$.

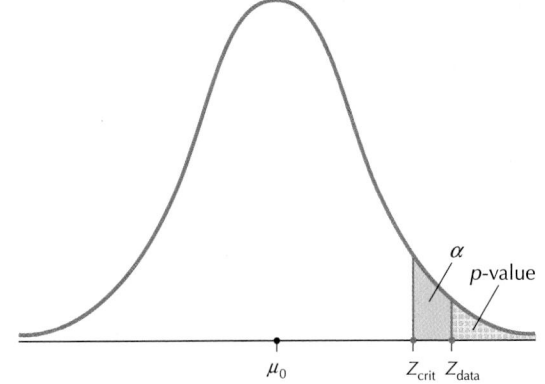

FIGURE 14b For a right-tailed test, $Z_{data} \geq Z_{crit}$ only when the *p*-value $\leq \alpha$.

how the *p*-value is determined by Z_{data}, and Z_{crit} is determined by α. In Figure 14a, when $Z_{data} < Z_{crit}$, it must also happen that the *p*-value $> \alpha$. In both cases we do not reject H_0. However, in Figure 14b, when $Z_{data} \geq Z_{crit}$, it also follows that the *p*-value is $\leq \alpha$. In both cases, we reject H_0. Thus, the *p*-value method and the critical-value method are equivalent.

4 Using Confidence Intervals for μ to Perform Two-Tailed Hypothesis Tests About μ

Consider a two-tailed hypothesis test for μ:

$$H_0 : \mu = \mu_0 \quad \text{versus} \quad H_a : \mu \neq \mu_0$$

and recall the $100(1 - \alpha)\%$ Z confidence interval for μ from Section 8.1:

$$\bar{x} \pm Z_{\alpha/2}\left(\sigma/\sqrt{n}\right)$$

Both inference methods are based on the Z statistic:

$$Z = \frac{\bar{x} - \mu}{\sigma/\sqrt{n}}$$

so it makes sense that the two-tailed hypothesis test and the confidence interval are equivalent.

Equivalence of a Two-Tailed Hypothesis Test and a Confidence Interval

- If a certain value for μ_0 lies *outside* the corresponding $100(1 - \alpha)\%$ Z confidence interval for μ, then the null hypothesis specifying this value for μ_0 would be *rejected* for level of significance α (see Figure 15).
- Alternatively, if a certain value for μ_0 lies *inside* the $100(1 - \alpha)\%$ Z confidence interval for μ, then the null hypothesis specifying this value for μ_0 would *not be rejected* for level of significance α.

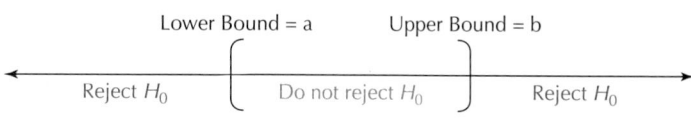

Lower Bound = a   Upper Bound = b

Reject H_0 Do not reject H_0 Reject H_0

FIGURE 15 Reject H_0 for values of μ_0 that lie outside confidence interval (a, b).

Table 7 shows the confidence levels and associated α levels of significance that will produce the equivalent inference.

Table 7 Confidence levels for equivalent α levels of significance

Confidence level	Level of significance α
90%	0.10
95%	0.05
99%	0.01

We may thus use a single confidence interval to test as many values of μ_0 as necessary.

EXAMPLE 16   Equivalence of two-tailed tests and confidence intervals

wavebreakmedia/ Shutterstock

Recall Example 4 from Section 8.1 (page 432), where we were 90% confident using a Z interval that the population mean score on the 2014 SAT Math test lies between 471.2 and 548.8. Test, using level of significance $\alpha = 0.10$, whether the population mean SAT Math test score differs from these values: (**a**) 470, (**b**) 510, (**c**) 550.

Solution

Once we have the 90% confidence interval, we may test as many possible values for μ_0 as necessary, as long as we use level of significance $\alpha = 0.10$ (see Table 7).

- If any values of μ_0 lie inside the confidence interval, that is, between 471.2 and 548.8, we will not reject H_0 for this value of μ_0.
- If any values of μ_0 lie outside the confidence interval, that is, either to the left of 471.2 or to the right of 548.8, we will reject H_0, as shown in Figure 16.

FIGURE 16
Reject H_0 for values of μ_0 that lie outside (471.2, 548.8).

Lower Bound = 471.2   Upper Bound = 548.8

Reject H_0 Do not reject H_0 Reject H_0

We set up the three two-tailed hypothesis tests as follows:

a. $H_0 : \mu = 470$ versus $H_a : \mu \neq 470$

b. $H_0 : \mu = 510$ versus $H_a : \mu \neq 510$

c. $H_0 : \mu = 550$ versus $H_a : \mu \neq 550$

To perform each hypothesis test, simply observe where each value of μ_0 falls on the number line shown in Figure 16. For example, in the first hypothesis test, the hypothesized value $\mu_0 = 470$ lies outside the interval (471.2, 548.8). Thus, we reject H_0. The three hypothesis tests are summarized here.

Value of μ_0	Form of hypothesis test, with $\alpha = 0.10$	Where μ_0 lies in relation to 90% confidence interval	Conclusion of hypothesis test
a. 470	$H_0 : \mu = 470$ vs. $H_a : \mu \neq 470$	Outside	Reject H_0
b. 510	$H_0 : \mu = 510$ vs. $H_a : \mu \neq 510$	Inside	Do not reject H_0
c. 550	$H_0 : \mu = 550$ vs. $H_a : \mu \neq 550$	Outside	Reject H_0

NOW YOU CAN DO
Exercises 41–46.

YOUR TURN
#7

For the Z interval from Example 16, test, using level of significance $\alpha = 0.10$, whether the population mean SAT Math test score differs from these values: (**a**) 548, (**b**) 477, (**c**) 549.

(The solutions are shown in Appendix A.)

Increasingly, technology is being used to perform statistical analysis, including hypothesis tests. Therefore, it is important to know how to read and interpret the software output from a hypothesis test.

EXAMPLE 17 Interpreting software output

Each of (**a**) and (**b**) represent software output from a *Z test* for μ. For each, examine the indicated software output, and provide the following steps:

Step 1 **State the hypotheses and the rejection rule.**

Step 2 **Calculate Z_{data}.**

Step 3 **Find the *p*-value.**

Step 4 **State the conclusion and interpretation.**

Let the level of significance be $\alpha = 0.05$ in each case.

a. TI-83/84 output for a *Z test* for μ, where μ represents the population mean length of laboratory mice (in cm)

b. Minitab output for a Z test for μ, where μ represents the population mean number of farmer's markets per county, nationwide.

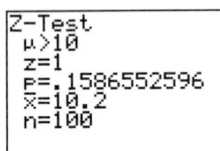

TI-83/84 output for part (**a**).

```
Test of μ = 2.3 vs ≠ 2.3
The assumed standard deviation = 5.7


Variable           N    Mean   StDev   SE Mean      95% CI        Z      P
Farmers Markets  3143   2.596   5.683    0.102   (2.397, 2.796)  2.91  0.004
```

Minitab output for part (**b**).

Solution

a. Interpreting the TI-83/84 output.

Step 1 **State the hypotheses and the rejection rule.**
In the TI-83/84 output, the "$\mu > 10$" indicates the *alternative* hypothesis. In other words, the hypotheses are:

$$H_0 : \mu = 10 \quad \text{versus} \quad H_a : \mu > 10$$

where μ represents the population mean length of laboratory mice (in cm). We will reject H_0 if the *p*-value is less than the level of significance $\alpha = 0.05$.

Step 2   **Find Z_{data}.**

The "z = 1" in the TI-83/84 output provides us the value of the test statistic, $Z_{data} = 1$.

Step 3   **Find the *p*-value.**

The "*p* = .1586552596" in the TI-83/84 output represents the *p*-value.

Step 4   **State the conclusion and interpretation.**

The *p*-value from Step 3 is not less than the level of significance $\alpha = 0.05$, so we do not reject H_0. There is insufficient evidence that the population mean length of laboratory mice is greater than 10 cm.

b.   Interpreting the Minitab output.

Step 1   **State the hypotheses and the rejection rule.**

The first line in the Minitab output is "Test of $\mu = 2.3$ vs $\neq 2.3$," which indicates a two-tailed test, as follows:

$$H_0 : \mu = 2.3 \quad \text{versus} \quad H_a : \mu \neq 2.3$$

where μ represents the population mean number of farmer's markets per county. We will reject H_0 if the *p*-value is less than the level of significance $\alpha = 0.05$.

Step 2   **Find Z_{data}.**

Under "*Z*" in the Minitab output is found "2.91," giving us $Z_{data} = 2.91$.

Step 3   **Find the *p*-value.**

Under "*p*" in the Minitab output is "0.004," representing our *p*-value.

Step 4   **State the conclusion and interpretation.**

The *p*-value (0.004) from Step 3 is less than the level of significance $\alpha = 0.05$, so we reject H_0. There is evidence that the population mean number of farmer's markets per county, nationwide, differs from 2.3.

STEP-BY-STEP TECHNOLOGY GUIDE: *Z* test for μ

We will use the birth weight data from Example 14 (page 512).

TI-83/84

If you have the data values:
Step 1   Enter the data into list **L1**.
Step 2   Press **STAT**, highlight **TESTS**.
Step 3   Press 1 (for **Z-Test**; see Figure 17a).
Step 4   For input (**Inpt**), highlight **Data**, and press **ENTER** (Figure 17b).
a.   For μ_0, enter the value of μ_0, **3200**.
b.   For σ, enter the value of σ, **528**.
c.   For **List**, press **2nd**, then **L1**.
d.   For **Freq**, enter 1.
e.   For μ, select the form of H_a. Here, we have a two-tailed test, so highlight $\neq \mu_0$ and press **ENTER**.
f.   Highlight **Calculate** and press **ENTER**. The results are shown in Figure 9 in Example 14.

If you have the summary statistics:
Step 1   Press **STAT**, highlight **TESTS**.
Step 2   Press 1 (for **Z-Test**; see Figure 17a).
Step 3   For input (**Inpt**), highlight **Stats** and press **ENTER** (Figure 17c).
a.   For μ_0, enter the value of μ_0, **3200**.
b.   For σ, enter the value of σ, **528**.
c.   For $\bar{x}$, enter the sample mean **3276**.
d.   For n, enter the sample size **44**.
e.   For μ, select the form of H_a. Here, we have a two-tailed test, so highlight $\neq \mu_0$ and press **ENTER**.
f.   Highlight **Calculate** and press **ENTER**. The results are shown in Figure 9 in Example 14.

FIGURE 17a

FIGURE 17b

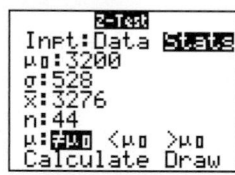

FIGURE 17c

EXCEL
JMP add-in for Excel.
There is a JMP add-in for Excel, which is activated when JMP is installed. This allows you to easily call upon JMP from within Excel, in order to conduct analyses that Excel itself does not provide. After installation of the JMP add-in for Excel, when you select **JMP > Data Table**, a new JMP workspace is created using the data from your Excel spreadsheet. Use the JMP add-in for Excel.

MINITAB
If you have the data values:
Step 1   Enter the data into column **C1**.
Step 2   Click **Stat > Basic Statistics > 1-Sample Z...**.
Step 3   Select **One or more samples, each in a column** from the drop-down menu. Click inside the box below the drop-down menu and select **C1**.
Step 4   Enter 528 as **Standard Deviation**.
Step 5   Select **Perform hypothesis test**. For **Hypothesized Mean**, enter 3200.
Step 6   Click **Options...**
a. Choose your **Confidence Level** as $100(1 - \alpha)$. Our level of significance α here is 0.10, so the confidence level is 90.0.
b. For **Alternative hypothesis**, select **Mean ≠ hypothesized mean** to symbolize the two-tailed test.
Step 7   Click **OK** and click **OK** again. The results are shown in Figure 10 in Example 14.

If you have the summary statistics:
Step 1   Click **Stat > Basic Statistics > 1-Sample Z...**.
Step 2   From the drop-down menu, select **Summarized Data**.
Step 3   Enter the **Sample Size** 44, **Sample Mean** 3276, and **Known standard deviation** 528.
Step 4   Select **Perform hypothesis test**, and enter the **Hypothesized mean** 3200.
Step 5   Click **Options...**
a. Choose your **Confidence Level** as $100(1 - \alpha)$. Our level of significance α here is 0.10, so the confidence level is 90.0.
b. For **Alternative hypothesis**, select **Mean ≠ hypothesized mean** to symbolize the two-tailed test.
Step 6   Click **OK** and click **OK** again. The results are shown in Figure 10 in Example 14.

JMP
If you have the data values:
Step 1   Click **File > New > Data Table**. Enter the data into **Column 1**.
Step 2   Click **Analyze > Distribution**. Click **Column 1** and then **Y, Columns**. Click **OK**.

Step 3   Click the red triangle beside **Column 1**, and select **Test Mean**. For **Specify Hypothesized Mean**, enter 3200. For **Enter True Standard Deviation to do z test rather than t test**, enter 528. Click **OK**. The results are shown in Figure 11 in Example 14.

CRUNCHIT!
If you have the data values:
Step 1   Click **File**, highlight **Load from Larose, Discostat3e > Chapter 9**, and click on **Example 03_14**.
Step 2   Click **Statistics**, highlight **z**, and select **1-sample**.
Step 3   With the **Columns** tab chosen, for **Sample** select **Birth weight (in grams)**. For **Standard Deviation**, enter 528.
Step 4   Select the **Hypothesis Test** tab. For **Mean under null hypothesis**, enter 3200. For **Alternative** select **Two-sided**. Then click **Calculate**.

If you have the summary statistics:
Step 1   Click **Statistics**, highlight **z**, and select **1-sample**.
Step 2   Choose the **Summarized** tab. For **n**, enter the sample size 44. For **Sample Mean**, enter 3276. For **Standard Deviation**, enter 528.
Step 3   Select the **Hypothesis Test** tab. For **Mean under null hypothesis**, enter 3200. For **Alternative**, select **Two-sided**. Then click **Calculate**.

Section 9.3 Summary

1. The *p*-value can be thought of as the probability of observing a sample statistic at least as extreme as the statistic in your sample if we assume that the null hypothesis is true. The rejection rule for the *p*-value method is to reject H_0 when the *p*-value $\leq \alpha$, the level of significance.

2. The *p*-value can be used to assess the strength of evidence against the null hypothesis.

3. The critical-value method and the *p*-value method are equivalent, and related in several ways.

4. We can use a single confidence interval for μ to help us perform any number of corresponding two-tailed hypothesis tests about μ.

Section 9.3 Exercises

CLARIFYING THE CONCEPTS

1. True or false: It is possible to get a *p*-value equal to 1.5. (p. 508)

2. State the rejection rule for the *p*-value method for performing the Z test for μ. (p. 509)

3. Explain why we might want to assess the strength of evidence against the null hypothesis, instead of delivering a simple "reject H_0 or do not reject H_0" conclusion. (p. 514)

4. What is the criterion for rejecting H_0 when using a confidence interval to perform a two-tailed hypothesis test for μ? (p. 517)

5. True or false: For a right-tailed test, when $Z_{data} < Z_{crit}$, the *p*-value is always $< \alpha$. (p. 516)

6. For (**a**)–(**c**), indicate whether or not the quantity represents a probability. (p. 516)

    a. Z_{data}

    b. *p*-value

    c. α

PRACTICING THE TECHNIQUES

✓ CHECK IT OUT!

To do	Check out	Topic
Exercises 7–20	Example 12	Calculating the *p*-value
Exercises 21–26	Example 13	One-tailed *Z* test for μ
Exercises 27–30	Example 14	Two-tailed *Z* test for μ
Exercises 31–40	Example 15	Assessing the strength of evidence against H_0
Exercises 41–46	Example 16	Equivalence of two-tailed tests and confidence intervals
Exercises 47–50	Example 17	Interpreting software output

For Exercises 7–40, assume that the conditions for performing the *Z* test are met.

For Exercises 7–17, find the *p*-value.

7. $H_0: \mu = 10$ vs. $H_a: \mu > 10, Z_{data} = 1.5$
8. $H_0: \mu = 10$ vs. $H_a: \mu > 10, Z_{data} = 2.5$
9. $H_0: \mu = 25$ vs. $H_a: \mu > 25, Z_{data} = 0.6$
10. $H_0: \mu = 25$ vs. $H_a: \mu > 25, Z_{data} = 1.2$
11. $H_0: \mu = 200$ vs. $H_a: \mu < 200, Z_{data} = -0.7$
12. $H_0: \mu = 200$ vs. $H_a: \mu < 200, Z_{data} = -1.27$
13. $H_0: \mu = 69$ vs. $H_a: \mu < 69, Z_{data} = -2.23$
14. $H_0: \mu = 69$ vs. $H_a: \mu < 69, Z_{data} = -2.55$
15. $H_0: \mu = 31$ vs. $H_a: \mu \neq 31, Z_{data} = 0.64$
16. $H_0: \mu = 31$ vs. $H_a: \mu \neq 31, Z_{data} = -0.64$
17. $H_0: \mu = 12$ vs. $H_a: \mu \neq 12, Z_{data} = 0$
18. Refer to Exercises 7–10. Explain what happens to the *p*-value for a right-tailed test as Z_{data} moves toward the right tail.
19. Refer to Exercises 11–14. Explain what happens to the *p*-value for a left-tailed test as Z_{data} moves toward the left tail.
20. Refer to Exercises 15 and 16. What can we say about the *p*-values of two two-tailed tests whose values of Z_{data} have the same absolute value?

For Exercises 21–30, perform the *Z* test for μ using level of significance $\alpha = 0.05$ by doing the following steps:

    a. State the hypotheses and the rejection rule.
    b. Calculate Z_{data}.
    c. Find the *p*-value.
    d. State the conclusion and the interpretation.

21. $H_0: \mu = 3.14$ vs. $H_a: \mu > 3.14, \bar{x} = 3.2, \sigma = 1, n = 100$
22. $H_0: \mu = 30$ vs. $H_a: \mu < 30, \bar{x} = 25, \sigma = 10, n = 16$
23. $H_0: \mu = -1.0$ vs. $H_a: \mu > -1.0, \bar{x} = 0, \sigma = 1, n = 400$
24. $H_0: \mu = 2000$ vs. $H_a: \mu > 2000, \bar{x} = 2050, \sigma = 200, n = 25$

25. $H_0: \mu = 500$ vs. $H_a: \mu < 500, \bar{x} = 450, \sigma = 100, n = 16$
26. $H_0: \mu = -32$ vs. $H_a: \mu > -32, \bar{x} = -30, \sigma = 40, n = 400$
27. $H_0: \mu = 10$ vs. $H_a: \mu \neq 10, \bar{x} = 10, \sigma = 5, n = 100$
28. $H_0: \mu = -5$ vs. $H_a: \mu \neq -5, \bar{x} = -5, \sigma = 1.5, n = 100$
29. $H_0: \mu = 0$ vs. $H_a: \mu \neq 0, \bar{x} = -0.12, \sigma = 0.4, n = 81$
30. $H_0: \mu = 46$ vs. $H_a: \mu \neq 46, \bar{x} = 47, \sigma = 15, n = 225$

For Exercises 31–40, use the indicated *p*-value to assess the strength of evidence against the null hypothesis, using Table 6.

31. *p*-value from Exercise 21
32. *p*-value from Exercise 22
33. *p*-value from Exercise 23
34. *p*-value from Exercise 24
35. *p*-value from Exercise 25
36. *p*-value from Exercise 26
37. *p*-value from Exercise 27
38. *p*-value from Exercise 28
39. *p*-value from Exercise 29
40. *p*-value from Exercise 30

For Exercises 41–46, a $100(1 - \alpha)\%$ *Z* confidence interval is given (see Section 8.1). Use the confidence interval to test, using level of significance α, whether μ differs from each of the indicated hypothesized values.

41. A 95% *Z* confidence interval for μ is $(-2.7, 6.9)$. Hypothesized values μ_0 are
    a. -3 **b.** -2 **c.** 0
    d. 5   **e.** 7
42. A 99% *Z* confidence interval for μ is $(45, 55)$. Hypothesized values μ_0 are
    a. 0   **b.** 44   **c.** 50
    d. 54   **e.** 56
43. A 90% *Z* confidence interval for μ is $(-10, -5)$. Hypothesized values μ_0 are
    a. -3 **b.** -8 **c.** -11
    d. 0   **e.** 7
44. A 95% *Z* confidence interval for μ is $(1024, 2056)$. Hypothesized values μ_0 are
    a. 1000   **b.** 2000   **c.** 3000
    d. 0   **e.** 1025
45. A 95% *Z* confidence interval for μ is $(0, 1)$. Hypothesized values μ_0 are
    a. 1.5   **b.** -1 **c.** 0.5
    d. 0.9   **e.** 1.2
46. A 95% *Z* confidence interval for μ is $(1.3275, 1.4339)$. Hypothesized values μ_0 are
    a. 1.3   **b.** 1.35   **c.** 1.4
    d. 1.45   **e.** 1.3275

For Exercises 47–50, software output from a *Z* test for μ is provided. For each, examine the indicated software output, and provide the following steps for level of significance $\alpha = 0.05$:

Step 1 State the hypotheses and the rejection rule.
Step 2 Find Z_{data}.
Step 3 Find the *p*-value.
Step 4 State the conclusion and interpretation.

47.
```
Z-Test
 μ<75
 z=-1.6
 p=.0547992894
 x̄=73
 n=64
```

48.
```
Z-Test
 μ≠2.5
 z=2
 p=.045500124
 x̄=2.6
 n=100
```

49.
```
Test of μ = 70 vs > 70
The assumed standard deviation = 240

Variable              N    Mean    StDev   SE Mean   95% Lower Bound    Z      P
Full Service Rest   3139   72.38   239.90    4.28             65.33   0.55   0.290
```

50.
```
Test of μ = 78 vs < 78
The assumed standard deviation = 237

Variable           N    Mean    StDev   SE Mean   95% Upper Bound    Z       P
Fast Food Rest   3139   68.91   237.17    4.23             75.86   -2.15   0.016
```

APPLYING THE CONCEPTS

For Exercises 51–58, do the following:
- **a.** State the hypotheses and the rejection rule.
- **b.** Calculate Z_{data}.
- **c.** Find the *p*-value.
- **d.** State the conclusion and the interpretation.

51. Car Insurance. Like sports cars? They can be expensive. Time.com/Money reports that in 2014 the mean annual car insurance premium for a Porsche Panamera Turbo-S was $3000. Suppose that a random sample of nine such Porsches taken this year has a mean car insurance premium of $3120. Assume $\sigma = \$600$, and assume that the distribution of premiums is normal. Test whether the population mean premium has increased, using level of significance $\alpha = 0.10$.

52. Mobile Apps. In 2014, Nielsen reported that young people ages 18–24 spent 37 hours per month using the apps on their mobile devices. Suppose that a random sample of 100 people ages 18–24 showed a sample mean of 40 hours. Assume $\sigma = 20$. Test whether the population mean number of hours using mobile apps by young people ages 18–24 has increased, at level of significance $\alpha = 0.05$.

53. Eating Trends. According to an NPD Group report, the mean number of meals prepared and eaten at home is less than 700 per year. Suppose that a random sample of 100 households showed a sample mean number of meals prepared and eaten at home of 650. Assume $\sigma = 25$. Test whether the population mean number of such meals is less than 700, using level of significance $\alpha = 0.10$.

54. DDT in Breast Milk. Researchers compared the amount of DDT in the breast milk of 12 Latina women in the Yakima Valley of Washington State with the amount of DDT in breast milk in the general U.S. population.[3] They measured the mean DDT level in the general population to be 47.2 parts per billion (ppb) and the mean DDT level in the 12 Latina women to be 219.7 ppb. Assume $\sigma = 36$ and a normally distributed population. Test whether the population mean DDT level in the breast milk of Latina women in the Yakima Valley is greater than that of the general population, using level of significance $\alpha = 0.01$.

55. Millennials' Income. The Pew Research Center reported in 2014 that Millennials (those ages 25–32) with Bachelor's degrees working full time are making on average $17,500 more per year than those Millennials with only a high school education ($45,500 vs. $28,000). Suppose that, in a random sample of 36 Millennials with Bachelor's degrees taken today, the sample mean income is $50,000. Assume $\sigma = \$18,000$. Test whether the population mean annual income has increased from its previous value of $45,500 using level of significance $\alpha = 0.05$.

56. Tree Rings. Do trees grow more quickly when they are young? The International Tree Ring Data Base collected data on a particular 440-year-old Douglas fir tree.[4] The mean annual ring growth in the tree's first 80 years of life was 1.4261 millimeters (mm). A random sample of size 100 taken from the tree's later years showed a sample mean growth of 0.56 mm per year. Assume $\sigma = 0.5$ mm and a normally distributed population. Test whether the population mean annual ring growth in the tree's later years is less than 1.4261 mm, using level of significance $\alpha = 0.05$.

57. Hybrid Vehicles. A study by Edmunds.com examined the time it takes for owners of hybrid vehicles to recoup their additional initial cost through reduced fuel consumption. Suppose that a random sample of nine hybrid cars showed a sample mean time of 2.1 years. Assume that the population is normal with $\sigma = 0.2$. Test, using level of significance $\alpha = 0.01$, whether the population mean time it takes owners of hybrid cars to recoup their initial cost is less than three years.

58. Americans' Height. Americans used to be, on average, the tallest people in the world. That is no longer the case, according to a study by Dr. Richard Steckel, professor of economics and anthropology at The Ohio State University. The Norwegians and Dutch are now the tallest, at 178 centimeters, followed by the Swedes at 177, and then the Americans, with a mean height of 175 centimeters (approximately 5 feet 9 inches). According to Dr. Steckel, "The average height of Americans has been pretty much stagnant for 25 years."[5] Suppose a random sample of 100 Americans taken this year shows a mean height of 174 centimeters, and we assume $\sigma = 10$ centimeters. Test, using level of significance $\alpha = 0.01$, whether the population mean height of Americans this year has changed from 175 centimeters.

For Exercises 59–66, use the *p*-value from the indicated exercise to assess the strength of evidence against the null hypothesis, using Table 6.

59. Car Insurance. Exercise 51.

60. Mobile Apps. Exercise 52.

61. Eating Trends. Exercise 53.

62. DDT in Breast Milk. Exercise 54.

63. Millennials' Income. Exercise 55.

64. Tree Rings. Exercise 56.

65. Hybrid Vehicles. Exercise 57.

66. Americans' Height. Exercise 58.

67. Advanced Placement Californians. The College Board reports that the mean score on all advanced placement tests taken in California in 2012 was 2.95. Suppose that a random sample of 49 test scores for tests taken by Californians this year is 2.95. Assume the population standard deviation is 0.21.

    a. Construct a 95% *Z* confidence interval for the population mean test score for this year. (*Hint:* See Section 8.1.)

    b. Use the confidence interval to test, at level of significance $\alpha = 0.05$, whether the population mean test score differs from the following amounts:

        i. 2.89   **ii.** 2.90

        iii. 3.00   **iv.** 3.01

68. Engineers' Starting Salary. The National Association of Colleges and Employers reported in 2013 that the college major with the highest mean starting salary was engineering, with $62,600. Suppose that a random sample of 36 engineering starting salaries is $62,600. Assume the population standard deviation is $10,000.

    a. Construct a 99% *Z* confidence interval for the population mean starting salary for engineers. (*Hint:* See Section 8.1.)

    b. Use the confidence interval to test, at level of significance $\alpha = 0.01$, whether the population mean starting salary for engineers differs from the following amounts:

        i. $66,000   **ii.** $58,000

        iii. $67,000   **iv.** $59,000

Health Care Premiums. Use the following information for Exercises 69–71. The National Conference of State Legislatures reports that the mean annual premium for employer-sponsored family health insurance coverage was $16,351 in 2014. A random sample of 100 such families showed a mean annual premium of $17,251. Assume $\sigma = \$5000$.

69. Test whether the population mean annual premium is greater than $16,351, using level of significance $\alpha = 0.05$.

WHAT IF **70.** *What if* the sample mean premium equaled some value larger than $17,251, while everything else stayed the same? Explain how this change would affect the following, if at all:

    a. The hypotheses

    b. Z_{crit}

    c. The critical region

    d. Z_{data}

    e. The conclusion

71. Test whether the population mean annual premium is greater than $16,351 using level of significance $\alpha = 0.01$. Compare your conclusion with the conclusion in Exercise 70. Suggest two possible methods to resolve this contradiction.

Mean Family Size. Use the following information for Exercises 72–74: According to the *Statistical Abstract of the United States,* the mean family size in 2010 was 3.14 persons, reflecting a slow decrease since 1980, when the mean family size was 3.29 persons. Has this trend continued to the present day? Suppose a random sample of 225 families taken this year yields a sample mean size of 3.05 persons, and suppose we assume that the population standard deviation of family sizes is 1 person.

 72. Test whether the population mean family size in America has decreased since 2010, using the *p*-value

method and level of significance $\alpha = 0.05$. (Try using the *p-value* applet to help you solve this problem.)

73. Refer to Exercise 72.

    a. What is the smallest *p*-value for which you will reject H_0?

    b. Which type of error is it possible that we are making—a Type I error or a Type II error? Which type of error are we certain we are not making?

    c. Suppose a newspaper headline referring to the study was "Mean Family Size Decreasing." Is the headline supported or not supported by the data and the hypothesis test?

WHAT IF **74.** Refer to Exercises 72 and 73, *What if* the 3.05 persons had been a typo, and the actual sample mean was 3.00 persons? How would this have affected the following?

    a. Z_{data}

    b. The *p*-value

    c. The conclusion

75. Women's Heart Rates. A random sample of 15 women produced the normal probability plot for their heart rates shown here. The sample mean was 75.6 beats per minute. Suppose the population standard deviation is known to be 9.

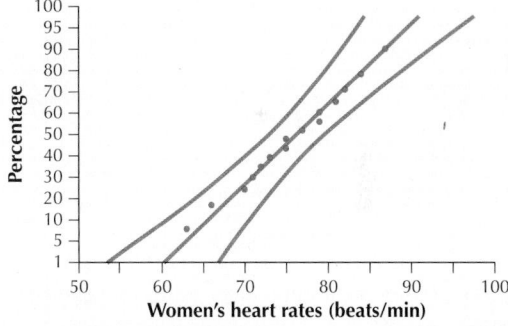

    a. Discuss the evidence for or against the normality assumption. Should we use the *Z* test? Why or why not?

    b. Assume that the plot does not contradict the normality assumption; test whether the population mean heart rate for all women *is less than* 78, using level of significance $\alpha = 0.05$.

    c. Test whether the population mean heart rate for all women *differs from* 78, using $\alpha = 0.05$.

76. Challenge Exercise. Refer to the previous exercise.

    a. Compare your conclusions from Exercises 75(**b**) and 75(**c**). Note that the conclusions differ, but the meanings of the hypotheses tested also differ. Combine the two conclusions into a single sentence. Do you find this sentence difficult to explain?

    b. Explain in your own words the difference between the hypotheses in Exercises 75(**b**) and 75(**c**). Also, explain how there could be evidence that the population mean heart rate is *less than 78 but not different from 78*.

    c. Assess the strength of the evidence against the null hypothesis for the hypothesis tests in Exercises 75(**b**) and 75(**c**).

77. Recognizing Extreme Values of Z_{data}. Try to put your calculator down for this exercise, and use your recognition of the extreme values of the standard normal distribution to solve it. Suppose we want to test whether the population mean test score differs from 70. The level of significance α is undisclosed, but lies somewhere between 0.01 and 0.10. Different samples provided each of the following different values of Z_{data}. Provide conclusions for each.

    a. $Z_{data} = 10$
    b. $Z_{data} = 0.1$
    c. $Z_{data} = -15$
    d. $Z_{data} = 0$

BRINGING IT ALL TOGETHER

Sodium in Breakfast Cereal. Use the following information for Exercises 78–83. A random sample of 23 breakfast cereals containing sodium had a mean sodium content per serving of 192.39 grams. Assume that the population standard deviation equals 50 grams. We are interested in whether the population mean sodium content per serving is less than 210 grams.

78. Based on the normal probability plot of the sodium content in the accompanying figure, should we proceed to apply the Z test? Why or why not?

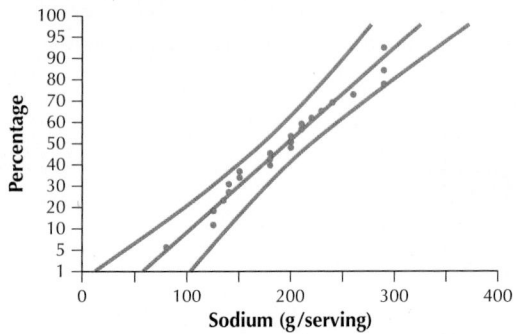

79. Assuming that the normal probability plot shows acceptable normality, we will test whether the population mean sodium content per serving is less than 210 grams, using level of significance $\alpha = 0.01$.

    a. State the hypotheses and the rejection rule. Make sure you state the meaning of μ.
    b. Calculate Z_{data}.
    c. Find the p-value. Draw a normal probability curve indicating Z_{data} and the p-value.
    d. State the conclusion and the interpretation.

80. For the p-value in Exercise 79, assess the strength of evidence against the null hypothesis.

81. Use the equivalence between two-tailed hypothesis tests and confidence intervals to perform a set of hypothesis tests, as follows.

    a. Construct a 95% Z confidence interval for the population mean sodium content. (*Hint:* See Section 8.1.)
    b. Use the confidence interval to test, at level of significance $\alpha = 0.05$, whether the population mean sodium content differs from the following amounts:
       i. 200   **ii.** 215
       iii. 170   **iv.** 160

WHAT IF ? **82.** Refer to the hypothesis test in Exercise 79. *What if* the population standard deviation of 50 grams had been a typo, and the actual population standard

deviation was smaller? How would this have affected the following?

    a. The standard deviation of the sampling distribution
    b. Z_{data}
    c. p-value
    d. The conclusion

WHAT IF ? **83.** Refer to the hypothesis test in Exercise 79. *What if* our level of significance α equaled 0.05 instead of 0.01?

    a. Perform the appropriate hypothesis test using the p-value method, but this time using level of significance $\alpha = 0.05$.
    b. Note that your conclusion differs from that obtained using level of significance $\alpha = 0.01$. Have the data changed? Why did your conclusion change?
    c. Suggest two alternatives for addressing the contradiction between Exercise 79 and Exercise 83(**a**).

WORKING WITH LARGE DATA SETS

Texas Towns. Work with the **Texas** data set for Exercises 84–86. **texas**

84. How many observations are in the data set? How many variables?

85. Use technology to explore the variable *tot_occ*, which lists the total occupied housing units for each county in Texas. Generate numerical summary statistics and graphs for the total occupied housing units. What is the sample mean? The sample standard deviation? Comment on the symmetry or skewness of the data set.

86. Suppose we are using the data in this data set as a sample of the total occupied housing units of all the counties in the southwestern United States, and let $\sigma = 88,400$. Use technology to test, at level of significance $\alpha = 0.05$, whether the population mean total occupied housing units for these counties differs from 40,000.

WORKING WITH LARGE DATA SETS

CASE STUDY **Chapter 9 Case Study: Clothing Store Sales.** Open the Chapter 9 Case Study data set, **Clothing Store**. Here, we will perform a hypothesis test for the population mean number of items purchased per customer. We will then see whether this hypothesis test made the correct decision. Use technology to do the following exercises. **clothingstore**

87. Obtain a random sample of size 100 from the data set.

88. The assistant manager wants to determine whether the mean number of items customers are buying is less than 20. Using your sample, perform the appropriate hypothesis test, using level of significance $\alpha = 0.05$. Assume $\sigma = 25$.

89. Using the same sample, test whether the population mean number of items purchased per customer is less than 18, using level of significance $\alpha = 0.05$. Assume $\sigma = 25$.

90. Find the actual value of the population mean number of items purchased per customer.

    a. Did your hypothesis test in Exercise 89 make the right decision? Explain.
    b. Discuss the decision your hypothesis test in Exercise 89 made, using the concept of "beyond a reasonable doubt."

9.4	*t* Test for the Population Mean

OBJECTIVES By the end of this section, I will be able to . . .

1   Perform the *t* test for the mean using the critical-value method.
2   Perform the *t* test for the mean using the *p*-value method.
3   Use confidence intervals to perform two-tailed hypothesis tests.

1 *t* Test for μ Using the Critical-Value Method

Note: Students may want to review the characteristics of the *t* distribution on page 450 of Chapter 8.

In many real-world scenarios, the value of the population standard deviation *s* is unknown. When this occurs, we should use neither the *Z* interval nor the *Z* test. Recall that in Section 8.2, we used the *t* distribution to find a confidence interval for the mean when σ was not known. The situation is similar for hypothesis testing.

Let $\bar{x}$ be the sample mean, μ be the unknown population mean, *s* be the sample standard deviation, and *n* be the sample size. The *t* statistic

$$t = \frac{\bar{x} - \mu}{s/\sqrt{n}}$$

with $n - 1$ degrees of freedom may be used when either the population is normal or the sample size is large. We call this *t* statistic t_{data} because its value depends largely on the sample data.

The test statistic used for the *t* test for the mean is

$$t_{\text{data}} = \frac{\bar{x} - \mu_0}{s/\sqrt{n}}$$

t_{data} represents the number of standard errors $\bar{x}$ lies above or below μ_0.

The degrees of freedom is a measure of how the *t* distribution changes as the sample size changes.

Extreme values of $\bar{x}$, that is, values of $\bar{x}$ that are significantly far from the hypothesized μ, will translate into extreme values of t_{data}. In other words, just as with Z_{data}, when $\bar{x}$ is far from μ_0, t_{data} will be far from 0. We answer the question "How extreme is extreme?" using the critical-value method by finding a *critical value* of *t*, called t_{crit}. This threshold value t_{crit} separates the values of t_{data} for which we reject H_0 (the *critical region*) from the values of t_{data} for which we will not reject H_0 (the *noncritical region*). Because a different *t* curve exists for every different sample size, you need to know the following to find the value of t_{crit}: (a) the form of the hypothesis test (right-tailed, left-tailed, or two-tailed), (b) the degrees of freedom (df = $n - 1$), and (c) the level of significance α.

---

t Test for the Population Mean μ: Critical-Value Method

When a random sample of size *n* is taken from a population, you can use the *t* test if either the population is normal or the sample size is large ($n \geq 30$).

Step 1   State the hypotheses.
Use one of the forms from Table 8. State the meaning of μ.

Step 2   Find t_{crit} and state the rejection rule.
Use Table D in the Appendix and Table 8.

Step 3   Calculate t_{data}.

$$t_{\text{data}} = \frac{\bar{x} - \mu_0}{s/\sqrt{n}}$$

Step 4   State the conclusion and the interpretation.
If t_{data} falls within the critical region, then reject H_0. Otherwise, do not reject H_0. Interpret your conclusion so that a nonspecialist can understand.

Table 8 contains the critical regions and rejection rules for the t test.

Table 8 Critical regions and rejection rules for various forms of the t test for μ

Form of test	Right-tailed test	Left-tailed test	Two-tailed test
Hypotheses	$H_0: \mu = \mu_0$ $H_a: \mu > \mu_0$ level of significance α	$H_0: \mu = \mu_0$ $H_a: \mu < \mu_0$ level of significance α	$H_0: \mu = \mu_0$ $H_a: \mu \neq \mu_0$ level of significance α
Critical region			
Rejection rule	Reject H_0 if $t_{data} \geq t_{crit}$	Reject H_0 if $t_{data} \leq -t_{crit}$	Reject H_0 if $t_{data} \geq t_{crit}$ or $t_{data} \leq -t_{crit}$

EXAMPLE 18 *t* Test for μ using critical-value method: Left-tailed test

We are interested in testing, using level of significance $\alpha = 0.05$, whether the mean age at onset of anorexia nervosa in young women has been decreasing. Assume that the previous mean age at onset was 15 years old. Data were gathered for a study of the onset age for this disorder.[6] From these data, a random sample (shown here) was taken of $n = 20$ young women who were admitted under this diagnosis to the Toronto Hospital for Sick Children. The Minitab descriptive statistics shown here indicate a sample mean age of $\bar{x} = 14.250$ years and a sample standard deviation of $s = 1.512$ years. If appropriate, perform the t test.

Age at onset of anorexia

14.50	15.75	14.17	14.00
14.67	17.25	11.00	16.00
14.50	15.17	12.00	13.00
13.00	13.50	15.42	16.08
14.00	12.58	13.50	14.92

Descriptive Statistics: Patient Age

Variable	Total Count	Mean	StDev
Patient Age	20	14.250	1.512

Descriptive statistics for anorexia data.

Solution

The sample size $n = 20$ is not large, so we need to verify normality. The normal probability plot of the ages at onset in Figure 18 indicates that the ages in the sample are normally distributed. We may proceed to perform the t test for the mean.

Step 1 **State the hypotheses.**
The key word "decreasing" guides us to state our hypotheses as follows:

$$H_0: \mu = 15 \quad \text{versus} \quad H_a: \mu < 15$$

where μ refers to the population mean age at onset.

Step 2 **Find t_{crit} and state the rejection rule.**
Our hypotheses from Step 1 indicate that we have a left-tailed test, meaning that the critical region represents an area in the left tail (see Figure 20, page 528). To find t_{crit},

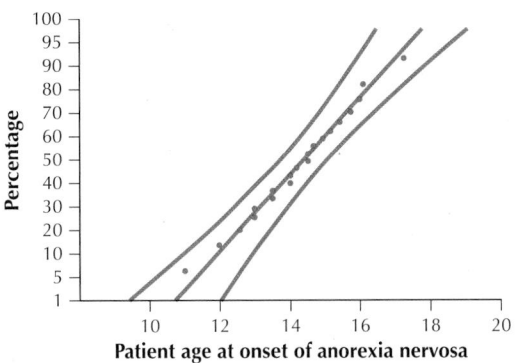

FIGURE 18
Normal probability plot for age at onset of anorexia nervosa.

we turn to the *t* table, an excerpt of which is shown in Figure 19. Because we have a one-tailed test, under "Area in one tail," select the column with our α value 0.05. Then choose the row with our df $= n - 1 = 20 - 1 = 19$, so that we get $t_{crit} = 1.729$. Because we have a left-tailed test, the rejection rule from Table 8 is "Reject H_0 if $t_{data} \leq -t_{crit}$"; that is, we will reject H_0 if $t_{data} \leq -1.729$.

FIGURE 19
Finding t_{crit} for a one-tailed test. For a two-tailed test, use "Area in two tails."

		0.10	0.05	**Area in one tail** 0.025
				Area in two tails
		0.20	0.10	0.05
df	1	3.078	6.314	12.706
	2	1.886	2.920	4.303
	3	1.638	2.353	3.182
	4	1.533	2.132	2.776
	5	1.476	2.015	2.571
	6	1.440	1.943	2.447
	7	1.415	1.895	2.365
	8	1.397	1.860	2.306
	9	1.383	1.833	2.262
	10	1.372	1.812	2.228
	11	1.363	1.796	2.201
	12	1.356	1.782	2.179
	13	1.350	1.771	2.160
	14	1.345	1.761	2.145
	15	1.341	1.753	2.131
	16	1.337	1.746	2.120
	17	1.333	1.740	2.110
	18	1.330	1.734	2.101
	19	1.328	**1.729**	2.093
	20	1.325	1.725	2.086

Step 3   **Calculate t_{data}.**
We have $n = 20$, $\bar{x} = 14.250$, and $s = 1.512$ years. Also, $\mu_0 = 15$, because this is the hypothesized value of μ stated in H_0. Therefore, our test statistic is

$$t_{data} = \frac{\bar{x} - \mu_0}{s/\sqrt{n}} = \frac{14.250 - 15}{1.512/\sqrt{20}} \approx -2.2183$$

Step 4   **State the conclusion and interpretation.**
The rejection rule from Step 2 says to reject H_0 if $t_{data} \leq -1.729$. From Step 3, we have $t_{data} = -2.2183$. Because -2.2183 is less than -1.729, our conclusion is to

reject H_0. If you prefer the graphical approach, consider Figure 20, which shows where t_{data} falls in relation to the critical region. Because $t_{data} = -2.2183$ falls within the critical region, our conclusion is to reject H_0. There is evidence at level of significance $\alpha = 0.05$ that the population mean age of onset has decreased from its previous level of 15 years.

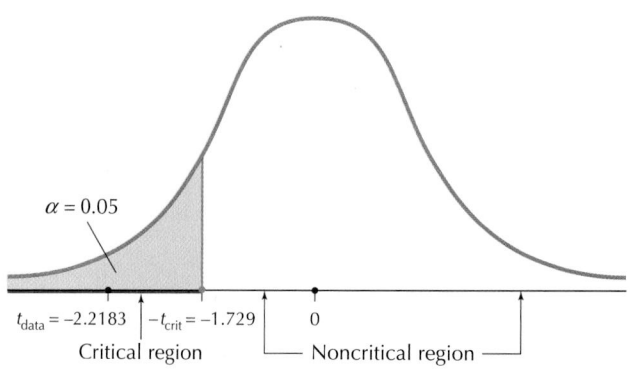

FIGURE 20 Our $t_{data} = -2.2183$ falls in the critical region.

NOW YOU CAN DO
Exercises 3–8.

a. For the data in Example 18, test, using level of significance $\alpha = 0.01$, whether the mean age at onset of anorexia nervosa in young women has been decreasing.

b. Discuss two possible resolutions to the contradiction between the conclusions in Example 18 and Part **(a)**.

(The solutions are shown in Appendix A.)

EXAMPLE 19   *t* Test for μ using critical-value method: Two-tailed test

CNN reported in 2014 that, even after factoring in part-time jobs, Americans worked an average of 38 hours per week. Suppose a social science researcher disputes this finding and is interested in testing whether the population mean number of hours worked per week differs from 38. A random sample of $n = 30$ working Americans yields a sample mean of $\bar{x} = 35.26$ hours worked, with a sample standard deviation of $s = 10$ hours. If the conditions are met, perform the appropriate hypothesis test using level of significance $\alpha = 0.10$.

Solution

Because $n = 30 \geq 30$, we may proceed with the *t* test.

Step 1   **State the hypotheses.**
The key words "differs from" indicate a two-tailed test, with $\mu_0 = 38$, because we are testing whether μ differs from 38. So our hypotheses are

$$H_0 : \mu = 38 \quad \text{versus} \quad H_a : \mu \neq 38$$

where μ represents the population mean number of hours worked per week.

Step 2   **Find t_{crit} and state the rejection rule.**
To find t_{crit} for a two-tailed test with level of significance $\alpha = 0.10$, we look in the 0.10 column in the "Area in two tails" section of Table D in the Appendix. The degrees of freedom df $= n - 1 = 29$ gives us $t_{crit} = 1.699$. From Table 8, the rejection rule is: "Reject H_0 if $t_{data} \geq 1.699$ or $t_{data} \leq -1.699$."

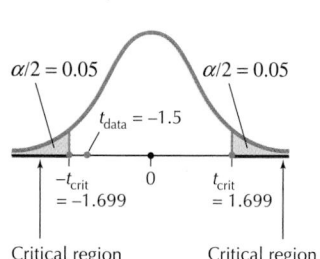

FIGURE 21 Critical region for two-tailed test.

Step 3 **Calculate t_{data}:**

$$t_{\text{data}} = \frac{\bar{x} - \mu_0}{s/\sqrt{n}} = \frac{35.26 - 38}{10/\sqrt{30}} \approx -1.5$$

Step 4 **State the conclusion and the interpretation.**

$t_{\text{data}} = -1.5$ is not ≥ 1.699 and it is not ≤ -1.699; therefore, we do not reject H_0. See Figure 21. There is insufficient evidence, at level of significance $\alpha = 0.10$, that the population mean number of hours worked per week differs from 38.

NOW YOU CAN DO
Exercises 9–14.

2 *t* Test for μ Using the *p*-Value Method

We may also use the *p*-value method for performing the *t* test for μ. The critical-value method and the *p*-value are equivalent, so they will provide identical conclusions.

---

t Test for the Population Mean μ: *p*-Value Method

When a random sample of size n is taken from a population, you can use the *t* test if either the population is normal or the sample size is large ($n \geq 30$).

Step 1 State the hypotheses and the rejection rule.

Use one of the forms from Table 9. State the meaning of μ. The rejection rule is "Reject H_0 if the *p*-value $\leq \alpha$."

Step 2 Calculate t_{data}.

$$t_{\text{data}} = \frac{\bar{x} - \mu_0}{s/\sqrt{n}}$$

Step 3 Find the *p*-value.

Either use technology to find the *p*-value or estimate the *p*-value using Table D, *t* Distribution, in the Appendix.

Step 4 State the conclusion and the interpretation.

If the *p*-value $\leq \alpha$, then reject H_0. Otherwise, do not reject H_0. Interpret your conclusion.

The definition of a *p*-value for a *t* test is similar to the *p*-value for a *Z* test. Unusual and extreme values of $\bar{x}$, and therefore of t_{data}, will have a small *p*-value, whereas values of $\bar{x}$ and t_{data} nearer to the center of the distribution will have a large *p*-value. Table 9 summarizes the definition of the *p*-value for *t* tests. Note that we will not be finding these *p*-values manually but will either (**a**) use a computer or calculator or (**b**) estimate them using the *t* table.

Table 9 *p*-Values for *t* tests

Form of test	Right-tailed test	Left-tailed test	Two-tailed test
Hypotheses	$H_0: \mu = \mu_0$ $H_a: \mu > \mu_0$ level of significance α	$H_0: \mu = \mu_0$ $H_a: \mu < \mu_0$ level of significance α	$H_0: \mu = \mu_0$ $H_a: \mu \neq \mu_0$ level of significance α
p-Value is tail area associated with t_{data}	$p\text{-value} = P(t > t_{\text{data}})$ Area to the right of t_{data}	$p\text{-value} = P(t < t_{\text{data}})$ Area to the left of t_{data}	$p\text{-value} = P(t > |t_{\text{data}}|) + P(t < -|t_{\text{data}}|)$ $= 2 \cdot P(t > |t_{\text{data}}|)$ Sum of the two tail areas

FIGURE 26 Mona Lisa's face
follows the golden ratio.

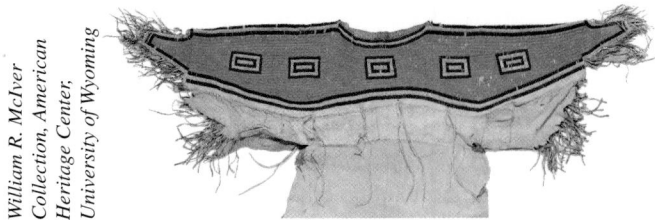

FIGURE 27 Beaded dress of Nahtoma, daughter of Chief Washakie of the Eastern Shoshone,
showing rectangles that may follow the golden ratio.

to consider whether the Shoshone beaded rectangles, such as those on this dress, follow the golden ratio. Table 10 contains the ratios of lengths to widths of 18 beaded
rectangles made by Shoshone artisans.[8] We will perform a hypothesis test to determine whether the population mean ratio of Shoshone beaded rectangles equals the
golden ratio of 1.618.

Table 10 Ratio of length to width of a sample of Shoshone beaded rectangles

1.44	1.75	1.64	1.66	1.64
1.51	1.34	1.53	1.74	1.81
1.45	1.49	1.63	1.49	
1.65	1.59	1.50	1.65	

Solution

The population standard deviation for such rectangles is unknown, so we must use a
t test instead of a Z test. Our sample size $n = 18$ is not large, so we must assess whether
the data are normally distributed. Figure 28 shows the normal probability plot, indicating acceptable support for the normality assumption. We proceed with the t test, using
level of significance $\alpha = 0.05$.

FIGURE 28
Normal probability plot.

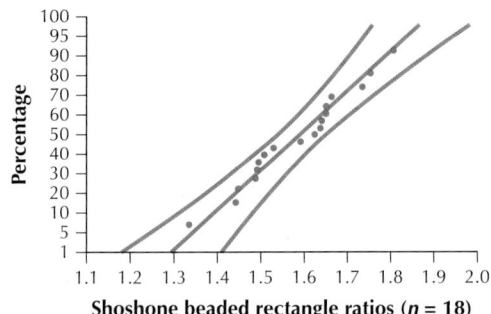

We use the TI-83/84 and CrunchIt! to perform this hypothesis test, using the Step-by-
Step Technology Guide at the end of this section.

Step 1   **State the hypotheses and the rejection rule.**
We are interested in whether the population mean length-to-width ratio of Shoshone
beaded rectangles *equals* the golden ratio of 1.618, so we perform a two-tailed test:

$$H_0 : \mu = 1.618 \quad \text{versus} \quad H_a : \mu \neq 1.618$$

where μ represents the population mean length-to-width ratio of Shoshone beaded
rectangles. We will reject H_0 if the p-value ≤ 0.05.

Step 2   **Find t_{data}.**
Using the statistics from Figure 29a, we have the test statistic

$$t_{data} = \frac{\bar{x} - \mu_0}{s/\sqrt{n}} = \frac{1.583888889 - 1.618}{0.1230083158/\sqrt{18}} \approx -1.176515482 \approx -1.1765$$

Step 3   **Find the *p*-value.**
From Figures 29a, 29b, and 30, we have

$$p\text{-value} = P(t > |-1.1765|) + P(t < -|-1.1765|) \approx 0.2556$$

Step 4   **State the conclusion and interpretation.**
Because *p*-value ≈ 0.2556 is *not* $\leq \alpha = 0.05$, we do *not* reject H_0. Thus, there is insufficient evidence, at level of significance $\alpha = 0.05$, that the population mean ratio differs from 1.618. In other words, the data do not reject the claim that Shoshone beaded rectangles follow the same golden ratio exhibited by the Parthenon and the *Mona Lisa*.

NOW YOU CAN DO
Exercises 21–26.

Null hypothesis:	Population mean = 1.618	
Alternative hypothesis:	Population mean is not 1.618	
n:	18	
Sample Mean:	1.584	
Standard Error:	0.02899	
df:	17	
t statistic:	-1.177	
P-value:	0.2556	

Form of H_a: ——→
t_{data} ——→
p-value ——→
Sample mean $\bar{x}$ ——→
Sample standard deviation *s* ——→
Sample size *n* ——→

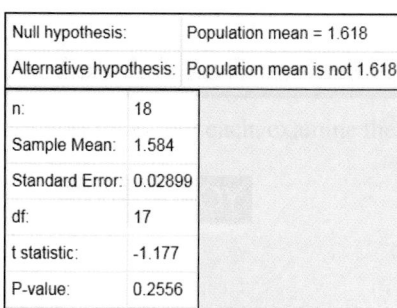

```
T-Test
μ≠1.618
t=-1.176515482
p=.255601045
x̄=1.583888889
Sx=.1230083158
n=18
```

FIGURE 29a TI-83/84 results.

FIGURE 29b CrunchIt! results.

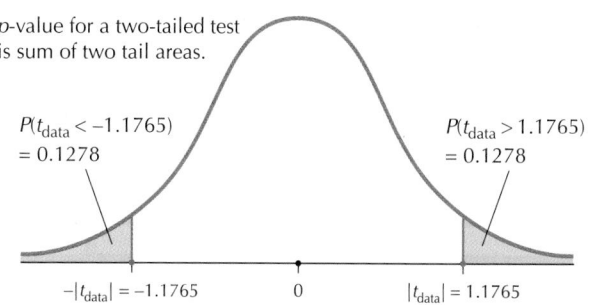

p-value for a two-tailed test
is sum of two tail areas.

$P(t_{data} < -1.1765)$
$= 0.1278$

$P(t_{data} > 1.1765)$
$= 0.1278$

$-|t_{data}| = -1.1765$ 0 $|t_{data}| = 1.1765$

FIGURE 30 *p*-Value for two-tailed *t* test is sum of two tail areas.

EXAMPLE 22   Estimating the *p*-value using the *t* table

Suppose we did not have access to technology. Estimate the *p*-value from Example 19 using the *t* table (Appendix Table D). For Example 19, our hypotheses are

$$H_0 : \mu = 38 \quad \text{versus} \quad H_a : \mu \neq 38$$

where μ represents the population mean number of hours worked per week. Our test statistic is $t_{data} = -1.5$.

Solution

For a two-tailed test, choose the row of the *t* table with the heading "Area in two tails." Then select the row in the table with the appropriate degrees of freedom, in this case

Solution

a.   Interpreting the SPSS output.

> Step 1   **State the hypotheses and the rejection rule.**
> In the SPSS output, the "Test Value = 40" indicates that $\mu_0 = 40$. Also, the "2-tailed" in the output indicates that we have a two-tailed test. Thus, our hypotheses are:

$$H_0 : \mu = 40 \quad \text{versus} \quad H_a : \mu \neq 40$$

> where μ represents the population mean number of orchard farms per county, nationwide. We will reject H_0 if the p-value is less than level of significance $\alpha = 0.10$.

> Step 2   **Find t_{data}.**
> Under the "t" in the SPSS is the value for t_{data}, -1.079.

> Step 3   **Find the p-value.**
> The abbreviation "Sig." stands for "Significance," which represents the p-value: 0.281.

> Step 4   **State the conclusion and the interpretation.**
> The p-value of 0.281 is not less than the level of significance $\alpha = 0.10$, so we do not reject H_0. There is insufficient evidence that the population mean number of orchard farms per county differs from 40.

b.   Interpreting the JMP output.

> Step 1   **State the hypotheses and the rejection rule.**
> In the JMP output, the "Hypothesized Value" indicates that $\mu_0 = 20$. Now, JMP is unusual in that it performs all three types of hypothesis test simultaneously: two-tailed, right-tailed, and left-tailed, as shown in the JMP output. It does not specify a particular form of the test. Let us use the right-tailed test for this example. Thus, our hypotheses are:

$$H_0 : \mu = 20 \quad \text{vs} \quad H_a : \mu > 20$$

> where μ represents the population mean number of grocery stores per county, nationwide. We will reject H_0 if the p-value is less than level of significance $\alpha = 0.10$.

> Step 2   **Find t_{data}.**
> Next to "Test Statistic" in the JMP output, we find the value of our test statistic, t_{data}, 0.3324.

> Step 3   **Find the p-value.**
> Here, we need to be careful, because JMP gives us three different p-values, depending on which form of the hypothesis test is performed. We chose the right-tailed test, so our p-value is next to "Prob > t": p-value = 0.3698, as indicated in the JMP output.

> Step 4   **State the conclusion and the interpretation.**
> The p-value of 0.3698 is not less than the level of significance $\alpha = 0.10$, so we do not reject H_0. There is insufficient evidence that the population mean number of grocery stores per county is greater than 20.

NOW YOU CAN DO
Exercises 37–40.

STEP-BY-STEP TECHNOLOGY GUIDE: *t* test for μ

We will use the golden ratio data from Example 21 (page 531).

TI-83/84

If you have the data values:
Step 1 Enter the data into list **L1**.
Step 2 Press **STAT**, highlight **TESTS**.
Step 3 Press **2** (for **T-Test**; see Figure 34a).
Step 4 For input (**Inpt**), highlight **Data** and press **ENTER** (Figure 34b).
a. For μ_0, enter the value of μ_0, 1.618.
b. For **List**, press **2nd**, then **L1**.
c. For **Freq**, enter 1.
d. For μ, select the form of H_a. Here, we have a two-tailed test, so highlight $\neq \mu_0$ and press **ENTER** (Figure 34b).
e. Highlight **Calculate** and press **ENTER**. The results are shown in Figure 29a in Example 21.

If you have the summary statistics:
Step 1 Press **STAT**, highlight **TESTS**.
Step 2 Press **2** (for **T-Test**; see Figure 34a).
Step 3 For input (**Inpt**), highlight **Stats** and press **ENTER** (Figure 34c).
a. For μ_0, enter the value of μ_0, 1.618.
b. For $\bar{x}$, enter the sample mean 1.583888889.
c. For **Sx**, enter the value of s, 0.1230083158.
d. For **n**, enter the sample size 18.
e. For μ, select the form of H_a. Here, we have a two-tailed test, so highlight $\neq \mu_0$ and press **ENTER** (Figure 34c).
f. Highlight **Calculate** and press **ENTER**. The results are shown in Figure 29a in Example 21.

FIGURE 34a

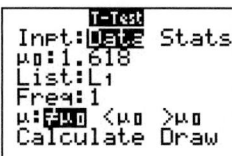

FIGURE 34b

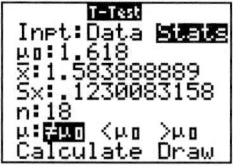

FIGURE 34c

EXCEL

Use the JMP add-in for Excel.

MINITAB

If you have the data values:
Step 1 Enter the data into column **C1**.
Step 2 Click **Stat > Basic Statistics > 1-Sample t…**.
Step 3 Select **One or more samples, each in a column** from the drop-down menu, click inside the box under the drop-down menu, and select **C1**.
Step 4 Select **Perform hypothesis test**. For **Hypothesized mean**, enter **1.618**.
Step 5 Click **Options…**
a. Choose your **Confidence Level** as $100(1 - \alpha)$. Our level of significance α here is 0.05, so the confidence level is 95.0.
b. Select **Mean ≠ hypothesized mean** for the **Alternative hypothesis**.
Step 6 Click **OK** and click **OK** again.

If you have the summary statistics:
Step 1 Click **Stat > Basic Statistics > 1-Sample t**.
Step 2 Select **Summarized Data** from the drop-down menu.
Step 3 Enter the **Sample size 18**, the **Sample mean 1.58**, and the **Standard deviation 0.123**. Check **Perform hypothesis test**, and enter a **Hypothesized mean** of **1.618**.
Step 4 Click **Options…**
a. Choose your **Confidence Level** as $100(1 - \alpha)$. Our level of significance α here is 0.05, so the confidence level is 95.0.
b. Select **Mean ≠ hypothesized mean** for the **Alternative hypothesis**.
Step 5 Click **OK** and click **OK** again.

SPSS

If you have the data values:
The following is for two-tailed tests only.
Step 1 Enter the data in the first column.
Step 2 Select **Analyze > Compare Means > One-sample T Test**. Move the variable to the **Test Variable(s)** box.

Step 3 Enter a **Test Value** of 1.618. Click **OK**. The *p*-value is found in the **Sig. (2-tailed)** box in the **One-Sample Test** table.

JMP

If you have the data values:

Step 1   Enter the data in the first column.
Step 2   Select **Analyze > Distribution**. Move the variable to the **Y, Columns** box. Click **OK**.
Step 3   Click the red triangle beside the variable name. Select **Test Mean**. For **Specify Hypothesized Mean**, enter 1.618. Click **OK**. The output is shown in Figure 34d.

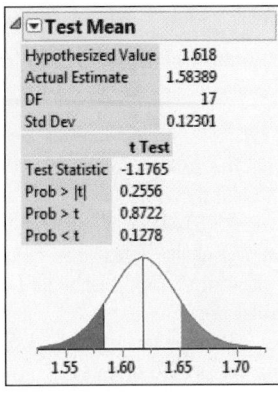

FIGURE 34d
JMP results.

CRUNCHIT!

If you have the data values:

Step 1   Click **File**, then highlight **Load from Larose, Discostat3e > Chapter 9** and click on **Example 04_20**.
Step 2   Click **Statistics**, highlight **t**, and select **1-sample**. With the **Columns** tab chosen, for **Sample** select **Ratio**.
Step 3   Select the **Hypothesis Test** tab. For **Mean under null hypothesis**, enter 1.618. For **Alternative**, select **Two-sided**. Then click **Calculate**. Results are in Figure 29b in Example 21.

If you have the summary statistics:

Step 1   Click **Statistics**, highlight **t**, and select **1-sample**.
Step 2   Choose the **Summarized** tab. For **n**, enter the sample size 18; for **Sample Mean** enter 1.583. For **Standard Deviation**, enter 0.123.
Step 3   Select the **Hypothesis Test** tab. For **Mean under null hypothesis**, enter 1.618. For **Alternative**, select **Two-sided**. Then click **Calculate**.

Section 9.4 Summary

1. The test statistic used for the t test for the mean is

$$t_{\text{data}} = \frac{\bar{x} - \mu_0}{s/\sqrt{n}}$$

with $n - 1$ degrees of freedom. The t test may be used under either of the following conditions: **(a)** the population is normal, or **(b)** the sample size is large ($n \geq 30$). For the critical-value method, we compare the values of t_{data} and t_{crit}. If t_{data} falls in the critical region, we reject H_0.

2. For the p-value method, we reject H_0 if the p-value $\leq \alpha$.

3. We may use $100(1 - \alpha)\%$ t confidence interval to perform two-tailed t tests at level of significance α for various values of μ_0.

Section 9.4 Exercises

CLARIFYING THE CONCEPTS

1. What assumption is required for performing the Z test that is not required for the t test? (p. 525)
2. What do we use to estimate the unknown population standard deviation σ? (p. 525)

PRACTICING THE TECHNIQUES

 **CHECK IT OUT!**

To do	Check out	Topic
Exercises 3–8	Example 18	One-tailed t test for μ: critical value method
Exercises 9–14	Example 19	Two-tailed t test for μ: critical value method
Exercises 15–20	Example 20	One-tailed t test for μ: p-value value method

To do	Check out	Topic
Exercises 21–26	Example 21	Two-tailed t test for μ: p-value method
Exercises 27–30	Example 22	Estimating the p-value for a t test for μ
Exercises 31–36	Example 23	Equivalence between confidence intervals and two-tailed t tests for μ
Exercises 37–40	Example 24	Interpreting software output

For Exercises 3–14, do the following:

a. State the hypotheses.
b. Calculate the t critical value t_{crit} and state the rejection rule. Also, sketch the critical region.
c. Find the test statistic t_{data}.
d. State the conclusion and the interpretation.

3. $H_0: \mu = 22$ vs. $H_a: \mu < 22, \bar{x} = 20, s = 4, n = 31,$ $\alpha = 0.05$

4. $H_0: \mu = 3$ vs. $H_a: \mu < 3, \bar{x} = 2, s = 1, n = 41,$ $\alpha = 0.10$

5. $H_0: \mu = 11$ vs. $H_a: \mu > 11, \bar{x} = 12, s = 3, n = 16,$ $\alpha = 0.01$, population is normal

6. $H_0: \mu = 80$ vs. $H_a: \mu > 80, \bar{x} = 82, s = 5, n = 9,$ $\alpha = 0.05$, population is normal

7. A random sample of size 25 from a normal population yields $\bar{x} = 104$ and $s = 10$. Researchers are interested in finding whether the population mean exceeds 100, using level of significance $\alpha = 0.01$.

8. A random sample of size 100 from a population with an unknown distribution yields a sample mean of -5 and a sample standard deviation of 5. Researchers are interested in finding whether the population mean is less than -4, using level of significance $\alpha = 0.05$.

9. $H_0: \mu = 102$ vs. $H_a: \mu \neq 102, \bar{x} = 106, s = 10, n = 81,$ $\alpha = 0.05$

10. $H_0: \mu = 95$ vs. $H_a: \mu \neq 95, \bar{x} = 99, s = 10, n = 31,$ $\alpha = 0.01$

11. $H_0: \mu = 1000$ vs. $H_a: \mu \neq 1000, \bar{x} = 975, s = 100,$ $n = 25, \alpha = 0.10$, population is normal

12. $H_0: \mu = -10$ vs. $H_a: \mu \neq -10, \bar{x} = -8, s = 5, n = 25,$ $\alpha = 0.05$, population is normal

13. A random sample of size 36 from a population with an unknown distribution yields $\bar{x} = 10$ and $s = 3$. Researchers are interested in finding whether the population mean differs from 9, using level of significance $\alpha = 0.10$.

14. A random sample of size 16 from a normal population yields $\bar{x} = 995$ and $s = 15$. Researchers are interested in finding whether the population mean differs from 1000, using level of significance $\alpha = 0.01$.

For Exercises 15–26, do the following:

    a. State the hypotheses and the rejection rule using the *p*-value method.

    b. Calculate the test statistic t_{data}.

    c. Find the *p*-value. (Use technology or estimate the *p*-value.)

    d. State the conclusion and the interpretation.

15. $H_0: \mu = 10$ vs. $H_a: \mu < 10, \bar{x} = 7, s = 5, n = 81,$ $\alpha = 0.01$

16. $H_0: \mu = 50$ vs. $H_a: \mu < 50, \bar{x} = 42, s = 8, n = 41,$ $\alpha = 0.05$

17. $H_0: \mu = 100$ vs. $H_a: \mu > 100, \bar{x} = 120, s = 50, n = 25,$ $\alpha = 0.10$, population is normal

18. $H_0: \mu = 3.0$ vs. $H_a: \mu > 3.0, \bar{x} = 3.2, s = 0.5,$ $n = 25, \alpha = 0.05$, population is normal

19. A random sample of size 400 from a population with an unknown distribution yields a sample mean of 230 and a sample standard deviation of 5. Researchers are interested in finding whether the population mean is greater than 200, using level of significance $\alpha = 0.05$.

20. A random sample of size 100 from a population with an unknown distribution yields $\bar{x} = 27$ and $s = 10$. Researchers are interested in finding whether the

population mean is less than 28, using level of significance $\alpha = 0.05$.

21. $H_0: \mu = 25$ vs. $H_a: \mu \neq 25, \bar{x} = 25, s = 1, n = 31,$ $\alpha = 0.01$

22. $H_0: \mu = 98.6$ vs. $H_a: \mu \neq 98.6, \bar{x} = 99, s = 10, n = 81,$ $\alpha = 0.05$

23. $H_0: \mu = 3.14$ vs. $H_a: \mu \neq 3.14, \bar{x} = 3.17, s = 0.5,$ $n = 9, \alpha = 0.10$, population is normal

24. $H_0: \mu = 2.72$ vs. $H_a: \mu \neq 2.72, \bar{x} = 2.57, s = 0.1,$ $n = 25, \alpha = 0.05$, population is normal

25. A random sample of size 9 from a normal population yields $\bar{x} = 1$ and $s = 0.5$. Researchers are interested in finding whether the population mean differs from 0, using level of significance $\alpha = 0.05$.

26. A random sample of size 16 from a normal population yields $\bar{x} = 2.2$ and $s = 0.3$. Researchers are interested in finding whether the population mean differs from 2.0, using level of significance $\alpha = 0.01$.

For Exercises 27–30, use the *t* table to estimate the *p*-value for the hypothesis tests in the indicated exercises.

27. Exercise 3

28. Exercise 4

29. Exercise 9

30. Exercise 10

For Exercises 31–36, a $100(1 - \alpha)\%$ *t* confidence interval is given. Use the confidence interval to test, using level of significance α, whether μ differs from each of the indicated hypothesized values.

31. A 95% *t* confidence interval for μ is (1, 4). Hypothesized values μ_0 are

    a. 0   **b.** 2   **c.** 5

32. A 99% *t* confidence interval for μ is (57, 58). Hypothesized values μ_0 are

    a. 55.5   **b.** 59.5   **c.** 57.5

33. A 90% *t* confidence interval for μ is $(-20, -10)$. Hypothesized values μ_0 are

    a. -21 **b.** -5 **c.** -12

34. A 95% *t* confidence interval for μ is (2010, 2015). Hypothesized values μ_0 are

    a. 2012   **b.** 2007   **c.** 2014

35. A 95% *t* confidence interval for μ is $(-1, 1)$. Hypothesized values μ_0 are

    a. 1.5   **b.** -1.5 **c.** 0

36. A 95% *t* confidence interval for μ is (19,570, 20,105). Hypothesized values μ_0 are

    a. 20,000   **b.** 21,000   **c.** 19,571

For Exercises 37–40, software output from a *t* test for μ is provided. For each, examine the indicated software output, and provide the following steps:

Step 1   State the hypotheses and the rejection rule.

Step 2   Find Z_{data}.

Step 3   Find the *p*-value.

Step 4   State the conclusion and interpretation.

Use level of significance $\alpha = 0.05$ for each hypothesis test.

37. TI-83/84 output

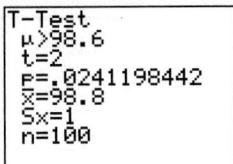

```
T-Test
 μ>98.6
 t=2
 p=.0241198442
 x̄=98.8
 Sx=1
 n=100
```

38. Minitab output

```
Test of μ = 46 vs < 46

Variable            N    Mean  StDev  SE Mean  95% Upper Bound    T      P
Farms with stands  3079  44.43  57.09    1.03            46.12  -1.52  0.064
```

39. SPSS output

One-Sample Test						
				Test Value = 20		
					95% Confidence Interval of the Difference	
	t	df	Sig. (2-tailed)	Mean Difference	Lower	Upper
Veggie_Farms	2.308	3177	.021	1.76432	.2657	3.2629

40. JMP output: Choose the left-tailed test.

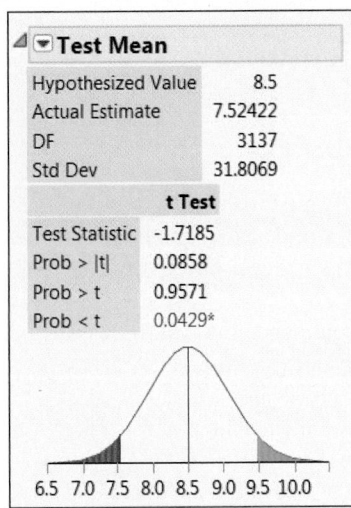

Test Mean	
Hypothesized Value	8.5
Actual Estimate	7.52422
DF	3137
Std Dev	31.8069
t Test	
Test Statistic	-1.7185
Prob > |t|	0.0858
Prob > t	0.9571
Prob < t	0.0429*

6.5 7.0 7.5 8.0 8.5 9.0 9.5 10.0

APPLYING THE CONCEPTS

41. Health Care Costs. The U.S. Agency for Healthcare Research and Quality (www.ahrq.gov) reports that, in 2010, the mean cost of a stay in the hospital for American women ages 18–44 was $15,200. A recent random sample of 400 hospital stays of women ages 18–44 showed a mean cost of $16,000, with a standard deviation of $5000. Test whether the population mean cost has increased since 2010, using level of significance $\alpha = 0.05$.

42. iPhone Apps. According to a 2010 Nielsen survey,[9] the mean number of apps downloaded by iPhone users is 40. Suppose a recent sample of 36 iPhone users downloaded an average of 45 apps, with a standard deviation of 24. Test whether the population mean number of apps is greater than 40, using level of significance $\alpha = 0.10$.

43. Facebook Friends. According to Facebook.com, the mean number of Facebook friends is 130. Suppose a sample of 100 Facebook users has a mean number of 110 Facebook friends, with a standard deviation of 50. Test whether the population mean number of Facebook friends is less than the reported 130, using level of significance $\alpha = 0.05$.

44. Small Business Employees. The U.S. Census Bureau reports that the average number of employees in a small business is 16.1. Suppose a sample of 49 small businesses showed a mean of 15 employees, with a standard deviation of 25. Test whether the population mean number of employees in a small business is different from the reported 16.1, using level of significance $\alpha = 0.01$.

Internet Response Times. Use the following information for Exercises 45 and 46: The Web site www.internettrafficreport.com monitors Internet traffic worldwide and reports on the response times of randomly selected servers.

45. On June 6, 2011, the Web site reported the following response times to Asia, in milliseconds:

165 175 2221 872 311 127 195 1801 769 225 261 249 421

We want to test whether the population mean response time is slower than 180 milliseconds, using a t test and level of significance $\alpha = 0.05$. A boxplot of the data is provided.

(*Hint:* The boxplot is right-skewed and the normal distribution is symmetric.) Can we proceed with the t test? Explain.

46. On June 6, 2011, the Web site reported the following response times to Asia, in milliseconds:

61 32 50 73 51 42 55 65 59 57 76 77 67 71

The normal probability plot of the data is also shown. We want to perform a t test.

   a. Are the conditions for performing the t test satisfied? Explain how you know.

   b. Test, using level of significance $\alpha = 0.05$, whether the population mean response time is less than 60 milliseconds.

   c. Explain why we can't use a Z test for this problem.

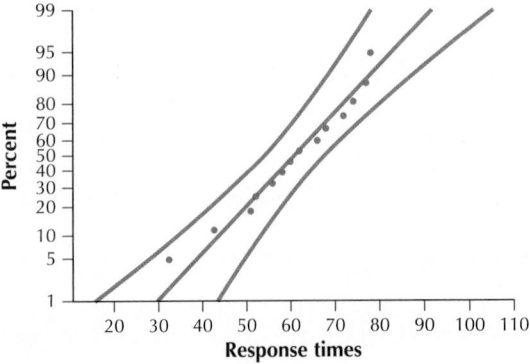

Deepwater Horizon **Cleanup Costs.** The following table represents the amount of money disbursed by BP to a random sample of six Florida counties, for cleanup of the *Deepwater Horizon* oil spill, in millions of dollars.[10] Use the following information for Exercises 47–49.

deepwaterclean

County	Cleanup costs ($ millions)
Broward	0.85
Escambia	0.70
Franklin	0.50
Pinellas	1.15
Santa Rosa	0.50
Walton	1.35

47. The normality of the data was confirmed in the Section 8.1 exercises. Test, at level of significance $\alpha = 0.10$, whether the population mean amount of cleanup money exceeds $500,000.

48. Answer the following:
   a. Repeat your test from Exercise 47, this time using level of significance $\alpha = 0.01$.
   b. How do you think we should resolve the apparent contradiction in Exercise 47 and part **(a)** of this exercise?
   c. Assess the strength of the evidence against the null hypothesis. Does this change depending on which level of α you use?

WHAT IF **49.** *What if* we changed μ_0 to some larger value (though still smaller than $\bar{x}$)? Otherwise, everything else remains unchanged. Describe how this change would affect the following, if at all:
   a. t_{data}
   b. t_{crit}
   c. The *p*-value
   d. The conclusion from Exercise 47
   e. The conclusion from Exercise 48(**a**)
   f. The strength of the evidence against the null hypothesis

50. Wii Game Sales. The following table represents the number of units sold in the United States for the week ending March 26, 2011, for a random sample of eight Wii games.[11] The normality of the data was confirmed in the Section 8.1 exercises.
   a. Construct and interpret a 95% *t* interval for the population mean number of units sold. (See Section 8.2.)
   b. Use your confidence interval to test, using level of significance $\alpha = 0.05$, whether μ differs from the following values:
      i.   30,000 units
      ii. 31,000 units
      iii. 0 units
      iv. 79,000 units   wiisales

Game	Units (1000s)	Game	Units (1000s)
Wii Sports Resort	65	Zumba Fitness	56
Super Mario All Stars	40	Wii Fit Plus	36
Just Dance 2	74	Michael Jackson	42
New Super Mario Bros.	16	Lego Star Wars	110

51. A Rainy Month in Georgia? The following table represents the total rainfall (in inches) for the month of February 2011 for a random sample of 10 locations in Georgia.[12] The normality was checked in the Section 8.1 exercises. Test whether the population mean amount of rainfall differs from 4 inches, using level of significance $\alpha = 0.10$.   georgiarain

Location	Rainfall (inches)	Location	Rainfall (inches)
Athens	4.72	Atlanta	4.25
Augusta	4.31	Cartersville	3.03
Dekalb	2.96	Fulton	4.36
Gainesville	4.06	Lafayette	3.75
Marietta	3.20	Rome	3.26

52. Electric Cars. The accompanying table shows the miles-per-gallon equivalent (MPGe) for five electric cars, as reported by www.hybridcars.com in 2014. Assume the data are drawn from a normal distribution. Test whether the population mean mileage is greater than 90 MPGe, using level of significance $\alpha = 0.10$.   electricmiles

Electric Vehicle	Mileage (MPGe)
Tesla Model S	89
Nissan Leaf	99
Ford Focus	105
Mitsubishi i-MiEV	112
Chevrolet Spark	119

BRINGING IT ALL TOGETHER

Community College Tuition. Use the following information for Exercises 53–63. The College Board reported that the mean tuition and fees at community colleges nationwide for the 2013–2014 academic year was $3264. Data were gathered on the total tuition and fees for a random sample of 10 community colleges in 2015. The normal probability plot is shown here.

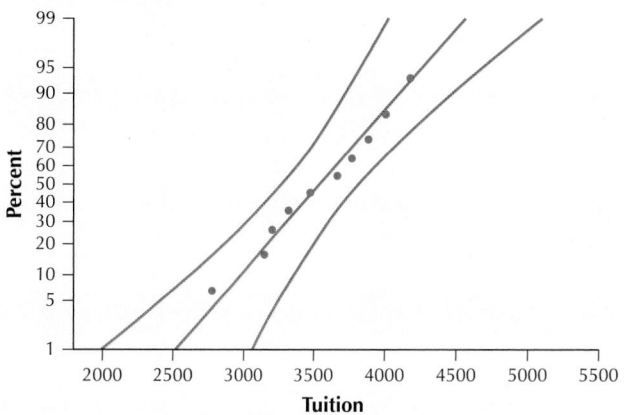

Normal probability plot.

53. Is it appropriate to apply the *t* test for the mean? Why or why not?

54. Find the *t* critical value for a right-tailed hypothesis test using level of significance $\alpha = 0.05$.

55. For the sample of 10 community colleges, the sample mean tuition and fees was $3541 with a sample standard deviation of $438. Test whether the population mean tuition and fees have increased using level of significance $\alpha = 0.05$.

56. Estimate the *p*-value for the hypothesis test in the previous exercise.

57. Assess the strength of evidence against the null hypothesis.

58. A 95% confidence interval for the population mean tuition and fees is given by lower bound = $3227, upper bound = $3854. Use this confidence interval to test, at level of significance $\alpha = 0.05$, whether μ differs from the following values.
    a. $4000
    b. $3500
    c. $3264

59. A data analyst, in attempting to use technology to test whether the population mean tuition and fees have increased, obtained the following Minitab output. However, it appears that the analyst asked for the wrong hypothesis test. How can you tell?

```
One-Sample T: Tuition

Test of μ = 3264 vs ≠ 3264


Variable   N   Mean  StDev  SE Mean     95% CI        T      P
Tuition    10  3541    438      139  (3227, 3854)  2.00  0.077
```

Minitab *t* test output.

60. How can we use the *p*-value on the Minitab printout to find the *p*-value needed for the right-tailed hypothesis test we performed in Exercise 55?

61. Suppose, when we did Exercise 55, we also asked for the wrong hypothesis test, and obtained this Minitab output. For level of significance $\alpha = 0.05$, what would have been our conclusion in that case? Why would this have been in error? What are some of the possible consequences of making an error of this sort?

62. Based on your experiences in these exercises, write a sentence about the importance of understanding the statistics behind the "point and click" power of statistical software.

63. **Challenge Exercise.** Note that we have concluded that there is insufficient evidence that the population mean cost has changed, but evidence exists that the population mean cost has increased. How can the mean cost have increased

without changing? Explain what is going on here, in terms of either critical regions or *p*-values.

WORKING WITH LARGE DATA SETS

New York Towns. Work with the **New York** data set for Exercises 64 and 65.  **newyork**

64. Use technology to find the summary statistics for the variable *tot_pop*, which lists the population for each of the towns and cities in New York with at least 1000 people.

65. Suppose we are using the data in this data set as a sample of the population of all the towns and cities in the northeastern United States with at least 1000 people. Use technology to test at level of significance $\alpha = 0.05$ whether the population mean of these towns differs from 50,000.

WORKING WITH LARGE DATA SETS

Fast Food versus Full Service Restaurants. Open the data set, **Restaurants.** Here, we will look at the variable *FFR per 1000*, which refers to the number of fast food restaurants per 1000 residents of the county. For example, a value of 0.6 would mean that there are 0.6 fast food restaurants per 1000 residents in the county. We will perform a hypothesis test for the population mean *FFR per 1000*. We will then see whether this hypothesis test made the correct decision. Use technology to do the following: **restaurants**

66. Obtain a random sample of size 100 from the data set.

67. Using your sample, test whether the population mean number of fast food restaurants per 1000 residents differs from 0.4, using level of significance $\alpha = 0.05$.

68. Find the actual value of the population mean number of fast food restaurants per county. Did your hypothesis test in Exercise 67 make the right decision? Explain.

WORKING WITH LARGE DATA SETS

Chapter 9 Case Study: Clothing Store Sales. Open the Chapter 9 Case Study data set, **Clothing Store.** Retail stores want you to come back again and again, with a short amount of time between purchases. The **Clothing Store** data set tracks the number of days since the last purchase for each customer. The marketing manager wants to make sure that the mean number of days since the last purchase is less than 150 days. Use technology to do the following: **clothingstore**

69. Obtain a random sample of size 100 from the data set.

70. Using your sample, test whether the population mean number of days since the last purchase is less than 150, using level of significance $\alpha = 0.05$.

Step 2   **Find t_{data}.**

Using the statistics from Figure 29a, we have the test statistic

$$t_{data} = \frac{\bar{x} - \mu_0}{s/\sqrt{n}} = \frac{1.583888889 - 1.618}{0.1230083158/\sqrt{18}} \approx -1.176515482 \approx -1.1765$$

Step 3   **Find the *p*-value.**

From Figures 29a, 29b, and 30, we have

$$p\text{-value} = P(t > |-1.1765|) + P(t < -|-1.1765|) \approx 0.2556$$

Step 4   **State the conclusion and interpretation.**

Because *p*-value ≈ 0.2556 is *not* $\leq \alpha = 0.05$, we do *not* reject H_0. Thus, there is insufficient evidence, at level of significance $\alpha = 0.05$, that the population mean ratio differs from 1.618. In other words, the data do not reject the claim that Shoshone beaded rectangles follow the same golden ratio exhibited by the Parthenon and the *Mona Lisa*.

NOW YOU CAN DO
Exercises 21–26.

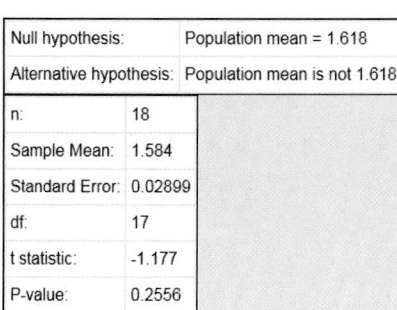

Null hypothesis:	Population mean = 1.618
Alternative hypothesis:	Population mean is not 1.618

n:	18
Sample Mean:	1.584
Standard Error:	0.02899
df:	17
t statistic:	-1.177
P-value:	0.2556

Form of H_a:
t_{data}
p-value
Sample mean $\bar{x}$
Sample standard deviation *s*
Sample size *n*

```
T-Test
µ≠1.618
t=-1.176515482
P=.255601045
x̄=1.583888889
Sx=.1230083158
n=18
```

FIGURE 29a TI-83/84 results.

FIGURE 29b Crunchlt! results.

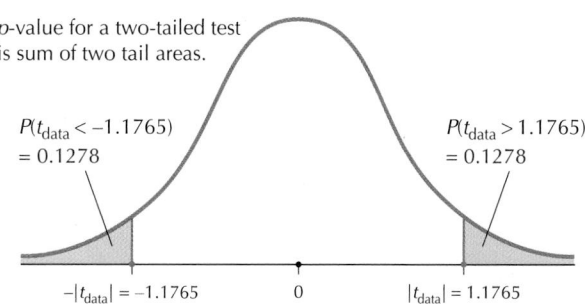

p-value for a two-tailed test is sum of two tail areas.

$P(t_{data} < -1.1765)$
$= 0.1278$

$P(t_{data} > 1.1765)$
$= 0.1278$

$-|t_{data}| = -1.1765$ 0 $|t_{data}| = 1.1765$

FIGURE 30 *p*-Value for two-tailed *t* test is sum of two tail areas.

EXAMPLE 22   Estimating the *p*-value using the *t* table

Suppose we did not have access to technology. Estimate the *p*-value from Example 19 using the *t* table (Appendix Table D). For Example 19, our hypotheses are

$$H_0: \mu = 38 \quad \text{versus} \quad H_a: \mu \neq 38$$

where μ represents the population mean number of hours worked per week. Our test statistic is $t_{data} = -1.5$.

Solution

For a two-tailed test, choose the row of the *t* table with the heading "Area in two tails." Then select the row in the table with the appropriate degrees of freedom, in this case df = $n - 1 = 30 - 1 = 29$. Note the *t*-values in this row: 1.311, 1.699, 2.045, 2.462, and 2.756. Think of these values as existing on a horizontal number line. We want to place our $t_{data} = -1.5$ somewhere on this number line, but all the *t*-values in the table are positive. Fortunately, because of the symmetry of the *t* distribution about zero, we may take $|t_{data}| = 1.5$. Now, where would $|t_{data}| = 1.5$ fit on this "number line"? Between 1.311 and 1.699, as indicated in Figure 31, an excerpt from the *t* table. Therefore, we may estimate the *p*-value to be between 0.20 and 0.10. In fact, the actual *p*-value for this problem is about 0.144 (see Figure 32), so that our estimate is confirmed.

NOW YOU CAN DO
Exercises 27–30.

		0.20	0.10	Area in two tails 0.05	0.02	0.01
df	29	1.311 $\quad |t_{data}| = 1.5$ 1.699		2.045	2.462	2.756

$|t_{data}| = 1.5$ lies between 1.311 and 1.699, so the *p*-value lies between 0.20 and 0.10.

FIGURE 31 Estimating the *p*-value using the *t* table.

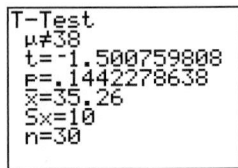

```
T-Test
μ≠38
t=-1.500759808
p=.1442278638
x̄=35.26
Sx=10
n=30
```

FIGURE 32 Actual *p*-value confirms the estimate.

YOUR TURN #10

Estimate the *p*-value for the hypothesis test in Example 18.

(The solution is shown in Appendix A.)

3 Using Confidence Intervals to Perform Two-Tailed *t* Tests

Just as we did for two-tailed *Z* tests in Section 9.3, we may use a $100(1 - \alpha)\%$ *t* confidence interval to perform a two-tailed *t* test with level of significance α for various hypothesized values of μ_0. The strategy is the same: if a certain value for μ_0 lies outside the $100(1 - \alpha)\%$ *t* confidence interval for μ, then the null hypothesis specifying this value for μ_0 would be rejected. Otherwise, it would not be rejected.

EXAMPLE 23 Using a confidence interval to perform two-tailed *t* tests

In Example 14 of Chapter 8 (page 452), we found the 99% *t*-confidence interval for μ, the population mean sodium content per serving of all breakfast cereals, to be the following:

Lower bound = 144.1 grams   Upper bound = 227.7 grams

Test, using level of significance $\alpha = 0.01$, whether the population mean amount of sodium differs from the following values: (**a**) 100 grams, (**b**) 170 grams, (**c**) 250 grams.

Solution

The key words "differs from" mean that we are using two-tailed tests. Then, for each hypothesized value of μ_0, we determine whether it falls inside or outside the given confidence interval.

a. $H_0 : \mu = 100$ versus $H_a : \mu \neq 100$

The confidence interval is (144.1, 227.7), and because $\mu_0 = 100$ lies outside the interval (see Figure 33), we reject H_0.

b. $H_0 : \mu = 170$ versus $H_a : \mu \neq 170$

$\mu_0 = 170$ lies inside the interval, so we do not reject H_0.

c. $H_0 : \mu = 250$ versus $H_a : \mu \neq 250$

$\mu_0 = 250$ lies outside the interval, so we reject H_0.

NOW YOU CAN DO
Exercises 31–36.

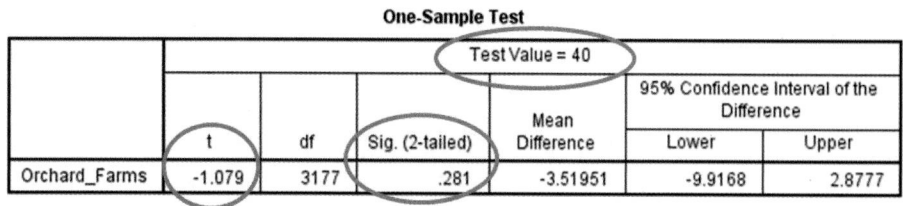

FIGURE 33 Reject H_0 for values of μ_0 that lie outside (144.1, 227.7).

In Section 9.3, we showed how to perform and interpret *Z* tests for μ using the TI-83/84 and Minitab software output. Here, in Section 9.4, we demonstrate how to perform and interpret *t* tests for μ using software output from SPSS and JMP.

EXAMPLE 24   Interpreting software output

Each of **(a)** and **(b)** represent software output from a *t* test for μ. For each, examine the indicated software output, and provide the following steps:

Step 1   **State the hypotheses and the rejection rule.**

Step 2   **Find t_{data}.**

Step 3   **Find the *p*-value.**

Step 4   **State the conclusion and the interpretation.**

Use level of significance $\alpha = 0.10$ for each hypothesis test.

a. SPSS output for a *t* test for μ, where μ represents the population mean number of orchard farms per county, nationwide.

One-Sample Test

| | \multicolumn{6}{c}{Test Value = 40} |
| | | | | | \multicolumn{2}{c}{95% Confidence Interval of the Difference} |
	t	df	Sig. (2-tailed)	Mean Difference	Lower	Upper
Orchard_Farms	-1.079	3177	.281	-3.51951	-9.9168	2.8777

b. JMP output for a *t* test for μ, where μ represents the population mean number of grocery stores per county, nationwide.

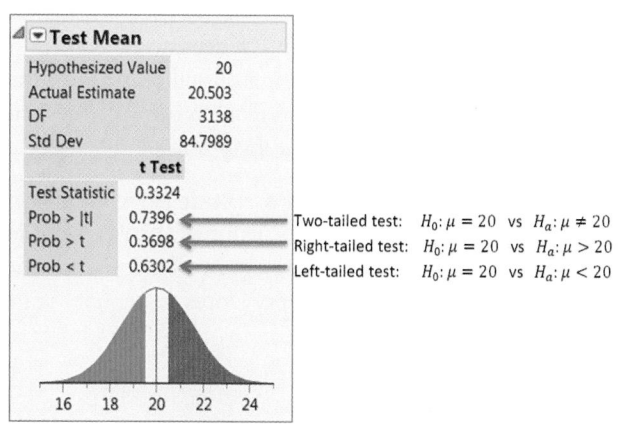

Solution

a. Interpreting the SPSS output.

Step 1 State the hypotheses and the rejection rule.
In the SPSS output, the "Test Value = 40" indicates that $\mu_0 = 40$. Also, the "2-tailed" in the output indicates that we have a two-tailed test. Thus, our hypotheses are:

$$H_0 : \mu = 40 \quad \text{versus} \quad H_a : \mu \neq 40$$

where μ represents the population mean number of orchard farms per county, nationwide. We will reject H_0 if the p-value is less than level of significance $\alpha = 0.10$.

Step 2 Find t_{data}.
Under the "t" in the SPSS is the value for t_{data}, -1.079.

Step 3 Find the p-value.
The abbreviation "Sig." stands for "Significance," which represents the p-value: 0.281.

Step 4 State the conclusion and the interpretation.
The p-value of 0.281 is not less than the level of significance $\alpha = 0.10$, so we do not reject H_0. There is insufficient evidence that the population mean number of orchard farms per county differs from 40.

b. Interpreting the JMP output.

Step 1 State the hypotheses and the rejection rule.
In the JMP output, the "Hypothesized Value" indicates that $\mu_0 = 20$. Now, JMP is unusual in that it performs all three types of hypothesis test simultaneously: two-tailed, right-tailed, and left-tailed, as shown in the JMP output. It does not specify a particular form of the test. Let us use the right-tailed test for this example. Thus, our hypotheses are:

$$H_0 : \mu = 20 \quad \text{vs} \quad H_a : \mu > 20$$

where μ represents the population mean number of grocery stores per county, nationwide. We will reject H_0 if the p-value is less than level of significance $\alpha = 0.10$.

Step 2 Find t_{data}.
Next to "Test Statistic" in the JMP output, we find the value of our test statistic, t_{data}, 0.3324.

Step 3 Find the p-value.
Here, we need to be careful, because JMP gives us three different p-values, depending on which form of the hypothesis test is performed. We chose the right-tailed test, so our p-value is next to "Prob > t": p-value = 0.3698, as indicated in the JMP output.

Step 4 State the conclusion and the interpretation.
The p-value of 0.3698 is not less than the level of significance $\alpha = 0.10$, so we do not reject H_0. There is insufficient evidence that the population mean number of grocery stores per county is greater than 20.

NOW YOU CAN DO
Exercises 37–40.

9.5   *Z* Test for the Population Proportion

OBJECTIVES By the end of this section, I will be able to . . .

1   Perform the *Z* test for *p* using the critical-value method.
2   Perform the *Z* test for *p* using the *p*-value method.
3   Use confidence intervals for *p* to perform two-tailed hypothesis tests about *p*.

1 The *Z* Test for *p* Using the Critical-Value Method

For example, if a baseball player has $x = 30$ hits in $n = 100$ at-bats, his batting average is $\hat{p} = x/n = 30/100 = 0.3$ (or .300).

Thus far, we have dealt with testing hypotheses about the population mean μ only. In this section, we will learn how to perform the *Z* test for the population proportion *p*. For our point estimate of the unknown population proportion *p*, we use the sample proportion $\hat{p} = x/n$, where *x* equals the number of successes.

Just as with the *Z* test for the mean, in the *Z* test for the proportion the null hypothesis will include a certain hypothesized value for the unknown parameter, which we call p_0. For example, the hypotheses for the two-tailed test have the following form:

$$H_0 : p = p_0 \quad \text{versus} \quad H_a : p \neq p_0$$

where p_0 represents a particular hypothesized value of the unknown population proportion *p*. For instance, if a researcher is interested in determining whether the population proportion of Americans who support increased funding for higher education differs from 50%, then $p_0 = 0.50$ and $q_0 = 1 - p_0 = 0.50$.

If we assume H_0 is correct, then the population proportion of successes is p_0. Then Facts 5 and 6 from Section 7.2 tell us that the sampling distribution of *p* has a mean of p_0 and the standard deviation

$$\sigma_{\hat{p}} = \sqrt{\frac{p \cdot q}{n}} = \sqrt{\frac{p_0 \cdot q_0}{n}}$$

because we claim in H_0 that $p = p_0$. Here, $\sigma_{\hat{p}}$ is called the **standard error of the proportion**. Fact 7 from Section 7.2 tells us that the sampling distribution of $\hat{p}$ is approximately normal whenever both of the following conditions are met: $n \cdot p \geq 5$ and $n \cdot q \geq 5$. This leads us to the following statement of the **essential idea about hypothesis testing for the proportion**.

> **The Essential Idea About Hypothesis Testing for the Proportion**
>
> When the sample proportion $\hat{p}$ is unusual or extreme in the sampling distribution of $\hat{p}$ that is based on the assumption that H_0 is correct, we reject H_0. Otherwise, there is insufficient evidence against H_0, and we should not reject H_0.

The remainder of this section explains the details of implementing hypothesis testing for the proportion. The critical-value method for the *Z* test for *p* is similar to that of the *Z* test for μ, in that we compare one *Z*-value (Z_{data}) with another *Z*-value (Z_{crit}). In this section, Z_{data} represents the number of standard errors ($\sigma_{\hat{p}}$) the sample proportion $\hat{p}$ lies above or below the hypothesized proportion p_0.

> The test statistic used for the *Z* test for the proportion is
>
> $$Z_{data} = \frac{\hat{p} - p_0}{\sqrt{\dfrac{p_0 \cdot q_0}{n}}}$$
>
> where $\hat{p}$ is the observed sample proportion of successes, p_0 is the value of *p* hypothesized in H_0, $q_0 = 1 - p_0$, and *n* is the sample size.

EXAMPLE 25 Calculating Z_{data} for the Z test for proportion

The NPD Group reported in 2013 that sales of the Chromebook accounted for 20% of the U.S. computer market. Suppose a random sample of $n = 400$ computers found 76 that were Chromebooks. We are interested in testing whether the population proportion of Chromebooks has changed from 20%.

a. Construct the hypotheses.

b. Calculate the test statistic Z_{data}.

Solution

The key words "has changed" indicate a two-tailed test. "Changed from what?" The hypothesized proportion $p_0 = 0.20$. The hypotheses are

$$H_0 : p = 0.20 \quad \text{versus} \quad H_a : p \neq 0.20$$

The sample proportion of Chromebooks is

$$\hat{p} = \frac{x}{n} = \frac{\text{number in sample that are Chromebooks}}{\text{sample size}} = \frac{76}{400} = 0.19$$

We then calculate the value of the test statistic Z_{data}:

$$Z_{data} = \frac{\hat{p} - p_0}{\sqrt{\dfrac{p_0 \cdot q_0}{n}}} = \frac{0.19 - 0.20}{\sqrt{\dfrac{0.20(0.80)}{400}}} = \frac{-0.01}{0.02} = -0.5$$

NOW YOU CAN DO
Exercises 7–14.

YOUR TURN

#11

For Example 25, suppose the sample found 50 of 400 computers that were Chromebooks. Calculate the test statistic Z_{data}.

(The solution is shown in Appendix A.)

To find the Z_{crit} critical values, the critical regions, or the rejection rules, you can use Table 11.

Table 11 Table of critical values Z_{crit} for common values of the level of significance α

	Form of Hypothesis Test		
Level of significance α	**Right-tailed** $H_0 : p = p_0$ $H_a : p > p_0$	**Left-tailed** $H_0 : p = p_0$ $H_a : p < p_0$	**Two-tailed** $H_0 : p = p_0$ $H_a : p \neq p_0$
0.10	$Z_{crit} = 1.28$	$Z_{crit} = -1.28$	$Z_{crit} = 1.645$
0.05	$Z_{crit} = 1.645$	$Z_{crit} = -1.645$	$Z_{crit} = 1.96$
0.01	$Z_{crit} = 2.33$	$Z_{crit} = -2.33$	$Z_{crit} = 2.58$
	α Noncritical region / Critical region	α Critical region / Noncritical region	$\alpha/2$ $\alpha/2$ Critical region / Noncritical region / Critical region
Rejection rule	Reject H_0 if $Z_{data} \geq Z_{crit}$	Reject H_0 if $Z_{data} \leq Z_{crit}$	Reject H_0 if $Z_{data} \leq -Z_{crit}$ or $Z_{data} \geq Z_{crit}$

Z Test for the Population Proportion p: Critical-Value Method

When a random sample of size n is taken from a population, you can use the Z test for the proportion if both of the normality conditions are satisfied:

$$n \cdot p_0 \geq 5 \quad \text{and} \quad n \cdot q_0 \geq 5$$

Step 1   State the hypotheses.
Use one of the forms from Table 11. State the meaning of p.

Step 2   Find Z_{crit} and state the rejection rule.
Use Table 11.

Step 3   Calculate Z_{data}.

$$Z_{data} = \frac{\hat{p} - p_0}{\sigma_{\hat{p}}} = \frac{\hat{p} - p_0}{\sqrt{\dfrac{p_0 \cdot q_0}{n}}}$$

Step 4   State the conclusion and the interpretation.
If Z_{data} falls in the critical region, then reject H_0. Otherwise, do not reject H_0. Interpret the conclusion so that a nonspecialist can understand.

EXAMPLE 26   Z test for p using the critical-value method

Refer to Example 25. Test whether the population proportion of Chromebook computers has changed from 20%, using the critical-value method and level of significance $\alpha = 0.10$.

Solution

First, we check that both of our normality conditions are met. From Example 25, we have $p_0 = 0.20$ and $n = 400$.

$$n \cdot p_0 = (400)(0.20) = 80 \geq 5 \quad \text{and} \quad n \cdot q_0 = (400)(0.80) = 320 \geq 5$$

The normality conditions are met and we may proceed with the hypothesis test.

As a check on your arithmetic, the two quantities you obtain when checking the normality conditions should add up to n. Here, $80 + 320 = 400 = n$.

Step 1   **State the hypotheses.**
From Example 25, our hypotheses are

$$H_0 : p = 0.20 \quad \text{versus} \quad H_a : p \neq 0.20$$

where p represents the population proportion of computers that are Chromebooks.

Step 2   **Find Z_{crit} and state the rejection rule.**
We have a two-tailed test, with $\alpha = 0.10$. This gives us our critical value $Z_{crit} = 1.645$. The rejection rule from Table 11 is: Reject H_0 if $Z_{data} \geq 1.645$ or $Z_{data} \leq -1.645$ (Figure 35).

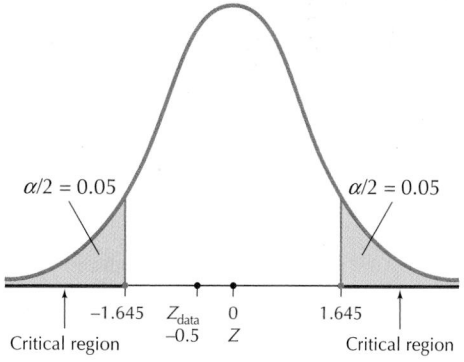

FIGURE 35
Z_{data} does not fall in the critical region.

Step 3   **Calculate Z_{data}.**
From Example 25, we have $Z_{data} = -0.5$

Step 4 **State the conclusion and the interpretation.**
The test statistic $Z_{data} = -0.5$ is not ≥ 1.645 and not ≤ -1.645. Thus, we do not reject H_0. There is insufficient evidence at level of significance $\alpha = 0.10$ that the population proportion of computers that are Chromebooks differs from 20%.

NOW YOU CAN DO
Exercises 15–18.

2 Z Test for p: The p-Value Method

The p-value method for the Z test for p is equivalent to the critical-value method. The p-values are defined similarly to those for the Z test for μ, as shown in Table 12.

Table 12 Finding the p-value depends on the form of the hypothesis test

Type of test	Right-tailed test	Left-tailed test	Two-tailed test
Hypotheses	$H_0: p = p_0$ $H_a: p > p_0$	$H_0: p = p_0$ $H_a: p < p_0$	$H_0: p = p_0$ $H_a: p \neq p_0$
p-Value is tail area associated with Z_{data}	p-value $= P(Z > Z_{data})$ Area to right of Z_{data}	p-value $= P(Z < Z_{data})$ Area to left of Z_{data}	p-value $= P(Z > \lvert Z_{data}\rvert)$ $+ P(Z < -\lvert Z_{data}\rvert)$ $= 2 \cdot P(Z > \lvert Z_{data}\rvert)$ Sum of the two tail areas.

Note that the p-value has precisely the same definition and behavior as in the Z test for the population mean μ. That is, the p-value is roughly a measure of how extreme your value of Z_{data} is and takes values between 0 and 1, with small values indicating extreme values of Z_{data}.

Developing Your Statistical Sense

The Difference Between the p-Value and the Population Proportion p

Be careful to distinguish between the p-value and the population proportion p. The latter represents the population proportion of successes for a binomial experiment and is a population parameter. The p-value is the probability of observing a value of Z_{data} at least as extreme as the Z_{data} actually observed. The p-value depends on the sample data, but the population proportion p does not depend on the sample data.

Z Test for the Population Proportion p: p-Value Method

When a random sample of size n is taken from a population, you can use the Z test for the proportion if both of the normality conditions are satisfied:

$$n \cdot p_0 \geq 5 \quad \text{and} \quad n \cdot q_0 \geq 5$$

Step 1 **State the hypotheses and the rejection rule.**
Use one of the forms from Table 12. State the meaning of *p*. State the rejection rule as "Reject H_0 when the *p*-value $\leq \alpha$."

Step 2 **Calculate Z_{data}.**

$$Z_{data} = \frac{\hat{p} - p_0}{\sqrt{\dfrac{p_0 \cdot q_0}{n}}}$$

Step 3 **Find the *p*-value.**
Either use technology to find the *p*-value, or calculate it using the form in Table 12 that corresponds to your hypotheses.

Step 4 **State the conclusion and the interpretation.**
If the *p*-value $\leq \alpha$, then reject H_0. Otherwise, do not reject H_0. Interpret your conclusion so that a nonspecialist can understand.

EXAMPLE 27   *Z* test for *p* using the *p*-value method

George Doyle/Punchstock

The National Transportation Safety Board publishes statistics on the number of automobile crashes that people in various age groups have. Young people ages 18–24 have an accident rate of 12%, meaning that on average 12 out of every 100 young drivers per year had an accident. A researcher claims that the population proportion of young drivers having accidents is greater than 12%. Her study examined 1000 young drivers ages 18–24 and found that 134 had an accident this year. Perform the appropriate hypothesis test using the *p*-value method with level of significance $\alpha = 0.05$.

Solution

First, we check that both of our normality conditions are met. We are interested in whether the proportion has increased from 12%, so we have $p_0 = 0.12$.

$$n \cdot p_0 = (1000)(0.12) = 120 \geq 5 \quad \text{and} \quad n \cdot q_0 = (1000)(0.88) = 880 \geq 5$$

The normality conditions are met and we may proceed with the hypothesis test.

Step 1   **State the hypotheses and the rejection rule.**
Our hypotheses are

$$H_0 : p = 0.12 \quad \text{versus} \quad H_a : p > 0.12$$

where *p* represents the population proportion of young people ages 18–24 who had an accident. We reject the null hypothesis if the *p*-value $\leq \alpha = 0.05$.

Step 2   **Calculate Z_{data}.**
Our sample proportion is $\hat{p} = 134/1000 = 0.134$. Because $p_0 = 0.12$, the standard error of $\hat{p}$ is

$$\sigma_{\hat{p}} = \sqrt{\frac{p_0 \cdot q_0}{n}} = \sqrt{\frac{(0.12)(0.88)}{1000}} \approx 0.0103$$

Thus, our test statistic is

We report Z_{data} to two decimal places to allow the use of the *Z* table to calculate the *p*-value.

$$Z_{data} = \frac{\hat{p} - p_0}{\sqrt{\dfrac{p_0 \cdot q_0}{n}}} = \frac{0.134 - 0.12}{\sqrt{\dfrac{(0.12)(0.88)}{1000}}} \approx 1.36$$

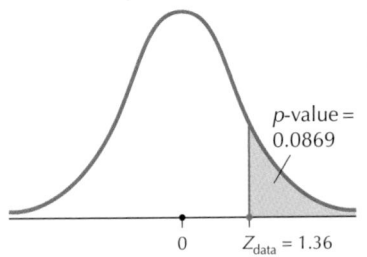

FIGURE 36 p-Value for a right-tailed test equals area to right of Z_{data}.

NOW YOU CAN DO

Exercises 19–22.

YOUR TURN

#12

That is, the sample proportion $\hat{p} = 0.134$ lies approximately 1.36 standard errors above the hypothesized proportion $p_0 = 0.12$.

Step 3   **Find the p-value.**

We have a right-tailed test, so our p-value from Table 12 is $P(Z > Z_{data})$. This is a Case 2 problem from Table 8 in Chapter 6 (page 355), where we find the tail area by subtracting the Z table area from 1 (Figure 36):

$$P(Z > Z_{data}) = P(Z > 1.36) = 1 - 0.9131 = 0.0869$$

Step 4   **State the conclusion and the interpretation.**

The p-value 0.0869 is not $\leq \alpha = 0.05$, so we do not reject H_0. There is insufficient evidence that the population proportion of young people ages 18–24 who had an accident has increased.

For Example 27, suppose 150 of the 1000 young drivers ages 18–24 had an accident this year. Now test whether the population proportion of young drivers who had an accident exceeds 0.12, using level of significance, $\alpha = 0.05$.

(The solution is shown in Appendix A.)

EXAMPLE 28   Performing the Z test for p using technology

A study reported that 1% of American Internet users who are married or in a long-term relationship met on a blind date or through a dating service.[13] A survey of 500 American Internet users who are married or in a long-term relationship found 8 who met on a blind date or through a dating service. If appropriate, test whether the population proportion has increased. Use the p-value method with level of significance $\alpha = 0.05$.

Solution

We have $p_0 = 0.01$ and $n = 500$. Checking the normality conditions, we have

$$n \cdot p_0 = (500)(0.01) = 5 \geq 5 \quad \text{and} \quad n \cdot q_0 = (500)(0.99) = 495 \geq 5$$

The normality conditions are met and we may proceed with the hypothesis test.

Step 1   **State the hypotheses and the rejection rule.**

Our hypotheses are

$$H_0 : p = 0.01 \quad \text{versus} \quad H_a : p > 0.01$$

where p represents the population proportion of American Internet users who are married or in a long-term relationship and who met on a blind date or through a dating service. We will reject H_0 if the p-value ≤ 0.05.

Step 2   **Calculate Z_{data}.**

We use the instructions supplied in the Step-by-Step Technology Guide on page 552. Figure 37 shows the TI-83/84 results from the Z test for p, Figure 38 shows the results from Minitab, and Figure 39 shows the results from CrunchIt!.

FIGURE 37
TI-83/84 results.

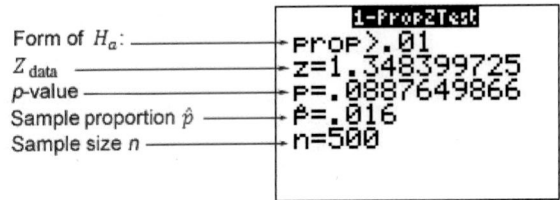

Form of H_a:
Z_{data}
p-value
Sample proportion $\hat{p}$
Sample size n

Note: Minitab, TI-83/84, and CrunchIt! round results to different numbers of decimal places.

```
Test of p = 0.01 vs p > 0.01
                                    95%
                                   Lower
  Sample  X    N   Sample p   Bound     Z-Value  P-Value
  1       8   500  0.016000   0.006770    1.35     0.089
          X    n     p̂      (not used)  Z_data   p-value
```

FIGURE 38 Minitab results.

Null hypothesis:	Proportion = 0.01
Alternative hypothesis:	Proportion > 0.01
n:	500
Successes:	8
p-hat:	0.01600
z statistic:	1.348
P-value:	0.08876

FIGURE 39 CrunchIt! results.

We have

$$Z_{\text{data}} = \frac{\hat{p} - p_0}{\sqrt{\dfrac{p_0 \cdot q_0}{n}}} = \frac{0.016 - 0.01}{\sqrt{\dfrac{(0.01)(0.99)}{500}}} \approx 1.348399725$$

which concurs with the TI-83/84 results in Figure 37.

Step 3   **Find the *p*-value.**
From Figures 37, 38, 39, and 40 we have

$$p\text{-value} = P(Z > 1.348399725) = 0.0887649866 \approx 0.08876$$

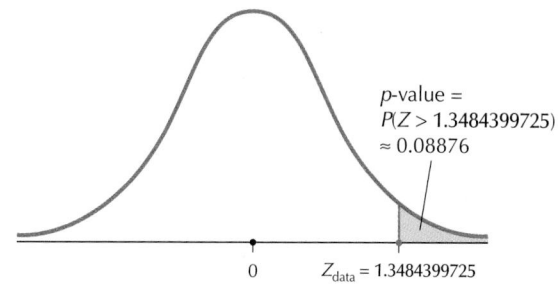

FIGURE 40
p-Value for a right-tailed test.

Step 4   **State the conclusion and interpretation.**
Because *p*-value ≈ 0.08876 is *not* $\leq \alpha = 0.05$, we do *not* reject H_0. There is insufficient evidence that the population proportion of American Internet users who are married or in a long-term relationship and who met on a blind date or through a dating service has increased.

3 Using Confidence Intervals for *p* to Perform Two-Tailed Hypothesis Tests About *p*

Just as for μ, we can use a $100(1 - \alpha)\%$ confidence interval for the population proportion *p* in order to perform a set of two-tailed hypothesis tests for *p*.

EXAMPLE 29   Using a confidence interval for *p* to perform two-tailed hypothesis tests about *p*

In 2013, Facebook reported that 73% of its users access Facebook using a mobile device. Suppose that a 95% confidence interval for the population of mobile accessers is (lower bound = 0.70, upper bound = 0.76). Use the confidence interval to test, using level of significance $\alpha = 0.05$, whether the population proportion differs from

a. 0.69

b. 0.72

c. 0.77

Solution

There is equivalence between a $100(1 - \alpha)\%$ confidence interval for p and a two-tailed test for p with level of significance α. Values of p_0 that lie outside the confidence interval lead to rejection of the null hypothesis, whereas values of p_0 within the confidence interval lead to not rejecting the null hypothesis. Figure 41 illustrates the 95% confidence interval for p.

FIGURE 41

Reject H_0 for values p_0 that lie outside the interval (0.70, 0.76).

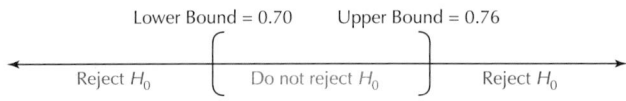

We want to perform the following two-tailed hypothesis tests:

a. $H_0 : p = 0.69$ versus $H_a : p \neq 0.69$

b. $H_0 : p = 0.72$ versus $H_a : p \neq 0.72$

c. $H_0 : p = 0.77$ versus $H_a : p \neq 0.77$

To perform each hypothesis test, simply observe where each value of p_0 falls on the number line. For example, in the first hypothesis test, the hypothesized value $p_0 = 0.69$ lies outside the interval (0.70, 0.76). Thus, we reject H_0. The three hypothesis tests are summarized here.

Value of p_0	Form of hypothesis test, with $\alpha = 0.05$	Where p_0 lies in relation to 95% confidence interval	Conclusion of hypothesis test
a. 0.69	$H_0 : p = 0.69$ $H_a : p \neq 0.69$	Outside	Reject H_0
b. 0.72	$H_0 : p = 0.72$ $H_a : p \neq 0.72$	Inside	Do not reject H_0
c. 0.77	$H_0 : p = 0.77$ $H_a : p \neq 0.77$	Outside	Reject H_0

NOW YOU CAN DO
Exercises 23–26.

EXAMPLE 30 Interpreting software output

Each of (**a**) and (**b**) represent software output from a Z test for p. For each, examine the indicated software output, and provide the following steps:

Step 1 **State the hypotheses and the rejection rule.**
Step 2 **Find Z_{data}.**
Step 3 **Find the p-value.**
Step 4 **State the conclusion and the interpretation.**

Use level of significance $\alpha = 0.05$ for each hypothesis test.

a. TI-83/84 output for a Z test for p, where p represents the population proportion of quiz questions answered correctly.

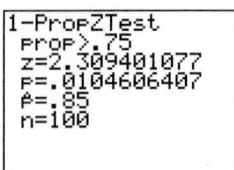

b. Minitab output for a *Z* test for *p*, where *p* represents the population proportion of counties having at least one specialty store.

```
Test of p = 0.62 vs p ≠ 0.62

Variable                    X     N  Sample p        95% CI         Z-Value  P-Value
SpecialtyStoresFlag      1986  3139  0.632686  (0.615821, 0.649550)    1.46    0.143
```

Solution

a. Interpreting the TI-83/84 output

Step 1   **State the hypotheses and the rejection rule.**
In the TI-83/84 output, the "prop > .75" tells us that we have a right-tailed test:

$$H_0 : p = 0.75 \quad \text{versus} \quad H_a : p > 0.75$$

where *p* represents the population proportion of quiz questions answered correctly. We will reject H_0 if the *p*-value is less than the level of significance $\alpha = 0.05$

Step 2   **Find Z_{data}.**
The "z = 2.309401077" in the TI-83/84 output gives us the value for Z_{data}.

Step 3   **Find the *p*-value.**
Here, we need to be a little bit careful, because there are two items containing *p* in the TI-83/84 output. *Don't pick $\hat{p}$*, which represents the sample proportion of successes. Instead, the *p*-value is given as "p = .0104606407."

Step 4   **State the conclusion and the interpretation.**
The *p*-value is less than the level of significance $\alpha = 0.05$, so we reject H_0. There is evidence that the population proportions of quiz questions answered correctly is greater than 0.75.

b. Interpreting the Minitab output

Step 1   **State the hypotheses and the rejection rule.**
The line "Test of $p = 0.62$ vs $p \neq 0.62$" tells us that we have the following two-tailed test.

$$H_0 : p = 0.62 \quad \text{versus} \quad H_a : p \neq 0.62$$

where *p* represents the population proportion of counties having at least one specialty store. We will reject H_0 if the *p*-value is less than the level of significance $\alpha = 0.05$.

Step 2   **Find Z_{data}.**
The "Z-value" of 1.46 in the Minitab output gives us the value for Z_{data}.

Step 3   **Find the *p*-value.**
Under "*P*-Value," Minitab gives us *p*-value = 0.143.

Step 4   **State the conclusion and the interpretation.**
The *p*-value is not less than the level of significance $\alpha = 0.05$, so we do not reject H_0. There is insufficient evidence that the population proportions of counties having at least one specialty store differs from 0.62.

NOW YOU CAN DO
Exercises 27–30.

STEP-BY-STEP TECHNOLOGY GUIDE: *Z* test for *p*

We will use the information from Example 28 (page 548).

TI-83/84

Step 1   Press **STAT**, highlight **TESTS**.
Step 2   Press **5** for **1-PropZTest** (see Figure 42a).
Step 3   For p_0, enter the value of p_0, 0.01.
Step 4   For **x**, enter the number of successes, 8.
Step 5   For **n**, enter the number of trials, 500.
Step 6   For **prop**, enter the form of H_a. Here, we have a right-tailed test, so highlight **> p_0** and press **ENTER** (see Figure 42b).
Step 7   Highlight **Calculate** and press **ENTER**. The results are shown in Figure 37 in Example 28.

```
EDIT CALC TESTS
1:Z-Test…
2:T-Test…
3:2-SampZTest…
4:2-SampTTest…
5:1-PropZTest…
6:2-PropZTest…
7↓ZInterval…
```

FIGURE 42a

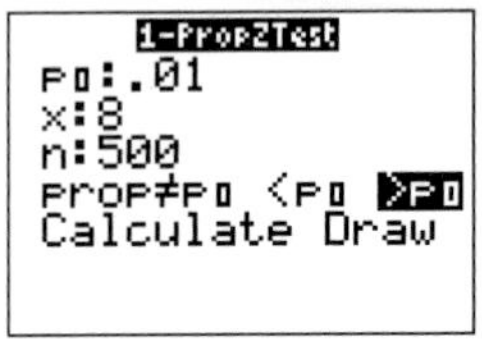

```
1-PropZTest
P0:.01
x:8
n:500
prop≠p0 <p0 >p0
Calculate Draw
```

FIGURE 42b

EXCEL

Use the JMP add-in for Excel.

MINITAB

If you have the summary statistics:
Step 1   Click Stat > Basic Statistics > 1 Proportion….
Step 2   Select Summarized Data from the drop-down menu.
Step 3   Enter the Number of events, 8 and the Number of trials, 500. Select Perform hypothesis test. For Hypothesized proportion, enter 0.01.
Step 4   Click Options.

a. Choose your **Confidence Level** as $100(1 - \alpha)$. Our level of significance α here is 0.05, so the confidence level is 95.0.
b. Select **Proportion > hypothesized proportion** for the Alternative hypothesis.
c. Select **Normal approximation** for Method.
Step 5   Click OK and click OK again. The results are shown in Figure 38 in Example 28.

CRUNCHIT!

If you have the summary statistics:
Step 1   Click **Statistics**, highlight **Proportion**, and select 1-sample.
Step 2   Choose the **Summarized** tab. For **n**, enter the number of trials 500; for **Successes**, enter 8.

Step 3   Select the **Hypothesis Test** tab. For **Proportion** under null hypothesis, enter 0.01.
Step 4   For **Alternative**, select **Greater than**. Then click **Calculate**. Results are shown in Figure 39 in Example 28.

Section 9.5 Summary

1. The test statistic used for the *Z* test for the proportion is

$$Z_{data} = \frac{\hat{p} - p_0}{\sqrt{\dfrac{p_0 \cdot q_0}{n}}}$$

where $\hat{p}$ is the observed sample proportion of successes, p_0 is the value of *p* hypothesized in H_0, $q_0 = 1 - p_0$, and *n* is the sample size. Z_{data} represents the number of standard deviations $(\sigma_{\hat{p}})$ the sample proportion $\hat{p}$ lies above or below the hypothesized proportion p_0. Extreme values of $\hat{p}$ will be

associated with extreme values of Z_{data}. The *Z* test for the proportion may be performed using either the *p*-value method or the critical-value method. For the critical-value method, we compare the values of Z_{data} and Z_{crit}. If Z_{data} falls in the critical region, we reject H_0.

2. For the *p*-value method, we reject H_0 if the *p*-value $\leq \alpha$.

3. We can use a single $100(1 - \alpha)\%$ confidence interval for *p* to help us perform any number of two-tailed hypothesis tests about *p* with level of significance α.

Section 9.5 Exercises

CLARIFYING THE CONCEPTS

1. What is the difference between $\hat{p}$ and *p*? (p. 543)
2. What are the conditions for the *Z* test for *p*? (p. 545)
3. Explain the essential idea about hypothesis testing for the proportion. (p. 543)

4. Explain what p_0 refers to. (p. 543)
5. What possible values can p_0 take? (p. 543)
6. What is the difference between *p* and a *p*-value? (p. 546)

PRACTICING THE TECHNIQUES

✓ CHECK IT OUT!

To do	Check out	Topic
Exercises 7–14	Example 25	Calculating Z_{data} for the Z test for proportion
Exercises 15–18	Example 26	Z test for p using the critical-value method
Exercises 19–22	Example 27	Z test for p using the p-value method
Exercises 23–26	Example 29	Using a confidence interval for p to perform two-tailed hypothesis tests about p
Exercises 27–30	Example 30	Interpreting software output

For Exercises 7–9, find the value of the test statistic Z_{data} for a right-tailed test with $p_0 = 0.4$.

7. A sample of size 50 yields 30 successes.

8. A sample of size 50 yields 40 successes.

9. A sample of size 50 yields 45 successes.

10. What kind of pattern do we observe in the value of Z_{data} for a right-tailed test as the number of successes becomes more extreme?

For Exercises 11–13, find the value of the test statistic Z_{data} for a two-tailed test with $p_0 = 0.5$.

11. A sample of size 80 yields 20 successes.

12. A sample of size 80 yields 30 successes.

13. A sample of size 80 yields 40 successes.

14. What kind of pattern do we observe in the value of Z_{data} as the sample proportion approaches p_0?

For Exercises 15–18, do the following:
   a. Check the normality conditions.
   b. State the hypotheses.
   c. Find Z_{crit} and the rejection rule.
   d. Calculate Z_{data}.
   e. Compare Z_{crit} with Z_{data}. State the conclusion and the interpretation.

15. Test whether the population proportion is less than 0.5. A random sample of size 225 yields 100 successes. Let level of significance $\alpha = 0.05$.

16. Test whether the population proportion differs from 0.3. A random sample of size 100 yields 25 successes. Let level of significance $\alpha = 0.01$.

17. Test whether the population proportion exceeds 0.6. A random sample of size 400 yields 260 successes. Let level of significance $\alpha = 0.05$.

18. Test whether p differs from 0.4. A random sample of size 900 yields 400 successes. Let level of significance $\alpha = 0.10$.

For Exercises 19–22, do the following:
   a. Check the normality conditions.
   b. State the hypotheses and the rejection rule for the p-value method, using level of significance $\alpha = 0.05$.
   c. Find Z_{data}.
   d. Find the p-value.
   e. Compare the p-value with level of significance $\alpha = 0.05$. State the conclusion and the interpretation.

19. Test whether the population proportion exceeds 0.4. A random sample of size 100 yields 44 successes.

20. Test whether the population proportion is less than 0.2. A random sample of size 400 yields 75 successes.

21. Test whether the population proportion differs from 0.5. A random sample of size 900 yields 475 successes.

22. Test whether the population proportion exceeds 0.9. A random sample of size 1000 yields 925 successes.

For Exercises 23–26, a $100(1 - \alpha)\%$ Z confidence interval for p is given. Use the confidence interval to test, using level of significance α, whether p differs from each of the indicated hypothesized values.

23. A 95% Z confidence interval for p is (0.1, 0.9). Hypothesized values p_0 are
   a. 0
   b. 1
   c. 0.5

24. A 99% Z confidence interval for p is (0.51, 0.52). Hypothesized values p_0 are
   a. 0.511
   b. 0.521
   c. 0.519

25. A 90% Z confidence interval for p is (0.1, 0.2). Hypothesized values p_0 are
   a. 0.09
   b. 0.9
   c. 0.19

26. A 95% Z confidence interval for p is (0.05, 0.95). Hypothesized values p_0 are
   a. 0.01
   b. 0.5
   c. 0.06

For Exercises 27–30, software output from a Z test for p is provided. For each, examine the indicated software output, and provide the following steps:

Step 1   State the hypotheses and the rejection rule.
Step 2   Find Z_{data}.
Step 3   Find the p-value.
Step 4   State the conclusion and the interpretation.

Use level of significance $\alpha = 0.05$ for each hypothesis test.

27. TI-83/84 output

```
1-PropZTest
 prop<.5
 z=-2.121320344
 p=.0169473661
 p̂=.425
 n=200
```

28. TI-83/84 output

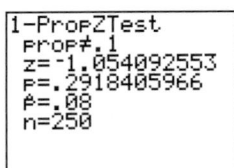

```
1-PropZTest
prop≠.1
z=-1.054092553
p=.2918405966
p̂=.08
n=250
```

29. Minitab output

```
Test of p = 0.05 vs p ≠ 0.05

Variable       X     N   Sample p       95% CI        Z-Value  P-Value
Web_buyer    220  5000   0.044000  (0.038315, 0.049685)   -1.95    0.052
```

30. Minitab output

```
Test of p = 0.4 vs p ≠ 0.4

Variable        X     N   Sample p       95% CI        Z-Value  P-Value
Credit_card  1973  5000   0.394600  (0.381052, 0.408148)   -0.78    0.436
```

APPLYING THE CONCEPTS

31. Facebook Not Cool Anymore? In 2014, Facebook reported that 23% of its users were ages 18–24, a decrease from the 2011 level of 30.9%. (Though this is partly due to more older people joining Facebook, there was nevertheless a loss of over 3 million users in this age group.) A survey of 500 randomly selected Facebook users showed that 100 of them were ages 18–24. If appropriate, test, using level of significance $\alpha = 0.10$, whether the population proportion of Facebook users who are ages 18–24 has decreased from 23%.

32. Twenty-Somethings. According to the U.S. Census Bureau, 7.1% of Americans living in 2004 were between the ages of 20 and 24. Suppose that a random sample of 400 Americans taken this year yields 35 between the ages of 20 and 24. If appropriate, test whether the population proportion of Americans ages 20–24 is different from 7.1%. Use level of significance $\alpha = 0.01$.

33. Nonmedical Pain Reliever Use. The National Survey on Drug Use and Health reported that 4.8% of persons ages 12 or older used a prescription pain reliever nonmedically.[14] Suppose that a random sample of 900 persons ages 12 or older found 54 who had used a prescription pain reliever nonmedically. If appropriate, test whether the population proportion has increased, using level of significance $\alpha = 0.01$.

34. Is This a Date, or What? In 2014, *USA Today* reported that 57% of 18- to 24-year-olds agreed that texting has made it more difficult to determine whether an outing is an actual date or not. Suppose that a recent random sample of 1000 18- to 24-year-olds found 500 who agreed with that sentiment. Test whether the population proportion of 18- to 24-year-olds who agree that texting has made it more difficult to determine whether an outing is an actual date has decreased, using level of significance $\alpha = 0.05$.

35. Mutual Fund Performance. *Business Insider* reported that 58% of mutual funds underperformed in 2013, against the Standard and Poor 500 benchmark. Suppose a random sample of 100 mutual funds showed 67 that had underperformed. Test whether the population proportion has increased, using level of significance $\alpha = 0.05$.

36. Affective Disorders Among Women. What do you think is the most common nonobstetric (not related to pregnancy) reason for hospitalization among 18- to 44-year-old American women? According to the U.S. Agency for Healthcare Research and Quality (www.ahrq.gov), this is the category of affective disorders, such as depression. In 2002, of hospitalizations among 18- to 44-year-old American women, 7% were for affective disorders. Suppose that a random sample taken this year of 1000 hospitalizations of 18- to 44-year-old women showed 80 admitted for affective disorders. We are interested in whether the population proportion of hospitalizations for affective disorders has changed since 2002. Test, using level of significance $\alpha = 0.10$.

37. Latino Household Income. The U.S. Census Bureau reported that 15.3% of Latino families had household incomes of at least $75,000. We are interested in whether the population proportion has changed, using the critical-value method and level of significance $\alpha = 0.01$. Suppose that a random sample of 100 Latino families reported 23 with household incomes of at least $75,000.

  a. Is it appropriate to perform the Z test for the proportion? Why or why not?

  b. Perform the appropriate hypothesis test.

38. Massively Online Dropouts? TechCrunch.com reported in 2014 that the completion rate for massively open online courses (MOOCs) is only 6.5%. Suppose a new study of 1000 randomly chosen students taking a MOOC found that 75 completed the course. Test, using level of significance $\alpha = 0.05$, whether the population proportion of students completing MOOCs has increased.

39. Living with the Parents. The National Association of Home Builders reported that 57% of young people ages 18–24 were living with their parents. Suppose a sample of 100 young people ages 18–24 showed 60 who were living with their parents. Test, using level of significance $\alpha = 0.10$, whether the population proportion has changed.

40. Living with the Parents. Refer to Exercise 39.

  a. Evaluate the strength of evidence against the null hypothesis.

  b. Suppose that we decide to perform the same Z test as Exercise 39, however, this time using a different method. (If you used the critical-value method earlier, use the *p*-value method, and vice-versa.) Without actually performing the test, what would the conclusion be and why?

  c. Would a 95% Z interval for *p* contain $p = 0.57$? Explain.

BRINGING IT ALL TOGETHER

Children and Environmental Tobacco Smoke at Home. Use the following information for Exercises 41–44. The Environmental Protection Agency reported that 11% of children age 6 and under were exposed to environmental tobacco smoke (ETS) at home on a regular basis (at least four times per week).[15] A random sample of 100 children age 6 and under showed that 6% of these children had been exposed to ETS at home on a regular basis.

41. Answer the following:

   a. Is it appropriate to perform the *Z* test for the proportion? Why or why not?

   b. Test at level of significance $\alpha = 0.05$ whether the population proportion of children age 6 and under exposed to ETS at home on a regular basis has decreased.

42. Refer to Exercise 41.

   a. Which is the only possible error you can be making here, a Type I or a Type II error? What are some consequences of this error?

   b. Suppose that a newspaper headline reported "Second-hand Smoke Prevalence Down." How would you respond? Does your inference support this headline?

43. Refer to your work in Exercise 41.

   a. Test at level of significance $\alpha = 0.10$ whether the population proportion of children age 6 and under exposed to ETS at home on a regular basis has decreased.

   b. How do you explain the different conclusions you got in the two hypothesis tests above?

   c. Evaluate the strength of evidence against the null hypothesis.

WHAT IF   **44.** Refer to Exercise 41. *What if* the sample proportion $\hat{p}$ decreased, but everything else stayed the same? Describe what would happen to the following, and why.

   a. $\sigma_{\hat{p}}$

   b. Z_{data}

   c. The *p*-value

   d. α

   e. The conclusion

Car Accidents Among Young Drivers. Use the following information for Exercises 45–47. The National Transportation Safety Board publishes statistics on the number of automobile crashes that people in various age groups have. Young people ages 18–24 have an accident rate of 12%, meaning that on average 12 out of every 100 young drivers per year had an accident. A researcher claims that the population proportion of young drivers having accidents is greater than 12%. Her study examined 1000 young drivers ages 19–24 and found that 134 had an accident this year.

45. Perform the appropriate hypothesis test with level of significance $\alpha = 0.05$.

WHAT IF   **46.** *What if* our sample size and the number of successes are doubled, so that $\hat{p}$ remains the same? Otherwise, everything else is the same. Describe how this change would affect the following.

   a. $\sigma_{\hat{p}}$

   b. Z_{data}

   c. The *p*-value

   d. α

   e. The conclusion

WHAT IF   **47.** *What if* the hypothesized proportion p_0 was no longer 0.12? Instead, p_0 takes some value between 0.12 and 0.134. Otherwise, everything else is the same as in the original example. Describe how this change would affect the following.

   a. $\sigma_{\hat{p}}$

   b. Z_{data}

   c. The *p*-value

   d. α

   e. The conclusion

WORKING WITH LARGE DATA SETS

Fast Food versus Full Service Restaurants. Open the data set, **Restaurants**. Here, we will look at the fast food restaurants, but in a different way. We will explore the proportion of counties that have no fast food restaurants at all. Use technology to do the following: **restaurants**

48. Obtain a random sample of size 100 from the data set.

49. Find the sample proportion of counties with zero fast food restaurants.

50. Using your sample, test whether the population proportion of counties with no fast food restaurants is less than 0.10, using level of significance $\alpha = 0.05$.

51. Find the actual value of the population proportion of counties with no fast food restaurants. Did your hypothesis test in Exercise 50 make the right decision? Explain.

WORKING WITH LARGE DATA SETS

Chapter 9 Case Study: Clothing Store Sales. Open the Chapter 9 Case Study data set, **Clothing Store**. Here, we will compare the proportions of buyers of blouses and buyers of suits. Use technology to do the following: **clothingstore**

52. Obtain a random sample of size 100 from the data set.

53. Find the sample proportion of customers who have bought a blouse (value = 1 for variable = *blouse*).

54. Find the sample proportion of customers who have bought a suit (value = 1 for variable = *suit*).

55. Using your sample, test whether the population proportion of customers who have bought a blouse exceeds 0.5, using level of significance $\alpha = 0.10$.

56. Explain why we need not perform a hypothesis test to determine whether the population proportion of customers who have bought a suit exceeds 0.5.

57. Using your sample, test whether the population proportion of customers who have bought a suit exceeds 0.1, using level of significance $\alpha = 0.10$.

9.6 Chi-Square Test for the Population Standard Deviation

OBJECTIVES By the end of this section, I will be able to . . .

1 Perform the χ^2 test for σ using the critical-value method.
2 Perform the χ^2 test for σ using the p-value method.
3 Use confidence intervals for σ to perform two-tailed hypothesis tests about σ.

1 χ^2 (Chi-Square) Test for σ using the Critical-Value Method

In Section 8.4, we used the χ^2 distribution to help us construct confidence intervals for the population variance and standard deviation. Here, in Section 9.6, we will use the χ^2 distribution to perform hypothesis tests about the population standard deviation σ. Why might we be interested in doing so? A pharmaceutical company that wants to ensure the safety of a particular new drug would perform statistical tests to make sure that the drug's effect was consistent and did not vary widely from patient to patient. The biostatisticians employed by the company would therefore perform a hypothesis test to make sure that the population standard deviation σ was not too large.

Under the assumption that $H_0 : \sigma = \sigma_0$ is true, the χ^2 statistic takes the following form:

$$\chi^2_{\text{data}} = \frac{(n-1)s^2}{\sigma_0^2}$$

For the hypothesis test about s, our test statistic is called χ^2_{data} because the values of $n - 1$ and s^2 come from the observed data. The test statistic χ^2_{data} takes a moderate value when the value of s^2 is moderate, assuming H_0 is true, and χ^2_{data} takes an extreme value when the value of s^2 is extreme, assuming H_0 is true. This leads us to the following.

Readers may want to review the characteristics of the χ^2 distribution in Section 8.4 on page 474.

> **The Essential Idea About Hypothesis Testing for the Standard Deviation**
>
> When the observed value of χ^2_{data} is unusual or extreme on the assumption that H_0 is true, we should reject H_0. Otherwise, there is insufficient evidence against H_0, and we should not reject H_0.

The remainder of Section 9.6 explains the details of implementing hypothesis testing for the standard deviation. The χ^2 test for σ may be performed using the p-value method or the critical-value method. We begin with the critical-value method.

> **χ^2 Test for σ: Critical-Value Method**
>
> This hypothesis test is valid only if we have a random sample from a normal population.
>
> **Step 1 State the hypotheses.**
> Use one of the forms in Table 13. State the meaning of σ.
>
> **Step 2 Find the χ^2 critical value or values and state the rejection rule.**
> Use Table 13.
>
> **Step 3 Calculate χ^2_{data}.**
> Either use technology to find the value of the test statistic χ^2_{data} or calculate the value of χ^2_{data} as follows:

$$\chi^2_{\text{data}} = \frac{(n-1)s^2}{\sigma_0^2}$$

which follows a χ^2 distribution with $n - 1$ degrees of freedom, and where s^2 represents the sample variance.

Step 4 State the conclusion and the interpretation.

If χ^2_{data} falls in the critical region, then reject H_0. Otherwise do not reject H_0. Interpret your conclusion so that a nonspecialist can understand.

The χ^2 critical values in the right-tailed, left-tailed, or two-tailed tests use the following notations: χ^2_α, $\chi^2_{1-\alpha}$, $\chi^2_{\alpha/2}$, and $\chi^2_{1-\alpha/2}$ (see Table 13). In each case, *the subscript indicates the area to the right* of the χ^2 critical value. Find these values just as you did in Section 8.4, using either technology or Table E, Chi-Square (χ^2) Distribution, in the Appendix.

Table 13 Critical values and rejection rules for the χ^2 test for σ

Right-tailed test	**Left-tailed test**	**Two-tailed test**
$H_0 : \sigma = \sigma_0$ $H_a : \sigma > \sigma_0$	$H_0 : \sigma = \sigma_0$ $H_a : \sigma < \sigma_0$	$H_0 : \sigma = \sigma_0$ $H_a : \sigma \neq \sigma_0$
Critical value: χ^2_α Reject H_0 if $\chi^2_{\text{data}} \geq \chi^2_\alpha$ level of significance α	Critical value: $\chi^2_{1-\alpha}$ Reject H_0 if $\chi^2_{\text{data}} \leq \chi^2_{1-\alpha}$ level of significance α	Critical values: $\chi^2_{\alpha/2}$ and $\chi^2_{1-\alpha/2}$ Reject H_0 if $\chi^2_{\text{data}} \geq \chi^2_{\alpha/2}$ or if $\chi^2_{\text{data}} \leq \chi^2_{1-\alpha/2}$ level of significance α

Right-tailed test: Reject H_0 if $\chi^2_{\text{data}} \geq \chi^2_\alpha$. Noncritical region / Critical region, χ^2_α.

Left-tailed test: Reject H_0 if $\chi^2_{\text{data}} \leq \chi^2_{1-\alpha}$. $\chi^2_{1-\alpha}$, Critical region / Noncritical region.

Two-tailed test: Reject H_0 if $\chi^2_{\text{data}} \leq \chi^2_{1-\alpha/2}$. Reject H_0 if $\chi^2_{\text{data}} \geq \chi^2_{\alpha/2}$. $\chi^2_{1-\alpha/2}$, $\chi^2_{\alpha/2}$. Critical region / Noncritical region / Critical region.

EXAMPLE 31 χ^2 test for σ using the critical-value method

carbonemissions8

Carbon emissions

State	Carbon emissions (millions of metric tons)
Florida	230.98
Kentucky	148.36
Missouri	135.54
New Hampshire	16.41
New Mexico	56.60
New York	166.32
Tennessee	105.73
Virginia	99.86

Source: U.S. Department of Energy.

The table contains the carbon emissions from all sources for a random sample of eight states. Test whether the population standard deviation σ of carbon emissions differs from 60 million metric tons, using level of significance $\alpha = 0.05$.

Solution

The normal probability plot indicates acceptable normality.

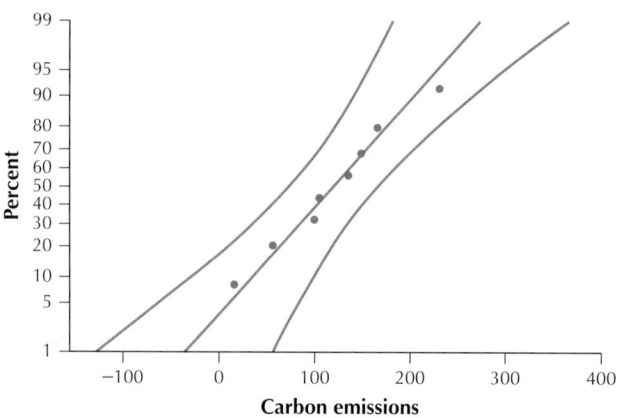

Step 1 State the hypotheses.

The phrase "differs from" indicates that we have a two-tailed test. The value $\sigma_0 = 60$ answers the question "Differs from what?" (Note that σ_0 is 60, and not 60,000,000 because the data are expressed in millions.) Thus, we have our hypotheses:

$$H_0 : \sigma = 60 \quad \text{versus} \quad H_a : \sigma \neq 60$$

where σ represents the population standard deviation of carbon emissions in millions of metric tons.

Step 2 Find the χ^2 critical values and state the rejection rule.

We have $n = 8$, so degrees of freedom $= n - 1 = 7$. Because α is given as 0.05, $\alpha/2 = 0.025$ and $1 - \alpha/2 = 0.975$. Then, from the χ^2 table (Appendix Table E), we have $\chi^2_{\alpha/2} = \chi^2_{0.025} = 16.013$, and $\chi^2_{1-\alpha/2} = \chi^2_{0.975} = 1.690$. We will reject H_0 if χ^2_{data} is either $\geq \chi^2_{\alpha/2} = 16.013$ or $\leq \chi^2_{1-\alpha/2} = 1.690$.

Step 3 Find χ^2_{data}.

The descriptive statistics in Figure 43 tell us that the sample variance is $s^2 = 4409.7$ million metric tons, squared.

FIGURE 43

Descriptive statistics from Minitab.

Descriptive Statistics: Carbon

Variable	Total Count	Mean	StDev	Variance
Carbon	8	120.0	66.4	4409.7

Thus, our test statistic is:

$$\chi^2_{\text{data}} = \frac{(n-1)s^2}{\sigma_0^2} = \frac{(8-1)(4409.7)}{60^2} \approx 8.57$$

Step 4 State the conclusion and the interpretation.

In Step 2, we said that we would reject H_0 if χ^2_{data} was either ≥ 16.013 or ≤ 1.690. Because $\chi^2_{\text{data}} = 8.57$ is neither ≥ 16.013 nor ≤ 1.690 (see Figure 44), we do not reject H_0. There is insufficient evidence at level of significance $\alpha = 0.05$ that the population standard deviation of the state carbon emissions differs from 60 million.

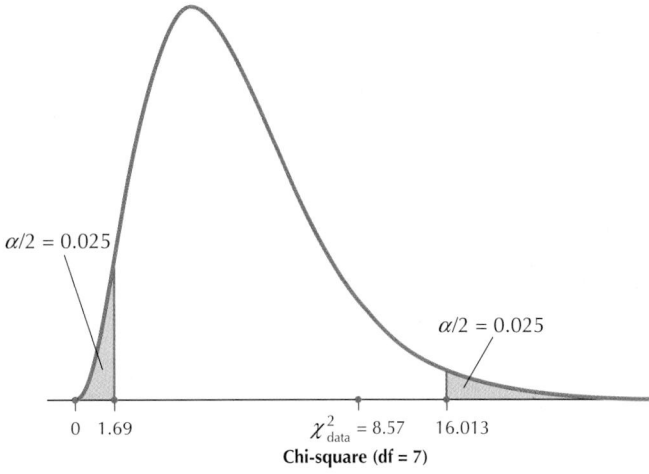

NOW YOU CAN DO
Exercises 7–12.

FIGURE 44 $\chi^2_{data} = 8.57$ does not fall in critical region, so do not reject H_0.

2 χ^2 Test for σ Using the p-Value Method

We may also use the p-value method to perform the χ^2 test for σ.

χ^2 Test for σ: p-Value Method

This hypothesis test is valid only if we have a random sample from a normal population.

Step 1 State the hypotheses and the rejection rule.
Use one of the forms in Table 14. State the rejection rule as "Reject H_0 when the p-value $\leq \alpha$." State the meaning of σ.

Step 2 Calculate χ^2_{data}.
Either use technology to find the value of the test statistic χ^2_{data} or calculate the value of χ^2_{data} as follows:

$$\chi^2_{data} = \frac{(n-1)s^2}{\sigma_0^2}$$

which follows a χ^2 distribution with $n-1$ degrees of freedom, and where s^2 represents the sample variance.

Step 3 Find the p-value.
Use Table 14.

Step 4 State the conclusion and the interpretation.
If the p-value $\leq \alpha$, then reject H_0. Otherwise, do not reject H_0. Interpret your conclusion so that a nonspecialist can understand.

Table 14 p-Value method for the χ^2 test for σ

Right-tailed test	Left-tailed test	Two-tailed test
$H_0 : \sigma = \sigma_0$ $H_a : \sigma > \sigma_0$	$H_0 : \sigma = \sigma_0$ $H_a : \sigma < \sigma_0$	$H_0 : \sigma = \sigma_0$ $H_a : \sigma \neq \sigma_0$

p-value $= P(\chi^2 > \chi^2_{data})$
 Area to right of χ^2_{data}
	p-value $= P(\chi^2 < \chi^2_{data})$
	 Area to left of χ^2_{data}
		If $P(\chi^2 > \chi^2_{data}) \leq 0.5$, then
		 a. χ^2_{data} is on the right side of the distribution
		 b. p-value $= 2 \cdot P(\chi^2 > \chi^2_{data})$
		 If $P(\chi^2 > \chi^2_{data}) > 0.5$, then
		 a. χ^2_{data} is on the left side of the distribution
		 b. p-value $= 2 \cdot P(\chi^2 < \chi^2_{data})$

EXAMPLE 32 χ^2 **test for** σ **using the** *p***-value method and technology**

The table contains the calories in ten entrée food items, courtesy of Food-A-Pedia. Test whether the population standard deviation is larger than 100 calories, using level of significance $\alpha = 0.05$.

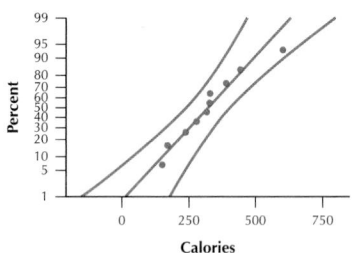
entreecalories

Entrée item	Calories	Entrée item	Calories
Grilled steak	387	Ground beef (95% lean, medium patty)	167
Fried steak	440	Grilled pork chop (large)	314
Breaded fried steak	600	Fried pork chop (large)	326
Ground beef (75% lean, medium patty)	234	Meat pizza, thin crust, large	325
1 large BBQ short rib with sauce	148	Fried catfish (breaded or battered)	276

Solution

The normal probability plot in Figure 45 indicates acceptable normality, allowing us to proceed with the hypothesis test.

Step 1 **State the hypotheses and the rejection rule.**
The phrase "larger than" indicates that we have a right-tailed test. The question "Larger than what?" tells us that $\sigma_0 = 100$, giving us

$$H_0 : \sigma = 100 \quad \text{versus} \quad H_a : \sigma > 100$$

We reject H_0 if the *p*-value $\leq \alpha = 0.05$.

Step 2 **Find** χ^2_{data}.
We use the Step-by-Step Technology Guide on page 562. The Minitab descriptive statistics in Figure 46 tell us that the sample variance is $s^2 = 17{,}742.5$ calories squared. Thus,

$$\chi^2_{\text{data}} = \frac{(n-1)s^2}{\sigma_0^2} = \frac{(10-1)(17{,}742.5)}{100^2} \approx 15.97$$

Step 3 **Find the** *p***-value.**
For our right-tailed test, Table 14 tells us that

$$p\text{-value} = P(\chi^2 > \chi^2_{\text{data}}) = P(\chi^2 > 15.97)$$

That is, the *p*-value is the area to the right of $\chi^2_{\text{data}} = 15.97$, as shown in Figure 47. To find the *p*-value, we use the instructions provided in the Step-by-Step Technology Guide provided at the end of this section. The TI-83/84 results shown in Figure 48 tell us that *p*-value $= P(\chi^2 > 15.97) = 0.0675107153 \approx 0.0675$.

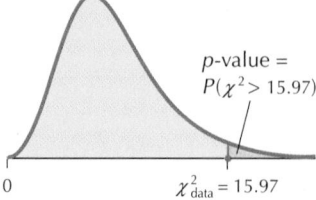

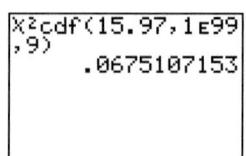

FIGURE 47 *p*-Value for χ^2 test

FIGURE 48 TI-83/84 results.

Percent (99, 95, 90, 80, 70, 60, 50, 40, 30, 20, 10, 5, 1) Calories (0, 250, 500, 750)

FIGURE 45 Normal probability plot of calories.

Descriptive Statistics: Calories

Variable	Total Count	Mean	StDev	Variance
Calories	10	321.7	133.2	17742.5

FIGURE 46 Calories descriptive statistics.

NOW YOU CAN DO
Exercises 13–18.

Step 4 **State the conclusion and the interpretation.**
Because *p*-value $= 0.0675$ is not $\leq \alpha = 0.05$, we do not reject H_0. There is insufficient evidence that the population standard deviation is greater than 100 calories.

3 Using Confidence Intervals for σ to Perform Two-Tailed Hypothesis Tests for σ

Suppose we have a $100(1 - \alpha)\%$ confidence interval for σ, of the form (lower bound, upper bound), and are interested in two-tailed hypothesis tests using level of significance α of the form:

$$H_0 : \sigma = \sigma_0 \quad \text{versus} \quad H_a : \sigma \neq \sigma_0$$

We will not reject H_0 for values of σ_0 that lie between the lower bound and upper bound of the confidence interval, and we will reject H_0 for values of σ_0 that lie outside this interval.

EXAMPLE 33   Using confidence intervals for σ to conduct two-tailed χ^2 tests for σ

A 95% confidence interval for the population mean sodium content of breakfast cereals, in milligrams (mg) per serving, is given by

$$(44.53 \text{ mg}, 81.50 \text{ mg})$$

Assume that the data are normally distributed. Test, using level of significance $\alpha = 0.05$, whether σ differs from the following.

a. 80 mg

b. 40 mg

Solution

a. For the hypothesis test $H_0 : \sigma = 80$ versus $H_a : \sigma \neq 80$, $\sigma_0 = 80$ lies between the lower bound 44.53 and the upper bound 81.50 of the confidence interval, and we therefore do not reject H_0. There is insufficient evidence that the population standard deviation of sodium content differs from 80 mg.

b. For the hypothesis test $H_0 : \sigma = 40$ versus $H_a : \sigma \neq 40$, $\sigma_0 = 40$ lies outside the confidence interval, and we therefore reject H_0. There is evidence that the population standard deviation of sodium content differs from 40 mg.

NOW YOU CAN DO
Exercises 19–22.

STEP-BY-STEP TECHNOLOGY GUIDE: Finding χ^2 *p*-value

We will use the information from Example 32 (page 559). The steps for finding the χ^2 critical values are given in the Step-by-Step Technology Guide at the end of Section 8.4 (pages 479–480).

TI-83/84
Step 1   Press **2nd > DISTR**, then χ^2 **cdf(**, and press **ENTER**.
Step 2   On the home screen, enter the value of χ^2_{data}, comma, 1E99, comma, degrees of freedom, close parenthesis, as shown

in Figure 48. (Remember that this "E" is inserted by pressing **2nd**, followed by the comma key.)
Step 3   Press **ENTER**. The results are shown in Figure 48 of Example 32.

EXCEL
Step 1   Select cell **A1**. Click the **Insert Function** icon f_x.
Step 2   For **Search for a Function**, type **chidist**, click **Go**, then click **OK**.

Step 3   For **X**, enter the value of χ^2_{data}, and for **Deg_freedom**, enter the degrees of freedom. Excel displays the *p*-value in the cell in the dialog box.

MINITAB

Step 1 Click on **Calc > Probability Distributions > Chi-Square**.
Step 2 Select **Cumulative probability**, and enter the **Degrees of freedom**.
Step 3 For **Input constant**, enter the value of χ^2_{data} and click **OK**.

Step 4 Minitab displays the area to the left of χ^2_{data} in the session window. To find the *p*-value, subtract this area from 1.

SPSS

Step 1 Enter a number into the first element of the spreadsheet.
Step 2 Select **Transform > Compute Variable…**. Enter the variable name under **Target Variable**. Select CDF and Noncentral CDF under **Function group**, and double-click **Cdf. Chisq** under **Functions and Special Variables**.

Step 3 Replace the two question marks in **Numeric Expression** with the value of χ^2_{data} and the degrees of freedom, respectively. Click **OK**, then **OK**. Minimize the output window.
Step 4 SPSS displays the area to the left of χ^2_{data} in the data spreadsheet. To find the *p*-value, subtract this area from 1.

JMP

Step 1 Click **File > New > Data Table**. Enter a number into the first element of Column 1. Select **Column 1** and click **Cols > Formula…**.
Step 2 Select **Probability** from the **Functions (grouped)** drop-down menu, then select **ChiSquare Distribution** from the list below the menu.

Step 3 In the formula, replace **x** with the value of χ^2_{data} and **DF** with the degrees of freedom. Click **OK**.
Step 4 JMP displays the area to the left of χ^2_{data} in the data spreadsheet. To find the *p*-value, subtract this area from 1.

CRUNCHIT!

Step 1 Click **Distribution Calculator** and select **Chi-square**.
Step 2 For **df** enter the degrees of freedom.

Step 3 Select the **Probability** tab. From the drop-down menu, select **>**. In the box, enter the value of χ^2_{data} and click **Calculate**.

Section 9.6 Summary

1. Under the assumption that H_0 is true, the χ^2 statistic takes the following form:

$$\chi^2_{\text{data}} = \frac{(n-1)s^2}{\sigma_0^2}$$

The hypothesis test about σ may be performed using the *p*-value method or the critical-value method. Either way, the test is valid only if we have a random sample from a normal population. The critical-value method compares χ^2_{data} with one or two critical values.

2. The *p*-value method compares the *p*-value to the level of significance α.

3. If we have a $100(1-\alpha)\%$ confidence interval for σ, of the form (lower bound, upper bound), and are interested in a two-tailed hypothesis test, using α, of the form

$$H_0 : \sigma = \sigma_0 \quad \text{versus} \quad H_a : \sigma \neq \sigma_0$$

we will not reject H_0 for values of σ_0 that lie between the lower bound and upper bound of the confidence interval. We will reject H_0 for values of σ_0 that lie outside this interval.

Section 9.6 Exercises

CLARIFYING THE CONCEPTS

1. Think of one instance where an analyst would be interested in performing a hypothesis test about the population standard deviation σ. (p. 556)
2. What is the difference between σ and σ_0? (p. 558)
3. Does it make sense to test whether $\sigma < 0$? Explain. (p. 556)

4. What condition must be fulfilled for us to perform a hypothesis test about σ? (p. 556)
5. Explain how we can use a confidence interval to determine significance. (p. 561)
6. In the previous exercise, what must be the relationship between α and the confidence level? (p. 561)

PRACTICING THE TECHNIQUES

✔ **CHECK IT OUT!**

To do	Check out	Topic
Exercises 7–12	Example 31	χ^2 test for σ: critical-value method
Exercises 13–18	Example 32	χ^2 test for σ: p-value method
Exercises 19–22	Example 33	Using confidence intervals for σ to conduct two-tailed χ^2 tests for σ

For Exercises 7–12, a random sample is drawn from a normal population. For each exercise, do the following:
a. State the hypotheses.
b. Find the χ^2 critical values and state the rejection rule.
c. Calculate χ^2_{data}.
d. Compare χ^2_{data} with the critical value or values. State the conclusion and interpretation.

7. Test whether $\sigma > 1$, using $\alpha = 0.05$. We have a sample of size $n = 21$ and a sample variance of $s^2 = 3$.
8. Test whether $\sigma < 5$, using $\alpha = 0.05$. We have a sample of size $n = 11$ and a sample variance of $s^2 = 25$.
9. Test whether $\sigma \neq 3$, using $\alpha = 0.05$. We have a sample of size $n = 16$ and a standard deviation of $s = 2.5$.
10. Test whether $\sigma > 10$, using $\alpha = 0.01$. We have a sample of size $n = 14$ and a standard deviation of $s = 12$.
11. Test whether $\sigma < 20$, using $\alpha = 0.10$. We have a sample of size $n = 8$ and a sample variance of $s^2 = 350$.
12. Test whether $\sigma \neq 5$, using $\alpha = 0.05$. We have a sample of size $n = 26$ with a standard deviation of $s = 5$.

For Exercises 13–18, a random sample is drawn from a normal population. Do the following:
a. State the hypotheses and the rejection rule.
b. Calculate χ^2_{data}. Draw a χ^2 distribution and indicate the location of χ^2_{data}.
c. Find the p-value and indicate the p-value in your distribution in **(b)**.
d. Compare the p-value with level of significance $\alpha = 0.05$. State the conclusion and interpretation.

13. We are testing whether $\sigma > 1$ and have a sample of size $n = 21$ with a sample variance of $s^2 = 3$.
14. We are testing whether $\sigma < 5$ and have a sample of size $n = 11$ with a sample variance of $s^2 = 25$.
15. We are testing whether $\sigma \neq 3$ and have a sample of size $n = 16$ with a standard deviation of $s = 2.5$.
16. We are testing whether $\sigma > 10$ and have a sample of size $n = 14$ with a standard deviation of $s = 12$.
17. We are testing whether $\sigma < 20$, and have a sample of size $n = 8$ and a sample variance of $s^2 = 350$.
18. We are testing whether $\sigma \neq 5$ and have a sample of size $n = 26$ with a standard deviation of $s = 5$.

For Exercises 19–22, a $100(1 - \alpha)\%$ χ^2 confidence interval for σ is given. Use the confidence interval to test, using level of significance α, whether σ differs from each of the indicated hypothesized values.

19. A 95% χ^2 confidence interval for σ is (1, 4). Hypothesized values σ_0 are
  a. 0
  b. 2
  c. 5
20. A 99% χ^2 confidence interval for σ is (10, 25). Hypothesized values σ_0 are
  a. 15
  b. 26
  c. 5
21. A 90% χ^2 confidence interval for σ is (100, 200). Hypothesized values σ_0 are
  a. 150
  b. 250
  c. 0
22. A 95% χ^2 confidence interval for σ is (127, 698). Hypothesized values σ_0 are
  a. 125
  b. 128
  c. 700

APPLYING THE CONCEPTS

23. Biomass Power Plants. Power plants around the country are retooling in order to consume biomass instead of or in addition to coal. The table contains a random sample of nine such power plants and the amount of energy generated in megawatts (MW) in 2014.[16] The normality was checked in the Section 8.4 exercises. Test whether the population standard deviation differs from 25 MW, using level of significance $\alpha = 0.05$. **biomass**

Company	Location	Capacity (MW)
Hoge Lumber Co.	New Knoxville, Ohio	3.7
Evergreen Clean Energy	Eagle, CO	12.0
GreenHunter Energy	Grapevine, TX	18.5
Covanta Energy Corporation	Niagara Falls, NY	30.0
Northwest Energy Systems Co.	Warm Springs, OR	37.0
Riverstone Holdings	Kenansville, NC	44.1
Lee County Solid Waste Authority	Ft. Myers, FL	57.0
Energy Investor Funds	Detroit, MI	68.0
Dominion Virginia Power	Hurt, VA	83.0

24. Carbon Emissions. The following table represents the carbon emissions (in millions of tons) from consumption of fossil fuels, for a random sample of five nations.[17] Test whether the population standard deviation is greater than 100 million tons, using level of significance $\alpha = 0.10$.

 **carbon**

Nation	Emissions (in millions of tons)
Brazil	361
Germany	844
Mexico	398
Great Britain	577
Canada	631

25. Calories in Breakfast Cereals. A random sample of six well-known breakfast cereals yielded the following calorie data. Can we perform a χ^2 test for the population standard deviation of the number of calories? Why or why not?

cerealcalories

Cereal	Calories
Apple Jacks	110
Cocoa Puffs	110
Mueslix	160
Cheerios	110
Corn Flakes	100
Shredded Wheat	80

26. *Deepwater Horizon* Cleanup Costs. The following table represents the amount of money disbursed by BP to a random sample of six Florida counties, for cleanup of the *Deepwater Horizon* oil spill, in millions of dollars.[18] The normality of the data was confirmed in the Section 8.1 exercises. Test whether the population standard deviation is less than $500,000, using level of significance $\alpha = 0.05$. deepwaterclean

County	Cleanup costs ($ millions)
Broward	0.85
Escambia	0.70
Franklin	0.50
Pinellas	1.15
Santa Rosa	0.50
Walton	1.35

27. Does Score Variability Differ by Gender? Recently, researchers have been examining the evidence for whether there is greater variability in boys' scores than girls' scores on cognitive abilities tests. For example, one study found that boys were overrepresented at both the top and the bottom of nonverbal reasoning tests and quantitative reasoning tests.[19] Suppose that the standard deviation for girls' scores is known to be 50 points for a particular test and that the population of all scores is normal. A random sample of 101 boys has a sample variance of 2600. Test whether the population standard deviation for boys exceeds 50 points, using level of significance $\alpha = 0.05$.

28. Heart Rate Variability. A reduction in heart rate variability is associated with elevated levels of stress, because the body continues to pump adrenaline after high-stress situations, even when at rest.[20] Suppose the standard deviation of heartbeats in the general population is 20 beats per minute, and that the population of heart rates is normal. A random sample of 50 individuals leading high-stress lives has a sample variance of 200 beats per minute. Test, using level of significance $\alpha = 0.05$, whether the population standard deviation for those leading high-stress lives is lower than that in the general population.

9.7 Probability of a Type II Error and the Power of a Hypothesis Test

OBJECTIVES By the end of this section, I will be able to . . .

1 Calculate the probability of a Type II error for a Z test for μ.
2 Compute the power of a Z test for μ and construct a power curve.

1 Probability of a Type II Error

In Section 9.1, we defined a Type II error as follows:

Type II error: not rejecting H_0 when H_0 is false

For example, the criminal trial scenario on page 494 had the following hypotheses:

H_0: defendant is not guilty   versus H_a: defendant is guilty

In this case, a Type II error was to find the defendant not guilty (not reject H_0) when in reality he did commit the crime (H_0 is false). In this section, we learn how to calculate the probability of making a Type II error for a Z test for μ, called β (beta), and to use the value of β to compute the power of a Z test for μ.

Calculating β, the Probability of a Type II Error

Use the following steps to calculate β, the probability of a Type II error.

Step 1 Recall that Z_{crit} divides the critical region from the noncritical region. Let $\bar{x}_{crit}$ be the value of the sample mean $\bar{x}$ associated with Z_{crit}. The following table shows how to calculate $\bar{x}_{crit}$ for the three forms of the hypothesis test.

Form of test			Value of $\bar{x}_{crit}$
Right-tailed	$H_0 : \mu = \mu_0$ vs.	$H_a : \mu > \mu_0$	$\bar{x}_{crit} = \mu_0 + Z_{crit} \cdot \dfrac{\sigma}{\sqrt{n}}$
Left-tailed	$H_0 : \mu = \mu_0$ vs.	$H_a : \mu < \mu_0$	$\bar{x}_{crit} = \mu_0 - Z_{crit} \cdot \dfrac{\sigma}{\sqrt{n}}$
Two-tailed	$H_0 : \mu = \mu_0$ vs.	$H_a : \mu \neq \mu_0$	$\bar{x}_{crit, lower} = \mu_0 - Z_{crit} \cdot \dfrac{\sigma}{\sqrt{n}}$
			$\bar{x}_{crit, upper} = \mu_0 + Z_{crit} \cdot \dfrac{\sigma}{\sqrt{n}}$

Here, μ_0 is the hypothesized value of the population mean, σ is the population standard deviation, and n is the sample size.

Step 2 Let μ_a represent a particular value for the population mean μ chosen from the values indicated in the alternative hypothesis H_a. Draw a normal curve centered at μ_a, with the value or values of $\bar{x}_{crit}$ from Step 1 indicated (see Example 34).

Step 3 Calculate β for the particular μ_a chosen using the following table.

Form of test			β = probability of Type II error
Right-tailed	$H_0 : \mu = \mu_0$ vs.	$H_a : \mu > \mu_0$	The area under the normal curve drawn in Step 2 to the left of $\bar{x}_{crit}$.
Left-tailed	$H_0 : \mu = \mu_0$ vs.	$H_a : \mu < \mu_0$	The area under the normal curve drawn in Step 2 to the right of $\bar{x}_{crit}$.
Two-tailed	$H_0 : \mu = \mu_0$ vs.	$H_a : \mu \neq \mu_0$	The area under the normal curve drawn in Step 2 between $\bar{x}_{crit, lower}$ and $\bar{x}_{crit, upper}$.

Let us illustrate the steps for calculating β, the probability of a Type II error, using an example.

EXAMPLE 34 Calculating β, the probability of a Type II error

ATM network operator Star Systems of San Diego reported that active users of debit cards used them an average of 11 times per month. Suppose we are interested in testing whether people use debit cards on average more than 11 times per month, using level of significance $\alpha = 0.01$. The hypotheses are

$$H_0 : \mu = 11 \quad \text{versus} \quad H_a : \mu > 11$$

where μ represents the population mean debit card usage per month. Suppose we have $n = 36$, $\bar{x} = 11.5$, and $\sigma = 3$, and from Table 4 (page 500) we have $Z_{\text{crit}} = 2.33$.

a. State what a Type II error would be in this case.

b. Let $\mu_a = 13$. That is, suppose the population mean debit card usage is actually 13 times per month. Calculate β, the probability of making a Type II error when $\mu_a = 13$.

Solution

a. We make a Type II error when we do not reject H_0 when H_0 is false. In this case, a Type II error would be to conclude that the population mean debit card usage was 11 times per month when in actuality it was more than 11 times per month.

b. We follow the steps for calculating β.

Step 1 We have a right-tailed test, so that

$$\bar{x}_{\text{crit}} = \mu_0 + Z_{\text{crit}} \cdot \frac{\sigma}{\sqrt{n}} = 11 + 2.33 \cdot \frac{3}{\sqrt{36}} = 12.165$$

Step 2 Figure 49 shows the normal curve centered at $\mu_a = 13$, with $\bar{x}_{\text{crit}} = 12.165$ labeled.

FIGURE 49
β probability of Type II error.

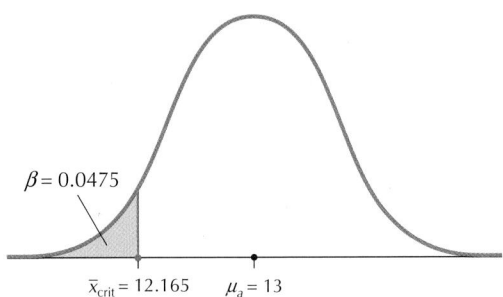

Step 3 The right-tailed test tells us that β equals the area under the normal curve drawn in Step 2 to the left of $\bar{x}_{\text{crit}} = 12.165$. This is the shaded area in Figure 49. Area represents probability, so we have

$$\beta = P(\bar{x} < 12.165) \text{ when } \mu_a = 13$$

Standardizing with $\mu_a = 13$, $\sigma = 3$, and $n = 36$:

$$\beta = P(\bar{x} < 12.165)$$

$$= P\left(Z < \frac{12.165 - 13}{3/\sqrt{36}}\right)$$

$$= P(Z < -1.67) = 0.0475$$

This is a Case 1 problem from Table 8 of Chapter 6 on page 355.

NOW YOU CAN DO
Exercises 5–16.

Thus, $\beta = 0.0475$. This represents the probability of making a Type II error, that is, of not rejecting the hypothesis that the population mean debit card usage is 11 times per month when in actuality it is 13 times per month.

2 Power of a Hypothesis Test

It is a correct decision to reject the null hypothesis when the null hypothesis is false. The probability of making this type of correct decision is called the **power of the test**.

> **Power of a Hypothesis Test**
> The **power of a hypothesis test** is the probability of rejecting the null hypothesis when the null hypothesis is false. Power is calculated as
> $$\text{power} = 1 - \beta$$

EXAMPLE 35 Power of a hypothesis test

Calculate the power, for the particular alternative value of the mean, of the hypothesis test in Example 34.

Solution

The probability of a Type II error was found in Example 34 to be $\beta = 0.0475$. Thus, the power of the hypothesis test is

$$\text{power} = 1 - \beta = 1 - 0.0475 = 0.9525$$

NOW YOU CAN DO
Exercises 17–28.

The probability of correctly rejecting the null hypothesis is 0.9525.

WHAT IF
?

What if Scenario: Type II Errors and Power of the Test

Suppose that we have the same hypothesis test from Example 34 and the same value $\bar{x}_{\text{crit}} = 12.165$. Now, *what if* we decrease μ_a such that it is less than 13 but still larger than 12.165? Describe what will happen to the following, and why:

a. The probability of a Type II error, β

b. The power of the test, $1 - \beta$

Solution

a. Consider Figure 50. The distribution of sample means remains centered at μ_a, so that a smaller μ_a will "slide" the normal curve toward the value of $\bar{x}_{\text{crit}} = 12.165$. This results in a larger area to the left of 12.165, as you can see by comparing Figure 50 with Figure 49. Therefore, a smaller μ_a leads to an *increase* in the probability of a Type II error, β.

b. As β increases, $1 - \beta$ decreases. Therefore, a smaller μ_a leads to a decrease in the power of the test.

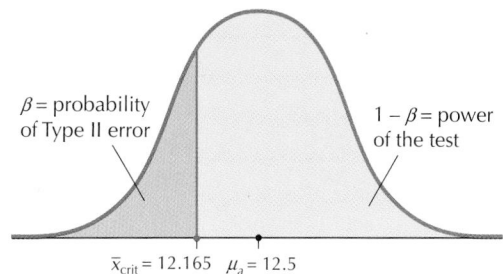

β = probability of Type II error

$1 - \beta$ = power of the test

$\bar{x}_{\text{crit}} = 12.165$ $\mu_a = 12.5$

FIGURE 50 Smaller μ_a leads to an increase in β.

A **power curve** plots the values for the power of the test versus the values of μ_a.

EXAMPLE 36 Power curve

a. Calculate the power of the hypothesis test from Example 34 for the following values of μ_a: 11.0, 11.5, 12.0, 12.165, 12.5, 13.5.

b. Construct the power curve by graphing the values for the power of the test on the vertical axis against the values of μ_a on the horizontal axis.

Solution

a. We have $\bar{x}_{\text{crit}} = 12.165$, $\sigma = 3$, and $n = 36$. The calculations are provided in the following table.

μ_a	Probability of Type II error: β	Power of the test: $1 - \beta$
11.0	$P\!\left(Z < \dfrac{12.165 - 11}{3/\sqrt{36}}\right) = P(Z < 2.33) = 0.9901$	$1 - 0.9901 = 0.0099$
11.5	$P\!\left(Z < \dfrac{12.165 - 11.5}{3/\sqrt{36}}\right) = P(Z < 1.33) = 0.9082$	$1 - 0.9082 = 0.0918$
12.0	$P\!\left(Z < \dfrac{12.165 - 12}{3/\sqrt{36}}\right) = P(Z < 0.33) = 0.6293$	$1 - 0.6293 = 0.3707$
12.165	$P\!\left(Z < \dfrac{12.165 - 12.165}{3/\sqrt{36}}\right) = P(Z < 0.00) = 0.5$	$1 - 0.5 = 0.5$
12.5	$P\!\left(Z < \dfrac{12.165 - 12.5}{3/\sqrt{36}}\right) = P(Z < -0.67) = 0.2514$	$1 - 0.2514 = 0.7486$
13.5	$P\!\left(Z < \dfrac{12.165 - 13.5}{3/\sqrt{36}}\right) = P(Z < -2.67) = 0.0038$	$1 - 0.0038 = 0.9962$

b. Figure 51 represents a power curve, because it plots the values for the power of the test on the vertical axis against the values of μ_a on the horizontal axis. Note that, as μ_a moves farther away from the hypothesized mean $\mu_0 = 11$, the power of the test increases. This is because it is more likely that the null hypothesis will be correctly rejected as the actual value of the mean μ_a gets farther away from the hypothesized value μ_0.

For completeness, we include the power for $\mu_a = 13$ from Example 35 in this power curve.

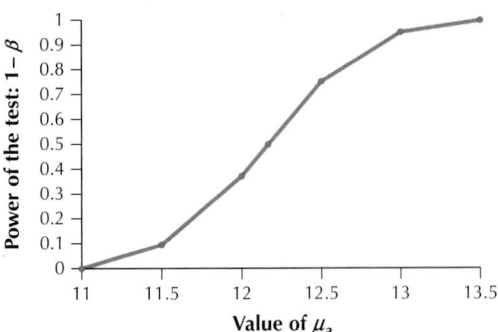

NOW YOU CAN DO
Exercises 29 and 30.

FIGURE 51 Power curve.

Section 9.7 Summary

1. We may calculate β, the probability of making a Type II error for a Z test for μ, given a particular alternative value for the population mean, μ_a.

2. The power of a hypothesis test is the probability of rejecting a false null hypothesis, and is calculated as *power* $= 1 - \beta$. We may then build the power curve by plotting the power against values of μ_a.

Section 9.7 Exercises

CLARIFYING THE CONCEPTS

1. Explain what a Type II error is. (p. 565)
2. Describe what $\bar{x}_{crit}$ is. (p. 565)
3. In words, what do we mean by the power of a hypothesis test? (p. 567)
4. How do we calculate the power of a test? (p. 567)

PRACTICING THE TECHNIQUES

 CHECK IT OUT!

To do	Check out	Topic
Exercises 5–16	Example 34	Calculating β, the probability of a Type II error
Exercises 17–28	Example 35	Power of a hypothesis test
Exercises 29–30	Example 36	Power curve

For Exercises 5–16, assume that the conditions for performing the Z test are met.

For Exercises 5–16, do the following:
a. Calculate the value or values of $\bar{x}_{crit}$.
b. Draw a normal curve, centered at μ_a, with the value or values of $\bar{x}_{crit}$ indicated.
c. Calculate β, the probability of a Type II error for that value of μ_a. Shade the corresponding area under the normal curve.

5. $H_0 : \mu = 50$ vs. $H_a : \mu > 50$, $\alpha = 0.10$, $\sigma = 4$, $n = 25$, $\mu_a = 51$
6. $H_0 : \mu = 50$ vs. $H_a : \mu > 50$, $\alpha = 0.10$, $\sigma = 4$, $n = 25$, $\mu_a = 52$
7. $H_0 : \mu = 50$ vs. $H_a : \mu > 50$, $\alpha = 0.10$, $\sigma = 4$, $n = 25$, $\mu_a = 53$
8. $H_0 : \mu = 50$ vs. $H_a : \mu > 50$, $\alpha = 0.10$, $\sigma = 4$, $n = 25$, $\mu_a = 54$
9. $H_0 : \mu = 50$ vs. $H_a : \mu > 50$, $\alpha = 0.10$, $\sigma = 4$, $n = 25$, $\mu_a = 55$
10. $H_0 : \mu = 50$ vs. $H_a : \mu > 50$, $\alpha = 0.10$, $\sigma = 4$, $n = 25$, $\mu_a = 56$
11. $H_0 : \mu = 100$ vs. $H_a : \mu < 100$, $\alpha = 0.05$, $\sigma = 12$, $n = 36$, $\mu_a = 96$
12. $H_0 : \mu = 100$ vs. $H_a : \mu < 100$, $\alpha = 0.05$, $\sigma = 12$, $n = 36$, $\mu_a = 94$
13. $H_0 : \mu = 100$ vs. $H_a : \mu < 100$, $\alpha = 0.05$, $\sigma = 12$, $n = 36$, $\mu_a = 92$

14. $H_0 : \mu = 100$ vs. $H_a : \mu < 100$, $\alpha = 0.05$, $\sigma = 12$, $n = 36$, $\mu_a = 90$
15. $H_0 : \mu = 100$ vs. $H_a : \mu < 100$, $\alpha = 0.05$, $\sigma = 12$, $n = 36$, $\mu_a = 88$
16. $H_0 : \mu = 100$ vs. $H_a : \mu < 100$, $\alpha = 0.05$, $\sigma = 12$, $n = 36$, $\mu_a = 86$

For Exercises 17–28, calculate the power of the hypothesis test for the indicated exercise.

17. Exercise 5
18. Exercise 6
19. Exercise 7
20. Exercise 8
21. Exercise 9
22. Exercise 10
23. Exercise 11
24. Exercise 12
25. Exercise 13
26. Exercise 14
27. Exercise 15
28. Exercise 16
29. Refer to Exercises 17–22. Construct the power curve for the given values of μ_a.
30. Refer to Exercises 23–28. Construct the power curve for the given values of μ_a.

APPLYING THE CONCEPTS

31. Stock Market. The *Statistical Abstract of the United States* reports that the mean daily number of shares traded on the New York Stock Exchange in 2009 was 2.9 billion. Let this value represent the hypothesized population mean, and assume that the population standard deviation equals 0.7 billion shares. Suppose that we have a random sample of 36 days from the present year, and we are interested in testing whether the population mean daily number of shares traded has increased, using level of significance $\alpha = 0.05$.
    a. Describe what a Type II error would mean in the context of this problem.
    b. What is the probability of making a Type II error when the actual mean number of shares traded equals the following values?
      i. 3.0 billion **ii.** 3.1 billion
      iii. 3.2 billion **iv.** 3.3 billion
    c. Calculate the power of the hypothesis test for the values of μ_a given in **(b)**.
    d. Construct the power curve for the values of μ_a given in **(b)**.

32. Credit Scores in Georgia. According to CreditReport.com, the mean credit score in Georgia in 2014 was 668. Suppose we have a recent random sample of 900 credit scores in Georgia, and assume that the population standard deviation is 150. We are interested in testing using level of significance $\alpha = 0.05$ whether the population mean credit score in Georgia has decreased since that time.

    a. Describe what a Type II error would mean in the context of this problem.

    b. What is the probability of making a Type II error when the actual mean credit score takes the following values?
      i. 650   **ii.** 645   **iii.** 640   **iv.** 635

    c. Calculate the power of the hypothesis test for the values of μ_a given in **(b)**.

    d. Construct the power curve for the values of μ_a given in **(b)**.

33. Accountants' Salaries. According to Salary.com, the mean salary for entry-level accountants in 2010 was $41,560. Let this value represent the hypothesized population mean, and assume that the population standard deviation equals $5000. Suppose we have a recent random sample of 100 entry-level accountants and want to test, using level of significance $\alpha = 0.05$, whether the population mean salary has changed since 2010.

    a. Describe what a Type II error would mean in the context of this problem.

    b. What is the probability of making a Type II error when the actual mean salary takes the following values?
      i. $42,000   **ii.** $43,000   **iii.** $44,000   **iv.** $45,000

    c. Calculate the power of the hypothesis test for the values of μ_a given in **(b)**.

    d. Construct the power curve for the values of μ_a given in **(b)**.

34. Price of Milk. The U.S. Bureau of Labor Statistics reports that the mean price for a gallon of milk in June 2014 was $3.63. Suppose that we have a random sample taken this year of 400 gallons of milk, and assume that the population standard deviation equals $1.00. We want to conduct a hypothesis test, using level of significance $\alpha = 0.01$, to investigate if the population mean price of milk this year has increased.

    a. Describe what a Type II error would mean in the context of this problem.

    b. What is the probability of making a Type II error when the actual mean price takes the following values?
      i. $3.70   **ii.** $3.90   **iii.** $4.10   **iv.** $4.30

    c. Calculate the power of the hypothesis test for the values of μ_a given in **(b)**.

    d. Construct the power curve for the values of μ_a given in **(b)**.

Chapter 9 Formulas and Vocabulary

SECTION 9.1
- **α (alpha)** (p. 495)
- **Alternative hypothesis** (p. 489)
- **β (beta)** (p. 495)
- **Hypothesis testing** (p. 488)
- **Level of significance** (p. 495)
- **Null hypothesis** (p. 489)
- **Statistical significance** (p. 492)
- **Type I error** (p. 494)
- **Type II error** (p. 494)

SECTION 9.2
- **Critical region** (p. 500)
- **Critical value, Z_{crit}** (p. 500)
- **Essential idea about hypothesis testing for the mean** (p. 498)
- **Level of significance α** (p. 500)
- **Noncritical region** (p. 500)
- **Test statistic** (p. 499)
- **Z_{data}** (p. 499).

$$Z_{data} = \frac{\bar{x} - \mu_0}{\sigma/\sqrt{n}}$$

SECTION 9.3
- **p-Value** (p. 508)
- **Rejection rule for performing a hypothesis test using the p-value method** (p. 509)

SECTION 9.4
- **Critical value, t_{crit}** (p. 525)
- **t_{data}** (p. 525).

$$t_{data} = \frac{\bar{x} - \mu_0}{s/\sqrt{n}}$$

SECTION 9.5
- **Essential idea about hypothesis testing for the proportion** (p. 543)
- **Standard error of the proportion** (p. 543)
- **Z_{data} for the hypothesis test for the population proportion** (p. 543).

$$Z_{data} = \frac{\hat{p} - p_0}{\sqrt{\dfrac{p_0(1 - p_0)}{n}}}$$

SECTION 9.6
- **χ^2_{data}** (p. 556).

$$\chi^2_{data} = \frac{(n - 1)s^2}{\sigma_0^2}$$

- **Essential idea about hypothesis testing for the standard deviation** (p. 556)

SECTION 9.7
- **Power curve** (p. 567)
- **Power of a hypothesis test** (p. 567)

Chapter 9 Review Exercises

SECTION 9.1

For Exercises 1–3, provide the null and alternative hypotheses.
1. Test whether $\mu < 12$.
2. Test whether $\mu > 10$.
3. Test whether μ is below zero.

For Exercises 4–6, do the following.
   a. Provide the null and alternative hypotheses.
   b. Describe the two ways a correct decision could be made.
   c. Describe what a Type I error would mean in the context of the problem.
   d. Describe what a Type II error would mean in the context of the problem.
4. **Household Size.** The U.S. Census Bureau reported (2010) that the mean household size is 2.58 persons. We conduct a hypothesis test to determine whether the population mean household size has changed.
5. **Speeding-Related Traffic Fatalities.** The National Highway Traffic Safety Administration reports that the mean number of speeding-related traffic fatalities over the Thanksgiving holiday period from 1994 to 2003 was 202.7. We conduct a hypothesis test to examine whether the population mean number of such fatalities has decreased.
6. **Salaries of Accounting Associate Professors.** Salary.com reports that the mean salary for accounting associate professors in 2014 was $94,000. A hypothesis test was conducted to determine if the population mean salary of accounting associate professors has increased.

SECTION 9.2

For Exercises 7–9, find the value of Z_{data}.
7. $\bar{x} = 59$, $\sigma = 10$, $n = 100$, $\mu_0 = 60$
8. $\bar{x} = 59$, $\sigma = 5$, $n = 100$, $\mu_0 = 60$
9. $\bar{x} = 59$, $\sigma = 1$, $n = 100$, $\mu_0 = 60$

For each of the following hypothesis tests in Exercises 10–12, do the following:
   a. Find the value of Z_{crit}.
   b. Find the critical-value rejection rule.
   c. Draw a standard normal curve and indicate the critical region.
   d. State the conclusion and interpretation.
10. $H_0 : \mu = \mu_0$ vs. $H_a : \mu \neq \mu_0$, $\alpha = 0.01$, $Z_{data} = -2.5$
11. $H_0 : \mu = \mu_0$ vs. $H_a : \mu > \mu_0$, $\alpha = 0.10$, $Z_{data} = 1.5$
12. $H_0 : \mu = \mu_0$ vs. $H_a : \mu > \mu_0$, $\alpha = 0.05$, $Z_{data} = -2.5$
13. **Credit Scores in Georgia.** According to CreditReport.com, the mean credit score in Georgia in 2014 was 668. A random sample of 144 Georgia residents this year shows a mean credit score of 650. Assume $\sigma = 50$. Perform a hypothesis test, using level of significance $\alpha = 0.05$, to determine if the population mean credit score in Georgia has decreased.
   a. State the hypotheses.
   b. Find the value of Z_{crit} and the rejection rule. Also, draw a standard normal curve, indicating the critical region.

   c. Calculate Z_{data}. Draw a standard normal curve showing Z_{crit}, the critical region, and Z_{data}.
   d. State the conclusion and the interpretation.

SECTION 9.3

For Exercises 14 and 15, perform the following steps:
   a. State the hypotheses and the rejection rule for the p-value method.
   b. Calculate Z_{data}.
   c. Find the p-value. Draw the standard normal curve, with Z_{data} and the p-value indicated on it.
   d. State the conclusion and the interpretation.
14. We are interested in testing at level of significance $\alpha = 0.05$ whether the population mean differs from 500. A random sample of size 100 is taken, with a mean of 520. Assume $\sigma = 50$.
15. We want to test, at level of significance $\alpha = 0.01$, whether the population mean is less than -10. A random sample of size 25 is taken from a normal population. The sample mean is -12. Assume $\sigma = 2$.
16. **Sleeping During the Full Moon.** A study found that subjects slept on average 20 minutes less on nights of the full moon than otherwise.[21] Other researchers disagree and are interested in testing whether the mean difference in sleep is less than 20 minutes. A random sample of 100 subjects showed that they slept, on average, 10 minutes less on nights of the full moon than otherwise. Assume the population standard deviation is 15 minutes. Perform the appropriate hypothesis test using level of significance $\alpha = 0.01$.

SECTION 9.4

For Exercises 17–19, find the critical value t_{crit} and sketch the critical region. Assume normality.
17. $H_0 : \mu = 100$, $H_a : \mu > 100$, $n = 8$, $\alpha = 0.10$
18. $H_0 : \mu = 100$, $H_a : \mu > 100$, $n = 8$, $\alpha = 0.05$
19. $H_0 : \mu = 100$, $H_a : \mu > 100$, $n = 8$, $\alpha = 0.01$
20. Describe what happens to the t critical value t_{crit} for right-tailed tests as α decreases.
21. A random sample of size 16 from a normal population yields a sample mean of 10 and a sample standard deviation of 3. Test whether the population mean differs from 9, using level of significance $\alpha = 0.10$.
22. A random sample of size 144 from an unknown population yields a sample mean of 45 and a sample standard deviation of 10. Test whether the population mean differs from 45, using level of significance $\alpha = 0.10$.

SECTION 9.5

For Exercises 23–25, do the following:
   a. Check the normality conditions.
   b. State the hypotheses.
   c. Find Z_{crit} and the rejection rule.
   d. Calculate Z_{data}.
   e. State the conclusion and the interpretation.

23. Test whether the population proportion exceeds 0.8. A random sample of size 1000 yields 830 successes. Let $\alpha = 0.10$.

24. Test whether the population proportion is below 0.2. A random sample of size 900 yields 160 successes. Let $\alpha = 0.05$.

25. Test whether the population proportion is not equal to 0.4. A random sample of size 100 yields 55 successes. Let $\alpha = 0.01$.

For Exercises 26 and 27, do the following:
   a. Check the normality conditions.
   b. State the hypotheses and the rejection rule for the p-value method, using level of significance $\alpha = 0.05$.
   c. Calculate Z_{data}.
   d. Calculate the p-value.
   e. State the conclusion and the interpretation.

26. Test whether the population proportion differs from 0.7. A random sample of size 144 yields 110 successes.

27. Test whether the population proportion is less than 0.25. A random sample of size 100 yields 25 successes.

28. DSL Internet Service. The U.S. Department of Commerce reports that 41.6% of Internet users preferred DSL as their method of service delivery.[22] A random sample of 1000 Internet users shows 350 who preferred DSL. If appropriate, test whether the population proportion who prefer DSL has decreased, using level of significance $\alpha = 0.05$.

SECTION 9.6

For Exercises 29 and 30, assume normality of the data, and do the following:
   a. State the hypotheses.
   b. Find the χ^2 critical value or values, and state the rejection rule.
   c. Find χ^2_{data}. Also, draw a χ^2 distribution and indicate χ^2_{data} and the χ^2 critical value or values.
   d. State the conclusion and the interpretation.

29. We are testing whether $\sigma > 6$ and have a random sample of size 20 with a standard deviation of $s = 9$. Let $\alpha = 0.05$.

30. We are testing whether $\sigma \neq 10$ and have a random sample of size 26 with a sample variance of 90. Let $\alpha = 0.05$.

For Exercises 31 and 32, assume normality of the data, and do the following:
   a. State the hypotheses and the p-value rejection rule for $\alpha = 0.05$.
   b. Find χ^2_{data}.
   c. Find the p-value. Also, draw a χ^2 distribution and indicate χ^2_{data} and the p-value.
   d. State the conclusion and the interpretation.

31. We are testing whether $\sigma < 35$ and have a random sample of size eight with a sample variance of 1200.

32. We are testing whether $\sigma \neq 50$ and have a random sample of size 26 with a standard deviation of $s = 45$.

SECTION 9.7

For Exercises 33–38, assume that the conditions for performing the Z test are met. Do the following:
   a. Calculate the value or values of $\bar{x}_{crit}$.
   b. Draw a normal curve, centered at μ_a, with the value or values of $\bar{x}_{crit}$ indicated.
   c. Calculate β, the probability of a Type II error for that value of μ_a. Shade the corresponding area under the normal curve.
   d. Calculate the power of the hypothesis test.

33. $H_0 : \mu = 100$ vs. $H_a : \mu \neq 100$, $\alpha = 0.01$, $\sigma = 15$, $n = 64$, $\mu_a = 103$

34. $H_0 : \mu = 100$ vs. $H_a : \mu \neq 100$, $\alpha = 0.01$, $\sigma = 15$, $n = 64$, $\mu_a = 106$

35. $H_0 : \mu = 100$ vs. $H_a : \mu \neq 100$, $\alpha = 0.01$, $\sigma = 15$, $n = 64$, $\mu_a = 109$

36. $H_0 : \mu = 100$ vs. $H_a : \mu \neq 100$, $\alpha = 0.01$, $\sigma = 15$, $n = 64$, $\mu_a = 112$

37. $H_0 : \mu = 100$ vs. $H_a : \mu \neq 100$, $\alpha = 0.01$, $\sigma = 15$, $n = 64$, $\mu_a = 115$

38. Refer to Exercises 33–37. Construct the power curve for the given values of μ_a.

Chapter 9 QUIZ

TRUE OR FALSE

1. True or false: It is possible that both the null and alternative hypotheses are correct at the same time.

2. True or false: The conclusion you draw from performing the critical-value method for the Z test is the same as the conclusion you draw from performing the p-value method for the Z test.

3. True or false: We do not need the estimated p-value method if we have access to a computer or calculator.

FILL IN THE BLANK

4. To reject H_0 when H_0 is true is a Type _____ error.

5. An extreme value of $\bar{x}$ is associated with a _____ p-value.

6. The rejection rule for performing a hypothesis test using the p-value method is to reject H_0 when the p-value is less than or equal to _____.

SHORT ANSWER

7. Under what conditions may we apply the Z test for the population proportion?

8. What does a small p-value indicate with respect to the null hypothesis? A large p-value?

9. Does the value of Z_{data} change when the form of the hypothesis test changes (for example, left-tailed instead of right-tailed)?

CALCULATIONS AND INTERPRETATIONS

10. ATM Fees. Do you hate paying the extra fees imposed by banks when withdrawing funds from an automated teller machine (ATM) not owned by your bank? The Federal Reserve System reported in 2010 that the mean such fee is $1.14. A random sample of 36 such transactions yielded a mean of $1.07 in extra fees. Suppose the population standard deviation of such extra fees is $0.25.

a. Test, using level of significance $\alpha = 0.05$, whether there has been a reduction in the population mean fee charged on such transactions.

b. Which type of error is it possible that we are making, a Type I error or a Type II error? Which type of error are we certain we are not making?

11. Alcohol-Related Fatal Car Accidents. The National Traffic Highway Safety Commission keeps statistics on the "mean years of potential life lost" in alcohol-related fatal automobile accidents. For males, the mean years of life lost is 32. That is, on average, males involved in fatal drinking-and-driving accidents had their lives cut short by 32 years. A random sample of 36 alcohol-related fatal accidents had a mean years of life lost of 33.8, with a standard deviation of 6 years.

a. Test whether the population mean years of life lost has changed, using a t test and level of significance $\alpha = 0.10$.

b. Assess the strength of the evidence against the null hypothesis.

12. Biomass Power Plants. Power plants around the country are retooling in order to consume biomass instead of or in addition to coal. The following table contains a random sample of 10 such power plants and the amount of biomass they consumed in 2006, in trillions of BTU (British thermal units).[23] Test whether the population standard deviation is greater than 2 trillion BTU using level of significance $\alpha = 0.05$.

Power plant	Location	Biomass consumed (trillions of BTU)
Georgia Pacific Naheola Mill	Choctaw, AL	13.4
Jefferson Smurfit Fernandina Beach	Nassau, FL	12.9
International Paper Augusta Mill	Richmond, GA	17.8
Gaylord Container Bogalusa	Washington, LA	15.1
Escanaba Paper Company	Delta, MI	19.5
Weyerhaeuser Plymouth NC	Martin, NC	18.6
International Paper	Georgetown, SC	13.8
Bowater Newsprint	McMinn, TN	10.6
Covington Facility	Covington, VA	12.7
Mosinee Paper	Marathon, WI	17.6

10

Two-Sample Inference

Introduction

This chapter examines differences in the characteristics of two populations. In Section 10.1, students construct confidence intervals and perform hypothesis tests to compare two dependent samples. Section 10.2 introduces inference techniques for comparing independent samples. Section 10.3 introduces the sampling distribution of the difference in independent sample proportions $\hat{p}_1 - \hat{p}_2$ and shows how to construct confidence intervals and hypothesis tests for $p_1 - p_2$. Finally, Section 10.4 covers inference for two independent standard deviations.

A new Chapter 10 Case Study, *Bank Loans*, examines a pair of large data sets, of 5000 observations each, of bank loan applicants. The data comes from *Data Mining and Predictive Analytics* by Daniel Larose and Chantal Larose, Wiley, 2015. The data are analyzed in Sections 10.2 and 10.3.

From the Author

Section 10.1 Inference for Mean Difference—Dependent Samples

- Starting in Section 10.1, we have moved coverage of hypothesis testing ahead of confidence intervals for the remainder of the book, in line with common practice.

- Use of a t confidence interval for μ_d to perform hypothesis tests about μ_d: Instructors may want to point out the similarity of the inference for μ_d to that of the t interval in Section 8.2 and the t test in Section 9.4.

Section 10.2 Inference for Two Independent Means

- Students often confuse μ_d with $\mu_1 - \mu_2$. Instructors may want to point out that μ_d represents the population mean of the paired differences, whereas $\mu_1 - \mu_2$ refers to the difference in population means.

- We use the terms "Welch's Hypothesis Test" and "Welch's Confidence Interval" for methods not using the pooled variance.

- Note that the Chapter 10 Case Study, *Bank Loans*, is used throughout the section.

Section 10.3 Inference for Two Independent Proportions

- Instructors may want to inform students, as a check on their calculations, that $\hat{p}_{\text{pooled}}$ must lie between $\hat{p}_1$ and $\hat{p}_2$.

Section 10.4 Inference for Two Independent Standard Deviations

- Instructors may find that they do not always have time to cover Section 10.4.

Teaching Tips

The following are motivating examples for the four sections. For Section 10.1, compare the running times of students on the track team at the beginning and at the end of the season. For

Section 10.2, compare the amount of time per week that men and women spend studying. For Section 10.3, compare the difference between the proportion of men and the proportion of women who have a certain characteristic, such as blue eyes or an education major.

For Section 10.1, dependent samples, emphasize that, after taking the differences, what we are left with is essentially a *single sample* of the differences. Therefore, the formulas are similar to the t intervals and t tests from Chapters 8 and 9, except that we are estimating the population mean of the differences.

For Section 10.2, independent samples, mention that $\bar{x}_1 - \bar{x}_2$ plays the same role that $\bar{x}$ played in the t inference from Chapters 8 and 9. For Section 10.3, mention that $\hat{p}_1 - \hat{p}_2$ plays the same role that $\hat{p}$ played in the Z interval for the proportion and the Z test for the proportion from Chapters 8 and 9, respectively.

In-Class Activities

1. **EESEE case study:** *Is Friday the 13th Unhealthy?* Let students conduct the projects listed at the end of this case study.

2. **Confidence Interval for the Difference in True Mean Height Between Men and Women**

   a. What do you think is the difference in true mean height between men and women?

   b. Have each student generate a random sample of 10 male students from your class, using the random generator from the Chapter 1 activities.

   c. Have each student generate a random sample of 10 female students from your class, using the random generator from the Chapter 1 activities.

   d. Calculate the mean of each of your samples in (**b**) and (**c**). Then calculate the difference in sample means.

   e. Give each student a sticky note to record his or her name and the difference in sample means calculated in (**d**).

   f. Have each student place his or her sticky note on the number line that the instructor has prepared on the blackboard. We thus have a collection of the differences in sample means.

   g. Using the evidence collected in (**f**), what would be a point estimate of the difference in true mean height for men and women?

   h. Is it plausible that the differences in sample means are normally distributed?

   i. Using his or her own sample, have each student calculate a 90% confidence interval for the difference in true mean heights between men and women.

   j. Are there any confidence intervals that contain the zero value? What would this indicate?

3. **Hypothesis Test for the Difference in Population Mean Heights of Men and Women**

   a. Suppose we believe that no difference exists between the population mean heights of men and women, and we want to test this hypothesis. State the appropriate hypotheses.

   b. Suppose we found the difference between sample mean heights of men and women to be 0.1 inch. Do you think this would be sufficient evidence to reject the null hypothesis that both sexes have the same mean height?

   c. Suppose we found the difference between sample mean heights of men and women to be 1 inch. Do you think this would be sufficient evidence to reject the null hypothesis that both sexes have the same mean height?

   d. Suppose we found the difference between sample mean heights of men and women to be 5 inches. Do you think this would be sufficient evidence to reject the null hypothesis that both sexes have the same mean height?

   e. Generate a random sample of 10 male students from your class, using either (**a**) technology or (**b**) the *Simple Random Sample* applet. Measure the heights of these students.

   f. Generate a random sample of 10 female students from your class, using either (**a**) technology or (**b**) the *Simple Random Sample* applet. Measure the heights of these students.

   g. Calculate the mean of each of your samples in (**e**) and (**f**). Then calculate the difference in sample means.

   h. Do you think your difference in sample means is evidence against or in favor of the null hypothesis?

   i. Each student, using his or her own sample, should perform a hypothesis test to determine whether or not the data are consistent with the hypothesis that no difference exists between mean heights of men and women. Use level of significance $\alpha = 0.10$. Will everyone in the class necessarily report the same conclusion? Why or why not?

   j. Use the data from the entire class to perform a hypothesis test to determine whether or not the data are consistent with the hypothesis that no difference exists between mean heights of men and women. Use level of significance $\alpha = 0.10$. Are you guaranteed to come up with the same conclusion as in (**i**)? Why or why not?

Supplements

- StatTutor 18.6, 18.7(a), (b), and (c); 19.1–19.8; and 21.1–21.6
- Stats@Work Simulations (*Note*: Not all parts of the following simulations deal with two population parameters.)
  - Which Method to Use? Hypothesis Testing: Means and Proportions; Gary Pop
  - Which Method to Use? Hypothesis Testing: Means and Proportions; Sam Sport
  - Which Method to Use? Hypothesis Testing: Means and Proportions; Justine Red
  - Which Method to Use? Hypothesis Testing: Means and Proportions; Mindy Admin
- EESEE case studies
  - Sleeping Patterns in Ducks (Questions 1 and 2 on estimation and hypothesis tests for population means)
  - Surgery in a Blanket (Questions 1, 2, 4, 5, 6, and 7)

- Keeping Balance as You Age (Questions 1 and 4. Students can be asked to compare the two groups through confidence intervals and hypothesis testing.)
- Is Caffeine Dependence Real? (Questions 1, 2, 3, 4, and 5)
- Checkmating and Reading Skills (Questions 1, 2, 3, and 5)
- Leave Survey After the Beep (Question 3)

Videos

- *Against All Odds*: *Inside Statistics:* www.learner.org/resources/series65.html
  - Program 27: Comparing Two Means
  - Program 28: Inference for Proportions

Web Sites

- The site Online Statistics: An Interactive Multimedia Course of Study has information on hypothesis testing and confidence intervals, and it has a link to the Rice Virtual Lab in Statistics with numerous simulation demonstrations and other topics related to statistics: http://onlinestatbook.com/index.html.
- This Web site lists other sites that do statistical calculations: http://statpages.org/.
- This Web site lists activities on topics relating to hypothesis testing: http://mathforum.org/mathtools/sitemap2/ps/.
- This site has several activities using Excel and Minitab: www.mathspace.com/NSF_ProbStat/Teaching_Materials/Primarily_Statistics.htm.

10

Two-Sample Inference

OVERVIEW

Ariel Skelley/Blend Images/Getty Images

Bank Loans

CASE STUDY

The Chapter 10 Case Study, *Bank Loans* examines a pair of large data sets, of 5000 observations each, of bank loan applicants: **BankLoan_Approved** and **BankLoan_Denied** (*Source: Data Mining and Predictive Analytics,*[1] by Daniel Larose and Chantal Larose, John Wiley and Sons, 2015). The variables examined in these data sets include *credit score, request amount, interest, debt-to-income ratio*, and whether the loan application was approved or not.

- In Section 10.2, we use the case study data for the following analysis:
  - We construct a two-sample confidence interval for the difference in mean credit score between loan applicants who were approved and those who were denied. We then use that confidence interval to conduct a series of hypothesis tests.
  - We conduct a pooled variance *t* test for the difference in mean debt-to-income ratio between those approved and those denied the loan. Then we construct a pooled variance *t* confidence interval for this difference.
  - In the Section 10.2 exercises, we examine whether a difference exists in the loan request amount between those approved and those denied the loan.
- Finally, in the Section 10.3 exercises, we examine whether a difference exists in the proportion of applicants with credit scores of 700 or higher between those approved for and those denied the loan.

THE BIG PICTURE

Where we are coming from and where we are headed . . .

- Thus far, our statistical inference has been limited to one population and one sample. In Chapter 8, we learned to construct confidence intervals, and in Chapter 9, we learned how to perform hypothesis tests, but all for a single population parameter.
- Here, in Chapter 10, "Two-Sample Inference," we perform inference on the differences in the parameters of two populations. For example, we may be interested in whether a difference exists in the population proportions of women and men who post personal information on the Internet.
- In Chapter 11, we will turn to inference methods for categorical data, such as contingency tables.

10.1 Inference for Mean Difference—Dependent Samples

OBJECTIVES By the end of this section, I will be able to . . .

1. Distinguish between independent samples and dependent samples.
2. Perform hypothesis tests for the population mean difference for dependent samples.
3. Construct and interpret confidence intervals for the population mean difference for dependent samples.
4. Use a t interval for μ_d to perform t tests about μ_d.

1 Independent Samples and Dependent Samples

Chapter 10 is about two-sample inference. The type of inference we apply depends on whether the data come from **independent samples** or **dependent samples**.

Independent Samples and Dependent Samples

Two samples are **independent** when the subjects selected for the first sample do not determine the subjects in the second sample. Two samples are **dependent** when the subjects in the first sample determine the subjects in the second sample. The data from dependent samples are called **matched-pair** or **paired** samples.

For example, suppose we are interested in comparing the heights of girl-boy fraternal twins. Selecting the girl twin for the first sample automatically results in the boy twin's being selected for the second sample. This is an example of dependent sampling, and the boy-girl pairs are called **matched-pair** samples or **paired** samples. However, suppose we are interested in comparing the heights of females and males in general. Then, if we took a random sample of 20 females at your school and another random sample of 20 males at your school, these samples would be **independent**, because the females selected in the first sample do not determine the males selected in the second sample.

EXAMPLE 1 Dependent or independent sampling?

Indicate whether each of the following experiments uses an independent or dependent sampling method:

a. A study was designed to compare the differences in price between name-brand merchandise and store-brand merchandise. Name-brand and store-brand items of the same size were purchased from each of the following six categories: paper towels, shampoo, cereal, ice cream, peanut butter, and milk.

b. A study was designed to compare traditional acupuncture with usual clinical care for a certain type of lower-back pain.[2] The 241 subjects suffering from persistent nonspecific lower-back pain were randomly assigned to receive either traditional acupuncture or the usual clinical care. The results were measured at 12 and 24 months.

Solution

a. For a given store, each name-brand item in the first sample is associated with exactly one store-brand item of that size in the second sample. Therefore, the items in the first sample determine the items in the second sample. This is an example of dependent sampling.

b. The subjects were randomly assigned to receive either of the two treatments. Thus, the subjects who received acupuncture did not determine those who received clinical care, and vice versa. This is an example of independent sampling.

Exercises 5–8.

2 Dependent Sample *t* Test for the Population Mean of the Differences

We begin with an example.

EXAMPLE 2   Finding the mean and standard deviation of the sample differences

Table 1 shows students' scores on two statistics quizzes. The "After" row (sample 1) contains scores after the students sought help in the Math Center, and the "Before" row (sample 2) shows scores before they had help. The observations are taken from the same students before and after they had help. Thus, sample 1 and sample 2 are dependent, matched-pair data.

Table 1 Statistics quiz scores of seven students before and after visiting the Math Center

Student	Ashley	Brittany	Chris	Dave	Emily	Fran	Greg
After (sample 1)	66	68	74	88	89	91	100
Before (sample 2)	50	55	60	70	75	80	88

a. Calculate the sample differences (after − before).

b. Explain the key idea behind dependent sampling.

c. Find the mean and standard deviation of the sample differences.

Solution

a. For each student, we subtract the "before" value from the "after" value. Notice that each student's score improved on the second quiz:

Ashley: $66 - 50 = 16$	Emily: $89 - 75 = 14$
Brittany: $68 - 55 = 13$	Fran: $91 - 80 = 11$
Chris: $74 - 60 = 14$	Greg: $100 - 88 = 12$
Dave: $88 - 70 = 18$	

b. The key idea behind dependent sampling is that we consider the set of these seven differences {16, 13, 14, 18, 14, 11, 12} as a sample, so that we can perform inference on these differences. In other words, we no longer have two samples. By matching the samples element by element and taking the difference, we have transformed two samples into one that is the sample of differences (Figure 1). We have already learned how to perform inference using a single sample, so the remainder of this section uses techniques you have used previously.

c. The Excel descriptive statistics show the mean and standard deviation of the differences, giving us

$$\bar{x}_d = 14 \quad \text{and} \quad s_d = 2.380476143$$

Differences	
Mean	14
Standard Error	0.899735411
Median	14
Mode	14
Standard Deviation	2.380476143

Excel descriptive statistics.

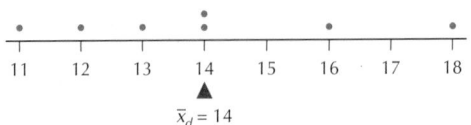

Difference in quiz scores (after – before)

FIGURE 1 Taking the differences reduces a two-sample problem to a single sample of differences.

The mean of the differences $\bar{x}_d = 14$ is shown as the balance point in Figure 1.

NOW YOU CAN DO
Exercises 9–14.

YOUR TURN
#1

Table 2 shows the change in English quiz scores for six students before and after getting help at the English Center. Calculate the mean $\bar{x}_d$ and the standard deviation s_d of the differences.

Table 2 English quiz scores

Student	Henrik	Ivana	Jen	Kayla	Luisa	Manuel
After	90	70	76	61	60	90
Before	92	70	75	60	58	86

(The solutions are shown in Appendix A.)

The sample of differences can be considered representative of the *population* of these differences, where the population represents all students who took statistics quizzes before and after visiting the Math Center. The sample mean difference $\bar{x}_d = 14$ is a point estimate of the *population mean difference* μ_d, which is the unknown mean difference in the (after − before) quiz scores for all students who visited the Math Center. Because μ_d is unknown, we need to perform hypothesis tests and construct confidence intervals to learn about its value.

Note that, in this book, μ_d always refers to sample 1 − sample 2—never sample 2 − sample 1. For example, μ_d represents the mean difference between the students' "after" scores and the "before" scores on the statistics quizzes in Table 1.

Paired Sample *t* Test for the Population Mean of the Differences μ_d: Critical-Value Method

For matched-pair data taken from dependent samples of two populations, find the differences to produce a random sample of the differences between the populations. You can use the *t* test whenever *either* of the following conditions is met:

- The population of differences is normal, or

- The sample size of differences is large ($n \geq 30$).

Step 1 State the hypotheses. Use one of the hypothesis test forms in Table 3. State the meaning of μ_d.

Step 2 Find t_{crit}, and state the rejection rule. To find t_{crit}, use the *t* table and degrees of freedom $n - 1$. To find the rejection rule, use Table 3.

Step 3 Calculate t_{data}.

$$t_{data} = \frac{\bar{x}_d}{s_d/\sqrt{n}}$$

which follows an approximate *t* distribution with degrees of freedom $n - 1$.

Step 4 State the conclusion and the interpretation. Compare t_{data} with t_{crit}.

Notice that we have only one sample of differences, so this procedure is very similar to the one-sample *t* test from Section 9.4.

Table 3 Critical regions and rejection rules for dependent sample *t* test

Form of test	Right-tailed test	Left-tailed test	Two-tailed test
Hypotheses	$H_0 : \mu_d = \mu_0$ $H_a : \mu_d > \mu_0$ level of significance α	$H_0 : \mu_d = \mu_0$ $H_a : \mu_d < \mu_0$ level of significance α	$H_0 : \mu_d = \mu_0$ $H_a : \mu_d \neq \mu_0$ level of significance α
Critical region	α 0 t_{crit} Noncritical region Critical region	α $-t_{crit}$ 0 Critical region Noncritical region	$\alpha/2$ $\alpha/2$ $-t_{crit}$ 0 t_{crit} Critical region Noncritical region Critical region
Rejection rule	Reject H_0 if $t_{data} \geq t_{crit}$	Reject H_0 if $t_{data} \leq -t_{crit}$	Reject H_0 if $t_{data} \geq t_{crit}$ or $t_{data} \leq -t_{crit}$

EXAMPLE 3   Paired *t* test using the critical-value method

For the Math Center data in Example 2, test, at level of significance $\alpha = 0.05$, whether the population mean μ_d of the differences in quiz scores (after − before) is greater than zero. Or, more informally, test whether the quiz scores after visiting the Math Center are larger on average than the quiz scores before visiting the Math Center.

Solution

The normal probability plot of the differences shown here shows acceptable normality, allowing us to proceed with the hypothesis test.

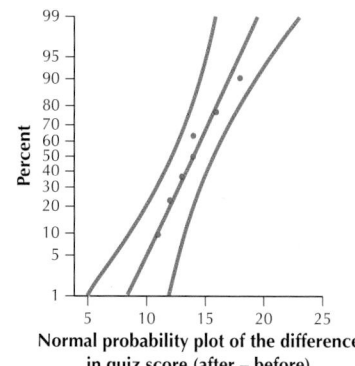

Normal probability plot of the difference in quiz score (after – before)

Step 1   **State the hypotheses.** "Greater than" implies that $\mu_d > 0$, leading to the hypotheses

$$H_0 : \mu_d = 0 \qquad \text{versus} \qquad H_a : \mu_d > 0$$

where μ_d represents the population mean difference in quiz scores after visiting the Math Center and before visiting the Math Center.

Step 2   **Find the critical value t_{crit} and state the rejection rule.** Use $n - 1$ degrees of freedom. Here $n = 7$, so df $= n - 1 = 6$. We have a right-tailed test with $\alpha = 0.05$, so we find our *t* critical value by choosing the column in the *t* table (Table D in the Appendix) with area 0.05 in one tail: $t_{crit} = 1.943$. The right-tailed test tells us that our rejection rule is to reject H_0 when t_{data} is greater than 1.943.

Step 3   **Find t_{data}.** We need to calculate $\bar{x}_d$ and s_d.
From Example 2, we have

$$\bar{x}_d = 14 \quad \text{and} \quad s_d = 2.380476143$$

This gives

$$t_{data} = \frac{\bar{x}_d}{s_d / \sqrt{n}} = \frac{14}{2.380476143 / \sqrt{7}} \approx 15.6$$

Step 4   **State the conclusion and the interpretation.** Because $t_{data} \approx 15.6$ is greater than $t_{crit} = 1.943$, we reject H_0. There is evidence that the population mean μ_d of the differences in quiz score (after − before) is greater than zero. That is, the quiz scores after visiting the Math Center are larger on average than the quiz scores before visiting the Math Center.

NOW YOU CAN DO
Exercises 15–17.

YOUR TURN #2

For the set of (after − before) English quiz score differences in Table 2 (page 578), test, at level of significance $\alpha = 0.10$, whether the population mean μ_d of the differences in quiz score (after − before) is greater than zero. (The normality of the data is fine, although you may check it with technology if you wish.)

(The solution is shown in Appendix A.)

The paired sample t test may also be performed using the p-value method.

Paired Sample t Test for the Population Mean of the Differences μ_d: p-Value Method

For matched-pair data taken from dependent samples of two populations, find the differences to produce a random sample of the differences between the populations. You can use the t test whenever *either* of the following conditions is met:

- The population of differences is normal, or
- The sample size of differences is large ($n \geq 30$).

Step 1 **State the hypotheses and the rejection rule.** Use one of the hypothesis test forms from Table 4 for a test at level of significance α. State the meaning of μ_d. *The rejection rule is: Reject H_0 if the p-value is less than α.*

Step 2 **Calculate t_{data}.**

$$t_{data} = \frac{\overline{x}_d}{s_d/\sqrt{n}}$$

which follows an approximate t distribution with degrees of freedom $n - 1$.

Step 3 **Find the p-value.** If you have access to technology, use it to find the p-value. Otherwise, calculate the p-value using one of the test forms in Table 4.

Step 4 **State the conclusion and the interpretation.** Compare the p-value with α.

Table 4 p-Values for dependent sample t tests

Form of test	Right-tailed test	Left-tailed test	Two-tailed test
Hypotheses	$H_0 : \mu_d = \mu_0$ $H_a : \mu_d > \mu_0$ p-value $= P(t > t_{data})$ Area to the right of t_{data}	$H_0 : \mu_d = \mu_0$ $H_a : \mu_d < \mu_0$ p-value $= P(t < t_{data})$ Area to the left of t_{data}	$H_0 : \mu_d = \mu_0$ $H_a : \mu_d \neq \mu_0$ p-value $= P(t > |t_{data}|) + P(t < -|t_{data}|)$ $= 2 \cdot P(t > |t_{data}|)$ Sum of the two tail areas
***p*-Value**			

EXAMPLE 4 Paired sample t test for μ_d: The p-value method

A study was performed to determine whether Reiki touch therapy was useful in the reduction of mean pain level in chronic pain sufferers, including cancer patients.[3] The pain level reported by a random sample of 13 patients before and after Reiki touch therapy is shown in Table 5. Test whether a mean reduction in pain level has occurred after the Reiki therapy, using level of significance $\alpha = 0.05$. In other words, test

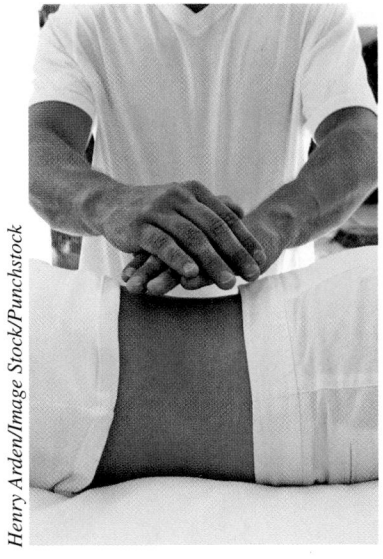

whether the population mean difference μ_d is less than zero, where μ_d is defined as the (after – before) difference in pain level.

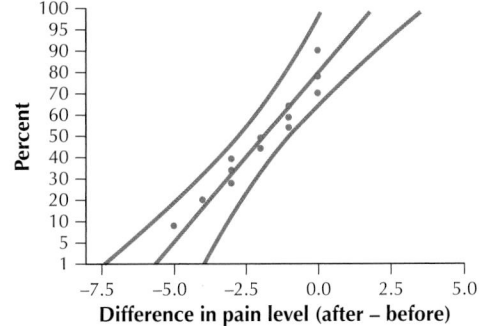 **reiki**

Table 5 Pain level reported by 13 patients before and after Reiki touch therapy

Patient	1	2	3	4	5	6	7	8	9	10	11	12	13
After	3	1	0	0	2	1	2	1	0	4	1	4	8
Before	6	2	2	3	3	4	2	5	1	6	6	4	8
Difference	−3	−1	−2	−3	−1	−3	0	−4	−1	−2	−5	0	0

Solution

For each patient, we subtract the "before" pain level from the "after" pain level to arrive at a set of $n = 13$ differences, highlighted in Table 5. The normal probability plot of the differences indicates acceptable normality, given the small sample size. The Minitab results from the t test are provided here.

```
Test of mu = 0 vs < 0
                                            95%
                                          Upper
Variable   N     Mean    StDev   SE Mean  Bound      T      P
[Diff]     13  -1.92308  1.60528  0.44522  -1.12956  -4.32  0.000
```

Step 1 **State the hypotheses and the rejection rule.** We are interested in testing whether a mean reduction in pain level occurred, which would mean that the mean pain level would be lower after the Reiki therapy than before the therapy. This implies that the population mean difference in pain level, $\mu_d =$ (after − before), is *less than* 0. Thus, from Table 4, the hypotheses are

$$H_0 : \mu_d = 0 \qquad H_a : \mu_d < 0$$

where μ_d represents the population mean difference in pain level. We will reject H_0 if the p-value < 0.05.

Step 2 **Find t_{data}.** As provided in the Minitab results,

$$t_{\text{data}} = \frac{\bar{x}_d}{s_d / \sqrt{n}} = \frac{-1.92308}{1.60528 / \sqrt{13}} \approx -4.32$$

which follows an approximate t distribution with degrees of freedom $n - 1 = 13 - 1 = 12$.

Step 3 **Find the p-value.** For a left-tailed test, the p-value is the area to the left of t_{data}. This area is essentially 0, as shown in Figure 2 and provided by Minitab,

$$P(t < t_{\text{data}}) = P(t < -4.32) \approx 0.000$$

FIGURE 2
The *p*-value =
$P(t < -4.32) \approx 0.000$.

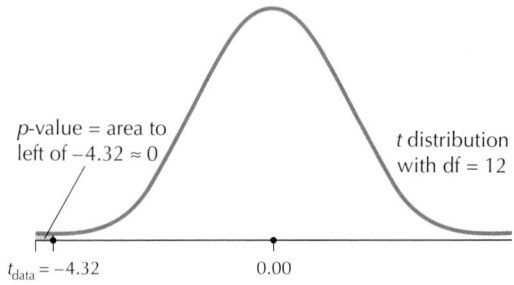

p-value = area to
left of $-4.32 \approx 0$

t distribution
with df = 12

$t_{data} = -4.32$ 0.00

Step 4 **State the conclusion and the interpretation.** Because *p*-value $\approx 0.000 \le$ $\alpha = 0.05$, we reject H_0. There is evidence that $\mu_d < 0$, thus the population mean difference in pain level (after $-$ before) is negative. That is, there is evidence, at level of significance $\alpha = 0.05$, that the Reiki touch therapy has worked to reduce the mean pain level for chronic pain sufferers.

NOW YOU CAN DO
Exercises 18–20.

3 *t* Confidence Intervals for the Population Mean Difference for Dependent Samples

Recall that in Section 8.2, we used the formula $\bar{x} \pm t_{\alpha/2}\left(s/\sqrt{n}\right)$ to calculate the *t* interval for the population mean μ. Here, to estimate the population mean of the differences μ_d, we use essentially the same formula, substituting $\bar{x}_d$ for $\bar{x}$ and s_d for s.

Confidence Interval for Population Mean Difference μ_d (Dependent Samples)

For matched-pair data taken from dependent samples of two populations, find the differences to produce a random sample of the differences between the populations. A $100(1 - \alpha)$% confidence interval for μ_d, the population mean of the differences, is given by

$$\text{lower bound} = \bar{x}_d - t_{\alpha/2}\left(\frac{s_d}{\sqrt{n}}\right) \qquad \text{upper bound} = \bar{x}_d + t_{\alpha/2}\left(\frac{s_d}{\sqrt{n}}\right)$$

where $\bar{x}_d$ and s_d represent the sample mean and sample standard deviation of the differences, respectively, of the set of *n* paired differences, $d_1, d_2, d_3, \ldots, d_n$, and where $t_{\alpha/2}$ is based on $n - 1$ degrees of freedom. This *t* interval applies whenever *either* of the following conditions is met:

• The population of differences is normal, or

• The sample size of differences is large ($n \ge 30$).

The $100(1 - \alpha)$% confidence interval for μ_d may also be expressed in the form

$$\bar{x}_d \pm t_{\alpha/2}\left(\frac{s_d}{\sqrt{n}}\right)$$

To construct this confidence interval, we need

$\bar{x}_d =$ mean of the differences of the two samples

$s_d =$ standard deviation of the differences of the two samples

$n =$ sample size of differences

$t_{\alpha/2} =$ critical value associated with confidence level $1 - \alpha$
          and degrees of freedom $n - 1$

EXAMPLE 5   *t* Confidence interval for μ_d

Use the "before" and "after" quiz scores from Table 1 to construct a 95% *t* confidence interval for the population mean of the differences in the statistics quiz scores.

Solution

The normality of the quiz scores was checked in Example 2. We ignore the original raw data (see Table 1) and concentrate only on the set of sample differences: {16, 13, 14, 18, 14, 11, 12}. From Example 2, we have

$$\bar{x}_d = 14 \quad \text{and} \quad s_d = 2.380476143 \approx 2.3805$$

For 95% confidence with $n - 1 = 6$ degrees of freedom, $t_{\alpha/2}$ equals 2.447 (see the *t* table in Appendix Table D). Using these values,

$$\text{lower bound} = \bar{x}_d - t_{\alpha/2}\left(s_d/\sqrt{n}\right)$$
$$= 14 - (2.447)\left(2.3805/\sqrt{7}\right)$$
$$\approx 14 - 2.2017 = 11.7983$$
$$\text{upper bound} = \bar{x}_d + t_{\alpha/2}\left(s_d/\sqrt{n}\right)$$
$$= 14 + (2.447)\left(2.3805/\sqrt{7}\right)$$
$$\approx 14 + 2.2017 = 16.2017$$

We are 95% confident that the population mean of the differences between quiz scores before and after visiting the Math Center lies between 11.7983 points and 16.2017 points. If no mean change in the quiz scores occurred, the difference would be 0, which is not in this confidence interval. Thus, we have evidence that the Math Center tutoring leads to a significant change in the mean quiz scores, with 95% confidence.

NOW YOU CAN DO
Exercises 21–26.

YOUR TURN
#3

For the set of (after − before) English quiz score differences in Table 2 (page 578), construct a 90% *t* confidence interval for the population mean of the differences in the English quiz scores.

(The solution is shown in Appendix A.)

4 Use a *t* Interval for μ_d to Perform *t* Tests About μ_d

Given a $100(1 - \alpha)\%$ *t* confidence interval for μ_d, we may perform two-tailed *t* tests for various values of μ_d, just as we did for the single sample case in Section 9.4. The methodology is the same: if a certain value for μ_d lies outside the $100(1 - \alpha)\%$ *t* confidence interval for μ_d, then the null hypothesis specifying this value for μ_d would be rejected. Otherwise it would not be rejected.

EXAMPLE 6   Using a *t* interval for μ_d to perform *t* tests about μ_d

Example 5 provided a 95% *t* confidence interval for the population mean of the differences between quiz scores before and after visiting the Math Center as (11.7983, 16.2017). Test, using level of significance $\alpha = 0.05$, whether the population mean of the differences between quiz scores before and after visiting the Math Center differs from these values: (**a**) 15 points, (**b**) 16 points, (**c**) 17 points.

Solution

We state the hypotheses and determine if each proposed value μ_0 lies inside or outside of the t confidence interval (11.7983, 16.2017).

a. $H_0 : \mu_d = 15$ versus $H_a : \mu_d \neq 15$

$\mu_0 = 15$ lies inside the interval (11.7983, 16.2017), so we do not reject H_0 (Figure 3).

b. $H_0 : \mu_d = 16$ versus $H_a : \mu_d \neq 16$

$\mu_0 = 16$ lies inside the interval, so we do not reject H_0.

c. $H_0 : \mu_d = 17$ versus $H_a : \mu_d \neq 17$

$\mu_0 = 17$ lies outside the interval, so we reject H_0.

NOW YOU CAN DO
Exercises 27–30.

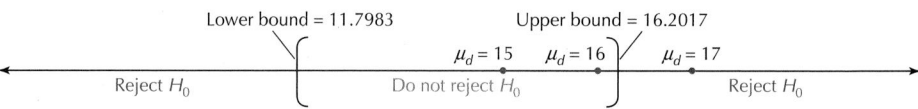

FIGURE 3 Reject H_0 for values of μ_d that lie outside the t confidence interval.

YOUR TURN #4

Use the 90% confidence interval you made for the population mean of the differences in the English quiz scores in the Your Turn #3 to test, using level of significance $\alpha = 0.10$, whether the population μ_d differs from these values: **(a)** 2 points, **(b)** 5.6 points, **(c)** 5.7 points.

(The solutions are shown in Appendix A.)

STEP-BY-STEP TECHNOLOGY GUIDE: Confidence Intervals and Hypothesis Tests for μ_d

TI-83/84

Hypothesis Test
(Example 3 is used to illustrate the procedure.)
Step 1   Enter samples 1 and 2 in lists L1 and L2. Press **2nd >** Quit.
Step 2   Type **(L1 − L2) STO→ L3** and press **ENTER** (Figure 4).
Step 3   Press **STAT** and highlight **TESTS**.
Step 4   For the hypothesis test, press **2** (for the **T-Test**). The T-Test menu appears.

Step 5   For input (**Inpt**), highlight **Data** and press **ENTER**. (If given the summary statistics for the differences, choose **Stats**.)
Step 6   For μ_0, enter the hypothesized value, 0. For **List**, press **2nd** then **L3**. For **Freq**, enter **1**. Choose the form of the hypothesis test, and press **ENTER** (Figure 5).
Step 7   When the cursor is over **Calculate**, make sure all your entries are correct, and press **ENTER**. The results are shown in Figure 6.

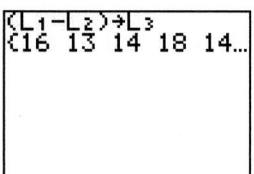

FIGURE 4

FIGURE 5

FIGURE 6

Confidence Interval
(Example 5 is used to illustrate the procedure.)
Step 1   Enter samples 1 and 2 in lists L1 and L2.
Step 2   Type **(L1 − L2) STO→ L3** and press **ENTER** (Figure 4).
Step 3   Press **STAT** and highlight **TESTS**.

Step 4   Press **8** (for the **TInterval**).
Step 5   For input (**Inpt**), highlight **Data** and press **ENTER**. (If given the summary statistics for the differences, choose **Stats**.)

Step 6 For **List**, press **2nd** then **L3**. For **Freq**, enter **1**. Enter the **C-Level** (confidence level, such as **0.95** for 95%), and press **ENTER** (Figure 7).

Step 7 Highlight **Calculate** and press **ENTER**. The results are shown in Figure 8.

```
  TInterval
Inpt:DATA Stats
List:L₃
Freq:1
C-Level:.95
Calculate
```

FIGURE 7

```
  TInterval
(11.798,16.202)
x̄=14
Sx=2.380476143
n=7
```

FIGURE 8

EXCEL

Hypothesis Test
Step 1 Enter samples 1 and 2 in columns A and B.
Step 2 Click **Data > Data Analysis > t-Test: Paired Two Sample for Means**, and click **OK**.

Step 3 For **Variable 1 Range**, highlight the cells for sample 1 in column A, and for **Variable 2 Range**, highlight the cells for sample 2 in column B.
Step 4 Enter the **Hypothesized Mean Difference** (usually 0), and enter a value for **alpha**. Then click **OK**.

MINITAB

Confidence Interval and Hypothesis Test
Step 1 Enter samples 1 and 2 in columns C1 and C2.
Step 2 Click **Stat > Basic Statistics > Paired t…**.
Step 3 Select **Each sample is in a column** from the drop-down menu. For **Sample 1**, enter **C1**, and for **Sample 2**, enter **C2**.

Step 4 Click **Options**.
a. For the confidence interval, specify the **Confidence Level**, then click **OK** twice.
b. For the hypothesis test, specify the form of the alternative **hypothesis**, then click **OK** twice.

SPSS

Confidence Interval and Two-Tailed Hypothesis Test
(Example 5 is used to illustrate the procedure.)
Step 1 Enter the data into the first two columns.
Step 2 Select **Analyze > Compare Means > Paired-Samples T Test…**.
Step 3 Move the first variable to Variable 1 in the **Paired Variables** table, and move the second variable to Variable 2.

Step 4 Click **Options…**, and enter the **Confidence Interval Percentage**. Click **Continue**, then click **OK**. The results are located in the Paired Samples Test table of the output window, shown in Figure 9.

		Paired Samples Test							
		Paired Differences							
					95% Confidence Interval of the Difference				
		Mean	Std. Deviation	Std. Error Mean	Lower	Upper	t	df	Sig. (2-tailed)
Pair 1	VAR00001 - VAR00002	14.00000	2.38048	.89974	11.79843	16.20157	15.560	6	.000

FIGURE 9 SPSS results.

JMP

Confidence Interval and Hypothesis Test
(Example 5 is used to illustrate the procedure.)
Step 1 Click **File > New > Data Table**. Enter the data into the first two columns.
Step 2 Select **Analyze > Matched Pairs**. Move both variables to the **Y, Paired Response** box. Click **OK**. The *t* test results and the 95% confidence interval are shown in Figure 10.
Step 3 To change the confidence level of the confidence interval, click the red triangle next to **Matched Pairs**, click **Set α level**, and select the desired level of α.

Column 2	68.2857	t-Ratio	-15.5601
Column 1	82.2857	DF	6
Mean Difference	-14	Prob > |t|	<.0001*
Std Error	0.89974	Prob > t	1.0000
Upper 95%	-11.798	Prob < t	<.0001*
Lower 95%	-16.202		
N	7		
Correlation	0.98657		

FIGURE 10 JMP results.

CRUNCHIT!

Paired *t* test and *t* interval for μ_d
We will use the data from Example 3.
Step 1   Click **File**, then highlight **Load from Larose, Discostat3e >**
Chapter 10, and click on **Example 01_03**.
Step 2   Click **Statistics**, highlight **t**, and select **Paired**. For **First Variable**, select **After**. For **Second Variable**, select **Before**.

a. For the hypothesis test, select the **Hypothesis Test** tab, enter the **Mean difference under null hypothesis**, choose the correct form of the **Alternative** hypothesis, and click **Calculate**.
b. For the confidence interval, select the **Confidence Interval** tab, enter the **Confidence Interval Level (%)**, and click **Calculate**.

Section 10.1 Summary

1. Two samples are independent when the subjects selected for the first sample do not determine the subjects in the second sample. Two samples are dependent when the subjects in the first sample determine the subjects in the second sample. The data from dependent samples are called matched-pair or paired samples. The key concept in this section is that we consider the differences of matched-pair data as a single sample and perform inference on this sample of differences.

2. The paired-sample *t* test for the population mean of the differences μ_d can be used either when the population is normally distributed or the sample size is large ($n \geq 30$).

The test may be performed using either the critical-value method or the *p*-value method.

3. A $100(1 - \alpha)\%$ confidence interval for μ_d, which is the population mean of the differences, is given by $\bar{x}_d \pm t_{\alpha/2}\left(s_d/\sqrt{n}\right)$, where $\bar{x}_d$ and s_d represent the sample mean and sample standard deviation of the differences, respectively, of the set of *n* paired differences, $d_1, d_2, d_3, \ldots, d_n$, and where $t_{\alpha/2}$ is based on $n - 1$ degrees of freedom.

4. This confidence interval may be used to conduct two-tailed hypothesis tests for μ_d.

Section 10.1 Exercises

CLARIFYING THE CONCEPTS

1. When are two samples considered independent? (p. 576)
2. When are two samples considered dependent? (p. 576)
3. What do we call the data obtained from dependent sampling? (p. 576)
4. How do we interpret the meaning of μ_d? (p. 578)

PRACTICING THE TECHNIQUES

 **CHECK IT OUT!**

To do	Check out	Topic
Exercises 5–8	Example 1	Dependent or independent sampling?
Exercises 9–14	Example 2	Calculating $\bar{x}_d$ and s_d
Exercises 15–17	Example 3	Paired *t* test for μ_d: critical-value method
Exercises 18–20	Example 4	Paired *t* test for μ_d: *p*-value method
Exercises 21–26	Example 5	*t* confidence interval for μ_d
Exercises 27–30	Example 6	Using a *t* interval for μ_d to perform *t* tests about μ_d

Determine whether the experiments in Exercises 5–8 represent an independent sampling method or a dependent sampling method. Explain your answer.

5. The Jacksonville Jaguars are interested in comparing the performance of their first-year players. For each player, a sample is taken of their games from their last year in college and compared with a sample of games taken from their first year in the pros.

6. For her senior project, an exercise science major takes a sample of females majoring in exercise science and a sample of females from her college who are not majoring in exercise science. She records the body mass index for each subject.

7. Before the first lecture, an algebra instructor gives a pretest to his students to determine the students' algebra readiness. At the end of the course, the instructor gives a post-test to the same students and compares the results with the pretest.

8. The sheriff's department takes a sample of vehicle speeds on a certain stretch of road and compares the results to a sample of vehicle speeds on a certain stretch of a different road. Both roads have the same posted speed limit.

In Exercises 9–14, assume that samples of differences are obtained through dependent sampling and follow a normal distribution. Calculate $\bar{x}_d$ and s_d.

9.

Subject	1	2	3	4	5
Sample 1	3.0	2.5	3.5	3.0	4.0
Sample 2	2.5	2.5	2.0	2.0	1.5

10.

Subject	1	2	3	4	5	6
Sample 1	10	12	9	14	15	8
Sample 2	8	11	10	12	14	9

11.

Subject	1	2	3	4	5	6	7
Sample 1	20	25	15	10	20	30	15
Sample 2	30	30	20	20	25	35	25

12.

Subject	1	2	3	4	5	6	7
Sample 1	1.5	1.8	2.0	2.5	3.0	3.2	4.0
Sample 2	1.0	1.7	2.1	2.0	2.7	2.9	3.3

13.

Subject	1	2	3	4	5	6	7	8
Sample 1	0	0.5	0.75	1.25	1.9	2.5	3.2	3.3
Sample 2	0.25	0.25	0.75	1.5	1.8	2.2	3.3	3.4

14.

Subject	1	2	3	4	5	6	7	8
Sample 1	105	88	103	97	115	125	122	92
Sample 2	110	95	108	97	116	127	125	95

15. For the data in Exercise 9, test whether $\mu_d > 0$, using the critical-value method and level of significance $\alpha = 0.05$.

16. For the data in Exercise 10, test whether $\mu_d \neq 0$, using the critical-value method and level of significance $\alpha = 0.01$.

17. For the data in Exercise 11, test whether $\mu_d < 0$, using the critical-value method and level of significance $\alpha = 0.10$.

18. For the data in Exercise 12, test whether $\mu_d > 0$, using the p-value method and level of significance $\alpha = 0.01$.

19. For the data in Exercise 13, test whether $\mu_d \neq 0$, using the p-value method and level of significance $\alpha = 0.05$.

20. For the data in Exercise 14, test whether $\mu_d < 0$, using the p-value method and level of significance $\alpha = 0.10$.

21. Using the data from Exercise 9, construct a 95% confidence interval for μ_d.

22. Using the data from Exercise 10, construct a 99% confidence interval for μ_d.

23. Using the data from Exercise 11, construct a 90% confidence interval for μ_d.

24. Using the data from Exercise 12, construct a 99% confidence interval for μ_d.

25. Using the data from Exercise 13, construct a 95% confidence interval for μ_d.

26. Using the data from Exercise 14, construct a 90% confidence interval for μ_d.

For Exercises 27–30 a $100(1 - \alpha)\%$ t confidence interval for μ_d is given. Use the confidence interval to test, using level of significance α, whether μ_d differs from each of the indicated hypothesized values.

27. A 95% t confidence interval for μ_d is $(-5, 5)$. Hypothesized values are
   a. 0
   b. -6
   c. 4

28. A 99% t confidence interval for μ_d is $(-10, -4)$. Hypothesized values are
   a. -12
   b. 0
   c. 4

29. A 90% t confidence interval for μ_d is $(10, 20)$. Hypothesized values are
   a. -10
   b. 25
   c. 0

30. A 95% t confidence interval for μ_d is $(0, 1)$. Hypothesized values are
   a. 0.41
   b. 0.29
   c. 1.23

APPLYING THE CONCEPTS

31. New Car Prices. Kelley's Blue Book (www.kbb.com) publishes data on new and used cars. The following table contains the fair market value for five new 2013 and 2014 vehicles (data recorded July 2014). We are interested in the difference in price between the 2013 models and the 2014 models. Assume that the population of price differences is normally distributed. ▓▓ carprice

   a. Find the mean of the differences, $\bar{x}_d$, and the standard deviation of the differences, s_d.
   b. Test whether 2014 models are on average more expensive, using level of significance $\alpha = 0.05$.

	Toyota Camry	Honda Civic	Ford F-150	Chevy Corvette	Tesla Model S
2014 (sample 1)	$20,672	$17,069	$24,362	$45,684	$68,738
2013 (sample 2)	$20,284	$16,499	$22,674	$44,021	$68,674

32. Does Friday the 13th Change Human Behavior? In Example 4 of Chapter 1, we discussed whether Friday the 13th changed human behavior. The researchers obtained data kept by the British Department of Transport on the traffic flow through certain junctions of the M25 motorway in England.[4]

   a. Find the mean of the differences, $\bar{x}_d$, and the standard deviation of the differences, s_d.
   b. Perform the appropriate hypothesis test for determining whether the mean traffic flow is smaller for Friday the 13th compared with Friday the 6th, using level of significance $\alpha = 0.10$. Assume normality. ▓▓ fridaythe13th

TABLE 6 Traffic through M25 junctions

Friday the 6th	Friday the 13th	Difference
139,246	138,548	698
134,012	132,908	1104
137,055	136,018	1037
133,732	131,843	1889
123,552	121,641	1911
121,139	118,723	2416
128,293	125,532	2761
124,631	120,249	4382
124,609	122,770	1839
117,584	117,263	321

33. High and Low Temperatures. The University of Waterloo Weather Station tracks the daily low and high temperatures in Waterloo, Ontario, Canada. The table contains a random sample of the daily high and low temperatures in degrees Celsius for 10 days in calendar year 2010. Assume that the temperature differences are normally distributed.

waterlootemp

Day	1	2	3	4	5	6	7	8	9	10
High	9.4	6.1	5.9	29.1	11.9	30.6	23.1	33.1	14.8	0.1
Low	0.8	−8.9	−1.3	19.3	6.7	21.5	10.5	18.7	7.4	−9.9

a. Find the mean of the differences, $\bar{x}_d$, and the standard deviation of the differences, s_d.
b. Construct and interpret a 95% confidence interval for μ_d, the population mean difference in temperature.

34. NASDAQ Stock Prices. The table provides the start-of-trading and end-of trading prices for the eight most active stocks on July 28, 2014. Assume that the differences are normally distributed. **nasdaq72814**

Stock	End-of-trading price	Start-of-trading price
Sirius XM	$3.38	$3.44
Apple	$99.02	$97.67
Facebook	$74.92	$75.19
Micron Technology	$31.98	$33.42
Dollar Tree	$54.87	$54.22
Intel	$34.23	$34.25
Microsoft	$43.97	$44.50
Cisco Systems	$25.92	$25.97

Source: NASDAQ.com.

a. Find the mean of the differences, $\bar{x}_d$, and the standard deviation of the differences, s_d.
b. Test whether the population mean difference in share prices differs from zero, using level of significance a $\alpha = 0.10$.

35. New Car Prices. Use the information in Exercise 31 to construct and interpret a 95% confidence interval for μ_d, the population mean difference in price.

36. Does Friday the 13th Change Human Behavior? Use the data from Exercise 32 to construct and interpret a 95% confidence interval for μ_d, the population mean difference in traffic for Friday the 13th and Friday the 6th.

37. High and Low Temperatures. Use the information in Exercise 33 for the following: Construct and interpret a 99% confidence interval for μ_d, the population mean difference in temperature.

38. NASDAQ Stock Prices. Use the information in Exercise 34 for the following:

a. Construct and interpret a 90% confidence interval for μ_d, the population mean difference in price.

b. Explain how your confidence interval supports your conclusion to the hypothesis test in Exercise 34.

BRINGING IT ALL TOGETHER

Mathematics Scores Worldwide. The National Center for Educational Statistics publishes the results from the Trends in International Math and Science Study (TIMSS). The table contains the 2007 and 2011 mean mathematics scores for eighth-graders from various countries. Assume that the population of score differences is normally distributed. Use this information for Exercises 39–44. **mathscores**

Eighth-grade math scores

Country	2007	2011
Singapore	593	611
Japan	570	570
Hong Kong	572	586
England	513	507
United States	508	509
Hungary	517	505
Italy	480	498
Russia	512	539
Ukraine	462	479
Australia	496	505
South Korea	597	613
Slovenia	501	505
Thailand	441	427
Norway	469	475
Indonesia	397	386

39. Explain why these are dependent samples and not independent samples.

40. Calculate the following statistics:
  a. $\bar{x}_d$
  b. s_d
  c. t_{data}

41. Using level of significance $\alpha = 0.05$, test whether the 2011 scores are higher than the 2007 scores, on average.

42. Construct a 95% confidence interval for μ_d, the population mean difference in score.

43. Use your confidence interval from the previous exercise to test, using level of significance $\alpha = 0.05$, whether the population mean difference equals the following values:
  a. 15 points
  b. 5 points
  c. 0 points

WHAT IF ? **44.** *What if* we added a certain constant number to each score in the table? How would this change affect the following?
  a. $\bar{x}_d$
  b. s_d
  c. t_{data}
  d. The conclusion

WORKING WITH LARGE DATA SETS

Phosphorus and Potassium. Use technology to solve Exercises 45–47. ᴴᵀᴴ nutrition

45. Open the **Nutrition** data set. Explore the variable *phosphor,* which lists the amount of phosphorus (in milligrams) for each food item. Generate numerical summary statistics and graphs for the amount of phosphorus in the food. What is the sample mean amount of phosphorus? The sample standard deviation?

46. Explore the variable *potass,* which lists the amount of potassium (in milligrams) for each food item. Generate numerical summary statistics and graphs for the amount of potassium in the food. What is the sample mean amount of potassium? The sample standard deviation?

47. Create a new variable in Excel or Minitab, **phos_pot,** which equals the amount of phosphorus minus the amount of potassium in each food item. Use a paired sample hypothesis test to test, at level of significance $\alpha = 0.05$, whether the population mean difference differs from 0.

WORKING WITH LARGE DATA SETS

Restaurants. Open the data set, **Restaurants**. Here, we will examine the difference in the number of fast food restaurants and the number of full service restaurants. Use technology to do the following: ᴴᵀᴴ restaurants

48. Obtain a random sample of size 100 from the data set.
49. For each county, compute the difference (number of fast food restaurants *minus* number of full service restaurants).
50. Test whether the population mean difference equals zero, using level of significance $\alpha = 0.05$.
51. Find the actual value of the population mean difference. Did your hypothesis test in Exercise 50 make the right decision? Explain.

10.2 Inference for Two Independent Means

OBJECTIVES By the end of this section, I will be able to . . .

1 Perform and interpret t tests about $\mu_1 - \mu_2$ using Welch's method.[5]
2 Compute and interpret t intervals for $\mu_1 - \mu_2$ using Welch's method.
3 Use confidence intervals for $\mu_1 - \mu_2$ to perform two-tailed t tests about $\mu_1 - \mu_2$.
4 Perform and interpret t tests and t intervals about $\mu_1 - \mu_2$ using the pooled variance method.
5 Apply Z tests and Z intervals for $\mu_1 - \mu_2$ when σ_1 and σ_2 are known.

1 Independent Sample t Test for $\mu_1 - \mu_2$

On page 178 in Chapter 3, we used boxplots to find evidence of a difference between male and female body temperature for a sample of 65 women and a sample of 65 men.[6] The summary statistics are shown in Table 7.

Table 7 Summary statistics for female versus male body temperatures in °F

Gender	Sample size	Sample mean body temperature	Sample standard deviation	Population mean body temperature
Females (sample 1)	$n_1 = 65$	$\bar{x}_1 = 98.394$	$s_1 = 0.743$	$\mu_1 = ?$
Males (sample 2)	$n_2 = 65$	$\bar{x}_2 = 98.105$	$s_2 = 0.699$	$\mu_2 = ?$

However, because the female subjects did not determine the male subjects, and vice versa, the 65 women and 65 men represent independent samples, so we cannot use the dependent sampling methods we learned in Section 10.1.

Note that for independent samples, we have two sample sizes, n_1 and n_2; two sample means, $\bar{x}_1$ and $\bar{x}_2$; two sample standard deviations, s_1 and s_2; and two unknown population means, μ_1 and μ_2. We are interested in the difference in the population means, so we consider the quantity

$$\mu_1 - \mu_2$$

The Difference Difference

A difference in interpretation exists between the quantity $\mu_1 - \mu_2$ and the quantity μ_d from Section 10.1. Here, $\mu_1 - \mu_2$ refers to the difference in population means, whereas μ_d represents the population mean of the paired differences.

In previous chapters, we used the statistic $\bar{x}$ to learn about the parameter μ. Here, we will use the statistic $\bar{x}_1 - \bar{x}_2$ to perform inference about the parameter $\mu_1 - \mu_2$, whose value is unknown. Note from Table 7 that the value of $\bar{x}_1 - \bar{x}_2$ for these samples is

$$\bar{x}_1 - \bar{x}_2 = 98.394 - 98.105 = 0.289$$

We use $\bar{x}_1 - \bar{x}_2 = 0.289$ as a *point estimate* of $\mu_1 - \mu_2$. If we repeat the experiment an infinite number of times, then the values of $\bar{x}_1 - \bar{x}_2$ will form a distribution called the **sampling distribution of** $\bar{x}_1 - \bar{x}_2$.

It is unlikely that the experimenter will have knowledge of both population standard deviations σ_1 and σ_2. Therefore, we use the estimates of σ_1 and σ_2 provided by the sample standard deviations s_1 and s_2. Recall from Section 8.2 that, when the population standard deviation σ is unknown, and if either the population is normally distributed or the sample size is large, the quantity

$$t = \frac{\bar{x} - \mu}{s/\sqrt{n}}$$

has a t distribution with $n - 1$ degrees of freedom. By analogy, we have the following sampling distribution.

Sampling Distribution of $\bar{x}_1 - \bar{x}_2$

When random samples are drawn independently from two populations with population means μ_1 and μ_2, and either (a) the two populations are normally distributed, or (b) the two sample sizes are large (at least 30), then the quantity

$$t = \frac{(\bar{x}_1 - \bar{x}_2) - (\mu_1 - \mu_2)}{\sqrt{\dfrac{s_1^2}{n_1} + \dfrac{s_2^2}{n_2}}}$$

approximately follows a t distribution with degrees of freedom equal to the smaller of $n_1 - 1$ and $n_2 - 1$, where $\bar{x}_1$ and s_1 represent the mean and standard deviation of the sample taken from population 1, and $\bar{x}_2$ and s_2 represent the mean and standard deviation of the sample taken from population 2.

This t statistic is called *Welch's approximate t*, after the twentieth-century English statistician Bernard Lewis Welch. Although there are other distributions that statisticians use to estimate the difference between two population means, we use this approximation because it is conservative and easy to calculate.

Researchers are often interested in testing whether the mean of one population is greater than, less than, or different from the mean of another population. Thus, we next learn how to perform hypothesis tests for the difference in population means $\mu_1 - \mu_2$. Usually, the most important hypothesized value for $\mu_1 - \mu_2$ is 0. Consider the two-tailed hypothesis test

$$H_0 : \mu_1 - \mu_2 = 0 \quad \text{versus} \quad H_a : \mu_1 - \mu_2 \neq 0$$

which is equivalent to

$$H_0 : \mu_1 = \mu_2 \quad \text{versus} \quad H_a : \mu_1 \neq \mu_2$$

In practice, the hypothesized difference between the two population means is nearly always $(\mu_1 - \mu_2)_0 = 0$. Thus, the test statistic takes the following form:

$$t_{\text{data}} = \frac{(\bar{x}_1 - \bar{x}_2) - 0}{\sqrt{\dfrac{s_1^2}{n_1} + \dfrac{s_2^2}{n_2}}} = \frac{(\bar{x}_1 - \bar{x}_2)}{\sqrt{\dfrac{s_1^2}{n_1} + \dfrac{s_2^2}{n_2}}}$$

Just as in Section 9.4, if t_{data} is extreme, then it represents evidence against the null hypothesis. The hypothesis test may be performed using either the critical-value method or the *p*-value method.

Welch's Hypothesis Test for the Difference in Two Population Means: Critical-Value Method

The hypothesis test applies whenever *either*

a. Both populations are normally distributed, or

b. Both samples are large, that is, $n_1 \geq 30$ and $n_2 \geq 30$.

Step 1   **State the hypotheses.**
Use one of the forms from Table 8. State the meaning of μ_1 and μ_2.

Step 2   **Find t_{crit} and state the rejection rule.**
To find t_{crit}, use the *t* table and degrees of freedom the *smaller* of $n_1 - 1$ and $n_2 - 1$. To find the rejection rule, use Table 8.

Step 3   **Calculate t_{data}.**

$$t_{\text{data}} = \frac{(\bar{x}_1 - \bar{x}_2)}{\sqrt{\dfrac{s_1^2}{n_1} + \dfrac{s_2^2}{n_2}}}$$

which follows an approximate *t* distribution with degrees of freedom the smaller of $n_1 - 1$ and $n_2 - 1$.

Step 4   **State the conclusion and the interpretation.**
Compare t_{data} with t_{crit}.

Table 8 Critical regions and rejection rules for *t* test for $\mu_1 - \mu_2$

Form of test	Right-tailed test	Left-tailed test	Two-tailed test
Hypotheses	$H_0 : \mu_1 = \mu_2$ $H_a : \mu_1 > \mu_2$ level of significance α	$H_0 : \mu_1 = \mu_2$ $H_a : \mu_1 < \mu_2$ level of significance α	$H_0 : \mu_1 = \mu_2$ $H_a : \mu_1 \neq \mu_2$ level of significance α
Critical region			
Rejection rule	Reject H_0 if $t_{\text{data}} \geq t_{\text{crit}}$	Reject H_0 if $t_{\text{data}} \leq -t_{\text{crit}}$	Reject H_0 if $t_{\text{data}} \geq t_{\text{crit}}$ or $t_{\text{data}} \leq -t_{\text{crit}}$

EXAMPLE 7   *t* Test for $\mu_1 - \mu_2$: Critical-value method

Using Table 7, test whether women's population mean body temperature differs from that of men, using the critical-value method and $\alpha = 0.05$.

Solution

Both sample sizes are large ($n_1 = n_2 = 65 \geq 30$), so we can perform the hypothesis test.

Step 1   **State the hypotheses.**
The key words "differs from" indicate a two-tail test:

$$H_0 : \mu_1 = \mu_2 \quad \text{versus} \quad H_a : \mu_1 \neq \mu_2$$

where μ_1 and μ_2 represent the population mean body temperature for women and men, respectively.

Step 2   **Find t_{crit} and state the rejection rule.**
The required degrees of freedom is the smaller of $n_1 - 1$ and $n_2 - 1$, which is $65 - 1 = 64$. Unfortunately, df $= 64$ is not in the t table in Appendix Table D, so we use the conservative df $= 60$. For $\alpha = 0.05$, this gives $t_{crit} = 2.000$. We have a two-tailed test, so Table 8 gives us the following rejection rule:

$$\text{Reject } H_0 \text{ if } t_{data} \geq 2.000 \text{ or } t_{data} \leq -2.000$$

Step 3   **Find t_{data}.**

$$t_{data} = \frac{(\bar{x}_1 - \bar{x}_2)}{\sqrt{\dfrac{s_1^2}{n_1} + \dfrac{s_2^2}{n_2}}} = \frac{(98.394 - 98.105)}{\sqrt{\dfrac{(0.743)^2}{65} + \dfrac{(0.699)^2}{65}}} \approx 2.28$$

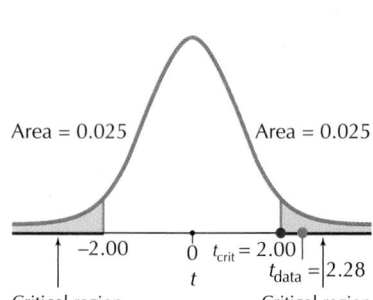

Area = 0.025   Area = 0.025

−2.00   0 t_{crit} = 2.00
                t t_{data} = 2.28
Critical region   Critical region

FIGURE 11 $t_{data} = 2.28$ falls within the critical region.

NOW YOU CAN DO
Exercises 3–6.

Step 4   **State the conclusion and the interpretation.**
The test statistic $t_{data} = 2.28$ is greater than $t_{crit} = 2.000$ (see Figure 11). We therefore reject H_0. There is evidence, at level of significance $\alpha = 0.05$, that the population mean body temperatures are not the same for women and men.

We may also use the p-value method to perform the independent sample t test for $\mu_1 - \mu_2$.

Welch's Hypothesis Test for the Difference in Two Population Means: p-Value Method

The hypothesis test applies whenever *either*

a. Both populations are normally distributed, or

b. Both samples are large, that is, $n_1 \geq 30$ and $n_2 \geq 30$.

Step 1   State the hypotheses and the rejection rule.
Use one of the forms from Table 9. State the meaning of μ_1 and μ_2. The rejection rule is: *Reject H_0 if the p-value is $\leq \alpha$.*

Step 2   Calculate t_{data}.

$$t_{data} = \frac{(\bar{x}_1 - \bar{x}_2)}{\sqrt{\dfrac{s_1^2}{n_1} + \dfrac{s_2^2}{n_2}}}$$

which follows an approximate t distribution with degrees of freedom the smaller of $n_1 - 1$ and $n_2 - 1$.

Step 3   Find the p-value.
Use technology or estimate using the t table.

Step 4   State the conclusion and the interpretation.
Compare the p-value with α.

Table 9 p-Values for t test for $\mu_1 - \mu_2$

Form of test	Right-tailed test	Left-tailed test	Two-tailed test
Hypotheses	$H_0 : \mu_1 = \mu_2$ $H_a : \mu_1 > \mu_2$ $p\text{-value} = P(t > t_{\text{data}})$ Area to the right of t_{data}	$H_0 : \mu_1 = \mu_2$ $H_a : \mu_1 < \mu_2$ $p\text{-value} = P(t < t_{\text{data}})$ Area to the left of t_{data}	$H_0 : \mu_1 = \mu_2$ $H_a : \mu_1 \neq \mu_2$ $p\text{-value} = P(t > |t_{\text{data}}|) + P(t < -|t_{\text{data}}|)$ $= 2 \cdot P(t > |t_{\text{data}}|)$ Sum of the two tail areas
p-Value			

(Right-tailed curve: p-value shaded area to the right of t_{data}, marked at 0, t_{data})

(Left-tailed curve: p-value shaded area to the left, marked at t_{data}, 0)

(Two-tailed curve: Sum of two areas is p-value, marked at $-|t_{\text{data}}|$, 0, $|t_{\text{data}}|$)

EXAMPLE 8 t Test for $\mu_1 - \mu_2$ using the p-value method

 bankloans

Bank Loans

Here, we do some analysis on our Chapter 10 Case Study data set: **Bank Loans**. When evaluating loan applicants for approval of a loan, banks examine several aspects of an applicant's financial history. Of particular importance is the applicant's credit score. Here, we will look for differences in the mean credit score between applicants who were approved and those who were denied. Independent samples of size 100 were taken from the **Bank Loans** data set, from among those approved and those denied the loan. The descriptive statistics for each group are shown in Table 10. Use the TI-83/84 or Excel, the p-value method, and level of significance $\alpha = 0.01$ to test whether the population mean credit score for successful loan applicants is greater than that for those who were denied the loan.

Note: Our degrees of freedom, the smaller of $n_1 - 1$ and $n_2 - 1$, is $100 - 1 = 99$. However, the TI-83/84 output on the next page shows df = 143.2490465. Why does the technology use different degrees of freedom than we do? Recall that we are using Welch's approximation to the t distribution. The TI-83/84, Excel, Minitab, and other technology calculate the degrees of freedom as follows:[7]

$$df = \frac{\left(\dfrac{s_1^2}{n_1} + \dfrac{s_2^2}{n_2}\right)^2}{\dfrac{\left(\dfrac{s_1^2}{n_1}\right)^2}{n_1 - 1} + \dfrac{\left(\dfrac{s_2^2}{n_2}\right)^2}{n_2 - 1}}$$

This provides a more accurate determination of the degrees of freedom than our method. However, our method is a conservative estimate that is easier to calculate, and it is recommended for hand calculations.

Table 10 Summary statistics for credit scores of loan applicants who were approved and those who were denied the loan

Loan Status	Sample size	Sample mean credit score	Sample standard deviation	Population mean credit score
Approved (sample 1)	$n_1 = 100$	$\bar{x}_1 = 708$	$s_1 = 34$	$\mu_1 = ?$
Denied (sample 2)	$n_2 = 100$	$\bar{x}_2 = 635$	$s_2 = 70$	$\mu_2 = ?$

Source: Data Mining and Predictive Analytics, by Daniel Larose and Chantal Larose, Wiley, 2015.

Solution

Because both samples are large ($n_1 = n_2 = 100 \geq 30$), we may proceed with the t test for $\mu_1 - \mu_2$.

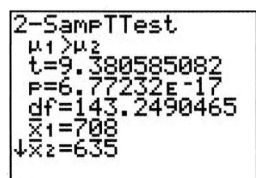

FIGURE 12 TI-83/84 output.

Remember that when the TI calculator gives you output such as "E-17," it is indicating that the decimal point needs to be moved 17 places to the left.

NOW YOU CAN DO
Exercises 7–10.

Step 1   **State the hypotheses and the rejection rule.**
The approved applicants represent sample 1, and we are interested in whether the mean credit score for the approved applicants is greater than that of the denied applicants. Thus, we have the following hypotheses:

$$H_0 : \mu_1 = \mu_2 \quad \text{versus} \quad H_a : \mu_1 > \mu_2$$

where μ_1 and μ_2 represent the population mean credit score for the approved and denied loan applicants, respectively. The rejection rule is to reject H_0 if p-value ≤ 0.01.

Step 2   **Find t_{data}.**
We use the instructions provided in the Step-by-Step Technology Guide at the end of this section. From Figure 12,

$$t_{\text{data}} = \frac{(\bar{x}_1 - \bar{x}_2)}{\sqrt{\dfrac{s_1^2}{n_1} + \dfrac{s_2^2}{n_2}}} \approx \frac{(708 - 635)}{\sqrt{\dfrac{34^2}{100} + \dfrac{70^2}{100}}} \approx 9.38$$

Step 3   **Find the p-value.**
From Figure 12,

$$p\text{-value} = P(t > t_{\text{data}}) = P(t > 9.38) \approx 0.0000000000000000677 \approx 0$$

Step 4   **State the conclusion and the interpretation.**
Our p-value of essentially zero is smaller than the level of significance 0.01. Therefore, we reject H_0. There is evidence that the population mean credit score for approved applicants is greater than that of the denied loan applicants. In fact, with a p-value so close to zero, the evidence is very strong indeed.

2 t Confidence Intervals for $\mu_1 - \mu_2$

Recall from Section 8.2 that to estimate the unknown population mean μ, we can use a t confidence interval:

$$\bar{x} \pm E = \bar{x} \pm t_{a/2} \left(s/\sqrt{n} \right)$$

where E is the **margin of error**. By analogy, here the t interval for $\mu_1 - \mu_2$ takes the following form.

Welch's Confidence Interval for $\mu_1 - \mu_2$

For two independent random samples taken from two populations with population means μ_1 and μ_2, a **100(1 − α)% confidence interval for $\mu_1 - \mu_2$** is given by

$$(\bar{x}_1 - \bar{x}_2) \pm t_{a/2} \sqrt{\frac{s_1^2}{n_1} + \frac{s_2^2}{n_2}}$$

where $\bar{x}_1$, s_1, and n_1 represent the mean, standard deviation, and sample size of the sample taken from population 1, $\bar{x}_2$, s_2, and n_2 represent the mean, standard deviation, and sample size of the sample taken from population 2, and $t_{a/2}$ is associated with the confidence level and degrees of freedom of the smaller of $n_1 - 1$ and $n_2 - 1$.

The t interval applies whenever *either* of the following conditions is met:

- Both populations are normally distributed, or
- Both sample sizes are large.

Margin of Error E

The margin of error for a 100(1 − α)% confidence interval for $\mu_1 - \mu_2$ is given by

$$E = t_{a/2} \cdot \sqrt{\frac{s_1^2}{n_1} + \frac{s_2^2}{n_2}}$$

Thus, the confidence interval for $\mu_1 - \mu_2$ takes the form $(\bar{x}_1 - \bar{x}_2) \pm E$.

This is a confidence interval for the difference in two population means, which is not the same as in Section 10.1, which was for the population mean of the differences of matched pairs. Here, we calculate the means of the samples and then compute the difference. In Section 10.1, we calculated the differences of sample values first and then computed the mean of these differences.

EXAMPLE 9   Confidence interval for $\mu_1 - \mu_2$

Bank Loans

Find a 90% confidence interval for the difference in population mean credit scores for those approved and those denied bank loans, using the data in Table 10.

Solution

Both sample sizes are large ($n_1 = n_2 = 100 \geq 30$), so we may construct the interval. For $t_{a/2}$, the required degrees of freedom is the smaller of $n_1 - 1$ and $n_2 - 1$, which is $100 - 1 = 99$. Because df = 99 is not listed in the t table, we use the next lower value listed as a conservative alternative: df = 90. For 90% confidence, then $t_{a/2} = 1.662$.

The margin of error is

$$E = t_{a/2} \cdot \sqrt{\frac{s_1^2}{n_1} + \frac{s_2^2}{n_2}} \approx (1.662) \cdot \sqrt{\frac{(34)^2}{100} + \frac{(70)^2}{100}} \approx 12.934$$

The 90% confidence interval is then

$$(\bar{x}_1 - \bar{x}_2) \pm E = (708 - 635) \pm 12.934 = (60.066, 85.934)$$

NOW YOU CAN DO
Exercises 11–16.

We are 90% confident that the difference in population mean credit scores $\mu_1 - \mu_2$ lies between 60.066 and 85.934. Because 0 is not contained in this interval, we may conclude that $\mu_1 \neq \mu_2$, just as we did in Example 8.

3 Using Confidence Intervals to Perform Hypothesis Tests

As in earlier sections, we may use a $100(1 - \alpha)\%$ t confidence interval for $\mu_1 - \mu_2$ to perform two-tailed t tests about $\mu_1 - \mu_2$.

Equivalence of a Two-Tailed t Test About $\mu_1 - \mu_2$ and a t Confidence Interval for $\mu_1 - \mu_2$

- If a certain value for $\mu_1 - \mu_2$ lies *outside* the corresponding $100(1 - \alpha)\%$ t confidence interval for $\mu_1 - \mu_2$, then the null hypothesis specifying this value would be *rejected* for level of significance α.
- Alternatively, if a certain value for $\mu_1 - \mu_2$ lies *inside* the $100(1 - \alpha)\%$ t confidence interval for $\mu_1 - \mu_2$, then the null hypothesis specifying this value would *not be rejected* for level of significance α.

EXAMPLE 10   Using a t confidence interval to perform a two-tailed t test about $\mu_1 - \mu_2$

Bank Loans

Use the 90% confidence interval for the difference in population mean credit scores $\mu_1 - \mu_2$ from Example 9 to test, using level of significance $\alpha = 0.10$, whether $\mu_1 - \mu_2$ differs from the following values:

a. 60   **b.** 70

Solution

The hypotheses for our two-sample t test look like this:

$$H_0 : \mu_1 = \mu_2 \quad \text{versus} \quad H_a : \mu_1 \neq \mu_2$$

which is equivalent to the following, if we subtract μ_2 from each side of the equations:

$$H_0 : \mu_1 - \mu_2 = 0 \quad \text{versus} \quad H_a : \mu_1 - \mu_2 \neq 0$$

a. Here, we replace zero with the value of 60 hypothesized for the difference in population means, obtaining

$$H_0 : \mu_1 - \mu_2 = 60 \quad \text{versus} \quad H_a : \mu_1 - \mu_2 \neq 60$$

The value of 60 lies outside the confidence interval (60.066, 85.934) from Example 9; therefore, we reject H_0.

b. Our hypotheses are

$$H_0 : \mu_1 - \mu_2 = 70 \quad \text{versus} \quad H_a : \mu_1 - \mu_2 \neq 70$$

NOW YOU CAN DO
Exercises 17–20.

The value of 70 lies inside the confidence interval (60.066, 85.934); therefore, we do not reject H_0.

4 t Inference for $\mu_1 - \mu_2$ Using Pooled Variance

Recall that the variance equals the square of the standard deviation.

An alternative method for t inference may be applied when the data analyst has reason to believe that $\sigma_1^2 = \sigma_2^2$, that is, the variances of the two populations are equal. A *pooled estimate* s_{pooled}^2 of the common variance $\sigma_1^2 = \sigma_2^2 = \sigma^2$ is used.

Pooled Estimate of the Common Variance σ^2

$$s_{pooled}^2 = \frac{(n_1 - 1)s_1^2 + (n_2 - 1)s_2^2}{n_1 + n_2 - 2}$$

Some statisticians think that the pooled variance method should be used sparingly.[8]

The conditions for performing t inference using pooled variance are the same as for Welch's method (page 591), with the additional condition that $\sigma_1^2 = \sigma_2^2$. The test statistic t_{data} for the pooled variance t test is then given by

$$t_{data} = \frac{(\bar{x}_1 - \bar{x}_2)}{\sqrt{s_{pooled}^2 \left(\frac{1}{n_1} + \frac{1}{n_2} \right)}}$$

We illustrate the pooled variance t test and the pooled variance t confidence interval using the following two examples.

EXAMPLE 11 Pooled variance t test

Bank Loans

Another indicator of financial health that banks look at when evaluating loan applicants is the *debt-to-income ratio,* which is defined as follows.

$$\text{Debt-to-income ratio} = \frac{\text{Applicant's total debt}}{\text{Applicant's annual income}}$$

For example, if you owed $5000 on your credit card and $10,000 on your car, and your annual income was $50,000, then your debt-to-income ratio would be ($5000 + $10,000)/$50,000 = 0.3.

Here, we examine whether the mean debt-to-income ratio differed between those approved and those denied the bank loans. Samples of size 100 were obtained from each group, with the summary statistics shown in Table 11.

Table 11 Summary statistics for debt-to-income ratios of loan applicants who were approved and those who were denied the loan

Loan status	Sample size	Sample mean debt-to-income ratio	Sample standard deviation	Population mean debt-to-income ratio
Approved (sample 1)	$n_1 = 100$	$\bar{x}_1 = 0.15$	$s_1 = 0.10$	$\mu_1 = ?$
Denied (sample 2)	$n_2 = 100$	$\bar{x}_2 = 0.20$	$s_2 = 0.15$	$\mu_2 = ?$

Use the critical-value method for the pooled variance t test to test whether the population mean debt-to-income ratio for those approved is less than that of those who were not approved for the loan. Assume $\sigma_1^2 = \sigma_2^2$ and use level of significance $\alpha = 0.05$.

Solution

Step 1 **State the hypotheses.**

$$H_0 : \mu_1 = \mu_2 \quad \text{versus} \quad H_a : \mu_1 < \mu_2$$

where μ_1 and μ_2 represent the population mean debt-to-income ratio for those approved and those denied the loan, respectively.

Step 2 **Find t_{crit}.**
The degrees of freedom for the pooled variance t test equals $n_1 + n_2 - 2 = 100 + 100 - 2 = 198$. Because df $= 198$ is not in the t table, we use the next lower value df $= 100$ instead, obtaining the critical value $t_{\text{crit}} = 1.984$. Reject H_0 if $t_{\text{data}} \leq -1.984$.

Step 3 **Calculate s_{pooled}^2 and t_{data}.**

$$s_{\text{pooled}}^2 = \frac{(n_1 - 1)s_1^2 + (n_2 - 1)s_2^2}{n_1 + n_2 - 2} = \frac{(100 - 1)0.10^2 + (100 - 1)0.15^2}{100 + 100 - 2} = 0.01625$$

Plugging this value into the following formula for the test statistic, we obtain

$$t_{\text{data}} = \frac{\bar{x}_1 - \bar{x}_2}{\sqrt{s_{\text{pooled}}^2 \left(\frac{1}{n_1} + \frac{1}{n_2}\right)}} = \frac{0.15 - 0.20}{\sqrt{0.01625\left(\frac{1}{100} + \frac{1}{100}\right)}} \approx -2.7735$$

Step 4 **Conclusion and interpretation.**
The test statistic $t_{\text{data}} \approx -2.7735$ is less than the critical value $t_{\text{crit}} = -1.984$. Therefore, we reject H_0. There is evidence that the population mean debt-to-income ratio for those whose loan was approved is less than that for those whose loan was not approved.

NOW YOU CAN DO
Exercises 21 and 22.

The pooled variance method may also be used to construct a t confidence interval for $\mu_1 - \mu_2$.

EXAMPLE 12   **Pooled variance *t* confidence interval for $\mu_1 - \mu_2$**

Bank Loans

Use the data from Example 11 to construct a 95% confidence interval for the difference in population mean debt-to-income ratio. Use the pooled variance method.

Solution

The $100(1 - \alpha)\%$ confidence interval for $\mu_1 - \mu_2$ using the pooled variance method is given by the following formula:

$$\bar{x}_1 - \bar{x}_2 \pm t_{a/2} \sqrt{s^2_{\text{pooled}} \left(\frac{1}{n_1} + \frac{1}{n_2} \right)}$$

where $t_{a/2}$ is found using $n_1 + n_2 - 2$ degrees of freedom. Similar to Example 11, we use df = 100 because df = 198 is not in the table. So, we have $t_{a/2} = 1.984$. Thus, our 95% confidence interval is:

$$0.15 - 0.20 \pm (1.984) \sqrt{0.01625 \left(\frac{1}{100} + \frac{1}{100} \right)} \approx -0.05 \pm 0.04 = (-0.09, -0.01)$$

NOW YOU CAN DO
Exercises 23 and 24.

We are 95% confident that the difference in population mean debt-to-income ratios, $\mu_1 - \mu_2$, lies between -0.09 and -0.01.

5 Z Inference for $\mu_1 - \mu_2$ When σ_1 and σ_2 Are Known

When the population standard deviations σ_1 and σ_2 are known, the data analyst may prefer to use *Z* inference for $\mu_1 - \mu_2$ because the margin of error for *Z* inference is smaller than for *t* inference. The conditions for performing *Z* inference for $\mu_1 - \mu_2$ are similar to Welch's method (page 591), with the additional condition that σ_1 and σ_2 are known. We illustrate the two-sample *Z* test and the *Z* confidence interval for $\mu_1 - \mu_2$ using the following two examples.

Do not use *Z* inference for $\mu_1 - \mu_2$ unless both σ_1 and σ_2 are known.

EXAMPLE 13   **Two-sample *Z* test**

A Kaiser Family Foundation report found that the mean amount of time that young people ages 8–18 spend talking on their cell phones is $\bar{x}_1 = 33$ minutes per day, whereas the mean amount of time spent watching TV shows on their cell phones is $\bar{x}_2 = 49$ minutes per day.[9] Assume that the sample sizes are $n_1 = 50$ and $n_2 = 40$, and that the population standard deviations are known to be $\sigma_1 = 15$ minutes and $\sigma_2 = 20$ minutes. Test, using the critical-value method and level of significance $\alpha = 0.05$, whether the population mean amount of time young people spending talking on their cell phones is less than the population mean amount of time they spend watching TV shows on their cell phones.

Solution

Step 1   **State the hypotheses.**

$$H_0 : \mu_1 = \mu_2 \quad \text{versus} \quad H_a : \mu_1 < \mu_2$$

where μ_1 and μ_2 represent the population mean amount of time young people spend talking and watching TV shows, respectively, on their cell phones.

Step 2 **Find Z_{crit}.**
From Table 4 in Chapter 9 (page 500), we have $Z_{crit} = -1.645$. Reject H_0 if $Z_{data} \leq -1.645$.

Step 3 **Calculate Z_{data}.**
The test statistic for the Z test for $\mu_1 - \mu_2$ takes the form

$$Z_{data} = \frac{\bar{x}_1 - \bar{x}_2}{\sqrt{\dfrac{\sigma_1^2}{n_1} + \dfrac{\sigma_2^2}{n_2}}} = \frac{33 - 49}{\sqrt{\dfrac{15^2}{50} + \dfrac{20^2}{40}}} \approx -4.202$$

Step 4 **Conclusion and interpretation.**
The test statistic $Z_{data} \approx -4.202$ is less than the critical value $Z_{crit} = -1.645$. Therefore, we reject H_0. There is evidence that the population mean amount of time young people spending talking on their cell phones is less than the population mean amount of time they spend watching TV shows on their cell phones.

NOW YOU CAN DO
Exercises 25 and 26.

When σ_1 and σ_2 are known, we can also construct a Z confidence interval for $\mu_1 - \mu_2$.

EXAMPLE 14 Z confidence interval for $\mu_1 - \mu_2$

Use the data from Example 13 to construct a 95% Z confidence interval for the difference in population mean amount of time spent using cell phones.

Solution

The $100(1 - \alpha)\%$ Z confidence interval for $\mu_1 - \mu_2$ is as follows:

$$\bar{x}_1 - \bar{x}_2 \pm Z_{a/2} \sqrt{\frac{\sigma_1^2}{n_1} + \frac{\sigma_2^2}{n_2}}$$

From Table 1 in Chapter 8 (page 432) we have $Z_{a/2} = 1.96$. Thus, our 95% confidence interval is

$$33 - 49 \pm (1.96) \sqrt{\frac{15^2}{50} + \frac{20^2}{40}} \approx -16 \pm 7.463 = (-23.463, -8.537)$$

We are 95% confident that the difference in population mean amounts of time spent by young people on cell phones talking and watching TV shows lies between -23.463 minutes and -8.537 minutes. In other words, we are 95% confident that young people spend between 8.537 minutes and 23.463 minutes longer watching TV shows on their cell phones rather than talking on their cell phones.

NOW YOU CAN DO
Exercises 27 and 28.

STEP-BY-STEP TECHNOLOGY GUIDE: Two-Sample t Test and Confidence Interval for $\mu_1 - \mu_2$

TI-83/84
Welch's t Test and Confidence Interval for $\mu_1 - \mu_2$
Data Option
Step 1 Enter one sample into List **L1** and the other sample into List **L2**.
Step 2 Press **STAT** and highlight **TESTS**.

Step 3 Press **4** (for the **2-Samp TTest**). The **2-Samp TTest** menu appears.
Step 4 For input (**INPT**), move the cursor over **Data** and press **ENTER**.
Step 5 For **List1** and **List2**, enter **L1** and **L2**.
Step 6 For **Freq1** and **Freq2**, enter **1**.

Step 7   For μ_1, choose the form of H_a.
Step 8   For **Pooled**, select **No** because we are not assuming the variances are equal, and do not need an estimate of the common variance (Figure 13).
Step 9   Highlight **Calculate** and press **ENTER**.

Stats Option.   (Example 8 is used to illustrate this method.) Here you enter the summary statistics.
Step 1   Press **STAT** and highlight **TESTS**.
Step 2   Press **4** (for the **2-Samp TTest**). The **2-Samp TTest** menu appears.

Step 3   For input (**Inpt**), move the cursor over **Stats** and press **ENTER**.
Step 4   For $\bar{x}_1$, enter **708**. For Sx_1, enter **34**. For n_1, enter **100**. For $\bar{x}_2$, enter **635**. For Sx_2, enter **70**. For n_2, enter **100** (Figure 14).
Step 5   For μ_1, choose the form of H_a. For Example 8, choose "$> \mu_2$" and press **ENTER**.
Step 6   For **Pooled**, press **No** (Figure 15).
Step 7   Highlight **Calculate** and press **ENTER**.

FIGURE 13   **FIGURE 14**   **FIGURE 15**

Welch's Two-Sample *t* interval for $\mu_1 - \mu_2$
Follow the same steps as for the *t* test, except select **0: 2-SampTInt**. Also, to select confidence level (**C-Level**), enter **0.95** for 95%.

Pooled Variance *t* Test and Confidence Interval for $\mu_1 - \mu_2$
Follow the same steps as for Welch's method, except select **Yes** for **Pooled** in Step 8.

***Z* Test for $\mu_1 - \mu_2$**
Data Option
Step 1   Enter the data into Lists **L1** and **L2**.
Step 2   Press **STAT** and highlight **TESTS**.
Step 3   Press **3** (for the **2-Samp Z Test**). The **2-Samp Z Test** menu appears.
Step 4   For input (**INPT**), move the cursor over **Data** and press **ENTER**.
Step 5   Enter the values for σ_1 and σ_2.
Step 6   For **List1** and **List2**, enter **L1** and **L2**.

Step 7   For **Freq1** and **Freq2**, enter **1**.
Step 8   For μ_1, choose the form of H_a and press **ENTER**.
Step 9   Highlight **Calculate** and press **ENTER**.

Stats Option
Step 1   Press **STAT** and highlight **TESTS**.
Step 2   Press **3** (for the **2-Samp Z Test**). The **2-Samp Z Test** menu appears.
Step 3   For input (**Inpt**), move the cursor over **Stats** and press **ENTER**.
Step 4   Enter the values for σ_1, σ_2, $s_{\bar{x}_1}$, n_1, $s_{\bar{x}_2}$, and n_2.
Step 5   For μ_1, choose the form of H_a and press **ENTER**.
Step 6   Highlight **Calculate** and press **ENTER**.

***Z* Confidence Interval for $\mu_1 - \mu_2$**
Follow the same steps as for the *Z* test, except select **9: 2-SampleZInt** at Step 2.

EXCEL

Welch's *t* Test for $\mu_1 - \mu_2$
Step 1   Enter Sample 1 and Sample 2 data into columns A and B, respectively.
Step 2   Select **Data > Data Analysis > *t*-Test: Two-Sample Assuming Unequal Variances**, and click **OK**.
Step 3   For the **Input**, select the cells in column A for the **Variable 1 Range** and the cells in column B for the **Variable 2 Range**.
Step 4   For the hypothesized mean difference, enter **0**, enter your value for **Alpha**, and click **OK**.

Pooled Variance *t* Test for $\mu_1 - \mu_2$
Follow the same steps as for Welch's method, except select **t-test: Two-Sample**

Assuming Equal Variances in Step 2.
***Z* Test for $\mu_1 - \mu_2$**
Step 1   Enter Sample 1 and Sample 2 data into columns A and B, respectively.
Step 2   Select **Data > Data Analysis > *Z*-Test: Two-Sample for Means**, and click **OK**.
Step 3   For the **Input**, select the cells in column A for the **Variable 1 Range** and the cells in column B for the **Variable 2 Range**. For the hypothesized mean difference, enter **0**.
Step 4   For **Variable 1 Variance (known)** and **Variable 2 Variance (known)**, enter the values for σ_1^2 and σ_2^2. Enter the value α for **ALPHA** and click **OK**.

MINITAB

Welch's *t* Test and Confidence Interval for $\mu_1 - \mu_2$
Step 1   Enter Sample 1 and Sample 2 data into columns C1 and C2, respectively.
Step 2   Click **Stat > Basic Statistics > 2-Sample t…**.

Step 3
a. If you have the data values, select **Each sample is in its own column**, and select **C1** and **C2** for **Sample 1** and **Sample 2**, respectively.

b. If you have the summary statistics, select **Summarized data** and enter the **sample size, mean,** and **standard deviation** for each of **Sample 1** and **Sample 2**.
Step 4 Click **Options**, enter your **Confidence level**, the **Hypothesized difference**, and select the form of the **Alternative hypothesis**.

Step 5 Click **OK** and click **OK** again.

Pooled Variance *t* Test and Confidence Interval for $\mu_1 - \mu_2$
Follow the same steps as for Welch's method, except select **Assume equal variances** at the end of Step 4.

SPSS

Welch's *t* Test and *t* Interval for $\mu_1 - \mu_2$
Step 1 Enter numeric group labels into the first column, and the corresponding data values into the second column.
Step 2 Select **Analyze > Compare Means > Independent-Samples T Test....**
Step 3 Put the variable with data values in the **Test Variable(s)** box, and the variable with group labels in the **Grouping Variable** box. Click **Define Groups...** and enter the values you used to define your groups. Click **Continue**.

Step 4 Click **Options...** and enter the **Confidence Interval Percentage**. Click **Continue**, and then click **OK**. The results are found in the **Independent Samples Test** box, in the **Equal variances not assumed** row.

Pooled Variance *t* Test and *t* Interval for $\mu_1 - \mu_2$
Step 1 Use the same steps as for Welch's *t* test and *t* interval. The results are found in the **Equal variances assumed** row of the **Independent Samples Test** box.

JMP

Welch's *t* Test and *t* Interval for $\mu_1 - \mu_2$
Step 1 Enter numeric group labels into **Column 1**, and the corresponding data values into **Column 2**. Right-click **Column 1**, click **Column Info...**, and change **Data Type** to **Character**. Click **OK**.
Step 2 Select **Analyze > Fit Y by X**. Put Column 1 in the in the **X, Factor** box, and Column 2 in the **Y, Response** box. Click **OK**.

Step 3 Click the red triangle beside **Oneway Analysis of Column 2 by Column 1**. Select *t* **Test**. To change the confidence level, select the red triangle, click **Set α level**, and select your desired α.

Pooled Variance *t* Test and *t* Interval for $\mu_1 - \mu_2$
Use the same steps as for Welch's *t* test and *t* interval, except choose **Means/Anova/Pooled *t*** during Step 3 and look at the *t* **Test** output.

CRUNCHIT!

Welch's *t* Test and *t* Interval for $\mu_1 - \mu_2$
Step 1 Click **Statistics**, highlight **t**, and select **2-sample**. Select the **Summarized** tab.
Step 2 Enter **n**, the **Sample Mean**, and the **Standard Deviation** for **Sample 1** and **Sample 2**.
Step 3
a. For the hypothesis test, select the **Hypothesis Test** tab, enter the **Difference of means under null hypothesis**, choose the correct form of the **Alternative** hypothesis, and click **Calculate**.
b. For the confidence interval, select the **Confidence Interval** tab, enter the **Confidence Interval Level (%)**, and click **Calculate**.

Pooled Variance *t* Test and *t* Interval for $\mu_1 - \mu_2$
Use the same steps as for Welch's *t* test and *t* interval, except make sure to check the **Pooled Variance** option in Step 2.

Z test for $\mu_1 - \mu_2$
Step 1 Click **Statistics**, highlight **z**, and select **2-sample**. Select the **Summarized** tab. Enter *n* and the Sample Mean for both **Sample 1** and **Sample 2**.
Step 2 Enter the population standard deviations for each sample under **StdDev of Group 1** and **StdDev of Group 2**. Enter the **Difference of means under null hypothesis**, choose the correct form of the **Alternative** hypothesis, and click **Calculate**.

Section 10.2 Summary

1. Section 10.2 examines inferential methods for $\mu_1 - \mu_2$, which is the difference between the means of two independent populations. Two-sample *t* tests may be performed using either the *p*-value method or the critical-value method.
2. $100(1 - \alpha)\%$ *t* confidence intervals for $\mu_1 - \mu_2$ are developed and illustrated.
3. The use of *t* confidence intervals for $\mu_1 - \mu_2$ to perform two-tailed *t* tests is illustrated.

4. The pooled variance method for *t* inference may be applied when the data analyst has reason to believe that the variances of the two populations are equal.
5. When the population standard deviations σ_1 and σ_2 are known, the data analyst may prefer to use *Z* inference for $\mu_1 - \mu_2$.

Section 10.2 Exercises

CLARIFYING THE CONCEPTS

1. What are the conditions that permit us to perform Welch's two-sample t test? (p. 592)
2. If a $100(1 - \alpha)\%$ confidence interval for $\mu_1 - \mu_2$ contains 0, then with level of significance α, what is our conclusion regarding the hypothesis that there is no difference in the population means? (p. 596)

PRACTICING THE TECHNIQUES

 CHECK IT OUT!

To do	Check out	Topic
Exercises 3–6	Example 7	t test for $\mu_1 - \mu_2$: critical-value method
Exercises 7–10	Example 8	t test for $\mu_1 - \mu_2$: p-value method
Exercises 11–16	Example 9	t confidence interval for $\mu_1 - \mu_2$
Exercises 17–20	Example 10	Equivalence between confidence intervals and t tests for $\mu_1 - \mu_2$
Exercises 21–22	Example 11	Pooled variance t test for $\mu_1 - \mu_2$
Exercises 23–24	Example 12	Pooled variance t confidence interval for $\mu_1 - \mu_2$
Exercises 25–26	Example 13	Z test for $\mu_1 - \mu_2$
Exercises 27–28	Example 14	Z confidence interval for $\mu_1 - \mu_2$

For Exercises 3–6, perform the indicated Welch's hypothesis test using the critical-value method. The summary statistics were taken from random samples that were drawn independently. For each exercise, follow these steps:
   a. State the hypotheses.
   b. Find the critical value t_{crit} and the rejection rule for this test.
   c. Calculate t_{data}.
   d. Compare t_{data} with t_{crit}. State and interpret your conclusion.
3. Test, at level of significance $\alpha = 0.10$, whether $\mu_1 \neq \mu_2$.

Sample 1	$n_1 = 36$	$\bar{x}_1 = 10$	$s_1 = 2$
Sample 2	$n_2 = 36$	$\bar{x}_2 = 8$	$s_2 = 2$

4. Test, at level of significance $\alpha = 0.05$, whether $\mu_1 > \mu_2$.

Sample 1	$n_1 = 64$	$\bar{x}_1 = 20$	$s_1 = 3$
Sample 2	$n_2 = 64$	$\bar{x}_2 = 18$	$s_2 = 2$

5. Test, at level of significance $\alpha = 0.01$, whether $\mu_1 < \mu_2$.

Sample 1	$n_1 = 100$	$\bar{x}_1 = 70$	$s_1 = 10$
Sample 2	$n_2 = 50$	$\bar{x}_2 = 80$	$s_2 = 12$

6. Test, at level of significance $\alpha = 0.05$, whether $\mu_1 > \mu_2$.

Sample 1	$n_1 = 60$	$\bar{x}_1 = 100$	$s_1 = 20$
Sample 2	$n_2 = 40$	$\bar{x}_2 = 90$	$s_2 = 10$

For Exercises 7–10, perform the indicated Welch's hypothesis test using the p-value method. The summary statistics were taken from random samples that were drawn independently. For each exercise follow these steps:
   a. State the hypotheses and the rejection rule.
   b. Calculate t_{data}.
   c. Find the p-value.
   d. Compare the p-value with level of significance α. State and interpret your conclusion.
7. Test, at level of significance $\alpha = 0.10$, whether $\mu_1 \neq \mu_2$.

Sample 1	$n_1 = 64$	$\bar{x}_1 = 0$	$s_1 = 3$
Sample 2	$n_2 = 49$	$\bar{x}_2 = 1$	$s_2 = 1$

8. Test, at level of significance $\alpha = 0.05$, whether $\mu_1 > \mu_2$.

Sample 1	$n_1 = 255$	$\bar{x}_1 = 103$	$s_1 = 17$
Sample 2	$n_2 = 400$	$\bar{x}_2 = 95$	$s_2 = 11$

9. Test, at level of significance $\alpha = 0.05$, whether $\mu_1 < \mu_2$.

Sample 1	$n_1 = 100$	$\bar{x}_1 = 50$	$s_1 = 10$
Sample 2	$n_2 = 100$	$\bar{x}_2 = 75$	$s_2 = 15$

10. Test, at level of significance $\alpha = 0.01$, whether $\mu_1 \neq \mu_2$.

Sample 1	$n_1 = 30$	$\bar{x}_1 = -10$	$s_1 = 5$
Sample 2	$n_2 = 30$	$\bar{x}_2 = -5$	$s_2 = 2$

For Exercises 11–16, do the following for the designated data:
   a. Provide the point estimate of $\mu_1 - \mu_2$.
   b. Calculate the margin of error for the confidence level indicated.
   c. Construct and interpret a t confidence interval for $\mu_1 - \mu_2$ with the confidence level indicated.
11. Data in Exercise 3, confidence level = 90%
12. Data in Exercise 4, confidence level = 95%
13. Data in Exercise 5, confidence level = 99%
14. Data in Exercise 6, confidence level = 95%
15. Data in Exercise 7, confidence level = 95%
16. Data in Exercise 8, confidence level = 90%

For Exercises 17–20 a $100(1 - \alpha)\%$ t confidence interval for $\mu_1 - \mu_2$ is given. Use the confidence interval to test, at level of significance α, whether $\mu_1 - \mu_2$ differs from each of the designated hypothesized values.

17. A 95% t confidence interval for $\mu_1 - \mu_2$ is (10, 15). Hypothesized values are
    a. 0   **b.** 12   **c.** 16
18. A 99% t confidence interval for $\mu_1 - \mu_2$ is (0, 100). Hypothesized values are
    a. 1   **b.** 99   **c.** 101
19. A 90% t confidence interval for $\mu_1 - \mu_2$ is (−10, 10). Hypothesized values are
    a. −10.1   **b.** −9.9   **c.** 0
20. A 95% t confidence interval for $\mu_1 - \mu_2$ is (−25, −15). Hypothesized values are
    a. −16   **b.** −26   **c.** 0

For Exercises 21–22, perform the indicated hypothesis test using the pooled variance method. The summary statistics were taken from random samples that were drawn independently. Assume $\sigma_1^2 = \sigma_2^2$.
21. Test, at level of significance $\alpha = 0.10$, whether $\mu_1 > \mu_2$.

Sample 1	$n_1 = 36$	$\bar{x}_1 = 54$	$s_1 = 10$
Sample 2	$n_2 = 36$	$\bar{x}_2 = 52$	$s_2 = 11$

22. Test, at level of significance $\alpha = 0.05$, whether $\mu_1 < \mu_2$.

Sample 1	$n_1 = 250$	$\bar{x}_1 = 3.0$	$s_1 = 0.25$
Sample 2	$n_2 = 150$	$\bar{x}_2 = 3.2$	$s_2 = 0.30$

For Exercises 23 and 24, construct a 95% confidence interval for $\mu_1 - \mu_2$ for the indicated data using the pooled variance method.
23. The data in Exercise 21
24. The data in Exercise 22

For Exercises 25 and 26, perform the indicated hypothesis test using the Z test. The summary statistics were taken from random samples that were drawn independently. Assume that σ_1 and σ_2 are known.
25. Test, at level of significance $\alpha = 0.05$, whether $\mu_1 > \mu_2$.

Sample 1	$n_1 = 49$	$\bar{x}_1 = 100$	$\sigma_1 = 1$
Sample 2	$n_2 = 36$	$\bar{x}_2 = 99$	$\sigma_2 = 2$

26. Test, at level of significance $\alpha = 0.10$, whether $\mu_1 < \mu_2$.

Sample 1	$n_1 = 64$	$\bar{x}_1 = 72$	$\sigma_1 = 3$
Sample 2	$n_2 = 100$	$\bar{x}_2 = 76$	$\sigma_2 = 5$

For Exercises 27 and 28, construct a 95% Z confidence interval for $\mu_1 - \mu_2$ for the indicated data.
27. The data in Exercise 25
28. The data in Exercise 26

APPLYING THE CONCEPTS

For Exercises 29–49, assume normality and use Welch's t test and t interval unless otherwise indicated.

29. **PC Sales.** A personal computer company launched an advertising campaign in the hopes of boosting sales. A random sample (sample 1) of 16 days before the advertising blitz showed mean sales of 120 computers per day with a standard deviation of 30. A random sample (sample 2) of 15 days after the advertisements appeared showed mean sales of 125 computers per day with a standard deviation of 35. If it is appropriate, test whether $\mu_1 < \mu_2$. If not, explain why not.
30. **Flight Delays.** The U.S. Customs and Border Protection Agency keeps track of the flight delays at all U.S. international airports. The summary statistics for the waiting times at Los Angeles International Airport (Tom Bradley Terminal) on July 1, 2014 at 5 P.M. and at 11 P.M. are provided in the following table.

Waiting Times at 5 P.M.	$n_1 = 1490$	$\bar{x}_1 = 27$ minutes	$s_1 = 20$ minutes
Waiting Times at 11 P.M.	$n_2 = 150$	$\bar{x}_2 = 11$ minutes	$s_2 = 15$ minutes

Test whether the mean waiting times at 5 P.M. are longer than those at 11 P.M., using level of significance $\alpha = 0.01$.
31. **Income in California Counties.** According to random samples taken in 2010 by the Bureau of Economic Analysis, the mean income for Sacramento County and Los Angeles County, California, was $31,987 and $33,179, respectively. Suppose the samples had the following sample statistics.

Sacramento County	$n_1 = 36$	$\bar{x}_1 = \$31{,}987$	$s_1 = \$5000$
Los Angeles County	$n_2 = 49$	$\bar{x}_2 = \$33{,}179$	$s_2 = \$6000$

   a. Provide the point estimate of the difference in population means $\mu_1 - \mu_2$.
   b. Calculate the margin of error for a confidence level of 95%.
   c. Construct and interpret a 95% confidence interval for $\mu_1 - \mu_2$.
   d. Test, at level of significance $\alpha = 0.05$, whether $\mu_1 < \mu_2$.
   e. Explain whether the confidence interval in (c) could have been used to perform the hypothesis test in (d). Why or why not?
32. **Math Scores.** The Institute of Educational Sciences published the results of the Trends in International Math and Science Study. In 2011, the sample mean mathematics scores for 8th-grade students from the United States and Hong Kong were 509 and 586, respectively. Suppose independent random samples are drawn from each population, and assume that the populations are normally distributed with the following summary statistics.

USA	$n_1 = 10$	$\bar{x}_1 = 509$	$s_1 = 80$
Hong Kong	$n_2 = 12$	$\bar{x}_2 = 586$	$s_2 = 70$

   a. Provide the point estimate of the difference in population means $\mu_1 - \mu_2$.

b. Calculate the margin of error for a confidence level of 90%.

c. Construct and interpret a 90% confidence interval for $\mu_1 - \mu_2$.

d. Test, at level of significance $\alpha = 0.01$, whether $\mu_1 < \mu_2$.

e. Provide two reasons why the confidence interval in (c) could not have been used to perform the hypothesis test in (d).

33. Children per Classroom. According to LocalSchoolDirectory.com, the sample mean number of children per teacher in the towns of Cupertino, California, and Santa Rosa, California, are 20.9 and 19.3, respectively. Suppose random samples of classrooms are taken from each county, with the following sample statistics.

Cupertino	$n_1 = 36$	$\bar{x}_1 = 20.9$	$s_1 = 5$
Santa Rosa	$n_2 = 64$	$\bar{x}_2 = 19.3$	$s_2 = 4$

a. Construct and interpret a 99% confidence interval for $\mu_1 - \mu_2$.

b. Use the confidence interval in (a) to test, at level of significance $\alpha = 0.01$, whether μ_1 differs from μ_2.

34. Property Taxes. Suppose you want to move to either a small town in Ohio (sample 1) or a small town in North Carolina (sample 2). You did some research on property taxes in each state and chose two random samples shown in the table. The data represent the property taxes in dollars for a residence assessed at $250,000. Test whether $\mu_1 \neq \mu_2$ using level of significance $\alpha = 0.05$. 🏢 **propertytax**

North Carolina		Ohio	
164	206	298	270
147	129	270	315
207	176	165	177
138	120	400	245
143	154	268	180
201	123	289	292
		285	291
		225	

35. Engineering Starting Salaries. The National Association of Colleges (NAC) reports mean starting salaries for college graduates in various disciplines. NAC reported that the mean salary for 2014 engineering graduates was $62,719, whereas the mean salary for 2013 engineering graduates was $62,535. Assume that these statistics were drawn from independent samples of size 100, each with a sample standard deviation of $10,000. Test whether there was a significant change in the salary of engineering graduates from 2013 to 2014, using level of significance $\alpha = 0.10$.

36. Park Usage. Suppose that planners for the town of The Woodlands, Texas, were interested in assessing usage of their parks. Random samples were taken of the number of daily visitors to Windvale Park and Cranebrook Park, with the statistics as reported here.

Windvale Park	$n_1 = 36$	$\bar{x}_1 = 110$	$s_1 = 60$
Cranebrook Park	$n_2 = 30$	$\bar{x}_2 = 150$	$s_2 = 75$

a. Construct and interpret a 95% confidence interval for $\mu_1 - \mu_2$.

b. Test at $\alpha = 0.05$ whether μ_1 is less than μ_2.

c. Explain whether the confidence interval in (a) could have been used to perform the hypothesis test in (b). Why or why not?

Coaching for the SAT. Use this information for Exercises 37–39. The College Board reports that a pretest and post-test study was done to investigate whether coaching had a significant effect on SAT scores. The improvement from pretest to post-test was 29 points for the coached sample of students, with a standard deviation of 59 points. For the uncoached students, the pretest to post-test improvement was 21 points with a standard deviation of 52 points.

37. Suppose we consider a sample of 100 students from each group. Perform a test, at level of significance $\alpha = 0.05$, to determine whether the population mean coached SAT pretest–post-test improvement is greater than that for the uncoached students.

38. Refer to Exercise 37.

a. Find a point estimate of the difference in population means.

b. Find a 99% confidence interval for the difference in population means.

c. Determine whether the population means differ, at level of significance $\alpha = 0.01$.

WHAT IF ? **39.** *What if* the sample sizes for each group were some number greater than $n = 100$?

a. How would this affect the width of the confidence interval in Exercise 38(b)? Is this good? Explain.

b. Would this change have any effect on our conclusion in the hypothesis test in Exercise 38(c)? Explain why or why not.

40. Nursing Support Services. A statistical study found that when nurses made home visits to pregnant teenagers to provide support services, discourage smoking, and otherwise provide care, the sample mean birth weight of the babies was higher for this treatment group (3285 grams) than for the control group (2922 grams) when the visits began before mid-gestation.[10] There were 21 patients in the treatment group and 11 in the control group. Suppose the birth weights for the babies in both groups follow a normal distribution. Assume that the population standard deviation in each sample is 500 grams.

a. Construct and interpret a 95% Z confidence interval for $\mu_1 - \mu_2$.

b. Test, at level of significance $\alpha = 0.05$, whether the population birth weight differs between the two groups. Use the Z test.

c. Assess the strength of evidence against the null hypothesis.

WHAT IF ? **41. Nursing Support Services.** Refer to Exercise 40. *What if* the birth weights of the babies in each group are the same certain amount greater? Explain how this would affect the following:

a. $\bar{x}_1 - \bar{x}_2$
b. t_{data}
c. *p*-value
d. conclusion

Zooplankton and Phytoplankton. Refer to the table below for Exercises 42 and 43. *Meta-analysis* refers to the statistical analysis of a set of similar research studies. In a meta-analysis, each data value represents an effect size calculated from the results of a particular study. The table contains effect sizes calculated in a meta-analysis for zooplankton and phytoplankton.[11] Not surprisingly, the paper found that zooplankton biomass was reduced by the introduction of zooplanktivorous fish (that is, fish that eat zooplankton), but that phytoplankton biomass increased, because there were now fewer zooplankton, and the zooplankton eat the phytoplankton. Such an effect is called a *trophic cascade*. See if you can replicate the scientists' results in the following exercises. **plankton**

Zooplankton		Phytoplankton	
−2.37	−3.00	10.61	3.04
−0.64	−0.68	2.97	0.65
−2.05	−1.39	1.58	2.55
−1.54	−0.64	2.55	1.05
−6.60	−3.88	5.67	2.11
0.26		1.57	

42. Use technology to construct a comparison dot plot or a comparison histogram of the zooplankton and the phytoplankton effect sizes. Is there evidence of a difference in mean effect size for the two groups?

43. Test whether the population means effect sizes differ, at level of significance $\alpha = 0.01$.

Studying Time for Biologists and Psychologists. Use the following information for Exercises 44–47: The National Survey of Student Engagement reported in 2012 that the mean number of hours spent studying per week for biology majors was 16.7, whereas the mean for psychology majors was 13.9. Assume that these are sample means from sample sizes of 100, and that the sample standard deviations were each 5 hours.

44. Calculate the margin of error for a 95% confidence interval for the difference in population mean studying times (Biology – Psychology).

45. Construct a 95% confidence interval for the difference in population mean studying times (Biology – Psychology).

46. Use the confidence interval to determine whether a significant difference exists between the population mean studying times.

47. Test, using level of significance $\alpha = 0.05$, whether the population mean studying times are equal. Does your conclusion agree with that in Exercise 46?

48. Does Multitasking Degrade Performance? The journal *Computers and Education* reported that multitasking during class may degrade student performance on assessments.[12] Students were randomly assigned into two groups: multitaskers and non-multitaskers. The multitaskers have their laptops open during class, doing things such as surfing Facebook and so forth. The following table provides the summary statistics for a class quiz for the two groups. Test whether the population mean score for the multitaskers is lower than that of the non-multitaskers, using level of significance $\alpha = 0.05$.

Multitaskers	$n_1 = 20$	$\bar{x}_1 = 55$	$s_1 = 11$
Non-multitaskers	$n_2 = 20$	$\bar{x}_2 = 66$	$s_2 = 12$

49. Does Multitasking Degrade Your Neighbor's Performance? Refer to the previous exercise. The same journal article reported that those sitting next to the multitaskers may have also suffered from degraded quiz performance, due to being distracted by their neighbor's laptop multitasking. The following table provides the summary statistics for a class quiz for the two groups: those sitting where they could see the multitasker's screen (multitasker's neighbors), and those who could not (multitasker's non-neighbors). Test whether the population mean score for the multitasker's neighbors is lower than that of the multitasker's non-neighbors, using level of significance $\alpha = 0.05$.

Multitasker's neighbors	$n_1 = 19$	$\bar{x}_1 = 56$	$s_1 = 12$
Multitasker's non-neighbors	$n_2 = 19$	$\bar{x}_2 = 73$	$s_2 = 12$

BRINGING IT ALL TOGETHER

Do Prior Student Evaluations Influence Students' Ratings of Professors? A study in 1950 reported that instructor reputation affected students' ratings of their instructors.[13] Towler and Dipboye uncovered experimental evidence in support of this phenomenon.[14] They randomly assigned to students one of two summaries of prior student evaluations: one for a "charismatic instructor" and the other for a "punitive instructor." The "charismatic" summary included such phrases as "always lively and stimulating in class" and "always approachable and treated students as individuals." The "punitive" summary included such phrases as "did not show an interest in students' progress" and "consistently seemed to grade students harder." All subjects were then shown the same 20-minute lecture video given by the same instructor. They were asked to rate the instructor using three questions, and a summary rating score was calculated. The summary statistics are shown below. Assume that both populations are normally distributed and that the samples are drawn independently. Were students' ratings influenced by the prior student evaluations? Find out by answering Exercises 50–58.

Reputation	Subjects	Sample mean rating	Sample standard deviation
Charismatic (sample 1)	$n_1 = 25$	$\bar{x}_1 = 2.613$	$s_1 = 0.533$
Punitive (sample 2)	$n_2 = 24$	$\bar{x}_2 = 2.236$	$s_2 = 0.543$

50. Test whether the population mean student evaluation rating for the charismatic group is greater than that of the punitive group, using a *t* test and level of significance $\alpha = 0.05$.

51. Construct a 95% *t* confidence interval for $\mu_1 - \mu_2$.

52. Recall we explored some equivalences between confidence intervals and hypothesis tests. Could we have used the confidence interval in Exercise 51 to perform the hypothesis test in Exercise 50? Explain.

53. Test whether the population mean student ratings differ, using a *t* test and level of significance $\alpha = 0.05$.

54. Use the confidence interval in Exercise 51 to conduct *t* tests, at level of significance $\alpha = 0.05$, for whether $\mu_1 - \mu_2$ takes the following values:
  a. 0
  b. 0.1
  c. 0.6
  d. 0.7

55. Do the following:
  a. Find the pooled variance.

  b. Test whether the population mean student ratings differ, using the pooled variance *t* test and level of significance $\alpha = 0.05$.
  c. Compare your results from (**b**) with Exercise 53.

56. Construct a 95% pooled variance *t* confidence interval for $\mu_1 - \mu_2$. Compare your results with Exercise 51.

For Exercises 57 and 58, assume that the given standard deviations are population standard deviations.

57. Use a *Z* test to test whether the population mean student ratings differ, using level of significance $\alpha = 0.05$. Compare your results to Exercises 53 and 55.

58. Construct a 95% *Z* confidence interval for $\mu_1 - \mu_2$. Compare your results with Exercises 51 and 56.

WORKING WITH LARGE DATA SETS

Case Study: Bank Loans. Open the Chapter 10 Case Study data sets, **BankLoans_Approved** and **BankLoans_Denied**. Here, we will examine whether a difference exists in the loan request amount between those approved for and those denied a loan. Use technology to do the following: ᎎᎎᎎ **bankloans_approved** ᎎᎎᎎ **bankloans_denied**

59. Obtain independent random samples of size 100, one each from the **BankLoans_Approved** data set and the **BankLoans_Denied** data set.

60. For each sample, find the summary statistics for the variable **Request Amount**, that is, the sample size, sample mean, and sample standard deviation.

61. Perform a *t* test, using level of significance $\alpha = 0.05$, to determine whether the population mean request amount differs between those approved for a loan and those denied the loan.

10.3 Inference for Two Independent Proportions

OBJECTIVES By the end of this section, I will be able to . . .

1 Perform and interpret *Z* tests for $p_1 - p_2$.
2 Compute and interpret *Z* intervals for $p_1 - p_2$.
3 Use *Z* intervals for $p_1 - p_2$ to perform two-tailed *Z* tests.

1 Independent Sample *Z* Tests for $p_1 - p_2$

So far in this chapter, we have learned how to perform inference about population *means*. In this section, we learn how to perform hypothesis tests and construct confidence intervals about the difference between two population *proportions*. Recall that the sample proportion of success $\hat{p} = x/n$ is the ratio of the number of successes x to the number of trials n in a binomial experiment.

In this section, we consider two independent samples, each of which yields a sample proportion: $\hat{p}_1 = x_1/n_1$ and $\hat{p}_2 = x_2/n_2$. For example, a recent survey found the sample proportion of males (sample 1) and females (sample 2) who agree that "technological changes will lead toward a future where people's lives are mostly better" to be

$$\hat{p}_1 = \frac{x_1}{n_1} = \frac{335}{500} = 0.67$$

and

$$\hat{p}_2 = \frac{x_2}{n_2} = \frac{255}{500} = 0.51$$

(See Example 15 for further details about these data.) Here, we are interested in performing inference for the difference in population proportions $p_1 - p_2$, such as the difference in the proportions of *all* males and females who think technological change will lead to a better future. We use the difference in sample proportions $\hat{p}_1 - \hat{p}_2$ as our point estimate of the difference in population proportions $p_1 - p_2$, which is unknown. And just as in earlier sections where we investigated the sampling distribution of $\bar{x}_1 - \bar{x}_2$ to perform inference on $\mu_1 - \mu_2$, here we use the sampling distribution of $\hat{p}_1 - \hat{p}_2$ to help us perform inference about $p_1 - p_2$.

Developing Your Statistical Sense

Independent Samples Only

The inferential methods of this section are reserved for *independent* samples only. An example of a problem that would not use the methods of this section is the following: In the latest poll, suppose 45% of the respondents supported the Democratic candidate and 45% supported the Republican one. Because each respondent had to choose between the Democratic candidate and the Republican candidate, their respective poll numbers are *not independent*.

The distribution of all possible values of $\hat{p}_1 - \hat{p}_2$ is called the **sampling distribution of $\hat{p}_1 - \hat{p}_2$**, with mean $p_1 - p_2$ and standard error

$$\sigma_{\hat{p}_1 - \hat{p}_2} = \sqrt{\frac{p_1(1 - p_1)}{n_1} + \frac{p_2(1 - p_2)}{n_2}}.$$

Let x_1 and x_2 denote the number of successes, and let $n_1 - x_1$ and $n_2 - x_2$ denote the number of failures in sample 1 and sample 2, respectively. The sampling distribution of $\hat{p}_1 - \hat{p}_2$ is approximately normal when the number of successes and the number of failures in each sample are each at least 5, that is, when $x_1 \geq 5$, $(n_1 - x_1) \geq 5$, $x_2 \geq 5$, and $(n_2 - x_2) \geq 5$. Let $q_1 = 1 - p_1$, $q_2 = 1 - p_2$, $\hat{q}_1 = 1 - \hat{p}_1$ and $\hat{q}_2 = 1 - \hat{p}_2$.

Sampling Distribution of $\hat{p}_1 - \hat{p}_2$

When two random samples are drawn independently from two populations, then the quantity

$$Z = \frac{(\hat{p}_1 - \hat{p}_2) - (p_1 - p_2)}{\sqrt{\frac{p_1 q_1}{n_1} + \frac{p_2 q_2}{n_2}}}$$

has an approximately standard normal distribution when the following conditions are satisfied:

$$x_1 \geq 5, \qquad (n_1 - x_1) \geq 5, \qquad x_2 \geq 5, \qquad (n_2 - x_2) \geq 5$$

and where $\hat{p}_1$ and n_1 represent the sample proportion and sample size of the sample taken from population 1 with population proportion p_1; $\hat{p}_2$ and n_2 represent the sample proportion and sample size of the sample taken from population 2 with population proportion p_2; and $q_1 = 1 - p_1$ and $q_2 = 1 - p_2$.

The three possible forms for the Z test for $p_1 - p_2$ are as follows:

$H_0: p_1 = p_2$	$H_a: p_1 > p_2$	Right-tailed test
$H_0: p_1 = p_2$	$H_a: p_1 < p_2$	Left-tailed test
$H_0: p_1 = p_2$	$H_a: p_1 \neq p_2$	Two-tailed test

The null hypothesis asserts that $H_0: p_1 = p_2$. We denote this *common population proportion* as p. The null hypothesis is assumed true, so the test statistic takes the following form:

$$Z_{\text{data}} = \frac{(\hat{p}_1 - \hat{p}_2) - (p_1 - p_2)}{\sqrt{\dfrac{p_1(1 - p_1)}{n_1} + \dfrac{p_2(1 - p_2)}{n_2}}} = \frac{(\hat{p}_1 - \hat{p}_2) - 0}{\sqrt{\dfrac{p_1(1 - p_1)}{n_1} + \dfrac{p_2(1 - p_2)}{n_2}}}$$

$$= \frac{(\hat{p}_1 - \hat{p}_2)}{\sqrt{\dfrac{p(1 - p)}{n_1} + \dfrac{p(1 - p)}{n_2}}} = \frac{(\hat{p}_1 - \hat{p}_2)}{\sqrt{p(1 - p)\left(\dfrac{1}{n_1} + \dfrac{1}{n_2}\right)}}$$

The common population proportion p is unknown, so we estimate it using the following **pooled estimate of p**:

$$\hat{p}_{\text{pooled}} = \frac{x_1 + x_2}{n_1 + n_2}$$

Note: As a check on your arithmetic, $\hat{p}_{\text{pooled}}$ must also lie between $\hat{p}_1$ and $\hat{p}_2$.

Substituting this into the formula for the test statistic gives

$$Z_{\text{data}} = \frac{(\hat{p}_1 - \hat{p}_2)}{\sqrt{\hat{p}_{\text{pooled}} \cdot (1 - \hat{p}_{\text{pooled}})\left(\dfrac{1}{n_1} + \dfrac{1}{n_2}\right)}}$$

Z_{data} measures the distance between the sample proportions. Extreme values of Z_{data} indicate evidence against the null hypothesis.

Hypothesis Test for the Difference in Two Population Proportions: Critical-Value Method

Suppose we have two independent random samples taken from two populations with population proportions p_1 and p_2, and the required conditions are met: $x_1 \geq 5$, $(n_1 - x_1) \geq 5$, $x_2 \geq 5$, and $(n_2 - x_2) \geq 5$.

Step 1 State the hypotheses.
Use one of the forms from Table 12 (page 609). State the meaning of p_1 and p_2.

Step 2 Find Z_{crit} and state the rejection rule.
Use Table 12 on page 609.

Step 3 Calculate Z_{data}

$$Z_{\text{data}} = \frac{\hat{p}_1 - \hat{p}_2}{\sqrt{\hat{p}_{\text{pooled}} \cdot (1 - \hat{p}_{\text{pooled}})\left(\dfrac{1}{n_1} + \dfrac{1}{n_2}\right)}}$$

where

$$\hat{p}_{\text{pooled}} = \frac{x_1 + x_2}{n_1 + n_2}$$

Z_{data} follows an approximately standard normal distribution if the required conditions are satisfied.

Step 4 State the conclusion and the interpretation.
Compare Z_{data} with Z_{crit}.

Table 12 Critical regions and rejection rules for Z test for $p_1 - p_2$

	Form of hypothesis test		
	Right-tailed	**Left-tailed**	**Two-tailed**
Level of significance α	$H_0 : p_1 = p_2$ $H_a : p_1 > p_2$	$H_0 : p_1 = p_2$ $H_a : p_1 < p_2$	$H_0 : p_1 = p_2$ $H_a : p_1 \neq p_2$
0.10 0.05 0.01	$Z_{crit} = 1.28$ $Z_{crit} = 1.645$ $Z_{crit} = 2.33$	$Z_{crit} = -1.28$ $Z_{crit} = -1.645$ $Z_{crit} = -2.33$	$Z_{crit} = 1.645$ $Z_{crit} = 1.96$ $Z_{crit} = 2.58$
Critical region	α 0 Z_{crit} Noncritical region Critical region	α $-Z_{crit}$ 0 Critical region Noncritical region	$\alpha/2$ $\alpha/2$ $-Z_{crit}$ 0 Z_{crit} Critical region Noncritical region Critical region
	Reject H_0 if $Z_{data} \geq Z_{crit}$	Reject H_0 if $Z_{data} \leq Z_{crit}$	Reject H_0 if $Z_{data} \leq -Z_{crit}$ or $Z_{data} \geq Z_{crit}$

EXAMPLE 15 Z test for $p_1 - p_2$ using the critical-value method

In April 2014, the Pew Research Center published a report called *U.S. Views of Technology and the Future*,[15] in which the results of a survey of Americans' views on the future of technology were examined. Among other questions, respondents were asked whether they agreed that "technological changes will lead toward a future where people's lives are mostly better." The results are shown in Table 13. Assume the samples are independent.

Table 13 Proportions of males and females who agree that technological change will lead to a better future

	Males	**Females**
Number agreeing	$x_1 = 335$	$x_2 = 255$
Sample size	$n_1 = 500$	$n_2 = 500$
Sample proportion	$\hat{p}_1 = x_1/n_1$ $= 335/500$ $= 0.67$	$\hat{p}_2 = x_2/n_2$ $= 255/500$ $= 0.51$
Population proportion	$p_1 = ?$	$p_2 = ?$

a. Find the point estimate of the difference in the population proportions of males and females, $\hat{p}_1 - \hat{p}_2$.

b. Compute the pooled estimate of the common proportion, $\hat{p}_{pooled}$.

c. Calculate the value of the test statistic Z_{data}.

d. Check whether the conditions for performing the Z test for $p_1 - p_2$ are met.

e. Test whether the population proportion of males who agree that technology will lead to a better future is greater than the population proportion of females who agree. Use the critical-value method at level of significance $\alpha = 0.01$.

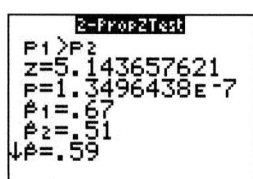

FIGURE 16 TI-83/84 results.

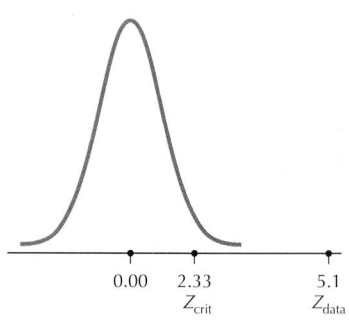

FIGURE 17 $Z_{data} = 5.1$ is extreme, leading to rejection of H_0.

NOW YOU CAN DO

Exercises 5–8.

Solution

a. The point estimate is $\hat{p}_1 - \hat{p}_2 = 0.67 - 0.51 = 0.16$

b. $\hat{p}_{pooled} = \dfrac{x_1 + x_2}{n_1 + n_2} = \dfrac{335 + 255}{500 + 500} = 0.59$

c. $Z_{data} = \dfrac{\hat{p}_1 - \hat{p}_2}{\sqrt{\hat{p}_{pooled} \cdot \left(1 - \hat{p}_{pooled}\right)\left(\dfrac{1}{n_1} + \dfrac{1}{n_2}\right)}} = \dfrac{0.67 - 0.51}{\sqrt{(0.59)(0.41)\left(\dfrac{1}{500} + \dfrac{1}{500}\right)}} \approx 5.1$

d. We check the conditions for performing the Z test for $p_1 - p_2$. We have: $x_1 = 335 \geq 5$, $x_2 = 255 \geq 5$, $n_1 - x_1 = 500 - 335 = 165 \geq 5$, and $n_2 - x_2 = 500 - 255 = 245 \geq 5$. We may thus proceed with the hypothesis test.

e. The Z test for $p_1 - p_2$ follows the steps below.

Step 1   State the hypotheses.
The key words "greater than," together with the fact that sample 1 represents the males, indicate that we have a right-tailed test:

$$H_0 : p_1 = p_2 \quad \text{versus} \quad H_a : p_1 > p_2$$

where p_1 and p_2 represent the population proportion of males and females, respectively, who agree that technology will lead to a better future.

Step 2   Find Z_{crit} and state the rejection rule.
For a right-tailed test with level of significance $\alpha = 0.01$, Table 12 gives us $Z_{crit} = 2.33$ and our rejection rule: Reject H_0 if $Z_{data} \geq 2.33$.

Step 3   Calculate Z_{data}.
From (**c**), we have $Z_{data} \approx 5.1$ (also see Figure 16).

Step 4   State the conclusion and the interpretation.
$Z_{data} \approx 5.1 \geq 2.33$; therefore, reject H_0 (see Figure 17). There is evidence at level of significance $\alpha = 0.01$ that the population proportion of males who agree that technology will lead to a better future is greater than the population proportion of females who agree.

We may also use the *p*-value method to perform the Z test for $p_1 - p_2$.

Hypothesis Test for the Difference in Two Population Proportions: *p*-Value Method

Suppose we have two independent random samples taken from two populations with population proportions p_1 and p_2, and the required conditions are met: $x_1 \geq 5$, $(n_1 - x_1) \geq 5$, $x_2 \geq 5$, and $(n_2 - x_2) \geq 5$.

Step 1   State the hypotheses and the rejection rule.
Use one of the forms from Table 12. State the meaning of p_1 and p_2. The rejection rule is: *Reject H_0 if the p-value $\leq \alpha$.*

Step 2   Calculate Z_{data}.

$$Z_{data} = \dfrac{\hat{p}_1 - \hat{p}_2}{\sqrt{\hat{p}_{pooled} \cdot \left(1 - \hat{p}_{pooled}\right)\left(\dfrac{1}{n_1} + \dfrac{1}{n_2}\right)}}$$

where $\hat{p}_{pooled} = \frac{x_1 + x_2}{n_1 + n_2}$. If the required conditions are satisfied, Z_{data} follows an approximately standard normal distribution.

Step 3   Find the *p*-value.
Either use technology or calculate the *p*-value using one of the forms in Table 14.

Step 4   State the conclusion and the interpretation.
Compare the *p*-value with α.

Table 14 *p*-Values for *Z* test for $p_1 - p_2$

Right-tailed test	Left-tailed test	Two-tailed test
$H_0 : p_1 = p_2$ $H_a : p_1 > p_2$	$H_0 : p_1 = p_2$ $H_a : p_1 < p_2$	$H_0 : p_1 = p_2$ $H_a : p_1 \neq p_2$
p-value = $P(Z > Z_{\text{data}})$ Area to right of Z_{data}	*p*-value = $P(Z < Z_{\text{data}})$ Area to left of Z_{data}	*p*-value = $P(Z > \lvert Z_{\text{data}} \rvert) +$ $P(Z < -\lvert Z_{\text{data}} \rvert) = 2 \cdot P(Z > \lvert Z_{\text{data}} \rvert)$ Sum of the two tail areas

EXAMPLE 16   *Z* Test for $p_1 - p_2$ using the *p*-value method

©Blend Images/Alamy

The General Social Survey tracks trends in American society through annual surveys. Married respondents were asked to characterize their feelings about being married. The results are shown here in a crosstabulation with gender. Test the hypothesis that the proportion of females who report being very happily married is smaller than the proportion of males who report being very happily married. Use the *p*-value method with level of significance $\alpha = 0.05$. **marriage**

	Very happy	Pretty happy/ Not too happy	Total
Female	257	166	423
Male	242	124	366
Total	499	290	789

Solution

From the crosstabulation, we assemble the statistics in Table 15 for the independent random samples of men and women.

Table 15 Sample statistics of very happily married respondents

	Sample size	Number very happy	Sample proportion very happy
Females (sample 1)	$n_1 = 423$	$x_1 = 257$	$\hat{p}_1 = \dfrac{x_1}{n_1} = \dfrac{257}{423} \approx 0.6076$
Males (sample 2)	$n_2 = 366$	$x_2 = 242$	$\hat{p}_2 = \dfrac{x_2}{n_2} = \dfrac{242}{366} \approx 0.6612$

We first check whether the conditions for the *Z* test are valid: $x_1 = 257 \geq 5$, $(n_1 - x_1) = (423 - 257) = 166 \geq 5$, $x_2 = 242 \geq 5$, and $(n_2 - x_2) = (366 - 242) = 124 \geq 5$. We can therefore proceed.

Step 1   **State the hypotheses and the rejection rule.**
We are interested in whether the proportion of females who report being very happily married *is smaller than* that of males and because the females represent sample 1, the hypotheses are

$$H_0 : p_1 = p_2 \qquad H_a : p_1 < p_2$$

where p_1 and p_2 represent the population proportions of all females and males, respectively, who report being very happily married. We will reject H_0 if the p-value $\leq a = 0.05$.

Step 2   Find Z_{data}.
First, use the data from Table 15 to find the value of $\hat{p}_{pooled}$.

$$\hat{p}_{pooled} = \frac{x_1 + x_2}{n_1 + n_2} = \frac{257 + 242}{423 + 366} \approx 0.63245$$

Then

$$Z_{data} = \frac{(0.6076 - 0.6612)}{\sqrt{0.63245 \cdot (1 - 0.63245)\left(\frac{1}{423} + \frac{1}{366}\right)}} \approx -1.56$$

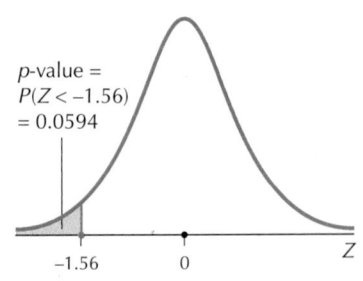

p-value =
$P(Z < -1.56)$
$= 0.0594$

FIGURE 18 p-Value for left-tailed Z test.

Note: When the p-value is close to α, many data analysts prefer to simply assess the strength of evidence against the null hypothesis using criteria such as those given in Table 6 in Chapter 9 (page 514).

Step 3   Find the p-value.
Because it is a left-tailed test, the p-value is given by Table 14 as $P(Z < Z_{data}) = P(Z < -1.56)$, as shown in Figure 18. This amounts to a Case 1 problem from Table 8 in Chapter 6 on page 357:

$$P(Z < -1.56) = 0.0594$$

Step 4   State the conclusion and the interpretation.
The p-value = 0.0594 is not less than or equal to $\alpha = 0.05$, so we do not reject H_0. There is insufficient evidence that the proportion of females who report being very happily married is smaller than the proportion of males who do so.

NOW YOU CAN DO
Exercises 9–12.

2   Independent Sample Z Interval for $p_1 - p_2$

We have learned how to perform Z tests for $p_1 - p_2$. Next, we learn how to use sample statistics to estimate $p_1 - p_2$ using a confidence interval.

Confidence Interval for $p_1 - p_2$

For two independent random samples taken from two populations with population proportions p_1 and p_2, a **100(1 − α)% confidence interval for $p_1 - p_2$** is given by

$$\hat{p}_1 - \hat{p}_2 \pm Z_{\alpha/2}\sqrt{\frac{\hat{p}_1 \cdot \hat{q}_1}{n_1} + \frac{\hat{p}_2 \cdot \hat{q}_2}{n_2}}$$

where $\hat{p}_1$ and n_1 represent the sample proportion and sample size of the sample taken from population 1 with population proportion p_1; $\hat{p}_2$ and n_2 represent the sample proportion and sample size of the sample taken from population 2 with population proportion p_2; $\hat{q}_1 = 1 - \hat{p}_1$ and $\hat{q}_2 = 1 - \hat{p}_2$, and the samples are drawn independently; and the following conditions are satisfied: $x_1 \geq 5$, $(n_1 - x_1) \geq 5$, $x_2 \geq 5$, and $(n_2 - x_2) \geq 5$.

Margin of Error E

The margin of error for a 100(1 − α)% confidence interval for $p_1 - p_2$ is given by

$$E = Z_{\alpha/2} \cdot \sqrt{\frac{\hat{p}_1 \cdot \hat{q}_1}{n_1} + \frac{\hat{p}_2 \cdot \hat{q}_2}{n_2}}$$

EXAMPLE 17   Z confidence interval for $p_1 - p_2$

Use the sample statistics from Example 15 to do the following:

a.   Calculate and interpret the margin of error E for confidence level 99%.

b.   Construct and interpret a 99% confidence interval for $p_1 - p_2$.

Solution

The conditions for the confidence interval are the same as for the hypothesis test and were checked in Example 15.

a. $\hat{q}_1 = 1 - \hat{p}_1 = 1 - 0.67 = 0.33$ $\qquad$ $\hat{q}_2 = 1 - \hat{p}_2 = 1 - 0.51 = 0.49$.
From Table 1 in Chapter 8 on page 432, the $Z_{\alpha/2}$ value for a 99% confidence level is 2.576. Therefore, the margin of error is

$$E = Z_{\alpha/2} \cdot \sqrt{\frac{\hat{p}_1 \cdot \hat{q}_1}{n_1} + \frac{\hat{p}_2 \cdot \hat{q}_2}{n_2}} = (2.576)\sqrt{\frac{(0.67)(0.33)}{500} + \frac{(0.51)(0.49)}{500}} \approx 0.079$$

The margin of error is 0.079, so we may estimate $p_1 - p_2$ to within 0.079 with 99% confidence.

b. The point estimate is $\hat{p}_1 - \hat{p}_2 = 0.67 - 0.51 = 0.16$. The 99% confidence interval is therefore

$$\hat{p}_1 - \hat{p}_2 \pm E = 0.16 \pm 0.079 = (0.081, 0.239)$$

We are 99% confident that the difference in population proportions of males and females who agree that technology will lead to a better future lies between 0.081 and 0.239.

NOW YOU CAN DO
Exercises 13–18.

3 Use Z Confidence Intervals to Perform Z Tests for $p_1 - p_2$

Given a $100(1 - \alpha)\%$ Z confidence interval for $\hat{p}_1 - \hat{p}_2$, we may perform two-tailed Z tests for various hypothesized values of $p_1 - p_2$. If a proposed value lies outside the $100(1 - \alpha)\%$ Z confidence interval for $p_1 - p_2$, then the null hypothesis specifying this value would be rejected. Otherwise, do not reject the null hypothesis.

EXAMPLE 18 Using a Z interval for $p_1 - p_2$ to perform Z tests about $p_1 - p_2$

This example asks whether $p_1 - p_2$ differs from (or is not equal to) a certain value, so we can use the Z confidence interval to test the hypotheses. Example 17 provided a 99% Z confidence interval for $p_1 - p_2$, the difference in population proportions of males and females who agree that technology will lead to a better future, as (0.081, 0.239). Test, using level of significance $\alpha = 0.01$, whether the $p_1 - p_2$ differs from these values: **(a)** 0.1, **(b)** 0.2, **(c)** 0.3.

Solution

a. $H_0 : p_1 - p_2 = 0.1$ versus $H_a : p_1 - p_2 \neq 0.1$.
The hypothesized value 0.1 lies outside the interval (0.081, 0.239), so we reject H_0.

b. $H_0 : p_1 - p_2 = 0.2$ versus $H_a : p_1 - p_2 \neq 0.2$.
The hypothesized value 0.2 lies inside the interval, so we do not reject H_0.

NOW YOU CAN DO
Exercises 19–22.

c. $H_0 : p_1 - p_2 = 0.3$ versus $H_a : p_1 - p_2 \neq 0.3$.
The hypothesized value 0.3 lies outside the interval, so we reject H_0.

STEP-BY-STEP TECHNOLOGY GUIDE: *Z* Test and *Z* Interval $p_1 - p_2$

(Example 15 is used to illustrate the procedure.)

TI-83/84

Z Test for $p_1 - p_2$

Step 1   Press **STAT** and highlight **TESTS**.
Step 2   Select **6** (for the **2-Prop ZTest**).
Step 3   For x_1, enter the number of successes in the first sample, 335.
Step 4   For n_1, enter the size of the first sample, 500.
Step 5   For x_2, enter the number of successes in the second sample, 255.
Step 6   For n_2, enter the size of the second sample, 500.
Step 7   For p_1, choose the form of the hypothesis test. For Example 15, choose $> p_2$ and press **ENTER** (Figure 19).
Step 8   Highlight **Calculate** and press **ENTER**. The results are shown in Figure 16 in Example 15.

Z Interval for $p_1 - p_2$

Follow the same steps as for the *Z* Test for $p_1 - p_2$, except "Select B: 2-PropZInt." Also, to select confidence level (**C-Level**), enter 0.95 for 95%, for example.

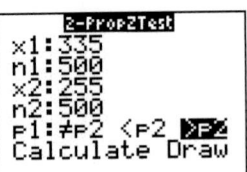

FIGURE 19

EXCEL

Use the JMP add-in for Excel.

MINITAB

Z Test and *Z* Interval for $p_1 - p_2$

Step 1   Click **Stat > Basic Statistics > 2 Proportions**....
Step 2   Select **Summarized data**.
Step 3   For the **Number of events** row, enter x_1 for **Sample 1** and x_2 for **Sample 2**.

Step 4   For the **Number of trials** row, enter n_1 for **Sample 1** and n_2 for **Sample 2**.
Step 5   Click **Options**..., enter the **Confidence level**, the **Hypothesized difference**, and select the form of the **Alternative hypothesis**. Under **Test method**, select **Use the pooled estimate of the proportion**. Then click **OK** twice.

CRUNCHIT!

Z test and *Z* Interval for $p_1 - p_2$

We will use the data from Example 15.
Step 1   Click **Statistics**, highlight **Proportion**, and select **2-sample**. Select the **Summarized** tab. For **Sample 1**, enter **n**: 500 and **Successes**: 335; for **Sample 2**, enter **n**: 500 and **Successes**: 255.

Step 2
a.   For the hypothesis test, select the **Hypothesis Test** tab, enter the **Difference of proportions under null hypothesis**, choose the correct form of the **Alternative** hypothesis, and click **Calculate**.
b.   For the confidence interval, select the **Confidence Interval** tab, enter the **Confidence Interval Level (%)**, and click **Calculate**.

Section 10.3 Summary

1. This section discusses inferential methods for $p_1 - p_2$, the difference between the proportions of two independent populations. Two-sample *Z* tests for $p_1 - p_2$ are discussed. These hypothesis tests may be performed using either the *p*-value method or the critical-value method.

2. $100(1 - \alpha)\%$ *Z* confidence intervals for $p_1 - p_2$ are developed and illustrated.
3. We may use *Z* confidence intervals for $p_1 - p_2$ to conduct two-tailed *Z* tests.

Section 10.3 Exercises

CLARIFYING THE CONCEPTS

1. $\hat{p}_{pooled}$ must always lie between which two quantities? (p. 608)
2. Does it make sense to use $\hat{p}_{pooled}$ when calculating confidence intervals for $p_1 - p_2$? Why or why not? (p. 612)

3. What does Z_{data} measure? What do extreme values of Z_{data} indicate? (p. 608)
4. What might we suggest if the *p*-value is very close to the level of significance α? (p. 612)

PRACTICING THE TECHNIQUES

✓ **CHECK IT OUT!**

To do	Check out	Topic
Exercises 5–8	Example 15	Z test for $p_1 - p_2$: critical-value method
Exercises 9–12	Example 16	Z test for $p_1 - p_2$: p-value method
Exercises 13–18	Example 17	Z confidence interval for $p_1 - p_2$
Exercises 19–22	Example 18	Equivalence between confidence intervals and Z tests for $p_1 - p_2$

The summary statistics in Exercises 5–7 and 9–11 were taken from random samples that were drawn independently. Let n_1 and n_2 denote the size of samples 1 and 2, respectively. Let x_1 and x_2 denote the number of successes in samples 1 and 2, respectively.

For Exercises 5–7, perform the indicated hypothesis test using the critical-value method. Answer (a)–(d) for each exercise.
   a. State the hypotheses and find the critical value Z_{crit} and the rejection rule.
   b. Calculate $\hat{p}_{pooled}$.
   c. Calculate Z_{data}.
   d. Compare Z_{data} with Z_{crit}. State and interpret your conclusion.

5. Test, at level of significance $\alpha = 0.10$, whether $p_1 \neq p_2$.

Sample 1	$n_1 = 100$	$x_1 = 80$
Sample 2	$n_2 = 40$	$x_2 = 30$

6. Test, at level of significance $\alpha = 0.05$, whether $p_1 < p_2$.

Sample 1	$n_1 = 10$	$x_1 = 5$
Sample 2	$n_2 = 12$	$x_2 = 5$

7. Test, at level of significance $\alpha = 0.01$, whether $p_1 > p_2$.

Sample 1	$n_1 = 200$	$x_1 = 60$
Sample 2	$n_2 = 250$	$x_2 = 40$

8. Refer to the data from Exercise 7. Test, at level of significance $\alpha = 0.01$, whether $p_1 \neq p_2$.

For Exercises 9–11, perform the indicated hypothesis test using the p-value method. Answer (a)–(e) for each exercise.
   a. State the hypotheses and the rejection rule.
   b. Calculate $\hat{p}_{pooled}$.
   c. Calculate Z_{data}.
   d. Calculate the p-value.
   e. Compare the p-value with α. State and interpret your conclusion.

9. Test, at level of significance $\alpha = 0.05$, whether $p_1 > p_2$.

Sample 1	$n_1 = 400$	$x_1 = 250$
Sample 2	$n_2 = 400$	$x_2 = 200$

10. Test, at level of significance $\alpha = 0.05$, whether $p_1 < p_2$.

Sample 1	$n_1 = 1000$	$x_1 = 490$
Sample 2	$n_2 = 1000$	$x_2 = 620$

11. Test, at level of significance $\alpha = 0.10$, whether $p_1 \neq p_2$.

Sample 1	$n_1 = 527$	$x_1 = 412$
Sample 2	$n_2 = 613$	$x_2 = 498$

12. Refer to the data from Exercise 11. Test, at level of significance $\alpha = 0.10$, whether $p_1 < p_2$.

For Exercises 13–18, refer to the indicated data to answer (a)–(d).
   a. We are interested in constructing a 95% confidence interval for $p_1 - p_2$. Is it appropriate to do so? Why or why not? If not appropriate, then do not perform (b)–(d).
   b. Provide the point estimate of the difference in population proportions $p_1 - p_2$.
   c. Calculate the margin of error for a confidence level of 95%. What does this number mean?
   d. Construct and interpret a 95% confidence interval for $p_1 - p_2$.
13. Data from Exercise 5
14. Data from Exercise 6
15. Data from Exercise 7
16. Data from Exercise 9
17. Data from Exercise 10
18. Data from Exercise 11

For Exercises 19–22, a $100(1 - \alpha)$% Z confidence interval for $p_1 - p_2$ is given. Use the confidence interval to test, using level of significance α, whether $p_1 - p_2$ differs from each of the indicated hypothesized values.
19. A 95% Z confidence interval for $p_1 - p_2$ is (0.5, 0.6). Hypothesized values are
   a. 0   **b.** 0.1   **c.** 0.57
20. A 99% Z confidence interval for $p_1 - p_2$ is (0.01, 0.99). Hypothesized values are
   a. 0.2   **b.** 0   **c.** 0.999
21. A 90% Z confidence interval for $p_1 - p_2$ is (0.1, 0.11). Hypothesized values are
   a. 0.151   **b.** 0.115   **c.** 0.105
22. A 95% Z confidence interval for $p_1 - p_2$ is (0.43, 0.57). Hypothesized values are
   a. 0.41   **b.** 0.51   **c.** 0.61

APPLYING THE CONCEPTS

23. Technological Change. Do attitudes about technological change differ between the young and their elders? The Pew report discussed earlier reported that 85 of 144 respondents 18 to 29 years old agreed that technology will lead to a better future, whereas 166 of 297 respondents 65 years old or older agreed.

a. Is it appropriate to perform the Z test for the difference in population proportions? Why or why not?

b. Clearly state the meaning of p_1 and p_2.

c. Test whether the proportions agreeing that technology will lead to a better future differ between the younger group and the older group, using level of significance $\alpha = 0.05$.

24. Medicare Recipients. The Centers for Medicare and Medicaid Services reported in 2004 that 3305 of the 50,350 Medicare recipients living in Alaska were age 85 or over, and 73,289 of the 754,642 Medicare recipients living in Arizona were age 85 or over.

a. Find a point estimate of the difference in population proportions.

b. Clearly state the difference in meaning between p_1 and $\hat{p}_1$.

c. Test whether the population proportions differ, using level of significance $\alpha = 0.05$.

25. Women's Ownership of Businesses. The U.S. Census Bureau tracks trends in women's ownership of businesses. A random sample of 100 Ohio businesses showed 34 that were woman-owned. A sample of 200 New Jersey businesses showed 64 that were woman-owned. Test whether the population proportions of female-owned businesses in Ohio is greater than that of New Jersey, using level of significance $\alpha = 0.10$.

26. Fetal Cells and Breast Cancer. A number of fetal stem cells may cross the placenta from the fetus to the mother during pregnancy and remain in the mother's tissue for decades. A recent study shows that the presence of fetal cells in the mother may offer some protection against the onset of breast cancer.[16] Of the 54 women in the study with breast cancer, 14 had fetal cells. Of the 45 women without breast cancer, 25 had fetal cells. Test whether the population proportions of women with fetal cells is lower among women with breast cancer compared with women without breast cancer, using level of significance $\alpha = 0.01$.

27. Technological Change. Refer to Exercise 23 to answer the following questions:

a. Construct and interpret a 95% confidence interval for the difference in population proportions.

b. Use the confidence interval from (a) to test, using level of significance $\alpha = 0.05$, whether the population proportions differ.

c. Does your conclusion from (b) agree with your conclusion from Exercise 23(c)?

28. Medicare Recipients. Refer to Exercise 24 to answer the following questions:

a. Construct and interpret a 95% confidence interval for the difference in population proportions.

b. Use the confidence interval from (a) to test, using level of significance $\alpha = 0.05$, whether the population proportions differ.

c. Does your conclusion from (b) agree with your conclusion from Exercise 24(c)?

29. Women's Ownership of Businesses. Refer to Exercise 25 to answer the following questions:

a. Construct and interpret a 90% confidence interval for the difference in population proportions.

b. Use the confidence interval from (a) to test, using level of significance $\alpha = 0.10$, whether the population proportions differ.

c. Explain whether or not we could use the confidence interval from part (b) to perform the hypothesis test in Exercise 25. Why or why not?

30. Fetal Cells and Breast Cancer. Refer to Exercise 26 to answer the following questions:

a. Construct and interpret a 99% confidence interval for the difference in population proportions.

b. Use the confidence interval from (a) to test, using level of significance $\alpha = 0.01$, whether the population proportions differ.

c. Explain whether or not we could use the confidence interval from (b) to perform the hypothesis test in Exercise 26. Why or why not?

31. Evidence for Alternative Medical Therapies? A company called QT, Inc., sells "ionized" bracelets, called Q-Ray bracelets, that it claims help to ease pain through balancing the body's flow of "electromagnetic energy." The Mayo Clinic decided to conduct a statistical experiment to determine whether the claims for the Q-Ray bracelets were justified.[17] At the end of 4 weeks, of the 305 subjects who wore the "ionized" bracelet, 236 (77.4%) reported improvement in their maximum pain index (where the pain was the worst). Of the 305 subjects who wore the placebo bracelet (a bracelet identical in every respect to the "ionized" bracelet except that there was no active ingredient—presumably, here, "ionization"), 234 (76.7%) reported improvement in their maximum pain index. Using level of significance $\alpha = 0.05$, test whether the population proportions reporting improvement differ between wearers of the ionized bracelet and wearers of the placebo bracelet.

BRINGING IT ALL TOGETHER

Males Listening to the Radio. Use the following information for Exercises 32–40: The Arbitron Corporation tracks trends in radio listening. In their publication *Radio Today*, Arbitron reported that 92% of 18- to 24-year-old males listen to the radio each week, whereas 87% of males 65 years and older listen to the radio each week. Suppose each sample size was 1000.

32. Is it appropriate to perform Z inference for the difference in population proportions? Why or why not?

33. Clearly describe what p_1 means and what p_2 means.

34. Explain what the difference is between p_1 and $\hat{p}_1$.

35. Calculate the margin of error for a 95% confidence interval for $p_1 - p_2$. Explain what this number means.

36. Construct and interpret a 95% confidence interval for $p_1 - p_2$.

37. Use the confidence interval from Exercise 36 to test, using level of significance $\alpha = 0.05$, whether $p_1 - p_2$ differs from the following:

    a. 0   **b.** 0.01   **c.** 0.05

38. Explain whether we could use the confidence interval from Exercise 36 to test whether the proportion of 18- to 24-year-old males who listen to the radio each week is greater than the proportion of males 65 years and older who do so. Why or why not?

39. Test, using level of significance $\alpha = 0.05$, whether the proportion of 18- to 24-year-old males who listen to the radio each week is greater than the proportion of males 65 years and older who do so.

WHAT IF **40.** *What if*, instead of 1000, each sample size was 100? How would this change affect each of the following measures?

    a. Margin of error in Exercise 35.
    b. *p*-value in Exercise 39.
    c. Conclusion of the hypothesis test in Exercise 39.

WORKING WITH LARGE DATA SETS

 **Case Study: Bank Loans.** Open the Chapter 10 Case Study data sets, **BankLoans_Approved**, and **BankLoans_Denied**. Here, we will examine whether a difference exists in the proportion of applicants with credit scores of 700 or higher between those approved for and those denied a loan. Use technology to do the following:

41. Obtain independent random samples of size 100, one each from the **BankLoans_Approved** data set and the **BankLoans_Denied** data set. ₥ **bankloans_approved** ₥ **bankloans_denied**

42. For each sample, calculate the proportion of applicants who have credit scores of 700 or higher.

43. Perform and interpret a Z test for the difference in the proportion of applicants who have credit scores of 700 or higher, using level of significance $\alpha = 0.05$.

44. Construct and interpret a 95% Z confidence interval for the difference in proportion of applicants who have credit scores of 700 or higher.

10.4 Inference for Two Independent Standard Deviations

OBJECTIVES By the end of this section, I will be able to . . .

1   Describe the characteristics of the *F* distribution and the *F* test for two population standard deviations.

2   Perform hypothesis tests for two population standard deviations using the critical-value method.

3   Perform hypothesis tests for two population standard deviations using the *p*-value method.

1 The *F* Distribution and the *F* Test

A. Barrington Brown/Science Source

Sir Ronald A. Fisher

In Sections 10.1–10.3, we were introduced to inference methods for comparing two population means and two population proportions. Here, we learn how to perform hypothesis tests regarding two population standard deviations. Wall Street investors are wary of excessive stock price variability. In this section, we will compare the variability of prices between two tech stocks, Google and Apple, using a new hypothesis test, called the *F* test. The *F* test will determine whether there is a significant difference in the variability of the stock prices, as measured by the respective population standard deviations. The *F* test is based on the *F* distribution, named in honor of the "grandfather of statistics," Sir Ronald A. Fisher.

Let population 1 be Google stock prices and population 2 be Apple stock prices. We can test whether Google's stock prices are more variable than those of Apple; that is, we may test whether the standard deviation of Google stock prices, σ_1, is greater than the standard deviation of Apple stock prices, σ_2. This gives us the following hypotheses for our *F* test:

$$H_0 : \sigma_1 = \sigma_2 \quad \text{versus} \quad H_a : \sigma_1 > \sigma_2$$

Table 16 provides the three possible forms of hypotheses available when performing the *F* test for comparing two population standard deviations for two populations with standard deviations σ_1 and σ_2, respectively.

Table 16 Three possible forms for the *F* test for comparing two population standard deviations

Form of test	Null hypothesis	Alternative hypothesis
Right-tailed test	$H_0 : \sigma_1 = \sigma_2$	$H_a : \sigma_1 > \sigma_2$
Left-tailed test	$H_0 : \sigma_1 = \sigma_2$	$H_a : \sigma_1 < \sigma_2$
Two-tailed test	$H_0 : \sigma_1 = \sigma_2$	$H_a : \sigma_1 \neq \sigma_2$

The requirements for performing the *F* test are the following:

1. We have two independent random samples taken from two populations.

2. The two populations are both normally distributed.

The test statistic for the *F* test is F_{data}, given as follows.

> **Test Statistic for the *F* Test for Comparing Two Population Standard Deviations**
>
> Suppose that the two population variances are equal, $\sigma_1^2 = \sigma_2^2$, and that we have independent random samples of size n_1 and n_2 from two normally distributed populations with sample variances s_1^2 and s_2^2, respectively. Then the test statistic for the *F* test
>
> $$F_{data} = \frac{s_1^2}{s_2^2}$$
>
> follows an *F* distribution with $n_1 - 1$ degrees of freedom in the numerator and $n_2 - 1$ degrees of freedom in the denominator.

Let's become better acquainted with the *F* distribution. Similar to the χ^2 distribution, the *F* distribution is right-skewed, never takes negative values, and has an infinite number of different *F* curves (Figure 20). The shape of the curve depends on degrees of freedom.

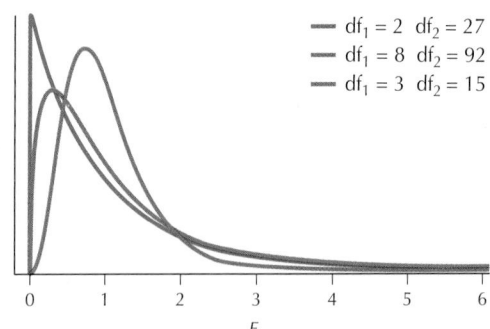

FIGURE 20 Shape of the *F* distribution for various degrees of freedom.

As noted, the *F* distribution resembles the χ^2 distribution. This is not surprising because the values of the *F* distribution represent ratios of two χ^2 distributions. Moreover, the *F* distribution has two different degrees of freedom, which we will call df_1 and df_2, derived from the degrees of freedom of the two χ^2 distributions represented in the ratio. Often, df_1 is called the *numerator degrees of freedom*, and df_2 is called the *denominator degrees of freedom*.

> **Properties of the *F* Curve**
> 1. The total area under the *F* curve equals 1.
> 2. The value of the *F* random variable is never negative, so the *F* curve starts at 0. However, it extends indefinitely to the right. The curve approaches but never quite meets the horizontal axis.
> 3. Because of the characteristics described in (2), the *F* curve is right-skewed.
> 4. There is a different *F* curve for each different pair of degrees of freedom: df_1 and df_2.

The F distribution is continuous, so we can find probabilities associated with values of F, and vice versa, just as we did with the normal, t, and χ^2 distributions. Just as for any continuous distribution, probability is represented by the area below the F curve above an interval.

2 Perform the *F* Test for Comparing Two Population Standard Deviations: Critical-Value Method

To perform the hypothesis tests in this section, as well as in Chapters 12 and 13, we need to find the critical values of an F distribution for a given level of significance α. For example, we may need to find the value of an F distribution that has area $\alpha = 0.05$ to the right of it, or we may need to find the value of an F distribution that has area $\alpha = 0.01$ to the left of it. To find these F critical values, we will work with the F tables (see Appendix Table F). The F tables are somewhat different from the other tables that we have worked with so far.

The notation

$$F_{\text{crit}} = F_{\alpha, n_1 - 1, n_2 - 1}$$

represents the critical value of the F distribution with $df_1 = n_1 - 1$ numerator degrees of freedom and $df_2 = n_2 - 1$ denominator degrees of freedom, with area α to the right of $F_{\alpha, n_1 - 1, n_2 - 1}$. For example, $F_{0.05, 15, 10}$ represents the value of the F distribution with $df_1 = 15$ and $df_2 = 10$, with area 0.05 to the right of it. Next, we learn how to find the F critical values using the F tables.

Procedure for Finding *F* Critical Value for a Given Area α to the Right of It

Suppose we have an F distribution with df_1 and df_2 degrees of freedom. To find the critical value F_{crit} that has area α to the right of it, do the following:

Step 1 Look across the top of the F table until you find your df_1. Then go down that column until you see your df_2 on the left.

Step 2 For each df_2 on the left, you will see a range of α values from 0.100 to 0.001. Choose the row next to df_2 that has your value of α. The F-value in that row and column is your value of F_{crit}.

Developing Your Statistical Sense

Note: When the degrees of freedom are not listed in the F table, we do not necessarily take the closest degrees of freedom we can find. This is because, sometimes, the closest degree of freedom is larger than the original, which leads to misleadingly overprecise results—a level of precision not warranted by the data. For example, this could lead us to find significance where none actually exists.

Degrees of Freedom Not Listed in the Table

Just as with the t table, not all the degrees of freedom are listed in the F table. If either of the degrees of freedom df_1 or df_2 are not listed in the F table, a conservative solution is to take the next smallest value for whichever of df_1 or df_2 is not listed. For example, suppose $df_1 = n_1 - 1 = 57 - 1 = 56$ and $df_2 = n_2 - 1 = 170 - 1 = 169$. Neither $df_1 = 56$ nor $df_2 = 169$ is given in the F table. Therefore, we set df_1 to be the next smallest value given in the table, $df_1 = 50$, and we set df_2 to be the next smallest value, $df_2 = 160$.

The F tables give only the F critical values for a given area α to the right. To find F critical values for a given area α to the left, we use the following property:

$$F_{1 - \alpha, n_1 - 1, n_2 - 1} = \frac{1}{F_{\alpha, n_2 - 1, n_1 - 1}}$$

In other words, the value from an F distribution with degrees of freedom $df_1 = n_1 - 1$ and $df_2 = n_2 - 1$ and area α to the *left* of it equals the reciprocal of the value from an F distribution with degrees of freedom $df_1 = n_2 - 1$ and $df_2 = n_1 - 1$ and area α to the *right* of it. Note that the two degrees of freedom get switched.

Procedure for Finding *F* Critical Value for a Given Area α to the Left of It

1. Switch the values of df_1 and df_2.
2. Find F_{α, n_2-1, n_1-1} using the *F* table.
3. Calculate $F_{crit} = F_{1-\alpha, n_1-1, n_2-1} = \dfrac{1}{F_{\alpha, n_2-1, n_1-1}}$

So, for example, to find the value of an *F* distribution with degrees of freedom $df_1 = 10$ and $df_2 = 15$ with area $\alpha = 0.05$ to the left of it, follow steps 1 and 2 above to find the value of an *F* distribution with $df_1 = 15$ and $df_2 = 10$ with area $\alpha = 0.05$ to the right of it. Then compute the reciprocal.

EXAMPLE 19   Finding critical values of the *F* distribution

Use the excerpt from the *F* distribution tables in Figure 21 to find the following critical values of the *F* distribution:

FIGURE 21
F table (excerpt).

	Area in right tail	**df_1** 1	2	3
		F distribution critical values		
1	0.100	39.86	49.59	53.59
	0.050	161.45	199.50	215.71
	0.025	647.79	799.50	864.16
	0.010	4052.20	4999.50	5403.40
	0.001	405284.00	500000.00	540379.00
6	0.100	3.78	3.46	3.29
	0.050	5.99	5.14	4.76
	0.025	8.81	7.26	6.60
	0.010	13.75	10.92	**9.78**
	0.001	35.51	27.00	23.70
7	0.100	3.59	3.26	3.07
	0.050	5.59	**4.74**	4.35
	0.025	8.07	6.54	5.89
	0.010	12.25	9.55	8.45
	0.001	29.25	21.69	18.77

a. Find the critical value with area $\alpha = 0.05$ to the right of it, for an *F* distribution with $df_1 = 2$ and $df_2 = 7$.

b. Find the critical value with area $\alpha = 0.01$ to the left of it, for an *F* distribution with $df_1 = 6$ and $df_2 = 3$.

Solution

a. Step 1.   Go across the top of the *F* table until we get to $df_1 = 2$, and go down that column until we see $df_2 = 7$ on the left.

Step 2.   Next to the 7 is a range of α values from 0.100 to 0.001. We choose the row with $\alpha = 0.05$. The *F*-value in that row and column is 4.74. Thus, our *F* critical value is

$$F_{crit} = F_{0.05, 2, 7} = 4.74$$

b. Step 1.   Switching the two degrees of freedom, we go across the top of the F table until we get to $df_1 = 3$. Then we go down that column until we see $df_2 = 6$ on the left.

Step 2.   Choose the row with $\alpha = 0.010 = 0.01$. The F-value in that row and column is 9.78. Thus, our F critical value is

$$F_{\text{crit}} = F_{0.99, 6, 3} = \frac{1}{F_{0.01, 6, 3}} = \frac{1}{9.78} = 0.1023$$

NOW YOU CAN DO
Exercises 9–20.

Now we use the critical values of the F distribution to help us with the critical-value method of performing the F test for comparing two population standard deviations. Later we show the steps for the p-value method, along with an example.

F Test for Comparing Two Population Standard Deviations: Critical-Value Method

Suppose we have two independent random samples of size n_1 and n_2 taken from two normally distributed populations, with population standard deviations σ_1 and σ_2, and sample standard deviations s_1 and s_2, respectively.

Step 1   **State the hypotheses.** Use one of the forms from Table 17. State the meaning of σ_1 and σ_2.

Step 2   **Find the critical value(s) and state the rejection rule.** Use Table 17 and the F tables.

Step 3   **Find F_{data}.**

$$F_{\text{data}} = \frac{s_1^2}{s_2^2}$$

follows an F distribution with $df_1 = n_1 - 1$ and $df_2 = n_2 - 1$.

Step 4   **State the conclusion and the interpretation.** Compare F_{data} with the F critical value from Table 17.

Table 17 Critical values, rejection rules, and rejection regions

	Right-tailed test	**Left-tailed test**	**Two-tailed test**
Hypotheses	$H_0: \sigma_1 = \sigma_2$ $H_a: \sigma_1 > \sigma_2$	$H_0: \sigma_1 = \sigma_2$ $H_a: \sigma_1 < \sigma_2$	$H_0: \sigma_1 = \sigma_2$ $H_a: \sigma_1 \neq \sigma_2$
Critical value(s)	F_{α, n_1-1, n_2-1} The F-value with area α to the right of it.	$F_{1-\alpha, n_1-1, n_2-1}$ The F-value with area α to the left of it.	$F_{1-\alpha/2, n_1-1, n_2-1}$ and $F_{\alpha/2, n_1-1, n_2-1}$ The F-value with area $\alpha/2$ to the left of it and the F-value with area $\alpha/2$ to the right of it.
Rejection rule	Reject H_0 if $F_{\text{data}} > F_{\alpha, n_1-1, n_2-1}$	Reject H_0 if $F_{\text{data}} < F_{1-\alpha, n_1-1, n_2-1}$	Reject H_0 if $F_{\text{data}} < F_{1-\alpha/2, n_1-1, n_2-1}$ or if $F_{\text{data}} > F_{\alpha/2, n_1-1, n_2-1}$
Critical region			

We now return to our example comparing the variability of stock prices between Google and Apple.

Table 18 shows independent samples from Google and Apple stock prices from July 2014 together with the sample sizes and sample standard deviations. Test, using the critical-value method, whether the standard deviation of Google stock prices σ_1 is greater than the standard deviation of Apple stock prices σ_2.

Table 18 Independent random samples of stock prices, July 2014

Google	Apple
574.79 590.76 583.04 593.06 580.82 599.02 579.55 587.78	93.52 94.03 95.39 96.45 95.60 99.02 97.03
$n_1 = 8$ $s_1 \approx 7.999701$	$n_2 = 7$ $s_2 \approx 1.862594$

Source: www.marketwatch.com/tools/quotes/historical.asp.

Solution

The normal probability plots in Figures 22a and 22b show acceptable normality for both samples. We may, therefore, proceed with the *F* test for comparing population standard deviations.

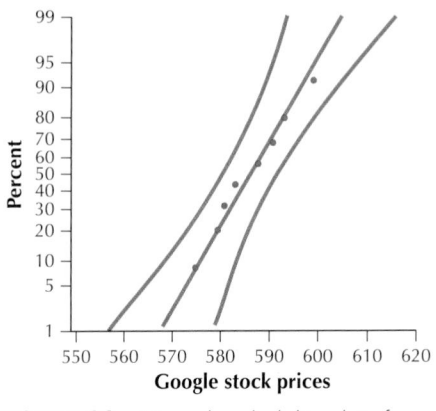

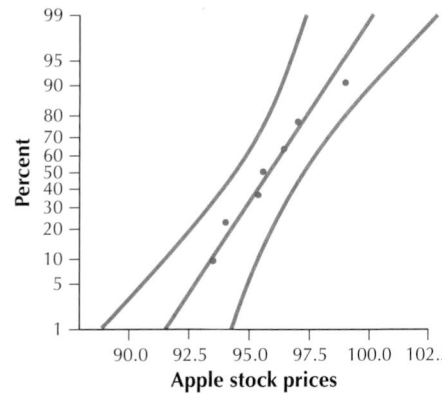

FIGURE 22a Normal probability plot of Google stock prices.

FIGURE 22b Normal probability plot of Apple stock prices.

Step 1   **State the hypotheses.** We are testing whether Google's stock prices are more variable than those of Apple. Thus, because Google represents population 1, we have the following hypotheses for our *F* test:

$$H_0 : \sigma_1 = \sigma_2 \quad \text{versus} \quad H_a : \sigma_1 > \sigma_2$$

where σ_1 represents the standard deviation of Google stock prices and σ_2 represents the standard deviation of Apple stock prices. Use level of significance $\alpha = 0.05$.

Step 2   **Find the critical value and state the rejection rule.** We have $df_1 = n_1 - 1 = 7$ and $df_2 = n_2 - 1 = 6$. From Table 17 and Appendix Table F, our critical value is the *F*-value with area $\alpha = 0.05$ to the right of it:

$$F_{crit} = F_{\alpha, n_1-1, n_2-1} = F_{0.05, 7, 6} = 4.21$$

Our rejection rule is, therefore, from Table 17: Reject H_0 if $F_{data} > 4.21$.

Step 3 Find F_{data}.

$$F_{data} = \frac{s_1^2}{s_2^2} = \frac{7.999701^2}{1.862594^2} \approx 18.45$$

follows an F distribution with $df_1 = n_1 - 1 = n_1 - 1 = 7$ and $df_2 = n_2 - 1 = n_2 - 1 = 6$.

Step 4 State the conclusion and the interpretation. Because $F_{data} \approx 18.45$ is greater than $F_{crit} = 4.21$, we reject H_0. There is evidence that the variability in Google stock prices is greater than the variability in Apple stock prices.

NOW YOU CAN DO
Exercises 21–26.

3 Perform the *F* Test for Comparing Two Population Standard Deviations: *p*-Value Method

We may also use the p-value method to perform the F test for comparing two population standard deviations. The requirements are the same.

F Test for Comparing Two Population Standard Deviations: *p*-Value Method

Suppose we have two independent random samples of size n_1 and n_2 taken from two normally distributed populations, with population standard deviations σ_1 and σ_2, and sample standard deviations s_1 and s_2, respectively.

Step 1 State the hypotheses and the rejection rule. Use one of the forms from Table 19. Clearly state the meaning of σ_1 and σ_2. The rejection rule is: *Reject H_0 if the p-value is less than α.*

Step 2 Find F_{data}.

$$F_{data} = \frac{s_1^2}{s_2^2}$$

follows an F distribution with $df_1 = n_1 - 1$ and $df_2 = n_2 - 1$.

Step 3 Find the *p*-value. Use technology and Table 19 to find the p-value.

Step 4 State the conclusion and the interpretation. Compare the p-value with α.

Table 19 *p*-Value for the *F* test for comparing two standard deviations

	Right-tailed test	**Left-tailed test**	**Two-tailed test**
Hypotheses	$H_0: \sigma_1 = \sigma_2$ $H_a: \sigma_1 > \sigma_2$	$H_0: \sigma_1 = \sigma_2$ $H_a: \sigma_1 < \sigma_2$	$H_0: \sigma_1 = \sigma_2$ $H_a: \sigma_1 \neq \sigma_2$
***p*-Value**	$P(F > F_{data})$ The area to the right of F_{data}. $p\text{-value} = P(F > F_{data})$	$P(F < F_{data})$ The area to the left of F_{data}. $p\text{-value} = P(F < F_{data})$ 	The *smaller* of: (i) $2 \cdot P(F > F_{data})$ and (ii) $2 \cdot P(F < F_{data})$.

We illustrate the p-value method with an example.

EXAMPLE 21 *F* Test for comparing two population standard deviations: *p*-value method

The Web site Medicare.gov publishes survey information on patient attitudes about their level of care. Table 20 shows the percentages of respondents taken from independent random samples of hospitals in Florida and Georgia, which reported that their nurses always communicated well.

Table 20 Independent random samples of hospital percentages

Florida	Georgia
67 66 70 70 72 73	72 75 78 73 68 71
63 69 65 68 65	82 72 77 75 73
$n_1 = 11$ $s_1 \approx 3.1305$	$n_2 = 11$ $s_2 \approx 3.8162$

Source: Medicare.gov.

Test whether there is a difference in variability between the two states, using $\alpha = 0.10$.

Solution

Because the normal probability plots in Figures 23a and 23b show acceptable normality, we may therefore proceed with the F test for comparing population standard deviations.

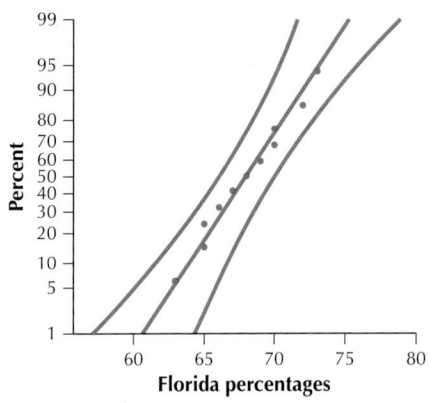

FIGURE 23a Normal probability plot of Florida percentages.

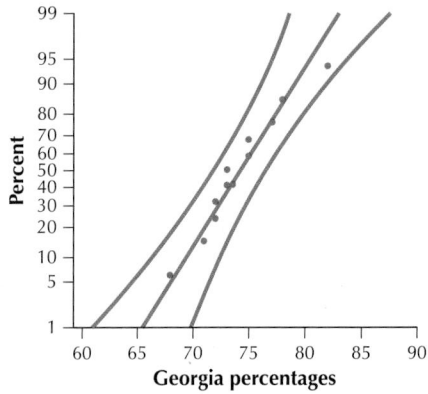

FIGURE 23b Normal probability plot of Georgia percentages.

Step 1   **State the hypotheses.** We are testing whether there is a difference in the standard deviation of the percentages for Florida (σ_1) and Georgia (σ_2). We therefore have a two-tailed test:

$$H_0 : \sigma_1 = \sigma_2 \quad \text{versus} \quad H_a : \sigma_1 \neq \sigma_2$$

where σ_1 and σ_2 represent the standard deviations of the percent of Florida and Georgia respondents, respectively, who reported that their nurses always communicated well. Use level of significance $\alpha = 0.10$.

Step 2   **Find F_{data}.**

$$F_{\text{data}} = \frac{s_1^2}{s_2^2} = \frac{3.1305^2}{3.8162^2} \approx 0.6729$$

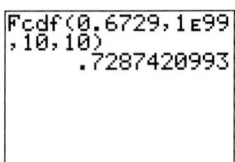

FIGURE 24a

```
Fcdf(0.6729,1E99
,10,10)
      .7287420993
```

follows an F distribution with $\text{df}_1 = n_1 - 1 = 10$ and $\text{df}_2 = n_2 - 1 = 10$.

Step 3   **Find the p-value.** Because we have a two-tailed test, Table 19 states that the p-value is

the smaller of **(i)** $2 \cdot P(F > F_{\text{data}})$ and **(ii)** $2 \cdot P(F < F_{\text{data}})$

Figure 24a shows the output from the TI-83/84, giving $P(F > F_{\text{data}})$ as the area under the F distribution curve between 0.6729 and infinity. This gives

$$2 \cdot P(F > F_{\text{data}}) = 2 \cdot 0.7287420993 = 1.457484199$$

```
Fcdf(0,0.6729,10
,10)
      .2712579007
```

FIGURE 24b

This *cannot* represent a valid p-value, because it is larger than 1. Figure 24b shows the output from the TI-83/84, giving $P(F < F_{\text{data}})$ as the area under the F distribution curve between 0 and 0.6729. Thus, our p-value is:

$$p\text{-value} = 2 \cdot P(F < F_{\text{data}}) = 2 \cdot P(F < 0.6729) = 2 \cdot 0.2712579007 \approx 0.5425$$

because this is smaller than 1.457484199.

SECTION 10.4

For Exercises 10 and 11, assume that the populations are normally distributed. Perform the indicated hypothesis tests using the critical-value method.

10. $H_0: \sigma_1 = \sigma_2$ vs. $H_a: \sigma_1 > \sigma_2$, $\alpha = 0.01$, $n_1 = 55$, $n_2 = 107$, $s_1 = 20.2$, $s_2 = 5.6$

11. $H_0: \sigma_1 = \sigma_2$ vs. $H_a: \sigma_1 < \sigma_2$, $\alpha = 0.05$, $n_1 = 125$, $n_2 = 27$, $s_1 = 5.7$, $s_2 = 10.2$

For Exercises 12 and 13, assume that the populations are normally distributed. Perform the indicated hypothesis tests using the *p*-value method for each exercise.

12. $H_0: \sigma_1 = \sigma_2$ vs. $H_a: \sigma_1 > \sigma_2$, $\alpha = 0.05$, $n_1 = 1050$, $n_2 = 250$, $s_1 = 10.4$, $s_2 = 4.6$

13. $H_0: \sigma_1 = \sigma_2$ vs. $H_a: \sigma_1 < \sigma_2$, $\alpha = 0.01$, $n_1 = 150$, $n_2 = 80$, $s_1 = 25.1$, $s_2 = 35.8$

Chapter 10 QUIZ

TRUE OR FALSE

1. True or false: In a dependent sampling method, the subjects in the first sample determine the subjects for selection in the second sample.

2. True or false: The pooled estimate of *p*, $\hat{p}_{pooled} = (x_1 + x_2)/(n_1 + n_2)$, always lies between $\hat{p}_1$ and $\hat{p}_2$.

3. True or false: The test statistic Z_{data} measures the size of the typical error in using $\hat{p}_1 - \hat{p}_2$ to estimate $p_1 - p_2$.

FILL IN THE BLANK

4. The conditions on paired sample data for performing a hypothesis test or constructing a confidence interval on paired sample data are that the population is _____ or the sample size is _____.

5. The notation *E* represents the _____ _____ _____ (three words).

6. _____ [notation] represents the sample mean of the set of *n* paired differences.

SHORT ANSWER

7. What is the notation used to indicate the difference in population means for two independent samples?

8. What statistic is used to estimate the common unknown population proportion?

9. If a $100(1 - \alpha)\%$ confidence interval for $\mu_1 - \mu_2$ contains 0, then with $100(1 - \alpha)\%$ confidence, what can you conclude about the difference in the population means?

CALCULATIONS AND INTERPRETATIONS

10. Trying to quit smoking? Butt-Enders, a cigarette dependence reduction program, claims to lower the average number of cigarettes smoked for its participants. A sample of 10 participants smoked the following numbers of cigarettes on a randomly chosen day before and after attending Butt-Enders. Assume that the differences are normally distributed.

Participant	1	2	3	4	5
Before	40	20	60	30	50
After	20	0	40	30	20
Participant	6	7	8	9	10
Before	60	20	40	30	20
After	60	20	20	0	20

a. Find a 90% confidence interval for the population mean difference in number of cigarettes smoked.

b. Use your confidence interval to test, at level of significance $\alpha = 0.10$, whether the population mean difference in number of cigarettes smoked differs from 0.

11. A family is trying to decide where to move. The choice has come down to Suburb A and Suburb B. A random sample of 40 households in Suburb A had a mean income of $50,000 and a standard deviation of $15,000. A random sample of 36 households in Suburb B had a mean income of $65,000 and a standard deviation of $20,000.

a. Test, at level of significance $\alpha = 0.05$, whether the population mean income in Suburb A is less than the population mean income in Suburb B.

b. Construct and interpret a 95% confidence interval for $\mu_1 - \mu_2$.

Use the following information for Exercises 12 and 13. A soft drink company recently performed a major overhaul of one of its bottling machines. Management is eager to determine whether the overhaul has resulted in an increase in productivity for the machine. One hundred "minute segments" are sampled at random from the updated machine (sample 1) and a machine that was not updated (sample 2), and the number of bottles processed is noted. The mean and standard deviation of the number of bottles processed by each machine is given in the table.

Updated machine	$n_1 = 100$	$\bar{x}_1 = 200$	$s_1 = 30$
Non-updated machine	$n_2 = 100$	$\bar{x}_2 = 190$	$s_2 = 25$

12. Construct and interpret a 95% confidence interval for $\mu_1 - \mu_2$.

13. Refer to the previous exercise.

a. Test, at level of significance $\alpha = 0.05$, whether μ_1 is greater than μ_2.

b. Explain whether the confidence interval in Exercise 12 could have been used to perform the hypothesis test in (**a**). Why or why not?

14. The U.S. Census Bureau reported that, for people 18 to 24 years old, the mean annual income for people who never married was $13,539 and for married people was $19,321. Suppose that this information came from a survey of 100 people from each group and that the sample standard deviations were $5000 for the people who never married and $8000 for the married people.

a. Test, at level of significance $\alpha = 0.10$, whether the population mean income for never-married people differs from that of married people.

b. If we construct a 90% confidence interval for $\mu_1 - \mu_2$, will the interval include 0? Explain why or why not.

c. Confirm your statements from (**b**).

11 Categorical Data Analysis

Introduction

Chapter 11 introduces the multinomial random variable. Students learn methods for performing hypothesis tests for this type of data using the χ^2 distribution.

Section 11.1 covers the multinomial random variable and the χ^2 goodness of fit test. In Section 11.2, students learn how to perform a χ^2 test for independence between two categorical variables and to test for differences in proportions among k populations.

From the Author

Section 11.1 χ^2 Goodness of Fit Test

- Instructors may want to review the chi-square distribution in Section 8.4 before starting Chapter 11.

- Instructors may want to use the first Developing Your Statistical Sense feature in this section to show how a goodness of fit test is a lot like trying on a new pair of gloves.

- Instructors may wish to emphasize that, in a goodness of fit test, we are comparing two sets of frequencies: observed frequencies and expected frequencies. Large differences between them will likely lead to rejection of the null hypothesis, as indicated by the What Results Might We Expect? feature in this section.

Section 11.2 χ^2 Tests for Independence and for Homogeneity of Proportions

- Instructors may point out that, even though the hypothesis tests and calculations in this section are different from those in Section 11.1, we are still comparing observed frequencies with expected frequencies.

- Students may note that, in a test for the homogeneity of proportions, the proportions usually do not sum to 1.

Teaching Tips

Starting Example: Collect data from the class to see whether a difference exists in the majors of men and women or to see whether a difference exists in class year (e.g., junior or senior) between men and women.

Extra Example: Using the chi-square goodness of fit test to confirm a sample is representative of a population.

In order to perform statistical inference, analysts need to use samples that are representative of their population. One way to confirm this for categorical data is to use a χ^2 (chi-square) goodness of fit test. Suppose a multinomial variable *Vehicle* takes the values *American*, *Japanese*, and *Other*, and suppose that we know that 40% of the individuals in the population are American, 35% are Japanese, and 25% report Others. We are taking a sample and would like to determine whether the

sample is representative of the population. We apply the χ^2 goodness of fit test as follows. The hypotheses for this χ^2 goodness of fit test would be as follows:

$$H_0 : p_{American} = 0.40, \quad p_{Japanese} = 0.35, \quad p_{Other} = 0.25$$

$$H_a : \text{At least one of the proportions in } H_0 \text{ is wrong.}$$

Our sample of size $n = 100$ yields the following *observed frequencies* (represented by the letter "O"):

$$O_{American} = 36, \quad O_{Japanese} = 35, \quad O_{Other} = 29$$

To determine whether these counts represent proportions that are significantly different from those expressed in H_0, we compare these observed frequencies with the *expected frequencies* that we would expect if H_0 were true. If H_0 were true, then we would expect 40% of our sample of 100 individuals to be American; that is, the expected frequency for *American* is:

$$E_{American} = n \cdot p_{American} = 100 \cdot 0.40 = 40$$

Similarly,

$$E_{Japanese} = n \cdot p_{Japanese} = 100 \cdot 0.35 = 35$$

$$E_{Ohter} = n \cdot p_{Other} = 100 \cdot 0.25 = 25$$

These frequencies are compared using the test statistic:

$$\chi^2_{data} = \sum \frac{(O - E)^2}{E}$$

Large differences between the observed and expected frequencies, and thus a large value for χ^2_{data}, will lead to a small *p*-value, and a rejection of the null hypothesis. The following table illustrates how the test statistic is calculated.

Calculating the test statistic χ^2_{data}

Vehicle	Observed frequency	Expected frequency	$\dfrac{(Obs - Exp)^2}{Exp}$
American	36	40	$\dfrac{(36 - 40)^2}{40} = 0.4$
Japanese	35	35	$\dfrac{(35 - 35)^2}{35} = 0$
Other	29	25	$\dfrac{(29 - 25)^2}{25} = 0.64$
			$\chi^2_{data} = 1.04$

The p-value is the area to the right of χ^2_{data} under the χ^2 curve with $k - 1$ degrees of freedom, where $k =$ the number of categories (here $k = 3$):

$$p\text{-value} = P(\chi^2 > \chi^2_{data}) = P(\chi^2 > 1.04) = 0.5945$$

Thus, there is no evidence that the observed frequencies represent proportions that differ significantly from those in the null hypothesis. In other words, our sample is representative of the population.

In-Class Activities

1. **Use the EESEE case study What Makes a Pre-Teen Popular?** Let students use some of the variables in the data set to construct contingency tables. Apply the χ^2 test for independence to the variables in the table.

2. **A χ^2 Goodness of Fit Test for the Proportions of Freshmen, Sophomores, Juniors, and Seniors at your College**

(*Note to instructor:* The small sample sizes involved in the following activities may not meet the assumptions for the χ^2 test. You can alleviate this by taking a larger sample or by collapsing the categories with the smaller expected frequencies.)

    a. What do you think are the proportions of freshmen, sophomores, juniors, and seniors at your college? Write out the hypotheses for the test to determine whether these proportions are correct.

    b. Use a sample of 25 students to test this hypothesis. What are the expected frequencies for each category? Do you have any data yet to evaluate your claim in (**a**)?

     c. Each student should generate a random sample of 25 students from the class, using either (**a**) technology or (**b**) the *Simple Random Sample* applet.

    d. For each student in your sample, ask which class he or she belongs to. Then add up the counts for each class. Which type of frequencies are these?

    e. You will use a level of significance of $\alpha = 0.10$ for this test. What is the critical value χ^2_{crit} for this test? Will it be the same for each student?

    f. Perform the χ^2 goodness of fit test with your sample data. Will the value of χ^2_{data} be the same for each student?

    g. Record your name and your value for the test statistic χ^2_{data} calculated in (**f**) on a sticky note.

    h. Place your sticky note on the number line that the instructor prepared on the blackboard, noting the location of the cutoff value χ^2_{crit}. We thus have a collection of values for the statistic χ^2_{data}, representing an approximation of the sampling distribution of χ^2_{data}. Can you make any comment on the shape of the distribution of χ^2_{data} values?

    i. What proportion of students rejected the null hypothesis?

    j. Reach a class consensus as to why most students either did or did not reject the null hypothesis.

Supplements

- StatTutor 6.1–6.2 and 23.2–23.10
- EESEE case studies
    - Alcoholism in Twins (Question 5 on χ^2 test)
    - Cancer and Power Lines (Questions 2, 3(b), 4(d))

Videos

- *Against All Odds: Inside Statistics:* www.learner.org/resources/series65.html
  - Program 29: Inference for Two-Way Tables

Web Sites

- The site Online Statistics: An Interactive Multimedia Course of Study has information on contingency tables and the χ^2 distribution, and it has a link to the Rice Virtual Lab in Statistics with simulation demonstrations and other topics related to statistics: http://onlinestatbook.com/index.html.

- This Web site lists other sites that do statistical calculations for contingency tables and the χ^2 distribution: http://statpages.org/.

- This site has an activity that deals with contingency tables and the χ^2 distribution: www.amstat.org/publications/jse/v6n3/smith.html.

11 Categorical Data Analysis

Susan Wides/Getty Images

Online Dating

CASE STUDY

The Pew Internet and American Life Project reports that about 16 million people, representing 11% of the American Internet-using public, have visited a dating Web site, and 37% of Internet users who are currently seeking partners have gone to a dating Web site.[1] In this chapter, we apply the concepts and methodologies of categorical data analysis to investigate online dating. In Section 11.2, we examine the following questions:

- When it comes to online dating, do men and women differ when it comes to reporting what kind of relationship they are in? In other words, do men and women report differently whether they are
  - in a committed relationship,
  - not in a committed relationship and not looking for a partner, or
  - not in a committed relationship, but looking for a partner?

- Also, when it comes to online dating, do men and women differ when it comes to how they self-report their physical appearance? In other words, do men and women self-report differently whether they are
  - very attractive
  - attractive
  - average, or
  - "prefer not to answer"?

THE BIG PICTURE

Where we are coming from and where we are headed . . .

- In Chapters 8–10, we learned how to perform inference for continuous variables.
- Here, in Chapter 11, we learn how to perform hypothesis tests for *multinomial data*, an extension of the binomial distribution. These methods rely on the χ^2 (chi-square) distribution, which we learned in Chapters 8 and 9.
- In later chapters, we will learn about *analysis of variance* and *regression*, which rank among the most commonly used data analytic methods in the world.

11.1 χ^2 Goodness of Fit Test

OBJECTIVES By the end of this section, I will be able to . . .

1 Explain what a multinomial random variable is and how to calculate expected frequencies.

2 Describe how a χ^2 goodness of fit test works.

3 Perform and interpret the results from the χ^2 goodness of fit test using the critical-value method and the p-value method.

According to the Adobe Digital Index, the market share for the leading Internet browsers (both desktop and mobile) in June 2014 was as follows: Google Chrome, 32%; Microsoft Internet Explorer, 31%; others, 37%. Change is rapid in the online environment. Have these market shares changed since June 2014? How would we go about performing a hypothesis test to determine whether market shares have changed significantly? In Section 11.1, we examine this question using a new type of hypothesis test called a χ^2 *goodness of fit test*. We begin by first considering a new type of random variable that is used to represent categorical data.

1 The Multinomial Random Variable

Recall from Chapter 1 that *categorical* (qualitative) variables take values that can be classified into categories. In Chapter 6, we considered binomial random variables, for which there are only two possible outcomes. Now, let's consider the following type of random variable, which can have more than two possible values.

> **Multinomial Random Variable**
>
> A random variable is *multinomial* if it satisfies each of the following conditions:
>
> - Each independent trial of the experiment has k possible outcomes, $k = 2, 3, 4, \ldots$
>
> - The ith outcome (category) occurs with probability p_i, where $i = 1, 2, \ldots, k$ (that is, p_i is the population proportion for category i).
>
> - $\sum p_i = 1$ (Law of Total Probability).
>
> Data from a **multinomial random variable** are said to follow a *multinomial distribution*.

Note: The binomial distribution may be considered a special case of the multinomial distribution, with $k = 2$.

For example, suppose 30% of the residents of a particular town are Democrats, 30% are Republicans, and 40% are Independents. If we select $n = 100$ residents at random, then the number of Democrats, Republicans, and Independents observed follows a multinomial distribution, with

$$p_{\text{Democrats}} = 0.30, \quad p_{\text{Republicans}} = 0.30, \quad p_{\text{Independents}} = 0.40,$$

and

$$\sum p_i = 0.30 + 0.30 + 0.40 = 1$$

EXAMPLE 1 Identifying a Multinomial Random Variable

For each of the following, determine whether the random variable is multinomial.

a. We select 10 students at random and define our random variable X to be the amount of time the student used a Web browser yesterday.

b. We select 10 students at random and define our random variable X to be the browser used most by the student the last time he or she was on the Internet, where the possible values are Google Chrome, Microsoft Internet Explorer, or Other.

Solution

a. The amount of time spent using a browser is a continuous random variable, not categorical. So, X cannot be multinomial.

NOW YOU CAN DO
Exercises 5–8.

b. The browser is categorical. We have $k = 3$ different categories, with the population proportions of the three categories adding up to one (see Example 2(**a**) below). Therefore, X is multinomial.

Next, recall from Section 6.2 that the formula for finding the expected value (mean) of a binomial random variable having n trials and probability of success p is

$$\text{expected value} = n \cdot p$$

For a multinomial random variable, the **expected frequency** of the ith category is

$$\text{expected frequency}_i = E_i = n \cdot p_i$$

where n represents the number of trials, and p_i represents the population proportion for the ith category.

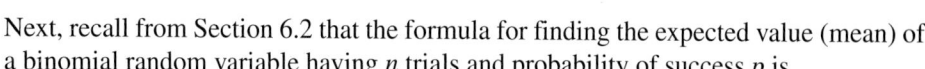

EXAMPLE 2   Finding the expected frequencies

According to the Adobe Digital Index, the market share for the leading Internet browsers (both desktop and mobile) in June 2014 was as shown in Table 1. Let $X = browser$ of a randomly selected Internet user.

a. Verify that $X = browser$ is a valid multinomial random variable.

b. Find the expected frequency for each category in a series of 200 trials.

Table 1 Distribution of browser market share

Browser	Relative frequency
Google Chrome	0.32
Microsoft Internet Explorer	0.31
Other	0.37

Solution

a. There are $k = 3$ possible outcomes: Google Chrome, Microsoft Internet Explorer, and Other. Assigning probabilities using the relative frequency method, we have the following hypothesized proportions for each browser:

$$p_{\text{Chrome}} = 0.32, \quad p_{\text{IE}} = 0.31, \quad p_{\text{Other}} = 0.37$$

and

$$\sum p_i = 0.32 + 0.31 + 0.37 = 1$$

Therefore, $X = browser$ is a valid multinomial random variable.

b. We have $n = 200$ trials (sample size $= 200$), so the expected frequencies are as provided in Table 2.

Table 2 Expected frequencies for browser preference in sample of size 100

Category	$Expected\ frequency_i = E_i = n \cdot p_i$
Google Chrome	$E_{Chrome} = 200 \cdot 0.32 = 64$
Microsoft Internet Explorer (IE)	$E_{IE} = 200 \cdot 0.31 = 62$
Other	$E_{Others} = 200 \cdot 0.37 = 74$

As a check on the calculations, we should have $\sum E_i = n$. In this case,

NOW YOU CAN DO
Exercises 9–12.

$$\sum E_i = 64 + 62 + 74 = 200 = n$$

YOUR TURN
#1

Publishers Weekly reported that, in 2014, the book format market share was as follows: paperbacks, 41%; hard covers, 34%; e-books, 13%; and all other formats, 12%. Suppose a survey was conducted this year of 2000 books purchased.

a. Verify that $X = book\ format$ is a valid multinomial random variable.

b. Find the expected frequency for each category.

(The solutions are shown in Appendix A.)

What Do These Expected Frequencies Mean?

Recall that the expected value of a random variable refers to the long-run mean of that random variable after an arbitrarily large number of trials. For example, if we repeatedly took samples of 200 Internet users and asked about browser preference, the mean number of persons who used Google Chrome would approach $E_{Chrome} = 64$ as we took more and more different samples, *if the proportions given in Table 1 are correct*. Similarly, because 31% of the entire population of Internet users use Microsoft IE, we would *expect* about 31% of any given sample of 200 Internet users to use Microsoft IE, because the sample is a subset of the population. This of course raises the question: Are the proportions in Table 1 still true? That is the type of question we will learn how to address next.

2 What Is a χ^2 Goodness of Fit Test?

Do the 2014 market shares still hold true today? In other words, has the distribution of the multinomial random variable *browser* given in Table 1 changed since June 2014? To determine this, we introduce a new type of hypothesis test, called a **χ^2 goodness of fit test**.

> **χ^2 Goodness of Fit Test**
> A **χ^2 goodness of fit test** is a hypothesis test used to determine whether a random variable follows a particular distribution. In a goodness of fit test, the hypotheses are
>
> H_0 : The random variable follows a particular distribution.
> H_a : The random variable does not follow the distribution specified in H_0.

For Example 2, the null hypothesis completely specifies each of the probabilities in the relative frequency distribution, as follows:

$$H_0 : p_{\text{Chrome}} = 0.32, \quad p_{\text{IE}} = 0.31, \quad p_{\text{Other}} = 0.37$$

The alternative hypothesis simply denies the claim made by the null hypothesis:

H_a : The random variable does not follow the distribution specified in H_0.

In other words, H_a claims that the browser market shares have changed since June 2014.

Developing Your Statistical Sense

Fitting the Model to the Data

Now, a goodness of fit test sounds like something you do in a clothing store dressing room. Actually, the analogy to clothes is rather appropriate. Suppose winter is coming and you are in the market for a new pair of gloves. You find one pair that is especially attractive, but the gloves don't fit your hands. What do you do? You reject the ill-fitting gloves and search for a new pair. In statistics, the gloves represent the models and your hands represent the actual "hard data" observed in the sample.

The null hypothesis H_0 represents what is called a *model,* a working theory of how the population proportions are distributed. Our working model of how the market shares are distributed is stated in the null hypothesis:

Model 1. $H_0 : p_{\text{Chrome}} = 0.32, \quad p_{\text{IE}} = 0.31, \quad p_{\text{Other}} = 0.37$

Of course, we could also try other models if we think the market has changed, such as the following:

Model 2. $H_0 : p_{\text{Chrome}} = 0.33, \quad p_{\text{IE}} = 0.33, \quad p_{\text{Other}} = 0.34$

Model 3. $H_0 : p_{\text{Chrome}} = 0.40, \quad p_{\text{IE}} = 0.30, \quad p_{\text{Other}} = 0.30$

In hypothesis testing, we "try on" only one model at a time.

In statistics, a goodness of fit test determines if the actual "hard data" observed in the sample are consistent with the proportions stated in the null hypothesis. Market researchers would collect data on the actual preferences of a sample of 100 real Internet users in order to determine whether or not the market shares have changed. The sample is summarized in a set of *observed frequencies* of Internet users who prefer the various browsers. The χ^2 goodness of fit test then *compares these observed frequencies with the expected frequencies* found in Example 2.

How a Goodness of Fit Test Works

The goodness of fit test is based on a comparison of the *observed frequencies* (sample data) with the *expected frequencies* when H_0 is true. That is, we compare what we actually see with what we would expect to see if H_0 were true. If the difference between the observed and expected frequencies is large, we reject H_0.

The difference between the observed and expected frequencies is measured by the test statistic, χ^2_{data}. As usual, it comes down to how large a difference is large.

Students may want to review the characteristics of the χ^2 distribution (Chapter 10, page 618) and the procedure for finding χ^2 critical values for a right-tailed test (Chapter 10, page 620).

Test Statistic for the χ^2 Goodness of Fit Test

For a multinomial random variable with k categories and n trials, let O_i represent the observed frequency for category i, and let E_i represent the expected frequency for category i. Then the test statistic for a goodness of fit test

$$\chi^2_{data} = \sum \frac{(O_i - E_i)^2}{E_i}$$

approximately follows a χ^2 (chi-square) distribution with $k - 1$ degrees of freedom (df), if the following conditions are satisfied:

a. None of the expected frequencies is less than 1.

b. At most, 20% of the expected frequencies are less than 5.

If the conditions are not satisfied, then it may be possible to combine two or more categories so that the conditions may then be fulfilled.

EXAMPLE 3   Calculating χ^2_{data}

Suppose the observed frequencies of browser preference in Table 3 come from a survey taken this year of 200 Internet users.

Table 3 Observed frequencies of browser preference in a sample of 200 Internet users

Browser	Observed frequency
Google Chrome	80
Microsoft Internet Explorer	62
Other	58

Calculate the test statistic χ^2_{data} by comparing the observed frequencies from Table 3 with the expected frequencies calculated in Table 2 of Example 2.

Solution

The observed frequencies O_i are found in Table 3, and the expected frequencies are given in Table 2. Table 4 then provides the quantities needed to calculate χ^2_{data}. Then

$$\chi^2_{data} = \sum \frac{(O_i - E_i)^2}{E_i} \approx 4 + 0 + 3.46 = 7.46$$

Table 4 Calculating χ^2_{data}

Category	p_i	O_i	E_i	$O_i - E_i$	$(O_i - E_i)^2$	$\dfrac{(O_i - E_i)^2}{E_i}$
Chrome	0.32	80	64	16	256	$\dfrac{(80-64)^2}{64} = 4$
IE	0.31	62	62	0	0	$\dfrac{(62-62)^2}{62} = 0$
Other	0.37	58	74	-16	-256	$\dfrac{(58-74)^2}{74} \approx 3.46$

NOW YOU CAN DO
Exercises 13–18.

YOUR TURN #2

Publishers Weekly reported that, in 2014, the book format market share was as follows: paperbacks, 41%; hard covers, 34%; e-books, 13%; and all other formats, 12%. Suppose a survey was conducted this year of 2000 books purchased, with the following book sales: 810 paperbacks, 680 hard covers, 280 e-books, and 230 others. Calculate the test statistic χ^2_{data}.

(The solution is shown in Appendix A.)

3 Performing the χ^2 Goodness of Fit Test

The χ^2 goodness of fit test may be performed using (a) the critical-value method or (b) the *p*-value method. We start with the critical value method.

> ### χ^2 Goodness of Fit Test: Critical-Value Method
>
> **Step 1 State the hypotheses and check the conditions.**
>
> - The null hypothesis states that the multinomial random variable follows a particular distribution.
> - The alternative hypothesis states that the random variable does not follow that distribution.
>
> The following conditions must be met:
>
> **a.** None of the expected frequencies is less than 1.
>
> **b.** At most, 20% of the expected frequencies are less than 5.
>
> The expected frequency for the *i*th category is $E_i = n \cdot p_i$, where *n* represents the number of trials and p_i represents the population proportion for the *i*th category.
>
> **Step 2 Find the χ^2 critical value, χ^2_{crit}, and state the rejection rule.** Use Table E in the Appendix. Reject H_0 if $\chi^2_{\text{data}} \geq \chi^2_{\text{crit}}$.
>
> **Step 3 Calculate χ^2_{data}.**
>
> $$\chi^2_{\text{data}} = \sum \frac{(O_i - E_i)^2}{E_i}$$
>
> where O_i = observed frequency, and E_i = expected frequency.
>
> **Step 4 State the conclusion and the interpretation.** Compare χ^2_{data} with χ^2_{crit}.

All hypothesis tests in this chapter are right-tailed tests, so that we need to find χ^2_{crit} for the area to the right of the critical value only.

EXAMPLE 4 Critical-value method for the χ^2 goodness of fit test

Test whether the Internet browser market shares from Example 2 have changed since June 2014, using level of significance $\alpha = 0.05$.

Solution

Step 1 **State the hypotheses and check the conditions.** The hypotheses are:

$$H_0: p_{\text{Chrome}} = 0.32, \quad p_{\text{IE}} = 0.31, \quad p_{\text{Other}} = 0.37$$

H_a : The random variable does not follow the distribution specified in H_0.

Checking the conditions, the expected frequencies from Table 2 are

$$E_{\text{Chrome}} = 64, \quad E_{\text{IE}} = 62, \quad E_{\text{Other}} = 74$$

Because none of these expected frequencies is less than 1, and none of the expected frequencies is less than 5, the conditions for performing the goodness of fit test are satisfied.

Step 2 **Find the χ^2 critical value, χ^2_{crit}, and state the rejection rule.** We have degrees of freedom $k - 1 = 3 - 1 = 2$ and $\alpha = 0.05$. Turning to the χ^2 table (Table E in the Appendix) in the column labeled $\chi^2_{0.05}$ and the row containing df = 2, we find $\chi^2_{\text{crit}} = \chi^2_{0.05} = 5.991$, as shown in Figure 1. The rejection rule is "Reject H_0 if $\chi^2_{\text{data}} \geq 5.991$."

FIGURE 1
Finding the χ^2 critical value for df = $k - 1 = 2$ and level of significance $\alpha = 0.05$.

Chi-square (χ^2) distribution

Area to the right of critical value

Degrees of freedom	0.995	0.99	0.975	0.95	0.90	0.10	0.05	0.025
1	—	—	0.001	0.004	0.016	2.706	3.841	5.024
2	0.010	0.020	0.051	0.103	0.211	4.605	5.991	7.378
3	0.072	0.115	0.216	0.352	0.584	6.251	7.815	9.348

Step 3   From Example 3, we have $\chi^2_{\text{data}} = 7.46$.

Step 4   **State the conclusion and the interpretation.** Compare χ^2_{data} with χ^2_{crit}. $\chi^2_{\text{data}} = 7.46$ is greater than $\chi^2_{\text{crit}} = 5.991$, as shown in Figure 2. Therefore, we reject H_0.

FIGURE 2
Reject H_0 when $\chi^2_{\text{data}} \geq \chi^2_{\text{crit}}$.

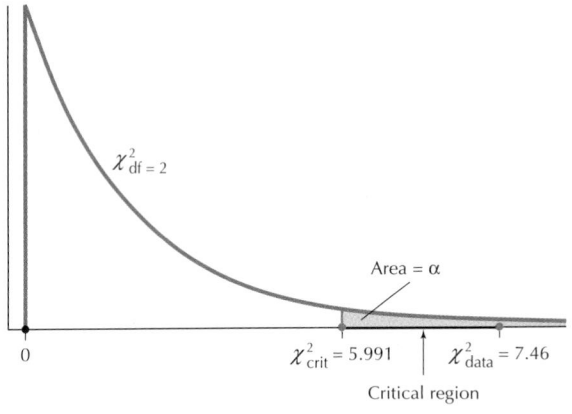

NOW YOU CAN DO
Exercises 19–22.

Evidence exists at level of significance $\alpha = 0.05$ that the random variable *browser* does not follow the distribution specified in H_0. In other words, evidence exists that the market shares for Internet browsers have changed.

YOUR TURN
#3

Test using level of significance $\alpha = 0.05$ whether the book format market shares have changed, using the information from Your Turn #1 on page 634 and Your Turn #2 on page 637.

(The solution is shown in Appendix A.)

Developing Your Statistical Sense

Be Careful How You Interpret the Conclusion

Note carefully what this conclusion says and what it doesn't say. The χ^2 goodness of fit test provides evidence that the random variable does not follow the distribution specified in H_0. In particular, the conclusion does *not* state, for example, that Chrome's proportion is significantly greater than it was in 2014. *Informally*, we can compare the observed frequency of 80 with the expected frequency of 64 for the Chrome browser and note that there appears to be evidence of an increase in market share for Chrome. But this is only informal and is not part of the hypothesis test. It is a common error in statistical analysis to form conclusions beyond what the hypothesis test is actually testing.

Next, we turn to the *p*-value method. The χ^2 goodness of fit test is a right-tailed test, so the *p*-value for the χ^2 statistic is defined as the area under the χ^2 curve to the right of the test statistic χ^2_{data}, as shown in Figure 3. That is,

$$p\text{-value} = P(\chi^2 > \chi^2_{data})$$

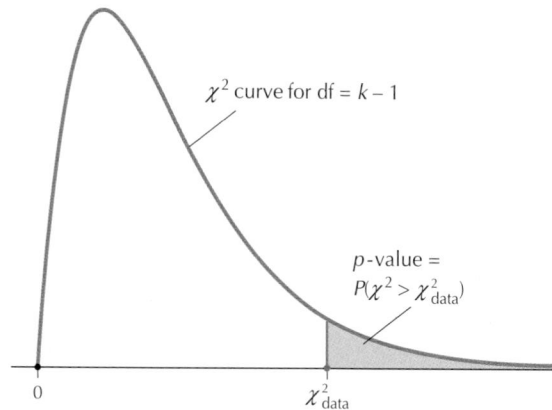

FIGURE 3 *p*-Value $= P(\chi^2 > \chi^2_{data})$.

χ^2 Goodness of Fit Test: *p*-Value Method

Step 1   State the hypotheses and the rejection rule. Check the conditions.

- The null hypothesis states that the multinomial random variable follows a particular distribution.
- The alternative hypothesis states that the random variable does not follow that distribution.
- Reject H_0 if the *p*-value $\le \alpha$.

The following conditions must be met:

a. None of the expected frequencies is less than 1.

b. At most, 20% of the expected frequencies are less than 5.

The expected frequency for the *i*th category is $E_i = n \cdot p_i$, where n represents the number of trials and p_i represents the population proportion for the *i*th category.

Step 2   Calculate χ^2_{data}.

$$\chi^2_{data} = \sum \frac{(O_i - E_i)^2}{E_i}$$

where O_i = observed frequency, and E_i = expected frequency.

Step 3   Find the *p*-value.

$$p\text{-value} = P(\chi^2 > \chi^2_{data}) \text{ (see Figure 3)}$$

Step 4   State the conclusion and the interpretation. Compare the *p*-value with α.

EXAMPLE 5   *p*-Value method for the χ^2 goodness of fit test using technology

Table 5 contains the distribution of violent crime in New York City in 2012.[2] Suppose that a random sample of 1000 violent crimes in New York City yielded the counts shown in Table 6. Test whether the population proportions have changed since 2012, using the *p*-value method and level of significance $\alpha = 0.05$.

Table 5   2012 violent crime in New York City

Murder	Rape	Robbery	Assault
0.01	0.04	0.35	0.60

Table 6 Sample of 1000 violent crimes in New York City this year

Murder	Rape	Robbery	Assault
6	50	350	594

Solution

Step 1 **State the hypotheses and the rejection rule. Check the conditions.**

$H_0: p_{\text{Murder}} = 0.01, \quad p_{\text{Rape}} = 0.04, \quad p_{\text{Robbery}} = 0.35, \quad p_{\text{Assault}} = 0.60$

H_a: The random variable does not follow the distribution specified in H_0.

Reject H_0 if the p-value ≤ 0.05.

What Results Might We Expect?

Before we do the formal hypothesis test, let's try to figure out what the conclusion might be. Figure 4 is a clustered bar graph (see Section 2.1) of the observed and expected frequencies for each of the four categories. If H_0 were true, then, for each category, we would expect the red bars (observed frequencies) and blue bars (expected frequencies) to have somewhat similar heights. In fact, the heights of the bars are fairly similar for all four categories, indicating not much difference between the crimes that were observed and the crimes that were expected. Thus, we might expect to *not* reject H_0.

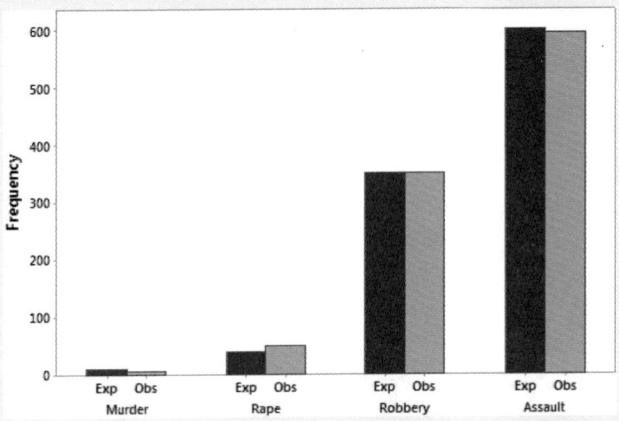

FIGURE 4 Graph indicates no evidence against H_0.

First, we need to find the expected frequencies. We have $n = 1000$, so the expected frequencies are as shown here.

Expected frequencies for violent crimes in a sample of size $n = 1000$

Category	*Expected frequency$_i$ = E_i = $n \cdot p_i$*
Murder	$E_{\text{Murder}} = 1000 \cdot 0.01 = 10$
Rape	$E_{\text{Rape}} = 1000 \cdot 0.04 = 40$
Robbery	$E_{\text{Robbery}} = 1000 \cdot 0.35 = 350$
Assault	$E_{\text{Assault}} = 1000 \cdot 0.60 = 600$

The expected frequency for a given cell is

$$\text{expected frequency} = \frac{(\text{row total}) \cdot (\text{column total})}{\text{grand total}}$$

CAUTION Do not include the row or column totals when counting the categories.

Step 2 **Find the critical value χ^2_{crit} and state the rejection rule.** Reject H_0 if $\chi^2_{data} \geq \chi^2_{crit}$. Use $(r - 1)(c - 1)$ degrees of freedom, where r is the number of categories in the row variable and c is the number of categories in the column variable.

Step 3 **Calculate χ^2_{data}.**

$$\chi^2_{data} = \sum \frac{(O_i - E_i)^2}{E_i}$$

where O_i = observed frequency and E_i = expected frequency for each cell.

Step 4 **State the conclusion and the interpretation.** Compare χ^2_{data} with χ^2_{crit}.

EXAMPLE 7   Performing the χ^2 test for independence using the critical-value method

Using Table 7, test whether *age group* is independent of *response*, using level of significance $\alpha = 0.05$.

Solution

Step 1   **State the hypotheses and check the conditions.**

H_0 : *Age group* and *response* are independent.

H_a : *Age group* and *response* are dependent.

We note from Table 8 that none of the expected frequencies are less than either 1 or 5. Therefore, the conditions are met, and we may proceed with the hypothesis test.

See Figure 1 (page 638) to review how to find χ^2_{crit}.

Step 2   **Find the critical value χ^2_{crit} and state the rejection rule.** The row variable, response, has three categories, so $r = 3$. The column variable, age group, has two categories, so $c = 2$. Thus,

$$\text{degrees of freedom} = (r - 1)(c - 1) = (3 - 1)(2 - 1) = 2$$

With level of significance $\alpha = 0.05$, this gives us $\chi^2_{crit} = 5.991$ from the χ^2 table. The rejection rule is therefore

$$\text{Reject } H_0 \text{ if } \chi^2_{data} \geq 5.991$$

Step 3   **Calculate χ^2_{data}.** The observed frequencies are found in Table 7 and the expected frequencies are found in Table 8. Then

$$\chi^2_{data} = \sum \frac{(O_i - E_i)^2}{E_i} = \frac{(180 - 204)^2}{204} + \frac{(330 - 306)^2}{306} + \frac{(378 - 325.2)^2}{325.2}$$

$$+ \frac{(435 - 487.8)^2}{487.8} + \frac{(42 - 70.8)^2}{70.8} + \frac{(135 - 106.2)^2}{106.2}$$

$$\approx 38.5192$$

Step 4   **State the conclusion and the interpretation.** Our χ^2_{data} of 38.5192 is greater than our χ^2_{crit} of 5.991 (see Figure 16), and so we reject H_0. The interpretation is: "There is evidence at level of significance $\alpha = 0.05$ that *age group* and *response* are dependent."

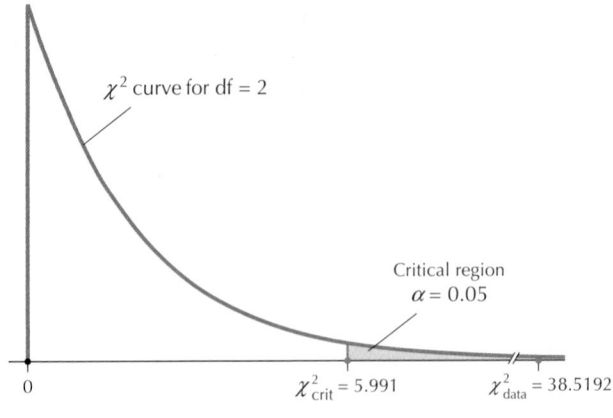

FIGURE 16
$\chi^2_{data} = 38.5192$ lies in the critical region.

NOW YOU CAN DO
Exercises 11–14.

χ^2 Test for Independence: p-Value Method

Step 1   State the hypotheses and the rejection rule. Check the conditions.

   H_0 : Variable A and Variable B are independent.
   H_a : Variable A and Variable B are dependent.

Reject H_0 if the p-value $\leq \alpha$.
   The following conditions must be met:

a.   None of the expected frequencies is less than 1.

b.   At most, 20% of the expected frequencies are less than 5.

The expected frequency for a given cell is

$$\text{expected frequency} = \frac{(\text{row total})(\text{column total})}{\text{grand total}}$$

Step 2   Calculate χ^2_{data}.

$$\chi^2_{data} = \sum \frac{(O_i - E_i)^2}{E_i}$$

where O_i = observed frequency and E_i = expected frequency for each cell.

Step 3   Find the p-value.

$$p\text{-value} = P(\chi^2 > \chi^2_{data})$$

Step 4   State the conclusion and the interpretation. Compare the p-value with α.

EXAMPLE 8 χ^2 test for independence using the p-value method and technology

 youngliving

The National Center for Health Statistics publishes information on the living arrangements of America's young people. Table 9 contains a random sample of 200 young people ages 1–24, indicating their gender and living arrangements. Test whether gender and living arrangement are independent, using the TI-83/84, Minitab, JMP, the p-value method, and level of significance $\alpha = 0.10$.

Table 9 Contingency table of living arrangements versus gender

Gender	Living arrangements			Total
	Living with parents	**Living with partner**	**All other arrangements**	
Female	51	22	28	101
Male	58	14	27	99
Total	109	36	55	200

Solution

Step 1 **State the hypotheses and the rejection rule. Check the conditions.**

H_0 : Gender and living arrangements are independent.

H_a : Gender and living arrangements are dependent.

Reject H_0 if the p-value ≤ 0.10.

Note that Minitab provides the expected counts (frequencies) below the observed counts. We can then verify that none of the expected frequencies is less than 1, and that none of the expected frequencies has a value less than 5.

Step 2 **Calculate χ^2_{data}.** We use the instructions found in the Step-by-Step Technology Guide at the end of this section. The TI-83/84 results in Figure 17 tell us that $\chi^2_{\text{data}} = 2.225723453$. The Minitab results in Figure 18 round this to "Pearson Chi-Square" $= \chi^2_{\text{data}} = 2.226$. The JMP results in Figure 19 ("Pearson") also round this to "ChiSquare" $= \chi^2_{\text{data}} = 2.226$.

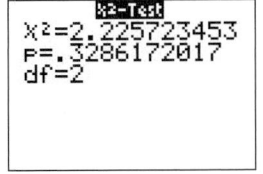

FIGURE 17 TI-83/84 χ^2 results.

```
Rows: Gender    Columns: Worksheet columns

               With       With
             Parents   Partner   Other   All

Female           51        22      28    101
              55.05     18.18   27.77

Male             58        14      27     99
              53.95     17.82   27.23

All             109        36      55    200

Cell Contents:        Count
                      Expected count


Pearson Chi-Square = 2.226, DF = 2, P-Value = 0.329
Likelihood Ratio Chi-Square = 2.241, DF = 2, P-Value = 0.326
```

FIGURE 18 Minitab χ^2 results.

Tests

N	DF	-LogLike	RSquare (U)
200	2	1.1203699	0.0056

Test	ChiSquare	Prob>ChiSq
Likelihood Ratio	2.241	0.3262
Pearson	2.226	0.3286

FIGURE 19 JMP χ^2 results.

Step 3 **Find the p-value.** From the TI-83/84 results in Figure 17, we have

$$p\text{-value} = P(\chi^2 > \chi^2_{\text{data}}) = 0.3286172017 \approx 0.329.$$

Step 4 **State the conclusion and the interpretation.** Because p-value ≈ 0.329 is not less than level of significance 0.10, we do not reject H_0. There is insufficient evidence that gender and living arrangements are dependent.

NOW YOU CAN DO
Exercises 15–18.

3 χ^2 Test for the Homogeneity of Proportions

Recall the two-sample Z test for $p_1 - p_2$ from Section 10.3, where we compared the proportions of two independent populations. When we extend that hypothesis test to k independent populations, we use a test statistic that follows a χ^2 distribution. Just as the null hypothesis for the two-sample test assumed no difference between the population proportions

- the null hypothesis for the k-sample test also assumes that all k proportions are equal, and

- the alternative hypothesis states that not all the population proportions are equal.

When performing the **test for the homogeneity of proportions**, we use the same steps as for the χ^2 test for independence.

**Developing Your
Statistical Sense**

Difference Between χ^2 Test for Homogeneity and χ^2 Test for Independence

The difference between the test for homogeneity of proportions and the test for independence has to do with how the data are collected. If a single sample is taken and two variables are measured, then the test for independence is appropriate.

If several (k) samples are taken and the sample proportion is measured for each sample, then the test for homogeneity of proportions is appropriate.

EXAMPLE 9 χ^2 Test for the homogeneity of proportions

The American Academy of Pediatrics recommends that children's TV-watching time be limited to two hours or less per day. Here, we examine whether a relationship exists between watching TV for more than two hours per day and being overweight. The National Center for Health Statistics conducted a survey of children 12–15 years old. Three random samples were taken, one sample of normal or underweight children, one sample of overweight children, and one sample of obese children. The surveys noted whether the children watched TV more than two hours per day. The results are shown in Table 10.

Test whether the population proportions of children watching more than two hours per day of TV are the same for the three weight statuses, using the *p*-value method, Minitab, and level of significance $\alpha = 0.05$.

tvandweight

Table 10   Numbers watching more than two hours of TV, for three weight statuses

	Normal or underweight	Overweight	Obese	Total
Number watching more than two hours of TV	140	44	82	266
Number watching two hours or less of TV	329	80	91	500
Total	469	124	173	766

Solution

The Minitab results are shown here in Figure 20.

FIGURE 20
Minitab results for the
χ^2 test for homogeneity.

```
Rows: C14    Columns: Worksheet columns

             Normal  Overweight   Obese  All

>2 Hours        140          44      82  266
             162.86       43.06   60.08

<=2 Hours       329          80      91  500
             306.14       80.94  112.92

All             469         124     173  766

Cell Contents:       Count
                     Expected count



Pearson Chi-Square = 17.207, DF = 2, P-Value = 0.000
Likelihood Ratio Chi-Square = 16.806, DF = 2, P-Value = 0.000
```

We use the same steps as for the χ^2 test for independence.

Step 1 **State the hypotheses and the rejection rule. Check the conditions.**

$H_0: p_{\text{Normal /Under}} = p_{\text{Overweight}} = p_{\text{Obese}}$

$H_a:$ Not all the proportions in H_0 are equal.

Reject H_0 if the p-value ≤ 0.05.

The expected frequencies are shown in Figure 20. None of them are less than either 1 or 5. Therefore, the conditions are met, and we may proceed with the hypothesis test.

Note: The conditions and the test statistic for the test for the homogeneity of proportions are the same as for the test for independence.

Step 2 **Find the test statistic χ^2_{data}.** χ^2_{data} is shown as "Pearson Chi-Square = 17.207." There are $r = 2$ rows and $c = 3$ columns, so the degrees of freedom are $(r - 1)(c - 1) = (2 - 1)(3 - 1) = 2$.

Step 3 **Find the p-value.** Minitab provides the p-value, which is essentially 0.000.

Step 4 **State the conclusion and the interpretation.** The p-value of 0.000 is less than $\alpha = 0.05$. We therefore reject H_0. Evidence exists, at level of significance $\alpha = 0.05$, that not all population proportions of watching TV more than two hours per day are equal.

NOW YOU CAN DO
Exercises 19–26.

Susan Wides/Getty Images

onlinedating

Online Dating

CASE STUDY

We look at two tests for independence in this Case Study. The first examines whether the type of relationship reported by respondents depends on the gender of the respondent. The second investigates whether the self-reported physical appearance of online daters depends on the person's gender.

Does the Reported Type of Relationship Depend on Gender?

The Pew Internet and American Life Project examined whether single men and women differed with respect to their current relationships. The observed frequencies are given in Table 11.

Table 11 Observed frequencies, online dating study

| | Gender | |
Type of relationship	Single men	Single women
In committed relationship	115	138
Not in committed relationship and not looking for partner	162	391
Not in committed relationship but looking for partner	89	54
Don't know/refused	19	18

We are interested in whether the type of relationship reported depends on the gender of the respondent. In other words, we will test whether the type of relationship is independent of gender. We will use the p-value method, with level of significance $\alpha = 0.05$, and we will follow the TI-83/84 instructions in the Step-by-Step Technology Guide on page 656 for the calculations.

What Results Might We Expect?

Table 11 and Figure 21 indicate that the proportion of men who are "looking" is greater than the proportion of women who are "looking." Similarly, the proportion of women who are "not looking" is greater than for men. This is evidence that the type of relationship depends on gender and that we might expect to reject the null hypothesis of independence.

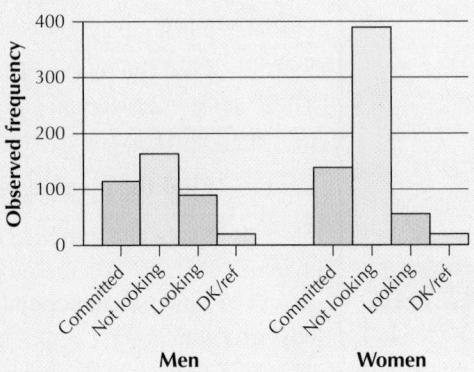

FIGURE 21 Graphical evidence indicates type of relationship depends on gender.

Step 1 **State the hypotheses and the rejection rule. Check the conditions.**

H_0 : *Type of relationship* and *gender* are independent.

H_a : *Type of relationship* and *gender* are dependent.

Reject H_0 if the p-value ≤ 0.05.

Figure 22 shows the expected frequencies, none of which are less than 5. Thus, the conditions are met.

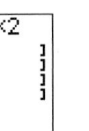

FIGURE 22 Expected frequencies.

Step 2 **Find χ^2_{data}.** The TI-83/84 results in Figure 23 tell us

$$\chi^2_{\text{data}} = 61.12955651$$

Step 3 **Find the p-value.** Figure 23 also gives us the p-value:

$$p\text{-value} = 3.372011\text{E-}13 \approx 0.0000000000003372011$$

FIGURE 23 χ^2 results on TI-83/84.

Step 4 **State the conclusion and the interpretation.** The p-value $\leq \alpha = 0.05$, so we reject H_0, as we expected. There is evidence that the type of relationship reported in the study depends on the gender of the respondent for level of significance $\alpha = 0.05$.

Does Self-Reported Physical Appearance of Online Daters Depend on Gender?

onlineappear

A Master's thesis from the Massachusetts Institute of Technology examined the characteristics and behavior of online daters.[6] Table 12 contains the self-reported physical appearance and gender of 52,817 users of an online dating service.

Table 12 Gender and self-reported physical appearance

| | Physical appearance | | | | |
	Very attractive	Attractive	Average	Prefer not to answer	Total
Female	3,113	16,181	6,093	3,478	28,865
Male	1,415	12,454	7,274	2,809	23,952
Total	4,528	28,635	13,367	6,287	52,817

Note from Table 12 that females seem to have higher proportions of those self-reporting as either attractive or very attractive, whereas males seem to have a higher proportion of those self-reporting as average. This is evidence that self-reported physical appearance does depend on gender and that we might expect to reject the null hypothesis of independence. We will test with the p-value method, using level of significance $\alpha = 0.01$, and Minitab. The hypotheses are

H_0 : *Self-reported physical appearance* and *gender* are independent.

H_a : *Self-reported physical appearance* and *gender* are dependent.

We reject H_0 if the p-value $\leq$ level of significance $\alpha = 0.01$.

The Minitab results in Figure 24 tell us

$$\chi^2_{\text{data}} = \text{``Chi-Sq''} = 847.702$$
$$p\text{-value} \approx 0$$

Figure 24 gives us the expected frequencies (highlighted in color), none of which are less than 5, allowing us to perform the hypothesis test. The p-value $\leq \alpha = 0.01$, so we reject H_0, as we expected. There is evidence at level of significance $\alpha = 0.01$ that the self-reported physical appearance depends on the gender of the online dater.

```
Expected counts are printed below observed counts
Chi-Square contributions are printed below expected counts

            VA        Att        Ave       PNTA   Total
     F     3113      16181       6093       3478   28865
        2474.60   15649.30    7305.19    3435.91
        164.698     18.065    201.147      0.516

     M     1415      12454       7274       2809   23952
        2053.40   12985.70    6061.81    2851.09
        198.480     21.770    242.406      0.621
 Total     4528      28635      13367       6287   52817

Chi-Sq = 847.702, DF = 3, P-Value = 0.000
```

FIGURE 24 Minitab results showing expected frequencies, χ^2_{data}, and the p-value.

STEP-BY-STEP TECHNOLOGY GUIDE: Test for Independence or Test for the Homogeneity of Proportions

We demonstrate using Example 8 (page 650).

TI-83/84

Entering Matrix Data
Step 1 Press **2nd**, then **MATRIX**.
Step 2 Highlight **EDIT**, and press **ENTER**.
Step 3 Set the dimensions of **MATRIX[A]** (number of rows × number of columns). Table 9 has 2 rows and 3 columns, so enter 2, press **ENTER**, enter 3, and press **ENTER**.
Step 4 Enter the first number in the first cell, **51**, and press **ENTER**.
Step 5 Continue entering the data row by row until the matrix is complete (Figure 25).

χ^2 **Test for Independence or Test for Homogeneity of Proportions**
Step 1 Enter the data into **Matrix[A]**.
Step 2 Press **STAT**, highlight **TESTS**, select χ^2 **Test**, and press **ENTER**.

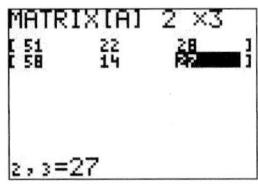

FIGURE 25

Step 3 Make sure that the Observed frequencies are put into **Matrix [A]**. The expected frequencies are automatically generated and put into **Matrix[B]**. Highlight **Calculate**, and press **ENTER**. The results are shown in Figure 17 in Example 8.
Step 4 To view the expected frequencies, press **2nd MATRIX**, highlight **EDIT**, and choose 2 for **Matrix[B]**.

EXCEL

Step 1 Enter the data from Table 9, *including row and column totals*, in cells A1 to D3.

Step 2 Use the JMP plug-in for Excel.

MINITAB

Step 1 Enter the observed frequencies from Table 9 into the Minitab worksheet, as shown here.
Step 2 Click **Stat > Tables > Chi-Square Test for Association....**
Step 3 Select **Summarized data in a two-way table**, then choose each of columns C2, C3, and C4 as the **Columns** containing the table.
Step 4 For **Labels for the table** (optional), select **Gender** for the **Rows** input. Then click **OK**. The results are shown in Figure 18 in Example 8.

C1-T	C2	C3	C4
Gender	With Parents	With Partner	Other
Female	51	22	28
Male	58	14	27

SPSS

Step 1 Enter the data as shown in Figure 26.

Arrangement	Gender	Count
With Parents	Female	51.00
With Parents	Male	58.00
With Partner	Female	22.00
With Partner	Male	14.00
Other	Female	28.00
Other	Male	27.00

FIGURE 26 SPSS data input.

Step 2 Click **Data > Weight Cases...**, select **Weight cases by**, and move the variable **Count** to **Frequency Variable**. Click **OK**.
Step 3 Click **Analyze > Descriptive Statistics > Crosstabs....** Move the variable **Gender** to **Row(s)**, and move **Arrangement** to **Column(s)**.

Step 4 Click **Statistics...**, and select **Chi-square**. Click **Continue**, then **OK**. The output is shown in Figure 27.

Chi-Square Tests

	Value	df	Asymp. Sig. (2-sided)
Pearson Chi-Square	2.226[a]	2	.329
Likelihood Ratio	2.241	2	.326
N of Valid Cases	200		

a. 0 cells (0.0%) have expected count less than 5. The minimum expected count is 17.82.

FIGURE 27 SPSS output.

JMP

Step 1 Click **File > New > Data Table**. Input the data in a similar way to Figure 26.
Step 2 Click **Analyze > Fit Y by X**. Move **Arrangement** to **Y, Response**, **Gender** to **X, Factor**, and **Count** to **Freq**. Click **OK**.

Step 3 The output is under the **Tests** heading, in the **Pearson** row. See Figure 19 in Example 8.

CRUNCHIT!

Step 1   Click File, highlight Load from Larose, Discostat3e > Chapter 11, and click on Example 11_08.
Step 2   Click Statistics and select Contingency table > with counts. For Row Variable select Gender. For Column Variable

select Arrangement. For Counts select Count. Then click Calculate.

Section 11.2 Summary

1. To determine whether two categorical variables are independent, using the data from a contingency table, we use a χ^2 test for independence. The hypotheses take the form
   H_0 : Variable A and Variable B are independent.
   H_a : Variable A and Variable B are dependent.
2. The χ^2 test for independence is performed using the critical-value method and the p-value method. The observed frequencies are compared with the expected frequencies on the assumption that H_0 is correct. Large differences lead to the rejection of the null hypothesis.

3. The k-sample test, called the test for the homogeneity of proportions, determines whether all k population proportions are equal. The result uses a test statistic that follows a χ^2 distribution. The null hypothesis for the k-sample test assumes that all k population proportions are equal. The alternative hypothesis states that not all the population proportions are equal. When performing the test for the homogeneity of proportions, the same steps are used as for the χ^2 test for independence.

Section 11.2 Exercises

CLARIFYING THE CONCEPTS

1. Explain what a contingency table is. (p. 646)
2. Explain in your own words what is meant by a test for independence. (p. 646)
3. What is the difference between the χ^2 test for homogeneity of proportions and the two-sample Z test for the difference in proportions from Chapter 10? (p. 651)
4. Explain how the expected frequencies are calculated without using the shortcut method. (p. 647)

PRACTICING THE TECHNIQUES

 CHECK IT OUT!

To do	Check out	Topic
Exercises 5–10	Example 6	Calculating expected frequencies
Exercises 11–14	Example 7	χ^2 test for independence: critical-value method
Exercises 15–18	Example 8	χ^2 test for independence: p-value method
Exercises 19–26	Example 9	χ^2 test for homogeneity of proportions

For Exercises 5–10, the observed frequencies are provided in a contingency table of two categorical variables. Find the expected frequencies, on the assumption that the variables are independent.

5.

	A1	A2
B1	10	20
B2	12	18

6.

	C1	C2
D1	50	100
D2	60	90

7.

	E1	E2	E3
F1	30	20	10
F2	35	24	8

8.

	G1	G2
H1	10	8
H2	8	10
H3	9	9

9.

	I1	I2	I3
J1	100	90	105
J2	50	60	55
J3	25	15	20

10.

	K1	K2	K3	K4
L1	40	70	90	100
L2	20	40	60	70
L3	30	65	65	70

For Exercises 11–14, test whether or not the variables are independent.
  a. State the hypotheses.
  b. Verify that the conditions for performing the χ^2 test for independence are met.
  c. Find χ^2_{crit} and state the rejection rule.
  d. Calculate χ^2_{data}.
  e. Compare χ^2_{data} with χ^2_{crit}. State the conclusion and the interpretation.

11. Exercise 5, level of significance $\alpha = 0.05$
12. Exercise 7, level of significance $\alpha = 0.10$
13. Exercise 9, level of significance $\alpha = 0.01$
14. Exercise 9, level of significance $\alpha = 0.10$

For Exercises 15–18, test whether or not the variables are independent.
  a. State the hypotheses and the rejection rule for the p-value method, and verify that the conditions for performing the χ^2 test for independence are met.
  b. Find χ^2_{data}.
  c. Calculate the p-value.
  d. Compare the p-value with α. State the conclusion and the interpretation.

15. Exercise 6, level of significance $\alpha = 0.05$
16. Exercise 8, level of significance $\alpha = 0.10$
17. Exercise 10, level of significance $\alpha = 0.01$
18. Exercise 10, level of significance $\alpha = 0.10$

For Exercises 19–22, test whether or not the proportions of successes are the same for all populations.
  a. State the hypotheses.
  b. Calculate the expected frequencies and verify that the conditions for performing the χ^2 test for homogeneity of proportions are met.
  c. Find χ^2_{crit} and state the rejection rule. Use level of significance $\alpha = 0.05$.
  d. Find χ^2_{data}.
  e. Compare χ^2_{data} with χ^2_{crit}. State the conclusion and the interpretation.

19.

	Sample 1	Sample 2	Sample 3
Successes	10	20	30
Failures	20	45	62

20.

	Sample 1	Sample 2	Sample 3
Successes	50	50	100
Failures	200	210	425

21.

	Sample 1	Sample 2	Sample 3	Sample 4
Successes	10	15	20	25
Failures	15	24	32	40

22.

	Sample 1	Sample 2	Sample 3	Sample 4
Successes	100	150	200	250
Failures	150	240	320	400

For Exercises 23–26, test whether or not the proportions of successes are the same for all populations.
  a. State the rejection rule for the p-value method using level of significance $\alpha = 0.05$, calculate the expected frequencies, and verify that the conditions for performing the χ^2 test for homogeneity of proportions are met.
  b. Find χ^2_{data}.
  c. Calculate the p-value.
  d. Compare the p-value with α. State the conclusion and the interpretation.

23.

	Sample 1	Sample 2	Sample 3
Successes	30	60	90
Failures	10	25	50

24.

	Sample 1	Sample 2	Sample 3
Successes	100	120	140
Failures	20	25	30

25.

	Sample 1	Sample 2	Sample 3	Sample 4
Successes	10	12	24	32
Failures	6	10	15	30

26.

	Sample 1	Sample 2	Sample 3	Sample 4
Successes	100	200	300	400
Failures	30	70	150	300

APPLYING THE CONCEPTS

27. **Email, Phone, or in Person?** What is the most effective way to handle a task at work: by email, by phone, or in person? Well, you probably say, it depends on the task. The Pew Internet and American Life Project Email at Work Survey surveyed 1000 randomly selected work email users, who chose the following methods as the best for handling certain work tasks. Test whether the proportions who favor email differ between the two tasks, using level of significance $\alpha = 0.05$ and the p-value method. **worktask**

Task	By email	By phone or in person
Edit or review documents	670	330
Arrange meetings or appointments	630	370

28. **Computer Usage and Weight in Children.** The National Center for Health Statistics conducted a survey of children 12–15 years old. Three random samples were taken, one sample of normal or underweight children, one sample of overweight children, and one sample of obese children. The surveys noted whether the children used a computer for more than two hours per day. The results are presented in the following table. Test whether the population proportions of children who use the computer for more than two hours per

day are the same for the three weight statuses, using level of significance $\alpha = 0.05$. 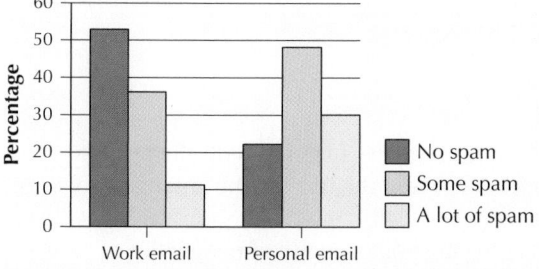 computerweight

	Normal or underweight	Overweight	Obese	Total
Using computer more than two hours per day	114	28	52	194
Using computer two hours or less per day	355	96	121	572
Total	469	124	173	766

29. Weather-Related Deaths. The Centers for Disease Control track the numbers of deaths due to weather-related causes. Is there is a difference in the pattern of deaths for young people and older people? The following table shows the number of deaths for three weather-related categories, for young people ages 15–24 and older people ages 75–84. Test, using level of significance $\alpha = 0.05$, whether cause of death and age group are independent. weatherdeaths

Weather-related cause of death

Age group	Heat-related	Cold-related	Floods/ storms/ lightning	Total
15–24	106	286	97	489
75–84	490	1010	53	1553
Total	596	1296	150	2042

30. Using Graphical Evidence. Sick of spam (unsolicited broadcast email)? Do you get more spam at your work, school, or home email address? The Pew Internet and American Life Project Email at Work Survey examined the proportion of spam in email users' work and home email accounts. Two random samples were used, one of work email and one of personal email. Using only the information in the clustered bar graph below, would you conclude that the proportion of those who report "a lot of spam" is the same for work email and personal email? Why?

31. Spam, Spam, Spam. Continue your work from the previous exercise. The following contingency table shows the actual percentages in the graph above based on samples of size 100 for each of work email and personal email. Test whether the proportions who report "a lot of spam" are the

same for work email and personal email, using level of significance $\alpha = 0.01$. Does your conclusion agree with your conjecture in the previous exercise?

	None	Some	A lot
Work email	53%	36%	11%
Personal email	22%	48%	30%

32. Gender Differences in Computer/Video/Online Gaming. The Pew Internet and American Life Project collected data on the College Students Gaming Survey. Among the questions they asked 1720 randomly selected college students was "Which one of the following do you play the most: video games, computer games, or online games?" The results are summarized by gender in the following contingency table. games

	Video games	Computer games	Internet games
Male	616	221	139
Female	198	372	174

a. Before you perform the hypothesis test, *what result might you expect*? Look over the data set carefully to see whether you can detect significant differences between the levels of the variables. Then see whether your hypothesis test bears out your intuition.
b. Test whether *gender* and *game type* are independent, using level of significance $\alpha = 0.01$.

33. Online Dating. A Pew Internet and American Life Project study reported that the proportion of urban residents who use online dating is 13%, whereas the proportion for suburban residents is 10% and the proportion for rural residents is 9%.[7] Test, using level of significance $\alpha = 0.05$, whether differences exist among the population proportions of residents from the three categories who use online dating. Assume that each sample size was 1000. (*Hint:* The null hypothesis assumes that all proportions are equal.)

WORKING WITH LARGE DATA SETS

Use Minitab or Excel for each of Exercises 34–38.

Goals of Middle School Students. Open the **Goals** data set. The subjects are students in grades 4, 5, and 6, from three school districts in Michigan. The students were asked which of the following was most important to them: good grades, athletic ability, or popularity. Information about the students' age, gender, race, and grade was also gathered, as well as whether their school was in an urban, suburban, or rural setting.[8] goals

34. How many observations are in the data set? How many variables?

35. Comparing gender and goals.
a. Looking at the data, do you think that boys and girls at this age differ in what is most important to them: grades, popularity, or sports? In other words, do you think that the variables *gender* and *goals* are dependent or independent?

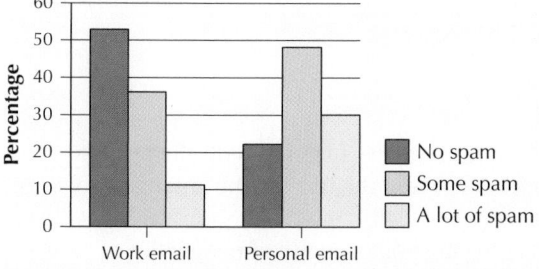

b. Perform the χ^2 test for independence, using level of significance $\alpha = 0.05$.

36. **Comparing gender and grade.**
   a. Looking at the data, do you think that the ratio of females to males differs significantly from grade to grade? In other words, do you think that the variables *gender* and *grade* are dependent or independent?
   b. Perform the χ^2 test for independence, using level of significance $\alpha = 0.05$.

37. **Comparing goals and school setting.**
   a. Looking at the data, do you think that the setting of the school (urban, suburban, or rural) affects the goals of the students? Or do you think that it has no effect? In other words, do you think that the variables *urb_rur* and *goals* are independent or dependent?
   b. Perform the χ^2 test for independence, using level of significance $\alpha = 0.10$.

38. **Comparing grades and goals.**
   a. One thing we know for sure is that, as students get older, they get more serious and grades get more important to them (don't they?). So we would expect that the variables *grade* and *goals* would be dependent, wouldn't we? Is this borne out by looking at the data?
   b. Perform the χ^2 test for independence, using level of significance $\alpha = 0.01$.

39. **1970 Military Draft.** Is there evidence that the 1970 military draft, conducted at the height of the Vietnam War, was not truly random? For this exercise, birth dates were ranked from 1 (for the first date drawn) to 366 (the last date drawn). In 1970, only those young men with birth date rankings up to 195 were eventually drafted. Because 195 of the 366 dates were "drafted," the overall proportion of "drafted dates" is $195/366 \approx 0.5328$. Assuming the draft was truly random, we do not expect the proportion of "drafted dates" to vary significantly from month to month. In other words, the proportion of "drafted dates" should be about the same for each of the 12 months. We therefore define a multinomial random variable *drafted,* with the $k = 12$ months as categories. The monthly counts of dates not drafted and drafted are provided here. (For example, for April, 12 dates out of 30 were chosen to be drafted.) Test whether the proportions of "drafted dates" are equal for all months, using level of significance $\alpha = 0.01$. **1970draft**

Month	Dates not drafted	Dates drafted	All
Jan.	17	14	31
Feb.	16	13	29
Mar.	21	10	31
Apr.	18	12	30
May	17	14	31
June	16	14	30
July	13	18	31
Aug.	12	19	31
Sept.	11	19	30
Oct.	17	14	31
Nov.	8	22	30
Dec.	5	26	31
All	171	195	366

40. **1971 Military Draft.** Criticism of the 1970 draft lottery led the U.S. Selective Service Bureau to focus on making sure that the 1971 draft lottery was truly random. Were their efforts successful? The results of the 1971 draft lottery are shown here (365 days). The Selective Service reports that all birth dates with a rank of 125 or less were chosen for the draft. Perform a χ^2 test for homogeneity of proportions to determine whether the population proportions of "drafted dates" per month were all equal, using level of significance $\alpha = 0.10$. **1971draft**

Month	Dates not drafted	Dates drafted
Jan.	19	12
Feb.	19	9
Mar.	21	10
Apr.	21	9
May	22	9
June	21	9
July	19	12
Aug.	18	13
Sept.	23	7
Oct.	19	12
Nov.	16	14
Dec.	22	9

Chapter 11 Formulas and Vocabulary

SECTION 11.1
- **Conditions for performing a goodness of fit test** (p. 636)
- χ^2 **Goodness of fit test** (p. 634)
- **Multinomial random variable** (p. 632)
- **Test statistic for the goodness of fit test** (p. 635).

$$\chi^2_{\text{data}} = \sum \frac{(O_i - E_i)^2}{E_i}$$

SECTION 11.2
- χ^2 **test for independence** (p. 647)

- **Conditions for performing both the test for independence and the test for the homogeneity of proportions** (p. 648)
- **Test for the homogeneity of proportions** (p. 651)
- **Test statistic for both the test for independence and the test for the homogeneity of proportions** (p. 648).

$$\chi^2_{\text{data}} = \sum \frac{(O_i - E_i)^2}{E_i}$$

Chapter 11 Review Exercises

SECTION 11.1

For Exercises 1–4, perform the χ^2 goodness of fit test.

1. Truck-Hauled Trade. According to the U.S. Census Bureau, 32% of North American international truck-hauled trade (in dollars) goes from the United States to Canada, 22% goes from the United States to Mexico, 31% goes from Canada to the United States, and 15% goes from Mexico to the United States. Suppose that a new survey showed that $25 billion went from the United States to Canada, $15 billion went from the United States to Mexico, $20 billion went from Canada to the United States, and $10 billion went from Mexico to the United States. Test whether the population proportions of truck-hauled trade have changed, using level of significance $\alpha = 0.05$.

2. Alcohol Abuse and Dependence in College. A report found that 25% of college students had abused alcohol in the last 12 months, whereas an additional 6% (not counted in the 25%) were alcohol-dependent.[9] Suppose that a new survey of 1000 randomly selected college students finds 275 who had abused alcohol in the last 12 months and an additional 50 (not counted in the 275) who are alcohol-dependent. Test whether the population proportions have changed, using level of significance $\alpha = 0.10$.

3. Truly Random Lottery Drawing? Have you ever wondered whether lottery drawings are truly random? For example, the accompanying histogram shows the frequencies of the third digit in the Maryland lottery's Pick 3 game (218 drawings from September 1989 to April 1990). In a Pick 3 game, you choose a three-digit number between 000 and 999, and if your number comes up, you win the cash prize. Notice that 1 appears as the third digit least of all the digits, and quite a bit less often than some of the other digits. Does the relative scarcity of 1s indicate that the system is flawed?

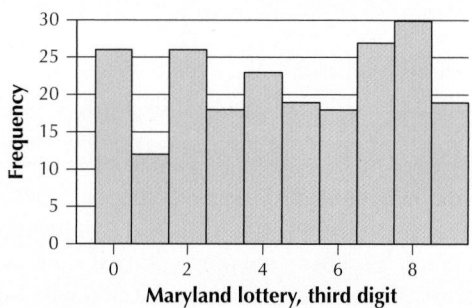

Frequency Histogram of third digits in Maryland lottery's Pick 3 game.

The relative frequency distribution of the third digit is shown in the following table. We would, of course, expect each digit to show up 10% of the time. Test whether the population proportions of digits are all 0.10, using level of significance $\alpha = 0.05$. **marylandlott**

Digit	Count	Percent
0	26	11.93
1	12	5.50
2	26	11.93
3	18	8.26
4	23	10.55
5	19	8.72
6	18	8.26
7	27	12.39
8	30	13.76
9	19	8.72
$N =$	218	

4. Alternative Medicine Use. A study examined the prevalence of alternative medicine usage by age group among persons with diabetes.[10] In the study, 5.7% of the subjects were ages 18–34 years, 20.7% were ages 35–49 years, 38.8% were ages 50–64 years, and 34.8% were age 65 or older. Suppose that a study conducted this year found that, of the 1000 randomly selected respondents with diabetes, 70 were 18–34 years old, 220 were 35–49 years old, 440 were 50–64 years old, and 270 were over age 65. Test, using level of significance $\alpha = 0.05$, whether the proportions have changed.

SECTION 11.2

5. Grades and the SAT. In its "Profile of College-Bound Seniors," the College Board provided the following data on high school grade point average and gender for the students taking the SAT exam. We are interested in testing whether the proportion of females is the same across the six grade categories. **highschoolgpa**

	High school grade point average					
Gender	A+	A	A−	B	C	D–F
Female	60	62	59	53	43	43
Male	40	38	41	47	57	57

a. Before you perform any calculations, *what result might we expect?* Examine the table carefully to see whether you can identify any differences in the proportions of females.

b. Test whether the proportion of females is the same across the six grade categories, using level of significance $\alpha = 0.05$.

6. Pregnancy and HIV Testing. A study examined the proportions of pregnant women in the United States who have had an HIV test in the past 12 months.[11] The proportions for the Northeast, Midwest, South, and West, resulting from four separate samples, were 56.8%, 49.3%, 58.5%, and 50.2%, respectively. Test whether the population proportions of pregnant women who have had an HIV test in the past 12 months are the same across all four regions, using level of significance $\alpha = 0.01$. Assume that each sample size equals 1000.

7. Radio-Listening Trends. The Arbitron Corporation tracks radio-listening trends among demographic groups. In a survey, 240 teens ages 12–17 listened to pop contemporary hit radio stations, whereas 170 teens listened to alternative radio stations. Also, 250 young adults ages 18–24 listened to pop contemporary hit radio stations, whereas 260 young adults listened to alternative radio stations. Test whether age and radio station type are independent, using level of significance $\alpha = 0.05$.

8. Happiness in Marriage. The General Social Survey tracks trends in American society. The accompanying crosstabulation shows the responses to a question that asked people to characterize their feelings about being married. Test whether happiness in marriage is independent of gender, using level of significance $\alpha = 0.05$.

🎴 **happymarriage**

Respondents' gender	Happiness in marriage			Total
	Very happy	Pretty happy	Not too happy	
Male	242	115	9	366
Female	257	149	17	423
Total	499	264	26	789

Chapter 11 QUIZ

TRUE OR FALSE

1. True or false: In a goodness of fit test, large differences between the observed frequencies and the expected frequencies lead to rejection of the null hypothesis.

2. True or false: In a test for independence, the degrees of freedom equals $k - 1$.

3. True or false: In the test for the homogeneity of proportions, the alternative hypothesis states that all the population proportions are different.

FILL IN THE BLANK

4. The conditions for performing a goodness of fit test are that none of the expected frequencies is less than _____, and, at most, 20% of the expected frequencies are less than _____.

5. In the test for the homogeneity of proportions, the null hypothesis states that all k population proportions are _____.

6. The _____ _____ [two words] of the ith category is given by the formula $n \cdot p_i$, where n is the number of trials and p_i is the population proportion for the ith category.

SHORT ANSWER

7. Name the two methods for performing the χ^2 goodness of fit test.

8. In the test for the homogeneity of proportions, which hypothesis states that not all population proportions are equal?

9. How does one calculate the degrees of freedom for the χ^2 test for independence?

CALCULATIONS AND INTERPRETATIONS

For Exercises 10–13, perform the χ^2 goodness of fit test, where the alternative hypothesis takes the following form:

> H_a: The random variable does not follow the distribution specified in H_0.

10. $H_0 : p_1 = 0.2, p_2 = 0.2, p_3 = 0.2, p_4 = 0.2, p_5 = 0.2$; $O_1 = 8, O_2 = 9, O_3 = 10, O_4 = 11, O_5 = 12; \alpha = 0.05$

11. $H_0 : p_1 = 0.3, p_2 = 0.25, p_3 = 0.20, p_4 = 0.15$, $p_5 = 0.06, p_6 = 0.04; O_1 = 50, O_2 = 40, O_3 = 30, O_4 = 20$, $O_5 = 10, O_6 = 10; \alpha = 0.05$

12. $H_0 : p_1 = 0.2, p_2 = 0.2, p_3 = 0.2, p_4 = 0.2, p_5 = 0.2$; $O_1 = 18, O_2 = 19, O_3 = 21, O_4 = 22, O_5 = 20; \alpha = 0.01$

13. $H_0 : p_1 = 0.3, p_2 = 0.25, p_3 = 0.20, p_4 = 0.15$, $p_5 = 0.06, p_6 = 0.04; O_1 = 65, O_2 = 55, O_3 = 30, O_4 = 25$, $O_5 = 15, O_6 = 10; \alpha = 0.05$

14. Illicit Drug Use Among Young People. Monitoring the Future (www.monitoringthefuture.org), at the University of Michigan, is an "an ongoing study of the behaviors, attitudes, and values of American secondary school students, college students, and young adults." They reported the lifetime prevalence of the use of any illicit drug among 8th-graders, 10th-graders, and 12th-graders, as shown in the table. 🎴 **druguse**

	8th-graders	10th-graders	12th-graders
Have used an illicit drug	3,655	6,527	7,461
Have never used an illicit drug	13,345	9,873	7,139

a. Before you perform the hypothesis test, *what result might you expect*? Look over the data set carefully to see whether you can detect significant differences between the levels of the variables. Then see whether your hypothesis test bears out your intuition.

b. Test, using level of significance $\alpha = 0.01$, for differences among the proportions of children in those grades who have ever used an illicit drug.

15. Sport Preference and Gender. A student group wants to determine whether or not gender and participatory sport preference are independent on campus. A survey of 200 randomly selected college students showed the following sport preferences. Test, at level of significance $\alpha = 0.05$, whether gender and sport preference are independent.

🎴 **sportsgender**

	Sport preference			
Gender	Basketball	Soccer	Swimming	Total
Female	30	20	50	100
Male	50	30	20	100
Total	80	50	70	200

16. Beef Cattle and Farm Size. The National Agricultural Statistics Service publishes data on farm products in the United States.[12] The accompanying table shows the number of beef cattle on smaller-scale operations (farms having fewer than 50 head) for three states. Three separate samples were taken. Test whether the proportions of cattle on smaller farms are the same across all three states, using level of significance $\alpha = 0.05$. cattlefarm

	Texas	Oklahoma	Pennsylvania
Beef cattle on smaller scale operations	103,000	3,600	11,400
Beef cattle on operations that are not smaller scale	28,000	44,400	600

12

Analysis of Variance

Introduction

Chapter 12 discusses analysis of variance (ANOVA), where we determine whether significant differences exist among the population means of several groups. Section 12.1 describes how ANOVA works and shows how to perform one-way ANOVA. In Section 12.2, we learn multiple comparisons, where we test for pairwise differences in means. Section 12.3 introduces the randomized block design, which can account for nuisance variables. Finally, Section 12.4 explains two-way ANOVA, where two factors can be tested at the same time.

From the Author

Section 12.1 One-Way Analysis of Variance (ANOVA)

- Instructors may ask students to carefully read Objective 1, How ANOVA Works, to get an intuitive sense of how ANOVA works by comparing variation among the groups to variation within the groups.

- Instructors may wish to review the F distribution in Section 10.4 at some point.

Section 12.2 Multiple Comparisons

- The Bonferroni method and Tukey's method are provided. Instructors may choose which method to teach.

- Students may wonder why they need to learn this material. The answer is that, even though ANOVA may reject the null hypothesis, we cannot determine which pairs of means are significantly different without performing a multiple comparisons test.

- Some instructors do not have time to teach all the calculations in this section, and prefer to simply work with the Minitab output to determine which pairs of means are significantly different.

Section 12.3 Randomized Block Design

- Why should students learn this material? Because randomized block design increases the ability of the ANOVA to reject the null hypothesis by accounting for a nuisance factor (or blocking factor).

- Note that, in Example 10, one-way ANOVA is unable to find significance, whereas in Example 11, the randomized block design does find significance for the same data.

Section 12.4 Two-Way ANOVA

- Why should students learn this material? Because two-way ANOVA allows us to account for the effects of two factors simultaneously. For example, we could test whether both *major* and *gender* have a significant effect on student GPA.

- Instructors may stress that, if interaction between the factors exists, then students should not perform the tests for Factor A or Factor B effects.

Teaching Tips

Starting Example: Using analysis of variance, see whether different fraternities or sororities have different GPAs. Other similar examples could be to explore whether men and women athletes have different GPAs or whether athletes on different sports teams have different GPAs.

In-Class Activities

1. Ask students to collect information on the number of pennies, nickels, dimes, and quarters a person may have in a wallet or purse. Data can be collected from approximately 50 people. Students can create a table with four columns to record the data set. They can construct boxplots for the four columns and discuss what is being revealed, and then use a one-way ANOVA to test whether a significant difference exists in the averages for the four categories and whether these results support their observations from the boxplot. Also, students can use the appropriate technology to test the assumptions for a one-factor (in this case, coins) experiment.

2. Analysis of Variance for the Mean Height of Freshmen, Sophomores, Juniors, and Seniors

You are going to test whether differences exist in the population mean height of freshmen, sophomores, juniors, and seniors. If the frequency of any class is too small, either eliminate that class or combine it with another class. However, do not go below three classes.

    a. State the appropriate hypotheses, and define your notation.

     b. Generate a sample of four students from each class, using either (a) technology or (b) the *Simple Random Sample* applet.

    c. Find out the height of each student in the sample and express the height in inches. Then find the sample mean height for each class.

    d. The sample means you calculated in (**c**) were not all the same. Is this sufficient evidence to reject the null hypothesis that the population means are all the same? Why or why not?

    e. Using the level of significance of $\alpha = 0.10$, what is the critical value F_{crit} for this test? Will it be the same for each student?

    f. Perform the analysis of variance with your sample data. Use $\alpha = 0.10$. Will the value of the test statistic F_{data} be the same for each student?

    g. Record your name and the value for the test statistic F_{data} calculated in (**f**) on a sticky note.

h. Place your sticky note on the number line that the instructor has prepared on the board, noting the location of the cutoff value F_{crit}. You now have a collection of the values for the statistic F_{data}. Can you make any comment on the shape of the distribution of F_{data} values?

i. What proportion of students rejected the null hypothesis?

j. Reach a class consensus as to why most students either did or did not reject the null hypothesis.

Optional Topic: ANOVA: The Estimated *P*-Value Method

If a computer or calculator is not available, we may still estimate the *p*-value for a given value of F_{data}, using the F table. Suppose we are interested in testing whether the true mean recovery times of three surgical procedures are equal. Six patients for each procedure are randomly selected from last year's surgeries. Thus we have: $k = 3$, $n_t = 6 + 6 + 6 = 18$, $df_1 = k - 1 = 3 - 1 = 2$, and $df_2 = n_t - k = 18 - 3 = 15$. When the results are in, suppose we find that $F_{data} = 3.2$. Consider Figure A here, excerpted from the F table, and Figure B.

	Area in Right Tail	1	2	3
	0.100	3.18	2.81	2.61
	0.050	4.75	3.89	3.49
12	0.025	6.55	5.10	4.47
	0.010	9.33	6.93	5.95
	0.001	18.64	12.97	10.80
	0.100	3.07	2.70	2.49
	0.050	4.54	3.68	3.29
15	0.025	6.20	4.77	4.15
	0.010	8.68	6.36	5.42
	0.001	16.59	11.34	9.34

FIGURE A Excerpt from the F Table

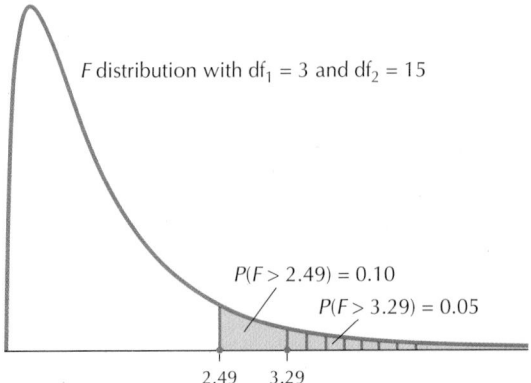

F distribution with $df_1 = 3$ and $df_2 = 15$

$P(F > 2.49) = 0.10$

$P(F > 3.29) = 0.05$

2.49   3.29

FIGURE B

The F-critical values for $df_1 = 3$ and $df_2 = 15$ are highlighted in Figure A, for various values of α. Each F-value has the indicated area α to the right. For example, $F = 2.49$ has area 0.10 to the right of it, and $F = 3.29$ has area 0.05 to the right of it (see Figure B). Now, our $F_{data} = 3.2$ falls between $F = 2.49$ and $F = 3.29$. Therefore, the area to the right of $F_{data} = 3.2$ lies between 0.010 and 0.05. Since $F_{data} = 3.2$ is closer to $F = 3.29$ than to $F = 2.49$, the estimated p-value for $F_{data} = 3.2$ is closer to 0.05 than to 0.10. By similar reasoning, for all values of F less than 2.49, the p-value will be larger than 0.10. For all values of F greater than 9.34, the p-value will be less than 0.001.

Supplements

- StatTutor 25.0–25.5
- EESEE case study:
  - Blinded Knee Doctors (Questions 5 and 6 on one-way ANOVA)

Applets

The *Analysis of Variance* applet is referenced in Chapter 12 for demonstrations and for Exercises 55 and 56 in Section 12.1.

Videos

- *Against All Odds: Inside Statistics:* www.learner.org/resources/series65.html
  - Program 31: One-Way ANOVA

Web Sites

- The site Online Statistics: An Interactive Multimedia Course of Study has information on ANOVA, and it has a link to the Rice Virtual Lab in Statistics with demonstrations and other topics related to statistics: http://onlinestatbook.com/index.html.
- This Web site lists other sites that do statistical calculations for ANOVA: http://statpages.org/.
- This site has an activity that deals with ANOVA and F tests: www.amstat.org/publications/jse/v6n3/smith.html.

12

Analysis of Variance

Elena Elisseeva/Superstock

Professors on Facebook

CASE STUDY

To improve communication with students, many instructors have been turning to computer-mediated interaction via online social networks. Online social networks serve as virtual meeting spots, where instructors and students can share information and interests outside the classroom. Do you think that the amount of information a professor posts about himself or herself, that is, *self-disclosure*, has an effect on the motivation, affective learning, classroom climate, or other important instructional issues? A recently published research paper asked precisely this question.[1] The researchers' primary research tool was *analysis of variance* (*ANOVA*), our topic here in Chapter 12. We investigate their results on page 678 in the Case Study "Professors on Facebook."

THE BIG PICTURE

Where we are coming from and where we are headed . . .

- In Chapter 10, we learned how to compare the population means of two populations.

- Here, in Chapter 12, we are introduced to *analysis of variance* (*ANOVA*), a way to compare the population means of several (more than two) different groups, and determine whether significant differences exist between the means. For example, are there significant differences in mean grade point average among the freshmen, sophomores, juniors, and seniors at your school?

- After Chapter 12, there are only two chapters left—Chapter 13, "Inference in Regression," where we extend our work from Chapter 4, "Correlation and Regression," and Chapter 14, "Nonparametric Statistics."

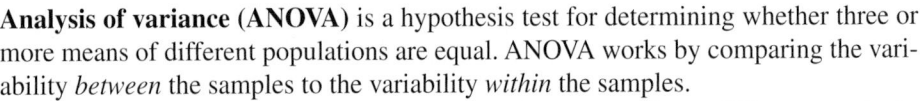

12.1 One-Way Analysis of Variance (ANOVA)

OBJECTIVES By the end of this section, I will be able to . . .

1 Explain how analysis of variance works.
2 Perform one-way analysis of variance.

1 How Analysis of Variance (ANOVA) Works

Analysis of variance (ANOVA) is a hypothesis test for determining whether three or more means of different populations are equal. ANOVA works by comparing the variability *between* the samples to the variability *within* the samples.

Suppose we are interested in determining whether significant differences exist in grade point averages (GPAs) among residents of three dormitories, A, B, and C. Table 1 displays three random samples of GPAs of 10 residents from each dormitory.

Table 1 Sample GPAs from Dorms A, B, and C

A	0.60	3.82	4.00	2.22	1.46	2.91	2.20	1.60	0.89	2.30
B	2.12	2.00	1.03	3.47	3.70	1.72	3.15	3.93	1.26	2.62
C	3.65	1.57	3.36	1.17	2.55	3.12	3.60	4.00	2.85	2.13

The sample mean GPA for Dormitory A is

$$\bar{x}_A = \frac{0.60 + 3.82 + 4.00 + 2.22 + 1.46 + 2.91 + 2.20 + 1.60 + 0.89 + 2.30}{10} = 2.2$$

Similarly, we can find the sample mean GPAs for the other dormitories: $\bar{x}_B = 2.5$ and $\bar{x}_C = 2.8$. We note that the sample means are not equal. The question is: Are the population means equal? Let μ_A, μ_B, and μ_C represent the population mean GPAs for Dormitories A, B, and C, respectively. We are interested in the following hypotheses, where μ_i represents the population mean GPA for dormitory *i*:

$$H_0 : \mu_A = \mu_B = \mu_C \quad \text{versus} \quad H_a : \text{not all the population means are equal}$$

Sufficient differences in the sample means would represent evidence that the population means were not equal. The question is: What represents "sufficiently" different? We need something to compare against, such as the *spread* of each sample. One measure of spread or variability is the *range:*

$$\text{range} = \max - \min$$

We have

$$\text{range (Dorm A)} = 4.00 - 0.60 = 3.40$$

$$\text{range (Dorm B)} = 3.93 - 1.03 = 2.90$$

$$\text{range (Dorm C)} = 4.00 - 1.17 = 2.83$$

These ranges are rather large spreads, and there is a considerable amount of overlap among the different dormitory GPAs, as shown in Figure 1.

Figure 1 shows the difference among the means for the three dorm GPAs compared with the spread of each dorm's GPAs, as measured by the range. The red triangles represent the sample means, $\bar{x}_A = 2.2$, $\bar{x}_B = 2.5$, and $\bar{x}_C = 2.8$. The spread of the sample means (shown by the red arrows) is much less than the spreads of the individual dorm GPAs (shown by the green arrows). Thus, the sample means $\bar{x}_A = 2.2$, $\bar{x}_B = 2.5$,

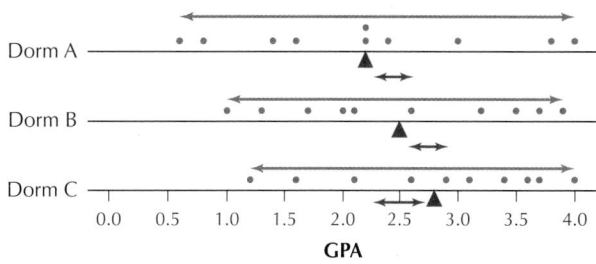

FIGURE 1 Comparison dotplot of GPAs for Dorms A, B, and C.

and $\bar{x}_C = 2.8$ are not sufficiently different when compared against the spread of the GPAs. This graph would therefore not provide evidence to reject the null hypothesis that the population mean GPAs are all equal.

Now we make a similar comparison for the GPAs for Dormitories D, E, and F in Table 2.

Table 2 Sample GPAs from Dorms D, E, and F

D	2.16	2.23	2.09	2.17	2.25	2.19	2.24	2.28	2.25	2.14
E	2.45	2.34	2.58	2.49	2.60	2.42	2.55	2.62	2.45	2.50
F	2.80	2.75	2.93	2.68	2.88	2.75	2.87	2.81	2.73	2.80

The sample mean GPAs for Dormitories D, E, and F are the *same* as those for Dormitories A, B, and C, respectively: $\bar{x}_D = 2.2$, $\bar{x}_E = 2.5$, and $\bar{x}_F = 2.8$. Again, we are interested in whether the population means are equal.

$$H_0 : \mu_D = \mu_E = \mu_F \quad \text{versus} \quad H_a : \text{not all the population means are equal}$$

Consider the comparison dotplot in Figure 2. There now seems to be better evidence for concluding that the three population means are not all equal. There is no overlap among the three samples because the spread *within* each dormitory is much smaller than for Dormitories A, B, and C.

$$\text{range (Dorm D)} = 2.28 - 2.09 = 0.19$$

$$\text{range (Dorm E)} = 2.62 - 2.34 = 0.28$$

$$\text{range (Dorm F)} = 2.93 - 2.68 = 0.25$$

Figure 2 shows the difference among the means for the three dorm GPAs compared with the range of each dorm's GPAs. The red triangles represent the sample means, $\bar{x}_D = 2.2$, $\bar{x}_E = 2.5$, and $\bar{x}_F = 2.8$. The spread of the sample means (red arrows) is much greater than the spreads of the individual dorm GPAs (green arrows). Thus, the

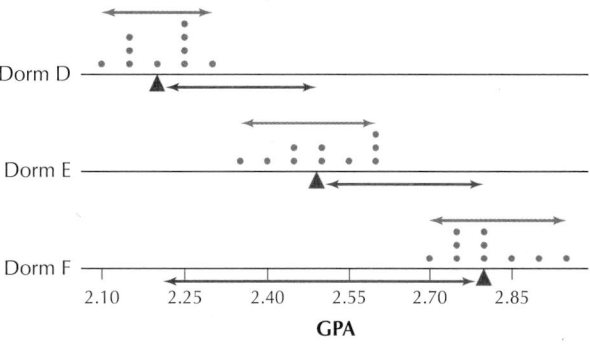

FIGURE 2 Comparison dotplot of GPAs for Dorms D, E, and F.

sample means $\bar{x}_D = 2.2$, $\bar{x}_E = 2.5$, and $\bar{x}_F = 2.8$ are sufficiently different when compared against the range of the GPAs. This graph would, therefore, provide some evidence to reject the null hypothesis that the population mean GPAs are all equal.

Note that we arrived at *opposite conclusions* for the two sets of dormitories, *even though the sample means of the first group are identical to the sample means of the second group.* Here is the key difference:

- The within-sample spreads of Dormitories A, B, and C are *large*. Compared to these large spreads, the difference in sample means did not seem large.

- The within-sample spreads of Dormitories D, E, and F are *small*. Compared to these small spreads, the difference in sample means did seem large.

These are the types of comparisons that the ANOVA method makes.

Instead of using the range as the measure of spread, analysis of variance uses the standard deviation of the individual samples. Recall that samples with larger spread have larger standard deviations, just as they have larger ranges.

Developing Your Statistical Sense

How Does Analysis of Variance Work?

The key to how analysis of variance works is the following comparison. Compare

a. the variability in the sample means—that is, how large the differences are between the sample means (indicated by the lengths of the red arrows in Figures 1 and 2)—with

b. the variability within each sample—that is, the within-sample spreads (indicated by the lengths of the green arrows in Figures 1 and 2).

When (**a**) is much larger than (**b**), this is evidence that the population means are not all equal and that we should reject the null hypothesis. Thus, our analysis depends on measuring variability—and hence the term *analysis of variance.*

Just as for hypothesis-testing procedures from previous chapters, analysis of variance can be performed only if certain requirements are met.

Requirements for Performing Analysis of Variance

1. Each of the k populations is normally distributed.

2. The variances (σ^2) of the populations are all equal.

3. The samples are independently drawn.

Note: In analysis of variance, the null hypothesis *always* states that all the population means are equal and the alternative hypothesis *always* states that not all the population means are equal. Note that H_a is *not* stating that the population means are all different. For H_a to be true, it is sufficient for a single population mean to be different, even though all the other population means may be equal.

Our hypotheses for testing for the equality of the population mean GPA for Dormitories A, B, and C are

$$H_0 : \mu_A = \mu_B = \mu_C \quad \text{versus} \quad H_a : \text{not all the population means are equal}$$

Let us stop for a moment to consider what these requirements and the hypotheses mean.

- If H_0 is true, then all three dormitories would have the same population mean GPA: $\mu_A = \mu_B = \mu_C = \mu$, where we denote the hypothesized common mean as μ.
- Requirement 1 states that each population is normally distributed.
- Requirement 2 states that all the population variances are equal. Let's call this common variance σ^2.

Putting all this together, H_0 assumes that the observations from each population come from the same normal distribution, with mean μ and variance σ^2.

Suppose we then take samples of size n from each group. Fact 3 in Chapter 7 states that the sampling distribution of $\bar{x}$ for a sample of size n taken from a normal population with mean μ and standard deviation σ (that is, variance σ^2) is also normal, with mean μ and standard deviation $\sigma/\sqrt{n}$ (that is, variance σ^2/n), as shown in Figure 3. Each dormitory's GPA is assumed (under H_0) to come from the same sampling distribution, so we would expect the sample means to be fairly close together.

On the other hand, if H_0 is not true, then not all the population means are equal (Figure 4). In this case, there is no sampling distribution common to all sample means, so we would not expect the sample means to be close together. Note in Figure 4 that each distribution nevertheless has the same shape (normal) and spread (that is, variance) because of the requirements.

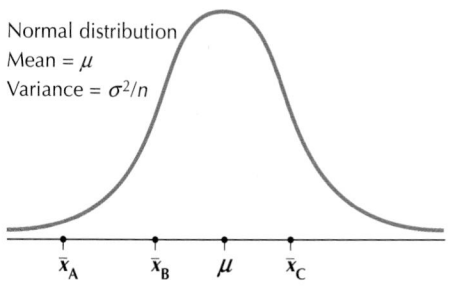

Normal distribution
Mean = μ
Variance = σ^2/n

FIGURE 3 Common sampling distribution when H_0 is true.

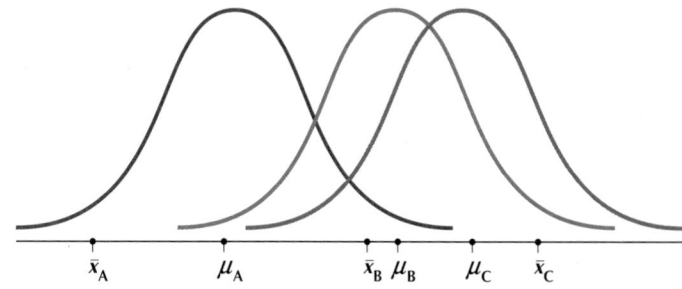

FIGURE 4 No common sampling distribution when H_0 is not true.

Note: Normal probability plots were introduced in Chapter 7.

Procedure for Verifying the Requirements for Analysis of Variance

Step 1 Normality. Check that the data from each group are normally distributed, using normality probability plots.

Step 2 Equal Variances. Compute the sample standard deviation for each group to verify that the largest standard deviation is not larger than twice the smallest standard deviation.

Step 3 Independence. Verify that the samples drawn from each group are independently drawn.

EXAMPLE 1   Verify the requirements for performing an analysis of variance

 dormitory

Verify the requirements for performing an analysis of variance using the hypotheses

$$H_0 : \mu_A = \mu_B = \mu_C \quad \text{versus} \quad H_a : \text{not all the population means are equal}$$

where μ_i represents the population mean GPA for Dormitory i, using data from Table 1.

Solution

Step 1 **Normality.** To verify that each of the $k = 3$ populations is normally distributed, we examine normal probability plots of each sample, shown in Figure 5. Each plot indicates acceptable normality.

FIGURE 5
Normal probability plots verify normality requirement.

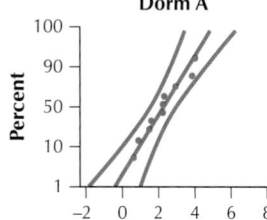

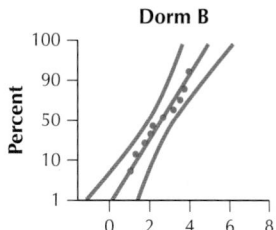

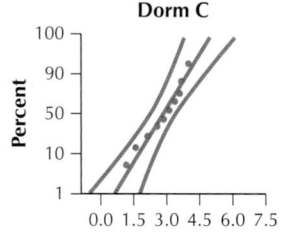

Step 2 **Equal Variances.** To find the standard deviation for Dorm A, we first find

$$\sum (x - \bar{x})^2 = (0.60 - 2.2)^2 + (3.82 - 2.2)^2 + (4.00 - 2.2)^2 + (2.22 - 2.2)^2$$
$$+ (1.46 - 2.2)^2 + (2.91 - 2.2)^2 + (2.20 - 2.2)^2 + (1.60 - 2.2)^2$$
$$+ (0.89 - 2.2)^2 + (2.30 - 2.2)^2$$
$$= 11.5626$$

Then

$$s_A = \sqrt{\frac{\sum (x - \bar{x})^2}{n - 1}} = \sqrt{\frac{11.5626}{10 - 1}} \approx 1.133460777$$

Note: We retain many decimal places when calculating s_A, s_B, and s_C because these values are used to calculate other quantities later on.

We similarly find $s_B \approx 1.030857248$ and $s_C \approx 0.9370284$. The largest, $s_A \approx 1.133460777$, is not larger than twice the smallest, $s_C \approx 0.9370284$. Thus, the equal variance requirement is satisfied.

Step 3 **Independence.** Because the students are randomly sampled from each dormitory, with the selection of students in one dormitory not affecting the selection of students sampled from the other dormitories, the independence assumption is also validated.

NOW YOU CAN DO
Exercises 7–10.

Note: This form for $\bar{\bar{x}}$ is a weighted mean with the weights being the sample sizes.

Assuming that H_0 is true, we estimate the common population mean μ using the **overall sample mean, $\bar{\bar{x}}$:**

$$\bar{\bar{x}} = \frac{(n_1 \bar{x}_1 + n_2 \bar{x}_2 + \cdots + n_k \bar{x}_k)}{n_t}$$

where there are k samples and n_t is the "total sample size" (sum of the k sample sizes). The overall sample mean $\bar{\bar{x}}$ is simply the mean of all the observations from all the samples. For the special case when all the sample sizes are equal, the overall sample mean $\bar{\bar{x}}$ is simply the mean of the k sample means,

$$\bar{\bar{x}} = \frac{(\bar{x}_1 + \bar{x}_2 + \cdots + \bar{x}_k)}{k}$$

EXAMPLE 2   Calculating the overall sample mean $\bar{\bar{x}}$

For the sample GPA data given in Table 1 for Dorms A, B, and C, calculate the overall sample mean, $\bar{\bar{x}}$.

Solution

We have $k = 3$ dormitories, with sample mean GPAs $\bar{x}_A = 2.2$, $\bar{x}_B = 2.5$, $\bar{x}_C = 2.8$. Also, $n_A = n_B = n_C = 10$, and $n_t = 10 + 10 + 10 = 30$. Thus,

$$\bar{\bar{x}} = \frac{10(2.2) + 10(2.5) + 10(2.8)}{30} = 2.5$$

All the sample sizes are equal, so we can also calculate $\bar{\bar{x}}$ as follows:

$$\bar{\bar{x}} = \frac{(2.2 + 2.5 + 2.8)}{3} = 2.5$$

NOW YOU CAN DO
Exercises 11–14.

$\bar{\bar{x}} = 2.5$ is the mean GPA for all 30 students from all three samples. We can use $\bar{\bar{x}}$ as our estimate of the common population mean μ assumed in H_0.

Recall that analysis of variance works by comparing the variability in the sample means to the variability within each sample. We use the following statistics to measure these variabilities.

The greater the distance between the sample means, the larger the MSTR.

> **The mean square treatment (MSTR)** measures the variability in the sample means. MSTR is the sample variance of the sample means, weighted by sample size.
>
> $$\text{MSTR} = \frac{\sum n_i (\bar{x}_i - \bar{\bar{x}})^2}{k - 1}$$
>
> where n_i and $\bar{x}_i$ are the sample size and mean of the ith sample, $\bar{\bar{x}}$ is the overall sample mean, and there are k populations.
>
> The **mean square error (MSE)** measures the variability within the samples. MSE is the mean of the sample variances, weighted by sample size.
>
> $$\text{MSE} = \frac{\sum (n_i - 1)s_i^2}{n_i - k}$$
>
> where n_i and s_i^2 are the sample size and variance of the ith sample, n_t is the total sample size, and there are k populations.

The larger the standard deviation of the k samples, the larger the MSE.

We compare MSTR to MSE by taking the ratio of these two quantities. This ratio MSTR/MSE follows the F distribution that we learned about in Section 10.4.

> The **test statistic for analysis of variance** is
>
> $$F_{\text{data}} = \frac{\text{MSTR}}{\text{MSE}}$$
>
> F_{data} measures the variability among the sample means, compared to the variability within the samples. F_{data} follows an F distribution with $df_1 = k - 1$ and $df_2 = n_t - k$, when the following requirements are met: (1) each of the k populations is normally distributed, (2) the variances of the populations are all equal, and (3) the samples are independently drawn.

The student may want to review the characteristics of the F distribution in Section 10.4.

The term *mean square* represents a weighted mean of quantities that are squared. Each mean square itself consists of two parts: the *sum of squares* in the numerator and the *degrees of freedom* in the denominator. The numerator for MSTR is called the **sum of squares treatment (SSTR)**, and the numerator for MSE is called the **sum of squares error (SSE)**.

$$\text{MSTR} = \frac{\text{sum of squares treatment}}{df_1} = \frac{\text{SSTR}}{df_1} = \frac{\sum n_i (\bar{x}_i - \bar{\bar{x}})^2}{k - 1}$$

$$\text{MSE} = \frac{\text{sum of squares error}}{df_2} = \frac{\text{SSE}}{df_2} = \frac{\sum (n_i - 1)s_i^2}{n_t - k}$$

The **total sum of squares (SST)** is found by adding SSTR and SSE:

$$\text{SST} = \text{SSTR} + \text{SSE}$$

The ANOVA table shown in Table 3 is a convenient way to display the various statistics calculated during an analysis of variance. Note that the quantities in the mean square column equal the ratio of the two columns to its left.

Table 3 ANOVA table

Source of variation	Sum of squares	Degrees of freedom	Mean square	F-test statistic
Treatment	SSTR	$df_1 = k - 1$	$MSTR = \dfrac{SSTR}{k - 1}$	$F_{data} = \dfrac{MSTR}{MSE}$
Error	SSE	$df_2 = n_t - k$	$MSE = \dfrac{SSE}{n_t - k}$	
Total	SST			

EXAMPLE 3 Constructing the ANOVA table

Use the summary statistics in Table 4 for the sample GPAs for Dorms A, B, and C to construct the ANOVA table.

Table 4 Summary statistics for sample GPAs for Dorms A, B, and C

	Dorm A	Dorm B	Dorm C
Mean	$\bar{x}_A = 2.2$	$\bar{x}_B = 2.5$	$\bar{x}_C = 2.8$
Standard deviation	$s_A \approx 1.133460777$	$s_B \approx 1.030857248$	$s_C \approx 0.9370284$
Sample size	$n_1 = 10$	$n_2 = 10$	$n_3 = 10$

Solution

We have $k = 3$ dormitories, and total sample size $n_t = 10 + 10 + 10 = 30$. Thus,

- $SSTR = \sum n_i (\bar{x}_i - \bar{\bar{x}})^2 = 10(2.2 - 2.5)^2 + 10(2.5 - 2.5)^2 + 10(2.8 - 2.5)^2$
$$= 10\left[(-0.3)^2 + (0)^2 + (0.3)^2\right] = 1.8$$

- $SSE \approx (10 - 1)(1.133460777)^2 + (10 - 1)(1.030857248)^2 + (10 - 1)(0.9370284)^2$
$$\approx 29.0288$$

- $SST = SSTR + SSE = 1.8 + 29.0288 = 30.8288$

- $MSTR = \dfrac{SSTR}{k - 1} = \dfrac{1.8}{3 - 1} = 0.9$

- $MSE = \dfrac{SSE}{n_t - k} = \dfrac{29.0288}{30 - 3} = 1.0751407407$

- $F_{data} = \dfrac{MSTR}{MSE} = \dfrac{0.9}{1.0751407407} = 0.8370997079 \approx 0.84$

NOW YOU CAN DO
Exercises 15–22.

We summarize these calculations in the following ANOVA table, with the results rounded for clarity.

Source of variation	Sum of squares	Degrees of freedom	Mean square	F-test statistic
Treatment	SSTR = 1.8	$df_1 = 3 - 1 = 2$	$MSTR = \dfrac{1.8}{2} = 0.9$	$F_{data} = \dfrac{0.9}{1.075} \approx 0.84$
Error	SSE = 29.0288	$df_2 = 30 - 3 = 27$	$MSE = \dfrac{29.0288}{27} \approx 1.075$	
Total	SST = 30.8288			

2 Performing One-Way ANOVA

Now that we know how it works, we next learn how to perform ANOVA.

One-Way Analysis of Variance

We have taken random samples from each of k populations and want to test whether the population means of the k populations are all equal.

Required conditions:

1. Each of the k populations is normally distributed.
2. The variances (σ^2) of the populations are all equal.
3. The samples are independently drawn.

Step 1   State the hypotheses, and state the rejection rule.

$$H_0 : \mu_1 = \mu_2 = \cdots = \mu_k \quad \text{versus} \quad H_a : \text{not all the population means are equal}$$

where the μ's represent the population mean from each population. The rejection rule is *Reject H_0 if the p-value $\leq \alpha$.*

Step 2   Calculate F_{data}.

$$F_{data} = \frac{MSTR}{MSE}$$

where

$$MSTR = \frac{\sum n_i(\bar{x}_i - \bar{\bar{x}})^2}{k - 1} \quad \text{and} \quad MSE = \frac{\sum (n_i - 1)s_i^2}{n_t - k}$$

F_{data} follows an F distribution with $df_1 = k - 1$ and $df_2 = n_t - k$ if the required conditions are satisfied, where n_t represents the total sample size.

Step 3   Find the p-value. Use technology to find the *p-value* $= P(F > F_{data})$, as shown in Figure 6.

Step 4   State the conclusion and the interpretation. Compare the *p*-value with α.

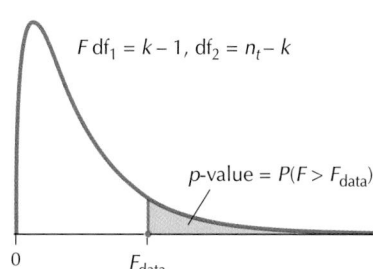

FIGURE 6
p-Value for the one-way ANOVA F test.

F df$_1$ = k − 1, df$_2$ = n_t − k

p-value = $P(F > F_{data})$

Remember: H_a is not stating that the population means are all different.

EXAMPLE 4   Performing one-way ANOVA using the p-value method

Test, using level of significance $\alpha = 0.05$, whether the population mean GPAs from Example 1 differ among the students in Dormitories A, B, and C.

> ◢ What Result Might We Expect?
>
> Recall that the comparison dotplot in Figure 1 (page 667) showed a large amount of overlap in the GPAs among dormitories A, B, and C. The large ranges illustrate the large within-dormitory spread of the GPAs for these dorms. When compared against this large within-sample variability, the variability in sample means may not seem large. Therefore, we might expect that the null hypothesis of no difference will *not be rejected.*

Solution

We already verified the requirements for performing the analysis of variance in Example 1.

Step 1 **State the hypotheses, and state the rejection rule.** Define the μ_i.

$$H_0 : \mu_A = \mu_B = \mu_C \quad \text{versus} \quad H_a : \text{not all the population means are equal}$$

where μ_i represents the population mean GPA of students from dormitory i. The rejection rule is *Reject* H_0 *if the* p-*value* $\le \alpha = 0.05$.

Step 2 **Calculate F_{data}.** From Example 3, we have MSTR = 0.9, MSE = 1.0751407407, and

$$F_{\text{data}} = \frac{\text{MSTR}}{\text{MSE}} = \frac{0.9}{1.0751407407} = 0.8370997079$$

F_{data} follows an F distribution with $df_1 = k - 1 = 3 - 1 = 2$ and $df_2 = n_t - k = 30 - 3 = 27$.

Step 3 **Find the p-value.** We use the instructions provided in the Step-by-Step Technology Guide at the end of this section (page 679). From Figures 7 and 8, we have

$$\text{p-value} = P(F > F_{\text{data}}) = P(F > 0.8370997079) = 0.4438929572 \approx 0.4439$$

> **CAUTION** When calculating the p-value for analysis of variance, always retain as many decimal places in the value of F_{data} as you can. This will make the p-value as accurate as possible. Rounding F_{data} too much will make the p-value less accurate.

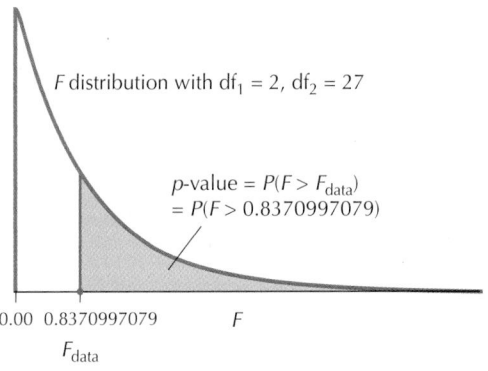

FIGURE 7 p-Value = $P(F > 0.8370997079)$.

FIGURE 8 TI-83/84 p-value.

Step 4 **State the conclusion and the interpretation.** Compare the p-value with α. The p-value of 0.4439 is not $\le \alpha = 0.05$, so we do not reject H_0. As expected, there is not enough evidence to conclude at level of significance $\alpha = 0.05$ that not all population mean GPAs are equal.

NOW YOU CAN DO
Exercises 23–28.

EXAMPLE 5 Performing one-way ANOVA using technology

© Redmond Durrell/Alamy

Researchers from the Institute for Behavioral Genetics at the University of Colorado investigated the effect that the enzyme protein kinase C (PKC) has on anxiety in mice. The genotype for a particular gene in a mouse (or a human) consists of two *alleles* (copies) of each chromosome, one each from the father and mother. The investigators in the study separated the mice into three groups. In Group 0, neither of the mice's alleles for PKC produced the enzyme. In Group 1, one of the two alleles for PKC produced the enzyme and the other did not. In Group 2, both PKC alleles produced the enzyme. To measure the anxiety in the mice, scientists measured the time (in seconds) the mice spent in the "open-ended" sections of an elevated maze. It was surmised that mice spending more time in open-ended sections exhibit decreased anxiety. The data are provided in Table 5. Use technology to test, at $\alpha = 0.01$, whether the population mean time spent in the open-ended sections of the maze was the same for all three groups.

micemaze

Table 5 Time spent in open-ended section of maze

Group 0		Group 1		Group 2	
15.8	14.4	5.2	7.6	10.6	9.2
16.5	25.7	8.7	10.4	6.4	14.5
37.7	26.9	0.0	7.7	2.7	11.1
28.7	21.7	22.2	13.4	11.8	3.5
5.8	15.2	5.5	2.2	0.4	8.0
13.7	26.5	8.4	9.5	13.9	20.7
19.2	20.5	17.2	0.0	0.0	0.0
2.5		11.9		16.5	

What Result Might We Expect?

Figure 9 shows a plot of the time in open-ended sections for the mice in the three groups. Note that the Group 1 and Group 2 mice spent on average about the same amount of time in the open-ended sections but that Group 0 spent on average somewhat more time in the open-ended sections. This would tend to suggest that the null hypothesis that all three population means are equal should be rejected. Remember that to reject H_0, it is sufficient for just one of the population means to be different.

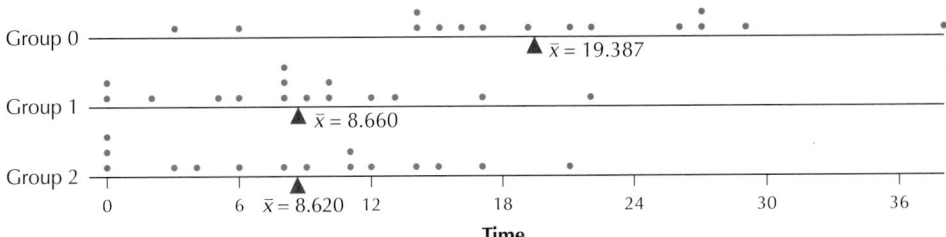

FIGURE 9 Evidence that the population mean of Group 0 is larger than the others.

Solution

We use the instructions provided in the Step-by-Step Technology Guide at the end of this section (page 679). We first verify whether the requirements are met.

- The normal probability plots in Figure 10 indicate acceptable normality.

- The group standard deviations are $s_0 \approx 9.0$, $s_1 \approx 6.0$, and $s_2 \approx 6.4$. Thus, the largest standard deviation is not greater than twice the smaller, which verifies the equal variances requirement.

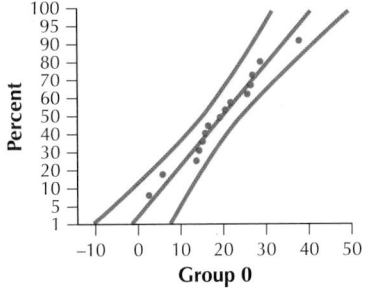
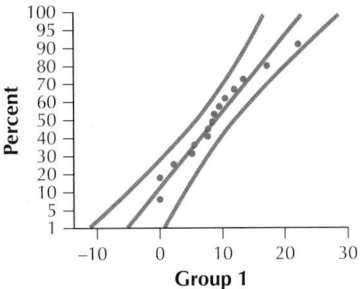
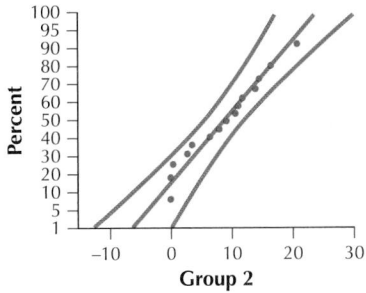

FIGURE 10 Normal probability plots.

- The selection of a mouse to a particular group did not affect the selection of mice to the other groups, so that the samples are independent.

Thus, we proceed with the one-way ANOVA.

$$H_0 : \mu_{\text{Group 0}} = \mu_{\text{Group 1}} = \mu_{\text{Group 2}}$$

H_a : not all the population means are equal

where the μ's represent the population mean time spent in the open-ended sections of the maze for each group.

Figure 11 contains the results from the TI-83/84, showing where each statistic corresponds to the ANOVA table structure in Table 3. We have $F_{\text{data}} = 10.906$, with a p-value of "1.5320224E-4" = 0.00015320224. This p-value is less than $\alpha = 0.01$, so we reject H_0. There is evidence at level of significance $\alpha = 0.01$ that the population mean times in the open-ended sections of the maze are not equal for all three groups.

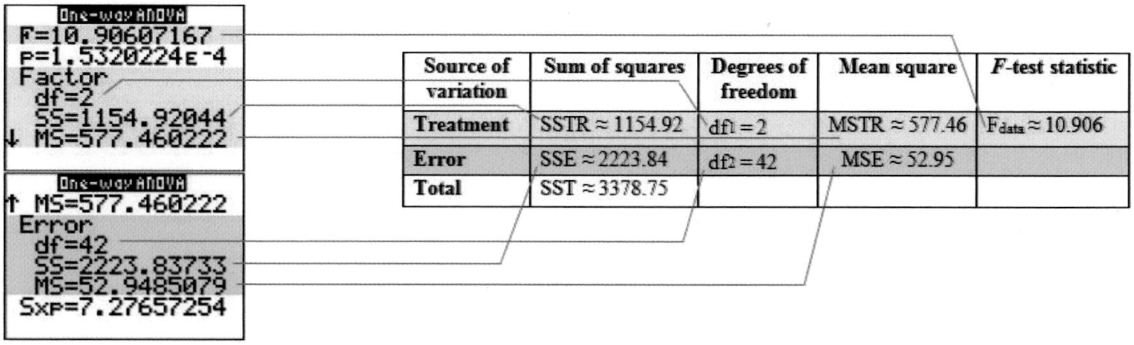

FIGURE 11 Correspondence between TI-83/84 ANOVA output and the ANOVA table.

Figure 12 contains the Excel ANOVA results, Figure 13 contains the Minitab ANOVA results, and Figure 14 contains the JMP ANOVA results. Values differ slightly due to rounding.

ANOVA						
Source of Variation	SS	df	MS	F	P-value	F crit
Between Groups	1154.92	2	577.4602	10.90607	0.000153	3.219938
Within Groups	2223.84	42	52.94851			
Total	3378.76	44				

FIGURE 12 Excel ANOVA results.

Source	DF	Adj SS	Adj MS	F-Value	P-Value
Factor	2	1155	577.46	10.91	0.000
Error	42	2224	52.95		
Total	44	3379			

FIGURE 13 Minitab ANOVA results.

△ Analysis of Variance

Source	DF	Sum of Squares	Mean Square	F Ratio	Prob > F
Group	2	1154.9204	577.460	10.9061	0.0002*
Error	42	2223.8373	52.949		
C. Total	44	3378.7578			

FIGURE 14 JMP results.

One-way ANOVA may also be conducted using the critical-value method. The conditions are the same as for the p-value method.

$\alpha_{EW} = 0.05$			
df	k	2	3
1		17.97	26.98
2		6.085	8.331
120		2.800	3.356

FIGURE 23 Finding the Tukey critical value q_{crit}.

When calculating the numerator of q_{data} for each pairwise comparison, be sure to subtract the smaller value of $\bar{x}$ from the larger value of $\bar{x}$, so that the value of q_{data} is positive.

NOW YOU CAN DO
Exercises 19–30.

degrees of freedom df $= n_t - k = 130 - 3 = 127$, and k = number of population means = 3. Using the table of Tukey critical values (Table G in the Appendix), we seek df = 127 on the left, but, when we don't find it, we conservatively choose df = 120. Then, in the column for $k = 3$, we find the Tukey critical value $q_{crit} = 3.356$ (Figure 23). The rejection rule for the Tukey method is "Reject H_0 if $q_{data} \geq q_{crit}$," that is, Reject H_0 if $q_{data} \geq 3.356$.

Step 3 **Calculate the Tukey test statistic q_{data} for each hypothesis test.** From Figure 18 on page 678, we get the sample means, the sample sizes, and the mean square error MSE = 168. Thus,

- Test 1:

$$q_{data} = \frac{\bar{x}_{High} - \bar{x}_{Medium}}{\sqrt{\frac{MSE}{2} \cdot \left(\frac{1}{n_{High}} + \frac{1}{n_{Medium}}\right)}} = \frac{81.09 - 79.36}{\sqrt{\left(\frac{168}{2}\right)\left(\frac{1}{43} + \frac{1}{44}\right)}} \approx 0.880$$

- Test 2:

$$q_{data} = \frac{\bar{x}_{High} - \bar{x}_{Low}}{\sqrt{\frac{MSE}{2} \cdot \left(\frac{1}{n_{High}} + \frac{1}{n_{Low}}\right)}} = \frac{81.09 - 70.63}{\sqrt{\left(\frac{168}{2}\right)\left(\frac{1}{43} + \frac{1}{43}\right)}} \approx 5.292$$

- Test 3:

$$q_{data} = \frac{\bar{x}_{Medium} - \bar{x}_{Low}}{\sqrt{\frac{MSE}{2} \cdot \left(\frac{1}{n_{Medium}} + \frac{1}{n_{Low}}\right)}} = \frac{79.36 - 70.63}{\sqrt{\left(\frac{168}{2}\right)\left(\frac{1}{44} + \frac{1}{43}\right)}} \approx 4.442$$

Step 4 **For each hypothesis test, state the conclusion and the interpretation.**

- Test 1: $q_{data} = 0.880$, which is not $\geq q_{crit} = 3.356$; therefore, do not reject H_0. There is insufficient evidence at the 0.05 level of significance that the population mean student motivation scores differ between professors having high and medium self-disclosure on Facebook.

- Test 2: $q_{data} = 5.292$, which is $\geq q_{crit} = 3.356$; therefore, reject H_0. There is evidence at the 0.05 level of significance that the population mean scores differ between high and low professor self-disclosure on Facebook.

- Test 3: $q_{data} = 4.442$, which is $\geq q_{crit} = 3.356$; therefore, reject H_0. There is evidence at the 0.05 level of significance that the population mean scores differ between medium and low professor self-disclosure on Facebook.

This set of three hypothesis tests has an experimentwise error rate $\alpha_{EW} = 0.05$.

3 Using Confidence Intervals to Perform Tukey's Test

Tukey's test for multiple comparisons may also be performed using confidence intervals and technology. Recall that when using confidence intervals for hypothesis tests, H_0 is rejected if the hypothesized value of the population mean does not fall inside the confidence interval.

Rejection Rule for Using Confidence Intervals to Perform Tukey's test

If a $100(1 - \alpha)\%$ confidence interval for $\mu_1 - \mu_2$ contains zero, then at level of significance α, we do not reject the null hypothesis $H_0: \mu_1 = \mu_2$. If the interval does not contain zero, then we do reject H_0.

We illustrate the concept of using confidence intervals to perform Tukey's test with an example using the Facebook data.

EXAMPLE 9   Using confidence intervals to perform Tukey's test

Use the 95% confidence intervals for the differences in population means provided by Minitab to perform Tukey's test for multiple comparisons on the Facebook data.

Solution

We use the steps in the Step-by-Step Technology Guide provided at the end of this section. Figure 24 contains the output from Minitab showing 95% confidence intervals for the differences in population means for the high, medium, and low professor disclosure levels. The output states that "Group = Low" is being subtracted from the other two groups, meaning that the first two confidence intervals are for $\mu_{Medium} - \mu_{Low}$ and $\mu_{High} - \mu_{Low}$. Later, "Group = Medium" is subtracted from the high group, indicating a confidence interval for $\mu_{High} - \mu_{Medium}$. The column headings "Lower" and "Upper" represent the lower and upper bounds of the confidence interval. Figure 25 shows the output from JMP, including 95% confidence intervals for the differences in population means. The output states that the second level listed is subtracted from the first, meaning that the first two confidence intervals are for $\mu_{High} - \mu_{Low}$ and $\mu_{Medium} - \mu_{Low}$. The columns "Lower CL" and "Upper CL" represent the lower and upper bounds of each confidence interval.

FIGURE 24
Using Minitab confidence intervals to perform Tukey's test.

```
Tukey 95% Simultaneous Confidence Intervals
All Pairwise Comparisons among Levels of Group

Individual confidence level = 98.06%

Group = Low subtracted from:

Group   Lower   Center   Upper    --+---------+---------+---------+-----
Medium   2.14    8.74    15.33                       (-------*-------)
High     3.84   10.47    17.09                       (-------*-------)
                                  --+---------+---------+---------+-----
                                  -7.0       0.0       7.0      14.0

Group = Medium subtracted from:

Group   Lower   Center   Upper    --+---------+---------+---------+-----
High    -4.86    1.73     8.32               (-------*-------)
                                  --+---------+---------+---------+-----
                                  -7.0       0.0       7.0      14.0
```

FIGURE 25
Using JMP confidence intervals to perform Tukey's test.

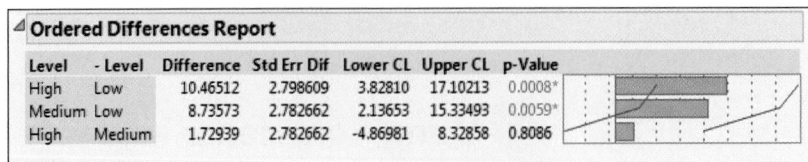

◢ **Ordered Differences Report**								
Level	**- Level**	**Difference**	**Std Err Dif**	**Lower CL**	**Upper CL**	**p-Value**		
High	Low	10.46512	2.798609	3.82810	17.10213	0.0008*		
Medium	Low	8.73573	2.782662	2.13653	15.33493	0.0059*		
High	Medium	1.72939	2.782662	-4.86981	8.32858	0.8086		

Thus, for our $c = 3$ hypothesis tests, we have

- Test 1: $H_0 : \mu_{Medium} = \mu_{Low}$ versus $H_a : \mu_{Medium} \neq \mu_{Low}$

  95% confidence interval for $\mu_{Medium} - \mu_{Low}$ is (2.14, 15.33), which does not contain zero, so we reject $H_0 : \mu_{Medium} = \mu_{Low}$ for level of significance $\alpha = 0.05$.

- Test 2: $H_0 : \mu_{High} = \mu_{Low}$ versus $H_a : \mu_{High} \neq \mu_{Low}$

  95% confidence interval for $\mu_{High} - \mu_{Low}$ is (3.84, 17.09), which does not contain zero, so we reject $H_0 : \mu_{High} = \mu_{Low}$ for level of significance $\alpha = 0.05$.

- Test 3: $H_0 : \mu_{\text{High}} = \mu_{\text{Medium}}$ versus $H_a : \mu_{\text{High}} \neq \mu_{\text{Medium}}$

  95% confidence interval for $\mu_{\text{High}} - \mu_{\text{Medium}}$ is $(-4.86, 8.32)$, which does contain zero, so we do not reject $H_0 : \mu_{\text{High}} = \mu_{\text{Medium}}$ for level of significance $\alpha = 0.05$.

Note that these conclusions are exactly the same as the conclusions from Example 8.

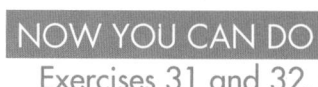

NOW YOU CAN DO
Exercises 31 and 32.

STEP-BY-STEP TECHNOLOGY GUIDE: Multiple Comparisons

Neither the TI-83/84 nor Excel performs multiple comparisons.

MINITAB

Step 1   Enter the data for ANOVA into columns C1, C2, etc., as shown in the Section 12.1 technology guide on page 679.
Step 2   Select **Stat > ANOVA > One way…**. Choose either Response data are in one column for all factor levels or Response data are in a separate column for each factor level,

depending on how you entered your data in Step 1. Enter the **response** and **factor** information as necessary.
Step 3   Click **Comparisons…**, check **Tukey**, and enter **Error rate for comparisons** (such as "5" for $\alpha_{EW} = 0.05$). Click **OK** twice. For the data in Example 9, this is shown in Figure 24.

SPSS

We demonstrate using the data from Example 9.
Step 1   Enter the **Motivation** data into the first column and a numeric code for the **Disclosure** groups in the second column. For example, let 2 represent High disclosure, 1 represent Medium disclosure, and 0 represent Low disclosure. Under the **Variable View** tab, set the decimals for the *Disclosure* variable to zero.

Step 2   Select **Analyze > Compare Means > One-way ANOVA…**. Move *Motivation* to **Dependent List**, and *Disclosure* to **Factor**.
Step 3   Click **Post Hoc…**, check **Tukey**, and enter a **Significance level** of 0.05. Click **Continue**, then **OK**. The output is in Figure 26.

Multiple Comparisons

Dependent Variable: Motivation
Tukey HSD

(I) Disclosure	(J) Disclosure	Mean Difference (I-J)	Std. Error	Sig.	95% Confidence Interval Lower Bound	Upper Bound
0	1	-8.73573*	2.78266	.006	-15.3348	-2.1366
	2	-10.46512*	2.79861	.001	-17.1020	-3.8282
1	0	8.73573*	2.78266	.006	2.1366	15.3348
	2	-1.72939	2.78266	.809	-8.3285	4.8697
2	0	10.46512*	2.79861	.001	3.8282	17.1020
	1	1.72939	2.78266	.809	-4.8697	8.3285

*. The mean difference is significant at the 0.05 level.

FIGURE 26 SPSS output.

JMP

We demonstrate using the data from Example 9.
Step 1   Select **File > New > Data Table**. Enter the **Motivation** values in the first column, and the **Disclosure** numeric labels in the second column. Right-click the *Disclosure* column heading, select **Column Info…**, and change **Data Type** to **Character**.

Step 2   Click **Analyze > Fit Y by X**. Move *Motivation* to **Y, Response** and *Disclosure* to **X, Factor**. Click **OK**.
Step 3   Click the red triangle beside "Oneway Analysis," and click **Compare Means > All pairs, Tukey HSD**. The output is shown in Figure 25 of Example 9.

Section 12.2 Summary

1. Once an ANOVA result has been found, significant multiple comparisons procedures are performed to determine which pairs of population means are significantly different. The Bonferroni method uses the Bonferroni adjustment, which multiplies the *p*-value of each pairwise hypothesis test by the number of comparisons being made.

2. Tukey's test may be performed using the critical-value method, where the test statistic q_{data} is compared to the critical value q_{crit} for each hypothesis test.
3. Tukey's test may also be performed using confidence intervals and technology.

Section 12.2 Exercises

CLARIFYING THE CONCEPTS

1. Explain why we cannot conclude, based on the ANOVA results alone, that one particular population mean differs from another. (p. 686)

2. Clarify why we do not perform multiple comparisons if we did not reject the null hypothesis in the original ANOVA. (p. 686)

3. What is a pairwise comparison? Explain how we calculate how many pairwise comparisons there are. (p. 686)

4. What do we mean by experimentwise error rate? (p. 686)

5. How does the Bonferroni adjustment correct for the greater size of the experimentwise error rate? (p. 686)

6. What do we do if the Bonferroni adjustment results in an adjusted p-value greater than 1? Why must we do this? (p. 686)

7. What are the requirements for performing multiple comparisons using either the Bonferroni method or Tukey's test? (pp. 687–688)

8. What is the rejection rule for using confidence intervals to perform Tukey's test? (p. 689)

PRACTICING THE TECHNIQUES

CHECK IT OUT!

To do	Check out	Topic
Exercises 9–18	Example 7	Bonferroni method of multiple comparisons
Exercises 19–30	Example 8	Tukey's test for multiple comparisons
Exercises 31–32	Example 9	Using confidence intervals to perform Tukey's test

For Exercises 9–14, use the summary statistics to calculate the value of the test statistic t_{data} for the Bonferroni method.

9. $\bar{x}_1 = 85, \bar{x}_2 = 75, MSE = 25, n_1 = 10, n_2 = 10$
10. $\bar{x}_1 = 85, \bar{x}_3 = 65, MSE = 25, n_1 = 10, n_3 = 10$
11. $\bar{x}_2 = 75, \bar{x}_3 = 65, MSE = 25, n_2 = 10, n_3 = 10$
12. $\bar{x}_1 = 100, \bar{x}_2 = 105, MSE = 100, n_1 = 8, n_2 = 8$
13. $\bar{x}_1 = 100, \bar{x}_3 = 85, MSE = 100, n_1 = 8, n_3 = 8$
14. $\bar{x}_2 = 105, \bar{x}_3 = 85, MSE = 100, n_2 = 8, n_3 = 8$

For Exercises 15 and 16, perform multiple comparisons using the Bonferroni method at level of significance $\alpha = 0.05$, for the data indicated. Assume the requirements are met. Do the following:

    a. For each hypothesis test, state the hypotheses and the rejection rule.

    b. Use the value of t_{data} from Exercises 9–14 for each hypothesis test.

    c. Find the Bonferroni-adjusted p-value for each hypothesis test.

    d. For each hypothesis test, state the conclusion and the interpretation.

15. The data from Exercises 9–11
16. The data from Exercises 12–14

For Exercises 17 and 18, assume the ANOVA requirements are met. If appropriate, perform multiple comparisons using the Bonferroni method with individual level of significance $\alpha = 0.01$ for the indicated data. If not appropriate to perform multiple comparisons, state why not.

17. The data from Exercise 19, Section 12.1
18. The data from Exercise 20, Section 12.1

For Exercises 19–24, use the summary statistics to calculate the value of the test statistic q_{data} for Tukey's test.

19. $\bar{x}_1 = 60, \bar{x}_2 = 61, MSE = 100, n_1 = 50, n_2 = 50$
20. $\bar{x}_1 = 60, \bar{x}_3 = 65, MSE = 100, n_1 = 50, n_3 = 50$
21. $\bar{x}_2 = 61, \bar{x}_3 = 65, MSE = 100, n_2 = 50, n_3 = 50$
22. $\bar{x}_1 = 220, \bar{x}_2 = 200, MSE = 450, n_1 = 25, n_2 = 25$
23. $\bar{x}_1 = 220, \bar{x}_3 = 210, MSE = 450, n_1 = 25, n_3 = 25$
24. $\bar{x}_2 = 200, \bar{x}_3 = 210, MSE = 450, n_2 = 25, n_3 = 25$

For Exercises 25 and 26, find the critical value q_{crit}.

25. Experimentwise error rate $\alpha_{EW} = 0.05, n_1 = 50, n_2 = 50, n_3 = 50$
26. Experimentwise error rate $\alpha_{EW} = 0.05, n_1 = 25, n_2 = 25, n_3 = 25$

For Exercises 27 and 28, perform multiple comparisons using Tukey's test at experimentwise error rate $\alpha_{EW} = 0.05$, for the data indicated. Assume the requirements are met. Do the following:

    a. For each hypothesis test, state the hypotheses.

    b. Use the value of q_{crit} from Exercises 25 or 26, and state the rejection rule.

    c. Use q_{data} from Exercises 19–24 for each hypothesis test.

    d. For each hypothesis test, state the conclusion and the interpretation.

27. The data from Exercises 19–21
28. The data from Exercises 22–24

For Exercises 29 and 30, assume the ANOVA requirements are met. If appropriate, perform multiple comparisons using Tukey's test with experimentwise error rate $\alpha_{EW} = 0.05$ for the indicated data. If not appropriate to perform multiple comparisons, state why not.

29. The data from Exercise 21, Section 12.1
30. The data from Exercise 22, Section 12.1

For Exercises 31 and 32, use the indicated output to perform Tukey's test, using the confidence interval method, to determine if significant pairwise differences exist among the population means.

31. Minitab output in Figure 27, for groups A, B, and C

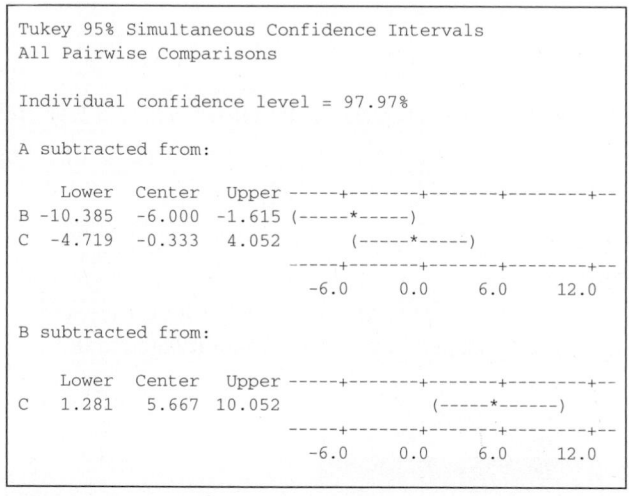

```
Tukey 95% Simultaneous Confidence Intervals
All Pairwise Comparisons

Individual confidence level = 97.97%

A subtracted from:

    Lower  Center  Upper  --+--------+--------+--------+----
B  -5.642  -2.667  0.309        (------*------)
C  -7.642  -4.667 -1.691  (------*------)
                          --+--------+--------+--------+----
                          -7.0     -3.5      0.0     3.5

B subtracted from:

    Lower  Center  Upper  --+--------+--------+--------+----
C  -4.976  -2.000  0.976           (------*------)
                          --+--------+--------+--------+----
                          -7.0     -3.5      0.0     3.5
```

FIGURE 27

32. Minitab output in Figure 28, for groups A, B, and C

```
Tukey 95% Simultaneous Confidence Intervals
All Pairwise Comparisons

Individual confidence level = 97.97%

A subtracted from:

    Lower   Center  Upper  -----+--------+--------+--------+--
B  -10.385  -6.000 -1.615  (-----*-----)
C   -4.719  -0.333  4.052        (-----*-----)
                           -----+--------+--------+--------+--
                           -6.0     0.0      6.0     12.0

B subtracted from:

    Lower  Center  Upper  -----+--------+--------+--------+--
C   1.281  5.667  10.052              (-----*------)
                          -----+--------+--------+--------+--
                          -6.0     0.0      6.0     12.0
```

FIGURE 28

APPLYING THE CONCEPTS

For Exercises 33–36, assume the ANOVA requirements are met. If appropriate, perform multiple comparisons using the Bonferroni method with individual level of significance $\alpha = 0.01$ for the indicated data. If not appropriate to perform multiple comparisons, state why not.

33. Student-Run Café Business. The data from Exercise 31, Section 12.1

34. Weight and Age. The data from Exercise 33, Section 12.1

35. Do Taller People Have Lower Voices? The data from Exercise 41, Section 12.1

36. Gas Mileage for European, Japanese, and American Cars. The data from Exercise 48, Section 12.1

For Exercises 37–40, assume the ANOVA requirements are met. If appropriate, perform multiple comparisons using Tukey's test with experimentwise error rate $\alpha_{EW} = 0.05$ for the indicated data. If not appropriate to perform multiple comparisons, state why not.

37. The Professionals versus the Darts. The data from Exercise 32, Section 12.1

38. The Full Moon and Emergency Room Visits. The data from Exercise 34, Section 12.1

39. Nutritional Rating of Breakfast Cereals. The data from Exercise 45, Section 12.1

40. Head Injuries and Vehicle Year. The data from Exercise 52, Section 12.1

12.3 Randomized Block Design

OBJECTIVE By the end of this section, I will be able to . . .

1 Explain the power of the randomized block design and perform a randomized block design ANOVA.

1 Randomized Block Design Explained and Performed

In the appropriate circumstances, we can use the **randomized block design** to improve the ability of the ANOVA to find significant differences among the treatment means. Suppose that a study was performed to determine whether significant differences existed among three types of educational methods: in-class learning, online learning, and the hybrid approach, which combines in-class and online learning. There were five courses taught in the study, and each course was taught using each method. For each course/method combination, the final exam average grade was calculated (see Table 6).

Table 6 Average final exam grade for five courses using three learning methods

		Learning method		
		In-class	**Online**	**Hybrid**
Course	**Statistics**	75	70	69
	English	80	77	82
	Psychology	72	69	72
	Biology	73	60	66
	Communication	88	82	85

The researchers are not interested in the differences among the courses, only the differences in population mean performance among the learning methods. Thus, our *factor of interest* is the treatment variable learning methods, and our **blocking factor** is the variable course.

Another common term for the blocking factor is *nuisance factor,* which clearly indicates we are not interested in it.

> A **blocking factor**, or **block**, is a variable that is not of primary interest to the researcher, but is included in the ANOVA in order to improve the ability of the ANOVA to find significant differences among the treatment means. In a **randomized block design ANOVA**, we test for differences among the treatment means, while accounting for the variability among the levels in the blocking factor.

Thus, in Table 6, the blocking factor *course* divides all the classes using a learning method into five blocks or courses.

We shall demonstrate the power of blocking by

1. first performing a usual one-way ANOVA testing for differences among the learning methods, and then

2. performing a *randomized block design* ANOVA, where we test for differences among the learning methods, while accounting for the variability among the courses (the blocks).

EXAMPLE 10   One-way ANOVA unable to find significance

 finalexam

Use technology to perform a one-way ANOVA, using level of significance $\alpha = 0.05$, to determine whether the population mean test grades for the three learning methods are all the same.

Solution

If we ignore the blocking factor *course,* Table 6 becomes Table 7.

Table 7 Average final exam grade for the three learning methods without using the blocking factor *course*

Learning method		
In-class	**Online**	**Hybrid**
75	70	69
80	77	82
72	69	72
73	60	66
88	82	85

Figure 29 shows acceptable normality, and the largest of the standard deviations in the ANOVA output in Figure 30 and Figure 31 is not larger than twice the smallest.

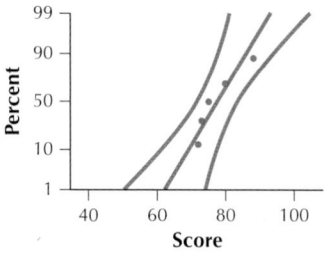 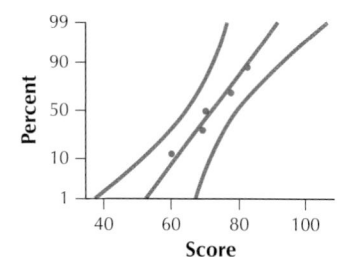 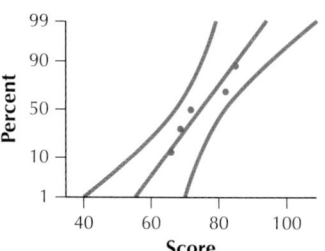

FIGURE 29 Normal probability plots for the three learning methods.

```
One-way ANOVA: Score versus Method

Analysis of Variance

Source   DF   Adj SS   Adj MS   F-Value   P-Value
Method    2    90.13    45.07      0.74     0.497
Error    12   729.20    60.77
Total    14   819.33


Model Summary

      S     R-sq   R-sq(adj)   R-sq(pred)
7.79530   11.00%      0.00%        0.00%


Means

Method   N    Mean   StDev      95% CI
1        5   77.60    6.58   (70.00, 85.20)
2        5   71.60    8.38   (64.00, 79.20)
3        5   74.80    8.29   (67.20, 82.40)
```

ANOVA

Grade

	Sum of Squares	df	Mean Square	F	Sig.
Between Groups	90.133	2	45.067	.742	.497
Within Groups	729.200	12	60.767		
Total	819.333	14			

Thus, we may proceed with the ANOVA. The hypotheses are

$$H_0: \mu_{\text{In-class}} = \mu_{\text{Online}} = \mu_{\text{Hybrid}} \quad \text{versus} \quad H_a: \text{not all the population means are equal}$$

Reject H_0 if the p-value ≤ 0.05.

The p-value from Figure 30 is 0.497, which is not ≤ 0.05; therefore, we do not reject H_0. There is insufficient evidence at level of significance $\alpha = 0.05$ that the population mean test performance differs among the three learning methods.

Next, we use the randomized block design to account for the variability among the courses.

Use technology and the randomized block design (RBD) to test for differences among the population mean test grades, at level of significance $\alpha = 0.05$.

Solution

Note that the *p*-value for the blocking factor is approximately zero, which means that the blocking factor is significant. But in the randomized block design we are not interested in the significance of the blocking factor. If we are interested in a second factor, we use *two-way ANOVA* in Section 4.

We use the Step-by-Step Technology Guide at the end of this section to run this analysis in Minitab and SPSS. The hypotheses and the rejection rule are the same for RBD as they are for one-way ANOVA:

$$H_0: \mu_{\text{In-class}} = \mu_{\text{Online}} = \mu_{\text{Hybrid}} \quad \text{versus} \quad H_a: \text{not all the population means are equal}$$

Reject H_0 if the *p*-value ≤ 0.05.

Figure 32 shows the Minitab output from the randomized block design, and Figure 33 shows the SPSS output. The factor of interest (that is, the treatment) is the learning method, whereas the blocking factor is the course. Note that there are two *p*-values, but we examine the *p*-value of the treatment only, not of the blocking factor.

FIGURE 32
Minitab output for randomized block design.

```
Analysis of Variance for Score

Source   DF      SS       MS      F      P
Course    4   677.33   169.33   26.12  0.000
Method    2    90.13    45.07    6.95  0.018
Error     8    51.87     6.48
Total    14   819.33
```

FIGURE 33
SPSS output for randomized block design.

Tests of Between-Subjects Effects

Dependent Variable: Score

Source	Type III Sum of Squares	df	Mean Square	F	Sig.
Model	84394.133[a]	7	12056.305	1859.584	.000
Method	90.133	2	45.067	6.951	.018
Course	677.333	4	169.333	26.118	.000
Error	51.867	8	6.483		
Total	84446.000	15			

a. R Squared = .999 (Adjusted R Squared = .999)

NOW YOU CAN DO
Exercises 5–10.

The *p*-value for the factor of interest from both Figure 32 and Figure 33 is 0.018, which is ≤ 0.05; therefore, we reject H_0. There is evidence at level of significance $\alpha = 0.05$ that the population mean test grade differs among the three learning methods.

Developing Your Statistical Sense

How Does Blocking Work?

Why did the randomized block design succeed in rejecting the null hypothesis, when the usual one-way ANOVA failed? From Section 1, we know that the following relationship exists among the three sums of squares:

$$SST = SSTR + SSE$$

Now, the sum of squares error (SSE) includes all unexplained variability, including random variation as well as the variability among the courses themselves. For example, note from Table 6 that the three values from the communication courses are all larger than the three values from the statistics courses. This represents variability, and becomes part of SSE when we perform the usual ANOVA. In other words, the variability in the courses makes SSE larger than it should be.

Why is a large SSE bad? Recall our F statistic from the ANOVA table (Table 3, page 672):

$$F = \frac{\text{MSTR}}{\text{MSE}} = \frac{\text{MSTR}}{\text{SSE}/\text{df}_2}$$

Recall that we reject H_0 when the p-value is small. Also note that

- the p-value is small only when F is large (see Figure 34); and
- F is large when SSE is small.

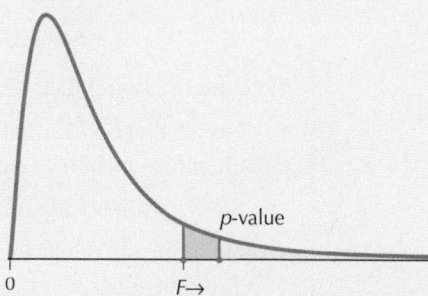

FIGURE 34 Larger $F \leftrightarrow$ smaller p-value.

So we want to keep SSE small (as we would any type of error). The randomized block design reduces SSE by extracting the *sum of squares blocks* (SSB), which represents the variability among the groups in the blocking factor, such as the variability among the courses. This is done as follows. The SSE from the original one-way ANOVA is separated into the sum of two quantities:

$$\text{SSE}_{\text{one-way ANOVA}} = \text{SSB} + \text{SSE}_{\text{RBD}}$$

Sums of squares are non-negative, so the new SSE_{RBD} from the RBD will probably be smaller than the original $\text{SSE}_{\text{one-way ANOVA}}$, and thus F will become larger, leading to easier rejection of H_0.

EXAMPLE 12   Demonstrating how RBD works

Demonstrate how randomized block design works, using the results from Examples 10 and 11.

Solution

Figure 30 from Example 10 provides us with

$$\text{SSTR} = 90.133, \text{SSE}_{\text{one-way ANOVA}} = 729.2, \text{ and SST} = 819.333$$

The RBD from Example 11 (Figure 32) separates $\text{SSE}_{\text{one-way ANOVA}}$ into the sum of the following two quantities:

$$\text{SSE}_{\text{one-way ANOVA}} = \text{SSB} + \text{SSE}_{\text{RBD}} = 677.333 + 51.867 = 729.2$$

The new sum of squares error SSE_{RBD} is small enough that the resulting new F statistic is large, leading to a rejection of the null hypothesis.

Let k and b represent the number of treatments and the number of blocks, respectively. Then, the ANOVA table for the randomized block design is as shown in Table 8.

Table 8 ANOVA table for randomized block design

Source	Sum of squares	Degrees of freedom	Mean square	F
Treatments	SSTR	$k - 1$	$\text{MSTR} = \dfrac{\text{SSTR}}{k - 1}$	$F = \dfrac{\text{MSTR}}{\text{MSE}}$
Blocks	SSB	$b - 1$	$\text{MSB} = \dfrac{\text{SSB}}{b - 1}$	
Error	SSE	$(k - 1)(b - 1)$	$\text{MSE} = \dfrac{\text{SSE}}{(k - 1)(b - 1)}$	
Total	SST	$n_t - 1$		

Note the following facts about the ANOVA table for randomized block design:

- Notice that SSTR, its degrees of freedom $k - 1$, and MSTR are all the same quantities as in the one-way ANOVA table (Table 3) on page 672.

- SSE_{RBD} is denoted simply as SSE.

- Quantities in the "Mean square" column equal the ratio of the quantities in the "Sum of squares" column divided by their respective degrees of freedom.

- We have SST = SSTR + SSB + SSE, and the three degrees of freedom values sum to $n_t - 1$.

- We are not interested in the blocks and thus the *mean square blocks MSB*, so there is no F statistic for blocks.

- In one-way ANOVA, the degrees of freedom for $\text{SSE}_{\text{one-way ANOVA}}$ is $n_t - k$. In RBD, this error degrees of freedom is partitioned into the degrees of freedom for SSB, $b - 1$, and the degrees of freedom for the new SSE, $(k - 1)(b - 1)$. An exercise in this section asks the student to show that these two degrees of freedom sum to $n_t - k$.

NOW YOU CAN DO
Exercises 11–13.

STEP-BY-STEP TECHNOLOGY GUIDE: Randomized Block Design

Neither the TI-83/84 nor Excel performs randomized block design analysis.

MINITAB

We shall use the data from Examples 10 and 11.
Step 1   Enter the score (response) data into column **C1**, the method (treatment) in column **C2**, and the course (block) in column **C3** (see Figure 35).
Step 2   Select **Stat > ANOVA > General Linear Model > Balanced ANOVA**.
Step 3   For **Responses**, select the score (response) variable; for **Factors**, enter the course and method variables. Click **OK**. Results are shown in Figure 32 of Example 11.

FIGURE 35
Entering the RBD data into Minitab.

Worksheet 1 ***			
↓	C1	C2-T	C3-T
	Score	Method	Course
1	75	In-class	Statistics
2	70	Online	Statistics
3	69	Hybrid	Statistics
4	80	In-class	English
5	77	Online	English
6	82	Hybrid	English
7	72	In-class	Psychology
8	69	Online	Psychology
9	72	Hybrid	Psychology
10	73	In-class	Biology
11	60	Online	Biology
12	66	Hybrid	Biology
13	88	In-class	Communication
14	82	Online	Communication
15	85	Hybrid	Communication

SPSS

Step 1   Enter the score (response) data into column **C1**, a numeric code for the method (treatment) in column **C2**, and a numeric code for the course (block) in column **C3**.
Step 2   Under Variable View, rename the characters **Score**, **Method**, and **Course**. Change the **Decimals** of **Method** and **Course** to 0 and the **Width** of **Method** and **Course** to 1.
Step 3   Select **Analyze > General Linear Model > Univariate....**

Step 4   For **Dependent Variable**, select *Score*. For **Fixed Factor(s)**, enter *Method* and *Course*.
Step 5   Click **Model...**, select **Custom**, choose **Main effects** from the **Build Term(s) Type:** drop-down menu, and move the *Course* and *Method* variables into the **Model** box. Deselect **Include intercept in model.** Click **Continue**, then **OK**. Results are shown in Figure 33 of Example 11.

JMP

Step 1   Enter the score (response) data into column **C1**, the method (treatment) in column **C2**, and the course (block) in column **C3** (see Figure 35).
Step 2   Select **Analyze > Fit Y by X**. Put *Score* in the **Y, Response** box, *Method* in the **X, Factor** box, and *Course* in the **Block** box. Click **OK**.
Step 3   Click the red triangle beside "Oneway Analysis," and select **Means/Anova**. The result is in the Analysis of Variance table, Figure 36.

◢ Analysis of Variance

Source	DF	Sum of Squares	Mean Square	F Ratio	Prob > F
Method	2	90.13333	45.067	6.9512	0.0178*
Course	4	677.33333	169.333	26.1183	0.0001*
Error	8	51.86667	6.483		
C. Total	14	819.33333			

FIGURE 36 JMP output for randomized block design.

Section 12.3 Summary

1. A blocking factor, or block, is a variable that is not of primary interest to the researcher, but is included in the ANOVA in order to improve the ability of the ANOVA to find significant differences among the treatment means. We perform a randomized block design ANOVA, where we test for differences among the factor of interest (treatment), while accounting for the variability among the blocks.

Section 12.3 Exercises

CLARIFYING THE CONCEPTS

1. Explain the difference between the variable of interest and the blocking factor. What is another term for the variable of interest? For the blocking factor? (p. 694)
2. What do we do in a randomized block design ANOVA? (p. 694)
3. Explain why a large SSE is bad if we want to reject the null hypothesis in ANOVA. (p. 697)
4. Explain how a larger *F* statistic leads to a more likely rejection of the null hypothesis in ANOVA. (p. 697)

PRACTICING THE TECHNIQUES

 CHECK IT OUT!

To do	Check out	Topic
Exercises 5–10	Example 11	Randomized block design (RBD)
Exercises 11–13	Example 12	Demonstrating how RBD works

For Exercises 5–8, use the indicated randomized block design ANOVA output to test, using level of significance $\alpha = 0.05$, whether the treatment means differ.
    a. Provide the hypotheses and the rejection rule.
    b. Find the relevant *p*-value; provide the conclusion and the interpretation.

5.

Source	DF	SS	MS	F	P
Treatmt	3	44.5	14.833	7.03	0.010
Block	3	402.5	134.167	63.55	0.000
Error	9	19.0	2.111		
Total	15	466.0			

6.

Source	DF	SS	MS	F	P
Treatmt	2	19.600	9.8000	2.64	0.132
Block	4	387.067	96.7667	26.04	0.000
Error	8	29.733	3.7167		
Total	14	436.400			

7.

Source	DF	SS	MS	F	P
Treatmt	3	2.27344	0.757813	5.03	0.009
Block	7	0.99219	0.141741	0.94	0.497
Error	21	3.16406	0.150670		
Total	31	6.42969			

8.

Source	DF	SS	MS	F	P
Treatmt	3	0.81250	0.270833	1.70	0.204
Block	6	1.33929	0.223214	1.40	0.269
Error	18	2.87500	0.159722		
Total	27	5.02679			

For Exercises 9 and 10, for the indicated partially completed randomized block ANOVA table, do the following:

a. Provide the hypotheses.
b. Complete the missing entries in the table.
c. Use technology to find the *p*-value. For level of significance $\alpha = 0.05$, provide the conclusion and the interpretation.

9.

Source	Sum of square	Degrees of freedom	Mean squares	F
Treatments	90	—	45	—
Blocks	50	—	—	
Error	—	8	7.5	
Total	—	15		

10.

Source	Sum of square	Degrees of freedom	Mean squares	F
Treatments	—	3	—	—
Blocks	—	6	40	
Error	216	—	—	
Total	636	—		

For Exercises 11 and 12, do the following:
a. Reconstruct the corresponding one-way ANOVA table.
b. Perform the one-way ANOVA using level of significance $\alpha = 0.05$. Provide the hypotheses, the conclusion, and the interpretation.
c. Is the conclusion from the one-way ANOVA different from that of the RBD analysis? Did the RBD help to find significance?

11. The randomized block design ANOVA table in Exercise 9
12. The randomized block design ANOVA table in Exercise 10
13. Verify that, in general, $n_t - k$ equals the sum of $b - 1$ and $(k - 1)(b - 1)$. (*Hint:* $n_t = k \cdot b$.)

APPLYING THE CONCEPTS

14. Vegetable Prices. Is there a difference in the mean price of vegetables from year to year? The table contains the price per pound (in dollars) for six different vegetables in June 2012, June 2013, and June 2014. Test, using the randomized block design at level of significance $\alpha = 0.05$, whether the population mean price is the same for all three years. **veggieprices**

	2012	2013	2014
Beans	1.44	1.40	1.48
Broccoli	1.68	1.73	1.77
Lettuce	0.90	0.93	1.12
Sweet Peppers	2.45	2.17	2.59
Potatoes	0.68	0.65	0.69
Tomatoes	1.49	1.47	1.68

Source: U.S. Department of Agriculture.

15. Eighth-Grade Math Scores Worldwide. Have worldwide eighth-grade math scores been changing? The table contains the TIMMS average math score for eighth-graders in five countries, for the years 1995, 2007, and 2011. Test, using the randomized block design at level of significance $\alpha = 0.10$, whether the population mean math score differs across the three years. **worldmath**

	1995	2007	2011
Singapore	609	593	611
Japan	581	570	570
Russia	524	512	539
USA	492	508	509
Sweden	540	491	484

Source: National Center for Education Statistics.

16. Health Insurance: Children Not Covered. The following table contains the number of children under 18 who are not covered by health insurance, for a sample of six states, for the years 2010, 2011, and 2012 (in thousands). Test, using the randomized block design at level of significance $\alpha = 0.10$, whether the population mean number of children not covered by health insurance is the same for all three years. **healthinschildren**

	2010	2011	2012
Alabama	103	86	94
Florida	566	516	527
Georgia	249	278	321
Michigan	122	123	78
New York	352	284	240
Virginia	157	111	108

Source: U.S. Census Bureau.

17. Vegetable Prices. Refer to Exercise 14.
a. Test whether the population mean price is the same for all three years, using a one-way ANOVA and level of significance $\alpha = 0.05$.
b. Compare your conclusions in Exercises 14 and 17(a). Did the randomized block design help us to find significance?

18. Eighth-Grade Math Scores Worldwide. Refer to Exercise 15.
a. Test whether the population mean math score differs across the three years, using a one-way ANOVA and level of significance $\alpha = 0.10$.
b. Compare your conclusions in Exercises 15 and 18(a). Did the randomized block design help us to reject the null hypothesis?

19. Health Insurance: Children Not Covered. Refer to Exercise 16.
a. Test whether the population mean number of children without health insurance is the same for all three years, using a one-way ANOVA and level of significance $\alpha = 0.10$.
b. Compare your conclusions in Exercises 16 and 19(a). Did the randomized block design help us to find significance?

20. Measures of Business Value. The following table contains three measures of the value of a business: the market capitalization, the firm value, and the business enterprise value (in millions of dollars). businessvalue

    a. Use one-way ANOVA to test, at level of significance $\alpha = 0.05$, whether the population mean values differ among the three measures.

    b. Notice how much some of the larger firms (like Caterpillar) differ from some of the smaller firms (like Abercrombie & Fitch) for all three measures. This indicates that there is *variability among the firms*. Let us account for this variability by using the firms as blocks and testing, using the randomized block design at level of significance $\alpha = 0.05$, whether the population mean values differ among the three measures.

    c. Did accounting for the variability among the firms help us to find significant differences among the three measures of firm value?

	Market cap	Firm value	Enterprise value
Abercrombie & Fitch	3.1	3.2	2.7
Caterpillar	36.3	71.9	69.1
CIGNA	10.0	12.1	10.7
Delta Air Lines	9.2	25.8	21.3
Nissan Motors	38.2	90.0	82.4
Repsol Petroleum	32.6	50.2	45.5

Source: U.S. Federal Trade Commission, 2009.

12.4 Two-Way ANOVA

OBJECTIVES By the end of this section, I will be able to . . .

1 Construct and interpret an interaction graph.
2 Perform a two-way ANOVA.

In Section 12.1, we analyzed the treatment means for a single factor in one-way ANOVA. Then, in Section 12.3, a second factor was introduced, but we were not interested in testing for this blocking factor (or nuisance factor). Finally, here in Section 12.4, we learn about *two-way ANOVA*, where we are interested in testing for the significance of two factors.

Consumer Reports publishes ratings of many different items, including cell phones and smartphones. Table 9 contains a random sample of the cell phone and smartphone ratings for three leading carriers, AT&T, T-Mobile, and Verizon, published November 15, 2010.

Table 9 Ratings of mobile phones showing type and carrier

Phone	Carrier	Type	Rating
Samsung Impression	AT&T	Cell	68
Pantech Breeze	AT&T	Cell	62
Samsung Gravity	T-Mobile	Cell	69
Sony Ericsson	T-Mobile	Cell	61
Nokia Twist	Verizon	Cell	66
LG Cosmos	Verizon	Cell	63
Apple iPhone 4	AT&T	Smart	76
Sony Experia	AT&T	Smart	68
Samsung Vibrant	T-Mobile	Smart	76
Blackberry Bold	T-Mobile	Smart	69
Motorola Droid	Verizon	Smart	75
Blackberry Storm	Verizon	Smart	70

Source: www.consumerreports.org/cro/electronics-computers/phones-mobile-devices/cell-phones-services/smart-phone-ratings/ratings-overview.htm.

Let us denote *carrier* as Factor A and *type* as Factor B, and rating as the response variable. Together, Factor A and Factor B are known as the *main effects*. We rearrange Table 9 into the 2 × 3 contingency table shown in Table 10. Each cell in the table contains two ratings of carrier/phone type combination as well as the mean of these ratings.

Table 10 *Consumer Reports* phone ratings, by type and carrier

		Factor A: Carrier		
		AT&T	**T-Mobile**	**Verizon**
Factor B: Type	Cell phone	68 62 Mean = 65	69 61 Mean = 65	66 63 Mean = 64.5
	Smartphone	76 68 Mean = 72	76 69 Mean = 72.5	75 70 Mean = 72.5

Informally, Table 10 shows that the cell means in a given row do not differ by much. For example, the means in the cell phone row are all roughly 65. In contrast, the cell means in a column do seem to differ somewhat. For example, the column AT&T has means 65 and 72. This hints that "Factor A: Carrier" may not be significant, whereas "Factor B: Type" may have a significant effect on ratings. But we need to perform the actual two-way ANOVA to determine this beyond a reasonable doubt.

1 | Constructing and Interpreting an Interaction Graph

Before we perform two-way ANOVA, however, we should ascertain whether factor interaction is present. If the two factors have substantial interaction, then we cannot safely draw conclusions about either factor. So it is important when performing two-way ANOVA to check for the presence of **interaction** between the factors.

> **Interaction** exists between two factors when the effect of one factor depends on the level of the other factor.

For example, suppose that the mean smartphone ratings for AT&T and T-Mobile were higher than their cell phone ratings, but for Verizon, the cell phone ratings were higher. This would be an example of interaction, because the effect of the type factor (cell phone versus smartphone ratings) depended on the level of the carrier factor.

We investigate the presence of interaction using an **interaction plot**.

> An **interaction plot** is a graphical representation of the cell means for each cell in the contingency table. To construct an interaction plot:
> 1. Compute the cell means for all cells.
> 2. Construct an $x - y$ plot (Cartesian plane). Label the horizontal axis for each level of Factor A. The vertical axis represents the response variable.
> 3. For the first level of Factor A, insert a point at a height representing the cell means for the response variable, for each level of Factor B. Then do this for the other levels of Factor A.
> 4. Connect points that have a common Factor B level.

EXAMPLE 13 Constructing an interaction plot

Construct the interaction plot for the data in Table 9.

Solution

We have already calculated the cell means in Table 10. We then use the instructions in the Step-by-Step Technology Guide to construct the interaction plot in Minitab, shown here in Figure 37.

FIGURE 37
Interaction plot for phone ratings.

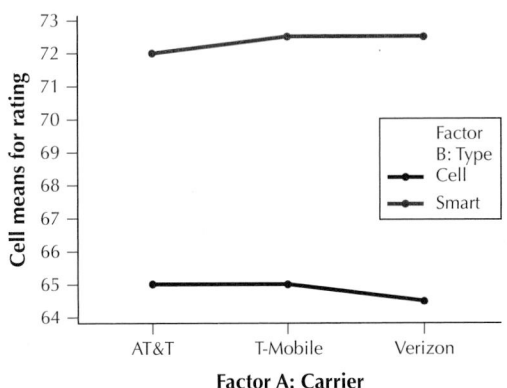

NOW YOU CAN DO
Exercises 9–16.

Figure 38 contains some typical interaction plots, showing (**a**) no interaction, (**b**) some interaction, and (**c**) significant interaction, with three levels for each of Factor A and Factor B. In an interaction plot, we are looking to see if the lines are parallel, or nearly parallel, or not at all parallel. Parallel lines indicate no interaction (Figure 38a); nearly parallel lines indicate some interaction (Figure 38b); lines that are not at all parallel indicate substantial interaction (Figure 38c).

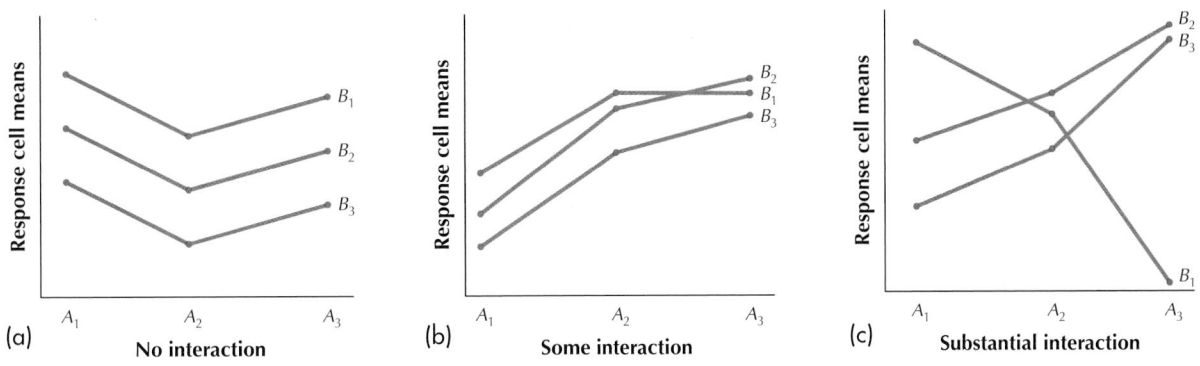

FIGURE 38 Examples of interaction plots.

2 Performing a Two-Way ANOVA

Next, we turn to performing the two-way ANOVA. The requirements are the same as for one-way ANOVA:

- Random samples are taken from each of k normally distributed populations.
- The variances (σ^2) of the populations are all equal.
- The samples are independently drawn.

These requirements may be checked as we did for the one-way ANOVA requirement.

Two-way ANOVA actually involves a series of three hypothesis tests:

1. Test for interaction between the factors

2. Test for Factor A effect

3. Test for Factor B effect

> **CAUTION** If *interaction between the factors occurs*, then we *cannot* draw conclusions about the main effects. If the test for interaction produces evidence that interaction is present, then *do not perform* the test for either Factor A or Factor B.

The steps involved in performing a two-way ANOVA are illustrated using the following example.

EXAMPLE 14 Performing a two-way ANOVA

 **cellsmartphone**

Perform a two-way ANOVA for the cell phone and smartphone ratings data in Table 9, using level of significance $\alpha = 0.05$. Assume the requirements are met.

Solution

The Minitab results are provided in Figure 39, and the SPSS results are provided in Figure 40.

FIGURE 39
Minitab two-way ANOVA results for the phone ratings data.

```
Two-way ANOVA: Rating versus Carrier, Type
Source        DF       SS        MS       F        P
Carrier        2     0.167     0.083    0.00    0.996
Type           1   168.750   168.750    8.20    0.029
Interaction    2     0.500     0.250    0.01    0.988
Error          6   123.500    20.583
Total         11   292.917
```

Tests of Between-Subjects Effects

Dependent Variable: Rating

Source	Type III Sum of Squares	df	Mean Square	F	Sig.
Corrected Model	169.417[a]	5	33.883	1.646	.280
Intercept	56444.083	1	56444.083	2742.223	.000
Carrier	.167	2	.083	.004	.996
Type	168.750	1	168.750	8.198	.029
Carrier * Type	.500	2	.250	.012	.988
Error	123.500	6	20.583		
Total	56737.000	12			
Corrected Total	292.917	11			

a. R Squared = .578 (Adjusted R Squared = .227)

FIGURE 40
SPSS two-way ANOVA results for the phone ratings data.

Step 1 **Test for interaction.** The hypotheses are:

H_0: There is no interaction between carrier (Factor A) and type (Factor B).
H_a: There is interaction between carrier and type.

The null hypothesis H_0 in the test for interaction always states that there is no interaction between the factors.

Reject H_0 if the *p*-value $\leq \alpha = 0.05$. The *p*-value in the Interaction row in Figure 39 is 0.988, which is not ≤ 0.05; therefore, do not reject H_0. There is insufficient evidence of interaction between carrier and type at level of significance $\alpha = 0.05$. Therefore, we may proceed with the tests for the main effects.

12.

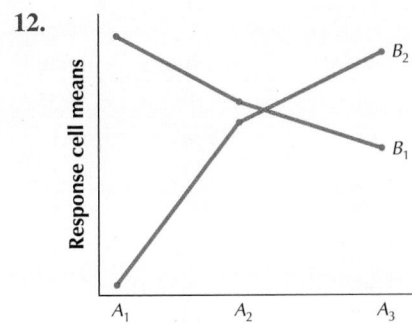

For the data in Exercises 13–16, draw an interaction plot and determine whether there exists no interaction, some interaction, or substantial interaction.

13.

Factor B	Factor A					
	A_1		A_2		A_3	
B_1	3.0	3.5	4.0	3.5	2.5	3.0
B_2	2.5	2.0	2.5	3.0	2.0	1.5

14.

Factor B	Factor A					
	A_1		A_2		A_3	
B_1	11	12	26	25	13	12
B_2	15	17	20	21	16	14

15.

Factor B	Factor A					
	A_1		A_2		A_3	
B_1	11	12	20	21	13	12
B_2	15	17	26	25	16	14

16.

Factor B	Factor A					
	A_1		A_2		A_3	
B_1	100	105	95	100	110	100
B_2	80	85	85	90	80	90

For Exercises 17–20, use level of significance $\alpha = 0.05$ to do the following for the indicated data:
  a. Test for interaction. Confirm that the result agrees with the interaction plot you constructed.
  b. If appropriate, test for the Factor A effect.
  c. If appropriate, test for the Factor B effect.
17. Data in Exercise 13
18. Data in Exercise 14
19. Data in Exercise 15
20. Data in Exercise 16

APPLYING THE CONCEPTS

21. Researchers at Harvard measured the cardiac output in liters per minute of patients involved in a cardiogenomics study. A random sample of eight patients is summarized here. Use level of significance $a = 0.05$ to do the following. **cardiac**
  a. Test for interaction.
  b. If appropriate, test whether cardiac output differs between those who have hypertension (high blood pressure) and those who do not.
  c. If appropriate, test whether cardiac output differs between the genders.

Factor B: Gender	Factor A: Hypertension			
	Yes		No	
Female	1.90	1.68	2.70	4.20
Male	2.60	2.20	4.30	5.50

22. Statistics and algebra instructors tested different types of problems for course assessment (multiple choice, short answer, and word problems) with the data provided here. Use level of significance $a = 0.05$ to do the following: **assessment**
  a. Test for interaction.
  b. If appropriate, test for the problem-type effect.
  c. If appropriate, test for the course effect.

Factor B: Course	Factor A: Problem type					
	Multiple choice		Short answer		Word problems	
Statistics	70	75	73	77	75	79
Algebra	68	76	75	74	80	77

23. The following table contains a random sample of breakfast cereals, along with their brands, whether or not they are high in fiber (at least 3 grams), and their nutritional rating. Use level of significance $\alpha = 0.05$ to perform the two-way ANOVA. **fibercereal**

Cereal	Brand	Fiber	Rating
Shredded Wheat	Kellogg's	High	68.2
All Bran	Kellogg's	High	59.4
Wheaties	General Mills	High	51.6
Total	General Mills	High	46.7
Nut'n Honey	Kellogg's	Low	29.9
Corn Flakes	Kellogg's	Low	45.9
Lucky Charms	General Mills	Low	26.7
Cocoa Puffs	General Mills	Low	22.7

24. A financial consultant is interested in differences in stock prices between industries and stock exchanges. She has collected a random sample of the stock prices of companies trading on the New York Stock Exchange and on NASDAQ, in the defense industry, the apparel industry, and the chemicals industry. Use level of significance $\alpha = 0.01$ to do the following: **stockindustry**

  a. Test for interaction.
  b. If appropriate, test for difference in mean stock price among the industries.
  c. If appropriate, test for difference in mean stock price between the exchanges.
  d. For the results in (**b**) and (**c**), would anything change if we used level of significance $\alpha = 0.05$ instead?

Company	Industry	Exchange	Stock price
Lockheed Martin	Defense	NYSE	75.35
General Dynamics	Defense	NYSE	68.17
Guess, Inc.	Apparel	NYSE	42.30
Van Heusen	Apparel	NYSE	40.68
Monsanto	Chemicals	NYSE	81.75
3M Company	Chemicals	NYSE	82.67
Argon ST	Defense	NASDAQ	21.71
Elbit Systems	Defense	NASDAQ	65.05
Iconix Group	Apparel	NASDAQ	12.67
Perry Ellis	Apparel	NASDAQ	15.06
Gulf Resources	Chemicals	NASDAQ	11.66
Hawkins, Inc.	Chemicals	NASDAQ	21.83

Source: Federal Trade Commission, November 17, 2009.

Chapter 12 Formulas and Vocabulary

SECTION 12.1
- **Analysis of variance (ANOVA)** (p. 666)
- F_{crit} (p. 677)
- F_{data} (p. 673).

$$F_{data} = \frac{MSTR}{MSE}$$

- **Hypotheses for analysis of variance** (p. 673).

$$H_0 : \mu_1 = \mu_2 = \cdots = \mu_k$$

versus

$$H_a : \text{not all the population means are equal}$$

- **Mean square error (MSE)** (p. 671).

$$MSE = \frac{\sum (n_i - 1) s_i^2}{n_t - k}$$

- **Mean square treatment (MSTR)** (p. 671).

$$MSTR = \frac{\sum n_i (\bar{x}_i - \bar{\bar{x}})^2}{k - 1}$$

- **Overall sample mean, $\bar{\bar{x}}$** (p. 670).

$$\bar{\bar{x}} = \frac{(n_1\bar{x}_1 + n_2\bar{x}_2 + \cdots + n_k\bar{x}_k)}{n_t}$$

- ***p*-Value** (p. 673)
- **Sum of squares error (SSE)** (p. 671).

$$SSE = \sum (n_i - 1) s_i^2$$

- **Sum of squares treatment (SSTR)** (p. 671).

$$SSTR = \sum n_i (\bar{x}_i - \bar{\bar{x}})^2$$

- **Total sum of squares (SST)** (p. 671).

$$SST = SSTR + SSE$$

SECTION 12.2
- **Bonferroni adjustment** (p. 686)
- **Experimentwise error rate α_{EW}** (p. 686)
- **Multiple comparisons** (p. 686)
- **Number of pairwise comparisons** (p. 686).

$$c = (_kC_2) = \frac{k!}{2!(k - 2)!}$$

- **Rejection rule for using confidence intervals to perform Tukey's test** (p. 689)
- **Test statistic for Bonferroni method** (p. 687).

$$t_{data} = \frac{\bar{x}_1 - \bar{x}_2}{\sqrt{MSE \cdot \left(\frac{1}{n_1} + \frac{1}{n_2}\right)}}$$

- **Test statistic for Tukey method** (p. 689).

$$q_{data} = \frac{\bar{x}_1 - \bar{x}_2}{\sqrt{\frac{MSE}{2} \cdot \left(\frac{1}{n_1} + \frac{1}{n_2}\right)}}$$

- **Tukey critical value q_{crit}** (p. 688)

SECTION 12.3
- **Blocking factor**, or **block** (p. 694)
- **Randomized block design** (p. 694)

SECTION 12.4
- **Interaction** (p. 702)
- **Interaction plot** (p. 702)

Chapter 12 Review Exercises

SECTION 12.1

1. For the following data, assume that the ANOVA assumptions are met, and calculate the measures in (**a**)–(**h**).

Sample A	Sample B	Sample C	Sample D
$\bar{x}_A = 0$	$\bar{x}_B = 10$	$\bar{x}_C = 20$	$\bar{x}_D = 10$
$s_A = 1.5$	$s_B = 2.25$	$s_C = 1.75$	$s_D = 2.0$
$n_A = 50$	$n_B = 100$	$n_C = 50$	$n_D = 100$

   a. df_1 and df_2
   b. $\bar{\bar{x}}$
   c. SSTR
   d. SSE
   e. SST
   f. MSTR
   g. MSE
   h. F_{data}

2. Construct the ANOVA table for the statistics in Exercise 1.

For Exercises 3–5, assume that the ANOVA assumptions are met and perform the appropriate analysis of variance using $a = 0.05$.

3. Differences in Medical Treatments. A psychologist is interested in investigating whether differences in mean client improvement exist for three medical treatments. Seven clients undergoing each medical treatment were asked to rate their level of satisfaction on a scale of 0 to 100. The data are provided in the following table. **ᴴⁱ medicaltreatmt**

Medical treatment 1	Medical treatment 2	Medical treatment 3
75	75	100
100	100	100
0	25	50
50	75	90
50	50	75
40	75	75
25	60	90

4. Customer Satisfaction. The district sales manager of a local chain store wants to determine whether significant differences exist in the mean customer satisfaction among the four franchise stores in her district. Customer satisfaction data were gathered over seven days at each of the four stores. The resulting data are summarized in Table 13.
ᴴⁱ customersatisfy

TABLE 13   Customer satisfaction in four stores

Store A	Store B	Store C	Store D
50	60	25	75
40	45	30	60
60	70	50	80
60	70	30	90
50	60	40	70
45	65	25	85
55	70	45	95
$\bar{x}_A = 51.43$	$\bar{x}_B = 62.86$	$\bar{x}_C = 35.00$	$\bar{x}_D = 79.29$
$s_A = 7.48$	$s_B = 9.06$	$s_C = 10.00$	$s_D = 12.05$

SECTION 12.2

For Exercises 5–7, use the summary statistics to calculate the value of the test statistic t_{data} for the Bonferroni method.
  5. $\bar{x}_1 = 50, \bar{x}_2 = 75$, MSE $= 1250$, $n_1 = 25$, $n_2 = 25$
  6. $\bar{x}_1 = 50, \bar{x}_3 = 65$, MSE $= 1250$, $n_1 = 25$, $n_3 = 25$
  7. $\bar{x}_2 = 75, \bar{x}_3 = 65$, MSE $= 1250$, $n_2 = 25$, $n_3 = 25$
8. Perform multiple comparisons using the Bonferroni method at level of significance $\alpha = 0.05$ for the data in Exercises 5–7. Assume the requirements are met. Do the following:
  a. For each hypothesis test, state the hypotheses and the rejection rule.
  b. Use the value of t_{data} from Exercises 5–7 for each hypothesis test.
  c. Find the Bonferroni-adjusted p-value for each hypothesis test.
  d. For each hypothesis test, state the conclusion and the interpretation.

For Exercises 9–11, use the summary statistics to calculate the value of the test statistic q_{data} for Tukey's test.
  9. $\bar{x}_1 = 200, \bar{x}_2 = 224$, MSE $= 14,400$, $n_1 = 100, n_2 = 100$
  10. $\bar{x}_1 = 200, \bar{x}_3 = 248$, MSE $= 14,400$, $n_1 = 100, n_2 = 100$
  11. $\bar{x}_2 = 224, \bar{x}_3 = 248$, MSE $= 14,400$, $n_1 = 100, n_2 = 100$
12. Find the Tukey critical value q_{crit} for experimentwise error rate $\alpha_{EW} = 0.05$, $n_1 = 100$, $n_2 = 100$, $n_3 = 100$.
13. Perform multiple comparisons using Tukey's test at experimentwise error rate $\alpha_{EW} = 0.05$, for the data in Exercises 9–11. Assume the requirements are met. Do the following.
  a. For each hypothesis test, state the hypotheses.
  b. Use the value of q_{crit} from Exercise 12, and state the rejection rule.
  c. Use q_{data} from Exercises 9–11 for each hypothesis test.
  d. For each hypothesis test, state the conclusion and the interpretation.

SECTION 12.3

14. For the partially completed randomized block ANOVA table, do the following:

  a. Provide the hypotheses.
  b. Complete the missing entries in the table.
  c. Use technology to find the p-value. For level of significance $a = 0.05$, provide the conclusion and the interpretation.

Source	Sum of squares	Degrees of freedom	Mean square	F
Treatments	—	—	—	2
Blocks	420	5	—	
Error	200	20	10	
Total	700	—		

15. The following table represents an excerpt from a U.S. Census Bureau report on the numbers of small businesses owned by females and males in four industries. Perform the randomized block design ANOVA using level of significance $\alpha = 0.05$. ▦ **genderindustry**

Block: Industry	Factor of interest: Gender of owner	
	Female	**Male**
Retail	9	11
Real estate	5	11
Health care	20	10
Entertainment	3	5

SECTION 12.4

For the data in Exercises 16 and 17, draw an interaction plot and determine whether there exists no interaction, some interaction, or substantial interaction.

16.

Factor B	Factor A					
	A_1		A_2		A_3	
B_1	100	95	120	115	85	80
B_2	120	125	150	155	100	95

17.

Factor B	Factor A					
	A_1		A_2		A_3	
B_1	10	8	3	2	12	9
B_2	3	4	11	7	5	2

For Exercises 18 and 19, use level of significance $\alpha = 0.05$ to do the following for the indicated data:

  a. Test for interaction. Confirm that the result agrees with the interaction plot you constructed.
  b. If appropriate, test for the Factor A effect.
  c. If appropriate, test for the Factor B effect.

18. Data in Exercise 16
19. Data in Exercise 17

Chapter 12 QUIZ

TRUE OR FALSE

1. True or false: The F curve is symmetric.
2. True or false: The total area under the F curve equals 1.
3. True or false: If we reject the null hypothesis in an ANOVA, we conclude that there is evidence that all the population means are different.

FILL IN THE BLANK

4. A weighted average (mean) of quantities that are squared is called a _____ _____ [two words].
5. The _____ _____ _____ [three words] measures the variability in the sample means.
6. The _____ _____ _____ [three words] measures the variability within the samples.

SHORT ANSWER

7. What do we use for an estimate of the overall population mean?
8. In the ANOVA table, what is the relationship between the quantities in the mean square column with respect to the quantities in the sum of squares column and the degrees of freedom column?

9. For analysis of variance, the p-value represents the area to the right of what?

CALCULATIONS AND INTERPRETATIONS

For Exercises 10–12, assume that the requirements for ANOVA have been met and test, using level of significance $\alpha = 0.05$, whether the population means are equal.

10. Gas Mileage and Number of Cylinders. When it comes to getting good gas mileage, does the number of cylinders in your engine make a difference? The following table provides the summary statistics regarding miles per gallon for 4-cylinder, 6-cylinder, and 8-cylinder cars.

	4 cylinders	**6 cylinders**	**8 cylinders**
n	199	83	103
$\bar{x}$	29.3	20.0	15.0
s	5.7	3.8	2.9

11. Hours Worked and Marital Status. The General Social Survey tracks demographic trends. Here, we are interested in whether the mean number of hours worked differs by marital status. The summary statistics are shown here.

	N	Mean	Std. Deviation
MARRIED	964	42.76	14.08
WIDOWED	72	40.13	14.28
DIVORCED	342	43.69	13.93
SEPARATED	79	41.66	15.71
NEVER MARRIED	478	41.03	14.03

12. Calories in Breakfast Cereals. A dietary researcher is interested in whether differences exist in the mean number of calories in breakfast cereals made by different manufacturers. The summary statistics for the samples from three manufacturers appear in the following table.

	Kellogg's	Quaker	Ralston Purina
n	23	8	8
$\bar{x}$	109	95	115
s	22	29	23

13

Inference in Regression

Introduction

Chapter 13 continues the development of the topic of regression analysis begun in Chapter 4, with emphasis on inference. Section 13.1 covers hypothesis tests and confidence intervals about the slope of the regression line. Section 13.2 introduces confidence intervals and prediction intervals. Finally, Section 13.3 covers multiple regression and model building.

From the Author

Section 13.1 Inference About the Slope of the Regression Line

- Instructors may want to review Section 4.2, "Introduction to Regression," and Section 4.3, "Further Topics in Regression Analysis," before discussing Section 13.1.

- Example 1 is a review of several of the regression topics covered in Sections 4.2 and 4.3. Note that the regression coefficients are fairly round numbers, allowing students to concentrate on the concepts and techniques.

- Students may not understand the difference between the regression equation and the population regression equation. The regression equation comes from a single sample, so, similar to $\bar{x}$, the values for b_0, and b_1 will vary from sample to sample. But the population regression equation comes from the entire population, so, similar to μ, the values for β_0 and β_1 are constants.

- Instructors may want to point out the importance of the value 0 for β_1. If $\beta_1 = 0$, then no linear relationship exists between x and y. If the confidence interval for β_1 contains zero, then, again, no linear relationship exists between x and y.

Section 13.2 Confidence Intervals and Prediction Intervals

- Students may confuse the confidence interval learned in this section with the confidence interval from Section 13.1. Point out that, in Section 13.1, the confidence interval was for the slope β_1, whereas in Section 13.2, the confidence interval is for the mean value of y for a given value of x.

- The calculations in this section are rather challenging. Instructors who are pressed for time may prefer to work with the confidence intervals and prediction intervals calculated by the software, and focus on the interpretation of the results.

Section 13.3 Multiple Regression

- Instructors may point out that multiple regression is simply an extension of the regression model from Section 13.1, to include more than one x variable. Many concepts in Section 13.3 are similar to those in Section 13.1.

- Because of the widespread use of multiple regression (including in my own consulting work), I was excited to write this section. If we can offer students a taste of model building, then some of them will surely catch "statistics fever" and go on to become data analysts like us!

Teaching Tips

Regression Review Example Using Technology

Students may benefit from an additional review example of the concepts from Chapter 4, this time using technology in the form of a Minitab printout. Data was collected on ten kayakers on the Shenandoah River in Virginia in 2015. Kayaking time in hours (x) and distance traveled in miles (y) were recorded. Regression of distance on time was performed, with the results shown in Figure A. Answer the following questions.

```
The regression equation is
Distance = 6.00 + 2.00 Time


Predictor    Coef   SE Coef      T       P
Constant   6.0000    0.9189    6.53   0.000
Time       2.0000    0.1667   12.00   0.000


S = 1.22474    R-Sq = 94.7%    R-Sq(adj) = 94.1%


Analysis of Variance

Source           DF       SS      MS       F       P
Regression        1   216.00  216.00  144.00   0.000
Residual Error    8    12.00    1.50
Total             9   228.00
```

FIGURE A Regression of distance on time for ten kayakers on the Shenandoah River.

1. Provide and interpret the slope b_1 of the regression equation.
2. Find and interpret the y intercept b_0 of the regression equation.
3. State and interpret the standard error of the estimate, s.
4. Find and interpret the coefficient of determination, r^2.
5. Predict the distance traveled by a kayaker who spent two hours kayaking.
6. Suppose the actual distance traveled by a particular kayaker who spent two hours kayaking was 9 miles. Calculate and interpret the prediction error.

Solution

1. The b_1 slope equals 2.0. This is interpreted as, for each additional hour of kayaking, the estimated increase in distance is two miles.

2. The y intercept b_0 equals 6.0. This means that, for a kayaker who spent zero hours kayaking, the estimated distance traveled is six miles. This does not make sense because it represents extreme extrapolation.

3. The standard error of the estimate is $s = 1.22474$. This is interpreted as the size of the typical error in prediction.

4. The value of r^2 is 0.947. That is, 94.7% of the variability in distance is accounted for by the linear relationship between distance and time.

5. The regression equation is $\hat{y} = 6 + 2(\text{time})$. For a time of 2 hours, we have the predicted distance traveled to be $\hat{y} = 6 + 2(2) = 10$ miles.

6. The prediction error is $y - \hat{y} = 9 - 10 = -1$ mile. The distance traveled is one mile less than predicted, given the time.

In-Class Activities

1. Ask students to collect data on two variables, such as the average daily temperature for their towns and the number of soft drinks they consumed during the observed day. This can be assigned to the students before the analysis is to be done in class, so that they can observe for a specified number of days (possibly 7). Because the students will be observing the same days, an average for the number of soft drinks consumed each day should be used as well as the average daily temperature. The activity should include scatterplots for the data and modeling of the data. Students can use the average number of soft drinks consumed as the y variable and the average daily temperature as the x variable. (*Note:* Use appropriate technology if available.)

2. Regression Activity for Distance and Time to Class

In this activity, we discuss whether the amount of time it takes a student to get to class depends on how far away the student lives.

 a. Generate a sample of 15 students from your class, using either (a) technology or (b) the *Simple Random Sample* applet. Each student's sample should be different.

b. Find out how long it takes each student in your sample to get to class and how far away the student lives. If the student lives on campus, obtain the data regarding the student's permanent residence (home mailing address).

    i. Construct a scatterplot of the time, in minutes, against the distance in miles. Will everyone's scatterplot look exactly the same? Will everyone's scatterplot look somewhat similar?

    ii. Draw a straight line that you think best approximates the relationship between time and distance.

    iii. Estimate the slope and y intercept of the line you drew in (**ii**).

    iv. Now calculate the regression equation using the data from your sample in (**a**). Draw the regression line on your scatterplot.

c. Explore the regression results.

    i. Clearly describe the meaning of the slope b_1.

    ii. Does the literal meaning of the y intercept b_0 make sense? What is its literal meaning? What is wrong with interpreting it this way?

    iii. Are your estimates in **b(iii)** close to the least-squares results in **b(iv)**?

    iv. Are time and distance positively correlated? Compute the correlation coefficient.

d. Make a prediction, and find the prediction error.

   i. Choose one student from your scatterplot and calculate the predicted time for the student. Then calculate the prediction error for the student.

   ii. What is the typical difference between the predicted amount of time and the actual amount of time? Do you want this statistic (*Hint: s*) to be large or small?

Supplements

- StatTutor 5.0–5.8, 24.1–24.8
- Stats@Work Simulations
    - Picture This, Scatterplots and Correlation; Jan Pepperoni
    - Picture This, Scatterplots and Correlation; Justine Red
- EESEE case studies
    - Brain Size and Intelligence (Questions 1, 2, 3, 4, and 5)
    - Is Old Faithful Faithful? (Questions 1, 2, 3, and 4)

Applets

This Web site has an applet for correlation and regression: http://wise.cgu.edu/.

Videos

- *Against All Odds: Inside Statistics:* www.learner.org/resources/series65.html
    - Program 11: Fitting Lines to Data
    - Program 12: Correlation
    - Program 30: Inference for Regression

Web Sites

- The site Online Statistics: An Interactive Multimedia Course of Study has information on regression and correlation, and it has a link to the Rice Virtual Lab in Statistics with simulation demonstrations and other topics related to statistics: http://onlinestatbook.com/index.html.
- This Web site lists other sites that do statistical calculations for correlation and regression analysis: http://statpages.org/.
- This site has an activity that deals with correlation and regression: http://www.amstat.org/publications/jse/v6n3/smith.html.

13

Inference in Regression

BEW Authors/AgeFotostock

How Fair Is the Scoring in Scrabble?

Scrabble[1] is a game in which the players choose letters randomly from a pool of letter tiles and take turns building words. Each letter tile is worth a specific number of points. For example, the letter A is worth 1 point, and the letter Z is worth 10 points. Do you think the way the various letters are valued is fair? Do you dread picking up the letter Z so much that you think it should be worth 25 points instead of 10? Do you think there are too many I's, perhaps, and not enough T's?

We can use *regression analysis,* the topic of this chapter, to help us find out why it often seems that there are too many I's and not enough T's. Regression can also help us understand why we would rather pick up an H tile instead of a V tile, even though each tile is worth 4 points. We can compare the distribution of letter points in Scrabble to the number of letter tiles in the game and to the frequency distribution of letters in the English language. To see more, check out this chapter's Case Study, *How Fair Is the Scoring in Scrabble?* (page 727).

THE BIG PICTURE

Where we are coming from and where we are headed . . .

- In Chapter 4, we were introduced to regression analysis using descriptive methods.
- Then, in Chapters 8–12, we learned about inference.
- Here, in Chapter 13, "Inference in Regression," we learn how to apply inference to regression analysis, such as the t test for the relationship between x and y.
- Where do you go from here? That's up to you. The world is overflowing with data and needs people like you who understand the statistical techniques and methods for uncovering the knowledge locked in the data. This course in statistics has tried to provide you with these tools. Here's hoping that your new expertise as a data analyst will help you achieve success in the Information Age.

13.1 Inference About the Slope of the Regression Line

OBJECTIVES By the end of this section, I will be able to . . .

1 Explain the regression model and the regression model assumptions.
2 Perform the hypothesis test for the slope β_1 of the population regression equation.
3 Construct confidence intervals for the slope β_1.
4 Use confidence intervals to perform the hypothesis test for the slope β_1.

1 The Regression Model and the Regression Assumptions

Before we learn about the regression model and assumptions, let us review the correlation and regression topics that we learned in Chapter 4. Recall that the regression line approximates the linear relationship between two continuous variables and is described by the regression equation $\hat{y} = b_1x + b_0$, where b_1 is the *slope* of the regression line, b_0 is the y *intercept*, x represents the *predictor variable*, y represents the *response variable*, and $\hat{y}$ represents the *estimated or predicted* y-*value*.

EXAMPLE 1   Review of regression topics

 **textms**

The Nielsen company has reported that the number of text messages that a person sends tends to decrease with age. Table 1 contains a random sample of 10 people, along with their age and the number of text messages they sent on the previous day.

a. Construct and interpret a scatterplot of the response variable y versus the predictor variable x.

b. Calculate and interpret the correlation coefficient r.

c. Compute the regression equation $\hat{y} = b_1x + b_0$. Interpret the meaning of the y intercept b_0 and the slope b_1 of the regression equation.

You may want to refer to Section 4.1 for **(a)** and **(b)**, and Section 4.2 for **(c)** and **(d)**.

d. Predict the number of text messages sent by a 20-year-old person, and calculate the prediction error (residual).

Table 1 Age and number of text messages

x = Age	y = Text messages	x = Age	y = Text messages
18	35	28	16
20	29	30	19
22	27	32	12
24	28	34	8
26	19	36	8

Solution

a. The number of messages depends on age, and not vice versa, so the predictor variable x is age and the response variable y is messages. Also, note that in **(d)**

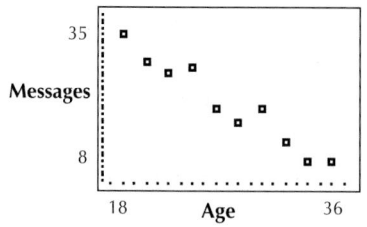

FIGURE 1 TI-83/84 scatterplot of messages versus age.

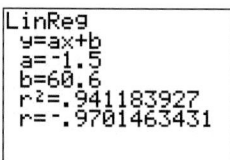

FIGURE 2 TI-83/84 correlation and regression results.

we are trying to predict the number of text messages, which tells us that messages is the response variable y because we never try to predict the known value of x. The TI-83/84 scatterplot is shown in Figure 1. As age increases, the number of messages tends to decrease.

b. Figure 2 shows the correlation coefficient $r \approx -0.9701$, calculated by the TI-83/84. Age and messages are negatively correlated. An increase in age is associated with a decrease in the number of messages.

c. Figure 2 shows that $a = b_1 = -1.5$ and $b = b_0 = 60.6$, and thus the regression equation is

$$\hat{y} = b_1 x + b_0 = (-1.5)(\text{age}) + 60.6$$

We can interpret b_0 and b_1 as follows:

- The y intercept $b_0 = 60.6$ is the estimated number of text messages sent by someone aged $x = 0$, which does not make sense because this value $x = 0$ lies far below the minimum value of x and therefore represents extreme extrapolation.

- The slope $b_1 = -1.5$ means there is an estimated *decrease* of 1.5 in the number of text messages for each additional year of age.

d. For a 20-year-old person, the estimated number of daily text messages is

$$\hat{y} = b_1 x + b_0 = (-1.5)(20) + 60.6 = 30.6$$

The actual number of text messages sent by our 20-year-old in Table 1 is $y = 29$. Our prediction from (c) is $\hat{y} = 30.6$. Thus, our prediction error (or residual) is: $(y - \hat{y}) = (29 - 30.6) = -1.6$. Our 20-year-old sent slightly fewer text messages than expected.

YOUR TURN #1

Age (x)	Score (y)
14	80
16	90
18	90
20	90
22	100

The table contains the age (x) and score in a video game (y) for a random sample of five young people.

a. Construct and interpret a scatterplot of the response variable y versus the predictor variable x.

b. Calculate and interpret the correlation coefficient r.

c. Compute the regression equation $\hat{y} = b_1 x + b_0$. Interpret the meaning of the y intercept b_0 and the slope b_1 of the regression equation.

d. Predict the score for a 22-year old person, and calculate the prediction error (residual).

(The solutions are shown in Appendix A.)

Example 1 and our work in Chapter 4 on regression represented descriptive statistics. Next, we turn to learning about *inference in regression*.

Note that the regression equation $\hat{y} = b_1 x + b_0 = (-1.5)(\text{age}) + 60.6$ depends on the sample. It is likely that a second sample will differ from the first, giving us a different regression line and different values for b_0 and b_1. In fact, for every different sample, b_0 and b_1 take different values because b_0 and b_1 are sample statistics. However, every sample comes from a population. We do not have data on the entire population, so we are not able to calculate the population regression equation. The y intercept β_0 and slope β_1 of the population regression equation are unknown population parameters, just as μ and p are parameters in other contexts. The values of β_0 and β_1 are unknown, so we need to perform inference to learn about them.

The **regression model** may be used to approximate the relationship between the predictor variable x and the response variable y for the *entire population* of (x, y) pairs.

Note that there is no "hat" on the y in the population regression equation because the equation represents a model of the relationship between the actual values of x and y, not an estimate of y.

Regression Model

The **population regression equation** is defined as

$$y = \beta_1 x + \beta_0 + \varepsilon$$

where β_0 is the y intercept of the population regression line, β_1 is the slope of the population regression line, and ε is the error term.

The 20-year-old in Table 1 sent 29 text messages. Suppose another 20-year-old sent 30 messages, so that both texters had age $x = 20$, but different values of y: $y = 29$ and $y = 30$. Then it would be impossible to draw a single regression line to pass through both ($x = 20$, $y = 29$) and ($x = 20$, $y = 30$). Thus, any linear approximation of the true relationship between x and y will introduce a certain amount of error. This is why the error term ε is needed.

Regression Model Assumptions

The regression model operates under a set of four assumptions that must be valid in order to perform the inference in this section.

Regression Model Assumptions

1. **Zero-mean assumption.** The error term ε is a random variable, with a mean of 0. That is, the expected value of the random variable ε is 0: $E(\varepsilon) = 0$.
2. **Constant variance assumption.** The variance of ε, which is denoted as σ^2, is the same regardless of the value of x.
3. **Independence assumption.** The values of ε are independent of each other.
4. **Normality assumption.** The error term ε is a normal random variable.

To summarize, for each value of x, the values of y come from a normally distributed population with a mean on the population regression line $E(y) = \beta_1 x + \beta_0$ and constant standard deviation σ^2. Figure 3 illustrates how y is distributed for each value of x. Note that each normal curve has the same shape, indicating constant variance for each x.

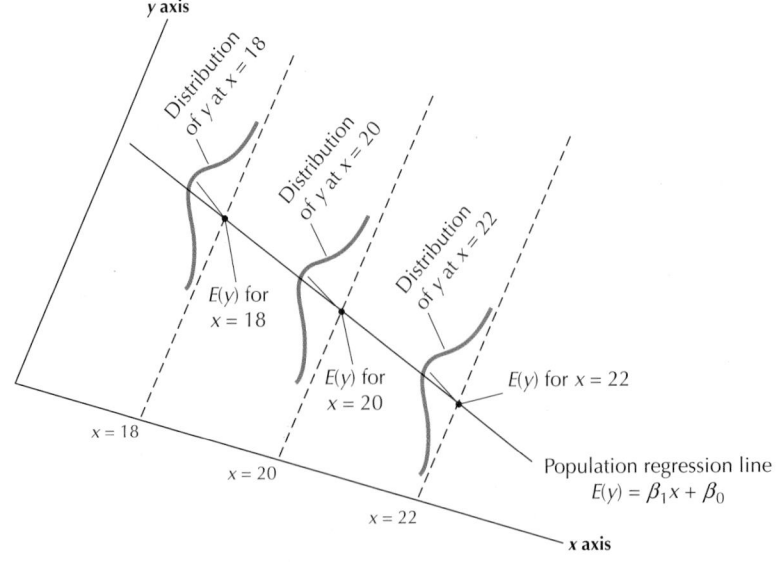

FIGURE 3 Illustrating the regression assumptions.

Verifying the Regression Assumptions

To check the regression model assumptions, we construct two graphs:

1. Scatterplot of the residuals (prediction errors, $y - \hat{y}$) against the fitted values (**fitted values** refers to the predicted values, $\hat{y}$)

2. Normal probability plot of the residuals

Figure 4 shows four types of patterns that might be observed in the residuals versus fitted values plots.

● Plot (a) is a "healthy" plot, displaying no noticeable patterns.

● Plot (b) is a curve, which indicates a violation of the independence assumption. Independence implies that knowing the value of a particular y does not help to predict the value of a different y. However, a curve suggests that knowing the value of a previous y helps in knowing the value of the next y.

● Plot (c) shows a "funnel" pattern, which contradicts the constant variance assumption. The residuals on the left are close together vertically (small variability), whereas the residuals on the right are far apart vertically (large variability).

● Plot (d) shows an increasing pattern that violates the zero-mean assumption. The residuals on the left are all below the midline, so $E(y) < \beta_1 x + \beta_0$, whereas the residuals on the right are all above the midline, so $E(y) > \beta_1 x + \beta_0$.

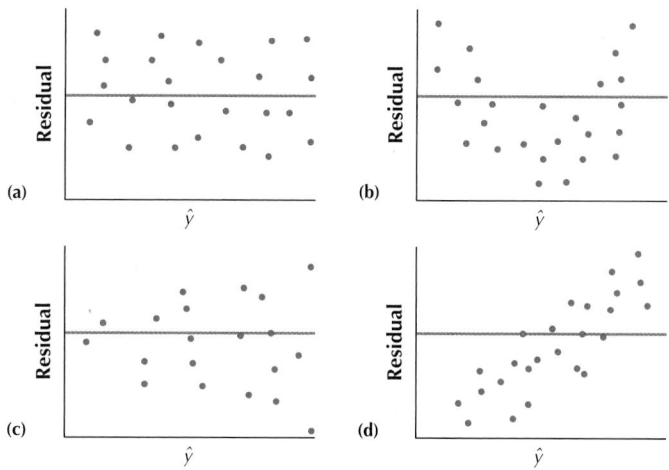

FIGURE 4 Patterns in the residuals versus predicted plots.

Developing Your Statistical Sense

Verifying the Regression Assumptions

With small data sets, it is difficult to ascertain whether or not patterns really exist. Be wary of seeing patterns where none exist. If one or more regression assumptions are violated, we should not proceed with inferential methods such as hypothesis tests or confidence intervals. However, even if one or more regression assumptions are violated, we can still report and interpret the descriptive regression statistics that we learned in Sections 4.2 and 4.3.

EXAMPLE 2 Calculating the residuals and verifying the regression assumptions

For the data in Example 1, do the following:

a. Calculate the residuals $y - \hat{y}$.

b. Verify the regression assumptions.

Solution

a. Table 2 contains the x and y data from Table 1, the fitted (predicted) values $\hat{y}$, and the residuals $y - \hat{y}$.

Table 2 Calculating the residuals

x = Age	y = Text messages	Fitted (predicted) values $\hat{y} = (-1.5)(\text{age}) + 60.6$	Residuals $y - \hat{y}$
18	35	33.6	1.4
20	29	30.6	−1.6
22	27	27.6	−0.6
24	28	24.6	3.4
26	19	21.6	−2.6
28	16	18.6	−2.6
30	19	15.6	3.4
32	12	12.6	−0.6
34	8	9.6	−1.6
36	8	6.6	1.4

b. The scatterplot in Figure 5 of the residuals versus fitted values shows no strong evidence of the unhealthy patterns shown in Figure 4. Thus, the independence assumption, the constant variance assumption, and the zero-mean assumption are verified. Also, the normal probability plot of the residuals in Figure 6 indicates no evidence of departures from normality in the residuals. Therefore, we conclude that the regression assumptions are verified.

NOW YOU CAN DO
Exercises 7–14.

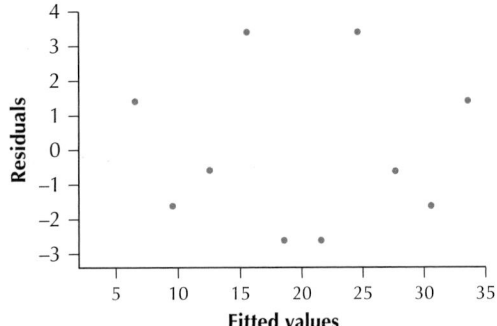

FIGURE 5 Scatterplot of residuals versus fitted values.

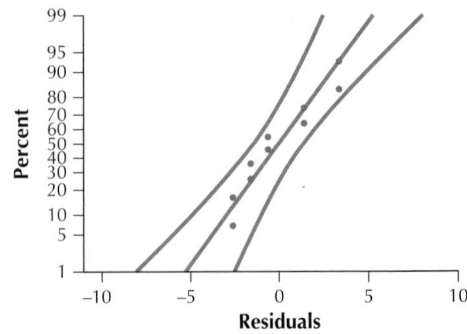

FIGURE 6 Normal probability plot of the residuals.

YOUR TURN
#2

For the data in the Your Turn #1 on page 717, do the following:

a. Calculate the residuals $y - \hat{y}$. for the regression of score on age.

b. Verify the regression assumptions.

(The solutions are shown in Appendix A.)

Once the regression assumptions have been verified, we may (a) perform hypothesis tests, and (b) construct confidence intervals for the population slope β_1.

2 Hypothesis Tests for Slope β_1

Suppose for a moment that, for the population regression equation $y = \beta_1 x + \beta_0 + \varepsilon$, the slope β_1 equals zero. Then the population regression equation would be

$$y = (0)x + \beta_0 + \varepsilon = \beta_0 + \varepsilon$$

That is,

- *If β_1 equals zero, then no relationship exists between x and y, because changing x in the equation $y = \beta_0 + \varepsilon$ does not affect y.*

- If β_1 equals any other value, then a linear relationship does exist between x and y.

This idea forms the basis for our inference in this section. To test whether a relationship exists between x and y, we begin with the hypothesis test to determine whether or not β_1 equals 0. The hypotheses are

$H_0 : \beta_1 = 0$ No linear relationship exists between x and y.
$H_a : \beta_1 \neq 0$ A linear relationship exists between x and y.

Assuming $H_0 : \beta_1 = 0$ is true, the test statistic t_{data} for this hypothesis test takes the following form.

CAUTION Here, s refers to the standard error of the estimate, *not the* sample standard deviation.

Test Statistic t_{data}

$$t_{\text{data}} = \frac{b_1 - \beta_1}{s/\sqrt{\sum(x - \bar{x})^2}} = \frac{b_1 - 0}{s/\sqrt{\sum(x - \bar{x})^2}} = \frac{b_1}{s/\sqrt{\sum(x - \bar{x})^2}}$$

where b_1 represents the slope of the regression line, $s = \sqrt{\frac{SSE}{n-2}}$ represents the standard error of the estimate (from Section 4.3), and $\sqrt{\sum(x - \bar{x})^2}$ is related to the sample variance of the x data (see page 229 in Section 4.3).

---

t_{data} consists of three quantities: b_1, s, and $\sqrt{\sum(x - \bar{x})^2}$. The next example shows how to calculate t_{data} by finding these three quantities.

EXAMPLE 3   Calculating t_{data}

Use the following steps to calculate the test statistic $t_{\text{data}} = \frac{b_1}{s/\sqrt{\sum(x - \bar{x})^2}}$ for the data in Table 2:

a. Find b_1, the slope of the regression line.

b. Calculate s, the standard error of the estimate.

c. Compute $\sqrt{\sum(x - \bar{x})^2}$, the numerator of the sample variance of the x data.

Solution

a. From Example 1, the slope of the regression line is $b_1 = -1.5$.

b. Recall from Section 4.3 (page 228) that

$$s = \sqrt{\frac{SSE}{n-2}} = \sqrt{\frac{\sum(y-\hat{y})^2}{n-2}} = \sqrt{\frac{\sum(\text{residual})^2}{n-2}}$$

is the *standard error of the estimate*. Squaring each residual from Table 2 gives us the squared residuals in Table 3, and the sum of squared residuals, or sum of squares error, equal to

$$SSE = \sum(y-\hat{y})^2 = 46.4$$

Then the standard error of the estimate is

$$s = \sqrt{\frac{SSE}{n-2}} = \sqrt{\frac{46.4}{8}} \approx 2.408318916.$$

All calculations up to the final result are expressed to nine decimal places.

Table 3 Calculating SSE

Residuals $y - \hat{y}$	Squared residuals $(y - \hat{y})^2$
1.4	1.96
−1.6	2.56
−0.6	0.36
3.4	11.56
−2.6	6.76
−2.6	6.76
3.4	11.56
−0.6	0.36
−1.6	2.56
1.4	1.96
	Sum = 46.4

c. To compute $\sum(x-\bar{x})^2$, we note from page 110 in Chapter 3 that the sample variance of x is

$$s_x^2 = \frac{\sum(x-\bar{x})^2}{n-1}$$

Multiplying each side of the equation by $n-1$, we obtain an equation for the quantity $\sum(x-\bar{x})^2$:

$$\sum(x-\bar{x})^2 = (n-1)\cdot s_x^2$$

The TI-83/84 output from Figure 7 shows that $s_x = 6.055300708$, and, because $n = 10$,

$$\sum(x-\bar{x})^2 = (n-1)\cdot s_x^2 = (9)(6.055300708)^2 = 330$$

Therefore,

$$t_{\text{data}} = \frac{b_1}{s/\sqrt{\sum(x-\bar{x})^2}} = \frac{-1.5}{2.408318916/\sqrt{330}} \approx -11.3$$

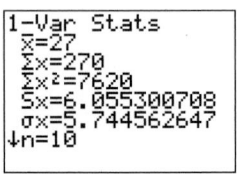

```
1-Var Stats
x̄=27
Σx=270
Σx²=7620
Sx=6.055300708
σx=5.744562647
↓n=10
```

FIGURE 7 Summary statistics for the x (age) data.

Now that we have t_{data}, we can perform the hypothesis test for the slope β_1, as the next example shows using the critical-value method.

EXAMPLE 4 **Hypothesis test for slope β_1 using the critical-value method**

Test whether a linear relationship exists between age and text messages, using the data from Table 1 at level of significance $\alpha = 0.01$.

Solution

The regression assumptions were shown to be valid in Example 2. We may thus proceed with the hypothesis test.

Step 1 **State the hypotheses.**

$H_0 : \beta_1 = 0$ No linear relationship exists between age and text messages.
$H_a : \beta_1 \neq 0$ A linear relationship exists between age and text messages.

Step 2 **Find the t critical value t_{crit} and the rejection rule.** To find t_{crit}, use the t distribution table (Table D in the Appendix) for a two-tailed test and degrees of freedom df $= n - 2$. The rejection rule for this two-tailed test is

$$\text{Reject } H_0 \text{ if } t_{\text{data}} \geq t_{\text{crit}} \text{ or } t_{\text{data}} \leq -t_{\text{crit}}$$

Here, $n = 10$, so df $= 8$. For level of significance $\alpha = 0.01$, the t table gives us $t_{\text{crit}} = 3.355$. We will reject H_0 if $t_{\text{data}} \geq 3.355$ or $t_{\text{data}} \leq -3.355$.

Step 3 **Calculate t_{data}.** From Example 3, we have

$$t_{\text{data}} = \frac{b_1}{s/\sqrt{\sum (x - \overline{x})^2}} \approx -11.3$$

NOW YOU CAN DO
Exercises 15–18.

Step 4 **State the conclusion and the interpretation.** Because $t_{\text{data}} \approx -11.3 \leq -3.335$, we reject H_0. There is evidence, at level of significance $\alpha = 0.01$, that $\beta_1 \neq 0$ and that a linear relationship exists between age and text messages.

The next example illustrates the steps for performing the hypothesis test for the slope β_1 using the p-value method.

EXAMPLE 5 **Hypothesis test for the slope β_1 using the p-value method and technology**

 shortmemory

In Section 4.3, we considered a study on short-term memory. Ten subjects were given a set of nonsense words to memorize within a certain amount of time and were later scored on the number of words they could remember. The results are repeated here in Table 4. Use the p-value method and technology to test, using level of significance $\alpha = 0.01$, whether a linear relationship exists between time and score.

Table 4

Time (x)	Score (y)
1	9
1	10
2	11
3	12
3	13
4	14
5	19
6	17
7	21
8	24

Solution

We begin by verifying the regression assumptions. The scatterplot of the residuals versus the fitted values in Figure 8 shows no strong evidence that the independence assumption, the constant variance assumption, or the zero-mean assumption is violated. Also, the normal probability plot of the residuals in Figure 9 offers evidence of the normality of the results. Therefore, we conclude that the regression assumptions are verified, and proceed with the hypothesis test.

Step 1 **State the hypotheses and the rejection rule.**

$H_0 : \beta_1 = 0$ No linear relationship exists between time and score.
$H_a : \beta_1 \neq 0$ A linear relationship exists between time and score.

Residual
2
1
0
−1
−2
10 12 14 16 18 20 22 24
Fitted value

FIGURE 8 Residuals versus fitted values plot.

Percent
100
95
90
80
70
60
50
40
30
20
10
5
1
−4 −3 −2 −1 0 1 2 3 4 5
Residuals

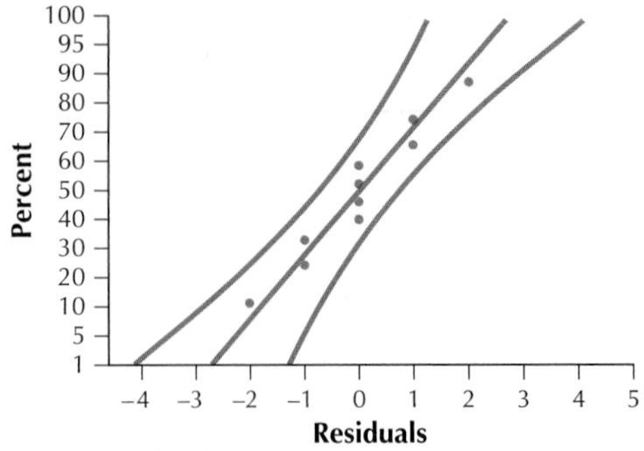

FIGURE 9 Normal probability plot of the residuals.

The rejection rule is: reject H_0 if the p-value ≤ 0.01.

Step 2 **Calculate t_{data}.**

$$t_{data} = \frac{b_1}{s/\sqrt{\sum(x-\bar{x})^2}}$$

From page 226 in Section 4.3, we have $b_1 = 2$. From Example 13 in Chapter 4 on page 228, we have

$$s = \sqrt{\frac{12}{8}} \approx 1.224744871$$

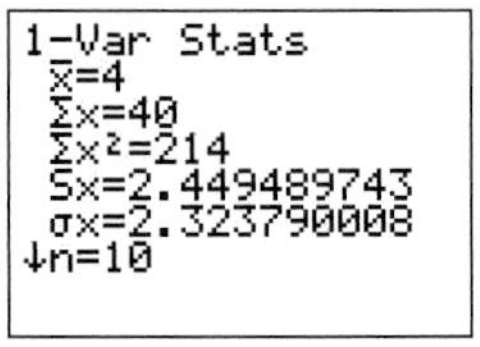

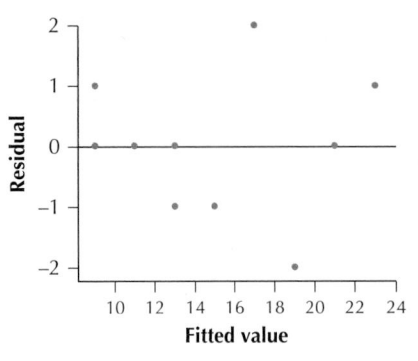

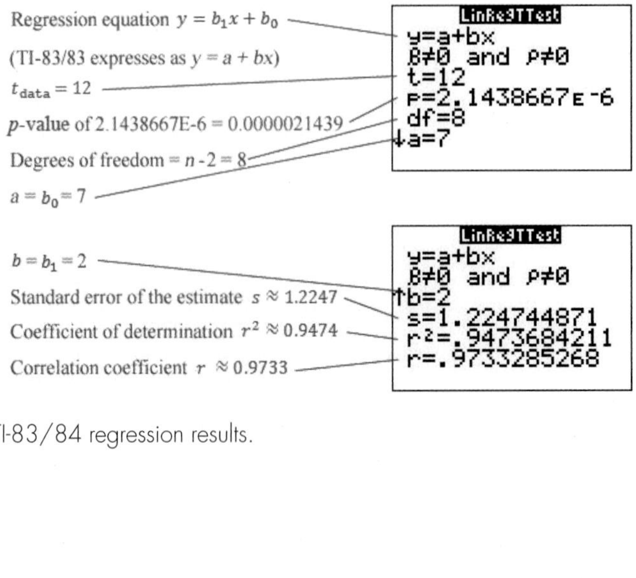

1-Var Stats
x̄=4
Σx=40
Σx²=214
Sx=2.449489743
σx=2.323790008
↓n=10

TI-83/84 summary statistics for x (time) data.

From the TI-83/84 summary statistics, we have the standard deviation of the x (time) data to be $s_x = 2.449489743$. Thus, using the relationship we learned in Example 3:

$$\sum(x-\bar{x})^2 = (n-1)\cdot s_x^2 = (9)2.449489743^2 = 54$$

Therefore,

$$t_{data} = \frac{b_1}{s/\sqrt{\sum(x-\bar{x})^2}} \approx \frac{2}{1.224744871/\sqrt{54}} = 12$$

Step 3 **Find the p-value.** For instructions, see the Step-by-Step Technology Guide on page 730. The regression results (including the p-value) for the TI-83/84, Excel, Minitab, and CrunchIt! are shown in Figures 10, 11, 12, and 13. (Differing results are due to rounding.)

Regression equation y = b₁x + b₀
(TI-83/83 expresses as y = a + bx)
t_data = 12
p-value of 2.1438667E-6 = 0.0000021439
Degrees of freedom = n - 2 = 8
a = b₀ = 7
LinRegTTest
y=a+bx
β≠0 and ρ≠0
t=12
p=2.1438667E-6
df=8
↓a=7
b = b₁ = 2
Standard error of the estimate s ≈ 1.2247
Coefficient of determination r² ≈ 0.9474
Correlation coefficient r ≈ 0.9733
LinRegTTest
y=a+bx
β≠0 and ρ≠0
↑b=2
s=1.224744871
r²=.9473684211
r=.9733285268

FIGURE 10 TI-83/84 regression results.

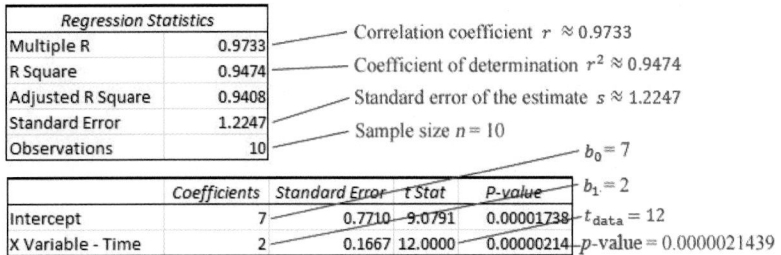

FIGURE 11 Excel regression results.

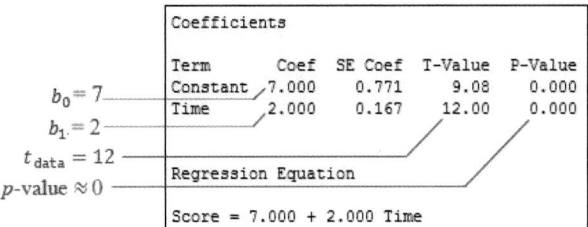

FIGURE 12 Minitab regression results.

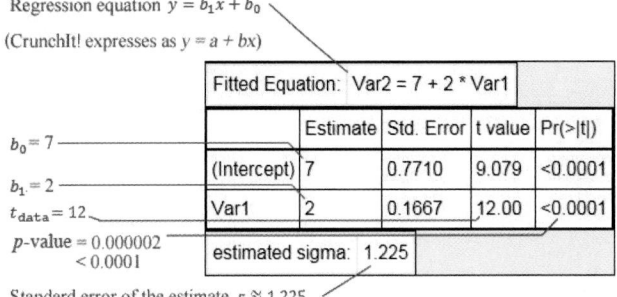

FIGURE 13 CrunchIt! regression results.

NOW YOU CAN DO
Exercises 19–22.

Step 4   The p-value of about 0.000 is $\leq \alpha = 0.01$, so we reject H_0. Evidence exists, at level of significance $\alpha = 0.01$, for a linear relationship between time and score.

YOUR TURN
#3

Recall the age and score data from the Your Turn #1 on page 717. Test, using level of significance $\alpha = 0.05$, whether a linear relationship exists between age and score.

(The solution is shown in Appendix A.)

3 Confidence Interval for Slope β_1

Recall that in Chapter 8 we constructed a confidence interval estimate for a population parameter, consisting of an interval of numbers that contain the parameter with a certain confidence level. Similarly, we can construct a confidence interval for the slope of the population regression equation β_1.

Confidence Interval for β_1

When the regression assumptions are met, a $100(1 - \alpha)\%$ confidence interval for β_1 is given by

$$b_1 \pm t_{\alpha/2} \cdot \frac{s}{\sqrt{\sum(x - \overline{x})^2}}$$

where b_1 is the point estimate of the slope β_1 of the population regression equation, s is the standard error of the estimate, and $t_{\alpha/2}$ has $n - 2$ degrees of freedom.

Margin of Error E

The margin of error for a $100(1 - \alpha)\%$ confidence interval for β_1 is given by

$$E = t_{\alpha/2} \cdot \frac{s}{\sqrt{\sum(x - \overline{x})^2}}$$

Thus, the confidence interval for β_1 takes the form $b_1 \pm E$.

EXAMPLE 6   Confidence interval for the slope β_1

Construct a 95% confidence interval for the slope β_1 of the population regression equation for the memory-test data in Example 5.

Solution

The regression assumptions were verified in Example 5, where we found:

- $b_1 = 2$,
- $s = 1.224744871$, and
- $\sum(x - \overline{x})^2 = 54$.

From the t table (Appendix Table D), we find that, for 95% confidence, $t_{\alpha/2}$ for $n - 2 = 10 - 2 = 8$ degrees of freedom is $t_{\alpha/2} = 2.306$. So, our margin of error E is

$$E = t_{\alpha/2} \cdot \frac{s}{\sqrt{\sum(x - \overline{x})^2}} = (2.306)\left(\frac{1.224744874}{\sqrt{54}}\right) \approx 0.3843$$

The 95% confidence interval for β_1 is then given by

$$b_1 \pm E = 2 \pm 0.3843 = (1.6157, 2.3843)$$

NOW YOU CAN DO
Exercises 23–30.

What Do These Numbers Mean?

- The margin of error $E = 0.3843$ means that, when we repeatedly take samples from this population, most of the time the sample estimate b_1 will be within $E = 0.3843$ of the unknown value of the slope β_1 of the population regression line.

- We are 95% confident that the interval $(1.6157, 2.3843)$ captures the slope β_1 of the population regression line.

- Because β_1 is the increase in memory-test score per added minute of memorization, *we are 95% confident that, for each additional minute of memorization, the increase in memory-test score will lie between 1.6157 and 2.3843 points.*

4 Using Confidence Intervals to Perform the *t* Test for the Slope β_1

As in earlier sections, we may use a $100(1 - \alpha)\%$ *t* confidence interval for the slope β_1 to perform the *t* test for β_1, which is a two-tailed test.

Equivalence of a Two-Tailed *t* Test About β_1 and a *t* Confidence Interval for β_1

- If a $100(1 - \alpha)\%$ *t* confidence interval for β_1 does not contain zero, then we would reject $H_0 : \beta_1 = 0$ for level of significance α, and conclude that a linear relationship exists between *x* and *y*.
- If a $100(1 - \alpha)\%$ *t* confidence interval for β_1 does contain zero, then we would not reject $H_0 : \beta_1 = 0$ for level of significance α.

EXAMPLE 7 Using confidence intervals to perform the *t* test for the slope β_1

a. Construct and interpret a 99% confidence interval for the slope β_1 for the text messaging data in Table 1.

b. Use the confidence interval in **(a)** to test whether a linear relationship exists between age and text messages, using level of significance $\alpha = 0.01$.

 **textms**

Solution

a. The regression assumptions were verified in Example 2. Also,

- In Example 1, we found $b_1 = -1.5$.
- In Example 3, we calculated $s = 2.408318916$, and $\Sigma(x - \bar{x})^2 = 330$.

From the *t* table, we find that, for 99% confidence, $t_{\alpha/2}$ for $n - 2 = 10 - 2 = 8$ degrees of freedom is $t_{\alpha/2} = 3.355$. So, our margin of error E is

$$E = t_{\alpha/2} \cdot \frac{s}{\sqrt{\sum(x - \bar{x})^2}} = (3.355)\left(\frac{2.408318916}{\sqrt{330}}\right) \approx 0.4448$$

The 99% confidence interval for β_1 is then given by

$$b_1 \pm E = -1.5 \pm 0.4448 = (-1.9448, -1.0552)$$

We are 99% confident that the interval $(-1.9448, -1.0552)$ captures the slope β_1 of the population regression line. That is, we are 99% confident that, for each additional year of age, the *decrease* in the number of text messages lies between 1.9448 and 1.0552.

b. The hypotheses are

$H_0 : \beta_1 = 0$ No linear relationship exists between age and text messages.

$H_a : \beta_1 \neq 0$ A linear relationship exists between age and text messages.

The confidence interval from **(a)** does not contain zero, so we may conclude that a linear relationship exists between age and text messages, at level of significance $\alpha = 0.01$.

NOW YOU CAN DO
Exercises 31–38.

Table 5 Frequency in English, frequency in Scrabble, and Scrabble point value of the letters in the alphabet

Letter	Rel. freq. in English language	Frequency in Scrabble	Point value in Scrabble	Letter	Rel. freq. in English language	Frequency in Scrabble	Point value in Scrabble
A	0.073	9	1	N	0.078	6	1
B	0.009	2	3	O	0.074	8	1
C	0.030	2	3	P	0.027	2	3
D	0.044	4	2	Q	0.003	1	10
E	0.130	12	1	R	0.077	6	1
F	0.028	2	4	S	0.063	4	1
G	0.016	3	2	T	0.093	6	1
H	0.035	2	4	U	0.027	4	1
I	0.074	9	1	V	0.013	2	4
J	0.002	1	8	W	0.016	2	4
K	0.003	1	5	X	0.005	1	8
L	0.035	4	1	Y	0.019	2	4
M	0.025	2	3	Z	0.001	1	10

scrabble

How Fair Is the Scoring in Scrabble?

In this Case Study, we consider the frequency and point values of Scrabble tiles. Table 5 shows the relative frequency in the English language, the frequency (number of tiles) in Scrabble, and the point value in Scrabble.

First of all, what is the relationship between the tile frequencies in Scrabble and the letter frequencies in the English language? Figure 14 shows a scatterplot of the tile frequencies in Scrabble against the letter frequencies in the English language. A positive relationship appears to exist between the two variables. That is, as the English frequencies increase, game frequencies also tend to increase.

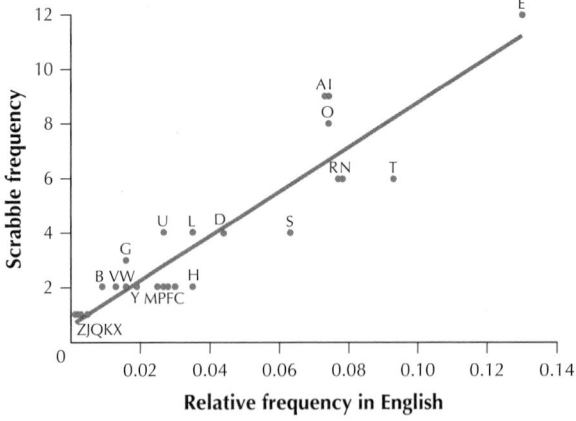

FIGURE 14 Scatterplot of Scrabble frequency versus English relative frequency of letters, with regression line.

Note that the letters above the regression line occur "too frequently" in the game, whereas the letters below the line occur "not frequently enough." Playing typical English words during a game of Scrabble would tend to leave you with a rack of letters similar to those above the regression line. Note that S is one of the letters that is rarer in the game than in the language.

Figure 15 displays the Minitab results from a regression of the tile frequencies against the English language relative frequencies. The regression equation is

$$\hat{y} = 81.5 \text{ (relative frequency in English)} + 0.636$$

The slope is positive, which concurs with the scatterplot in Figure 14.

```
The regression equation is
Scrabble Freq = 0.636 + 81.5 Rel Freq Eng

Predictor         Coef  SE Coef       T       P
Constant        0.6362   0.3484    1.83   0.080
Rel Freq Eng    81.458    6.856   11.88   0.000


S = 1.16096    R-Sq = 85.5%    R-Sq(adj) = 84.9%


Analysis of Variance

Source            DF       SS       MS        F       P
Regression         1   190.27   190.27   141.17   0.000
Residual Error    24    32.35     1.35
Total             25   222.62
```

FIGURE 15 Minitab regression output.

Next, we turn to the hypothesis test:

$H_0 : \beta_1 = 0$ No linear relationship exists between Scrabble frequency and English relative frequency.

$H_a : \beta_1 \neq 0$ A linear relationship exists between Scrabble frequency and English relative frequency.

The 0.000 in red represents the *p*-value for the *t* test. This *p*-value is smaller than any α, so we reject the null hypothesis that no linear relationship exists between the game frequencies and the English frequencies. Does the model fit the data well? The coefficient of determination r^2 is 0.855, which is good, and the correlation coefficient *r* equals 0.924, which indicates that the variables are positively correlated.

But the fit really could be better. Look at the value of *s*, the standard error of the estimate: $s \approx 1.16$. This means that, given the English language frequency of a letter, the estimate of the tile frequency will typically differ from the actual tile frequency by more than one tile.

Next, what is the relationship between the Scrabble point values and the English relative frequencies? Figure 16 shows a scatterplot of these two variables. The first thing you might notice about this relationship is that it is not linear. Therefore, it would not be appropriate to perform linear regression on this data set.

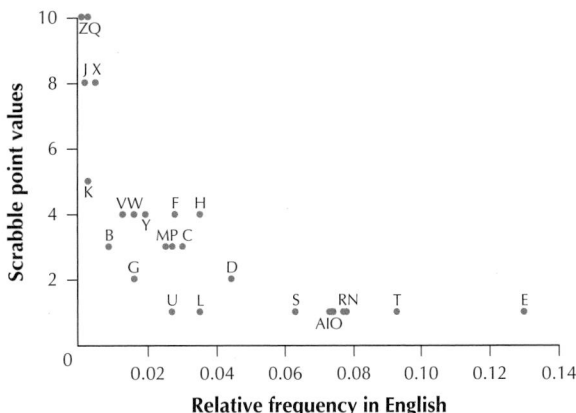

FIGURE 16 Scatterplot of Scrabble frequency versus English relative frequency. Linear regression would not be appropriate.

We can, nevertheless, make some descriptive remarks.

- What is a "good" Scrabble tile to pick up? In the best case, it would be a letter with high English frequency worth lots of Scrabble game points. Unfortunately, the two do not go together. The high-frequency letters such as E and T have low point values, and the high point-value letters such as Q and Z have low frequencies. But we can still make comparisons.

- Which would you rather pick up, a D or a G? It would seem that D would be preferable because it has the same point value as G with much higher English frequency.

- Which do you prefer between J and X? They are worth the same points, but X has a higher frequency in English, so it is easier to make words with it in the game. The letter H seems to have a good combination of high points and moderate frequency.

STEP-BY-STEP TECHNOLOGY GUIDE: Regression Analysis

Data from Example 5 (page 723) are used to illustrate the steps.

TI-83/84
Step 1 Enter the **X (Time)** data in **L1** and the **Y (Score)** data in **L2**.
Step 2 Press **STAT**, highlight **CALC**, and press **4** to choose **LinReg(ax+b)**. On the home screen, the following command appears: **LinReg(ax + b)**.
Step 3 Press **ENTER**. The output shows y = ax + b, a = 2, b = 7. The TI-83/84 denotes the slope β_1 as **a** and the y intercept b_0 as **b**. Thus, the TI-83/84 is telling you that the estimated regression equation is $\hat{y} = 2x + 7$.
Step 4 Now Press **STAT** again and highlight **TESTS**.
Step 5 Press the **down arrow** key until **LinRegTTest** is highlighted.

Step 6 Press **ENTER**. The LinRegTTest menu appears.
Step 7 For **Xlist**, enter **L1** (or whichever list you entered the X data in).
Step 8 For **Ylist**, enter **L2** (or whichever list you entered the Y data in).
Step 9 For **Freq**, enter **1**, and for β & p highlight "≠ 0" and press **ENTER**.
Step 10 Move the cursor over **Calculate**, make sure all your entries are correct, and press **ENTER**. The results are as shown in Figure 10 (page 724).

EXCEL
Step 1 Enter the "Time" variable in column **A** and the "Score" variable in column **B**.
Step 2 Click on **Data > Data Analysis > Regression** and click **OK**.
Step 3 For **Input Y Range**, select cells **B1–B10**. For **Input X Range**, select cells **A1–A10**. Make sure **Labels** is unchecked.

Step 4 If you want to verify the regression assumptions, then select **Residual Plots** and **Normal Probability Plots**.
Step 5 Click **OK**. The results are as shown in Figure 11 (page 725).

MINITAB
Step 1 Enter the "Time" variable in **C1** and the "Score" variable in **C2**.
Step 2 Click on **Stat > Regression > Regression > Fit Regression Model**.
Step 3 Move "Score" to the **Responses** box, and "Time" to the **Continuous predictors** box.

Step 4 If you want to verify the regression assumptions, click the button labeled **Graphs…** and select **Four in One**.
Step 5 Click **OK** twice. The results are as shown in Figure 12 (page 725).

SPSS
Step 1 Input the time and score data into the first two columns. Under the **Variable View** tab, rename the columns *Time* and *Score*.
Step 2 Select **Analyze > Regression > Linear…**.
Step 3 Choose *Score* as the **Dependent** variable and *Time* as the **Independent** variable.

Step 4 If you want to verify the regression assumptions, click **Plots…** and select **Normal probability plot**. Click **Continue**.
Step 5 Click **OK**.

JMP

Step 1   Click **File > New > Data Table**. Input the time and score data into the first two columns. Rename the columns *Time* and *Score*.
Step 2   Select **Analyze > Fit Y by X**. Choose *Score* as the **Y, Response** and *Time* as the **X, Factor**. Click **OK**.

Step 3   Click the red triangle beside "Bivariate Fit of Score by Time," and select **Fit Line**. The desired output is in the **Parameter Estimates** table.
Step 4   If you want to verify the regression assumptions, click the red triangle beside **Linear Fit**, and select **Plot Residuals**.

CRUNCHIT!

Step 1   Input the time and score data into the first two columns.
Step 2   Select **Statistics > Regression > Simple Linear**. Choose *Var2* as the **Dependent Variable** and *Var1* as the **Independent Variable**.

Step 3
a. If you want to verify the regression assumptions, select **Residuals Plot** as the **Display** and click **Calculate**.
b. To observe the test statistics and p-values, select **Numeric Results** as the **Display** and click **Calculate**. Results are shown in Figure 13 of Example 5.

Section 13.1   Summary

1. This section examines inferential methods for regression analysis. The regression model, or the (population) regression equation, is $y = \beta_1 x + \beta_0 + \varepsilon$, where β_0 is the y intercept of the population regression line, β_1 is the slope of the population regression line, and ε is the error term.

2. A hypothesis test may be performed to determine whether a linear relationship exists between x and y.
3. We can construct confidence intervals for the true value of the population regression slope β_1 because it is unknown.

Section 13.1   Exercises

CLARIFYING THE CONCEPTS

1. What is the difference between the regression equation (calculated using the sample) and the population regression equation? (p. 718)
2. What are the four regression model assumptions? (p. 718)
3. How do we go about verifying the regression model assumptions? (p. 719)
4. What is the difference between b_0 and b_1 on the one hand and β_0 and β_1 on the other hand? (p. 718)
5. What does it mean for the relationship between x and y when β_1 equals 0? (p. 721)
6. What is the difference between s and s_x? (p. 722)

PRACTICING THE TECHNIQUES

 CHECK IT OUT!

To do	Check out	Topic
Exercises 7–14	Example 2	Calculating the residuals and verifying the regression assumptions
Exercises 15–18	Example 4	Test for the slope β_1: critical-value method
Exercises 19–22	Example 5	Test for the slope β_1: p-value method
Exercises 23–30	Example 6	Confidence interval for the slope β_1
Exercises 31–38	Example 7	Using confidence intervals to perform the t test for the slope β_1

For Exercises 7–14, you are given the regression equation.
   a. Calculate the predicted values.
   b. Compute the residuals.
   c. Construct a scatterplot of the residuals versus the predicted values.
   d. Use technology to construct a normal probability plot of the residuals.
   e. Verify that the regression assumptions are valid.

7.

x	y
1	15
2	20
3	20
4	25
5	25

$\hat{y} = 2.5x + 13.5$

8.

x	y
0	10
5	20
10	45
15	50
20	75

$\hat{y} = 3.2x + 8$

9.

x	y
-5	0
-4	8
-3	8
-2	16
-1	16

$\hat{y} = 4x + 21.6$

10.

x	y
-3	-5
-1	-15
1	-20
3	-25
5	-30

$\hat{y} = -3x - 16$

11.

x	y
10	100
20	95
30	85
40	85
50	80

$\hat{y} = -0.5x + 104$

12.

x	y
0	11
20	11
40	16
60	21
80	26

$\hat{y} = 0.2x + 9$

13.

x	y
1	1
2	1
3	2
4	3
5	3

$\hat{y} = 0.6x + 0.2$

14.

x	y
1	6
2	5
2	4
2	3
3	2

$\hat{y} = -2x + 8$

For Exercises 15–18, follow these steps. Assume that the regression model assumptions are valid.

    a. Find t_{crit} for a two-tailed test with level of significance $\alpha = 0.05$ and df $= n - 2$.

    b. Calculate s.

    c. Compute $\Sigma(x - \bar{x})^2$.

    d. Calculate t_{data}.

    e. Perform the hypothesis test for the linear relationship between x and y, using the critical-value method and $\alpha = 0.05$.

15. Data in Exercise 7, where $b_1 = 2.5$

16. Data in Exercise 8, where $b_1 = 3.2$

17. Data in Exercise 9, where $b_1 = 4.0$

18. Data in Exercise 10, where $b_1 = -3$

For Exercises 19–22, follow these steps. Assume that the regression model assumptions are valid.

    a. Calculate s.

    b. Compute $\Sigma(x - \bar{x})^2$.

    c. Calculate t_{data}.

    d. Find p-value $2 \cdot P(t > |t_{data}|)$.

    e. Perform the hypothesis test for the linear relationship between x and y using the p-value method and level of significance $\alpha = 0.05$.

19. Data in Exercise 11, where $b_1 = -0.5$

20. Data in Exercise 12, where $b_1 = 0.2$

21. Data in Exercise 13, where $b_1 = 0.6$

22. Data in Exercise 14, where $b_1 = -2$

For Exercises 23–30, follow these steps. Assume that the regression model assumptions are valid.

    a. Find $t_{\alpha/2}$ for a 95% confidence interval for β_1.

    b. Find the margin of error E.

    c. Construct a 95% confidence interval for β_1.

23. Data in Exercise 7

24. Data in Exercise 8

25. Data in Exercise 9

26. Data in Exercise 10

27. Data in Exercise 11

28. Data in Exercise 12

29. Data in Exercise 13

30. Data in Exercise 14

For Exercises 31–38, using the confidence interval from the indicated exercise, perform the t test for β_1 at level of significance $\alpha = 0.05$.

31. Exercise 23

32. Exercise 24

33. Exercise 25

34. Exercise 26

35. Exercise 27

36. Exercise 28

37. Exercise 29

38. Exercise 30

APPLYING THE CONCEPTS

For Exercises 39–46, assume the regression requirements are met. Test for the linear relationship between x and y, using level of significance $\alpha = 0.05$.

39. Volume and Weight. The following table contains the volume (*x*, in cubic meters) and weight (*y*, in kilograms) of five randomly chosen packages shipped to a local college.

 volweight

Volume (x)	Weight (y)
4	10
8	16
12	25
16	30
20	35

40. Family Size and Pets. The number of family members (*x*) in a random sample taken from a suburban neighborhood, along with the number of pets (*y*) belonging to each family are shown in the following table. familypet

Family size (x)	Pets (y)
2	1
3	2
4	2
5	3
6	3

41. World Temperatures. Listed in the following table are the low (*x*) and high (*y*) temperatures for a particular day, measured in degrees Fahrenheit, for a random sample of cities worldwide. worldtemp

City	Low (x)	High (y)
Kolkata	57	77
London	36	45
Montreal	7	21
Rome	39	55
San Juan	70	83
Shanghai	34	45

42. Video Game Sales. The Chapter 1 Case Study looked at video game sales for the top 30 video games. The following table contains the total sales (*y*, in game units) and weeks on the top 30 list (*x*) of five randomly chosen video games.

Video game	Weeks (x)	Total sales (y)
Super Mario Bros. U for WiiU	78	1,690,689
NBA 2K14 for PS4	27	608,899
Battlefield 4 for PS3	29	911,687
Titanfall for Xbox One	10	1,150,856
Yoshi's New Island for 3DS	10	172,680

 videogamereg

43. Darts and the Dow Jones. The following table contains a random sample of eight days from the Chapter 3 Case Study data set, indicating the stock market gain or loss for the portfolio chosen by the random darts (*y*), as well as the Dow Jones Industrial Average gain or loss for that day (*x*).

 dartsdjia

Darts (y)	DJIA (x)
−27.4	−12.8
18.7	9.3
42.2	8
−16.3	−8.5
11.2	15.8
28.5	10.6
1.8	11.5
16.9	−5.3

44. Age and Height. The following table provides a random sample from the Chapter 4 Case Study data set **Body Females**, showing the age (*x*) and height (*y*) of the eight women. ageheight

Age (x)	Height (y)
40	63.5
28	63
25	64.4
34	63
26	63.8
21	68
19	61.8
24	69

45. Gardasil Shots and Age. The accompanying table shows a random sample of 10 patients from the Chapter 5 Case Study data set, **Gardasil**, including the age of the patient (*x*) and the number of shots taken by the patient (*y*).

 gardasilreg

Age (x)	Shots (y)
13	3
21	3
16	3
17	2
17	3
18	1
25	2
15	3
12	1
16	1

46. NCAA Power Ratings. The accompanying table shows the top 10 teams' winning percentage (*x*) and power rating (*y*) for the 2013–2014 NCAA basketball season, according to www.teamrankings.com.

Team	Winning proportion (x)	Power rating (y)
Florida	0.923	121.2
Wichita State	0.971	119.1
Arizona	0.868	118.8
Louisville	0.838	117.9
Connecticut	0.800	117.2
Virginia	0.811	116.8
Wisconsin	0.789	116.6
Villanova	0.853	116.4
Michigan State	0.763	115.9
Michigan	0.757	115.9

ncaa2014

For Exercises 47–54, do the following for the indicated data:
  a. Calculate the margin of error E for a 95% confidence interval for β_1.
  b. Construct a 95% confidence interval for β_1.
  c. Interpret the confidence interval.

47. Volume and Weight. Data from Exercise 39
48. Family Size and Pets. Data from Exercise 40
49. World Temperatures. Data from Exercise 41
50. Video Game Sales. Data from Exercise 42
51. Darts and the Dow Jones. Data from Exercise 43
52. Age and Height. Data from Exercise 44
53. Gardasil Shots and Age. Data from Exercise 45
54. NCAA Power Ratings. Data from Exercise 46

55. Saturated Fat and Calories. The table contains the calories and saturated fat in a sample of 10 food items.
  a. Construct a 90% confidence interval for the slope of the regression line, for the regression of calories on saturated fat.
  b. Using your confidence interval, conclude whether a linear relationship exists between calories and saturated fat, with level of significance $\alpha = 0.10$.

satfatreg

Food item	Calories	Grams of saturated fat
Chocolate bar (1.45 oz)	215.66	6.9618
Meat & veggie pizza, big slice (1/8 lg pizza)	363.81	5.6472
New England clam chowder (cup)	148.80	1.8600
Baked chicken drumstick (no skin, medium size)	75.24	0.6424
Curly fries, deep-fried (4 ounces)	276.21	3.1752
Wheat bagel (large)	374.66	0.2751
Chicken curry (cup)	146.32	1.5930
Cake doughnut hole (one)	58.94	0.5068
Rye bread (slice)	67.34	0.1638
Raisin Bran cereal (cup)	194.59	0.3355

Source: Food-A-Pedia.

56. Engine Displacement and Gas Mileage. The table provides the engine displacement (size, in liters) and the city MPG (miles per gallon) gas mileage of a random sample of 12 vehicles .
  a. Construct a 95% confidence interval for the slope of the regression line, for the regression of city MPG on engine displacement.
  b. Using your confidence interval, conclude whether a linear relationship exists between city MPG and engine displacement, with level of significance $\alpha = 0.05$. **displacement**

Vehicle	Engine displacement	City MPG
GMC Yukon Denali	6.2	13
Ford E350 Wagon	5.4	11
BMW 435i Coupe	3.0	20
Land Rover Range Rover	5.0	13
Infiniti Q50a	3.7	19
Dodge Journey	3.6	17
Jaguar XF	5.0	15
Dodge Challenger	6.4	14
Toyota Highlander Hybrid	3.5	28
Mercedes-Benz S 550	4.7	17
Ford Fiesta	1.6	29
Hyundai Elantra	2.0	24

Batting Average and Runs Scored. The table shows the top 10 hitters in Major League Baseball for 2013. We are interested in estimating the number of runs scored (y) by using the player's batting average (x). Use this information for Exercises 57–60. **batters2013**

Batter	Team	Runs scored	Batting average
Miguel Cabrera	Tigers	103	0.348
Mike Trout	Angels	109	0.323
Matt Carpenter	Cardinals	126	0.318
Andrew McCutchen	Pirates	97	0.317
Paul Goldschmidt	Diamondbacks	103	0.302
Josh Donaldson	Athletics	89	0.301
Chris Davis	Orioles	103	0.286
Carlos Gomez	Brewers	80	0.284
Manny Machado	Orioles	88	0.283
Evan Longoria	Rays	91	0.269

57. Assess the regression assumptions. Is it okay to proceed with the regression?
58. Regress runs scored on batting average. Test for the significance of the linear relationship, using level of significance $\alpha = 0.10$.
59. Find the residual for Matt Carpenter. What is unusual about Matt Carpenter in this regression?

60. Test for the significance of the linear relationship, using level of significance $\alpha = 0.05$ Compare your conclusion with the earlier regression, using level of significance $\alpha = 0.05$. How do you suggest we resolve this dilemma?

BRINGING IT ALL TOGETHER

SAT Reading and Math Scores. Use this information for Exercises 61–65. The table shows the SAT scores for five states as reported by the College Board. We are interested in whether a linear relationship exists between the SAT Reading score (x) and the SAT Math score (y).

statesat

State	SAT Reading (x)	SAT Math (y)
New York	497	510
Connecticut	515	515
Massachusetts	518	523
New Jersey	501	514
New Hampshire	522	521

61. What Result Might We Expect? Consider the accompanying scatterplot of Math score versus Reading score. Is there evidence for or against the null hypothesis that no linear relationship exists? Explain.

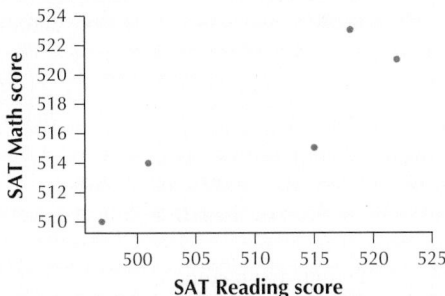

62. Consider the following graphics. Is there strong evidence that the regression assumptions are violated?

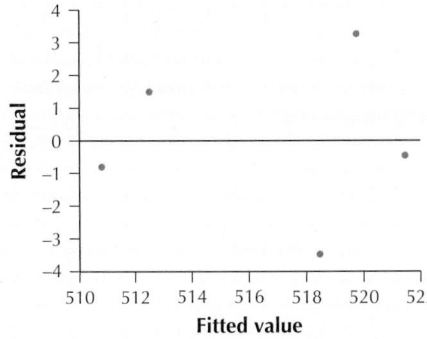

Plot of residuals versus fitted values.

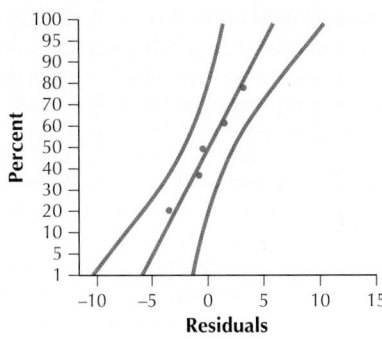

Normality plot of residuals.

63. Test, using level of significance $\alpha = 0.10$, whether a linear relationship exists between the SAT reading score and the SAT Math score.

64. Construct and interpret a 90% confidence interval for a slope β_1.

65. Do your inferences in Exercises 63 and 64 agree with each other? Explain.

WHAT IF ? **66. Challenge Exercise.** Suppose we have a regression equation whose slope was not significant (that is, the null hypothesis was not rejected). *What if* we add five new data values to the original data set, and all five data values are identical, $(\bar{x}, \bar{y})$? How and why will this affect the following statistics? Will the statistic increase, decrease, or remain unchanged, or is there insufficient information to determine? (*Hint:* The data point $(\bar{x}, \bar{y})$ always lies on the estimated regression line.)

   a. n **d.** SSR
   b. SSE   **e.** MSE
   c. SST   **f.** MSR

67. Challenge Exercise. Refer to Exercise 66. How and why will the change affect the following items?

   a. t_{data} **d.** p-value
   b. r^2 **e.** conclusion
   c. s

WHAT IF ? **68. Challenge Exercise.** Suppose a regression analysis of y on x was found to be significant (that is, the null hypothesis was rejected). *What if* we get 10 new data values, all with different values of x and *all of which can be found on the estimated regression line of the original model*? How and why will this change affect the following statistics? Will the statistic increase, decrease, or remain unchanged, or is there insufficient information to determine?

   a. n **d.** SSR
   b. SSE   **e.** MSE
   c. SST   **f.** MSR

69. Refer to Exercise 68. How and why will this change affect the following measures?

   a. t_{data} **d.** p-value
   b. r^2 **e.** conclusion
   c. s

WHAT IF ? **70. Challenge Exercise.** Suppose a regression analysis of y on x was found to be significant (that is, the null hypothesis was rejected) and the slope $b_1 > 0$. Consider the observation (max x, y), which represents the (x, y)

data value for the maximum value of x in the data set. Suppose the residual for (max x, y) is negative. *What if* we increase max x by an arbitrary amount c so that the new data value is (max $x + c$, y)? (All other data values in the data set are unchanged.) How will this increase affect the following measures? Will they increase, decrease, or remain unchanged, or is there insufficient information to determine the effect?

a. n	**e.** MSE
b. SSE	**f.** MSR
c. SST	**g.** F
d. SSR	

71. Challenge Exercise. Refer to Exercise 70. How and why will the change affect the following measures?

a. t_{data}	**d.** p-value
b. r^2	**e.** conclusion
c. s	

WORKING WITH LARGE DATA SETS

For Exercises 72–74, use technology to solve the following problems:

a. Verify the regression model assumptions.
b. Construct and interpret a 95% confidence interval for β_1.
c. Based on the confidence interval constructed in (**b**), would you expect the hypothesis test to reject the null hypothesis that $\beta_1 = 0$?
d. Test, at $\alpha = 0.05$, whether a linear relationship exists between x and y.

72. Open the **Darts** data set, which we used for the Chapter 3 Case Study. Use the Dow Jones Industrial Average (x) to estimate the pros' performance (y). **darts**

73. Open the **Nutrition** data set. Estimate the number of calories per gram (y) using the amount of fat per gram (x). **nutrition**

74. Open the **Pulse and Temp** data set. Estimate body temperature (y) using heart rate (x). **pulseandtemp**

Use technology for Exercises 75–78. Open the **Crash** data set, which contains information about the severity of injuries sustained by crash dummies when the National Transportation Safety Board crashed automobiles into a wall at 35 miles per hour. **crash**

75. The variable *head_inj* contains a measure of the severity of the head injury sustained by the dummies. The variable *chest_in* is a measure of the severity of the chest injury suffered by the crash dummies.

a. Would you expect a linear relationship to exist between the severity of head injuries and chest injuries?
b. Construct a scatterplot of the *head_inj* against *chest_in*. Describe the relationship between the variables.
c. If we were to perform a regression analysis using these two variables, is it clear which of the two variables we should label as the predictor and which we should label as the response? Explain.

76. Perform a regression of the head injury severity (y) on the chest injury severity (x).

a. What is the regression equation? Write it out in words and numbers, so that a nonstatistician would understand it.
b. Perform the appropriate hypothesis test, using level of significance $\alpha = 0.01$.
c. Clearly interpret the meaning of the slope estimate b_1.
d. Construct and interpret a 99% confidence interval for the true slope β_1 of the relationship between severity of head injury and severity of chest injury. How does your confidence interval support your conclusion in (**b**)?

77. The variable *lleg_inj* contains a measure of the severity of the injury sustained by the dummies' left legs. The variable *weight* contains the weight of the vehicles.

a. Would you expect a linear relationship to exist between the severity of left leg injuries and the weight of the vehicles?
b. Construct a scatterplot of the *lleg_inj* against *weight*. Describe the relationship between the variables.
c. If we were to perform a regression analysis using these two variables, is it clear which of the two variables we should label as the predictor and which we should label as the response? Explain.

78. Perform a regression of the left leg injury severity (y) on the vehicle weight (x).

a. What is the regression equation? Write it out in words and numbers, so that a nonstatistician would understand it.
b. Is the relationship significant? Perform the appropriate hypothesis test, using level of significance $\alpha = 0.5$.
c. Construct and interpret a 95% confidence interval for the true slope β_1 of the relationship between vehicle weight and severity of left leg injury. How does your confidence interval support your conclusion in (**b**)?

13.2 Confidence Intervals and Prediction Intervals

OBJECTIVES By the end of this section, I will be able to . . .

1   Construct confidence intervals for the mean value of y for a given value of x.
2   Construct prediction intervals for a randomly chosen value of y for a given value of x.

1 Construct Confidence Intervals for the Mean Value of y for a Given x

Suppose a cell phone company wants to increase sales to 20-year-olds, and therefore wants to estimate the number of text messages sent by 20-year-old men and women. In earlier sections, we learned how to calculate the predicted value $\hat{y}$ by using the regression equation $\hat{y} = b_1 x + b_0$. For example, in Example 1 (page 716), we saw that the predicted number of text messages for the 20-year-old texter is $\hat{y} = (-1.5)(20) + 60.6 = 30.6$ messages. However, this estimate of 30.6 messages is only a point estimate, similar to using the sample mean $\bar{x}$ alone to estimate the unknown population mean μ. We learned in Section 8.1 about the pitfalls of point estimates, such as the lack of a measure of confidence associated with the estimate. So, just as we learned about confidence intervals for μ in Chapter 8, in this section we will learn how to construct:

●   confidence intervals for the mean value of y for a given value of x, and

●   prediction intervals for a randomly chosen value of y for a given value of x.

Refer to Figure 3 (reproduced here from page 719), which illustrates the regression assumptions. Note that, for $x = 20$, there is assumed to be an entire population of y-values. The cell phone company is interested in this population, and wants to construct a confidence interval for the mean number of text message sent by all 20-year-olds.

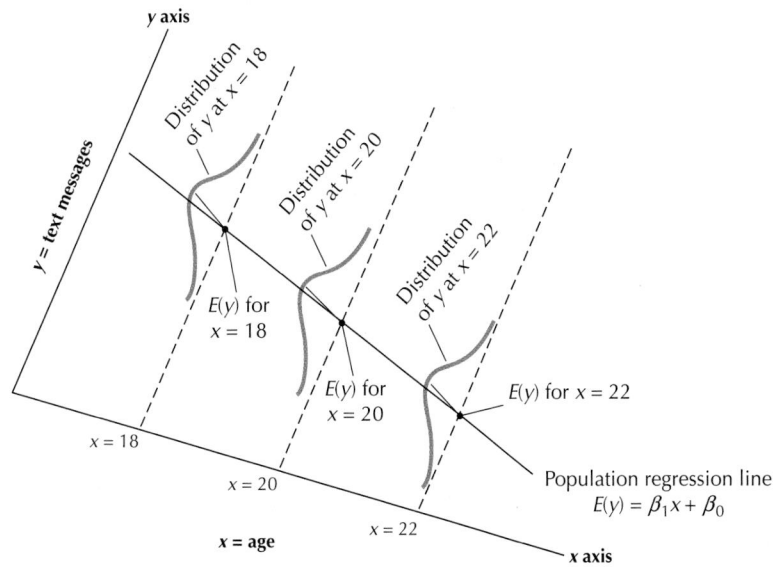

FIGURE 3 For $x = 20$ years old, there is an entire population of y = number of text messages.

Confidence Interval for the Mean Value of y for a Given x

A $100(1 - \alpha)\%$ confidence interval for the mean response, that is, for the population mean of all values of y, given a value of x, may be constructed using the following lower and upper bounds:

$$\text{Lower bound: } \hat{y} - t_{\alpha/2} \cdot s \sqrt{\frac{1}{n} + \frac{(x^* - \bar{x})^2}{\sum (x_i - \bar{x})^2}}$$

$$\text{Upper bound: } \hat{y} + t_{\alpha/2} \cdot s \sqrt{\frac{1}{n} + \frac{(x^* - \bar{x})^2}{\sum (x_i - \bar{x})^2}}$$

where x^* represents the given value of the predictor variable x, $s = \sqrt{\frac{MSE}{n-2}}$ represents the standard error of the estimate, n is the number of observations, $\bar{x}$ is the sample mean of the x-values, and $t_{\alpha/2}$ is the critical value with $n - 2$ degrees of freedom. The requirements for the confidence interval are met if the regression assumptions are met.

Let's look at an example of constructing and interpreting such a confidence interval.

EXAMPLE 8   Constructing and interpreting a confidence interval for the mean value of y for a given x

Construct and interpret a 95% confidence interval for the population mean number of text messages sent for all 20-year-olds.

Solution

Example 2 showed that the regression assumptions for this data set are met, thus allowing us to construct this confidence interval.

- Using our given value $x^* = 20$, we calculate the point estimate as follows: $\hat{y} = (-1.5)(20) + 60.6 = 30.6$.

- The data set in Table 1 shows that there are $n = 10$ people, so that the degrees of freedom for the t critical value is $n - 2 = 8$. From the t table (Table D in the Appendix), the t critical value for confidence level 95% is $t_{\alpha/2} = 2.306$.

- In Example 3, we calculated $\sum (x - \bar{x})^2 = 330$ and the standard error of the estimate to be $s \approx 2.4083$.

- The Minitab output in Figure 17 shows that $\bar{x} = 27$ years of age.

Thus,

$$\text{Lower bound: } \hat{y} - t_{\alpha/2} \cdot s \sqrt{\frac{1}{n} + \frac{(x^* - \bar{x})^2}{\sum (x_i - \bar{x})^2}}$$

$$\approx 30.6 - (2.306) \cdot 2.4083 \sqrt{\frac{1}{10} + \frac{(20 - 27)^2}{330}} \approx 27.8317$$

$$\text{Upper bound: } \hat{y} + t_{\alpha/2} \cdot s \sqrt{\frac{1}{n} + \frac{(x^* - \bar{x})^2}{\sum (x_i - \bar{x})^2}}$$

$$\approx 30.6 + (2.306) \cdot 2.4083 \sqrt{\frac{1}{10} + \frac{(20 - 27)^2}{330}} \approx 33.3683$$

We are 95% confident that the population mean number of text messages for all 20-year-olds lies between 27.8317 and 33.3683 messages.

Descriptive Statistics: age

Variable	Mean
age	27.00

FIGURE 17 Minitab shows $\bar{x} = 27$.

 CAUTION Extrapolation (page 215) continues to be a danger for confidence intervals and prediction intervals. For example, it would not have been wise to ask for a confidence interval for the population mean number of text messages sent by all 50-year-old people because $x = 50$ lies outside the range of x.

NOW YOU CAN DO
Exercises 3–8.

2 Construct Prediction Intervals for an Individual Value of y for a Given x

The formula for constructing a prediction interval for an individual random value of y is similar to that for the confidence interval for the mean y, except that the margin of error is larger. We talk about why this is so later in this section.

> ### Prediction Interval for a Randomly Selected Value of y for a Given x
>
> A $100(1 - \alpha)\%$ prediction interval for a randomly selected value of y given a value of x may be constructed using the following lower and upper bounds:
>
> $$\text{Lower bound: } \hat{y} - t_{\alpha/2} \cdot s \sqrt{1 + \frac{1}{n} + \frac{(x^* - \bar{x})^2}{\sum (x_i - \bar{x})^2}}$$
>
> $$\text{Upper bound: } \hat{y} + t_{\alpha/2} \cdot s \sqrt{1 + \frac{1}{n} + \frac{(x^* - \bar{x})^2}{\sum (x_i - \bar{x})^2}}$$
>
> where x^* represents the given value of the predictor variable x, $s = \sqrt{\frac{MSE}{n-2}}$ represents the standard error of the estimate, n is the number of observations, $\bar{x}$ is the sample mean of the x values, and $t_{\alpha/2}$ is the critical value with $n - 2$ degrees of freedom. The requirements for the confidence interval are met if the regression assumptions are met.

EXAMPLE 9   Constructing and interpreting a prediction interval for a randomly selected value of y for a given x

Construct and interpret a 95% prediction interval for the number of text messages sent by a randomly selected 20-year-old.

Solution

The requirements for the prediction interval are the same as for the confidence interval, so we know from Example 8 that the requirements for constructing the prediction interval are met. Using the values obtained in Example 8, we calculate the lower and upper bounds as follows:

$$\text{Lower bound: } \hat{y} - t_{\alpha/2} \cdot s \sqrt{1 + \frac{1}{n} + \frac{(x^* - \bar{x})^2}{\sum (x_i - \bar{x})^2}}$$

$$\approx 30.6 - (2.306) \cdot 2.4083 \sqrt{1 + \frac{1}{10} + \frac{(20 - 27)^2}{330}} \approx 24.3947$$

$$\text{Upper bound: } \hat{y} + t_{\alpha/2} \cdot s \sqrt{1 + \frac{1}{n} + \frac{(x^* - \bar{x})^2}{\sum (x_i - \bar{x})^2}}$$

$$\approx 30.6 + (2.306) \cdot 2.4083 \sqrt{1 + \frac{1}{10} + \frac{(20 - 27)^2}{330}} \approx 36.8053$$

NOW YOU CAN DO
Exercises 9–14.

We are 95% confident that the number of text messages sent by a randomly selected 20-year-old lies between 24.3947 and 36.8053.

Developing Your Statistical Sense

Why Are Prediction Intervals Wider?

The prediction interval (24.3947, 36.8053) from Example 9 is *wider* than the corresponding confidence interval from Example 8. This is true in general, but why? Note that the only difference in the formulas for the prediction interval, compared to the confidence interval, is the "1 +" inside the radical in the formula for the prediction interval. This ensures that the prediction interval always has a larger margin of error than the confidence interval for a given x. This larger margin of error reflects the greater variability of individual responses compared to the mean response.

To see how individual values have greater variability than their means, consider that you would probably stand a better chance of guessing the class mean height to within 2 inches instead of the height of a randomly chosen student. This is because the variability of the randomly selected y is larger than that of the mean y. In the one-sample case, this is because the standard deviation σ of a particular value of x is always larger than the standard deviation of the sample mean $\bar{x}$, which is $\sigma_{\bar{x}} = \sigma/\sqrt{n}$. Similarly, the prediction interval for a randomly chosen y is wider than the confidence interval for the mean value of y, for a given value of x.

Finally, we illustrate the use of technology to construct confidence intervals and prediction intervals.

EXAMPLE 10   Confidence intervals and prediction intervals using technology

Use Minitab and JMP to construct a 95% confidence interval for the population mean number of text messages sent by all 20-year-olds, and to construct a 95% prediction interval for the number of text messages sent by a randomly selected 20-year-old.

Solution

We use the instructions provided in the Step-by-Step Technology Guide below. Figure 18 shows the Minitab results, and Figure 19 shows the JMP results.

```
Prediction for Messages

Regression Equation

Messages = 60.60 - 1.500 Age


Variable  Setting
Age            20


 Fit   SE Fit      95% CI            95% PI
30.6  1.20050  (27.8316, 33.3684)  (24.3947, 36.8053)
```

FIGURE 18 Minitab results for confidence interval and prediction interval.

Age	Messages	Lower 95% Mean Messages	Upper 95% Mean Messages	Lower 95% Indiv Messages	Upper 95% Indiv Messages
20	•	27.831630634	33.368369366	24.394658079	36.805341921

FIGURE 19 JMP results for confidence interval and prediction interval.

STEP-BY-STEP TECHNOLOGY GUIDE: Confidence Intervals and Prediction Intervals

MINITAB
Step 1 Enter the x data into column **C1** and the y data into column **C2**.
Step 2 Select **Stat > Regression > Regression > Fit Regression Model....**

Step 3 For **Responses**, select *Messages*. For **Continuous predictors**, select *Age*. Click **OK**.
Step 4 Select **Stat > Regression > Regression > Predict....**
Step 5 Enter **20** in the first element of the **Age** column. Click **OK**. The results are shown in Figure 18.

SPSS
Step 1 Enter the x and y data into the first two columns.
Step 2 Select **Analyze > Regression > Linear....**
Step 3 For **Dependent**, select the y variable. For **Independent(s)**, select the x variable.
Step 4 Click **Save...**, and select **Mean** and **Individual** under **Prediction Intervals**. Click **Continue**, then **OK**.
Step 5 Return to the **Data window** and the **Data View** tab. The new variables *LMCI_1* and *UMCI_1* represent confidence

intervals for the corresponding x value, and the new variables *LICI_1* and *UICI_1* represent prediction intervals for the corresponding x value.
Step 6 To make confidence or prediction intervals for x values not in the dataset, add the value (for example, **21**) in the first empty element of the x column, and repeat Steps 2 through 4. Look for the variables *LMCI_2* and *UMCI_2*, and *LICI_2* and *UICI_2*.

JMP
Step 1 Enter the x and y data into the first two columns. Rename the columns *Age* and *Messages*.
Step 2 Select **Analyze > Fit Y by X**. For **Y, Response**, select *Messages*. For **X, Factor**, select *Age*. Click **OK**.
Step 3 Click the red triangle beside "Bivariate Fit of Messages by Age" and select **Fit Line**.
Step 4 Click the red triangle beside "Linear Fit" and select **Mean Confidence Limit Formula** and then **Indiv Confidence Limit**

Formula. Confidence intervals and prediction intervals for all existing *Age* values are supplied.
Step 5 To make confidence or prediction intervals for *Age* values not in the dataset, add the value (for example, **21**) in the first empty element of the *Age* column, and hit **ENTER**. The intervals will automatically appear.

CRUNCHIT!
Step 1 Enter the x and y data into the first two columns.
Step 2 Select **Statistics**, highlight **Regression**, and select **Simple Linear**.

Step 3 For **Dependent Variable**, select *Var2*. For **Independent Variable**, select *Var1*. For **Predict (optional)**, enter the value (for example **21**). Click **Calculate**.

Section 13.2 Summary

1. A confidence interval may be constructed for the mean response y for a given value of x. The confidence interval takes the form *point estimate ± margin of error*, where the point estimate is $\hat{y}$.

2. A prediction interval may be constructed for a random value of y for a given value of x. This prediction interval is similar to the confidence interval for the mean response y, except that the margin of error is always larger. Thus, the prediction interval is always wider than the confidence interval.

Section 13.2 Exercises

CLARIFYING THE CONCEPTS

1. Explain the difference between the confidence interval and the prediction interval we learned about in this section. (pp. 738, 739)
2. True or false: For a given value of x, the confidence interval is always wider than the prediction interval. (p. 740)

PRACTICING THE TECHNIQUES

 **CHECK IT OUT!**

To do	Check out	Topic
Exercises 3–8	Example 8	Constructing and interpreting a confidence interval for the mean value of y for a given x
Exercises 9–14	Example 9	Constructing and interpreting a prediction interval for a randomly selected value of y for a given x

For Exercises 3–8, use the data provided to construct a 95% confidence interval for the mean value of y for the given value of x. The regression equation is provided; assume the regression assumptions are met.

3.

x	y
1	15
2	20
3	20
4	25
5	25

$\hat{y} = 2.5x + 13.5; x = 3$

4.

x	y
0	10
5	20
10	45
15	50
20	75

$\hat{y} = 3.2x + 8; x = 10$

5.

x	y
-5	0
-4	8
-3	8
-2	16
-1	16

$\hat{y} = 4x + 21.6; x = -4$

6.

x	y
-3	-5
-1	-15
1	-20
3	-25
5	-30

$\hat{y} = -3x - 16; x = 3$

7.

x	y
10	100
20	95
30	85
40	85
50	80

$\hat{y} = -0.5x + 104; x = 10$

8.

x	y
0	11
20	11
40	16
60	21
80	26

$\hat{y} = 0.2x + 9; x = 20$

For Exercises 9–14, use the data from the indicated exercise to construct a 95% prediction interval for a randomly chosen value of y for the given value of x. Assume the regression assumptions are met.

9. The data in Exercise 3
10. The data in Exercise 4
11. The data in Exercise 5
12. The data in Exercise 6
13. The data in Exercise 7
14. The data in Exercise 8

APPLYING THE TECHNIQUES

For Exercises 15–24, use the data and the regression equations that you calculated in the Section 13.1 exercises.

15. Volume and Weight. For the data from Exercise 39 in Section 13.1, do the following:
   a. Predict the weight of a package that has a volume of 4 cubic meters.
   b. Construct a 95% confidence interval for the population mean weight of all packages that have a volume of 4 cubic meters.

16. Family Size and Pets. For the data from Exercise 40 in Section 13.1, do the following:
   a. Predict the population mean number of pets for a family of size 5.
   b. Construct a 95% confidence interval for the population mean number of pets for all families of size 5.

17. World Temperatures. For the data from Exercise 41 in Section 13.1, do the following:
   a. Predict the population mean high temperature for a city with a low temperature of 30 degrees.
   b. Construct a 99% confidence interval for the population mean high temperature for all cities with a low temperature of 30 degrees.

18. Video Game Sales. For the data from Exercise 42 in Section 13.1, do the following:
   a. Predict the population mean total sales for a game that has been on the top 30 list for 20 weeks.
   b. Construct a 99% confidence interval for the population mean total sales for all games that have been on the top 30 list for 20 weeks.

19. Volume and Weight. Refer to the data from Exercise 15.
   a. Construct a 95% prediction interval for the weight of a randomly selected package that has a volume of 4 cubic meters.
   b. Compare the intervals from Exercise 15(**b**) and Exercise 19(**a**). Which interval is wider, and why?

20. Family Size and Pets. Refer to the data from Exercise 16.
   a. Construct a 95% prediction interval for the number of pets for a randomly selected family of size 5.
   b. Compare the intervals from Exercise 16(**b**) and Exercise 20(**a**). Which interval is wider, and why?

21. World Temperatures. Refer to the data from Exercise 17.
   a. Construct a 90% prediction interval for the high temperature for a randomly chosen city with a low temperature of 30 degrees.

b. Suppose we were asked to provide a prediction interval for the high temperature for a randomly chosen city with a low temperature of zero degrees. Explain what the danger would be in constructing such a prediction interval.

22. **Video Game Sales.** Refer to the data from Exercise 18.
   a. Construct a 99% prediction interval for the total sales for a randomly chosen video game that has been on the top list for 20 weeks.
   b. Suppose we were asked to provide a prediction interval for the total sales for a randomly chosen video game that has been on the top list for 104 weeks. Explain what the danger would be in constructing such a prediction interval.

23. **Working with Large Data Sets.** Open the **Pulse and Temp** data set. Select the females data only. Use technology to do the following: **pulseandtemp**

a. Construct a 95% confidence interval for the population mean body temperature for all females with a heart rate of 72.
b. Construct a 95% prediction interval for the body temperature for a randomly selected female with a heart rate of 72.

24. **Working with Large Data Sets.** Open the **Nutrition** data set. Use technology to do the following:

nutrition

a. Construct a 95% confidence interval for the population mean number of calories per gram (y) for all foods with the amount of fat per gram (x) of 0.5.
b. Construct a 95% prediction interval for the number of calories per gram for a randomly selected food with the amount of fat per gram (x) of 0.5.

13.3 Multiple Regression

OBJECTIVES By the end of this section, I will be able to . . .

1. Find the multiple regression equation, interpret the multiple regression coefficients, and use the multiple regression equation to make predictions.
2. Calculate and interpret the adjusted coefficient of determination.
3. Perform the F test for the overall significance of the multiple regression.
4. Conduct t tests for the significance of individual predictor variables.
5. Explain the use and effect of dummy variables in multiple regression.
6. Apply the strategy for building a multiple regression model.

1 Finding the Multiple Regression Equation, Interpreting the Coefficients, and Making Predictions

Regression analysis using a single y variable and a single x variable is called simple linear regression.

Thus far, we have examined the relationship between the response variable y and a single predictor variable x. In our data-filled world, however, we often encounter situations where we can use more than one x variable to predict the y variable. This is called **multiple regression**.

> **Multiple regression** describes the linear relationship between one response variable y and more than one predictor variable, x_1, x_2, x_3, ... The **multiple regression equation** is an extension of the regression equation
>
> $$\hat{y} = b_0 + b_1 x_1 + b_2 x_2 + \cdots b_k x_k$$
>
> where k represents the number of x variables in the equation, and b_0, b_1, b_2, . . . b_k represent the **multiple regression coefficients**.

The interpretation of the regression coefficients is similar to the interpretation of the slope b_1 in simple linear regression, except that we also state that the other x variables are held constant. The interpretation of the y intercept b_0 is similar to the simple linear regression case. The next example illustrates the multiple regression equation, and shows how to interpret the multiple regression coefficients.

EXAMPLE 11 Multiple regression equation, coefficients, and prediction

breakfastcereals

The data set **Breakfast Cereals** includes several predictor variables and one response variable, y = (nutritional) rating.

a. Use technology to find the multiple regression equation for predicting y = rating, using x_1 = fiber and x_2 = sugar. State the equation with a sentence.

b. State the values of the multiple regression coefficients.

c. Interpret the multiple regression coefficients for using x_1 = fiber and x_2 = sugar.

d. Use the multiple regression equation to predict the rating of a breakfast cereal with 5 mg of fiber and 10 mg of sugar.

Solution

Using the instructions in the Step-by-Step Technology Guide at the end of this section, we open the **Breakfast Cereals** data set and perform a multiple regression of y = rating on x_1 = fiber and x_2 = sugar. Note that this does not represent extrapolation, as there are cereals in the data set that have either zero grams of fiber (such as Cap'n Crunch) or zero grams of sugar (such as Cream of Wheat).

When we perform a multiple regression of one variable on *(or* against *or* versus*) a set of other variables, the first variable is always the* y *variable, and the set of variables following the word* on *are the* x *variables.*

a. A partial Minitab printout is shown in Figure 20. A partial SPSS printout is in Figure 21. The multiple regression equation is

$$\hat{y} = b_0 + b_1 x_1 + b_2 x_2$$
$$= 52.22 + 2.869(\text{fiber}) - 2.246(\text{sugar})$$

The estimated nutritional rating equals 52.22 points plus 2.869 times the number of grams of fiber minus 2.246 times the number of grams of sugar.

Regression Analysis: rating versus fiber, sugar

Analysis of Variance

Source	DF	Adj SS	Adj MS	F-Value	P-Value
Regression	2	12251.6	6125.80	165.13	0.000
fiber	1	3482.8	3482.77	93.88	0.000
sugar	1	7134.0	7134.05	192.31	0.000
Error	74	2745.2	37.10		
Lack-of-Fit	44	1962.1	44.59	1.71	0.063
Pure Error	30	783.1	26.10		
Total	76	14996.8			

Model Summary

S	R-sq	R-sq(adj)	R-sq(pred)
6.09075	81.69%	81.20%	80.15%

Regression Equation

rating = 52.22 + 2.869 fiber - 2.246 sugar

FIGURE 20 Multiple regression equation in Minitab.

Model		Unstandardized Coefficients	
		B	Std. Error
1	(Constant)	52.217	1.541
	Fiber	2.869	.296
	Sugar	-2.246	.162

FIGURE 21 Multiple regression equation in SPSS.

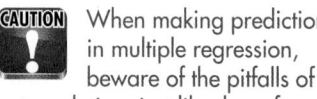
When making predictions in multiple regression, beware of the pitfalls of extrapolation, just like those for simple linear regression. Further, in multiple regression, the values for all *predictor variables must lie within their respective ranges. Otherwise, the prediction represents extrapolation, and it may be misleading.*

b. The values of the multiple regression coefficients are $b_0 = 52.22$, $b_1 = 2.869$, and $b_2 = -2.246$.

c. The multiple regression coefficients are interpreted as follows:

- $b_0 = 52.22$ (y intercept). The estimated nutritional rating when there are zero grams of fiber and zero grams of sugar is 52.22.

- $b_1 = 2.869$. For every increase of one gram of fiber, the estimated increase in nutritional rating is 2.869 points, *when the amount of sugar is held constant*.

- $b_2 = -2.246$. For every increase of one gram of sugar, the estimated *decrease* in nutritional rating is 2.246 points, *when the amount of fiber is held constant*.

So far, all of our x variables have been continuous. But what if we want to include a categorical variable as a predictor?

5 Dummy Variables in Multiple Regression

The data set **Pulse and Temp** contains the heart rate, body temperature, and sex of 130 men and women. We want to use x_1 = heart rate and x_2 = sex to predict y = body temperature. However, sex is a categorical variable, so we must *recode* the values of x_2 as follows:

$$x_2: \text{Let "female"} = 1 \text{ and let "male"} = 0$$

The variable x_2 is called a **dummy variable**, because it recodes the values of the binomial (categorical) variable *sex* into values of 0 and 1.

> A **dummy variable** is a predictor variable used to recode a binomial categorical variable in regression, and taking values 0 or 1.

This recoding will provide us with two different regression equations, one for the females ($x_2 = 1$) and one for the males ($x_2 = 0$), shown here:

- Females: $\hat{y} = b_0 + b_1 x_1 + b_2 x_2 = b_0 + b_1 x_1 + b_2(1) = (b_0 + b_2) + b_1 x_1$
- Males: $\hat{y} = b_0 + b_1 x_1 + b_2 x_2 = b_0 + b_1 x_1 + b_2(0) = b_0 + b_1 x_1$

Note that these two regression equations have the *same slope* b_1, but different y intercepts. The females have y intercept $(b_0 + b_2)$, whereas the males have y intercept b_0. See Figure 29 on page 750. The *difference in* y *intercepts* is $(b_0 + b_2) - b_0 = b_2$, which is the coefficient of the dummy variable x_2. Let us illustrate with an example.

EXAMPLE 15 Dummy variables in multiple regression

a. Verify that the regression assumptions are met.

b. Perform a multiple regression of y = body temperature on x_1 = heart rate and x_2 = sex, using level of significance $\alpha = 0.05$. Find the two regression equations, one for females and the other for males.

c. Construct a scatterplot of y = body temperature versus x_1 = heart rate, using different-shaped points to show the different sexes. Place the two regression equations on the scatterplot.

d. Interpret the coefficient of the dummy variable x_2 = sex.

Solution

a. Figures 26 and 27 contain no evidence of unhealthy patterns. We therefore conclude that the regression assumptions are verified.

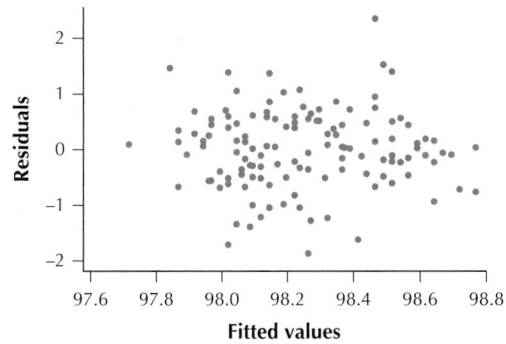

FIGURE 26 Scatterplot of residuals versus fitted values.

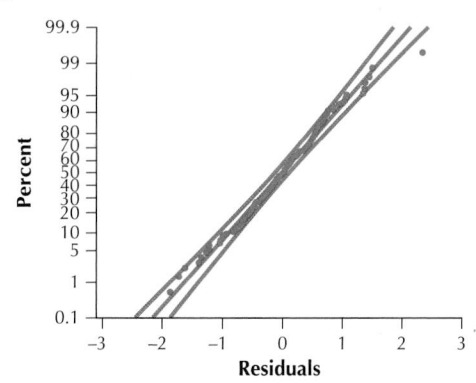

FIGURE 27 Normal probability plot of the residuals.

b. Figure 28 contains the multiple regression results. The *p*-value for the *F* test is 0.001, which is $\leq \alpha = 0.05$, so we conclude that the overall regression is significant. The regression results tell us that $b_0 = 96.251$, $b_1 = 0.02526$, and $b_2 = 0.270$. Thus, our two regression equations are:

- Females: $\hat{y} = (b_0 + b_2) + b_1 x_1 = (96.251 + 0.27) + 0.02526 x_1$
$$= 96.521 + 0.02526 x_1$$

- Males: $\hat{y} = b_0 + b_1 x_1 = 96.251 + 0.02526 x_1$

Regression Analysis: Body Temp versus Heart Rate, Sex

Analysis of Variance

Source	DF	Adj SS	Adj MS	F-Value	P-Value
Regression	2	6.813	3.4065	6.92	0.001
Heart Rate	1	4.092	4.0922	8.31	0.005
Sex	1	2.353	2.3535	4.78	0.031
Error	127	62.533	0.4924		
Lack-of-Fit	49	18.859	0.3849	0.69	0.920
Pure Error	78	43.674	0.5599		
Total	129	69.346			

Model Summary

S	R-sq	R-sq(adj)	R-sq(pred)
0.701701	9.82%	8.40%	5.85%

Regression Equation

Body Temp = 96.251 + 0.02526 Heart Rate + 0.270 Sex

FIGURE 28 Results for multiple regression of y = body temperature on x_1 = heart rate and x_2 = sex.

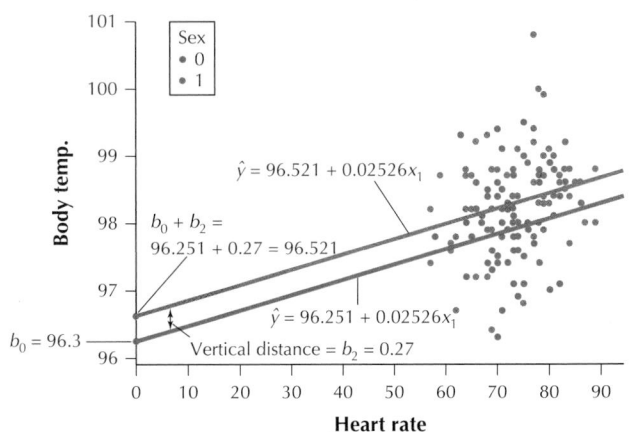

FIGURE 29 Scatterplot showing parallel regression lines when using dummy variables.

c. Figure 29 contains the scatterplot of y = body temperature versus x_1 = heart rate, with the orange dots representing females and the blue dots representing males. The regression lines are shown, orange for females, blue for males. *Note that the regression lines are parallel, because they each have the same slope $b_1 = 0.02521$.* So the only difference in the lines is the y intercepts.

- For females, the y intercept is $b_0 + b_2 = 96.251 + 0.27 = 96.521$.
- For males, the y intercept is simply $b_0 = 96.251$.

d. The vertical distance between the parallel regression lines equals $b_2 = 0.27$, as shown in Figure 29. Thus, we interpret the coefficient b_2 of the dummy variable x_2 as the estimated increase in y = body temperature for those observations with $x_2 = 1$ (females), as compared to those with $x_2 = 0$ (males), when heart rate is held constant. That is, for the same heart rate, females have a body temperature that is higher than that of males, by an estimated 0.27 degrees.

NOW YOU CAN DO
Exercise 29.

6 Strategy for Building a Multiple Regression Model

In order to bring together all you have learned of multiple regression, we now present a general strategy for building a multiple regression model.

Strategy for Building a Multiple Regression Model

Step 1 The *F* Test.
Construct the multiple regression equation using all relevant predictor variables. Apply the *F* test for the significance of the overall regression, in order to make sure that a linear relationship exists between the response y and at least one of the predictor variables.

We eliminate only one variable at a time. It may happen that eliminating one nonsignificant variable will nudge a second, formerly nonsignificant, variable into significance.

Step 2   The t Tests.
Perform the t tests for the individual predictors. If at least one of the predictors is not significant (that is, its p-value is greater than level of significance α), then eliminate the x variable with the largest p-value from the model. Ignore the p-value of β_0. *Repeat Step 2 until all remaining predictors are significant.*

Step 3   Verify the Assumptions.
For your *final model*, verify the regression assumptions.

Step 4   Report and Interpret Your Final Model.
a.   Provide the multiple regression equation for your final model.
b.   Interpret the multiple regression coefficients so that a nonstatistician could understand.
c.   Report and interpret the standard error of the estimate s and the adjusted coefficient of determination R^2_{adj}.

We illustrate this strategy, known as **backward stepwise regression**, in the following example.

EXAMPLE 16   Strategy for building a multiple regression model

 baseball2013

Ronald C. Modra/Sports Imagery/Getty Images

The author of this book first became interested in the field of statistics through the enjoyment of sports statistics, especially baseball, which is packed with interesting statistics. Today, professional sports teams are seeking competitive advantage through the analysis of data and statistics, such as *Sabermetrics* (Society of American Baseball Research, www.sabr.org), as shown in the motion picture *Moneyball*.

Suppose a baseball researcher is interested in predicting y = runs scored, using the data set **Baseball 2013** and the following predictor variables:

- x_1 = Hits, the number of hits (of all kinds) the player makes
- x_2 = Doubles, the number of doubles the player makes
- x_3 = Triples, the number of triples the player makes
- x_4 = Home Runs, the number of home runs the player makes
- x_5 = RBIs, the number of runs batted in (runs scored by other players, but caused by this player)
- x_6 = Walks, the number of walks issued to the player
- x_7 = Batting Average, the number of hits divided by the number of at-bats
- x_8 = Red Sox, a dummy variable equal to 1 if the player plays for the Boston Red Sox, and 0 otherwise

Use the Strategy for Building a Multiple Regression Model to build the best multiple regression model for predicting the number of runs scored using these predictor variables, at level of significance $\alpha = 0.05$.

Solution

The data set **Baseball 2013** contains the batting statistics of the $n = 448$ players in Major League Baseball who had at least 100 at-bats during the 2013 season (*Source:* www.seanlahman.com/baseball-archive/statistics).

Step 1   **The F Test.** Figure 30 shows the Minitab results of a regression of y = runs scored on the set of predictor variables $x_1, x_2, x_3, \ldots, x_8$. The *p-value for the F test is significant, so we know that a linear relationship exists between y* = runs scored and at least one of the x variables.

Step 2   **The t test (the first time).** In Figure 30, the p-value for Batting Average is greater than level of significance $\alpha = 0.05$. We therefore eliminate the Batting Average from the model. Perhaps surprisingly, a player's batting average is evidently not helpful in predicting the number of runs that player will score when all other predictors are held constant.

```
Analysis of Variance

Source        DF  Adj SS   Adj MS  F-Value  P-Value       Step 1:
Regression     8  264398  33049.8   782.22    0.000        p-value for F test
  Hits         1    8394   8393.9   198.67    0.000
  Doubles      1     341    340.9     8.07    0.005
  Triples      1    2521   2520.9    59.66    0.000
  Home Runs    1    2428   2428.4    57.47    0.000
  RBI          1     262    261.6     6.19    0.013
  Walks        1    3629   3629.3    85.90    0.000
  Bat Ave      1      14     14.1     0.33    0.563
  RedSox       1     319    319.1     7.55    0.006
Error        439   18548     42.3
Total        447  282947


Model Summary

      S   R-sq  R-sq(adj)  R-sq(pred)
6.50012  93.44%    93.33%      93.05%


Coefficients

Term        Coef   SE Coef  T-Value  P-Value    VIF
Constant    0.14      2.54     0.05    0.957
Hits      0.3225    0.0229    14.09    0.000  12.75
Doubles   0.2092    0.0736     2.84    0.005   6.39
Triples    1.429     0.185     7.72    0.000   1.40
Home Runs 0.6827    0.0901     7.58    0.000   6.00
RBI      -0.1064    0.0428    -2.49    0.013  13.32
Walks     0.2292    0.0247     9.27    0.000   2.78
Bat Ave     -6.6      11.4    -0.58    0.563   1.67       Step 2:
RedSox      5.19      1.89     2.75    0.006   1.07       p-value for t test
```

FIGURE 30 Step 1: *F* test is significant.

```
Analysis of Variance

Source        DF  Adj SS   Adj MS  F-Value  P-Value
Regression     7  264384  37769.2   895.27    0.000
  Hits         1    9144   9144.2   216.75    0.000
  Doubles      1     346    345.8     8.20    0.004
  Triples      1    2521   2521.0    59.76    0.000
  Home Runs    1    2433   2433.3    57.68    0.000
  RBI          1     259    259.1     6.14    0.014
  Walks        1    3858   3858.2    91.45    0.000
  RedSox       1     309    308.5     7.31    0.007
Error        440   18563     42.2
Total        447  282947


Model Summary

      S   R-sq  R-sq(adj)  R-sq(pred)
6.49520  93.44%    93.34%      93.07%


Coefficients

Term         Coef   SE Coef  T-Value  P-Value    VIF
Constant   -1.283     0.650    -1.97    0.049
Hits       0.3182    0.0216    14.72    0.000  11.40
Doubles    0.2105    0.0735     2.86    0.004   6.38
Triples     1.429     0.185     7.73    0.000   1.40
Home Runs  0.6833    0.0900     7.59    0.000   6.00
RBI       -0.1059    0.0427    -2.48    0.014  13.31
Walks      0.2319    0.0242     9.56    0.000   2.68
RedSox       5.08      1.88     2.70    0.007   1.05
```

FIGURE 31 All *x* variables are significant; we have our final model.

Step 2 **The *t* test (the second time).** We repeat Step 2 as long as there are *x* variables with *p*-values greater than level of significance $\alpha = 0.05$. Figure 31 shows the results of performing the multiple regression of *y* = Runs Scored on all the *x* variables except Batting Average. No further variables have *p*-values below 0.05; therefore, no further variables are excluded from the model. In other words, we have our final model.

Step 3 **Verify the assumptions.** For our final model, we now verify the regression assumptions. Figures 32 and 33 show no patterns for the bulk of the data that would indicate a violation of the regression assumptions. We therefore conclude that the regression assumptions are verified.

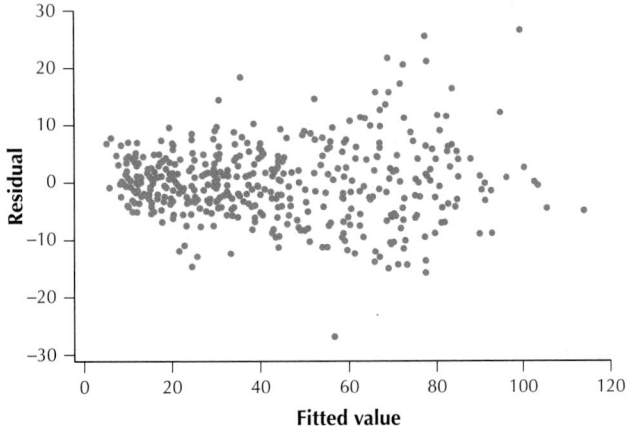

FIGURE 32 Scatterplot of residuals versus fitted values.

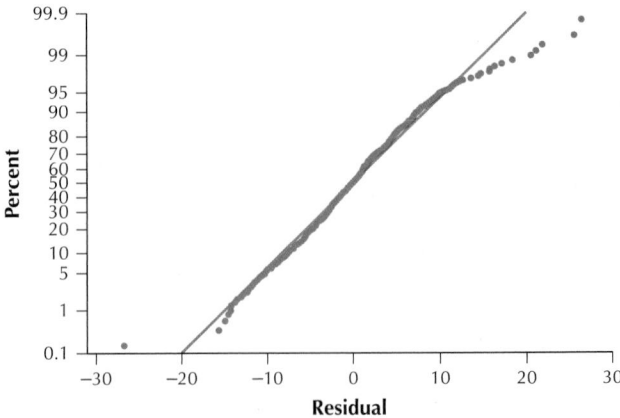

FIGURE 33 Normal probability plots of the residuals.

Step 4 **Report and interpret your final model.**

a. The multiple regression equation for the final model is shown here.

$$\hat{y} = -1.283 + 0.3182 \text{ Hits} + 0.2105 \text{ Doubles} + 1.429 \text{ Triples} + 0.6833 \text{ Home Runs} - 0.1059 \text{ RBIs} + 0.2319 \text{ Walks} + 5.08 \text{ Red Sox}$$

b. We interpret the coefficient for Hits, and leave to the exercises the interpretation of the other multiple regression coefficients. "For each additional hit that a player makes, the estimated increase in the number of runs that player will score is 0.3182, when all the other x variables are held constant."

c. The standard error of the estimate for the final model is $s = 6.4952 \approx 6.5$. That is, using the multiple regression equation in **(a)**, the size of the typical prediction error will be about 6.5 runs. The value of the adjusted coefficient of determination is $R^2_{adj} = 93.07\%$. In other words, 93.07% of the variability in the number of runs scored is accounted for by this multiple regression equation.

NOW YOU CAN DO
Exercises 24–26
and 31–33.

STEP-BY-STEP TECHNOLOGY GUIDE: Multiple Regression

TI-83/84
The TI-83/84 calculators do not perform the calculations for multiple regression.

EXCEL
Step 1   Load the response variable y into column **A**, and the predictor variables into columns **B**, **C**, and so on.
Step 2   Click on **Data**, then **Data Analysis**, then **Regression**, and click **OK**.
Step 3   For **Input Y Range**, select the cells in column **A**. For **Input X Range**, select the cells in columns **B**, **C**, etc.

Step 4   If you want to verify the regression assumptions, select **Residual Plots** and **Normal Probability Plots**.
Step 5   Click **OK**.

Note: Excel does not construct prediction intervals and confidence intervals for multiple regression.

MINITAB
Step 1   Load the response variable y into **C1** and the predictor variables into **C2**, **C3**, and so on.
Step 2   Click on **Stat > Regression > Regression > Fit Regression Model...** .
Step 3   Select **C1** as your **Responses**, and select **C2**, **C3**, etc. as your **Continuous predictors**.

Step 4   If you want to verify the regression assumptions, click the button labeled **Graphs...**, select **Four in One**, and click **OK**.
Step 5   Click **OK**.

Note: See the Guide on page 741 for instructions on how to construct confidence intervals and prediction intervals for multiple regression.

SPSS
Step 1   Load the response variable y into **VAR00001** and the predictor variables into **VAR00002**, **VAR00003**, and so on.
Step 2   Select **Analyze > Regression > Linear...** .
Step 3   Move the y variable to **Dependent**, and the x variables to **Independent(s)**.

Step 4   If you would like to verify the regression assumptions, click the button labeled **Plots...**, select **Normal Probability Plot**, and click **Continue**. Click **OK**.

Note: See the Guide on page 741 for instructions on how to construct confidence intervals and prediction intervals for multiple regression.

JMP
Step 1   Load the response variable y into **Column 1** and the predictor variables into **Column 2**, **Column 3**, and so on.
Step 2   Select **Analyze > Fit Model**.
Step 3   Move the y variable to **Y**. Select the x variables and click **Add** under **Construct Model Effects**. Click **Run**.

Step 4   If you would like to verify the regression assumptions, observe the **Residual by Predicted Plot** under **Effect Tests**.

Note: See the Guide on page 741 for instructions on how to construct confidence intervals and prediction intervals for multiple regression.

CRUNCHIT!
Step 1   Enter data into the spreadsheet.
Step 2   Select **Statistics**, highlight **Regression**, and select **Multiple Linear**.

Step 3   Select the **Dependent Variable** from the drop-down menu, and check the **Independent Variables**. Click **Calculate**.

Note: CrunchIt! does not construct prediction intervals and confidence intervals for multiple regression.

Section 13.3 Summary

1. Multiple regression describes the linear relationship between one response variable y and more than one predictor variable, $x_1, x_2, x_3, \ldots$. The multiple regression equation is an extension of the regression equation: $\hat{y} = b_0 + b_1 x_1 + b_2 x_2 + \cdots + b_k x_k$ where k represents the number of x variables in the equation, and $b_1, b_2, b_3, \cdots b_k$ represent the multiple regression coefficients.

2. The multiple coefficient of determination R^2 represents the proportion of the variability in the response y that is explained by the multiple regression equation. The adjusted coefficient of determination R^2_{adj} adjusts the value of R^2 as a penalty for having too many unhelpful x variables in the equation.

3. The multiple regression model is an extension of the regression model from Section 13.1. The population multiple regression equation is $y = \beta_0 + \beta_1 x_1 + \beta_2 x_2 + \cdots + \beta_k x_k + \varepsilon$. The F test is performed to assess the significance of the overall model.

4. To determine whether a particular x variable has a significant linear relationship with the response variable y, we perform the t test for the significance of that x variable. One may perform as many such t tests as there are x variables in the model, which is k assuming the overall F test is significant.

5. Dummy variables are 0/1 variables that allow, via recoding, categorical variables to be included in the multiple regression model.

6. The Strategy for Building a Multiple Regression Model brings together all we have learned about multiple regression modeling.

Section 13.3 Exercises

CLARIFYING THE CONCEPTS

1. Write the multiple regression equation for $k = 3$ predictor variables. (p. 743)
2. Which is preferable, R^2 or R^2_{adj}, and why? (p. 745)
3. Which test do we perform if we want to determine whether our multiple regression is useful? (p. 746)
4. If we conclude from the F test that our multiple regression is useful, is it still possible that one of the β's equals zero? Explain. (p. 746)
5. Explain the difference between the F test and the t test we learned in this section. (p. 747)
6. How many t tests may we perform for a multiple regression model? (p. 747)
7. How do we interpret the coefficient for a dummy variable. (*Hint:* Consider Figure 29.) (p. 750)
8. What are the four steps of the Strategy for Building a Multiple Regression Model. (p. 750)

PRACTICING THE TECHNIQUES

 CHECK IT OUT!

To do	Check out	Topic
Exercises 9–16	Example 11	Multiple regression equation, coefficients, and predictions
Exercises 17–20	Example 12	Calculating and interpreting the adjusted coefficient of determination R^2_{adj}
Exercises 21–22 and 27–28	Example 13	F test for the overall significance of the multiple regression
Exercises 23 and 30	Example 14	Performing a set of t tests for the significance of a set of individual x variables

To do	Check out	Topic
Exercise 29	Example 15	Dummy variables in multiple regression
Exercises 24–26 and 31–33	Example 16	Strategy for building a multiple regression model

Use the following information for Exercises 9–12: A multiple regression model has been produced for a set of $n = 20$ observations with multiple regression equation $\hat{y} = 10 + 5x_1 + 8x_2$, with multiple coefficient of determination $R^2 = 0.5$. Assume the regression assumptions are met.

9. Interpret the value of the coefficient for x_1.
10. Explain what the value of b_2 means.
11. Interpret the coefficients b_0, b_1, and b_2.
12. Find point estimates of y for the following values of x_1 and x_2:
    a. $x_1 = 6, x_2 = 4$
    b. $x_1 = 10, x_2 = 8$

Use the following information for Exercises 13–16: A multiple regression model has been produced for a set of $n = 50$ observations with multiple regression equation $\hat{y} = 0.5 - 0.1x_1 + 0.9x_2$, with multiple coefficient of determination $R^2 = 0.75$. Assume the regression assumptions are met.

13. Interpret the value of the coefficient for x_1.
14. Explain what the value of b_2 means.
15. Interpret the coefficients b_0, b_1, and b_2.
16. Find point estimates of y for the following values of x_1 and x_2:
    a. $x_1 = 5, x_2 = 6$
    b. $x_1 = 4, x_2 = 3$
17. For the data in Exercises 9–12, how should the value of R^2 be interpreted?
18. Calculate R^2_{adj} for the data in Exercises 9–12.
19. For the data in Exercises 13–16, how should the value of R^2 be interpreted?
20. Calculate R^2_{adj} for the data in Exercises 13–16.

Use the following data set for Exercises 21–26.

y	x_1	x_2	x_3
0.6	1	10	1.3
4.0	2	10	−3.2
3.2	3	8	−1.0
9.0	4	8	0.9
1.8	5	6	−2.5
8.4	6	6	0.9
9.8	7	4	1.0
10.4	8	4	2.0
8.8	9	2	0.2
14.7	10	2	−2.2

21. Perform the multiple regression of y on x_1, x_2, and x_3, and write the multiple regression equation.
22. Assume the regression assumptions are met. Perform the F test for the significance of the overall regression, using level of significance $\alpha = 0.05$. Do the following:
 a. State the hypotheses and the rejection rule.
 b. Find the F statistic and the p-value.
 c. State the conclusion and interpretation.
23. Perform the t test for the significance of the individual predictor variables, using level of significance $\alpha = 0.05$. Do the following:
 a. For each hypothesis test, state the hypotheses and the rejection rule.
 b. For each hypothesis test, find the t statistic and the p-value.
 c. For each hypothesis test, state the conclusion and interpretation.
24. Identify any predictors that have corresponding p-values greater than the level of significance $\alpha = 0.05$. Of these, discard the variable with the largest p-value. Then redo Exercise 23, omitting this predictor. Repeat if necessary.
25. Verify the regression assumptions for your final model from Exercise 24.
26. Report and interpret your final model from Exercise 24, by doing the following:
 a. Provide the multiple regression equation for your final model.
 b. Interpret the multiple regression coefficients so that a nonstatistician could understand.
 c. Report and interpret the standard error of the estimate s and the adjusted coefficient of determination R^2_{adj}.

Use the following data set for Exercises 27–33. Note that x_3 is a dummy variable.

y	x_1	x_2	x_3
−0.7	2	0.1	0
6.4	4	−2.5	1
2.8	6	2.7	0
9.4	8	2.8	1
8.6	10	−1.6	0
13.1	12	1.0	1
12.2	14	−1.4	0
19.1	16	−0.5	1
18.8	18	1.0	0
23.2	20	−2.3	1

27. Perform the multiple regression of y on x_1, x_2, and x_3, and write the multiple regression equation.
28. Assume the regression assumptions are met. Perform the F test for the significance of the overall regression, using level of significance $\alpha = 0.01$. Do the following:
 a. State the hypotheses and the rejection rule.
 b. Find the F statistic and the p-value.
 c. State the conclusion and interpretation.
29. Interpret the coefficient for the dummy variable.
30. Perform the t test for the significance of the individual predictor variables, using level of significance $\alpha = 0.01$. Do the following:
 a. For each hypothesis test, state the hypotheses and the rejection rule.
 b. For each hypothesis test, find the t statistic and the p-value.
 c. For each hypothesis test, state the conclusion and interpretation.
31. Identify any predictors that have corresponding p-values greater than the level of significance $\alpha = 0.01$. Of these, discard the variable with the largest p-value. Then redo Exercise 30, omitting this predictor. Repeat if necessary.
32. Verify the regression assumptions for your final model from Exercise 31.
33. Report and interpret your final model from Exercise 31, by doing the following:
 a. Provide the multiple regression equation for your final model.
 b. Interpret the multiple regression coefficients so that a nonstatistician could understand.
 c. Report and interpret the standard error of the estimate s, and the adjusted coefficient of determination R^2_{adj}.

APPLYING THE CONCEPTS

For Exercises 34–39, apply the Strategy for Building a Multiple Regression Model by performing the following steps, using level of significance $\alpha = 0.05$:
 a. *Step 1* Perform the F test for significance of the overall regression.
 b. *Step 2* Perform the t tests for the individual predictors. If at least one of the predictors is not significant, then eliminate the x variable with the largest p-value from the model. Repeat Step 2 until all remaining predictors are significant.
 c. *Step 3* Verify the assumptions.
 d. *Step 4* Report and interpret your final model. Report and interpret the coefficients, the standard error of the estimate s, and the adjusted coefficient of determination R^2_{adj}.
34. Best Places for Dating. Sperling's Best Places published the list of best places for dating in America for 2010. Table 6 shows the top 10 places, along with the overall dating score (y) and a set of predictor variables.

TABLE 6 Best places for dating in America

City	y = Overall dating score	Percentage 18–24 years old	Percentage 18–24 years and single	Online dating score
Austin	100.0	13.40%	81.20%	77.8
Colorado Springs	88.7	10.50%	74.20%	88.9
San Diego	84.0	11.30%	79.40%	77.4
Raleigh	80.7	11.60%	82.90%	79.2
Seattle	78.7	9.00%	83.90%	100.0
Charleston	78.7	11.20%	82.70%	66.9
Norfolk	77.0	11.20%	75.60%	82.9
Ann Arbor	75.5	12.90%	90.30%	51.1
Springfield	75.2	11.70%	89.80%	63.5
Honolulu	75.2	10.10%	82.30%	50.2

 bestdating

35. Ease of Doing Business. Doing Business (www.doingbusiness.org) publishes statistics on how easy or difficult different countries make it to do business. Table 7 shows the top 12 countries for ease of doing business, with y = easiness score along with a set of predictor variables.

bestbusiness

TABLE 7 Best countries for ease of doing business

Country	Easiness score	Starting a business	Employing workers	Paying taxes
Singapore	100	10	1	5
New Zealand	99	1	14	12
United States	98	6	1	46
Hong Kong	97	15	20	3
Denmark	96	16	10	13
U.K.	95	8	28	16
Ireland	94	5	38	6
Canada	93	2	18	28
Australia	92	3	8	48
Norway	91	33	99	18
Iceland	90	17	62	32
Japan	89	64	17	112

36. Virginia Weather. Table 8 contains data on weather in a sample of cities in the state of Virginia. We are interested in predicting y = heating degree days using the other predictor variables.

vaweather

TABLE 8 Data on the weather in Virginia

City	Heating degree-days	Avg. Jan. temp.	Avg. July temp.	Cooling degree-days
Alexandria	4055	34.9	79.2	1531
Arlington	4055	34.9	79.2	1531
Blacksburg	5559	30.9	71.1	533
Charlottesville	4103	35.5	76.9	1212
Chesapeake	3368	40.1	79.1	1612
Danville	3970	36.6	78.8	1418
Hampton	3535	39.4	78.5	1432
Harrisonburg	5333	30.5	73.5	758
Leesburg	5031	31.5	75.2	911
Lynchburg	4354	34.5	75.1	1075
Manassas	4925	31.7	75.7	1075
Newport News	3179	41.2	80.3	1682
Norfolk	3368	40.1	79.1	1612
Petersburg	3334	39.7	79.6	1619
Portsmouth	3368	40.1	79.1	1612
Richmond	3919	36.4	77.9	1435
Roanoke	4284	35.8	76.2	1134
Suffolk	3467	39.6	78.5	1427
Virginia Beach	3336	40.7	78.8	1482

Source: National Oceanic and Atmospheric Administration.

37. Health Insurance Coverage. We are interested in estimating y = the number of people covered by health insurance, using x_1 = the number of adults not covered and x_2 = the number of children not covered. Use the data in Table 9, containing a random sample of U.S. states. All data are in thousands. healthinsurance

TABLE 9 Health insurance coverage

State	Persons covered	Adults not covered	Children not covered
Alabama	3,843	689	82
Arizona	4,958	1,311	283
Colorado	3,977	826	176
Georgia	7,688	1,659	314
Illinois	10,867	1,776	302
Kentucky	3,467	639	98
Maryland	4,836	776	137
Massachusetts	5,678	657	103
Michigan	8,928	1,043	116
Minnesota	4,675	475	104
Missouri	5,028	772	127
New Jersey	7,319	1,341	277
North Carolina	7,266	1,585	307
Ohio	10,181	1,138	157
Pennsylvania	11,108	1,237	203
South Carolina	3,553	672	112
Tennessee	5,111	809	94
Virginia	6,532	1,006	185
Washington	5,572	746	105
Wisconsin	4,995	481	63

38. Regression in Accounting. We are interested in estimating y = current ratio using x_1 = price–earnings ratio, x_2 = total assets, and x_3 = total liabilities. Use the data in Table 10, containing a random sample of large technology companies in 2010. Total assets and total liabilities are in billions of dollars. **accounting**

TABLE 10 Accounting data for large technology companies

Company	Current ratio	Price–earnings ratio	Assets	Liabilities
Microsoft	1.82	12.51	77.9	38.3
Intel	2.79	18.44	53.1	11.4
Dell	1.28	10.95	33.7	28.0
Apple	1.88	24.57	53.9	26.0
Google	10.62	18.87	40.5	4.5

Source: Lexis-Nexis.

   a. Build the final multiple regression model using level of significance $\alpha = 0.10$.

   b. Comment on your results from **(a)**.

   c. Redo your work from **(a)**, this time using level of significance $\alpha = 0.13$.

   d. Report and interpret your final model from **(c)**.

39. Blood Pressure. Open the data set **Systolic**. We are interested in estimating y = systolic blood pressure, based on the other predictor variables. **systolic**

40. Baseball. In Example 16, interpret the coefficients for Triples, Hits, Home Runs, RBIs, Walks, and Yankees.

Your Best Model. Work with the **Nutrition** data sets for Exercises 41 and 42. **nutrition**

41. Use technology to apply the Strategy for Building a Multiple Regression Model, using level of significance $\alpha = 0.05$, for predicting the number of calories, with the following x-variables: protein, fat, saturated fat, cholesterol, carbohydrates, calcium, phosphorous, iron, potassium, sodium, thiamin, niacin, and ascorbic acid.

42. Write a summary to interpret each regression coefficient, and comment on which variables are the most important for predicting the number of calories.

Chapter 13 Formulas and Vocabulary

SECTION 13.1
- **Confidence interval for slope β_1** (p. 726).

$$b_1 \pm t_{\alpha/2} \cdot \frac{s}{\sqrt{\sum (x - \bar{x})^2}}$$

- **Fitted values** (p. 719)
- **Margin of error E** (p. 726)
- **Population regression equation** (p. 718)
- **Regression model** (p. 718)
- **Regression model assumptions** (p. 718)
- **Test statistic t_{data}** (p. 721).

$$t_{data} = \frac{b_1}{s/\sqrt{\sum (x - \bar{x})^2}}$$

SECTION 13.2
- **Confidence interval for the mean of the values of y for a given x** (p. 738)
- **Prediction interval for a randomly selected value of y for a given x** (p. 739)

SECTION 13.3
- **Adjusted coefficient of determination R^2_{adj}** (p. 745).

$$R^2_{adj} = 1 - (1 - R^2)\left(\frac{n - 1}{n - k - 1}\right)$$

- **Backward, stepwise regression** (p. 751)
- **Dummy variable** (p. 749)
- **Multiple coefficient of determination R^2** (p. 745).

$$R^2 = \frac{SSR}{SST}$$

- **Multiple regression** (p. 743)
- **Multiple regression coefficients** (p. 743)
- **Multiple regression equation** (p. 743)
- **Multiple regression model** (p. 746)
- **Population multiple regression equation** (p. 746)
- **Strategy for building a multiple regression model** (p. 750)

Chapter 13 Review Exercises

SECTION 13.1
For Exercises 1–3, test whether a linear relationship exists between x and y, using level of significance $\alpha = 0.05$.

  1. Education and Earnings. The U.S. Census Bureau reports the mean annual earnings of American citizens according to the number of years of education. We are interested in the relationship between earnings (y, in thousands of dollars) and years of education (x).

 **eduearn**

Education (x)	Annual earnings (y)
8	18.6
10	18.9
12	27.3
13	29.7
14	34.2
16	51.2
18	60.4

2. High School GPA and College GPA. The college admissions office wants to determine if a relationship exists between the high school grade point average (x) and the grade point average of first-year college students (y), using the data in the following table. ⩍⩍⩍ **gpa**

Student	High school GPA (x)	First-year college GPA (y)
1	2.4	2.6
2	2.5	1.9
3	2.9	2.7
4	2.7	2.5
5	3.0	2.4
6	3.5	2.9
7	3.0	2.7
8	3.6	3.1
9	3.4	3.0
10	3.9	3.3

3. Used Cars: Price versus Age. Do you think you can predict the price of a used car based on how old it is? The table shows the age (x, in years) and the price (y, in thousands of dollars) of 10 previously owned vehicles of the same make and model. ⩍⩍⩍ **ageprice**

Car	Age (x)	Price (y)
1	1	18.0
2	2	16.0
3	3	15.5
4	4	13.5
5	4	14.5
6	5	10.5
7	5	12.0
8	6	9.5
9	7	8.5
10	8	7.0

For Exercises 4–6, construct and interpret a 95% confidence interval for β_1.
  4. Data in Exercise 1
  5. Data in Exercise 2
  6. Data in Exercise 3

SECTION 13.2
For Exercises 7–9, do the following, for the indicated data:
  a. Find the point estimate of y, for the given x.
  b. Calculate and interpret a 95% confidence interval for the mean value of y for the given x.
  c. Compute and interpret a 95% prediction interval for a randomly chosen value of y for the given x.
  7. Data in Exercise 1, for 10 years of education
  8. Data in Exercise 2, for a high school GPA of 3.0
  9. Data in Exercise 3, for a used car that is eight years old

SECTION 13.3
Use the following data set for Exercises 10–13.

y	x_1	x_2	x_3
18.7	2	100	4.1
18.4	4	90	5.0
21.8	6	90	3.8
22.0	8	70	5.2
25.2	10	70	3.8
25.7	12	50	5.3
26.9	14	50	5.4
28.3	16	30	5.4
28.6	18	30	4.3
31.8	20	10	4.9

10. Assume the regression assumptions are met. Perform the F test for the significance of the overall regression, using level of significance $\alpha = 0.05$. Do the following:
  a. State the hypotheses and the rejection rule.
  b. Find the F statistic and the p-value.
  c. State the conclusion and interpretation.
11. Perform the t test for the significance of the individual predictor variables, using level of significance $\alpha = 0.05$. Do the following:
  a. For each hypothesis test, state the hypotheses and the rejection rule.
  b. For each hypothesis test, find the t statistic and the p-value.
  c. For each hypothesis test, state the conclusion and interpretation.
12. Identify any predictors that have corresponding p-values greater than the level of significance $\alpha = 0.05$. Of these, discard the variable with the largest p-value. Then redo Exercise 11, omitting this predictor. Repeat if necessary.
13. Verify the regression assumptions for your final model from Exercise 12.

Chapter 13 **QUIZ**

TRUE OR FALSE

1. True or false: We may check all the regression assumptions using a scatterplot of the residuals versus the fitted values.

2. True or false: When $\beta_1 = 0$, no relationship exists between the predictor variable and the response variable.

3. True or false: For a given value of x, the prediction interval for a randomly chosen value of y is always wider than the confidence interval for the mean value of y.

FILL IN THE BLANK

4. Two other terms for the fitted values of a regression are the **(a)** _____ and the **(b)** _____ values.

5. The point estimate of the slope _____ of the population regression equation in Section 13.1 is

_____.

6. A predictor variable used to recode a binomial categorical variable in regression, and taking values 0 or 1, is called a _____ variable.

SHORT ANSWER

7. What are the four regression assumptions?

8. What other form of inference is the $100(1 - \alpha)\%$ confidence interval for β_1 equivalent to?

9. Why is R^2_{adj} preferable to R^2 in multiple regression?

CALCULATIONS AND INTERPRETATIONS

10. Men's Heights and Weights. The university medical unit is collecting data on the heights and weights of the male students on campus. A random sample of 10 male students showed the following heights (in inches, x) and weights (in pounds, y). Test for a linear relationship between height and weight, using level of significance $\alpha = 0.01$. **heightweight**

Student	Height (x)	Weight (y)
1	66	150
2	68	145
3	69	160
4	70	165
5	70	165
6	71	180
7	72	175
8	72	180
9	73	195
10	75	210

11. For the data in Problem 10, answer the following:
   a. Will a 99% confidence interval for β_1 include zero? Why or why not?
   b. Verify your statement in **(a)** by constructing and interpreting a 99% confidence interval for β_1.

12. Using the data in Problem 10, do the following:
   a. Find the point estimate of the height of a man 70 inches tall.
   b. Calculate and interpret a 95% confidence interval for the mean height of all men 70 inches tall.
   c. Compute and interpret a 95% prediction interval for a randomly chosen man 70 inches tall.

For Exercise 13, refer to the data in Review Exercises 10–13 (page 758).

13. Report and interpret your final model from Exercise 12, by doing the following:
   a. Provide the multiple regression equation for your final model.
   b. Interpret the multiple regression coefficients so that a nonstatistician could understand.
   c. Report and interpret the standard error of the estimate s and the adjusted coefficient of determination R^2_{adj}.

Chapter 1

YOUR TURN #1

Section 1.2, page 9

1. The variable *type* takes the following values: adventure, platform, sports, shooter, action, role-playing, and racing.

2. The observation for *Spiderman* 2 for PS4 is as follows:

Game	Platform	Studio	Type	Sales for Week	Sales Total	Weeks on List
Spiderman 2 for PS4	PS4	Activision	Action	6510	49,292	3

YOUR TURN #2

Section 1.2, page 10

1. Since Sirius XM is a stock, it is an element.

2. Another name for the variable *Exchange* would be *Market*.

YOUR TURN #3

Section 1.2, page 10

1. Because the number of medals won is finite, the variable *number of medals won* is discrete.

2. Because the time to run a 100-meter dash can take on an infinite number of values, the variable *racing time in the 100-meter dash* is continuous.

YOUR TURN #4

Section 1.2, page 11

1. Exchange represents nominal data, because the data cannot be ordered in a natural or obvious way. Also, no arithmetic can be performed on exchange.

2. Last price represents ratio data. Here division makes sense. A last price of $20 is twice a last price of $10.

YOUR TURN #5

Section 1.2, page 13

1. Since Florida has more than 3 counties, this represents a sample.

2. Since we are talking about all of the counties in Florida, this represents a population.

YOUR TURN #6

Section 1.2, page 13

1. Since this represents a sample, the most expensive hotel in the 3 counties represents a statistic.

2. Since this represents a population, the most expensive hotel in Florida represents a parameter.

YOUR TURN #7

Section 1.2, page 15

1. The average grade on the first quiz is a descriptive statistic, since the average grade is only for your class. However, no inference is made regarding a larger population.

2. Jessica won 2 out of 10 games of ping pong to her friend Lu Li. Therefore she won $(2/10) \cdot 100\% = 20\%$ of her ping pong games to Lu Li. This represents a statistic. She then used this statistic to perform statistical inference to conclude that she will only win 20% of her games of ping pong to Lu Li.

YOUR TURN #8

Section 1.3, page 23

1. Systematic sample: Bill Gates, Charles Koch, Jim Walton, Michael Bloomberg, Larry Page, Jacqueline Mars, George Soros

2. Systematic sample: Larry Ellison, Alice Walton, Larry Page, Carl Icahn

YOUR TURN #9

Section 1.3, page 25

1. Since you used a random sample from your school, this is not convenience sampling.

2. Since you obtain data from your 5 closest friends at school, this is convenience sampling. You are choosing a sample that is convenient for you.

YOUR TURN #10

Section 1.3, page 26

1. Convenience sampling: you are choosing a sample convenient for you.

2. Systematic Sampling: where every kth student is taken, with $k = 10$.

3. Cluster sampling: (a) the population was divided into clusters (classes), (b) a random sample of one cluster (class) is taken, and (c) every student in that cluster (class) is selected.

4. Stratified sampling: (a) the population was divided into subgroups (nursing majors and all others), and (b) a random sample from each of the groups was drawn.

5. Random sampling: the two names were selected randomly.

Chapter 2

YOUR TURN #1

Section 2.1, page 41

1.

Borough	Frequency
Brooklyn	5
Manhattan	7

2.

Violation type	Frequency
Cell phone	4
Safety belt	3
Speeding	2
Disobey sign	3

YOUR TURN #2

Section 2.1, page 42

1.

Borough	Relative frequency
Brooklyn	$5/12 = 0.42$
Manhattan	$7/12 = 0.58$

2.

Violation type	Relative frequency
Cell phone	$4/12 = 0.33$
Safety belt	$3/12 = 0.25$
Speeding	$2/12 = 0.17$
Disobey sign	$3/12 = 0.25$

YOUR TURN #3

Section 2.1, page 43

1.
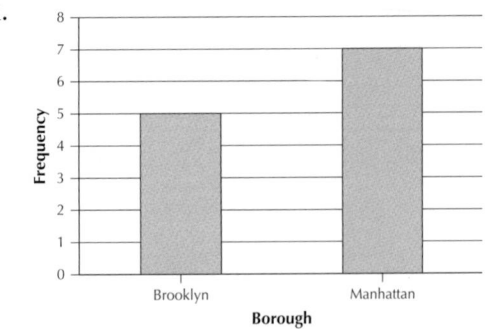

2.
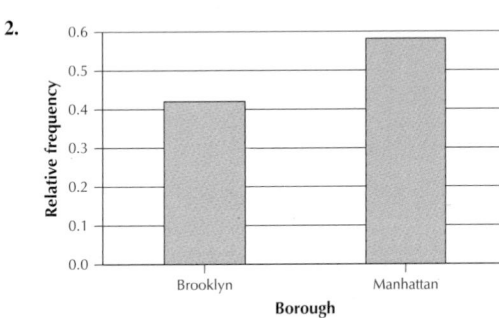

3.
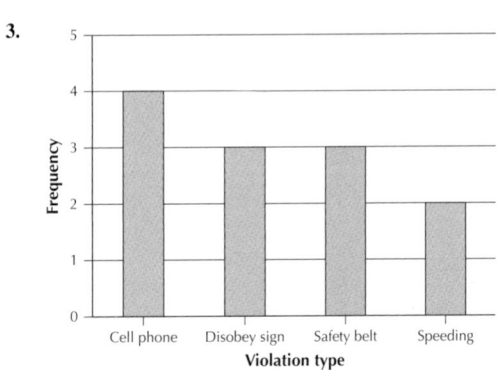

4.
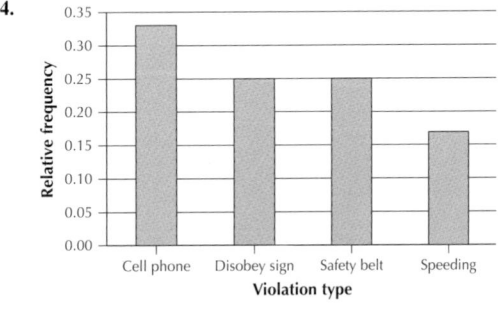

YOUR TURN #4

Section 2.1, page 47

	Violation Type				
Borough	**Cell phone**	**Disobey sign**	**Safety belt**	**Speeding**	**Total**
Brooklyn	2	1	0	2	5
Manhattan	2	2	3	0	7
Total	4	3	3	2	12

YOUR TURN #5

Section 2.1, page 48

(a)

SAT Writing score	Frequency
200−390	347,920
400−590	1,015,121
600−800	297,006
Total	1,660,047

(b)
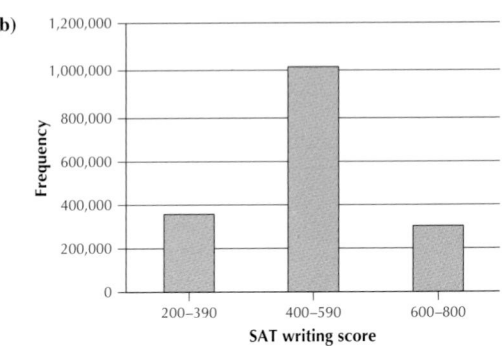

YOUR TURN #6

Section 2.1, page 49

1.
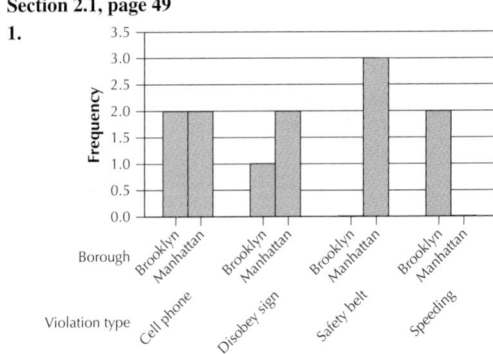

2.
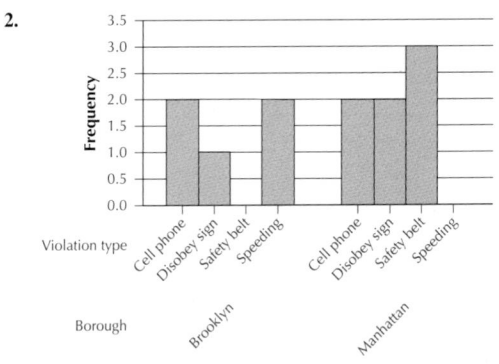

YOUR TURN #7

Section 2.2, pages 64–65

Class boundaries: 1 to < 3.5, 3.5 to < 6, 6 to < 8.5, 8.5 to <11, 11 to < 13.5

Class: $x =$	Frequency	Relative frequency
1 to < 3.5	2	2/20 = 0.1
3.5 to < 6	5	5/20 = 0.25
6 to < 8.5	7	7/20 = 0.35
8.5 to < 11	3	3/20 = 0.15
11 to < 13.5	3	3/20 = 0.15
Total	**20**	**20/20 = 1.00**

YOUR TURN #8

Section 2.2, page 66

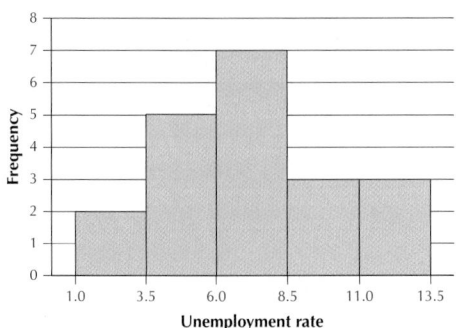

YOUR TURN #9

Section 2.2, page 68

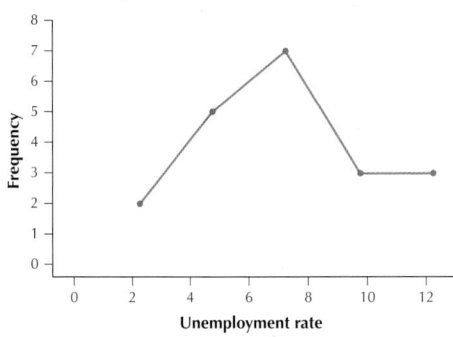

YOUR TURN #10

1.
```
 2 | 0
 3 | 4 7
 4 | 1 8 9
 5 | 7
 6 | 2 2 3 4 7
 7 | 0 0
 8 | 8 8
 9 |
10 | 2
11 | 5
12 | 0 3
```

2.
```
 2 | 0
 2 |
 3 | 4
 3 | 7
 4 | 1
 4 | 8 9
 5 |
 5 | 7
 6 | 2 2 3 4
 6 | 7
 7 | 0 0
 7 |
 8 |
 8 | 8 8
 9 |
 9 |
10 | 2
10 |
11 |
11 | 5
12 | 0 3
```

YOUR TURN #11

Section 2.2, page 70

```
|·    ·  · ·  ·  · ·· ··     ·      · ·  ·|
  2.8   4.2   5.6   7.0   8.4   9.8   11.2  12.6
              Unemployment rate
```

YOUR TURN #12

Section 2.2, page 72

1. $2 + 5 = 7$

2. No. The class boundaries are from 6 to 8.5.

3. 3

YOUR TURN #13

Section 2.3, page 88

Unemployment rate	Frequency	Relative frequency	Cumulative frequency	Cumulative relative frequency
$1 \le x < 3.5$	2	2/20 = 0.10	2	0.10
$3.5 \le x < 6$	5	5/20 = 0.25	2 + 5 = 7	0.10 + 0.25 = 0.35
$6 \le x < 8.5$	7	7/20 = 0.35	2 + 5 + 7 = 14	0.10 + 0.25 + 0.35 = 0.70
$8.5 \le x < 11$	3	3/20 = 0.15	2 + 5 + 7 + 3 = 17	0.10 + 0.25 + 0.35 + 0.15 = 0.85
$11 \le x < 13.5$	3	3/20 = 0.15	2 + 5 + 7 + 3 + 3 = 20	0.10 + 0.25 + 0.35 + 0.15 + 0.15 = 1.00
Total	**20**	**20/20 = 1.00**		

YOUR TURN #14

Section 2.3, page 89

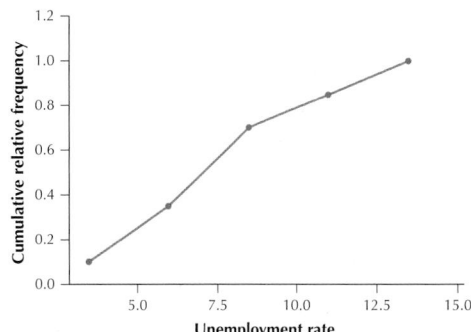

Chapter 3

YOUR TURN #1

Section 3.1, page 109

Mean number of tropical storms =

$$\frac{10 + 15 + 16 + 9 + 19 + 19 + 19 + 14}{8} = 15.125$$

The population mean number of tropical storms is 15.125 storms.

YOUR TURN #2

Section 3.1, page 110

(a) The sample mean would be lower than \$337.50 without the Sony Xperia Z2 since the price of this phone is \$600, which is higher than the prices of the other phones.

(b) $\bar{x} = \dfrac{\sum x}{n} = \dfrac{200 + 250 + 300}{3} = \$250$. Yes, the sample mean of \$250 is lower than \$337.50.

YOUR TURN #3

Section 3.1, page 115

The sample consists of the number of tropical storms in the years 2006, 2008, 2010, and 2012: 10, 16, 19, 19

The mean is $\bar{x} = \dfrac{\sum x}{n} = \dfrac{10 + 16 + 19 + 19}{4} = 16$.

The data set is already in ascending order.

Because $n = 4$ is even, the median is the mean of the two data values that lie on either side of the $\left(\dfrac{n+1}{2}\right)^{\text{th}} = \left(\dfrac{4+1}{2}\right)^{\text{th}} = 2.5$th position. That is, the median is the mean of the 2nd and 3rd data values, 16 and 19. Splitting the difference between these two, we get median number of tropical storms $= \dfrac{16+19}{2} = 17.5$ storms.

Since 19 storms appears two times and no other number appears more than once, the mode is 19.

YOUR TURN #4

Section 3.2, page 128

1.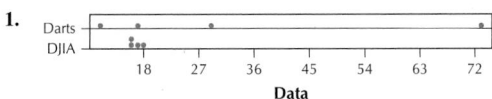

2. Darts has the larger range.

3. $\text{range}_{\text{Darts}} = $ largest value $-$ smallest value $= 72.9 - 11.2 = 61.7$
   $\text{range}_{\text{DJIA}} = $ largest value $-$ smallest value $= 17.7 - 15.8 = 1.9$

As we expected, the range for Darts is indeed larger than the range for DJIA.

YOUR TURN #5

Section 3.2, page 131

1. $\mu = \dfrac{\sum x}{N} = \dfrac{7.8 + 1.9 + 14.9 + 1.5 + 2.7 + 1.6}{6} = \$5.07$ billion

2.

x	$x - \mu$	$(x - \mu)^2$
7.8	$7.8 - 5.07 = 2.73$	7.4529
1.9	$1.9 - 5.07 = -3.17$	10.0489
14.9	$14.9 - 5.07 = 9.83$	96.6289
1.5	$1.5 - 5.07 = -3.57$	12.7449
2.7	$2.7 - 5.07 = -2.37$	5.6169
1.6	$1.6 - 5.07 = -3.47$	12.0409
		$\sum (x - \mu)^2 = 144.5334$

From the table, $\sum (x - \mu)^2 = 144.5334$. Therefore, the variance is

$$\sigma^2 = \dfrac{\sum(x - \mu)^2}{6} = \dfrac{144.5334}{6} = 24.0889$$

YOUR TURN #6

Section 3.2, page 132

From Your Turn #5 on p. 131, the population variance is $\sigma^2 = 24.0889$. Therefore the population standard deviation is

$$\sigma = \sqrt{\sigma^2} = \sqrt{24.0889} = 4.9080$$

YOUR TURN #7

Section 3.2, page 134

From Table 14 on p. 130, the CDC provided $1.9 million to Maine, $1.5 to New Hampshire, and $1.6 million to Vermont to fight HIV/AIDS. Therefore our sample data is: $1.9 million, $1.5 million, $1.6 million

1. Smaller. The sample contains the 3 smallest numbers in the population.

2. First we need to find the sample mean.

$$\bar{x} = \dfrac{\sum x}{n} = \dfrac{1.5 + 1.9 + 1.6}{3} = \$1.67 \text{ million}$$

x	$x - \bar{x}$	$(x - \bar{x})^2$
1.5	$1.5 - 1.67 = -0.17$	0.0289
1.9	$1.9 - 1.67 = 0.23$	0.0529
1.6	$1.6 - 1.67 = -0.07$	0.0049
		$\sum (x - \bar{x})^2 = 0.0867$

From the table, $\sum (x - \bar{x})^2 = 0.0867$. Therefore the sample variance is
$s^2 = \dfrac{\sum(x - \bar{x})^2}{n-1} = \dfrac{0.0867}{3-1} = 0.04335$ billion of dollars squared.

3. From Problem 2 the sample variance is $s^2 = 0.04335$ billion of dollars squared. Therefore the sample standard deviation is $s = \sqrt{s^2} = \sqrt{0.04335} = \$0.2082$ billion.

4. For this sample of CDC funding to fight HIV/AIDS in the northeastern United States, the typical difference between a state's funding and the mean funding is $0.2082 billion.

YOUR TURN #8

Section 3.2, page 137

1. $\mu - 1\sigma = 70 - 1 \cdot 5 = 65$ mph and $\mu + 1\sigma = 70 + 1 \cdot 5 = 75$ mph Therefore the percentage of vehicle speeds that lie between 65 mph and 75 mph is the percentage of vehicle speeds that lie within 1 standard deviation of the mean. Thus, from the Empirical Rule, approximately 68% of vehicle speeds lie between 65 mph and 75 mph.

2. From Problem 1, 65 mph lies 1 standard deviation below the mean. Therefore the Empirical Rule tells us that $\frac{1}{2}(100\% - 68\%) = 16\%$ of all vehicle speeds are at most 65 mph.

YOUR TURN #9

Section 3.2, page 139

Since $\mu - 2\sigma = 130 - 2 \cdot 10 = 110$ and $\mu + 2\sigma = 130 + 2 \cdot 10 = 150$, $k = 2$. Therefore Chebyshev's Rule tells us that at least $\left(1 - \frac{1}{k^2}\right)100\% = \left(1 - \frac{1}{2^2}\right)100\% = \left(\frac{3}{4}\right)100\% = 75\%$ of the systolic blood pressure readings will lie between 110 and 150.

YOUR TURN #10

Section 3.3, page 149

The data values are 90, 70, and 85. The weights are 50, 20, and 30. The course weighted mean is then calculated as follows:

$$\bar{x} = \dfrac{\sum(w \cdot x)}{\sum w} = \dfrac{(50)(90) + (20)(70) + (30)(85)}{50 + 20 + 30} = \dfrac{8450}{100} = 84.5$$

YOUR TURN #11

Section 3.4, page 156

Note here that we have population values, with $\mu = \$96$ and $\sigma = \$40$.

1. The amount Austin spent on video games is $x = \$136$.

   The z-score $= \dfrac{\text{data value} - \text{mean}}{\text{standard deviation}} = \dfrac{x - \mu}{\sigma} = \dfrac{136 - 96}{40} = 1$.

2. The amount Brian spent on music downloads is $x = \$16$.

   The z-score $= \dfrac{\text{data value} - \text{mean}}{\text{standard deviation}} = \dfrac{x - \mu}{\sigma} = \dfrac{16 - 96}{40} = -2$.

3. The amount Courtney spent on gifts for her friends is $x = \$256$.

   The z-score $= \dfrac{\text{data value} - \text{mean}}{\text{standard deviation}} = \dfrac{x - \mu}{\sigma} = \dfrac{256 - 96}{40} = 4$.

YOUR TURN #12

Section 3.4, page 157

Note here that we have population values, with $\mu = \$96$ and $\sigma = \$40$.

1. David's z-score was -1.5. Therefore he spent

   $x = z\text{-score} \cdot \sigma + \mu = (-1.5) \cdot (\$40) + \$96 = \$36$

2. Emily's z-score was 2.5. Therefore she spent

   $x = z\text{-score} \cdot \sigma + \mu = (2.5) \cdot (\$40) + \$96 = \$196$

3. Frances's z-score was 0. Therefore she spent

   $x = z\text{-score} \cdot \sigma + \mu = (0) \cdot (\$40) + \$96 = \$96$

YOUR TURN #13

Section 3.4, page 158

Gisele used a tablet. Note here that we have population values, with $\mu = \$96$ and $\sigma = \$40$. The amount Gisele spent on an online holiday shopping order is $x = \$120$.

The $z\text{-score} = \dfrac{\text{data value} - \text{mean}}{\text{standard deviation}} = \dfrac{x - \mu}{\sigma} = \dfrac{120 - 96}{40} = 0.6.$

Hong used a cell phone. Note here that we have population values, with $\mu = \$85$ and $\sigma = \$40$. The amount Hong spent on an online holiday shopping order is $x = \$120$.

The $z\text{-score} = \dfrac{\text{data value} - \text{mean}}{\text{standard deviation}} = \dfrac{x - \mu}{\sigma} = \dfrac{120 - 85}{40} = 0.875.$

Since the z-score for Hong is larger than the z-score for Gisele, Hong spent more relative to his group.

YOUR TURN #14

Section 3.4, page 159

1. Austin's z-score is 1, which lies in the range, $-2 < z\text{-score} < 2$. Therefore the amount Austin spent on video games is not considered unusual.

2. Brian's z-score is -2, which lies in the range, $-3 < z\text{-score} \leq -2$. Therefore the amount Brian spent on music downloads may be considered moderately unusual.

3. Courtney's z-score is 4, which is ≥ 3. Therefore the amount Courtney spent on gifts for her friends may be considered an outlier.

YOUR TURN #15

Section 3.4, page 161

Position	1	2	3	4	5	6	7	8	9
Rating	1.9	2.5	3.6	4.2	5.4	5.7	7.1	8.7	9.3

Since there are 9 numbers, $n = 9$. Since we want the number corresponding to the 20th percentile, $p = 20$. Thus $i = \left(\frac{p}{100}\right) n = \left(\frac{20}{100}\right) 9 = 1.8$.

Here, i is not an integer, so round i up to 2. The 20th percentile is the number in position 2, which is 2.5. Thus, the 20th percentile is 2.5.

YOUR TURN #16

Section 3.4, page 162

Position	1	2	3	4	5	6	7	8	9
Rating	1.9	2.5	3.6	4.2	5.4	5.7	7.1	8.7	9.3

Here, $x = 9.0$. Eight of the movies have a ranking of 9.0 or below, so the percentile rank of a movie with a rating of 9.0 or below is

Percentile rank of data value $(x = 9.0) =$

$\dfrac{\text{number of values in data set} \leq 9.0}{\text{total number of values is data set}} \cdot 100\% = \dfrac{8}{9} \cdot 100\% = 89\%.$

Thus a rating of 9 represents the 89th percentile of movie ratings.

YOUR TURN #17

Section 3.4, page 164

Position	1	2	3	4	5	6	7	8	9
Rating	1.9	2.5	3.6	4.2	5.4	5.7	7.1	8.7	9.3

Here, $n = 9$. To find Q1, plug $p = 25$ into $i = \left(\frac{p}{100}\right) n$, where $n = 9$. We get $i = \left(\frac{p}{100}\right) n = \left(\frac{25}{100}\right) 9 = 2.25$. Here, i is not an integer so round i up to 3. The 25th percentile is the number in position 3, which is 3.6. Thus Q1, the 25th percentile, is 3.6.

To find Q2 = the median, plug $p = 50$ into $i = \left(\frac{p}{100}\right) n$, where $n = 9$. We get $i = \left(\frac{p}{100}\right) n = \left(\frac{50}{100}\right) 9 = 4.5$. Here, i is not an integer so round i up to 5. The 50th percentile is the number in position 5, which is 5.4. Thus Q2 = the median, the 50th percentile, is 5.4.

To find Q3, plug $p = 75$ into $i = \left(\frac{p}{100}\right) n$, where $n = 9$. We get $i = \left(\frac{p}{100}\right) n = \left(\frac{75}{100}\right) 9 = 6.75$. Here, i is not an integer so round i up to 7. The 75th percentile is the number in position 7, which is 7.1. Thus Q3, the 75th percentile, is 7.1.

YOUR TURN #18

Section 3.4, page 166

From Your Turn #17 on p. 164, Q1 = 3.6 and Q3 = 7.1. Thus IQR = Q3 $-$ Q1 = 7.1 $-$ 3.6 = 3.5.

YOUR TURN #19

Section 3.5, page 172

Position	1	2	3	4	5	6	7	8	9
Rating	1.9	2.5	3.6	4.2	5.4	5.7	7.1	8.7	9.3

From Your Turn #17, p. 164, Q1 = 3.6, the median = Q2 = 5.4, and Q3 = 7.1. From the table, the minimum is 1.9 and the maximum is 9.3. Thus, the five-number summary is:

1. Minimum = 1.9
2. First quartile, Q1 = 3.6
3. Median = Q2 = 5.4
4. Third quartile, Q3 = 7.1
5. Maximum = 9.3

YOUR TURN #20

Section 3.5, page 175

From Your Turn #19 on p. 172, the five-number summary is:

1. Minimum = 1.9
2. First quartile, Q1 = 3.6
3. Median = Q2 = 5.4
4. Third quartile, Q3 = 7.1
5. Maximum = 9.3

The IQR = Q3 $-$ Q1 = 7.1 $-$ 3.6 = 3.5.

The lower fence = Q1 $-$ 1.5(IQR) = 3.6 $-$ 1.5(3.5) = -1.65.

The upper fence = Q3 + 1.5(IQR) = 7.1 + 1.5(3.5) = 12.35.

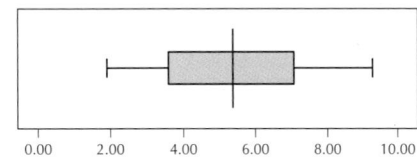

YOUR TURN #21

Section 3.5, page 178

From Your Turn # 20 on p. 175:

The lower fence = Q1 − 1.5(IQR) = 3.6 − 1.5(3.5) = −1.65.

The upper fence = Q3 + 1.5(IQR) = 7.1 + 1.5(3.5) = 12.35.

Thus, for this data set, a data value would be an outlier if it is −1.65 or less or 12.35 or more. Since none of the data values are −1.65 or less or 12.35 or more there are no outliers in this data set.

Chapter 4

YOUR TURN #1

Section 4.1, page 189

Because the response variable depends on the predictor variable, and because the grade on an exam depends in part on the number of hours spent studying for the exam, the number of hours spent studying for the exam is the predictor (x) variable and the grade on the exam is the response (y) variable.

YOUR TURN #2

Section 4.1, page 190

1. Because the response variable depends on the predictor variable, and because the weight of a person depends in part on the height of the person, height is the predictor (x) variable and weight is the response (y) variable.

2.

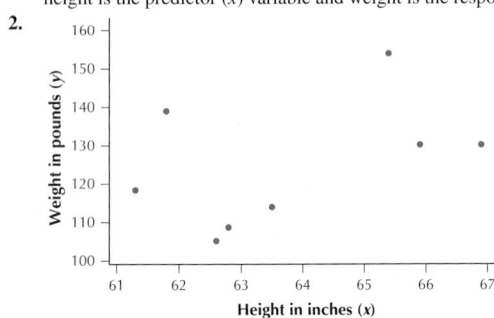

YOUR TURN #3

Section 4.1, page 192

Height and weight have a positive linear relationship.

YOUR TURN #4

Section 4.1, page 194

Step 1

$$\bar{x} = \frac{\sum x}{n} = \frac{63.5 + 65.9 + 62.8 + 61.8 + 61.3 + 66.9 + 62.6 + 65.4}{8}$$
$$= 63.775 \text{ inches}$$

$$\bar{y} = \frac{\sum y}{n} = \frac{113.8 + 130.1 + 108.5 + 138.9 + 118.2 + 130.1 + 104.9 + 153.9}{8}$$
$$= 124.8 \text{ pounds}$$

Step 2

x	y	$(x - \bar{x})$	$(x - \bar{x})^2$	$(y - \bar{y})$	$(y - \bar{y})^2$	$(x - \bar{x})(y - \bar{y})$
63.5	113.8	−0.275	0.075625	−11	121	3.025
65.9	130.1	2.125	4.515625	5.3	28.09	11.2625
62.8	108.5	−0.975	0.950625	−16.3	265.69	15.8925
61.8	138.9	−1.975	3.900625	14.1	198.81	−27.8475
61.3	118.2	−2.475	6.125625	−6.6	43.56	16.335
66.9	130.1	3.125	9.765625	5.3	28.09	16.5625
62.6	104.9	−1.175	1.380625	−19.9	396.01	23.3825
65.4	153.9	1.625	2.640625	29.1	846.81	47.2875
			$\sum(x - \bar{x})^2 =$		$\sum(y - \bar{y})^2 =$	$\sum(x - \bar{x})(y - \bar{y})$
			29.355		**1928.06**	**= 105.9**

Step 3

$$s_x = \sqrt{\frac{\sum (x - \bar{x})^2}{n - 1}} = \sqrt{\frac{29.355}{8 - 1}} \approx 2.047821142$$

$$s_y = \sqrt{\frac{\sum (y - \bar{y})^2}{n - 1}} = \sqrt{\frac{1928.06}{8 - 1}} \approx 16.59629907$$

Step 4

$$r = \frac{\sum(x - \bar{x})(y - \bar{y})}{(n - 1)s_x s_y} = \frac{105.9}{(8 - 1)(2.047821142)(16.59629907)}$$
$$\approx 0.4451379745 \approx 0.4451$$

YOUR TURN #5

Section 4.1, page 198

From Your Turn #4, p. 194, $r \approx 0.4451$. Since r is positive, we would therefore say that height and weight are positively correlated. As height increases, weight also tends to increase.

YOUR TURN #6

Section 4.2, page 211

(a) From Your Turn #4, page 194:

Step 1   Calculate the respective sample means $\bar{x}$ and $\bar{y}$. We have already done this in Your Turn #4 (page 194): $\bar{x} = 63.775$ inches and $\bar{y} = 124.8$ pounds.

Step 2   Calculate the respective sample standard deviations s_x and s_y. We have already done this in Your Turn #4 (page 194): $s_x \approx 2.047821142$ $s_y \approx 16.59629907$.

Step 3   Find the correlation coefficient r. We have already done this in Your Turn #4 (page 194): $r \approx 0.4451379745$.

Step 4   Combine the statistics from Steps 2 and 3 to calculate b_1:

$$b_1 = r \cdot \frac{s_y}{s_x} = 0.4451379745 \cdot \frac{16.59629907}{2.047821142} \approx 3.607562595 \approx 3.6076.$$

Step 5   Use the statistics from steps 1–4 to calculate b_0:

$$b_0 = \bar{y} - (b_1 \cdot \bar{x}) = 124.8 - (3.607562595)(63.775) = -105.2723045$$
$$\approx -105.2723$$

(b) $\hat{y} = 3.6076x - 105.2723$

(c) Because y and x represent weight and height, respectively, this regression equation is read as follows: "The estimated weight of a woman is 3.6076 times her height minus 105.2723 pounds."

YOUR TURN #7

Section 4.2, page 211

(a) For a woman with height 0 inches, her estimated weight is −105.2723 pounds.

(b) For each increase of 1 inch to a woman's height, her estimated weight increases by 3.6076 pounds.

YOUR TURN #8

Section 4.2, page 213

In Your Turn #7 (page 211), we calculated the regression equation to be $\hat{y} = 3.6076x - 105.2723$. Therefore, for a woman who is $x = 63.5$ inches tall, her estimated weight is $\hat{y} = 3.6076x - 105.2723 = 3.6076(63.5) - 105.2723 = 123.8103$ pounds.

YOUR TURN #9

Section 4.2, page 215

In Your Turn #8 (page 213), the estimated weight for a woman who is $x = 63.5$ inches tall is $\hat{y} = 123.8103$ pounds. From Table 2 on page 190, the actual weight of the woman who is $x = 63.5$ inches tall is $y = 113.8$ pounds. Therefore the prediction error is $y - \hat{y} = 113.8 - 123.8103 = -10.0103$ pounds. Since the prediction error is negative, her actual weight lies below the regression line.

YOUR TURN #10

Section 4.2, page 216

From Table 2, the smallest value of x is 61.3 inches and the largest is 66.9 inches, so estimates for any value of x between 61.3 inches and 66.9 inches, inclusive, would not represent extrapolation.

(a) For $x = 65$ inches, $\hat{y} = 3.6076x - 105.2723 = 3.6076(65) - 105.2722 = 129.2217$ pounds. Because $x = 65$ inches lies between 61.3 inches and 66.9 inches, inclusive, this estimate does not represent extrapolation.

(b) For $x = 70$ inches, $\hat{y} = 3.6076x - 105.2722 = 3.6076(70) - 105.2723 = 147.2597$ pounds. Because $x = 70$ inches does not lie between 61.3 inches and 66.9 inches, this estimate represents extrapolation.

Chapter 5

YOUR TURN #1

Section 5.1, page 243

(a) The chance of a recession near zero means that the probability of a recession is near zero. Therefore it is very unlikely that a recession will occur.

(b) The chance of satisfaction with the purchase of a new car is 100% means the probability that you will be satisfied with the purchase of a new car is 1. Therefore it is certain that you will be satisfied with the purchase of a new car.

YOUR TURN #2

Section 5.1, page 245

(a) A total of 52 outcomes is included in the sample space, so $N(S) = 52$. Let E be the event a heart is drawn. Then E consists of the 13 hearts $\{2\heartsuit, 3\heartsuit, 4\heartsuit, 5\heartsuit, 6\heartsuit, 7\heartsuit, 8\heartsuit, 9\heartsuit, 10\heartsuit, J\heartsuit, Q\heartsuit, K\heartsuit, A\heartsuit\}$ so $N(E) = 13$. Therefore the probability of drawing a heart is $P(E) = \frac{N(E)}{N(S)} = \frac{13}{52} = \frac{1}{4}$.

(b) A total of 52 outcomes is included in the sample space, so $N(S) = 52$. Let F be the event a black card is drawn. Then F consists of the 26 black cards $\{2\spadesuit, 3\spadesuit, 4\spadesuit, 5\spadesuit, 6\spadesuit, 7\spadesuit, 8\spadesuit, 9\spadesuit, 10\spadesuit, J\spadesuit, Q\spadesuit, K\spadesuit, A\spadesuit, 2\clubsuit, 3\clubsuit, 4\clubsuit, 5\clubsuit, 6\clubsuit, 7\clubsuit, 8\clubsuit, 9\clubsuit, 10\clubsuit, J\clubsuit, Q\clubsuit, K\clubsuit, A\clubsuit\}$ so $n(F) = 26$. Therefore the probability of drawing a black card is $P(F) = \frac{N(F)}{N(S)} = \frac{26}{52} = \frac{1}{2}$.

YOUR TURN #3

Section 5.1, page 245

(a) Since only a roll of 6 wins, the other 5 possible rolls $\{1, 2, 3, 4, 5\}$ don't win. Therefore the probability of not winning is $\frac{5}{6}$.

(b) There are 6 possible rolls. Only one of them is a 5. Therefore the probability of rolling a 5 is $\frac{1}{6}$.

YOUR TURN #4

Section 5.1, page 246

(a)

(b) {HHH, HHT, HTH, HTT, THH, THT, TTH, TTT}

YOUR TURN #5

Section 5.1, page 247

(a) Let E be the event toss two heads. Then $E = \{HH\}$, so $N(E) = 1$. Therefore $P(E) = \frac{N(E)}{N(S)} = \frac{1}{4}$.

(b) Let F be the event toss two tails. Then $F = \{TT\}$, so $N(F) = 1$. Therefore $P(F) = \frac{N(F)}{N(S)} = \frac{1}{4}$.

YOUR TURN #6

Section 5.1, page 247

(a) Let E denote the event roll a sum of 7. Then E consists of the outcomes $\{(1, 6), (2, 5), (3, 4), (4, 3), (5, 2), (6, 1)\}$, so $N(E) = 6$. Therefore $P(E) = \frac{N(E)}{N(S)} = \frac{6}{36} = \frac{1}{6}$.

(b) Let F denote the event roll a sum of 12. Then F consists of the outcome $\{(6, 6)\}$, so $N(F) = 1$. Therefore $P(F) = \frac{N(F)}{N(S)} = \frac{1}{36}$.

(c) Let G denote the event roll a sum of 2. Then G consists of the outcome $\{(1, 1)\}$, so $N(G) = 1$. Therefore $P(G) = \frac{N(G)}{N(S)} = \frac{1}{36}$.

YOUR TURN #7

Section 5.1, page 250

(a) Define D: Patient did not complete treatment.

(b) Relative frequency of $D = \frac{\text{frequency of } D}{\text{number of trials in the experiment}} = \frac{944}{1413} \approx 0.6681$.

(c) $P(D) \approx$ relative frequency of $D \approx 0.6681$.

YOUR TURN #8

Section 5.2, page 260

From Example 7 on page 247, $N(S) = 36$.

(a) B consists of the outcomes $\{(2, 6), (3, 5), (4, 4), (5, 3), (6, 2)\}$, so $N(B) = 5$. Therefore $P(B) = \frac{N(B)}{N(S)} = \frac{5}{36}$.

(b) B^C consists of all of the outcomes in the sample space that are not in B. Therefore B consists of the outcomes $\{(1,1), (1, 2), (1, 3), (1, 4), (1, 5), (1, 6), (2, 1), (2, 2), (2, 3), (2, 4), (2, 5), (3, 1), (3, 2), (3, 3), (3, 4), (3, 6), (4, 1), (4, 2), (4, 3), (4, 5), (4, 6), (5, 1), (5, 2), (5, 4), (5, 5), (5, 6), (6, 1), (6, 3), (6, 4), (6, 5), (6, 6)\}$, so $N(B^C) = 31$. Therefore $P(B^C) = \frac{N(B^C)}{N(S)} = \frac{31}{36}$.

YOUR TURN #9

Section 5.2, page 262

(a) The set (M and N) is the set of outcomes that are common to both M and N. The green cell belongs to both the *Male* column and the *Person did not survive* row. The green cell represents (M and N), which includes the 1364 males who did not survive.

	Female	Male	Total
Did not survive	126	1364	1490
Survived	344	367	711
Total	470	1731	2201

(b) The set (M or N) is the set of outcomes that are in M or N or both. The green cells are the people that were either male or did not survive or both. Therefore (M or N) is represented by the green cells.

	Female	Male	Total
Did not survive	126	1364	1490
Survived	344	367	711
Total	470	1731	2201

YOUR TURN #10

Section 5.2, page 263

(a) Here we seek $P(M)$. There were 1731 males out of a total of 2201 passengers aboard the *Titanic*. Therefore $P(M) = \frac{1731}{2201}$.

(b) We are looking for $P(N)$. Of the 2201 passengers onboard, 1490 of them did not survive. Therefore $P(N) = \frac{1490}{2201}$.

(c) Those who were both male and did not survive represent (M and N). From Your Turn #9, p. 262, there are 1364 out of the 2201 passengers who were male and did not survive. Therefore $P(M \text{ and } N) = \frac{1364}{2201}$.

(d) Here we seek $P(M \text{ or } N)$. From the Addition Rule, $P(M \text{ or } N) = P(M) + P(N) - P(M \text{ and } N) = \frac{1731}{2201} + \frac{1490}{2201} - \frac{1364}{2201} = \frac{1857}{2201}$.

YOUR TURN #11

Section 5.2, page 265

Here we seek $P(\text{On Campus or Off Campus})$. Of the 19,375 students at the university, 2608 live on campus and 9911 live off campus. Thus $P(\text{On Campus}) = \frac{2608}{19,375}$ and $P(\text{Off Campus}) = \frac{9911}{19,375}$. Since no one is living both on campus and off campus at the same time, the events *On Campus* and *Off Campus* are mutually exclusive. From the Addition Rule for Mutually Exclusive Events, $P(\text{On Campus or Off Campus}) = P(\text{On Campus}) + P(\text{Off Campus}) = \frac{2,608}{19,735} + \frac{9,911}{19,735} = \frac{12,519}{19,735}$

YOUR TURN #12

Section 5.3, page 274

(a) The event did not respond to the marketing campaign is the complement of the event responded to the marketing campaign. Therefore R^C denotes the event did not respond to the marketing campaign. $P(R^C) = \frac{N(R^C)}{N(S)} = \frac{240}{288} = 0.8333$. We could have also used the formula for the probability of a complement. This gives us $P(R^C) = 1 - P(R) = 1 - 0.1667 = 0.8333$.

(b) The event did not have a credit card on file is the complement of the event has a credit card on file. Therefore C^C denotes the event does not have a credit card on file. Then $P(R^C \text{ given } C^C) = P(R^C|C^C) = \frac{N(R^C \text{ and } C^C)}{N(C^C)} = \frac{161}{178} = 0.9043$.

YOUR TURN #13

Section 5.3, page 275

	Female	Male	Total
Did not survive	126	1364	1490
Survived	344	367	711
Total	470	1731	2201

Define the following events:

M: Person is male.
N: Person did not survive.

Then $P(M) = \frac{N(M)}{N(S)} = \frac{1731}{2201} \approx 0.79$ and $P(M|N) = \frac{N(M \text{ and } N)}{N(N)} = \frac{1364}{1490} \approx 0.92$. Since $P(M) \neq P(M|N)$, M and N are not independent.

YOUR TURN #14

Section 5.3, page 277

Define the following events:

F: Roll a 6 on the first toss.
D: Roll a 6 on the second toss.

Then $P(F) = \frac{1}{6}$. Since the second toss of the die is not affected by the first toss of the die, F and D are independent. Therefore $P(D|F) = P(D) = \frac{1}{6}$. Using the Multiplication Rule we have $P(F \text{ and } D) = P(F) \cdot P(D|F) = P(F) \cdot P(D) = \left(\frac{1}{6}\right) \cdot \left(\frac{1}{6}\right) = \frac{1}{36}$.

YOUR TURN #15

Section 5.3, page 277

$P(\text{not getting red on the first spin}) = P(A^C) = 1 - P(A) = 1 - \frac{18}{38} = \frac{20}{38} = P(\text{not getting red on the second spin}) = P(B^C)$. Using the Multiplication Rule for Independent Events, we get $P(A^C \text{ and } B^C) = P(A^C) \cdot P(B^C) = \left(\frac{20}{38}\right)\left(\frac{20}{38}\right) \approx 0.2770$.

YOUR TURN #16

Section 5.3, p. 278

Define the following events:

G: Observe a heart on the first draw.
H: Observe a heart on the second draw.

Since we are sampling with replacement, G and H are independent. Then $P(G) = P(H) = \frac{13}{52} = \frac{1}{4}$. Using the Multiplication Rule for Independent Events we get $P(G \text{ and } H) = P(G) \cdot P(H) = \left(\frac{1}{4}\right)\left(\frac{1}{4}\right) = \frac{1}{16} = 0.0625$.

YOUR TURN #17

Section 5.3, page 279

Define the following events:

G: Observe a heart on the first draw.
H: Observe a heart on the second draw.

Since we are sampling without replacement, G and H are dependent. Then $P(G) = \frac{13}{52} = \frac{1}{4}$. After drawing a heart from the deck there are 51 cards left and 12 hearts left. Thus $P(H|G) = \frac{12}{51} = \frac{4}{17}$. Using the Multiplication Rule we get $P(G \text{ and } H) = P(G) \cdot P(H|G) = \left(\frac{1}{4}\right)\left(\frac{4}{17}\right) = \frac{1}{17} \approx 0.0588$.

YOUR TURN #18

Section 5.3, page 281

$P(S_1 \text{ and } S_2 \text{ and } S_3 \text{ and } S_4 \text{ and } S_5 \text{ and } S_6 \text{ and } S_7 \text{ and } S_8 \text{ and } S_9 \text{ and } S_{10}) = P(S_1) \cdot P(S_2) \cdot P(S_3) \cdot P(S_4) \cdot P(S_5) \cdot P(S_6) \cdot P(S_7) \cdot P(S_8) \cdot P(S_9) \cdot P(S_{10})$
$= (0.24)(0.24)(0.24)(0.24)(0.24)(0.24)(0.24)(0.24)(0.24)(0.24)$
$= (0.24)^{10} \approx 0.0000006340$.

YOUR TURN #19

Section 5.3, page 281

$P(\text{At least 1 of the 4 Americans smokes})$

$= 1 - P(\text{None of the 4 Americans smoke}) = 1 - P(\text{First one doesn't smoke and Second one doesn't smoke and Third one doesn't smoke and Fourth one doesn't smoke})$

$= 1 - P(\text{First one doesn't smoke}) \cdot P(\text{Second one doesn't smoke}) \cdot P(\text{Third one doesn't smoke}) \cdot P(\text{Fourth one doesn't smoke}) = 1 - (0.76)^4 \approx 0.6664$.

YOUR TURN #20

Section 5.3, page 283

Define the following event:

H: Used hospital-based insurance

Then using Bayes' Rule we get

$$P(H \text{ given } C) = \frac{P(H) \cdot P(C \text{ given } H)}{P(H) \cdot P(C \text{ given } H) + P(H^C) \cdot P(C \text{ given } H^C)}$$

$$= \frac{\left(\frac{84}{1413}\right)\left(\frac{39}{84}\right)}{\left(\frac{84}{1413}\right)\left(\frac{39}{84}\right) + \left(\frac{1329}{1413}\right)\left(\frac{430}{1329}\right)} = \frac{\frac{39}{1413}}{\frac{39}{1413} + \frac{430}{1413}} = \frac{39}{469} \approx 0.08$$

YOUR TURN #21

Section 5.4, page 293

Once again no one can finish in more than one place. Therefore there is no repetition. Thus there are $6 \cdot 5 \cdot 4 \cdot 3 = 360$ possible sets of trophy winners.

YOUR TURN #22

Section 5.4, page 294

$10 \cdot 9 \cdot 8 \cdot 7 \cdot 6 \cdot 5 \cdot 4 \cdot 3 \cdot 2 \cdot 1 = 3{,}628{,}800$

YOUR TURN #23

Section 5.4, page 294

(a) $9! = 9 \cdot 8 \cdot 7 \cdot 6 \cdot 5 \cdot 4 \cdot 3 \cdot 2 \cdot 1 = 362{,}880$

(b) $10! = 10 \cdot 9 \cdot 8 \cdot 7 \cdot 6 \cdot 5 \cdot 4 \cdot 3 \cdot 2 \cdot 1 = 3{,}628{,}800$, as in Your Turn #22.

YOUR TURN #24

Section 5.4, page 295

$10 \cdot 9 \cdot 8 \cdot 7 \cdot 6 = 30{,}240$

YOUR TURN #25

Section 5.4, page 296

(a) $_5P_3 = \dfrac{5!}{(5-3)!} = \dfrac{5 \cdot 4 \cdot 3 \cdot 2!}{2!} = 5 \cdot 4 \cdot 3 = 60$

(b) $_6P_4 = \dfrac{6!}{(6-4)!} = \dfrac{6 \cdot 5 \cdot 4 \cdot 3 \cdot 2!}{2!} = 6 \cdot 5 \cdot 4 \cdot 3 = 360$

(c) $_2P_2 = \dfrac{2!}{(2-2)!} = \dfrac{2 \cdot 1}{0!} = 2 \cdot 1 = 2$

YOUR TURN #26

Section 5.4, page 297

$_{10}C_2 = \dfrac{10!}{2!(10-2)!} = \dfrac{10!}{2!8!} = \dfrac{10 \cdot 9 \cdot 8!}{2 \cdot 1 \cdot 8!} = \dfrac{90}{2} = 45$

YOUR TURN #27

Section 5.4, page 298

(a) $_5C_4 = \dfrac{5!}{4!(5-4)!} = \dfrac{5!}{4!1!} = \dfrac{5 \cdot 4!}{4! \cdot 1!} = \dfrac{5}{1} = 5$

(b) $_5C_3 = \dfrac{5!}{3!(5-3)!} = \dfrac{5!}{3!2!} = \dfrac{5 \cdot 4 \cdot 3!}{3! \cdot 2 \cdot 1} = \dfrac{20}{2} = 10$

(c) $_5C_2 = \dfrac{5!}{2!(5-2)!} = \dfrac{5!}{2!3!} = \dfrac{5 \cdot 4 \cdot 3!}{2 \cdot 1 \cdot 3!} = \dfrac{20}{2} = 10$

Chapter 6

YOUR TURN #1

Section 6.1, page 312

(a) Your best friend's height is something that is measured, not counted. Therefore your best friend's height is a continuous variable. Possible values for your best friend's height in feet are $3 \le H \le 7$.

(b) The number of cats you have is something you count. Therefore the number of cats you have is a discrete variable. Possible values are $\{0,1,2,3, \dots\}$.

YOUR TURN #2

Section 6.1, page 314

All of the probabilities $P(X)$ are between 0 and 1. However, the sum of the probabilities is $\sum P(X) = 0.25 + 0.30 + 0.30 + 0.20 = 1.05$, which is not equal to 1. Therefore this is not a valid discrete probability distribution.

YOUR TURN #3

Section 6.1, page 315

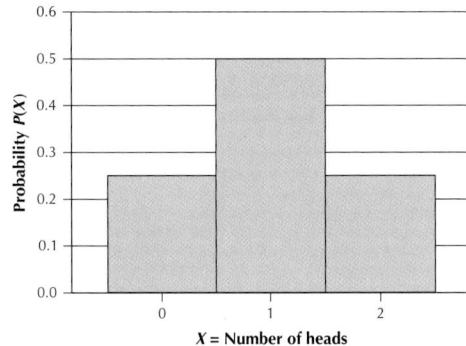

YOUR TURN #4

Section 6.1, page 315

(a) $P(X = 1\ \text{point}\ or\ X = 3\ \text{points}) = P(X = 1) + P(X = 3) = 0.24 + 0.5 = 0.74$.

(b) The outcomes $X = 1$ and $X = 3$ are mutually exclusive. Therefore, $P(X = 1\ \text{point}\ and\ X = 3\ \text{points}) = 0$.

(c) The phrase *at most* means "that many or fewer." Thus, $P(X \le 3) = P(X = 3\ \text{points}\ or\ X = 1\ \text{point}\ or\ X = 0\ \text{points}) = P(X = 3) + P(X = 1) + P(X = 0) = 0.5 + 0.24 + 0.26 = 1$.

(d) The phrase *at least* means "that many or more." Thus, $P(X \ge 3) = P(X = 3) = 0.5$.

YOUR TURN #5

Section 6.1, page 317

X = Number of heads	$P(X)$	$X \cdot P(X)$
0	$\frac{1}{4} = 0.25$	$0 \cdot 0.25 = 0$
1	$\frac{1}{2} = 0.50$	$1 \cdot 0.50 = 0.50$
2	$\frac{1}{4} = 0.25$	$2 \cdot 0.25 = 0.50$
Total		$\mu = \sum[X \cdot P(X) = 1]$

Therefore the mean number of heads is 1.

YOUR TURN #6

Section 6.1, page 318

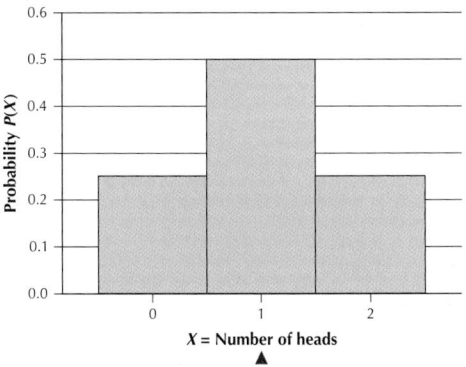

Balance point at mean = 1

YOUR TURN #7

Section 6.1, page 318

The largest probability in the table is $P(X = 1)$ and the largest bar in the graph in Your Turn #6 is for $X = 1$. Therefore the most likely number of heads is $X = 1$.

YOUR TURN #8

Section 6.2, page 331

$n = 3, p = 0.5, q = 1 - p = 1 - 0.5 = 0.5$

(a) $P(X = 0) = (_3C_0)(0.5)^0(0.5)^{3-0} = (1)(1)(0.125) = 0.125$

(b) $P(X = 1) = (_3C_1)(0.5)^1(0.5)^{3-1} = (3)(0.5)(0.25) = 0.375$

(c) $P(X \leq 1) = P(X = 0) + P(X = 1) = 0.500$

YOUR TURN #9

Section 6.2, page 334

$n = 10, p = 0.5$

(a) From the binomial table:

$P(X \leq 5) = P(X = 0) + P(X = 1) + P(X = 2) + P(X = 3) + P(X = 4) + P(X = 5) = 0.0010 + 0.0098 + 0.0439 + 0.1172 + 0.2051 + 0.2461 = 0.6231.$

From the TI-84 Plus: 0.623046875

(b) From the binomial table:

$P(X < 6) = P(X = 0) + P(X = 1) + P(X = 2) + P(X = 3) + P(X = 4) + P(X = 5) = 0.0010 + 0.0098 + 0.0439 + 0.1172 + 0.2051 + 0.2461 = 0.6231.$

From the TI-84 Plus: 0.623046875

(c) From the binomial table:

$P(0 \leq X \leq 5) = P(X = 0) + P(X = 1) + P(X = 2) + P(X = 3) + P(X = 4) + P(X = 5) = 0.0010 + 0.0098 + 0.0439 + 0.1172 + 0.2051 + 0.2461 = 0.6231.$

From the TI-84 Plus: 0.623046875

(d) From the binomial table:

$P(0 < X < 5) = P(X = 1) + P(X = 2) + P(X = 3) + P(X = 4)$
$= 0.0098 + 0.0439 + 0.1172 + 0.2051 = 0.376.$

From the TI-84 Plus: 0.3759765625

YOUR TURN #10

Section 6.2, page 335

$n = 50, p = 0.6, q = 1 - p = 1\sigma$

(a) $\mu = n \cdot p = (50)(0.6) = 30$

(b) $\sigma^2 = n \cdot p \cdot q = (50)(0.6)(0.4) = 12$

$\sigma = \sqrt{\sigma^2} = \sqrt{12} = 3.46101615 \approx 3.4641$

(c) We use the Z-score method. The Z-score for $X = 36$ is
$z = \frac{X - \mu}{\sigma} = \frac{36 - 30}{3.4641} \approx 1.73$ since $-2 < Z < 2$, $X = 36$ is neither an outlier nor unusual.

YOUR TURN #11

Section 6.3, page 344

(a) Here, $X = 10$, so the probability that $X = 10$ is

$P(10) = \frac{10^{10} \cdot e^{-10}}{10!} \approx 0.1251.$

(b) Here, $X = 12$, so the probability that $X = 12$ is

$P(12) = \frac{10^{12} \cdot e^{-10}}{12!} \approx 0.0948.$

YOUR TURN #12

Section 6.3, page 345

(a) Mean $= \mu = 10$, Variance $= \sigma^2 = 10$, Standard deviation $= \sigma = \sqrt{\mu} = \sqrt{10} \approx 3.1623.$

(b) A data value farther than 2 standard deviations from the mean is considered moderately unusual. The number of customers to a boutique shop that lie 2 standard deviations above and below the mean are calculated as follows:

$$\mu + 2\sigma \approx 10 + 2(3.1623) = 16.3246$$

$$\mu - 2\sigma \approx 10 - 2(3.1623) = 3.6754$$

Thus if there were 3 or less customers to the boutique shop this would be considered moderately unusual. Similarly, if there were 17 or more customers to the boutique shop this would also be considered moderately unusual.

YOUR TURN #13

Section 6.4, page 351

Area $=$ base $\times$ height $= (8 - 4) \times 0.1 = 4 \times 0.1 = 0.4$

YOUR TURN #14

Section 6.4, page 356

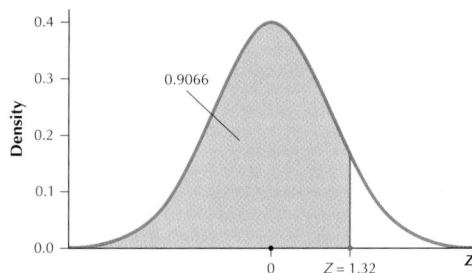

YOUR TURN #15

Section 6.4, page 357

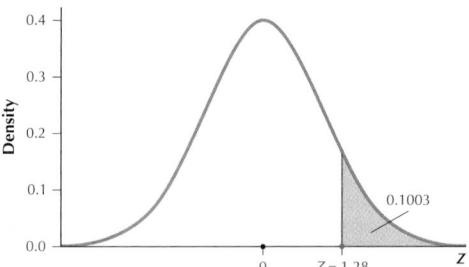

$1 - 0.8997 = 0.1003.$

YOUR TURN #16

Section 6.4, page 360

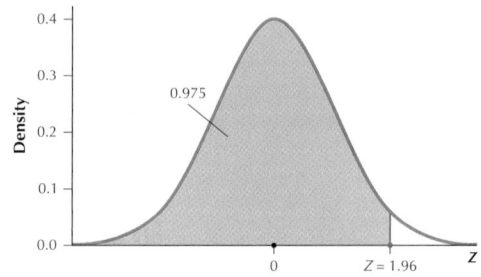

YOUR TURN #17

Section 6.4, page 361

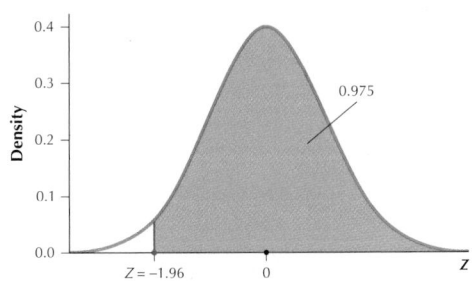

$1 - 0.975 = 0.025.$

YOUR TURN #18

Section 6.4, page 362

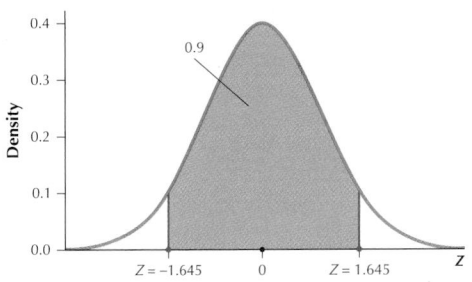

$Z = -1.645$ and $Z = 1.645.$

YOUR TURN #19

Section 6.5, page 370

We want $P(X > 600)$. Thus we want to find the area shaded in the graph below.

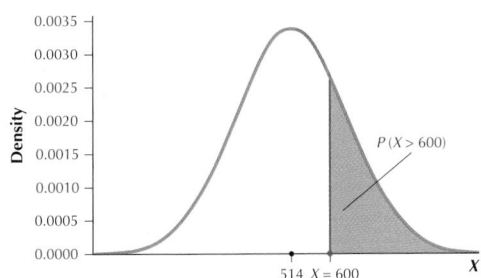

Standardizing we get $P(X > 600) = P\left(\dfrac{X - \mu}{\sigma} > \dfrac{600 - 514}{118}\right) \approx P(Z > 0.73)$

$1 - 0.7673 = 0.2327.$

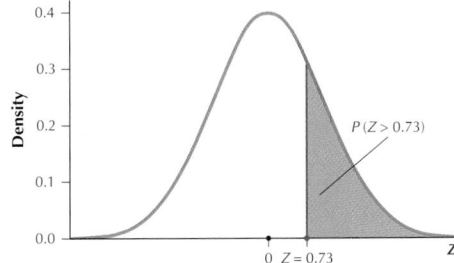

YOUR TURN #20

Section 6.5, page 372

We want to find $P(305 < X < 605)$.

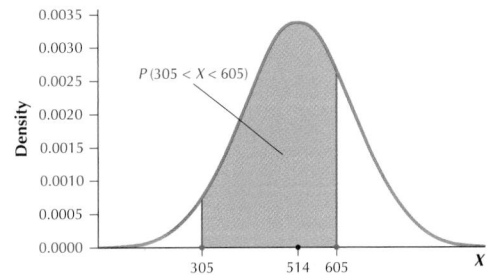

Standardizing we get

$P(305 < X < 605) = P\left(\dfrac{305 - 514}{118} < \dfrac{X - 514}{118} < \dfrac{605 - 514}{118}\right)$

$\approx P(-1.77 < Z < 0.77) = 0.7794 - 0.0384 = 0.741.$

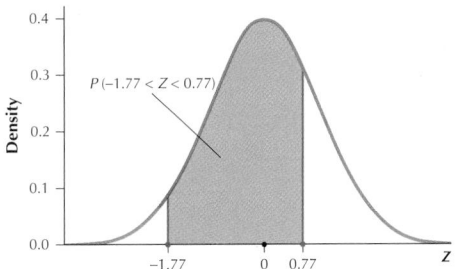

YOUR TURN #21

Section 6.5, page 373

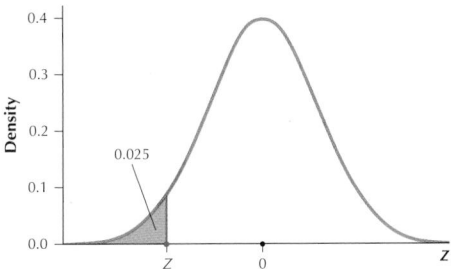

From the table $Z = -1.96$. Then $X = Z\sigma + \mu = -1.96(118) + 514 = 282.72.$

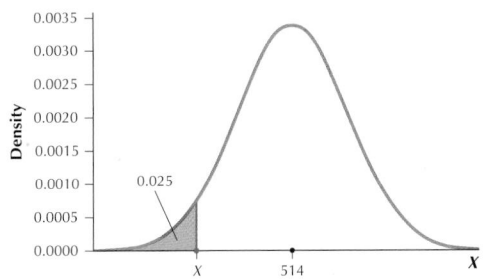

YOUR TURN #22

Section 6.5, page 374

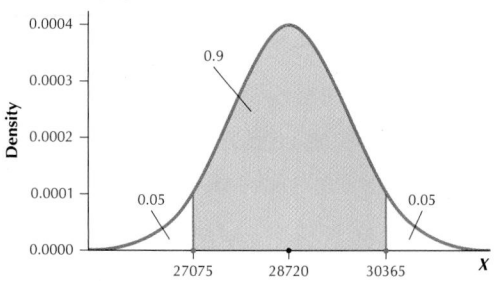

From the table, $Z_1 = -1.645$ and $Z_2 = 1.645$. Then $X_1 = Z_1\sigma + \mu = (-1.645)(\$1000) + \$28,720 = \$27,075$ and $X_2 = Z_2\sigma + \mu = (1.645)(\$1000) + \$28,720 = \$30,365$.

YOUR TURN #23

Section 6.6, page 389

$$P(X_{binomial} < 12) \approx P(Y_{normal} < 11.5) \approx P\left(\frac{Y_{normal} - 12.8}{3.2} < \frac{11.5 - 12.8}{3.2}\right)$$

$$\approx P(Z < -0.41) = 0.3409.$$

Chapter 7

YOUR TURN #1

Section 7.1, page 399

We have $\mu_{\bar{x}} = \mu = 9.3$. $n = 900$, so $\sigma_{\bar{x}} = \frac{\sigma}{\sqrt{n}} = \frac{1}{\sqrt{900}} \approx 0.0333$.

YOUR TURN #2

Section 7.1, page 403

We want to find $P(\bar{x} < 75)$.

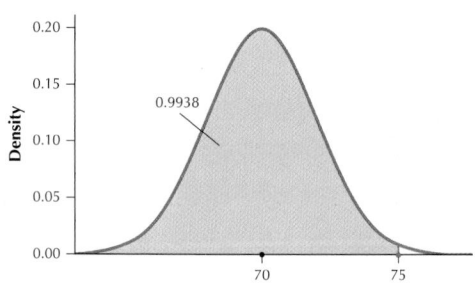

From Example 5, $\mu_{\bar{x}} = \mu = 70$ page likes and $\sigma_{\bar{x}} = \frac{\sigma}{\sqrt{n}} = \frac{10}{\sqrt{25}} = 2$ page likes.

Therefore $P(\bar{x} < 75) = P\left(\frac{\bar{x} - \mu}{\sigma_{\bar{x}}} < \frac{75 - \mu}{\sigma_{\bar{x}}}\right) = P\left(\frac{\bar{x} - \mu}{\sigma/\sqrt{n}} < \frac{75 - \mu}{\sigma/\sqrt{n}}\right)$

$= P\left(\frac{\bar{x} - 70}{10/\sqrt{25}} < \frac{75 - 70}{10/\sqrt{25}}\right) = P\left(\frac{\bar{x} - 70}{2} < \frac{75 - 70}{2}\right) = P(Z < 2.5) = 0.9938.$

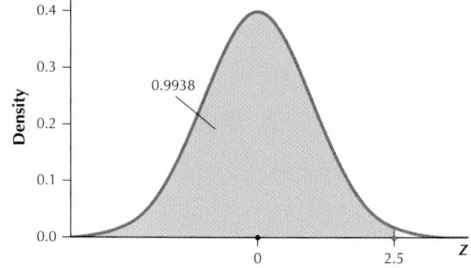

YOUR TURN #3

Section 7.1, page 405

From Example 5, $\mu_{\bar{x}} = \mu = 70$ page likes and $\sigma_{\bar{x}} = \frac{\sigma}{\sqrt{n}} = \frac{10}{\sqrt{25}} = 2$ page likes.

We want the 2.5th percentile, so we look for 0.0250 on the inside of the Z table. Working backward from 0.0250 we find that $Z = -1.96$. Thus $\bar{x} = Z \cdot \sigma_{\bar{x}} + \mu = (-1.96)(2) + 70 = 66.08$.

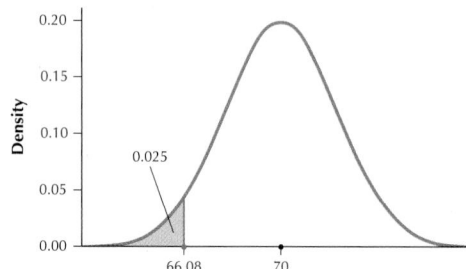

YOUR TURN #4

Section 7.1, page 406

From Example 7, $\mu_{\bar{x}} = 20$ and $\sigma_{\bar{x}} = 1$. We want to find $P(\bar{x} < 18.5)$.

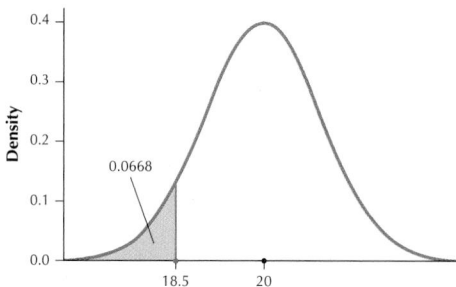

Standardizing, we get

$$Z = \frac{18.5 - 20}{1} = -1.50.$$

Therefore $P(\bar{x} < 18.5) = P(Z < -1.50) = 0.0668.$

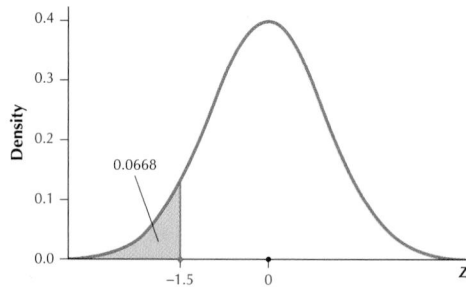

YOUR TURN #5

Section 7.1, page 407

From Example 7, $\mu_{\bar{x}} = 20$ and $\sigma_{\bar{x}} = 1$. We want the middle 90% of the sample means, so we will use our calculators to find the sample mean $\bar{x}_1$ with 5% to the left of it and the sample mean $\bar{x}_2$ with 95% to the left of it. The calculator gives us $\bar{x}_1 \approx 18.36$ mpg and $\bar{x}_2 = 21.64$ mpg.

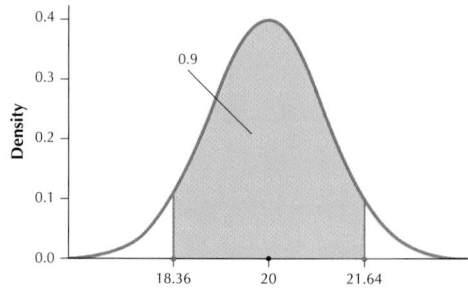

YOUR TURN #6

Section 7.2, page 415

The survey sample size is $n = 3058$, and the number of successes is $X = 1009$. We calculate

$$\hat{p} = \frac{X}{n} = \frac{1009}{3058} \approx 0.33.$$

YOUR TURN #7

Section 7.2, page 416

From Example 11, $p = 0.08$. $n = 400$. Then $\mu_{\hat{p}} = p = 0.08$ and

$$\sigma_{\hat{p}} = \sqrt{\frac{p \cdot q}{n}} = \sqrt{\frac{0.08 \cdot (1 - 0.08)}{400}} = \sqrt{0.000184} \approx 0.01356.$$

YOUR TURN #8

Section 7.2, page 420

We want to find $P(\hat{p} < 0.04)$. From Example 13, $p = 0.043$. From Example 13 (a) the minimum sample size required to produce a sampling distribution of $\hat{p}$ that is approximately normal is 117 vehicles. Since $n = 225$ vehicles is greater than 117 vehicles, the sampling distribution of $\hat{p}$ is approximately normal. Then $\mu_{\hat{p}} = p = 0.043$ and

$$\sigma_{\hat{p}} = \sqrt{\frac{p \cdot q}{n}} = \sqrt{\frac{0.043 \cdot (1 - 0.043)}{225}} \approx \sqrt{0.0001828933333} \approx 0.01352.$$

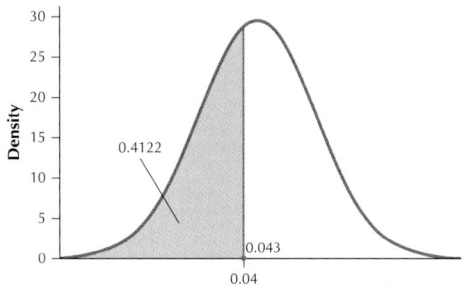

Notice from the graph that if we use technology to find $P(\hat{p} < 0.04)$ we get

$$P(\hat{p} < 0.04) = 0.4122.$$

We standardize as follows:

$$Z = \frac{0.04 - \mu_{\hat{p}}}{\sigma_{\hat{p}}} = \frac{0.04 - 0.043}{0.01352} \approx -0.22$$

Then $P(\hat{p} < 0.04) = P(Z < -0.22) = 0.4129$.

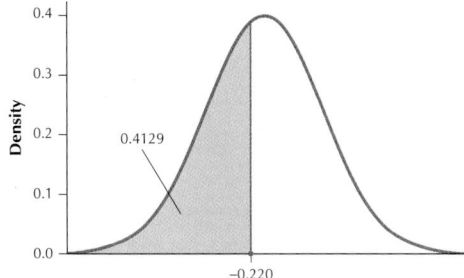

YOUR TURN #9

Section 7.2, page 421

From Example 13, $\mu_{\hat{p}} = 0.043$ and $\sigma_{\hat{p}} = 0.01875$. The middle 90% lies between the 5th and the 95th percentile. Using the calculator we get the 5th percentile ≈ 0.01216 and the 95th percentile ≈ 0.07384. Therefore the middle 90% lies between 0.01216 and 0.07384.

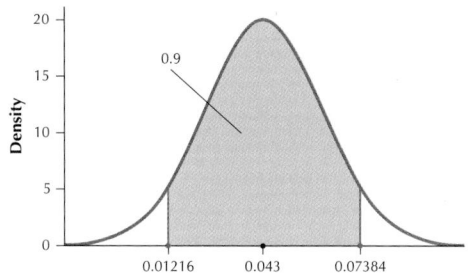

Chapter 8

YOUR TURN #1

Section 8.1, page 429

(a) The sample mean yield is calculated as

$$\bar{x} = \frac{\sum x}{n} = \frac{36 + 68 + 65 + 70 + 63}{5} = 60.4.$$

(b) The point estimate of μ, the unknown nationwide mean winter wheat yield for all 50 states, is 60.4 bushels per acre.

YOUR TURN #2

Section 8.1, page 431

(a) Since the population is not normally distributed and the sample size is less than 30, the Z interval may not be used.

(b) Since the sample size is greater than or equal to 30 and σ is known, the Z interval may be used.

YOUR TURN #3

Section 8.1, page 432

(a) Table 1 gives us $Z_{\alpha/2} = 2.576$.

(b) Table 1 gives us $Z_{\alpha/2} = 1.645$.

YOUR TURN #4

Section 8.1, page 433

Because the population is normal and the population standard deviation σ is known, the requirements for the Z interval are met:

lower bound $= \bar{x} - Z_{\alpha/2}(\sigma/\sqrt{n})$ upper bound $= \bar{x} + Z_{\alpha/2}(\sigma/\sqrt{n})$

We are given $\bar{x} = 510$, $\sigma = 118$, and $n = 25$. From Table 1, we have $Z_{\alpha/2} = 1.96$. Thus,

lower bound $= 510 - 1.96(118/\sqrt{25}) = 463.74$

upper bound $= 510 + 1.96(118/\sqrt{25}) = 556.26$

Thus we are 95% confident that the population mean score on the 2014 Mathematics SAT test lies between 463.74 and 556.26.

YOUR TURN #5

Section 8.1, page 434

The formula for the confidence interval is given by

lower bound $= \bar{x} - Z_{\alpha/2}(\sigma/\sqrt{n})$

upper bound $= \bar{x} + Z_{\alpha/2}(\sigma/\sqrt{n})$

We are given $\bar{x} = 20.71$, $\sigma = 5.637$, and $n = 100$. From Table 1, we have $Z_{\alpha/2} = 2.576$. Plugging into the formula:

lower bound $= 20.71 - 2.576(5.637/\sqrt{100}) \approx 20.71 - 1.45 = 19.26$

upper bound $= 20.71 + 2.576(5.637/\sqrt{100}) \approx 20.71 + 1.45 = 22.16$

We are 99% confident that μ, the population mean city MPG for all motor vehicles, lies between 19.26 mpg and 22.16 mpg.

YOUR TURN #6

Section 8.1, page 441

"Within \$100" means that the margin of error E is \$100, and 1.96 is the value associated with 95% confidence. Substituting into the formula for sample size we get:

$$n = \left(\frac{Z_{\alpha/2} \cdot \sigma}{E}\right)^2 = \left(\frac{1.96 \cdot \$5000}{\$100}\right)^2 = 9604$$

YOUR TURN #7

Section 8.2, page 451

First we need to find our degrees of freedom, df $= n - 1 = 20 - 1 = 19$. Then, using the table for a 90% confidence interval, $t_{\alpha/2} = 1.729$.

YOUR TURN #8

Section 8.2, page 452

(a) The sample size is not large (n is not ≥ 30) and we are not told that the population is normal. Therefore, the conditions are not met for the t interval for μ. It is not okay to construct the t interval.

(b) We are not told that the population is normal. However, the sample size is large (n is ≥ 30). Therefore, the conditions are met for the t interval for μ. It is okay to construct the t interval.

YOUR TURN #9

Section 8.2, page 453

The value of $t_{\alpha/2}$ for 95% confidence and 15 degrees of freedom is 2.131. A 95% confidence interval for μ is given by the interval

lower bound $= \bar{x} - t_{\alpha/2}(s/\sqrt{n})$, upper bound $= \bar{x} + t_{\alpha/2}(s/\sqrt{n})$

From Example 14, $\bar{x} = 185.9$, $s = 56.8$, and $n = 16$, From the table, $t_{\alpha/2} = 2.131$. Substituting, we get

lower bound $= 185.9 - (2.131)(56.8/\sqrt{16}) = 185.9 - 30.3 = 159.6$

upper bound $= 185.9 + (2.131)(56.8/\sqrt{16}) = 185.9 + 30.3 = 216.2$

We are 95% confident that μ, the population mean sodium content per serving of all breakfast cereals, lies between 155.6 grams and 216.2 grams.

YOUR TURN #10

Section 8.2, page 454

From Your Turn #9, we have:

$$E = (2.131)(56.8/\sqrt{16}) = 30.3$$

The margin of error for mean sodium content is 30.3 grams. We can estimate the population mean sodium content per serving of all breakfast cereals to within 30.3 grams with 95% confidence.

YOUR TURN #11

Section 8.3, page 463

(a) $\hat{p} = \dfrac{x}{n} = \dfrac{50}{100} = 0.5$

The point estimate of the proportion p is 0.5.

(b) $\hat{p} = \dfrac{x}{n} = \dfrac{90}{160} = 0.5625$

The point estimate of the proportion p is 0.5625.

YOUR TURN #12

Section 8.3, page 465

(a) There are $x = 50$ successes, which is ≥ 5 and there are $n - x = 100 - 50 = 50$ failures, which is also ≥ 5. The conditions for constructing the Z interval for p have been met.

From Table 1, the confidence level of 95% gives $Z_{\alpha/2} = 1.96$. Thus, the confidence interval is

lower bound $= \hat{p} - Z_{\alpha/2}\sqrt{\dfrac{\hat{p} \cdot \hat{q}}{n}} = 0.5 - 1.96\sqrt{\dfrac{0.5(0.5)}{100}}$

$= 0.5 - 1.96(0.05) = 0.5 - 0.098 = 0.402$

upper bound $= \hat{p} + Z_{\alpha/2}\sqrt{\dfrac{\hat{p} \cdot \hat{q}}{n}} = 0.5 + 1.96\sqrt{\dfrac{0.5(0.5)}{100}}$

$= 0.5 + 1.96(0.05) = 0.5 + 0.098 = 0.598$

We are 95% confident that the population proportion lies between 0.402 and 0.598.

(b) There are $x = 90$ successes, which is ≥ 5 and there are $n - x = 160 - 90 = 70$ failures, which is also ≥ 5. The conditions for constructing the Z interval for p have been met.

From Table 1, the confidence level of 95% gives $Z_{\alpha/2} = 1.96$. Thus, the confidence interval is

lower bound $= \hat{p} - Z_{\alpha/2}\sqrt{\dfrac{\hat{p} \cdot \hat{q}}{n}} = 0.5625 - 1.96\sqrt{\dfrac{0.5625(0.4375)}{160}}$

$= 0.5625 - 1.96(0.0392184387) = 0.5625 - 0.0768681399$

$= 0.4856318601 \approx 0.4856$

upper bound $= \hat{p} + Z_{\alpha/2}\sqrt{\dfrac{\hat{p} \cdot \hat{q}}{n}} = 0.5625 + 1.96\sqrt{\dfrac{0.5625(0.4375)}{160}}$

$= 0.5625 + 1.96(0.0392184387) = 0.5625 + 0.0768681399$

$= 0.6393681399 \approx 0.6394$

We are 95% confident that the population proportion lies between 0.4856 and 0.6394.

YOUR TURN #13

Section 8.3, page 467

(a) The margin of error is

$$E = Z_{\alpha/2}\sqrt{\dfrac{\hat{p} \cdot \hat{q}}{n}} = 1.96\sqrt{\dfrac{0.5(0.5)}{100}} = 1.96(0.05) = 0.098$$

We can estimate the population proportion to within 0.098 with 95% confidence.

(b) The margin of error is

$$E = Z_{\alpha/2}\sqrt{\dfrac{\hat{p} \cdot \hat{q}}{n}} = 1.96\sqrt{\dfrac{0.5625(0.4375)}{160}} = 0.0768681399 \approx 0.0769$$

We can estimate the population proportion to within 0.0769 with 95% confidence.

YOUR TURN #14

Section 8.3, page 469

The required sample size is

$$n = \hat{p} \cdot \hat{q}\left(\frac{Z_{\alpha/2}}{E}\right)^2 = 0.37(0.63)\left(\frac{2.576}{0.03}\right)^2 \approx 1718.665984$$

Rounding up, this gives us a minimum required sample size of 1719.

YOUR TURN #15

Section 8.3, page 469

The required sample size is

$$n = \left[\frac{0.5 Z_{\alpha/2}}{E}\right]^2 = \left[\frac{(0.5)(1.96)}{0.05}\right]^2 = 384.16$$

Rounding up, this gives us a minimum required sample size of 385.

YOUR TURN #16

Section 8.4, page 476

For a 95% confidence interval,

$$(1 - \alpha) = 0.95 \qquad \frac{\alpha}{2} = \frac{0.05}{2} = 0.025 \qquad 1 - \frac{\alpha}{2} = 1 - 0.025 = 0.975$$

So we are seeking (1) $\chi^2_{0.975}$, the critical value with area $1 - \frac{\alpha}{2} = 0.975$ to the right of it and (2) $\chi^2_{0.025}$, the critical value with area $\frac{\alpha}{2} = 0.025$ to the right of it. Because $n = 20$, the degrees of freedom is $n - 1 = 20 - 1 = 19$. From the table, $\chi^2_{0.975} = 8.907$ and $\chi^2_{0.025} = 32.852$.

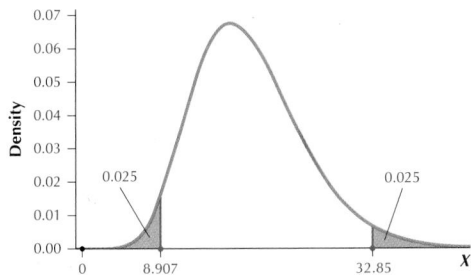

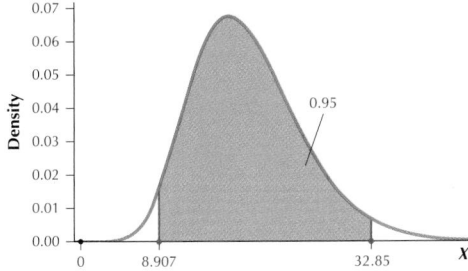

Chapter 9

YOUR TURN #1

Section 9.1, page 492

Step 1   **Search the word problem for certain key English words and select the appropriate symbol.**

The problem uses the word "increased" which means "is greater than." Thus, we will use a hypothesis that contains the > symbol.

Step 2   **Determine the form of the hypothesis.**

From Table 1, we see that the > symbol means that we use a right-tail test:

$$H_0 : \mu = \mu_0 \qquad \text{versus} \qquad H_a : \mu > \mu_0$$

Step 3   **Find the value for μ_0 and write your hypotheses.**

The alternative hypothesis H_a states that the mean monthly amount of time that iPhone and Android users spend using the apps on their devices has increased from some value μ_0. Increased from what? 30 hours per month. Write the two hypotheses with $\mu_0 = 30$.

$$H_0 : \mu = 30 \qquad \text{versus} \qquad H_a : \mu > 30$$

YOUR TURN #2

Section 9.1, page 495

(a) A Type I error would be to reject H_0 when H_0 is true. This would be to conclude that μ is less than 7 when, in reality, $\mu = 7$. In other words, a Type I error would be to conclude that the population mean of the pair of dice tossed in Example 1 is less than 7 when, in reality, it is equal to 7.

  A Type II error occurs when we do not reject H_0 when H_0 is false. This would be to conclude that μ is equal to 7 when, in reality, it is less than 7.

(b) A Type I error would be to reject H_0 when H_0 is true. This would be to conclude that μ had decreased when, in reality, it had stayed the same. In other words, a Type I error would be to conclude that the population mean rainfall in Arizona had decreased when, in reality, it had not decreased.

  A Type II error occurs when we do not reject H_0 when H_0 is false. This would be to conclude that μ had stayed the same when, in reality, it had decreased.

YOUR TURN #3

Section 9.2, page 499

Now $n = 36$, but $\bar{x} = 480$, $\sigma = 670$, and $\mu_0 = 413$ have all stayed the same.

Therefore,

$$z_{\text{data}} = \frac{\bar{x} - \mu_0}{\sigma/\sqrt{n}} = \frac{480 - 413}{670/\sqrt{36}} = 0.6$$

YOUR TURN #4

Section 9.2, page 501

(a) We have a right-tailed test and level of significance $\alpha = 0.10$, so Table 4 tells us that the critical value is $Z_{\text{crit}} = 1.28$.

(b)

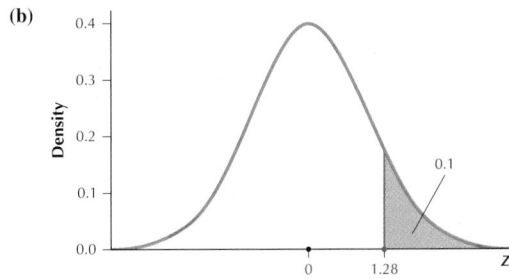

YOUR TURN #5

Section 9.3, page 509

(a) We have a right-tailed test, so that the p-value is the area in the right tail:

$$p\text{-value} = P(Z > Z_{\text{data}}) = P(Z > 0.5)$$

The Z table gives the probability for $P(Z < 0.5)$. Thus,

$$p\text{-value} = P(Z > 0.5) = 1 - P(Z < 0.5) = 1 - 0.6915 = 0.3085$$

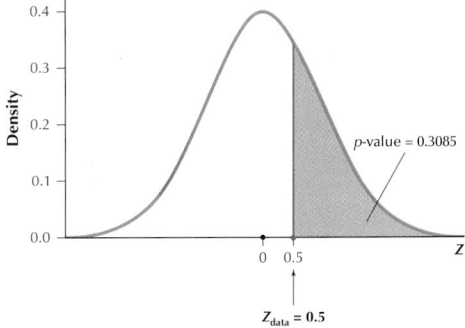

(b) We have a left-tailed test, so that the p-value is the area in the left tail:
$$p\text{-value} = P(Z < Z_{\text{data}}) = P(Z < -1.2) = 0.1151$$

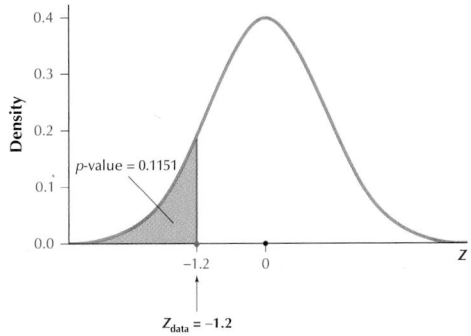

(c) Here, we have a two-tailed test, so that the p-value is the sum of the area in the two tails:

p-value $= P(Z < -|Z_{data}|) + P(Z > |Z_{data}|) = P(Z < -|-0.1|) + P(Z > |-0.1|)$

$= P(Z < -0.1) + P(Z > 0.1) = 0.4602 + 0.4602 = 0.9204$

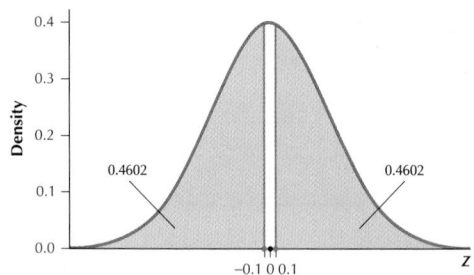

YOUR TURN #6

Section 9.3, page 515

(a) The p-value of 0.1587 implies that there is no evidence against the null hypothesis that the population mean equals 3.0.

(b) The p-value of 0.0735 implies that there is moderate evidence against the null hypothesis that the population mean equals 10.

(c) The p-value of 0.0456 implies that there is solid evidence against the null hypothesis that the population mean equals 100.

YOUR TURN #7

Section 9.3, page 518

Value of μ_0	Form of hypothesis test, with $\alpha = 0.10$	Where μ_0 lies in relation to 90% confidence interval	Conclusion of hypothesis test
a. 548	$H_0 : \mu = 548$ vs. $H_a : \mu \neq 548$	Inside	Do not reject H_0
b. 477	$H_0 : \mu = 477$ vs. $H_a : \mu \neq 477$	Inside	Do not reject H_0
c. 549	$H_0 : \mu = 549$ vs. $H_a : \mu \neq 549$	Outside	Reject H_0

YOUR TURN #8

Section 9.4, page 528

(a) Step 1 **State the hypotheses.**

From Example 18 the hypotheses are

$$H_0 : \mu = 15 \quad \text{versus} \quad H_a : \mu < 15$$

where μ refers to the population mean age at onset.

Step 2 **Find t_{crit} and state the rejection rule.**

We still have a left-tailed test and df = 19, but now $\alpha = 0.01$. From the table $t_{crit} = 2.539$. Therefore we will reject H_0 if $t_{data} \leq -2.539$.

Step 3 **Calculate t_{data}.**

From Example 18 $t_{data} = -2.2183$.

Step 4 **State the conclusion and interpretation.**

The rejection rule from Step 2 says to reject H_0 if $t_{data} \leq -2.539$. From Step 3, we have $t_{data} = -2.2183$. Because -2.2183 is not less than -2.539, our conclusion is to not reject H_0. There is insufficient evidence at level of significance $\alpha = 0.01$ that the population mean age of onset has decreased from its previous level of 15 years.

(b) In Example 18, $\alpha = 0.05$ and the conclusion is to reject H_0. In (a), $\alpha = 0.01$ and the conclusion is to not reject H_0. This is because the cutoff for rejecting the null hypothesis at $\alpha = 0.01 = -2.539 < t_{data} = -2.2183 \leq -1.729 = $ the cutoff for rejecting the null hypothesis at $\alpha = 0.05$.

From the "Developing Your Statistical Sense" box on p. 515, there are two alternatives available in situations like this.

1. Turn to a direct assessment of the strength of evidence against the null hypothesis.

2. Obtain more data, perhaps through a call for further research.

YOUR TURN #9

Section 9.4, page 531

(a) The hypotheses do not depend on α. Therefore, a change in α will not affect the hypotheses.

(b) The test statistic t_{data} does not depend on α. Therefore, a change in α will not affect t_{data}.

(c) The p-value does not depend on α. Therefore, a change in α will not affect the p-value.

(d) The conclusion does depend on α. Decreasing α from $\alpha = 0.05$ to $\alpha = 0.01$ changes the rejection rule from reject H_0 if the p-value $\leq \alpha = 0.05$ to reject H_0 if the p-value $\leq \alpha = 0.01$. The p-value $= 0.0228$ is not ≤ 0.01. Therefore we do not reject H_0. Thus the conclusion has changed from reject H_0 if $\alpha = 0.05$ to do not reject H_0 if $\alpha = 0.01$.

YOUR TURN #10

Section 9.4, page 534

From Example 18, our hypotheses are

$$H_0 : \mu = 15 \quad \text{versus} \quad H_a : \mu < 15.$$

Our test statistic is $t_{data} = -2.2183$ and df = 19.

We have a left-tailed test, which is a one-tailed test. Since the table only has positive values of t, we will look up $|t_{data}| = |-2.2183| = 2.2183$ in the table.

		Area in one tail					
		0.10	0.05	0.025		0.01	0.005
df	**19**	1.328	1.729	2.093	$|t_{data}| = 2.2183$	2.539	2.861

From the table we see that $2.093 < |t_{data}| = 2.2183 < 2.593$. Therefore $0.01 < p$-value < 0.025.

YOUR TURN #11

Section 9.5, page 544

The sample proportion of Chromebooks is

$$\hat{p} = \frac{x}{n} = \frac{\text{number in sample that are Chromebooks}}{\text{sample size}} = \frac{50}{400} = 0.125.$$

We then calculate the value of the test statistic Z_{data}:

$$Z_{data} = \frac{\hat{p} - p_0}{\sqrt{\dfrac{p_0 \cdot q_0}{n}}} = \frac{0.125 - 0.20}{\sqrt{\dfrac{0.20(0.80)}{400}}} = \frac{-0.075}{0.02} = -3.75$$

YOUR TURN #12

Section 9.5, page 548

Step 1 **State the hypotheses and the rejection rule.**

The hypotheses are

$$H_0 : p = 0.12 \quad \text{versus} \quad H_a : p > 0.12$$

where p represents the population proportion of young people ages 18–24 who had an accident. We reject H_0 if the p-value $\leq \alpha = 0.05$.

Step 2 **Calculate Z_{data}.**

Our sample proportion is $\hat{p} = 150/1000 = 0.15$.

Thus, our test statistic is

$$Z_{\text{data}} = \frac{\hat{p} - p_0}{\sqrt{\dfrac{p_0 \cdot q_0}{n}}} = \frac{0.15 - 0.12}{\sqrt{\dfrac{0.12(0.88)}{1000}}} \approx 2.92$$

Step 3 **Find the *p*-value.**

We have a right-tailed test, so the *p*-value is

$$p\text{-value} = P(Z > Z_{\text{data}}) = P(Z > 2.92) = 1 - P(Z < 2.92)$$
$$= 1 - 0.9982 = 0.0018$$

Step 4 **State the conclusion and the interpretation.**

The *p*-value 0.0018 is $\leq \alpha = 0.05$, so we reject H_0. There is evidence that the population proportion of young people ages 18–24 who had an accident has increased.

Chapter 10

YOUR TURN #1

Section 10.1, page 578

For each student, we subtract the "Before" value from the "After" value.

Henrik: $90 - 92 = -2$

Ivana: $70 - 70 = 0$

Jen: $76 - 75 = 1$

Kayla: $61 - 60 = 1$

Luisa: $60 - 58 = 2$

Manuel: $90 - 86 = 4$

We now consider the set of these six differences $\{-2, 0, 1, 1, 2, 4\}$ as a sample.

Descriptive Statistics: Differences

```
Variable      N   N*   Mean   StDev
Differences   6   0    1.000  2.000
```

From the Minitab output we see that $\bar{x}_d = 1$ and $s_d = 2$.

YOUR TURN #2

Section 10.1, page 580

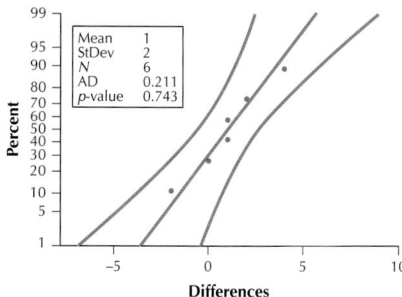

From the normal probability plot we see that we have acceptable normality.

Step 1 **State the hypotheses.**

The hypotheses are

$$H_0 : \mu_d = 0 \quad \text{versus} \quad H_a : \mu_d > 0$$

where μ_d represents the population mean difference in English quiz scores.

Step 2 **Find the critical value t_{crit} and state the rejection rule.**

We have a right-tailed test with area in one tail equal to $\alpha = 0.10$ and degrees of freedom equal to df $= n - 1 = 6 - 1 = 5$. Therefore our critical value from the table is $t_{\text{crit}} = 1.476$. Therefore our rejection rule is that we will reject H_0 when t_{data} is greater than or equal to 1.476.

Step 3 **Find t_{data}.**

From Your Turn #1, $\bar{x}_d = 1$ and $s_d = 2$. Therefore

$$t_{\text{data}} = \frac{\bar{x}_d}{s_d / \sqrt{n}} = \frac{1}{2 / \sqrt{6}} \approx 1.22$$

Step 4 **State the conclusion and the interpretation.**

Since $t_{\text{data}} \approx 1.22$ is not greater than or equal to 1.476, we do not reject H_0. There is insufficient evidence at level of significance $\alpha = 0.10$ that the population mean difference of English quiz scores is greater than 0.

YOUR TURN #3

Section 10.1, page 583

The normality was checked in Your Turn #2. From Your Turn #1, $\bar{x}_d = 1$ and $s_d = 2$. For a 90% confidence interval with degrees of freedom equal to df $= n - 1 = 6 - 1 = 5$, $t_{\alpha/2} = 2.015$.

Using these values,

$$\text{lower bound} = \bar{x}_d - t_{\alpha/2}(s_d / \sqrt{n})$$
$$= 1 - (2.015)(2/\sqrt{6})$$
$$\approx 1 - 1.6452 = -0.6452$$
$$\text{upper bound} = \bar{x}_d + t_{\alpha/2}(s_d / \sqrt{n})$$
$$= 1 + (2.015)(2/\sqrt{6})$$
$$\approx 1 + 1.6452 = 2.6452$$

We are 90% confident that the population mean of the differences between English quiz scores before and after visiting the English Center lies between -0.6452 point and 2.6452 points.

YOUR TURN #4

Section 10.1, page 584

From Your Turn #3, our 90% confidence interval is $(-0.6452, 2.6432)$.

(a) $H_0 : \mu_d = 2$ vs. $H_a : \mu_d \neq 2$

$\mu_0 = 2$ lies inside the interval, so we do not reject H_0.

(b) $H_0 : \mu_d = 5.6$ vs. $H_a : \mu_d \neq 5.6$

$\mu_0 = 5.6$ lies outside the interval, so we reject H_0.

(c) $H_0 : \mu_d = 5.7$ vs. $H_a : \mu_d \neq 5.7$

$\mu_0 = 5.7$ lies outside the interval, so we reject H_0.

Chapter 11

YOUR TURN #1

Section 11.1, page 634

(a) There are $k = 4$ possible outcomes: paperbacks, hardcovers, e-Books, and all other formats. Assigning probabilities using the relative frequency method, we have the following hypothesized proportions for each book format:

$$p_{\text{paperback}} = 0.41, \; p_{\text{hardcover}} = 0.34, \; p_{\text{e-Book}} = 0.13, \; p_{\text{all other formats}} = 0.12$$

and

$$\sum p_i = 0.41 + 0.34 + 0.13 + 0.12 = 1.$$

Therefore, $X = $ *book format* is a valid multinomial random variable.

(b) We have $n = 2000$ books (sample size $= 2000$), so the expected frequencies are as provided in the following table.

Category	Expected frequency$_i = E_i = n \cdot p_i$
Paperback	$E_{\text{paperback}} = 2000 \cdot 0.41 = 820$
Hardcover	$E_{\text{hardcover}} = 2000 \cdot 0.34 = 680$
e-Book	$E_{\text{e-Book}} = 2000 \cdot 0.13 = 260$
All other formats	$E_{\text{all other formats}} = 2000 \cdot 0.12 = 240$

YOUR TURN #2

Section 11.1, page 637

Category	p_i	O_i	E_i	$O_i - E_i$	$(O_i - E_i)^2$	$\dfrac{(O_i - E_i)^2}{E_i}$
Paperback	0.41	810	820	−10	100	$\dfrac{(810 - 820)^2}{820} \approx 0.122$
Hardcover	0.34	680	680	0	0	$\dfrac{(680 - 680)^2}{680} = 0$
e-Book	0.13	280	260	20	400	$\dfrac{(280 - 260)^2}{260} \approx 1.538$
All other formats	0.12	230	240	−10	100	$\dfrac{(230 - 240)^2}{240} \approx 0.417$

Then

$$\chi^2_{data} = \sum \frac{(O_i - E_i)^2}{E_i} \approx 0.122 + 0 + 1.538 + 0.417 = 2.077$$

YOUR TURN #3

Section 11.1, page 638

Step 1 **State the hypotheses and check the conditions.** The hypotheses are:

H_0: $p_{paperback} = 0.41$, $p_{hardcover} = 0.34$, $p_{e\text{-}Book} = 0.13$, $p_{all\ other\ formats} = 0.12$

H_a: The random variable does not follow the distribution specified in H_0.

Checking the conditions, the expected frequencies from Your Turn #1 are

$$E_{paperback} = 820, \ E_{hardcover} = 680, \ E_{e\text{-}Book} = 260, \ E_{all\ other\ formats} = 240$$

Because none of these expected frequencies is less than one, and none of the expected frequencies is less than five, the conditions for performing the goodness of fit test are met.

Step 2 **Find the χ^2 critical value, χ^2_{crit}, and state the rejection rule.**

We have degrees of freedom $k - 1 = 4 - 1 = 3$ and $\alpha = 0.05$. Then, from the table, $\chi^2_{crit} = \chi^2_{0.05} = 7.815$. The rejection rule is "Reject H_0 if $\chi^2_{data} \geq 7.815$."

Step 3 **Find χ^2_{data}.**

From Your Turn #2, $\chi^2_{data} \approx 2.077$

Step 4 **State the conclusion and interpretation.**

Compare χ^2_{data} with χ^2_{crit}. $\chi^2_{data} \approx 2.077$ is not $\geq$ to $\chi^2_{crit} = 7.815$. Therefore, we do not reject H_0. There is insufficient evidence that the variable *book format* does not follow the distribution specified in H_0. In other words, there is insufficient evidence that the book format market shares have changed.

Chapter 12

There are no Your Turn exercises in this chapter.

Chapter 13

YOUR TURN #1

Section 13.1, page 717

(a)

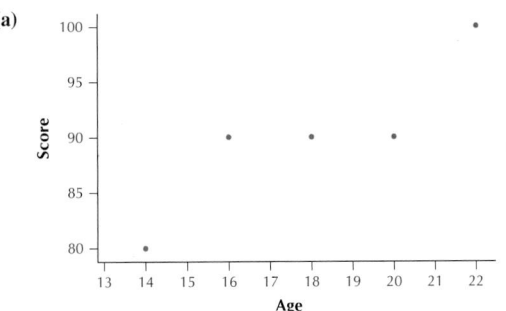

As age increases, the score on the video game tends to increase.

(b) As calculated by the TI-84, $r = 0.894427191 \approx 0.8944$. Age and score are positively correlated. An increase in age is associated with an increase in the score on the video game.

(c) As calculated by the TI-84, $\hat{y} = 2x + 54$. The slope $b_1 = 2$ means there is an estimated increase of 2 in the score on the video game for each one year increase in age. The y-intercept $b_0 = 54$ is the estimated score on the video game of someone aged $x = 0$.

(d) For a 22-year-old person, the estimated score on the video game is $\hat{y} = 2x + 54 = 2(22) + 54 = 98$.

The actual score of the 22-year-old person in the sample is $y = 100$. Thus, our prediction error is $y - \hat{y} = 100 - 98 = 2$. The 22-year-old person scored slightly higher than expected.

YOUR TURN #2

Section 13.1, page 721

(a)

x = Age	y = score on video game	Fitted (predicted) values $\hat{y} = 2(age) + 54$	Residuals $y - \hat{y}$
14	80	82	−2
16	90	86	4
18	90	90	0
20	90	94	−4
22	100	98	2

(b)

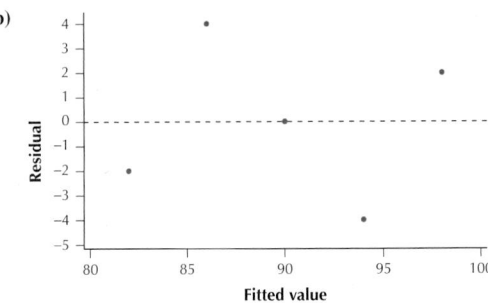

The scatterplot of the residuals versus the fitted values shows no evidence of the unhealthy patterns shown in Figure 4. Thus, the independence assumption, the constant variance assumption, and the zero-mean assumption are verified.

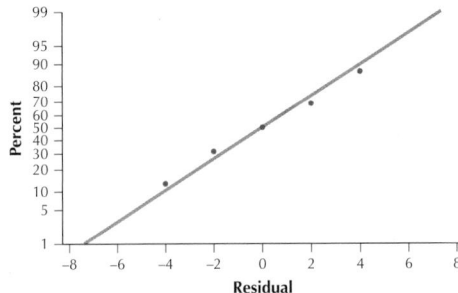

Also, the normal probability plot of the residuals indicates no evidence of departures from normality in the residuals. Therefore, we conclude that the regression assumptions are verified.

YOUR TURN #3

Section 13.1, page 725

Hypothesis test using the *p*-value method.

Step 1 **State the hypotheses and the rejection rule.**

$H_0: \beta_1 = 0$ No linear relationship exists between age and score.

$H_a: \beta_1 \neq 0$ A linear relationship exists between age and score.

Reject H_0 if the *p*-value $\leq \alpha = 0.05$.

Step 2 **Calculate t_{data}.**

$$t_{data} = \frac{b_1}{s/\sqrt{\sum(x - \bar{x})^2}}$$

From Your Turn #1, $b_1 = 2$. From the TI-84, $s_x^2 = 10$. Therefore,

$$\sum(x - \bar{x})^2 = (n - 1)s_x^2 = (5 - 1)10 = 40.$$

SSE $= \sum(y - \hat{y})^2$. The residuals $y - \hat{y}$ were calculated in Your Turn #2 (a). They are $-2, 4, 0, -4$, and 2. Therefore,

$$SSE = \sum(y - \hat{y})^2 = (-2)^2 + (4)^2 + (0)^2 + (-4)^2 + (2)^2$$

$$= 4 + 16 + 0 + 16 + 4 = 40. \text{ Then}$$

$$s = \sqrt{\frac{SSE}{n - 2}} = \sqrt{\frac{40}{5 - 2}} = 3.651483717.$$

Thus

$$t_{data} = \frac{b_1}{s/\sqrt{\sum(x - \bar{x})^2}} = \frac{2}{3.65148317/\sqrt{40}} = 3.464101615 \approx 3.4641.$$

Step 3 **Find the p-value.**

Using the TI-84 we get p-value $= 0.0405193264$.

Step 4 **State the conclusion and the interpretation.**

The p-value $= 0.0405193264$ is $\leq \alpha = 0.05$, so we reject H_0. Evidence exists, at level of significance $\alpha = 0.05$, for a linear relationship between age and score.

Hypothesis test using the critical value method.

Step 1 **State the hypotheses** $H_0 : \beta_1 = 0$ No linear relationship exists between age and score.

$H_a : \beta_1 \neq 0$ A linear relationship exists between age and score.

Step 2 **Find the critical value t_{crit} and the rejection rule.**

The degrees of freedom are df $= n - 2 = 5 - 2 = 3$. We have a two-tailed test with $\alpha = 0.05$. From the table, $t_{crit} = 3.182$. Therefore the rejection rule is Reject H_0 if $t_{data} \leq -3.182$ or $t_{data} \geq 3.182$.

Step 3 **Calculate t_{data}.**

$$t_{data} = \frac{b_1}{s/\sqrt{\sum(x - \bar{x})^2}}$$

From Your Turn #1, $b_1 = 2$. From the TI-84, $s_x^2 = 10$. Therefore,

$$\sum(x - \bar{x})^2 = (n - 1)s_x^2 = (5 - 1)10 = 40.$$

SSE $= \sum(y - \hat{y})^2$. The residuals $y - \hat{y}$ were calculated in Your Turn #2 (a). They are $-2, 4, 0, -4$, and 2. Therefore,

$$SSE = \sum(y - \hat{y})^2 = (-2)^2 + (4)^2 + (0)^2 + (-4)^2 + (2)^2$$

$$= 4 + 16 + 0 + 16 + 4 = 40. \text{ Then}$$

$$s = \sqrt{\frac{SSE}{n - 2}} = \sqrt{\frac{40}{5 - 2}} = 3.651483717.$$

Thus

$$t_{data} = \frac{b_1}{s/\sqrt{\sum(x - \bar{x})^2}} = \frac{2}{3.65148317/\sqrt{40}} = 3.464101615 \approx 3.4641.$$

Step 4 **State the conclusion and the interpretation.**

$t_{data} \approx 3.4641$ is ≥ 3.182, so we reject H_0. Evidence exists, at level of significance $\alpha = 0.05$, for a linear relationship between age and score.

Tables Appendix

Table A Random numbers

10480	15011	01536	02011	81647	91646	67179	14194	62590	36207	20969	99570	91291	90700
22368	46573	25595	85393	30995	89198	27982	53402	93965	34095	52666	19174	39615	99505
24130	48360	22527	97265	76393	64809	15179	24830	49340	32081	30680	19655	63348	58629
42167	93093	06243	61680	07856	16376	39440	53537	71341	57004	00849	74917	97758	16379
37570	39975	81837	16656	06121	91782	60468	81305	49684	60672	14110	06927	01263	54613
77921	06907	11008	42751	27756	53498	18602	70659	90655	15053	21916	81825	44394	42880
99562	72905	56420	69994	98872	31016	71194	18738	44013	48840	63213	21069	10634	12952
96301	91977	05463	07972	18876	20922	94595	56869	69014	60045	18425	84903	42508	32307
89579	14342	63661	10281	17453	18103	57740	84378	25331	12566	58678	44947	05584	56941
85475	36857	43342	53988	53060	59533	38867	62300	08158	17983	16439	11458	18593	64952
28918	69578	88231	33276	70997	79936	56865	05859	90106	31595	01547	85590	91610	78188
63553	40961	48235	03427	49626	69445	18663	72695	52180	20847	12234	90511	33703	90322
09429	93969	52636	92737	88974	33488	36320	17617	30015	08272	84115	27156	30613	74952
10365	61129	87529	85689	48237	52267	67689	93394	01511	26358	85104	20285	29975	89868
07119	97336	71048	08178	77233	13916	47564	81056	97735	85977	29372	74461	28551	90707
51085	12765	51821	51259	77452	16308	60756	92144	49442	53900	70960	63990	75601	40719
02368	21382	52404	60268	89368	19885	55322	44819	01188	65255	64835	44919	05944	55157
01011	54092	33362	94904	31273	04146	18594	29852	71585	85030	51132	01915	92747	64951
52162	53916	46369	58586	23216	14513	83149	98736	23495	64350	94738	17752	35156	35749
07056	97628	33787	09998	42698	06691	76988	13602	51851	46104	88916	19509	25625	58104
48663	91245	85828	14346	09172	30168	90229	04734	59193	22178	30421	61666	99904	32812
54164	58492	22421	74103	47070	25306	76468	26384	58151	06646	21524	15227	96909	44592
32639	32363	05597	24200	13363	38005	94342	28728	35806	06912	17012	64161	18296	22851
29334	27001	87637	87308	58731	00256	45834	15398	46557	41135	10367	07684	36188	18510
02488	33062	28834	07351	19731	92420	60952	61280	50001	67658	32586	86679	50720	94953
81525	72295	04839	96423	24878	82651	66566	14778	76797	14780	13300	87074	79666	95725
29676	20591	68086	26432	46901	20849	89768	81536	86645	12659	92259	57102	80428	25280
00742	57392	39064	66432	84673	40027	32832	61362	98947	96067	64760	64584	96096	98253
05366	04213	25669	26422	44407	44048	37937	63904	45766	66134	75470	66520	34693	90449
91921	26418	64117	94305	26766	25940	39972	22209	71500	64568	91402	42416	07844	69618
00582	04711	87917	77341	42206	35126	74087	99547	81817	42607	43808	76655	62028	76630
00725	69884	62797	56170	86324	88072	76222	36086	84637	93161	76038	65855	77919	88006
69011	65797	95876	55293	18988	27354	26575	08625	40801	59920	29841	80150	12777	48501
25976	57948	29888	88604	67917	48708	18912	82271	65424	69774	33611	54262	85963	03547
09763	83473	73577	12908	30883	18317	28290	35797	05998	41688	34952	37888	38917	88050
91567	42595	27958	30134	04024	86385	29880	99730	55536	84855	29080	09250	79656	73211
17955	56349	90999	49127	20044	59931	06115	20542	18059	02008	73708	83517	36103	42791
46503	18584	18845	49618	02304	51038	20655	58727	28168	15475	56942	53389	20562	87338
92157	89634	94824	78171	84610	82834	09922	25417	44137	48413	25555	21246	35509	20468
14577	62765	35605	81263	39667	47358	56873	56307	61607	49518	89656	20103	77490	18062
98427	07523	33362	64270	01638	92477	66969	98420	04880	45585	46565	04102	46880	45709
34914	63976	88720	82765	34476	17032	87589	40836	32427	70002	70663	88863	77775	69348
70060	28277	39475	46473	23219	53416	94970	25832	69975	94884	19661	72828	00102	66794
53976	54914	06990	67245	68350	82948	11398	42878	80287	88267	47363	46634	06541	97809
76072	29515	40980	07391	58745	25774	22987	80059	39911	96189	41151	14222	60697	59583
90725	52210	83974	29992	65831	38857	50490	83765	55657	14361	31720	57375	56228	41546
64364	67412	33339	31926	14883	24413	59744	92351	97473	89286	35931	04110	23726	51900
08962	00358	31662	25388	61642	34072	81249	35648	56891	69352	48373	45578	78547	81788
95012	68379	93526	70765	10593	04542	76463	54328	02349	17247	28865	14777	62730	92277
15664	10493	20492	38391	91132	21999	59516	81652	27195	48223	46751	22923	32261	85653

Courtesy W. H. Beyer.

Table B Binomial distribution

n	X	0.10	0.15	0.20	0.25	0.30	0.35	0.40	0.45	0.50
2	0	0.8100	0.7225	0.6400	0.5625	0.4900	0.4225	0.3600	0.3025	0.2500
	1	0.1800	0.2550	0.3200	0.3750	0.4200	0.4550	0.4800	0.4950	0.5000
	2	0.0100	0.0225	0.0400	0.0625	0.0900	0.1225	0.1600	0.2025	0.2500
3	0	0.7290	0.6141	0.5120	0.4219	0.3430	0.2746	0.2160	0.1664	0.1250
	1	0.2430	0.3251	0.3840	0.4219	0.4410	0.4436	0.4320	0.4084	0.3750
	2	0.0270	0.0574	0.0960	0.1406	0.1890	0.2389	0.2880	0.3341	0.3750
	3	0.0010	0.0034	0.0080	0.0156	0.0270	0.0429	0.0640	0.0911	0.1250
4	0	0.6561	0.5220	0.4096	0.3164	0.2401	0.1785	0.1296	0.0915	0.0625
	1	0.2916	0.3685	0.4096	0.4219	0.4116	0.3845	0.3456	0.2995	0.2500
	2	0.0486	0.0975	0.1536	0.2109	0.2646	0.3105	0.3456	0.3675	0.3750
	3	0.0036	0.0115	0.0256	0.0469	0.0756	0.1115	0.1536	0.2005	0.2500
	4	0.0001	0.0005	0.0016	0.0039	0.0081	0.0150	0.0256	0.0410	0.0625
5	0	0.5905	0.4437	0.3277	0.2373	0.1681	0.1160	0.0778	0.0503	0.0312
	1	0.3280	0.3915	0.4096	0.3955	0.3602	0.3124	0.2592	0.2059	0.1562
	2	0.0729	0.1382	0.2048	0.2637	0.3087	0.3364	0.3456	0.3369	0.3125
	3	0.0081	0.0244	0.0512	0.0879	0.1323	0.1811	0.2304	0.2757	0.3125
	4	0.0004	0.0022	0.0064	0.0146	0.0284	0.0488	0.0768	0.1128	0.1562
	5		0.0001	0.0003	0.0010	0.0024	0.0053	0.0102	0.0185	0.0312
6	0	0.5314	0.3771	0.2621	0.1780	0.1176	0.0754	0.0467	0.0277	0.0156
	1	0.3543	0.3993	0.3932	0.3560	0.3025	0.2437	0.1866	0.1359	0.0938
	2	0.0984	0.1762	0.2458	0.2966	0.3241	0.3280	0.3110	0.2780	0.2344
	3	0.0146	0.0415	0.0819	0.1318	0.1852	0.2355	0.2765	0.3032	0.3125
	4	0.0012	0.0055	0.0154	0.0330	0.0595	0.0951	0.1382	0.1861	0.2344
	5	0.0001	0.0004	0.0015	0.0044	0.0102	0.0205	0.0369	0.0609	0.0938
	6			0.0001	0.0002	0.0007	0.0018	0.0041	0.0083	0.0156
7	0	0.4783	0.3206	0.2097	0.1335	0.0824	0.0490	0.0280	0.0152	0.0078
	1	0.3720	0.3960	0.3670	0.3115	0.2471	0.1848	0.1306	0.0872	0.0547
	2	0.1240	0.2097	0.2753	0.3115	0.3177	0.2985	0.2613	0.2140	0.1641
	3	0.0230	0.0617	0.1147	0.1730	0.2269	0.2679	0.2903	0.2918	0.2734
	4	0.0026	0.0109	0.0287	0.0577	0.0972	0.1442	0.1935	0.2388	0.2734
	5	0.0002	0.0012	0.0043	0.0115	0.0250	0.0466	0.0774	0.1172	0.1641
	6		0.0001	0.0004	0.0013	0.0036	0.0084	0.0172	0.0320	0.0547
	7				0.0001	0.0002	0.0006	0.0016	0.0037	0.0078
8	0	0.4305	0.2725	0.1678	0.1001	0.0576	0.0319	0.0168	0.0084	0.0039
	1	0.3826	0.3847	0.3355	0.2670	0.1977	0.1373	0.0896	0.0548	0.0312
	2	0.1488	0.2376	0.2936	0.3115	0.2965	0.2587	0.2090	0.1569	0.1094
	3	0.0331	0.0839	0.1468	0.2076	0.2541	0.2786	0.2787	0.2568	0.2188
	4	0.0046	0.0185	0.0459	0.0865	0.1361	0.1875	0.2322	0.2627	0.2734
	5	0.0004	0.0026	0.0092	0.0231	0.0467	0.0808	0.1239	0.1719	0.2188
	6		0.0002	0.0011	0.0038	0.0100	0.0217	0.0413	0.0703	0.1094
	7			0.0001	0.0004	0.0012	0.0033	0.0079	0.0164	0.0313
	8					0.0001	0.0002	0.0007	0.0017	0.0039

Note: Blank entries indicate a binomial probability of less than 0.00005.

(Continued)

Table B Binomial distribution (*continued*)

n	X	0.10	0.15	0.20	0.25	0.30	0.35	0.40	0.45	0.50
						p				
9	0	0.3874	0.2316	0.1342	0.0751	0.0404	0.0207	0.0101	0.0046	0.0020
	1	0.3874	0.3679	0.3020	0.2253	0.1556	0.1004	0.0605	0.0339	0.0176
	2	0.1722	0.2597	0.3020	0.3003	0.2668	0.2162	0.1612	0.1110	0.0703
	3	0.0446	0.1069	0.1762	0.2336	0.2668	0.2716	0.2508	0.2119	0.1641
	4	0.0074	0.0283	0.0661	0.1168	0.1715	0.2194	0.2508	0.2600	0.2461
	5	0.0008	0.0050	0.0165	0.0389	0.0735	0.1181	0.1672	0.2128	0.2461
	6	0.0001	0.0006	0.0028	0.0087	0.0210	0.0424	0.0743	0.1160	0.1641
	7			0.0003	0.0012	0.0039	0.0098	0.0212	0.0407	0.0703
	8				0.0001	0.0004	0.0013	0.0035	0.0083	0.0176
	9						0.0001	0.0003	0.0008	0.0020
10	0	0.3487	0.1969	0.1074	0.0563	0.0282	0.0135	0.0060	0.0025	0.0010
	1	0.3874	0.3474	0.2684	0.1877	0.1211	0.0725	0.0403	0.0207	0.0098
	2	0.1937	0.2759	0.3020	0.2816	0.2335	0.1757	0.1209	0.0763	0.0439
	3	0.0574	0.1298	0.2013	0.2503	0.2668	0.2522	0.2150	0.1665	0.1172
	4	0.0112	0.0401	0.0881	0.1460	0.2001	0.2377	0.2508	0.2384	0.2051
	5	0.0015	0.0085	0.0264	0.0584	0.1029	0.1536	0.2007	0.2340	0.2461
	6	0.0001	0.0012	0.0055	0.0162	0.0368	0.0689	0.1115	0.1596	0.2051
	7		0.0001	0.0008	0.0031	0.0090	0.0212	0.0425	0.0746	0.1172
	8			0.0001	0.0004	0.0014	0.0043	0.0106	0.0229	0.0439
	9					0.0001	0.0005	0.0016	0.0042	0.0098
	10							0.0001	0.0003	0.0010
12	0	0.2824	0.1422	0.0687	0.0317	0.0138	0.0057	0.0022	0.0008	0.0002
	1	0.3766	0.3012	0.2062	0.1267	0.0712	0.0368	0.0174	0.0075	0.0029
	2	0.2301	0.2924	0.2835	0.2323	0.1678	0.1088	0.0639	0.0339	0.0161
	3	0.0853	0.1720	0.2362	0.2581	0.2397	0.1954	0.1419	0.0923	0.0537
	4	0.0213	0.0683	0.1329	0.1936	0.2311	0.2367	0.2128	0.1700	0.1208
	5	0.0038	0.0193	0.0532	0.1032	0.1585	0.2039	0.2270	0.2225	0.1934
	6	0.0005	0.0040	0.0155	0.0401	0.0792	0.1281	0.1766	0.2124	0.2256
	7		0.0006	0.0033	0.0115	0.0291	0.0591	0.1009	0.1489	0.1934
	8		0.0001	0.0005	0.0024	0.0078	0.0199	0.0420	0.0762	0.1208
	9			0.0001	0.0004	0.0015	0.0048	0.0125	0.0277	0.0537
	10					0.0002	0.0008	0.0025	0.0068	0.0161
	11						0.0001	0.0003	0.0010	0.0029
	12								0.0001	0.0002
15	0	0.2059	0.0874	0.0352	0.0134	0.0047	0.0016	0.0005	0.0001	
	1	0.3432	0.2312	0.1319	0.0668	0.0305	0.0126	0.0047	0.0016	0.0005
	2	0.2669	0.2856	0.2309	0.1559	0.0916	0.0476	0.0219	0.0090	0.0032
	3	0.1285	0.2184	0.2501	0.2252	0.1700	0.1110	0.0634	0.0318	0.0139
	4	0.0428	0.1156	0.1876	0.2252	0.2186	0.1792	0.1268	0.0780	0.0417
	5	0.0105	0.0449	0.1032	0.1651	0.2061	0.2123	0.1859	0.1404	0.0916
	6	0.0019	0.0132	0.0430	0.0917	0.1472	0.1906	0.2066	0.1914	0.1527
	7	0.0003	0.0030	0.0138	0.0393	0.0811	0.1319	0.1771	0.2013	0.1964
	8		0.0005	0.0035	0.0131	0.0348	0.0710	0.1181	0.1647	0.1964
	9		0.0001	0.0007	0.0034	0.0116	0.0298	0.0612	0.1048	0.1527
	10			0.0001	0.0007	0.0030	0.0096	0.0245	0.0515	0.0916
	11				0.0001	0.0006	0.0024	0.0074	0.0191	0.0417
	12					0.0001	0.0004	0.0016	0.0052	0.0139
	13						0.0001	0.0003	0.0010	0.0032
	14								0.0001	0.0005
	15									

Note: Blank entries indicate a binomial probability of less than 0.00005.

Table B Binomial distribution (*continued*)

						p				
n	*X*	**0.10**	**0.15**	**0.20**	**0.25**	**0.30**	**0.35**	**0.40**	**0.45**	**0.50**
18	0	0.1501	0.0536	0.0180	0.0056	0.0016	0.0004	0.0001		
	1	0.3002	0.1704	0.0811	0.0338	0.0126	0.0042	0.0012	0.0003	0.0001
	2	0.2835	0.2556	0.1723	0.0958	0.0458	0.0190	0.0069	0.0022	0.0006
	3	0.1680	0.2406	0.2297	0.1704	0.1046	0.0547	0.0246	0.0095	0.0031
	4	0.0700	0.1592	0.2153	0.2130	0.1681	0.1104	0.0614	0.0291	0.0117
	5	0.0218	0.0787	0.1507	0.1988	0.2017	0.1664	0.1146	0.0666	0.0327
	6	0.0052	0.0301	0.0816	0.1436	0.1873	0.1941	0.1655	0.1181	0.0708
	7	0.0010	0.0091	0.0350	0.0820	0.1376	0.1792	0.1892	0.1657	0.1214
	8	0.0002	0.0022	0.0120	0.0376	0.0811	0.1327	0.1734	0.1864	0.1669
	9		0.0004	0.0033	0.0139	0.0386	0.0794	0.1284	0.1694	0.1855
	10		0.0001	0.0008	0.0042	0.0149	0.0385	0.0771	0.1248	0.1669
	11			0.0001	0.0010	0.0046	0.0151	0.0374	0.0742	0.1214
	12				0.0002	0.0012	0.0047	0.0145	0.0354	0.0708
	13					0.0002	0.0012	0.0045	0.0134	0.0327
	14						0.0002	0.0011	0.0039	0.0117
	15							0.0002	0.0009	0.0031
	16								0.0001	0.0006
	17									0.0001
	18									
20	0	0.1216	0.0388	0.0115	0.0032	0.0008	0.0002			
	1	0.2702	0.1368	0.0576	0.0211	0.0068	0.0020	0.0005	0.0001	
	2	0.2852	0.2293	0.1369	0.0669	0.0278	0.0100	0.0031	0.0008	0.0002
	3	0.1901	0.2428	0.2054	0.1339	0.0716	0.0323	0.0123	0.0040	0.0011
	4	0.0898	0.1821	0.2182	0.1897	0.1304	0.0738	0.0350	0.0139	0.0046
	5	0.0319	0.1028	0.1746	0.2023	0.1789	0.1272	0.0746	0.0365	0.0148
	6	0.0089	0.0454	0.1091	0.1686	0.1916	0.1712	0.1244	0.0746	0.0370
	7	0.0020	0.0160	0.0545	0.1124	0.1643	0.1844	0.1659	0.1221	0.0739
	8	0.0004	0.0046	0.0222	0.0609	0.1144	0.1614	0.1797	0.1623	0.1201
	9	0.0001	0.0011	0.0074	0.0271	0.0654	0.1158	0.1597	0.1771	0.1602
	10		0.0002	0.0020	0.0099	0.0308	0.0686	0.1171	0.1593	0.1762
	11			0.0005	0.0030	0.0120	0.0336	0.0710	0.1185	0.1602
	12			0.0001	0.0008	0.0039	0.0136	0.0355	0.0727	0.1201
	13				0.0002	0.0010	0.0045	0.0146	0.0366	0.0739
	14					0.0002	0.0012	0.0049	0.0150	0.0370
	15						0.0003	0.0013	0.0049	0.0148
	16							0.0003	0.0013	0.0046
	17								0.0002	0.0011
	18									0.0002
	19									
	20									

Note: Blank entries indicate a binomial probability of less than 0.00005.

Table B Binomial distribution (*continued*)

						p				
n	X	**0.55**	**0.60**	**0.65**	**0.70**	**0.75**	**0.80**	**0.85**	**0.90**	**0.95**
2	0	0.2025	0.1600	0.1225	0.0900	0.0625	0.0400	0.0225	0.0100	0.0025
	1	0.4950	0.4800	0.4550	0.4200	0.3750	0.3200	0.2550	0.1800	0.0950
	2	0.3025	0.3600	0.4225	0.4900	0.5625	0.6400	0.7225	0.8100	0.9025
3	0	0.0911	0.0640	0.0429	0.0270	0.0156	0.0080	0.0034	0.0010	0.0001
	1	0.3341	0.2880	0.2389	0.1890	0.1406	0.0960	0.0574	0.0270	0.0071
	2	0.4084	0.4320	0.4436	0.4410	0.4219	0.3840	0.3251	0.2430	0.1354
	3	0.1664	0.2160	0.2746	0.3430	0.4219	0.5120	0.6141	0.7290	0.8574
4	0	0.0410	0.0256	0.0150	0.0081	0.0039	0.0016	0.0005	0.0001	
	1	0.2005	0.1536	0.1115	0.0756	0.0469	0.0256	0.0115	0.0036	0.0005
	2	0.3675	0.3456	0.3105	0.2646	0.2109	0.1536	0.0975	0.0486	0.0135
	3	0.2995	0.3456	0.3845	0.4116	0.4219	0.4096	0.3685	0.2916	0.1715
	4	0.0915	0.1296	0.1785	0.2401	0.3164	0.4096	0.5220	0.6561	0.8145
5	0	0.0185	0.0102	0.0053	0.0024	0.0010	0.0003	0.0001		
	1	0.1128	0.0768	0.0488	0.0284	0.0146	0.0064	0.0022	0.0005	
	2	0.2757	0.2304	0.1811	0.1323	0.0879	0.0512	0.0244	0.0081	0.0011
	3	0.3369	0.3456	0.3364	0.3087	0.2637	0.2048	0.1382	0.0729	0.0214
	4	0.2059	0.2592	0.3124	0.3601	0.3955	0.4096	0.3915	0.3281	0.2036
	5	0.0503	0.0778	0.1160	0.1681	0.2373	0.3277	0.4437	0.5905	0.7738
6	0	0.0083	0.0041	0.0018	0.0007	0.0002	0.0001			
	1	0.0609	0.0369	0.0205	0.0102	0.0044	0.0015	0.0004	0.0001	
	2	0.1861	0.1382	0.0951	0.0595	0.0330	0.0154	0.0055	0.0012	0.0001
	3	0.3032	0.2765	0.2355	0.1852	0.1318	0.0819	0.0415	0.0146	0.0021
	4	0.2780	0.3110	0.3280	0.3241	0.2966	0.2458	0.1762	0.0984	0.0305
	5	0.1359	0.1866	0.2437	0.3025	0.3560	0.3932	0.3993	0.3543	0.2321
	6	0.0277	0.0467	0.0754	0.1176	0.1780	0.2621	0.3771	0.5314	0.7351
7	0	0.0037	0.0016	0.0006	0.0002	0.0001				
	1	0.0320	0.0172	0.0084	0.0036	0.0013	0.0004	0.0001		
	2	0.1172	0.0774	0.0466	0.0250	0.0115	0.0043	0.0012	0.0002	
	3	0.2388	0.1935	0.1442	0.0972	0.0577	0.0287	0.0109	0.0026	0.0002
	4	0.2918	0.2903	0.2679	0.2269	0.1730	0.1147	0.0617	0.0230	0.0036
	5	0.2140	0.2613	0.2985	0.3177	0.3115	0.2753	0.2097	0.1240	0.0406
	6	0.0872	0.1306	0.1848	0.2471	0.3115	0.3670	0.3960	0.3720	0.2573
	7	0.0152	0.0280	0.0490	0.0824	0.1335	0.2097	0.3206	0.4783	0.6983
8	0	0.0017	0.0007	0.0002	0.0001					
	1	0.0164	0.0079	0.0033	0.0012	0.0004	0.0001			
	2	0.0703	0.0413	0.0217	0.0100	0.0038	0.0011	0.0002		
	3	0.1719	0.1239	0.0808	0.0467	0.0231	0.0092	0.0026	0.0004	
	4	0.2627	0.2322	0.1875	0.1361	0.0865	0.0459	0.0185	0.0046	0.0004
	5	0.2568	0.2787	0.2786	0.2541	0.2076	0.1468	0.0839	0.0331	0.0054
	6	0.1569	0.2090	0.2587	0.2965	0.3115	0.2936	0.2376	0.1488	0.0515
	7	0.0548	0.0896	0.1373	0.1977	0.2670	0.3355	0.3847	0.3826	0.2793
	8	0.0084	0.0168	0.0319	0.0576	0.1001	0.1678	0.2725	0.4305	0.6634

Note: Blank entries indicate a binomial probability of less than 0.00005.

Table B Binomial distribution (*continued*)

		\(p \)								
n	*X*	**0.55**	**0.60**	**0.65**	**0.70**	**0.75**	**0.80**	**0.85**	**0.90**	**0.95**
9	0	0.0008	0.0003	0.0001						
	1	0.0083	0.0035	0.0013	0.0004	0.0001				
	2	0.0407	0.0212	0.0098	0.0039	0.0012	0.0003			
	3	0.1160	0.0743	0.0424	0.0210	0.0087	0.0028	0.0006	0.0001	
	4	0.2128	0.1672	0.1181	0.0735	0.0389	0.0165	0.0050	0.0008	
	5	0.2600	0.2508	0.2194	0.1715	0.1168	0.0661	0.0283	0.0074	0.0006
	6	0.2119	0.2508	0.2716	0.2668	0.2336	0.1762	0.1069	0.0446	0.0077
	7	0.1110	0.1612	0.2162	0.2668	0.3003	0.3020	0.2597	0.1722	0.0629
	8	0.0339	0.0605	0.1004	0.1556	0.2253	0.3020	0.3679	0.3874	0.2985
	9	0.0046	0.0101	0.0207	0.0404	0.0751	0.1342	0.2316	0.3874	0.6302
10	0	0.0003	0.0001							
	1	0.0042	0.0016	0.0005	0.0001					
	2	0.0229	0.0106	0.0043	0.0014	0.0004	0.0001			
	3	0.0746	0.0425	0.0212	0.0090	0.0031	0.0008	0.0001		
	4	0.1596	0.1115	0.0689	0.0368	0.0162	0.0055	0.0012	0.0001	
	5	0.2340	0.2007	0.1536	0.1029	0.0584	0.0264	0.0085	0.0015	0.0001
	6	0.2384	0.2508	0.2377	0.2001	0.1460	0.0881	0.0401	0.0112	0.0010
	7	0.1665	0.2150	0.2522	0.2668	0.2503	0.2013	0.1298	0.0574	0.0105
	8	0.0763	0.1209	0.1757	0.2335	0.2816	0.3020	0.2759	0.1937	0.0746
	9	0.0207	0.0403	0.0725	0.1211	0.1877	0.2684	0.3474	0.3874	0.3151
	10	0.0025	0.0060	0.0135	0.0282	0.0563	0.1074	0.1969	0.3487	0.5987
12	0	0.0001								
	1	0.0010	0.0003	0.0001						
	2	0.0068	0.0025	0.0008	0.0002					
	3	0.0277	0.0125	0.0048	0.0015	0.0004	0.0001			
	4	0.0762	0.0420	0.0199	0.0078	0.0024	0.0005	0.0001		
	5	0.1489	0.1009	0.0591	0.0291	0.0115	0.0033	0.0006		
	6	0.2124	0.1766	0.1281	0.0792	0.0401	0.0155	0.0040	0.0005	
	7	0.2225	0.2270	0.2039	0.1585	0.1032	0.0532	0.0193	0.0038	0.0002
	8	0.1700	0.2128	0.2367	0.2311	0.1936	0.1329	0.0683	0.0213	0.0021
	9	0.0923	0.1419	0.1954	0.2397	0.2581	0.2362	0.1720	0.0852	0.0173
	10	0.0339	0.0639	0.1088	0.1678	0.2323	0.2835	0.2924	0.2301	0.0988
	11	0.0075	0.0174	0.0368	0.0712	0.1267	0.2062	0.3012	0.3766	0.3413
	12	0.0008	0.0022	0.0057	0.0138	0.0317	0.0687	0.1422	0.2824	0.5404
15	0									
	1	0.0001								
	2	0.0010	0.0003	0.0001						
	3	0.0052	0.0016	0.0004	0.0001					
	4	0.0191	0.0074	0.0024	0.0006	0.0001				
	5	0.0515	0.0245	0.0096	0.0030	0.0007	0.0001			
	6	0.1048	0.0612	0.0298	0.0116	0.0034	0.0007	0.0001		
	7	0.1647	0.1181	0.0710	0.0348	0.0131	0.0035	0.0005		
	8	0.2013	0.1771	0.1319	0.0811	0.0393	0.0138	0.0030	0.0003	
	9	0.1914	0.2066	0.1906	0.1472	0.0917	0.0430	0.0132	0.0019	
	10	0.1404	0.1859	0.2123	0.2061	0.1651	0.1032	0.0449	0.0105	0.0006
	11	0.0780	0.1268	0.1792	0.2186	0.2252	0.1876	0.1156	0.0428	0.0049

Note: Blank entries indicate a binomial probability of less than 0.00005.

(Continued)

Table B Binomial distribution (*continued*)

n	X	0.55	0.60	0.65	0.70	0.75	0.80	0.85	0.90	0.95
	12	0.0318	0.0634	0.1110	0.1700	0.2252	0.2501	0.2184	0.1285	0.0307
	13	0.0090	0.0219	0.0476	0.0916	0.1559	0.2309	0.2856	0.2669	0.1348
	14	0.0016	0.0047	0.0126	0.0305	0.0668	0.1319	0.2312	0.3432	0.3658
	15	0.0001	0.0005	0.0016	0.0047	0.0134	0.0352	0.0874	0.2059	0.4633
18	0									
	1									
	2	0.0001								
	3	0.0009	0.0002							
	4	0.0039	0.0011	0.0002						
	5	0.0134	0.0045	0.0012	0.0002					
	6	0.0354	0.0145	0.0047	0.0012	0.0002				
	7	0.0742	0.0374	0.0151	0.0046	0.0010	0.0001			
	8	0.1248	0.0771	0.0385	0.0149	0.0042	0.0008	0.0001		
	9	0.1694	0.1284	0.0794	0.0386	0.0139	0.0033	0.0004		
	10	0.1864	0.1734	0.1327	0.0811	0.0376	0.0120	0.0022	0.0002	
	11	0.1657	0.1892	0.1792	0.1376	0.0820	0.0350	0.0091	0.0010	
	12	0.1181	0.1655	0.1941	0.1873	0.1436	0.0816	0.0301	0.0052	0.0002
	13	0.0666	0.1146	0.1664	0.2017	0.1988	0.1507	0.0787	0.0218	0.0014
	14	0.0291	0.0614	0.1104	0.1681	0.2130	0.2153	0.1592	0.0700	0.0093
	15	0.0095	0.0246	0.0547	0.1046	0.1704	0.2297	0.2406	0.1680	0.0473
	16	0.0022	0.0069	0.0190	0.0458	0.0958	0.1723	0.2556	0.2835	0.1683
	17	0.0003	0.0012	0.0042	0.0126	0.0338	0.0811	0.1704	0.3002	0.3763
	18		0.0001	0.0004	0.0016	0.0056	0.0180	0.0536	0.1501	0.3972
20	0									
	1									
	2									
	3	0.0002								
	4	0.0013	0.0003							
	5	0.0049	0.0013	0.0003						
	6	0.0150	0.0049	0.0012	0.0002					
	7	0.0366	0.0146	0.0045	0.0010	0.0002				
	8	0.0727	0.0355	0.0136	0.0039	0.0008	0.0001			
	9	0.1185	0.0710	0.0336	0.0120	0.0030	0.0005			
	10	0.1593	0.1171	0.0686	0.0308	0.0099	0.0020	0.0002		
	11	0.1771	0.1597	0.1158	0.0654	0.0271	0.0074	0.0011	0.0001	
	12	0.1623	0.1797	0.1614	0.1144	0.0609	0.0222	0.0046	0.0004	
	13	0.1221	0.1659	0.1844	0.1643	0.1124	0.0545	0.0160	0.0020	
	14	0.0746	0.1244	0.1712	0.1916	0.1686	0.1091	0.0454	0.0089	0.0003
	15	0.0365	0.0746	0.1272	0.1789	0.2023	0.1746	0.1028	0.0319	0.0022
	16	0.0139	0.0350	0.0738	0.1304	0.1897	0.2182	0.1821	0.0898	0.0133
	17	0.0040	0.0123	0.0323	0.0716	0.1339	0.2054	0.2428	0.1901	0.0596
	18	0.0008	0.0031	0.0100	0.0278	0.0669	0.1369	0.2293	0.2852	0.1887
	19	0.0001	0.0005	0.0020	0.0068	0.0211	0.0576	0.1368	0.2702	0.3774
	20			0.0002	0.0008	0.0032	0.0115	0.0388	0.1216	0.3585

Note: Blank entries indicate a binomial probability of less than 0.00005.

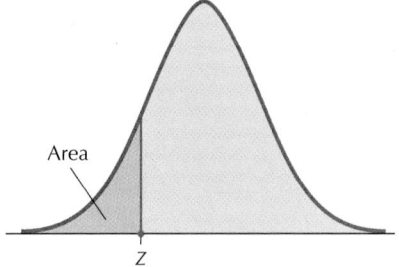

Area

Z

Table C Standard normal distribution

Z	0.00	0.01	0.02	0.03	0.04	0.05	0.06	0.07	0.08	0.09
-3.4	0.0003	0.0003	0.0003	0.0003	0.0003	0.0003	0.0003	0.0003	0.0003	0.0002
-3.3	0.0005	0.0005	0.0005	0.0004	0.0004	0.0004	0.0004	0.0004	0.0004	0.0003
-3.2	0.0007	0.0007	0.0006	0.0006	0.0006	0.0006	0.0006	0.0005	0.0005	0.0005
-3.1	0.0010	0.0009	0.0009	0.0009	0.0008	0.0008	0.0008	0.0008	0.0007	0.0007
-3.0	0.0013	0.0013	0.0013	0.0012	0.0012	0.0011	0.0011	0.0011	0.0010	0.0010
-2.9	0.0019	0.0018	0.0018	0.0017	0.0016	0.0016	0.0015	0.0015	0.0014	0.0014
-2.8	0.0026	0.0025	0.0024	0.0023	0.0023	0.0022	0.0021	0.0021	0.0020	0.0019
-2.7	0.0035	0.0034	0.0033	0.0032	0.0031	0.0030	0.0029	0.0028	0.0027	0.0026
-2.6	0.0047	0.0045	0.0044	0.0043	0.0041	0.0040	0.0039	0.0038	0.0037	0.0036
-2.5	0.0062	0.0060	0.0059	0.0057	0.0055	0.0054	0.0052	0.0051	0.0049	0.0048
-2.4	0.0082	0.0080	0.0078	0.0075	0.0073	0.0071	0.0069	0.0068	0.0066	0.0064
-2.3	0.0107	0.0104	0.0102	0.0099	0.0096	0.0094	0.0091	0.0089	0.0087	0.0084
-2.2	0.0139	0.0136	0.0132	0.0129	0.0125	0.0122	0.0119	0.0116	0.0113	0.0110
-2.1	0.0179	0.0174	0.0170	0.0166	0.0162	0.0158	0.0154	0.0150	0.0146	0.0143
-2.0	0.0228	0.0222	0.0217	0.0212	0.0207	0.0202	0.0197	0.0192	0.0188	0.0183
-1.9	0.0287	0.0281	0.0274	0.0268	0.0262	0.0256	0.0250	0.0244	0.0239	0.0233
-1.8	0.0359	0.0351	0.0344	0.0336	0.0329	0.0322	0.0314	0.0307	0.0301	0.0294
-1.7	0.0446	0.0436	0.0427	0.0418	0.0409	0.0401	0.0392	0.0384	0.0375	0.0367
-1.6	0.0548	0.0537	0.0526	0.0516	0.0505	0.0495	0.0485	0.0475	0.0465	0.0455
-1.5	0.0668	0.0655	0.0643	0.0630	0.0618	0.0606	0.0594	0.0582	0.0571	0.0559
-1.4	0.0808	0.0793	0.0778	0.0764	0.0749	0.0735	0.0721	0.0708	0.0694	0.0681
-1.3	0.0968	0.0951	0.0934	0.0918	0.0901	0.0885	0.0869	0.0853	0.0838	0.0823
-1.2	0.1151	0.1131	0.1112	0.1093	0.1075	0.1056	0.1038	0.1020	0.1003	0.0985
-1.1	0.1357	0.1335	0.1314	0.1292	0.1271	0.1251	0.1230	0.1210	0.1190	0.1170
-1.0	0.1587	0.1562	0.1539	0.1515	0.1492	0.1469	0.1446	0.1423	0.1401	0.1379
-0.9	0.1841	0.1814	0.1788	0.1762	0.1736	0.1711	0.1685	0.1660	0.1635	0.1611
-0.8	0.2119	0.2090	0.2061	0.2033	0.2005	0.1977	0.1949	0.1922	0.1894	0.1867
-0.7	0.2420	0.2389	0.2358	0.2327	0.2296	0.2266	0.2236	0.2206	0.2177	0.2148
-0.6	0.2743	0.2709	0.2676	0.2643	0.2611	0.2578	0.2546	0.2514	0.2483	0.2451
-0.5	0.3085	0.3050	0.3015	0.2981	0.2946	0.2912	0.2877	0.2843	0.2810	0.2776
-0.4	0.3446	0.3409	0.3372	0.3336	0.3300	0.3264	0.3228	0.3192	0.3156	0.3121
-0.3	0.3821	0.3783	0.3745	0.3707	0.3669	0.3632	0.3594	0.3557	0.3520	0.3483
-0.2	0.4207	0.4168	0.4129	0.4090	0.4052	0.4013	0.3974	0.3936	0.3897	0.3859
-0.1	0.4602	0.4562	0.4522	0.4483	0.4443	0.4404	0.4364	0.4325	0.4286	0.4247
-0.0	0.5000	0.4960	0.4920	0.4880	0.4840	0.4801	0.4761	0.4721	0.4681	0.4641

(Continued)

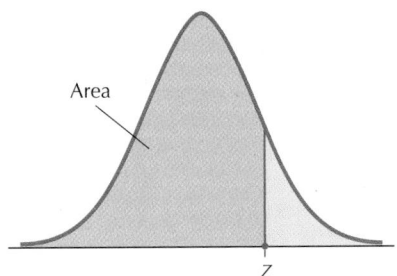

Area

Z

Table C Standard normal distribution (*continued*)

Z	0.00	0.01	0.02	0.03	0.04	0.05	0.06	0.07	0.08	0.09
0.0	0.5000	0.5040	0.5080	0.5120	0.5160	0.5199	0.5239	0.5279	0.5319	0.5359
0.1	0.5398	0.5438	0.5478	0.5517	0.5557	0.5596	0.5636	0.5675	0.5714	0.5753
0.2	0.5793	0.5832	0.5871	0.5910	0.5948	0.5987	0.6026	0.6064	0.6103	0.6141
0.3	0.6179	0.6217	0.6255	0.6293	0.6331	0.6368	0.6406	0.6443	0.6480	0.6517
0.4	0.6554	0.6591	0.6628	0.6664	0.6700	0.6736	0.6772	0.6808	0.6844	0.6879
0.5	0.6915	0.6950	0.6985	0.7019	0.7054	0.7088	0.7123	0.7157	0.7190	0.7224
0.6	0.7257	0.7291	0.7324	0.7357	0.7389	0.7422	0.7454	0.7486	0.7517	0.7549
0.7	0.7580	0.7611	0.7642	0.7673	0.7704	0.7734	0.7764	0.7794	0.7823	0.7852
0.8	0.7881	0.7910	0.7939	0.7967	0.7995	0.8023	0.8051	0.8078	0.8106	0.8133
0.9	0.8159	0.8186	0.8212	0.8238	0.8264	0.8289	0.8315	0.8340	0.8365	0.8389
1.0	0.8413	0.8438	0.8461	0.8485	0.8508	0.8531	0.8554	0.8577	0.8599	0.8621
1.1	0.8643	0.8665	0.8686	0.8708	0.8729	0.8749	0.8770	0.8790	0.8810	0.8830
1.2	0.8849	0.8869	0.8888	0.8907	0.8925	0.8944	0.8962	0.8980	0.8997	0.9015
1.3	0.9032	0.9049	0.9066	0.9082	0.9099	0.9115	0.9131	0.9147	0.9162	0.9177
1.4	0.9192	0.9207	0.9222	0.9236	0.9251	0.9265	0.9279	0.9292	0.9306	0.9319
1.5	0.9332	0.9345	0.9357	0.9370	0.9382	0.9394	0.9406	0.9418	0.9429	0.9441
1.6	0.9452	0.9463	0.9474	0.9484	0.9495	0.9505	0.9515	0.9525	0.9535	0.9545
1.7	0.9554	0.9564	0.9573	0.9582	0.9591	0.9599	0.9608	0.9616	0.9625	0.9633
1.8	0.9641	0.9649	0.9656	0.9664	0.9671	0.9678	0.9686	0.9693	0.9699	0.9706
1.9	0.9713	0.9719	0.9726	0.9732	0.9738	0.9744	0.9750	0.9756	0.9761	0.9767
2.0	0.9772	0.9778	0.9783	0.9788	0.9793	0.9798	0.9803	0.9808	0.9812	0.9817
2.1	0.9821	0.9826	0.9830	0.9834	0.9838	0.9842	0.9846	0.9850	0.9854	0.9857
2.2	0.9861	0.9864	0.9868	0.9871	0.9875	0.9878	0.9881	0.9884	0.9887	0.9890
2.3	0.9893	0.9896	0.9898	0.9901	0.9904	0.9906	0.9909	0.9911	0.9913	0.9916
2.4	0.9918	0.9920	0.9922	0.9925	0.9927	0.9929	0.9931	0.9932	0.9934	0.9936
2.5	0.9938	0.9940	0.9941	0.9943	0.9945	0.9946	0.9948	0.9949	0.9951	0.9952
2.6	0.9953	0.9955	0.9956	0.9957	0.9959	0.9960	0.9961	0.9962	0.9963	0.9964
2.7	0.9965	0.9966	0.9967	0.9968	0.9969	0.9970	0.9971	0.9972	0.9973	0.9974
2.8	0.9974	0.9975	0.9976	0.9977	0.9977	0.9978	0.9979	0.9979	0.9980	0.9981
2.9	0.9981	0.9982	0.9982	0.9983	0.9984	0.9984	0.9985	0.9985	0.9986	0.9986
3.0	0.9987	0.9987	0.9987	0.9988	0.9988	0.9989	0.9989	0.9989	0.9990	0.9990
3.1	0.9990	0.9991	0.9991	0.9991	0.9992	0.9992	0.9992	0.9992	0.9993	0.9993
3.2	0.9993	0.9993	0.9994	0.9994	0.9994	0.9994	0.9994	0.9995	0.9995	0.9995
3.3	0.9995	0.9995	0.9995	0.9996	0.9996	0.9996	0.9996	0.9996	0.9996	0.9997
3.4	0.9997	0.9997	0.9997	0.9997	0.9997	0.9997	0.9997	0.9997	0.9997	0.9998

Table D *t*-Distribution

		Confidence level				
		80%	90%	95%	98%	99%
		Area in one tail				
		0.10	0.05	0.025	0.01	0.005
		Area in two tails				
		0.20	0.10	0.05	0.02	0.01
df	1	3.078	6.314	12.706	31.821	63.657
	2	1.886	2.920	4.303	6.965	9.925
	3	1.638	2.353	3.182	4.541	5.841
	4	1.533	2.132	2.776	3.747	4.604
	5	1.476	2.015	2.571	3.365	4.032
	6	1.440	1.943	2.447	3.143	3.707
	7	1.415	1.895	2.365	2.998	3.499
	8	1.397	1.860	2.306	2.896	3.355
	9	1.383	1.833	2.262	2.821	3.250
	10	1.372	1.812	2.228	2.764	3.169
	11	1.363	1.796	2.201	2.718	3.106
	12	1.356	1.782	2.179	2.681	3.055
	13	1.350	1.771	2.160	2.650	3.012
	14	1.345	1.761	2.145	2.624	2.977
	15	1.341	1.753	2.131	2.602	2.947
	16	1.337	1.746	2.120	2.583	2.921
	17	1.333	1.740	2.110	2.567	2.898
	18	1.330	1.734	2.101	2.552	2.878
	19	1.328	1.729	2.093	2.539	2.861
	20	1.325	1.725	2.086	2.528	2.845
	21	1.323	1.721	2.080	2.518	2.831
	22	1.321	1.717	2.074	2.508	2.819
	23	1.319	1.714	2.069	2.500	2.807
	24	1.318	1.711	2.064	2.492	2.797
	25	1.316	1.708	2.060	2.485	2.787
	26	1.315	1.706	2.056	2.479	2.779
	27	1.314	1.703	2.052	2.473	2.771
	28	1.313	1.701	2.048	2.467	2.763
	29	1.311	1.699	2.045	2.462	2.756
	30	1.310	1.697	2.042	2.457	2.750
	31	1.309	1.696	2.040	2.453	2.744
	32	1.309	1.694	2.037	2.449	2.738
	33	1.308	1.692	2.035	2.445	2.733
	34	1.307	1.691	2.032	2.441	2.728
	35	1.306	1.690	2.030	2.438	2.724
	36	1.306	1.688	2.028	2.435	2.719
	37	1.305	1.687	2.026	2.431	2.715
	38	1.304	1.686	2.024	2.429	2.712
	39	1.304	1.685	2.023	2.426	2.708
	40	1.303	1.684	2.021	2.423	2.704
	50	1.299	1.676	2.009	2.403	2.678
	60	1.296	1.671	2.000	2.390	2.660
	70	1.294	1.667	1.994	2.381	2.648
	80	1.292	1.664	1.990	2.374	2.639
	90	1.291	1.662	1.987	2.368	2.632
	100	1.290	1.660	1.984	2.364	2.626
	1000	1.282	1.646	1.962	2.330	2.581
	z	1.282	1.645	1.960	2.326	2.576

Table E Chi-square (χ^2) distribution

	Area to the right of critical value									
Degrees of freedom	**0.995**	**0.99**	**0.975**	**0.95**	**0.90**	**0.10**	**0.05**	**0.025**	**0.01**	**0.005**
1	—	—	0.001	0.004	0.016	2.706	3.841	5.024	6.635	7.879
2	0.010	0.020	0.051	0.103	0.211	4.605	5.991	7.378	9.210	10.597
3	0.072	0.115	0.216	0.352	0.584	6.251	7.815	9.348	11.345	12.838
4	0.207	0.297	0.484	0.711	1.064	7.779	9.488	11.143	13.277	14.860
5	0.412	0.554	0.831	1.145	1.610	9.236	11.071	12.833	15.086	16.750
6	0.676	0.872	1.237	1.635	2.204	10.645	12.592	14.449	16.812	18.548
7	0.989	1.239	1.690	2.167	2.833	12.017	14.067	16.013	18.475	20.278
8	1.344	1.646	2.180	2.733	3.490	13.362	15.507	17.535	20.090	21.955
9	1.735	2.088	2.700	3.325	4.168	14.684	16.919	19.023	21.666	23.589
10	2.156	2.558	3.247	3.940	4.865	15.987	18.307	20.483	23.209	25.188
11	2.603	3.053	3.816	4.575	5.578	17.275	19.675	21.920	24.725	26.757
12	3.074	3.571	4.404	5.226	6.304	18.549	21.026	23.337	26.217	28.299
13	3.565	4.107	5.009	5.892	7.042	19.812	22.362	24.736	27.688	29.819
14	4.075	4.660	5.629	6.571	7.790	21.064	23.685	26.119	29.141	31.319
15	4.601	5.229	6.262	7.261	8.547	22.307	24.996	27.488	30.578	32.801
16	5.142	5.812	6.908	7.962	9.312	23.542	26.296	28.845	32.000	34.267
17	5.697	6.408	7.564	8.672	10.085	24.769	27.587	30.191	33.409	35.718
18	6.265	7.015	8.231	9.390	10.865	25.989	28.869	31.526	34.805	37.156
19	6.844	7.633	8.907	10.117	11.651	27.204	30.144	32.852	36.191	38.582
20	7.434	8.260	9.591	10.851	12.443	28.412	31.410	34.170	37.566	39.997
21	8.034	8.897	10.283	11.591	13.240	29.615	32.671	35.479	38.932	41.401
22	8.643	9.542	10.982	12.338	14.042	30.813	33.924	36.781	40.289	42.796
23	9.260	10.196	11.689	13.091	14.848	32.007	35.172	38.076	41.638	44.181
24	9.886	10.856	12.401	13.848	15.659	33.196	36.415	39.364	42.980	45.559
25	10.520	11.524	13.120	14.611	16.473	34.382	37.652	40.646	44.314	46.928
26	11.160	12.198	13.844	15.379	17.292	35.563	38.885	41.923	45.642	48.290
27	11.808	12.879	14.573	16.151	18.114	36.741	40.113	43.194	46.963	49.645
28	12.461	13.565	15.308	16.928	18.939	37.916	41.337	44.461	48.278	50.993
29	13.121	14.257	16.047	17.708	19.768	39.087	42.557	45.722	49.588	52.336
30	13.787	14.954	16.791	18.493	20.599	40.256	43.773	46.979	50.892	53.672
40	20.707	22.164	24.433	26.509	29.051	51.805	55.758	59.342	63.691	66.766
50	27.991	29.707	32.357	34.764	37.689	63.167	67.505	71.420	76.154	79.490
60	35.534	37.485	40.482	43.188	46.459	74.397	79.082	83.298	88.379	91.952
70	43.275	45.442	48.758	51.739	55.329	85.527	90.531	95.023	100.425	104.215
80	51.172	53.540	57.153	60.391	64.278	96.578	101.879	106.629	112.329	116.321
90	59.196	61.754	65.647	69.126	73.291	107.565	113.145	118.136	124.116	128.299
100	67.328	70.065	74.222	77.929	82.358	118.498	124.342	129.561	135.807	140.169

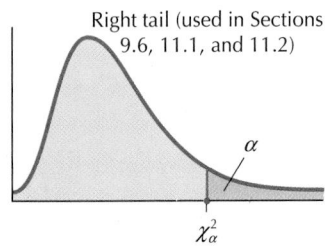

Right tail (used in Sections 9.6, 11.1, and 11.2)

α

χ^2_α

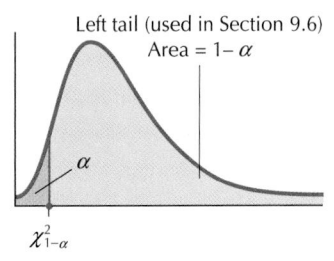

Left tail (used in Section 9.6)

Area = $1 - \alpha$

α

$\chi^2_{1-\alpha}$

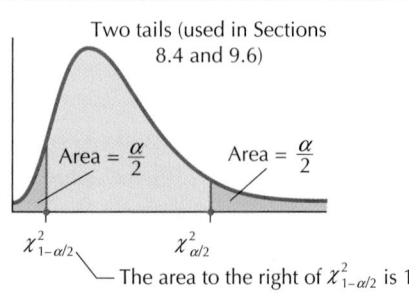

Two tails (used in Sections 8.4 and 9.6)

Area = $\dfrac{\alpha}{2}$ Area = $\dfrac{\alpha}{2}$

$\chi^2_{1-\alpha/2}$ $\chi^2_{\alpha/2}$

The area to the right of $\chi^2_{1-\alpha/2}$ is $1 - \dfrac{\alpha}{2}$.

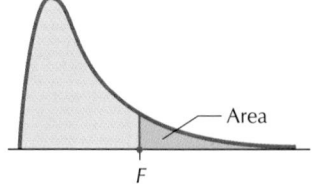
Area
F

Table F *F*-Distribution critical values

	Area in right tail	df₁							
		1	**2**	**3**	**4**	**5**	**6**	**7**	**8**
1	0.100	39.86	49.59	53.59	55.83	57.24	58.20	58.91	59.44
	0.050	161.45	199.50	215.71	224.58	230.16	233.99	236.77	238.88
	0.025	647.79	799.50	864.16	899.58	921.85	937.11	948.22	956.66
	0.010	4052.20	4999.50	5403.40	5624.60	5763.60	5859.00	5928.40	5981.10
	0.001	405284.00	500000.00	540379.00	562500.00	576405.00	585937.00	592873.00	598144.00
2	0.100	8.53	9.00	9.16	9.24	9.29	9.33	9.35	9.37
	0.050	18.51	19.00	19.16	19.25	19.30	19.33	19.35	19.37
	0.025	38.51	39.00	39.17	39.25	39.30	39.33	39.36	39.37
	0.010	98.50	99.00	99.17	99.25	99.30	99.33	99.36	99.37
	0.001	998.50	999.00	999.17	999.25	999.30	999.33	999.36	999.37
3	0.100	5.54	5.46	5.39	5.34	5.31	5.28	5.27	5.25
	0.050	10.13	9.55	9.28	9.12	9.01	8.94	8.89	8.85
	0.025	17.44	16.04	15.44	15.10	14.88	14.73	14.62	14.54
	0.010	34.12	30.82	29.46	28.71	28.24	27.91	27.67	27.49
	0.001	167.03	148.50	141.11	137.10	134.58	132.85	131.58	130.62
4	0.100	4.54	4.32	4.19	4.11	4.05	4.01	3.98	3.95
	0.050	7.71	6.94	6.59	6.39	6.26	6.16	6.09	6.04
	0.025	12.22	10.65	9.98	9.60	9.36	9.20	9.07	8.98
	0.010	21.20	18.00	16.69	15.98	15.52	15.21	14.98	14.80
	0.001	74.14	61.25	56.18	53.44	51.71	50.53	49.66	49.00
5	0.100	4.06	3.78	3.62	3.52	3.45	3.40	3.37	3.34
	0.050	6.61	5.79	5.41	5.19	5.05	4.95	4.88	4.82
	0.025	10.01	8.43	7.76	7.39	7.15	6.98	6.85	6.76
	0.010	16.26	13.27	12.06	11.39	10.97	10.67	10.46	10.29
	0.001	47.18	37.12	33.20	31.09	29.75	28.83	28.16	27.65
6	0.100	3.78	3.46	3.29	3.18	3.11	3.05	3.01	2.98
	0.050	5.99	5.14	4.76	4.53	4.39	4.28	4.21	4.15
	0.025	8.81	7.26	6.60	6.23	5.99 '	5.82	5.70	5.60
	0.010	13.75	10.92	9.78	9.15	8.75	8.47	8.26	8.10
	0.001	35.51	27.00	23.70	21.92	20.80	20.03	19.46	19.03
7	0.100	3.59	3.26	3.07	2.96	2.88	2.83	2.78	2.75
	0.050	5.59	4.74	4.35	4.12	3.97	3.87	3.79	3.73
	0.025	8.07	6.54	5.89	5.52	5.29	5.12	4.99	4.90
	0.010	12.25	9.55	8.45	7.85	7.46	7.19	6.99	6.84
	0.001	29.25	21.69	18.77	17.20	16.21	15.52	15.02	14.63
8	0.100	3.46	3.11	2.92	2.81	2.73	2.67	2.62	2.59
	0.050	5.32	4.46	4.07	3.84	3.69	3.58	3.50	3.44
	0.025	7.57	6.06	5.42	5.05	4.82	4.65	4.53	4.43
	0.010	11.26	8.65	7.59	7.01	6.63	6.37	6.18	6.03
	0.001	25.41	18.49	15.83	14.39	13.48	12.86	12.40	12.05

df₂

(Continued)

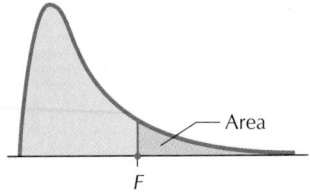

Area

F

Table F F-Distribution critical values (continued)

					df$_1$				
	Area in right tail	**9**	**10**	**15**	**20**	**30**	**60**	**120**	**1000**
1	0.100	59.86	60.19	61.22	61.74	62.26	62.79	63.06	63.30
	0.050	240.54	241.88	245.95	248.01	250.10	252.20	253.25	254.19
	0.025	963.28	968.63	984.87	993.10	1001.4	1009.8	1014.0	1017.7
	0.010	6022.5	6055.8	6157.3	6208.7	6260.6	6313.0	6339.4	6362.7
	0.001	602284.0	605621.0	615764.0	620908.0	626099.0	631337.0	633972.0	636301.0
2	0.100	9.38	9.39	9.42	9.44	9.16	9.47	9.48	9.49
	0.050	19.38	19.40	19.43	19.45	19.46	19.48	19.49	19.49
	0.025	39.39	39.40	39.43	39.45	39.46	39.48	39.49	39.50
	0.010	99.39	99.40	99.43	99.45	99.47	99.48	99.49	99.50
	0.001	999.39	999.40	999.43	999.45	999.47	999.48	999.49	999.50
3	0.100	5.24	5.23	5.20	5.18	5.17	5.15	5.14	5.13
	0.050	8.81	8.79	8.70	8.66	8.62	8.57	8.55	8.53
	0.025	14.47	14.42	14.25	14.17	14.08	13.99	13.95	13.91
	0.010	27.35	27.23	26.87	26.69	26.50	26.32	26.22	26.14
	0.001	129.86	129.25	127.37	126.42	125.45	124.47	123.97	123.53
4	0.100	3.94	3.92	3.87	3.84	3.82	3.79	3.78	3.76
	0.050	6.00	5.96	5.86	5.80	5.75	5.69	5.66	5.63
	0.025	8.90	8.84	8.66	8.56	8.46	8.36	8.31	8.26
	0.010	14.66	14.55	14.20	14.02	13.84	13.65	13.56	13.47
	0.001	48.47	48.05	46.76	46.10	45.43	44.75	44.40	44.09
5	0.100	3.32	3.30	3.24	3.21	3.17	3.14	3.12	3.11
	0.050	4.77	4.74	4.62	4.56	4.50	4.43	4.40	4.37
	0.025	6.68	6.62	6.43	6.33	6.23	6.12	6.07	6.02
	0.010	10.16	10.05	9.72	9.55	9.38	9.20	9.11	9.03
	0.001	27.24	26.92	25.91	25.39	24.87	24.33	24.06	23.82
6	0.100	2.96	2.94	2.87	2.84	2.80	2.76	2.74	2.72
	0.050	4.10	4.06	3.94	3.87	3.81	3.74	3.70	3.67
	0.025	5.52	5.46	5.27	5.17	5.07	4.96	4.90	4.86
	0.010	7.98	7.87	7.56	7.40	7.23	7.06	6.97	6.89
	0.001	18.69	18.41	17.56	17.12	16.67	16.21	15.98	15.77
7	0.100	2.72	2.70	2.63	2.59	2.56	2.51	2.49	2.47
	0.050	3.68	3.64	3.51	3.44	3.38	3.30	3.27	3.23
	0.025	4.82	4.76	4.57	4.47	4.36	4.25	4.20	4.15
	0.010	6.72	6.62	6.31	6.16	5.99	5.82	5.74	5.66
	0.001	14.33	14.08	13.32	12.93	12.53	12.12	11.91	11.72
8	0.100	2.56	2.54	2.46	2.42	2.38	2.34	2.32	2.30
	0.050	3.39	3.35	3.22	3.15	3.08	3.01	2.97	2.93
	0.025	4.36	4.30	4.10	4.00	3.89	3.78	3.73	3.68
	0.010	5.91	5.81	5.52	5.36	5.20	5.03	4.95	4.87
	0.001	11.77	11.54	10.84	10.48	10.11	9.73	9.53	9.36

df$_2$

Table F *F*-Distribution critical values (*continued*)

df_2	Area in right tail	df_1 = 1	2	3	4	5	6	7	8	9	10
	0.100	3.36	3.01	2.81	2.69	2.61	2.55	2.51	2.47	2.44	2.42
	0.050	5.12	4.26	3.86	3.63	3.48	3.37	3.29	3.23	3.18	3.14
9	0.025	7.21	5.71	5.08	4.72	4.48	4.32	4.20	4.10	4.03	3.96
	0.010	10.56	8.02	6.99	6.42	6.06	5.80	5.61	5.47	5.35	5.26
	0.001	22.86	16.39	13.90	12.56	11.71	11.13	10.70	10.37	10.11	9.89
	0.100	3.29	2.92	2.73	2.61	2.52	2.46	2.41	2.38	2.35	2.32
	0.050	4.96	4.10	3.71	3.48	3.33	3.22	3.14	3.07	3.02	2.98
10	0.025	6.94	5.46	4.83	4.47	4.24	4.07	3.95	3.85	3.78	3.72
	0.010	10.04	7.56	6.55	5.99	5.64	5.39	5.20	5.06	4.94	4.85
	0.001	21.04	14.91	12.55	11.28	10.48	9.93	9.52	9.20	8.96	8.75
	0.100	3.18	2.81	2.61	2.48	2.39	2.33	2.28	2.24	2.21	2.19
	0.050	4.75	3.89	3.49	3.26	3.11	3.00	2.91	2.85	2.80	2.75
12	0.025	6.55	5.10	4.47	4.12	3.89	3.73	3.61	3.51	3.44	3.37
	0.010	9.33	6.93	5.95	5.41	5.06	4.82	4.64	4.50	4.39	4.30
	0.001	18.64	12.97	10.80	9.63	8.89	8.38	8.00	7.71	7.48	7.29
	0.100	3.07	2.70	2.49	2.36	2.27	2.21	2.16	2.12	2.09	2.06
	0.050	4.54	3.68	3.29	3.06	2.90	2.79	2.71	2.64	2.59	2.54
15	0.025	6.20	4.77	4.15	3.80	3.58	3.41	3.29	3.20	3.12	3.06
	0.010	8.68	6.36	5.42	4.89	4.56	4.32	4.14	4.00	3.89	3.80
	0.001	16.59	11.34	9.34	8.25	7.57	7.09	6.74	6.47	6.26	6.08
	0.100	2.97	2.59	2.38	2.25	2.16	2.09	2.04	2.00	1.96	1.94
	0.050	4.35	3.49	3.10	2.87	2.71	2.60	2.51	2.45	2.39	2.35
20	0.025	5.87	4.46	3.86	3.51	3.29	3.13	3.01	2.91	2.84	2.77
	0.010	8.10	5.85	4.94	4.43	4.10	3.87	3.70	3.56	3.46	3.37
	0.001	14.82	9.95	8.10	7.10	6.46	6.02	5.69	5.44	5.24	5.08
	0.100	2.92	2.53	2.32	2.18	2.09	2.02	1.97	1.93	1.89	1.87
	0.050	4.24	3.39	2.99	2.76	2.60	2.49	2.40	2.34	2.28	2.24
25	0.025	5.69	4.29	3.69	3.35	3.13	2.97	2.85	2.75	2.68	2.61
	0.010	7.77	5.57	4.68	4.18	3.85	3.63	3.46	3.32	3.22	3.13
	0.001	13.88	9.22	7.45	6.49	5.89	5.46	5.15	4.91	4.71	4.56
	0.100	2.81	2.41	2.20	2.06	1.97	1.90	1.84	1.80	1.76	1.73
	0.050	4.03	3.18	2.79	2.56	2.40	2.29	2.20	2.13	2.07	2.03
50	0.025	5.34	3.97	3.39	3.05	2.83	2.67	2.55	2.46	2.38	2.32
	0.010	7.17	5.06	4.20	3.72	3.41	3.19	3.02	2.89	2.78	2.70
	0.001	12.22	7.96	6.34	5.46	4.90	4.51	4.22	4.00	3.82	3.67
	0.100	2.76	2.36	2.14	2.00	1.91	1.83	1.78	1.73	1.69	1.66
	0.050	3.94	3.09	2.70	2.46	2.31	2.19	2.10	2.03	1.97	1.93
100	0.025	5.18	3.83	3.25	2.92	2.70	2.54	2.42	2.32	2.24	2.18
	0.010	6.90	4.82	3.98	3.51	3.21	2.99	2.82	2.69	2.59	2.50
	0.001	11.50	7.41	5.86	5.02	4.48	4.11	3.83	3.61	3.44	3.30
	0.100	2.73	2.33	2.11	1.97	1.88	1.80	1.75	1.70	1.66	1.63
	0.050	3.89	3.04	2.65	2.42	2.26	2.14	2.06	1.98	1.93	1.88
200	0.025	5.10	3.76	3.18	2.85	2.63	2.47	2.35	2.26	2.18	2.11
	0.010	6.76	4.71	3.88	3.41	3.11	2.89	2.73	2.60	2.50	2.41
	0.001	11.15	7.15	5.63	4.81	4.29	3.92	3.65	3.43	3.26	3.12
	0.100	2.71	2.31	2.09	1.95	1.85	1.78	1.72	1.68	1.64	1.61
	0.050	3.85	3.00	2.61	2.38	2.22	2.11	2.02	1.95	1.89	1.84
1000	0.025	5.04	3.70	3.13	2.80	2.58	2.42	2.30	2.20	2.13	2.06
	0.010	6.66	4.63	3.80	3.34	3.04	2.82	2.66	2.53	2.43	2.34
	0.001	10.89	6.96	5.46	4.65	4.14	3.78	3.51	3.30	3.13	2.99

(Continued)

Table F F-Distribution critical values (*continued*)

					df$_1$					
	Area in right tail	12	15	20	25	30	40	50	60	120	1000
9	0.100	2.38	2.34	2.30	2.27	2.25	2.23	2.22	2.21	2.18	2.16
	0.050	3.07	3.01	2.94	2.89	2.86	2.83	2.80	2.79	2.75	2.71
	0.025	3.87	3.77	3.67	3.60	3.56	3.51	3.47	3.45	3.39	3.34
	0.010	5.11	4.96	4.81	4.71	4.65	4.57	4.52	4.48	4.40	4.32
	0.001	9.57	9.24	8.90	8.69	8.55	8.37	8.26	8.19	8.00	7.84
10	0.100	2.28	2.24	2.20	2.17	2.16	2.13	2.12	2.11	2.08	2.06
	0.050	2.91	2.85	2.77	2.73	2.70	2.66	2.64	2.62	2.58	2.54
	0.025	3.62	3.52	3.42	3.35	3.31	3.26	3.22	3.20	3.14	3.09
	0.010	4.71	4.56	4.41	4.31	4.25	4.17	4.12	4.08	4.00	3.92
	0.001	8.45	8.13	7.80	7.60	7.47	7.30	7.19	7.12	6.94	6.78
12	0.100	2.15	2.10	2.06	2.03	2.01	1.99	1.97	1.96	1.93	1.91
	0.050	2.69	2.62	2.54	2.50	2.47	2.43	2.40	2.38	2.34	2.30
	0.025	3.28	3.18	3.07	3.01	2.96	2.91	2.87	2.85	2.79	2.73
	0.010	4.16	4.01	3.86	3.76	3.70	3.62	3.57	3.54	3.45	3.37
	0.001	7.00	6.71	6.40	6.22	6.09	5.93	5.83	5.76	5.59	5.44
15	0.100	2.02	1.97	1.92	1.89	1.87	1.85	1.83	1.82	1.79	1.76
	0.050	2.48	2.40	2.33	2.28	2.25	2.20	2.18	2.16	2.11	2.07
	0.025	2.96	2.86	2.76	2.69	2.64	2.59	2.55	2.52	2.46	2.40
	0.010	3.67	3.52	3.37	3.28	3.21	3.13	3.08	3.05	2.96	2.88
	0.001	5.81	5.54	5.25	5.07	4.95	4.80	4.70	4.64	4.47	4.33
20	0.100	1.89	1.84	1.79	1.76	1.74	1.71	1.69	1.68	1.64	1.61
	0.050	2.28	2.20	2.12	2.07	2.04	1.99	1.97	1.95	1.90	1.85
	0.025	2.68	2.57	2.46	2.40	2.35	2.29	2.25	2.22	2.16	2.09
	0.010	3.23	3.09	2.94	2.84	2.78	2.69	2.64	2.61	2.52	2.43
	0.001	4.82	4.56	4.29	4.12	4.00	3.86	3.77	3.70	3.54	3.40
25	0.100	1.82	1.77	1.72	1.68	1.66	1.63	1.61	1.59	1.56	1.52
	0.050	2.16	2.09	2.01	1.96	1.92	1.87	1.84	1.82	1.77	1.72
	0.025	2.51	2.41	2.30	2.23	2.18	2.12	2.08	2.05	1.98	1.91
	0.010	2.99	2.85	2.70	2.60	2.54	2.45	2.40	2.36	2.27	2.18
	0.001	4.31	4.06	3.79	3.63	3.52	3.37	3.28	3.22	3.06	2.91
50	0.100	1.68	1.63	1.57	1.53	1.50	1.46	1.44	1.42	1.38	1.33
	0.050	1.95	1.87	1.78	1.73	1.69	1.63	1.60	1.58	1.51	1.45
	0.025	2.22	2.11	1.99	1.92	1.87	1.80	1.75	1.72	1.64	1.56
	0.010	2.56	2.42	2.27	2.17	2.10	2.01	1.95	1.91	1.80	1.70
	0.001	3.44	3.20	2.95	2.79	2.68	2.53	2.44	2.38	2.21	2.05
100	0.100	1.61	1.56	1.49	1.45	1.42	1.38	1.35	1.34	1.28	1.22
	0.050	1.85	1.77	1.68	1.62	1.57	1.52	1.48	1.45	1.38	1.30
	0.025	2.08	1.97	1.85	1.77	1.71	1.64	1.59	1.56	1.46	1.36
	0.010	2.37	2.22	2.07	1.97	1.89	1.80	1.74	1.69	1.57	1.45
	0.001	3.07	2.84	2.59	2.43	2.32	2.17	2.08	2.01	1.83	1.64
200	0.100	1.58	1.52	1.46	1.41	1.38	1.34	1.31	1.29	1.23	1.16
	0.050	1.80	1.72	1.62	1.56	1.52	1.46	1.41	1.39	1.30	1.21
	0.025	2.01	1.90	1.78	1.70	1.64	1.56	1.51	1.47	1.37	1.25
	0.010	2.27	2.13	1.97	1.87	1.79	1.69	1.63	1.58	1.45	1.30
	0.001	2.90	2.67	2.42	2.26	2.15	2.00	1.90	1.83	1.64	1.43
1000	0.100	1.55	1.49	1.43	1.38	1.35	1.30	1.27	1.25	1.38	1.08
	0.050	1.76	1.68	1.58	1.52	1.47	1.41	1.36	1.31	1.24	1.11
	0.025	1.96	1.85	1.72	1.64	1.58	1.50	1.45	1.41	1.29	1.13
	0.010	2.20	2.06	1.90	1.79	1.72	1.61	1.54	1.50	1.35	1.16
	0.001	2.77	2.54	2.30	2.14	2.02	1.87	1.77	1.69	1.49	1.22

df$_2$

Table G Critical values for Tukey's test

$\alpha_{EW} = 0.05$

df	k: 2	3	4	5	6	7	8	9	10	k: 11	12	13	14	15	16	17	18	19
1	17.97	26.98	32.82	37.08	40.41	43.12	15.40	47.36	19.07	50.59	51.96	53.20	54.33	55.36	56.32	57.22	58.04	58.83
2	6.085	8.331	9.798	10.88	11.74	12.44	13.03	13.54	13.99	14.39	14.75	15.08	15.38	15.65	15.91	16.14	16.37	16.37
3	4.501	5.910	6.825	7.502	8.037	8.478	8.853	9.177	9.462	9.717	9.946	10.15	10.35	10.53	10.69	10.84	10.98	11.11
4	3.927	5.040	5.757	6.287	6.707	7.053	7.347	7.602	7.826	8.027	8.208	8.373	8.525	8.664	8.794	8.914	9.028	9.134
5	3.635	4.602	5.218	5.673	6.033	6.330	6.582	6.802	6.995	7.168	7.324	7.466	7.596	7.717	7.828	7.932	8.030	8.122
6	3.461	4.339	4.896	5.305	5.628	5.895	6.122	6.319	6.493	6.649	6.789	6.917	7.034	7.143	7.244	7.338	7.426	7.508
7	3.344	4.165	4.681	5.060	5.359	5.606	5.815	5.998	6.158	6.302	6.431	6.550	6.658	6.759	6.852	6.939	7.020	7.097
8	3.261	4.041	4.529	4.886	5.167	5.399	5.597	5.767	5.918	6.054	6.175	6.287	6.389	6.483	6.571	6.653	6.729	6.802
9	3.199	3.949	4.415	4.756	5.024	5.244	5.432	5.595	5.739	5.867	5.983	6.089	6.186	6.276	6.359	6.437	6.510	6.579
10	3.151	3.877	4.327	4.654	4.912	5.124	5.305	5.461	5.599	5.722	5.833	5.935	6.028	6.114	6.194	6.269	6.339	6.405
11	3.113	3.820	4.256	4.574	4.823	5.028	5.202	5.353	5.487	5.605	5.713	5.811	5.901	5.984	6.062	6.134	6.202	6.265
12	3.082	3.773	4.199	4.508	4.751	4.950	5.119	5.265	5.395	5.511	5.615	5.710	5.798	5.878	5.953	6.023	6.089	6.151
13	3.055	3.735	4.151	4.453	4.690	4.885	5.049	5.192	5.318	5.431	5.533	5.625	5.711	5.789	5.862	5.931	5.995	6.055
14	3.033	3.702	4.111	4.407	4.639	4.829	4.990	5.131	5.254	5.364	5.463	5.554	5.637	5.714	5.786	5.852	5.915	5.974
15	3.014	3.674	4.076	4.367	4.595	4.782	4.940	5.077	5.198	5.306	5.404	5.493	5.574	5.649	5.720	5.785	5.846	5.904
16	2.998	3.649	4.046	4.333	4.557	4.741	4.897	5.031	5.150	5.256	5.352	5.439	5.520	5.593	5.662	5.727	5.786	5.843
17	2.984	3.628	4.020	4.303	4.524	4.706	4.858	4.991	5.108	5.212	5.307	5.392	5.471	5.544	5.612	5.675	5.734	5.790
18	2.971	3.609	3.997	4.277	4.495	4.673	4.824	4.956	5.071	5.174	5.267	5.352	5.429	5.501	5.568	5.630	5.688	5.743
19	2.960	3.593	3.977	4.253	4.469	4.645	4.794	4.924	5.038	5.140	5.231	5.315	5.391	5.462	5.528	5.589	5.647	5.701
20	2.950	3.578	3.958	4.232	4.445	4.620	4.768	4.896	5.008	5.108	5.199	5.282	5.357	5.427	5.493	5.553	5.610	5.663
24	2.919	3.532	3.901	4.166	4.373	4.541	4.684	4.807	4.915	5.012	5.099	5.179	5.251	5.319	5.381	5.439	5.494	5.545
30	2.888	3.486	3.845	4.102	4.302	4.464	4.602	4.720	4.824	4.917	5.001	5.077	5.147	5.211	5.271	5.327	5.379	5.429
40	2.858	3.442	3.791	4.039	4.232	4.389	4.521	4.635	4.735	4.824	4.904	4.977	5.044	5.106	5.163	5.216	5.266	5.313
60	2.829	3.399	3.737	3.977	4.163	4.314	4.441	4.550	4.646	4.732	4.808	4.878	4.942	5.001	5.056	5.107	5.154	5.199
120	2.800	3.356	3.685	3.917	4.096	4.241	4.363	4.468	4.560	4.641	4.714	4.781	4.842	4.898	4.950	4.998	5.044	5.086
∞	2.772	3.314	3.633	3.858	4.030	4.170	4.286	4.387	4.474	4.552	4.622	4.685	4.743	4.796	4.845	4.891	4.934	4.974

(*Continued*)

Table G Critical values for Tukey's test (continued)

$\alpha_{EW} = 0.05$

df	k: 20	22	24	26	28	30	32	34	36	k: 38	40	50	60	70	80	90	100
1	59.56	60.91	62.12	63.22	64.23	65.15	66.01	66.81	67.56	68.26	68.92	71.73	73.97	75.82	77.40	78.77	79.98
2	16.77	17.13	17.45	17.75	18.02	18.27	18.50	18.72	18.92	19.11	19.28	20.05	20.66	21.16	21.59	21.96	22.29
3	11.24	11.47	11.68	11.87	12.05	12.21	12.36	12.50	12.63	12.75	12.87	13.36	13.76	14.08	14.36	14.61	14.82
4	9.233	9.418	9.584	9.736	9.875	10.00	10.12	10.23	10.34	10.44	10.53	10.93	11.24	11.51	11.73	11.92	12.09
5	8.208	8.368	8.512	8.643	8.764	8.875	8.979	9.075	9.165	9.250	9.330	9.674	9.949	10.18	10.38	10.54	10.69
6	7.587	7.730	7.861	7.979	8.088	8.189	8.283	8.370	8.452	8.529	8.601	8.913	9.163	9.370	9.548	9.702	9.839
7	7.170	7.303	7.423	7.533	7.634	7.728	7.814	7.895	7.972	8.043	8.110	8.400	8.632	8.824	8.989	9.133	9.261
8	6.870	6.995	7.109	7.212	7.307	7.395	7.477	7.554	7.625	7.693	7.756	8.029	8.248	8.430	8.586	8.722	8.843
9	6.644	6.763	6.871	6.970	7.061	7.145	7.222	7.295	7.363	7.428	7.488	7.749	7.958	8.132	8.281	8.410	8.526
10	6.467	6.582	6.686	6.781	6.868	6.948	7.023	7.093	7.159	7.220	7.279	7.529	7.730	7.897	8.041	8.166	8.276
11	6.326	6.436	6.536	6.628	6.712	6.790	6.863	6.930	6.994	7.053	7.110	7.352	7.546	7.708	7.847	7.968	8.075
12	6.209	6.317	6.414	6.503	6.585	6.660	6.731	6.796	6.858	6.916	6.970	7.205	7.394	7.552	7.687	7.804	7.909
13	6.112	6.217	6.312	6.398	6.478	6.551	6.620	6.684	6.744	6.800	6.854	7.083	7.267	7.421	7.552	7.667	7.769
14	6.029	6.132	6.224	6.309	6.387	6.459	6.526	6.588	6.647	6.702	6.754	6.979	7.159	7.309	7.438	7.550	7.650
15	5.958	6.059	6.149	6.233	6.309	6.379	6.445	6.506	6.564	6.618	6.669	6.888	7.065	7.212	7.339	7.449	7.546
16	5.897	5.995	6.084	6.166	6.241	6.310	6.374	6.434	6.491	6.544	6.594	6.810	6.984	7.128	7.252	7.360	7.457
17	5.842	5.940	6.027	6.107	6.181	6.249	6.313	6.372	6.427	6.479	6.529	6.741	6.912	7.054	7.176	7.283	7.377
18	5.794	5.890	5.977	6.055	6.128	6.195	6.258	6.316	6.371	6.422	6.471	6.680	6.848	6.989	7.109	7.213	7.307
19	5.752	5.846	5.932	6.009	6.081	6.147	6.209	6.267	6.321	6.371	6.419	6.626	6.792	6.930	7.048	7.152	7.244
20	5.714	5.807	5.891	5.968	6.039	6.104	6.165	6.222	6.275	6.325	6.373	6.576	6.740	6.877	6.994	7.097	7.187
24	5.594	5.683	5.764	5.838	5.906	5.968	6.027	6.081	6.132	6.181	6.226	6.421	6.579	6.710	6.822	6.920	7.008
30	5.475	5.561	5.638	5.709	5.774	5.833	5.889	5.941	5.990	6.037	6.080	6.267	6.417	6.543	6.650	6.744	6.827
40	5.358	5.439	5.513	5.581	5.642	5.700	5.753	5.803	5.849	5.893	5.934	6.112	6.255	6.375	6.477	6.566	6.645
60	5.241	5.319	5.389	5.453	5.512	5.566	5.617	5.664	5.708	5.750	5.789	5.958	6.093	6.206	6.303	6.387	6.462
120	5.126	5.200	5.266	5.327	5.382	5.434	5.481	5.526	5.568	5.607	5.644	5.802	5.929	6.035	6.126	6.205	6.275
∞	5.012	5.081	5.144	5.201	5.253	5.301	5.346	5.388	5.427	5.463	5.498	5.646	5.764	5.863	5.947	6.020	6.085

Answers to Exercises and Chapter Quizzes

Chapter 1

Section 1.1
1. 1994 and 1995
2. No
3. (a) About 36,000,000 (b) About 7600
4. (a) About 23,000,000 (b) About 900
5. About 5400
6. About 3500
7. Eiler, Junction
8. Eiler, Happy Camp Complex

Section 1.2
1. Answers will vary.
2. Elements
3. A *qualitative variable* is usually classified into categories; a *quantitative variable* takes on numerical values.
4. Categorical variable
5. True
6. A population is the collection of all elements (persons, items, or data) of interest in a particular study. A sample is a subset of the population from which the information is collected.
7. A *statistic* is a characteristic of a sample.
8. The value of a parameter is constant but usually unknown. The value of a statistic may vary from sample to sample but is usually known.
9. Collections of data from every element in the population
10. False
11. Teams: Dragonborn, Sprites, Enchanters, Trolls
12. Captain's gender, wins, rank, winning percentage
13. (a) Male, female (b) Female, 9, 2, 0.600
14. Quantitative variables: wins, winning percentage; Qualitative variables: captain's gender, rank
15. Discrete: wins; Continuous: winning percentage
16. Captain's gender is nominal, wins is ratio, rank is ordinal, winning percentage is ratio
17. The players: Miguel Cabrera, Michael Cuddyer, Joe Mauer, Michael Trout, Chris Johnson
18. Team, batting average, hits, rank, year of birth
19. (a) Detroit Tigers, Colorado Rockies, Minnesota Twins, Los Angeles Angels, Atlanta Braves (b) Detroit Tigers, 0.348, 193, 1, 1983
20. Quantitative: batting average, hits, year of birth; Qualitative: team, rank
21. Discrete: hits, year of birth; Continuous: batting average
22. Team is nominal, batting average is ratio, hits is ratio, rank is ordinal, year of birth is interval
23. Schools: University of Phoenix, Devry University, ITT Technical Institute, Penn State University, Kaplan University
24. State, school type, recipients, total loan amount ($ millions)
25. (a) Proprietary and public (b) PA, Public, 42,011, $151 million
26. Quantitative variables: recipients, total loan amount ($ millions); Qualitative variables: state, school type
27. Discrete: recipients, total loan amount ($ millions)
28. State is nominal, school type is nominal, recipients is ratio, total loan amount ($ millions) is ratio
29. (a) Quantitative, discrete (b) Interval
30. (a) Qualitative (b) Nominal
31. (a) Quantitative, discrete (b) Ratio
32. (a) Quantitative, discrete (b) Interval
33. (a) Quantitative, discrete (b) Ratio
34. (a) Qualitative (b) Nominal
35. (a) Qualitative (b) Ordinal
36. (a) Quantitative, discrete (b) Ratio
37. (a) Qualitative (b) Nominal
38. (a) Quantitative, discrete (b) Ratio
39. (a) Qualitative (b) Nominal
40. (a) Quantitative, discrete (b) Ratio
41. (a) Quantitative, continuous (b) Ratio
42. (a) Qualitative (b) Nominal
43. Population

44. Sample
45. Sample
46. Parameter
47. Statistic
48. Statistic
49. Descriptive statistics
50. Statistical inference
51. Statistical inference
52. Population: all home sales in Tarrant County, Texas; Sample: 100 home sales selected
53. Population: all veterans returning from war; Sample: the 20 veterans selected.
54. Population: all 4-H clubs in Maricopa County, Arizona; Sample: ten selected 4-H clubs.
55. Population: all older women; Sample: the physical therapist's 10 selected patients.
56. Population: all students at Portland Community College; Sample: 50 selected Portland Community College students.
57. Population: all companies that recently underwent a merger; Sample: the 50 selected companies that recently underwent a merger.
58. Descriptive statistics; the variable describes a sample.
59. Inferential statistics; the sample was used to draw a conclusion about the entire population.
60. Statistical inference; the sample was used to draw a conclusion about the entire population.
61. Descriptive statistics; the variable describes a sample.
62. Descriptive statistics; the variable describes a sample.
63. Inferential statistics; the sample was used to draw a conclusion about the entire population.
64. Descriptive statistics; the variable describes a sample.
65. (a) Elements: Endangered species pygmy rabbit, Florida panther, red wolf, and West Indian manatee; Variables: year listed as endangered, estimated number remaining, and range. (b) Qualitative variables: range; Quantitative variables: year listed as endangered and estimated number remaining. (c) Year listed as endangered—discrete, estimated number remaining—discrete. (d) Year listed as endangered—interval; estimated number remaining—ratio, range—nominal. (e) 1973, 50, Florida.
66. (a) Elements: Companies City of Santa Monica, St. John's Health Center, The Macerich Company, Fremont General Corp., and Entravision Corp.; Variables: Employees and industry. (b) Qualitative variables: Industry; Quantitative variables: Employees. (c) Employees—discrete (d) Employees—ratio; Industry—nominal (e) 1892, government
67. (a) Elements: States Texas, Missouri, Minnesota, Ohio, and South Dakota; Variables: proportion of GE corn and most prevalent type. (b) Qualitative variables: most prevalent type; quantitative variables: proportion of GE corn (c) proportion of GE corn—ratio; most prevalent type—nominal (d) proportion of GE corn—continuous (e) 89%, herbicide-tolerant
68. (a) Elements: Hospital names Hardy Wilson, Humphreys County, Jefferson County, Lackey Memorial, Leake Memorial, Madison County, Monfort Jones, and Rankin Medical Center; Variables: beds, city, and zip (b) Qualitative variables: city and zip; Quantitative variables: beds (c) Beds—discrete (d) Beds—ratio; city—nominal; zip—nominal (e) 134, Brandon, 39042
69. (a) Elements: Hospitals Briarcliff Manor, Buchanan, Cortlandt, Croton-on-Hudson, Mount Pleasant, Ossining 1, Ossining 2, Peekskill, Pleasantville, and Sleepy Hollow; Variables: births and average maternal age (b) Qualitative variables: There are no qualitative variables; Quantitative variables: births and average maternal age (c) Births—discrete; average maternal age—continuous (d) Births—ratio; average maternal age—ratio (e) 134, 29.2
70. (a) Elements: Commodities oil, gold, and wheat; Variables: price per share and percent change (b) Qualitative variables: There are no qualitative variables; Quantitative variables: price per share and percent change (c) Price per share—discrete; percent change—continuous (d) Price per share—ratio; percent change—ratio (e) $1243.62, −0.110%
71. (a) Elements are the tornado names: Tri-State, Natchez, St. Louis, Tupelo, Gainesville; Variables: deaths, year (b) Quantitative variables: deaths, year (c) Discrete: deaths, year (d) Deaths is ratio, year is interval (e) 255, 1896

72. **(a)** Sample **(b)** No, these companies are relatively large and there are probably many more small companies than larger companies.

73. **(a)** Sample **(b)** No, only the five tornadoes with the highest death toll are included.

74. They compared the average lifetime of a sample of their own light bulb to the reported average lifetimes of other current models of light bulbs.

75. **(a)** Statistic **(b)** An estimate of the average lifetime of all new light bulbs is the average lifetime of the sample of 100 light bulbs, which is 2000 hours.

76. **(a)** The elements are the institutions: Ashford University, Arizona State University, Liberty University, Miami Dade College, Lone Star College System **(b)** State, enrollment, rank **(c)** State, rank **(d)** Enrollment **(e)** State is nominal, enrollment is ratio, rank is ordinal

77. **(a)** Sample **(b)** No, only the five campuses with the largest enrollment are included. **(c)** Arizona, 72,254, 2

78. Platform, studio, type

79. Sales for week, sales total, weeks on list

80. Discrete

81. Sample, only the 30 best-selling video games are included.

82. Sample; statistic

83. Platform, studio, type, weeks on list

84. No

85. Sales for week, sales total

86. No. The variables platform, studio, and type are qualitative and there is no natural ordering for their values. The variables sales for week, sales total, and weeks on list have values that can be divided and have natural zeros.

87. Descriptive statistics

88. Statistical inference

Section 1.3

1. Convenience sampling usually only includes a select group of people. For example, surveying people at a mall on a workday during working hours would probably include few if any people who work full time.

2. Answers will vary.

3. Answers will vary; could have chosen a random sample of houses and apartments and surveying the people door to door, for instance.

4. No.

5. A sample for which every element has an equal chance of being included.

6. Observational study—observes association between explanatory and response variables; experimental study—investigates how varying the explanatory variable affects the response by placing subjects into treatment and control groups.

7. Answers will vary.

8. Answers will vary.

9. Answers will vary.

10. Answers will vary.

11. Illinois, Iowa, Michigan State, Nebraska, Ohio State, Perdue

12. Arkansas, Georgia, Mississippi, South Carolina, Vanderbilt

13. Alabama, Georgia, Mississippi State, Texas A&M

14. California, Oregon State, USC, Washington State

15. Answers will vary.

16. Answers will vary.

17. Answers will vary.

18. Answers will vary.

19. Answers will vary.

20. Answers will vary.

21. No

22. No

23. Cluster sampling

24. Systematic sampling

25. Convenience sampling

26. Cluster sampling

27. Target population: All college students; Potential population: All students working out at the gymnasium on the Monday night Brandon was there.

28. Yes. Students working out at the gymnasium are more likely to be physically fit than the rest of the students.

29. Target population: All small businesses; Potential population: Small businesses near the state university.

30. Yes. Businesses near the state university are more likely to employ college students than businesses farther away.

31. Vague terminology

32. Leading question

33. Neither simple nor clear

34. Asking two questions in one

35. **(a)** Observational **(b)** response variable: how often they attend religious services; predictor variable: whether or not the family is large (at least four children)

36. **(a)** Observational **(b)** response variable: stock price; predictor variable: whether or not the company gives large bonuses to CEOs

37. **(a)** Experimental **(b)** response variable: performance of the electronics equipment; predictor variable: whether or not a piece of equipment has a new computer processor

38. **(a)** Experimental **(b)** response variable: whether or not the person's blood pressure is lowered; predictor variable: whether or not the person is taking the new drug

39. Level of insect damage to crops

40. Whether or not the new pesticide was used

41. The new pesticide

42. The traditional pesticide

43. LDL cholesterol level in the bloodstream

44. Whether person is given new medication or placebo

45. New medication

46. Placebo

47. **(a)** Randomization is present for the 100 randomly assigned subjects but not for the subject with high LDL cholesterol levels. **(b)** The sample of 100 people is probably enough replication.

48. **(a)** Randomization is present **(b)** Two subjects each is probably insufficient replication to uncover any strong statistical results.

49. Experiment

50. Observational study

51. **(a)** Answers will vary **(b)** No. Every possible sample of 5 video games has the same chance of being selected. **(c)** No. Every possible sample of 5 video games has the same chance of being selected. Some of the samples will contain the video game and some won't. **(d)** Answers will vary; answers will vary

52. Minecraft for PS3, Titanfall for Xbox One, Titanfall for Xbox 360, Super Luigi U for Wii U, Battlefield 4 for Xbox 360, Battlefield 4 for PS3, Yoshi's New Island for 3DS, Mario Kart 7 for 3DS

53. Answers will vary

54. Answers will vary

55. Answers will vary. For instance, the poll by Ann Landers was extremely biased. Only people who read the Ann Landers column and felt strongly about the poll responded to this poll. The *Newsday* poll was done professionally, and therefore the sample used was more likely to be representative of the population.

56. Target population: all high schools in New England; potential population: all high schools in greater Boston. Potential population is not a random sample of all high schools in New England; dropout rate for target population may differ from potential population

57. Target population: all the people living in Chicago; potential population: people who have phones and who have their phone number listed in the Chicago phone directory. Reasons for potential selection bias vary; for instance, many of the people living below the poverty level in Chicago may not have phones.

58. Desired response type is open to interpretation: preference or yes/no.

59. The survey question is a leading question.

60. No, the conclusion leaves out the details of the survey question.

61. **(a)** No, different random samples may contain different lowest stock prices. **(b)** Answers will vary. **(c)** No; see answer (a). **(d)** Answers will vary.

62. **(a)** No; no. We don't know which two companies will be selected. **(b)–(c)** Answers will vary

63. It is a variable that may vary from sample to sample.

64. Predictor variable: patient diet, Mediterranean or Western; response variable: risk for a second heart attack.

65. **(a)** Forcing the parents of a treatment group to smoke tobacco will increase the occurrence of respiratory illnesses of their children. This is not very ethical. **(b)** Observational study

66. It is unethical.

67. **(a)** The 305 subjects that wore the placebo bracelet **(b)** The subjects were randomly assigned to wear either the placebo bracelet or the ionized bracelet. **(c)** There are 305 subjects in both the treatment and the control groups.
68. **(a)** Type of bracelet worn, placebo or ionized **(b)** Wearing the ionized bracelet **(c)** Measure of pain.
69. This study is an experimental study because the subjects were randomly assigned to either a treatment or a control.

Chapter 1 Review

1. **(a)** Make/Models: Chevrolet, Corvette, Ferrari 458 Italia, Honda CR-Z, Jaguar F Convertible, Porsche Boxster S **(b)** Cylinders, transmission, combined mileage
2. **(a)** Transmission **(b)** Cylinders, combined mileage **(c)** Cylinders is ratio, transmission is nominal, combined mileage is ratio
3. 8, manual, 21
4. **(a)** Elements are the states: California, Texas, New York, Florida, Illinois; Variables: population (1960, in 1000's), population (2013, in 1000's), increase **(b)** Quantitative **(c)** 4,952, 19,953, 14,601 **(d)** Largest: California, Texas, Florida; Smallest: New York, Illinois
5. **(a)** Turn on all 1 million light bulbs, leave them all on until they burn out, compute the average of all bulb lifetimes. **(b)** Use the average lifetime of a sample to estimate the average lifetime of the population.
6. **(a)** All registered voters in the United States. **(b)** People on the lists of people who owned cars and had telephones **(c)** All people on the lists of people who owned cars and had telephones **(d)** Not similar; answers will vary.
7. **(a)** Observational study **(b)** It would be impractical to randomly reassign people to a statistics class after classes have started.
8. **(a)** Replication **(b)** Surveying only four dentists is not likely to get a sample representative of the population of all dentists.
9. Observational; it would be impossible to randomly assign a child to come from a single-parent family or a two-parent family.

Chapter 1 Quiz

1. False
2. False
3. collecting
4. observation
5. sample
6. Observational study
7. Experimental study
8. Predictor variable: drug given to an elderly patient with Alzheimer's, new or placebo; response variable: whether or not the patient's Alzheimer's symptoms are reduced.
9. **(a)** All statistics students **(b)** The students in the statistics class who were selected for the sample **(c)** Left-handed or not; qualitative **(d)** No; not likely to be very far away from the population proportion since enrollment in a specific statistics class is not dependent on being left-handed or not.
10. Different people have different interpretations of the words *often, occasionally, sometimes,* and *seldom.*

Chapter 2

Section 2.1

1. We use graphical and tabular form to summarize data in order to organize it in a format where we can better assess the information. If we just report the raw data, it may be extremely difficult to extract the information contained in the data.
2. A frequency distribution is a listing of all the values with the corresponding frequencies for each value of a variable; a relative frequency distribution lists the relative frequencies for the values.
3. True.
4. False
5. The sample size, *n*.
6. 1.00
7. The row totals; the column totals
8. The grand total; the sample size
9. True
10. False

11.

Vote	Frequency
Republican	15
Democrat	2

12.

Vote	Relative frequency
Republican	15/17 = 0.88
Democrat	2/17 = 0.12

13.

Size	Frequency
Small	9
Medium	6
Large	2

14.

Size	Relative frequency
Small	9/17 = 0.53
Medium	6/17 = 0.35
Large	2/17 = 0.12

15.

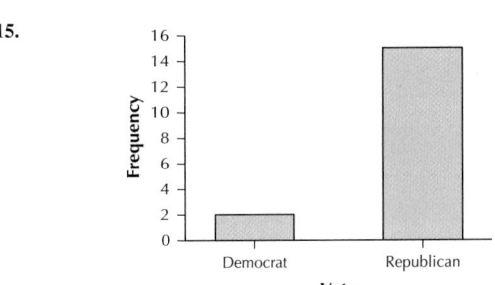

16.

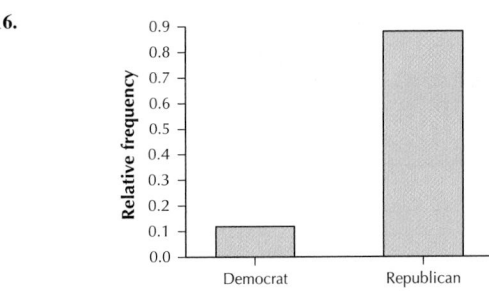

17.

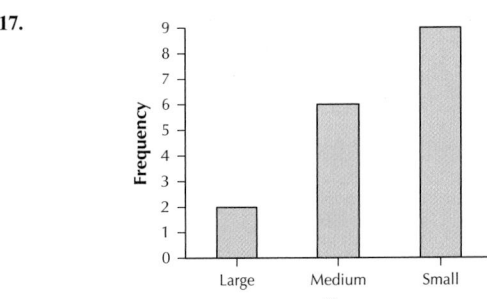

18.

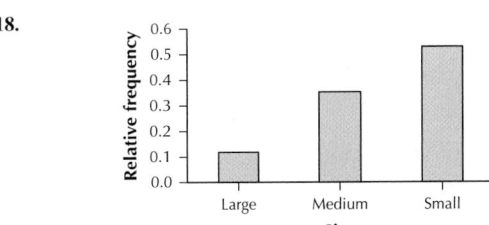

19.

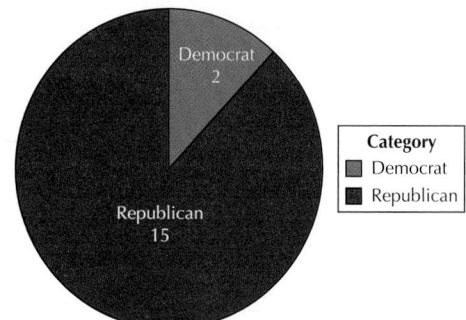

20.

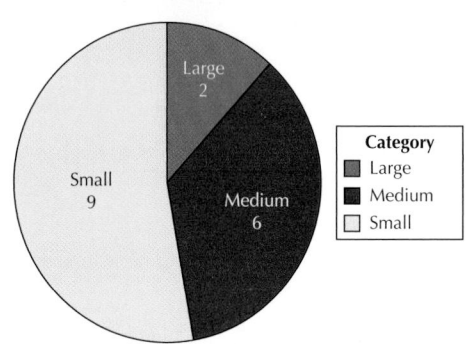

21.

	Large	Medium	Small	Total
Democrat	2	0	0	2
Republican	0	6	9	15
Total	2	6	9	17

22. The Democrats carried the two large counties in Nevada. The large counties have more people than the small and medium counties. Since the electoral votes are based on who wins the popular vote, it is possible enough people voted Democrat in the two large counties for the Democrats to win the popular vote.

23.

Gender	Frequency
Male	16
Female	4

24.

Gender	Relative frequency
Male	16/20 = 0.8
Female	4/20 = 0.2

25.

Survived	Frequency
Yes	9
No	11

26.

Survived	Relative frequency
Yes	9/20 = 0.45
No	11/20 = 0.55

27.

Class	Frequency
1st class	2
2nd class	4
3rd class	5
Crew	9

28.

Class	Relative frequency
1st class	2/20 = 0.1
2nd class	4/20 = 0.2
3rd class	5/20 = 0.25
Crew	9/20 = 0.45

29.

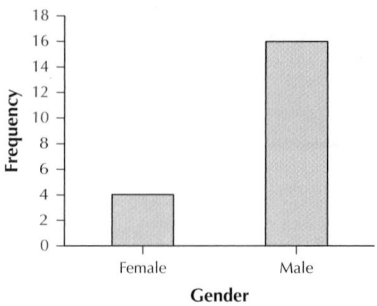

30.

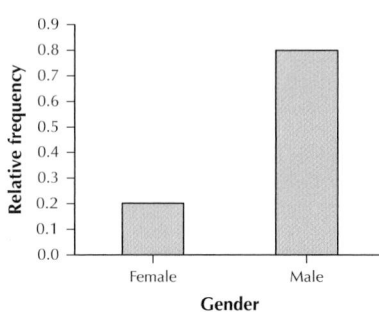

31.

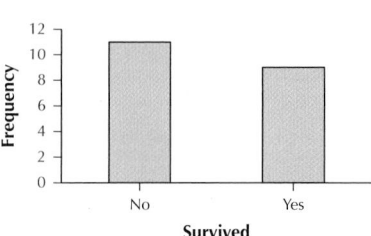

32.

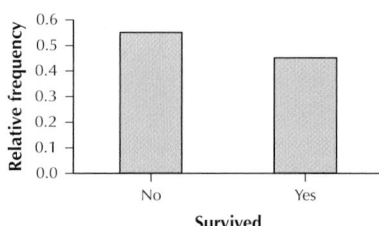

33.

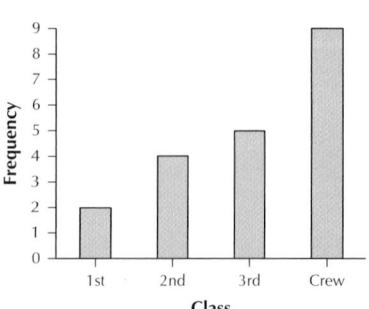

34.

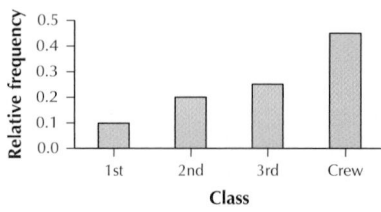

35.

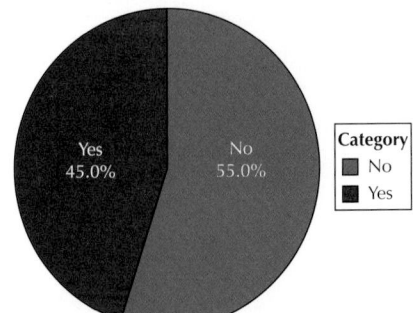

36.

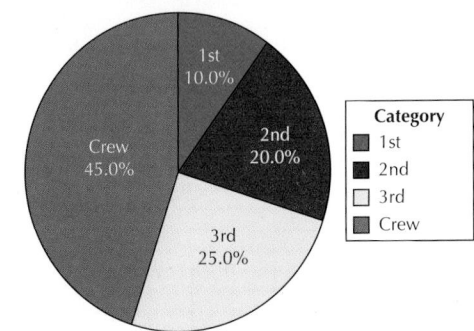

37.

	1st	2nd	3rd	Crew	Total
Female	1	1	2	0	4
Male	1	3	3	9	16
Total	2	4	5	9	20

38. None

39. Crew. In 1912 women did not work as crew members on a ship. Only the people who could afford first class bought tickets for both men and women.

40. 1st

41.

	No	Yes	Total
Female	1	3	4
Male	10	6	16
Total	11	9	20

42. 0.75

43. 0.375

44. No. We need to know the number of females total, the number of males total, the number of females that survived, and the number of males that survived to calculate this information. This information is only available in the crosstabulation.

45.

	No	Yes	Total
1st	0	2	2
2nd	2	2	4
3rd	2	3	5
Crew	7	2	9
Total	11	9	20

46. 1

47. 0.5, 0.6, 0.22

48. All of the first class passengers survived. Only some of the second and third class passengers survived. Very few crew members survived.

49.

Gender	Frequency
Female	20
Male	21

50.

Gender	Relative frequency
Female	20/41 = 0.49
Male	21/41 = 0.51

51.

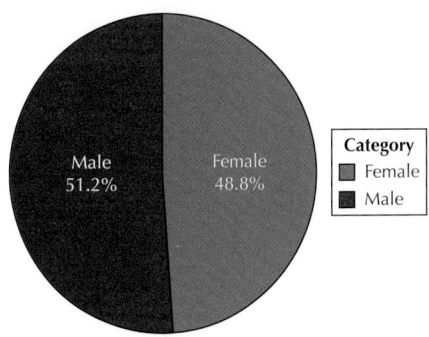

52.

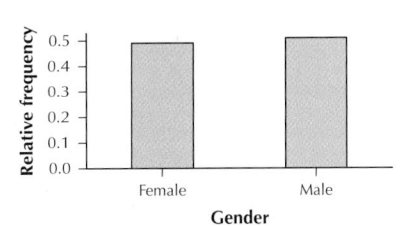

53.

Sport	Frequency
Alpine skiing	21
Figure skating	9
Ice hockey	44

54.

Sport	Relative frequency
Alpine skiing	21/74 = 0.28
Figure skating	9/74 = 0.12
Ice hockey	44/74 = 0.59

55.

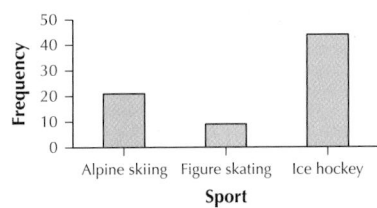

56.

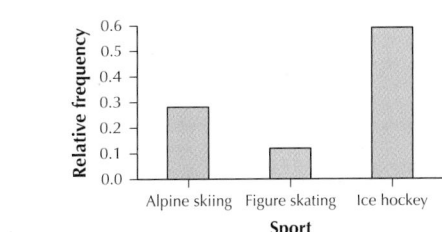

57.

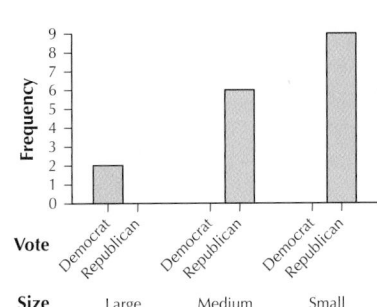

58.

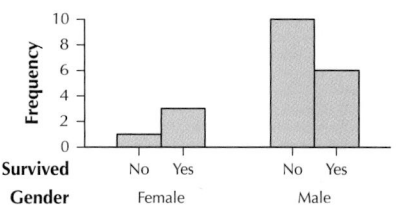

59.

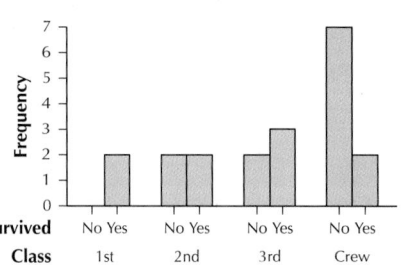

60.

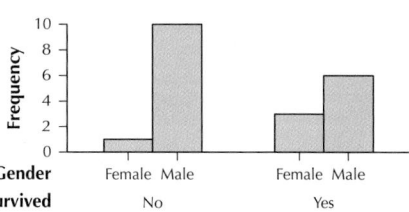

61. (a)–(b)

Continent	Frequency	Relative frequency
Africa	1	0.10
Asia	5	0.50
Europe	1	0.10
North America	2	0.20
South America	1	0.10

(c)

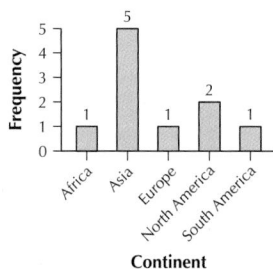

(d)

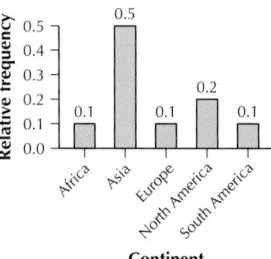

(e)

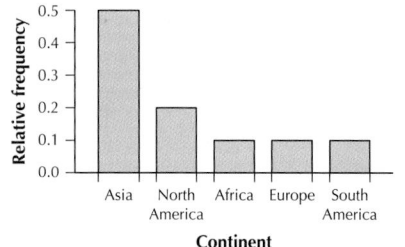

(f)

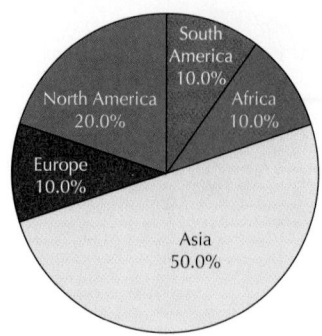

62. (a)–(b)

Climate	Frequency	Relative frequency
Arid	5	0.50
Temperate	4	0.40
Tropical	1	0.10

(c)

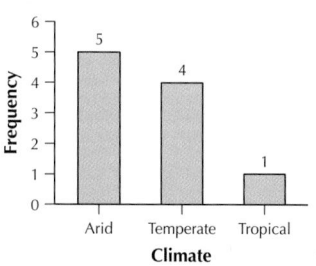

(d)

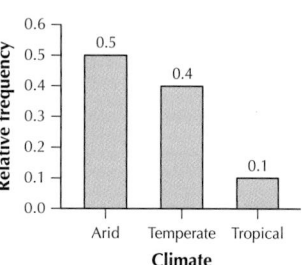

(e)

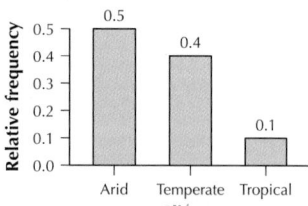

(f)

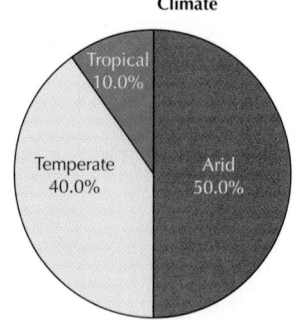

63. (a)–(b)

Main use	Frequency	Relative frequency
Industry	2	0.20
Irrigation	6	0.60
Not reported	2	0.20

(c)

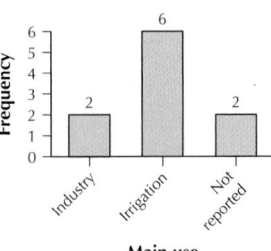

(d)

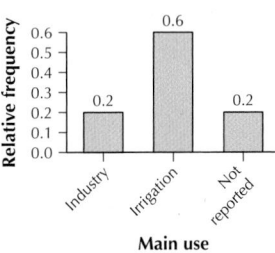

(e)

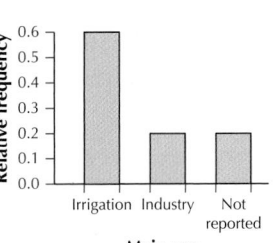

(f)

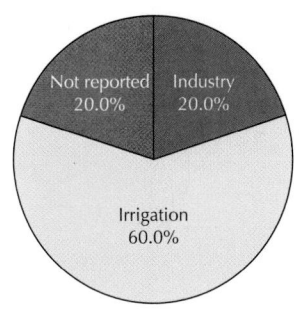

64. The countries are the elements.

65.

	Arid	Temperate	Tropical	Total
Africa	0	0	1	1
Asia	4	1	0	5
Europe	0	1	0	1
North America	0	2	0	2
South America	1	0	0	1
Total	5	4	1	10

66.

	Industry	Irrigation	Not reported	Total
Africa	0	1	0	1
Asia	0	4	1	5
Europe	0	0	1	1
North America	2	0	0	2
South America	0	1	0	1
Total	2	6	2	10

67.

	Industry	Irrigation	Not reported	Total
Arid	0	5	0	5
Temperate	2	0	2	4
Tropical	0	1	0	1
Total	2	6	2	10

68.

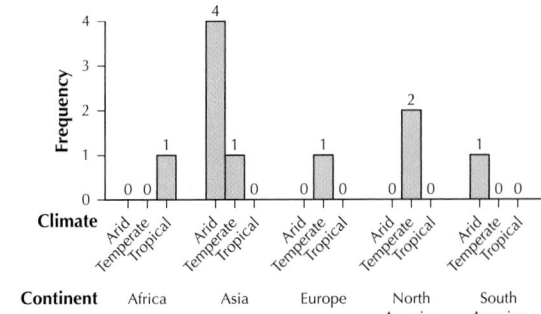

69.

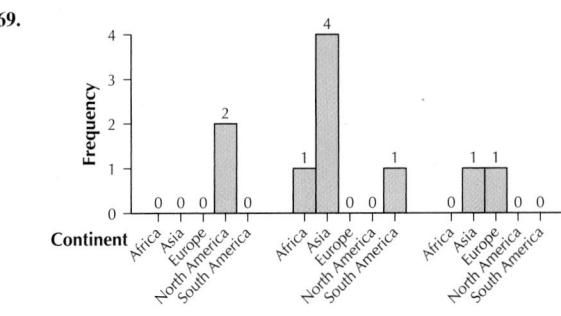

70.

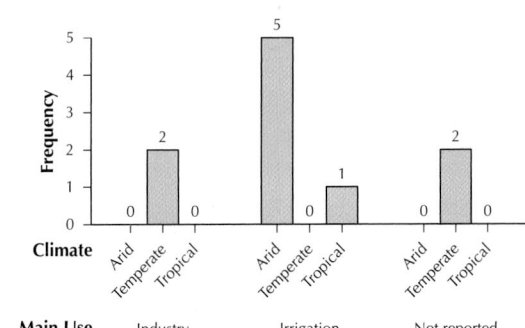

71.

```
     Type   Count   Percent
   Action       9     30.00
Adventure       2      6.67
 Platform       3     10.00
   Racing       2      6.67
Role-Playing    1      3.33
  Shooter      10     33.33
   Sports       3     10.00
      N=        30
```

Shooter, role-playing

72.

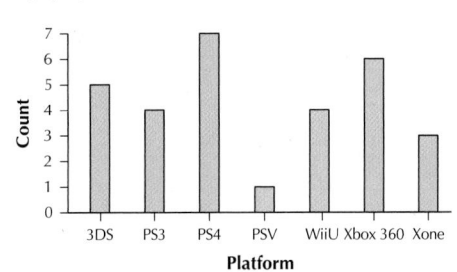

PS4

73.

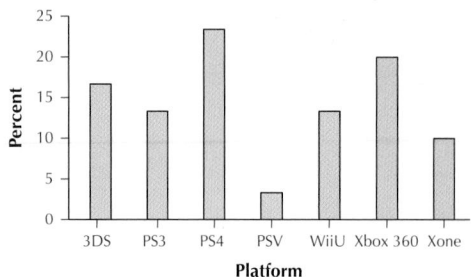

Percent within all data.

23%

74.

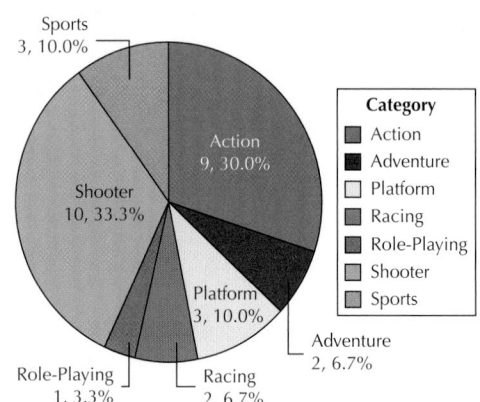

(b)

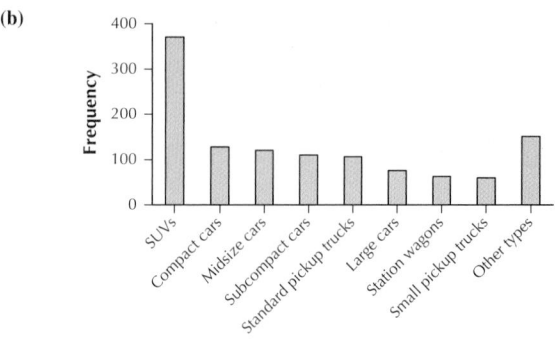

(c)

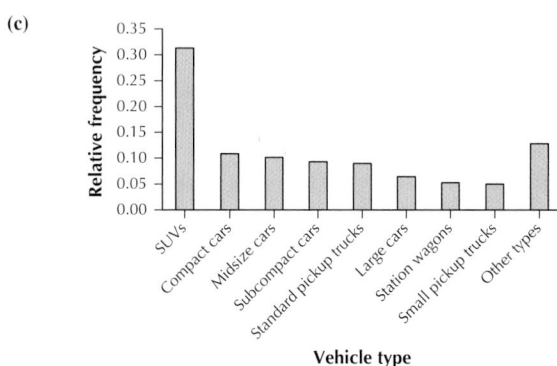

(d)

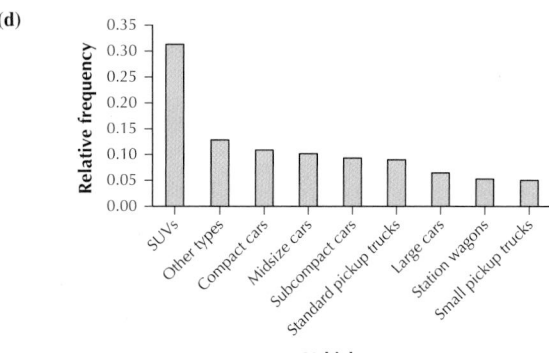

75.

	3DS	PS3	PS4	PSV	WiiU	360	Xone	All
Action	2	1	3	0	2	1	0	9
Adventure	0	1	0	0	0	1	0	2
Platform	1	0	0	0	2	0	0	3
Racing	1	0	0	0	0	0	1	2
Role-Playing	1	0	0	0	0	0	0	1
Shooter	0	2	2	1	0	3	2	10
Sports	0	0	2	0	0	1	0	3
All	5	4	7	1	4	6	3	30

76. PS4, 360

77. No. There are actually two categorical variables—level of education and whether or not the person owns a cell phone. The percents are percents of each category of level of education who own cell phones and not the percent of the whole group who own cell phones.

78. No. There are actually two categorical variables—level of education and whether or not the person owns a cell phone. The percents are percents of each category of level of education who own cell phones and not the percent of the whole group who own cell phones. Also, the categories are arranged in order from lowest percent to highest percent and not in order from highest percent to lowest percent.

79. (a) Several times a day; 43.4% (b) Every few weeks; 5.1%

80. (a) 12.1% (b) 12.1% (c) 121

81. (a) Fractures; 26% (b) Traumatic brain injury; 9% (c) Yes. It would have to be one of the injuries included in the category "Other injuries."

82. (a) 25% (b) 16%

83. (a) Relative frequency distribution of *vehicle type*

Vehicle type	Relative frequency
SUVs	0.3130
Compact cars	0.1083
Midsize cars	0.1015
Subcompact cars	0.0931
Standard pickup trucks	0.0897
Large cars	0.0643
Station wagons	0.0525
Small pickup trucks	0.0499
Other types	0.1277
Total	1.0000

(e)

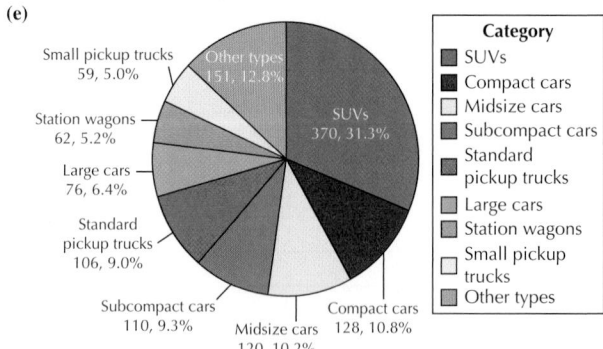

84. The sample may be a representative sample of all U.S. teenagers if one assumes that teenagers of all walks of life visit this particular Web site and participate in the survey. If only teenagers of a particular social class visit this Web site, then although the sample is "large" it will not be a representative sample. Bias may be introduced if only teenagers of a particular social standing, region, ethnicity, and so on visit this Web site.

85. A pie chart cannot be constructed for the data since the percentages for the different classes do not add up to 100%. Respondents could select more than one category.

86.

Response	Frequency
Yes	1860
No	2700
I've never thought about it.	1440

87.

Response	Relative frequency
Yes	0.31
No	0.45
I've never thought about it	0.24

88.

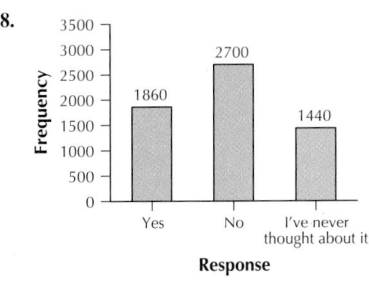

89. Relative frequencies are expressed as percentages.

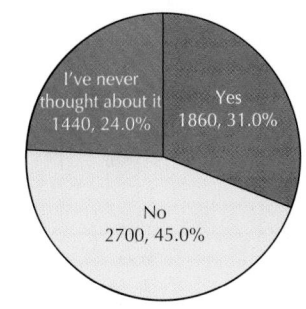

90. (a) Virgo (b) Aries

91. (a)

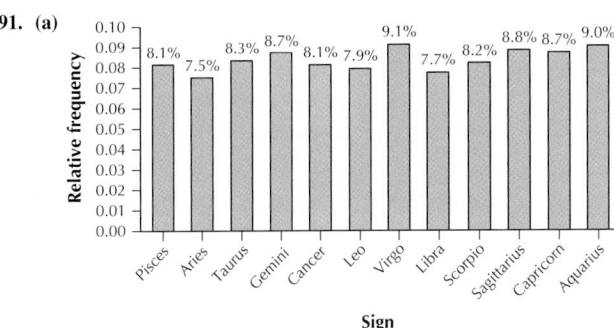

(b)

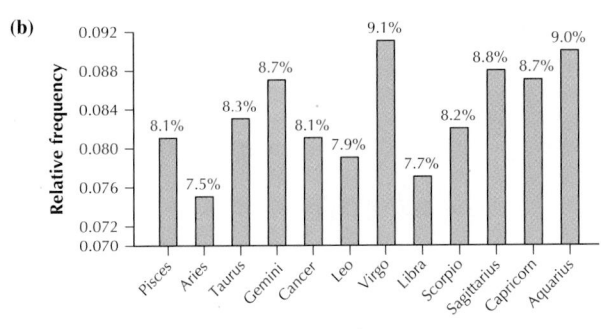

The graph in (b) uses an adjusted scale, which is misleading. Use this graph to magnify the small variability in percentages.

92. Virgo occurs the most and Aries occurs the least.

Sign	Relative frequency	Frequency
Pisces	0.081	118
Aries	0.075	110
Taurus	0.083	121
Gemini	0.087	127
Cancer	0.081	118
Leo	0.079	116
Virgo	0.091	133
Libra	0.077	113
Scorpio	0.082	120
Sagittarius	0.088	129
Capricorn	0.087	127
Aquarius	0.090	132

93.

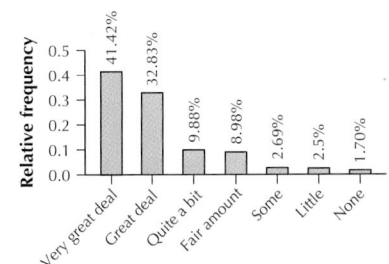

Categories are qualitative; interpretation will vary.

94. Yes.

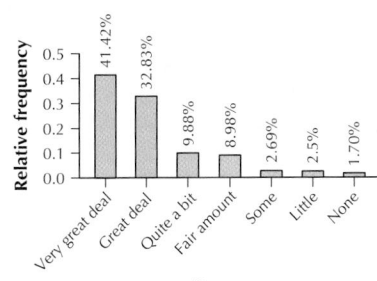

95. No.

96.

	Beverages	Food	Household	Tobacco	All
Glass	2	0	0	0	2
Metal	0	0	2	0	2
Paper	0	0	1	3	4
Plastic	1	2	1	0	4
All	3	2	4	3	12

97.

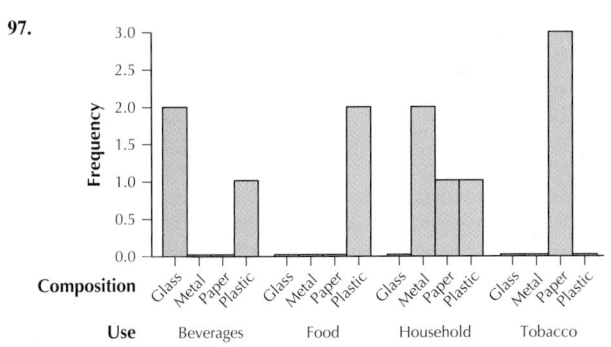

98. The composition of most of the litter from tobacco use is paper, the composition of most of the litter from food is plastic, the composition of most of the litter from beverages is glass or plastic, and the composition of most of the litter from household use is metal, paper, or plastic.

99. Petit larceny

100. Approximately 84,000.

101. Yes. The categories are arranged in order from highest frequency to lowest frequency.

102. 0.355, 0.253

103. (a)

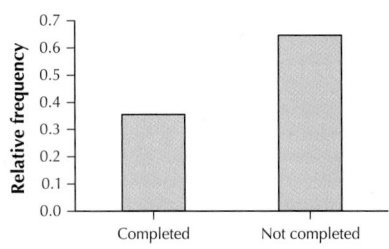

(b)

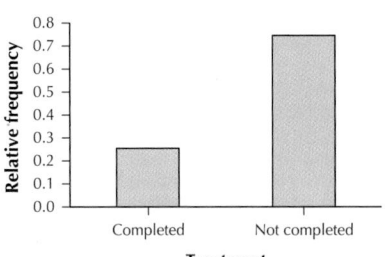

104. Real disparity

105. Missing values are in red

"How much do you enjoy shopping?"	Gender		
	Male	Female	Total
A lot	388	950	1338
Some	582	673	1255
Only a little	662	497	1159
Not at all	497	220	717
Don't know/refused	20	25	45
Total	2149	2365	4514

106.

"How much do you enjoy shopping?"	Gender		
	Male	Female	Total
A lot	29.00%	71.00%	100%
Some	46.37%	53.63%	100%
Only a little	57.12%	42.88%	100%
Not at all	69.32%	30.68%	100%
Don't know/refused	44.44%	55.56%	100%
Total	47.61%	52.39%	100%

107. (a) Women (b) Women (c) Men (d) Men

108.

Gender	Frequency
Male	2149
Female	2365

109.

How much do you enjoy shopping?	Frequency
A lot	1338
Some	1255
Only a little	1159
Not at all	717
Don't know	45

110.

Gender	Relative frequency
Male	0.4761
Female	0.5239

111.

How much do you enjoy shopping?	Relative frequency
A lot	0.2964
Some	0.2780
Only a little	0.2568
Not at all	0.1588
Don't know	0.0100

112.

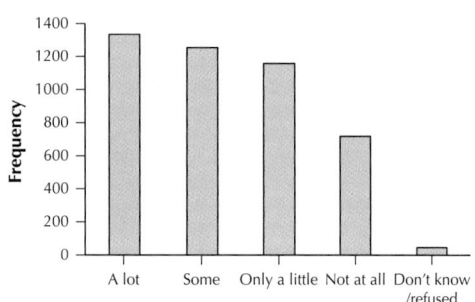

113.

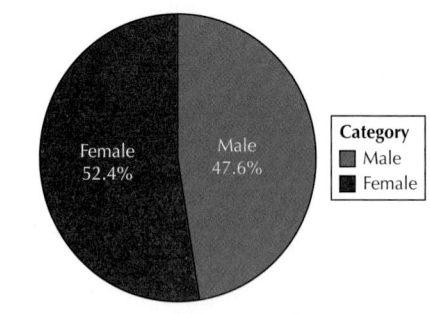

114.

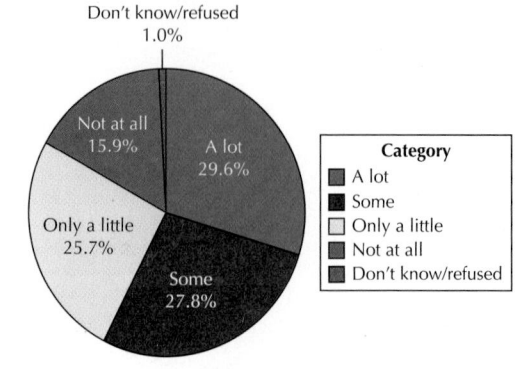

115.

116.

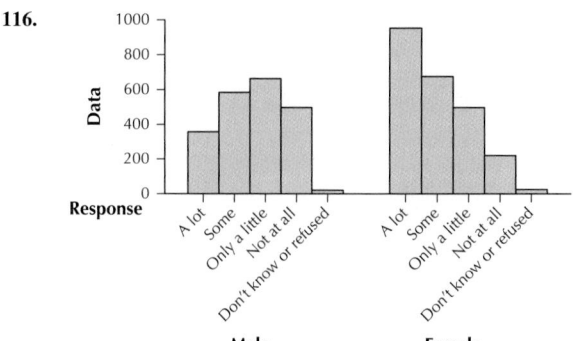

117.

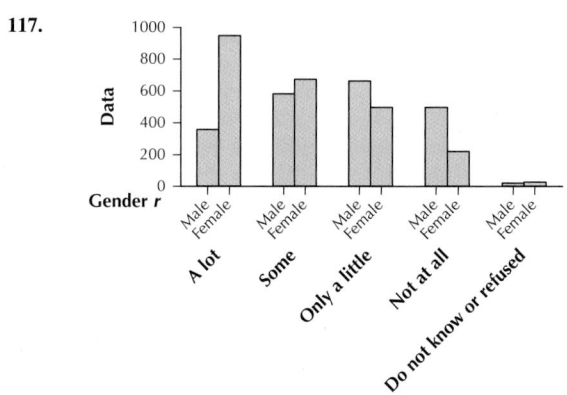

118. 0.5239
119. **(a)** Frequencies would double **(b)** Stay the same **(c)** Stay the same
120. **(a)** Girls: 0.525; boys: 0.475

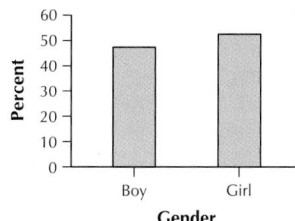

(b) Grades: 51.67%; popular: 29.50%; sports: 18.83%

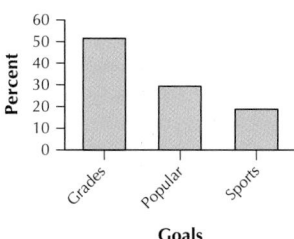

121. **(a)**

Tally for Discrete Variables: GENDER	
GENDER	Percent
boy	47.49
girl	52.51

(b)

Tally for Discrete Variables: GOALS	
GOALS	Percent
Grades	51.67
Popular	29.50
Sports	18.83

122. There are 51 observations and 8 variables.
123. Qualitative: State; Quantitative: Tot_hhld, Fam_tpc, Fam_mpc, Fam_fpc, Nfm_tpc, Nfm_lpc, and Ave_size.
124. Each relative frequency in the relative frequency distribution would be the same, each slice in the pie chart would be the same size, and each rectangle in the bar graph would be the same height.
125.

(a)

Tally for Discrete Variables: Class		
Class	Count	Percent
1st	325	14.77
2nd	285	12.95
3rd	706	32.08
Crew	885	40.21
N=	2201	

(b)

Tally for Discrete Variables: Age		
Age	Count	Percent
Adult	2092	95.05
Child	109	4.95
N=	2201	

(c)

Tally for Discrete Variables: Sex		
Sex	Count	Percent
Female	470	21.35
Male	1731	78.65
N=	2201	

(d)

Tally for Discrete Variables: Survived_1		
Survived_1	Count	Percent
No	1490	67.70
Yes	711	32.30
N=	2201	

126. **(a)**

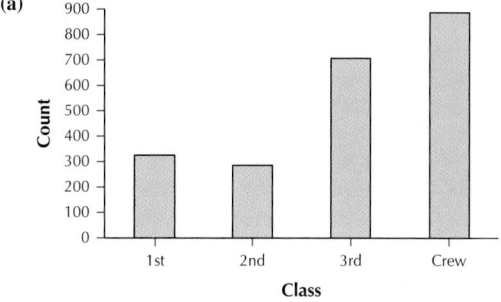

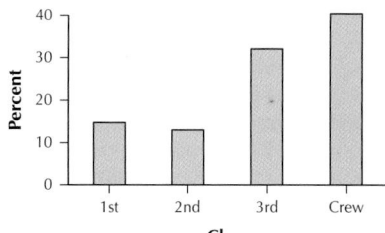

Percent within all data.

(b)

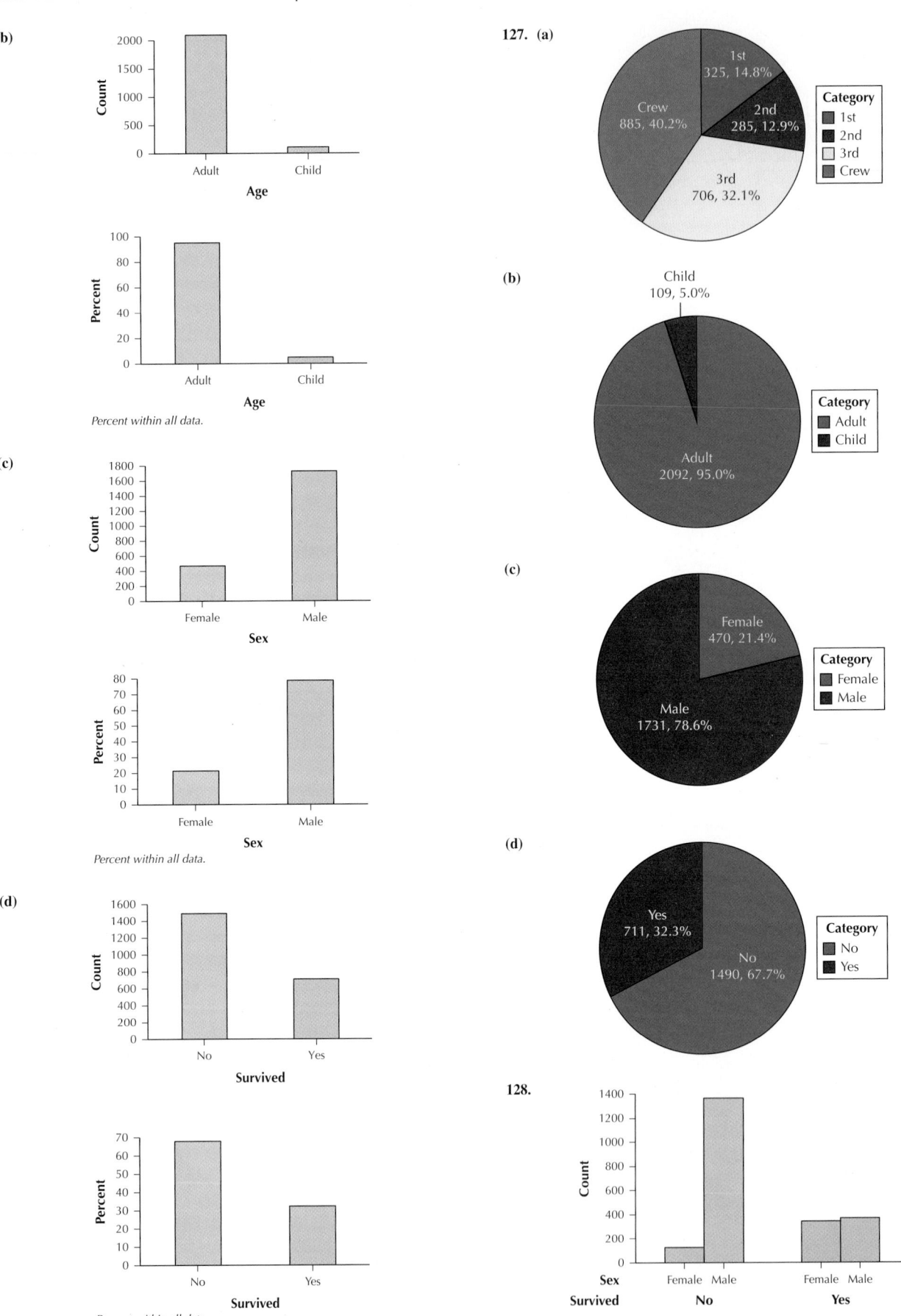

127. (a)

A much higher proportion of females survived than males.

129.

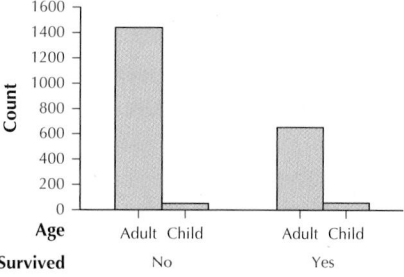

A child.

130.

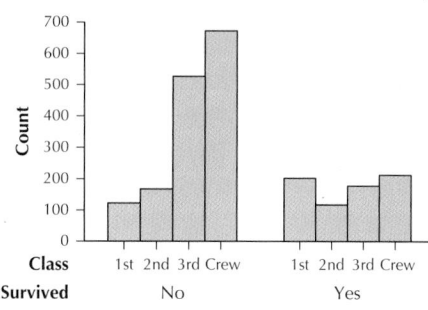

First class.

131. (a) and **(b)**

Class	Frequency	Relative frequency
Freshman	5	0.25
Sophomore	5	0.25
Junior	5	0.25
Senior	5	0.25

132. Answers will vary.

133. Answers will vary.

134.

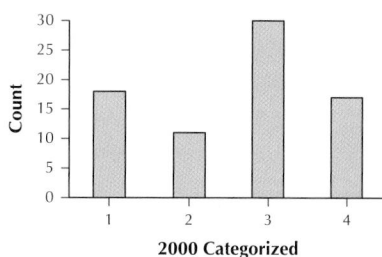

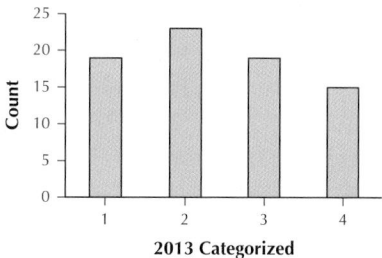

135. In 2000 there were more precincts that had category 3 and 4 numbers of petit larcenies than in 2013. In 2000 there were fewer precincts that had category 1 and 2 numbers of petit larcenies than in 2013. Good news.

136.

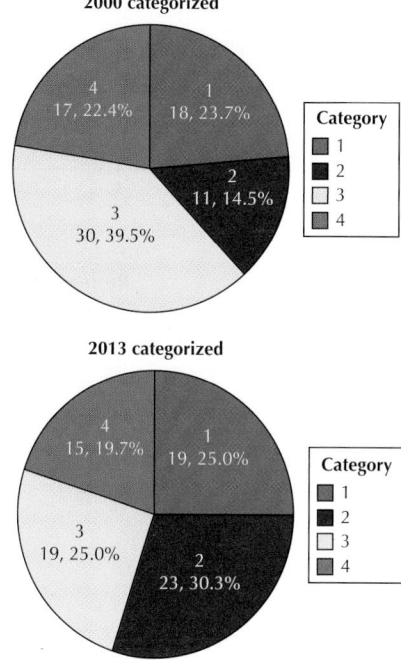

137. Pie charts, since they show each category as a percent of the whole.

Section 2.2

1. Both: frequency distribution, relative frequency distribution; quantitative data only: histograms, frequency polygons, stem-and-leaf displays, dotplot.

2. Answers will vary.

3. Between 5 and 20

4. We would be unable to use such summary information in constructing histograms.

5. Answers will vary.

6. Answers will vary; of a symmetric shape; circle; of a nonsymmetric shape; scalene triangle.

7. Answers will vary.

8. False.

9 and **10.**

Number of nominations	Frequency	Relative frequency
2	5	5/13 = 0.3846
3	3	3/13 = 0.2308
4	1	1/13 = 0.0769
5	1	1/13 = 0.0769
6	1	1/13 = 0.0769
7	1	1/13 = 0.0769
8	1	1/13 = 0.0769
Total	13	13/13 = 1.0000

11 and **12.**

Grand slams	Frequency	Relative frequency
14	4	4/18 = 0.2222
15	2	2/18 = 0.1111
16	3	3/18 = 0.1667
17	3	3/18 = 0.1667
18	2	2/18 = 0.1111
19	1	1/18 = 0.0556
21	1	1/18 = 0.0556
23	1	1/18 = 0.0556
24	1	1/18 = 0.0556
Total	18	18/18 = 1.0000

13. and **14.**

Nominations	Frequency	Relative frequency
0–3	8	8/13 = 0.6154
4–6	3	3/13 = 0.2308
7–9	2	2/13 = 0.1538
Total	13	13/13 = 1.0000

15. and **16.**

Grand slams	Frequency	Relative frequency
11–15	6	6/18 = 0.3333
16–20	9	9/18 = 0.5000
21–25	3	3/18 = 0.1667
Total	18	18/18 = 1.0000

17. 1 to < 2, 2 to < 3, 3 to < 4, 4 to < 5, 5 to < 6
18. 1, 2, 3, 4, 5, 6

19. and **20.**

Exports to	Frequency	Relative frequency
1 to < 2	3	3/9 = 0.3333
2 to < 3	1	1/9 = 0.1111
3 to < 4	2	2/9 = 0.2222
4 to < 5	2	2/9 = 0.2222
5 to < 6	1	1/9 = 0.1111
Total	9	9/9 = 1.0000

21. 2 to < 4, 4 to < 6, 6 to < 8, 8 to < 10, 10 to < 12
22. 2, 4, 6, 8, 10, 12
23. and **24.**

Imports from	Frequency	Relative frequency
2 to < 4	4	4/9 = 0.4444
4 to < 6	3	3/9 = 0.3333
6 to < 8	0	0/9 = 0.0000
8 to < 10	0	0/9 = 0.0000
10 to < 12	2	2/9 = 0.2222
Total	9	9/9 = 1.0000

25. 2 to < 3.5, 3.5 to < 5, 5 to < 6.5, 6.5 to < 8, 8 to < 9.5, 9.5 to < 11, 11 to < 12.5
26. 2, 3.5, 5, 6.5, 8, 9.5, 11, 12.5
27. and **28.**

Imports from	Frequency	Relative frequency
2 to < 3.5	2	2/9 = 0.2222
3.5 to < 5	4	4/9 = 0.4444
5 to < 6.5	1	1/9 = 0.1111
6.5 to < 8	0	0/9 = 0.0000
8 to < 9.5	0	0/9 = 0.0000
9.5 to < 11	1	1/9 = 0.1111
11 to < 12.5	1	1/9 = 0.1111
Total	9	9/9 = 1.0000

29. −6 to < −5, −5 to < −4, −4 to < −3, −3 to < −2, −2 to < −1, −1 to < 0, 0 to < 1, 1 to < 2

30. −6, −5, −4, −3, −2, −1, 0, 1, 2
31. and **32.**

Trade balance	Frequency	Relative frequency
−6 to < −5	2	2/9 = 0.2222
−5 to < −4	0	0/9 = 0.0000
−4 to < −3	0	0/9 = 0.0000
−3 to < −2	1	1/9 = 0.1111
−2 to < −1	4	4/9 = 0.4444
−1 to < 0	0	0/9 = 0.0000
0 to < 1	1	1/9 = 0.1111
1 to < 2	1	1/9 = 0.1111
Total	9	9/9 = 1.0000

33. 0 to < 25, 25 to < 50, 50 to < 75, 75 to < 100, 100 to < 125, 125 to < 150, 150 to < 175, 175 to < 200, 200 to < 225
34. 0, 25, 50, 75, 100, 125, 150, 175, 200, 225
35. and **36.**

Motor vehicle theft rate	Frequency	Relative frequency
0 to < 25	1	1/20 = 0.05
25 to < 50	6	6/20 = 0.30
50 to < 75	4	4/20 = 0.20
75 to < 100	3	3/20 = 0.15
100 to < 125	3	3/20 = 0.15
125 to < 150	0	0/20 = 0.00
150 to < 175	2	2/20 = 0.10
175 to < 200	0	0/20 = 0.00
200 to < 225	1	1/20 = 0.05
Total	20	20/20 = 1.00

37. 20 to < 45, 45 to < 70, 70 to < 95, 95 to < 120, 120 to < 145, 145 to < 170, 170 to < 195, 195 to < 220
38. 20, 45, 70, 95, 120, 145, 170, 195, 220
39. and **40.**

Motor vehicle theft rate	Frequency	Relative frequency
20 to < 45	5	5/20 = 0.25
45 to < 70	6	6/20 = 0.30
70 to < 95	3	3/20 = 0.15
95 to < 120	3	3/20 = 0.15
120 to < 145	0	0/20 = 0.00
145 to < 170	1	1/20 = 0.05
170 to < 195	1	1/20 = 0.05
195 to < 220	1	1/20 = 0.05
Total	20	20/20 = 1.00

41.

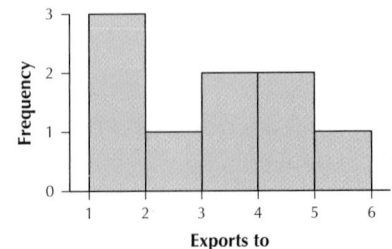

42.

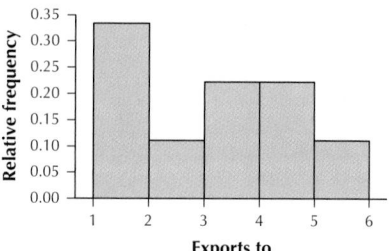

48.

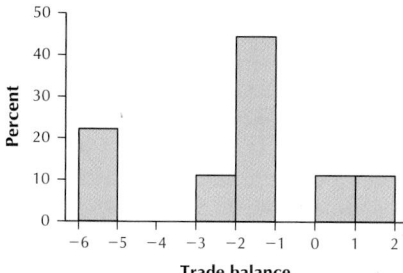

43.

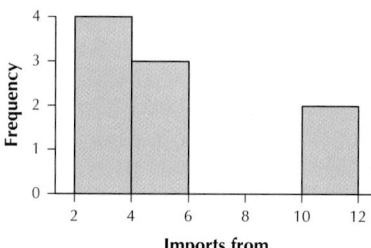

49.

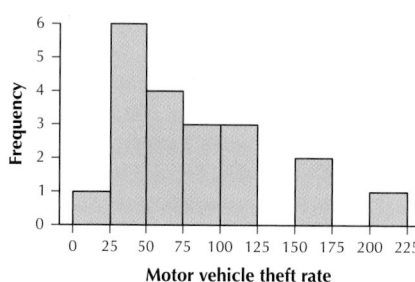

44.

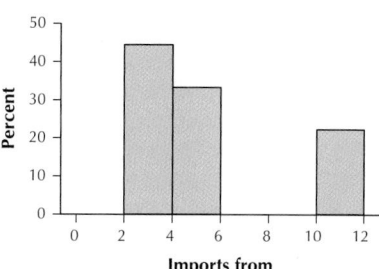

50.

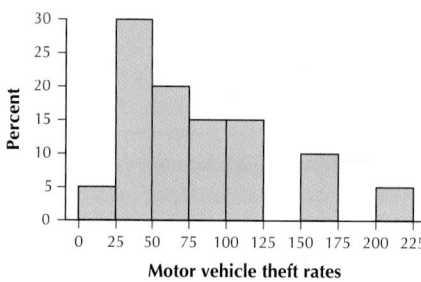

45.

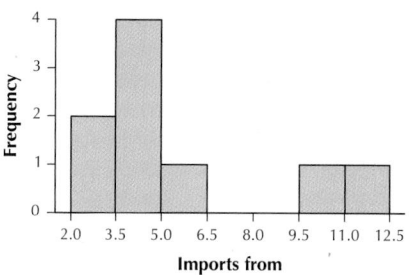

51.

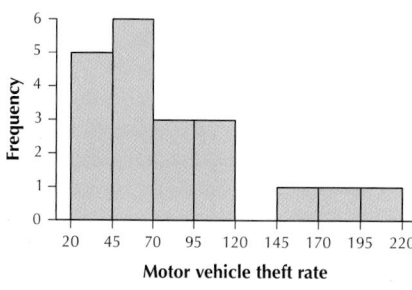

46.

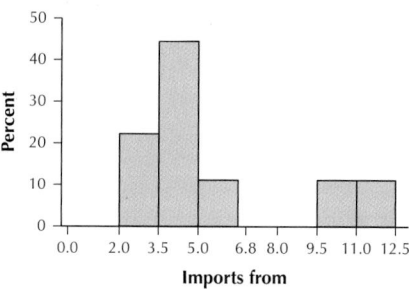

52.

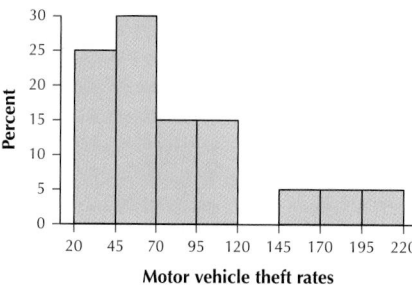

47.

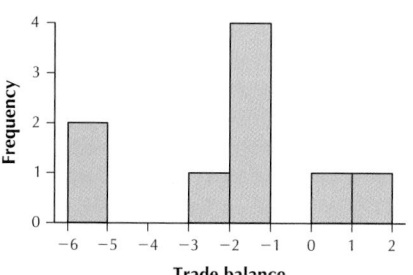

53.

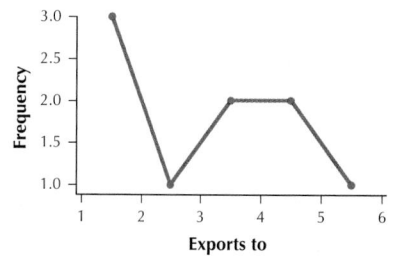

54.

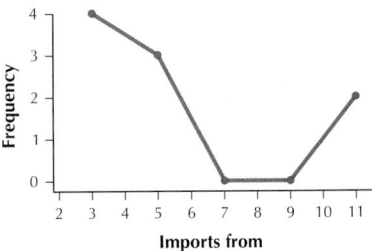

55.

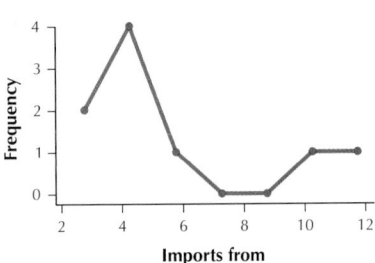

56.

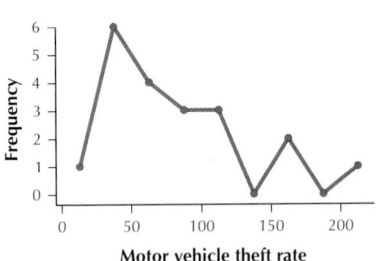

57.

```
1 | 279
2 | 8
3 | 58
4 | 45
5 | 6
```

58.

```
 2 | 5
 3 | 257
 4 | 04
 5 | 6
 6 |
 7 |
 8 |
 9 |
10 | 0
11 | 3
```

59.

```
-5 | 66
-4 |
-3 |
-2 | 4
-1 | 8832
 0 | 0
 1 | 0
```

60.

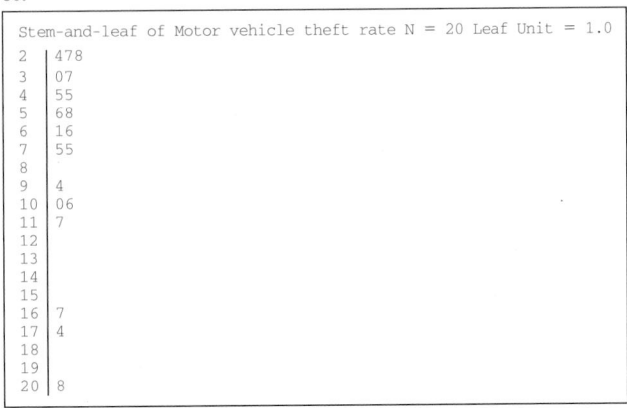

61.

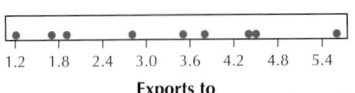

62.

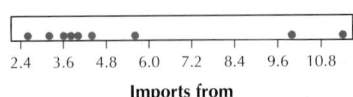

63.

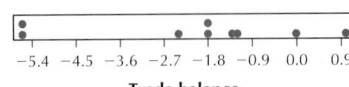

64.

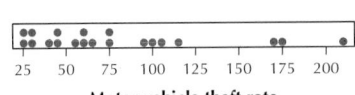

65.

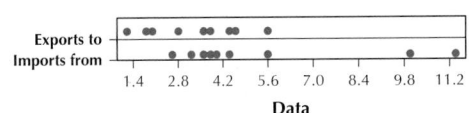

66.

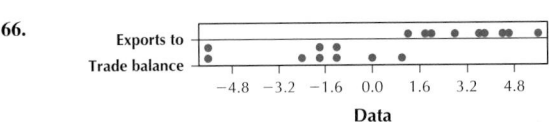

67. 0.75, 2.25, 3.75, 5.25, 6.75, 8.25, 9.75, 11.25
68. 48, 0.48
69. 2, 0.02
70. 50
71. 2, 6, 10, 14, 18, 22, 26, 30
72. 20 and 24
73. No. The histogram does not tell us the actual quiz scores. Yes.
74. 0.03
75. 56.25, 58.75, 61.25, 63.75, 66.25, 68.75, 71.25, 73.75, 76.25
76. 65 inches and 67.5 inches
77. The scale on the x-axis would shift up by 1 inch.
78. The scale on the y-axis, the heights and widths of the rectangles, the class width, and the number of rectangles
79. Right-skewed
80. Left-skewed
81. Symmetric
82. Right-skewed
83. (a)

Fruit cups sold per day	Frequency	Relative frequency
0	8	8/47 = 0.1702
1	12	12/47 = 0.2553
2	17	17/47 = 0.3617
3	6	6/47 = 0.1277
4	4	4/47 = 0.0851
Total	47	47/47 = 1.0000

(b)

Fruit cups sold per day	Frequency	Relative frequency
0–1	20	20/47 = 0.4255
2–3	23	23/47 = 0.4894
4–5	4	4/47 = 0.0851
Total	47	47/47 = 1.0000

(c) 0.2128, (a), Since 2 and 3 fruit cups sold per day are grouped together, we don't know how many of the days had 3 fruit cups sold that day.

84. (a)

Number of sandwiches sold	Frequency	Relative frequency
0	1	1/47 = 0.0213
1	2	2/47 = 0.0426
2	5	5/47 = 0.1064
3	7	7/47 = 0.1489
4	10	10/47 = 0.2128
5	3	3/47 = 0.0638
6	8	8/47 = 0.1702
7	4	4/47 = 0.0851
8	6	6/47 = 0.1277
9	1	1/47 = 0.0213
Total	47	47/47 = 1.0000

(b)

Number of sandwiches sold	Frequency	Relative frequency
0–2	8	8/47 = 0.1702
3–5	20	20/47 = 0.4255
6–8	18	18/47 = 0.3830
9–11	1	1/47 = 0.0213
Total	47	47/47 = 1.0000

(c) No. More numbers at the lower end of the distributions.
(d) 0.5957
85.

(a) and **(b)**

Frauds	Frequency	Relative frequency
0 to < 40	1	1/23 = 0.0435
40 to < 80	8	8/23 = 0.3478
80 to < 120	8	8/23 = 0.3478
120 to < 160	2	2/23 = 0.0870
160 to < 200	2	2/23 = 0.0870
200 to < 240	1	1/23 = 0.0435
240 to < 280	1	1/23 = 0.0435
Total	23	23/23 = 1.0000

(c)

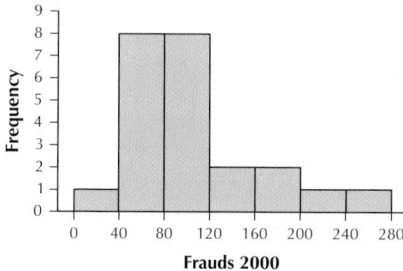

(d)

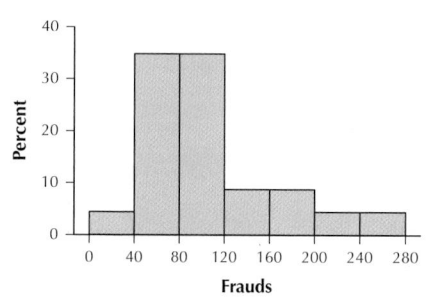

86.
(a), (b), (c)

Frauds	Frequency	Relative frequency
0 to < 40	8	8/23 = 0.3478
40 to < 80	13	13/23 = 0.5652
80 to < 120	1	1/23 = 0.0435
120 to < 160	1	1/23 = 0.0435
160 to < 200	0	0/23 = 0.0000
200 to < 240	0	0/23 = 0.0000
240 to < 280	0	0/23 = 0.0000
Total	23	23/23 = 1.0000

(d)

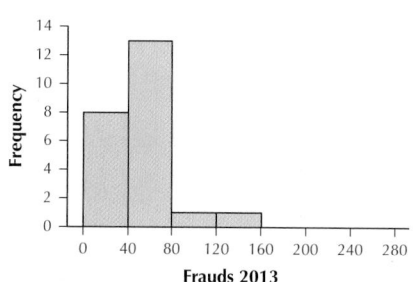

(e)

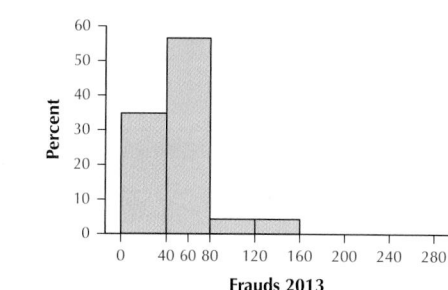

(f) The number of fraud cases in Brooklyn has decreased. This is good news.
87. (a) 0.0870 **(b)** 0.6522 **(c)** 0.3478, 1 − 0.6522 = 0.3478
88. (a) 20, 60, 100, 140, 180, 220, 260

(b)

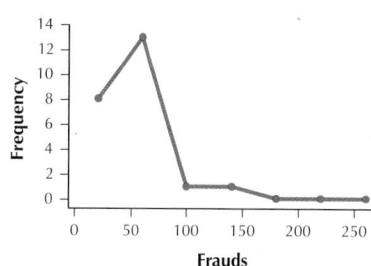

(c)

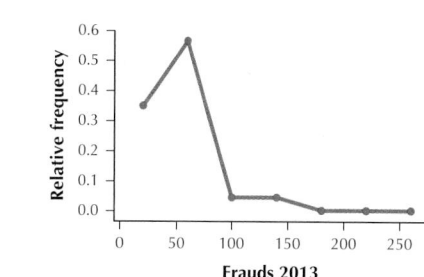

(d) 0.0435

89. (a)

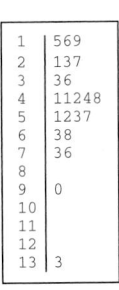

```
 1 | 569
 2 | 137
 3 | 36
 4 | 11248
 5 | 1237
 6 | 38
 7 | 36
 8 |
 9 | 0
10 |
11 |
12 |
13 | 3
```

(b)

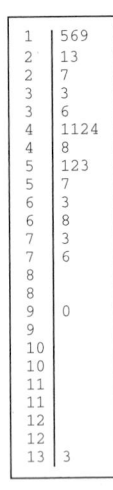

```
 1 | 569
 2 | 13
 2 | 7
 3 | 3
 3 | 6
 4 | 1124
 4 | 8
 5 | 123
 5 | 7
 6 | 3
 6 | 8
 7 | 3
 7 | 6
 8 |
 8 |
 9 | 0
 9 |
10 |
10 |
11 |
11 |
12 |
12 |
13 | 3
```

(c) 9. Could not have used histogram because the histogram does not contain the actual data.

90. (a)

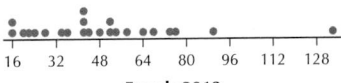

Frauds 2013

(b)

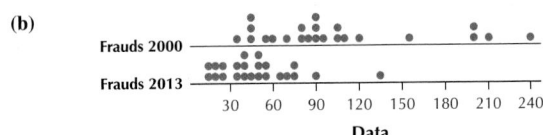

Data

(c) There were less fraud cases in 2013 than in 2000.

91. (a), (b), (c)

Coffee sold per day	Frequency	Relative frequency
0 to < 8	7	7/47 = 0.1489
8 to < 16	7	7/47 = 0.1489
16 to < 24	10	10/47 = 0.2128
24 to < 32	15	15/47 = 0.3191
32 to < 40	6	6/47 = 0.1277
40 to < 48	1	1/47 = 0.0213
48 to < 56	1	1/47 = 0.0213
Total	47	47/47 = 1.0000

(d)

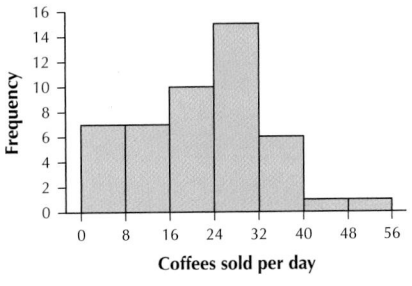

(e)

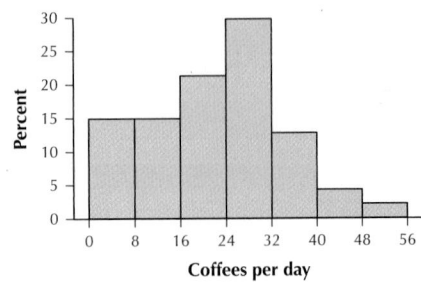

92.

(a), (b), (c)

Sodas sold per day	Frequency	Relative frequency
10 to < 15	6	6/47 = 0.1277
15 to < 20	4	4/47 = 0.0851
20 to < 25	8	8/47 = 0.1702
25 to < 30	6	6/47 = 0.1277
30 to < 35	8	8/47 = 0.1702
35 to < 40	6	6/47 = 0.1277
40 to < 45	2	2/47 = 0.0426
45 to < 50	3	3/47 = 0.0638
50 to < 55	3	3/47 = 0.0638
55 to < 60	1	1/47 = 0.0213
Total	47	47/47 = 1.0000

(d)

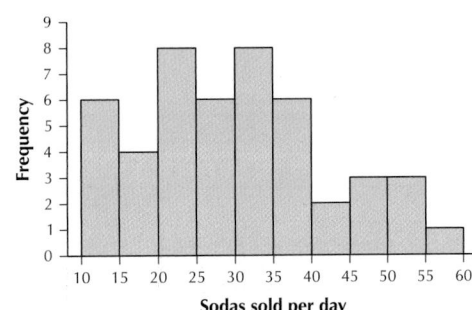

(e)

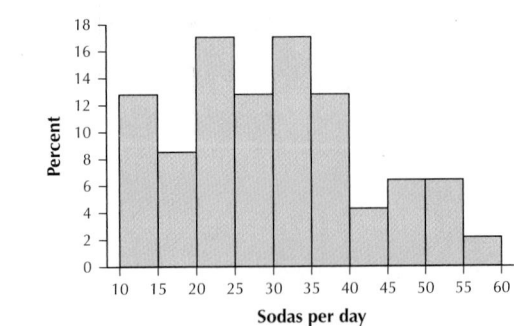

93. (a) 0.0426 **(b)** 0.1702 **(c)** 0.8298; 1 − 0.1702 = 0.9298 **(d)** 0.5319
94. (a) 0.3191 **(b)** 0.2128 **(c)** 0.7872, 1 − 0.2128 = 0.7872 **(d)** 0.5957
95. (a) 4, 12, 20, 28, 36, 44, 52

(b)

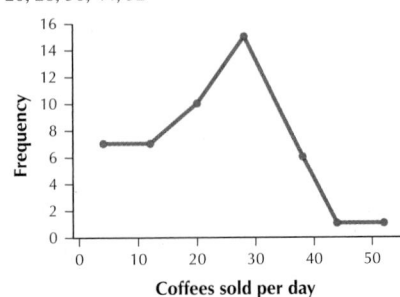

(c)

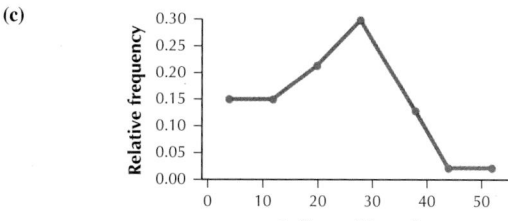

(d) 0.3404
96. **(a)** 12.5, 17.5, 22.5, 27.5, 32.5, 37.5, 42.5, 47.5, 52.5, 57.5
(b)

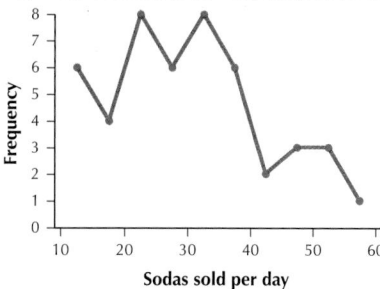

(c)

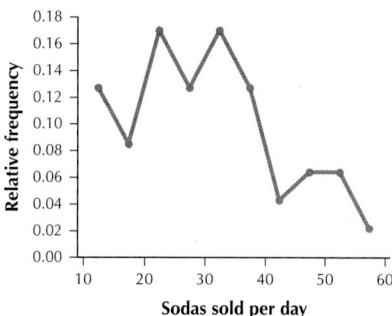

(d) 0.6170
97. **(a)**

```
0 | 344445688
1 | 01134668
2 | 00112334455677789
3 | 0011233455
4 | 118
```

(b)

```
0 | 34444
0 | 5688
1 | 01134
1 | 668
2 | 001123344
2 | 55677789
3 | 00112334
3 | 55
4 | 11
4 | 8
```

(c)

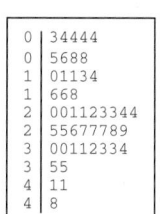

(d) 15. No, the stem-and-leaf display contains the actual data.
98. **(a)**

```
1 | 1233345579
2 | 00234444566679
3 | 01133334556679
4 | 23558
5 | 0045
```

(b)

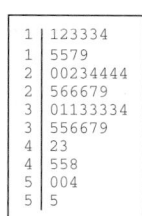

```
1 | 123334
1 | 5579
2 | 00234444
2 | 566679
3 | 01133334
3 | 556679
4 | 23
4 | 558
5 | 004
5 | 5
```

(c)

(d) 20. No, the stem-and-leaf display contains the actual data.
99. **(a)** Right-skewed, tail on right
(b)

(c) No. Histogram does not contain actual data.
100. **(a)** 2000 **(b)** 1000 **(c)** 1000 to 3000 **(d)** 17,000 to 19,000
101. **(a)** Approximately 72 **(b)** Approximately 2 **(c)** Approximately 18
102. **(a)** and **(b)**

Continent	Frequency	Relative frequency
Africa	1	0.1
Asia	5	0.5
Europe	1	0.1
North America	2	0.2
South America	1	0.1

(c) We cannot construct because the variable is qualitative. **(d)** We cannot construct because the variable is qualitative. **(e)** We cannot construct because the variable is qualitative.
103. **(a)** Divide the frequency values by the total frequency—classes not affected **(b)** change the scale along the relative frequency (vertical) axis by multiplying the relative frequency values by the total frequency—shape of distribution not affected **(c)** 19
104. **(a)** 2 **(b)** 0.1053 **(c)** 5 **(d)** 0.2632
105. **(a)** 0 **(b)** 0 **(c)** $25 to $27.5 has the largest relative frequency, 4/19 = 0.2105. **(d)** 3 **(e)** 0
106. **(c)** Tending to be left-skewed
107. Data set: 23 24 25 26 27 28 28 29 30 31 31 32 32 32 33 35 36 37 39 40
108.

Classes	Frequency	Relative frequency
22–25	3	3/20 = 0.15
26–29	5	5/20 = 0.25
30–33	7	7/20 = 0.35
34–37	3	3/20 = 0.15
38–41	2	2/20 = 0.10

109. Histogram with five classes

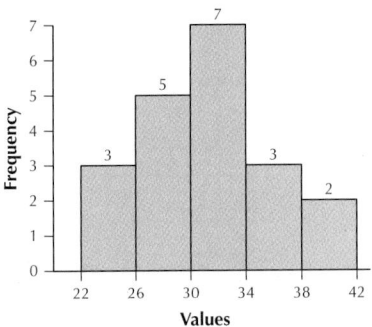

110.

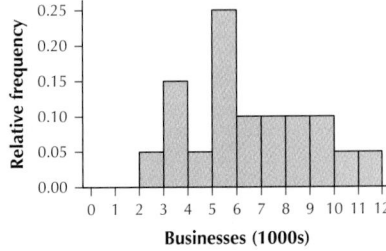

111. (a) 15 (b) 37.5 (c) 52.5 (d) 67.5 to 82.5 (e) 22.5 to 37.5
112. (a) Approximately 33 (b) Approximately 9 (c) No, because this interval does not contain only values in the 90s.
113. Fairly symmetrical
114. 62%; middle
115. $n \approx 690$.
116. No. We need the original data to construct a stem-and-leaf display. Since the data are grouped with a class width of 2, we only know how many observations are in each class, not the actual numbers.
117. (a) and (b)

Businesses (1000s)	Frequency	Relative frequency
2 to < 3	1	1/20 = 0.05
3 to < 4	3	3/20 = 0.15
4 to < 5	1	1/20 = 0.05
5 to < 6	5	5/20 = 0.25
6 to < 7	2	2/20 = 0.10
7 to < 8	2	2/20 = 0.10
8 to < 9	2	2/20 = 0.10
9 to < 10	2	2/20 = 0.10
10 to < 11	1	1/20 = 0.05
11 to < 12	1	1/20 = 0.05
Total	20	20/20 = 1.00

(c)

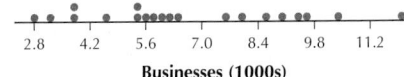

118. (a)

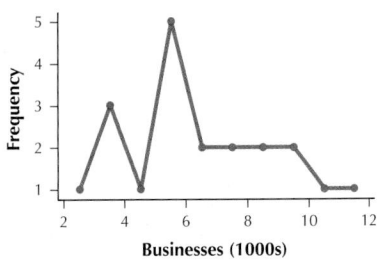

(b)

(c)

```
2  | 7
3  | 188
4  | 6
5  | 44679
6  | 13
7  | 59
8  | 69
9  | 35
10 | 3
11 | 9
```

119.

	Dotplot	Histogram	Stem-and-leaf	Frequency polygon
(a) Symmetry and skewness	Appropriate to use for small ranges of data	Appropriate to use	Appropriate to use for small ranges of data	Appropriate to use
(b) Construct using pencil and paper	Easily done for small ranges of data	Easily done for small ranges of data	Easily done for small ranges of data	Easily done for small ranges of data
(c) Retain complete knowledge of the data	Appropriate	Appropriate only if the data are ungrouped	Appropriate	Appropriate only if the data are ungrouped
(d) Presentation in front of non-statisticians	Appropriate	Appropriate	Appropriate	Appropriate

120. (a) The scale on the x-axis would be 1000 less. Everything else would stay the same. (b) The scale on the x-axis would be 1000 less. Everything else would stay the same. (c) All of the numbers would be 1000 less. (d) The scale on the x-axis would be 1000 less. Everything else would stay the same.
121. 961; 25
122. The histogram is right-skewed.

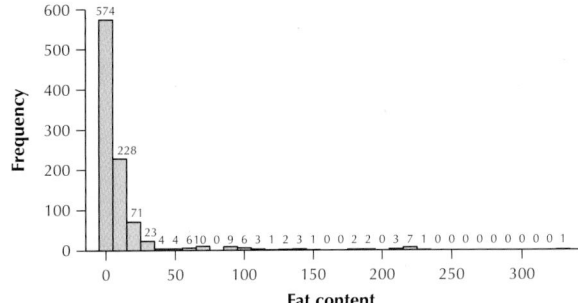

123. Yes; fats and oils.
124. The histogram is right-skewed.

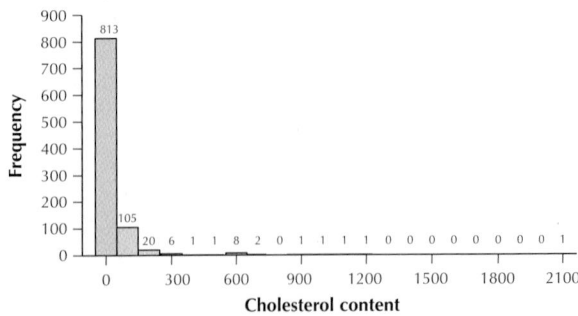

125. One whole cheesecake (2053 grams of cholesterol)
126. (a) 9 (b) 0.30
127. (a) 2 (b) 4.00, 4.30
128. (a) Decreases (b) Increases
129. (a) Increases (b) Decreases
130. (a) 3 (b) 57. There are 57 observations in the data set.
131. (a) 6 (b) 57 (c) Split stems; answers will vary
132. Answers will vary.
133. Answers will vary.

134.

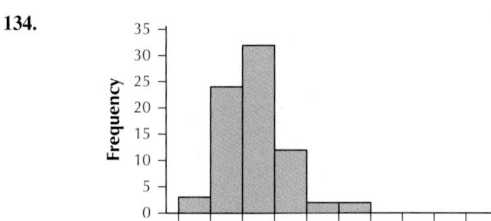

135.

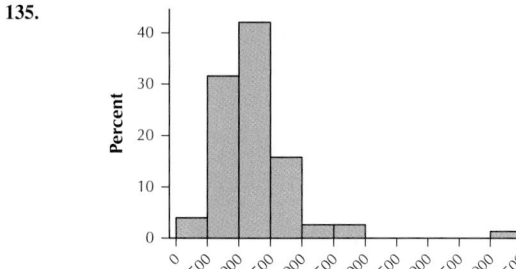

2000

136.

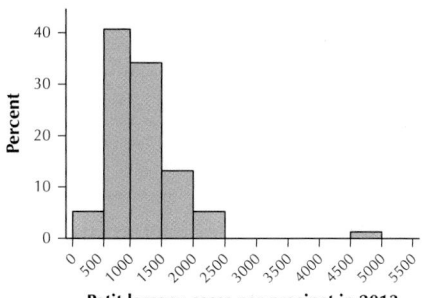

Petit larceny cases per precinct in 2013

137. The number of petit larceny cases appears to be decreasing. This is good news.
138. Precinct 14, midtown south precinct
139.

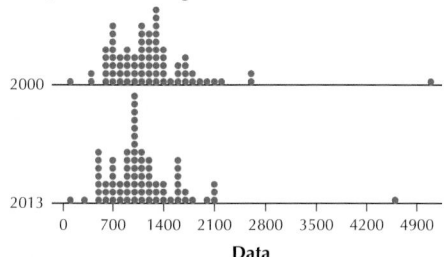

The number of petit larcenies appears to be decreasing.
140. For 2000

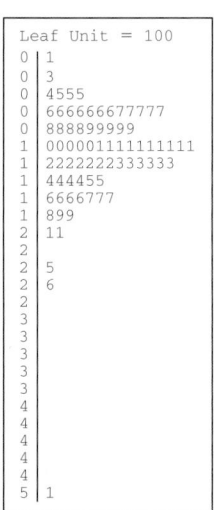

```
Leaf Unit = 100
0 | 1
0 | 3
0 | 4555
0 | 666666677777
0 | 888899999
1 | 000001111111111
1 | 2222222333333
1 | 444455
1 | 6666777
1 | 899
2 | 11
2 |
2 | 5
2 | 6
2 |
3 |
3 |
3 |
3 |
3 |
3 |
4 |
4 |
4 |
4 |
4 |
4 |
5 | 1
```

For 2013

```
Stem-and-leaf of 2013 N = 76
Leaf Unit = 10
 1 | 1
 2 | 9
 3 |
 4 | 57
 5 | 0013468
 6 | 49
 7 | 023447
 8 | 2256788
 9 | 236678999
10 | 00023445778
11 | 11456678
12 | 058
13 | 27
14 | 13
15 | 359
16 | 0034
17 | 02
18 | 1
19 |
20 | 179
21 | 2
22 |
23 |
24 |
25 |
26 |
27 |
28 |
29 |
30 |
31 |
32 |
33 |
34 |
35 |
36 |
37 |
38 |
39 |
40 |
41 |
42 |
43 |
44 |
45 | 5
```

The number of petit larceny cases appears to be decreasing.
141.

	Dotplot	Histogram	Stem-and-leaf
(a) Symmetry and skewness	Appropriate to use for small ranges of data	Appropriate to use	Appropriate to use for small ranges of data
(b) Construct using pencil and paper	Easily done for small ranges of data	Easily done for small ranges of data	Easily done for small ranges of data
(c) Retain complete knowledge of the data	Appropriate	Appropriate only if the data are ungrouped	Appropriate
(d) Presentation in front of non-statisticians	Appropriate	Appropriate	Appropriate

Section 2.3
1. A frequency distribution gives the frequency counts for each class (grouped or ungrouped). A cumulative frequency distribution gives the number of values that are less than or equal to the upper limit of a given class for grouped data or it gives the number of values that are less than or equal to a given number for ungrouped data.
2. A cumulative frequency distribution gives the number of values that are less than or equal to the upper limit of a given class for grouped data or it gives the number of values that are less than or equal to a given number for ungrouped data. A cumulative relative frequency distribution gives the proportion of values that are less than or equal to the upper limit of a given class for grouped data or it gives the proportion of values that are less than or equal to a given number for ungrouped data.
3. Ogive.

4. Plot the upper class limits along the x axis and the cumulative frequency or cumulative relative frequency along the y axis and connect the points with line segments.

5. Time series data.

6. Plot the values of the variable along the y axis and the time points (hours, days, months, years) along the x axis. Connect these points with line segments.

7 and 8.

Age	Frequency (millions)	Relative frequency	Cumulative frequency (millions)	Cumulative relative frequency
$15 \le x < 35$	22.7	0.24	22.7	0.24
$35 \le x < 45$	22.2	0.24	22.7 + 22.2 = 44.9	0.24 + 0.24 = 0.48
$45 \le x < 55$	25.8	0.27	22.7 + 22.2 + 25.8 = 70.7	0.24 + 0.24 + 0.27 = 0.75
$55 \le x < 65$	23.2	0.25	22.7 + 22.2 + 25.8 + 23.2 = 93.9	0.24 + 0.24 + 0.27 + 0.25 = 1.00

9. **(a)** and **(b)**

Carbon emissions	Frequency	Relative frequency	Cumulative frequency	Cumulative relative frequency
$50 \le x < 60$	2	2/20 = 0.10	2	0.10
$60 \le x < 70$	5	5/20 = 0.25	2 + 5 = 7	0.10 + 0.25 = 0.35
$70 \le x < 80$	2	2/20 = 0.10	2 + 5 + 2 = 9	0.10 + 0.25 + 0.10 = 0.45
$80 \le x < 90$	2	2/20 = 0.10	2 + 5 + 2 + 2 = 11	0.10 + 0.25 + 0.10 + 0.10 = 0.55
$90 \le x < 100$	6	6/20 = 0.30	2 + 5 + 2 + 2 + 6 = 17	0.10 + 0.25 + 0.10 + 0.10 + 0.30 = 0.85
$100 \le x < 110$	2	2/20 = 0.10	2 + 5 + 2 + 2 + 6 + 2 = 19	0.10 + 0.25 + 0.10 + 0.10 + 0.30 + 0.10 = 0.95
$110 \le x < 120$	1	1/20 = 0.05	2 + 5 + 2 + 2 + 6 + 2 + 1 = 20	0.10 + 0.25 + 0.10 + 0.10 + 0.30 + 0.10 + 0.05 = 1.00
Total	20	20/20 = 1.00		

10. **(a)** and **(b)**

Unemployment rate	Frequency	Relative frequency	Cumulative frequency	Cumulative relative frequency
$1 \le x < 3.5$	2	2/20 = 0.10	2	0.10
$3.5 \le x < 6$	5	5/20 = 0.25	2 + 5 = 7	0.10 + 0.25 = 0.35
$6 \le x < 8.5$	7	7/20 = 0.35	2 + 5 + 7 = 14	0.10 + 0.25 + 0.35 = 0.70
$8.5 \le x < 11$	3	3/20 = 0.15	2 + 5 + 7 + 3 = 17	0.10 + 0.25 + 0.35 + 0.15 = 0.85
$11 \le x < 13.5$	3	3/20 = 0.15	2 + 5 + 7 + 3 + 3 = 20	0.10 + 0.25 + 0.35 + 0.15 + 0.15 = 1.00
Total	20	20/20 = 1.00		

11. **(a)** and **(b)**

Dangerous weapons cases	Frequency	Relative frequency	Cumulative frequency	Cumulative relative frequency
$0 \le x < 20$	3	3/20 = 0.15	3	0.15
$20 \le x < 40$	5	5/20 = 0.25	8	0.40
$40 \le x < 60$	4	4/20 = 0.20	12	0.60
$60 \le x < 80$	2	2/20 = 0.10	14	0.70
$80 \le x < 100$	5	5/20 = 0.25	19	0.95
$100 \le x < 120$	1	1/20 = 0.05	20	1.00
Total	20	1.00		

12. **(a)** and **(b)**

Frauds	Frequency	Relative frequency	Cumulative frequency	Cumulative relative frequency
$0 \le x < 20$	3	3/23 = 0.1304	3	0.1304
$20 \le x < 40$	5	5/23 = 0.2174	8	0.3478
$40 \le x < 60$	9	9/23 = 0.3913	17	0.7391
$60 \le x < 80$	4	4/23 = 0.1739	21	0.9130
$80 \le x < 100$	1	1/23 = 0.0435	22	0.9565
$100 \le x < 120$	0	0/23 = 0.0000	22	0.9565
$120 \le x < 140$	1	1/23 = 0.0435	23	1.0000
Total	23	23/23 = 1.0000		

13. **(a)**

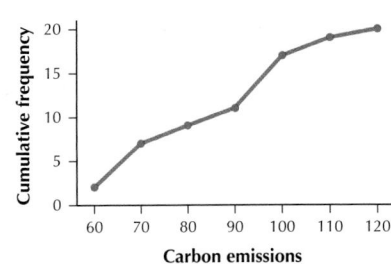

(b)

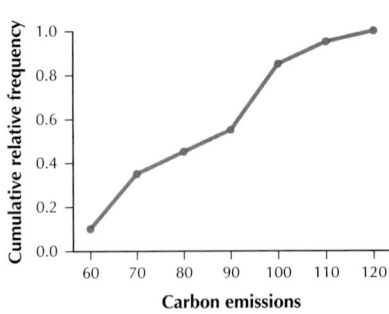

14. **(a)**

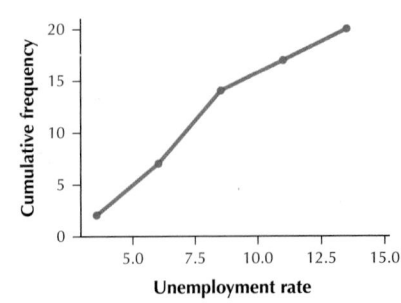

(b)

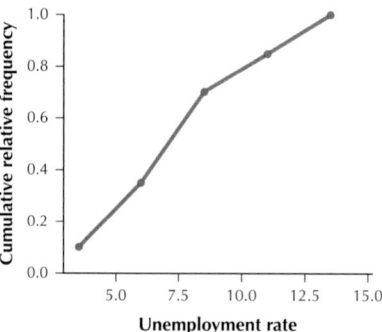

15. (a)

(b)

16. (a)

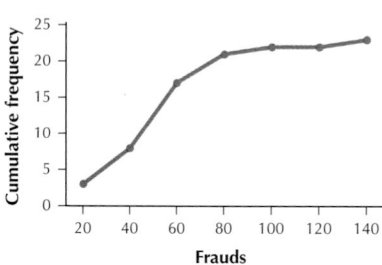

(b)

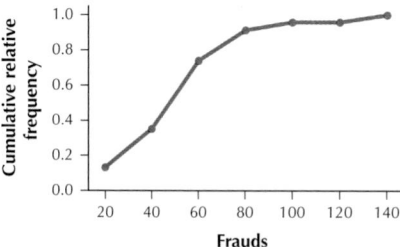

17. (a)

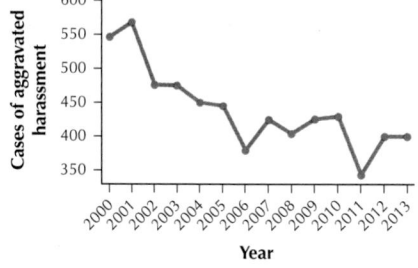

(b) Generally decreasing

18. (a)

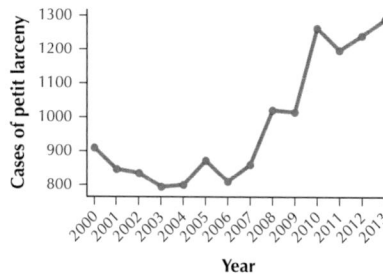

(b) Increasing in recent years.
19. (a) 0.8 **(b)** 2.39 **(c)** 1.99
20. (a) Approximately 120 **(b)** Approximately 300 **(c)** Approximately 67
21. (a)

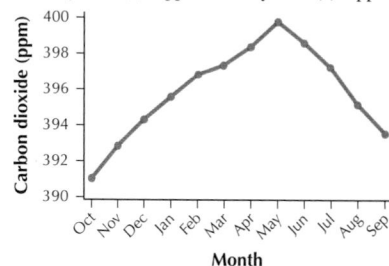

(b) The level of carbon dioxide increases from October to May and decreases from May to September.
22.

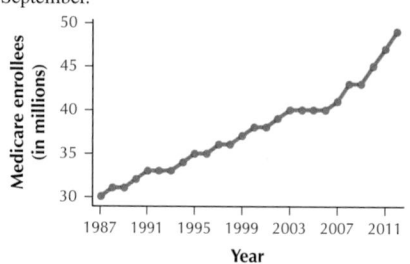

23. 2009
24.

Agricultural exports (in billions of dollars)	Frequency	Cumulative frequency
$0−$1.9	3	3
$2.0−$3.9	9	12
$4.0−$5.9	6	18
$6.0−$7.9	1	19
$8.0−$9.9	0	19
$10.0−$11.9	0	19
$12.0−$13.9	1	20
Total	20	

(a) 12 **(b)** 18 **(c)** 2
25.

Agricultural exports (in billions of dollars)	Frequency	Relative frequency	Cumulative relative frequency
$0−$1.9	3	0.15	0.15
$2.0−$3.9	9	0.45	0.60
$4.0−$5.9	6	0.30	0.90
$6.0−$7.9	1	0.05	0.95
$8.0−$9.9	0	0	0.95
$10.0−$11.9	0	0	0.95
$12.0−$13.9	1	0.05	1.00
Total	20	1.00	

(a) 0.60 **(b)** 0.90 **(c)** 0.10

26.

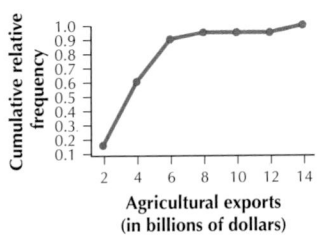

27. (a)

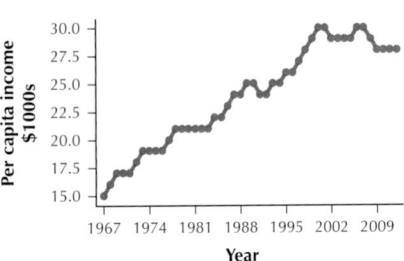

(b) 2008

28. (a)

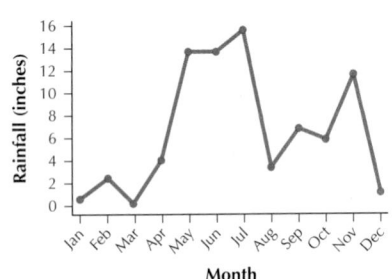

(b) Summer

29. The only change would be that the entire time series graph would shift up 3 units. The horizontal scale would stay the same and the shape of the graph would stay the same.

30. (a)

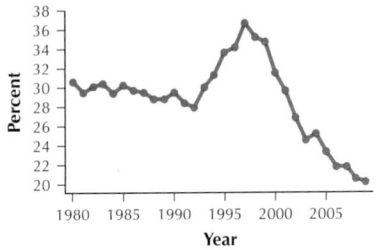

(b) The percent of 12th-graders who smoked cigarettes was relatively constant from 1980 to 1992, it increased from 1992 to 1997, and then it decreased from 1997 to 2009.

31.

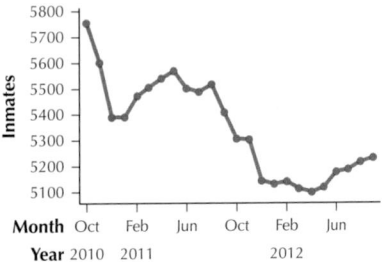

32. PS4
33. 3DS

34.

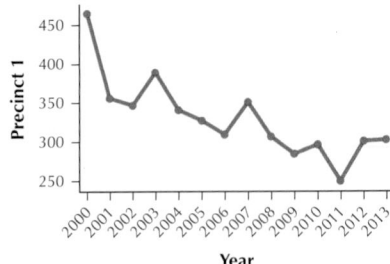

The number of assaults appears to be decreasing.

35 and 36.

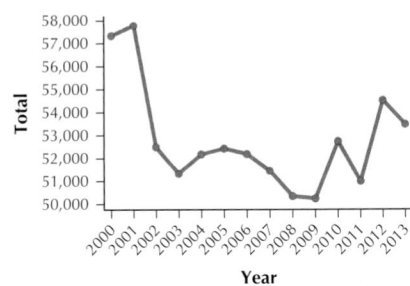

Section 2.4

1. Answers will vary.
2. False
3. Figure 54 visually reinforces the magnitude of the differences.
4. By factoring the "number of cars per country," we can compute the number of cars stolen per 1000 cars in each population. This would give more useful statistics than the per capita statistics.
5. Table 44 gives the actual number of cars stolen.
6. (a) Bar graph **(b)** Omitting the zero on the relevant scale; biased distortion or embellishment; inaccuracy in relative lengths of bars in a bar chart. **(c)** We can use a Pareto chart or pie chart.
7. (a) Biased distortion or embellishment; omitting the zero on the relevant scales; inaccuracy in relative lengths of bars in a bar chart. **(b)** A Pareto chart or pie chart can be used.
8. (a) Biased distortion or embellishment; omitting the zero on the relevant scales; inaccuracy in relative lengths of bars in a bar chart; careless combination of categories. **(b)** We can use a Pareto chart. Note: we cannot split the categories.
9. (a) The number of people living with AIDS is increasing.
(b) Using two dimensions (area) to emphasize a one-dimensional difference.
(c)

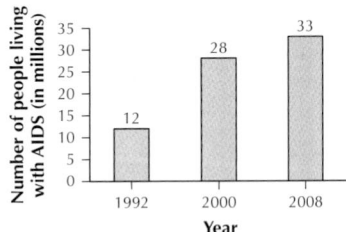

10. (a) Omitting the zero on the vertical scale.
(b)

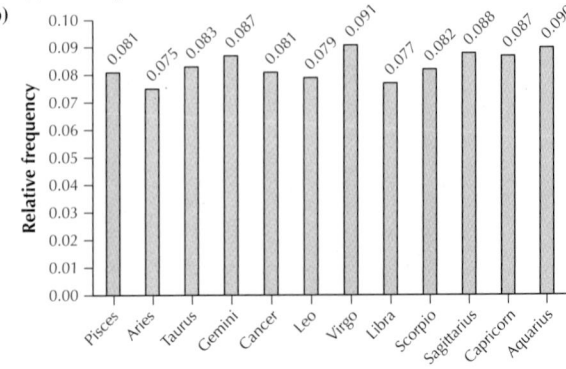

11. (a)

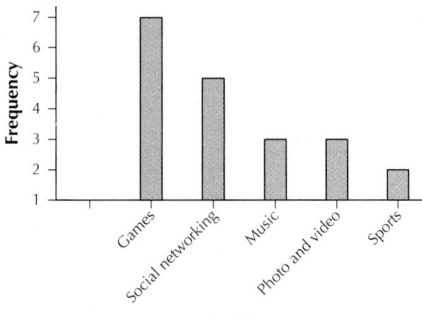

(b) Manipulating the scale, omitting the 0 on the vertical scale

(c)

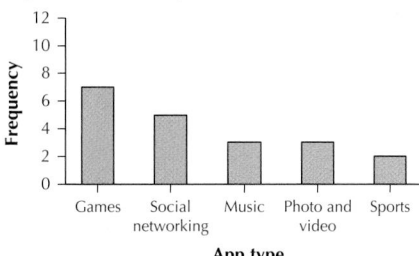

(d) Manipulating the scale

12. (a)

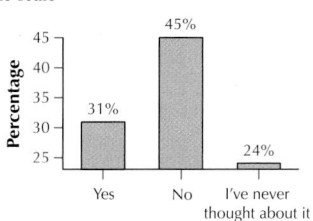

(b)

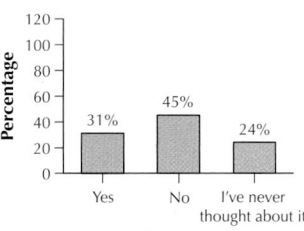

(c)

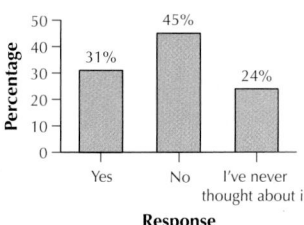

13. (a)

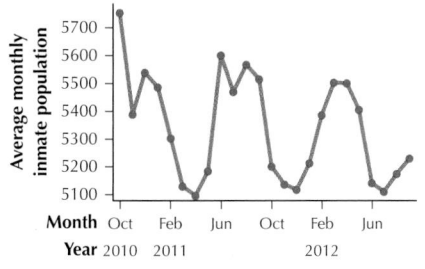

(b)

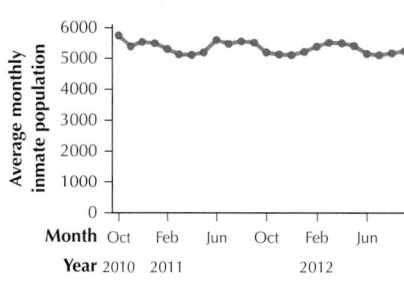

14. Answers will vary.
15. Answers will vary.
16. Answers will vary.

Chapter 2 Review

1. No, because the variable is categorical.

2.

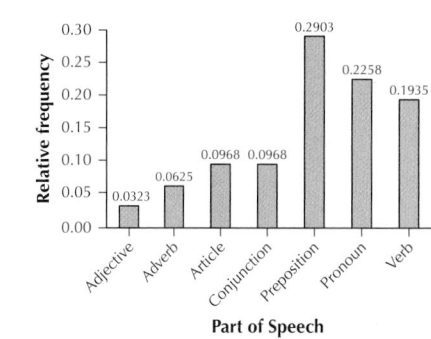

3. and 4.

Part of speech	Frequency	Relative frequency
Adjective	1	0.0323
Adverb	2	0.0645
Article	3	0.0968
Conjunction	3	0.0968
Preposition	9	0.2903
Pronoun	7	0.2258
Verb	6	0.1935
Total	31	31/31 = 1.000

5.

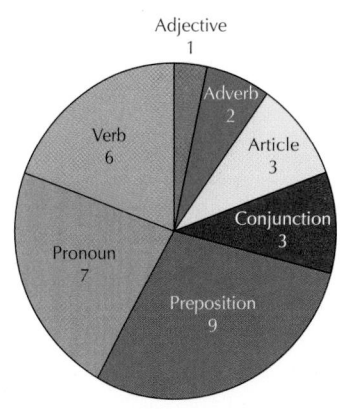

6.

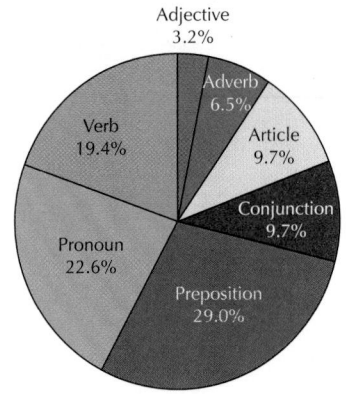

7. 0.6612

8. 0.6076

9. 0.0246

10. 0.0402

11. Answers will vary. May have clustered bar graph by happiness of marriage or clustered bar graph by sex.

12.

Chips	Frequency	Relative frequency
0 to < 7	12	0.2553
7 to < 14	28	0.5957
14 to < 21	5	0.1064
21 to < 28	2	0.0426
Total	47	1.0000

13.

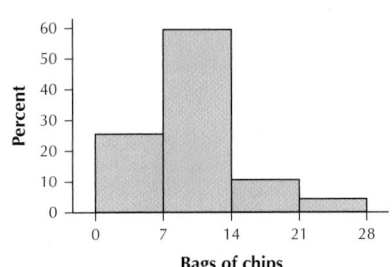

14. 4.26%

15, 16, 17.

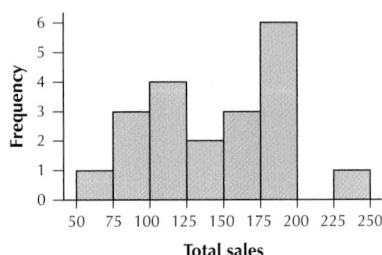

18. 5%

19. 0.5

20. 45%

21. (a) and (b)

Frauds	Frequency	Relative frequency	Cumulative frequency	Cumulative relative frequency
0 to < 40	8	8/23 = 0.3478	8	0.3478
40 to < 80	13	13/23 = 0.5652	21	0.9130
80 to < 120	1	1/23 = 0.0435	22	0.9565
120 to < 160	1	1/23 = 0.0435	23	1.0000
Total	23	1.0000		

22. (a)

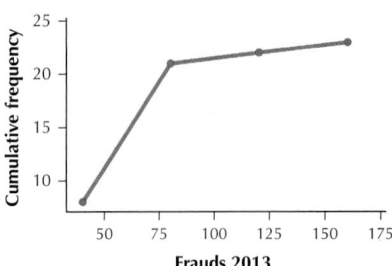

(b)

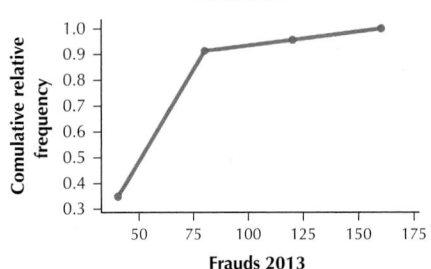

23. (a)

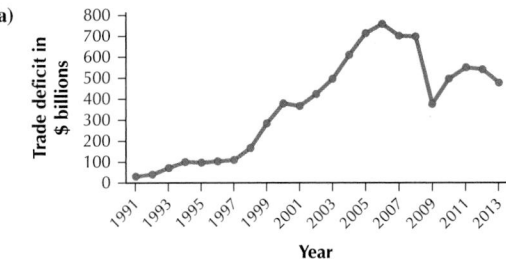

(b) The trade deficit generally increased from 1991 to 2006, then it generally decreased from 2006 until 2009, then it increased from 2009 until 2011, and then it decreased from 2011 to 2013.

24. (a)

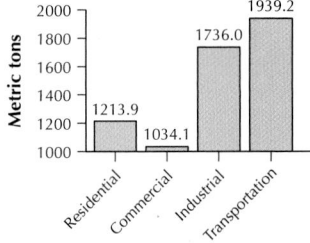

(b) Manipulating the scale.

(c)

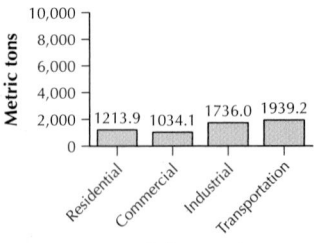

(d) Manipulating the scale.

(e)

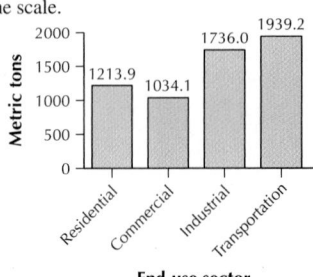

Chapter 2 Quiz
1. False
2. True
3. sample size
4. frequency distribution
5. Symmetric
6. Right skewed

7 – 10.

Life Expectancy	Frequency	Relative frequency	Cumulative frequency	Cumulative relative frequency
$40 \leq x < 50$	1	0.1	1	0.1
$50 \leq x < 60$	0	0	1	0.1
$60 \leq x < 70$	3	0.3	4	0.4
$70 \leq x < 80$	3	0.3	7	0.7
$80 \leq x < 90$	3	0.3	10	1.0
Total	10	1.0		

11.

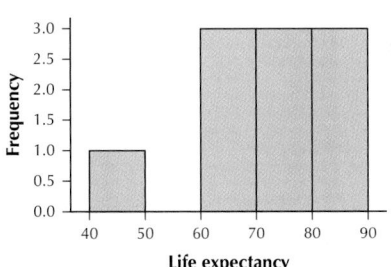

12.

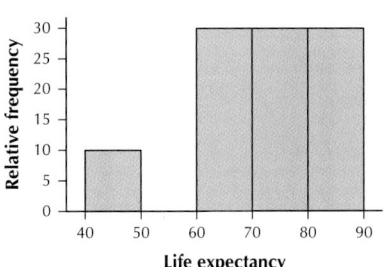

13.

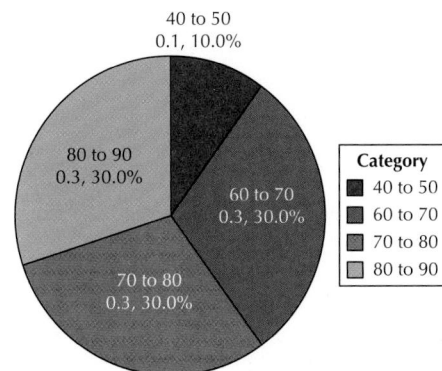

14.

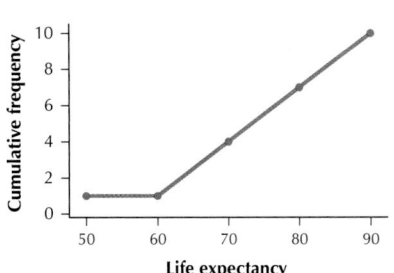

15.

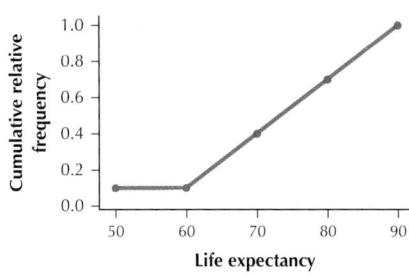

Chapter 3

Section 3.1
1. A value that locates the center of the data set.
2. Mean; answers will vary.
3. Because the mean depends in part on the sum of all data values, an outlier will skew the mean (pull it in one direction or another). Since the median simply depends on position in an ordered list, it is not sensitive to outliers.
4. Mean, the median, and the mode
5. Sample size (n)
6. Population size (N)
7. $\sum x$
8. μ
9. $\bar{x}$
10. Median
11. Mode
12. $\bar{x}$
13. (a) 6 (b) $0.83 billion
14. (a) 5 (b) 77.4 wins
15. (a) 10 (b) 118.37 motor vehicles stolen per 100,000 residents
16. (a) 10 (b) 516.9
17. (a) 10 (b) 1758.2 petit larceny cases
18. (a) 10 (b) 165.4 criminal trespass cases
19. (a) 3 (b) $1.33 billion
20. (a) 3 (b) 80 wins
21. (a) 3 (b) 158.67 motor vehicles stolen per 100,000 residents
22. (a) 4 (b) 548.25
23. (a) 5 (b) 2110 petit larceny cases
24. (a) 4 (b) 200.25 criminal trespass cases
25. $3.5 billion, larger than $1.33 billion
26. 65 wins, less than 80 wins
27. 369, larger than 158.67
28. 438.6, smaller than 548.25
29. 5091.67 petit larceny cases, larger than 2110 petit larceny cases
30. 460.2 criminal trespass cases, larger than 200.25 criminal trespass cases
31. (a) $1.4 billion (b) $1.9 billion
32. (a) 78 wins (b) 74.5 wins
33. (a) 167.8 motor vehicles stolen per 100,000 residents (b) 187.9 motor vehicles stolen per 100,000 residents
34. (a) 554 (b) 552
35. (a) 1607 petit larceny cases (b) 1810.5 petit larceny cases
36. (a) 176.5 criminal trespass cases (b) 233 criminal trespass cases
37. 24 dangerous weapons cases
38. 4 multi-platinum singles
39. 6.2
40. Games
41. Mean < Median < Mode
42. Mode < Median < Mean
43. Mean = Median = Mode
44. Mode < Median < Mean
45. (a) 4 (b) 7.75 wins
46. (a) 6 (b) 5.5 electoral votes
47. (a) 2 (b) 8 wins
48. (a) 3 (b) 6 electoral votes
49. (a) 0.92 million units (b) 0.9 million units
50. (a) 30.8 weeks (b) 27 weeks
51. (a) 9.45 (b) 14.05
52. (a) 3.575 (b) 8.65

53. (a) 27.125 years (b) 25.5 years
54. (a) 64.5625 inches (b) 63.65 inches
55. (a) 192.2 calories (b) 172 calories
56. (a) 2.12 grams (b) 1.1 grams
57. 9
58. −$2.08 billion
59. −$1.8 billion
60. −$5.6 billion and −$1.8 billion
61. (a) 6 cylinders (b) 5 cylinders (c) 4 cylinders
62. The mode. It represents the "typical" number of cylinders.
63. (a) 3.317 liters (b) 2.75 liters (c) No mode
64. (a) 24 mpg (b) 21.5 mpg (c) 18 mpg
65. USA
66. (a) 16.6 (b) 17 (c) 5 and 15
67. (a) Mean = 18.8 (b) Median = 19 (c) No mode
68. (a) Mean = 59.8 (b) Median = 57.5 (c) Mode = 55
69. $7.83, 7.83, no mode, A typical book on the best-sellers list would cost less than $14.00 since the mean and median are all less than $14.
70. (a) $17.83 (b) The new mean is $10 more than the original mean. (c) If a number d is added to each value of a data set, the mean of the resulting data set will be equal to the mean of the original data set plus d.
71. (a) $39.13 (b) The new mean is 5 times the original mean. (c) If each value of a data set is multiplied by a number m, the mean of the resulting data set will be the mean of the original data set times m.
72. (a) No mode (b) George R. R. Martin
73. It does not make sense to find the mean and the median of the variable author because the variable author is a categorical variable.
74. Cell phone; no
75. Manhattan
76. No. The mean and the median can only be calculated for numerical data and the variables *violation type* and *borough* are categorical variables.
77. Mean = 2013.5, Median = 2014, Mode = 2015
78. (a) 0 years (b) Mean = 1.5 years, Median = 1 year
79. Mean = 11.5 years, Median = 11 years, Mode = 10 years
80. $6.12
81. $21.07
82. $25
83. 2 Mb
84. 25
85. 16
86. (a) n = 59 (b) Sample mean = 45.46 (c) Sample median = 42; (d) Highest ratings = 93.7, lowest ratings = 22.4
87. The mean is larger than the median, which implies the distribution is positively skewed.
88. New mean = 40.46; new median = 37; new mode = old mode − 5. Each statistic will be affected equally.
89. The mean, median and mode will all be halved. Each statistic will be affected equally.
90. (a) Female (b) Approximately 73 (c) Approximately 74 (d) Female; yes
91. (a) 73 (b) 76 (c) Females; yes
92. (a) 78 (b) 79 (c) Females; yes
93. (a) Decrease (b) Unchanged (c) Unchanged
94. (a) 6.615 thousand businesses (b) 6.51875 thousand businesses (c) 6.4917 thousand businesses
95. (a) Smaller (b) Larger (c) Same
96. (a) $7.60 (b) 2013
97. 2.190 mph
98. Since she walked the first mile at a speed of 5 mph, her time for walking the first mile was $\frac{1}{5}$ hour. Similarly, her time to walk the second, third, fourth, and fifth miles was $\frac{1}{4}, \frac{1}{3}, \frac{1}{2}$, and $\frac{1}{1}$ hour, respectively. Thus the total time it took Emily to walk 5 miles is $\frac{1}{5} + \frac{1}{4} + \frac{1}{3} + \frac{1}{2} + \frac{1}{1} = \sum \frac{1}{x}$ hours. Therefore her average speed is $\frac{\text{distance}}{\text{time}} = \frac{n}{\sum \frac{1}{x}}$, which is the harmonic mean. The arithmetic mean is just the average of the 5 rates.
99. Geometric mean = 1.0664, 6.64% growth
100. Answers will vary.
101. Answers will vary.
102. Answers will vary.
103. Answers will vary.
104. (a) Answers will vary. (b) The mean increases. (c) The median remains the same.

105. Answers will vary.
106. Mean = 12,102, Median = 8256
107. The mean
108. Mean = 11,253, Median = 8189
109. Mean = 1,562,798, Median = 876,193
110. Grand Theft Auto V for Xbox 360, Mean = 1,360,928, Median = 840,698, the mean
111. (a) PS4 (b) Nintendo (c) Shooter
112. Mean = 27.70, Median = 27, Modes: 2, 3, 26, 27
113. (a) Increase by x (b) Increase by x (c) Increase by x

Section 3.2

1. Deviation for a data value gives the distance the value is from the mean.
2. A distance from the mean that is representative for that data set
3. Benefit—simple to calculate, Drawbacks—quite sensitive to extreme values, does not use all of the data values.
4. False
5. Benefit—uses all of the numbers in a data set. Drawback—can be time-consuming to calculate.
6. Population variance
7. False
8. Never
9. When all of the data values are the same
10. When the distribution is bell-shaped.
11. (a) 2.2 (b) $0.83 billion (c) −$0.63 billion, −$0.53 billion, −$0.53 billion, −$0.43 billion, $0.57 billion, $1.57 billion (d) 0.6556 billions of dollars squared (e) $0.8097 billion dollars
12. (a) 45 wins (b) 77.4 wins (c) 18.6 wins, 13.6 wins, 0.6 win, −6.4 wins, −26.4 wins (d) 253.84 wins squared (e) 15.93 wins
13. (a) 142 motor vehicles stolen per 100,000 people (b) 118.37 motor vehicles stolen per 100,000 people (c) 89.63, 55.73, 49.43, −1.17. −12.37, −18.17, −24.27, −43.17, −43.27, −52.37 motor vehicles stolen per 100,000 people (d) 2113.4821 motor vehicles stolen per 100,000 people squared (e) 45.97 motor vehicles stolen per 100,000 people
14. (a) 201 (b) 516.9 (c) 61.1, 54.1, 41.1, 39.1, 35.1, 10.1, 3.1, −6.9, −96.9, −139.9 (d) 4023.09 (e) 63.43
15. (a) 3967 petit larceny cases (b) 1758.2 petit larceny cases (c) 255.8, −470.2, −203.2, −1174.2, −151.2, −763.2, 335.8, 2792.8, −935.2, 312.8 petit larceny cases (d) 1,119,682.96 petit larceny cases squared (e) 1058.15 petit larceny cases
16. (a) 266 criminal trespass cases (b) 165.4 criminal trespass cases (c) −57.4, −60.4, −52.4, 67.6, 53.6, 41.6, −30.4, 174.6, −91.4, −45.4 criminal trespass cases (d) 6068.64 criminal trespass cases squared (e) 77.90 criminal trespass cases
17. (a) 1.2133 billion dollars squared (b) $1.10 billion (c) For this sample of state export data, the typical difference between a state's export amount and the mean export amount is $1.10 billion.
18. (a) 103 wins squared (b) 10.15 wins (c) For this sample of baseball teams, the typical difference between a baseball team's number of wins and the mean number of wins is 10.15 wins.
19. (a) 2967.7733 motor vehicles stolen per 100,000 people squared (b) 54.48 motor vehicles stolen per 100,000 people (c) For this sample of motor vehicle theft data, the typical difference between a country's motor vehicle theft rate and the mean motor vehicle theft rate is 54.48 motor vehicles stolen per 100,000 people.
20. (a) 206.9167 (b) 14.38 (c) For this sample of science scores, the typical difference between a country's average science score and the mean science score is 14.38.
21. (a) 2,046,275 petit larceny cases squared (b) 1430.48 petit larceny cases (c) For this sample of New York precincts, the typical difference between a precinct's number of petit larceny cases and the mean number of petit larceny cases is 1430.48 petit larceny cases.
22. (a) 11,850.9167 criminal trespass cases squared (b) 108.86 criminal trespass cases (c) For this sample of New York precincts, the typical difference between a precinct's number of criminal trespass cases and the mean number of criminal trespass cases is 108.86 criminal trespass cases.
23. About 68%
24. About 95%
25. About 99.7%
26. About 16%
27. About 68%
28. About 2.5%
29. About 2.5%

30. About 95%
31. At least 75%
32. At least 88.89%
33. At least 93.75%
34. At least 91.84%
35. At least 93.75%
36. At least 88.89%
37. At least 55.56%
38. Can't do since $k = 1$.
39. (i)—(d); (ii)—(b); (iii)—(c); (iv)—(a)
40. (i)—(b); (ii)—(d); (iii)—(a); (iv)—(c)
41. (a) 1.5 million game units (b) 0.327 million game units squared (c) 0.5718 million game units
42. (a) 68 weeks (b) 777.7 weeks squared (c) 27.89 weeks
43. (a) 69.6 (b) 522.7286 (c) 22.86
44. (a) 28.6 (b) 115.2393 (c) 10.73
45. (a) 21 years (b) 47.5536 years squared (c) 6.90 years
46. (a) 7.2 inches (b) 6.5370 inches squared (c) 2.56 inches
47. (a) 316 calories (b) 13,520.1778 calories squared (c) 116.28 calories
48. (a) 6.8 grams (b) 5.8507 grams squared (c) 2.42 grams
49. No, standard deviation
50. (a) No (b) No (c) Yes
51. At least 60.94%
52. 76.67%, Yes
53. The darts
54. Yes
55. About 68%
56. No. The Empirical Rule does not tell us anything about what percent of the data values lies within 0.5 standard deviation of the mean.
57. (a) No (b) No (c) Yes
58. At least 55.56%
59. (a) Yes (b) Yes (c) Yes
60. About 95%
61. (a) No (b) No (c) Yes
62. At least 55.56%
63. (a) 8 cylinders (b) 9.6 cylinders^2 (c) 3.098 cylinders
64. (a) 4.9 liters (b) 3.078 liters^2 (c) 1.754 liters
65. (a) 30 mpg (b) 116 mpg^2 (c) 10.770 mpg
66. Range for Colony A = 73; range for Colony B = 91 (a) Colony B (b) Colony B
67. Colony A = 21.91, Colony B = 26.35 (a) Colony B (b) Colony B; yes (c) Colony B, because it has the larger standard deviation
68. (a) 60.24 wins squared (b) 7.76 wins
69. (a) 16.33 wins squared (b) 4.04 wins
70. (a) Boston Red Sox and Toronto Blue Jays. The number of wins for these two teams are the farthest apart (b) Baltimore Orioles and New York Yankees, both teams have the same number of wins
71. (a) About 68% (b) About 95% (c) Can't do, the Empirical Rule does not say what percent of the data values lie within 2.67 standard deviations of the mean.
72. (a) About 68% (b) About 95% (c) About 97.5%
73. (a) Can't do (b) At least 75% (c) At least 85.94%
74. (a) Can't do (b) At least 75% (c) Can't do

75.

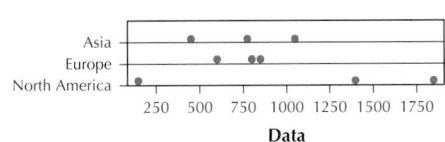

North America, because the dots for North America are more spread out than the dots for Asia and Europe.
76. Asia: Range = 591 watts per person, Variance = 87,651 watts per person squared Europe; Range = 239 watts per person, Variance = 15,582.33 watts per person squared; North America: Range = 1740 watts per person, Variance = 810,500.33 watts per person squared. Yes.
77. (a) North America (b) Europe
78. (a) Reduce them. (b) Range = 469 watts per person, Variance = 109,980.5 watts per person squared, Standard deviation = 331.63 watts per person. Yes.
79. (a) Increase because the contribution of a value of 84 to the computation is greater than the contribution of a value of 72. (b) 7.27; yes.

80. (a) Range = 15; standard deviation = 5.48. (b) Adding a positive constant to each value in a data set will not change the value of the original range or standard deviation.
81. (a) Range = 12; standard deviation = 4.9 (b) Multiplying each value in a data set by a constant will increase the value of the original range or standard deviation by a factor of the constant.
82. (a) Cylinders: CV = 51.64%; Engine size: CV = 52.89%; City mpg: CV = 44.88% (b) Engine size; City mpg
83. Asia: 39.32%, Europe: 16.37%, North America: 79.34% Yes
84. (a) Cylinders: MAD = 2; Engine size: MAD = 1.189; City mpg: MAD = 8.333 (b) City mpg; Engine size
85. (a) Asia: 204 watts per person; Europe: 93.56 watts per person; North America: 669.11 watts per person (b) North America, yes
86. (a) Skewness = 0; symmetric (b) Skewness = 3; right−skewed (c) Skewness = −3; left−skewed (d) Skewness = −1.5; left-skewed (e) Skewness = 0; symmetric (f) Skewness = 0.6; right-skewed
87. 0
88. (a) Pros: Skewness = 0.182; Darts: Skewness = 0.196; DJIA: Skewness = −0.077 (b) Pros and Darts are slightly right-skewed, DJIA are slightly left-skewed.
89. (a)

	Range	Sample variance	Sample standard deviation	Coefficient of variation	Mean absolute deviation
Cylinders	8	9.6	3.098	51.64%	2
Engine size	4.9	3.078	1.754	52.89%	1.189
City mpg	30	116	10.770	44.88%	8.333

(b) City mpg: Range, Sample variance, Sample standard deviation, Mean absolute deviation. Engine size: Coefficient of variation; Cylinders: None of them.
90. (a) Asia: 296.06 watts per person; Europe: 124.83 watts per person; North America: 900.28 watts per person
(b)

Continent	Range	Sample variance	Sample standard deviation	Coefficient of variation	Mean absolute deviation
Asia	591	87,651	296.06	39.32	204
Europe	239	15,582.33	124.83	16.37	93.56
North America	1740	810,500.33	900.28	79.34	669.11

(c) Yes (d) less
91. Answers will vary.
92. Answers will vary.
93. Answers will vary.
94. Answers will vary.
95. Answers will vary.
96.

```
Variable   Mean    StDev
PROS       10.95   22.25
Variable   Mean    StDev
DARTS      4.52    19.39
Variable   Mean    StDev
DJIA       6.793   8.031
```

97.

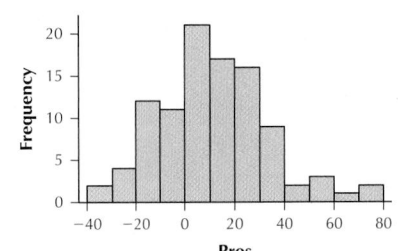

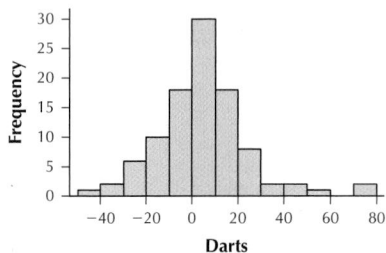

Darts

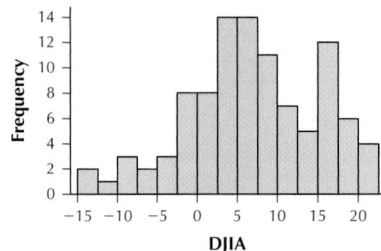

DJIA

98. (a)
$\mu - 1\sigma = 10.95 - 1 \cdot 22.25 = -11.3, \mu + 1\sigma = 10.95 + 1 \cdot 22.25 = 33.2$
$\mu - 2\sigma = 10.95 - 2 \cdot 22.25 = -33.55, \mu + 2\sigma = 10.95 + 2 \cdot 22.25 = 55.45$
$\mu - 3\sigma = 10.95 - 3 \cdot 22.25 = -55.8, \mu + 3\sigma = 10.95 + 3 \cdot 22.25 = 77.7$
(b) About 68% of the stock returns lie between -11.3 and 33.2. About
95% of the stock returns lie between -33.55 and 55.45. About 99.7% of
the stock returns lie between -55.8 and 77.7. **(c)** 70% of the stock returns
lie between -11.3 and 33.2. 96% of the stock returns lie between -33.55
and 55.45. 100% of the stock returns lie between -55.8 and 77.7. **(d)** The
actual percentages that lie within each interval are close to the approximate
percentages given by the Empirical Rule.
99. (a)
$\mu - 1\sigma = 4.52 - 1 \cdot 19.39 = -14.87, \mu + 1\sigma = 4.52 + 1 \cdot 19.39 = 23.91$
$\mu - 2\sigma = 4.52 - 2 \cdot 19.39 = -34.26, \mu + 2\sigma = 4.52 + 2 \cdot 19.39 = 43.3$
$\mu - 3\sigma = 4.52 - 3 \cdot 19.39 = -53.65, \mu + 3\sigma = 4.52 + 3 \cdot 19.39 = 62.69$
(b) About 68% of the stock returns lie between -14.87 and 23.91. About
95% of the stock returns lie between -34.26 and 43.3. About 99.7% of the
stock returns lie between -53.65 and 62.69. **(c)** 76% of the stock returns
lie between -14.87 and 23.91. 94% of the stock returns lie between -34.26
and 43.3. 98% of the stock returns lie between -53.65 and 62.69. **(d)** The
actual percentages that lie within each interval are close to the approximate
percentages given by the Empirical Rule.
100. (a)
$\mu - 1\sigma = 6.793 - 1 \cdot 8.031 = -1.238$
$\mu + 1\sigma = 6.793 + 1 \cdot 8.031 = 14.824$
$\mu - 2\sigma = 6.793 - 2 \cdot 8.031 = -9.269$
$\mu + 2\sigma = 6.793 + 2 \cdot 8.031 = 22.855$
$\mu - 3\sigma = 6.793 - 3 \cdot 8.031 = -17.3$
$\mu + 3\sigma = 6.793 + 3 \cdot 8.031 = 30.886$
(b) About 68% of the stock returns lie between -1.238 and 14.824. About
95% of the stock returns lie between -9.269 and 22.855. About 99.7% of the
stock returns lie between -17.3 and 30.886. **(c)** 66% of the stock returns lie
between -1.238 and 14.824. 96% of the stock returns lie between -9.269
and 22.855. 100% of the stock returns lie between -17.3 and 30.886. **(d)** The
actual percentages that lie within each interval are close to the approximate
percentages given by the Empirical Rule.

Section 3.3
1. These formulas will provide only estimates because we will not know the
exact data values.
2. The weighted mean is the sum of the products of the data points with their
respective weights, divided by the sum of the weights.
3. The mean
4. 70
5. 80.5
6. 22.8571
7. 2.45
8. 82.5

9. (a)–(c)

Class	Frequency f	Midpoint x	Product $f \cdot x$
$0 \le GPA < 1.0$	2	0.5	1.0
$1.0 \le GPA < 2.0$	10	1.5	15.0
$2.0 \le GPA < 3.0$	13	2.5	32.5
$3.0 \le GPA < 4.0$	5	3.5	17.5
Total	$\sum f = 30$		$\sum(f \cdot x) = 66$

(d) $\bar{x} = \dfrac{\sum(f \cdot x)}{\sum f} = \dfrac{66}{30} = 2.2$

10. (a)–(c)

Class	Frequency f	Midpoint x	Product $f \cdot x$
$-10 \le$ golf score < -5	3	-7.5	-22.5
$-5 \le$ golf score < 0	7	-2.5	-17.5
$0 \le$ golf score < 5	7	2.5	17.5
$5 \le$ golf score < 10	3	7.5	22.5
Total	$\sum f = 20$		$\sum(f \cdot x) = 0$

(d) $\bar{x} = \dfrac{\sum(f \cdot x)}{\sum f} = \dfrac{0}{20} = 0$

11. (a)–(c)

Class	Frequency f	Midpoint x	Product $f \cdot x$
$0 \le$ score < 2	10	1	10
$2 \le$ score < 4	20	3	60
$4 \le$ score < 6	30	5	150
$6 \le$ score < 8	20	7	140
$8 \le$ score < 10	10	9	90
Total	$\sum f = 90$		$\sum(f \cdot x) = 450$

(d) $\bar{x} = \dfrac{\sum(f \cdot x)}{\sum f} = \dfrac{450}{90} = 5$

12. (a)–(c)

Class	Frequency f	Midpoint x	Product $f \cdot x$
$0 \le$ grade < 50	5	25	125
$50 \le$ grade < 70	10	60	600
$70 \le$ grade < 80	15	75	1125
$80 \le$ grade < 90	20	85	1700
$90 \le$ grade < 100	20	95	1900
Total	$\sum f = 70$		$\sum(f \cdot x) = 5450$

(d) $\bar{x} = \dfrac{\sum(f \cdot x)}{\sum f} = \dfrac{5450}{70} = 77.86$

13. (a)–(c)

Class	Frequency f	Midpoint x	Product $f \cdot x$
$0 \le$ cost < 5	100	2.5	250
$5 \le$ cost < 10	150	7.5	1,125
$10 \le$ cost < 15	200	12.5	2,500
$15 \le$ cost < 20	250	17.5	4,375
$20 \le$ cost < 30	300	25	7,500
$30 \le$ cost < 50	350	40	14,000
$50 \le$ cost < 100	400	75	30,000
$100 \le$ cost < 200	450	150	67,500
Total	$\sum f = 2200$		$\sum(f \cdot x) = 127,250$

(d) $\bar{x} = \dfrac{\sum(f \cdot x)}{\sum f} = \dfrac{127,250}{2200} = 57.84$

14. (a)–(c)

Class	Frequency f	Midpoint x	Product $f \cdot x$
$0 \leq$ cash < 10	15	5	75
$10 \leq$ cash < 20	10	15	150
$20 \leq$ cash < 30	5	25	125
$30 \leq$ cash < 40	4	35	140
$40 \leq$ cash < 50	4	45	180
$50 \leq$ cash < 75	2	62.5	125
$75 \leq$ cash < 100	1	87.5	87.5
$100 \leq$ cash < 200	1	150	150
Total	$\sum f = 42$		$\sum(f \cdot x) = 1032.5$

(d) $\bar{x} = \dfrac{\sum(f \cdot x)}{\sum f} = \dfrac{1032.5}{42} = 24.58$

15. Variance: $s^2 = 0.6767$; Standard deviation: $s = 0.8226$
16. Variance: $s^2 = 21.25$; Standard deviation: $s = 4.6098$
17. Variance: $s^2 = 5.3333$; Standard deviation: $s = 2.3094$
18. Variance: $s^2 = 345.4082$; Standard deviation: $s = 18.5852$
19. Variance: $s^2 = 2672.3270$; Standard deviation: $s = 51.6946$
20. Variance: $s^2 = 746.1062$; Standard deviation: $s = 27.3149$
21. (a)

Age	Frequency	Midpoints
0–4.99	63,422	2.5
5–17.99	240,629	11.5
18–64.99	540,949	41.5

(b) Estimated mean $= 30.0298$ years **(c)** Estimated standard deviation $= 15.455909$ years; estimated variance $= 238.8851$ years squared
22. (a)

Class (1000s)	Housing units	Midpoints (1000s)
$0 \leq$ value < 50	5,430	25
$50 \leq$ value < 100	90,605	75
$100 \leq$ value < 150	90,620	125
$150 \leq$ value < 200	54,295	175
$200 \leq$ value < 300	34,835	250
$300 \leq$ value < 500	15,770	400
$500 \leq$ value < 1000	5.595	750

(b) Estimated mean $= \$158,079.25$ **(c)** Estimated standard deviation $= \$116,223.13$; estimated variance $= 13,507,815,862.68$ (dollars squared)
23. Estimated mean $= 135.5224$; estimated standard deviation $= 95.6874$
24. 82.8%
25. $58.72
26. $70,940
27. If $w_i = 1$ for all i, then the weighted mean formula will be equivalent to the formula for the sample mean.
28. 8, 23.5, 38.5, 53.5, 76, 105.5
29.

Delay (minutes)	Frequency f	Midpoint x	Product $f \cdot x$
0 to < 16	665	8	5,320
16 to < 31	551	23.5	12,948.5
31 to < 46	497	38.5	19,134.5
46 to < 61	399	53.5	21,346.5
61 to < 91	355	76	26,980
91 to < 120	27	105.5	2,848.5
Total	$\sum f = 2494$		$\sum(f \cdot x) = 88,578$

30. $\bar{x} = \dfrac{\sum(f \cdot x)}{\sum f} = \dfrac{88578}{2494} = 35.52$

31.

Delay (minutes)	Midpoint x	Frequency f	$\bar{x}$	$x - \bar{x}$	$(x - \bar{x})^2 \cdot f$
0 to < 16	8	665	35.52	-27.52	503,638.016
16 to < 31	23.5	551	35.52	-12.02	79,608.7004
31 to < 46	38.5	497	35.52	2.98	4,413.5588
46 to < 61	53.5	399	35.52	17.98	128,988.8796
61 to < 91	76	355	35.52	40.48	581,713.792
91 to < 120	105.5	27	35.52	69.98	132,224.4108
Total		$\sum f = 2494$			$\sum(x - x)^2 \cdot f =$ 1,430,587.3576

32. 573.6116
33. 23.9502
34.
Professionals

Class	Frequency
$-50 \leq$ price change < -25	3
$-25 \leq$ price change < 0	26
$0 \leq$ price change < 25	43
$25 \leq$ price change < 50	22
$50 \leq$ price change < 75	5
$75 \leq$ price change < 100	1
Total	100

Darts

Class	Frequency
$-50 \leq$ price change < -25	4
$-25 \leq$ price change < 0	33
$0 \leq$ price change < 25	53
$25 \leq$ price change < 50	7
$50 \leq$ price change < 75	3
$75 \leq$ price change < 100	0
Total	100

DJIA

Class	Frequency
$-50 \leq$ price change < -25	0
$-25 \leq$ price change < 0	19
$0 \leq$ price change < 25	81
$25 \leq$ price change < 50	0
$50 \leq$ price change < 75	0
$75 \leq$ price change < 100	0
Total	100

35.
Professionals: 13.25
Darts: 5.5
DJIA: 7.75
36.
Professionals:
Variance: 555.69, Standard deviation: 23.57
Darts:
Variance: 376, Standard deviation: 19.39
DJIA:
Variance: 96.188, Standard deviation: 9.808

37.

Variable	Mean	StDev	Variance
PROS	10.95	22.25	494.91
Variable	Mean	StDev	Variance
DARTS	4.52	19.39	375.91
Variable	Mean	StDev	Variance
DJIA	6.793	8.031	64.505

Professionals:
Mean: $10.95 - 13.25 = -2.3$
Variance: $494.91 - 555.69 = -60.78$
Standard deviation: $22.25 - 23.57 = -1.32$
Darts:
Mean: $4.52 - 5.5 = -0.98$
Variance: $375.91 - 376 = -0.09$
Standard deviation: $19.39 - 19.39 = 0$
DJIA:
Mean: $6.793 - 7.75 = -0.957$
Variance: $64.505 - 96.188 = -31.683$
Standard deviation: $8.031 - 9.808 = -1.777$
38. 3,944,153
39. 37.30 years
40. 22.70 years
41. 14.6 years and 60 years
42. 62.86%

Section 3.4

1. Positive z-score: the data value is above the mean. Negative z-score: the data value is below the mean. z-score of zero: the data value is equal to the mean.
2. Answers will vary.
3. No more than 100% of the data values can be less than or equal to any value in the data set.
4. It is possible for the 1st percentile to equal the 99th percentile if all of the data values are the same.
5. A percentile is a data value while a percentile rank is a percentage.
6. False
7. 2
8. 0.5
9. -1.0
10. -4.33
11. 1
12. -2
13. 4
14. -0.5
15. 1
16. 3.5
17. **(a)** The standard deviation, which is 10 mg/dl. **(b)** 2.5 **(c)** Chelsea's blood sugar level lies 2.5 standard deviations above the mean blood sugar level of 100 mg/dl.
18. **(a)** -1.5 **(b)** David's blood sugar level lies 1.5 standard deviations below the mean blood sugar level of 100 mg/dl.
19. 100
20. 175
21. 130
22. 25
23. 150
24. 50
25. 87.5
26. 100
27. 119.6
28. 74.24
29. 80.4
30. 125.76
31. Elizabeth: $z = 1$; Fiona: $z = 2$; Fiona did better.
32. Juan: $z = 1$; Luis: $z = 1$; they both did the same.
33. Moderately unusual
34. Not unusual
35. Not unusual
36. Outlier
37. Not unusual
38. Moderately unusual
39. Outlier
40. Not unusual

41. Not unusual
42. Outlier
43. Moderately unusual
44. Not unusual
45. 28
46. 20
47. 31
48. 30
49. 21
50. 31
51. 225
52. 125
53. 260
54. 140
55. 80
56. 290
57. 92nd percentile
58. 100th percentile
59. 8th percentile
60. 50th percentile
61. 75th percentile
62. 83rd percentile
63. 6th percentile
64. 100th percentile
65. 94th percentile
66. 19th percentile
67. 81st percentile
68. 75th percentile
69. 23.5
70. 25.5
71. 28
72. 25.5, the same
73. 137.5
74. 190
75. 225
76. 190, the same
77. 4.5. The middle 50%, or middle half, of the highway MPG data ranged over 4.5 miles per gallon.
78. 87.5. The middle 50%, or middle half, of the sodium content of breakfast cereals data ranged over 87.5 milligrams per serving.
79. **(a)** -0.79 **(b)** 1.79 **(c)** -1.65 **(d)** 0.93
80. **(a)** 109.167 calories **(b)** 120.812 calories **(c)** 97.522 calories
(d) 114.989 calories
81. No outliers
82. **(a)** 105 calories **(b)** 110 calories **(c)** 115 calories **(d)** 130 calories
83. **(a)** 17% **(b)** 92% **(c)** 75% **(d)** 25%
84. **(a)** 105 calories **(b)** 110 calories **(c)** 115 calories **(d)** 10 calories
85. The middle 50%, or half, of the number of calories in 12 breakfast cereals ranges over 10 calories.
86. **(a)** 90 calories **(b)** Bran Chex and Bran Flakes
87. **(a)** 2.87 **(b)** -0.91 **(c)** -0.89 **(d)** 1.11
88. **(a)** 5.073 million **(b)** 15.151 million **(c)** -5.005 million **(d)** 8.433 million
89. Echinacea with 14.7 million users is moderately unusual.
90. **(a)** Valerian with 2.1 million **(b)** Ginseng with 8.8 million **(c)** Saw Palmetto with 2 million **(d)** Echinacea with 14.7 million
91. **(a)** 100% **(b)** 7% **(c)** 93% **(d)** 13%
92. **(a)** Bee pollen with 2.8 million **(b)** Fish oil with 4.2 million **(c)** Garlic with 7.1 million **(d)** 4.3 million
93. The middle 50%, or half, of the usage of dietary supplements ranges over 4.3 million.
94. **(a)** 7.7 million **(b)** Ginkgo biloba, Ginseng, Echinacea
95. **(a)** Right-skewed **(b)** Greater than, since the distribution is right-skewed. **(c)** Answers will vary.
96. All numbers between Q1 and Q2 would be the same; median line and line for Q1 would be the same.
97. Right-skewed with a few values much larger than the rest; median line of box plot closer to the line for Q3 than the line for Q1.
98. Can't do. The 50th percentile of a data set can never be smaller than the 25th percentile.
99. Not possible. Q1, the 25th percentile, will always be less than or equal to Q3, the 75th percentile. Thus the IQR = Q3 − Q1 is always greater than or equal to zero.
100. 10%

101. 20%

102. 0.9%

103. 0.1%

104. No. 100 followers is between 98 followers, which is the 60th percentile, and 154 followers, which is the 70th percentile Therefore between 60 and 70 percent of Twitter accounts have 100 or fewer followers.

105. 95th percentile

106.

Variable	Median
PROS	9.60

Variable	Median
DARTS	3.25

Variable	Median
DJIA	7.000

The median increase in stock prices for the stocks randomly selected by darts is 3.25%. This is lower than the median increase in stock prices of the stocks selected by the professionals (9.60%) or the stocks in the DJIA (7.000%). Therefore, the relative value of the medians says that using darts is the worst method of picking stocks.

107.

Variable	Mean	StDev	Median
PROS	10.95	22.25	9.60

z-score $= -0.0607$

Variable	Mean	StDev	Median
DARTS	4.52	19.39	3.25

z-score $= -0.0655$

Variable	Mean	StDev	Median
DJIA	6.793	8.031	7.000

z-score $= 0.0258$

For the pros and the darts data the median is below the mean. For the DJIA data the median is above the mean.

108. (a) Pros: 55.45; Darts: 43.3; DJIA: 22.855 **(b)** Pros: -33.35; Darts: -34.26; DJIA: -9.269

109. About 95%

110. Pros: Q1 $= -5.85$, Q3 $= 26.55$; Darts: Q1 $= -6.25$, Q3 $= 14.35$; DJIA: Q1 $= 1.55$, Q3 $= 13.2$

111. Pros: IQR $= 32.4$

For the professionals, the middle 50%, or middle half, of the change in stocks data ranged over 32.4%.

Darts: IQR $= 20.60$

For using darts, the middle 50%, or middle half, of the change in stocks data ranged over 20.60%.

DJIA: IQR $= 11.65$

For the DJIA, the middle 50%, or middle half, of the change in stocks data ranged over 11.65%.

112.

Variable	StDev	Range
PROS	22.25	112.80

IQR $= 32.4$

Variable	StDev	Range
DARTS	19.39	115.90

IQR $= 20.60$

Variable	StDev	Range
DJIA	8.031	35.600

IQR $= 11.65$

The IQR, range, and standard deviation all agree that stocks in the DJIA have the least amount of variability.

However, the IQR and the standard deviation indicate that the stocks chosen by the professionals have the most variability while the range indicates that the stocks chosen by the darts has the most variability.

113. (a) -0.6673 **(b)** 0.3203 **(c)** 1.5615

114. (a) -0.10 fatality per 100,000 people **(b)** 2.15 fatalities per 100,000 people **(c)** 3.65 fatalities per 100,000 people

115. No outliers

116. These are the largest and smallest numbers in the data set.

117. (a) 1.445 fatalities per 100,000 people **(b)** 1.66 fatalities per 100,000 people **(c)** 0.90 fatality per 100,000 people

118. (a) 17th percentile **(b)** 50th percentile **(c)** 100th percentile

119. (a) 0.90 fatality per 100,000 people **(b)** 1.445 fatalities per 100,000 people **(c)** 1.66 fatalities per 100,000 people **(d)** 0.76 fatality per 100,000 people

120. The middle 50%, or middle half, of the change in the fatality rate data ranged over 0.76 fatality per 100,000 people.

Section 3.5

1. False

2. The boxplot is a graphical display of the five-number summary.

3. (a) The median will be about the same distance from Q1 and Q3, and the upper and lower whiskers will be about the same length. **(b)** The median is closer to Q1 than to Q3, and the upper whisker is much longer than the lower whisker. **(c)** The median is closer to Q3 than to Q1, and the lower whisker is much longer than the upper whisker.

4. Never

5. Any data value located 1.5 (IQR) or more below Q1 or 1.5 (IQR) or more above Q3 is considered an outlier.

6. The z-score is sensitive to the outliers themselves while the IQR is not. The IQR method is more robust.

7. Q1 $=$ \$200, Q2 $=$ median $=$ \$225, Q3 $=$ \$450

8. Minimum $=$ \$200, Q1 $=$ \$200, Q2 $=$ median $=$ \$225, Q3 $=$ \$450, Maximum $=$ \$800

9. \$250

10.

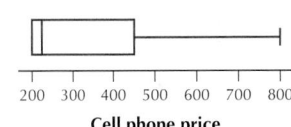

Cell phone price

11. No

12. No

13. Q1 $= 2325$, Q2 $=$ Median $= 2500$, Q3 $= 2775$

14. Minimum $= 1750$, Q1 $= 2325$, Q2 $=$ Median $= 2500$, Q3 $= 2775$, Maximum $= 3100$

15. 450

16.

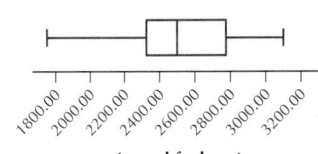

Annual fuel cost

17. No

18. No

19. Q1 $= 55$, Q2 $=$ median $= 101$, Q3 $= 166$

20. Minimum $= 32$, Q1 $= 55$, Q2 $=$ median $= 101$, Q3 $= 166$, Maximum $= 451$

21. 111

22.

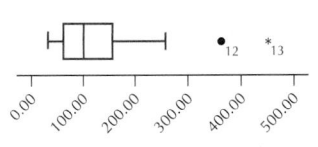

Criminal trespass cases

23. Not an outlier

24. Outlier

25. (a) Right-skewed **(b)** Minimum $= 0$, Q1 $= 1$, Q2 $=$ median $= 3$, Q3 $= 7.5$, maximum $= 12$

26. (a) Symmetric **(b)** Minimum $= 1$, Q1 $= 2$, Q2 $=$ median $= 3$, Q3 $= 4$, maximum $= 5$

27. (a) Right-skewed (b) Minimum = 5, Q1 = 10, Q2 = median = 15, Q3 = 25, maximum = 45

28. (a) Symmetric (b) Minimum = 15, Q1 = 20, Q2 = median = 25, Q3 = 30, maximum = 35

29. x

30. x

31. Minimum = 3.38, Q1 = 25.28, Q2 = median = 38.28, Q3 = 65.28, Maximum = 95.18

32. IQR = 40. The middle 50%, or middle half, of the stock price data ranged over $40.

33. No outliers.

34.

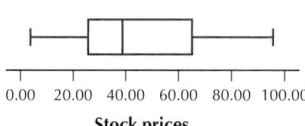

Stock prices

35. Minimum = −0.36, Q1 = −0.01, Q2 = median = 0.09, Q3 = 0.15, Maximum = 1.09

36. 0.16

37. −0.36, 0.41, and 1.09 are outliers.

38.

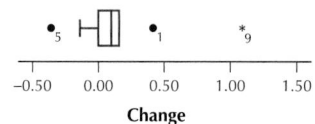

Change

39. Min = 2,000,000; Q1 = 2,800,000; median = 4,200,000; Q3 = 7,100,000; max = 14,700,000

40. IQR = 4,300,000, which is the spread of the middle 50% of the data.

41. Q1 − 1.5 * IQR = −3.65 and Q3 + 1.5 * IQR = 13.55. Usage of 14,700,000 is the only outlier.

42.

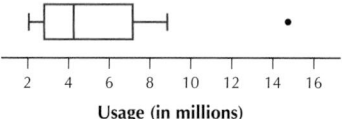

Usage (in millions)

43. Mean: 5,073,300; standard deviation: 3,359,300

44. The z-score is 2.87, so echinacea usage is moderately unusual but not an outlier. The robust method indicates that echinacea usage is an outlier.

45. Honda: Minimum = 17, Q1 = 18, Q2 = median = 23, Q3 = 31, Maximum = 42; Lexus: Minimum = 15, Q1 = 18, Q2 = median = 19.5, Q3 = 23, Maximum = 24

46.

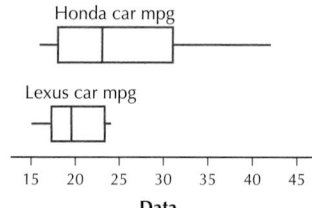

Data

47. The distribution of the mpg for the Honda cars is right-skewed. The distribution for the mpg for the Lexus cars is right-skewed.

48. Honda: Mean greater than median; Lexus: Mean greater than median.

49. Honda: Mean = 25.14 mpg; Lexus: Mean = 19.83 mpg. Yes

50. The location of the box for the Lexus cars starts a little to the left of the location of the box for the Honda cars. The box for the Honda cars extends further to the right than the box for the Lexus cars. Honda. Yes

51. Honda: IQR = 13 mpg; Lexus: IQR = 5 mpg. The IQR for the Honda cars is larger than the IQR for the Lexus cars. This indicates that the data for the Honda cars has a larger spread than the data for the Lexus cars.

52. Honda

53. No outliers in either group.

54. (a) 961 (b) 25

55. Mean = 1.784 mg, standard deviation = 3.138 mg, min = 0.000 mg, Q1 = 0.300 mg, median = 0.800 mg, Q3 = 1.700 mg, max = 37.600 mg. Range = 37.600 mg – 0.000 mg = 37.600 mg. IQR = 1.700 mg − 0.300 mg = 1.400 mg

56. Homemade dark fruitcake (1 cake)

57. The boxplot is very right-skewed.

58. Pros:

Variable	Minimum	Q1	Median	Q3	Maximum
PROS	−37.80	−5.85	9.60	26.55	75.00

Darts:

Variable	Minimum	Q1	Median	Q3	Maximum
DARTS	−43.00	−6.25	3.25	14.35	72.90

DJIA:

Variable	Minimum	Q1	Median	Q3	Maximum
DJIA	−13.100	1.55	7.000	13.2	22.500

59.

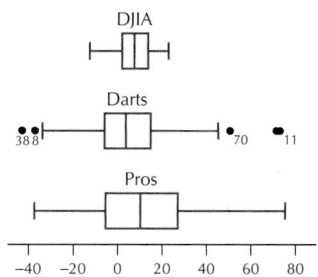

Pros has the greatest variability, DJIA has the smallest.

60.

Variable	StDev	Range
PROS	22.25	112.80

Variable	StDev	Range
DARTS	19.39	115.90

Variable	StDev	Range
DJIA	8.031	35.600

Yes.

61. Pros and DJIA

62. 5; IQR = Q3 − Q1 = 14.35 − (−6.25) = 20.6. The lower fence = Q1 −1.5(IQR) = −6.25 − 1.5(20.6) = −37.15. The upper fence = Q3 + 1.5(IQR) = 14.35 + 1.5(20.6) = 45.25. Thus, for this data set, a data value would be an outlier if it is −37.15 or less or 45.25 or more. Since −43.0 and −37.3 are ≤ −37.15 and 72.9, 71.3, and 50.5 are all ≥ 45.25 these five numbers are outliers in the data set.

63.

Variable	StDev	Range
DARTS	4.52	19.39

z-score for $x = −43.0$: −2.4507
Moderately unusual
z-score for $x = −37.3$: −2.1568
Moderately unusual
z-score for $x = 72.9$: 3.5266
Outlier
z-score for $x = 71.3$: 3.4440
Outlier
z-score for $x = 50.5$: 2.3713
Moderately unusual

Chapter 3 Review

1. $5.07 million

2. $2.3 million

3. Mean = $13.21 million, Median = $2.7 million. The mean. The mean is affected by extreme values more than the median is.

4. 109.17 calories
5. 110 calories
6. 110 calories
7. The mode, since the value with the largest frequency is unaffected by the deletion of values 90 or less.
8. Each will be increased by 10 calories.
9. $13.4 million
10. Connecticut: $2.73 million, Maine: $-$3.17 million, Massachusetts: $9.83 million, New Hampshire: $-$3.57 million, Rhode Island: $-$2.37 million, Vermont: $-$3.47 million.
11. No. It is always 0.
12. $\sum(x - \mu)^2 = 144.5334$ million dollars squared

$\sigma^2 = \frac{\sum(x - \mu)^2}{N} = \frac{144.5334}{6} = 24.0889$ million dollars squared
13. $4.9080 million.
14. For this population of states in New England, the typical difference between a state's CDC funding and the mean CDC funding is $4.9080 million.
15. 11.6450 calories
16. 85.88 calories, 132.46 calories
17. 85.88 calories, 132.46 calories
18. Mean = 396.8; range = 803
19. About 300.
20. 276.2
21. Standard deviation = 276.2. **(a)** 24 **(b)** The frequency counts for the syllables typically differ from the mean of 396.8 by only 276.2.
22. 24.45 years
23. 2.89 years
24. 21.56 years and 27.34 years
25. 59.48%. The distribution is not bell-shaped.
26. 16.5
27. 34.5
28. 59.5
29. 0.7088
30. 1.4358
31. -1.7145
32. No outliers, no moderately unusual observations
33. 30th percentile
34. 90th percentile
35. 80th percentile
36. Q1 = 25, Q2 = 34.5, Q3 = 48
37. IQR = 23, which is the spread of the middle 50% of the data set
38. No outliers
39. **(a)** Min = 8, Q1 = 25, median = 34.5, Q3 = 48, max = 60
(b)

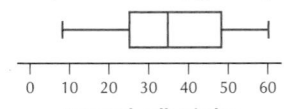

Ragweed pollen index

(c) Close to symmetric, slightly right-skewed. **(d)** The mean should be close to the median or a little above the median. **(e)** Mean = 36.30, standard deviation = 16.51. The value of the mean is slightly above the median of 34.50.
40. No outliers, yes
41. Unchanged; robustness

Chapter 3 Quiz
1. False
2. False
3. False
4. outlier
5. center
6. mean
7. Robust measures
8. Mode
9. Zero
10. Class midpoint
11. 3.31
12. **(a)** 1.5 **(b)** -1 **(c)** 1 **(d)** -1.5 **(e)** 0
13. **(a)** 60 **(b)** 68% **(c)** No, Chebyshev's Rule can't be used for $k = 1$. **(d)** 16%
14. **(a)** 501.5 **(b)** 512 **(c)** 518
15. IQR = 16.5

16. Min = 499, Q1 = 501.5, median = 512, Q3 = 518, max = 523.
17. Q1 $-$ 1.5 * IQR = 476.75 and Q3 + 1.5 * IQR = 542.75. All the SAT scores lie between 476.75 and 542.75, so there are no outliers.
18.

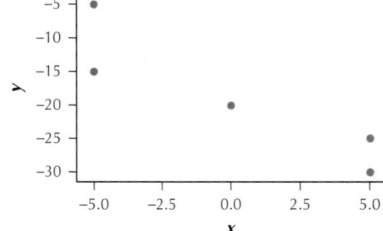

SAT 1 math score

Chapter 4
Section 4.1
1. Scatterplot
2. It measures the strength and direction of the linear relationship between two variables; r.
3. Between -1 and 1, inclusive
4. **(a)** The variables are positively correlated. **(b)** The variables are negatively correlated. **(c)** The variables are not correlated.
5. Often, the value of the x variable can be used to predict or estimate the value of the y variable.
6. They increase.
7. They decrease.
8. No. It means that x and y don't have a linear relationship. They may have another type of relationship.
9. Predictor (x): Height
   Response (y): Weight
10. Predictor (x): The number of days absent from class
   Response (y): Course grade
11. Predictor (x): The number of hours spent on the repair
   Response (y): The cost of a repair job
12. Predictor (x): The amount of rain at a baseball stadium that day
   Response (y): Attendance at a baseball game
13. **(a)**

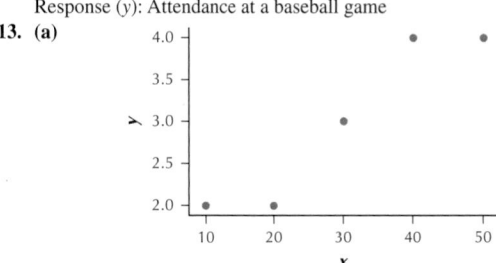

(b) The variables x and y have a positive linear relationship. **(c)** $r = 0.9487$
(d) This value of r is very close to the maximum value $r = 1$. We would therefore say that x and y are positively correlated. As x increases, y also tends to increase.
14. **(a)**

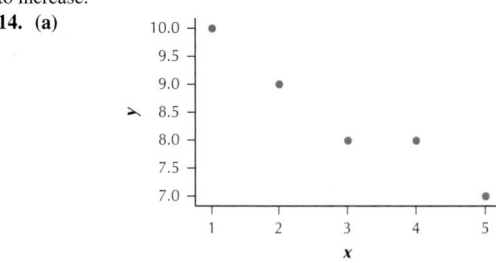

(b) The variables x and y have a negative linear relationship. **(c)** $r = -0.9707$
(d) This value of r is very close to the minimum value $r = -1$. We would therefore say that x and y are negatively correlated. As x increases, y tends to decrease.
15. **(a)**

(b) The variables x and y have a negative linear relationship. **(c)** $r = -0.9098$ **(d)** This value of r is very close to the minimum value $r = -1$. We would therefore say that x and y are negatively correlated. As x increases, y tends to decrease.

16. (a)

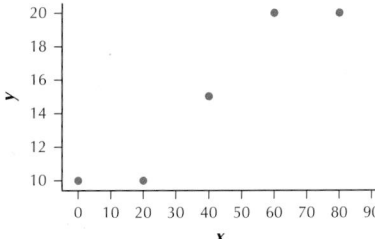

(b) The variables x and y have a positive linear relationship. **(c)** $r = 0.9487$ **(d)** This value of r is very close to the maximum value $r = 1$. We would therefore say that x and y are positively correlated. As x increases, y also tends to increase.

17. (a)

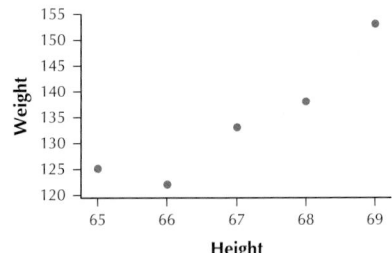

(b) The variables height and weight have a positive linear relationship. **(c)** $r = 0.9274$ **(d)** This value of r is very close to the maximum value $r = 1$. We would therefore say that height and weight are positively correlated. As height increases, weight also tends to increase.

18. (a)

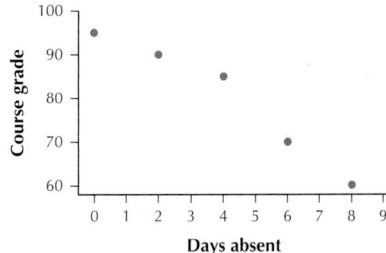

(b) The number of days absent from class and the course grade have a negative linear relationship. **(c)** $r = -0.9762$ **(d)** This value of r is very close to the minimum value $r = -1$. We would therefore say that the number of days absent and the course grade are negatively correlated. As the number of days absent increases, the course grade tends to decrease.

19. (a)

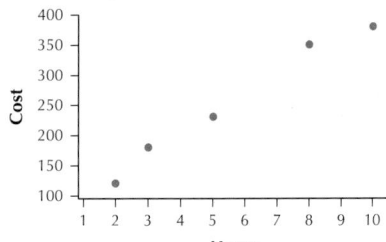

(b) The hours a repair job takes and the cost of the repair job have a positive linear relationship. **(c)** $r = 0.9896$ **(d)** This value of r is very close to the maximum value $r = 1$. We would therefore say that hours worked and cost are positively correlated. As the number of hours worked increases, the cost also tends to increase.

20. (a)

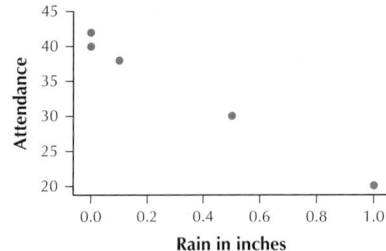

(b) The amount of rain and the attendance at a baseball game have a negative linear relationship. **(c)** $r = -0.9959$ **(d)** This value of r is very close to the minimum value $r = -1$. We would therefore say that rain and attendance are negatively correlated. As the amount of rain increases, attendance tends to decrease.

21. (a) Strong negative linear relationship **(b)** They decrease.
22. (a) Perfect positive linear relationship **(b)** They increase.
23. (a) Moderate positive linear relationship **(b)** They increase.
24. (a) Moderate negative linear relationship **(b)** They decrease.
25. (a) Perfect negative linear relationship **(b)** They decrease.
26. (a) Nonlinear relationship but no linear relationship **(b)** First they decrease, then they increase.
27. i
28. ii
29. iii
30. iv
31. (a) (1,1), (2,3), (3,3), (4,4), (5,6), (6,6), (7,7), (8,7), (9,9), (10,11)
(b) Minitab: Pearson correlation of x and y = 0.978. TI-83/84: r = 0.9781316853
32. (a) (1,3), (2,2), (3,2), (4,3), (5,3), (6,3), (7,2), (8,3), (9,3), (10,2) **(b)** Minitab: Correlations: x, y; Pearson correlation of x and y = 0.000. TI-83/84: $r = 0$
33. (a) (1,7), (2,8), (3,7), (4,6), (5,6), (6,5), (7,6), (8,5), (9,7), (10,6) **(b)** Minitab: Pearson correlation of x and y = -0.522. TI-83/84: $r = -0.5222329679$
34. (a) (1,11), (2,9), (3,7), (4,7), (5,6), (6,6), (7,4), (8,3), (9,3), (10,1)
(b) Minitab: Correlations: x, y; Pearson correlation of x and y = -0.978. TI-83/84: $r = -0.9781316853$

35. (a)

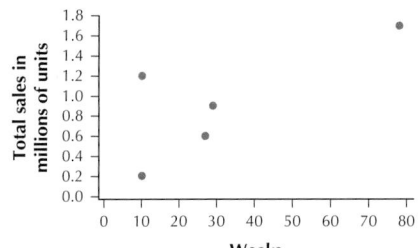

(b) The number of weeks on the top 30 list and the total sales in millions of game units have a positive linear relationship. **(c)** $r = 0.7403$ **(d)** This value of r is close to the maximum value $r = 1$. We would therefore say that weeks and total sales are positively correlated. As the number of weeks on the top 30 lists increases, the total sales also tends to increase.

36. (a)

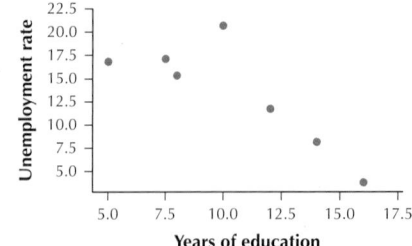

(b) The years of education and the unemployment rate have a negative linear relationship. **(c)** $r = -0.8261$ **(d)** This value of r is very close to the minimum value $r = -1$. We would therefore say that years of education and unemployment rate are negatively correlated. As the amount of number of years of education increases, the unemployment rate tends to decrease.

37. (a)

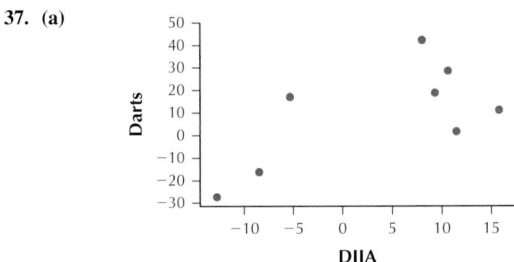

(b) The change in price of the stocks in the DJIA and the change in price of the stocks in the portfolio selected by the darts have a positive linear relationship. **(c)** $r = 0.6639$ **(d)** This value of r is close to the maximum value $r = 1$. We would therefore say that DJIA and Darts are positively correlated. As the change in stock prices of the stocks in the DJIA increases, the change in stock prices selected by darts also tends to increase.

38. (a)

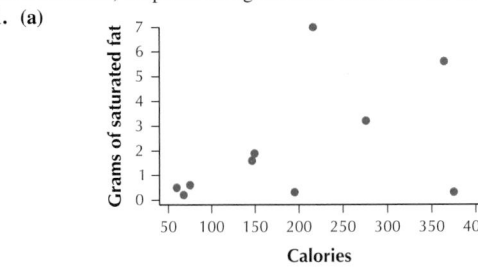

(b) The age and height of a person have a negative linear relationship. **(c)** $r = -0.3003$ **(d)** This value of r is negative. We would therefore say that age and height are negatively correlated. As age increases, height tends to decrease.

39. (a)

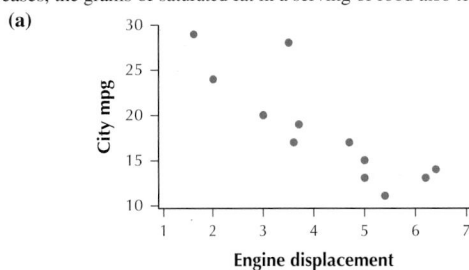

(b) Age and number of shots have no apparent relationship. **(c)** $r = 0.0641$ **(d)** This value of r is very close to the value $r = 0$. We would therefore say that there is no linear relationship between age and shots.

40. (a)

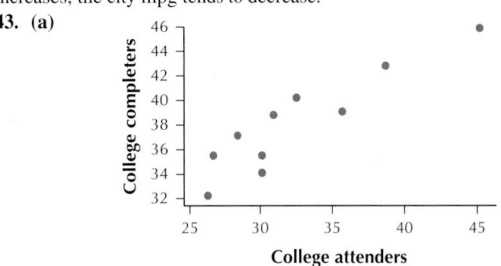

(b) The winning proportion and the power rating of a baseball team have a positive linear relationship. **(c)** $r = 0.8328$ **(d)** This value of r is very close to the maximum value $r = 1$. We would therefore say that winning proportion and power rating are positively correlated. As the winning proportion of games increases, the power rating of a team also tends to increase.

41. (a)

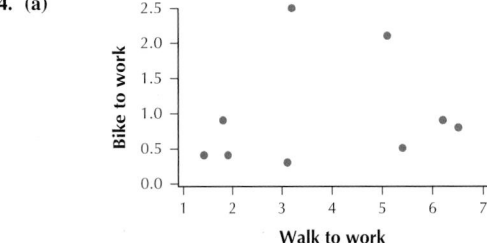

(b) The number of calories in a serving of food and the grams of saturated fat in a serving of food have a positive linear relationship. **(c)** $r = 0.4482$ **(d)** This value of r is positive. We would therefore say that the number of calories in a serving of food and the grams of saturated fat in a serving of food are positively correlated. As the number of calories in a serving of food increases, the grams of saturated fat in a serving of food also tends to increase.

42. (a)

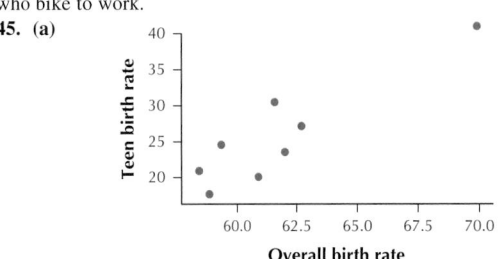

(b) The engine displacement of a car and the city mpg of a car have a negative linear relationship. **(c)** $r = -0.8497$ **(d)** This value of r is very close to the minimum value $r = -1$. We would therefore say that engine displacement and city mpg are negatively correlated. As the engine displacement (size) increases, the city mpg tends to decrease.

43. (a)

(b) The percent of residents of a state who attend college and the percent of residents of a state who graduate from college have a positive linear relationship. **(c)** $r = 0.9196$ **(d)** This value of r is very close to the maximum value $r = 1$. We would therefore say that college attenders and college completers are positively correlated. As the number of college attenders increases, the number of college completers also tends to increase.

44. (a)

(b) The percent of residents of a city who walk to work and the percent of residents of a city who bike to work have no apparent relationship. **(c)** $r = 0.2388$ **(d)** This value of r is very close to the value $r = 0$. We would therefore say that there is no linear relationship between the percentage of people who walk to work and the percentage of people who bike to work.

45. (a)

(b) The overall birth rate and the teenage birth rate have a positive linear relationship. **(c)** $r = 0.9059$ **(d)** This value of r is very close to the maximum value $r = 1$. We would therefore say that overall birth rate and teen birth rate are positively correlated. As the overall birth rate increases, the teen birth rate also tends to increase.

46. (a)

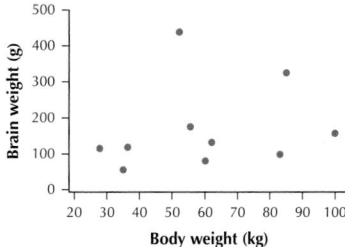

(b) The body weight of an animal and the brain weight of an animal have no apparent relationship. **(c)** $r = 0.2067$ **(d)** This value of r is very close to the value $r = 0$. We would therefore say that there is no linear relationship between body weight and brain weight.

47. (a)

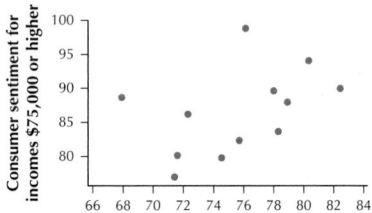

(b) The consumer sentiment for incomes under \$75,000 and the consumer sentiment for incomes \$75,000 or higher have a positive linear relationship. **(c)** $r = 0.4296$ **(d)** This value of r is positive. We would therefore say that the consumer sentiment for incomes under \$75,000 and the consumer sentiment for incomes \$75,000 or higher are positively correlated. As the consumer sentiment for incomes under \$75,000 increases, the consumer sentiment for incomes \$75,000 or higher also tends to increase.

48. (a)

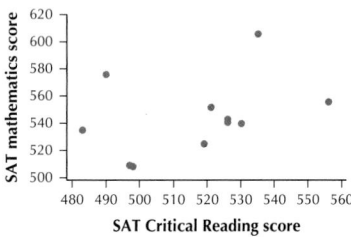

(b) The SAT Critical Reading score and the SAT Mathematics score have a positive linear relationship. **(c)** $r = 0.3715$ **(d)** This value of r is positive. We would therefore say that SAT Critical Reading score and SAT Mathematics score are positively correlated. As the SAT Critical Reading score increases, the SAT Mathematics score also tends to increase.

49. (a)

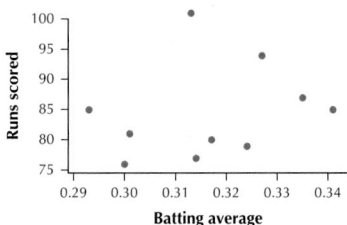

(b) The batting average of a baseball player and the number of runs scored by a baseball player have no apparent relationship. **(c)** $r = 0.2332$ **(d)** This value of r is very close to the value $r = 0$. We would therefore say that there is no linear relationship between batting average and runs scored.

50.

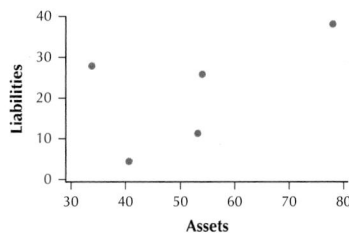

Assets and liabilities have a positive linear relationship.
51. $r = 0.5554$; This value of r is positive. We would therefore say that assets and liabilities are positively correlated. As assets increase, liabilities also tend to increase.

52.

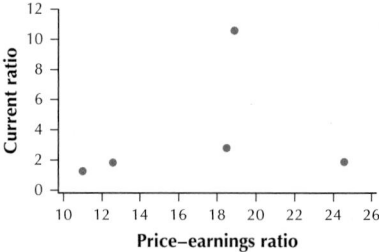

Price–earnings ratio and current ratio have no apparent relationship.
53. $r = 0.2443$; This value of r is very close to the value $r = 0$. We would therefore say that there is no linear relationship between price–earnings ratio and current ratio.
54. Microsoft: 39.6; Intel: 41.7; Dell: 5.7; Apple: 27.9; Google: 36
55.

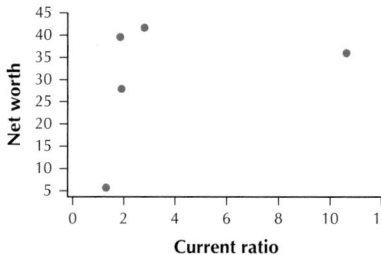

The current ratio of a company and the company's net worth have a positive linear relationship.
56. $r = 0.3284$; This value of r is positive. We would therefore say that their current ratio and net worth are positively correlated. As the current ratio increases, the company's net worth also tends to increase.

57. (a)

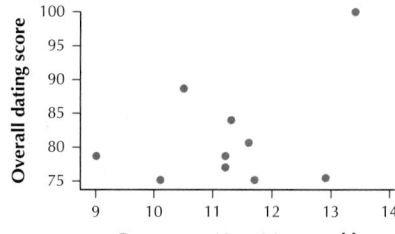

The percentage of 18- to 24-year-olds in a city and the overall dating score in a city have a positive linear relationship.

(b)

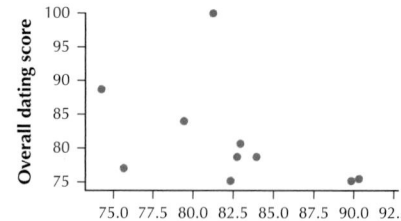

The percentage of a city's residents who are 18- to 24-year-olds and single and the overall dating score of a city have a negative linear relationship.

(c)

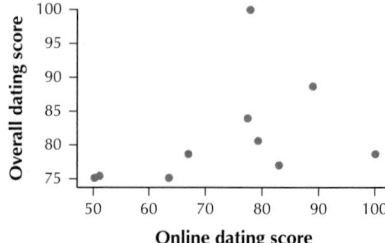

The online dating score of a city and the overall dating score of a city have a positive linear relationship.

58. (a) $r = 0.3947$. This value of r is positive. We would therefore say that the percentage of 18- to 24-year-olds in a city and the overall dating score of the city are positively correlated. As the percentage of 18- to 24-year-olds in a city increases, the overall dating score of the city also tends to increase. **(b)** $r = -0.4288$. This value of r is negative. We would therefore say that the percentage of 18- to 24-year-olds who are single in a city and the overall dating score of a city are negatively correlated. As the percentage of 18- to 24-year-olds who are single in a city increases, the overall dating score of a city tends to decrease. **(c)** $r = 0.3940$. This value of r is positive. We would therefore say that the online dating score of a city and the overall dating score of a city are positively correlated. As the online dating score of a city increases, the overall dating score of a city also tends to increase.

59. Percentage of 18- to 24-year-olds who are single

60.

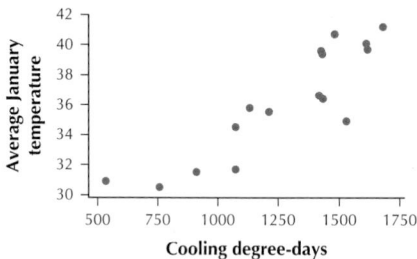

Cooling degree-days and the average January temperature have a positive linear relationship.

61. Positive. The scatterplot indicates a positive linear relationship between cooling-degree days and the average January temperature.

62. $r = 0.8636$. This value of r is very close to the maximum value $r = 1$. We would therefore say that cooling degree-days and average January temperature are positively correlated. As the number of cooling degree-days increases, the average January temperature also tends to increase.

63.

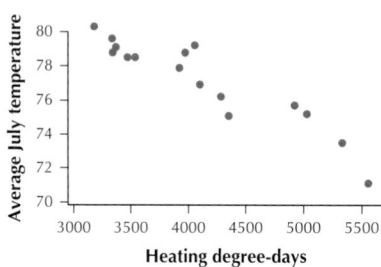

Heating degree-days and the average July temperature have a negative linear relationship.

64. $r = -0.9182$. This value of r is very close to the minimum value $r = -1$. We would therefore say that heating degree-days and the average July temperature are negatively correlated. As the number of heating degree-days increases, the average July temperature tends to decrease.

65. The relationship between the average July temperatures and the heating degree-days is stronger.

66.

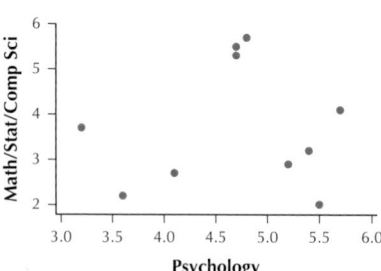

67. No.

68. $r = 0.0724$. Yes.

69.

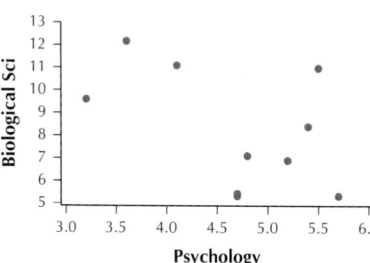

70. Yes.

71. $r = -0.4858$. Yes.

72.

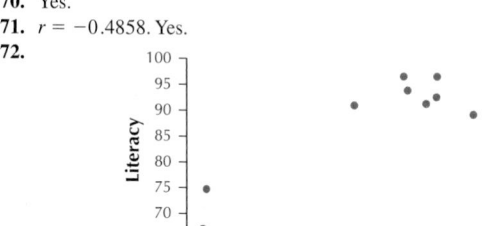

Economy and literacy have a positive linear relationship.

73. $r = 0.8811$. This value of r is very close to the maximum value $r = 1$. We would therefore say that economy and literacy are positively correlated. As the economy increases, literacy also tends to increase.

74.

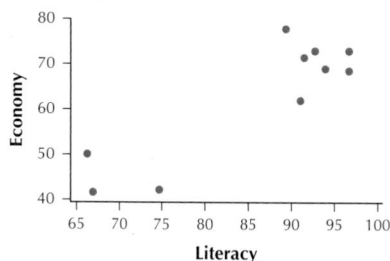

Literacy and economy have a positive linear relationship. Both scatterplots have a group of 3 dots in the lower left hand corner and a group of 7 dots in the upper right hand corner. The patterns made by the dots are different.

75. Yes. $r = 0.8811$. Yes.

76.

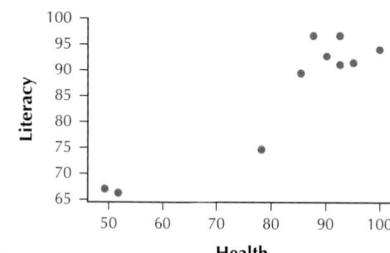

Health and literacy have a positive linear relationship.

77. $r = 0.9309$. This value of r is very close to the maximum value $r = 1$. We would therefore say that health and literacy are positively correlated. As health increases, literacy also tends to increase.

78. $r = 0.9309$

79.

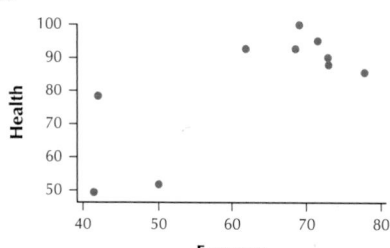

Economy and health have a positive linear relationship.

80. $r = 0.7598$. This value of r is very close to the maximum value $r = 1$. We would therefore say that economy and health are positively correlated. As the economy increases, health also tends to increase.

81. $r = 0.7598$

82. They are the same.

83. $r = 0.9434$

84. (a)

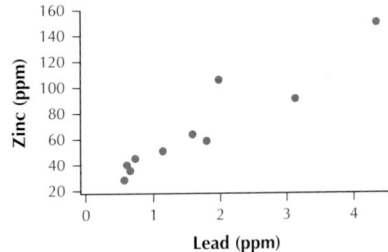

(b) Positive **(c)** Positive

85. (a) $r = 0.9405$ **(b)** Yes. **(c)** This value of r is very close to the maximum value $r = 1$. We would therefore say that lead and zinc are positively correlated. As the lead content in the fish increases, the zinc content in the fish also tends to increase.

86. Yes.

87. (a)

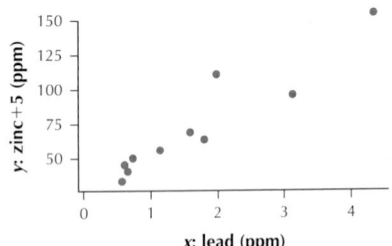

Everything is the same except the dots are shifted up 5 ppm. **(b)** $r = 0.9405$ **(c)** They are the same. **(d)** When a constant is added to each y-data value the correlation coefficient stays the same.

88. (a) Everything will be the same except the dots will be shifted right the number of units that was added on to each x-data value. **(b)** None. Any time an x appears in the formula for the correlation coefficient r it appears as "$x - \bar{x}$." Since x and $\bar{x}$ increase by the same amount, the differences will remain the same. **(c)** When a constant is added to each x-data value, the correlation coefficient remains the same.

89. Answers will vary.

90. Answers will vary.

91. (a)–(c) Answers will vary.

92. (a) Positively correlated **(b)** Negatively correlated **(c)** Not correlated

93. (a)–(c) Answers will vary.

94.

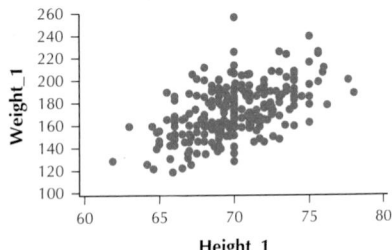

95. Positive

96. $r = 0.5351$

97. This value of r is positive. We would therefore say that height and weight are positively correlated. As height increases, weight also tends to increase.

98. (a) Both scatterplots show a weak positive linear relationship between height and weight. **(b)** Both relationships are positive linear relationships. **(c)** Both correlation coefficients are around 0.5. The correlation coefficient for men is a little above 0.5 and the correlation coefficient for women is a little below 0.5.

Section 4.2

1. To approximate the relationship between two numerical variables using the regression line and the regression equation

2. $\hat{y} = b_1 x + b_0$, where $b_1 = \dfrac{\sum(x - \bar{x})(y - \bar{y})}{\sum(x - \bar{x})^2}$ and $b_0 = \bar{y} - (b_1 \cdot \bar{x})$

3. We can find the predicted value of y by plugging a given value of x into the regression equation and simplifying.

4. y is the actual value of the response variable for a given x, and $\hat{y}$ is the value of the response variable predicted from the regression equation.

5. Extrapolation is the process of making predictions based on x-values that are beyond the range of the x-values in our data set.

6. The slope of the regression line and the correlation coefficient always have the same sign.

7. Negative

8. Positive

9. Positive

10. Negative

11. Negative

12. Near 0

13. (a) $b_1 = 0.34$, $b_0 = -1.5$, $\hat{y} = 0.34x - 1.5$ **(b)** For each increase of 1 unit, the estimated value of y increases by 0.34 unit. When $x = 0$ the estimated value of y is -1.5.

14. (a) $b_1 = -2$, $b_0 = 56$, $\hat{y} = -2x + 56$ **(b)** For each increase of 1 unit, the estimated value of y decreases by 2 units. When $x = 0$ the estimated value of y is 56.

15. (a) $b_1 = 4$, $b_0 = 31.6$, $\hat{y} = 4x + 31.6$ **(b)** For each increase of 1 unit, the estimated value of y increases by 4 units. When $x = 0$ the estimated value of y is 31.6.

16. (a) $b_1 = -3$, $b_0 = -36$, $\hat{y} = -3x - 36$ **(b)** For each increase of 1 unit, the estimated value of y decreases by 3 units. When $x = 0$ the estimated value of y is -36.

17. (a) $b_1 = 0.0114$, $b_0 = 7.4667$, $\hat{y} = 0.0114x + 7.4667$ **(b)** For each increase of 1 unit, the estimated value of y increases by 0.0114 unit. When $x = 0$ the estimated value of y is 7.4667.

18. (a) $b_1 = 0$, $b_0 = 9$, $\hat{y} = 0x + 9 = 9$ **(b)** For each increase of 1 unit, the estimated value of y remains 9. When $x = 0$ the estimated value of y is 9.

19. (a) $b_1 = 0$, $b_0 = 12.5$, $\hat{y} = 0x + 12.5 = 12.5$ **(b)** For each increase of 1 unit, the estimated value of y remains 12.5. When $x = 0$ the estimated value of y is 12.5.

20. (a) $b_1 = -1.4081$, $b_0 = 54.1914$, $\hat{y} = -1.4081x + 54.1914$ **(b)** For each increase of 1 unit, the estimated value of y decreases by 1.4081 units. When $x = 0$ the estimated value of y is 54.1914.

21. (a) $b_1 = 7.2$, $b_0 = -348.2$, $\hat{y} = 7.2x - 348.2$ **(b)** For each increase of 1 inch, the estimated weight increases by 7.2 pounds. When $x = 0$ inches, the estimated weight is -348.2 pounds.

22. (a) $b_1 = -4.5$, $b_0 = 98$, $\hat{y} = -4.5x + 98$ **(b)** For each increase of 1 one day absent, the estimated grade decreases by 4.5 points. When $x = 0$ days absent, the estimated grade is 98.

23. (a) $b_1 = 32.61$, $b_0 = 69.38$, $\hat{y} = 32.61x + 69.38$ **(b)** For each increase of 1 hour of labor, the estimated cost of the repairs increases by \$32.61. When $x = 0$ hours of labor, the estimated cost of the repairs is \$69.38.

24. (a) $b_1 = -20.86$, $b_0 = 40.67$, $\hat{y} = -20.86x + 40.67$ **(b)** For each increase of 1 inch of rain, the estimated attendance decreases by 20.86 thousand. When $x = 0$ inches of rain, the estimated attendance is 40.67 thousand.

25. (a) $\hat{y} = 8.7$ **(b)** $y - \hat{y} = 0.3$. The data point lies above the regression line, so the actual y-value of 9 is greater than the predicted y-value of 8.7. **(c)** Does not represent extrapolation

26. (a) $\hat{y} = 52$ **(b)** $y - \hat{y} = 8$. The data point lies above the regression line, so the actual y-value of 60 is greater than the predicted y-value of 52. **(c)** Does not represent extrapolation

27. (a) $\hat{y} = 11.6$ **(b)** $y - \hat{y} = -1.6$. The data point lies below the regression line, so the actual y-value of 10 is less than the predicted y-value of 11.6. **(c)** Does not represent extrapolation

28. (a) $\hat{y} = -45$ **(b)** $y - \hat{y} = 0$. The data point lies on the regression line, so the actual y-value of -45 is equal to the predicted y-value of -45. **(c)** Does not represent extrapolation

29. (a) $\hat{y} = 7.4667$ **(b)** $y - \hat{y}$ can't be found **(c)** Represents extrapolation

30. (a) $\hat{y} = 9$ **(b)** $y - \hat{y}$ can't be found **(c)** Represents extrapolation

31. (a) $\hat{y} = 12.5$ **(b)** $y - \hat{y} = -7.5$. The data point lies below the regression line, so the actual y-value of 5 is less than the predicted y-value of 12.5. **(c)** Does not represent extrapolation

32. (a) $\hat{y} = 47.1509$ **(b)** $y - \hat{y} = -0.1509$. The data point lies below the regression line, so the actual y-value of 47 is less than the predicted y-value of 47.1509. **(c)** Does not represent extrapolation

33. (a) $\hat{y} = 141.4$ pounds **(b)** $y - \hat{y} = -3.4$ pounds. The data point lies below the regression line, so the actual weight of 138 pounds is less than the predicted weight of 141.4 pounds. **(c)** Does not represent extrapolation

34. **(a)** $\hat{y} = 53$ **(b)** $y - \hat{y}$ can't be found **(c)** Represents extrapolation

35. **(a)** $\hat{y} = \$13,113.38$ **(b)** $y - \hat{y}$ can't be found **(c)** Represents extrapolation

36. **(a)** $\hat{y} = 35.46$ thousand people **(b)** $y - \hat{y}$ can't be found **(c)** Does not represent extrapolation

37. **(a)** $b_1 = 0.02$, $b_0 = 0.45$, $\hat{y} = 0.02x + 0.45$ **(b)** The predicted total sales in millions of game units of a video game is 0.02 times the number of weeks on the top 30 list plus 0.45 million game units. **(c)** For each increase of 1 week on the top 30 list the predicted number of game units of that game sold increases by 0.02 million units. **(d)** The predicted number of game units for a game that has been on the top 30 list for $x = 0$ weeks is 0.45 million units.

38. **(a)** $b_1 = -1.24$, $b_0 = 26.19$, $\hat{y} = -1.24x + 26.19$ **(b)** The estimated unemployment rate is -1.24 times the number of years of education plus 26.19. **(c)** For each increase of 1 year of education, the estimated unemployment rate decreases by 1.24%. **(d)** When the number of years of education equals 0, the estimated unemployment rate is 26.19%.

39. **(a)** $b_1 = 1.41$, $b_0 = 4.39$, $\hat{y} = 1.41x + 4.39$ **(b)** The predicted gain or loss in one day by the portfolio selected by the darts is 1.41 times the gain or loss of the DJIA plus \$4.39. **(c)** For each increase of \$1 in the DJIA the predicted value of the portfolio selected by the darts increases by \$1.41. **(d)** The predicted loss or gain in one day by the portfolio selected by the darts for a day when the gain or loss in the DJIA is $x = \$0$ is \$4.39.

40. **(a)** $b_1 = -0.11$, $b_0 = 67.58$, $\hat{y} = -0.11x + 67.58$ **(b)** The predicted height of a person is -0.11 times their age in years plus 67.58 inches. **(c)** For each increase of 1 year of age the predicted height decreases by 0.11 inch. **(d)** The predicted height of a person who is $x = 0$ years old is 67.58 inches.

41. **(a)** $b_1 = 0.02$, $b_0 = 1.93$, $\hat{y} = 0.02x + 1.93$ **(b)** The predicted number of shots that a person will get is 0.02 times the person's age in years plus 1.93. **(c)** For each increase of 1 year of age the predicted number of shots that a person will get increases by 0.02. **(d)** The predicted number of shots that a person who is $x = 0$ years old will get is 1.93.

42. **(a)** $b_1 = 20.432$, $b_0 = 100.472$, $\hat{y} = 20.432x + 100.472$ **(b)** The predicted power rating of a team is 20.432 times the winning proportion plus 100.472 **(c)** For each increase in the winning proportion of 1 the predicted power rating increases by 20.432. **(d)** The predicted power rating of a team with a winning proportion of $x = 0$ is 100.472.

43. **(a)** $b_1 = 0.01$, $b_0 = 0.33$, $\hat{y} = 0.01x + 0.33$ **(b)** The predicted number of grams of saturated fat in a food item is 0.01 times the number of calories in the food item plus 0.33 gram. **(c)** For each increase of 1 calorie in a food item the predicted number of grams of saturated fat increases by 0.01 gram. **(d)** The predicted number of grams of saturated fat in a food item with $x = 0$ calories is 0.33 gram.

44. **(a)** $b_1 = -3.28$, $b_0 = 32.03$, $\hat{y} = -3.28x + 32.03$ **(b)** The predicted city mpg of a car is -3.28 times the engine displacement (size) plus 32.03 mpg. **(c)** For each increase of 1 liter in the engine displacement (size) the city mpg decreases by 3.28 mpg. **(d)** The predicted city mpg of a car with engine displacement $x = 0$ liters is 32.03 mpg.

45. **(a)** $b_1 = 0.64$, $b_0 = 17.26$, $\hat{y} = 0.64x + 17.26$ **(b)** The predicted percentage of college students in a state who have completed their college degrees is 0.46 times the percentage of people who have attended college plus 17.26%. **(c)** For each increase of 1% in the percent of people in a state who have attended college the predicted percentage of college students who have completed their degree increases by 0.64%. **(d)** The predicted percent of college students who will complete their degree in a state with $x = 0\%$ of its residents attending college is 17.26%

46. **(a)** $b_1 = 0.093$, $b_0 = 0.581$, $\hat{y} = 0.093x + 0.581$ **(b)** The predicted percent of Americans in a city who bike to work is 0.093 times the percent of Americans in the city who walk to work plus 0.581 percent. **(c)** For each increase of 1% of Americans who walk to work in a city the predicted percent of Americans in that city who bike to work increases by 0.093% **(d)** The predicted percent of Americans who bike to work in a city that has $x = 0\%$ of its citizens who walk to work is 0.581%.

47. **(a)** $b_1 = 1.83$, $b_0 = -87.2$, $\hat{y} = 1.83x - 87.2$ **(b)** The predicted teenage birth rate of a state is 1.83 times the overall birth rate of the state minus 87.2 live births per 1000 women aged 15–19. **(c)** For each increase of 1 live birth per 1000 women the predicted teenage birth rate increases by 1.83 live births per 1000 women aged 15–19. **(d)** The predicted teenage birth rate for a state with an overall birth rate of $x = 0$ live births per 1000 women is -87.2 live births per 1000 women aged 15–19.

48. **(a)** $b_1 = 1.048$ $b_0 = 107.355$, $\hat{y} = 1.048x + 107.355$. **(b)** The estimated brain weight of a mammal is 107.355 grams plus 1.048 times the body

weight. **(c)** The estimated brain weight increases by 1.048 grams (slope b_1) for each 1 kilogram increase in body weight. **(d)** The y intercept $b_0 = 107.355$ is the predicted brain weight in grams of a mammal with a body weight of 0 kilograms.

49. **(a)** $b_1 = 0.641$, $b_0 = 38.1$, $\hat{y} = 0.641x + 38.1$ **(b)** The predicted consumer sentiment for incomes \$75,000 or higher is 0.641 times the consumer sentiment for incomes under \$75,000 plus 38.1. **(c)** For each increase of 1 in the consumer sentiment for incomes under \$75,000 the predicted consumer sentiment for incomes \$75,000 or higher increases by 0.641. **(d)** The predicted consumer sentiment for incomes \$75,000 or higher in a month when consumer incomes for under \$75,000 is $x = 0$ is 38.1.

50. **(a)** $b_1 = 0.479$, $b_0 = 297$, $\hat{y} = 0.479x + 297$ **(b)** The predicted SAT Mathematics score is 0.479 times the SAT Critical Reading score plus 297. **(c)** For each increase of 1 point in the SAT Critical Reading score the predicted SAT Mathematics score increases by 0.479. **(d)** The predicted SAT Mathematics score for a SAT Critical Reading score of $x = 0$ is 297.

51. **(a)** $b_1 = 118$, $b_0 = 47.2$, $\hat{y} = 118x + 47.2$ **(b)** The predicted number of runs scored by a player is 118 times the player's batting average plus 47.2 runs. **(c)** For each increase of 1 in a player's batting average the predicted number of runs scored increases by 118 runs. **(d)** The predicted number of runs scored by a player with a batting average of $x = 0$ is 47.2.

52. **(a)** 0.99 million game units **(b)** 0.65 million game units **(c)** -0.39 million game units, below, prediction error is negative **(d)** Titanfall for Xbox One: 0.55, above the regression line. Yoshi's New Island for 3DS: -0.45, below the regression line. Titanfall for Xbox One lies above Yoshi's New Island for 3DS.

53. **(a)** 13.79 **(b)** 7.59 **(c)** 20 years is outside of the range of the data set. **(d)** The result in part (a) is the predicted unemployment rate for individuals with 10 years of education while 20.6 is the actual unemployment rate for individuals with 10 years of education. 6.81, above the regression line. The observed unemployment rate of 20.6 is greater than the predicted unemployment rate of 13.79 for 10 years of education.

54. **(a)** $\hat{y} = 15.67$ **(b)** Not appropriate. $x = 20$ lies outside the range of the given x-values, which is from -12.8 to 15.8 **(c)** -9.71 **(d)** 26.53. The actual gain or loss of $y = 42.2$ in one day by the portfolio selected by the darts is greater than the predicted gain or loss in one day of $\hat{y} = 15.67$.

55. **(a)** $\hat{y} = 63.18$ inches **(b)** No. A newborn baby will not be 67.58 inches tall. **(c)** It is misleading to use the regression equation to predict the height of a 50-year-old person. An age of $x = 50$ years old is outside the range of the given x-values which is from 19 to 40 years old. **(d)** The height in (a) is the predicted height of a 40-year-old while the first height in the table is the actual height of a 40-year-old. **(e)** 0.32 inch. The actual height of the 40-year-old of $y = 63.5$ inches is greater than the predicted height of $\hat{y} = 63.18$ inches.

56. **(a)** 212.155 g. **(b)** No, there are no mammals with a body weight of 0 kg. **(c)** Dangerous; a mammal with a body weight of 10 g is outside of the range of the data set so the predicted brain weight would represent extrapolation. **(d)** The brain weight of 157 g is the actual brain weight of a mammal with a body weight of 100 kg, and the brain weight of 212.155 g is the predicted brain weight of a mammal with a body weight of 100 kg. **(e)** -55.155 g. The observed brain weight of 157 g for a mammal with a body weight of 100 kg is less than the predicted brain weight of 212.155 g.

57. **(a)** Decrease. Since all of the x-values are decreased $\sum x$ will also decrease. Therefore, $\bar{x} = \frac{\sum x}{n}$ will decrease. **(b)** No change. The y-values aren't changed. **(c)** Increase if $b_1 \cdot \bar{x} > 0$, no change if $b_1 \cdot \bar{x} = 0$, and decrease if $b_1 \cdot \bar{x} < 0$. Since $\bar{y}$ and b_1 stay the same and $\bar{x}$ decreases, $b_0 = \bar{y} - (b_1 \cdot \bar{x})$ will increase if $b_1 \cdot \bar{x} > 0$, no change if $b_1 \cdot \bar{x} = 0$, and decrease if $b_1 \cdot \bar{x} < 0$. **(d)** No change. Since r, s_x, and s_y all stay the same, $b_1 = r \cdot \frac{s_y}{s_x}$ stays the same. **(e)** No change. The correlation coefficient is $r = \frac{\sum(x - \bar{x})(y - \bar{y})}{(n-1)s_x s_y}$ where $s_x = \sqrt{\frac{\sum(x - \bar{x})^2}{n-1}}$ and $s_y = \sqrt{\frac{\sum(y - \bar{y})^2}{n-1}}$. Since the y-values aren't changed, s_y and $(y - \bar{y})$ aren't changed. Since x and $\bar{x}$ are decreased by the same amount, s_x and $(x - \bar{x})$ aren't changed. Since the number of data values hasn't changed, n hasn't changed. Therefore r does not change.

58. $b_1 = 2.4$, $b_0 = -60$, $\hat{y} = 2.4x - 60$

59. **(a)** 10.5. In a state with 0 households, 10.5% of the households are headed by women. Since all states have households, the value $x = 0$ would not occur. **(b)** This estimate would be considered extrapolation since the value of $x = 0$ is outside the range of x-values in the data set. **(c)** 0.000000282. For each increase of one household, the percentage of households headed by women increases by 0.000000282. **(d)** (Percentage of households headed by women) = 10.5 + 0.000000282 (Total number of households). The estimated percentage of households headed by women equals 10.5 plus 0.000000282 times the total number of households. **(e)** Positive since the slope is positive.

60. (a) Then State A has 0.282% more households headed by women than State B. **(b)** Then State C has 1.41% fewer households headed by women than State D.
61. (a) 12.474% **(b)** $x = 100{,}000$ is not in the range of the x-values in the data set.
62. Negative
63. Curved
64. No; curved
65. (a) Darts is the y variable and DJIA is the x variable. **(b)** -2.49 **(c)** When the Dow Jones Industrial Average changes 0%, the portfolios chosen by the Darts decreases by 2.49%. Since the Dow Jones Industrial Average is based on a different set of stocks than the stock portfolio selected by the Darts, this situation makes sense. **(d)** The estimate in (c) would not be considered extrapolation since a value of $x = 0$ lies within the range of the x-values of the data set.
66. (a) 1.032. The estimated percent change in the stock prices selected by the Darts increases by 1.032% for each unit percent increase in the Dow Jones Industrial Average. **(b)** Darts = $-2.49 + 1.032$ DJIA. The estimated percent change in the stock prices selected by the Darts is equal to -2.49 plus 1.032 times the percent change in the Dow Jones Industrial Average. **(c)** Positive, since the slope of the regression line is positive.
67. (a) The percent change in the stocks selected by the Darts for Contest A is 1.032 (10) = 10.32% more than the percent change in the stocks selected by the Darts for Contest B. **(b)** The percent change in the stocks selected by the Darts for Contest C is 1.032 (5) = 5.16% less than the percent change in the stocks selected by the Darts for Contest D.
68. (a) 20.214% **(b)** -12.81% **(c)** $x = -22$ is not in the range of the x-values of the data set.
69.

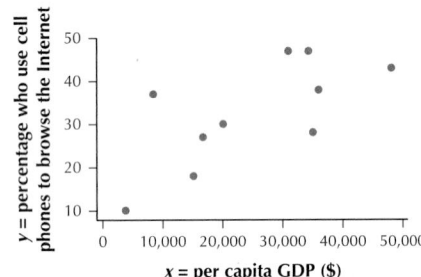

Per capita GDP ($) and percentage who use their cell phones to browse the Internet have a positive linear relationship.
70. Positive
71. $r = 0.6958$
72. $b_1 = 0.000604$, $b_0 = 17.50$, $\hat{y} = 0.000604x + 17.50$
The estimated percentage of cell phone users who use their cell phones to browse the Internet is 0.000604 times that country's per capita GDP in US dollars plus 17.50 percent.
73. For each increase of $1 in the country's per capita GDP the estimated percentage of cell phone users who use their cell phones to browse the Internet increases by 0.000604 percent.
The estimated percentage of cell phone users who use their cell phones to browse the Internet for a country with a per capita GDP of $x = \$0$ is 17.50%. This represents extrapolation.
74. 46.58%
75. USA. -3.58. The actual percentage of cell phone users who use their cell phones to browse the Internet for the USA of 43% lies below the estimated percentage of cell phone users who use their cell phones to browse the Internet for the USA of 46.58%.
76. (a) 37,000,000 **(b)** 7800
77. (a) 23,000,000 **(b)** 900
78. 5600
79. 3500
80. (a) It decreases.
(b)

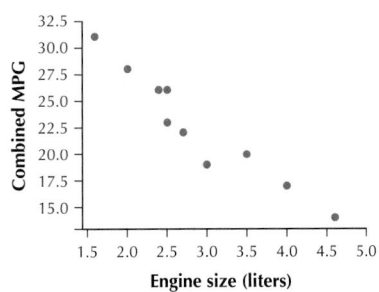

(c) Engine size and combined mpg have a negative linear relationship. Yes.
81. (a) Negative; **(b)** -0.9; the dots lie in a pattern that is close to a straight line **(c)** Negative; the dots lie in a pattern that is close to a straight line with a negative slope.
82. (a) $r = -0.9585$; yes **(b)** As the engine size of a car increases the combined mpg tends to decrease.
83. (a) $b_1 = -5.49$; yes **(b)** $b_0 = 38.41$ **(c)** The slope of $b_1 = -5.49$ means that the combined mpg will decrease by 5.49 mpg for each 1-liter increase in engine size. The y intercept of $b_0 = 38.41$ is the predicted combined mpg for an engine size of 0 liters.
84. (a) 21.94 mpg **(b)** -2.94, below. The observed combined mpg of 19 for the Chevrolet Equinox is less than the predicted combined mpg of 21.94.
85. Answers will vary.
86. Answers will vary.
87. Answers will vary.
88. Answers will vary.
89. Weight = $-96.2 + 3.541$. Height $\hat{y} = 3.541x - 96.2$
90. For each increase of 1 inch in height the estimated weight of a woman increases by 3.541 pounds.
The estimated weight of a woman who is $x = 0$ inches tall is -96.2 pounds.
91. 128.6535 pounds
92. -14.8535 inches
93. (a) 133.965 pounds **(b)** 151.67 pounds
94. Weight = $-134.7 + 4.386$ Height. $\hat{y} = 4.386x - 134.7$
95. For each increase of 1 inch in height the estimated weight of a man increases by 4.386 pounds.
The estimated weight of a man who is $x = 0$ inches tall is -134.7 pounds.
96. 165.741 pounds
97. -21.141 pounds
98. (a) 128.46 pounds, represents extrapolation **(b)** 150.39 pounds
99. Hip girth = $43.80 + 0.7428$ Waist girth
The predicted hip girth of a woman is 0.7428 times her waist girth (in centimeters) plus 43.80 centimeters.
100. For each increase of 1 centimeter in waist girth the estimated hip girth of a woman increases by 0.7428 centimeter.
The estimated hip girth of a woman who has a waist girth of $x = 0$ centimeters is 43.80 centimeters.
101. Hip girth = $49.82 + 0.5671$ Waist girth
The predicted hip girth of a man is 0.5671 times his waist girth (in centimeters) plus 49.82 centimeters.
102. For each increase of 1 centimeter in waist girth the estimated hip girth of a man increases by 0.5671 centimeter.
The estimated hip girth of a man who has a waist girth of $x = 0$ centimeters is 49.82 centimeters.
103. The slopes and y-intercepts for both are positive. The slopes for both are less than 1 and the y-intercepts for both are in the forties. The slope is larger for women and the y-intercept is larger for men.

Section 4.3
1. The standard error of the estimate s is a measure of the size of the typical difference between the predicted value of y and the observed value of y.
2. Out of all possible straight lines, the least-squares criterion chooses the line with the smallest SSE.
3. SSE measures the prediction errors. SSE is the sum of the squared prediction errors. Since we want our prediction errors to be small, we want SSE to be as small as possible.
4. It measures the amount of improvement in the accuracy of our estimate when using the regression equation compared with relying only on the y-values and ignoring the x information. We want SSR to be large, since SSR measures the amount of improvement in the accuracy of our estimate when using the regression equation compared with relying only on the y-values and ignoring the x information.
5. Measure of the variability in y. The variance s^2 of the y's.
6. A value of r^2 close to 1 indicates that the regression equation fits the data extremely well. A value of r^2 close to 0 indicates that the regression equation fits the data extremely poorly.
7. No
8. No, we don't have enough information to determine the sign of r.
9. 64% of the variability in the variable y is accounted for by the linear relationship between x and y.
10. False

11. (a) 0.2 **(b)** 0.316228. The typical difference between the predicted value of y and the actual observed value of y is 0.316228. **(c)** 58 **(d)** 57.8 **(e)** 0.9966. Therefore, 99.66% of the variability in y is accounted for by the linear relationship between y and x. **(f)** 0.9983

12. (a) 120 **(b)** 7.74597. The typical difference between the predicted value of y and the actual observed value of y is 7.74597. **(c)** 200 **(d)** 80 **(e)** 0.4000. Therefore, 40% of the variability in y is accounted for by the linear relationship between y and x. **(f)** -0.6325

13. (a) 19.20 **(b)** 2.52982. The typical difference between the predicted value of y and the actual observed value of y is 2.52982. **(c)** 179.20 **(d)** 160 **(e)** 0.8929. Therefore, 89.29% of the variability in y is accounted for by the linear relationship between y and x. **(f)** 0.9449

14. (a) 10 **(b)** 1.82574 The typical difference between the predicted value of y and the actual observed value of y is 1.82574. **(c)** 370 **(d)** 360 **(e)** 0.9730. Therefore, 97.30% of the variability in y is accounted for by the linear relationship between y and x. **(f)** -0.9864

15. (a) 1.27619 **(b)** 0.564843. The typical difference between the predicted value of y and the actual observed value of y is 0.564843. **(c)** 1.33333 **(d)** 0.05714 **(e)** 0.0429. Therefore, 4.29% of the variability in y is accounted for by the linear relationship between y and x. **(f)** 0.2071

16. (a) 0 **(b)** 0. The typical difference between the predicted value of y and the actual observed value of y is 0. **(c)** 0 **(d)** 0 **(e)** Undefined **(f)** Undefined

17. (a) 250 **(b)** 6.45497. The typical difference between the predicted value of y and the actual observed value of y is 6.45497. **(c)** 250 **(d)** 0 **(e)** 0. Therefore, 0% of the variability in y is accounted for by the linear relationship between y and x. **(f)** 0

18. (a) 195 **(b)** 5.70050. The typical difference between the predicted value of y and the actual observed value of y is 5.70050. **(c)** 5598 **(d)** 5403 **(e)** 0.9652. Therefore, 96.52% of the variability in y is accounted for by the linear relationship between y and x. **(f)** -0.9824

19. (a) 84.40 **(b)** 5.30409 pounds. The typical difference between the predicted value of y = weight and the actual observed value of y = weight is 5.30409 pounds. **(c)** 602.80 **(d)** 518.40 **(e)** 0.8600. Therefore, 86.00% of the variability in weight is accounted for by the linear relationship between y = weight and x = height. **(f)** 0.9274

20. (a) 40.00 **(b)** 3.65148. The typical difference between the predicted value of y = course grade and the actual observed value of y = course grade is 3.65148. **(c)** 850.00 **(d)** 810.00 **(e)** 0.9529. Therefore, 95.29% of the variability in y = course grade is accounted for by the linear relationship between y = course grade and x = days absent. **(f)** -0.9762

21. (a) 1012 **(b)** 18.3662. The typical difference between the predicted value of y = cost and the actual observed value of y = cost is \$18.3662. **(c)** 49080 **(d)** 48068 **(e)** 0.9794. Therefore, 97.94% of the variability in y = cost is accounted for by the linear relationship between y = cost and x = hours. **(f)** 0.9896

22. (a) 2.652 **(b)** 0.940285. The typical difference between the predicted value of y = attendance and the actual observed value of y = attendance is 0.940285. **(c)** 328 **(d)** 325.348 **(e)** 0.9919. Therefore, 99.19% of the variability in y = attendance is accounted for by the linear relationship between y = attendance and x = rain. **(f)** -0.9959

23. (a) 0.59123 **(b)** 0.443933. The typical difference between the predicted number of total sales in millions and the actual observed number of total sales in millions of units is 0.443933 million units. **(c)** 1.30800 **(d)** 0.71677 **(e)** 0.5480. Therefore, 54.80% of the variability in total sales in millions of units is accounted for by the linear relationship between total sales in millions of units and weeks on top 30 list. **(f)** 0.7403. This value of r is close to the maximum value $r = 1$. We would therefore say that the total sales in millions of units and weeks in the top 30 are positively correlated. As the number of weeks in the top 30 increases, the total sales in millions of units also tends to increase.

24. (a) 64.72 **(b)** 3.59790. The typical difference between the predicted unemployment rate and the actual observed unemployment rate is 3.59790. **(c)** 203.82 **(d)** 139.10 **(e)** 0.6824. Therefore, 68.24% of the variability in unemployment rate is accounted for by the linear relationship between the unemployment rate and the years of education. **(f)** -0.8261. This value of r is very close to the minimum value $r = -1$. We would therefore say that the unemployment rate and the number of years of education are negatively correlated. As the number of years of education increases, the unemployment rate tends to decrease.

25. (a) 2046 **(b)** 18.4665. The typical difference between the predicted value of the change in the stocks in the portfolio of stocks selected by the darts and the actual observed value of the change in the stocks in the

portfolio of stocks selected by the darts is 18.4665. **(c)** 3659 **(d)** 1613 **(e)** 0.4408. Therefore, 44.08% of the variability in the change in the stocks in the portfolio of stocks selected by the darts is accounted for by the linear relationship between the change in the stocks in the portfolio of stocks selected by the darts and the change in the stocks in the DJIA. **(f)** 0.6639. This value of r is positive. We would therefore say that the change in the stocks in the portfolio of stocks selected by the darts and the change in the stocks in the DJIA are positively correlated. As the change in the stocks in the DJIA increases, the change in the stocks in the portfolio of stocks selected by the darts also tends to increase.

26. (a) 41.632 **(b)** 2.63414 inches. The typical difference between the predicted height and the actual observed height is 2.63414 inches. **(c)** 45.759 **(d)** 4.127 **(e)** 0.0902. Therefore, 9.02% of the variability in height is accounted for by the linear relationship between height and age. **(f)** -0.3003. This value of r is negative. We would therefore say that height and age are negatively correlated. As age increases, height tends to decrease.

27. (a) 7.56875 **(b)** 0.972674 shot. The typical difference between the predicted number of shots and the actual observed number of shots is 0.972674. **(c)** 7.6 **(d)** 0.03125 **(e)** 0.0041. Therefore, 0.41% of the variability in the number of shots is accounted for by the linear relationship between the number of shots and age. **(f)** 0.0640. This value of r is close to $r = 0$. We would therefore say that there is no linear relationship between the number of shots a child gets and the child's age.

28. (a) 7.891 **(b)** 0.993181. The typical difference between the predicted power rating of a baseball team and the actual observed power rating of a baseball team is 0.993181. **(c)** 25.756 **(d)** 17.865 **(e)** 0.6936. Therefore, 69.36% of the variability in the power rating of a baseball team is accounted for by the linear relationship between the power rating of a baseball team and the winning percentage of the baseball team. **(f)** 0.8328. This value of r is very close to the maximum value $r = 1$. We would therefore say that the power rating of a baseball team and the winning percentage of the baseball team are positively correlated. As the winning percentage increases, the power rating also tends to increase.

29. (a) 42.08 **(b)** 2.29340 grams. The typical difference between the predicted amount of saturated fat and the actual observed amount of saturated fat is 2.29340 grams. **(c)** 52.66 **(d)** 10.58 **(e)** 0.2009. Therefore, 20.09% of the variability in the amount of saturated fat is accounted for by the linear relationship between the amount of saturated fat and the calories per serving. **(f)** 0.4482. This value of r is positive. We would therefore say that the number of calories per serving and the number of grams of saturated fat per serving are positively correlated. As the number of calories in a serving of food increases, the number of grams of saturated fat also tends to increase.

30. (a) 107.510 **(b)** 3.27888 mpg. The typical difference between the predicted gas mileage and the actual observed gas mileage is 3.27888 mpg. **(c)** 386.667 **(d)** 279.156 **(e)** 0.7220. Therefore, 72.20% of the variability in the gas mileage is accounted for by the linear relationship between the gas mileage and the engine displacement. **(f)** -0.8497. This value of r is very close to the minimum value $r = -1$. We would therefore say that the gas mileage and the engine displacement are negatively correlated. As the engine displacement of a car increases, the gas mileage of the car tends to decrease.

31. (a) 23.800 **(b)** 1.72483%. The typical difference between the predicted percent of a state's population that completes college and the actual observed percent of the state's population that completes college is 1.74283%. **(c)** 154.156 **(d)** 130.356 **(e)** 0.8456. Therefore, 84.56% of the variability in the percent of a state's population that completes college is accounted for by the linear relationship between the percent of a state's population that completes college and the percent of a state's population that attends college. **(f)** 0.9196. This value of r is very close to the maximum value $r = 1$. We would therefore say that the percent of a state's population that completes college and the percent of a state's population that attends college are positively correlated. As the percent of a state's population that attends college increases, the percent of a state's population that completes college tends to increase.

32. (a) 4.97525 **(b)** 0.788611%. The typical difference between the predicted percent of a city's population that bikes to work and the actual observed percent of the city's population that bikes to work is 0.788611%. **(c)** 5.27600 **(d)** 0.30075 **(e)** 0.0570. Therefore, 5.70% of the variability in the percent of a city's population that bikes to work is accounted for by the linear relationship between the percent of a city's population that bikes to work and the percent of a city's population that walks to work. **(f)** 0.2387. This value of r is close to $r = 0$. We would therefore say that the percent of a city's population that bikes to work and the percent of a city's population that walks to work have no apparent linear relationship.

33. (a) 68.77 (b) 3.38560. The typical difference between the predicted teen birth rate and the actual observed teen birth rate is 3.38560. (c) 383.40 (d) 314.63 (e) 0.8206. Therefore, 82.06% of the variability in the teen birth rate is accounted for by the linear relationship between the teen birth rate and the overall birth rate. (f) 0.9059. This value of r is very close to the maximum value $r = 1$. We would therefore say that the teen birth rate and the overall birth rate are positively correlated. As the overall birth rate increases, the teen birth rate also tends to increase.

34. (a) 124,645 (b) 124.822 grams. The typical difference between the predicted brain weight of an animal and the actual observed brain weight of an animal is 124.822 grams. (c) 130,210 (d) 5566 (e) 0.0427. Therefore, 4.27% of the variability in the brain weight of an animal is accounted for by the linear relationship between the brain weight of the animal and the body weight of the animal. (f) 0.2067 This value of r is close to the value $r = 0$. We would therefore say that the brain weight of an animal and the body weight of the animal have no linear relationship.

35. (a) 358.82 (b) 5.99013. The typical difference between the predicted consumer sentiment for incomes $75,000 or higher and the actual observed consumer sentiment for incomes $75,000 or higher is 5.99013. (c) 440.01 (d) 81.19 (e) 0.1845. Therefore, 18.45% of the variability in the consumer sentiment for incomes $75,000 or higher is accounted for by the linear relationship between the consumer sentiment for incomes $75,000 or higher and the consumer sentiment for incomes under $75,000. (f) 0.4295. This value of r is positive. We would therefore say that the consumer sentiment for incomes $75,000 or higher and the consumer sentiment for incomes under $75,000 are positively correlated. As the consumer sentiment for incomes under $75,000 increases, the consumer sentiment for incomes $75,000 or higher also tends to increase.

36. (a) 6948.36 (b) 27.7856. The typical difference between the SAT Mathematics score and the actual observed SAT Mathematics score is 27.7856. (c) 8060.55 (d) 1112.18 (e) 0.1380. Therefore, 13.80% of the variability the SAT Mathematics score is accounted for by the linear relationship between the SAT Mathematics score and the SAT Critical Reading score. (f) 0.3715. This value of r is positive. We would therefore say that the SAT Mathematics score and the SAT Critical Reading score are positively correlated. As the SAT Critical Reading score increases, the SAT Mathematics score also tends to increase.

37. (a) 530.02 (b) 8.13958. The typical difference between the predicted batting average of a player and the actual observed batting average of a player is 8.13958. (c) 560.50 (d) 30.48 (e) 0.0544. Therefore, 5.44% of the variability in the runs scored is accounted for by the linear relationship between the batting averages and the runs scored. (f) 0.2332. This value of r is close to the value $r = 0$. We would therefore say that the batting averages and the runs scored have no linear relationship.

38.

Company	Net worth
Microsoft	39.6
Intel	41.7
Dell	5.7
Apple	27.9
Google	36

39. $\hat{y} = 0.088x + 1.03$; $b_1 = 0.088$ means that for each increase of $1 billion in the net worth of a company the current ratio increases by 0.088. $b_0 = 1.03$ means that the predicted current ratio of a company with a net worth of $x = $0 is 1.03.

40. $\hat{y} = 0.175x + 0.69$; $b_1 = 0.175$ means that for each increase of 1 in the price–earnings ratio of a company the current ratio increases by 0.175. $b_0 = 0.69$ means that the predicted current ratio of a company with a price–earnings ratio of $x = 0$ is 0.69.

41. (a) 10.79% (b) 5.97%

42. (a) 4.27357 (b) 4.38743

43. Net worth

44. 15.35382.

45. (a) y = overall dating score = 54.0 + 2.43 Percentage 18–24 years old (b) y = overall dating score = 134.5 − 0.646 Percentage 18–24 who are single (c) y = overall dating score = 67.1 + 0.194 Online dating score

46. (a) 15.58% (b) 18.39% (c) 15.52%

47. (a) 7.63309 (b) 7.50504 (c) 7.63561

48. Percentage 18–24 who are single

49. Heating degree-days = 11264 − 198.1 Average January temperatures

50. $b_1 = -198.1$ means that for each increase of 1°F in the average January temperature the number of heating degree-days decreases by 198.1.

51. 164.753

52. 95.19%

53. Cooling degree-days = −8999 + 133.22 Average July temperatures

54. $b_1 = 133.22$ means that for each increase of 1°F in the average July temperature the predicted number of cooling degree-days increases by 133.22.

55. 55.3145

56. 97.27%

57. (a) $x = 10$ years of education; $y = 20.6$ = unemployment rate. It doesn't follow the trend of the higher the number of years of education, the lower the unemployment rate. (b) Since $r^2 = 0.6824$, 68.24% of the variability in the variable y = unemployment rate is accounted for by the linear relationship between x = years of education and y = unemployment rate. Hence the statement is not true. (c) Since the absolute values of the residuals for 5, 10, and 16 years of education are more than 1%, this claim is not always true. (d) Since $b_1 = -1.24$, we can say that each additional year of education drops the predicted unemployment rate by 1.24%.

58. The y intercept would increase and the slope would increase in absolute value (decrease since it is negative). This is because an increase in the y-value of the point closest to the y axis would pull the left side of the regression line higher up, increasing the y intercept and making the slope more negative.

59. SST = 228, SSR = 216

60. (a)

$x =$ Engine size (liters)	$y =$ Combined (city/highway) gas mileage (MPG)	$\hat{y} = -5.49x + 38.41$	$(y - \hat{y})$	$(y - \hat{y})^2$
1.6	31	29.626	1.374	1.887876
2.0	28	27.43	0.57	0.3249
2.5	26	24.685	1.315	1.729225
2.4	26	25.234	0.766	0.586756
2.5	23	24.685	−1.685	2.839225
2.7	22	23.587	−1.587	2.518569
3.5	20	19.195	0.805	0.648025
3.0	19	21.94	−2.94	8.6436
4.0	17	16.45	0.55	0.3025
4.6	14	13.156	0.844	0.712336
				SSE = 20.193012

(b) SSE is the sum of the squared residuals. Since we know that $\hat{y} = -5.49x + 38.41$ is the regression line, according to the least-squares criterion, no other possible straight line would result in a smaller SSE. (c) Chevrolet Equinox. It has much less combined mpg than expected.

61. (a) $s = 1.5887$ (b) If we know car's engine size (x), then our estimate of the combined mpg will typically differ from the actual mpg by 1.5887 mpg.

62. (a) $s^2 = 27.6$, SST = 248.4 (b) SSR = 228.206988. SSR measures the amount of improvement in the accuracy of our estimates using the regression equation compared with relying only on the y-values and ignoring the x information. (c) $r^2 = 0.9187$; 91.87% of the variability in the variable y = combined mpg is accounted for by the linear relationship between x = engine size and y = combined mpg.

63. (a) $r = -0.9585$ (b) This value of r is very close to the minimum value $r = -1$. We would therefore say that the gas mileage and the engine displacement are negatively correlated. As the engine displacement of a car increases, the gas mileage of the car tends to decrease.

64. Since $(\bar{x}, \bar{y})$ is on the regression line, the slope and the y intercept would remain the same.

65. The slope would increase and the y intercept would decrease.

66. 38.1744186 mpg

67. Since $b_1 = 0$, the regression equation is $\hat{y} = \bar{y} = 25.02$. Thus $\hat{y} - \bar{y} = 0$ for all of the $\hat{y} - \bar{y}$'s. Hence SSR = 0, so SSR would decrease. Since SST = SSR + SSE and SSR = 0, SSE = SST = 358.5622634, so both SSE and SST increase. Since the regression line doesn't include any information from the x-values, SSR = 0, $r^2 = 0$, and $r = 0$. Since SSR = 0, $r^2 = \frac{SSR}{SST}$, and $r = \sqrt{r^2}$, $r^2 = 0$ and $r = 0$. Since $s = \sqrt{\frac{SSE}{n-2}} = 6.6948$ and SSE increases, s increases.

68. (a)

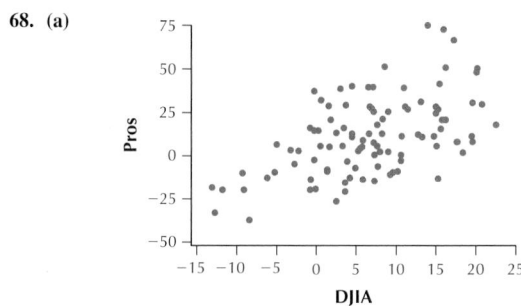

(b) $\hat{y} = 1.49x + 0.83$. The estimated increase (in percent) in the Pros stock portfolio equals 1.49 times the increase in the DJIA plus 0.83. **(c)** $r^2 = 0.289$, so 28.9% of the variability in the Pros price increase is accounted for by the linear relationship between the Pros price increase and the DJIA. **(d)** $s = 18.8545$. The typical difference between the predicted Pros price increase and the actual Pros price increase is 18.8545%. **(e)** $r = \sqrt{r^2} = \sqrt{0.289} = 0.5376$.

69. (a)

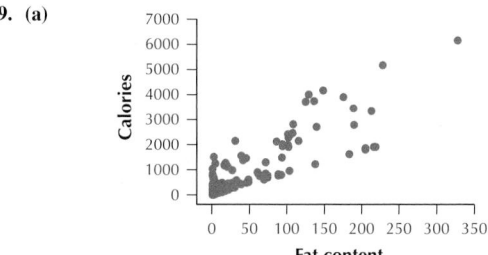

(b) $\hat{y} = 8.12x + 1.28$. The estimated calories per gram equals 8.12 times the amount of fat per gram, plus 1.28. **(c)** $r^2 = 0.736$, so 73.6% of the variability in the number of calories per gram is accounted for by the linear relationship between calories per gram and fat per gram. **(d)** $s = 0.9944$. The typical difference between the predicted number of calories per gram and the actual calories per gram is 0.9944. **(e)** $r = \sqrt{r^2} = \sqrt{0.736} = 0.8579$

70. (a)

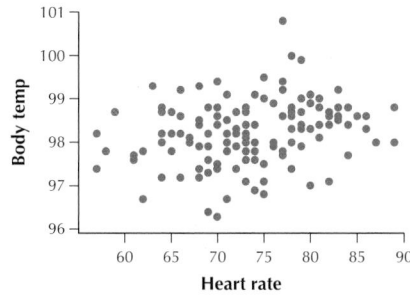

(b) $\hat{y} = 0.0263x + 96.3$. The estimated body temperature equals 0.0263 times the heart rate, plus 96.3. **(c)** $r^2 = 0.064$, so 6.4% of the variability in body temperature is accounted for by the linear relationship between body temperature and heart rate. **(d)** $s = 0.7120$. The typical difference between the predicted body temperature and the actual body temperature is 0.7120.
(e) $r = \sqrt{r^2} = \sqrt{0.064} = 0.2530$
71. Answers will vary.
72. Answers will vary.
73. Answers will vary.
74. Answers will vary.
75. Answers will vary.
76. Answers will vary.
77. (a) Answers will vary. **(b)** That your adjusted line matches the least squares regression line **(c)** Yes
78. Answers will vary.

Regression Analysis: Weight versus Height

Analysis of Variance

Source	DF	Adj SS	Adj MS	F-Value	P-Value
Regression	1	21559	21559.0	58.65	0.000
Height	1	21559	21559.0	58.65	0.000
Error	258	94835	367.6		
Lack-of-Fit	71	27586	388.5	1.08	0.336
Pure Error	187	67248	359.6		
Total	259	116394			

Model Summary

S	R-sq	R-sq(adj)	R-sq(pred)
19.1723	18.52%	18.21%	17.26%

Coefficients

Term	Coef	SE Coef	T-Value	P-Value	VIF
Constant	-96.2	30.0	-3.20	0.002	
Height	3.541	0.462	7.66	0.000	1.00

Regression Equation

Weight = -96.2 + 3.541 Height

Fits and Diagnostics for Unusual Observations

Obs	Weight	Fit	Resid	Std Resid		
10	165.80	127.19	38.61	2.02	R	
16	109.80	109.13	0.67	0.04		X
24	193.60	149.85	43.75	2.30	R	
28	148.40	154.45	-6.05	-0.32		X
57	133.80	155.16	-21.36	-1.13		X
65	176.40	128.25	48.15	2.52	R	
99	147.30	154.45	-7.15	-0.38		X
102	161.40	114.44	46.96	2.48	R	
112	231.90	144.54	87.36	4.58	R	
123	98.80	112.31	-13.51	-0.71		X
160	186.30	130.37	55.93	2.92	R	
173	180.30	158.70	21.60	1.15		X
210	178.40	133.91	44.49	2.32	R	
219	190.30	133.91	56.39	2.95	R	
227	229.50	133.91	95.59	5.00	R	
230	177.50	126.83	50.67	2.65	R	
235	180.30	130.37	49.93	2.61	R	
247	103.20	112.67	-9.47	-0.50		X
252	166.40	126.83	39.57	2.07	R	
255	169.30	123.29	46.01	2.41	R	

R Large residual
X Unusual X

Regression Analysis: Weight_1 versus Height_1

Analysis of Variance

Source	DF	Adj SS	Adj MS	F-Value	P-Value
Regression	1	37835	37834.6	98.30	0.000
Height_1	1	37835	37834.6	98.30	0.000
Error	245	94299	384.9		
Lack-of-Fit	64	25305	395.4	1.04	0.416
Pure Error	181	68994	381.2		
Total	246	132133			

Model Summary

S	R-sq	R-sq(adj)	R-sq(pred)
19.6187	28.63%	28.34%	27.54%

Coefficients

Term	Coef	SE Coef	T-Value	P-Value	VIF
Constant	-134.7	31.0	-4.35	0.000	
Height_1	4.386	0.442	9.91	0.000	1.00

Regression Equation

Weight_1 = -134.7 + 4.386 Height_1

Fits and Diagnostics for Unusual Observations

Obs	Weight_1	Fit	Resid	Std Resid		
80	200.40	205.71	-5.31	-0.28		X
106	128.70	136.85	-8.15	-0.42		X
120	127.90	172.38	-44.48	-2.27	R	
124	256.60	172.38	84.22	4.30	R	
127	188.50	207.47	-18.97	-0.99		X
141	239.40	194.31	45.09	2.32	R	
159	225.50	185.54	39.96	2.05	R	
161	211.40	163.60	47.80	2.44	R	
167	201.50	161.41	40.09	2.05	R	
169	213.40	172.38	41.02	2.10	R	
198	205.50	160.09	45.41	2.32	R	
209	159.40	141.67	17.73	0.92		X
221	226.00	172.38	53.62	2.74	R	

R Large residual
X Unusual X

79. 19.1723 pounds. The typical difference between the predicted weight of a woman and the actual weight of a woman is 19.1723 pounds.
80. 0.1852. Therefore, 18.52% of the variability in women's weights is accounted for by the linear relationship between women's weights and women's heights.
81. 0.4303
82. 19.6187 pounds. The typical difference between the predicted weight of a man and the actual weight of a man is 19.6187 pounds.

83. 0.2863. Therefore, 28.63% of the variability in men's weights is accounted for by the linear relationship between men's weights and men's heights.
84. 0.5351
85. **(a)** The values of s for men and women are both between 19 and 20 but are not the same. **(b)** The values of r^2 for men and women are both low but the value of r^2 for men is higher than the value of r^2 for women.
86. Bicep girth $= 3.05 + 0.4378$ Thigh girth. $\hat{y} = 0.4378x + 3.05$. The predicted bicep girth of a woman is 0.4378 times her thigh girth plus 3.05. $b_1 = 0.4378$ means that for each increase of 1 centimeter in a woman's thigh girth the predicted bicep girth increases by 0.4378 centimeter. $b_0 = 3.05$ means that the predicted bicep girth of a woman with a thigh girth of $x = 0$ centimeters is 3.05 centimeters.
87. 1.79882 centimeters. The typical difference between the predicted bicep girth of a woman and the actual bicep girth of a woman is 1.79882 centimeters.
88. 0.5612. Therefore, 56.12% of the variability in women's bicep girths is accounted for by the linear relationship between women's bicep girths and women's thigh girths.
89. Bicep girth_1 $= 8.12 + 0.4651$ Hip girth_1. $\hat{y} = 0.4651x + 8.12$. The predicted bicep girth of a man is 0.4651 times his thigh girth plus 8.12. $b_1 = 0.4651$ means that for each increase of 1 centimeter in a man's thigh girth the predicted bicep girth increases by 0.4651 centimeter. $b_0 = 8.12$ means that the predicted bicep girth of a man with a thigh girth of $x = 0$ centimeters is 8.12 centimeters.
90. 2.23852 centimeters. The typical difference between the predicted bicep girth of a man and the actual bicep girth of a man is 2.23852 centimeters.
91. 0.4388. Therefore, 43.88% of the variability in men's bicep girths is accounted for by the linear relationship between men's bicep girths and men's thigh girths.
92. **(a)** The slopes for men and women are both small and positive. The slope for women is larger than the slope for men. **(b)** The y-intercepts for men and women are both positive and under 10. The y-intercept for men is larger than the one for women. **(c)** The s for both men and women is around 2. The s for men is larger than the s for women. **(d)** The r^2 for women is larger than the r^2 for men.

Chapter 4 Review

1.

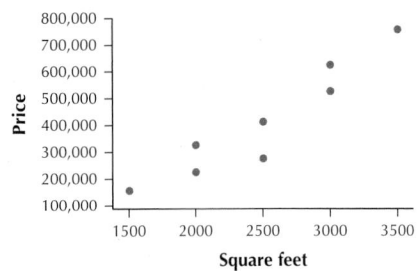

2. Positive.
3. As the number of square feet increases, the price also tends to increase.
4. 0.9434.
5. This value of r is very close to the maximum value $r = 1$. We would therefore say that price and square feet are positively correlated. As square feet increases, price also tends to increase.
6. $b_1 = 300.0$, $b_0 = -337,500$, $\hat{y} = 300.0x - 337,500$
7. The predicted price of a house is 300.0 times the number of square feet minus $337,500.
8. $b_1 = 300.0$ means that for each increase of 1 square foot of space the predicted price of a house increases by $300.0.
9. $b_0 = -337,500$ means that the predicted price for a home with $x = 0$ square feet is $-\$337,500$.
10. $712,500
11. $43,750. The actual observed price of house with 3500 square feet lies above the predicted price of a house with 3500 square feet.
12. $1,012,500. This represents extrapolation.
13. 33,359,375,000
14. $74,564.7. The typical difference between the predicted price of a house and the actual price is $74,564.7.
15. SST $= 3.03359 \times 10^{11}$, SSR $= 2.70000 \times 10^{11}$
16. 0.8900. Therefore, 89.00% of the variability in house prices is accounted for by the linear relationship between house prices and square feet.

17. 0.9434. This value of r is very close to the maximum value $r = 1$. We would therefore say that price and square feet are positively correlated. As square feet increases, price also tends to increase.

Chapter 4 Quiz

1. False
2. False
3. estimate
4. unit
5. extrapolation
6. negative
7.

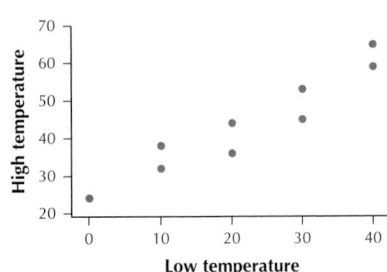

8. Positive
9. $\hat{y} = 0.900x + 24.00$
10. SSR $= 1260.00$, SSE $= 132.00$, SST $= 1392.00$
11. 4.34248°F. The typical difference between the predicted high temperature and the actual high temperature is 4.34248°F.
12. 0.9052. Therefore, 90.52% of the variability in high temperatures is accounted for by the linear relationship between high temperatures and low temperatures.
13. 0.9514. This value of r is close to the maximum value $r = 1$. We would therefore say that high temperatures and low temperatures are positively correlated. As the low temperature increases, the high temperature also tends to increase.
14. 0°F

Chapter 5

Section 5.1

1. Answers will vary; chance, likelihood.
2. A numerical value for probability is more definite and is interpreted the same way by everyone.
3. Answers will vary.
4. Classical method, relative frequency method, and subjective method
5. The experiment has equally likely outcomes.
6. When we have prior knowledge about the relative frequency of an outcome
7. We consider all available information, tempered by our experience and intuition, and then assign a probability value that expresses our estimate of the likelihood that the outcome will occur.
8. The classical and relative frequency methods
9. First find out how many students are at your college and then find out how many of them like hip-hop music. Then calculate the relative frequency of students who like hip-hop music. Use the relative frequency method.
10. **(a)** Outcome or event is very unlikely. **(b)** Outcome or event cannot occur. **(c)** Outcome or event is nearly certain to occur. **(d)** Outcome or event is certain to occur. It's "a sure thing."
11. Not a probability model. The probability for males is negative and the probabilities don't add up to 1.
12. Not a probability model. The probability for tenors is negative and the probabilities don't add up to 1.
13. Not a probability model. The probabilities don't add up to 1.
14. Not a probability model. The probability for a math major is negative.
15. Probability model.
16. Probability model.
17. "The chances of recovering from the surgery is near 100%" means that the probability of recovering from the surgery is near 1. Therefore it is almost certain that the person will recover from the surgery.
18. No chance means probability is equal to 0. Therefore if the student's grade does not improve, the student is definitely not getting the new video game system as a birthday present.
19. "The chances are high that the blue chip stock will gain in value this year" means that the probability that the blue chip stock will gain in value this year is near 1. Therefore it is almost certain that the blue chip stock will gain in value this year.

20. The candidate's chances of winning the election are about 2%. This means that the probability that the candidate will win the election is about 0.02. This is a very small number. Therefore it is very unlikely that the candidate will win the election.

21. $\frac{1}{13}$

22. $\frac{1}{4}$

23. $\frac{1}{52}$

24. $\frac{1}{2}$

25. $\frac{1}{6}$

26. $\frac{1}{2}$

27. $\frac{2}{3}$

28. $\frac{1}{6}$

29. $\frac{1}{3}$

30. 0

31.

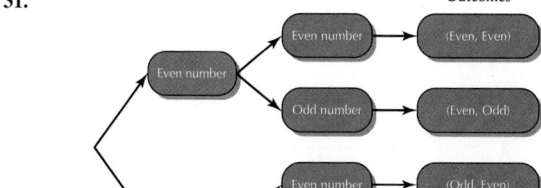

32. {(Even, Even), (Even, Odd), (Odd, Even), (Odd, Odd)}

33. Let L = tossing a number less than 4 and G = tossing a number greater than or equal to 4.

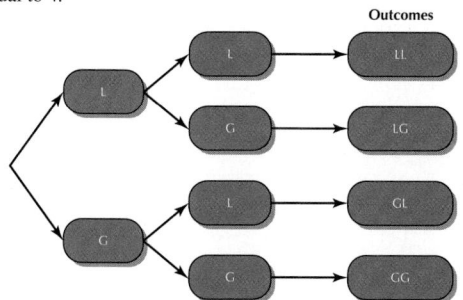

34. {LL, LG, GL, GG}

35.

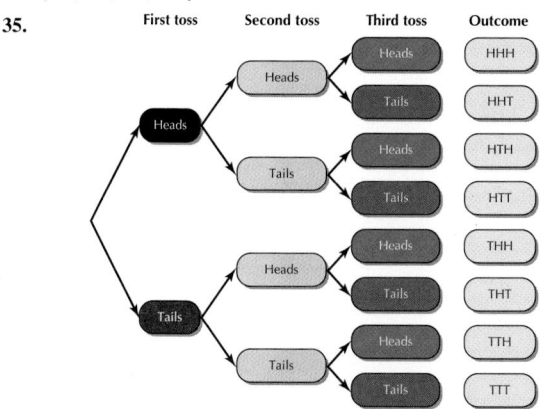

36. {HHH, HHT, HTH, HTT, THH, THT, TTH, TTT}

37. We can follow the branches to get all possible outcomes.

38. Choose a branch for the first toss, then choose a branch for the second toss, then choose a branch for the third toss. This will give one of the possible outcomes. If we follow all possible choices of branches for each toss, we get all the outcomes in the sample space.

39. $\frac{1}{4}$

40. $\frac{1}{2}$

41. $\frac{1}{4}$

42.

Number of heads	Probability
0	$\frac{1}{4}$
1	$\frac{1}{2}$
2	$\frac{1}{4}$

43. $\frac{1}{9}$

44. 0

45. $\frac{1}{18}$

46. $\frac{1}{6}$

47. 0

48.

Sum of dice	Probability
2	1/36
3	2/36 = 1/18
4	3/36 = 1/12
5	4/36 = 1/9
6	5/36
7	6/36 = 1/6
8	5/36
9	4/36 = 1/9
10	3/36 = 1/12
11	2/36 = 1/18
12	1/36

49. Sum of 7

50. Sum of 2 and sum of 12

51. $\frac{7}{20}$

52. $\frac{1}{5}$

53. $\frac{1}{5}$

54. $\frac{1}{4}$

55. Relative frequency method

56.

Hot caffeinated beverage	Probability
Regular coffee	$\frac{7}{20}$
Latte	$\frac{1}{5}$
Cappuccino	$\frac{1}{5}$
Tea	$\frac{1}{4}$

57. $\frac{2}{5}$

58. $\frac{2}{5}$

59. $\frac{1}{5}$

60.

Where student lives	Probability
On campus	$\frac{2}{5}$
With family off campus	$\frac{2}{5}$
In an apartment off campus	$\frac{1}{5}$

61.

Favorite color	Probability
Red	$\frac{1}{4}$
Blue	$\frac{1}{4}$
Green	$\frac{1}{5}$
Black	$\frac{1}{10}$
Violet	$\frac{1}{10}$
Yellow	$\frac{1}{10}$

62.

Favorite season	Probability
Summer	$\frac{7}{20}$
Spring	$\frac{7}{20}$
Autumn	$\frac{1}{4}$
Winter	$\frac{1}{20}$

63. (a)

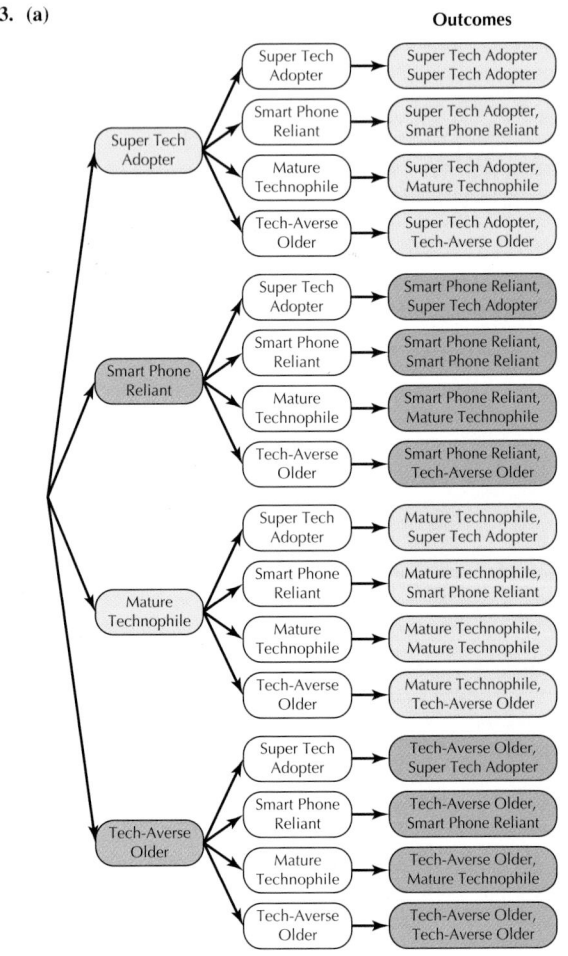

(b) {{Super Tech Adopter, Super Tech Adopter}, {Super Tech Adopter, Smart Phone Reliant}, {Super Tech Adopter, Mature Technophile}, {Super Tech Adopter, Tech-Averse Older}, {Smart Phone Reliant, Super Tech Reliant}, {Smart Phone Reliant, Smart Phone Reliant}, {Smart Phone Reliant, Mature Technophile}, {Smart Phone Reliant, Tech-Averse Older}, {Mature Technophile, Super Tech Adopter}, {Mature Technophile, Smart Phone

Reliant}, {Mature Technophile, Mature Technophile}, {Mature Technophile, Tech-Averse Older}, {Tech-Averse Older, Super Tech Adopter}, {Tech-Averse Older, Smart Phone Reliant}, {Tech-Averse Older, Mature Technophile}. {Tech-Averse Older, Tech-Averse Older}

64. (a)

Gender	Relative frequency
Male	0.5333
Female	0.4556
Unknown	0.0111

(b)

Gender	Probability $\approx$ Relative frequency
Male	0.5333
Female	0.4556
Unknown	0.0111

(c)

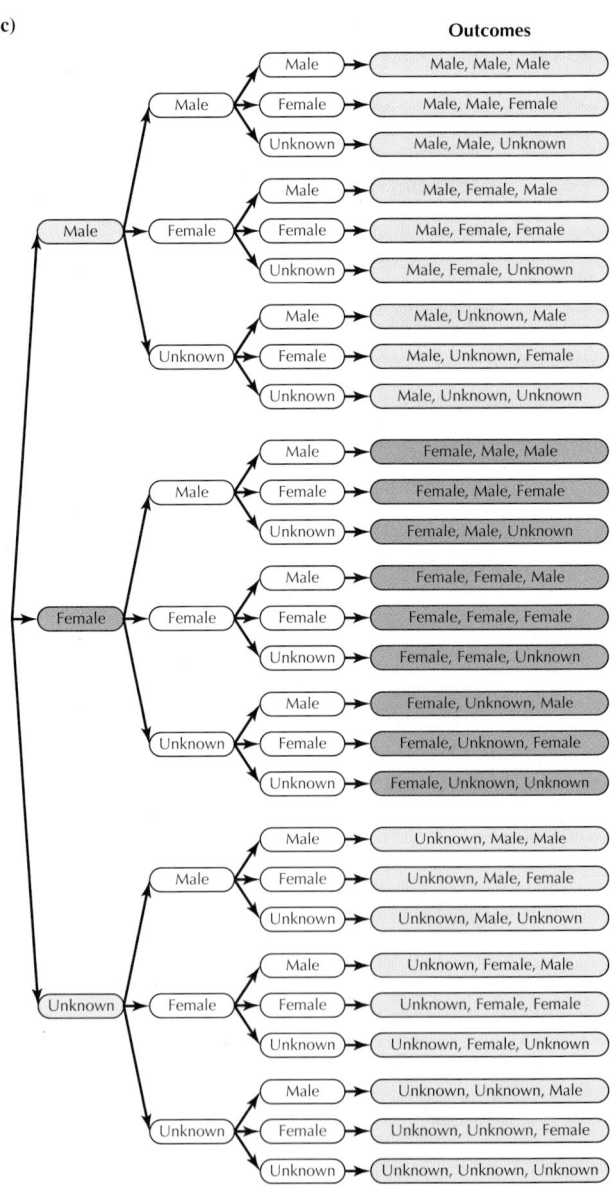

(d) {MMM, MMF, MMU, MFM, MFF, MFU, MUM, MUF, MUU, FMM, FMF, FMU, FFM, FFF, FFU, FUM, FUF, FUU, UMM, UMF, UMU, UFM, UFF, UFU, UUM, UUF, UUU}

65. (a) $\frac{7}{72}$ **(b)** $\frac{65}{72}$ **(c)** Relative frequency method

66. (a) and **(b)** Answers will vary.

67. (a)

	Frequency	Relative frequency
Girls	18	$18/44 = 0.4091$
Boys	26	$26/44 = 0.5909$
Total	44	$44/44 = 1.0000$

(b)

Outcome	Probability
Girl	$18/44 = 0.4091$
Boy	$26/44 = 0.5909$

68. (a) $\frac{1}{52}$ **(b)** $3.85

69. (a) $\frac{1}{6}$ **(b)** $\frac{5}{6}$ **(c)** $1.67

70. (a) No, because the outcomes of sharing a social media profile with your partner and not sharing a social media profile with your partner are not equally likely **(b)** $\frac{14}{127}$; relative frequency method

71. $1/8 = 0.125$

72. $3/8 = 0.375$

73. $3/8 = 0.375$

74. $1/8 = 0.125$

75.

Number of heads	Probability
0	$1/8 = 0.125$
1	$3/8 = 0.375$
2	$3/8 = 0.375$
3	$1/8 = 0.125$

76. The classical method.

77. $\frac{1}{8}$

78. $\frac{7}{8}$

79. $\frac{3}{8}$

80. $\frac{5}{8}$

81. (a) Answers will vary. **(b)** As the sample size increases, the relative frequencies approach the probabilities.

82. The sum of the percents is greater than 100%.

83. (a)

Type of music	Probability
Hip-hop/rap	0.27
Pop	0.23
Rock/Punk	0.17
Alternative	0.07
Christian/gospel	0.06
R&B	0.06
Country	0.05
Techno/house	0.04
Jazz	0.01
Other	0.04

(b) Yes. **(c)** Answers will vary. **(d)** As the sample size increases, the relative frequencies approach the probabilities.

84. (a) $(1/2)^8$ **(b)** $(1/2)^9$ **(c)** $1/2$ **(d)** $(1/2)^k$

85. (a) Greater **(b)** Greater

86.

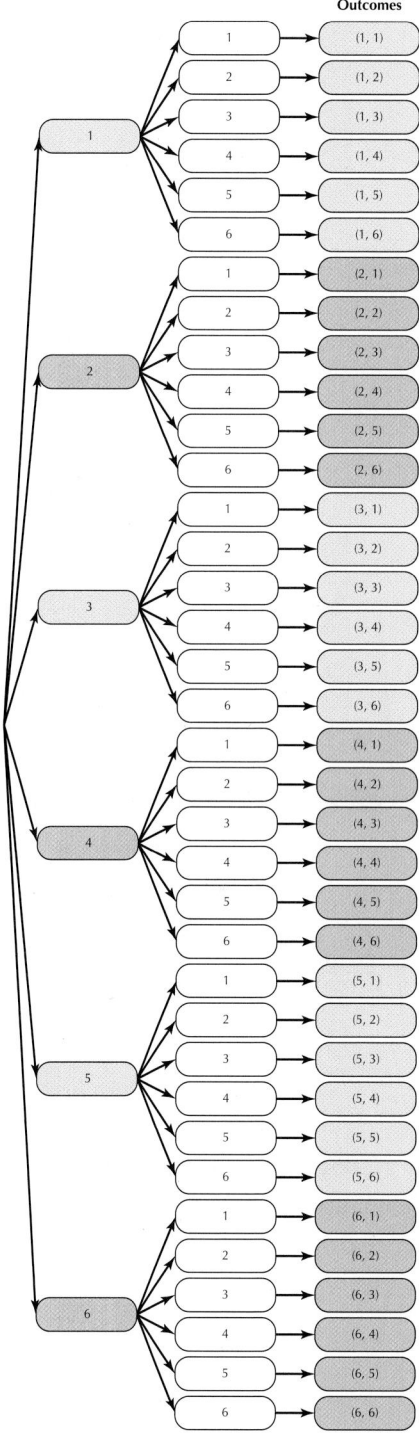

87.
$$\begin{Bmatrix} (1,1),(1,2),(1,3),(1,4),(1,5),(1,6),(2,1),(2,2),(2,3),(2,4),(2,5),(2,6), \\ (3,1),(3,2),(3,3),(3,4),(3,5),(3,6),(4,1),(4,2),(4,3),(4,4),(4,5),(4,6), \\ (5,1),(5,2),(5,3),(5,4),(5,5),(5,6),(6,1),(6,2),(6,3),(6,4),(6,5),(6,6), \end{Bmatrix}$$

Example 5.5, Experiment is tossing 2 fair dice.

88. Events can consist of more than one outcome, but outcomes can't consist of more than one event.

89. 1/36. Classical probability method; have the sample space but no actual data and can assume outcomes are equally likely

90. 1/9. Classical probability method; have the sample space but no actual data and can assume outcomes are equally likely

91. (a) 1/9 **(b)** 4/9 **(c)** 4/9 **(d)** 5/9 **(e)** 8/9

92. Sony, MS, Nintendo, Electronic Arts, Activision, Focus, Take-Two

93. $\dfrac{3}{20}$

94.

Studio	Probability ≈ Relative frequency
Sony	$\dfrac{3}{20}$
MS	$\dfrac{1}{10}$
Nintendo	$\dfrac{1}{5}$
Electronic Arts	$\dfrac{3}{20}$
Activision	$\dfrac{1}{5}$
Focus	$\dfrac{1}{20}$
Take-Two	$\dfrac{3}{20}$

Relative frequency method

95. All probabilities are between 0 and 1 and the sum of the probabilities is 1.

96. (a)

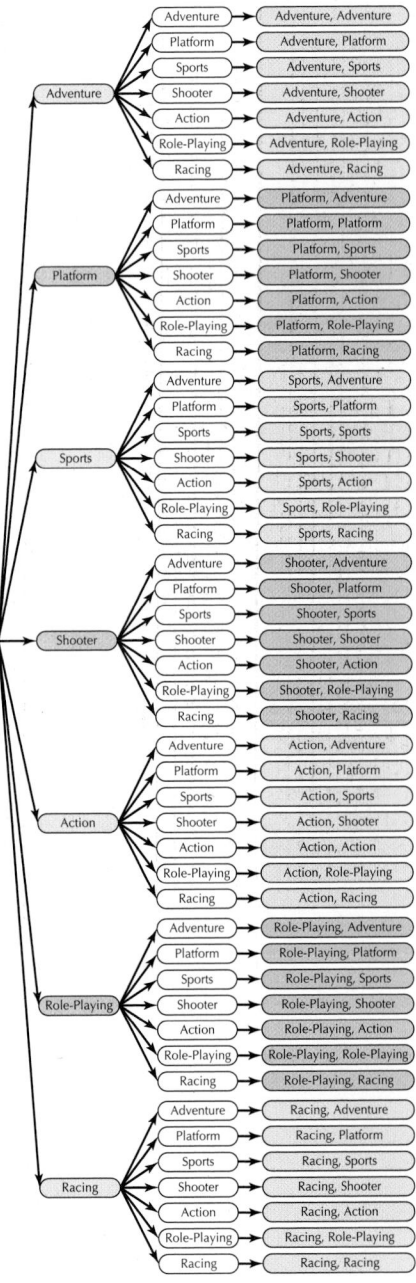

(b) {{Adventure, Adventure}, {Adventure, Platform}, {Adventure, Sports}, {Adventure, Shooter}, {Adventure, Action}, {Adventure, Role-Playing}, {Adventure, Racing}, {Platform, Adventure}, {Platform, Platform}, {Platform, Sports}, {Platform, Shooter}, {Platform, Action}, {Platform, Role-Playing}, {Platform, Racing}, {Sports, Adventure}, {Sports, Platform}, {Sports, Sports}, {Sports, Shooter}, {Sports, Action}, {Sports, Role-Playing}, {Sports, Racing}, {Shooter, Adventure}, {Shooter, Platform}, {Shooter, Sports}, {Shooter, Shooter}, {Shooter, Action}, {Shooter, Role-Playing}, {Shooter, Racing}, {Action, Adventure}, {Action, Platform}, {Action, Sports}, {Action, Shooter}, {Action, Action}, {Action, Role-Playing}, {Action, Racing}, {Role-Playing, Adventure}, {Role-Playing, Platform}, {Role-Playing, Sports}, {Role-Playing, Shooter}, {Role-Playing, Action}, {Role-Playing, Role-Playing}, {Role-Playing, Racing}, {Racing, Adventure}, {Racing, Platform}, {Racing, Sports}, {Racing, Shooter}, {Racing, Action}, {Racing, Role-Playing}, {Racing, Racing}}

97.

Type	Relative frequency
Adventure	$\dfrac{1}{10}$
Platform	$\dfrac{1}{10}$
Sports	$\dfrac{1}{20}$
Shooter	$\dfrac{2}{5}$
Action	$\dfrac{1}{4}$
Role-playing	$\dfrac{1}{20}$
Racer	$\dfrac{1}{20}$

98.

Type	Probability ≈ Relative frequency
Adventure	$\dfrac{1}{10}$
Platform	$\dfrac{1}{10}$
Sports	$\dfrac{1}{20}$
Shooter	$\dfrac{2}{5}$
Action	$\dfrac{1}{4}$
Role-playing	$\dfrac{1}{20}$
Racer	$\dfrac{1}{20}$

All probabilities are between 0 and 1 and the sum of the probabilities is 1.

99. (a)–(d) Answers will vary.

100. As the sample size increases, the relative frequencies approach the probabilities.

101.

Clinic location	Frequency
Suburban	963
Urban	450
Total	1413

102.

Clinic location	Relative frequency ≈ Probability
Suburban	0.6815
Urban	0.3185

103. 0.3185; Relative frequency method

104.

Insurance Type	Frequency
Hospital-based	84
Med assistance	275
Military	331
Private payer	723
Total	1413

105.

Insurance Type	Relative frequency ≈ Probability
Hospital-based	0.0594
Med assistance	0.1946
Military	0.2343
Private payer	0.5117

106. 0.5117; Relative frequency method

Section 5.2

1. Two events are mutually exclusive if they have no outcomes in common.

2. It contains no outcomes.

3. It is all of the outcomes in each of the events. There are no outcomes in both.

4. Yes

5. You are more likely to select a male than a male football player. All male football players are males, but most males are not football players. Therefore, there are many more males than male football players at any college or university.

6. Answers will vary; $1 - P$ (will rain); they are complementary events.

7. $\frac{5}{6}$

8. $\frac{5}{6}$

9. $\frac{1}{2}$

10. $\frac{2}{3}$

11. $\frac{1}{2}$

12. $\frac{1}{2}$

13. $\{J\clubsuit, J\spadesuit\}$

14. $\{J\spadesuit\}$

15. $\{2\spadesuit, 3\spadesuit, 4\spadesuit, 5\spadesuit, 6\spadesuit, 7\spadesuit, 8\spadesuit, 9\spadesuit, 10\spadesuit, J\spadesuit, Q\spadesuit, K\spadesuit, A\spadesuit\}$

16. $\{J\diamondsuit, J\heartsuit, J\spadesuit, J\clubsuit, 2\spadesuit, 3\spadesuit, 4\spadesuit, 5\spadesuit, 6\spadesuit, 7\spadesuit, 8\spadesuit, 9\spadesuit, 10\spadesuit, Q\spadesuit, K\spadesuit, A\spadesuit, 2\clubsuit, 3\clubsuit, 4\clubsuit, 5\clubsuit, 6\clubsuit, 7\clubsuit, 8\clubsuit, 9\clubsuit, 10\clubsuit, Q\clubsuit, K\clubsuit, A\clubsuit\}$

17. $\{J\diamondsuit, J\heartsuit, J\spadesuit, J\clubsuit, 2\spadesuit, 3\spadesuit, 4\spadesuit, 5\spadesuit, 6\spadesuit, 7\spadesuit, 8\spadesuit, 9\spadesuit, 10\spadesuit, Q\spadesuit, K\spadesuit, A\spadesuit\}$

18. $\{2\spadesuit, 3\spadesuit, 4\spadesuit, 5\spadesuit, 6\spadesuit, 7\spadesuit, 8\spadesuit, 9\spadesuit, 10\spadesuit, J\spadesuit, Q\spadesuit, K\spadesuit, A\spadesuit, 2\clubsuit, 3\clubsuit, 4\clubsuit, 5\clubsuit, 6\clubsuit, 7\clubsuit, 8\clubsuit, 9\clubsuit, 10\clubsuit, J\clubsuit, Q\clubsuit, K\clubsuit, A\clubsuit\}$

19. $\frac{1}{26}$

20. $\frac{1}{52}$

21. $\frac{1}{4}$

22. $\frac{7}{13}$

23. $\frac{4}{13}$

24. $\frac{1}{2}$

25. 1

26. 0

27. $\frac{57}{2201}$

28. $\frac{763}{2201}$

29. $\frac{52}{2201}$

30. $\frac{1547}{2201}$

31. 0

32. $\frac{654}{2201}$

33. $\frac{2149}{2201}$

34. $\frac{1438}{2201}$

35. $\frac{2144}{2201}$

36. Child

37. $\frac{1}{2}$

38. $\frac{1}{2}$

39. $\frac{3}{5}$

40. $\frac{2}{5}$

41. $\frac{2}{5}$

42. $\frac{1}{5}$

43. $\frac{9}{10}$

44. $\frac{3}{5}$

45. $\frac{2}{3}$

46. $\frac{5}{12}$

47. $\frac{1}{4}$

48. $\frac{1}{6}$

49. (a) 0.5 (b) 0.53

50. (a) 0.69 (b) 0.41

51. (a) 0.72 (b) 0.28

52. (a) $\frac{49}{90}$ (b) 1 (c) 1

53. $\frac{2}{9}$

54. $\frac{4}{13}$

55. $0.92

56. $\frac{7}{13}$

57. $1.62

58. (a) 0.5326 (b) 0.5700 (c) 0.3305 (d) 0.7721 (e) 0.4674 (f) 0.4300

59. (a) 0.5465 (b) 0.5422 (c) 0.3064 (d) 0.7823 (e) 0.7952

60.

	Regular	Premium	Total
Compact	2	1	3
Midsize	2	2	4
Large	1	2	3
Total	5	5	10

61. (a) $\frac{3}{10}$ (b) $\frac{7}{10}$ (c) $\frac{2}{5}$ (d) $\frac{3}{5}$

62. (a) $\frac{1}{2}$ (b) $\frac{1}{2}$ (c) $\frac{1}{10}$ (d) $\frac{1}{5}$

63. (a) $\frac{1}{5}$ (b) $\frac{1}{5}$ (c) $\frac{1}{5}$ (d) $\frac{1}{10}$

64. (a) $\frac{7}{10}$ (b) $\frac{3}{5}$ (c) $\frac{3}{10}$ (d) $\frac{9}{10}$ (e) $\frac{3}{5}$

65. $\frac{1}{10}$

66. $\frac{3}{10}$

67. $\frac{3}{5}$

68. $\frac{4}{5}$

69. $\frac{1}{10}$

70. 0

71. $\frac{1}{5}$

72. $\frac{2}{5}$

73. $\frac{2}{5}$

74.

Letter	Probability
A	0.073
B	0.009
C	0.030
D	0.044
E	0.130
F	0.028
G	0.016
H	0.035
I	0.074
J	0.002
K	0.003
L	0.035
M	0.025
N	0.078
O	0.074
P	0.027
Q	0.003
R	0.077
S	0.063
T	0.093
U	0.027
V	0.013
W	0.016
X	0.005
Y	0.019
Z	0.001

75. 0.13

76. 0.378

77. 0.622

78. Consonant

79. 4

80. Less likely to guess a correct letter in the puzzle

81. T, N, R, S, D

82. 0.5729

83. 0.4271

84. 0.2434

85. 0.1255

86. 0.1179

87. 0.6908

88. 0.5526

89. 0.7566

90. 0.4474

91. 0.3092

92. 0.8821

93. 0.8745

94. (a) Minecraft for PS3, MLB 14 The Show for PS4, inFamous: Second Son for PS4 (b) Kirby: Triple Deluxe for 3DS, Pokemon X/Y for 3DS, Super Luigi U for WiiU, Super Mario Brothers U for WiiU

95.

Studio	Probability
Sony	$\frac{3}{20}$
MS	$\frac{1}{10}$
Nintendo	$\frac{1}{5}$
Electronic Arts	$\frac{3}{20}$
Activision	$\frac{1}{5}$
Focus	$\frac{1}{20}$
Take-Two	$\frac{3}{20}$

96. (a) Not made by Sony (b) Not made by Nintendo

97. (a) $\frac{17}{20}$ (b) $\frac{4}{5}$

98. 1

99. (a) inFamous: Second Son for PS4 (b) Minecraft for PS3, MLB 14 The Show for PS4, inFamous: Second Son for PS4, Super Mario Brothers U for WiiU, Grand Theft Auto V for PS3, Grand Theft Auto V for Xbox 360, Bound by Flame for PS4 (c) Kirby: Triple Deluxe for 3DS, Super Luigi U for WiiU (d) Kirby: Triple Deluxe for 3DS, Pokemon X/Y for 3DS, Super Luigi U for WiiU, Super Mario Brothers U for WiiU (e) Super Mario Brothers U for WiiU (f) Kirby: Triple Deluxe for 3DS, Pokemon X/Y for 3DS, Super Luigi U for WiiU, Super Mario Brothers U for WiiU, inFamous: Second Son for PS4, Grand Theft Auto V for PS3, Grand Theft Auto V for Xbox 360, Bound by Flame for PS4

100. (a) $\frac{1}{20}$ (b) $\frac{7}{20}$ (c) $\frac{1}{10}$ (d) $\frac{1}{5}$ (e) $\frac{1}{20}$ (f) $\frac{2}{5}$

101. (a) None (b) Minecraft for PS3, MLB 14 The Show for PS4, inFamous: Second Son for PS4, Kirby: Triple Deluxe for 3DS, Pokemon X/Y for 3DS, Super Luigi U for WiiU, Super Mario Brothers U for WiiU (c) None (d) inFamous: Second Son for PS4, Grand Theft Auto V for PS3, Grand Theft Auto V for Xbox 360, Bound by Flame for PS4, Kirby: Triple Deluxe for 3DS, Super Luigi U for WiiU, Super Mario Brothers U for WiiU

102. (a) 0 (b) $\frac{7}{20}$ (c) 0 (d) $\frac{7}{20}$

103.

Had Medical Assistance	Frequency	Relative Frequency
No	1138	0.8054
Yes	275	0.1946
Total	1413	1.0000

Completed Medical Treatment	Frequency	Relative Frequency
No	944	0.6681
Yes	469	0.3319
Total	1413	1.0000

(a) 0.1946

(b) 0.3319

104.

	Did not complete treatment	Completed treatment	Total
Did not have medical assistance	724	414	1138
Had medical assistance	220	55	275
Total	944	469	1413

105. (a) 0.0389 (b) 0.1557 (c) 0.2930 (d) 0.5124

106. No. Patients who had medical assistance and completed the treatment compared to patients who had medical assistance and did not complete the treatment:

$$\frac{0.0389}{0.1557} \approx 0.25$$

Therefore, the proportion of people with medical assistance who completed the treatment was approximately 0.25 $\left(\text{or } \frac{1}{4}\right)$ of the proportion of people with medical assistance who did not complete the treatment. Patients who did not have medical assistance and completed the treatment compared to patients who did not have medical assistance and did not complete the treatment:

$$\frac{0.2930}{0.5124} \approx 0.57$$

Therefore, the proportion of people who did not have medical assistance who completed the treatment is approximately 0.57 $\left(\text{or } \frac{5}{7}\right)$ of the proportion of people who did not have medical assistance who did not complete the treatment.

107. (a) 0.4876 (b) 1

Section 5.3

1. (a) Yes. (b) The probability of winning the football game depends on whether or not the star quarterback can play in the game.

2. Answers will vary.

3. For $P(A \mid B)$, we assume that the event B has occurred, and now need to find the probability of event A, given event B. On the other hand, for $P(A \cap B)$, we do not assume that event B has occurred, and instead need to determine the probability that both events occurred.

4. Answers will vary.

5. Answers will vary.

6. $A \cap B$ is a subset of A, $P(A \cap B) \leq P(A)$, and $A \cap B$ is a subset of B, $P(A \cap B) \leq P(B)$. Therefore, $P(A \cap B)$ cannot equal 0.30 if $P(A) = 0.25$ and $P(B) = 0.25$.

7. (a) Independent; sampling with replacement (b) Dependent; sampling without replacement

8. (a) Independent. The probability of selecting any particular card is not influenced by the outcome of the coin toss. The probability of heads or tails is not influenced by which card is selected. (b) Dependent, since we are sampling without replacement.

9. $\frac{1}{2}$

10. $\frac{1}{6}$

11. $\frac{1}{3}$

12. $\frac{1}{4}$

13. $\frac{1}{4}$

14. $\frac{1}{2}$

15. $\frac{3}{4}$

16. $\frac{1}{2}$

17. $\frac{1}{2}$

18. $\frac{1}{4}$

19. $\frac{1}{2}$

20. $\frac{1}{2}$

21. Endangered

22. Africa

23. $\frac{1}{3}$

24. $\frac{1}{3}$

25. $\frac{2}{3}$

26. $\frac{2}{3}$

27. $\frac{1}{3}$

28. $\frac{1}{3}$

29. 1

30. $\frac{2}{3}$

31. $\frac{2}{3}$

32. 0

33. Not independent

34. Not independent

35. Not independent

36. Not independent

37. Not independent

38. Not independent

39. Independent

40. Independent

41. Independent

42. Independent

43. Not independent

44. Not independent

45. Not independent

46. Not independent

47. Not independent

48. Not independent

49. It is not true.

50. The probabilities in Exercises 33–38.

51. It is not true.

52. It is true.

53. 0.33725

54. 0.184184

55. 0.27

56. 0.05

57. 0.125

58. 0.24

59. 0.1296

60. 0.1368

61. 0.1368

62. 0.1444

63. 0.36

64. 0.38

65. $P(\text{F1 and F2}) = (0.36)(0.36)$ whereas $P(\text{F2 given F1}) = 0.36$; sampling less than 1% of population so 1% Guideline applies.

66. $P(\text{F1 and M2}) = (0.36)(0.38)$ whereas $P(\text{M2 given F1}) = 0.38$; sampling less than 1% of population so 1% Guideline applies.

67. 0.25

68. 0.0625

69. 0.125

70. 0.125

71. 0.2451

72. 0.0588

73. 0.1176

74. 0.0129

75. (a) A sample of size $n = 2$ is $\frac{2}{1,500,000} \cdot 100\% \approx 0.00013\%$ of the population, which is less than 1%. Therefore the 1% Guideline applies.

(b) A sample of size $n = 3$ is $\frac{3}{1,500,000} \cdot 100\% = 0.0002\%$ of the population, which is less than 1%. Therefore the 1% Guideline applies. **(c)** A sample of size $n = 5$ is $\frac{5}{1,500,000} \cdot 100\% \approx 0.00033\%$ of the population, which is less than 1%. Therefore the 1% Guideline applies.

76. 0.2304
77. 0.1106
78. 0.0255
79. Not independent
80. Not independent
81. Not independent
82. Not independent
83. Not independent
84. Not independent
85. Not independent
86. Not independent
87. Not independent
88. Not independent
89. Not independent
90. No; they are mutually exclusive
91. No; they are mutually exclusive
92. No; they are mutually exclusive
93. $\frac{1}{8}$
94. $\frac{1}{16}$
95. $\frac{1}{32}$
96. $\frac{1}{1024}$
97. $\frac{7}{8}$
98. $\frac{15}{16}$
99. $\frac{31}{32}$
100. $\frac{1023}{1024}$
101. 0.75
102. 0.875
103. 0.9375
104. 0.96875
105. 0.375
106. 0.3488
107. 0.2941
108. 0.9574
109. **(a)** 0.16 **(b)** 0.064 **(c)** 0.01024 **(d)** 0.8704
110. **(a)** 0.36 **(b)** 0.216 **(c)** 0.936
111. **(a)** 0.6 **(b)** 5/9 **(c)** 1/2
112. **(a)** 0.4770 **(b)** 0.5230 **(c)** 0.5646 **(d)** 0.4354
113. **(a)** 0.4900 **(b)** 0.4600 **(c)** 0.5100 **(d)** 0.5400
114. **(a)** Not independent **(b)** Not independent **(c)** Not independent **(d)** Not independent
115. **(a)** 0.3307 **(b)** 0.5444
116. **(a)** 0.1984 **(b)** 0.6767 **(c)** 0.6693 **(d)** 0.4556
117. **(a)** 0.3644 **(b)** 0.6356 **(c)** 0.2903 **(d)** 0.7097
118. Not independent
119. **(a)** 0.5361 **(b)** 0.4639 **(c)** 0.0330
120. **(a)** 0.0215 **(b)** 0.0114
121. Not independent. $P(F) \cdot P(Not) = (0.5361) \cdot (0.0330) = 0.0176913 \neq 0.0215 = P(F \text{ and } Not)$ and $P(M) \cdot P(Not) = (0.4639) \cdot (0.0330) = 0.0153087 \neq 0.0114 = P(M \text{ and } Not)$
122. **(a)** Without replacement; the only way to make sure that we sample two different computers is to sample without replacement. **(b)** 1/10 **(c)** 1/11 **(d)** 109/110 **(e)** 1/110
123. Either reject the batch if at least one computer is defective or increase the sample size.
124. **(a)** 0.2583 **(b)** 0.3772 **(c)** 0.3319
125. **(a)** 0.0750 **(b)** 0.1423
126. **(a)** 0.2904 **(b)** 0.3771
127. **(a)** Not independent **(b)** Not independent

128.

	Did not complete treatment	Completed treatment	Total
11–17 years old	454	247	701
18–26 years old	490	222	712
Total	944	469	1413

129. **(a)** 0.3319 **(b)** 0.4961 **(c)** 0.5039 **(d)** 0.1748 **(e)** 0.1571 **(f)** 0.3524 **(g)** 0.3118
130. **(a)** Not independent **(b)** Not independent
131. 11–17 years old; true variation

Section 5.4

1. Tree diagram
2. Multiply all the numbers from 5 down to 1 together. $5! = 5 \cdot 4 \cdot 3 \cdot 2 \cdot 1 = 120$
3. In a permutation, order is important. In a combination, order is not important.
4. No; this symbol represents the number of ways to select and order 9 objects from 8 different objects, which is not possible.
5. Answers will vary.
6. **(a)** Classical **(b)** All outcomes in the sample space are equally likely.
7.

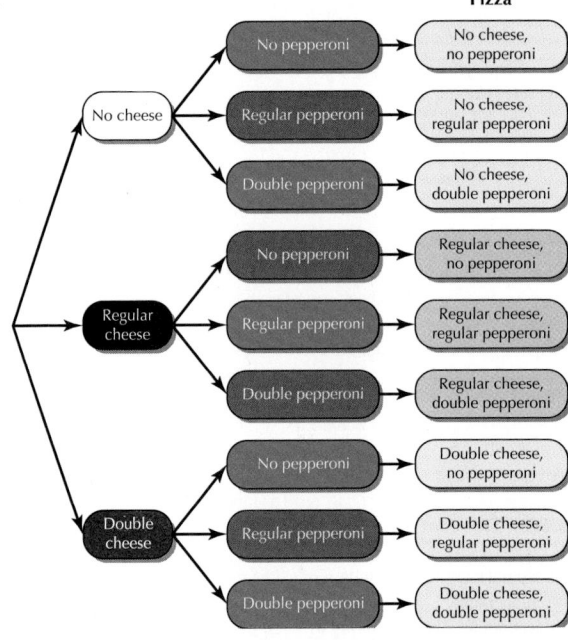

8.

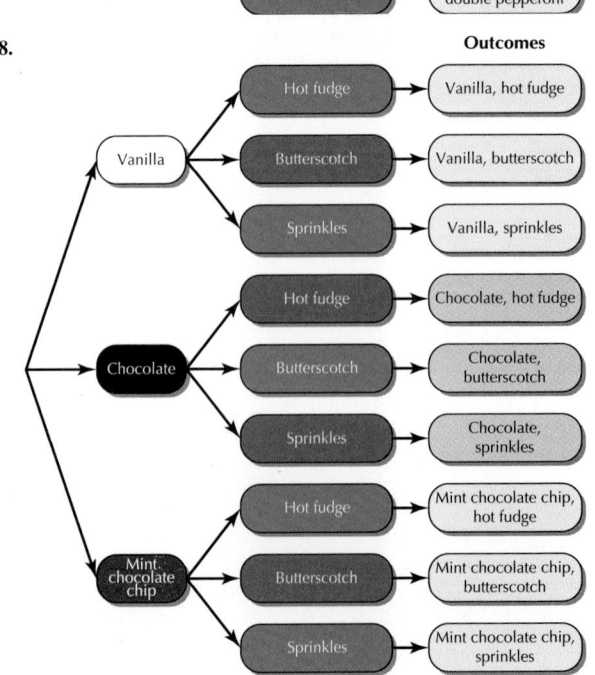

9.

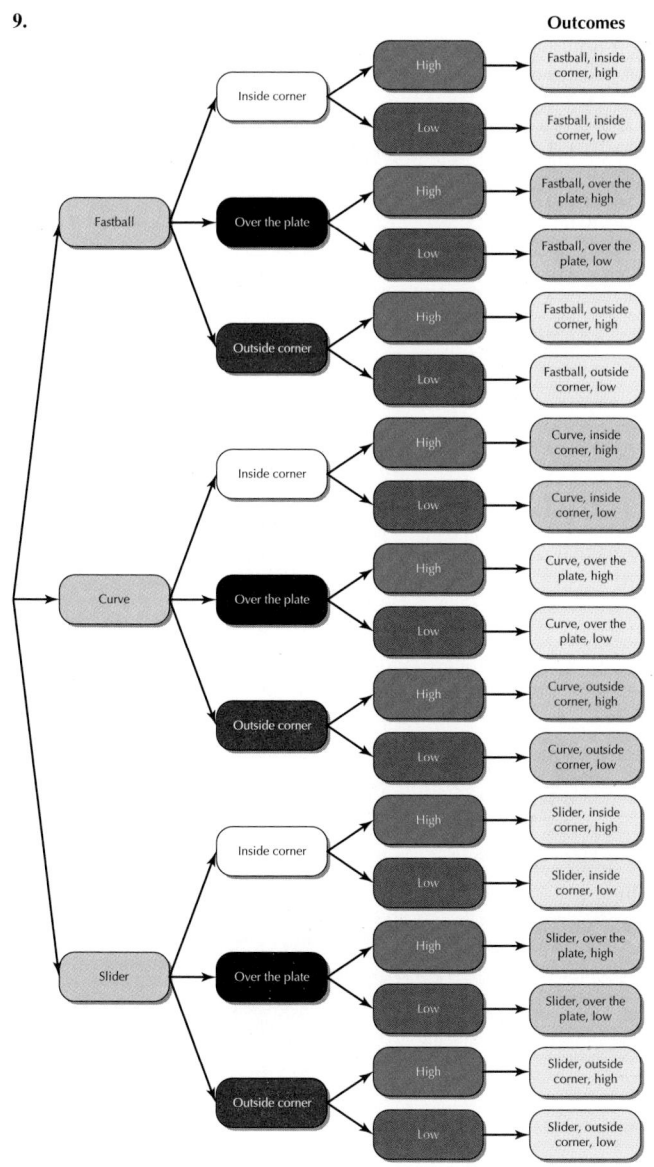

Outcomes

10. See the Solutions Manual.
11. 26^4
12. $3^5 = 243$
13. 20
14. 15
15. 24
16. 720
17. 720
18. 362,880
19. 1
20. 39,916,800
21. 1
22. 1,307,674,368,000
23. 12
24. 360
25. 210
26. 840
27. 6720
28. 336
29. 100
30. 1
31. 93,326,215,443,944,152,681,699,238,856,266,700,490,715,968,264,381, 621,468,592,963,895,217,599,993,229,915,608,941,463,976,156,518,286,253, 697,920,827,223,758,251,185,210,916,864,000,000,000,000,000,000,000,000

32. 93,326,215,443,944,152,681,699,238,856,266,700,490,715,968,264, 381,621,468,592,963,895,217,599,993,229,915,608,941,463,976,156,518, 286,253,697,920,827,223,758,251,185,210,916,864,000,000,000,000,000, 000,000,000
33. 35
34. 35
35. 165
36. 55
37. 11
38. 1
39. 1
40. 100
41. $5!/(2!1!1!1!) = 60$
42. $9!/(3!2!1!1!1!1!) = 30,240$
43. $_7C_3 = 7!/(3! \cdot 4!) = 7!/(4! \cdot 3!) = {}_7C_4$
44. $_{11}C_8 = 11!/(8! \cdot 3!) = 11!/(3! \cdot 8!) = {}_{11}C_3$
45. {Amy, Bob, Chris}, {Amy, Chris, Bob}, {Bob, Amy, Chris}, {Bob, Chris, Amy}, {Chris, Amy, Bob}, {Chris, Bob, Amy}, {Amy, Bob, Danielle}, {Amy, Danielle, Bob}, {Bob, Amy, Danielle}, {Bob, Danielle, Amy}, {Danielle, Amy, Bob}, {Danielle, Bob, Amy}, {Amy, Chris, Danielle}, {Amy, Danielle, Chris}, {Chris, Amy, Danielle}, {Chris, Danielle, Amy}, {Danielle, Amy, Chris}, {Danielle, Chris, Amy}, {Bob, Chris, Danielle}, {Bob, Danielle, Chris}, {Chris, Bob, Danielle}, {Chris, Danielle, Bob}, {Danielle, Bob, Chris}, {Danielle, Chris, Bob}. $_4P_3 = 24$
46. {Amy, Bob, Chris}, {Amy, Bob, Danielle}, {Amy, Chris, Danielle}, {Bob, Chris, Danielle}, $_4C_3 = 4$
47. {Amy, Bob, Chris}, {Amy, Chris, Bob}, {Chris, Amy, Bob}, {Chris, Bob, Amy}, {Bob, Amy, Chris}, and {Bob, Chris, Amy} are all different permutations but the same combination.
48. 3!
49. $r!$
50. **(a)**

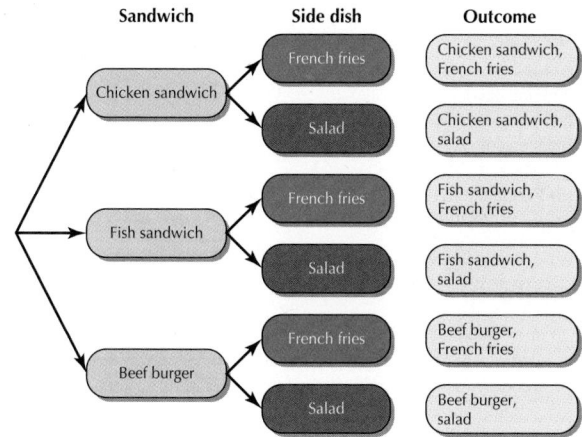

Sandwich   Side dish   Outcome

(b) 6
51. **(a)** See Solutions Manual. **(b)** 18
52. 576
53. 3,628,800
54. 120
55. 720
56. 360
57. 20
58. 30
59. 300
60. 455
61. 20
62. 1
63. 184,756
64. 4,989,600
65. 6720
66. **(a)** Combination. We are interested only in which two PDAs are selected, not in the order in which they are selected. **(b)** 3 **(c)** 0.01

Chapter 5 Review

1. 3/8
2. 1/2
3. 0
4. 3/8
5. 1/2
6. (a) 0.75 (b) The relative frequency method was used because there is prior knowledge of the relative frequencies of the sonnets.
7. (a) 0.213 (b) 0.656 (c) 0 (d) 0.376
8. (a) 0.9269 (b) 0 (c) 1
9. 0
10. (a) 1/2 (b) 1/3
11. (a) 1/6 (b) 1/6
12. (a) 5/18 (b) 5/12
13. Men's TV channel, since $P(\text{Dog} \mid \text{Male}) = 5/12 > 5/18 = P(\text{Dog} \mid \text{Female})$
14. 34,650
15. 60
16. (a) Combination; order is not important. (b) 3 (c) 0.0196.

Chapter 5 Quiz

1. False
2. True
3. 0, 1
4. or, and
5. 0.5
6. 1
7. With replacement
8. Intersection of A and B.
9. (a) 1/9 (b) 8/9 (c) 5/18 (d) 1/18 (e) 1/3
10. 0.2
11. 0.2125
12. (a) 1/4 (b) 3/13 (c) 1/13 (d) 1/2 (e) 1/52 (f) 1/26
13. Not independent
14. Not independent
15. Not independent
16. Not independent
17. Not independent
18. Independent
19. 4
20. (a) Permutation; the order in which the numbers are selected is important. (b) 6840 (c) 1/6840

Chapter 6

Note to instructors and students: Some answers may vary slightly depending on whether you round at intermediate steps or wait until you get the final answer to round. Also, different software and different forms of technology may give slightly different answers.

Section 6.1

1. Answers will vary.
2. Your individual height is not a random variable. But if your height were "drawn from a hat" consisting of the heights of randomly selected people, your height would be considered a random variable.
3. Discrete: takes finite or a countable number of values that can be graphed as separate points on the number line; continuous takes infinitely many values that form an interval on the number line.
4. A discrete random variable can take either a finite or a countable number of values. Each value can be graphed as a separate point on the number line with space between each point. A probability distribution of a discrete random variable provides all the possible values that the random variable can assume, together with the probability associated with each value.
5. $\sum P(X) = 1$ and $0 \leq P(X) \leq 1$.
6. The mean μ of a discrete random variable represents the mean result when the experiment is repeated an indefinitely large number of times. It is the mean of the results from the population of all possible repetitions of the experiment. The sample mean x is simply the arithmetic mean of a finite set of sample values.
7. Discrete

8. Continuous
9. Continuous
10. Continuous
11. Discrete
12. Discrete
13. {0, 1, 2, 3, 4, 5, 6, 7, 8, 9, 10, 11, 12, 13, 14, 15}
14. {0, 1, 2, 3, 4, 5, 6, 7, 8, 9, 10}
15. {0, 1, 2, 3, 4}
16. {0, 1, 2, 3}

17. (a)

X = Number of CDs	0	1	2	3	4
$P(X)$	0.06	0.24	0.38	0.22	0.10

(b)

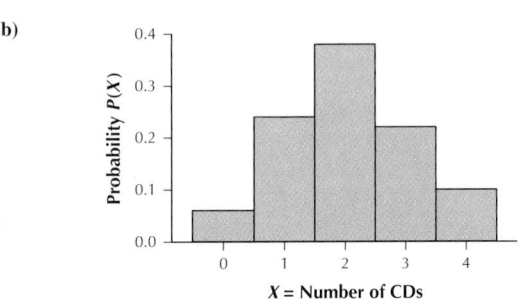

18. (a)

X = Number of goals	0	1	2	3
$P(X)$	0.25	0.35	0.25	0.15

(b)

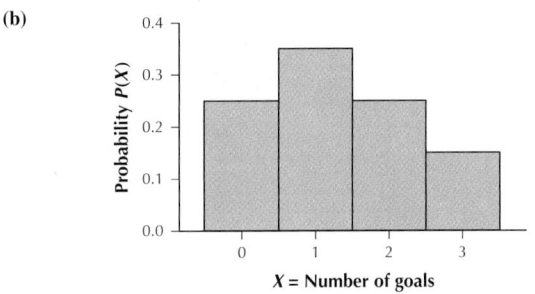

19. (a)

X = Money gained	−$10,000	$10,000	$50,000
$P(X)$	1/3	1/2	1/6

(b)

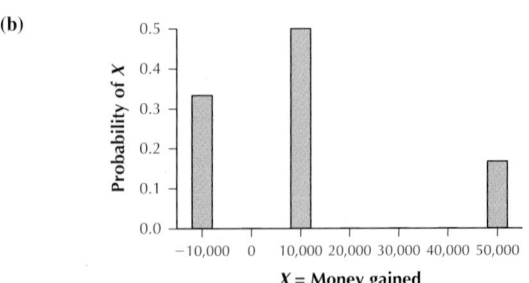

20. (a)

X = Number of pets	0	1	2	3
$P(X)$	0.5	0.3	0.1	0.1

(b)

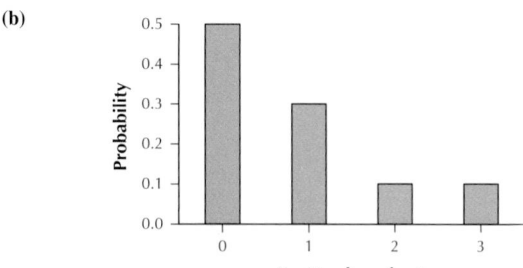

21. (a)

X = Number of games in the series	P(X)
4	0.2083
5	0.2083
6	0.2917
7	0.2917
Total	1.0000

(b)

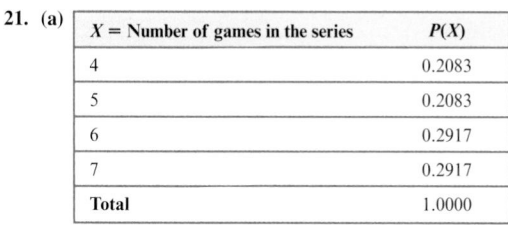

22. (a)

X = Number of SAT subject tests	P(X)
1	0.1306
2	0.4640
3	0.4054
Total	1.0000

(b)

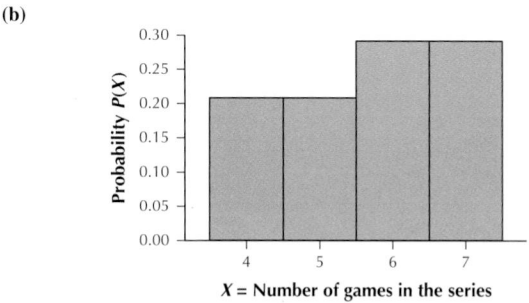

23. (a)

X = Number of years of high school math	P(X)
4	0.7465
3	0.2103
2	0.0302
1	0.0131
Total	1.0001

(b)

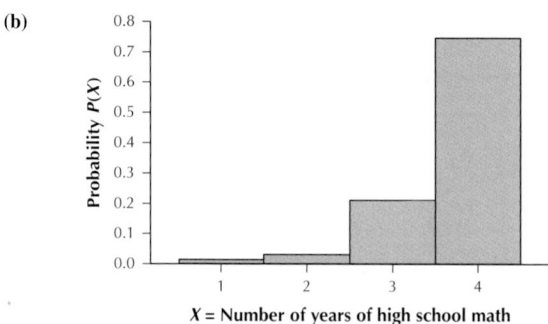

24. (a)

X = Number of major hurricanes	P(X)
0	0.08
1	0.16
2	0.28
3	0.12
4	0.08
5	0.16
6	0.08
7	0.04
Total	1.00

(b)

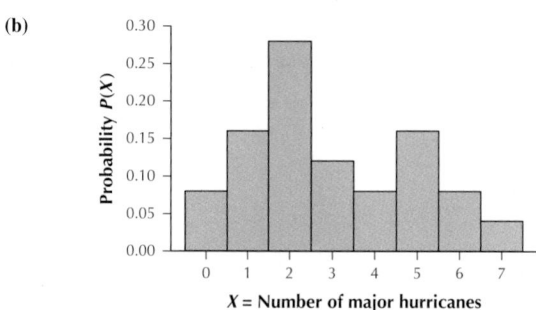

25. Not a valid probability distribution. The probabilities don't add up to 1.
26. Valid probability distribution.
27. Not a valid probability distribution. $P(X = 1)$ is negative.
28. Not a valid probability distribution. $P(X = 100,000)$ is larger than 1.
29. 0.5833
30. 0.7083
31. 0.7917
32. 0.2917
33. 0.8694
34. 0.5946
35. 0.5946
36. 0
37. 0.9567
38. 0
39. 0.9567
40. 0.7465
41. 0.24
42. 0.44
43. 0
44. 0.92
45. (a) 2.06 CDs (b) 2 CDs (c) 2.06 CDs
46. (a) 1.3 goals (b) 1 goal (c) 1.3 goals
47. (a) $10,000 (b) $10,000 (c) $10,000
48. (a) 0.8 pet (b) 0 pets (c) 0.8 pet
49. (a) 5.6667 games (b) 6 and 7 games (c) 5.6667 games
50. (a) 2.2748 SAT subject tests (b) 2 SAT subject tests (c) 2.2748 SAT subject tests
51. (a) 3.6901 years (b) 4 years (c) 3.6901 years
52. (a) 2.96 hurricanes (b) 2 hurricanes (c) 2.96 hurricanes
53. (a) 1.0964 CDs squared (b) 1.0471 CDs (c) No outliers or unusual data values.

54. (a) 1.01 goals squared **(b)** 1.0050 goals **(c)** No outliers or unusual data values

55. (a) 400,000,000 dollars squared **(b)** $20,000 **(c)** $50,000 is moderately unusual.

56. (a) 0.96 pet squared **(b)** 0.9798 pet **(c)** No outliers; 3 pets is moderately unusual

57. (a) 1.22217776 games squared **(b)** 1.1055 games **(c)** No outliers or unusual data values.

58. (a) 0.4605 SAT subject test squared **(b)** 0.6786 SAT subject test **(c)** No outliers or unusual data values.

59. (a) 0.35154784 year squared **(b)** 0.5929 year **(c)** 1 year of high school math is considered an outlier; 2 years of high school math is considered moderately unusual

60. (a) 3.7184 hurricanes squared **(b)** 1.9283 hurricanes **(c)** No outliers; 7 major hurricanes is considered moderately unusual.

61. (a) A faculty member from all degree-granting institutions in the United States is randomly selected. **(b)** The number of classes taught by a faculty member is counted.

(c)

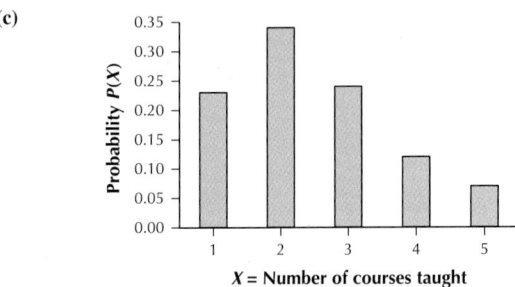

(d) 0.43 **(e)** 0.19 **(f)** 2 classes

62. (a) A Florida resident is randomly selected. **(b)** The number of vehicles a Florida resident owns is counted.

(c)

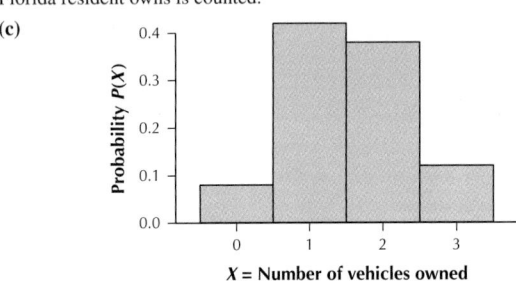

(d) 0.88 **(e)** Most likely number of vehicles is 1; least likely number of vehicles is 0.

63. (a)

X = Number of vehicles involved	P(X)
1	0.0808
2	0.8839
3	0.0306
4	0.0034
5	0.0011
8	0.0003
Total	1.0001

(b) 0.0354 **(c)** 0 **(d)** 0.9953 **(e)** 0.8839

64. (a)

X = Number of people injured	P(X)
0	0.8494
1	0.1296
2	0.0160
3	0.0028
4	0.0008
5	0.0006
6	0.0003
7	0.0006
Total	1.0001

(b) 0.8494 **(c)** 0.1506 **(d)** 0.9950 **(e)** 0.1296

65. (a) The expected number of classes taught is 2.46. **(b)** $\sigma^2 = 1.3684$ classes squared; $\sigma = 1.1698$ classes. **(c)** It is unusual to teach 5 classes.

66. (a) The expected number of vehicles a Florida resident owns is 1.54 **(b)** $\sigma^2 = 0.6484$ vehicle squared; $\sigma = 0.8052$ vehicle. **(c)** It is not unusual for a Florida resident to own 0 vehicles.

67. (a) The expected number of vehicles involved in a crash is 1.9619 vehicles. **(b)** $\sigma^2 = 0.14384839$ vehicle squared; $\sigma = 0.3793$ vehicle. **(c)** 4, 5, and 8 vehicles are outliers. 1 vehicle and 3 vehicles are moderately unusual.

68. (a) The expected number of people injured in a crash is 0.1818. **(b)** $\sigma^2 = 0.241563716$ people squared; $\sigma = 0.4915$ people. **(c)** 2, 3, 4, 5, 6, and 7 people injured are all outliers. None of the values are moderately unusual.

69.

X = Sum of dice	2	3	4	5	6	7	8	9	10	11	12
P(X)	1/36	1/18	1/12	1/9	5/36	1/6	5/36	1/9	1/12	1/18	1/36

70. The mean is about 7.

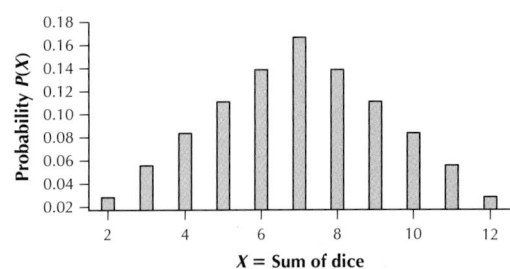

71. $\mu = 7$. The estimate is equal to the actual value. If we were to consider tossing two dice an infinite number of times, the mean sum of the dice would be 7.

72. $\sigma = 2.4152$

73. Z-score is -2.07; moderately unusual. By symmetry, so is $X = 12$.

74. No. The mean is 2 but the most likely value is 0.

X	0	2	8
P(X)	0.6	0.2	0.2

75. Symmetric, one mode

76. (a) Increases by k. **(b)** Remains unchanged.

77. (a) An NFL team is selected at random. **(b)** The number of games a team has won is counted.

(c)

X = Number of games won	P(X)
2	0.0313
3	0.0313
4	0.1563
5	0.0313
6	0.0313
7	0.1250
8	0.2188
9	0.0313
10	0.0625
11	0.1250
12	0.0938
13	0.0625
Total	1.0004

(d)
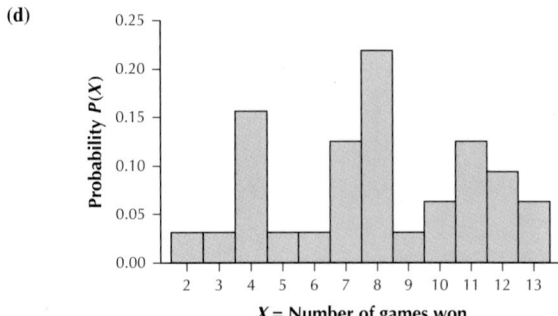

(e) 0.2502 **(f)** 8 games won
78. A student who took the AP Statistics Exam is selected at random.
79. There are finitely many possible exam scores.

80.

X = AP Statistics exam score	P(X)
1	0.2339
2	0.1906
3	0.2445
4	0.1972
5	0.1337
Total	0.9999

81. All probabilities are between 0 and 1 and the probabilities add up to 1.

82.
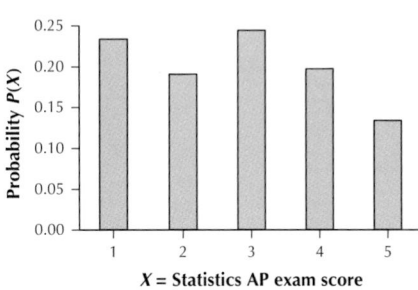

83. **(a)** 0.7661 **(b)** 0.7661 **(c)** 0.4245 **(d)** 0.4245 **(e)** Same; Same; X is a
discrete random variable.
84. 2.8063
85. 3
86. 2.8063
87. $\sigma^2 = 1.82165775$; $\sigma = 1.3497$.
88. No outliers or unusual values.

89.

X = Number of cylinders	Relative frequency
4	0.4075
6	0.3637
8	0.2287
Total	0.9999

90.

X = Number of cylinders	Probability ≈ relative frequency
4	0.4075
6	0.3637
8	0.2287
Total	0.9999

91.
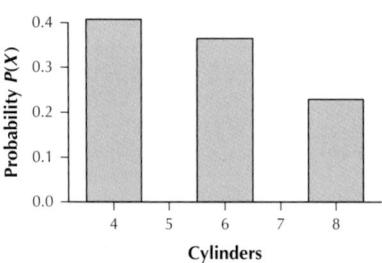

92. 5.6418 cylinders
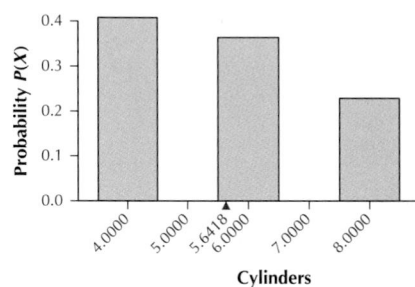

93. 4 cylinders

94.

X = Number of gears	P(X)
1	0.0447
4	0.0289
5	0.0771
6	0.5188
7	0.1604
8	0.1621
9	0.0079
Total	0.9999

95.
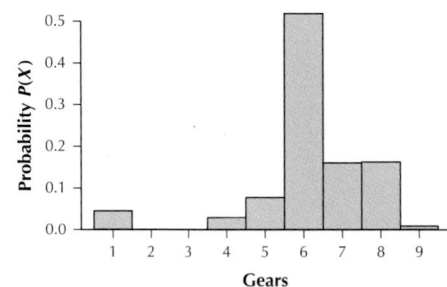

96. 6.1493 gears
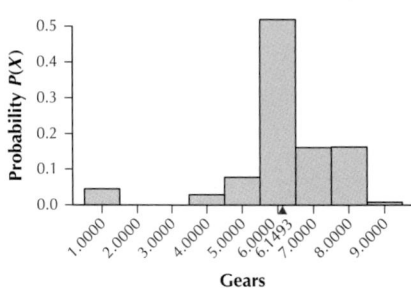

97. $\sigma^2 = 2.17140951$ gears squared; $\sigma = 1.4736$ gears.
98. 1 gear is considered an outlier

Section 6.2

1. **(i)** Each trial of the experiment has only two possible mutually exclusive outcomes (or is defined in such a way that the number of outcomes is reduced to two). One outcome is denoted a success and the other a failure. **(ii)** There is a fixed number of trials, known in advance of the experiment. **(iii)** The experimental outcomes are independent of each other. **(iv)** The probability of observing a success remains the same from trial to trial.

2. A *success* denotes simply the outcomes we are interested in, without necessarily implying that the outcome is desirable.

3. If you perform an experiment n times, you can't have more than n successes. For example, if you flip a coin 10 times you can't get 11 heads.

4. $P(X) = (_nC_X) (P(\text{success}))^{\text{number of successes}} (P(\text{failures}))^{\text{number of failures}}$

5. Not binomial; the events "Person A comes to party" and "Person B comes to party" may not be independent.

6. Not binomial; more than two possible total number of spots.

7. Binomial, X = number of correct answers, $n = 8$, $p = 1/4 = 0.25$, $1 - p = 3/4 = 0.75$

8. Binomial; $n = 3$, X = number of sixes, $p = 1/6$, $1 - p = 5/6$

9. Not binomial; not a fixed number of trials

10. Binomial, X = number of queens, $n = 4$, $p = 4/52 = 1/13$, $1 - p = 48/52 = 12/13$

11. Not binomial, trials are not independent, sample is more than 1% of the population.

12. Not binomial, more than 2 possible outcomes.

13. Binomial; $n = 2$, X = number of games won, $p = 0.25$, $1 - p = 0.75$

14. Not binomial; not a fixed number of trials.

15. 0.3955
16. 0.2373
17. 0.1172
18. 0.0439
19. 0.2301
20. 0.3766
21. 0.6328
22. 0.7627
23. 0.1611
24. 0
25. 0.8891
26. 0.1109
27. 0.9744
28. 0.9957
29. 0.421875
30. 0.421875
31. 0.140625
32. 0.984375
33. 0.578125
34. 0.984375
35. 0.0015609375
36. 0.0015925781
37. 0.7320941406
38. 0.9695300781
39. 0.2679058594
40. 0.0304382812
41. 0.59049
42. 0.40951
43. 0.99144
44. 0
45. 0.004096
46. 1
47. 0.451584
48. 0.13824

49. **(a)** 0.75 business major; If we take an infinite number of random samples of size 3 from all of the Masters degrees granted in 2012 and calculated the mean number of business majors in the samples it would be 0.75 business major. **(b)** 0.5625 business major squared **(c)** 0.75 business major

50. **(a)** 0.3 child. If we take an infinite number of random samples of size 4 from all American children aged 6–17 and calculated the mean number of children taking medication for emotional or behavioral problems the mean would be 0.3 child. **(b)** 0.2775 child squared **(c)** 0.5268 child

51. **(a)** 0.5 American 25–29 years old. If we take an infinite number of random samples of size 5 of Americans aged 25–29 and calculated the mean

number of them living alone it would be 0.5 American. **(b)** 0.45 American squared **(c)** 0.6708 American

52. **(a)** 2.4 18- to 29-year-olds. If we take an infinite number of samples of size 6 of 18–29 year olds and calculated the mean number who had Twitter accounts it would be 2.4 18- to 29-year-olds. **(b)** 1.44 18- to 29-year-olds squared **(c)** 1.2 18- to 29-year-olds

53. **(a)**

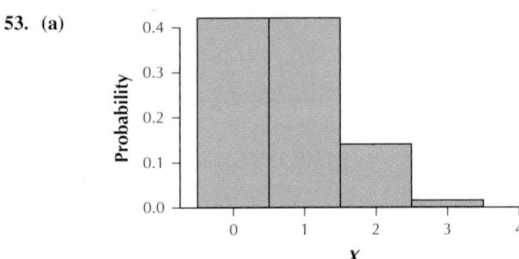

(b) 0 and 1 business major

54. **(a)**

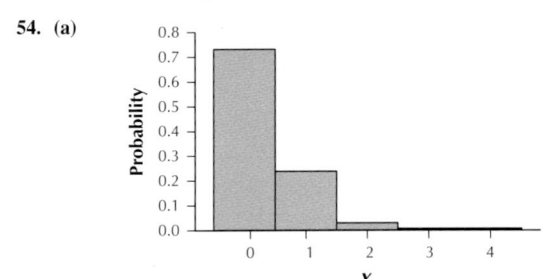

(b) 0 6- to 17-year-olds

55. **(a)**

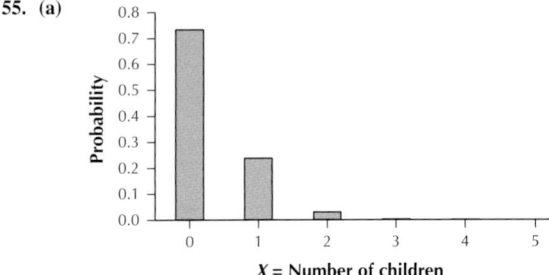

(b) 0 25- to 29-year-olds

56. **(a)**

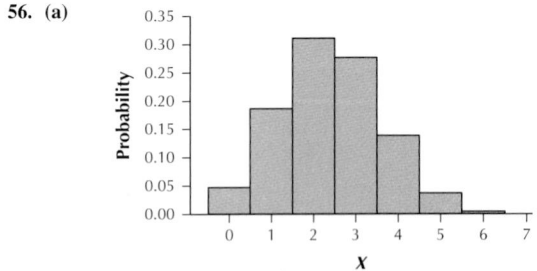

(b) 2 18- to 29-year-olds

57. **(a)** It fulfills the requirements: (i) There are only two possible outcomes for each trial: correct answer or incorrect answer. (ii) We know in advance that the quiz will have 5 questions. (iii) Since you are randomly guessing the answer to each question, the trials are independent. (iv) Since each question has 4 responses, the probability of guessing correctly remains the same from question to question. **(b)** $n = 5$, $p = 1/4 = 0.25$ **(c)** 0.1035 **(d)** 0.8965

58. **(a)** 0.1171 **(b)** 0.0005 **(c)** 0.4565 (TI-83/84: 0.4566)

59. **(a)** 0.2054 **(b)** 0.4101 **(c)** 0.6189

60. **(a)** $p = 0.22$ is not in the table. **(b)** 0.2457 **(c)** 0.5726

61. **(a)** $\mu = 1.25$ correct answers. If we repeat this experiment an infinite number of times, record the number of correct answers for each quiz taken, and take the mean of all of the quizzes, the mean number of correct answers will equal $\mu = 1.25$. $\sigma^2 = 0.9375$ correct answer squared, $\sigma = 0.9682$ correct answer. **(b)** Five correct answers is considered an outlier; 4 correct answers is considered moderately unusual. **(c)** Mode is 1 correct answer.

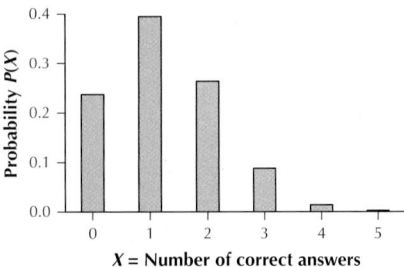

(d) 0.3955

62. (a) $\mu = 8$ women. If we repeat this experiment an infinite number of times, record the number of women in each sample, and take the mean of all the samples, the mean number of women will equal $\mu = 8$. $\sigma^2 = 4.8$ women squared, $\sigma = 2.1909$ women **(b)** $Z = -0.9129$, not unusual **(c)** 8 women in management positions **(d)** 0.1797

63. (a) $\mu = 4.92$ telephone users who have abandoned their landline; $\sigma^2 = 2.9028$ telephone users who have abandoned their landline squared; $\sigma = 1.7038$ telephone users who have abandoned their landline. If we take infinitely many random samples of telephone users of size 12 and calculate the mean number of telephone users who have abandoned their landline it would be 4.92. **(b)** Moderately unusual

64. (a) $\mu = 3.3$ 25- to 34-year-olds who have used online dating; $\sigma^2 = 2.574$ 25- to 34-year-olds who have used online dating squared; $\sigma = 1.6044$ 25- to 34-year-olds who have used online dating; If we take infinitely many random samples of size 15 of 25- to 34-year-olds and calculate the mean number of 25- to 34-year-olds who have used online dating it would be 3.3. **(b)** Moderately unusual

65. (a) $\mu = 0.6$ woman. If we repeat this experiment an infinite number of times, record the number of women affected by a depressive disorder for each sample, and take the mean of all the samples, the mean number of women living with a depressive disorder will equal $\mu = 0.6$. **(b)** Not possible. The expected value of X is not an integer. **(c)** $X = 0$ women

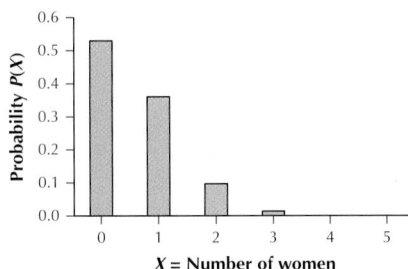

(d) 0.5277

66. (a) $\mu = 0.33$ man. If we repeat this experiment an infinite number of times, record the number of men affected by a depressive disorder for each sample, and take the mean of all of the samples, the mean number of men living with a depressive disorder will equal $\mu = 0.33$. **(b)** Not possible. The expected value of Y is not an integer. **(c)** $Y = 0$ men

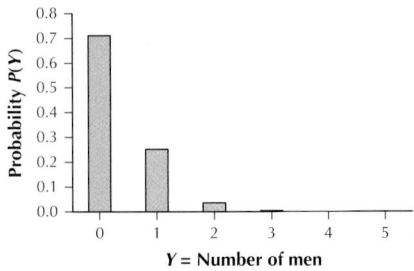

(d) 0.7108

67. (a) 90 students **(b)** 90 students **(c)** 90 students
68. (a) 0.7 **(b)** 0.21 **(c)** 0.063
69. (a) 1287/2,598,960 ≈ 0.0005 **(b)** 27,885/2,598,960 ≈ 0.0107 **(c)** 29,172/2,598,960 ≈ 0.0112 **(d)** 211,926/2,598,960 ≈ 0.0815 **(e)** 2,357,862/2,598,960 ≈ 0.9072
70. (a) 0.0784 **(b)** 0.0588 **(c)** 0.0588
71. (i) Either a new job is provided by a small company or it is not provided by a small company. These are the only two possible outcomes and they are mutually exclusive.
   (ii) It is known in advance that exactly 10 new jobs will be selected.
   (iii) The sample is random, so the outcomes are independent.
   (iv) The sample is quite small compared to the size of the population, so that the probability that a new job was provided by a small business remains the same from job to job.
72. (a) 0.2503 **(b)** 0.2816
73. (a) 0.2440 **(b)** 0.2440 **(c)** 0.0197 **(d)** 0.0197; X is a discrete variable.
74. (a) 7.5 jobs **(b)** 1.875 jobs squared **(c)** 1.3693 jobs
If we take infinitely many samples of size 10 small businesses and calculated the mean and the standard deviation they would be equal to 7.5 jobs and 1.3693 jobs, respectively.
75. Not possible because $\mu = 7.5$ jobs is not a whole number.

76.

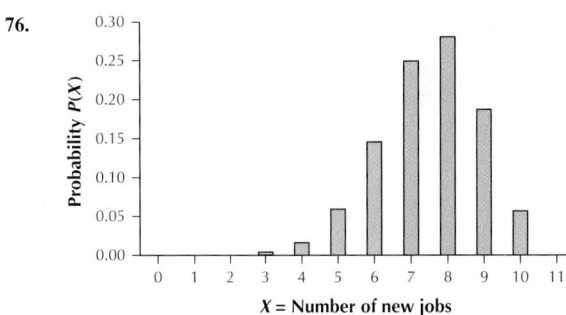

77. 8 jobs
78. 0.2816
79. 0, 1, 2, and 3 jobs are outliers. 4 jobs is moderately unusual.
80. 0.1639
81. 16.39 cars
82. 3.7018 cars
83. Answers will vary
84. Answers will vary.

Section 6.3

1. When observing the number of occurrences of an event within a fixed interval of space and time
2. The number of occurrences of the event in the interval
3. The occurrences must be random, independent, and uniformly distributed over the given interval.
4. $n \geq 100$ and $np \leq 10$
5. Does not follow a Poisson distribution
6. Does not follow a Poisson distribution
7. Follows a Poisson distribution
8. Does not follow a Poisson distribution
9. 0.1126
10. 0.1251
11. 0.6472
12. 0.3528
13. 0.2381
14. 0.7619
15. (a) 4.4721 **(b)** 11.0558, 28.9442, any values less than or equal to 11 or greater than or equal to 29 will be considered moderately unusual
16. (a) 5 **(b)** 15, 35, any values less than or equal to 15 or greater than or equal to 35 will be considered moderately unusual
17. (a) 3.1623 **(b)** 3.6754, 16.3246, any values less than or equal to 3 or greater than or equal to 17 will be considered moderately unusual
18. (a) 1.4142 **(b)** −0.8284, 4.8284, any values that are greater than or equal to 5 will be considered moderately unusual
19. (a) $n = 200 \geq 100$ and $np = 10 \leq 10$ **(b)** 0.0948
20. (a) $n = 120 \geq 100$ and $np = 3.6 \leq 10$ **(b)** 0.1912

21. (a) $n = 200 \geq 100$ and $np = 8 \leq 10$ (b) 0.0993
22. (a) $n = 120 \geq 100$ and $np = 6 \leq 10$ (b) 0.1606
23. (a) 0.0399 (b) 0.0360 (c) 0.0395
24. (a) 0.0967 (b) 0.1224 (c) 0.0116 (d) 0.0033 (e) 0.9851
25. (a) n = 247,000,000 ≥ 100 and $np = 5.2858 \leq 10$ (b) 5.2858
(c) 0.0051 (d) 0.9949 (e) 0.1026
26. (a) n = 142 ≥ 100 and $np = 5.68 \leq 10$ (b) 5.68 (c) 0.0551 (d) 0.9966
27. (a) 0.1749 (b) 0.1684 (c) 0.0056 (d) 0
28. (a) 0.0827 (b) 0.0803 (c) 0 (d) 0
29. (a) 0.2046 (b) 0.1615 (c) 0.0224 (d) 0.5265
30. (a) $n = 200$ so $n \geq 100$, $np = (200)(0.01) = 2 \leq 10$ (b) $\mu = 2$
(c) 0.2707 (d) 0.8647
31. (a) $n = 100 \geq 100$ and $np = 10 \leq 10$ (b) 10 (c) 0.1251 (d) 0.3336
32. (a) (i) 0, 10, 11, 12; (ii) 13, 14, 15, . . . (b) (i) 9, 10, 11, 12, 13,
33, 34, 35, 36, 37; (ii) 0, 1, 2, 3, 4, 5, 6, 7, 8, 38, 39, 40, . . . (c) (i) 8, 9; (ii) 10,
11, 12, . . . (d) (i) 5, 6 (ii) 7, 8, 9, . . .

Section 6.4

1. The probability that X equals some particular value is zero.
2. X values
3. Area under the normal distribution curve above an interval.
4. Any real number; any positive real number
5. False
6. No difference. They are the same number.
7. 0
8. 1
9. True
10. True
11. 0.5
12. 0.5
13. 0.65
14. 0.2
15. 0.01
16. 0
17. 0.5
18. 1
19. 0.1
20. 0.6
21. A has mean 10; B has mean 25. The peak of a normal curve is at the
mean; from the graphs we see that the mean of A is less than the mean of B.
22. A: standard deviation 3; B: standard deviation 6. B is more spread out
than A, so B has a larger standard deviation than A.
23. $\mu = 0, \sigma = 1$
24. $\mu = 100, \sigma = 5$
25. $\mu = -10, \sigma = 10$
26. $\mu = 1000, \sigma = 100$
27. $\mu = 100, \sigma = 12.5$
28. $\mu = 10, \sigma = 2$
29. $\mu = 0, \sigma = 2$
30. $\mu = 10, \sigma = 2$
31. 0
32. 0.5
33. 0.5
34. They are both equal to 0.5.
35. Less than 0.5. Since $X = 4285$ is greater than the mean of 3285 and the
area to the right of $\mu = 3285$ is 0.5, the area to the right of $X = 4285$ is less
than the area to the right of $X = 3285$.
36. Greater than 0.5. Since $X = 4285$ is greater than the mean of 3285 and the
area to the left of $\mu = 3285$ is 0.5, the area to the left of $X = 4285$ is greater
than the area to the left of $X = 3285$.

37. (a)
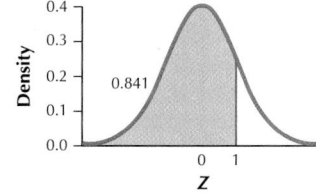

(b) 0.8413

38. (a)
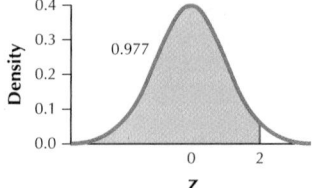

(b) 0.9772
39. (a)
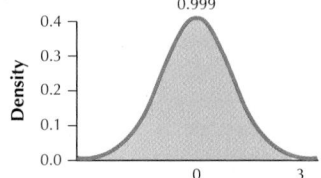

(b) 0.9987
40. (a)
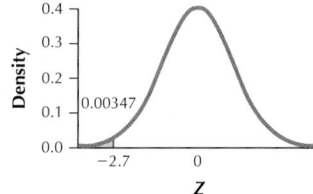

(b) 0.6915
41. (a)
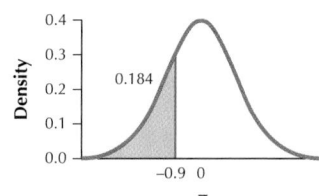

(b) 0.0035
42. (a)
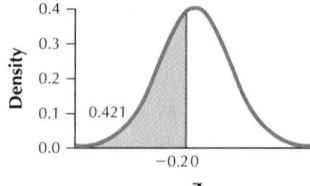

(b) 0.1841
43. (a)
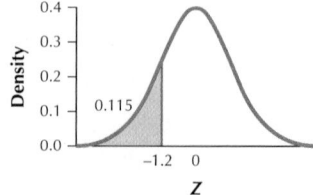

(b) 0.4207
44. (a)

(b) 0.1151

45. (a)

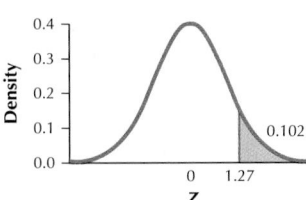

(b) 0.1020

46. (a)

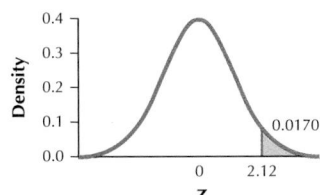

(b) 0.0170

47. (a)

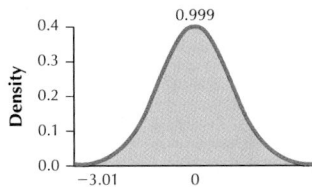

(b) 0.9987

48. (a)

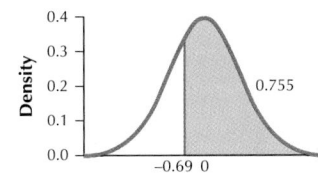

(b) 0.7549

49. (a)

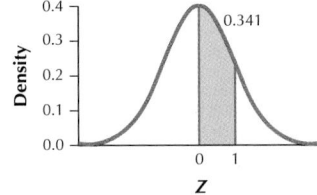

(b) 0.3413

50. (a)

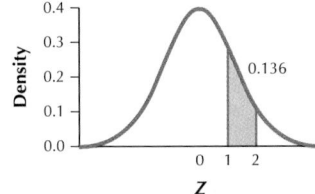

(b) 0.1359

51. (a)

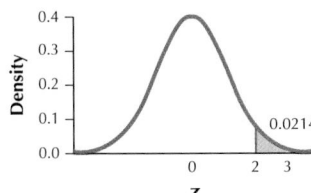

(b) 0.0214

52. (a)

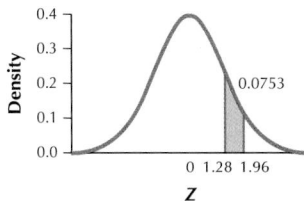

(b) 0.0753

53. (a)

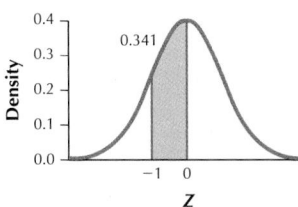

(b) 0.3413

54. (a)

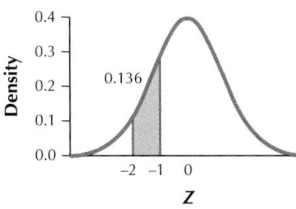

(b) 0.1359

55. (a)

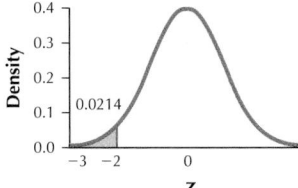

(b) 0.0214

56. (a)

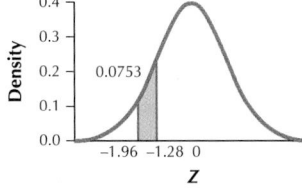

(b) 0.0753

57. (a)

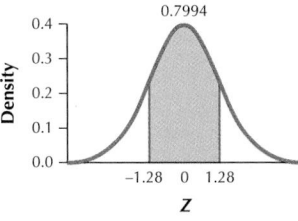

(b) 0.7994

58. (a)

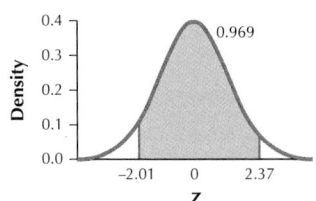

(b) 0.9689

59. (a)

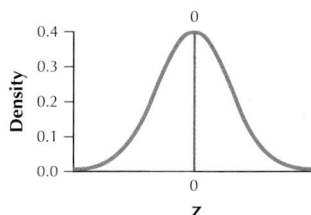

(b) 0
60. (a)

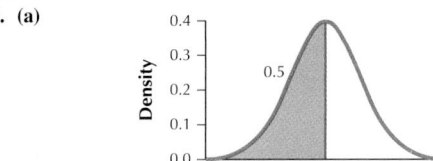

(b) 0.5
61. (a)

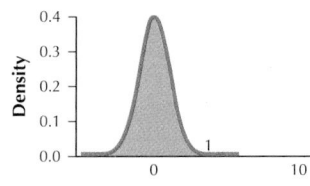

(b) 1
62. (a)

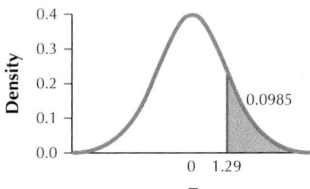

(b) 0.0985
63. (a)

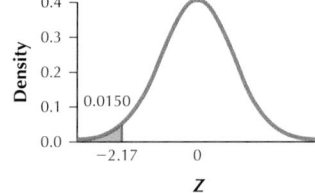

(b) 0.0150
64. (a)

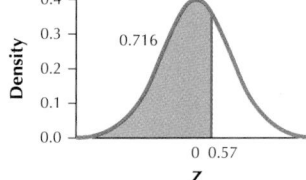

(b) 0.7157
65. (a)

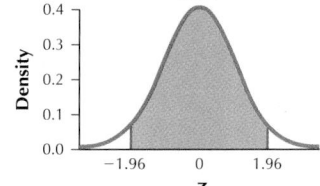

(b) 0.9500

66. (a)

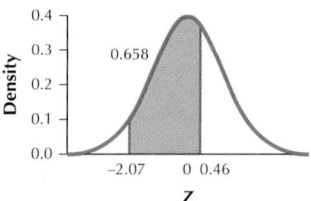

(b) 0.6580
67. (a)

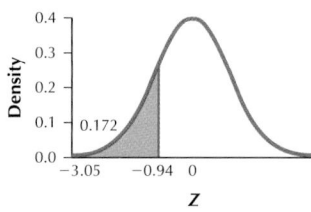

(b) 0.1725
68. (a)

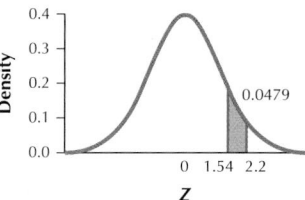

(b) 0.0479
69. (a)

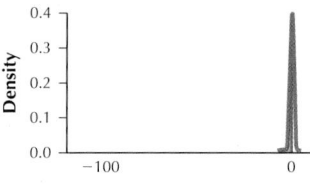

(b) 0.5000
70. (a)

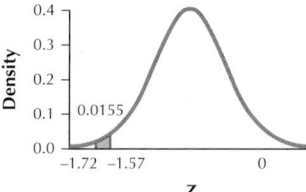

(b) 0.0155
71.

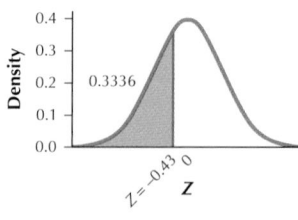

$Z = -0.43$
72.

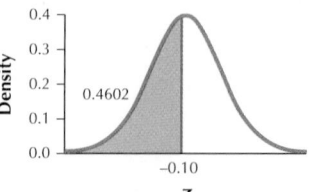

$Z = -0.10$

73.

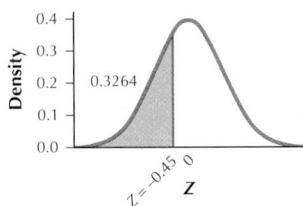

$Z = -0.45$
74.

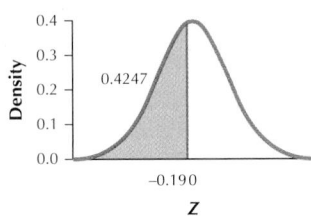

$Z = -0.19$
75. 1.65 (TI-83/84: 1.645)

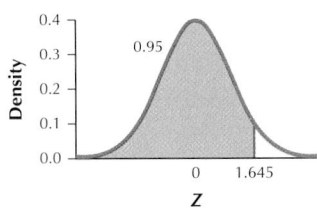

76. 1.96

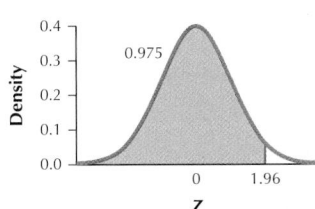

77. 2.05

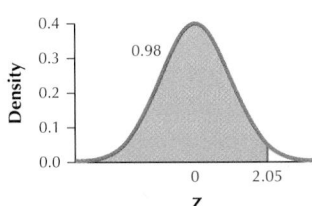

78. 2.33

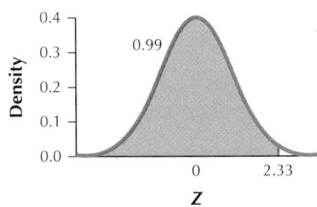

79.

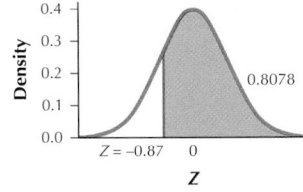

$Z = -0.87$

80.

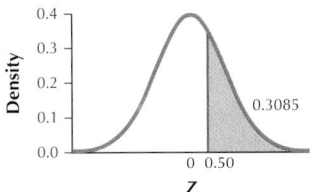

$Z = 0.50$
81.

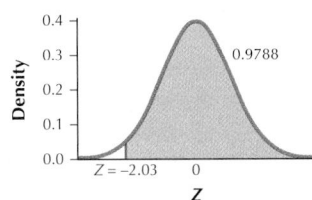

$Z = -2.03$
82.

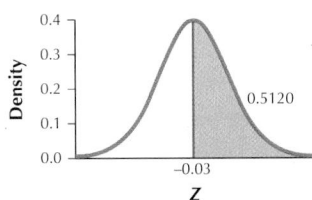

$Z = -0.03$
83. -1.28

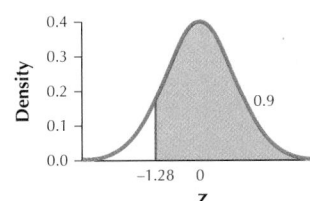

84. -1.96

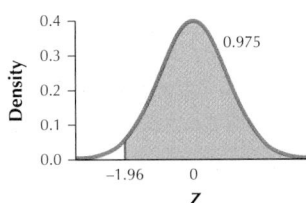

85. -3.036 (Using the table, both -3.03 and -3.04 have area to the left of them equal to 0.0012 and area to the right of them as 0.9988.)

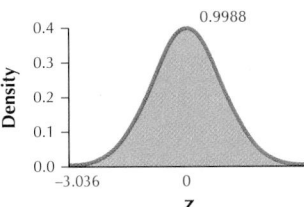

86. -3.54

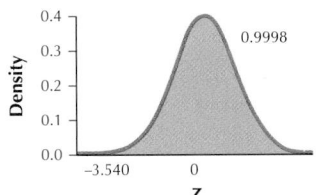

87. -1.28 and 1.28
88. -1.96 and 1.96
89. -2.33 and 2.33
90. -1.44 and 1.44
91. $Z = 0$
92. 0.67
93. $Z = 2.58$
94. -2.58
95. **(a)** 0.25 **(b)** 0.25 **(c)** 0 **(d)** 0. Area underneath the curve for a single value of X is the area of a line which is 0.
96. **(a)** 0.3 **(b)** 0.65 **(c)** 0 **(d)** 0; Area underneath the curve for a single value of X is the area of a line that is 0.

97.

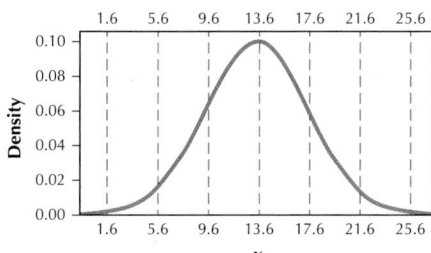

98.

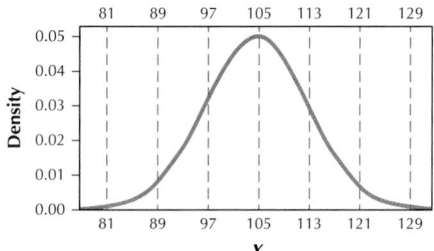

99.

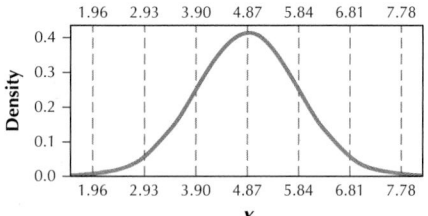

100.

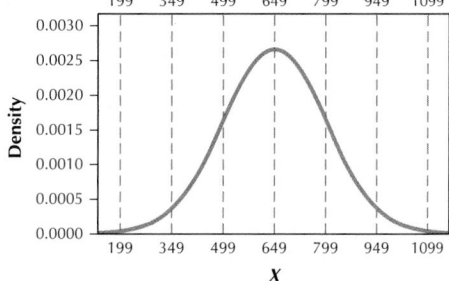

101. **(a)** 9.5 minutes **(b)** 9 minutes **(c)** 9.75 minutes **(d)** 0.5 minute **(e)** 1 minute **(f)** 0.25 minute
102. **(a)** The distribution is symmetric, so the balance point (mean) is the midpoint of the bottom side of the rectangle. **(b)** The distribution is rectangular shaped, so a line connecting the midpoints of the top side and bottom divides the rectangle; 50% of the area of the rectangle is left of this line and 50% of the area of the rectangle is right of this line. Thus the

midpoint of the bottom side of the rectangle is the median. From part (a), this is also the mean of the distribution.
103. 0.9750
104. 0.8997
105. 0.4821
106. 0.4641
107. 0.0179
108. 0.0446
109. 0.8020
110. 0.1359
111. 0.1832
112. 0.1253
113. 0.9641
114. 0.4772
115. 0.8413
116. 0.9332
117. -0.13
118. -1.28
119. The area between $Z = -2$ and $Z = 2$ is 0.9544. By the Empirical Rule, the area between $Z = -2$ and $Z = 2$ is about 0.95.
120. The area between $Z = -3$ and $Z = 3$ is 0.9974. By the Empirical Rule, the area between $Z = -3$ and $Z = 3$ is about 0.997.
121. **(a)** 0.0668 **(b)** 0.9332 **(c)** 0.8664
122. **(a)** 0.9965 **(b)** 0.0035 **(c)** 0.9930
123. $Z = -2.58$ and $Z = 2.58$.
124. $Z = -1.645$ and $Z = 1.645$
125. -0.67; 0; 0.67

126.

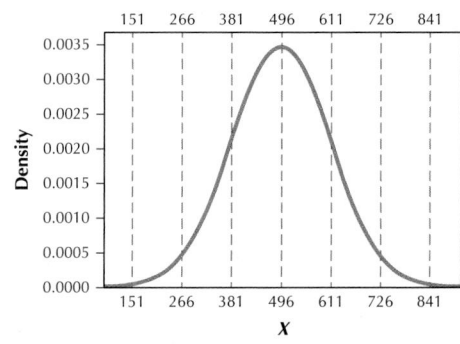

127. The peak would still be at $\mu = 496$ but since the standard deviation is larger the curve would be flatter and more spread out.
128. The peak would be shifted up to $\mu = 600$ but since the standard deviation is the same the curve would be just as spread out as the original curve.
129. They are both equal to 0.5. The property that the mean equals the median.
130. $P(X > 611)$ is larger. From the graph in Exercise 126 the area to the right of $X = 611$ is larger than the area to the right of $X = 726$.
131. $P(X > 611) = P(Z > 1) = 0.1587$ and $P(X > 726) = P(Z > 2) = 0.0228$; yes

Section 6.5

1. To standardize things means to make them all the same, uniform, or equivalent. To standardize a normal random variable X, we transform X into the standard normal random variable Z using the formula $Z = \frac{X - \mu}{\sigma}$. We do this so that we can use the standard normal table to find the probabilities.
2. The Z-value is for the standard normal distribution and we want the value for the original normal distribution. For the graphs for Problems 3–15 see the Solutions Manual.
3. 0.5
4. 0.1587
5. 0.8413
6. 0.0062
7. 0.0062
8. 0.8413
9. 0.9332
10. 0.8400
11. 0.8400

12. 0.0214 using technology or 0.0215 using the table
13. 0.0049
14. 0.3413
15. 86.45; see Solutions Manual for graph.
16. 53.55

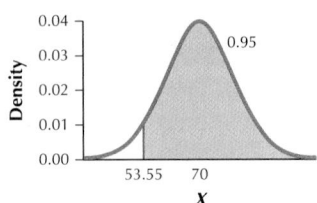

17. 89.6

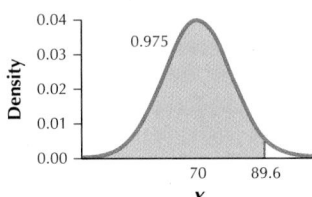

18. 50.4

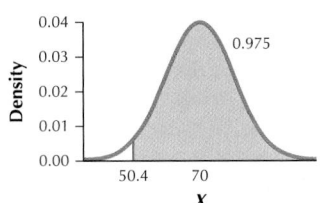

19. 46.7

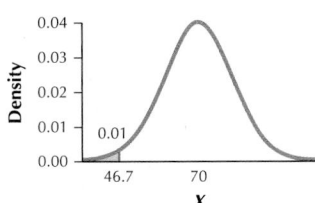

20. 93.3

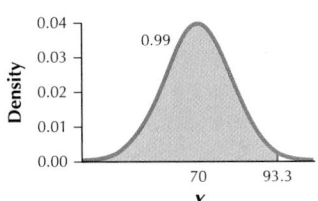

21. 44.2

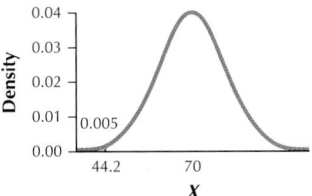

22. 95.8

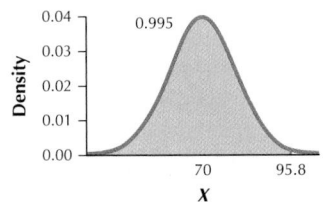

23. 53.55 and 86.45; see Solutions Manual for graph.
24. 50.4 and 89.6; see Solutions Manual for graph.
25. 46.7 and 93.3

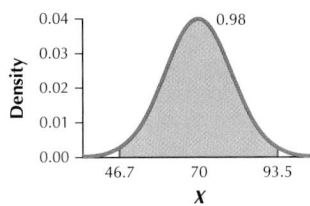

26. 44.2 and 95.8; see Solutions Manual for graph.
27. No
28. Yes
29. Yes
30. No
31. (a) 0.5 (b) 0.1587 (c) 0.1359
32. (a) 0.1038 (b) 0.1075 (c) 0.7887
33. (a) 0.1423 (b) 0.1423 (c) 26.67% (d) $X = 27.6$ mph (e) Z-score is -2.27; moderately unusual
34. (a) 0.0668 (b) 0.0228 (c) 0.9104 (d) 749.5 (e) Yes; z-score $= 3.5 \geq 3$.
35. (a) 18.29 ounces (b) 11.71 ounces (c) 11.71 ounces and 18.29 ounces
36. (a) \$171.40 (b) \$78.20 (c) \$78.20 and \$171.40
37. (a) 21.28 mph (TI-83/84: 21.29 mph) (b) 5.92 mph (TI-83/84: 5.91 mph) (c) 5.92 (5.91) mph and 21.28 (21.29) mph
38. (a) 402.25 (b) 895.75 (c) 402.25 and 895.75: yes, use the answers to (a) and (b) (d) 355 and 943 (e) Farther apart; The farther apart X_1 and X_2 are, the larger the percentage between them.
39. (a) 0.0179 (b) 0.5111 (c) 4.9 days; mean = median (d) Yes; z-score $= 3.1 \geq 3$
40. (a) 0.6915 (b) 3.66 million tobacco-related deaths (TI-83/84: 3.65 million) (c) The 25th percentile is $5 - 3.66 = 1.34$ $(5 - 3.65 = 1.35)$ million tobacco-related deaths below the mean. By symmetry, the 75th percentile is 1.34 (1.35) million tobacco-related deaths above the mean. Therefore, the 75th percentile is $5 + 1.34 = 6.34$ $(5 + 1.35 = 6.35)$ million tobacco-related deaths. (d) Z-score is 1.5; not unusual
41. (a) 0.2643 (b) 0.8643 (c) \$84.36 (d) \$125.64 (e) \$84.36 and \$125.64; yes, use the answers to (c) and (d)
42. (a) -1.04 calories per gram (b) No, food cannot have negative calories per gram. (c) Yes (d) The distribution of the number of calories per gram is right-skewed, so it is not normal.
43. For the graphs see the Solutions Manual. (a) 18.29 ounces (b) 11.71 ounces (c) \$171.40 (d) 21.28 mph (e) 5.92 mph and 21.28 mph
44. (a) 0.0322 (b) 0.0287 (c) A higher proportion of females than males will win the writing scholarship. This makes sense given that females have a higher average on the Writing SAT test than males.
45. (a) 0.2033 (b) 0.2389 (c) A higher proportion of males will be identified as at-risk writers than females. This makes sense given that females have a higher average on the Writing SAT test than males.
46. 273.48 and 712.52
47. 256.6 and 707.4
48. (a) Decrease (b) Increase (c) Increase

Section 6.6
1. For certain values of n and p, it may be inconvenient to calculate probabilities for the binomial distribution. For example, if $n = 100$ and $p = 0.5$, it may be tedious to calculate $P(X > 57)$, which, in the absence of technology, would involve 43 applications of the binomial probability formula.
2. $n \cdot p \geq 5$ and $n \cdot q \geq 5$
3. Met
4. Unmet
5. Unmet
6. Met
7. Unmet
8. Met
9. 0.1272
10. 0.5636
11. 0.4364
12. 0.5636
13. 0.4364

14. 0.5704
15. 0.3616
16. 0.4660
17. 0.7764
18. 0.8577
19. 0.6772
20. 0.1423
21. 0.0853
22. 0.2981
23. 0.0992
24. 0.2168
25. (a) 0.4761 (b) 0.5239 (c) 0.4761 (d) 0.5223
26. (a) 0.1251 (b) 0.1251 (c) 0.8749 (d) 0.1054
27. (a) 0.0037 (b) 1 (c) 0.5517 (d) 0
28. (a) 0.1350 (b) 0.5675 (c) 0.4325 (d) 0.5675 (e) 0.4325
29. (a) No (b) The normal distribution is not a good approximation to the binomial distribution, so not appropriate. (c) $n \cdot p = 2 < 5$, so the conditions are not met.

Chapter 6 Review

1. (a) 0.15 (b) 0.60

2. (a)

X	0	2	50	2000	1,500,000
P(X)	0.9746625583	0.02381	0.001495	0.0000323	0.0000001417

(b) $E(X) = \$0.40$; The expected winnings are $0.60 less than the $1 it costs to play.
3. (a) 0.1887 (b) 0.8113 (c) 0.7880
4. (a) 0.1285 (b) 0.7385 (c) Mean $\mu = 13.5$ Americans, variance $\sigma^2 = 1.35$ Americans squared, standard deviation = 1.1619 Americans. If we take an infinite number of samples of Americans of size $n = 15$ and calculate the mean number of Americans who said they were hearing mostly bad news about the price of gasoline it would be $\mu = 13.5$.
5. 0.1755
6. 0.4913
7. 0.1247
8. 0.6160
9. (a) 0.3169 (b) 0.1920 (c) 0.7703 (d) 0.2297
10. 0
11. 0.5
12. Less than 0.5. Since the area to the right of the mean $\mu = 106$ mm is 0.5 and $X = 110$ mm is greater than the mean $\mu = 106$ mm, the area to the right of $X = 110$ mm is less than the area to the right of the mean $\mu = 106$.
13. About 0.68
14. (a)

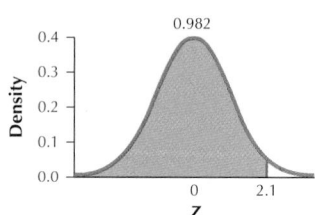

(b) 0.9821
15. (a)

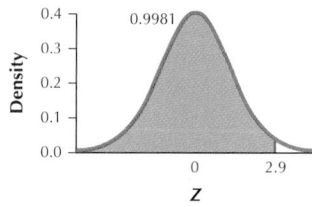

(b) 0.9981
16. (a)

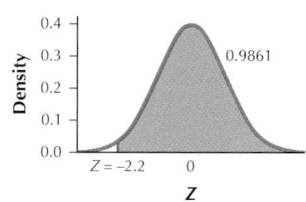

(b) 0.9861
17. (a)

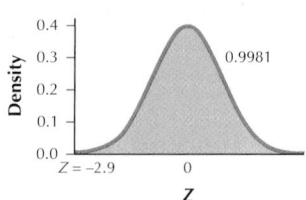

(b) 0.9981
18. (a)

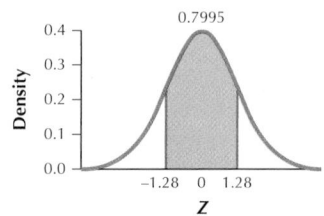

(b) 0.7995
19. (a)

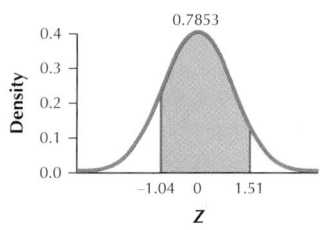

(b) 0.7853
20. (a) 0.0013 (b) 0.4287 (c) $X = 78.910$ deaths (d) The Z-score for $X = 60$ deaths is 1. Since $-2 < 1 < 2$, 60 people in the United States who are killed by drunk driving in a particular day is not unusual.
21. (a) 0.0062 (b) 0.4938 (c) 136.55 (d) Yes; z-score = $3.5 \geq 3$.
22. (a) 0.1192 (b) 0.0003 (c) 0.4404 (d) 0.5596 (e) 0.4404 (f) 0.6157

Chapter 6 Quiz

1. True
2. False
3. False
4. 0.5
5. 0
6. 0
7. discrete
8. binomial
9. $\mu = 0, \sigma = 1$
10. (a) 0.0962 (b) 19 CEOs (c) $\mu = 19$ CEOs, $Var(X) = 15.39$, $SD(X) =$ 3.9230. The expected number of CEOs who drive luxury cars in a random sample of 100 CEOs is 19. (d) Z-score is 5.35; unusual
11. (a) 0.1003 (b) 33.22% (c) $4329.50 (d) Z-score is -2.05; moderately unusual
12. (a) 0.7967 (b) 0.0188 (c) 0.2967 (d) 72.46 mph
(e) Yes; Z-score = $-3.325 \leq -3$.

Chapter 7

Note to instructors and students: Some answers may vary slightly depending on whether you round at intermediate steps or wait until you get the final answer to round. Also, different software and different forms of technology may give slightly different answers.

Section 7.1

1. It consists of the sample statistics of all possible samples of size n from the population. Sampling distributions can tell us about the expected location and variability of a statistic.
2. For a given sample size n, it is normal with mean μ and standard deviation $\sigma_{\bar{x}} = \sigma/\sqrt{n}$.

3. True

4. It becomes approximately normal $(\mu, \sigma/\sqrt{n})$.

5. $n = 30$

6. Normal, approximately normal, or unknown

7. 4 times as large

8. The population is either non-normal or of unknown distribution and the sample size is less than 30.

9. (a) 10 **(b)** 0.4

(c) and **(d)**

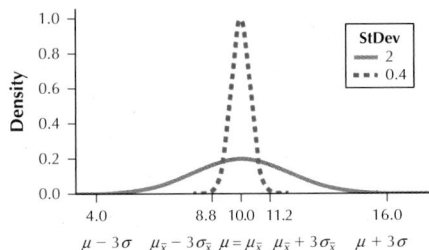

10. (a) 10 **(b)** 0.2

(c) and **(d)**

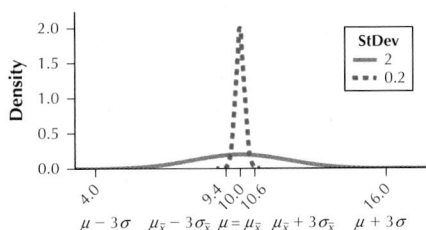

11. (a) 0 **(b)** 0.3333

(c) and **(d)**

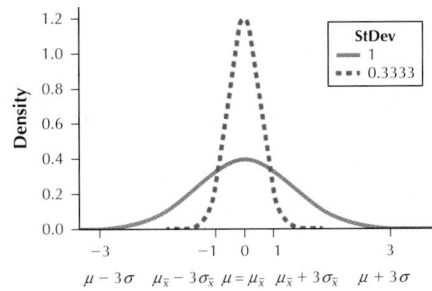

12. (a) 0 **(b)** 0.2

(c) and **(d)**

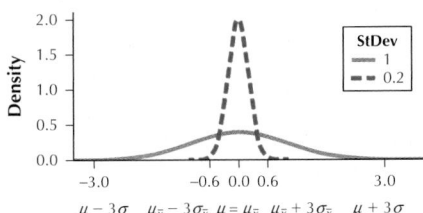

13. (a) 100 **(b)** 1

(c) and **(d)**

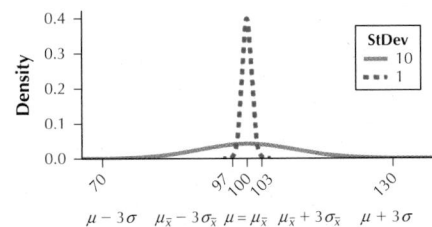

14. (a) 100 **(b)** 0.5

(c) and **(d)**

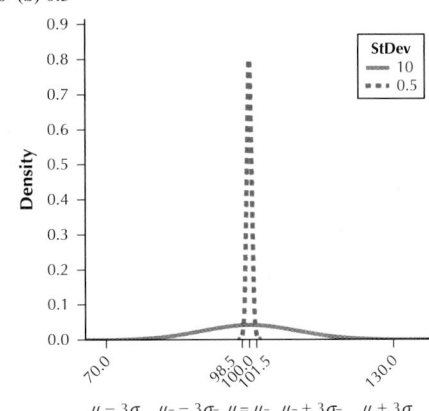

15. (a) 75 **(b)** 2.5

(c) and **(d)**

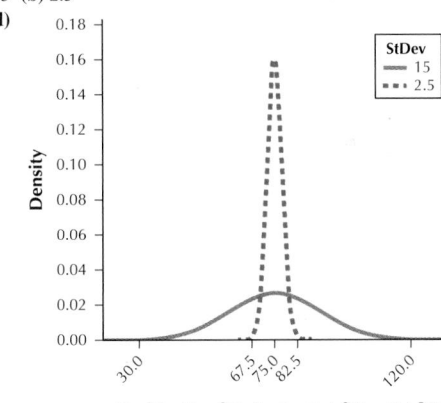

16. (a) 75 **(b)** 3

(c) and **(d)**

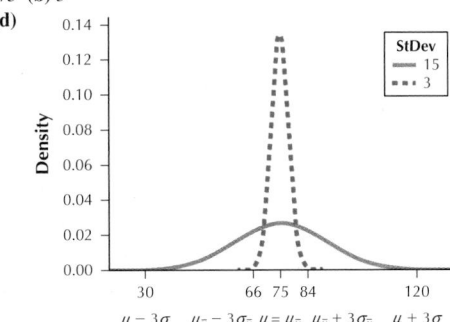

17. Approximately normal

18. Unknown

19. Unknown

20. Approximately normal

21. Unknown

22. Approximately normal

23. Normal

24. Approximately normal

25. Unknown

26. Approximately normal

27. $\mu_{\bar{x}} = 100$, $\sigma_{\bar{x}} = 3$

28.

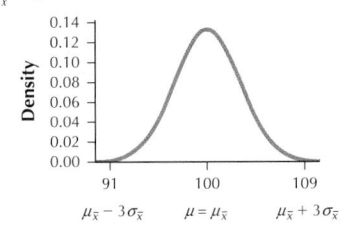

29.

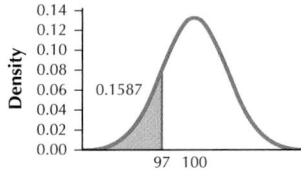

0.1587

30.

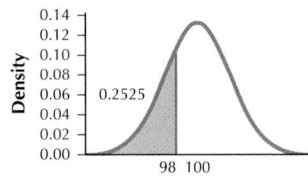

Larger; greater than; 0.2525; yes

31. 0.2525
32. 0.4950
33. $\mu_{\bar{x}} = 0$, $\sigma_{\bar{x}} = 0.1$

34.

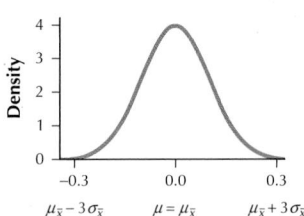

35.

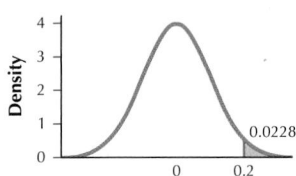

0.0228

36.

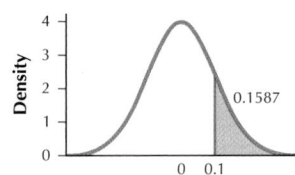

0.1587

37. 0.1587
38. 0.6826
39. $\mu_{\bar{x}} = 10$, $\sigma_{\bar{x}} = 1$

40.

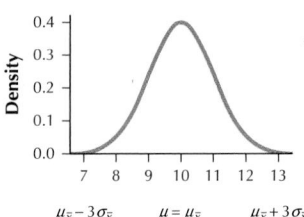

41. 11.65
42. 8.35
43. 11.96
44. 8.04
45. 8.35 and 11.65
46. 8.04 and 11.96
47. $\mu_{\bar{x}} = 100$, $\sigma_{\bar{x}} = 3$

48.

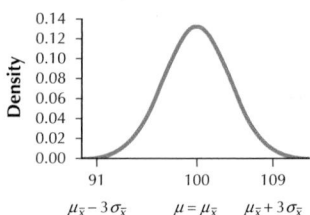

49. 104.95
50. 95.05
51.

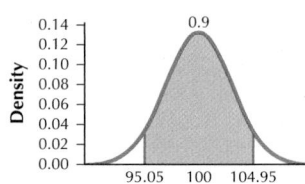

95.05 and 104.95

52. 0.10
53. $\mu_{\bar{x}} = 0$, $\sigma_{\bar{x}} = 0.25$

54.

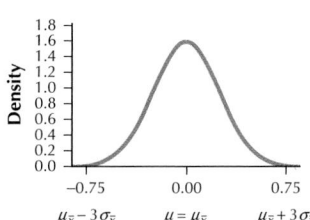

55. 0.49
56. −0.49
57.

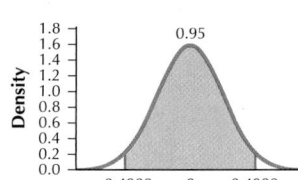

−0.49 and 0.49

58. 0.05
59. $\mu_{\bar{x}} = 2.5$, $\sigma_{\bar{x}} = 0.05$

60.

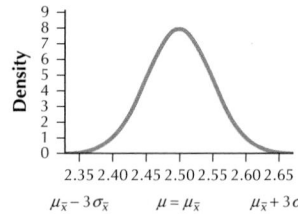

61.

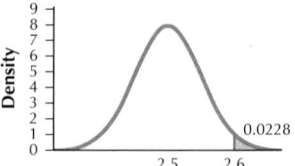

0.0228

62.

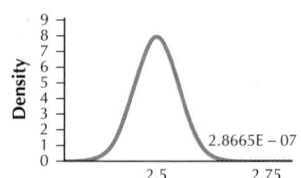

Approximately 0.

63.

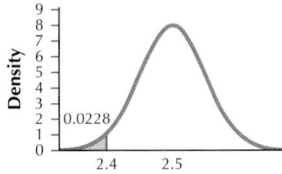

0.0228

64.

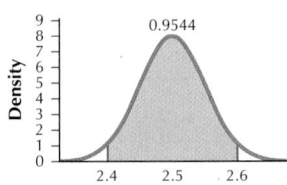

0.9544

65. $\mu_{\bar{x}} = -5, \sigma_{\bar{x}} = 0.5$

66.

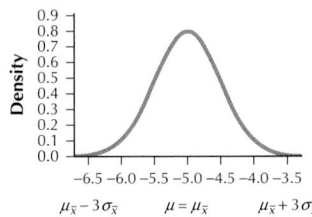

67.

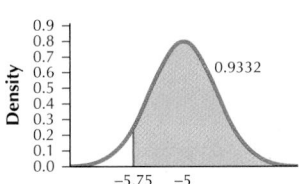

0.9332

68.

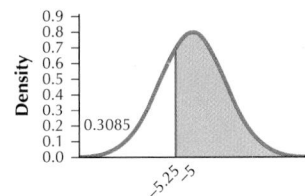

0.3085

69.

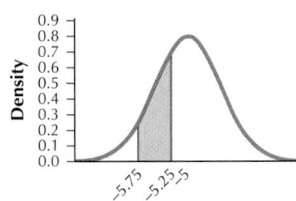

70. 0.2417
71. $\mu_{\bar{x}} = 80, \sigma_{\bar{x}} = 1$
72.

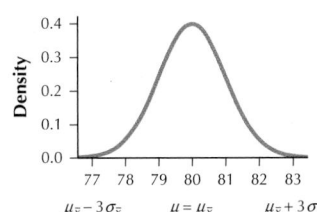

73. 82.58
74. 77.42

75.

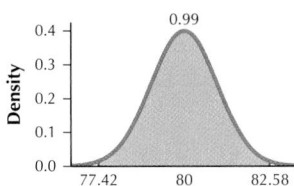

77.42 and 82.58
76. 0.01
77. $\mu_{\bar{x}} = 5, \sigma_{\bar{x}} = 0.25$
78.

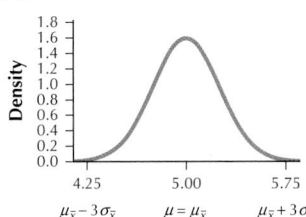

79.

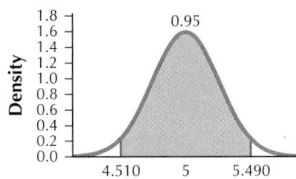

4.51 and 5.49
80. 0.05
81. Not possible; variable not normally distributed and sample size less than 30
82. 9.51 and 10.49
83. \$3.755 and \$4.245
84. Not possible; variable not normally distributed and sample size less than 30
85. Not possible; variable not normally distributed and sample size less than 30
86. \$76,080 and \$83,920
87. 76.04 and 79.96
88. Not possible; variable not normally distributed and sample size less than 30
89. Not possible; variable not normally distributed and sample size less than 30
90. 53.04 and 56.96
91. (a) $\mu_{\bar{x}} = 31.4$ micrograms, $\sigma_{\bar{x}} = 3.55$ micrograms **(b)** 0.1587 **(c)** 0.0228
92. (a) $\mu_{\bar{x}} = \$29,400, \sigma_{\bar{x}} = \$3,000$ **(b)** 0.1587 **(c)** 0.0668
93. (a) Approximately 0 **(b)** Approximately 1 **(c)** Approximately 0
94. (a) 0.0013 **(b)** Approximately 0 **(c)** 0.9987
95. (a) 37.2575 micrograms **(b)** 25.5425 micrograms **(c)** 25.5425 micrograms and 37.2575 micrograms
96. (a) \$35,280 **(b)** \$23,520 **(c)** \$23,520 and \$35,280
97. (a) 61.05 and 71.35
(b)

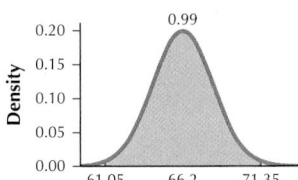

98. (a) \$97,437 and \$102,563
(b)

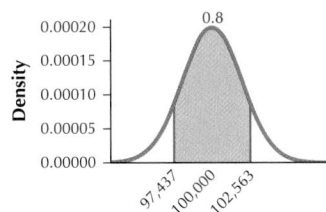

99. (a) 0.0918 (TI-83/84: 0.0912) (b) 0.8164 (TI-83/84: 0.8176)
100. (a) 0.7995 (b) 0.1256 (TI-83/84: 0.1255)
101. (a) 0.0026 (b) 0.4974
102. (a) 214.34 (b) 189.66
103. (a) 41.87 (b) 35.33

(c)

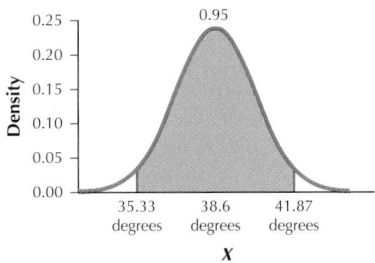

104. (a) 111.1 computers (b) 136.9 computers
(c)

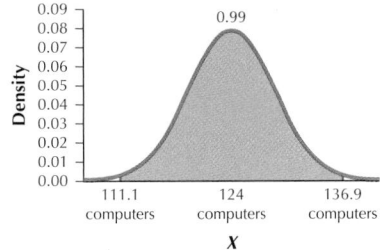

105. (a) 0.9544 (b) 397 ppm; in a normal distribution the median equals the mean. (c) 384.12 ppm and 409.88 ppm
106. (a) 0.6826 (b) 6.608 bits and 7.392 bits
107. In favor of normality. All of the points are between the curved lines and most of the points are close to the center line.
108. 0.0359
109. 0 (TI-83/84: 0.0001591)
110. Against normality. There are several points outside of the curved lines and most of the points are close to the upper curved line.
111. Not possible. The variable is not normally distributed and the sample size is less than 30.
112. 0.9544 (TI-83/84: 0.9545)
113. (a) Remain unchanged. From Fact 1 in Section 7.1, $\mu_{\bar{x}} = \mu$. Thus $\mu_{\bar{x}}$ does not depend on the sample size n. (b) Decrease. Since $\sigma_{\bar{x}} = \sigma/(\sqrt{n})$, an increase in the sample size n results in a decrease in $\sigma_{\bar{x}}$. (c) Insufficient information to tell. If $\bar{x} > 50{,}000$, then $\bar{x} - 50{,}000 > 0$. Since $\sigma_{\bar{x}}$ decreases and is positive, $Z = (x - \mu_{\bar{x}})/\sigma_{\bar{x}}$ will increase. If $\bar{x} = 50{,}000$, then $\bar{x} - 50{,}000 = 0$. Thus $Z = (x - \mu_{\bar{x}})/\sigma_{\bar{x}}$ will remain 0. If $\bar{x} < 50{,}000$, then $\bar{x} - 50{,}000 < 0$. Since $\sigma_{\bar{x}}$ decreases and is positive, $Z = (x - \mu_{\bar{x}})/\sigma_{\bar{x}}$ will decrease.
(d) Increase. From part (c), $Z = (x - \mu_{\bar{x}})/\sigma_{\bar{x}} = (60{,}000 - 50{,}000)/\sigma_{\bar{x}}$ will increase and $Z = (x - \mu_{\bar{x}})/\sigma_{\bar{x}} = (40{,}000 - 50{,}000)/\sigma_{\bar{x}}$ will decrease. Thus the area between these two values will increase. Since $P(\$40{,}000 < \bar{x} < \$60{,}000)$ is the area between these two values of Z, $P(\$40{,}000 < \bar{x} < \$60{,}000)$ will increase.
114. 0.0456
115. (a) 0.1067 gram (b) 1.067 grams (c) About 0.997
116. (a) 0.9803 (b) 0.9803, 0.0197 (c) The values found in (b) favor the Master of the Mint.
117. (a) 0.0002 (b) 0.0002, 0.9998 (c) The value found in the original case study in the text favors the Master of the Mint.
118. $\mu = 127.751946$ grams
119. (a) $n = 2$

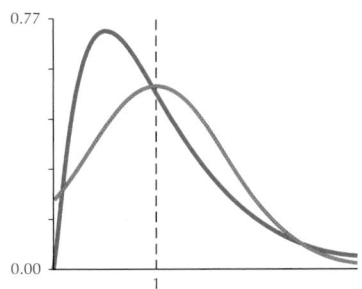

(b) $n = 5$

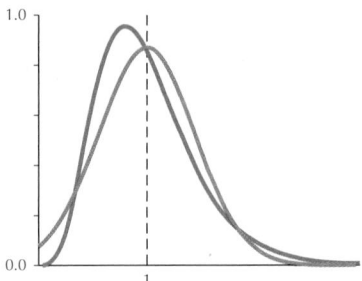

(c) $n = 30$

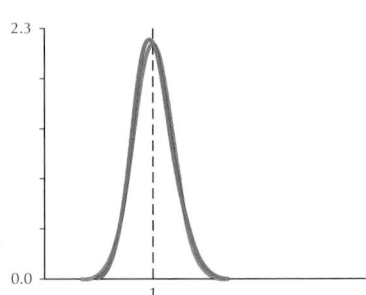

120. $n = 30$
121. 0.4522
122. (a) 514 (b) 29 (c) Normal with mean $\mu_{\bar{x}} = 514$ and $\sigma_{\bar{x}} = 29$; Case 1
123. (a) 0.3156 (b) Means are less variable than individual values.
124. (a) 588.7 (b) 439.3
125. (a) Increase (b) Remain the same (c) Increase (d) Still normal with $\mu_{\bar{x}} = 514$ but $\sigma_{\bar{x}}$ would increase
126. (a) Increase (b) Increase (c) Decrease

Section 7.2
1. If we take a sample of size n, the sample proportion $\hat{p}$ is $\hat{p} = x/n$, where x represents the number of individuals in the sample that have the particular characteristic. Examples will vary.
2. The mean of the sampling distribution of $\hat{p}$ is $\mu_{\hat{p}} = p$.
3. $\sigma_{\hat{p}} = \sqrt{p \cdot (1 - p)/n}$
4. Both (1) $np \geq 5$ and (2) $n(1 - p) \geq 5$.
5. It decreases by a factor of $1/\sqrt{2} \approx 0.7071$.
6. (a) $\hat{p} = 1/4 = 0.25$ (b) $\hat{p} = 1/3 = 0.3333$ (c) $\hat{p} = 1/2 = 0.5$
(d) $\hat{p} = 1/1000 = 0.001$
7. (a) 0.5 (b) 0.025 (c) Approximately normal
8. (a) 0.5 (b) 0.1118 (c) Approximately normal
9. (a) 0.05 (b) 0.01541 (c) Approximately normal
10. (a) 0.05 (b) 0.02179 (c) Approximately normal
11. (a) 0.3 (b) 0.1323 (c) Unknown
12. (a) 0.3 (b) 0.1025 (c) Approximately normal
13. (a) 0.002 (b) 0.000999 (c) Unknown
14. (a) 0.002 (b) 0.0008935 (c) Approximately normal
15. (a) 0.02 (b) 0.008854 (c) Approximately normal
16. (a) 0.02 (b) 0.009899 (c) Unknown
17. (a) 0.01 (b) 0.004450 (c) Approximately normal
18. (a) 0.01 (b) 0.009950 (c) Unknown
19. 25
20. 34
21. 50
22. 100
23. 500
24. 5000
25. 0.0668
26. Not possible since $n \cdot p = (8)(0.5) = 4 < 5$.
27. Not possible since $n \cdot p = (36)(0.03) = 1.08 < 5$.
28. 0.9525
29. 0.0531
30. 0.0293
31. Not possible since $n \cdot q = (225)(0.02) = 4.50 < 5$.

32. 0.4562

33. 0.6628

34. 0.5372

35. Not possible since $n \cdot q = (225)(0.01) = 2.25 < 5$.

36. Not possible since $n \cdot q = (225)(0.01) = 2.25 < 5$.

37. 0.1477

38. 0.2523

39. (a) $\mu_{\hat{p}} = 0.25$, $\sigma_{\hat{p}} \approx 0.0722$ (b) Approximately normal (0.25, 0.0722) (c) 0.4443 (TI-83/84: 0.4449)

40. (a) $\mu_{\hat{p}} = 0.75$ and $\sigma_{\hat{p}} \approx 0.06124$ (b) Approximately normal (c) 0.9929

41. (a) $\mu_{\hat{p}} = 0.75$, $\sigma_{\hat{p}} \approx 0.0968$ (b) 0.7324 (TI-83/84: 0.7323) (c) 0.0959 (TI-83/84: 0.0954)

42. (a) $\mu_{\hat{p}} = 0.840$ and $\sigma_{\hat{p}} = 0.022913$ (b) 0.1904 (c) 0.6173

43. (a) 0.1312, 0.3688

(b)

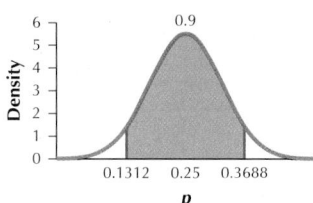

(c) For $\hat{p} = 2/36 \approx 0.0556$, $Z = -2.69$. Thus $\hat{p} = 2/36$ is considered moderately unusual. (d) Sample proportions between 0 and 0.0334 inclusive and between 0.4666 and 1 inclusive would be considered outliers.

44. (a) 0.63 and 0.87

(b)

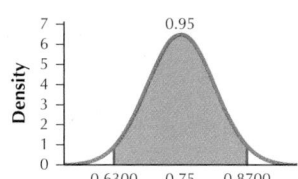

(c) z-score $= -4.08 \le -3$. Therefore a 50% success rate from the free-throw line would be considered an outlier or "poor shooting" by LeBron's standards. (d) z-score $= 4.08 \ge 3$. Therefore a 100% success rate from the free-throw line would be considered an outlier or "hot shooting" by LeBron's standards.

45. (a) 0.5003, 0.9997 (TI-83/84: 0.5007, 0.9993)

(b)

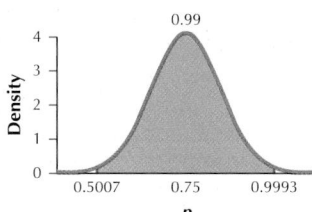

(c) For $\hat{p} = 14/20 = 0.7$, $Z = -0.5165$. Thus $\hat{p} = 0.7$ is neither moderately unusual nor an outlier.

46. (a) 0.7951 and 0.8849

(b)

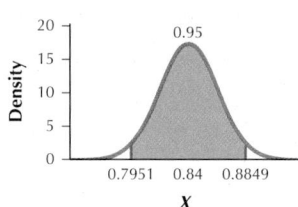

(c) 0.1914 (d) No.

47. (a) Remains the same since $\mu_{\hat{p}} = p$ does not depend on n.

(b) Decrease. Since the sample size n is in the denominator of $\sigma_{\hat{p}} = \sqrt{\dfrac{p \cdot q}{n}}$,

$\sigma_{\hat{p}}$ decreases as the sample size n increases. (c) Decrease. Standardizing we get $Z = \dfrac{0.86 - \mu_{\hat{p}}}{\sigma_{\hat{p}}} = \dfrac{0.86 - 0.840}{\sigma_{\hat{p}}} = \dfrac{0.02}{\sigma_{\hat{p}}}$. From (b), $\sigma_{\hat{p}}$ decreases as the

sample size n increases. Therefore $Z = \dfrac{0.02}{\sigma_{\hat{p}}}$ increases as the sample size n

increases. Therefore $P(\hat{p} > 0.86) = P\left(Z > \dfrac{0.02}{\sigma_{\hat{p}}}\right)$ decreases.

(d) Increase. Standardizing we get $Z = \dfrac{0.82 - \mu_{\hat{p}}}{\sigma_{\hat{p}}} = \dfrac{0.82 - 0.840}{\sigma_{\hat{p}}} = \dfrac{-0.02}{\sigma_{\hat{p}}}$

and $Z = \dfrac{0.86 - \mu_{\hat{p}}}{\sigma_{\hat{p}}} = \dfrac{0.86 - 0.840}{\sigma_{\hat{p}}} = \dfrac{0.02}{\sigma_{\hat{p}}}$. From (b), $\sigma_{\hat{p}}$

decreases as the sample size n increases. Therefore $Z = \dfrac{-0.02}{\sigma_{\hat{p}}}$

decreases and $Z = \dfrac{0.02}{\sigma_{\hat{p}}}$ increases as the sample size n increases.

Thus $P(0.82 < \hat{p} < 0.86) = P\left(\dfrac{-0.02}{\sigma_{\hat{p}}} < Z < \dfrac{0.02}{\sigma_{\hat{p}}}\right)$ increases as

the sample size n increases. (e) Decrease. Standardizing we get

$Z = \dfrac{0.82 - \mu_{\hat{p}}}{\sigma_{\hat{p}}} = \dfrac{0.82 - 0.840}{\sigma_{\hat{p}}} = \dfrac{-0.02}{\sigma_{\hat{p}}}$. From (b), $\sigma_{\hat{p}}$ decreases as the

sample size n increases. Therefore $Z = \dfrac{-0.02}{\sigma_{\hat{p}}}$ decreases as the sample size

n increases. Thus $P(\hat{p} < 0.82) = P\left(Z < \dfrac{-0.02}{\sigma_{\hat{p}}}\right)$ decreases as the sample

size n increases. (f) Increase. The 2.5th percentile is found by the formula $\hat{p}_1 = (-1.96)\sigma_{\hat{p}} + \mu_{\hat{p}}$. From (a) $\mu_{\hat{p}}$ remains the same as the sample size n increases and from (b) $\sigma_{\hat{p}}$ decreases as the sample size n increases. Therefore $\hat{p}_1 = (-1.96)\sigma_{\hat{p}} + \mu_{\hat{p}}$ increases as the sample size n increases. (g) Decrease. The 97.5th percentile is found by the formula $\hat{p}_2 = (1.96)\sigma_{\hat{p}} + \mu_{\hat{p}}$. From (a) $\mu_{\hat{p}}$ remains the same as the sample size n increases and from (b) $\sigma_{\hat{p}}$ decreases as the sample size n increases. Therefore $\hat{p}_2 = (1.96)\sigma_{\hat{p}} + \mu_{\hat{p}}$ decreases as the sample size n increases.

48. (a) $\mu_{\hat{p}} = 0.65$, $\sigma_{\hat{p}} \approx 0.0477$ (b) $\mu_{\hat{p}} = 0.41$, $\sigma_{\hat{p}} \approx 0.0492$

49. (a) 0.5 (b) 0 (c) 0 (d) 0.5

50. (a) 0.5565, 0.7435 (b) 0.3136, 0.5064

51. The results of Exercises 49 and 50 do not support this claim. The 97.5th percentile for the males is less than the 2.5th percentile for the females. Also $P(p < 0.41)$ and $P(p > 0.65)$ are both very different for males and females.

Chapter 7 Review

1. $\mu_{\bar{x}} = 100$, $\sigma_{\bar{x}} = 3$

2. $\mu_{\bar{x}} = 100$, $\sigma_{\bar{x}} = 2.5$

3. $\mu_{\bar{x}} = 100$, $\sigma_{\bar{x}} \approx 2.1429$

4. $\mu_{\bar{x}} = 0$, $\sigma_{\bar{x}} = 0.5$

5. $\mu_{\bar{x}} = 0$, $\sigma_{\bar{x}} = 0.25$

6. Not possible. Variable not normally distributed and sample size is less than 30.

7. 0.4525

8. (a) 48.18 years (b) 39.82 years (c) 39.82 years and 48.18 years

(d)

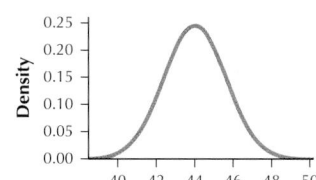

9. Not possible since $n \cdot q = (40)(0.1) = 4 < 5$.

10. 0.3192

11. 0.9753

12. 0.9762

13. (a) 0.1093 (b) 0.3907 (c) 0.8907 (d) 0.0665 and 0.1735

Chapter 7 Quiz

1. True

2. False

3. Sampling error

4. Approximately normal

5. No

6. $np \geq 5$ and $n(1 - p) \geq 5$
7. (a) 0.1587 (b) 0.9500 (c) 0.1056
8. (a) 43.29 grams (b) 36.71 grams (c) 36.71 grams and 43.29 grams
9. (a) 0.0668 (b) 0.2643
10. (a) \$27,115.88 (b) \$27,815.88
11. (a) 0.5 (b) 0.4013 (c) 0.0417, 0.0903

Chapter 8

Section 8.1

1. A range of values is more likely to contain μ than a point estimate is to be exactly equal to μ. We have no measure of confidence that our point estimate is close to μ. A confidence level for a confidence interval means that if we take sample after sample for a very long time, then in the long run, the percent of intervals that will contain the population mean μ will equal the confidence level.
2. $\bar{x} + Z_{\alpha/2}(\sigma/\sqrt{n})$, and $\bar{x} - Z_{\alpha/2}(\sigma/\sqrt{n}), \bar{x} + Z_{\alpha/2}(\sigma/\sqrt{n}))$
Point estimate $\pm$ margin of error and (point estimate – margin of error, point estimate + margin of error)
3. We are 95% confident that the population mean football score lies between 15 and 25.
4. True
5. $\bar{x} \pm E$ is shorthand for writing the two values $\bar{x} - E$ and $\bar{x} + E$. $\pm$ is shorthand notation for writing two numbers.
6. *A confidence interval estimate* of a parameter consists of an interval of numbers generated by a point estimate, together with an associated *confidence level* specifying the probability that the interval contains the parameter.
7. (a) $Z_{\alpha/2}$ increases. (b) Since the confidence level is $(1 - \alpha) \times 100\%$, as the confidence level increases, $1 - \alpha$ increases. Thus α and $\alpha/2$ will decrease. Since $\alpha/2$ is the area underneath the standard normal curve to the right of $Z_{\alpha/2}$, a decrease in $\alpha/2$ will result in an increase in $Z_{\alpha/2}$.
8. Increase the sample size.
9. Increases, Decreases
10. Decreases, Increases
11. 4
12. 16
13. 13
14. 104
15. No
16. Yes
17. Yes
18. Yes
19. Yes
20. No
21. $Z_{\alpha/2} = 2.576$
22. $Z_{\alpha/2} = 1.96$
23. $Z_{\alpha/2} = 1.96$
24. $Z_{\alpha/2} = 1.96$
25. $Z_{\alpha/2} = 1.645$
26. $Z_{\alpha/2} = 2.576$
27. (a) 2.5 (b) 1.96 (c) (70.1, 79.9). We are 95% confident that the population mean μ lies between 70.1 and 79.9.
28. (a) 2 (b) 1.96 (c) (96.08, 103.92). We are 95% confident that the population mean μ lies between 96.08 and 103.92.
29. (a) 2.6667 (b) 1.96 (c) (14.7733, 25.2267). We are 95% confident that the population mean μ lies between 14.7733 and 25.2267.
30. (a) 2 (b) 1.96 (c) (0.08, 7.92). We are 95% confident that the population mean μ lies between 0.08 and 7.92.
31. (a) 1 (b) 1.96 (c) (18.04, 21.96). We are 95% confident that the population mean μ lies between 18.04 and 21.96.
32. (a) 0.1 (b) 1.96 (c) (0.304, 0.696). We are 95% confident that the population mean μ lies between 0.304 and 0.696.
33. (a) 0.1 (b) 1.96 (c) $(-5.196, -4.804)$. We are 95% confident that the population mean μ lies between -5.196 and -4.804.
34. (a) 0.1875 (b) 1.96 (c) (26.6325, 27.3675). We are 95% confident that the population mean μ lies between 26.6325 and 27.3675.
35. (a) 4.9 (b) We can estimate the population mean μ to within 4.9 with 95% confidence.
36. (a) 3.92 (b) We can estimate the population mean μ to within 3.92 with 95% confidence.

37. (a) 5.2267 (b) We can estimate the population mean μ to within 5.2267 with 95% confidence.
38. (a) 3.92 (b) We can estimate the population mean μ to within 3.92 with 95% confidence.
39. (a) 1.96 (b) We can estimate the population mean μ to within 1.96 with 95% confidence.
40. (a) 0.196 (b) We can estimate the population mean μ to within 0.196 with 95% confidence.
41. (a) 0.196 (b) We can estimate the population mean μ to within 0.196 with 95% confidence.
42. (a) 0.3675 (b) We can estimate the population mean μ to within 0.3675 with 95% confidence.
43. (a) (95.1, 104.9) (b) We are 95% confident that the population mean μ lies between 95.1 and 104.9. (c) 4.9 (d) We can estimate the population mean μ to within 4.9 with 95% confidence.
44. (a) (93.56, 106.44) (b) We are 99% confident that the population mean μ lies between 93.56 and 106.44. (c) 6.44 (d) We can estimate the population mean μ to within 6.44 with 99% confidence.
45. (a) (2.6807, 2.8193) (b) We are 95% confident that the population mean μ lies between 2.6807 and 2.8193. (c) 0.0693 (d) We can estimate the population mean μ to within 0.0693 with 95% confidence.
46. (a) (2.6918, 2.8082) (b) We are 90% confident that the population mean μ lies between 2.6918 and 2.8082. (c) 0.0582 (d) We can estimate the population mean μ to within 0.0582 with 90% confidence.
47. 1
48. 2
49. 7
50. It increases by a factor of 4.
51. 3
52. 4
53. 7
54. It increases.
55. (a) (9.342, 10.658) (b) (9.216, 10.784) (c) (8.9696, 11.0304) (d) It increases.
56. (a) (48.712, 51.288) (b) (49.02, 50.98) (c) (49.1775, 50.8225) (d) It decreases.
57. (a) 69 gallons (b) 3.65 gallons (c) $Z_{\alpha/2} = 1.96$ (d) $n \geq 30$ and the population standard deviation σ is known. (e) (61.84, 76.16). We are 95% confident that μ lies between 61.84 gallons and 76.16 gallons.
58. (a) $\bar{x} = 2$ billion shares (b) 83,333,333.33 shares (c) $Z_{\alpha/2} = 1.96$ (d) $n \geq 30$ and the population standard deviation σ is known. (e) (1836.667, 2163.333); TI-83/84: (1836.7, 2163.3). We are 95% confident that the population mean number of shares traded daily on the New York Stock Exchange lies between 1836.667 (1836.7) million shares and 2163.333 (2163.3) million shares.
59. (a) 107 seconds (b) 19.5 seconds (c) $Z_{\alpha/2} = 1.96$ (d) $n \geq 30$ and the population standard deviation σ is known. (e) (68.78, 145.22). We are 95% confident that the true mean length of time that boys remain engaged with a science exhibit at a museum μ lies between 68.78 seconds and 145.22 seconds.
60. (a) \$177 (b) 25 (c) 1.645 (d) $n \geq 30$ and the population standard deviation σ is known. (e) (135.87, 218.13). We are 90% confident that μ lies between \$135.87 and \$218.13.
61. (a) 7.16 gallons. We can estimate μ to within 7.16 gallons with 95% confidence. (b) 3 (c) 62
62. (a) 163,333,333.33 shares. We can estimate μ to within 163,333,333.33 shares with 95% confidence. (b) 97 (c) 9604 days, approximately 26.31 years
63. (a) $E = 38.22$ seconds. We can estimate μ, the mean length of time that boys remain engaged with a science exhibit at a museum, to within 38.22 seconds with 95% confidence. (b) 59 (c) 5844 days
64. (a) \$41.13. We can estimate μ to within \$41.13 with 90% confidence. (b) 35 (c) 865
65. (a) (2.3040, 2.6960) (b) We are 95% confident that the population mean μ lies between 2.3040 and 2.6960. (c) 0.1960 (d) We can estimate the population mean μ to within 0.1960 with 95% confidence.
66. (a) (4.9497, 5.3103) (b) We are 95% confident that the population mean μ lies between 4.9497 and 5.3103. (c) 0.1803 (d) We can estimate the population mean μ to within 0.1803 with 95% confidence.
67. (a) (22.97, 23.22) (b) We are 95% confident that the population mean μ lies between 22.97 and 23.22. (c) 0.125 (d) We can estimate the population mean μ to within 0.125 with 95% confidence.

68. (a) (5.929361, 7.914795) (b) We are 95% confident that the population mean μ lies between 5.929361 and 7.914795. (c) 0.992717 (d) We can estimate the population mean μ to within 0.992717 with 95% confidence.
69. (a) The normal probability plot indicates an acceptable level of normality.

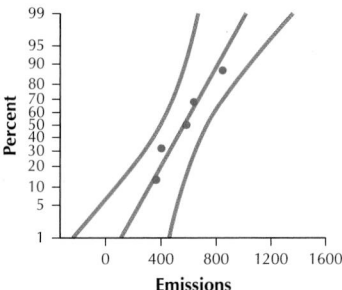

Emissions

(b) (415.067, 709.333); TI-83/84: (415.08, 709.32). We are 90% confident that the population mean carbon emissions lies between 415.067 (415.08) million tons and 709.333 (709.32) million tons. (c) $E = 147.133$ million tons. We can estimate the population mean emissions level of all nations to within 147.133 million tons with 90% confidence. (d) 44 nations
70. (a) The normal probability plot indicates an acceptable level of normality.

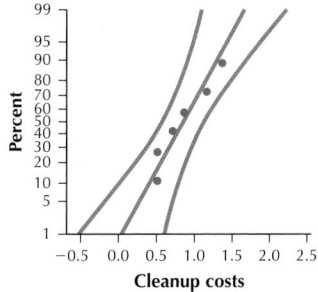

Cleanup costs

(b) (0.562, 1.122). We are 95% confident that the population mean cleanup costs in Florida counties for the BP oil spill lies between 0.562 million dollars and 1.122 million dollars. (c) $E = 0.280$ million dollars. We can estimate the population mean cleanup costs in Florida counties for the BP oil spill to within 0.280 million dollars with 95% confidence. (d) 189 counties
71. (a) The normal probability plot indicates an acceptable level of normality.

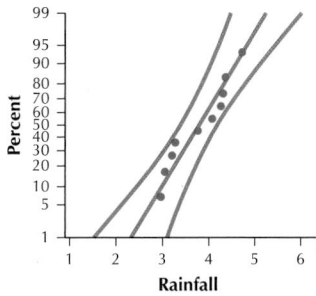

Rainfall

(b) (3.393, 4.187). We are 95% confident that the population mean rainfall in Georgia lies between 3.393 inches and 4.187 inches. (c) $E = 0.397$ inch. We can estimate the population mean rainfall in Georgia to within 0.397 inch with 95% confidence. (d) 158 locations
72. (a) 7 bits (b) 2 bits (c) $Z_{\alpha/2} = 1.96$ (d) 97 (e) 385
73. (a) $E = 1.07$ miles. We can estimate μ, the mean commuting distance, to within 1.07 miles with 95% confidence. (b) (8.86, 11.00). We are 95% confident that the true mean commuting distance μ lies between 8.86 miles and 11.00 miles.

74. (a) $E = 3.70$ ng/g. We can estimate μ, the mean concentration of the herbicide dicamba in Iowa homes, to within 3.70 ng/g with 95% confidence. (b) (176.30, 183.70). We are 95% confident that the true mean concentration of the herbicide dicamba in Iowa homes μ lies between 176.30 ng/g and 183.70 ng/g. (c) 113 homes. This is only one more home than the original sample size. The margin of error for the 95% confidence interval is 3.7041 homes, which is close to 3.7 homes. (d) 194 homes
75. (a) $Z_{\alpha/2} = 1.96$ (b) 5.91 (c) 3.1948. We can estimate the true mean quality of life score within 3.1948 with 95% confidence. (d) 9.78
76. (a) $Z_{\alpha/2} = 1.96$ (b) 144 steps (c) 132 steps. We can estimate the true mean number of fewer steps for each additional hour of television viewing within 132 steps with 95% confidence. (d) 673.5
77. (a) $n \geq 30$ and the population standard deviation σ is known. (b) (0.902, 1.098) (c) We are 95% confident that μ, the population mean lead contamination for all trout on the Spokane River, lies between 0.902 ppm and 1.098 ppm.
78. (a) Unchanged. (b)–(d) Decrease (e) Unchanged
79. (a) Since σ is a population characteristic, it stays constant and is unaffected by a decrease in confidence level. (b) The quantity $\sigma/\sqrt{n}$ is unaffected by a decrease in confidence level. (c) A decrease in the confidence level will result in a decrease in $Z_{\alpha/2}$. The width of the confidence interval $= 2 \cdot E = 2 \cdot Z_{\alpha/2}(\sigma/\sqrt{n})$. Thus a decrease in $Z_{\alpha/2}$ will result in a decrease in the width of the confidence interval. (d) The quantity $\bar{x}$ depends only on the sample taken, so it will remain unaffected by a decrease in the confidence level. (e) A decrease in the confidence level will result in a decrease in $Z_{\alpha/2}$. Since the margin of error is $E = Z_{\alpha/2}(\sigma/\sqrt{n})$, a decrease in $Z_{\alpha/2}$ will result in a decrease in the margin of error.

80.

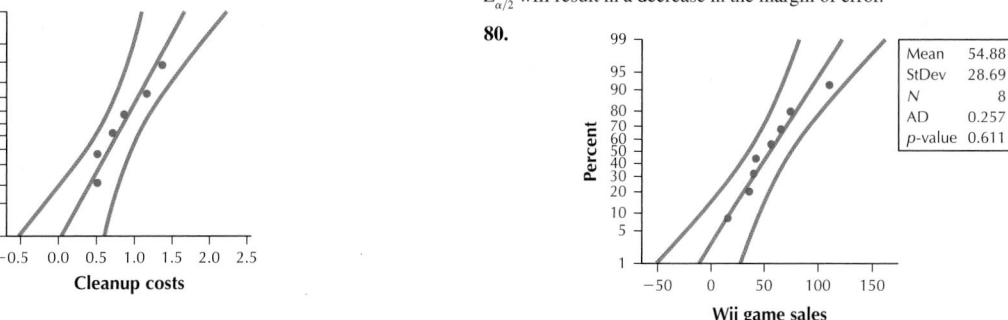

Wii game sales

All points lie within the curved lines so the normality is acceptable.
81. The distribution is approximately normal and σ is known.
82. 2.576
83. (27.6, 82.2). We are 99% confident that μ, the population mean Wii game sales, lies between 27.6 thousand units and 82.2 thousand units.
84. 27.3 thousand units. We can estimate μ, the population mean Wii game sales, to within 27.3 thousand units with 99% confidence.
85. 239 games
86. 4969 small firms
87. See the histogram in the answer for exercise 90.

88.

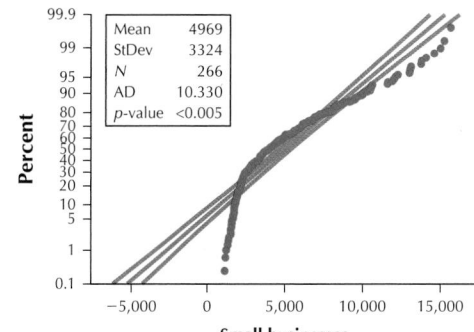

Small businesses

Since the majority of the points lie outside of the curved lines, the normality assumption is not valid.
89. (3189, 9209). We are 95% confident that the average number of small firms per metropolitan area lies between 3189 and 9209.

90.

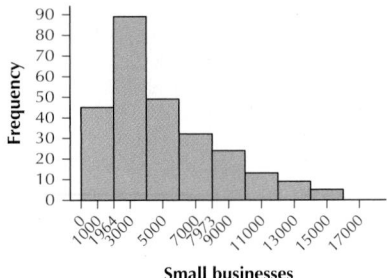

Small businesses

91. Answers will vary.
92. (a) 1.28 (b) 1.44 (c) 2.33
93. Answers will vary.
94. Answers will vary.
95. Answers will vary.
96. Answers will vary.
97. 0.90

Section 8.2

1. In most real-world problems, the population standard deviation σ is unknown, so we can't use the Z interval.
2. Yes. With even moderate sample sizes, reporting the t interval rather than the Z interval may offer peace of mind to the data analyst.
3. The t curve approaches closer and closer to the Z curve.
4. $E = t_{\alpha/2}(s/\sqrt{n})$
5. (a) 1.725 (b) 2.086 (c) 2.845
6. (a) 2.228 (b) 2.086 (c) 2.042
7. (a) For a given sample size, the value of $t_{\alpha/2}$ increases as the confidence level increases. (b) As the value of $t_{\alpha/2}$ increases, the confidence interval becomes wider. With a wider confidence interval you can be more confident that your confidence interval contains μ.

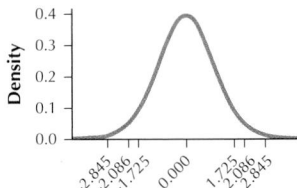

8. (a) For a given confidence level, as the sample size increases, the value of $t_{\alpha/2}$ decreases. (b) This is because as the sample size increases, the degrees of freedom increases, so the t distribution is approaching the Z distribution. Therefore the area in the middle is increasing, so the values of $-t_{\alpha/2}$ and $t_{\alpha/2}$ don't need to be as far away from 0 to bound 95% of the area.

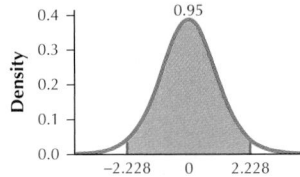

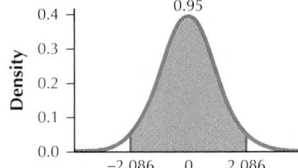

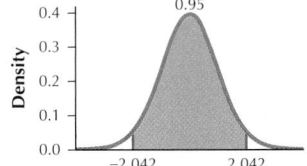

9. The sample size is not large (n is not ≥ 30) and we are not told that the population is normal. Therefore, the conditions are not met for the t interval for μ. It is not okay to construct the t interval.
10. The sample size is large (n is ≥ 30) and we are told that the population is normal. Therefore, the conditions are met for the t interval for μ. It is okay to construct the t interval.
11. We are not told that the population is normal. However, the sample size is large (n is ≥ 30). Therefore, the conditions are met for the t interval for μ. It is okay to construct the t interval.
12. We are not told that the population is normal. However, the sample size is large (n is ≥ 30). Therefore, the conditions are met for the t interval for μ. It is okay to construct the t interval.
13. (a) $\bar{x} = 4$, $s = 1$ (b) $t_{\alpha/2} = 2.776$ (c) (2.759, 5.241). We are 95% confident that the population mean μ lies between 2.759 and 5.241.
14. (a) $\bar{x} = 16$, $s = 2$ (b) $t_{\alpha/2} = 2.776$ (c) (13.517, 18.483). We are 95% confident that the population mean μ lies between 13.517 and 18.483.
15. (a) $\bar{x} = 13$, $s = 3$ (b) $t_{\alpha/2} = 2.776$ (c) (9.276, 16.724). We are 95% confident that the population mean μ lies between 9.276 and 16.724.
16. (a) $\bar{x} = 104$, $s = 4$ (b) $t_{\alpha/2} = 2.776$ (c) (99.034, 108.966). We are 95% confident that the population mean μ lies between 99.034 and 108.966.
17. (a) $t_{\alpha/2} = 1.753$ (b) We are told that the population is normal.
(c) (7.8, 12.2); We are 90% confident that the population mean μ lies between 7.8 and 12.2.
(d)

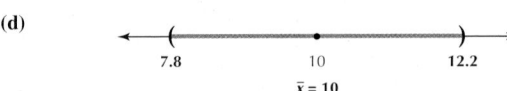

18. (a) $t_{\alpha/2} = 2.131$ (b) We are told that the population is normal.
(c) (20.0, 24.0). We are 95% confident that the population mean μ lies between 20.0 and 24.0.
(d)

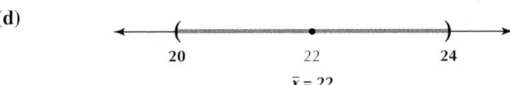

19. (a) $t_{\alpha/2} = 3.355$ (b) We are told that the population is normal.
(c) (43.3, 56.7). We are 99% confident that the population mean μ lies between 43.3 and 56.7.
(d)

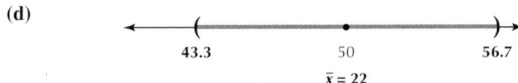

20. (a) $t_{\alpha/2} = 2.064$ (b) We are told that the population is normal.
(c) (−3.3, 3.3). We are 95% confident that the population mean μ lies between −3.3 and 3.3.
(d)

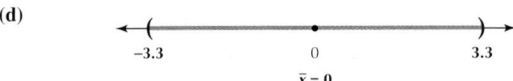

21. (a) $t_{\alpha/2} = 1.984$ (b) We are told that the population is normal.
(c) (−20.6, −19.4). We are 95% confident that the population mean μ lies between −20.6 and −19.4.
(d)

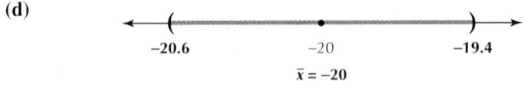

22. (a) $t_{\alpha/2} = 1.662$ (b) We are told that the population is normal.
(c) (−2.5, 2.5). We are 90% confident that the population mean μ lies between −2.5 and 2.5
(d)

23. (a) $t_{\alpha/2} = 1.660$ (b) We are not told that the population is normal. However, the sample size is large (n is ≥ 30). (c) (98.3, 101.7). We are 90% confident that the population mean μ lies between 98.3 and 101.7.

(d)

$\bar{x} = 100$

24. **(a)** $t_{\alpha/2} = 1.987$ **(b)** We are not told that the population is normal. However, the sample size is large (n is ≥ 30). **(c)** (248.0, 252.0). We are 95% confident that the population mean μ lies between 248.0 and 252.0.

(d)

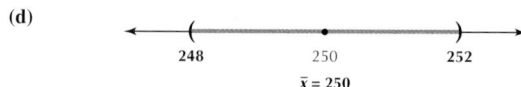

$\bar{x} = 250$

25. **(a)** $t_{\alpha/2} = 1.660$ **(b)** We are not told that the population is normal. However, the sample size is large (n is ≥ 30). **(c)** (34.4, 35.6). We are 90% confident that the population mean μ lies between 34.4 and 35.6.

(d)

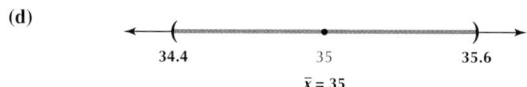

$\bar{x} = 35$

26. **(a)** $t_{\alpha/2} = 2.626$ **(b)** We are not told that the population is normal. However, the sample size is large (n is ≥ 30). **(c)** (40.7, 43.3). We are 99% confident that the population mean μ lies between 40.7 and 43.3.

(d)

$\bar{x} = 42$

27. **(a)** $t_{\alpha/2} = 2.000$ **(b)** We are not told that the population is normal. However, the sample size is large (n is ≥ 30). **(c)** ($-21, -19$). We are 95% confident that the population mean μ lies between -21 and -19.

(d)

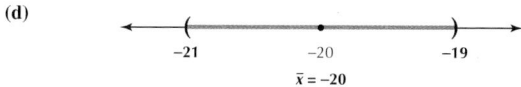

$\bar{x} = -20$

28. **(a)** $t_{\alpha/2} = 1.984$ **(b)** We are not told that the population is normal. However, the sample size is large (n is ≥ 30). **(c)** ($-1, 1$). We are 95% confident that the population mean μ lies between -1 and 1.

(d)

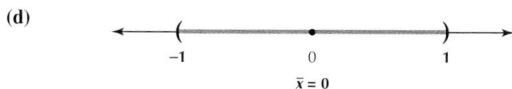

$\bar{x} = 0$

29. $E = 2.2$. We can estimate the population mean μ to within 2.2 with 90% confidence.

30. $E = 2.0$. We can estimate the population mean μ to within 2.0 with 95% confidence.

31. $E = 6.7$. We can estimate the population mean μ to within 6.7 with 99% confidence.

32. $E = 3.3$. We can estimate the population mean μ to within 3.3 with 95% confidence.

33. $E = 0.6$. We can estimate the population mean μ to within 0.6 with 95% confidence.

34. $E = 2.5$. We can estimate the population mean μ to within 2.5 with 90% confidence.

35. $E = 1.7$. We can estimate the population mean μ to within 1.7 with 90% confidence.

36. $E = 2.0$. We can estimate the population mean μ to within 2.0 with 95% confidence.

37. $E = 0.6$. We can estimate the population mean μ to within 0.6 with 90% confidence.

38. $E = 1.3$. We can estimate the population mean μ to within 1.3 with 99% confidence.

39. $E = 1$. We can estimate the population mean μ to within 1 with 95% confidence.

40. $E = 1$. We can estimate the population mean μ to within 1 with 95% confidence.

41. **(a)** We are not told that the population is normal. However, the sample size is large (n is ≥ 30). **(b)** $t_{\alpha/2} = 2.042$ **(c)** (3.7, 5.9). We are 95% confident that μ, the population mean length of stay in hospital for all heart attack victims, lies between 3.7 days and 5.9 days.

42. **(a)** We are not told that the population is normal. However, the sample size is large (n is ≥ 30). **(b)** $t_{\alpha/2} = 1.684$ **(c)** (21,422, 31,942). We are 90% confident that μ, the population mean student loan amount for all students, lies between \$21,422 and \$31,942.

43. **(a)** We are not told that the population is normal. However, the sample size is large (n is ≥ 30). **(b)** $t_{\alpha/2} = 1.987$ **(c)** (80.5, 84.5). We are 95% confident that μ, the population mean consumer sentiment for all consumers, lies between 80.5 and 84.5.

44. **(a)** We are not told that the population is normal. However, the sample size is large (n is ≥ 30). **(b)** $t_{\alpha/2} = 1.660$ **(c)** (53,932, 58,888). We are 90% confident that μ, the population mean teacher salary for all teachers, lies between \$53,932 and \$58,888.

45. **(a)** $E = 1.1$ days. We can estimate the population mean length of stay in hospital of all heart attack victims to within 1.1 days with 95% confidence. **(b)** It decreases.

46. **(a)** $E = \$5,260$. We can estimate the population mean student loan amount for all students to within \$5,260 with 90% confidence. **(b)** It decreases.

47. **(a)** $E = 2$. We can estimate the population mean consumer sentiment for all consumers to within 2 with 95% confidence. **(b)** Decrease the confidence level or increase the sample size. Increase the sample size. The only way to have both high confidence and a tight interval is to boost the sample size.

48. **(a)** $E = \$2478$. We can estimate the population mean teacher salary for all teachers to within \$2478 with 90% confidence. **(b)** Decrease the confidence level or increase the sample size. Increase the sample size. The only way to have both high confidence and a tight interval is to boost the sample size.

49. **(a)** (72.2, 77.8) **(b)** We are 95% confident that μ, the population mean exam score for all exam takers, lies between 72.2 and 77.8. **(c)** 2.8 **(d)** We can estimate the population mean exam score for all exam takers to within 2.8 with 95% confidence.

50. **(a)** (462.72, 501.43) **(b)** We are 95% confident that μ, the population mean total net sales for a clothing store, lies between \$462.72 and \$501.43 **(c)** \$19.36 **(d)** We can estimate the population mean total net sales for a clothing store to within \$19.36 with 95% confidence.

51. **(a)** (20.3, 23.3) **(b)** We are 95% confident that μ, the population mean number of vegetable farms per county for all counties nationwide, lies between 20.3 vegetable farms per county and 23.3 vegetable farms per county. **(c)** 1.5 vegetable farms per county **(d)** We can estimate the population mean number of vegetable farms per county for all counties nationwide to within 1.5 vegetable farms per county with 95% confidence.

52. **(a)** (18, 23) **(b)** We are 90% confident that μ, the population mean number of grocery stores per county for all counties nationwide, lies between 18 grocery stores per county and 23 grocery stores per county. **(c)** 2.5 grocery stores per county **(d)** We can estimate the population mean number of grocery stores per county for all counties nationwide to within 2.5 grocery stores per county with 90% confidence.

53. **(a)** (22,400, 25,630) **(b)** We are 95% confident that μ, the population mean number of small businesses per city for all cities nationwide, lies between 22,400 small businesses per city and 25,630 small businesses per city. **(c)** 1615 small businesses per city **(d)** We can estimate the population mean number of small businesses per city for all cities nationwide to within 1615 small businesses per city with 95% confidence.

54. **(a)** (2.30, 2.80) **(b)** We are 95% confident that μ, the population mean amount of protein for all breakfast cereals, lies between 2.30 grams and 2.80 grams. **(c)** 0.25 gram **(d)** We can estimate the population mean amount of protein for all breakfast cereals to within 0.25 gram with 95% confidence.

55. **(a)** (6.08, 7.77) **(b)** We are 90% confident that μ, the population mean amount of sugar for all breakfast cereals, lies between 6.08 grams and 7.77 grams. **(c)** 0.84 gram **(d)** We can estimate the population amount of sugar for all breakfast cereals to within 0.84 gram with 90% confidence.

56. **(a)** (22.28, 23.91) **(b)** We are 99% confident that μ, the population mean length of fourth graders' feet for all fourth graders, lies between 22.28 cm and 23.91 cm. **(c)** 0.815 cm **(d)** We can estimate the population mean length of fourth graders' feet for all fourth graders to within 0.815 cm with 99% confidence.

57. **(a)** Acceptable normality **(b)** (320.31, 804.09). We are 95% confident that μ, the population mean carbon emissions for all nations, lies between 320.31 million tons and 804.09 million tons. **(c)** $E = 241.89$ million tons. We can estimate the population mean carbon emissions for all nations to within 241.89 million tons with 95% confidence. **(d)** Decrease the confidence level or increase the sample size. Increase the sample size. The only way to have both high confidence and a tight interval is to boost the sample size.

58. (a) Acceptable normality **(b)** (0.48, 1.21). We are 95% confident that μ, the population mean clean up cost, lies between $0.48 million and $1.21 million. **(c)** $E = \$0.37$ million. We can estimate the population mean clean up cost to within $0.37 million with 95% confidence. **(d)** Decrease the confidence level or increase the sample size. Increase the sample size. The only way to have both high confidence and a tight interval is to boost the sample size.

59. (a) Acceptable normality **(b)** (31, 79). We are 95% confident that μ, the population mean number of Wii game units sold per week for all weeks, lies between 31 thousand units and 79 thousand units. **(c)** $E = 24$ thousand units. We can estimate the population mean number of Wii game units sold per week for all weeks to within 24 thousand units with 95% confidence. **(d)** Decrease the confidence level or increase the sample size. Increase the sample size. The only way to have both high confidence and a tight interval is to boost the sample size.

60. (a)

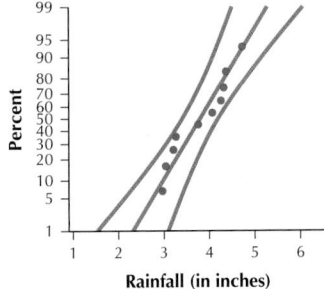

Acceptable normality. **(b)** (3.42, 4.16). We are 90% confident that μ, the population mean rainfall in Georgia, lies between 3.42 inches and 4.16 inches. **(c)** $E = 0.37$ inch. We can estimate the population mean rainfall in Georgia to within 0.37 inch with 90% confidence. **(d)** Decrease the confidence level or increase the sample size. Increase the sample size. The only way to have both high confidence and a tight interval is to boost the sample size.

61. (a)

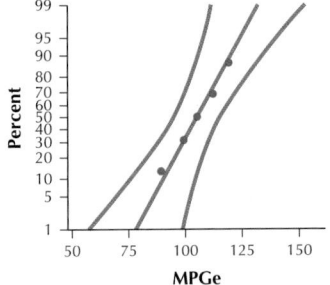

Acceptable normality. **(b)** $t_{\alpha/2} = 2.132$ **(c)** $E = 11.0$ MPGe. We can estimate the population mean mileage to within 11.0 MPGe with 90% confidence. **(d)** (93.8, 115.8). We are 90% confident that μ, the population mean mileage, lies between 93.8 MPGe and 115.8 MPGe.

62. (a)

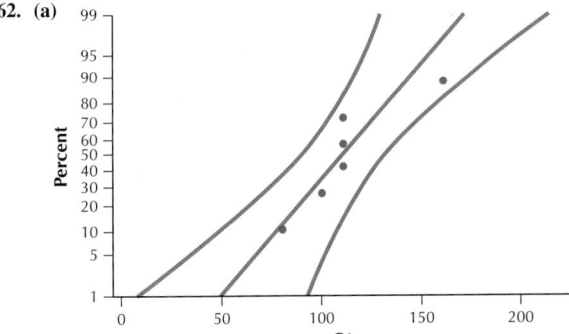

(b) Yes, the points do not appear to lie in a straight line. **(c)** Since the data do not appear to be normal, Case 1 does not apply. Since the sample size of $n = 6$ is small ($n < 30$), Case 2 does not apply. Thus a t interval cannot be used.

63. (a) $t_{\alpha/2} = 1.699$ **(b)** $E = 0.9$ miles. We can estimate the population mean commuting distance to within 0.9 miles with 90% confidence. **(c)** (9.0, 10.8). We are 90% confident that μ, the population mean commuting distance, lies between 9.0 miles and 10.8 miles.
64. (a) Decrease **(b)** Decrease **(c)** Decrease
65. The graph is symmetric about the middle value with the values with the highest frequency in the middle. This indicates that the normality assumption is valid. Since the normality assumption appears to be valid and σ is unknown, Case 1 applies, so we can use the t interval.
66. $E = 184$ cigarettes. We can estimate the population mean number of cigarettes smoked per capita to within 184 cigarettes with 90% confidence.
67. (2208, 2576). We are 90% confident that μ, the population mean number of cigarettes smoked per capita, lies between 2208 cigarettes and 2576 cigarettes.
68. The data appear to be highly left-skewed and therefore not acceptably normal. Since the population is not normally distributed, Case 1 does not apply. Since the sample size of $n = 20$ is small ($n < 30$), Case 2 does not apply.
69.

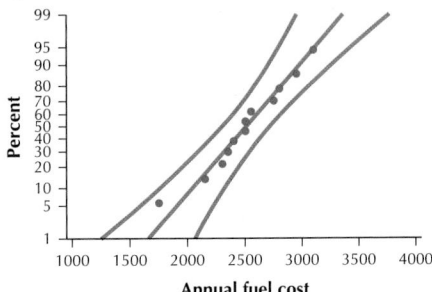

Acceptable normality.
70. $\bar{x} = \$2508$
71. $s = \$366.1$
72. $t_{\alpha/2} = 1.796$
73. (2318.2, 2697.8). We are 90% confident that μ, the population mean annual fuel cost, lies between $2318.2 and $2697.8.
74. $E = \$189.8$. We can estimate the population mean annual fuel cost to within $189.8 with 90% confidence.
75. Answers will vary.
76. No.
77. Answers will vary.
78. Answers will vary.
79. Answers will vary.
80. 0.90

Section 8.3
1. No, unless there is some reason to suspect that the value of p has changed.
2. National pollsters almost always use 95% as their confidence level and plus or minus 3 percentage points as their margin of error.
3. 0.36
4. 0.2
5. 0.5844
6. 0.6121
7. (a) $Z_{\alpha/2} = 1.96$ **(b)** Conditions are met. **(c)** (0.3005, 0.4195)
(d)

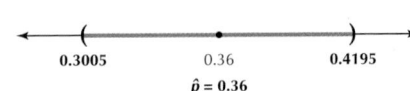

8. (a) $Z_{\alpha/2} = 1.96$ **(b)** Conditions are not met. **(c)** Can't do. **(d)** Can't do.
9. (a) $Z_{\alpha/2} = 1.96$ **(b)** Conditions are not met. **(c)** Can't do. **(d)** Can't do.
10. (a) $Z_{\alpha/2} = 2.576$ **(b)** Conditions are met. **(c)** (0.097, 0.303)
(d)

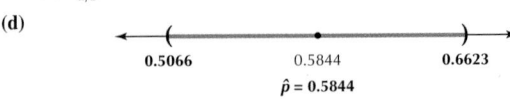

11. (a) $Z_{\alpha/2} = 1.96$ **(b)** Conditions are not met. **(c)** Can't do. **(d)** Can't do
12. (a) $Z_{\alpha/2} = 1.96$ **(b)** Conditions are met. **(c)** (0.5066, 0.6623)
(d)

13. (a) $Z_{\alpha/2} = 1.96$ (b) Conditions are met. (c) (0.5494, 0.6748)

(d)

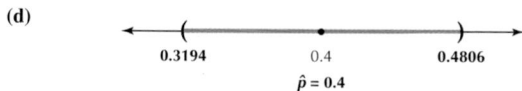

$\hat{p} = 0.6121$

14. (a) $Z_{\alpha/2} = 1.96$ (b) Conditions are not met. (c) Can't do. (d) Can't do.
15. (a) $Z_{\alpha/2} = 1.645$ (b) Conditions are met. (c) (0.3194, 0.4806)

(d)

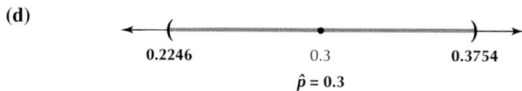

$\hat{p} = 0.4$

16. (a) $Z_{\alpha/2} = 1.645$ (b) Conditions are met. (c) (0.2246, 0.3754)

(d)

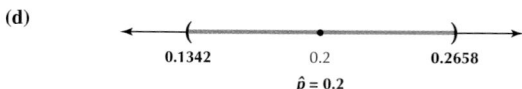

$\hat{p} = 0.3$

17. (a) $Z_{\alpha/2} = 1.645$ (b) Conditions are met. (c) (0.1342, 0.2658)

(d)

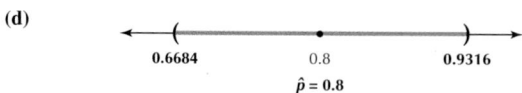

$\hat{p} = 0.2$

18. (a) $Z_{\alpha/2} = 1.645$ (b) Conditions are met. (c) (0.6684, 0.9316)

(d)

$\hat{p} = 0.8$

19. (a) $Z_{\alpha/2} = 1.96$ (b) Conditions are met. (c) (0.6432, 0.9568)

(d)

$\hat{p} = 0.8$

20. (a) $Z_{\alpha/2} = 2.576$ (b) Conditions are met. (c) (0.5939, 1.0061)

(d)

$\hat{p} = 0.8$

21. 0.0806
22. 0.0754
23. 0.0658
24. (a) If the sample size and the confidence level stay the same, then the margin of error decreases as the sample proportion decreases. (b) It decreases.
25. 0.1316
26. 0.1568
27. 0.2061
28. (a) If the sample size and the sample proportion stay the same, then the margin of error increases as the confidence level increases. (b) It increases.
29. (a) 0.3099 (b) 0.098 (c) 0.0310 (d) 0.0098
30. (a) 0.0588 (b) 0.0784 (c) 0.0898 (d) 0.0960 (e) 0.098
31. (a) Since the margin of error is $E = Z_{\alpha/2} \cdot \sqrt{\hat{p}(1 - \hat{p})/n}$, an increase in the sample size while $\hat{p}$ remains constant results in a decrease in the margin of error. (b) Since the width of the confidence interval is $2\,E$, an increase in the sample size while $\hat{p}$ remains constant results in a decrease in the width of the confidence interval.
32. (a) Since the margin of error is $E = Z_{\alpha/2} \cdot \sqrt{\hat{p}(1 - \hat{p})/n}$, the margin of error increases as the sample proportion approaches 0.5 while the sample size remains constant. (b) Increases
33. 897
34. 897
35. 385
36. 43
37. 5
38. Decreases
39. 752
40. 1068
41. 1844

42. 4269
43. 17,074
44. 68,296
45. Increases
46. Increases by about a factor of 4.
47. (a) $Z_{\alpha/2} = 1.96$ (b) Conditions are met. (c) (0.1923, 0.2487). We are 95% confident that p, the population proportion of millennials who are married, lies between 0.1923 and 0.2487.

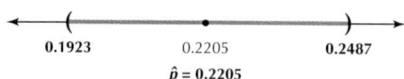

$\hat{p} = 0.2205$

48. (a) $Z_{\alpha/2} = 1.645$ (b) Conditions are met. (c) (0.3002, 0.4598). We are 90% confident that p, the population proportion of Minnesota residents who go fishing, lies between 0.3002 and 0.4598.

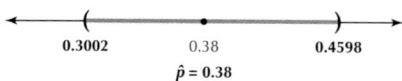

$\hat{p} = 0.38$

49. (a) $Z_{\alpha/2} = 2.576$ (b) Conditions are met. (c) (0.7332, 0.9268). We are 99% confident that p, the population proportion of college females who agree that heavier drinking occurs on spring break trips than is typically found on campus, lies between 0.7332 and 0.9268.

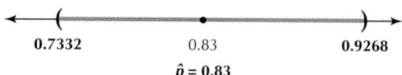

$\hat{p} = 0.83$

50. (a) $Z_{\alpha/2} = 1.645$ (b) Conditions are met. (c) (0.3745, 0.4255). We are 90% confident that p, the population proportion of NASCAR racing attendees who own a pickup truck, lies between 0.3745 and 0.4255.

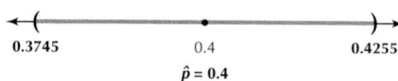

$\hat{p} = 0.4$

51. (a) 0.0282 (b) We can estimate the population proportion of millennials who are married to within 0.0282 with 95% confidence.
52. (a) 0.0798 (b) We can estimate the population proportion of Minnesota residents who go fishing to within 0.0798 with 90% confidence.
53. (a) 0.0968 (b) We can estimate the population proportion of college females who agree that heavier drinking occurs on spring break trips than is typically found on campus to within 0.0968 with 99% confidence.
54. (a) 0.0255 (b) We can estimate the population proportion of NASCAR attendees who own a pickup truck to within 0.0255 with 90% confidence.
55. (a) (0.61018, 0.78982) (b) We are 95% confident that p, the population proportion of times the weather forecaster correctly predicted rain, lies between 0.61018 and 0.78982. (c) 0.08982 (d) We can estimate the population proportion of times the weather forecaster correctly predicted rain to within 0.08982 with 95% confidence.
56. (a) (0.381017, 0.408306) (b) We are 95% confident that p, the population proportion of clothing store purchases made with a credit card, lies between 0.381017 and 0.408306. (c) 0.0136445 (d) We can estimate the population proportion of clothing store purchases made with a credit card to within 0.0136445 with 95% confidence.
57. (a) (0.0383, 0.0497) (b) We are 95% confident that p, the population proportion of clothing store purchases made online, lies between 0.0383 and 0.0497. (c) 0.0057 (d) We can estimate the population proportion of clothing store purchases made online to within 0.0057 with 95% confidence.
58. (a) (0.317472, 0.339328) (b) We are 90% confident that p, the population proportion of clothing store purchases made using a coupon, lies between 0.317472 and 0.339328. (c) 0.010928 (d) We can estimate the population proportion of clothing store purchases made using a coupon to within 0.010928 with 90% confidence.
59. (a) 0.0387 (b) We can estimate the population proportion of Hawaii residents who are thriving to within 0.0387 with 99% confidence. (c) (0.6163, 0.6937) (d) We are 99% confident that p, the population proportion of Hawaii residents who are thriving, lies between 0.6163 and 0.6937.
60. 11,009
61. (0.5565, 0.7435)

62. **(a)–(c)** Increases
63. **(a)** Decrease **(b)** Unchanged **(c)** Decrease
64. **(a)** Decrease **(b)** Decrease **(c)** Decrease
65. $\hat{p} = 0.4630$
66. Yes.
67. $Z_{\alpha/2} = 1.96$
68. 0.1330. We can estimate the population proportion of workers reporting skin rashes to within 0.1330 with 95% confidence.
69. 0.4630 ± 0.1330
70. (0.3300, 0.5960). We are 95% confident that p, the population proportion of workers reporting skin rashes, lies between 0.3300 and 0.5960.
71. 54. Same margin of error and same confidence level yields the same sample size.
72. Larger. Depending on whether $\hat{p}$ is known, $n = \hat{p} \cdot \hat{q} \left(\dfrac{Z_{\alpha/2}}{E} \right)^2$ or $n = \left[\dfrac{0.5 Z_{\alpha/2}}{E} \right]^2$. An increase in in the confidence level will result in an increase in $Z_{\alpha/2}$. In either case $Z_{\alpha/2}$ is in the numerator, so n will increase. 94
73. We have $n\hat{p} = 40(0.975) = 39 \geq 5$ but $n(1 - \hat{p}) = 40(1 - 0.975) = 1 < 5$. Thus we cannot use the Z interval for p.
74. (0.7198, 0.8694)
75. **(a)** Decrease in $Z_{\alpha/2}$ from 1.96 to 1.645. **(b)** Decrease in the margin of error from 0.0748 to 0.0628. **(c)** Decrease in the width of the confidence interval from 0.1496 to 0.1256.
76. Answers will vary.
77. Answers will vary.
78. Answers will vary.
79. Answers will vary.
80. It would increase.
81. Answers will vary.
82. 0.1639
83. Answers will vary.
84. Answers will vary.
85. 0.1770. Answers will vary.

Section 8.4

1. The population must be normal.
2. The difference between σ^2 and s^2 is that σ^2 is the population variance and s^2 is the sample variance.
3. To use this method, the distribution has to be symmetric and the X^2 curve is not symmetric.
4. Answers will vary.
5. False. The X^2 curve is *not* symmetric. It is right-skewed.
6. True
7. True
8. True
9. $\chi^2_{0.95} = 26.509$ and $\chi^2_{0.05} = 55.758$.
[Using Minitab: $\chi^2_{0.95} = 33.9303$ and $\chi^2_{0.05} = 66.3386$]
10. $\chi^2_{0.975} = 24.433$ and $\chi^2_{0.025} = 59.342$.
[Using Minitab: $\chi^2_{0.975} = 31.5549$ and $\chi^2_{0.025} = 70.2224$]
11. $\chi^2_{0.995} = 20.707$ and $\chi^2_{0.005} = 66.766$.
[Using Minitab: $\chi^2_{0.995} = 27.2493$ and $\chi^2_{0.005} = 78.230$]
12. $\chi^2_{0.975} = 8.907$ and $\chi^2_{0.025} = 32.852$.
13. $\chi^2_{0.975} = 12.401$ and $\chi^2_{0.025} = 39.364$.
14. $\chi^2_{0.975} = 16.047$ and $\chi^2_{0.025} = 45.722$.
15. $\chi^2_{1 - \alpha/2}$ decreases and $\chi^2_{\alpha/2}$ increases.
16. Both $\chi^2_{1 - \alpha/2}$ and $\chi^2_{\alpha/2}$ increase.
17. (21.87, 35.80) [Using Minitab: (20.1, 32.1)]
18. (20.95, 37.70) [Using Minitab: (19.3, 33.7)]
19. (19.29, 41.81) [Using Minitab: (17.8, 37.2)]
20. (4.68, 5.98) [Using Minitab: (4.48, 5.67)]
21. (4.58, 6.14) [Using Minitab: (4.39, 5.81)]
22. (4.39, 6.47) [Using Minitab: (4.22, 6.10)]
23. Lower bound decreases while the upper bound increases.
24. Lower bound decreases while the upper bound increases.
25. (15.86, 45.18)
26. (20.75, 58.07) [Using Minitab: (16.8, 41.2)]
27. (20.64, 50.14) [Using Minitab: (17.4, 38.8)]
28. (3.98, 6.72)
29. (4.56, 7.62) [Using Minitab: (4.10, 6.42)]
30. (4.54, 7.08) [Using Minitab: (4.18, 6.23)]

31. Lower bound increases while the upper bound decreases.
32. Both increase, then decrease.
33. **(a)** Acceptable normality. **(b)** $\chi^2_{0.975} = 2.180$ and $\chi^2_{0.025} = 17.535$ **(c)** (319.75, 2571.91). We are 95% confident that the population variance σ^2 lies between 319.75 megawatts squared and 2571.91 megawatts squared. **(d)** (17.88, 50.71). We are 95% confident that the population standard deviation σ lies between 17.88 megawatts and 50.71 megawatts.
34. **(a)** $\chi^2_{0.975} = 0.484$ and $\chi^2_{0.025} = 11.143$ **(b)** (13623.51, 313650.41) [Using Minitab: (13, 623, 313, 379).] We are 95% confident that the population variance σ^2 lies between $13,623.51$ millions of tons squared and $313,650.41$ millions of tons squared.
35. **(a)** Megawatts squared **(b)** Megawatts **(c)** Megawatts
36. **(a)** Millions of tons squared **(b)** No **(c)** Millions of tons **(d)** (116.72, 560.05). We are 95% confident that the population standard deviation σ lies between 116.72 million tons and 5060.05 million tons.
37. No; not normally distributed
38. (0.234, 0.728). We are 90% confident that the population standard deviation σ lies between \$0.234 million and \$0.728 million.
39. (16.86, 76.33). We are 99% confident that the population standard deviation σ lies between 16.86 thousand units and 76.33 thousand units.

Chapter 8 Review

1. **(a)** 5 **(b)** $Z_{\alpha/2} = 1.96$ **(c)** $E = 9.8$. We can estimate the population mean μ to within 9.8 with 95% confidence. **(d)** (90.2, 109.2). We are 95% confident that the population mean μ lies between 90.2 and 109.2.
2. **(a)** 1 **(b)** $Z_{\alpha/2} = 1.96$ **(c)** $E = 1.96$. We can estimate the population mean μ to within 1.96 with 95% confidence. **(d)** (98.04, 101.96). We are 95% confident that the population mean μ lies between 98.04 and 101.96.
3. **(a)** 7 points **(b)** 0.2981 point **(c)** 1.645 **(d)** 0.4904 point. We can estimate μ to within 0.4904 point with 90% confidence. **(e)** (6.5096, 7.4904). We are 90% confident that the true mean increase in IQ points for all children after listening to a Mozart piano sonata for about 10 minutes μ lies between 6.5106 points and 7.4904 points.
4. 385
5. 139
6. 16
7. 68
8. Not appropriate. We are told that the population is non-normal and the sample size is small (n is < 30).
9. (40.3, 43.7)
10. (41.2, 42.8)
11. (18.91, 59.60)
12. **(a)** $Z_{\alpha/2} = 1.96$ **(b)** $x = 5$ is ≥ 5 and $n - x = 10 - 5 = 5$ is ≥ 5. **(c)** $E = 0.310$. We can estimate the population proportion to within 0.310 with 95% confidence. **(d)** (0.190, 0.810). We are 95% confident that the population proportion p lies between 0.190 and 0.810.
13. **(a)** $Z_{\alpha/2} = 1.96$ **(b)** $x = n\hat{p} = (500)(0.01) = 5$ is ≥ 5 and $n - x = 500 - 5 = 495$ is ≥ 5. **(c)** $E = 0.0087$. We can estimate the population proportion to within 0.0087 with 95% confidence. **(d)** (0.0013, 0.0187). We are 95% confident that the population proportion p lies between 0.0013 and 0.0187.
14. **(a)** 0.0583. We can estimate p to within $E = 0.0583$ with 95% confidence. **(b)** (0.7117, 0.8283). We are 95% confident that the true proportion of all emergency room patients mentioning MDMA (Ecstasy) as a factor in their admission who are age 25 and under lies between 0.7117 and 0.8283.
15. 239
16. 16
17. 2
18. 267
19. 385
20. 601
21. (224.00, 366.63)
22. (214.54, 386.06)
23. (14.97, 19.15)
24. (14.65, 19.65)
25. Lower bound = 30.537, upper bound = 104.367. We are 95% confident that σ, the population standard deviation of total union membership per state, lies between 30.537 and 104.367 thousand.

Chapter 8 Quiz

1. False
2. True
3. 4
4. less
5. α is a probability.
6. Either the population is normal or the sample size is large ($n \geq 30$).
7. (a) $E = \$653.33$. We can estimate μ, the population mean cost of college education per year, to within $\$653.33$ with 95% confidence. (b) (29,846.67, 31,153.33). We are 95% confident that μ, the population mean cost of college education per year, lies between $\$29,846.67$ and $\$31,153.33$.
8. (a) $E = 164.5$ pounds. We can estimate μ, the population mean femur load in a frontal crash for the passenger in a Chevrolet Equinox SUV, to within 164.5 pounds with 90% confidence.
(b) (838.5, 1167.5). We are 95% confident that μ, the population mean femur load in a frontal crash for the passenger in a Chevrolet Equinox SUV, lies between 838.5 pounds and 1167.5 pounds.
9. (a) $E = 0.0386$. We can estimate p, the true proportion of all Québecois who favor independence for the Province of Quebec, to within 0.0386 with 99% confidence. (b) (0.3014, 0.3786)
10. (a) (1.34, 6.43). We are 95% confident that the population standard deviation σ lies between 1.34 hours and 6.43 hours. (b) (2.68, 12.86). We are 95% confident that the population standard deviation σ lies between 2.68 hours and 12.86 hours.
11. 752

Chapter 9

Section 9.1

1. The null hypothesis is assumed to be true unless the sample evidence indicates that the alternative hypothesis is true instead. It represents what has been tentatively assumed about the value of the parameter. It is the status quo hypothesis. The alternative hypothesis represents an alternative claim about the value of the parameter. The researcher concludes that the alternative hypothesis is true only if the evidence provided by the sample data indicates that it is true.
2. μ_0 is a specified value for the unknown mean μ.
3.

Form	Null hypothesis		Alternative hypothesis
1	$H_0 : \mu = \mu_0$	vs.	$H_a : \mu > \mu_0$
2	$H_0 : \mu = \mu_0$	vs.	$H_a : \mu < \mu_0$
3	$H_0 : \mu = \mu_0$	vs.	$H_a : \mu \neq \mu_0$

4. (1) Finding the defendant guilty when in reality he did not commit the crime (a Type I error) and (2) finding the defendant not guilty when in reality he did commit the crime (a Type II error).
5. A Type I error occurs when one rejects H_0 when H_0 is true. A Type II error occurs when one does not reject H_0 when H_0 is false.
6. The two ways of making the correct decision are to not reject H_0 when H_0 is true and to reject H_0 when H_0 is false.
7. No. It depends on how many standard deviations the sample mean of 90 is below the population mean of 100 and the level of significance of the test.
8. False
9. Invalid. Statistics like $\bar{x}$ should not be used in the hypotheses. So the correct form is: $H_0 : \mu = 100$ versus $H_a : \mu < 100$
10. Valid.
11. Invalid. The equal sign always goes in H_0. So the correct form is: $H_0 : \mu = 3.14$ versus $H_a : \mu > 3.14$
12. Invalid. The same value for μ_0 should go in H_0 and H_a. So one possible correct form is: $H_0 : \mu = 2$ versus $H_a : \mu > 2$
13. $H_0 : \mu = 79$ versus $H_a : \mu > 79$
14. $H_0 : \mu = 50$ versus $H_a : \mu < 50$
15. $H_0 : \mu = 75$ versus $H_a : \mu \neq 75$
16. $H_0 : \mu = 12,500$ versus $H_a : \mu \neq 12,500$
17. $H_0 : \mu = 1000$ versus $H_a : \mu \neq 1000$
18. $H_0 : \mu = 32$ versus $H_a : \mu > 32$
19. (a) $H_0 : \mu = 10$ versus $H_a : \mu < 10$ (b) Conclude that the population mean achievement gap has decreased when, in reality, it has remained the

same. (c) Conclude that the population mean achievement gap has remained the same when, in reality, it has decreased.
20. (a) $H_0 : \mu = 1000$ versus $H_a : \mu > 1000$ (b) Conclude that the population mean throughout has increased since the manager has come on the job when, in reality, it has remained the same. (c) Conclude that the population mean throughout has remained the same since the manager has come on the job when, in reality, it has increased.
21. (a) $H_0 : \mu = 20,000$ versus $H_a : \mu < 20,000$ (b) Conclude that the population mean cost of the new treatment has decreased from the cost of the previous treatment when, in reality, it has remained the same. (c) Conclude that the population mean cost of the new treatment has remained the same as the cost of the previous treatment when, in reality, it has decreased.
22. (a) $H_0 : \mu = 5$ versus $H_a : \mu \neq 5$ (b) Conclude that the population mean time of the 40-yard dash has changed when, in reality, it has remained the same. (c) Conclude that the population mean time of the 40-yard dash has remained the same when, in reality, it has changed.
23. (a) $H_0 : \mu = 18$ versus $H_a : \mu > 18$ (b) Conclude that the population mean number of hours nursing majors study per week is greater than 18 hours when, in reality, it is greater than 18 hours. Conclude that the population mean number of hours nursing majors study per week is equal to 18 hours when, in reality, it is equal to 18 hours. (c) Conclude that the population mean number of hours nursing majors study per week is greater than 18 hours when, in reality, it is equal to 18 hours. (d) Conclude that the population mean number of hours nursing majors study per week is equal to 18 hours when, in reality, it is greater than 18 hours.
24. (a) $H_0 : \mu = 700$ versus $H_a : \mu < 700$ (b) Conclude that the population mean number of times consumers dine out is less than 700 per year when, in reality, it is less than 700 times per year. Conclude that the population mean number of times consumers dine out is equal to 700 times per year when, in reality, it is equal to 700 times per year. (c) Conclude that the population mean number of times consumers dine out is less than 700 times per year when, in reality, it is equal to 700 times per year. (d) Conclude that the population mean number of times consumers dine out is equal to 700 times per year when, in reality, it is less than 700 times per year.
25. (a) $H_0 : \mu = 3$ versus $H_a : \mu < 3$ (b) Conclude that the population mean number years it takes for owners of hybrid vehicles to recoup their initial increased cost through reduced fuel consumption is less than 3 years when, in reality, it is less than 3 years. Conclude that the population mean number years it takes for owners of hybrid vehicles to recoup their initial increased cost through reduced fuel consumption is equal to 3 years when, in reality, it is equal to 3 years. (c) Conclude that the population mean number years it takes for owners of hybrid vehicles to recoup their initial increased cost through reduced fuel consumption is less than 3 years when, in reality, it is equal to 3 years. (d) Conclude that the population mean number years it takes for owners of hybrid vehicles to recoup their initial increased cost through reduced fuel consumption is equal to 3 years when, in reality, it is less than 3 years.
26. (a) $H_0 : \mu = 339.1$ vs. $H_a : \mu \neq 339.1$ (b) (1) The mean number of fatal and injury collisions is different from 339.1 per year when the population mean number of fatal and injury collisions is actually different than 339.1 per year and (2) the average number of fatal and injury collisions is equal to 339.1 per year when in actuality the population mean number of fatal and injury collisions is equal to 339.1 per year. (c) The mean number of fatal and injury collisions is different from 339.1 per year when the population mean number of fatal and injury collisions is actually equal to 339.1 per year.
(d) The mean number of fatal and injury collisions is equal to 339.1 per year when actually the population mean number of fatal and injury collisions is not equal to 339.1 per year.
27. (a) $H_0 : \mu = 3.24$ vs. $H_a : \mu > 3.24$ (b) Conclude that the mean is greater than $\$3.24$ when it actually is greater than $\$3.24$, and conclude that the mean is equal to $\$3.24$ when it actually is equal to $\$3.24$. (c) Concluding that the mean is greater than $\$3.24$ when it actually is equal to $\$3.24$
(d) Concluding that the mean is equal to $\$3.24$ when it actually is greater than $\$3.24$
28. (a) $H_0 : \mu = 175$ vs. $H_a : \mu \neq 175$ (b) (1) The mean height of Americans this year has changed from 175 centimeters when it actually is different from 175 centimeters and (2) the mean height of Americans has not changed from 175 centimeters when it actually is equal to 175 centimeters. (c) The mean height of Americans this year has changed from 175 centimeters when it actually is equal to 175 centimeters. (d) The mean height of Americans has not changed from 175 centimeters when it actually is not equal to 175 centimeters.

29. (a) $H_0 : \mu = 350$ vs. $H_a : \mu \neq 350$ **(b)** Conclude that the population mean amount of caffeine in a 16-ounce Starbucks Park Place brewed coffee has changed from 350 milligrams when, in reality, it has changed from 350 milligrams. Conclude that the population mean amount of caffeine in a 16-ounce Starbucks Park Place brewed coffee has not changed from 350 milligrams when, in reality, it has not changed from 350 milligrams. **(c)** Conclude that the population mean amount of caffeine in a 16-ounce Starbucks Park Place brewed coffee has changed from 350 milligrams when, in reality, it has not changed from 350 milligrams. **(d)** Conclude that the population mean amount of caffeine in a 16-ounce Starbucks Park Place brewed coffee has not changed from 350 milligrams when, in reality, it has changed from 350 milligrams.

30. (a) $H_0 : \mu = 28.6$ vs. $H_a : \mu > 28.6$ **(b)** Conclude that the population mean number of apps used per month by iPhone and Android users has increased from its 2014 level of 28.6 when, in reality, it has increased. Conclude that the population mean number of apps used per month by iPhone and Android users has stayed the same as its 2014 level of 28.6 when, in reality, it has stayed the same. **(c)** Conclude that the population mean number of apps used per month by iPhone and Android users has increased from its 2014 level of 28.6 when, in reality, it has stayed the same. **(d)** Conclude that the population mean number of apps used per month by iPhone and Android users has stayed the same as its 2014 level of 28.6 when, in reality, it has increased.

31. (a) $H_0 : \mu = 5$ vs. $H_a : \mu > 5$ **(b)** Conclude that the population mean number of visits that customers make to the store in the six-month period is greater than 5 when, in reality, it is greater than 5. Conclude that the population mean number of visits that customers make to the store in the six-month period is equal to 5 when, in reality, it is equal to 5. **(c)** Conclude that the population mean number of visits that customers make to the store in the six-month period is greater than 5 when, in reality, it is equal to 5. **(d)** Conclude that the population mean number of visits that customers make to the store in the six-month period is equal to 5 when, in reality, it is greater than 5.

32. (a) $H_0 : \mu = 413$ vs. $H_a : \mu > 413$ **(b)** Conclude that the population mean total sales is greater than $413 per customer when, in reality, it is greater than $413. Conclude that the population mean total sales is equal to $413 per customer when, in reality, it is equal to $413. **(c)** Conclude that the population mean total sales is greater than $413 per customer when, in reality, it is equal to $413. **(d)** Conclude that the population mean total sales is equal to $413 per customer when, in reality, it is greater than $413.

33. (a) $H_0 : \mu = 20$ vs. $H_a : \mu < 20$ **(b)** Conclude that the population mean number of items that customers are buying is less than 20 when, in reality, it is less than 20. Conclude that the population mean number of items that customers are buying is equal to 20 when, in reality, it is equal to 20. **(c)** Conclude that the population mean number of items that customers are buying is less than 20 when, in reality, it is equal to 20. **(d)** Conclude that the population mean number of items that customers are buying is equal to 20 when, in reality, it is less than 20.

34. (a) $H_0 : \mu = 0.75$ vs. $H_a : \mu \neq 0.75$ **(b)** Concluding that the population mean number of coupons used differs from 0.75 when, in reality, it differs from 0.75. Concluding that the population mean number of coupons used is equal to 0.75 when, in reality, it is equal to 0.75. **(c)** Concluding that the population mean number of coupons used differs from 0.75 when, in reality, it is equal to 0.75. **(d)** Concluding that the population mean number of coupons used is equal to 0.75 when, in reality, it differs from 0.75.

35. (a) $H_0 : \mu = 150$ vs. $H_a : \mu < 150$ **(b)** Concluding that the population mean number of days since purchase is less than 150 when, in reality, it is less than 150. Concluding that the population mean number of days since purchase is equal to 150 when, in reality, it is equal to 150. **(c)** Concluding that the population mean number of days since purchase is less than 150 when, in reality, it is equal to 150. **(d)** Concluding that the population mean number of days since purchase is equal to 150 when, in reality, it is less than 150.

Section 9.2
1. When the observed value of $\bar{x}$ is unusual or extreme in the sampling distribution of $\bar{x}$ that assumes H_0 is true, we should reject H_0. Otherwise, we should not reject H_0.
2. The number of standard deviations that the sample mean $\bar{x}$ lies above or below the hypothesized value of the mean μ_0 from H_0.
3. A statistic generated from a data set for the purpose of testing a statistical hypothesis

4. The critical region consists of the range of values of the test statistic Z_{data} for which we reject the null hypothesis. The noncritical region consists of the range of values of the test statistic Z_{data} for which we do not reject the null hypothesis.
5. The value of Z that separates the critical region from the noncritical region.
6. There is evidence at level of significance α that the population mean is less than 5.
7. The critical region for a right-tailed test lies in the right (upper) tail.
8. True
9. $Z_{data} = 1$
10. $Z_{data} = 1.5$
11. $Z_{data} = 1$
12. $Z_{data} = 4$
13. $Z_{data} = -1$
14. $Z_{data} = -3$
15. $Z_{data} = -2$
16. $Z_{data} = -3$
17. $Z_{data} = 1$
18. $Z_{data} = -1$
19. $Z_{data} = -1$
20. $Z_{data} = 1$
21. It increases.
22. It decreases.
23. (a) $Z_{crit} = 1.28$

(b)

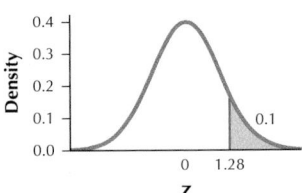

(c) Reject H_0 if $Z_{data} \geq 1.28$.
24. (a) $Z_{crit} = 1.645$

(b)

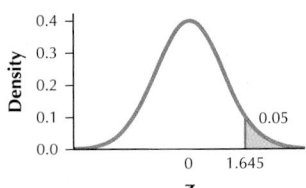

(c) Reject H_0 if $Z_{data} \geq 1.645$.
25. (a) $Z_{crit} = 1.645$

(b)

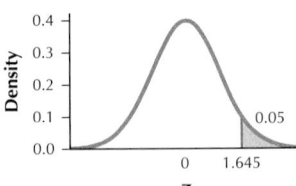

(c) Reject H_0 if $Z_{data} \geq 1.645$.
26. (a) $Z_{crit} = 2.33$

(b)

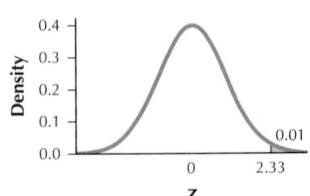

(c) Reject H_0 if $Z_{data} \geq 2.33$.

27. **(a)** $Z_{crit} = -1.645$

(b)

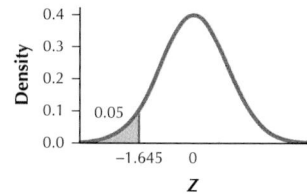

(c) Reject H_0 if $Z_{data} \leq -1.645$.

28. **(a)** $Z_{crit} = -2.33$

(b)

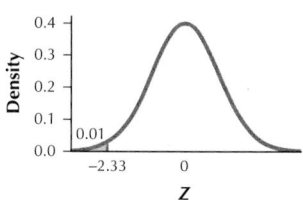

(c) Reject H_0 if $Z_{data} \leq -2.33$.

29. **(a)** $Z_{crit} = -1.28$

(b)

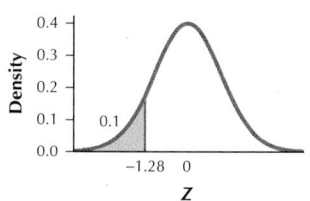

(c) Reject H_0 if $Z_{data} \leq -1.28$.

30. **(a)** $Z_{crit} = -1.645$

(b)

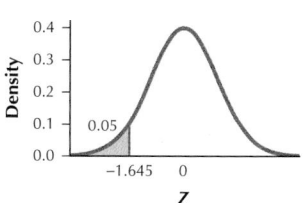

(c) Reject H_0 if $Z_{data} \leq -1.645$.

31. **(a)** $Z_{crit} = 1.96$

(b)

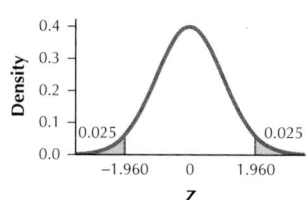

(c) Reject H_0 if $Z_{data} \leq -1.96$ or $Z_{data} \geq 1.96$.

32. **(a)** $Z_{crit} = 2.58$

(b)

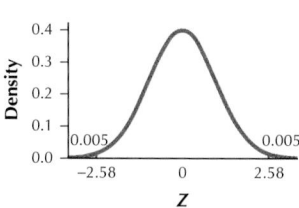

(c) Reject H_0 if $Z_{data} \leq -2.58$ or $Z_{data} \geq 2.58$.

33. **(a)** $Z_{crit} = 1.645$

(b)

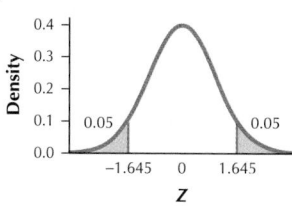

(c) Reject H_0 if $Z_{data} \leq -1.645$ or $Z_{data} \geq 1.645$.

34. **(a)** $Z_{crit} = 1.96$

(b)

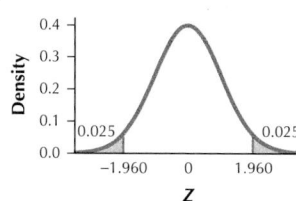

(c) Reject H_0 if $Z_{data} \leq -1.96$ or $Z_{data} \geq 1.96$.

35. **(a)** Z_{crit} increases. **(b)** The critical region decreases.

36. **(a)** Z_{crit} decreases. **(b)** The critical region decreases.

37. **(a)** $H_0 : \mu = 75$ vs. $H_a : \mu > 75$ **(b)** $Z_{crit} = 1.28$. Reject H_0 if $Z_{data} \geq 1.28$. **(c)** $Z_{data} = 1$ **(d)** $Z_{data} = 1$ is not ≥ 1.28. Therefore we do not reject H_0. There is insufficient evidence at level of significance $\alpha = 0.10$ that the population mean is greater than 75.

38. **(a)** $H_0 : \mu = 75$ vs. $H_a : \mu > 75$ **(b)** $Z_{crit} = 1.645$. Reject H_0 if $Z_{data} \geq 1.645$. **(c)** $Z_{data} = 1.5$ **(d)** $Z_{data} = 1.5$ is not ≥ 1.645. Therefore we do not reject H_0. There is insufficient evidence at level of significance $\alpha = 0.05$ that the population mean is greater than 75.

39. **(a)** $H_0 : \mu = 50$ vs. $H_a : \mu > 50$ **(b)** $Z_{crit} = 1.645$. Reject H_0 if $Z_{data} \geq 1.645$. **(c)** $Z_{data} = 1$ **(d)** $Z_{data} = 1$ is not ≥ 1.645. Therefore we do not reject H_0. There is insufficient evidence at level of significance $\alpha = 0.05$ that the population mean is greater than 50.

40. **(a)** $H_0 : \mu = 50$ vs. $H_a : \mu > 50$ **(b)** $Z_{crit} = 2.33$. Reject H_0 if $Z_{data} \geq 2.33$. **(c)** $Z_{data} = 4$ **(d)** $Z_{data} = 4$ is ≥ 2.33. Therefore we reject H_0. There is evidence at level of significance $\alpha = 0.01$ that the population mean is greater than 50.

41. **(a)** $H_0 : \mu = 98.6$ vs. $H_a : \mu < 98.6$ **(b)** $Z_{crit} = -1.645$. Reject H_0 if $Z_{data} \leq -1.645$. **(c)** $Z_{data} = -1$ **(d)** $Z_{data} = -1$ is not ≤ -1.645. Therefore we do not reject H_0. There is insufficient evidence at level of significance $\alpha = 0.05$ that the population mean is less than 98.6.

42. **(a)** $H_0 : \mu = 98.6$ vs. $H_a : \mu < 98.6$ **(b)** $Z_{crit} = -2.33$. Reject H_0 if $Z_{data} \leq -2.33$. **(c)** $Z_{data} = -3$ **(d)** $Z_{data} = -3$ is ≤ -2.33. Therefore we reject H_0. There is evidence at level of significance $\alpha = 0.01$ that the population mean is less than 98.6.

43. **(a)** $H_0 : \mu = 20$ vs. $H_a : \mu < 20$ **(b)** $Z_{crit} = -1.28$. Reject H_0 if $Z_{data} \leq -1.28$. **(c)** $Z_{data} = -2$ **(d)** $Z_{data} = -2$ is ≤ -1.28. Therefore we reject H_0. There is evidence at level of significance $\alpha = 0.10$ that the population mean is less than 20.

44. **(a)** $H_0 : \mu = 20$ vs. $H_a : \mu < 20$ **(b)** $Z_{crit} = -1.645$. Reject H_0 if $Z_{data} \leq -1.645$. **(c)** $Z_{data} = -3$ **(d)** $Z_{data} = -3$ is ≤ -1.645. Therefore we reject H_0. There is evidence at level of significance $\alpha = 0.05$ that the population mean is less than 20.

45. **(a)** $H_0 : \mu = 1000$ vs. $H_a : \mu \neq 1000$ **(b)** $Z_{crit} = 1.96$. Reject H_0 if $Z_{data} \leq -1.96$ or $Z_{data} \geq 1.96$. **(c)** $Z_{data} = 1$ **(d)** $Z_{data} = 1$ is not ≤ -1.96 and not ≥ 1.96. Therefore we do not reject H_0. There is insufficient evidence at level of significance $\alpha = 0.05$ that the population mean differs from 1000.

46. **(a)** $H_0 : \mu = 1000$ vs. $H_a : \mu \neq 1000$ **(b)** $Z_{crit} = 2.58$. Reject H_0 if $Z_{data} \leq -2.58$ or $Z_{data} \geq 2.58$. **(c)** $Z_{data} = -1$ **(d)** $Z_{data} = -1$ is not ≤ -2.58 and not ≥ 2.58. Therefore we do not reject H_0. There is insufficient evidence at level of significance $\alpha = 0.01$ that the population mean differs from 1000.

47. **(a)** $H_0 : \mu = 2.5$ vs. $H_a : \mu \neq 2.5$ **(b)** $Z_{crit} = 1.645$. Reject H_0 if $Z_{data} \leq -1.645$ or $Z_{data} \geq 1.645$. **(c)** $Z_{data} = -1$ **(d)** $Z_{data} = -1$ is not ≤ -1.645 and not ≥ 1.645. Therefore we do not reject H_0. There is insufficient evidence at level of significance $\alpha = 0.10$ that the population mean differs from 2.5.

48. **(a)** $H_0 : \mu = 2.5$ vs. $H_a : \mu \neq 2.5$ **(b)** $Z_{crit} = 1.96$. Reject H_0 if $Z_{data} \leq -1.96$ or $Z_{data} \geq 1.96$. **(c)** $Z_{data} = 1$ **(d)** $Z_{data} = 1$ is not ≤ -1.96 and

not ≥ 1.96. Therefore we do not reject H_0. There is insufficient evidence at level of significance $\alpha = 0.05$ that the population mean differs from 2.5.
49. **(a)** $H_0 : \mu = 80$ vs. $H_a : \mu > 80$ **(b)** $Z_{crit} = 1.645$. Reject H_0 if $Z_{data} \geq 1.645$. **(c)** $Z_{data} = 1$.

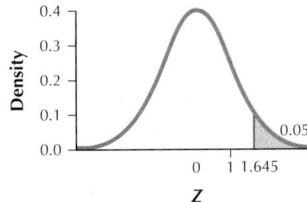

(d) Since $Z_{data} = 1$ is not ≥ 1.645, the conclusion is do not reject H_0. There is insufficient evidence at the 0.05 level of significance that the population mean number of connections to community pages, groups, and events is greater than 80.
50. **(a)** $H_0 : \mu = 18$ vs. $H_a : \mu > 18$ **(b)** $Z_{crit} = 1.645$. Reject H_0 if $Z_{data} \geq 1.645$. **(c)** $Z_{data} = 1$

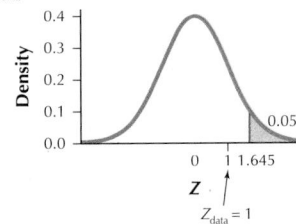

(d) $Z_{data} = 1$ is not ≥ 1.645. Therefore we do not reject H_0. There is insufficient evidence at level of significance $\alpha = 0.05$ that the population mean number of hours that nursing majors study per week is greater than 18.
51. **(a)** $H_0 : \mu = 60$ vs. $H_a : \mu \neq 60$ **(b)** $Z_{crit} = 2.58$. Reject H_0 if $Z_{data} \leq -2.58$ or $Z_{data} \geq 2.58$ **(c)** $Z_{data} = 3$

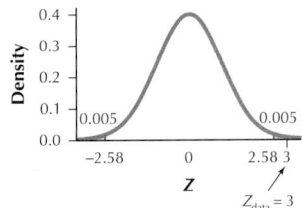

(d) $Z_{data} = 3$ is ≥ 2.58. Therefore we reject H_0. There is evidence at level of significance $\alpha = 0.01$ that the population mean number of text messages sent per day for young people ages $12-17$ differs from 60.
52 **(a)** $H_0 : \mu = 35$ vs. $H_a : \mu < 35$ **(b)** $Z_{crit} = -1.645$. Reject H_0 if $Z_{data} \leq -1.645$. **(c)** $Z_{data} = -1$

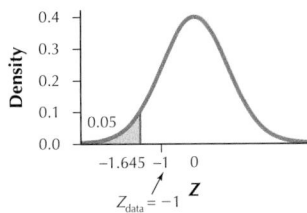

(d) $Z_{data} = -1$ is not ≤ -1.645. Therefore we do not reject H_0. There is insufficient evidence at level of significance $\alpha = 0.05$ that the population mean age of video gamers is less than 35.
53. **(a)** $H_0 : \mu = 3.70$ vs. $H_a : \mu > 3.70$ **(b)** $Z_{crit} = 1.645$. Reject H_0 if $Z_{data} \geq 1.645$. **(c)** $Z_{data} = 2$.

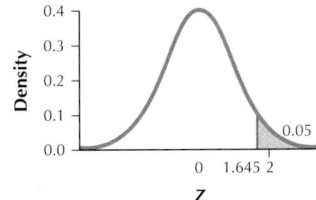

(d) Since $Z_{data} = 2$ is ≥ 1.645, the conclusion is reject H_0. There is evidence at the 0.05 level of significance that the population mean price of regular gasoline is greater than \$3.70 per gallon. Therefore we can conclude at the 0.05 level of significance that the population mean price for a gallon of regular gasoline has risen since June 2011.
54. **(a)** $H_0 : \mu = 93$ vs. $H_a : \mu \neq 93$ **(b)** $Z_{crit} = 1.645$. Reject H_0 if $Z_{data} \leq -1.645$ or $Z_{data} \geq 1.645$. **(c)** $Z_{data} = 2$

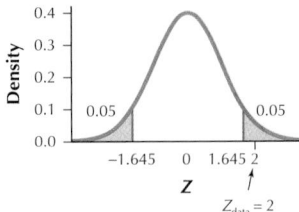

(d) $Z_{data} = 2$ is ≥ 1.645. Therefore we reject H_0. There is evidence at level of significance $\alpha = 0.10$ that the population mean amount of time smartphone owners spend using shopping apps on their smartphones differs from 93 minutes.
55. **(a)** $H_0 : \mu = 175$ vs. $H_a : \mu \neq 175$ **(b)** $Z_{crit} = 1.645$. Reject H_0 if $Z_{data} \leq -1.645$ or if $Z_{data} \geq 1.645$. **(c)** $Z_{data} = 8$.

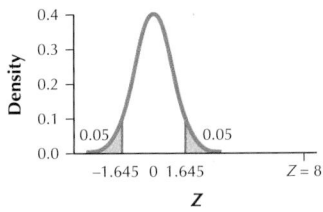

(d) Since $Z_{data} = 8$ is ≥ 1.645, the conclusion is reject H_0. There is evidence at the 0.10 level of significance that the population mean height of Americans has changed from 175 centimeters.
56. **(a)** $H_0 : \mu = 3.34$ vs. $H_a : \mu > 3.34$ **(b)** $Z_{crit} = 1.645$. Reject H_0 if $Z_{data} \geq 1.645$. **(c)** $Z_{data} = 2$.

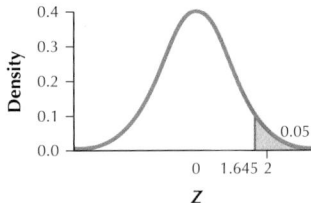

(d) Since $Z_{data} = 2$ is ≥ 1.645, the conclusion is reject H_0. There is evidence at the 0.05 level of significance that the population mean price for milk this year is greater than \$3.34 per gallon. Therefore we can conclude that the population mean price of milk this year has increased from the 2011 value.
57. **(a)** Since the sample size of $n = 100$ is large ($n \geq 30$), Case 2 applies, so it is appropriate to apply the Z test. **(b)** Even though the sample mean $\bar{x} = 6.2$ cents per mile is greater than the hypothesized mean $\mu_0 = 5.9$ cents per mile, this is not enough by itself to reject the null hypothesis. It also depends on the variability of the data and on α. **(c)** Since $\bar{x} = 6.2$ cents per mile is 2 standard errors above $\mu_0 = 5.9$ cents per mile, it is mildly extreme.
58. **(a)** $H_0 : \mu = 5.9$ versus $H_a : \mu > 5.9$ **(b)** $Z_{crit} = 1.645$ Reject H_0 if $Z_{data} \geq 1.645$ **(c)** $Z_{data} = 2$ **(d)** Since $Z_{data} = 2$ is ≥ 1.645, we reject H_0. There is evidence at the 0.05 level of significance that the population mean cost of operating a vehicle in the United States is greater than 5.9 cents per mile.
59. The histogram indicates that the data are extremely right-skewed and therefore not normally distributed. Thus Case 1 does not apply. Since the sample size of $n = 16$ is small ($n < 30$), Case 2 does not apply. Thus it is not appropriate to apply the Z test.

60.

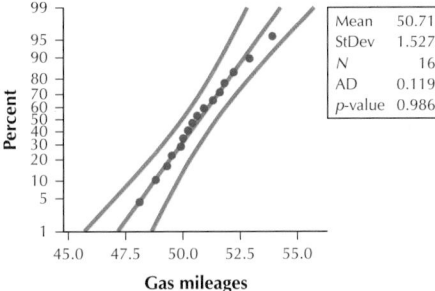

Mean	50.71
StDev	1.527
N	16
AD	0.119
p-value	0.986

Yes. Acceptable normality and σ is known.

61. $H_0 : \mu = 50$ vs. $H_a : \mu > 50$ where μ is the population mean city/highway combined gas mileage for the Toyota Prius hybrid car.

62. $Z_{crit} = 1.28$. Reject H_0 if $Z_{data} \geq 1.28$.

63. $Z_{data} = 1.425$.

64. $Z_{data} = 1.425$ is ≥ 1.28. Therefore we reject H_0. There is evidence at level of significance $\alpha = 0.10$ that the population mean city/highway combined gas mileage for the Toyota Prius hybrid car is greater than 50 mpg.

65. (a) Increase **(b)** Stay the same **(c)** Stay the same **(d)** Increase **(e)** Stay the same **(f)** Stay the same

66. $Z_{crit} = 1.645$. Reject H_0 if $Z_{data} \geq 1.645$. $Z_{data} = 1.425$ is not ≥ 1.645. Therefore we do not reject H_0. There is insufficient evidence at level of significance $\alpha = 0.05$ that the population mean city/highway combined gas mileage for the Toyota Prius hybrid car is greater than 50 mpg.

67. Turn to a direct assessment of the strength of evidence against the null hypothesis or obtain more data.

68. (a) Graphs will vary

```
Descriptive Statistics: SODIUM

Variable     N     N*     Mean    SE Mean   StDev   Minimum    Q1
SODIUM      961     0     320.4    20.2     625.7     0.0      7.0

          Median     Q3    Maximum
           75.0    313.5   8142.0
```

(b)

```
One-Sample Z: SODIUM

Test of mu = 280 vs > 280
The assumed standard deviation = 625

Variable     N    Mean   StDev   SE Mean   95% Lower Bound
SODIUM      961   320.4  625.7    20.2          287.3

              Z      P
            2.01   0.022
```

$H_0 : \mu = 280$ vs. $H_a : \mu > 280$. $Z_{crit} = 1.645$. Reject H_0 if $Z_{data} \geq 1.645$. $Z_{data} = 2.01$. Since $Z_{data} = 2.01$ is ≥ 1.645, the conclusion is reject H_0. There is evidence at the 0.05 level of significance that the population mean amount of sodium is greater than 280 mg.

(c)

```
One-Sample Z: SODIUM

Test of mu = 290 vs > 290
The assumed standard deviation = 625

Variable     N    Mean   StDev   SE Mean   95% Lower Bound
SODIUM      961   320.4  625.7    20.2          287.3

              Z      P
            1.51   0.065
```

$H_0 : \mu = 290$ vs. $H_a : \mu > 290$. $Z_{crit} = 1.645$. Reject H_0 if $Z_{data} \geq 1.645$. $Z_{data} = 1.51$. Since $Z_{data} = 1.51$ is not ≥ 1.645, the conclusion is do not reject H_0. There is insufficient evidence at the 0.05 level of significance that the population mean amount of sodium is greater than 290 mg.

69. Answers will vary.

70. $H_0 : \mu = 0.75$ versus $H_a : \mu \neq 0.75$; $Z_{crit} = 1.96$. Reject H_0 if $Z_{data} \leq -1.96$ or $Z_{data} \geq 1.96$. The answers to the rest of the hypothesis test will vary.

71. 0.7724 coupon

Section 9.3

1. False

2. Reject H_0 if the p-value $\leq \alpha$, the level of significance. Otherwise, do not reject H_0.

3. It gives us extra information about whether H_0 was barely rejected or not rejected or whether it was a no-brainer decision to reject or not reject H_0.

4. If a certain value for μ_0 lies outside the corresponding $100(1 - \alpha)\%$ Z confidence interval for μ, then the null hypothesis specifying this value for μ_0 would be rejected for level of significance α. Alternatively, if a certain value for μ_0 lies inside the corresponding $100(1 - \alpha)\%$ Z confidence interval for μ, then the null hypothesis specifying this value for μ_0 would not be rejected for level of significance α.

5. False

6. (a) Not a probability **(b)** A probability **(c)** A probability

7. 0.0668

8. 0.0062

9. 0.2743

10. 0.1151

11. 0.2420

12. 0.1020

13. 0.0129

14. 0.0054

15. 0.5222

16. 0.5222

17. 1

18. It decreases.

19. It decreases.

20. They are the same.

21. (a) $H_0 : \mu = 3.14$ vs. $H_a : \mu > 3.14$. We will reject H_0 if the p-value $\leq \alpha = 0.05$. **(b)** $Z_{data} = 0.6$ **(c)** p-value $= 0.2743$ **(d)** The p-value $= 0.2743$ is not ≤ 0.05. Therefore we do not reject H_0. There is insufficient evidence at level of significance $\alpha = 0.05$ that the population mean is greater than 3.14.

22. (a) $H_0 : \mu = 30$ vs. $H_a : \mu < 30$. We will reject H_0 if the p-value $\leq \alpha = 0.05$. **(b)** $Z_{data} = -2$ **(c)** p-value $= 0.0228$ **(d)** The p-value $= 0.0228$ is ≤ 0.05. Therefore we reject H_0. There is evidence at level of significance $\alpha = 0.05$ that the population mean is less than 30.

23. (a) $H_0 : \mu = -1.0$ vs. $H_a : \mu > -1.0$ We will reject H_0 if the p-value $\leq \alpha = 0.05$. **(b)** $Z_{data} = 20$ **(c)** p-value ≈ 0 **(d)** The p-value ≈ 0 is ≤ 0.05. Therefore we reject H_0. There is evidence at level of significance $\alpha = 0.05$ that the population mean is greater than -1.0.

24. (a) $H_0 : \mu = 2000$ vs. $H_a : \mu > 2000$. We will reject H_0 if the p-value $\leq \alpha = 0.05$. **(b)** $Z_{data} = 1.25$ **(c)** p-value $= 0.1056$ **(d)** The p-value $= 0.1056$ is not ≤ 0.05. Therefore we do not reject H_0. There is insufficient evidence at level of significance $\alpha = 0.05$ that the population mean is greater than 2000.

25. (a) $H_0 : \mu = 500$ vs. $H_a : \mu < 500$. We will reject H_0 if the p-value $\leq \alpha = 0.05$. **(b)** $Z_{data} = -2$ **(c)** p-value $= 0.0228$ **(d)** The p-value $= 0.0228$ is ≤ 0.05. Therefore we reject H_0. There is evidence at level of significance $\alpha = 0.05$ that the population mean is less than 500.

26. (a) $H_0 : \mu = -32$ vs. $H_a : \mu > -32$. We will reject H_0 if the p-value $\leq \alpha = 0.05$. **(b)** $Z_{data} = 1$ **(c)** p-value $= 0.1587$ **(d)** The p-value $= 0.1587$ is not ≤ 0.05. Therefore we do not reject H_0. There is insufficient evidence at level of significance $\alpha = 0.05$ that the population mean is greater than -32.

27. (a) $H_0 : \mu = 10$ vs. $H_a : \mu \neq 10$. We will reject H_0 if the p-value $\leq \alpha = 0.05$. **(b)** $Z_{data} = 0$ **(c)** p-value $= 1$ **(d)** The p-value $= 1$ is not ≤ 0.05. Therefore we do not reject H_0. There is insufficient evidence at level of significance $\alpha = 0.05$ that the population mean is not equal to 10.

28. (a) $H_0 : \mu = -5$ vs. $H_a : \mu \neq -5$. We will reject H_0 if the p-value $\leq \alpha = 0.05$. **(b)** $Z_{data} = 0$ **(c)** p-value $= 1$ **(d)** The p-value $= 1$ is not ≤ 0.05. Therefore we do not reject H_0. There is insufficient evidence at level of significance $\alpha = 0.05$ that the population mean is not equal to -5.

29. (a) $H_0 : \mu = 0$ vs. $H_a : \mu \neq 0$. We will reject H_0 if the p-value $\leq \alpha = 0.05$.

(b) $Z_{\text{data}} = -2.7$ **(c)** p-value $= 0.0070$ **(d)** The p-value $= 0.0070$ is ≤ 0.05. Therefore we reject H_0. There is evidence at level of significance $\alpha = 0.05$ that the population mean is not equal to 0.

30. (a) $H_0 : \mu = 46$ vs. $H_a : \mu \neq 46$. We will reject H_0 if the p-value $\leq \alpha = 0.05$. **(b)** $Z_{\text{data}} = 1$ **(c)** p-value $= 0.3174$ **(d)** The p-value $= 0.3174$ is not ≤ 0.05. Therefore we do not reject H_0. There is insufficient evidence at level of significance $\alpha = 0.05$ that the population mean is not equal to 46.

31. The p-value of 0.2743 implies that there is no evidence against the null hypothesis that the population mean equals 3.14.

32. The p-value of 0.0228 implies that there is solid evidence against the null hypothesis that the population mean equals 30.

33. The p-value of approximately 0 implies that there is extremely strong evidence against the null hypothesis that the population mean equals -1.0.

34. The p-value of 0.1056 implies that there is slight evidence against the null hypothesis that the population mean equals 2000.

35. The p-value of 0.0228 implies that there is solid evidence against the null hypothesis that the population mean equals 500.

36. The p-value of 0.1587 implies that there is no evidence against the null hypothesis that the population mean equals -32.

37. The p-value of 1 implies that there is no evidence against the null hypothesis that the population mean equals 10.

38. The p-value of 1 implies that there is no evidence against the null hypothesis that the population mean equals -5.

39. The p-value of 0.007 implies that there is very strong evidence against the null hypothesis that the population mean equals 0.

40. The p-value of 0.3174 implies that there is no evidence against the null hypothesis that the population mean equals 46.

41.

	Value of μ_0	Form of hypothesis test, with $\alpha = 0.05$	Where μ_0 lies in relation to 95% confidence interval	Conclusion of hypothesis test
a.	-3	$H_0 : \mu = -3$ vs. $H_a : \mu \neq -3$	Outside	Reject H_0
b.	-2	$H_0 : \mu = -2$ vs. $H_a : \mu \neq -2$	Inside	Do not reject H_0
c.	0	$H_0 : \mu = 0$ vs. $H_a : \mu \neq 0$	Inside	Do not reject H_0
d.	5	$H_0 : \mu = 5$ vs. $H_a : \mu \neq 5$	Inside	Do not reject H_0
e.	7	$H_0 : \mu = 7$ vs. $H_a : \mu \neq 7$	Outside	Reject H_0

42.

	Value of μ_0	Form of hypothesis test, with $\alpha = 0.01$	Where μ_0 lies in relation to 99% confidence interval	Conclusion of hypothesis test
a.	0	$H_0 : \mu = 0$ vs. $H_a : \mu \neq 0$	Outside	Reject H_0
b.	44	$H_0 : \mu = 44$ vs. $H_a : \mu \neq 44$	Outside	Reject H_0
c.	50	$H_0 : \mu = 50$ vs. $H_a : \mu \neq 50$	Inside	Do not reject H_0
d.	54	$H_0 : \mu = 54$ vs. $H_a : \mu \neq 54$	Inside	Do not reject H_0
e.	56	$H_0 : \mu = 56$ vs. $H_a : \mu \neq 56$	Outside	Reject H_0

43.

	Value of μ_0	Form of hypothesis test, with $\alpha = 0.10$	Where μ_0 lies in relation to 90% confidence interval	Conclusion of hypothesis test
a.	-3	$H_0 : \mu = -3$ vs. $H_a : \mu \neq -3$	Outside	Reject H_0
b.	-8	$H_0 : \mu = -8$ vs. $H_a : \mu \neq -8$	Inside	Do not reject H_0
c.	-11	$H_0 : \mu = -11$ vs. $H_a : \mu \neq -11$	Outside	Reject H_0
d.	0	$H_0 : \mu = 0$ vs. $H_a : \mu \neq 0$	Outside	Reject H_0
e.	7	$H_0 : \mu = 7$ vs. $H_a : \mu \neq 7$	Outside	Reject H_0

44.

	Value of μ_0	Form of hypothesis test, with $\alpha = 0.05$	Where μ_0 lies in relation to 95% confidence interval	Conclusion of hypothesis test
a.	1000	$H_0 : \mu = 1000$ vs. $H_a : \mu \neq 1000$	Outside	Reject H_0
b.	2000	$H_0 : \mu = 2000$ vs. $H_a : \mu \neq 2000$	Inside	Do not reject H_0
c.	3000	$H_0 : \mu = 3000$ vs. $H_a : \mu \neq 3000$	Outside	Reject H_0
d.	0	$H_0 : \mu = 0$ vs. $H_a : \mu \neq 0$	Outside	Reject H_0
e.	1025	$H_0 : \mu = 1025$ vs. $H_a : \mu \neq 1025$	Inside	Do not reject H_0

45.

	Value of μ_0	Form of hypothesis test, with $\alpha = 0.05$	Where μ_0 lies in relation to 95% confidence interval	Conclusion of hypothesis test
a.	1.5	$H_0 : \mu = 1.5$ vs. $H_a : \mu \neq 1.5$	Outside	Reject H_0
b.	-1	$H_0 : \mu = -1$ vs. $H_a : \mu \neq -1$	Outside	Reject H_0
c.	0.5	$H_0 : \mu = 0.5$ vs. $H_a : \mu \neq 0.5$	Inside	Do not reject H_0
d.	0.9	$H_0 : \mu = 0.9$ vs. $H_a : \mu \neq 0.9$	Inside	Do not reject H_0
e.	1.2	$H_0 : \mu = 1.2$ vs. $H_a : \mu \neq 1.2$	Outside	Reject H_0

46.

	Value of μ_0	Form of hypothesis test, with $\alpha = 0.05$	Where μ_0 lies in relation to 95% confidence interval	Conclusion of hypothesis test
a.	1.3	$H_0 : \mu = 1.3$ vs. $H_a : \mu \neq 1.3$	Outside	Reject H_0
b.	1.35	$H_0 : \mu = 1.35$ vs. $H_a : \mu \neq 1.35$	Inside	Do not reject H_0
c.	1.4	$H_0 : \mu = 1.4$ vs. $H_a : \mu \neq 1.4$	Inside	Do not reject H_0
d.	1.45	$H_0 : \mu = 1.45$ vs. $H_a : \mu \neq 1.45$	Outside	Reject H_0
e.	1.3275	$H_0 : \mu = 1.3275$ vs. $H_a : \mu \neq 1.3275$	Outside	Reject H_0

47. *Step 1* $H_0 : \mu < 75$ vs. $H_a : \mu < 75$
We will reject H_0 if the p-value $\leq \alpha = 0.05$.
Step 2 $Z_{\text{data}} = -1.6$
Step 3 p-value $= 0.0547992894$
Step 4 The p-value $= 0.0547992894$ is not ≤ 0.05. Therefore we do not reject H_0. There is insufficient evidence at level of significance $\alpha = 0.05$ that the population mean is less than 75.

48. *Step 1* $H_0 : \mu = 2.5$ vs. $H_a : \mu \neq 2.5$
We will reject H_0 if the p-value $\leq \alpha = 0.05$.
Step 2 $Z_{\text{data}} = 2$
Step 3 p-value $= 0.04550124$
Step 4 The p-value $= 0.04550124$ is ≤ 0.05. Therefore we reject H_0. There is evidence at level of significance $\alpha = 0.05$ that the population mean is not equal to 2.5.

49. *Step 1* $H_0 : \mu = 70$ vs. $H_a : \mu > 70$
We will reject H_0 if the p-value $\leq \alpha = 0.05$.
Step 2 $Z_{\text{data}} = 0.55$
Step 3 p-value $= 0.290$
Step 4 The p-value $= 0.290$ is not ≤ 0.05. Therefore we do not reject H_0. There is insufficient evidence at level of significance $\alpha = 0.05$ that the population mean is greater than 70.

50. *Step 1* $H_0 : \mu = 78$ vs. $H_a : \mu < 78$
We will reject H_0 if the p-value $\leq \alpha = 0.05$.
Step 2 $Z_{\text{data}} = -2.15$
Step 3 p-value $= 0.016$
Step 4 The p-value $= 0.016$ is ≤ 0.05. Therefore we reject H_0. There is evidence at level of significance $\alpha = 0.05$ that the population mean is less than 78.

51. (a) $H_0 : \mu = 3000$ vs. $H_a : \mu > 3000$. We will reject H_0 if the p-value $\leq \alpha = 0.10$. (b) $Z_{data} = 0.6$ (c) p-value $= 0.2743$ (d) The p-value $= 0.2743$ is not ≤ 0.10. Therefore we do not reject H_0. There is insufficient evidence at level of significance $\alpha = 0.10$ that the population mean annual car insurance premium for a Porsche Panamera Turbo-S has increased from $3000.

52. (a) $H_0 : \mu = 37$ vs. $H_a : \mu > 37$. We will reject H_0 if the p-value $\leq \alpha = 0.05$. (b) $Z_{data} = 1.5$ (c) p-value $= 0.0668$ (d) The p-value $= 0.0668$ is not ≤ 0.05. Therefore we do not reject H_0. There is insufficient evidence at level of significance $\alpha = 0.05$ that the population mean number of hours using mobile apps by young people ages $18-24$ has increased from 37 hours.

53. (a) $H_0 : \mu = 700$ vs. $H_a : \mu < 700$. Reject H_0 if the p-value ≤ 0.10. (b) -20 (c) ≈ 0 (d) Since the p-value $\leq \alpha$, reject H_0. There is evidence that the population mean number of meals prepared and eaten at home is less than 700.

54. (a) $H_0 : \mu = 47.2$ vs. $H_a : \mu > 47.2$. Reject H_0 if the p-value ≤ 0.01. (b) 16.60 (c) p-value ≈ 0. (d) Since the p-value $\leq \alpha$, reject H_0. There is evidence that the population mean DDT level in the breast milk of Hispanic women in the Yakima valley is greater than 47.2 parts per billion.

55. (a) $H_0 : \mu = 45{,}500$ vs. $H_a : \mu > 45{,}500$. We will reject H_0 if the p-value $\leq \alpha = 0.05$. (b) $Z_{data} = 1.5$ (c) p-value $= 0.0668$ (d) The p-value $= 0.0668$ is not ≤ 0.05. Therefore we do not reject H_0. There is insufficient evidence at level of significance $\alpha = 0.05$ that the population mean annual income of Millennials with Bachelor's degrees working full time has increased from $45,500.

56. (a) $H_0 : \mu = 1.4261$ vs. $H_a : \mu < 1.4261$. Reject H_0 if the p-value ≤ 0.05. (b) -17.32 (c) ≈ 0 (d) Since the p-value $\leq \alpha$, reject H_0. There is evidence that the population mean annual ring growth in the tree's later years is less than 1.4261 mm per year.

57. (a) $H_0 : \mu = 3$ vs. $H_a : \mu < 3$. Reject H_0 if the p-value ≤ 0.01. (b) -13.5 (c) p-value ≈ 0 (d) Since the p-value $\leq \alpha$, reject H_0. There is evidence that the population mean time hybrid cars take to recoup their initial cost is less than 3 years.

58. (a) $H_0 : \mu = 175$ vs. $H_a : \mu \neq 3$. Reject H_0 if the p-value ≤ 0.01. (b) -1 (c) 0.3174 (d) Since the p-value is not less than or equal to $\alpha = 0.01$, do not reject H_0. There is insufficient evidence that the population mean height of Americans this year has changed from 175 centimeters.

59. The p-value of 0.2743 implies that there is no evidence against the null hypothesis that the population mean equals $3000.

60. The p-value of 0.0668 implies that there is moderate evidence against the null hypothesis that the population mean equals 37 hours.

61. The p-value of approximately 0 implies that there is extremely strong evidence against the null hypothesis that the population mean equals 700 meals.

62. The p-value of approximately 0 implies that there is extremely strong evidence against the null hypothesis that the population mean equals 47.2 parts per billion.

63. The p-value of 0.0668 implies that there is moderate evidence against the null hypothesis that the population mean equals $45,500.

64. The p-value of approximately 0 implies that there is extremely strong evidence against the null hypothesis that the population mean equals 1.4261 millimeters.

65. The p-value of approximately 0 implies that there is extremely strong evidence against the null hypothesis that the population mean equals 3 years.

66. The p-value of 0.3174 implies that there is no evidence against the null hypothesis that the population mean equals 175 centimeters.

67. (a) (2.89, 3.01)

(b)

	Value of μ_0	Form of hypothesis test, with $\alpha = 0.05$	Where μ_0 lies in relation to 95% confidence interval	Conclusion of hypothesis test
(i)	2.89	$H_0 : \mu = 2.89$ vs. $H_a : \mu \neq 2.89$	Outside	Reject H_0
(ii)	2.90	$H_0 : \mu = 2.90$ vs. $H_a : \mu \neq 2.90$	Inside	Do not reject H_0
(iii)	3.00	$H_0 : \mu = 3.00$ vs. $H_a : \mu \neq 3.00$	Inside	Do not reject H_0
(iv)	3.01	$H_0 : \mu = 3.01$ vs. $H_a : \mu \neq 3.01$	Outside	Reject H_0

68. (a) (58,300, 66,900)

(b)

	Value of μ_0	Form of hypothesis test, with $\alpha = 0.01$	Where μ_0 lies in relation to 99% confidence interval	Conclusion of hypothesis test
(i)	$66,000	$H_0 : \mu = 66{,}000$ vs. $H_a : \mu \neq 66{,}000$	Inside	Do not reject H_0
(ii)	$58,000	$H_0 : \mu = \$58{,}000$ vs. $H_a : \mu \neq \$58{,}000$	Outside	Reject H_0
(iii)	$67,000	$H_0 : \mu = 67{,}000$ vs. $H_a : \mu \neq 67{,}000$	Outside	Reject H_0
(iv)	$59,000	$H_0 : \mu = 59{,}000$ vs. $H_a : \mu \neq 59{,}000$	Inside	Do not reject H_0

69. $H_0 : \mu = 16{,}351$ vs. $H_a : \mu > 16{,}351$. We will reject H_0 if the p-value $\leq \alpha = 0.05$. $Z_{data} = 1.8$. p-value $= 0.0359$. The p-value $= 0.0359$ is ≤ 0.05. Therefore we reject H_0. There is evidence at level of significance $\alpha = 0.05$ that the population mean annual premium for employer-sponsored family health insurance coverage has increased from $16,351.

70. (a) Remain the same (b) Remain the same (c) Remain the same (d) Increase (e) Remain the same.

71. $H_0 : \mu = 16{,}351$ vs. $H_a : \mu > 16{,}351$. We will reject H_0 if the p-value $\leq \alpha = 0.01$. $Z_{data} = 1.8$. p-value $= 0.0359$. The p-value $= 0.0359$ is not ≤ 0.01. Therefore we do not reject H_0. There is insufficient evidence at level of significance $\alpha = 0.01$ that the population mean annual premium for employer-sponsored family health insurance coverage has increased from $16,351. Turn to a direct assessment of the strength of evidence against the null hypothesis or obtain more data.

72. $H_0 : \mu = 3.14$ vs. $H_a : \mu < 3.14$. $Z_{data} = -1.35$. The p-value is 0.0885. Since the p-value $= 0.0885$ is not ≤ 0.05, we therefore do not reject H_0. There is insufficient evidence at level of significance $\alpha = 0.05$ that the population mean family size is less than 3.14 persons.

73. (a) 0.0885 (b) Type II error; Type I error (c) This headline is not supported by the data and our hypothesis test.

74. (a) Decrease from -1.35 to -2.1 (b) Decrease from 0.0885 to 0.0179 (c) Since the p-value is less than α, we reject H_0. There is evidence that the true mean family size in America is less than 3.14 persons.

75. (a) Yes, the normal probability plot indicates acceptable normality. (b) $H_0 : \mu = 78$ vs. $H_a : \mu < 78$. Reject H_0 if p-value ≤ 0.05. $Z_{data} = -1.03$. p-value $= 0.1515$. Since p-value $= 0.1515$ is not $\leq \alpha = 0.05$, do not reject H_0. There is insufficient evidence at level of significance $\alpha = 0.05$ that the population mean heart rate for all women is less than 78 beats per minute. (c) $H_0 : \mu = 78$ vs. $H_a : \mu \neq 78$. Reject H_0 if p-value ≤ 0.05. $Z_{data} = -1.03$. p-value $= 0.303$. Since p-value $= 0.303$ is not $\leq \alpha = 0.05$, do not reject H_0. There is insufficient evidence that the population mean heart rate for all women is different from 78 beats per minute.

76. (a) There is insufficient evidence that the true mean heart rates for all women is less than 78 beats per minute and there is insufficient evidence that the true mean heart rate for all woman is different than 78 beats per minute. (b) The p-value for (c) is twice the p-value in (b). If α is between these two p-values, then the conclusion for the one-tailed test will be "Reject H_0" and the conclusion for the two-tailed test will be "Do not reject H_0." (c) There is no evidence against the null hypothesis in (b) and (c).

77. (a) Reject H_0. (b) Do not reject H_0. (c) Reject H_0. (d) Do not reject H_0.

78. Yes, the normal probability plot indicates acceptable normality.

79. (a) $H_0 : \mu = 210$ vs. $H_a : \mu < 210$, where μ refers to the population mean sodium content per serving of breakfast cereal. We will reject H_0 if the p-value $\leq \alpha = 0.01$. (b) $Z_{data} = -1.69$. (c) p-value $= 0.0455$

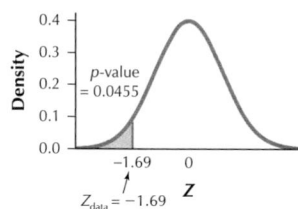

(d) The p-value $= 0.0455$ is not ≤ 0.01. Therefore we do not reject H_0.

There is insufficient evidence at level of significance $\alpha = 0.01$ that the population mean sodium content per serving of breakfast cereal is less than 210 grams.
80. The p-value of 0.0455 implies that there is solid evidence against the null hypothesis that the population mean equals 210 grams.
81. (a) (171.96, 212.82)
(b)

	Value of μ_0	Form of hypothesis test, with $\alpha = 0.05$	Where μ_0 lies in relation to 95% confidence interval	Conclusion of hypothesis test
(i)	200	$H_0: \mu = 200$ vs. $H_a: \mu \neq 200$	Inside	Do not reject H_0
(ii)	215	$H_0: \mu = 215$ vs. $H_a: \mu \neq 215$	Outside	Reject H_0
(iii)	170	$H_0: \mu = 170$ vs. $H_a: \mu \neq 170$	Outside	Reject H_0
(iv)	160	$H_0: \mu = 160$ vs. $H_a: \mu \neq 160$	Outside	Reject H_0

82. (a) Decrease **(b)** Decrease **(c)** Decrease **(d)** Depends on new value of σ.
83. (a) $H_0: \mu = 210$ vs. $H_a: \mu < 210$. Reject H_0 if p-value ≤ 0.05. $Z_{data} = -1.69$. p-value $= 0.0455$. Since p-value $= 0.0455$ is $\leq \alpha = 0.05$, reject H_0. There is evidence that the population mean sodium content per serving of breakfast cereal is less than 210 grams. **(b)** The data have not changed; α was increased to a value greater than the p-value. **(c)** Report the p-value and assess the strength of the evidence against the null hypothesis; obtain more data.
84. 254 observations, 6 variables
85. $\bar{x} = 23,901$, $s = 88,421$. The distribution is right-skewed.
86. Critical-value method: $H_0: \mu = 40,000$ vs. $H_a: \mu \neq 40,000$. For a two-tailed test with level of significance $\alpha = 0.05$, $Z_{crit} = 1.96$. Reject H_0 if $Z_{data} \leq -1.96$ or $Z_{data} \geq 1.96$. From the Minitab output, $Z_{data} = -2.90$. Since $Z_{data} = -2.90$ is ≤ -1.96, we reject H_0. There is evidence that the population mean total occupied housing units for these counties in Texas differs from 40,000. p-value method: $H_0: \mu = 40,000$ vs. $H_a: \mu \neq 40,000$. Reject H_0 if p-value ≤ 0.05. From the Minitab output, $Z_{data} = -2.90$. From the Minitab output, p-value $= 0.004$. Since p-value $= 0.004$ is ≤ 0.05, we reject H_0. There is evidence that the population mean total occupied housing units for these counties in Texas differs from 40,000.
87. Answers will vary.
88. $H_0: \mu = 20$ vs. $H_a: \mu < 20$. We will reject H_0 if the p-value $\leq \alpha = 0.05$. Answers will vary for the rest of the problem.
89. $H_0: \mu = 18$ vs. $H_a: \mu < 18$. We will reject H_0 if the p-value $\leq \alpha = 0.05$. Answers will vary for the rest of the problem.
90. 17.4784 items **(a)** Answers will vary. **(b)** Answers will vary.

Section 9.4
1. The population standard deviation σ is known.
2. Sample standard deviation s
3. (a) $H_0: \mu = 22$ vs. $H_a: \mu < 22$ **(b)** $t_{crit} = -1.697$. Reject H_0 if $t_{data} \leq -1.697$.

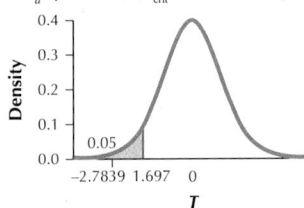

(c) $t_{data} = -2.7839$ **(d)** Since $t_{data} = -2.7839$ is ≤ -1.697, the conclusion is reject H_0. There is evidence at the 0.05 level of significance that the population mean is less than 22.
4. (a) $H_0: \mu = 3$ vs. $H_a: \mu < 3$ **(b)** $t_{crit} = -1.303$. Reject H_0 if $t_{data} \leq -1.303$.

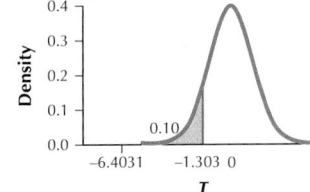

(c) $t_{data} = -6.4031$ **(d)** Since $t_{data} = -6.4031$ is ≤ -1.303, the conclusion is reject H_0. There is evidence at the 0.10 level of significance that the population mean is less than 3.
5. (a) $H_0: \mu = 11$ vs. $H_a: \mu < 11$ **(b)** $t_{crit} = 2.602$. Reject H_0 if $t_{data} \geq 2.602$.

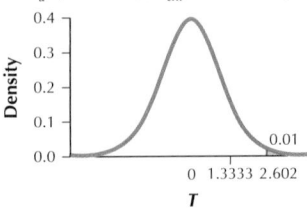

(c) $t_{data} = 1.3333$ **(d)** Since $t_{data} = 1.3333$ is not ≥ 2.602, the conclusion is do not reject H_0. There is insufficient evidence at the 0.01 level of significance that the population mean is greater than 11.
6. (a) $H_0: \mu = 80$ vs. $H_a: \mu > 80$ **(b)** $t_{crit} = 1.860$. Reject H_0 if $t_{data} \geq 1.860$.

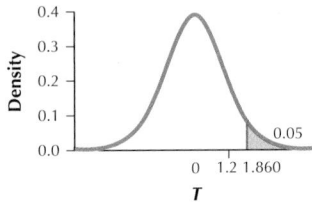

(c) $t_{data} = 1.2$ **(d)** Since $t_{data} = 1.2$ is not ≥ 1.860, the conclusion is do not reject H_0. There is insufficient evidence at the 0.05 level of significance that the population mean is greater than 80.
7. (a) $H_0: \mu = 100$ vs. $H_a: \mu > 100$ **(b)** $t_{crit} = 2.492$. Reject H_0 if $t_{data} \geq 2.492$.

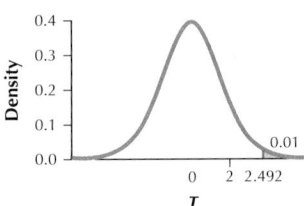

(c) $t_{data} = 2$ **(d)** Since $t_{data} = 2$ is not ≥ 2.492, the conclusion is do not reject H_0. There is insufficient evidence at the 0.01 level of significance that the population mean is greater than 100.
8. (a) $H_0: \mu = -4$ vs. $H_a: \mu < -4$ **(b)** $t_{crit} = -1.662$. Reject H_0 if $t_{data} \leq -1.662$.

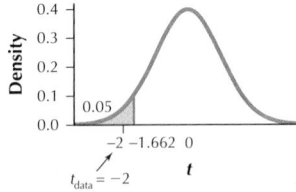

(c) $t_{data} = -2$ **(d)** Since $t_{data} = -2$ is ≤ -1.662, the conclusion is reject H_0. There is evidence at the 0.05 level of significance that the population mean is less than -4.
9. (a) $H_0: \mu = 102$ vs. $H_a: \mu \neq 102$ **(b)** $t_{crit} = 1.990$. Reject H_0 if $t_{data} \leq -1.990$ or if $t_{data} \geq 1.990$.

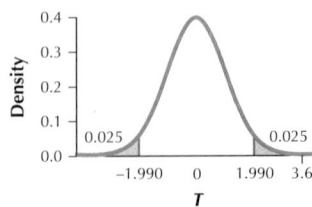

(c) $t_{data} = 3.6$ **(d)** Since $t_{data} = 3.6$ is ≥ 1.990, the conclusion is reject H_0. There is evidence at the 0.05 level of significance that the population mean differs from 102.

10. (a) $H_0 : \mu = 95$ vs. $H_a : \mu \neq 95$ **(b)** $t_{crit} = 2.750$. Reject H_0 if $t_{data} \leq -2.750$ or if $t_{data} \geq 2.750$.

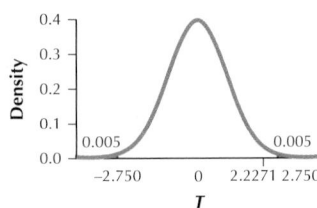

(c) $t_{data} = 2.2271$ **(d)** Since $t_{data} = 2.2271$ is not ≤ -2.750 and not ≥ 2.750, the conclusion is do not reject H_0. There is insufficient evidence at the 0.01 level of significance that the population mean differs from 95.

11. (a) $H_0 : \mu = 1000$ vs. $H_a : \mu \neq 1000$ **(b)** $t_{crit} = 1.711$. Reject H_0 if $t_{data} \leq -1.711$ or if $t_{data} \geq 1.711$.

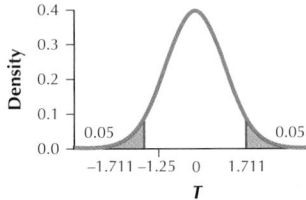

(c) $t_{data} = -1.25$ **(d)** Since $t_{data} = -1.25$ is not ≤ -1.711 and not ≥ 1.711, the conclusion is do not reject H_0. There is insufficient evidence at the 0.10 level of significance that the population mean differs from 1000.

12. (a) $H_0 : \mu = -10$ vs. $H_a : \mu \neq -10$ **(b)** $t_{crit} = 2.064$. Reject H_0 if $t_{data} \leq -2.064$ or if $t_{data} \geq 2.064$.

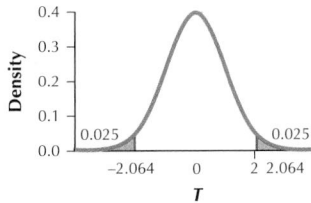

(c) $t_{data} = 2$ **(d)** Since $t_{data} = 2$ is not ≤ -2.064 and not ≥ 2.064, the conclusion is do not reject H_0. There is insufficient evidence at the 0.05 level of significance that the population mean differs from -10.

13. (a) $H_0 : \mu = 9$ vs. $H_a : \mu \neq 9$ **(b)** $t_{crit} = 1.690$. Reject H_0 if $t_{data} \leq -1.690$ or if $t_{data} \geq 1.690$.

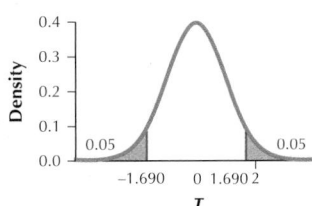

(c) $t_{data} = 2$ **(d)** Since $t_{data} = 2$ is ≥ 1.690, the conclusion is reject H_0. There is evidence at the 0.10 level of significance that the population mean differs from 9.

14. (a) $H_0 : \mu = 1000$ vs. $H_a : \mu \neq 1000$ **(b)** $t_{crit} = 2.947$. Reject H_0 if $t_{data} \leq -2.947$ or if $t_{data} \geq 2.947$.

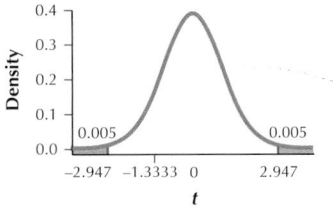

(c) $t_{data} = -1.3333$ **(d)** Since $t_{data} = -1.3333$ is not ≤ -2.947 and not ≥ 2.947, the conclusion is do not reject H_0. There is insufficient

evidence at the 0.01 level of significance that the population mean differs from 1000.

15. (a) $H_0 : \mu = 10$ vs. $H_a : \mu < 10$. Reject H_0 if the p-value $\leq \alpha = 0.01$ **(b)** $t_{data} = -5.4$ **(c)** 0 **(d)** Since the p-value $= 0$ is $\leq \alpha = 0.01$, the conclusion is reject H_0. There is evidence at the 0.01 level of significance that the population mean is less than 10.

16. (a) $H_0 : \mu = 50$ vs. $H_a : \mu < 50$. Reject H_0 if the p-value $\leq \alpha = 0.05$ **(b)** $t_{data} = -6.4031$ **(c)** 0 **(d)** Since the p-value $= 0$ is $\leq \alpha = 0.05$, the conclusion is reject H_0. There is evidence at the 0.05 level of significance that the population mean is less than 50.

17. (a) $H_0 : \mu = 100$ vs. $H_a : \mu > 100$. Reject H_0 if the p-value $\leq \alpha = 0.10$ **(b)** $t_{data} = 2$ **(c)** 0.0285 **(d)** Since the p-value $= 0.0285$ is $\leq \alpha = 0.10$, the conclusion is reject H_0. There is evidence at the 0.10 level of significance that the population mean is greater than 100.

18. (a) $H_0 : \mu = 3.0$ vs. $H_a : \mu > 3.0$. Reject H_0 if the p-value $\leq \alpha = 0.05$ **(b)** $t_{data} = 2$ **(c)** 0.0285 **(d)** Since the p-value $= 0.0285$ is $\leq \alpha = 0.05$, the conclusion is reject H_0. There is evidence at the 0.05 level of significance that the population mean is greater than 3.0.

19. (a) $H_0 : \mu = 200$ vs. $H_a : \mu > 200$. Reject H_0 if the p-value $\leq \alpha = 0.05$ **(b)** $t_{data} = 120$ **(c)** 0 **(d)** Since the p-value $= 0$ is $\leq \alpha = 0.05$, the conclusion is reject H_0. There is evidence at the 0.05 level of significance that the population mean is greater than 200.

20. (a) $H_0 : \mu = 28$ vs. $H_a : \mu < 28$. Reject H_0 if the p-value $\leq \alpha = 0.05$ **(b)** $t_{data} = -1$ **(c)** 0.1599 **(d)** Since the p-value $= 0.1599$ is not $\leq \alpha = 0.05$, the conclusion is do not reject H_0. There is insufficient evidence at the 0.05 level of significance that the population mean is less than 28.

21. (a) $H_0 : \mu = 25$ vs. $H_a : \mu \neq 25$. Reject H_0 if the p-value $\leq \alpha = 0.01$ **(b)** $t_{data} = 0$ **(c)** 1 **(d)** Since the p-value $= 1$ is not $\leq \alpha = 0.01$, the conclusion is do not reject H_0. There is insufficient evidence at the 0.01 level of significance that the population mean differs from 25.

22. (a) $H_0 : \mu = 98.6$ vs. $H_a : \mu \neq 98.6$. Reject H_0 if the p-value $\leq \alpha = 0.05$ **(b)** $t_{data} = 0.36$ **(c)** 0.7198 **(d)** Since the p-value $= 0.7918$ is not $\leq \alpha = 0.05$, the conclusion is do not reject H_0. There is insufficient evidence at the 0.05 level of significance that the population mean differs from 98.6.

23. (a) $H_0 : \mu = 3.14$ vs. $H_a : \mu \neq 3.14$. Reject H_0 if the p-value $\leq \alpha = 0.10$ **(b)** $t_{data} = 0.18$ **(c)** 0.8616 **(d)** Since the p-value $= 0.8616$ is not $\leq \alpha = 0.10$, the conclusion is do not reject H_0. There is insufficient evidence at the 0.10 level of significance that the population mean differs from 3.14.

24. (a) $H_0 : \mu = 2.72$ vs. $H_a : \mu \neq 2.72$. Reject H_0 if the p-value $\leq \alpha = 0.05$ **(b)** $t_{data} = -7.5$ **(c)** 0 **(d)** Since the p-value $= 0$ is $\leq \alpha = 0.05$, the conclusion is reject H_0. There is evidence at the 0.05 level of significance that the population mean differs from 2.72.

25. (a) $H_0 : \mu = 0$ vs. $H_a : \mu \neq 0$. Reject H_0 if the p-value $\leq \alpha = 0.05$ **(b)** $t_{data} = 6$ **(c)** 0.0003 **(d)** Since the p-value $= 0.0003$ is $\leq \alpha = 0.05$, the conclusion is reject H_0. There is evidence at the 0.05 level of significance that the population mean differs from 0.

26. (a) $H_0 : \mu = 2.0$ vs. $H_a : \mu \neq 2.0$. Reject H_0 if the p-value $\leq \alpha = 0.01$ **(b)** $t_{data} = 2.6667$ **(c)** 0.0176 **(d)** Since the p-value $= 0.0176$ is not $\leq \alpha = 0.01$, the conclusion is do not reject H_0. There is insufficient evidence at the 0.01 level of significance that the population mean differs from 2.0.

27. p-value < 0.005

28. p-value < 0.005

29. p-value < 0.01

30. $0.02 < p$-value < 0.05

31.

	Value of μ_0	Form of hypothesis test, with $\alpha = 0.05$	Where μ_0 lies in relation to 95% confidence interval (1, 4)	Conclusion of hypothesis test
(a)	0	$H_0 : \mu = 0$ vs. $H_a : \mu \neq 0$	Outside	Reject H_0
(b)	2	$H_0 : \mu = 2$ vs. $H_a : \mu \neq 2$	Inside	Do not reject H_0
(c)	5	$H_0 : \mu = 5$ vs. $H_a : \mu \neq 5$	Outside	Reject H_0

32.

	Value of μ_0	Form of hypothesis test, with $\alpha = 0.01$	Where μ_0 lies in relation to 99% confidence interval (57, 58)	Conclusion of hypothesis test
(a)	55.5	$H_0 : \mu = 55.5$ vs. $H_a : \mu \neq 55.5$	Outside	Reject H_0
(b)	59.5	$H_0 : \mu = 59.5$ vs. $H_a : \mu \neq 59.5$	Outside	Reject H_0
(c)	57.5	$H_0 : \mu = 57.5$ vs. $H_a : \mu \neq 57.5$	Inside	Do not reject H_0

33.

	Value of μ_0	Form of hypothesis test, with $\alpha = 0.10$	Where μ_0 lies in relation to 90% confidence interval $(-20, -10)$	Conclusion of hypothesis test
(a)	-21	$H_0 : \mu = -21$ vs. $H_a : \mu \neq -21$	Outside	Reject H_0
(b)	-5	$H_0 : \mu = -5$ vs. $H_a : \mu \neq -5$	Outside	Reject H_0
(c)	-12	$H_0 : \mu = -12$ vs. $H_a : \mu \neq -12$	Inside	Do not reject H_0

34.

	Value of μ_0	Form of hypothesis test, with $\alpha = 0.05$	Where μ_0 lies in relation to 95% confidence interval (2010, 2015)	Conclusion of hypothesis test
(a)	2012	$H_0 : \mu = 2012$ vs. $H_a : \mu \neq 2012$	Inside	Do not reject H_0
(b)	2007	$H_0 : \mu = 2007$ vs. $H_a : \mu \neq 2007$	Outside	Reject H_0
(c)	2014	$H_0 : \mu = 2014$ vs. $H_a : \mu \neq 2014$	Inside	Do not reject H_0

35.

	Value of μ_0	Form of hypothesis test, with $\alpha = 0.05$	Where μ_0 lies in relation to 95% confidence interval $(-1, 1)$	Conclusion of hypothesis test
(a)	1.5	$H_0 : \mu = 1.5$ vs. $H_a : \mu \neq 1.5$	Outside	Reject H_0
(b)	-1.5	$H_0 : \mu = -1.5$ vs. $H_a : \mu \neq -1.5$	Outside	Reject H_0
(c)	0	$H_0 : \mu = 0$ vs. $H_a : \mu \neq 0$	Inside	Do not reject H_0

36.

	Value of μ_0	Form of hypothesis test, with $\alpha = 0.05$	Where μ_0 lies in relation to 95% confidence interval (19,570, 20,105)	Conclusion of hypothesis test
(a)	20,000	$H_0 : \mu = 20,000$ vs. $H_a : \mu \neq 20,000$	Inside	Do not reject H_0
(b)	21,000	$H_0 : \mu = 21,000$ vs. $H_a : \mu \neq 21,000$	Outside	Reject H_0
(c)	19,571	$H_0 : \mu = 19,571$ vs. $H_a : \mu \neq 19,571$	Inside	Do not reject H_0

37. *Step 1* $H_0 : \mu = 98.6$ vs. $H_a : \mu > 98.6$
We will reject H_0 if the p-value $\leq \alpha = 0.05$.
Step 2 $t_{data} = 2$
Step 3 p-value $= 0.0241198442$
Step 4 The p-value $= 0.0241198442$ is ≤ 0.05. Therefore we reject H_0. There is evidence at level of significance $\alpha = 0.05$ that the population mean is greater than 98.6.

38. *Step 1* $H_0 : \mu = 46$ vs. $H_a : \mu < 46$
We will reject H_0 if the p-value $\leq \alpha = 0.05$.
Step 2 $t_{data} = -1.52$
Step 3 p-value $= 0.064$
Step 4 The p-value $= 0.064$ is not ≤ 0.05. Therefore we do not reject H_0. There is insufficient evidence at level of significance $\alpha = 0.05$ that the population mean is less than 46.

39. *Step 1* $H_0 : \mu = 20$ vs. $H_a : \mu \neq 20$
We will reject H_0 if the p-value $\leq \alpha = 0.05$.
Step 2 $t_{data} = 2.308$
Step 3 p-value $= 0.021$
Step 4 The p-value $= 0.021$ is ≤ 0.05. Therefore we reject H_0. There is evidence at level of significance $\alpha = 0.05$ that the population mean is not equal to 20.

40. *Step 1* $H_0 : \mu = 8.5$ vs. $H_a : \mu < 8.5$
We will reject H_0 if the p-value $\leq \alpha = 0.05$.
Step 2 $t_{data} = -1.7185$
Step 3 p-value $= 0.0429$
Step 4 The p-value $= 0.0429$ is ≤ 0.05. Therefore we reject H_0. There is evidence at level of significance $\alpha = 0.05$ that the population mean is less than 8.5.

41. Critical-value method: $H_0 : \mu = 15,200$ vs. $H_a : \mu > 15,200$. $t_{crit} = 1.660$. Reject H_0 if $t_{data} \geq 1.660$. $t_{data} = 3.2$. Since $t_{data} = 3.2$ is ≥ 1.660, the conclusion is reject H_0. There is evidence at the 0.05 level of significance that the population mean cost of a stay in the hospital for women aged 18–44 is greater than $15,200. Therefore we can conclude at level of significance 0.05 that the population mean cost of a stay in the hospital for American women aged 18–24 has increased since 2010. p-value method: $H_0 : \mu = 15,200$ vs. $H_a : \mu > 15,200$. Reject H_0 if the p-value $\leq \alpha = 0.05$. $t_{data} = 3.2$. p-value $= 0.0007$. Since the p-value $= 0.0007$ is $\leq \alpha = 0.05$, the conclusion is reject H_0. There is evidence at the 0.05 level of significance that the population mean cost of a stay in the hospital for women aged 18–44 is greater than $15,200. Therefore we can conclude at level of significance 0.05 that the population mean cost of a stay in the hospital for American women aged 18–24 has increased since 2010.

42. Critical-value method: $H_0 : \mu = 40$ vs. $H_a : \mu > 40$. $t_{crit} = 1.306$. Reject H_0 if $t_{data} \geq 1.306$. $t_{data} = 1.25$. Since $t_{data} = 1.25$ is not ≥ 1.306, the conclusion is do not reject H_0. There is insufficient evidence at the 0.10 level of significance that the population mean number of apps downloaded by iPhone users is greater than 40. p-value method: $H_0 : \mu = 40$ vs. $H_a : \mu > 40$. Reject H_0 if the p-value $\leq \alpha = 0.10$. $t_{data} = 1.25$. p-value $= 0.1098$. Since the p-value $= 0.1098$ is not $\leq \alpha = 0.10$, the conclusion is do not reject H_0. There is insufficient evidence at the 0.10 level of significance that the population mean number of apps downloaded by iPhone users is greater than 40.

43. Critical-value method: $H_0 : \mu = 130$ vs. $H_a : \mu < 130$. $t_{crit} = -1.662$. Reject H_0 if $t_{data} \leq -1.662$. $t_{data} = -4$. Since $t_{data} = -4$ is ≤ -1.662, the conclusion is reject H_0. There is evidence at the 0.05 level of significance that the population mean number of Facebook friends is less than 130. p-value method: $H_0 : \mu = 130$ vs. $H_a : \mu < 130$. Reject H_0 if p-value $\leq \alpha = 0.05$. $t_{data} = -4$. p-value $= 0$. Since the p-value $= 0$ is $\leq \alpha = 0.05$, the conclusion is reject H_0. There is evidence at the 0.05 level of significance that the population mean number of Facebook friends is less than 130.

44. Critical-value method: $H_0 : \mu = 16.1$ vs. $H_a : \mu \neq 16.1$. $t_{crit} = 2.704$. Reject H_0 if $t_{data} \leq -2.704$ or if $t_{data} \geq 2.704$. $t_{data} = -0.308$. Since $t_{data} = -0.308$ is not ≤ -2.704 and not ≥ 2.704, the conclusion is do not reject H_0. There is insufficient evidence at the 0.01 level of significance that the population mean number of employees in a small business is different from 16.1. p-value method: $H_0 : \mu = 16.1$ vs. $H_a : \mu \neq 16.1$. Reject H_0 if p-value $\leq \alpha = 0.01$. $t_{data} = -0.308$. p-value $= 0.7594$. Since the p-value $= 0.7594$ is not $\leq \alpha = 0.01$, the conclusion is do not reject H_0. There is insufficient evidence at the 0.01 level of significance that the population mean number of employees in a small business is different from 16.1.

45. No. The distribution of the variable is not normal and the sample size is less than 30.

46. (a) Yes. Most of the points in the normal probability plot lie close to the center line and all of the points lie between the two curved lines. (b) Critical-value method: $H_0 : \mu = 60$ vs. $H_a : \mu < 60$. $t_{crit} = -1.771$. Reject H_0 if $t_{data} \leq -1.771$. $t_{data} = -0.0816$. Since $t_{data} = -0.0816$ is not ≤ -1.771, the conclusion is do not reject H_0. There is insufficient evidence at the 0.05 level of significance that the population mean response time to Asia is less than 60 milliseconds. p-value method: $H_0 : \mu = 60$ vs. $H_a : \mu < 60$. Reject H_0 if p-value $\leq \alpha = 0.05$. $t_{data} = -0.0816$. p-value = 0.4681. Since the p-value = 0.4681 is not $\leq \alpha = 0.05$, the conclusion is do not reject H_0. There is insufficient evidence at the 0.05 level of significance that the population mean response time to Asia is less than 60 milliseconds. (c) The population standard deviation is unknown.

47. *Step 1* $H_0 : \mu = 500{,}000$ vs. $H_a : \mu > 500{,}000$
We will reject H_0 if the p-value $\leq \alpha = 0.10$.
Step 2 $t_{data} = 2.4018$
Step 3 p-value $= 0.0307414454$
Step 4 The p-value $= 0.0307414454$ is ≤ 0.10. Therefore we reject H_0. There is evidence at level of significance $\alpha = 0.10$ that the population mean amount of cleanup money is greater than \$500,000.

48. (a) *Step 1* $H_0 : \mu = 500{,}000$ vs. $H_a : \mu > 500{,}000$
We will reject H_0 if the p-value $\leq \alpha = 0.01$.
Step 2 $t_{data} = 2.4018$
Step 3 p-value $= 0.0307414454$
Step 4 The p-value $= 0.0307414454$ is not ≤ 0.01. Therefore we do not reject H_0. There is insufficient evidence at level of significance $\alpha = 0.01$ that the population mean amount of cleanup money is greater than \$500,000.
(b) Turn to a direct assessment of the strength of evidence against the null hypothesis or obtain more data. (c) The p-value of 0.0307414454 implies that there is solid evidence against the null hypothesis that the population mean equals \$500,000.

49. (a) Decrease (b) Stay the same (c) Increase (d) Depends on new value of μ_0 (e) Stay the same (f) Depends on new value of μ_0

50. (a) (30.9, 78.9). We are 95% confident that the population mean number of game units sold lies between 30.9 thousand units and 78.9 thousand units.

(b)

	Value of μ_0	Form of hypothesis test, with $\alpha = 0.05$	Where μ_0 lies in relation to 95% confidence interval	Conclusion of hypothesis test
(i)	30,000	$H_0 : \mu = 30{,}000$ vs. $H_a : \mu \neq 30{,}000$	Outside	Reject H_0
(ii)	31,000	$H_0 : \mu = 31{,}000$ vs. $H_a : \mu \neq 31{,}000$	Inside	Do not reject H_0
(iii)	0	$H_0 : \mu = 0$ vs. $H_a : \mu \neq 0$	Outside	Reject H_0
(iv)	79,000	$H_0 : \mu = 79{,}000$ vs. $H_a : \mu \neq 79{,}000$	Outside	Reject H_0

51. *Step 1* $H_0 : \mu = 4$ vs. $H_a : \mu \neq 4$
We will reject H_0 if the p-value $\leq \alpha = 0.10$.
Step 2 $t_{data} = -1.044$
Step 3 p-value $= 0.3237340082$
Step 4 The p-value $= 0.3237340082$ is not ≤ 0.10. Therefore we do not reject H_0. There is insufficient evidence at level of significance $\alpha = 0.10$ that the population mean amount of rainfall differs from 4 inches.

52. *Step 1* $H_0 : \mu = 90$ vs. $H_a : \mu > 90$
We will reject H_0 if the p-value $\leq \alpha = 0.10$.
Step 2 $t_{data} = 2.8567$
Step 3 p-value $= 0.0230404904$
Step 4 The p-value $= 0.0230404904$ is ≤ 0.10. Therefore we reject H_0. There is evidence at level of significance $\alpha = 0.10$ that the population mean mileage is greater than 90 MPGe.

53. Yes. The normal probability plot indicates acceptable normality.

54. $t_{crit} = 1.833$

55. *Step 1* $H_0 : \mu = 3264$ vs. $H_a : \mu > 3264$
Step 2 $t_{crit} = 1.833$. Reject H_0 if $t_{data} \geq 1.833$.
Step 3 $t_{data} = 2.000$

Step 4 $t_{data} = 2.000$ is ≥ 1.833. Therefore we reject H_0. There is evidence at level of significance $\alpha = 0.05$ that the population mean tuition and fees for community colleges has increased from the previous level of \$3264.

56. $0.025 < p$-value < 0.05

57. There is solid evidence against the null hypothesis that the population mean is \$3264.

58.

	Value of μ_0	Form of hypothesis test, with $\alpha = 0.05$	Where μ_0 lies in relation to 95% confidence interval	Conclusion of hypothesis test
a.	\$4000	$H_0 : \mu = \$4000$ vs. $H_a : \mu \neq \$4000$	Outside	Reject H_0
b.	\$3500	$H_0 : \mu = \$3500$ vs. $H_a : \mu \neq \$3500$	Inside	Do not reject H_0
c.	\$3264	$H_0 : \mu = \$3264$ vs. $H_a : \mu \neq \$3264$	Inside	Do not reject H_0

59. The line "Test of $\mu = 3264$ vs. $\neq 3264$."

60. The p-value needed for the right-tailed test is one half of the p-value given on the Minitab printout.

61. Do not reject H_0. Fees have increased. Answers will vary.

62. Answers will vary.

63. Since the p-value for the two-tailed test is twice the p-value for the one-tailed test, it is possible to conclude that there is insufficient evidence that the population mean cost has changed, but there is evidence that the population mean cost has increased if α is between the two p-values.

64.

```
Descriptive Statistics: TOT_POP

Variable    N   N*    Mean   SE Mean   StDev   Minimum    Q1
Median    Q3   Maximum
TOT_POP   790   0   18305      9284  260938      1000  1901
4013   9059  7322564
```

65.

```
One-Sample T: TOT_POP

Test of mu = 50000 vs not = 50000

Variable    N    Mean    StDev  SE Mean       95% CI       T
TOT_POP   790   18305   260938     9284  (81, 36529)   -3.41
P
0.001
```

66. Answers will vary.

67. $H_0 : \mu = 0.4$ vs. $H_a : \mu \neq 0.4$. Rest of answer will vary.

68. 0.561675 per 1000 per county. Answers will vary.

69. Answers will vary.

70. $H_0 : \mu = 150$ vs. $H_a : \mu < 150$. Rest of answer will vary.

Section 9.5

1. $\hat{p}$ is the sample proportion and p is population proportion.

2. Both $np_0 \geq 5$ and $n(1 - p_0) \geq 5$

3. Answers will vary.

4. It is the hypothesized value of the population proportion.

5. Between 0 and 1 inclusive: $0 \leq p_0 \leq 1$

6. p represents the population proportion of successes for a binomial experiment and is a population parameter. The p-value is roughly a measure of how extreme the value of Z_{data} is.

7. 2.8868

8. 5.7735

9. 7.2169

10. It increases.

11. -4.47

12. -2.24

13. 0

14. As $\hat{p}$ approaches p_0, the value of Z_{data} approaches 0.

15. (a) We have $np_0 = 225(0.5) = 112.5 \geq 5$ and $n(1 - p_0) = 225(1 - 0.5) = 112.5 \geq 5$, so we can use the Z test for proportions. (b) $H_0 : p = 0.5$ vs. $H_a : p < 0.5$ (c) $Z_{crit} = -1.645$. Reject H_0 if $Z_{data} \leq -1.645$. (d) -1.67 (e) Since $Z_{data} \leq -1.645$, we reject H_0. There is evidence that the population proportion is less than 0.5.

16. (a) $np_0 = 100(0.3) = 30 \geq 5$ and $n(1 - p_0) = 100(1 - 0.3) = 70 \geq 5$, so we may use the Z test for proportions. (b) $H_0 : p = 0.3$ vs. $H_a : p \neq 0.3$

(c) $Z_{crit} = 2.58$. Reject H_0 if $Z_{data} \leq -2.58$ or $Z_{data} \geq 2.58$. (d) $Z_{data} = -1.09$.
(e) Since Z_{data} is not ≤ -2.58 and Z_{data} is not ≥ 2.58, we do not reject H_0. There is insufficient evidence that the population proportion is not equal to 0.3.
17. (a) We have $np_0 = 400(0.6) = 240 \geq 5$ and $n(1 - p_0) = 400(1 - 0.6) = 160 \geq 5$, so we can use the Z test for proportions. (b) $H_0 : p = 0.6$ vs. $H_a : p > 0.6$
(c) $Z_{crit} = 1.645$. Reject H_0 if $Z_{data} \geq 1.645$. (d) 2.04 (e) Since $Z_{data} \geq 1.645$, we reject H_0. There is evidence that the population proportion is greater than 0.6.
18. (a) $np_0 = 900(0.4) = 360 \geq 5$ and $n(1 - p_0) = 900(1 - 0.4) = 540 \geq 5$, so the normality conditions are met. (b) $H_0 : p = 0.4$ vs. $H_a : p \neq 0.4$
(c) $Z_{crit} = 1.645$. Reject H_0 if $Z_{data} \leq -1.645$ or if $Z_{data} \geq 1.645$. (d) $\hat{p} = 400/900 = 0.4444444$, $Z_{data} = 2.7217$ (e) Since $Z_{data} = 2.7217$ is ≥ 1.645, the conclusion is reject H_0. There is evidence at the 0.10 level of significance that the population proportion differs from 0.4.
19. (a) We have $np_0 = 100(0.4) = 40 \geq 5$ and $n(1 - p_0) = 100(1 - 0.4) = 60 \geq 5$, so we can use the Z test for proportions. (b) $H_0 : p = 0.4$ vs. $H_a : p > 0.4$. Reject H_0 if the p-value ≤ 0.05. (c) 0.82 (d) p-value $= 0.2061$ (e) Since the p-value is not ≤ 0.05, we do not reject H_0. There is insufficient evidence that the population proportion is greater than 0.4.
20. (a) We have $np_0 = 400(0.2) = 80 \geq 5$ and $n(1 - p_0) = 400(1 - 0.2) = 320 \geq 5$, so we may use the Z test for proportions. (b) $H_0 : p = 0.2$ vs. $H_a : p < 0.2$. Reject H_0 if the p-value ≤ 0.05. (c) $Z_{data} = -0.63$ (d) p-value $= 0.2643$ (e) Since the p-value is not ≤ 0.05, we do not reject H_0. There is insufficient evidence that the population proportion is less than 0.2.
21. (a) We have $np_0 = 900(0.5) = 450 \geq 5$ and $n(1 - p_0) = 900(1 - 0.5) = 450 \geq 5$, so we may use the Z test for proportions. (b) $H_0 : p = 0.5$ vs. $H_a : p \neq 0.5$ Reject H_0 if the p-value ≤ 0.05. (c) 1.67 (d) p-value $= 0.095$
(e) Since the p-value is not ≤ 0.05, we do not reject H_0. There is insufficient evidence that the population proportion is not equal to 0.5.
22. (a) $np_0 = 1000(0.9) = 900 \geq 5$ and $n(1 - p_0) = 1000(1 - 0.9) = 100 \geq 5$, so we may use the Z test for proportions. (b) $H_0 : p = 0.9$ vs. $H_a : p > 0.9$. Reject H_0 if p-value ≤ 0.05. (c) $Z_{data} = 2.64$ (d) p-value $= 0.0041$ (e) Since p-value ≤ 0.05, we reject H_0. There is evidence that the population proportion is greater than 0.9.

23.

	Value of p_0	Form of hypothesis test, with $\alpha = 0.05$	Where p_0 lies in relation to 95% confidence interval (0.1, 0.9)	Conclusion of hypothesis test
(a)	0	$H_0 : p = 0$ vs. $H_a : p \neq 0$	Outside	Reject H_0
(b)	1	$H_0 : p = 1$ vs. $H_a : p \neq 1$	Outside	Reject H_0
(c)	0.5	$H_0 : p = 0.5$ vs. $H_a : p \neq 0.5$	Inside	Do not reject H_0

24.

	Value of p_0	Form of hypothesis test, with $\alpha = 0.01$	Where p_0 lies in relation to 99% confidence interval (0.51, 0.52)	Conclusion of hypothesis test
(a)	0.511	$H_0 : p = 0.511$ vs. $H_a : p \neq 0.511$	Inside	Do not reject H_0
(b)	0.521	$H_0 : p = 0.521$ vs. $H_a : p \neq 0.521$	Outside	Reject H_0
(c)	0.519	$H_0 : p = 0.519$ vs. $H_a : p \neq 0.519$	Inside	Do not reject H_0

25.

	Value of p_0	Form of hypothesis test, with $\alpha = 0.10$	Where p_0 lies in relation to 90% confidence interval (0.1, 0.2)	Conclusion of hypothesis test
(a)	0.09	$H_0 : p = 0.09$ vs. $H_a : p \neq 0.09$	Inside	Reject H_0
(b)	0.9	$H_0 : p = 0.9$ vs. $H_a : p \neq 0.9$	Outside	Reject H_0
(c)	0.19	$H_0 : p = 0.19$ vs. $H_a : p \neq 0.19$	Inside	Do not reject H_0

26.

	Value of p_0	Form of hypothesis test, with $\alpha = 0.05$	Where p_0 lies in relation to 95% confidence interval (0.05, 0.95)	Conclusion of hypothesis test
(a)	0.01	$H_0 : p = 0.01$ vs. $H_a : p \neq 0.01$	Outside	Reject H_0
(b)	0.5	$H_0 : p = 0.5$ vs. $H_a : p \neq 0.5$	Inside	Do not reject H_0
(c)	0.06	$H_0 : p = 0.06$ vs. $H_a : p \neq 0.06$	Inside	Do not reject H_0

27. *Step 1* $H_0 : p = 0.5$ vs. $H_a : p < 0.5$
We reject H_0 if the p-value $\leq \alpha = 0.05$.
Step 2 $Z_{data} = -2.12$
Step 3 p-value $= 0.0169473661$
Step 4 The p-value 0.0169473661 is $\leq \alpha = 0.05$, so we reject H_0. There is evidence at level of significance $\alpha = 0.05$ that the population proportion is less than 0.5.

28. *Step 1* $H_0 : p = 0.1$ vs. $H_a : p \neq 0.1$
We reject H_0 if the p-value $\leq \alpha = 0.05$.
Step 2 $Z_{data} = -1.05$
Step 3 p-value $= 0.2918405966$
Step 4 The p-value 0.2918405966 is not $\leq \alpha = 0.05$, so we do not reject H_0. There is insufficient evidence at level of significance $\alpha = 0.05$ that the population proportion is not equal to 0.1.

29. *Step 1* $H_0 : p = 0.05$ vs. $H_a : p \neq 0.05$.
We reject H_0 if the p-value $\leq \alpha = 0.05$.
Step 2 $Z_{data} = -1.95$
Step 3 p-value $= 0.052$
Step 4 The p-value 0.052 is not $\leq \alpha = 0.05$, so we do not reject H_0. There is insufficient evidence at level of significance $\alpha = 0.05$ that the population proportion is not equal to 0.05.

30. *Step 1* $H_0 : p = 0.4$ vs. $H_a : p \neq 0.4$
We reject H_0 if the p-value $\leq \alpha = 0.05$.
Step 2 $Z_{data} = -0.78$
Step 3 p-value $= 0.436$
Step 4 The p-value 0.436 is not $\leq \alpha = 0.05$, so we do not reject H_0. There is insufficient evidence at level of significance $\alpha = 0.05$ that the population proportion is not equal to 0.4.

31. Both $n \cdot p_0 = 500(0.23) = 115 \geq 5$ and $n \cdot q_0 = 500(1 - 0.23) = 500$ $(0.77) = 385 \geq 5$, so the conditions for using the Z test for proportions are met.
Step 1 $H_0 : p = 0.23$ vs. $H_a : p < 0.23$
We reject H_0 if the p-value $\leq \alpha = 0.10$.
Step 2 $Z_{data} = -1.59$
Step 3 p-value $= 0.0559$
Step 4 The p-value 0.0559 is $\leq \alpha = 0.10$, so we reject H_0. There is evidence at level of significance $\alpha = 0.10$ that the population proportion of Facebook users ages 18–24 has decreased from 23%.

32. Both $n \cdot p_0 = 400(0.071) = 28.4 \geq 5$ and $n \cdot q_0 = 400(1 - 0.071) = 400$ $(0.929) = 371.6 \geq 5$, so the conditions for using the Z test for proportions are met.
Step 1 $H_0 : p = 0.071$ vs. $H_a : p \neq 0.071$
We reject H_0 if the p-value $\leq \alpha = 0.01$.
Step 2 $Z_{data} = 1.28$
Step 3 p-value $= 0.2006$
Step 4 The p-value 0.2006 is not $\leq \alpha = 0.01$, so we do not reject H_0. There is insufficient evidence at level of significance $\alpha = 0.01$ that the population proportion of Americans ages 20–24 differs from 7.1%.

33. Both $n \cdot p_0 = 900(0.048) = 43.2 \geq 5$ and $n \cdot q_0 = 400(1 - 0.048) = 900$ $(0.952) = 856.8 \geq 5$, so the conditions for using the Z test for proportions are met.
Step 1 $H_0 : p = 0.048$ vs. $H_a : p > 0.048$
We reject H_0 if the p-value $\leq \alpha = 0.01$.
Step 2 $Z_{data} = 1.68$
Step 3 p-value $= 0.0465$
Step 4 The p-value 0.0465 is not $\leq \alpha = 0.01$, so we do not reject H_0. There is insufficient evidence at level of significance $\alpha = 0.01$ that the population proportion of persons ages 12 or older who used a prescription pain reliever nonmedically has increased from 4.8%.

34. Both $n \cdot p_0 = 1000(0.57) = 570 \geq 5$ and $n \cdot q_0 = 1000(1 - 0.57) = 1000$ $(0.43) = 430 \geq 5$, so the conditions for using the Z test for proportions are met.

Step 1 $H_0 : p = 0.57$ vs. $H_a : p < 0.57$
We reject H_0 if the p-value $\leq \alpha = 0.05$.

Step 2 $Z_{data} = -4.47$

Step 3 p-value ≈ 0

Step 4   The p-value ≈ 0 is $\leq \alpha = 0.05$, so we reject H_0. There is evidence at level of significance $\alpha = 0.05$ that the population proportion of 18- to 24-year-olds who agree that texting has made it more difficult to determine whether an outing is an actual date has decreased from 57%.

35. Both $n \cdot p_0 = 100(0.58) = 58 \geq 5$ and $n \cdot q_0 = 100(1 - 0.58) = 100(0.42)$ $= 42 \geq 5$, so the conditions for using the Z test for proportions are met.

Step 1 $H_0 : p = 0.58$ vs. $H_a : p > 0.58$
We reject H_0 if the p-value $\leq \alpha = 0.05$.

Step 2 $Z_{data} = 1.82$

Step 3 p-value $= 0.0341$

Step 4   The p-value 0.0341 is $\leq \alpha = 0.05$, so we reject H_0. There is evidence at level of significance $\alpha = 0.05$ that the population proportion of mutual funds that underperform has increased from 58%.

36. $np_0 = 1000(0.07) = 70 \geq 5$ and $n(1 - p_0) = 1000(1 - 0.07) = 930 \geq$ 5, so we may use the Z test for proportions. $H_0 : p = 0.07$ vs. $H_a : p \neq 0.07$ Reject H_0 if p-value ≤ 0.10. $Z_{data} = 1.24$. p-value $= 0.2150$. Since p-value is not ≤ 0.10, we do not reject H_0. There is insufficient evidence that the population proportion of hospitalizations of 18- to 44-year-old American women for affective disorders is not equal to 0.07.

37. (a) Yes. We have $np_0 = 100(0.153) = 15.3 \geq 5$ and $n(1 - p_0) = 100$ $(1 - 0.153) = 84.7 \geq 5$. **(b)** $H_0 : p = 0.153$ vs. $H_a : p \neq 0.153$ Reject H_0 if p-value ≤ 0.01. $Z_{data} = 2.14$. p-value $= 0.0324$. Since the p-value is not ≤ 0.01, we do not reject H_0. There is insufficient evidence that the population proportion of Hispanic families that had a household income of at least \$75,000 is not equal to 0.153.

38. Both $n \cdot p_0 = 1000(0.065) = 65 \geq 5$ and $n \cdot q_0 = 1000(1 - 0.065) = 1000$ $(0.935) = 935 \geq 5$, so the conditions for using the Z test for proportions are met.

Step 1 $H_0 : p = 0.065$ vs. $H_a : p > 0.065$
We reject H_0 if the p-value $\leq \alpha = 0.05$.

Step 2 $Z_{data} = 1.28$

Step 3 p-value $= 0.1003$

Step 4   The p-value 0.1003 is not $\leq \alpha = 0.05$, so we do not reject H_0. There is insufficient evidence at level of significance $\alpha = 0.05$ that the population proportion of students completing MOOCs has increased from 6.5%.

39. Both $n \cdot p_0 = 100(0.57) = 57 \geq 5$ and $n \cdot q_0 = 100(1 - 0.57) = 100(0.43) = 43 \geq 5$, so the conditions for using the Z test for proportions are met.

Step 1 $H_0 : p = 0.57$ vs. $H_a : p \neq 0.57$
We reject H_0 if the p-value $\leq \alpha = 0.10$.

Step 2 $Z_{data} = 0.61$

Step 3 p-value $= 0.5418$

Step 4   The p-value 0.5418 is not $\leq \alpha = 0.10$, so we do not reject H_0. There is insufficient evidence at level of significance $\alpha = 0.10$ that the population proportion of young people ages 18–24 who are living with their parents differs from 57%.

40. (a) No evidence. **(b)** Yes. The two methods are equivalent. **(c)** Yes. The conclusion is do not reject H_0.

41. (a) We have $np_0 = 100(0.11) = 11 \geq 5$ and $n(1 - p_0) = 100(1 - 0.11) = 89 \geq 5$, so we can use the Z test for proportions. **(b)** $H_0 : p = 0.11$ vs. $H_a : p < 0.11$. $Z_{crit} = -1.645$. Reject H_0 if $Z_{data} \leq -1.645$. $Z_{data} = -1.60$. Since Z_{data} is not ≤ -1.645, we do not reject H_0. There is insufficient evidence that the population proportion of children age 6 and under exposed to ETS at home on a regular basis is less than 0.11.

42. (a) Type II; answers will vary. **(b)** Since we did not reject H_0, our hypothesis test does not support this headline.

43. (a) Since the p-value ≤ 0.10, we reject H_0. There is evidence that the population proportion of children age 6 and under exposed to environmental tobacco smoke at home on a regular basis is less than 0.11. **(b)** The difference is because we changed the value of α and not because we used different methods for the two different hypothesis tests. **(c)** Since $0.05 \leq p$-value ≤ 0.10, there is mild evidence against the null hypothesis that the population proportion of children age 6 and under exposed to environmental tobacco smoke at home on a regular basis is greater than or equal to 0.11.

44. (a) Unchanged **(b)–(c)** Decrease **(d)** Unchanged **(e)** Depends on new value of $\hat{p}$.

45. Both $n \cdot p_0 = 1000(0.12) = 120 \geq 5$ and $n \cdot q_0 = 1000(1 - 0.12) = 1000$ $(0.88) = 880 \geq 5$, so the conditions for using the Z test for proportions are met.

Step 1 $H_0 : p = 0.12$ vs. $H_a : p > 0.12$
We reject H_0 if the p-value $\leq \alpha = 0.05$.

Step 2 $Z_{data} = 1.36$

Step 3 p-value $= 0.0869$

Step 4   The p-value 0.0869 is not $\leq \alpha = 0.05$, so we do not reject H_0. There is insufficient evidence at level of significance $\alpha = 0.05$ that the population proportion of young people ages 18–24 who have had an accident has increased from 12%.

46. (a) Decreases **(b)** Increases **(c)** Decreases **(d)** Stays the same. **(e)** Depends on new sample size.

47. (a) Stays the same. **(b)** Decreases **(c)** Increases **(d)** Stays the same **(e)** Stays the same

48. Answers will vary.

49. Answers will vary.

50. $H_0 : p = 0.10$ vs. $H_a : p < 0.10$
The rest of the answer will vary.

51. 0.0507. Answers will vary.

52. Answers will vary.

53. Answers will vary.

54. Answers will vary.

55. $H_0 : p = 0.5$ vs. $H_a : p > 0.5$
The rest of the answer will vary.

56. Most of the data values are 0, representing the customer did not buy a suit.

57. $H_0 : p = 0.1$ vs. $H_a : p > 0.1$
The rest of the answer will vary.

Section 9.6

1. Answers will vary.

2. σ is the actual population standard deviation and σ_0 is the hypothesized population standard deviation.

3. No, σ will never be less than 0.

4. The population must be normal.

5. Answers will vary.

6. The confidence level is $(1 - \alpha)100\%$.

7. (a) $H_0 : \sigma = 1$ vs. $H_a : \sigma > 1$ **(b)** $\chi^2_\alpha = \chi^2_{0.05} = 31.410$. Reject H_0 if $\chi^2_{data} \geq 31.410$. **(c)** $\chi^2_{data} = 60$ **(d)** Since $\chi^2_{data} \geq 31.410$, reject H_0. There is evidence that the population standard deviation is greater than 1.

8. (a) $H_0 : \sigma = 5$ vs. $H_a : \sigma < 5$ **(b)** $\chi^2_{1-\alpha} = \chi^2_{0.95} = 3.940$. Reject H_0 if $\chi^2_{data} \leq 3.940$. **(c)** 10 **(d)** Since χ^2_{data} is not ≤ 3.940, we do not reject H_0. There is insufficient evidence that the population standard deviation is less than 5.

9. (a) $H_0 : \sigma = 3$ vs. $H_a : \sigma \neq 3$ **(b)** $\chi^2_{\alpha/2} = \chi^2_{0.025} = 27.488$ and $\chi^2_{1-\alpha/2} = \chi^2_{0.975} = 6.262$. Reject H_0 if $\chi^2_{data} \leq 6.262$ or $\chi^2_{data} \geq 27.488$. **(c)** $\chi^2_{data} = 10.417$ **(d)** Since χ^2_{data} is not ≤ 6.262 and χ^2_{data} is not ≥ 27.488, we do not reject H_0. There is insufficient evidence that the population standard deviation is different from 3.

10. (a) $H_0 : \sigma = 10$ vs. $H_a : \sigma > 10$ **(b)** $\chi^2_\alpha = \chi^2_{0.01} = 27.688$. Reject H_0 if $\chi^2_{data} \geq 27.688$. **(c)** 18.72 **(d)** Since χ^2_{data} is not ≥ 27.688, we do not reject H_0. There is insufficient evidence that the population standard deviation is greater than 10.

11. (a) $H_0 : \sigma = 20$ vs. $H_a : \sigma < 20$ **(b)** $\chi^2_{1-\alpha} = \chi^2_{0.90} = 2.833$. Reject H_0 if $\chi^2_{data} \leq 2.833$. **(c)** $\chi^2_{data} = 6.125$ **(d)** Since χ^2_{data} is not ≤ 2.833, we do not reject H_0. There is insufficient evidence that the population standard deviation is less than 20.

12. (a) $H_0 : \sigma = 5$ vs. $H_a : \sigma \neq 5$ **(b)** $\chi^2_{\alpha/2} = \chi^2_{0.025} = 40.646$ and $\chi^2_{1-\alpha/2} = \chi^2_{0.975} = 13.120$. Reject H_0 if $\chi^2_{data} < 13.120$ or $\chi^2_{data} \geq 40.646$. **(c)** 25 **(d)** Since χ^2_{data} is not ≤ 13.120 and χ^2_{data} is not ≥ 40.646, we do not reject H_0. There is insufficient evidence that the population standard deviation is different from 5.

13. (a) $H_0 : \sigma = 1$ versus $H_a : \sigma > 1$ Reject H_0 if the p-value $\leq \alpha = 0.05$. **(b)** $\chi^2_{data} = 60$

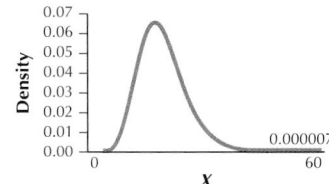

(c) p-value $= 7.121750863 \times 10^{-6}$ **(d)** Since the p-value ≤ 0.05, we reject H_0. There is evidence that the population standard deviation is greater than 1.

14. (a) $H_0 : \sigma = 5$ vs. $H_a : \sigma < 5$. Reject H_0 if the p-value $\le \alpha = 0.05$.
(b) $\chi^2_{\text{data}} = 10$

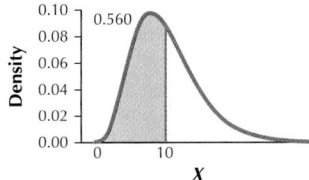

(c) p-value $= 0.5595067149$ **(d)** Since the p-value is not ≤ 0.05, we do not reject H_0. There is insufficient evidence that the population standard deviation is less than 5.

15. (a) $H_0 : \sigma = 3$ vs. $H_a : \sigma \ne 3$. Reject H_0 if the p-value $\le \alpha = 0.05$.
(b) $\chi^2_{\text{data}} = 10.417$

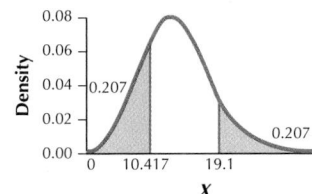

(c) p-value $= 0.4145552434$ **(d)** Since the p-value is not ≤ 0.05, we do not reject H_0. There is insufficient evidence that the population standard deviation is different from 3.

16. (a) $H_0 : \sigma = 10$ vs. $H_a : \sigma > 10$. Reject H_0 if the p-value $\le \alpha = 0.05$.
(b) $\chi^2_{\text{data}} = 18.72$

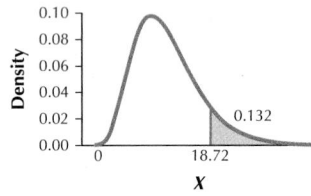

(c) p-value $= 0.1320442389$ **(d)** Since the p-value is not ≤ 0.05, we do not reject H_0. There is insufficient evidence that the population standard deviation is greater than 10.

17. (a) $H_0 : \sigma = 20$ vs. $H_a : \sigma < 20$. Reject H_0 if the p-value $\le \alpha = 0.05$.
(b) $\chi^2_{\text{data}} = 6.125$

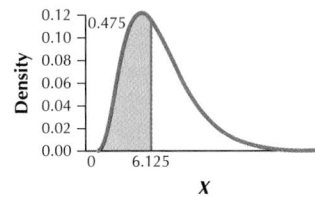

(c) p-value $= 0.4747679539$ **(d)** Since the p-value is not ≤ 0.05, we do not reject H_0. There is insufficient evidence that the population standard deviation is less than 20.

18. (a) $H_0 : \sigma = 5$ vs. $H_a : \sigma \ne 5$. Reject H_0 if the p-value $\le \alpha = 0.05$.
(b) $\chi^2_{\text{data}} = 25$

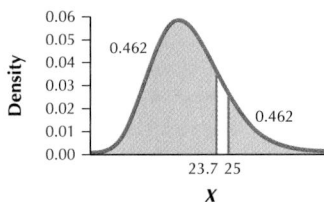

(c) p-value $= 0.9247473258$ **(d)** Since the p-value is not ≤ 0.05, we do not reject H_0. There is insufficient evidence that the population standard deviation is different than 5.

19.

	Value of σ_0	Form of hypothesis test, with $\alpha = 0.05$	Where σ_0 lies in relation to 95% confidence interval (1, 4)	Conclusion of hypothesis test
(a)	0	$H_0 : \sigma = 0$ vs. $H_a : \sigma \ne 0$	Outside	Reject H_0
(b)	2	$H_0 : \sigma = 2$ vs. $H_a : \sigma \ne 2$	Inside	Do not reject H_0
(c)	5	$H_0 : \sigma = 5$ vs. $H_a : \sigma \ne 5$	Outside	Reject H_0

20.

	Value of σ_0	Form of hypothesis test, with $\alpha = 0.01$	Where σ_0 lies in relation to 99% confidence interval (10, 25)	Conclusion of hypothesis test
(a)	15	$H_0 : \sigma = 15$ vs. $H_a : \sigma \ne 15$	Inside	Do not reject H_0
(b)	26	$H_0 : \sigma = 26$ vs. $H_a : \sigma \ne 26$	Outside	Reject H_0
(c)	5	$H_0 : \sigma = 5$ vs. $H_a : \sigma \ne 5$	Outside	Reject H_0

21.

	Value of σ_0	Form of hypothesis test, with $\alpha = 0.10$	Where σ_0 lies in relation to 90% confidence interval (100, 200)	Conclusion of hypothesis test
(a)	150	$H_0 : \sigma = 150$ vs. $H_a : \sigma \ne 150$	Inside	Do not reject H_0
(b)	250	$H_0 : \sigma = 250$ vs. $H_a : \sigma \ne 250$	Outside	Reject H_0
(c)	0	$H_0 : \sigma = 0$ vs. $H_a : \sigma \ne 0$	Outside	Reject H_0

22.

	Value of σ_0	Form of hypothesis test, with $\alpha = 0.05$	Where σ_0 lies in relation to 95% confidence interval (127, 698)	Conclusion of hypothesis test
(a)	125	$H_0 : \sigma = 125$ vs. $H_a : \sigma \ne 125$	Outside	Reject H_0
(b)	128	$H_0 : \sigma = 128$ vs. $H_a : \sigma \ne 128$	Inside	Do not reject H_0
(c)	700	$H_0 : \sigma = 700$ vs. $H_a : \sigma \ne 700$	Outside	Reject H_0

23. *Step 1* $H_0 : \sigma = 25$ vs. $H_a : \sigma \ne 25$
Step 2 $\chi^2_{0.975} = 2.180$ and $\chi^2_{0.025} = 17.535$. Therefore we reject H_0 if $\chi^2_{\text{data}} \le \chi^2_{0.975} = 2.180$ or $\chi^2_{\text{data}} \ge \chi^2_{0.025} = 17.535$.
Step 3 $\chi^2_{\text{data}} = 8.97$
Step 4 $\chi^2_{\text{data}} = 8.97$ is not $\le \chi^2_{0.975} = 2.180$ and is not $\ge \chi^2_{0.025} = 17.535$, so we do not reject H_0. There is insufficient evidence at level of significance $\alpha = 0.05$ that the population standard deviation differs from 25 MW.

24. *Step 1* $H_0 : \sigma = 100$ vs. $H_a : \sigma > 100$
Step 2 $\chi^2_{0.10} = 7.779$. Therefore we reject H_0 if $\chi^2_{\text{data}} \ge \chi^2_{0.10} = 7.779$.
Step 3 $\chi^2_{\text{data}} = 15.18$
Step 4 $\chi^2_{\text{data}} = 15.18$ is $\ge \chi^2_{0.10} = 7.779$, so we reject H_0. There is evidence at level of significance $\alpha = 0.10$ that the population standard deviation is greater than 100 million tons.

25. No, population is not normal.

26. *Step 1* $H_0 : \sigma = 500{,}000$ vs. $H_a : \sigma < 500{,}000$
Step 2 $\chi^2_{0.95} = 1.145$. Therefore we reject H_0 if $\chi^2_{\text{data}} \le \chi^2_{0.95} = 1.145$.
Step 3 $\chi^2_{\text{data}} = 2.43$
Step 4 $\chi^2_{\text{data}} = 2.43$ is not $\le \chi^2_{0.95} = 1.145$, so we do not reject H_0. There is insufficient evidence at level of significance $\alpha = 0.05$ that the population standard deviation is less than \$500,000.

27. *p*-value method: $H_0 : \sigma = 50$ vs. $H_a : \sigma > 50$. Reject H_0 if *p*-value. ≤ 0.05. $\chi^2_{data} = 104$. *p*-value $= 0.3721497012$. Since *p*-value is not ≤ 0.05, we do not reject H_0. There is insufficient evidence that the population standard deviation of test scores for boys is greater than 50 points.

Critical-value method: $H_0 : \sigma = 50$ vs. $H_a : \sigma > 50$. $\chi^2_\alpha = \chi^2_{0.05} = 124.342$. Reject H_0 if $\chi^2_{data} \geq 124.342$. $\chi^2_{data} = 104$. Since χ^2_{data} is not ≥ 124.342, we do not reject H_0. There is insufficient evidence that the population standard deviation of test scores for boys is greater than 50 points.

28. $H_0 : \sigma = 20$ vs. $H_a : \sigma < 20$. Reject H_0 if *p*-value ≤ 0.05. $\chi^2_{data} = 24.5$. *p*-value $= 0.0013251562$. Since *p*-value ≤ 0.05, we reject H_0. There is evidence that the population standard deviation of heart rates of people leading high-stress lives is less than 20 beats per minute.

Section 9.7

1. A Type II error is not rejecting H_0 when H_0 is false.
2. $\bar{x}_{crit}$ is the value of the sample mean $\bar{x}$ associated with Z_{crit}.
3. The probability of rejecting H_0 when H_0 is false.
4. $1 - \beta$, where β is the probability of a Type II error.
5. (a) 51.024
(b)

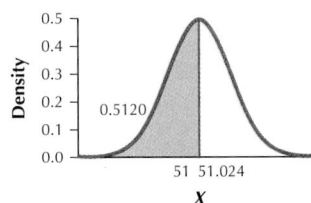

(c) 0.5120
6. (a) 51.024
(b)

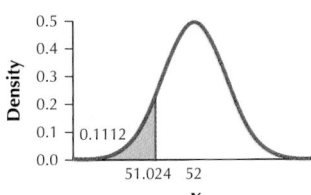

(c) 0.1112
7. (a) 51.024
(b)

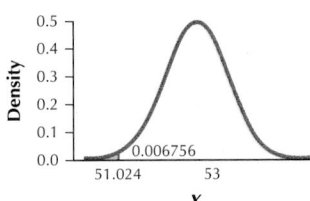

(c) 0.0068
8. (a) 51.024
(b)

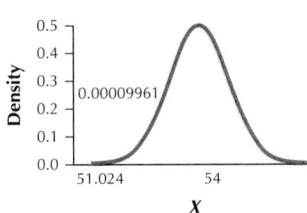

(c) TI-83/84: 0.00009964
9. (a) 51.024
(b)

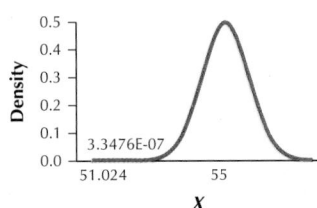

(c) TI-83/84: 0.0000003353

10. (a) 51.024
(b)

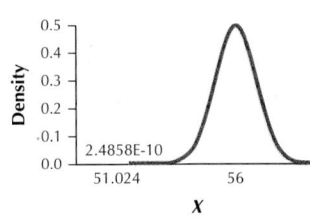

(c) TI-83/84: 0.0000000002496
11. (a) 96.71
(b)

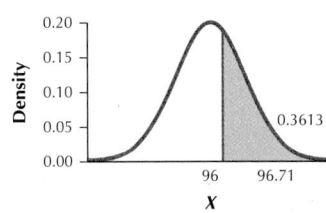

(c) 0.3613
12. (a) 96.71
(b)

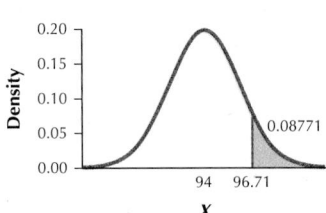

(c) 0.0877
13. (a) 96.71
(b)

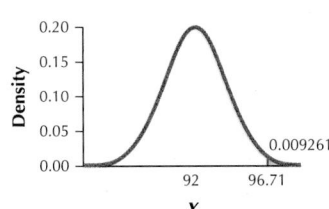

(c) 0.0093
14. (a) 96.71
(b)

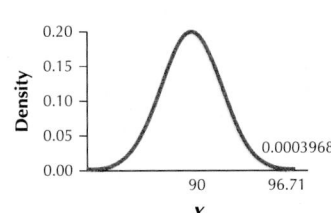

(c) 0.0003969
15. (a) 96.71
(b)

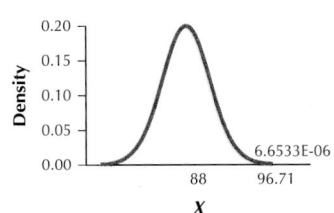

(c) 0.000006658

16. (a) 96.71

(b)

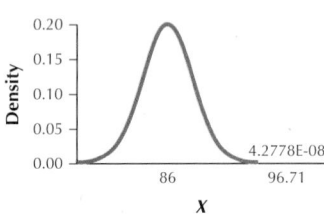

(c) 0.00000004287
17. 0.4880
18. 0.8888
19. 0.9932
20. 0.99990036
21. 0.9999996647
22. 0.9999999997504
23. 0.6387
24. 0.9123
25. 0.9907
26. 0.9996031
27. 0.999993342
28. 0.99999995713

29.

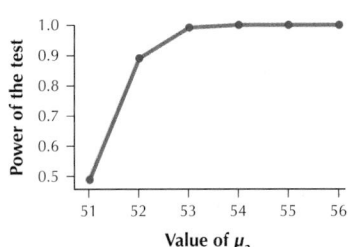

30.

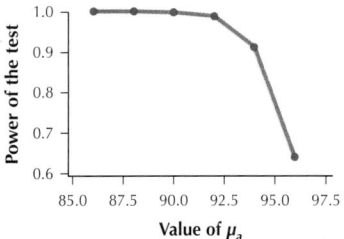

31. (a) Concluding that the population mean daily number of shares traded equals 2.9 billion shares when, in reality, it has increased from 2.9 billion shares.

(b) and (c)

	μ_a	Probability of Type 11 error: β	Power of Test: $1 - \beta$
(i)	3.0 billion	$P\left(Z < \dfrac{3.09 - 3.0}{0.7/\sqrt{36}}\right) =$ $P(Z < 0.77) = 0.7794$	$1 - 0.7794 = 0.2206$
(ii)	3.1 billion	$P\left(Z < \dfrac{3.09 - 3.1}{0.7/\sqrt{36}}\right) =$ $P(Z < -0.09) = 0.4641$	$1 - 0.4641 = 0.5359$
(iii)	3.2 billion	$P\left(Z < \dfrac{3.09 - 3.2}{0.7/\sqrt{36}}\right) =$ $P(Z < -0.94) = 0.1736$	$1 - 0.1736 = 0.8264$
(iv)	3.3 billion	$P\left(Z < \dfrac{3.09 - 3.3}{0.7/\sqrt{36}}\right) =$ $P(Z < -1.8) = 0.0359$	$1 - 0.0359 = 0.9641$

(d)

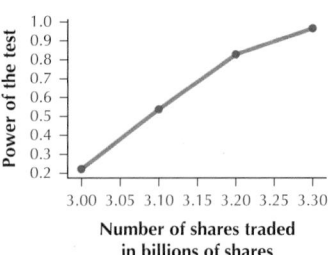

32. (a) Concluding that the population mean credit score in Georgia equals 668 when, in reality, it has decreased from 668.

(b) and (c)

	μ_a	Probability of Type 11 error: β	Power of Test: $1 - \beta$
(i)	650	$P\left(Z > \dfrac{659.775 - 650}{150/\sqrt{900}}\right) =$ $P(Z > 1.96) = 0.0250$	$1 - 0.0250 = 0.9750$
(ii)	645	$P\left(Z > \dfrac{659.775 - 645}{150/\sqrt{900}}\right) =$ $P(Z > 2.96) = 0.0015$	$1 - 0.0015 = 0.9985$
(iii)	640	$P\left(Z > \dfrac{659.775 - 640}{150/\sqrt{900}}\right) =$ $P(Z > 3.96) = 0.00004$	$1 - 0.00004 = 0.99996$
(iv)	635	$P\left(Z > \dfrac{659.775 - 635}{150/\sqrt{900}}\right) =$ $P(Z > 4.96) = 0$	$1 - 0 = 1$

(d)

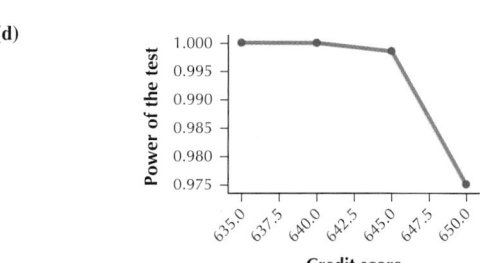

33. (a) Concluding that the population mean salary for accountants is equal to $41,560 when, in reality, it has changed from $41,560.

(b) and (c)

	μ_a	Probability of Type 11 error: β	Power of Test: $1 - \beta$
(i)	$42,000	$P\left(\dfrac{40,580 - 42,000}{5000/\sqrt{100}} < Z < \dfrac{42,540 - 42,000}{5000/\sqrt{100}}\right)$ $= P(-2.84 < Z < 1.08) =$ $0.8599 - 0.0023 = 0.8576$	$1 - 0.8576 = 0.1424$
(ii)	$43,000	$P\left(\dfrac{40,580 - 43,000}{5000/\sqrt{100}} < Z < \dfrac{42,540 - 43,000}{5000/\sqrt{100}}\right)$ $= P(-4.84 < Z < -0.92) =$ $0.1788 - 0 = 0.1788$	$1 - 0.1788 = 0.8212$
(iii)	$44,000	$P\left(\dfrac{40,580 - 44,000}{5000/\sqrt{100}} < Z < \dfrac{42,540 - 44,000}{5000/\sqrt{100}}\right)$ $= P(-6.84 < Z < -2.92) =$ $0.0018 - 0 = 0.0018$	$1 - 0.0018 = 0.9982$
(iv)	$45,000	$P\left(\dfrac{40,580 - 45,000}{5000/\sqrt{100}} < Z < \dfrac{42,540 - 45,000}{5000/\sqrt{100}}\right)$ $= P(-8.84 < Z < -4.92) = 0 - 0 = 0$	$1 - 0 = 1$

(d)

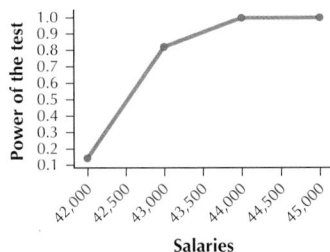

34. (a) Concluding that the population mean price of a gallon of milk is $3.63 when it is, in reality, greater than $3.63.

(b) and **(c)**

	μ_a	Probability of Type 11 error: β	Power of Test: $1 - \beta$
(i)	$3.70	$P\left(Z < \dfrac{3.75 - 3.70}{1.00/\sqrt{400}}\right) =$ $P(Z < 1.00) = 0.8413$	$1 - 0.8413 = 0.1587$
(ii)	$3.90	$P\left(Z < \dfrac{3.75 - 3.90}{1.00/\sqrt{400}}\right) =$ $P(Z < -3.00) = 0.0013$	$1 - 0.0013 = 0.9987$
(iii)	$4.10	$P\left(Z < \dfrac{3.75 - 4.10}{1.00/\sqrt{400}}\right) =$ $P(Z < -7.00) = 0$	$1 - 0 = 1$
(iv)	$4.30	$P\left(Z < \dfrac{3.75 - 4.30}{1.00/\sqrt{400}}\right) =$ $P(Z < -11.00) = 0$	$1 - 0 = 1$

(d)

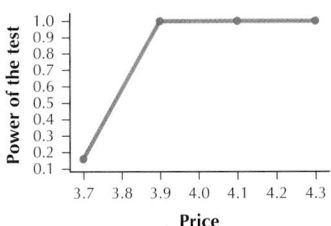

Chapter 9 Review

1. $H_0 : \mu = 12$ vs. $H_a : \mu < 12$

2. $H_0 : \mu = 10$ vs. $H_a : \mu > 10$

3. $H_0 : \mu = 0$ vs. $H_a : \mu < 0$

4. (a) $H_0 : \mu = 2.58$ vs. $H_a : \mu \neq 2.58$ **(b)** Concluding that the population mean has changed from 2.58 when it actually is different from 2.58, and concluding that the population mean has not changed from 2.58 when it actually is not different from 2.58 **(c)** Concluding that the population mean has changed from 2.58 when it hasn't **(d)** Concluding that the population mean has not changed from 2.58 when it actually has changed from 2.58

5. (a) $H_0 : \mu = 202.7$ vs. $H_a : \mu < 202.7$ **(b)** We conclude that (1) the population mean number of speeding-related fatalities is less than 202.7 when it actually is and (2) the mean number of speeding-related fatalities is not less than 202.7 when it actually is. **(c)** The population mean number of speeding-related fatalities is less than 202.7 when it actually is not less than 202.7. **(d)** The population mean number of speeding-related fatalities is not less than 202.7 when it actually is less than 202.7.

6. (a) $H_0 : \mu = 94,000$ vs. $H_a : \mu > 94,000$ **(b)** Concluding that the population mean salary is greater than $94,000 when it actually is, and concluding that the population mean is not greater than 94,000 when it actually isn't. **(c)** Concluding that the population mean is greater than $94,000 when it actually isn't. **(d)** Concluding that the population mean is not greater than $94,000 when it actually is.

7. -1

8. -2

9. -10

10. (a) 2.58 **(b)** Reject H_0 if $Z_{data} \leq -2.58$ or $Z_{data} \geq 2.58$.

(c)

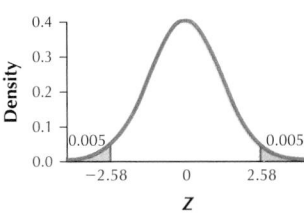

(d) Since Z_{data} is not ≤ -2.58 and Z_{data} is not ≥ 2.58, we do not reject H_0. There is insufficient evidence that the population mean is different from μ_0.

11. (a) 1.28 **(b)** Reject H_0 if $Z_{data} \geq 1.28$.

(c)

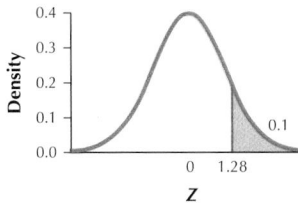

(d) Since $Z_{data} \geq 1.28$, we reject H_0. There is evidence that the population mean is greater than μ_0.

12. (a) 1.645 **(b)** Reject H_0 if $Z_{data} \geq 1.645$.

(c)

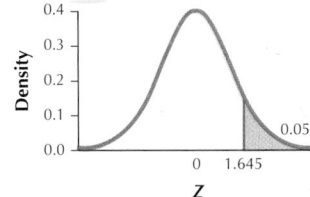

(d) Since Z_{data} is not ≥ 1.645, we do not reject H_0. There is insufficient evidence that the population mean is greater than μ_0.

13. (a) $H_0 : \mu = 668$ vs. $H_a : \mu < 668$ **(b)** -1.645; reject H_0 if $Z_{data} \leq -1.645$ **(c)** $Z_{data} = -4.52$

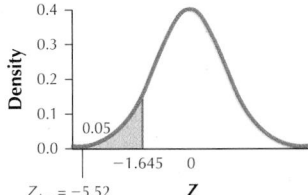

(d) Since $Z_{data} \leq -1.645$, we reject H_0. There is evidence that the population mean credit score in Georgia is less than 668.

14. (a) $H_0 : \mu = 500$ vs. $H_a : \mu \neq 500$. Reject H_0 if the p-value ≤ 0.05. **(b)** 4 **(c)** $6.337206918 \times 10^{-5}$

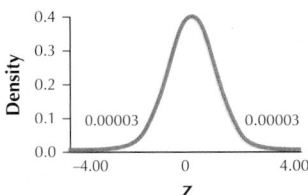

(d) Since the p-value ≤ 0.05, reject H_0. There is evidence that the population mean is different than 500.

15. (a) $H_0 : \mu = -10$ vs. $H_0 : \mu < -10$. Reject H_0 if the p-value ≤ 0.01. **(b)** -5 **(c)** p-value $= 2.87105 \times 10^{-7}$

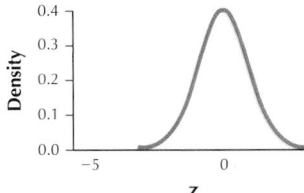

(d) Since the p-value ≤ 0.01, reject H_0. There is evidence that the population mean is less than -10.

16. *Step 1* $H_0 : \mu = 20$ vs. $H_0 : \mu < 20$
We will reject H_0 if the p-value $\leq \alpha = 0.01$.
Step 2 $Z_{\text{data}} = -6.67$
Step 3 p-value ≈ 0
Step 4 The p-value ≈ 0 is ≤ 0.01. Therefore we reject H_0.
There is evidence at level of significance $\alpha = 0.01$ that the population mean difference in sleep is less than 20 minutes.

17. $t_{\text{crit}} = 1.415$

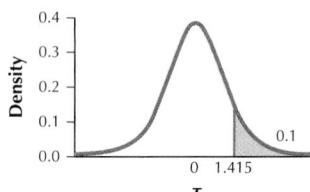

18. $t_{\text{crit}} = 1.895$

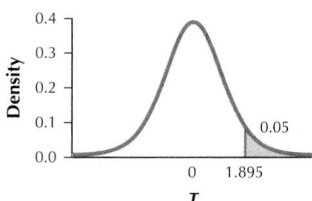

19. $t_{\text{crit}} = 2.998$

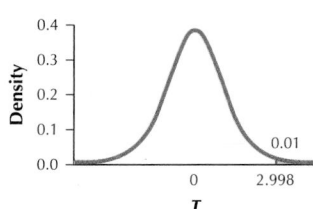

20. For right-tailed tests, t_{crit} increases as α decreases.
21. $H_0 : \mu = 9$ vs. $H_a : \mu \neq 9$. $t_{\text{crit}} = 1.753$. Reject H_0 if $t_{\text{data}} \leq -1.753$ or $t_{\text{data}} \geq 1.753$. $t_{\text{data}} = 1.33$. Since t_{data} is not ≤ -1.753 and t_{data} is not ≥ 1.753, we do not reject H_0. There is insufficient evidence that the population mean is different from 9.
22. $H_0 : \mu = 45$ vs. $H_a : \mu \neq 45$. Reject H_0 if p-value ≤ 0.10. $t_{\text{data}} = 0$. p-value $= 1$ Since p-value is not ≤ 0.10, we do not reject H_0. There is insufficient evidence that the population mean differs from 45.
23. **(a)** We have $np_0 = 100(0.8) = 800 \geq 5$ and $n(1 - p_0) = 1000(1 - 0.8)$ $= 200 \geq 5$. **(b)** $H_0 : p = 0.8$ vs. $H_a : p > 0.8$ **(c)** $Z_{\text{crit}} = 1.28$. Reject H_0 if $Z_{\text{data}} \geq 1.28$. **(d)** $Z_{\text{data}} = 2.37$ **(e)** Since $Z_{\text{data}} \geq 1.28$, we reject H_0. There is evidence that the population proportion is greater than 0.8.
24. **(a)** We have $np_0 = 900(0.2) = 180 \geq 5$ and $n(1 - p_0) = 900(1 - 0.2)$ $= 720 \geq 5$. **(b)** $H_0 : p = 0.2$ vs. $H_a : p < 0.2$ **(c)** 1.645; reject H_0 if $Z_{\text{data}} \leq -1.645$. **(d)** -1.67 **(e)** Since $Z_{\text{data}} \leq -1.645$, we reject H_0. There is evidence that the population proportion is less than 0.2.
25. **(a)** We have $np_0 = 100(0.4) = 40 \geq 5$ and $n(1 - p_0) = 100(1 - 0.4)$ $= 60 \geq 5$. **(b)** $H_0 : p = 0.4$ vs. $H_a : p \neq 0.4$ **(c)** $Z_{\text{crit}} = 2.58$. Reject H_0 if $Z_{\text{data}} \leq -2.58$ or $Z_{\text{data}} \geq 2.58$. **(d)** $Z_{\text{data}} = 3.06$ **(e)** Since $Z_{\text{data}} \geq 2.58$, we reject H_0. There is evidence that the population proportion is not equal to 0.4.
26. **(a)** We have $np_0 = 144(0.7) = 100.8 \geq 5$ and $n(1 - p_0) = 144(1 - 0.7)$ $= 43.2 \geq 5$. **(b)** $H_0 : p = 0.7$ vs. $H_a : p \neq 0.7$. Reject H_0 if the p-value ≤ 0.05. **(c)** $Z_{\text{data}} = 1.67$ **(d)** p-value $= 0.095$ **(e)** Since the p-value is not ≤ 0.05, we do not reject H_0. There is insufficient evidence that the population proportion is different than 0.7.
27. **(a)** We have $np_0 = 100(0.25) = 25 \geq 5$ and $n(1 - p_0) = 100(1 - 0.25)$ $= 75 \geq 5$. **(b)** $H_0 : p = 0.25$ vs. $H_a : p < 0.25$. Reject H_0 if the p-value ≤ 0.05. **(c)** 0 **(d)** 0.5 **(e)** Since the p-value is not ≤ 0.05, we do not reject H_0. There is insufficient evidence that the population proportion is less than 0.25.
28. $np_0 = 1000(0.416) = 416 \geq 5$ and $n(1 - p_0) = 1000(1 - 0.416) = 584 \geq 5$. Thus we can use the Z test. $H_0 : p = 0.416$ vs. $H_a : p < 0.416$. $Z_{\text{crit}} = -1.645$.

Reject H_0 if $Z_{\text{data}} \leq -1.645$. $Z_{\text{data}} \leq -4.23$. Since $Z_{\text{data}} \leq -1.645$, we reject H_0. There is evidence that the population proportion who prefer DSL is less than 0.416.
29. **(a)** $H_0 : \sigma = 6$ vs. $H_a : \sigma > 6$ **(b)** $\chi_a^2 = \chi_{0.05}^2 = 30.144$. Reject H_0 if $\chi_{\text{data}}^2 \geq 30.144$. **(c)** $\chi_{\text{data}}^2 = 42.75$

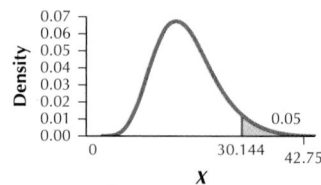

(d) Since $\chi_{\text{data}}^2 \geq 30.144$, we reject H_0. There is evidence that the population standard deviation is greater than 6.
30. **(a)** $H_0 : \sigma = 10$ vs. $H_a : \sigma \neq 10$ **(b)** $\chi_{\alpha/2}^2 = \chi_{0.05}^2 = 40.646$ and $\chi_{1-\alpha/2}^2 = \chi_{0.975}^2 = 13.120$. Reject H_0 if $\chi_{\text{data}}^2 \geq 40.646$ or $\chi_{\text{data}}^2 \leq 13.120$. **(c)** 22.5

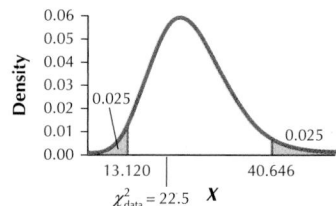

(d) Since χ_{data}^2 is not ≤ 13.120 and χ_{data}^2 is not ≥ 40.646, we do not reject H_0. There is insufficient evidence that the population standard deviation differs from 10.
31. **(a)** $H_0 : \sigma = 35$ vs. $H_a : \sigma < 35$. Reject H_0 if the p-value ≤ 0.05.
(b) 6.857 **(c)** p-value $= 0.5560805474$

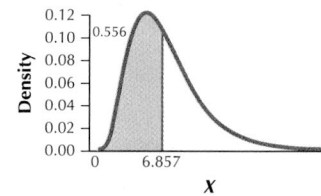

(d) Since the p-value is not ≤ 0.05, we do not reject H_0. There is insufficient evidence that the population standard deviation is less than 35.
32. **(a)** $H_0 : \sigma = 50$ vs. $H_a : \sigma \neq 50$. Reject H_0 if the p-value ≤ 0.05.
(b) 20.25 **(c)** 0.5328172862

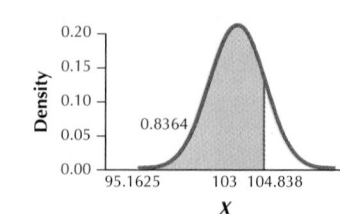

(d) Since the p-value is not ≤ 0.05, we do not reject H_0. There is insufficient evidence that the population standard deviation is not equal to 50.
33. **(a)** $\bar{x}_{\text{critical, lower}} = 95.1625$, $\bar{x}_{\text{critical, upper}} = 104.8375$

(b)

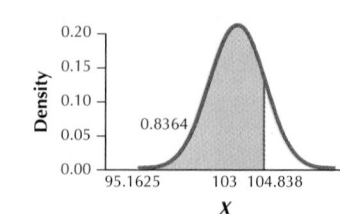

(c) TI-83/84: 0.8364 **(d)** 0.1636

Section 10.3

1. $\hat{p}_1$ and $\hat{p}_2$

2. No, because the confidence interval is not based on the assumption that $p_1 = p_2$.

3. Z_{data} measures the standardized distance between sample proportions. Extreme values of Z_{data} indicate evidence against the null hypothesis.

4. We can either assess the strength of the evidence against the null hypothesis or conduct further studies involving larger samples.

5. (a) $H_0 : p_1 = p_2$ vs. $H_a : p_1 \neq p_2$; $Z_{crit} = 1.645$. Reject H_0 if $Z_{data} \leq -1.645$ or $Z_{data} \geq 1.645$. (b) 0.7857 (c) 0.65 (d) Since $Z_{data} \geq -1.645$ and $Z_{data} \leq 1.645$, we do not reject H_0. There is insufficient evidence that the population proportion from Population 1 is different from the population proportion from Population 2.

6. (a) $H_0 : p_1 = p_2$ vs. $H_a : p_1 \leq p_2$; $Z_{crit} = -1.645$. Reject H_0 if $Z_{data} \leq -1.645$. (b) $\hat{p}_{pooled} = 0.45454545$ (c) $Z_{data} \approx 0.39$ (d) $Z_{data} \approx 0.39$ is not ≤ -1.645. Therefore we do not reject H_0. There is insufficient evidence at level of significance $\alpha = 0.05$ that the population proportion for Population 1 is less than the population proportion for Population 2.

7. (a) $H_0 : p_1 = p_2$ vs. $H_a : p_1 > p_2$. $Z_{crit} = 2.33$. Reject H_0 if $Z_{data} \geq 2.33$ (b) $\hat{p}_{pooled} = 100/450 \approx 0.2222$. (c) $Z_{data} = 3.550$. (d) Since $Z_{data} = 3.550$ is ≥ 2.33, we reject H_0. There is evidence at the $\alpha = 0.01$ level of significance that the population proportion of Population 1 is greater than the population proportion of Population 2.

8. (a) $H_0 : p_1 = p_2$ vs. $H_a : p_1 \neq p_2$. $Z_{crit} = 2.58$. Reject H_0 if $Z_{data} \leq -2.58$ or $Z_{data} \geq 2.58$ (b) $\hat{p}_{pooled} = 100/450 \approx 0.2222$. (c) $Z_{data} \approx 3.550$. (d) Since $Z_{data} \approx 3.550$ is ≥ 2.58, we reject H_0. There is evidence at the $\alpha = 0.01$ level of significance that the population proportion of Population 1 is different from the population proportion of Population 2.

9. (a) $H_0 : p_1 = p_2$ vs. $H_a : p_1 > p_2$. Reject H_0 if p-value ≤ 0.05. (b) $\hat{p}_{pooled} = 450/800 = 0.5625$. (c) $Z_{data} = 3.563$. (d) p-value $= 0.0002$ (e) Since p-value $= 0.0002$ is ≤ 0.05, we reject H_0. There is evidence at the $\alpha = 0.05$ level of significance that the population proportion of Population 1 is greater than the population proportion of Population 2.

10. (a) $H_0 : p_1 = p_2$ vs. $H_a : p_1 < p_2$. Reject H_0 if p-value ≤ 0.05. (b) $\hat{p}_{pooled} = 1110/2000 = 0.555$. (c) $Z_{data} = -5.849$. (d) p-value ≈ 0 (e) Since p-value ≈ 0 is ≤ 0.05, we reject H_0. There is evidence at the $\alpha = 0.05$ level of significance that the population proportion of Population 1 is less than the population proportion of Population 2.

11. (a) $H_0 : p_1 = p_2$ vs. $H_a : p_1 \neq p_2$. Reject H_0 if p-value ≤ 0.10. (b) $\hat{p}_{pooled} = 910/1140 \approx 0.7982$. (c) $Z_{data} \approx -1.284$. (d) p-value ≈ 0.1991 (e) Since p-value ≈ 0.1991 is not ≤ 0.10, we do not reject H_0. There is insufficient evidence at the $\alpha = 0.10$ level of significance that the population proportion of Population 1 is different from the population proportion of Population 2.

12. (a) $H_0 : p_1 = p_2$ vs. $H_a : p_1 < p_2$. Reject H_0 if p-value ≤ 0.10. (b) $\hat{p}_{pooled} = 910/1140 \approx 0.7982$. (c) $Z_{data} \approx -1.284$. (d) p-value $= 0.0996$ (e) Since p-value $= 0.0996$ is ≤ 0.10, we reject H_0. There is evidence at the $\alpha = 0.10$ level of significance that the population proportion of Population 1 is less than the population proportion of Population 2.

13. (a) $x_1 = 80 \geq 5$, $n_1 - x_1 = 20 \geq 5$, $x_2 = 30 \geq 5$, and $n_2 - x_2 = 10 \geq 5$, so it is appropriate. (b) 0.05 (c) 0.1554. The point estimate $\hat{p}_1 - \hat{p}_2$ will lie within $E = 0.1554$ of the difference in population proportions $p_1 - p_2$ 95% of the time. (d) $(-0.1054, 0.2054)$. We are 95% confident that the difference in population proportions lies between -0.1054 and 0.2054.

14. (a) $x_1 = 5 \geq 5$, $n_1 - x_1 = 5 \geq 5$, $x_2 = 5 \geq 5$, and $n_2 - x_2 = 7 \geq 5$, so it is appropriate. (b) 0.0833 (c) 0.41694. The point estimate $\hat{p}_1 - \hat{p}_2$ will lie within $E = 0.41694$ of the difference in population proportions $p_1 - p_2$ 95% of the time. (d) $(-0.3336, 0.50028)$. We are 95% confident that the difference in population proportions lies between -0.3336 and 0.50028.

15. (a) $x_1 = 60 \geq 5$, $n_1 - x_1 = 140 \geq 5$, $x_2 = 40 \geq 5$, and $n_2 - x_2 = 210 \geq 5$, so it is appropriate. (b) 0.14 (c) 0.078. The point estimate $\hat{p}_1 - \hat{p}_2$ will lie within $E = 0.078$ of the difference in population proportions $p_1 - p_2$ 95% of the time. (d) $(0.062, 0.218)$. We are 95% confident that the difference in population proportions lies between 0.062 and 0.218.

16. (a) Yes, because $x_1 = 250 \geq 5$, $n_1 - x_1 = 150 \geq 5$, $x_2 = 200 \geq 5$, and $n_2 - x_2 = 200 \geq 5$. (b) 0.125 (c) 0.0682. The point estimate will lie within 0.0682 of the difference in population proportions $p_1 - p_2$ 95% of the time. (d) $(0.0568, 0.1932)$. We are 95% confident that the difference in population proportions lies between 0.0568 and 0.1932.

17. (a) $x_1 = 490 \geq 5$, $n_1 - x_1 = 510 \geq 5$, $x_2 = 620 \geq 5$, and $n_2 - x_2 = 380 \geq 5$, so it is appropriate. (b) -0.13 (c) 0.0432. The point estimate

$\hat{p}_1 - \hat{p}_2$ will lie within $E = 0.0432$ of the difference in population proportions $p_1 - p_2$ 95% of the time. (d) $(-0.1732, -0.0868)$. We are 95% confident that the difference in population proportions lies between -0.1732 and -0.0868.

18. (a) Yes, because $x_1 = 412 \geq 5$, $n_1 - x_1 = 115 \geq 5$, $x_2 = 498 \geq 5$, and $n_2 - x_2 = 115 \geq 5$. (b) -0.0306 (c) 0.04689. The point estimate will lie within 0.04689 of the difference in population proportions $p_1 - p_2$ 95% of the time. (d) $(-0.0775, 0.01628)$. We are 95% confident that the difference in population proportions lies between -0.0775 and 0.01628.

19. (a) $H_0 : p_1 - p_2 = 0$ vs. $H_a : p_1 - p_2 \neq 0$. The hypothesized value of 0 lies outside the interval (0.5, 0.6), so we reject H_0 at the $\alpha = 0.05$ level of significance. (b) $H_0 : p_1 - p_2 = 0.1$ vs. $H_a : p_1 - p_2 \neq 0.1$. The hypothesized value of 0.1 lies outside the interval (0.5, 0.6), so we reject H_0 at the $\alpha = 0.05$ level of significance. (c) $H_0 : p_1 - p_2 = 0.57$ vs. $H_a : p_1 - p_2 \neq 0.57$. The hypothesized value of 0.57 lies inside the interval (0.5, 0.6), so we do not reject H_0 at the $\alpha = 0.05$ level of significance.

20. (a) $H_0 : p_1 - p_2 = 0.2$ vs. $H_a : p_1 - p_2 \neq 0.2$. The hypothesized value of 0.2 lies inside of the interval (0.01, 0.99), so we do not reject H_0 at the $\alpha = 0.01$ level of significance. (b) $H_0 : p_1 - p_2 = 0$ vs. $H_a : p_1 - p_2 \neq 0$. The hypothesized value of 0 lies outside of the interval (0.01, 0.99), so we reject H_0 at the $\alpha = 0.01$ level of significance. (c) $H_0 : p_1 - p_2 = 0.999$ vs. $H_a : p_1 - p_2 \neq 0.999$. The hypothesized value of 0.999 lies outside of the interval (0.01, 0.99), so we reject H_0 at the $\alpha = 0.01$ level of significance.

21. (a) $H_0 : p_1 - p_2 = 0.151$ vs. $H_a : p_1 - p_2 \neq 0.151$. The hypothesized value of 0.151 lies outside of the interval (0.1, 0.11), so we reject H_0 at the $\alpha = 0.10$ level of significance. (b) $H_0 : p_1 - p_2 = 0.115$ vs. $H_a : p_1 - p_2 \neq 0.115$. The hypothesized value of 0.115 lies outside of the interval (0.1, 0.11), so we reject H_0 at the $\alpha = 0.10$ level of significance. (c) $H_0 : p_1 - p_2 = 0.105$ vs. $H_a : p_1 - p_2 \neq 0.105$. The hypothesized value of 0.105 lies inside of the interval (0.1, 0.11), so we do not reject H_0 at the $\alpha = 0.10$ level of significance.

22. (a) $H_0 : p_1 - p_2 = 0.41$ vs. $H_a : p_1 - p_2 \neq 0.41$. The hypothesized value of 0.41 lies outside of the interval (0.43, 0.57), so we reject H_0 at the $\alpha = 0.05$ level of significance. (b) $H_0 : p_1 - p_2 = 0.51$ vs. $H_a : p_1 - p_2 \neq 0.51$. The hypothesized value of 0.51 lies inside of the interval (0.43, 0.57), so we do not reject H_0 at the $\alpha = 0.05$ level of significance. (c) $H_0 : p_1 - p_2 = 0.61$ vs. $H_a : p_1 - p_2 \neq 0.61$. The hypothesized value of 0.61 lies outside of the interval (0.43, 0.57), so we reject H_0 at the $\alpha = 0.05$ level of significance.

23. (a) $x_1 = 85 \geq 5$, $n_1 - x_1 = 144 - 85 = 59 \geq 5$, $x_2 = 166 \geq 5$, and $n_2 - x_2 = 297 - 166 = 131 \geq 5$. Therefore it is appropriate to perform the Z test for the difference in population proportions. (b) p_1 refers to the population proportion of people 18–29 years old who agree that technology will lead to a better future and p_2 refers to the population proportion of people 65 years old or older who agree that technology will lead to a better future. (c) $H_0 : p_1 = p_2$ vs. $H_a : p_1 \neq p_2$. Reject H_0 if the p-value $\leq \alpha = 0.05$. $Z_{data} \approx 0.62$; p-value $= 0.5352$ (TI-84: 0.5329195849). The p-value $= 0.5352$ is not $\leq \alpha = 0.05$, so we do not reject H_0. There is insufficient evidence at level of significance $\alpha = 0.05$ that the population proportions agreeing that technology will lead to a better future differ between the younger group and the older group.

24. (a) $\hat{p}_1 = x_1/n_1 = 3{,}305/50{,}350 \approx 0.0656$ and $\hat{p}_2 = x_2/n_2 = 73{,}289/754{,}642 \approx 0.0971$. $\hat{p}_1 - \hat{p}_2 \approx -0.0315$ (b) p_1 is the population proportion of Medicare recipients living in Alaska who are age 85 or over and $\hat{p}_1$ is the sample proportion of Medicare recipients living in Alaska who are age 85 or over. (c) Critical-value method: $H_0 : p_1 = p_2$ vs. $H_a : p_1 \neq p_2$. $Z_{crit} = 1.96$. Reject H_0 if $Z_{data} \leq -1.96$ or if $Z_{data} \geq 1.96$. $\hat{p}_{pooled} = 76{,}594/804{,}992 \approx 0.0951$. $Z_{data} = -23.329$ (TI-83/84: -23.307). Since $Z_{data} = -23.329$ is ≤ -1.96, we reject H_0. There is evidence at the $\alpha = 0.05$ level of significance that the population proportion of Medicare recipients living in Alaska who are age 85 or over differs from the population proportion of Medicare recipients living in Arizona who are age 85 or older. p-value method: $H_0 : p_1 = p_2$ vs. $H_a : p_1 \neq p_2$. Reject H_0 if the p-value ≤ 0.05. $\hat{p}_{pooled} = 76{,}594/804{,}992 \approx 0.0951$. $Z_{data} = -23.329$ (TI-83/84: -23.307). p-value ≈ 0. Since the p-value ≈ 0 is ≤ 0.05, we reject H_0. There is evidence at the $\alpha = 0.05$ level of significance that the population proportion of Medicare recipients living in Alaska who are age 85 or over differs from the population proportion of Medicare recipients living in Arizona who are age 85 or older.

25. $x_1 = 34 \geq 5$, $(n_1 - x_1) = 66 \geq 5$, $x_2 = 64 \geq 5$, and $(n_2 - x_2) = 136 \geq 5$. Therefore it is appropriate to perform the Z test for the difference in population proportions. p_1 is the population proportion of Ohio businesses that are owned by women and p_2 is the population proportion of New Jersey businesses that are owned by women. Critical-value method: $H_0 : p_1 = p_2$ vs. $H_a : p_1 > p_2$. $Z_{crit} = 1.28$. Reject H_0 if $Z_{data} \geq 1.28$. $\hat{p}_{pooled} = 98/300 \approx 0.3267$. $Z_{data} = 0.348$. Since $Z_{data} = 0.348$ is not ≥ 1.28, we do not reject H_0. There is insufficient evidence at the $\alpha = 0.10$ level of significance that the population proportion of Ohio businesses that are owned by women is greater than the population proportion of New Jersey businesses that are owned by women. p-value method: $H_0 : p_1 = p_2$ vs. $H_a : p_1 > p_2$. Reject H_0 if the p-value ≤ 0.10. $\hat{p}_{pooled} = 98/300 \approx 0.3267$. $Z_{data} = 0.348$. p-value $= 0.3639$. Since the p-value $= 0.3639$ is not ≤ 0.10, we do not reject H_0. There is insufficient evidence at the $\alpha = 0.10$ level of significance that the population proportion of Ohio businesses that are owned by women is greater than the population proportion of New Jersey businesses that are owned by women.

26. (a) $\hat{p}_1 = x_1/n_1 = 14/54 \approx 0.2593$ and $\hat{p}_2 = X_2/n_2 = 25/45 \approx 0.5556$. $\hat{p}_1 - \hat{p}_2 \approx -0.2963$ (b) p_1 is the population proportion of women with breast cancer who have fetal cells and $\hat{p}_1$ is the sample proportion of women with breast cancer who have fetal cells. (c) Critical-value method: $H_0 : p_1 = p_2$ vs. $H_a : p_1 < p_2$. $Z_{crit} = -2.33$. Reject H_0 if $Z_{data} \leq -2.33$. $\hat{p}_{pooled} = 39/99 \approx 0.3939$. $Z_{data} = -3.004$. Since $Z_{data} = -3.004$ is ≤ -2.33, we reject H_0. There is evidence at the $\alpha = 0.01$ level of significance that the population proportion of women with breast cancer who have fetal cells is less than the population proportion of women without breast cancer with fetal cells. p-value method: $H_0 : p_1 = p_2$ vs. $H_a : p_1 < p_2$. Reject H_0 if the p-value ≤ 0.01. $\hat{p}_{pooled} = 39/99 \approx 0.3939$. $Z_{data} = -3.004$. p-value $= 0.0013$. Since the p-value $= 0.0013$ is ≤ 0.01, we reject H_0. There is evidence at the $\alpha = 0.01$ level of significance that the population proportion of women with breast cancer who have fetal cells is less than the population proportion of women without breast cancer with fetal cells.

27. (a) TI-84 $(-0.0668, 0.1295)$. We are 95% confident that the difference in the population proportions of people 18–29 years old and people age 65 years or older who agree that technology will lead to a better future lies between -0.0668 and 0.1295. (b) $H_0 : p_1 = p_2$ vs. $H_a : p_1 \neq p_2$; 0 lies in the interval so we do not reject H_0 (c) Yes.

28. (a) $(-0.0341, -0.0289)$. TI-83/84: $(-0.0337, -0.0292)$. We are 95% confident that the difference of the population proportion of Medicare recipients living in Alaska who are age 85 and older and the population proportion of Medicare recipients living in Arizona who are age 85 and older lies between $-0.0341(-0.0337)$ and $-0.0289(-0.0292)$. (b) $H_0 : p_1 = p_2$ vs. $H_a : p_1 \neq p_2$. Our hypothesized value of 0 lies outside of the interval in (a), so we reject H_0. There is evidence that the population proportion of Medicare recipients living in Alaska who are age 85 and older differs from the population proportion of Medicare recipients living in New Jersey who are age 85 and older. (c) Yes, it agrees.

29. (a) $(-0.0745, 0.1145)$. TI-83/84: $(-0.0749, 0.1150)$. We are 90% confident that the difference of the population proportion of Ohio businesses that are owned by women and the population proportion of New Jersey businesses that are owned by women lies between $-0.0745(-0.0749)$ and $0.1145(0.1150)$. (b) $H_0 : p_1 = p_2$ vs. $H_a : p_1 \neq p_2$. Our hypothesized value of 0 lies inside the interval in (a), so we do not reject H_0. There is insufficient evidence that the population proportion of Ohio businesses that are owned by women differs from the population proportion of New Jersey businesses that are owned by women. (c) No, it is a one-sided test and confidence intervals can only be used to perform two-sided tests.

30. (a) $(-0.5507, -0.0419)$. TI-83/84: $(-0.5412, -0.0513)$. We are 99% confident that the difference of the population proportion of women with breast cancer who have fetal cells and the population proportion of women without breast cancer who have fetal cells lies between $-0.5507(-0.5412)$ and $-0.0419(-0.0513)$. (b) $H_0 : p_1 = p_2$ vs. $H_a : p_1 \neq p_2$. The hypothesized value of 0 lies outside in (a), so we reject H_0. There is evidence that the population proportion of women with breast cancer who have fetal cells differs from the population proportion of women without breast cancer who have fetal cells. (c) No, it is a one-sided test and confidence intervals can only be used to perform two-sided tests.

31. $H_0 : p_1 = p_2$ vs. $H_a : p_1 \neq p_2$. Reject H_0 if the p-value ≤ 0.05. $\hat{p}_{pooled} = 0.7705$. $Z_{data} = 0.19$. p-value $= 0.8473$. Since the p-value is not $0 \leq 0.05$, we do not reject H_0. There is insufficient evidence that the proportion of the people who wore the ionized bracelets who reported improvement in their

maximum pain index is different from the proportion of the people who wore the placebo bracelets who reported improvement in their maximum pain index.

32. $x_1 = 920 \geq 5$, $n_1 - x_1 = 80 \geq 5$, $x_2 = 870 \geq 5$, and $n_2 - x_2 = 130 \geq 5$, so it is appropriate.

33. $p_1 =$ the population proportion of 18- to 24-year-old males who listen to the radio each week and $p_2 =$ the population proportion of males age 65 or older who listen to the radio each week.

34. p_1 is the population proportion of 18- to 24-year-old males who listen to radio each week and $\hat{p}_1$ is the sample proportion of 18- to 24-year-old males who listen to radio each week.

35. 0.02678. The point estimate of the difference in the population proportion of 18- to 24-year-old males who listen to the radio each week and the population proportion of males 65 years and older who listen to the radio each week will lie within $E = 0.002678$ of the difference in population proportions $p_1 - p_2$ 95% of the time.

36. $(0.02322, 0.07678)$. We are 95% confident that the difference in the population proportion of 18- to 24-year-old males who listen to the radio each week and the population proportion of males 65 years and older who listen to the radio each week lies between 0.02322 and 0.07678.

37. (a) $H_0 : p_1 - p_2 = 0$ vs. $H_a : p_1 - p_2 \neq 0$. The hypothesized value of 0 does not lie in the interval from Exercise 36, so we reject H_0. There is evidence that the difference in the population proportion of 18- to 24-year-old males who listen to the radio each week and the population proportion of males 65 years and older who listen to the radio each week differs from 0. (b) $H_0 : p_1 - p_2 = 0.01$ vs. $H_a : p_1 - p_2 \neq 0.01$. The hypothesized value of 0.01 does not lie in the interval from Exercise 36, so we reject H_0. There is evidence that the difference in the population proportion of 18- to 24-year-old males who listen to the radio each week and the population proportion of males 65 years and older who listen to the radio each week differs from 0.01. (c) $H_0 : p_1 - p_2 = 0.05$ vs. $H_a : p_1 - p_2 \neq 0.05$. The hypothesized value of 0.05 lies in the interval from Exercise 36, so we do not reject H_0. There is insufficient evidence that the difference in the population proportion of 18- to 24-year-old males who listen to the radio each week and the population proportion of males 65 years and older who listen to the radio each week differs from 0.05.

38. We can't because it is a one-tailed test and confidence intervals can only be used to perform a two-tailed test.

39. Critical-value method: $H_0 : p_1 = p_2$ vs. $H_a : p_1 > p_2$. $Z_{crit} = 1.645$. Reject H_0 if $Z_{data} \geq 1.645$. $\hat{p}_{pooled} = 1790/2000 = 0.895$. $Z_{data} = 3.647$. Since $Z_{data} = 3.647$ is ≥ 1.645, we reject H_0. There is evidence at the $\alpha = 0.05$ level of significance that the population proportion of 18- to 24-year-old males who listen to the radio each week is greater than the population proportion of males 65 years and older who listen to the radio each week. p-value method: $H_0 : p_1 = p_2$ vs. $H_a : p_1 > p_2$. Reject H_0 if the p-value ≤ 0.05. $\hat{p}_{pooled} = 1790/2000 = 0.895$. $Z_{data} = 3.647$. p-value $= 0.00013$. Since the p-value 0.00013 is ≤ 0.05, we reject H_0. There is evidence at the $\alpha = 0.05$ level of significance that the population proportion of 18- to 24-year-old males who listen to the radio each week is greater than the population proportion of males 65 years and older who listen to the radio each week.

40. (a) $E = 0.085$. It would increase. (b) It would increase. $Z_{data} = 1.153$. p-value $= 0.1245$. (c) It would change to "Do not reject H_0" since the p-value $= 0.1245 \geq 0.05$.

41. Answers will vary.

42. Answers will vary.

43. $H_0 : p_1 = p_2$ vs. $H_a : p_1 \neq p_2$
Rest of answer will vary.

44. Answers will vary.

Section 10.4

1. Two population standard deviations

2. False

3. True

4. $F_{data} = s_1^2/s_2^2$

5. An F distribution with $n_1 - 1$ degrees of freedom in the numerator and $n_2 - 1$ degrees of freedom in the denominator.

6. False

7. The F-value with $df_1 = n_1 - 1$ numerator degrees of freedom and $df_2 = n_2 - 1$ denominator degrees of freedom, with area α to the right of F_{α, n_1-1, n_2-1}.

8. Switch the values of df_1 and df_2. Find F_{α, n_2-1, n_1-1} using the F table. Calculate $F_{crit} = F_{1-\alpha, n_1-1, n_2-1} = 1/F_{\alpha, n_2-1, n_1-1}$.
9. $F_{\alpha, n_1-1, n_2-1} = F_{0.05, 10, 9} = 3.14$.
10. $F_{\alpha, n_1-1, n_2-1} = F_{0.10, 10, 9} = 2.42$.
11. $F_{\alpha, n_1-1, n_2-1} = F_{0.01, 5, 4} = 15.52$.
12. $F_{\alpha, n_1-1, n_2-1} = F_{0.01, 5, 20} = 4.10$.
13. $F_{1-\alpha, n_1-1, n_2-1} = F_{0.95, 20, 3} = 0.3226$.
14. $F_{1-\alpha, n_1-1, n_2-1} = F_{0.99, 20, 3} = 0.2024$.
15. $F_{1-\alpha, n_1-1, n_2-1} = F_{0.95, 3, 20} = 0.1155$.
16. $F_{1-\alpha, n_1-1, n_2-1} = F_{0.99, 3, 20} = 0.0375$.
17. $F_{\alpha/2, n_1-1, n_2-1} = F_{0.025, 15, 6} = 5.27$, $F_{1-\alpha/2, n_1-1, n_2-1} = F_{0.975, 15, 6} = 0.2933$.
18. $F_{\alpha/2, n_1-1, n_2-1} = F_{0.05, 30, 100} = 1.57$, $F_{1-\alpha/2, n_1-1, n_2-1} = F_{0.95, 30, 100} = 0.5495$.
19. $F_{\alpha/2, n_1-1, n_2-1} = F_{0.025, 6, 15} = 3.41$, $F_{1-\alpha/2, n_1-1, n_2-1} = F_{0.975, 15, 6} = 0.1898$.
20. $F_{\alpha/2, n_1-1, n_2-1} = F_{0.05, 60, 100} = 1.45$, $F_{1-\alpha/2, n_1-1, n_2-1} = F_{0.95, 60, 100} = 0.6329$.
21. **(a)** $H_0 : \sigma_1 = \sigma_2$ vs. $H_a : \sigma_1 > \sigma_2$ **(b)** $F_{crit} = F_{\alpha, n_1-1, n_2-1} = F_{0.05, 10, 6} = 4.06$. Reject H_0 if $F_{data} \geq 4.06$. **(c)** $F_{data} = 4$. **(d)** Since $F_{data} = 4$ is not ≥ 4.06, we do not reject H_0. There is insufficient evidence at the $\alpha = 0.05$ level of significance that the population standard deviation of Population 1 is greater than the population standard deviation of Population 2.
22. **(a)** $H_0 : \sigma_1 = \sigma_2$ vs. $H_a : \sigma_1 < \sigma_2$ **(b)** $F_{crit} = F_{1-\alpha, n_1-1, n_2-1} = F_{0.95, 15, 7} = 0.3690$. Reject H_0 if $F_{data} \leq 0.3690$. **(c)** $F_{data} = 0.4444$. **(d)** Since $F_{data} = 0.4444$ is not ≤ 0.3690, we do not reject H_0. There is insufficient evidence at the $\alpha = 0.05$ level of significance that the population standard deviation of Population 1 is less than the population standard deviation of Population 2.
23. **(a)** $H_0 : \sigma_1 = \sigma_2$ vs. $H_a : \sigma_1 < \sigma_2$ **(b)** $F_{crit} = F_{1-\alpha, n_1-1, n_2-1} = F_{0.90, 9, 20} = 0.4348$. Reject H_0 if $F_{data} \leq 0.4348$. **(c)** $F_{data} = 0.5102$. **(d)** Since $F_{data} = 0.5102$ is not ≤ 0.4348, we do not reject H_0. There is insufficient evidence at the $\alpha = 0.10$ level of significance that the population standard deviation of Population 1 is less than the population standard deviation of Population 2.
24. **(a)** $H_0 : \sigma_1 = \sigma_2$ vs. $H_a : \sigma_1 > \sigma_2$ **(b)** $F_{crit} = F_{\alpha, n_1-1, n_2-1} = F_{0.10, 30, 6} = 2.80$. Reject H_0 if $F_{data} \geq 2.80$. **(c)** $F_{data} = 4$. **(d)** Since $F_{data} = 4$ is ≥ 2.80, we reject H_0. There is evidence at the $\alpha = 0.10$ level of significance that the population standard deviation of Population 1 is greater than the population standard deviation of Population 2.
25. **(a)** $H_0 : \sigma_1 = \sigma_2$ vs. $H_a : \sigma_1 \neq \sigma_2$ **(b)** $F_{crit} = F_{\alpha/2, n_1-1, n_2-1} = F_{0.025, 5, 12} = 3.89$ and $F_{crit} = F_{1-\alpha/2, n_1-1, n_2-1} = F_{0.975, 5, 12} = 0.1511$. Reject H_0 if $F_{data} \geq 3.89$ or if $F_{data} \leq 0.1511$. **(c)** $F_{data} = 5.686$. **(d)** Since $F_{data} = 5.686$ is ≥ 3.89, we reject H_0. There is evidence at the $\alpha = 0.05$ level of significance that the population standard deviation of Population 1 differs from the population standard deviation of Population 2.
26. **(a)** $H_0 : \sigma_1 = \sigma_2$ vs. $H_a : \sigma_1 \neq \sigma_2$ **(b)** $F_{crit} = F_{\alpha/2, n_1-1, n_2-1} = F_{0.025, 20, 6} = 5.17$ and $F_{crit} = F_{1-\alpha/2, n_1-1, n_2-1} = F_{0.975, 20, 6} = 0.3195$. Reject H_0 if $F_{data} \geq 5.17$ or if $F_{data} \leq 0.3195$. **(c)** $F_{data} = 1.168$. **(d)** Since $F_{data} = 1.168$ is not ≥ 5.17 and is not ≤ 0.3195, we do not reject H_0. There is insufficient evidence at the $\alpha = 0.05$ level of significance that the population standard deviation of Population 1 differs from the population standard deviation of Population 2.
27. **(a)** $H_0 : \sigma_1 = \sigma_2$ vs. $H_a : \sigma_1 > \sigma_2$. Reject H_0 if the p-value ≤ 0.01. **(b)** $F_{data} = 2.778$ **(c)** p-value $= 0.0420$. **(d)** Since the p-value $= 0.0420$ is not ≤ 0.01, we do not reject H_0. There is insufficient evidence at the $\alpha = 0.01$ level of significance that the population standard deviation of Population 1 is greater than the population standard deviation of Population 2.
28. **(a)** $H_0 : \sigma_1 = \sigma_2$ vs. $H_a : \sigma_1 < \sigma_2$. Reject H_0 if the p-value ≤ 0.01. **(b)** $F_{data} = 0.3265$ **(c)** p-value $= 0.0272$. **(d)** Since the p-value $= 0.\ 0272$ is not ≤ 0.01, we do not reject H_0. There is insufficient evidence at the $\alpha = 0.01$ level of significance that the population standard deviation of Population 1 is less than the population standard deviation of Population 2.
29. **(a)** $H_0 : \sigma_1 = \sigma_2$ vs. $H_a : \sigma_1 < \sigma_2$. Reject H_0 if the p-value ≤ 0.05. **(b)** $F_{data} = 0.2899$ **(c)** p-value $= 0.0090$. **(d)** Since the p-value $= 0.0090$ is ≤ 0.05, we reject H_0. There is evidence at the $\alpha = 0.05$ level of significance that the population standard deviation of Population 1 is less than the population standard deviation of Population 2.
30. **(a)** $H_0 : \sigma_1 = \sigma_2$ vs. $H_a : \sigma_1 > \sigma_2$. Reject H_0 if the p-value ≤ 0.05. **(b)** $F_{data} = 3.906$ **(c)** p-value ≈ 0. **(d)** Since the p-value ≈ 0 is ≤ 0.05, we reject H_0. There is evidence at the $\alpha = 0.05$ level of significance that the population standard deviation of Population 1 is greater than the population standard deviation of Population 2.
31. **(a)** $H_0 : \sigma_1 = \sigma_2$ vs. $H_a : \sigma_1 \neq \sigma_2$. Reject H_0 if the p-value ≤ 0.10. **(b)** $F_{data} = 1$ **(c)** p-value ≈ 1. **(d)** Since the p-value ≈ 1 is not ≤ 0.10, we do not reject H_0. There is insufficient evidence at the $\alpha = 0.10$ level of significance that the population standard deviation of Population 1 is different from the population standard deviation of Population 2.
32. **(a)** $H_0 : \sigma_1 = \sigma_2$ vs. $H_a : \sigma_1 \neq \sigma_2$. Reject H_0 if the p-value ≤ 0.10. **(b)** $F_{data} = 0.5373$ **(c)** p-value $= 0.3056$. **(d)** Since the p-value $= 0.3056$ is not ≤ 0.10, we do not reject H_0. There is insufficient evidence at the $\alpha = 0.10$ level of significance that the population standard deviation of Population 1 is different from the population standard deviation of Population 2.
33. Critical-value method: $H_0 : \sigma_1 = \sigma_2$ vs. $H_a : \sigma_1 < \sigma_2$. $F_{crit} = F_{1-\alpha, n_1-1, n_2-1} = F_{0.95, 1159, 1570} = 0.9009$. Reject H_0 if $F_{data} \leq 0.9009$. $F_{data} = 0.5890$. Since $F_{data} = 0.5890$ is ≤ 0.9009, we reject H_0. There is evidence at the $\alpha = 0.05$ level of significance that the population standard deviation of the males' BMI is less than the population standard deviation of the females' BMI. p-value method: $H_0 : \sigma_1 = \sigma_2$ vs. $H_a : \sigma_1 < \sigma_2$. Reject H_0 if the p-value ≤ 0.05. $F_{data} = 0.5890$. p-value $= 0$. Since the p-value $= 0$ is ≤ 0.05, we reject H_0. There is evidence at the $\alpha = 0.05$ level of significance that the population standard deviation of the males' BMI is less than the population standard deviation of the females' BMI.
34. Critical-value method: $H_0 : \sigma_1 = \sigma_2$ vs. $H_a : \sigma_1 > \sigma_2$. $F_{crit} = F_{\alpha, n_1-1, n_2-1} = F_{0.10, 42, 246} = 1.34$. Reject H_0 if $F_{data} \geq 1.34$. $F_{data} = 1.8595$. Since $F_{data} = 1.8595$ is ≥ 1.34, we reject H_0. There is evidence at the $\alpha = 0.10$ level of significance that the population standard deviation of the number of cigarettes smoked per day of the Morning High group is greater than the population standard deviation of the number of cigarettes smoked per day of the Flatline group. p-value method: $H_0 : \sigma_1 = \sigma_2$ vs. $H_a : \sigma_1 > \sigma_2$. Reject H_0 if the p-value ≤ 0.10. $F_{data} = 1.8595$. p-value $= 0.0021$. Since the p-value $= 0.0021$ is ≤ 0.10, we reject H_0. There is evidence at the $\alpha = 0.10$ level of significance that the population standard deviation of the number of cigarettes smoked per day of the Morning High group is greater than the population standard deviation of the number of cigarettes smoked per day of the Flatline group.
35. Critical-value method: $H_0 : \sigma_1 = \sigma_2$ vs. $H_a : \sigma_1 \neq \sigma_2$. $F_{crit} = F_{\alpha/2, n_1-1, n_2-1} = F_{0.05, 84, 340} = 1.39$ and $F_{crit} = F_{1-\alpha/2, n_1-1, n_2-1} = F_{0.95, 84, 340} = 0.6623$. Reject H_0 if $F_{data} \leq 0.6623$ or if $F_{data} \geq 1.39$. $F_{data} = 0.8292$. Since $F_{data} = 0.8292$ is not ≤ 0.6623 and not ≥ 1.39, we do not reject H_0. There is insufficient evidence at the $\alpha = 0.10$ level of significance that the population standard deviation of the IQ scores of children with autism differs from the population standard deviation of the IQ scores of children with Asperger's syndrome. p-value method: $H_0 : \sigma_1 = \sigma_2$ vs. $H_a : \sigma_1 \neq \sigma_2$. Reject H_0 if the p-value ≤ 0.10. $F_{data} = 0.8292$. p-value $= 0.3031$. Since the p-value $= 0.3031$ is not ≤ 0.10, we do not reject H_0. There is insufficient evidence at the $\alpha = 0.10$ level of significance that the population standard deviation of the IQ scores of children with autism differs from the population standard deviation of the IQ scores of children with Asperger's syndrome.
36. Critical-value method: $H_0 : \sigma_1 = \sigma_2$ vs. $H_a : \sigma_1 < \sigma_2$. $F_{crit} = F_{1-\alpha, n_1-1, n_2-1} = F_{0.95, 99, 99} = 0.6329$. Reject H_0 if $F_{data} \leq 0.6329$. $F_{data} = 0.1479$. Since $F_{data} = 0.1479$ is ≤ 0.6329, we reject H_0. There is evidence at the $\alpha = 0.05$ level of significance that the population standard deviation of the blood pressures of adults aged 30–44 is less than the population standard deviation of the blood pressures of adults aged 70–79. p-value method: $H_0 : \sigma_1 = \sigma_2$ vs. $H_a : \sigma_1 < \sigma_2$. Reject H_0 if the p-value ≤ 0.05. $F_{data} = 0.1479$. p-value ≈ 0. Since the p-value ≈ 0 is ≤ 0.05, we reject H_0. There is evidence at the $\alpha = 0.05$ level of significance that the population standard deviation of the blood pressures of adults aged 30–44 is less than the population standard deviation of the blood pressures of adults aged 70–79.
37. **(a)** No **(b)** Critical-value method: $H_0 : \sigma_1 = \sigma_2$ vs. $H_a : \sigma_1 > \sigma_2$. $F_{crit} = F_{\alpha, n_1-1, n_2-1} = F_{0.01, 13, 13} = 4.16$. Reject H_0 if $F_{data} \geq 4.16$. $F_{data} = 3173.62408/1.164810989 \approx 2724.5829$. Since $F_{data} \approx 2724.5829$ is ≥ 4.16, we reject H_0. There is evidence at the $\alpha = 0.01$ level of significance that the population standard deviation of gold prices is greater than the population standard deviation of silver prices. p-value method: $H_0 : \sigma_1 = \sigma_2$ vs. $H_a : \sigma_1 > \sigma_2$. Reject H_0 if the p-value ≤ 0.01. $F_{data} = 3173.62408/1.164810989 \approx 2724.5829$. p-value ≈ 0. Since the p-value ≈ 0 is ≤ 0.01, we reject H_0. There is evidence at the $\alpha = 0.01$ level of significance that the population standard deviation of gold prices is greater than the population standard deviation of silver prices.

38. **(a)** Different. The box and the whiskers of the boxplot for Honda cars are much longer than the box and whiskers of the boxplot for Lexus cars. **(b)** Critical-value method: $H_0 : \sigma_1 = \sigma_2$ vs. $H_a : \sigma_1 \neq \sigma_2$. $F_{crit} = F_{\alpha/2, n_1 - 1, n_2 - 1}$ = $F_{0.05, 6, 5}$ = 4.95 and $F_{crit} = F_{1 - \alpha/2, n_1 - 1, n_2 - 1}$ = $F_{0.95, 6, 5}$ = 0.2278. Reject H_0 if $F_{data} \leq 0.2278$ or if $F_{data} \geq 4.95$. $F_{data} = 76.47619048 / 10.96666667 \approx 6.9735$. Since $F_{data} \approx 6.9735$ is ≥ 4.95, we reject H_0. There is evidence at the $\alpha = 0.10$ level of significance that the population standard deviation of the combined mpg of Honda cars differs from the population standard deviation of the combined mpg of Lexus cars. p-value method: $H_0 : \sigma_1 = \sigma_2$ vs. $H_a : \sigma_1 \neq \sigma_2$. Reject H_0 if the p-value ≤ 0.10. $F_{data} = 76.47619048 / 10.96666667 \approx 6.9735$. p-value = 0.0501. Since the p-value = 0.0501 is ≤ 0.10, we reject H_0. There is evidence at the $\alpha = 0.10$ level of significance that the population standard deviation of the combined mpg of Honda cars differs from the population standard deviation of the combined mpg of Lexus cars.

Chapter 10 Review

1. **(a)** $\bar{x}_d = -2.6875$, $s_d = 1.6146$ **(b)** $(-4.0376, -1.3374)$
2. $H_0 : \mu_d = 0$ vs. $H_a : \mu_d < 0$. $t_{crit} = -1.895$. Reject H_0 if $t_{data} \leq -1.895$. $t_{data} = -4.708$. Since $t_{data} \leq -1.895$, we reject H_0. There is evidence that the population mean of the differences is less than 0.
3. $H_0 : \mu_d = 0$ vs. $H_a : \mu_d < 0$. Reject H_0 if p-value < 0.05. $t_{data} = -4.708$. p-value = 0.0010939869. Since the p-value ≤ 0.05, we reject H_0. There is evidence that the population mean of the differences is less than 0.
4. Since both sample sizes are large, $n_1 \geq 30$ and $n_2 \geq 30$, it is appropriate to construct a 95% confidence interval for $\mu_1 - \mu_2$.
5. 0.1
6. 0.005684
7. (0.094, 0.106). We are 95% confident that the interval captures the difference in population means.
8. **(a)** $H_0 : \mu_1 = \mu_2$ vs. $H_a : \mu_1 > \mu_2$. $t_{crit} = 1.306$. Reject H_0 if $t_{data} \geq 1.306$. $t_{data} = 5.117$. Since $t_{data} \geq 1.306$, we reject H_0. There is evidence that the population mean salary of young persons with a college degree is greater than the population mean salary of young persons without a college degree. **(b)** (3348.77, 6651.23). We are 90% confident that the difference in the population mean salaries of young persons with and without college degrees lies between \$3348.77 and \$6651.23.
9. **(a)** $(-0.0715, 0.03212)$ **(b)** Since the p-value = 0.3278 is not ≤ 0.01, we do not reject H_0. There is insufficient evidence that the population proportion of new mothers in Florida who took their babies in for a checkup within one week of delivery is different from the proportion of new mothers in North Carolina who took their babies in for a checkup within one week of delivery.
10. $H_0 : \sigma_1 = \sigma_2$ vs. $H_a : \sigma_1 > \sigma_2$. $F_{crit} = F_{\alpha, n_1 - 1, n_2 - 1}$ = $F_{0.01, 54, 106}$ = 1.74. Reject H_0 if $F_{data} \geq 1.74$. $F_{data} = 13.011$. Since $F_{data} = 13.011$ is ≥ 1.74, we reject H_0. There is evidence at the $\alpha = 0.01$ level of significance that the population standard deviation of Population 1 is greater than the population standard deviation of Population 2.
11. $H_0 : \sigma_1 = \sigma_2$ vs. $H_a : \sigma_1 < \sigma_2$. $F_{crit} = F_{1-\alpha, n_1 - 1, n_2 - 1}$ = $F_{0.95, 124, 26}$ = 0.6173. Reject H_0 if $F_{data} \leq 0.6173$. $F_{data} = 0.3123$. Since $F_{data} = 0.3123$ is ≤ 0.6173, we reject H_0. There is evidence at the $\alpha = 0.05$ level of significance that the population standard deviation of Population 1 is less than the population standard deviation of Population 2.
12. $H_0 : \sigma_1 = \sigma_2$ vs. $H_a : \sigma_1 > \sigma_2$. Reject H_0 if the p-value ≤ 0.05. $F_{data} = 5.112$. p-value ≈ 0. Since the p-value ≈ 0 is ≤ 0.05, we reject H_0. There is evidence at the $\alpha = 0.05$ level of significance that the population standard deviation of Population 1 is greater than the population standard deviation of Population 2.
13. $H_0 : \sigma_1 = \sigma_2$ vs. $H_a : \sigma_1 < \sigma_2$. Reject H_0 if the p-value ≤ 0.01. $F_{data} = 0.4916$. p-value = 0.0001. Since the p-value = 0.0001 is ≤ 0.01, we reject H_0. There is evidence at the $\alpha = 0.01$ level of significance that the population standard deviation of Population 1 is less than the population standard deviation of Population 2.

Chapter 10 Quiz
1. True
2. True
3. False
4. normal; large (greater than or equal to 30)
5. margin of error
6. $\bar{x}_d$
7. $\mu_1 - \mu_2$

8. $\hat{p}_{pooled}$
9. No difference
10. **(a)** (6.6680, 21.3320) **(b)** Since 0 does not lie in the confidence interval, we reject H_0. There is evidence that the population mean difference in the number of cigarettes smoked before and after attending Butt-Enders is different from 0.
11. **(a)** Critical-value method: $H_0 : \mu_1 = \mu_2$ vs. $H_a : \mu_1 < \mu_2$. $t_{crit} = -1.690$. Reject H_0 if $t_{data} \leq -1.690$. $t_{data} = -3.667$. Since $t_{data} = -3.667$ is ≤ -1.690, we reject H_0. There is evidence at the $\alpha = 0.05$ level of significance that the population mean income in Suburb A is less than the population mean income in Suburb B. p-value method: $H_0 : \mu_1 = \mu_2$ vs. $H_a : \mu_1 < \mu_2$. Reject H_0 if p-value ≤ 0.05. $t_{data} = -3.667$. p-value = 0.0004. Since p-value = 0.0004 is ≤ 0.05, we reject H_0. There is evidence at the $\alpha = 0.05$ level of significance that the population mean income in Suburb A is less than the population mean income in Suburb B. **(b)** $(-23, 304.69, -6, 695.31)$. We are 95% confident that the interval captures the difference of the population mean income of Suburb A and the population mean income of Suburb B.
12. (2.2406, 17.7594). We are 95% confident that the interval captures the difference of the population mean number of bottles processed by the updated machine and the population mean number of bottles processed by the non-updated machine.
13. **(a)** Since $t_{data} \geq 1.662$, we reject H_0. There is evidence that the population mean number of bottles processed by the updated machine is greater than the population mean number of bottles processed by the non-updated machine. **(b)** Since confidence intervals can be used only to perform two-tailed tests and the hypothesis test in (a) is a one-tailed test, the confidence interval in Exercise 12 cannot be used to perform the hypothesis test in (a).
14. **(a)** $H_0 : \mu_1 = \mu_2$ vs. $H_a : \mu_1 \neq \mu_2$. $t_{crit} = 1.662$. Reject H_0 if $t_{data} \leq -1.662$ or $t_{data} \geq 1.662$. $t_{data} = -6.129$. Since $t_{data} \leq -1.662$, we reject H_0. There is evidence that the population mean income of people 18 to 24 years old who never married is different from the population mean income of people 18 to 24 years old who are married. **(b)** No, the conclusion of the two-tailed hypothesis test for $\alpha = 0.10$ is "Reject H_0." **(c)** $(-\$7349.928, -\$4214.072)$. The confidence interval does not include 0.

Chapter 11

Section 11.1
1. (1) Each independent trial of the experiment has k possible outcomes, $k = 2, 3, \ldots$ (2) The ith outcome (category) occurs with probability p_i, where $i = 1, 2, \ldots, k$ (3)

$$\sum_{i=1}^{k} p_i = 1$$

2. Answers will vary.
3. It is the long-run mean of that random variable after an arbitrarily large number of trials.
4. H_0: The random variable follows a particular distribution. H_a: The random variable does not follow the distribution specified in H_0.
5. Multinomial
6. Not multinomial
7. Multinomial
8. Not multinomial
9. **(a)** $E_1 = 50, E_2 = 25, E_3 = 25$ **(b)** Conditions are met.
10. **(a)** $E_1 = 2, E_2 = 3, E_3 = 4, E_4 = 1$ **(b)** χ^2 goodness of fit conditions are not met.
11. **(a)** $E_1 = 90, E_2 = 5, E_3 = 4, E_4 = 1$ **(b)** Conditions are not met.
12. **(a)** $E_1 = 80, E_2 = 70, E_3 = 20, E_4 = 20, E_5 = 10$. **(b)** χ^2 goodness of fit conditions are met.
13. 0.667
14. 3.5
15. 7.333
16. 2.183
17. 17.667
18. 8.8535
19. **(a)** $E_1 = 40, E_2 = 30, E_3 = 30$; conditions are met. **(b)** $\chi^2_{crit} = \chi^2_{0.05} = 5.991$. Reject H_0 if $\chi^2_{data} \geq 5.991$. **(c)** 4.167 **(d)** Since χ^2_{data} is not ≥ 5.991, we do not reject H_0. There is insufficient evidence that the random variable does not follow the distribution specified in H_0.
20. **(a)** $E_1 = 30, E_2 = 30, E_3 = 30$. Since none of the expected frequencies is less than 1 and none of the expected frequencies is less than 5, the conditions

for performing the χ^2 goodness of fit test are met. **(b)** $\chi^2_{crit} = \chi^2_{0.01} = 9.210$. Reject H_0 if $\chi^2_{data} \geq 9.210$. **(c)** $\chi^2_{data} = 6.667$ **(d)** Since χ^2_{data} is not ≥ 9.210, we do not reject H_0. There is insufficient evidence that the random variable does not follow the distribution specified in H_0.

21. (a) $E_1 = 80, E_2 = 70, E_3 = 20, E_4 = 20, E_5 = 10$; conditions are met. **(b)** $\chi^2_{crit} = \chi^2_{0.10} = 7.779$. Reject H_0 if $\chi^2_{data} \geq 7.779$. **(c)** 6.607 **(d)** Since χ^2_{data} is not ≥ 7.779, we do not reject H_0. There is insufficient evidence that the random variable does not follow the distribution specified in H_0.

22. (a) $E_1 = 60, E_2 = 40, E_3 = 40, E_4 = 40, E_5 = 20$. Since none of the expected frequencies is less than 1 and none of the expected frequencies is less than 5, the conditions for performing the χ^2 goodness of fit test are met. **(b)** $\chi^2_{crit} = \chi^2_{0.05} = 9.488$. Reject H_0 if $\chi^2_{data} \geq 9.488$. **(c)** $\chi^2_{data} = 0.8$ **(d)** Since χ^2_{data} is not ≥ 9.488, we do not reject H_0. There is insufficient evidence that the random variable does not follow the distribution specified in H_0.

23. (a) Reject H_0 if the p-value ≤ 0.05. $E_1 = 50, E_2 = 50$; conditions are met. **(b)** 4 **(c)** p-value $= 0.0455$. **(d)** Since the p-value ≤ 0.05, we reject H_0. There is evidence that the random variable does not follow the distribution specified in H_0.

24. (a) Reject H_0 if the p-value ≤ 0.10. $E_1 = 50, E_2 = 25, E_3 = 25$. Since none of the expected frequencies is less than 1 and none of the expected frequencies is less than 5, the conditions for performing the χ^2 goodness of fit test are met. **(b)** 0.24 **(c)** p-value $= 0.8869$. **(d)** Since the p-value is not ≤ 0.10, we do not reject H_0. There is insufficient evidence that the random variable does not follow the distribution specified in H_0.

25. (a) Reject H_0 if the p-value ≤ 0.10. $E_1 = 100, E_2 = 50, E_3 = 30, E_4 = 20$; conditions are met. **(b)** 6.083 **(c)** p-value $= 0.1076$. **(d)** Since the p-value is not ≤ 0.10, we do not reject H_0. There is insufficient evidence that the random variable does not follow the distribution specified in H_0.

26. (a) Reject H_0 if the p-value ≤ 0.05. $E_1 = 80, E_2 = 40, E_3 = 40, E_4 = 20$ $E_5 = 20$. Since none of the expected frequencies is less than 1 and none of the expected frequencies is less than 5, the conditions for performing the χ^2 goodness of fit test are met. **(b)** 8.125 **(c)** p-value $= 0.0871$. **(d)** Since the p-value is not ≤ 0.05, we do not reject H_0. There is insufficient evidence that the random variable does not follow the distribution specified in H_0.

27. $H_0 : p_{\text{Barnes and Noble}} = 0.23, p_{\text{Amazon}} = 0.20, p_{\text{Others}} = 0.57$. H_a: The random variable does not follow the distribution specified in H_0. Checking the conditions, the expected frequencies are $E_{\text{Barnes and Noble}} = 230, E_{\text{Amazon}} = 200, E_{\text{Others}} = 570$. Because none of these expected frequencies is less than one, and none of the expected frequencies is less than five, the conditions for performing the goodness of fit test are met. $\chi^2_{crit} = 5.991$. Reject H_0 if $\chi^2_{data} \geq 5.991$. $\chi^2_{data} \approx 6.088$. Compare χ^2_{data} with χ^2_{crit}. $\chi^2_{data} \approx 6.088$ is $\geq \chi^2_{crit} = 5.991$. Therefore, we reject H_0. There is evidence that the variable *seller* does not follow the distribution specified in H_0. In other words, there is evidence that the market share for sales of children's books has changed.

28. $H_0 : p_{\text{believe}} = 0.78, p_{\text{not sure}} = 0.12, p_{\text{did not believe}} = 0.10$. H_a: The random variable does not follow the distribution specified in H_0. $E_{\text{believe}} = 780, E_{\text{not sure}} = 120, E_{\text{did not believe}} = 100$. Since none of the expected frequencies is less than 1 and none of the expected frequencies is less than 5, the conditions for performing the χ^2 goodness of fit test are met. $\chi^2_{crit} = \chi^2_{0.05} = 5.991$. Reject H_0 if $\chi^2_{data} \geq 5.991$. $\chi^2_{data} = 11.885$. Since $\chi^2_{data} \geq 5.991$, we reject H_0. There is evidence that the population proportions of people who either believe, are not sure, or don't believe in angels have changed.

29. $H_0 : p_{\text{Windows 7}} = 0.51, p_{\text{Windows XP}} = 0.25, p_{\text{Windows 8}} = 0.13, p_{\text{Others}} = 0.11$. H_a: The random variable does not follow the distribution specified in H_0. Checking the conditions, the expected frequencies are $E_{\text{Windows 7}} = 5,100$, $E_{\text{Windows XP}} = 2,500, E_{\text{Windows 8}} = 1,300, E_{\text{Others}} = 1,100$, Because none of these expected frequencies is less than one, and none of the expected frequencies is less than five, the conditions for performing the goodness of fit test are met. $\chi^2_{crit} = 7.815$. Reject H_0 if $\chi^2_{data} \geq 7.815$. $\chi^2_{data} \approx 1583.875$. Compare χ^2_{data} with χ^2_{crit}. $\chi^2_{data} \approx 1583.875$ is $\geq \chi^2_{data} = 7.815$. Therefore, we reject H_0. There is evidence that the variable *operating system* does not follow the distribution specified in H_0. In other words, there is evidence that the desktop operating systems market share has changed.

30. $H_0 : p_{\text{Less than high school}} = 0.08, p_{\text{High school diploma}} = 0.23, p_{\text{Some college}} = 0.32$, $p_{\text{Bachelor's degree}} = 0.24, p_{\text{Graduate or professional degree}} = 0.13$. H_a: The random variable does not follow the distribution specified in H_0. $E_{\text{Less than high school}} = 16$, $E_{\text{High school diploma}} = 46, E_{\text{Some college}} = 64, E_{\text{Bachelor;s degree}} = 48$,

$E_{\text{Graduate or professional degree}} = 26$. Since none of the expected frequencies is less than 1 and none of the expected frequencies is less than 5, the conditions for performing the χ^2 goodness of fit test are met. $\chi^2_{crit} = \chi^2_{0.05} = 9.488$. Reject H_0 if $\chi^2_{data} \geq 9.488$. $\chi^2_{data} = 3.980$. Since χ^2_{data} is not ≥ 9.488, we do not reject H_0. There is insufficient evidence that the distribution of education levels has changed.

31. $H_0 : p_{\text{fast food}} = 0.30, p_{\text{food courts}} = 0.46, p_{\text{restaurants}} = 0.24$. H_a: The random variable does not follow the distribution specified in H_0. $E_{\text{fast food}} = 30, E_{\text{food courts}} = 46, E_{\text{restaurants}} = 24$. Since none of the expected frequencies is less than 1 and none of the expected frequencies is less than 5, the conditions for performing the χ^2 goodness of fit test are met. $\chi^2_{crit} = \chi^2_{0.10} = 4.605$. Reject H_0 if $\chi^2_{data} \geq 4.605$. $\chi^2_{data} = 1.371$. Since χ^2_{data} is not ≥ 4.605, we do not reject H_0. There is insufficient evidence that the random variable does not follow the distribution specified in H_0.

32. $H_0 : p_{\text{phip}} = 0.30, p_{\text{mm}} = 0.556, p_{\text{other}} = 0.144$. H_a: The random variable does not follow the distribution specified in H_0. $E_{\text{phip}} = 300, E_{\text{mm}} = 556$, $E_{\text{other}} = 144$. Since none of the expected frequencies is less than 1 and none of the expected frequencies is less than 5, the conditions for performing the χ^2 goodness of fit test are met. $\chi^2_{crit} = \chi^2_{0.05} = 5.991$. Reject H_0 if $\chi^2_{data} \geq 5.991$. $\chi^2_{data} = 14.224$. Since $\chi^2_{data} \geq 5.991$, we reject H_0. There is evidence that the population proportions of minority patients who suffered spinal cord injuries, who had a private health insurance provider, Medicare, Medicaid, or other arrangements, have changed.

33. $H_0 : p_{\text{pizza}} = 0.25, p_{\text{cheeseburger}} = 0.25, p_{\text{quiche}} = 0.25, p_{\text{sushi}} = 0.25$. H_a: The random variable does not follow the distribution specified in H_0. $E_{\text{pizza}} = 125$, $E_{\text{cheeseburger}} = 125, E_{\text{quiche}} = 125, E_{\text{sushi}} = 125$. Since none of the expected frequencies is less than 1 and none of the expected frequencies is less than 5, the conditions for performing the χ^2 goodness of fit test are met. $\chi^2_{crit} = \chi^2_{0.01} = 11.345$. Reject H_0 if $\chi^2_{data} \geq 11.345$. $\chi^2_{data} = 377.2$. Since $\chi^2_{data} \geq 11.345$, we reject H_0. There is evidence that there is a difference in student preference among the four entries.

34. $H_0 : p_{\text{Comcast}} = 0.07, p_{\text{AT\&T}} = 0.07, p_{\text{TimeWarner}} = 0.05, p_{\text{Verizon}} = 0.05$ $p_{\text{Others}} = 0.76$ H_a: The random variable does not follow the distribution specified in H_0. Checking the conditions, the expected frequencies are $E_{\text{Comcast}} = 700, E_{\text{AT\&T}} = 700, E_{\text{TimeWarner}} = 500, E_{\text{Verizon}} = 500$, $E_{\text{Others}} = 7600$. Because none of these expected frequencies is less than one, and none of the expected frequencies is less than five, the conditions for performing the goodness of fit test are met. $\chi^2_{crit} = 13.277$ Reject H_0 if $\chi^2_{data} \geq 13.277$. $\chi^2_{data} \approx 218.797$. Compare χ^2_{data} with χ^2_{crit}. $\chi^2_{data} \approx 218.797$ is $\geq \chi^2_{crit} = 13.277$. Therefore, we reject H_0. There is evidence that the variable *Internet service provider* does not follow the distribution specified in H_0. In other words, there is evidence that the worldwide market share for the Internet service provider market has changed.

35. (a) Stay the same **(b)** Stay the same **(c)** Increases **(d)** Decreases **(e)** Stays the same

Section 11.2

1. Tabular summary of the relationship between two categorical variables

2. Answers will vary.

3. The two-sample Z test for the difference in proportions from Chapter 10 is for comparing proportions of two independent populations, and the χ^2 test for homogeneity of proportions is for comparing proportions of k independent populations.

4. Under the assumption that the variables are independent, $P(A \cap B) = P(A) \cdot P(B) = \left(\frac{\text{row total}}{n}\right) \cdot \left(\frac{\text{column total}}{n}\right)$. Thus the expected frequency of A and B is $n \cdot P(A \cap B) = n \cdot \left(\frac{\text{row total}}{n}\right) \cdot \left(\frac{\text{column total}}{n}\right)$.

5.

	A1	A2	Total
B1	11	19	30
B2	11	19	30
Total	22	38	60

6.

	C1	C2	Total
D1	55	95	150
D2	55	95	150
Total	110	190	300

7.

	E1	E2	E3	Total
F1	30.71	20.79	8.50	60
F2	34.29	23.21	9.50	67
Total	65	44	18	127

8.

	G1	G2	Total
H1	9	9	18
H2	9	9	18
H3	9	9	18
Total	27	27	54

9.

	I1	I2	I3	Total
J1	99.2788	93.6058	102.1154	295
J2	55.5288	52.3558	57.1154	165
J3	20.1923	19.0385	20.7692	60
Total	174.9999	165.0001	180	520

10.

	K1	K2	K3	K4	Total
L1	37.5	72.92	89.58	100	300
L2	23.75	46.18	56.74	63.33	190
L3	28.75	55.90	68.68	76.67	230
Total	90	175	215	240	720

11. (a) H_0: Variable A and Variable B are independent. H_a: Variable A and Variable B are dependent.
(b)

	A1	A2	Total
B1	11	19	30
B2	11	19	30
Total	22	38	60

Since none of the expected frequencies is less than 1 and none of the expected frequencies is less than 5, the conditions for performing the χ^2 test for independence are met. (c) 3.841. Reject H_0 if $\chi^2_{\text{data}} \geq 3.841$. (d) 0.2871 (e) Since χ^2_{data} is not ≥ 3.841, we do not reject H_0. There is insufficient evidence that variable A and variable B are dependent.
12. (a) H_0: Variables E and F are independent. H_a: Variables E and F are dependent.
(b)

	E1	E2	E3	Total
F1	30.71	20.79	8.50	60
F2	34.29	23.21	9.50	67
Total	65	44	18	127

Since none of the expected frequencies is less than 1 and none of the expected frequencies is less than 5, the conditions for performing the χ^2 test for independence are met. (c) $\chi^2_{\text{crit}} = \chi^2_{0.10} = 4.605$. Reject H_0 if $\chi^2_{\text{data}} \geq 4.605$. (d) 0.590 (e) Since χ^2_{data} is not ≥ 4.605, we do not reject H_0. There is insufficient evidence that variable E and variable F are dependent.
13. (a) H_0: Variable I and Variable J are independent. H_a: Variable I and Variable J are dependent.
(b)

	I1	I2	I3	Total
J1	99.2788	93.6058	102.1154	295
J2	55.5288	52.3558	57.1154	165
J3	20.1923	19.0385	20.7692	60
Total	174.9999	165.0001	180	520

Since none of the expected frequencies is less than 1 and none of the expected frequencies is less than 5, the conditions for performing the χ^2 test for

independence are met. (c) 13.277. Reject H_0 if $\chi^2_{\text{data}} \geq 13.277$. (d) 4.000
(e) Since χ^2_{data} is not ≥ 13.277, we do not reject H_0. There is insufficient evidence that variable I and variable J are dependent.
14. (a) H_0: Variables I and J are independent. H_a: Variables I and J are dependent.
(b)

	I1	I2	I3	Total
J1	99.2788	93.6058	102.1154	295
J2	55.5288	52.3558	57.1154	165
J3	20.1923	19.0385	20.7692	60
Total	174.9999	165.0001	180	520

Since none of the expected frequencies is less than 1 and none of the expected frequencies is less than 5, the conditions for performing the χ^2 test for independence are met. (c) $\chi^2_{\text{crit}} = \chi^2_{0.10} = 7.779$. Reject H_0 if $\chi^2_{\text{data}} \geq 7.779$. (d) 4.000 (e) Since χ^2_{data} is not ≥ 7.779, we do not reject H_0. There is insufficient evidence that variable I and variable J are dependent.
15. (a) H_0: Variable C and Variable D are independent. H_a: Variable C and Variable D are dependent. Reject H_0 if the p-value ≤ 0.05.

	C1	C2	Total
D1	55	95	150
D2	55	95	150
Total	110	190	300

Since none of the expected frequencies is less than 1 and none of the expected frequencies is less than 5, the conditions for performing the χ^2 test for independence are met. (b) 1.4354 (c) p-value = 0.2309 (d) Since the p-value is not ≤ 0.05, we do not reject H_0. There is insufficient evidence that variable C and variable D are dependent.
16. (a) H_0: Variables G and H are independent. H_a: Variables G and H are dependent.

	G1	G2	Total
H1	9	9	18
H2	9	9	18
H3	9	9	18
Total	27	27	54

Since none of the expected frequencies is less than 1 and none of the expected frequencies is less than 5, the conditions for performing the χ^2 test for independence are met. Reject H_0 if the p-value ≤ 0.10. (b) 0.444
(c) p-value = 0.8007 (d) Since the p-value is not ≤ 0.10, we do not reject H_0. There is insufficient evidence that variables G and H are dependent.
17. (a) H_0: Variable K and Variable L are independent. H_a: Variable K and Variable L are dependent. Reject H_0 if p-value ≤ 0.01.

	K1	K2	K3	K4	Total
L1	37.5	72.92	89.58	100	300
L2	23.75	46.18	56.74	63.33	190
L3	28.75	55.90	68.68	76.67	230
Total	90	175	215	240	720

Since none of the expected frequencies is less than 1 and none of the expected frequencies is less than 5, the conditions for performing the χ^2 test for independence are met. (b) $\chi^2_{\text{data}} = 4.906$ (c) p-value = 0.5560 (d) Since the p-value is not ≤ 0.01, we do not reject H_0. There is insufficient evidence that variable K and variable L are dependent.
18. (a) H_0: Variables K and L are independent. H_a: Variables K and L are dependent.

	K1	K2	K3	K4	Total
L1	37.5	72.92	89.58	100	300
L2	23.75	46.18	56.74	63.33	190
L3	28.75	55.90	68.68	76.67	230
Total	90	175	215	240	720

Since none of the expected frequencies is less than 1 and none of the expected frequencies is less than 5, the conditions for performing the χ^2 test for independence are met. Reject H_0 if the p-value ≤ 0.10. **(b)** $\chi^2_{data} = 4.906$ **(c)** p-value $= 0.5560$ **(d)** Since the p-value is not ≤ 0.10, we do not reject H_0. There is insufficient evidence that variables K and L are dependent.

19. (a) $H_0: p_1 = p_2 = p_3$. H_a: Not all the proportions in H_0 are equal.

(b)

	Sample 1	Sample 2	Sample 3	Total
Successes	9.63	20.86	29.52	60.01
Failures	20.37	44.14	62.48	126.99
Total	30	65	92	187

Since none of the expected frequencies is less than 1 and none of the expected frequencies is less than 5, the conditions for performing the χ^2 test for homogeneity of proportions are met. **(c)** 5.991. Reject H_0 if $\chi^2_{data} \geq 5.991$. **(d)** 0.0847 **(e)** Since χ^2_{data} is not ≥ 5.991, we do not reject H_0. There is insufficient evidence that not all the proportions in H_0 are equal.

20. (a) $H_0: p_1 = p_2 = p_3$. H_a: Not all the proportions in H_0 are equal.

(b)

	Sample 1	Sample 2	Sample 3	Total
Successes	48.3092	50.2415	101.4493	200
Failures	201.6908	209.7585	423.5507	835
Total	250	260	525	1035

Since none of the expected frequencies is less than 1 and none of the expected frequencies is less than 5, the conditions for performing the χ^2 test for homogeneity of proportions are met. **(c)** $\chi^2_{crit} = \chi^2_{0.05} = 5.991$. Reject H_0 if $\chi^2_{data} \geq 5.991$. **(d)** 0.100 **(e)** Since χ^2_{data} is not ≥ 5.991, we do not reject H_0. There is insufficient evidence that not all the proportions in H_0 are equal.

21. (a) $H_0: p_1 = p_2 = p_3 = p_4$. H_a: Not all the proportions in H_0 are equal.

(b)

	Sample 1	Sample 2	Sample 3	Sample 4	Total
Successes	9.67	15.08	20.11	25.14	70
Failures	15.33	23.92	31.89	39.86	111
Total	25	39	52	65	181

Since none of the expected frequencies is less than 1 and none of the expected frequencies is less than 5, the conditions for performing the χ^2 test for homogeneity of proportions are met. **(c)** 7.815. Reject H_0 if $\chi^2_{data} \geq 7.815$. **(d)** 0.0215 **(e)** Since χ^2_{data} is not ≥ 7.815, we do not reject H_0. There is insufficient evidence that not all the proportions in H_0 are equal.

22. (a) $H_0: p_1 = p_2 = p_3 = p_4$. H_a: Not all the proportions in H_0 are equal.

(b)

	Sample 1	Sample 2	Sample 3	Sample 4	Total
Successes	96.6851	150.8287	201.1050	251.3812	700
Failures	153.3149	239.1713	318.8950	398.6188	1110
Total	250	390	520	650	1810

Since none of the expected frequencies is less than 1 and none of the expected frequencies is less than 5, the conditions for performing the χ^2 test for homogeneity of proportions are met. **(c)** $\chi^2_{crit} = \chi^2_{0.05} = 7.815$. Reject H_0 if $\chi^2_{data} \geq 7.815$. **(d)** 0.2150 **(e)** Since χ^2_{data} is not ≥ 7.815, we do not reject H_0. There is insufficient evidence that not all the proportions in H_0 are equal.

23. (a) $H_0: p_1 = p_2 = p_3$. H_a: Not all the proportions in H_0 are equal. Reject H_0 if the p-value ≤ 0.05.

	Sample 1	Sample 2	Sample 3	Total
Successes	27.17	57.74	95.09	180
Failures	12.83	27.26	44.91	85
Total	40	85	140	265

Since none of the expected frequencies is less than 1 and none of the expected frequencies is less than 5, the conditions for performing the χ^2 test for homogeneity of proportions are met. **(b)** 2.0468 **(c)** p-value $= 0.3594$.

(d) Since the p-value is not ≤ 0.05, we do not reject H_0. There is insufficient evidence that not all the proportions in H_0 are equal.

24. (a) $H_0: p_1 = p_2 = p_3$. H_a: Not all the proportions in H_0 are equal. Reject H_0 if the p-value ≤ 0.05.

	Sample 1	Sample 2	Sample 3	Total
Successes	99.3103	120	140.6897	360
Failures	20.6897	25	29.3103	75
Total	120	145	170	435

Since none of the expected frequencies is less than 1 and none of the expected frequencies is less than 5, the conditions for performing the χ^2 test for homogeneity of proportions are met. **(b)** 0.047 **(c)** p-value $= 0.9766$. **(d)** Since the p-value is not ≤ 0.05, we do not reject H_0. There is insufficient evidence that not all the proportions in H_0 are equal.

25. (a) $H_0: p_1 = p_2 = p_3 = p_4$. H_a: Not all the proportions in H_0 are equal. Reject H_0 if the p-value ≤ 0.05.

	Sample 1	Sample 2	Sample 3	Sample 4	Total
Successes	8.98	12.35	21.88	34.79	78
Failures	7.02	9.65	17.12	27.21	61
Total	16	22	39	62	139

Since none of the expected frequencies is less than 1 and none of the expected frequencies is less than 5, the conditions for performing the χ^2 test for homogeneity of proportions are met. **(b)** 1.263 **(c)** p-value $= 0.7379$. **(d)** Since the p-value is not ≤ 0.05, we do not reject H_0. There is insufficient evidence that not all the proportions in H_0 are equal.

26. (a) $H_0: p_1 = p_2 = p_3 = p_4$. H_a: Not all the proportions in H_0 are equal. Reject H_0 if the p-value ≤ 0.05.

	Sample 1	Sample 2	Sample 3	Sample 4	Total
Successes	83.8710	174.1935	290.3226	451.6129	1000
Failures	46.1290	95.8065	159.6774	248.3871	550
Total	130	270	450	700	1550

Since none of the expected frequencies is less than 1 and none of the expected frequencies is less than 5, the conditions for performing the χ^2 test for homogeneity of proportions are met. **(b)** 37.048 **(c)** p-value ≈ 0. **(d)** Since the p-value ≤ 0.05, we reject H_0. There is evidence that not all the proportions in H_0 are equal.

27. $H_0: p_{edit} = p_{arrange}$. H_a: Not all the proportions in H_0 are equal. Reject H_0 if p-value ≤ 0.05. Since none of the expected frequencies is less than 1 and none of the expected frequencies is less than 5, the conditions for performing the χ^2 test for homogeneity of proportions are met. $\chi^2_{data} = 3.5165$. p-value $= 0.0608$. Since the p-value is not ≤ 0.05, we do not reject H_0. There is insufficient evidence that the proportions who favor email differ between the two tasks.

28. $H_0: p_{Normal\ or\ underweight} = p_{Overweight} = p_{Obese}$. H_a: Not all of the proportions in H_0 are equal.

```
Chi-Square Test for Association: Uses the computer, Worksheet columns

Rows: Uses the computer   Columns: Worksheet columns

                    Normal or
                    underweight  Overweight   Obese   All
>2 hours per day         114          28        52    194
                      118.78       31.40     43.81

<=2 hours per dat        355          96       121    572
                      350.22       92.60    129.19

All                      469         124       173    766

Cell Contents:       Count
                     Expected count

Pearson Chi-Square = 2.800, DF = 2, P-Value = 0.247
Likelihood Ratio Chi-Square = 2.739, DF = 2, P-Value = 0.254
```

From the Minitab output above, none of these expected frequencies is less than one, and none of the expected frequencies is less than five. Therefore, the conditions for performing the χ^2 test for homogeneity of proportions are met.

Reject H_0 if the p-value $\leq \alpha = 0.05$. $\chi^2_{\text{data}} = 2.800$. p-value $= 0.247$. The p-value $= 0.247$ is not less than or equal to $\alpha = 0.05$. Therefore, we do not reject H_0. Insufficient evidence exists, at level of significance $\alpha = 0.05$, that the population proportions of children who use the computer more than two hours per day are not the same for the three weight statuses.

29. H_0: *Cause of death* and *age group* are independent. H_a: *Cause of death* and *age group* are dependent.

```
Chi-Square Test for Association: Age group, Worksheet columns

Rows: Age group   Columns: Worksheet columns

                  Heat-related  Cold-related  Floods/Storms/Lightening   All

15 to 24          106           286                                97    489
                  142.7         310.4                              35.9

75 to 84          490           1010                               53   1553
                  453.3         985.6                             114.1

All               596           1296                              150   2042

Cell Contents:    Count
                  Expected count

Pearson Chi-Square = 151.500, DF = 2, P-Value = 0.000
Likelihood Ratio Chi-Square = 127.302, DF = 2, P-Value = 0.000
```

From the Minitab output above, none of these expected frequencies is less than one, and none of the expected frequencies is less than five. Therefore, the conditions for performing the χ^2 test for independence are met. Reject H_0 if the p-value $\leq \alpha = 0.05$. $\chi^2_{\text{data}} = 151.500$. p-value ≈ 0. The p-value ≈ 0 is less than or equal to $\alpha = 0.05$. Therefore, we reject H_0. Evidence exists, at level of significance $\alpha = 0.05$, that the variables *Cause of death* and *age group* are dependent.

30. No. For the work email group the dark green bar is longer, and for the personal email group the light green bar is longer. This means that no spam is more common for the work email group and that some spam is more common for the personal email group.

31. H_0: $p_{\text{work}} = p_{\text{personal}}$. H_a: Not all the proportions in H_0 are equal. Reject H_0 if p-value ≤ 0.01. Since none of the expected frequencies is less than 1 and none of the expected frequencies is less than 5, the conditions for performing the χ^2 test for homogeneity of proportions are met. $\chi^2_{\text{data}} = 23.3325$. p-value ≈ 0. Since p-value ≤ 0.01, we reject H_0. There is evidence that the population proportions who report "a lot of spam" are not the same for work email and personal email. Yes.

32. (a) The type of game with the highest frequency for males is video games and the type of game with the highest frequency for females is computer games. We see some evidence that the most frequently played type of game depends in part on gender and that the two variables may not be independent. We thus might expect to reject H_0. **(b)** Since p-value ≤ 0.01, we reject H_0. There is evidence that *gender* and *type of game* are dependent.

33. H_0: $p_{\text{urban}} = p_{\text{suburban}} = p_{\text{rural}}$. H_a: Not all the proportions in H_0 are equal. Reject H_0 if p-value ≤ 0.05. Since none of the expected frequencies is less than 1 and none of the expected frequencies is less than 5, the conditions for performing the χ^2 test for homogeneity of proportions are met. $\chi^2_{\text{data}} = 9.095$. p-value $= 0.0106$. Since p-value ≤ 0.05, we reject H_0. There is evidence that the population proportions of residents from the three categories who use online dating are not all the same.

34. 478 observations, 11 variables.

35. (a) Dependent **(b)** Since the p-value ≈ 0, p-value ≤ 0.05. Thus we reject H_0. There is evidence that *gender* and *goals* are dependent.

36. (a) Independent **(b)** Since the p-value $= 0.841$, the p-value is not ≤ 0.05. Thus we do not reject H_0. There is insufficient evidence that *gender* and *grade* are dependent.

37. (a) Dependent **(b)** Since the p-value ≈ 0.001, p-value ≤ 0.10. Thus we reject H_0. There is evidence that *urb_rural* and *goals* are dependent.

38. (a) Yes; yes **(b)** H_0: *Grade* and *goals* are independent. H_a: *Grade* and *goals* are dependent. Reject H_0 if p-value ≤ 0.01. Since the p-value $= 0.859$, the p-value is not ≤ 0.01. Thus we do not reject H_0. There is insufficient evidence that *grade* and *goals* are dependent.

39. H_0: $p_{\text{Jan}} = p_{\text{Feb}} = p_{\text{Mar}} = p_{\text{Apr}} = p_{\text{May}} = p_{\text{June}} = p_{\text{July}} = p_{\text{Aug}} = p_{\text{Sept}} = p_{\text{Oct}} = p_{\text{Nov}} = p_{\text{Dec}}$. H_a: Not all the proportions in H_0 are equal. Reject H_0 if p-value ≤ 0.01. $\chi^2_{\text{data}} = 30.2538$. p-value $= 0.001$. Since p-value ≤ 0.01, we reject H_0. There is evidence that the population proportion of "drafted dates" is not equal for all months.

40. H_0: $p_{\text{Jan}} = p_{\text{Feb}} = p_{\text{Mar}} = p_{\text{Apr}} = p_{\text{May}} = p_{\text{June}} = p_{\text{July}} = p_{\text{Aug}} = p_{\text{Sept}} = p_{\text{Oct}} = p_{\text{Nov}} = p_{\text{Dec}}$. H_a: Not all the proportions in H_0 are equal. Since none of the

expected frequencies is less than 1 and none of the expected frequencies is less than 5, the conditions for performing the χ^2 test for homogeneity of proportions are met. $\chi^2_{\text{crit}} = \chi^2_{0.10} = 17.275$. Reject H_0 if $\chi^2_{\text{data}} \geq 17.275$. $\chi^2_{\text{data}} = 6.617$. Since χ^2_{data} is not ≥ 17.275, we do not reject H_0. There is insufficient evidence that not all the proportions in H_0 are equal.

Chapter 11 Review

1. H_0: $p_{\text{US Can}} = 0.32$, $p_{\text{US Mex}} = 0.22$, $p_{\text{Can US}} = 0.31$, $p_{\text{Mex US}} = 0.15$. H_a: The random variable does not follow the distribution specified in H_0. $E_{\text{US Can}} = \$22.4$ billion, $E_{\text{US Mex}} = \$15.4$ billion, $E_{\text{Can US}} = \$21.7$ billion, $E_{\text{Mex US}} = \$10.5$ billion. Since none of the expected frequencies is less than 1 and none of the expected frequencies is less than 5, the conditions for performing the χ^2 goodness of fit test are met. Reject H_0 if p-value ≤ 0.05. $\chi^2_{\text{data}} = 0.469$. p-value $= 0.9256$. Since p-value is not ≤ 0.05, we do not reject H_0. There is insufficient evidence that the population proportions of truck-hauled trade have changed.

2. H_0: $p_{\text{abused alcohol}} = 0.25$, $p_{\text{alcohol dependent}} = 0.06$, $p_{\text{other}} = 0.69$. H_a: The random variable does not follow the distribution specified in H_0. $E_{\text{abused alcohol}} = 250$, $E_{\text{alcohol dependent}} = 60$, $E_{\text{other}} = 690$. Since none of the expected frequencies is less than 1 and none of the expected frequencies is less than 5, the conditions for performing the χ^2 goodness of fit test are met. Reject H_0 if p-value ≤ 0.10. $\chi^2_{\text{data}} = 4.493$. p-value $= 0.1057817985$. Since the p-value is not ≤ 0.10, we do not reject H_0. There is insufficient evidence that the population proportions have changed.

3. H_0: $p_0 = 0.10$, $p_1 = 0.10$, $p_2 = 0.10$, $p_3 = 0.10$, $p_4 = 0.10$, $p_5 = 0.10$, $p_6 = 0.10$, $p_7 = 0.10$, $p_8 = 0.10$, $p_9 = 0.10$. H_a: Not all the proportions in H_0 are equal. Reject H_0 if p-value ≤ 0.05. $E_0 = 21.8$, $E_1 = 21.8$, $E_2 = 21.8$, $E_3 = 21.8$, $E_4 = 21.8$, $E_5 = 21.8$, $E_6 = 21.8$, $E_7 = 21.8$, $E_8 = 21.8$, $E_9 = 21.8$. Since none of the expected frequencies is less than 1 and none of the expected frequencies is less than 5, the conditions for performing the χ^2 goodness of fit test are met. Reject H_0 if p-value ≤ 0.05. $\chi^2_{\text{data}} = 12.4587$. p-value $= 0.1886664729$. Since the p-value is not ≤ 0.05, we do not reject H_0. There is insufficient evidence that the population proportions of digits are not all 0.10.

4. H_0: $p_{18-34} = 0.057$, $p_{35-49} = 0.207$, $p_{50-64} = 0.388$, $p_{\text{over 65}} = 0.348$. H_a: The random variable does not follow the distribution specified in H_0. $E_{18-34} = 57$, $E_{35-49} = 207$, $E_{50-64} = 388$, $E_{\text{over 65}} = 348$. Since none of the expected frequencies is less than 1 and none of the expected frequencies is less than 5, the conditions for performing the χ^2 goodness of fit test are met. $\chi^2_{\text{crit}} = \chi^2_{0.05} = 7.815$. Reject H_0 if $\chi^2_{\text{data}} \geq 7.815$. $\chi^2_{\text{data}} = 28.233$. Since $\chi^2_{\text{data}} \geq 7.815$, we reject H_0. There is evidence that the proportions have changed.

5. (a) A higher proportion of the females with high GPAs take the SAT exam than the proportion of the females with lower GPAs. **(b)** H_0: $p_{A+} = p_A = p_{A-} = p_B = p_C = p_{D/F}$. H_a: Not all the proportions in H_0 are equal. Reject H_0 if p-value ≤ 0.05. Since none of the expected frequencies is less than 1 and none of the expected frequencies is less than 5, the conditions for performing the χ^2 test for homogeneity of proportions are met. $\chi^2_{\text{data}} = 14.6786$. p-value $= 0.0118277763$. Since the p-value ≤ 0.05, we reject H_0. There is evidence that the proportion of females is not all the same across the six grade categories.

6. H_0: $p_{\text{Northeast}} = p_{\text{Midwest}} = p_{\text{South}} = p_{\text{West}}$. H_a: Not all the proportions in H_0 are equal. Reject H_0 if p-value ≤ 0.01. Since none of the expected frequencies is less than 1 and none of the expected frequencies is less than 5, the conditions for performing the χ^2 test for homogeneity of proportions are met. $\chi^2_{\text{data}} = 25.846$. p-value ≈ 0. Since the p-value ≤ 0.01, we reject H_0. There is evidence that the population proportions of pregnant women who have had an HIV test in the past 12 months are not the same across all four regions.

7. H_0: Age and radio station type are independent. H_a: Age and radio station type are not independent. Reject H_0 if p-value ≤ 0.05. Since none of the expected frequencies is less than 1 and none of the expected frequencies is less than 5, the conditions for performing the χ^2 test for independence are met. $\chi^2_{\text{data}} = 8.269$. p-value $= 0.0040$. Since p-value ≤ 0.05, we reject H_0. There is evidence that age and radio station type are not independent.

8. H_0: *Happiness in marriage* and *gender* are independent. H_a: *Happiness in marriage* and *gender* are not independent. Since none of the expected frequencies is less than 1 and none of the expected frequencies is less than 5, the conditions for performing the χ^2 test for independence are met. $\chi^2_{\text{crit}} = \chi^2_{0.05} = 5.991$. Reject H_0 if $\chi^2_{\text{data}} \geq 5.991$. $\chi^2_{\text{data}} = 3.190$. Since χ^2_{data} is not ≤ 5.991, we do not reject H_0. There is insufficient evidence that *happiness in marriage* and *gender* are not independent.

Chapter 11 Quiz

1. True
2. False
3. False
4. 1, 5
5. equal
6. expected frequency
7. The critical value method and the *p*-value method.
8. H_a, the alternative hypothesis
9. Degrees of freedom $= (r - 1)(c - 1)$, where $r =$ the number of categories in the row variable and $c =$ the number of categories in the column variable.
10. $E_1 = 10$, $E_2 = 10$, $E_3 = 10$, $E_4 = 10$, $E_5 = 10$. Conditions are met. $\chi^2_{\text{crit}} = 9.488$. Reject H_0 if $\chi^2_{\text{data}} \geq 9.488$. $\chi^2_{\text{data}} = 1$. Since χ^2_{data} is not ≥ 9.488, we do not reject H_0. There is insufficient evidence that the random variable does not follow the distribution specified in H_0.
11. $E_1 = 48$, $E_2 = 40$, $E_3 = 32$, $E_4 = 24$, $E_5 = 9.6$, $E_6 = 6.4$. Conditions are met. $\chi^2_{\text{crit}} = 11.071$. Reject H_0 if $\chi^2_{\text{data}} \geq 11.071$. $\chi^2_{\text{data}} = 2.917$. Since χ^2_{data} is not ≥ 11.071, we do not reject H_0. There is insufficient evidence that the random variable does not follow the distribution specified in H_0.
12. $E_1 = 20$, $E_2 = 20$, $E_3 = 20$, $E_4 = 20$, $E_5 = 20$. Conditions are met. $\chi^2_{\text{crit}} = 13.277$. Reject H_0 if $\chi^2_{\text{data}} \geq 13.277$. $\chi^2_{\text{data}} = 0.5$. Since χ^2_{data} is not ≥ 13.277, we do not reject H_0. There is insufficient evidence that the random variable does not follow the distribution specified in H_0.
13. $E_1 = 60$, $E_2 = 50$, $E_3 = 40$, $E_4 = 30$, $E_5 = 12$, $E_6 = 8$. Conditions are met. $\chi^2_{\text{crit}} = 11.071$. Reject H_0 if $\chi^2_{\text{data}} \geq 11.071$. $\chi^2_{\text{data}} = 5.5$. Since χ^2_{data} is not ≥ 11.071, we do not reject H_0. There is insufficient evidence that the random variable does not follow the distribution specified in H_0.
14. (a) The higher the grade level, the higher the proportion of students who have used an illicit drug. **(b)** $H_0: p_{\text{8th-graders}} = p_{\text{10th-graders}} = p_{\text{12th-graders}}$. H_a: Not all the proportions in H_0 are equal. Reject H_0 if *p*-value ≤ 0.01. Since none of the expected frequencies is less than 1 and none of the expected frequencies is less than 5, the conditions for performing the χ^2 test for homogeneity of proportions are met. $\chi^2_{\text{data}} = 3060.142231$. *p*-value ≈ 0. Since the *p*-value ≤ 0.01, we reject H_0. There is evidence that the proportions of children in those grades that have ever used an illicit drug are not all the same.
15. H_0: *Gender* and *sport preference* are independent. H_a: *Gender* and *sport preference* are not independent. Reject H_0 if *p*-value ≤ 0.05. Since none of the expected frequencies is less than 1 and none of the expected frequencies is less than 5, the conditions for performing the χ^2 test for independence are met. $\chi^2_{\text{data}} = 19.857$. *p*-value $= 0.00004876$. Since *p*-value ≤ 0.05, we reject H_0. There is evidence that *gender* and *sport preference* are not independent.
16. $H_0: p_{\text{Texas}} = p_{\text{Oklahoma}} = p_{\text{Pennsylvania}}$. H_a: Not all the proportions in H_0 are equal. Reject H_0 if *p*-value ≤ 0.05. Since none of the expected frequencies is less than 1 and none of the expected frequencies is less than 5, the conditions for performing the χ^2 test for homogeneity of proportions are met. $\chi^2_{\text{data}} = 81,246.70747$. *p*-value ≈ 0. Since *p*-value ≤ 0.01, we reject H_0. There is evidence that the population proportions of cattle on smaller farms are not the same across all three states.

Chapter 12

Section 12.1

1. No. If the sample sizes are not all the same, then we need to calculate the overall sample mean by calculating the weighted mean of the sample means where the weights are the sample sizes.
2. MSTR measures the variability in the sample means. MSE measures the variability within the samples.
3. Answers will vary.
4. Each of the k populations is normally distributed, the variances of the populations are all equal, and the samples are independently drawn.
5. Against.
6. False. It means that at least one of the population means is different from the rest.

7. (a)

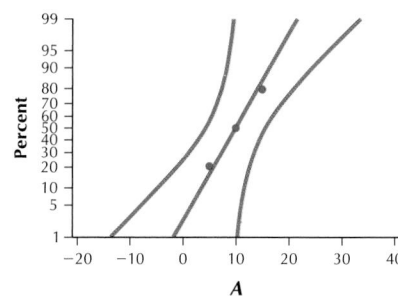

A

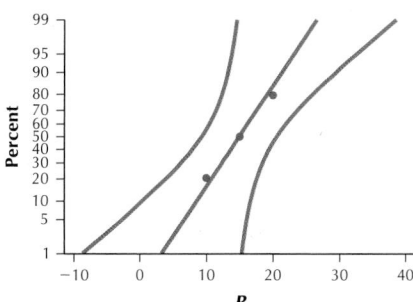

B

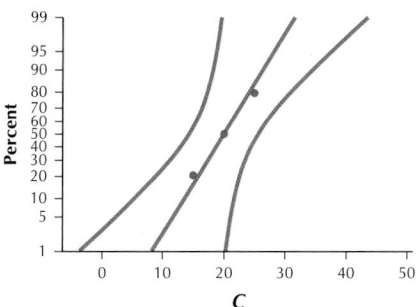

C

All three graphs show acceptable normality.

(b)

Means			
Factor	N	Mean	StDev
A	3	10.00	5.00
B	3	15.00	5.00
C	3	20.00	5.00

All three standard deviations are equal to 5, so all of them are less than $2(5.00) = 10.00$. Therefore, the equal variance requirement is satisfied.

8. (a)

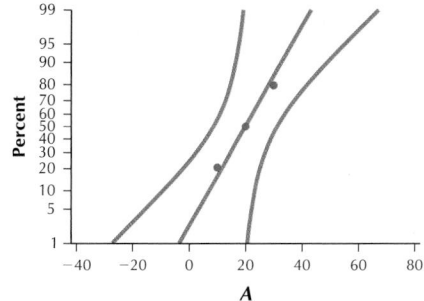

A

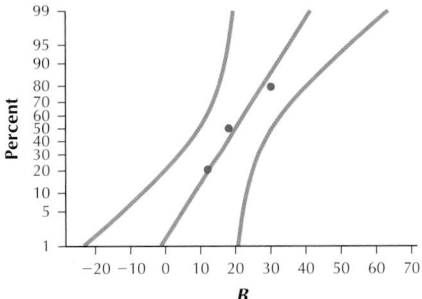

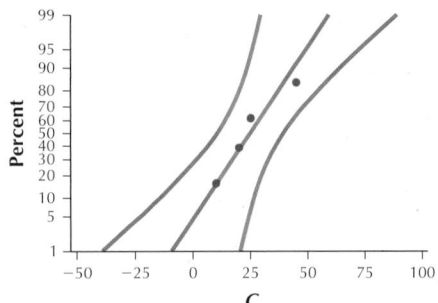

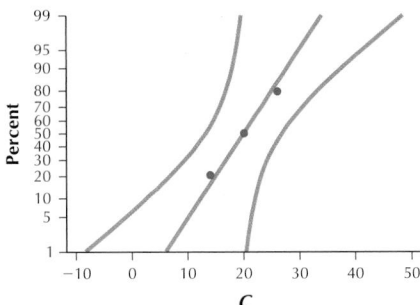

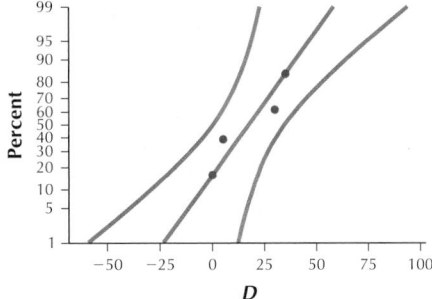

All four graphs show acceptable normality.

All three graphs show acceptable normality.

(b)

Factor	N	Mean	StDev
A	3	20.00	10.00
B	3	20.00	9.17
C	3	20.00	6.00

(b)

Means			
Factor	N	Mean	StDev
A	4	13.75	16.01
B	4	52.50	17.08
C	4	25.00	14.72
D	4	17.50	17.56

The smallest standard deviation is 6.00, which is the standard deviation for group C. Twice this standard deviation is 12. The other two standard deviations are less than 12, so the equal variance requirement is satisfied.

The smallest standard deviation is 14.72 which is the standard deviation for group C. Twice 14.72 is 29.44. All other standard deviations are less than 29.44 so the equal variance requirement is satisfied.

9. (a)

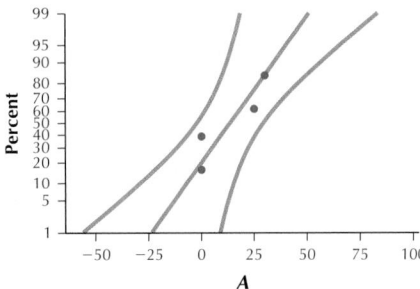

10. (a)

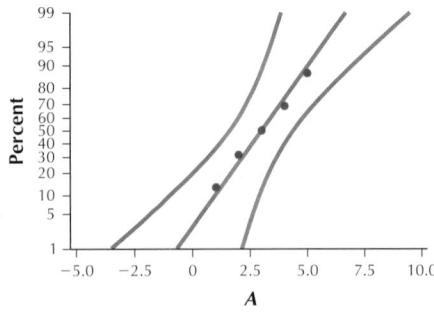

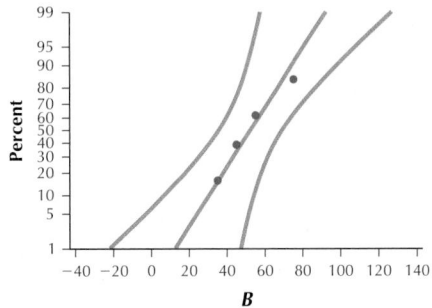

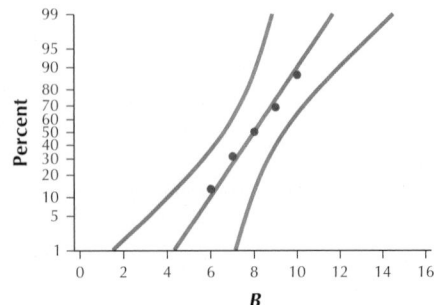

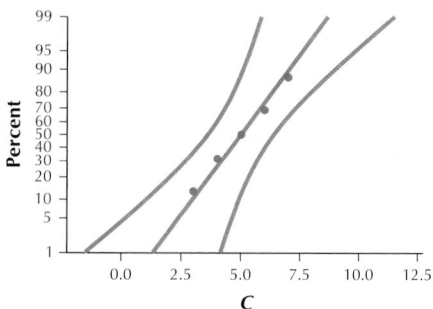

C

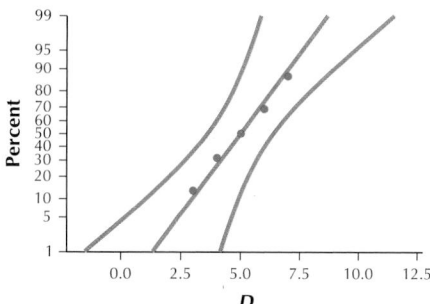

D

All four graphs show acceptable normality.

(b)

```
Means

Factor  N  Mean   StDev
A       5  3.000  1.581
B       5  8.000  1.581
C       5  5.000  1.581
D       5  5.000  1.581
```

All four standard deviations are equal to 1.581, so all of them are less than $2(1.581) = 3.162$. Therefore, the equal variance requirement is satisfied.

11. $\bar{\bar{x}} = 15$

12. $\bar{\bar{x}} = 20$

13. $\bar{\bar{x}} = 27.1875$

14. $\bar{\bar{x}} = 5.25$

15. **(a)** **(i)** SSTR = 150 **(ii)** SSE = 150 **(iii)** SST = 300 **(iv)** $df_1 = 2$ **(v)** $df_2 = 6$ **(vi)** MSTR = 75 **(vii)** MSE = 25

(b)

```
Analysis of Variance

Source  DF  Adj SS  Adj MS  F-Value  P-Value
Factor  2   150.0   75.00   3.00     0.125
Error   6   150.0   25.00
Total   8   300.0
```

16. **(a)** **(i)** SSTR = 0 **(ii)** SSE = 440 **(iii)** SST = 440 **(iv)** $df_1 = 2$ **(v)** $df_2 = 6$ **(vi)** MSTR = 0 **(vii)** MSE = 73.3333

(b)

```
Analysis of Variance

Source  DF  Adj SS   Adj MS   F-Value  P-Value
Factor  2   0.000    0.0000   0.00     1.000
Error   6   440.000  73.3333
Total   8   440.000
```

17. **(a)** **(i)** SSTR = 3680 **(ii)** SSE = 3219 **(iii)** SST = 6898 **(iv)** $df_1 = 3$ **(v)** $df_2 = 12$ **(vi)** MSTR = 1226.6 **(vii)** MSE = 268.2

(b)

```
Analysis of Variance

Source  DF  Adj SS  Adj MS  F-Value  P-Value
Factor  3   3680    1226.6  4.57     0.023
Error   12  3219    268.2
Total   15  6898
```

18. **(a)** **(i)** SSTR = 63.75 **(ii)** SSE = 40 **(iii)** SST = 103.75 **(iv)** $df_1 = 3$ **(v)** $df_2 = 16$ **(vi)** MSTR = 21.250 **(vii)** MSE = 2.5

(b)

```
Analysis of Variance

Source  DF  Adj SS  Adj MS  F-Value  P-Value
Factor  3   63.75   21.250  8.50     0.001
Error   16  40.00   2.500
Total   19  103.75
```

19. **(a)** **(i)** SSTR = 40 **(ii)** SSE = 12 **(iii)** SST = 52 **(iv)** $df_1 = 2$ **(v)** $df_2 = 12$ **(vi)** MSTR = 20 **(vii)** MSE = 1

(b)

Source of variation	Sum of squares	Degrees of freedom	Mean square	F
Treatments	40	2	20	20
Error	12	12	1	
Total	52			

20. **(a)** **(i)** SSTR = 100 **(ii)** SSE = 16 **(iii)** SST = 116 **(iv)** $df_1 = 3$ **(v)** $df_2 = 16$ **(vi)** MSTR = 33.3333 **(vii)** MSE = 1

(b)

Source of variation	Sum of squares	Degrees of freedom	Mean square	F
Treatments	100	3	33.3333	33.3333
Error	16	16	1	
Total	116			

21. **(a)** **(i)** SSTR = 395,833.3333 **(ii)** SSE = 15,748 **(iii)** SST = 411,581.3333 **(iv)** $df_1 = 3$ **(v)** $df_2 = 596$ **(vi)** MSTR = 131,944.4444 **(vii)** MSE = 26.4228

(b)

Source of variation	Sum of squares	Degrees of freedom	Mean square	F
Treatments	395,833.3333	3	131,944.4444	4993.5794
Error	15,748	594	26.4228	
Total	411,581.3333			

22. **(a)** **(i)** SSTR = 10,000 **(ii)** SSE = 1157.50 **(iii)** SST = 11,157.50 **(iv)** $df_1 = 3$ **(v)** $df_2 = 296$ **(vi)** MSTR = 3,333.3333 **(vii)** MSE = 3.9105

(b)

Source of variation	Sum of squares	Degrees of freedom	Mean square	F
Treatments	10,000	3	3,333.3333	852.4118
Error	1,157.50	296	3.9105	
Total	11,157.50			

23. $H_0: \mu_A = \mu_B = \mu_C$ versus H_a: Not all the population means are equal.
Reject H_0 if the p-value $\le \alpha = 0.05$.
$F_{data} = 3.00$.
p-value $= 0.125$.
The p-value $= 0.125$ is not $\le \alpha = 0.05$. Therefore we do not reject H_0. There is not enough evidence to conclude at level of significance $\alpha = 0.05$ that the population means are not all equal.

24. $H_0: \mu_A = \mu_B = \mu_C$ versus H_a: Not all the population means are equal.
Reject H_0 if the p-value $\le \alpha = 0.05$.
$F_{data} = 0.00$.
p-value $= 1.000$.
The p-value $= 1.000$ is not $\le \alpha = 0.05$. Therefore we do not reject H_0. There is not enough evidence to conclude at level of significance $\alpha = 0.05$ that the population means are not all equal.

25. $H_0: \mu_A = \mu_B = \mu_C = \mu_D$ versus H_a: Not all the population means are equal.
Reject H_0 if the p-value $\le \alpha = 0.05$.
$F_{data} = 4.57$.
p-value $= 0.023$.
The p-value $= 0.023$ is $\le \alpha = 0.05$. Therefore we reject H_0. There is evidence at level of significance $\alpha = 0.05$ that the population means are not all equal.

26. $H_0: \mu_A = \mu_B = \mu_C = \mu_D$ versus H_a: Not all the population means are equal.
Reject H_0 if the p-value $\le \alpha = 0.05$.

$F_{data} = 8.50$.

p-value $= 0.001$.

The p-value $= 0.001$ is $\leq \alpha = 0.05$. Therefore we reject H_0. There is evidence at level of significance $\alpha = 0.05$ that the population means are not all equal.

27. $H_0 : \mu_A = \mu_B = \mu_C$

H_a: Not all the population means are equal.

Reject H_0 if the p-value $\leq \alpha = 0.05$.

$F_{data} = 20$.

p-value $= 0.0002$.

The p-value $= 0.0002$ is $\leq \alpha = 0.05$. Therefore we reject H_0. There is evidence at level of significance $\alpha = 0.05$ that the population means are not all equal.

28. $H_0 : \mu_A = \mu_B = \mu_C = \mu_D$;

H_a: Not all the population means are equal.

Reject H_0 if the p-value $\leq \alpha = 0.05$.

$F_{data} = 33.3333$.

p-value ≈ 0.

The p-value ≈ 0 is $\leq \alpha = 0.05$. Therefore we reject H_0. There is evidence at level of significance $\alpha = 0.05$ that the population means are not all equal.

29. $H_0 : \mu_A = \mu_B = \mu_C = \mu_D$;

H_a: Not all the population means are equal.

$F_{crit} = 3.815$ Reject H_0 if $F_{data} \geq 3.815$;

$F_{data} = 4793.5794$;

$F_{data} = 4793.5794$ is ≥ 3.815. Therefore we reject H_0. There is evidence at level of significance $\alpha = 0.01$ that the population means are not all equal.

30. $H_0 : \mu_A = \mu_B = \mu_C = \mu_D$;

H_a: Not all the population means are equal.

$F_{crit} = 3.815$. Reject H_0 if $F_{data} \geq 3.815$.

$F_{data} = 852.4118$;

$F_{data} = 852.4118$ is ≥ 3.815. Therefore we reject H_0. There is evidence at level of significance $\alpha = 0.01$ that the population means are not all equal.

31. (a)

Means			
Factor	N	Mean	StDev
Wraps	5	14.20	6.61
Muffins	5	5.60	3.65
Chips	5	9.00	4.47

The smallest standard deviation is 3.65, which is the standard deviation for muffins. Twice 3.65 is 7.30. All other standard deviations are less than 7.30, so the equal variance requirement is satisfied.

(b) (i) df$_1 = 2$, df$_2 = 12$ **(ii)** $\bar{\bar{x}} = 9.6$ **(iii)** SSTR $= 187.6$ **(iv)** SSE $= 308$ **(v)** SST $= 495.6$ **(vi)** MSTR $= 93.80$ **(vii)** MSE $= 25.67$ **(viii)** $F_{data} = 3.65$

(c)

Analysis of Variance					
Source	DF	Adj SS	Adj MS	F-Value	P-Value
Factor	2	187.6	93.80	3.65	0.058
Error	12	308.0	25.67		
Total	14	495.6			

(d) $H_0 : \mu_{Wraps} = \mu_{Muffins} = \mu_{Chips}$;

H_a: Not all the population means are equal.

Reject H_0 if the p-value $\leq \alpha = 0.05$.

$F_{data} = 3.65$.

p-value $= 0.058$.

The p-value $= 0.058$ is not $\leq \alpha = 0.05$. Therefore we do not reject H_0. There is not enough evidence to conclude at level of significance $\alpha = 0.05$ that the population means of the number of different types of food items are not all equal.

32. (a)

Means			
Factor	N	Mean	StDev
Pros	5	16.76	11.52
Darts	5	10.88	8.92
DJIA	5	7.42	6.95

The smallest standard deviation is 6.95, which is the standard deviation for the DJIA. Twice 6.95 is 13.9. All other standard deviations are less than 13.9, so the equal variance requirement is satisfied.

(b) (i) df$_1 = 2$, df$_2 = 12$ **(ii)** $\bar{\bar{x}} = 11.69$ **(iii)** SSTR $= 223$ **(iv)** SSE $= 1042.3$ **(v)** SST $= 1265.3$ **(vi)** MSTR $= 111.48$ **(vii)** MSE $= 86.86$ **(viii)** $F_{data} = 1.28$

(c)

Analysis of Variance					
Source	DF	Adj SS	Adj MS	F-Value	P-Value
Factor	2	223.0	111.48	1.28	0.313
Error	12	1042.3	86.86		
Total	14	1265.3			

(d) $H_0 : \mu_{Darts} = \mu_{Pros} = \mu_{DJIA}$;

H_a: Not all the population means are equal.

Reject H_0 if the p-value $\leq \alpha = 0.05$.

$F_{data} = 1.28$.

p-value $= 0.313$.

The p-value $= 0.313$ is not $\leq \alpha = 0.05$. Therefore we do not reject H_0. There is not enough evidence to conclude at level of significance $\alpha = 0.05$ that the population means of the daily stock market returns are not all equal.

33. (a)

Means			
Factor	N	Mean	StDev
Younger	5	132.00	13.91
Middle	5	114.26	17.14
Older	5	135.2	25.9

The smallest standard deviation is 13.91, which is the standard deviation for the younger. Twice 13.91 is 27.82. All other standard deviations are less than 27.82, so the equal variance requirement is satisfied.

(b) (i) df$_1 = 2$, df$_2 = 12$ **(ii)** $\bar{\bar{x}} = 127.17$ **(iii)** SSTR $= 1276$ **(iv)** SSE $= 4640$ **(v)** SST $= 5916$ **(vi)** MSTR $= 637.8$ **(vii)** MSE $= 386.7$ **(viii)** $F_{data} = 1.65$

(c)

Analysis of Variance					
Source	DF	Adj SS	Adj MS	F-Value	P-Value
Factor	2	1276	637.8	1.65	0.233
Error	12	4640	386.7		
Total	14	5916			

(d) $H_0 : \mu_{Younger} = \mu_{Middle} = \mu_{Older}$;

H_a: Not all the population means are equal.

Reject H_0 if the p-value $\leq \alpha = 0.05$.

$F_{data} = 1.65$.

p-value $= 0.233$.

The p-value $= 0.233$ is not $\leq \alpha = 0.05$. Therefore we do not reject H_0. There is not enough evidence to conclude at level of significance $\alpha = 0.05$ that the population mean weights of the different age groups are not all equal.

34. Using the TI-83/84: Before: $\bar{x} = 10.91666667$, $s^2 = 13.10333333$, $s = 3.619852667$. During: $\bar{x} = 13.41666667$, $s^2 = 30.26515152$, $s = 5.501377238$. After: $\bar{x} = 11.45833333$, $s^2 = 9.668106061$, $s = 3.109357821$. **(a)** The largest sample standard deviation $s_{during} = 5.501377238$ is not more than twice the smallest sample standard deviation $2 \cdot s_{after} = 2(3.109357821) = 6.218715642$.

(b) (i) df$_1 = 2$, df$_2 = 33$ **(ii)** $\bar{\bar{x}} = 11.93055556$ **(iii)** SSTR $= 41.51388896$ **(iv)** SSE $= 583.4025$ **(v)** SST $= 624.916389$ **(vi)** MSTR $= 20.75694448$ **(vii)** MSE $= 17.67886364$ **(viii)** $F_{data} = 1.74110786$

(c)

Source of variation	Sum of squares	Degrees of freedom	Mean square	F
Treatment	41.51388896	2	20.75694448	1.74110786
Error	583.4025	33	17.67886364	
Total	624.916389			

(d) $H_0 : \mu_{before} = \mu_{during} = \mu_{after}$ versus H_a: Not all the population means are equal. Reject H_0 if $F_{data} \geq 3.39$. $F_{data} = 1.74110786$. Since $F_{data} = 1.74110786$ is not ≥ 3.39, we do not reject H_0. There is insufficient evidence that not all the population mean numbers of emergency room visits are equal.

35. (a) Missing values are in red.

Source of variation	Sum of squares	Degrees of freedom	Mean square	F
Treatments	120	6	20	4
Error	315	63	5	
Total	435			

(b) $H_0 : \mu_1 = \mu_2 = \mu_3 = \mu_4 = \mu_5 = \mu_6 = \mu_7$. H_a: Not all the population means are equal. Reject H_0 if p-value ≤ 0.05. $F_{data} = 4$ p-value $= 0.0018680725$.

Since the p-value ≤ 0.05, we reject H_0. There is evidence that not all the population means are equal.

36. (a) Missing values are in red.

Source of variation	Sum of squares	Degrees of freedom	Mean square	F
Treatment	60	2	30	5
Error	90	15	6	
Total	150			

(b) $H_0: \mu_1 = \mu_2 = \mu_3$. H_a: Not all the population means are equal. $F_{crit} = 6.36$. Reject H_0 if $F_{data} \geq 6.36$. $F_{data} = 5$. Since F_{data} is not ≥ 6.36, we do not reject H_0. There is insufficient evidence that not all the population means are equal.

37. (a) Missing values are in red.

Source of variation	Sum of squares	Degrees of freedom	Mean square	F-test statistic
Treatments	SSTR = 40	$df_1 = 4$	MSTR = 10	$F_{data} = 1.0$
Error	SSE = 400	$df_2 = 40$	MSE = 10	
Total	SST = 440			

(b) $H_0: \mu_1 = \mu_2 = \mu_3 = \mu_4 = \mu_5$. H_a: Not all the population means are equal. $F_{crit} = 2.09$ Reject H_0 if $F_{data} \geq 2.09$ $F_{data} = 1.0$. Since F_{data} is not ≥ 2.09 we do not reject H_0. There is insufficient evidence that not all the population means are equal.

38. (a) Missing values are in red.

Source of variation	Sum of squares	Degrees of freedom	Mean square	F-test statistic
Treatments	SSTR = 96	$df_1 = 2$	MSTR = 48	$F_{data} = 2.0$
Error	SSE = 480	$df_2 = 20$	MSE = 24	
Total	SST = 576			

(b) $H_0: \mu_1 = \mu_2 = \mu_3$. H_a: Not all the population means are equal. Reject H_0 if the p-value ≤ 0.05. $F_{data} = 2.0$; p-value $= 0.1615$. Since the p-value is not ≤ 0.05, we do not reject H_0. There is insufficient evidence that not all the population means are equal.

39. (a) $H_0: \mu_{Females} = \mu_{Males}$, H_a: The two population means are not equal. $\mu_{Females} =$ the population mean heart rate for females; $\mu_{Males} =$ the population mean heart rate for males. Reject H_0 if the p-value ≤ 0.05. $F_{data} = 4.896939413$, p-value $= 0.0287$. Since the p-value is ≤ 0.05, we reject H_0. There is evidence that the population mean heart rates are not equal. **(b)** Inference for Two Independent Means, Section 10.2.

40. No evidence

41. Since the p-value ≈ 0 is ≤ 0.01, we reject H_0. There is evidence that not all the population mean heights are equal.

42. (a) Since one of the boxplots overlaps only 2 of the 5 other boxplots, the comparison boxplot of the nutritional ratings may be considered as evidence against the null hypothesis that all population mean nutritional ratings were equal. **(b)** Rejected

43. Since $S_{QuakerOats} = 10.603 \leq 11.018 = 2 S_{Nabisco}$, the equal variance requirement is satisfied.

44. Yes, since the cereals were selected randomly from the manufacturers so that the selection of a cereal from one manufacturer did not affect the selection of cereals from other manufacturers.

45. $H_0: \mu_{GeneralMills} = \mu_{Kelloggs} = \mu_{Nabisco} = \mu_{Post} = \mu_{QuakerOats} = \mu_{RalstonPurina}$. H_a: Not all the population means are equal. Reject H_0 if the p-value ≤ 0.05. $F_{data} = 15.71$; p-value ≈ 0. Since the p-value is ≤ 0.05, we reject H_0. There is evidence that the population mean nutritional ratings of the cereals made by the different manufacturers are not all equal.

46. (a) (i) df_1 and df_2 would stay the same **(ii)** $\bar{\bar{x}}$ would increase **(iii)** SSTR would increase **(iv)** SSE would increase **(v)** SST would increase **(vi)** MSTR would increase **(vii)** MSE would increase **(viii)** F_{data} would decrease **(ix)** the p-value would increase
(b) F_{data} would decrease and the p-value would increase. Since the conclusion is do not reject H_0, an increase in the p-value would not change the conclusion.

47. (a) (i) df_1 and df_2 would stay the same **(ii)** $\bar{\bar{x}}$ would increase **(iii)** SSTR would decrease **(iv)** SSE would stay the same **(v)** SST would decrease **(vi)** MSTR would decrease **(vii)** MSE would stay the same **(viii)** F_{data} would decrease **(ix)** the p-value would increase. **(b)** F_{data} would decrease and the p-value would increase. Therefore, the conclusion would still be do not reject H_0.

48. (a) Since the box part of the boxplot for the gas mileage of automobiles manufactured in the United States does not overlap the other boxplots, the conclusion might be to reject H_0. **(b)** $H_0: \mu_{Europe} = \mu_{Japan} = \mu_{USA}$. H_a: Not all of the population means are equal. Reject H_0 if the p-value ≤ 0.01. $F_{data} = 96.6250761$; p-value $= 8.53843292 \times 10^{-35}$. Since the p-value is ≤ 0.01, we reject H_0. There is evidence that not all the population mean gas mileages are equal. **(c)** Yes

49. (a) (i) (25.4805, 29.7255) **(ii)** (28.6366, 32.2654) **(iii)** (18.9526, 21.1134) **(b)** The confidence interval for the population mean gas mileage of American cars does not overlap the other two confidence intervals. This is evidence against the null hypothesis that all the population means are equal.

50. (a)–(b) No change **(c)** Increase **(d)** No change **(e)–(f)** Increase **(g)** No change **(h)** Increase **(i)** Decrease **(j)** No change

51. $H_0: \mu_1 = \mu_2 = \mu_3 = \mu_4 = \mu_5 = \mu_6 = \mu_7 = \mu_8$. H_a: Not all the population means are equal. Reject H_0 if the p-value ≤ 0.05. 8.27 p-value ≈ 0. Since the p-value is ≤ 0.05, we reject H_0. There is evidence that not all the population means are equal.

```
Source     DF        SS         MS        F      P
SIZE2       7   10879518    1554217    8.27   0.000
Error     330   62003922     187891
Total     337   72883441
```

52. $H_0: \mu_{87} = \mu_{88} = \mu_{89} = \mu_{90} = \mu_{91}$. H_a: Not all the population means are equal. Reject H_0 if the p-value ≤ 0.05. $F_{data} = 1.34$. p-value $= 0.255$. Since the p-value is not ≤ 0.05, we do not reject H_0. There is insufficient evidence that not all the population means are equal.

Source	df	SS	MS	F	P
Year	4	1155151	288788	1.34	0.255
Error	333	71728290	215400		
Total	337	72883441			

53. $H_0: \mu_1 = \mu_2 = \mu_3 = \mu_4$. H_a: Not all the population means are equal. Reject H_0 if the p-value ≤ 0.01. 3.12. p-value $= 0.026$. Since the p-value is not ≤ 0.01, we do not reject H_0. There is insufficient evidence that not all the population means are equal.

```
Source     DF        SS        MS       F      P
PROTECT2    3      844.4     281.5    3.12   0.026
Error     334    30168.7      90.3
Total     337    31013.1
```

54. $H_0: \mu_1 = \mu_2 = \mu_3 = \mu_4$. H_a: Not all the population means are equal. Reject H_0 if the p-value ≤ 0.05. $F_{data} = 3.12$. p-value $= 0.026$. Since the p-value is ≤ 0.05, we reject H_0. There is evidence that not all the population means are equal.

Source	df	SS	MS	F	P
Protect2	3	844.4	281.5	3.12	0.026
Error	334	30168.7	90.3		
Total	337	31013.1			

55. (a) Decreases **(b)** Since all of the sample means are changed until they are about the same, SSTR and MSTR should decrease to small numbers. Since the variability of each sample stays the same, MSE stays the same. Thus $F_{data} = \frac{MSTR}{MSE}$ decreases

56. (a) Decreases **(b)** Decreases. Since the within-sample variability is increased, SSE and MSE are increased. Since the sample means remain the same, SSTR and MSTR remain unchanged. Thus $F_{data} = \frac{MSTR}{MSE}$ decreases.

Section 12.2

1. ANOVA only tests for whether or not all of the population means are equal. It does not test for which pairs of means are not equal.

2. If the ANOVA does not indicate any differences in the means, then there is no point in performing multiple comparisons to test which pairs of means are different.

3. It is performing a hypothesis test to determine whether two means are different. $C = {}_kC_2 = k!/[2! \, (k-2)!]$, where k is the number of populations.

4. The probability of making at least one Type I error in the set of hypothesis tests when performing multiple comparisons

5. By multiplying the p-value of each pairwise hypothesis test by the number of comparisons being made

6. Set it equal to 1. The p-value is a probability and probabilities cannot be greater than 1.

7. The requirements for ANOVA have been met and the null hypothesis that the population means are all equal has been rejected.

8. If a $100(1 - \alpha)\%$ confidence interval for $\mu_1 - \mu_2$ contains zero, then at level of significance α we do not reject the null hypothesis $H_0 : \mu_1 = \mu_2$. If the interval does not contain zero, then we do reject H_0.

9. $t_{\text{data}} = 4.472$

10. $t_{\text{data}} = 8.944$

11. $t_{\text{data}} = 4.472$

12. $t_{\text{data}} = -1$

13. $t_{\text{data}} = 3$

14. $t_{\text{data}} = 4$

15. **(a)** Test 1: $H_0 : \mu_1 = \mu_2$ vs. $H_a : \mu_1 \neq \mu_2$; Test 2: $H_0 : \mu_1 = \mu_3$ vs. $H_a : \mu_1 \neq \mu_3$; Test 3: $H_0 : \mu_2 = \mu_3$ vs. $H_a : \mu_2 \neq \mu_3$. For each hypothesis test, reject H_0 if the Bonferroni-adjusted p-value $\leq \alpha = 0.05$. **(b)** Test 1: $t_{\text{data}} = 4.472$; Test 2: $t_{\text{data}} = 8.944$; Test 3: $t_{\text{data}} = 4.472$ **(c)** Test 1: Bonferroni-adjusted p-value $= 0.0004$; Test 2: Bonferroni-adjusted p-value $= 0$; Test 3: Bonferroni-adjusted p-value $= 0.0004$ **(d)** Test 1: The adjusted p-value $= 0.0004$, which is ≤ 0.05; therefore we reject H_0. There is evidence at the $\alpha = 0.05$ level of significance that the population mean of Population 1 differs from the population mean of Population 2. Test 2: The adjusted p-value $= 0$, which is ≤ 0.05; therefore we reject H_0. There is evidence at the $\alpha = 0.05$ level of significance that the population mean of Population 1 differs from the population mean of Population 3. Test 3: The adjusted p-value $= 0.0004$, which is ≤ 0.05; therefore we reject H_0. There is evidence at the $\alpha = 0.05$ level of significance that the population mean of Population 2 differs from the population mean of Population 3.

16. **(a)** Test 1: $H_0 : \mu_1 = \mu_2$ vs. $H_a : \mu_1 \neq \mu_2$; Test 2: $H_0 : \mu_1 = \mu_3$ vs. $H_a : \mu_1 \neq \mu_3$; Test 3: $H_0 : \mu_2 = \mu_3$ vs. $H_a : \mu_2 \neq \mu_3$. For each hypothesis test, reject H_0 if the Bonferroni-adjusted p-value $\leq \alpha = 0.05$. **(b)** Test 1: $t_{\text{data}} = -1$; Test 2: $t_{\text{data}} = 3$; Test 3: $t_{\text{data}} = 4$ **(c)** Test 1: Bonferroni-adjusted p-value $= 0.9861$; Test 2: Bonferroni-adjusted p-value $= 0.0205$; Test 3: Bonferroni-adjusted p-value $= 0.0019$ **(d)** Test 1: The adjusted p-value $= 0.9861$, which is not ≤ 0.05; therefore we do not reject H_0. There is insufficient evidence at the $\alpha = 0.05$ level of significance that the population mean of Population 1 differs from the population mean of Population 2. Test 2: The adjusted p-value $= 0.0205$, which is ≤ 0.05; therefore we reject H_0. There is evidence at the $\alpha = 0.05$ level of significance that the population mean of Population 1 differs from the population mean of Population 3. Test 3: The adjusted p-value $= 0.0019$, which is ≤ 0.05; therefore we reject H_0. There is evidence at the $\alpha = 0.05$ level of significance that the population mean of Population 2 differs from the population mean of Population 3.

17. Test 1: $H_0 : \mu_A = \mu_B$ vs. $H_a : \mu_A \neq \mu_B$; Test 2: $H_0 : \mu_A = \mu_C$ vs. $H_a : \mu_A \neq \mu_C$; Test 3: $H_0 : \mu_B = \mu_C$ vs. $H_a : \mu_B \neq \mu_C$. For each hypothesis test, reject H_0 if the Bonferroni-adjusted p-value $\leq \alpha = 0.01$. Test 1: $t_{\text{data}} = -3.162$; Test 2: $t_{\text{data}} = 3.162$; Test 3: $t_{\text{data}} = 6.325$. Test 1: Bonferroni-adjusted p-value $= 0.0246$; Test 2: Bonferroni-adjusted p-value $= 0.0246$; Test 3: Bonferroni-adjusted p-value $= 0.0001$. Test 1: The adjusted p-value $= 0.0246$, which is not ≤ 0.01; therefore we do not reject H_0. There is insufficient evidence at the $\alpha = 0.01$ level of significance that the population mean of Population A differs from the population mean of Population B. Test 2: The adjusted p-value $= 0.0246$, which is not ≤ 0.01; therefore we do not reject H_0. There is insufficient evidence at the $\alpha = 0.01$ level of significance that the population mean of Population A differs from the population mean of Population C. Test 3: The adjusted p-value $= 0.0001$, which is ≤ 0.01; therefore we reject H_0. There is evidence at the $\alpha = 0.01$ level of significance that the population mean of Population B differs from the population mean of Population C.

18. See Solutions Manual.

19. $q_{\text{data}} = 0.707$

20. $q_{\text{data}} = 3.536$

21. $q_{\text{data}} = 2.828$

22. $q_{\text{data}} = 4.714$

23. $q_{\text{data}} = 2.357$

24. $q_{\text{data}} = 2.357$

25. $q_{\text{crit}} = 3.356$

26. $q_{\text{crit}} = 3.399$

27. **(a)** Test 1: $H_0 : \mu_1 = \mu_2$ vs. $H_a : \mu_1 \neq \mu_2$; Test 2: $H_0 : \mu_1 = \mu_3$ vs. $H_a : \mu_1 \neq \mu_3$; Test 3: $H_0 : \mu_2 = \mu_3$ vs. $H_a : \mu_2 \neq \mu_3$. **(b)** $q_{\text{crit}} = 3.356$. Reject H_0 if $q_{\text{data}} \geq 3.356$; **(c)** Test 1: $q_{\text{data}} = 0.707$; Test 2: $q_{\text{data}} = 3.536$; Test 3: $q_{\text{data}} = 2.828$ **(d)** Test 1: $q_{\text{data}} = 0.707$, which is not ≥ 3.356; therefore we do not reject H_0. There is insufficient evidence at the $\alpha = 0.05$ level of significance that the population mean of Population 1 differs from the population mean of Population 2. Test 2: $q_{\text{data}} = 3.536$, which is ≥ 3.356; therefore we reject H_0. There is evidence at the $\alpha = 0.05$ level of significance that the population mean of Population 1 differs from the population mean of Population 3. Test 3: $q_{\text{data}} = 2.828$, which is not ≥ 3.356; therefore we do not reject H_0. There is insufficient evidence at the $\alpha = 0.05$ level of significance that the population mean of Population 2 differs from the population mean of Population 3.

28. **(a)** Test 1: $H_0 : \mu_1 = \mu_2$ vs. $H_a : \mu_1 \neq \mu_2$; Test 2: $H_0 : \mu_1 = \mu_3$ vs. $H_a : \mu_1 \neq \mu_3$; Test 3: $H_0 : \mu_2 = \mu_3$ vs. $H_a : \mu_2 \neq \mu_3$. **(b)** $q_{\text{crit}} = 3.399$. Reject H_0 if $q_{\text{data}} \geq 3.399$; **(c)** Test 1: $q_{\text{data}} = 4.714$; Test 2: $q_{\text{data}} = 2.357$; Test 3: $q_{\text{data}} = 2.357$ **(d)** Test 1: $q_{\text{data}} = 4.714$, which is ≥ 3.399; therefore we reject H_0. There is evidence at the $\alpha = 0.05$ level of significance that the population mean of Population 1 differs from the population mean of Population 2. Test 2: $q_{\text{data}} = 2.357$, which is not ≥ 3.399; therefore we do not reject H_0. There is insufficient evidence at the $\alpha = 0.05$ level of significance that the population mean of Population 1 differs from the population mean of Population 3. Test 3: $q_{\text{data}} = 2.357$, which is not ≥ 3.399; therefore we do not reject H_0. There is insufficient evidence at the $\alpha = 0.05$ level of significance that the population mean of Population 2 differs from the population mean of Population 3.

29. Test 1: $H_0 : \mu_A = \mu_B$ vs. $H_a : \mu_A \neq \mu_B$; Test 2: $H_0 : \mu_A = \mu_C$ vs. $H_a : \mu_A \neq \mu_C$; Test 3: $H_0 : \mu_A = \mu_D$ vs. $H_a : \mu_A \neq \mu_D$; Test 4: $H_0 : \mu_B = \mu_C$ vs. $H_a : \mu_B \neq \mu_C$; Test 5: $H_0 : \mu_B = \mu_D$ vs. $H_a : \mu_B \neq \mu_D$; Test 6: $H_0 : \mu_C = \mu_D$ vs. $H_a : \mu_C \neq \mu_D$; $q_{\text{crit}} = 3.685$. For each hypothesis test, reject H_0 if $q_{\text{data}} \geq 3.685$. Test 1: $q_{\text{data}} = 53.484$, which is ≥ 3.685; therefore we reject H_0. There is evidence at the $\alpha = 0.05$ level of significance that the population mean of Population A differs from the population mean of Population B. Test 2: $q_{\text{data}} = 112.755$, which is ≥ 3.685; therefore we reject H_0. There is evidence at the $\alpha = 0.05$ level of significance that the population mean of Population A differs from the population mean of Population C. Test 3: $q_{\text{data}} = 175.069$, which is ≥ 3.685; therefore we reject H_0. There is evidence at the $\alpha = 0.05$ level of significance that the population mean of Population A differs from the population mean of Population D. Test 4: $q_{\text{data}} = 63.926$, which is ≥ 3.685; therefore we reject H_0. There is evidence at the $\alpha = 0.05$ level of significance that the population mean of Population B differs from the population mean of Population C. Test 5: $q_{\text{data}} = 133.711$, which is ≥ 3.685; therefore we reject H_0. There is evidence at the $\alpha = 0.05$ level of significance that the population mean of Population B differs from the population mean of Population D. Test 6: $q_{\text{data}} = 72.783$, which is ≥ 3.685; therefore we reject H_0. There is evidence at the $\alpha = 0.05$ level of significance that the population mean of Population C differs from the population mean of Population D.

30. See Solutions Manual.

31. Test 1: $H_0 : \mu_A = \mu_B$ vs. $H_a : \mu_A \neq \mu_B$. 95% confidence interval for $\mu_B - \mu_A$ is $(-5.642, 0.309)$, which does contain zero, so we do not reject $H_0 : \mu_A = \mu_B$ for level of significance $\alpha = 0.05$. Test 2: $H_0 : \mu_A = \mu_C$ vs. $H_a : \mu_A \neq \mu_C$. 95% confidence interval for $\mu_C - \mu_A$ is $(-7.642, -1.691)$, which does not contain zero, so we reject $H_0 : \mu_A = \mu_C$ for level of significance $\alpha = 0.05$. Test 3: $H_0 : \mu_B = \mu_C$ vs. $H_a : \mu_B \neq \mu_C$. 95% confidence interval for $\mu_C - \mu_B$ is $(-4.976, 0.976)$, which does contain zero, so we do not reject $H_0 : \mu_B = \mu_C$ for level of significance $\alpha = 0.05$.

32. Test 1: $H_0 : \mu_A = \mu_B$ vs. $H_a : \mu_A \neq \mu_B$. 95% confidence interval for $\mu_B - \mu_A$ is $(-10.385, -1.615)$, which does not contain zero, so we reject $H_0 : \mu_A = \mu_B$ for level of significance $\alpha = 0.05$. Test 2: $H_0 : \mu_A = \mu_C$ vs. $H_a : \mu_A \neq \mu_C$. 95% confidence interval for $\mu_C - \mu_A$ is $(-4.719, 4.052)$, which does contain zero, so we do not reject $H_0 : \mu_A = \mu_C$ for level of significance $\alpha = 0.05$. Test 3: $H_0 : \mu_B = \mu_C$ vs. $H_a : \mu_B \neq \mu_C$. 95% confidence interval for $\mu_C - \mu_B$ is $(1.281, 10.052)$, which does not contain zero, so we reject $H_0 : \mu_B = \mu_C$ for level of significance $\alpha = 0.05$.

33. Not appropriate since the conclusion is do not reject H_0.

34. Not appropriate since the conclusion is do not reject H_0.

35. See Solutions Manual.

36. Test 1: $H_0: \mu_{Europe} = \mu_{Japan}$ vs. $H_a: \mu_{Europe} \neq \mu_{Japan}$; Test 2: $H_0: \mu_{Europe} = \mu_{USA}$ vs. $H_a: \mu_{Europe} \neq \mu_{USA}$; Test 3: $H_0: \mu_{Japan} = \mu_{USA}$ vs. $H_a: \mu_{Japan} \neq \mu_{USA}$. For each hypothesis test, reject H_0 if the Bonferroni-adjusted p-value $\leq \alpha = 0.01$. Test 1: $t_{data} = -2.692$; Test 2: $t_{data} = 8.635$; Test 3: $t_{data} = 12.589$. Test 1: The Bonferroni-adjusted p-value $= 0.0222$, which is not ≤ 0.01; therefore we do not reject H_0. There is insufficient evidence at the $\alpha = 0.01$ level of significance that the population mean gas mileage of cars manufactured in Europe differs from the population mean gas mileage of cars manufactured in Japan. Test 2: The Bonferroni-adjusted p-value $= 0$, which is ≤ 0.01; therefore we reject H_0. There is evidence at the $\alpha = 0.01$ level of significance that the population mean gas mileage of cars manufactured in Europe differs from the population mean gas mileage of cars manufactured in the United States. Test 3: The Bonferroni-adjusted p-value $= 0$, which is ≤ 0.01; therefore we reject H_0. There is evidence at the $\alpha = 0.01$ level of significance that the population mean gas mileage of cars manufactured in Japan differs from the population mean gas mileage of cars manufactured in the United States.

37. Not appropriate since the conclusion is do not reject H_0.

38. The conclusion was do not reject H_0, so it is not appropriate to perform multiple comparisons.

39. See Solutions Manual.

40. See Solutions Manual.

Section 12.3

1. The variable of interest is the variable we are interested in studying, and the blocking factor is a variable that is not of primary interest to the researcher but is included in the ANOVA in order to improve the ability of the ANOVA to find significant differences among the treatment means. Treatment. Nuisance factor.

2. In a randomized block design ANOVA, we test for differences among the treatment means, while accounting for the variability among the levels in the blocking factor.

3. We reject H_0 when the p-value is small. The smaller SSE is, the larger F is and therefore the smaller the p-value is.

4. The larger F is, the smaller the p-value is. Smaller p-values will more likely lead us to reject the null hypothesis.

5. (a) $H_0: \mu_1 = \mu_2 = \mu_3 = \mu_4$ vs. H_a: Not all of the population means are equal. Reject H_0 if p-value ≤ 0.05. (b) The p-value $= 0.010$, which is ≤ 0.05; therefore we reject H_0. There is evidence at level of significance $\alpha = 0.05$ that the population means are not all equal.

6. (a) $H_0: \mu_1 = \mu_2 = \mu_3$ vs. H_a: Not all of the population means are equal. Reject H_0 if p-value ≤ 0.05. (b) The p-value $= 0.132$, which is not ≤ 0.05; therefore we do not reject H_0. There is insufficient evidence at level of significance $\alpha = 0.05$ that the population means are not all equal.

7. (a) $H_0: \mu_1 = \mu_2 = \mu_3 = \mu_4$ vs. H_a: Not all of the population means are equal. Reject H_0 if p-value ≤ 0.05. (b) The p-value $= 0.009$, which is ≤ 0.05; therefore we reject H_0. There is evidence at level of significance $\alpha = 0.05$ that the population means are not all equal.

8. (a) $H_0: \mu_1 = \mu_2 = \mu_3 = \mu_4$ vs. H_a: Not all of the population means are equal. Reject H_0 if p-value ≤ 0.05. (b) The p-value $= 0.204$, which is not ≤ 0.05; therefore we do not reject H_0. There is insufficient evidence at level of significance $\alpha = 0.05$ that the population means are not all equal.

9. (a) $H_0: \mu_1 = \mu_2 = \mu_3$ vs. H_a: Not all of the population means are equal. Reject H_0 if p-value ≤ 0.05. (b) Answers to missing values are in red.

Source	Sum of squares	Degrees of freedom	Mean square	F
Treatments	90	2	45	6
Blocks	50	5	10	
Error	60	8	7.5	
Total	200	15		

(c) The p-value $= 0.0256$, which is ≤ 0.05; therefore we reject H_0. There is evidence at level of significance $\alpha = 0.05$ that the population means are not all equal.

10. (a) $H_0: \mu_1 = \mu_2 = \mu_3 = \mu_4$ vs. H_a: Not all of the population means are equal. Reject H_0 if p-value ≤ 0.05. (b) Answers to missing values are in red.

Source	Sum of squares	Degrees of freedom	Mean square	F
Treatments	180	3	60	5
Blocks	240	6	40	
Error	216	18	12	
Total	636	27		

(c) The p-value $= 0.0107$, which is ≤ 0.05; therefore we reject H_0. There is evidence at level of significance $\alpha = 0.05$ that the population means are not all equal.

11. (a)

Source	Sum of squares	Degrees of freedom	Mean square	F
Treatments	90	2	45	5.3182
Error	110	13	8.4615	
Total	200	15		

(b) $H_0: \mu_1 = \mu_2 = \mu_3$ vs. H_a: Not all of the population means are equal. Reject H_0 if p-value ≤ 0.05. The p-value $= 0.0205$, which is ≤ 0.05; therefore we reject H_0. There is evidence at level of significance $\alpha = 0.05$ that the population means are not all equal. (c) The conclusions are the same for both tests. No.

12. (a)

Source	Sum of squares	Degrees of freedom	Mean square	F
Treatments	180	3	60	3.1579
Error	456	24	19	
Total	636	27		

(b) $H_0: \mu_1 = \mu_2 = \mu_3 = \mu_4$ vs. H_a: Not all of the population means are equal. Reject H_0 if p-value ≤ 0.05. The p-value $= 0.0431$, which is ≤ 0.05; therefore we reject H_0. There is evidence at level of significance $\alpha = 0.05$ that the population means are not all equal. (c) The conclusions are the same for both tests. No.

13. $b - 1 + (k - 1)(b - 1) = b - 1 + k \cdot b - k - b + 1 = k \cdot b - k = n_T - k$

14.
```
Analysis of Variance

Source      DF   Adj SS    Adj MS   F-Value  P-Value
Year         2   0.08448   0.04224    6.00    0.019
Vegetable    5   5.44478   1.08896  154.56    0.000
Error       10   0.07046   0.00705
Total       17   5.59971
```

$H_0: \mu_{2012} = \mu_{2013} = \mu_{2014}$ vs. H_a: Not all the population means are equal. Reject H_0 if the p-value $\leq \alpha = 0.05$.
$F_{data} = 6.00$.
p-value $= 0.019$.
The p-value $= 0.019$ is $\leq \alpha = 0.05$. Therefore we reject H_0. There is evidence at level of significance $\alpha = 0.05$ that not all of the population mean prices of the vegetables are the same for all three years.

15.
```
Analysis of Variance

Source      DF   Adj SS    Adj MS   F-Value  P-Value
Year         2    519.6    259.8     0.96     0.423
Country      4  24201.1   6050.3    22.35     0.000
Error        8   2165.7    270.7
Total       14  26886.4
```

$H_0: \mu_{1995} = \mu_{2007} = \mu_{2011}$ vs. H_a: Not all the population means are equal. Reject H_0 if the p-value $\leq \alpha = 0.10$.
$F_{data} = 0.96$.
p-value $= 0.423$.
The p-value $= 0.423$ is not $\leq \alpha = 0.10$. Therefore we do not reject H_0. There is not enough evidence to conclude at level of significance $\alpha = 0.10$ that not all of the population mean math scores are the same for all three years.

16.

```
Analysis of Variance

Source   DF   Adj SS   Adj MS   F-Value   P-Value
Year      2     3137     1568      1.54     0.262
State     5   432650    86530     84.75     0.000
Error    10    10211     1021
Total    17   445998
```

H_0: $\mu_{2010} = \mu_{2011} = \mu_{2012}$ vs. H_a: Not all the population means are equal.
Reject H_0 if the p-value $\le \alpha = 0.10$.
$F_{data} = 1.54$.
p-value $= 0.262$.
The p-value $= 0.262$ is not $\le \alpha = 0.10$. Therefore we do not reject H_0. There is not enough evidence to conclude at level of significance $\alpha = 0.10$ that not all of the population mean number of children not covered by health insurance are the same for all three years.

17. (a)

```
Analysis of Variance

Source   DF     Adj SS    Adj MS   F-Value   P-Value
Year      2    0.08448   0.04224     0.11     0.892
Error    15    5.51523   0.36768
Total    17    5.59971
```

H_0: $\mu_{2012} = \mu_{2013} = \mu_{2014}$ vs. H_a: Not all the population means are equal.
Reject H_0 if the p-value $\le \alpha = 0.05$.
$F_{data} = 0.11$.
p-value $= 0.892$.
The p-value $= 0.892$ is not $\le \alpha = 0.05$. Therefore we do not reject H_0. There is not enough evidence to conclude at level of significance $\alpha = 0.05$ that not all of the population mean prices of the vegetables are the same for all three years.
(b) Yes.

18. (a)

```
Analysis of Variance

Source   DF     Adj SS   Adj MS   F-Value   P-Value
Year      2      519.6    259.8     0.12     0.890
Error    12    26366.8   2197.2
Total    14    26886.4
```

H_0: $\mu_{1995} = \mu_{2007} = \mu_{2011}$ vs. H_a: Not all the population means are equal.
Reject H_0 if the p-value $\le \alpha = 0.10$.
$F_{data} = 0.12$.
p-value $= 0.890$.
The p-value $= 0.890$ is not $\le \alpha = 0.10$. Therefore we do not reject H_0. There is not enough evidence to conclude at level of significance $\alpha = 0.10$ that not all of the population mean math scores are the same for all three years.
(b) No.

19. (a)

```
Analysis of Variance

Source   DF   Adj SS   Adj MS   F-Value   P-Value
Year      2     3137     1568      0.05     0.948
Error    15   442861    29524
Total    17   445998
```

H_0: $\mu_{2010} = \mu_{2011} = \mu_{2012}$ vs. H_a: Not all the population means are equal.
Reject H_0 if the p-value $\le \alpha = 0.10$.
$F_{data} = 0.05$.
p-value $= 0948$.
The p-value $= 948$ is not $\le \alpha = 0.10$. Therefore we do not reject H_0. There is not enough evidence to conclude at level of significance $\alpha = 0.10$ that not all of the population mean number of children not covered by health insurance are the same for all three years.
(b) No.

20. (a) H_0: $\mu_{MarketCap} = \mu_{FirmValue} = \mu_{EnterpriseValue}$ vs. H_a: Not all of the population means are equal. Reject H_0 if p-value ≤ 0.05. The p-value $= 0.435$, which is not ≤ 0.05; therefore we do not reject H_0. There is insufficient evidence at level of significance $\alpha = 0.05$ that the population mean values differ among the three measures. **(b)** H_0: $\mu_{MarketCap} = \mu_{FirmValue} = \mu_{EnterpriseValue}$ vs. H_a: Not all of the population means are equal. Reject H_0 if p-value ≤ 0.05. The p-value $= 0.019$, which is ≤ 0.05; therefore we reject H_0. There is evidence at level of significance $\alpha = 0.05$ that the population mean values differ among the three measures. **(c)** Yes

Section 12.4

1. We are interested in testing for the significance of both factors and we need to test for interaction of the two factors.
2. The main effects
3. Factor interaction
4. Answers will vary.
5. It is a graphical representation of the cell means for each cell in the contingency table. It is used to investigate the presence of interaction between the two factors.
6. (a) Significant interaction **(b)** No interaction **(c)** Some interaction
7. Test for interaction between the factors, test for Factor A effect, Test for Factor B effect
8. Do not perform the test for either Factor A or Factor B. If there is an interaction between factors, then we cannot draw conclusions about the main effects.
9. Significant interaction
10. No interaction
11. Some interaction
12. Significant interaction
13. The lines are parallel, so there is no interaction between Factor A and Factor B.

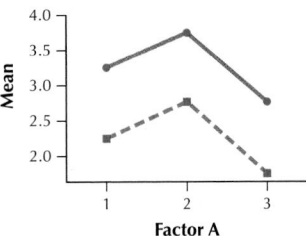

14. There appears to be significant interaction between Factor A and Factor B.

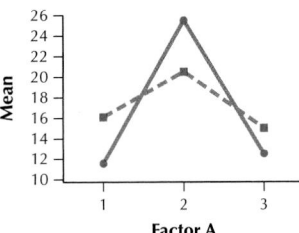

15. The lines are nearly parallel, so there is no interaction between Factors A and B.

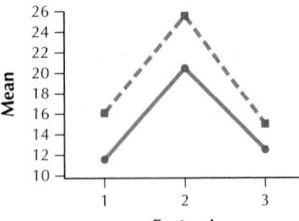

16. The graphs don't intersect, but they are not parallel. Therefore there may be a slight interaction between Factors A and B.

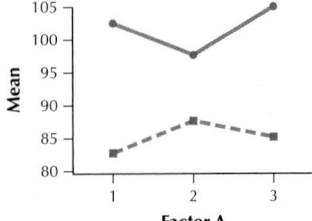

17. (a) H_0: There is no interaction between carrier (Factor A) and type (Factor B). H_a: There is interaction between carrier (Factor A) and type (Factor B). Reject H_0 if the p-value ≤ 0.05. The p-value $= 1.00$, which is not ≤ 0.05; therefore we do not reject H_0. There is insufficient evidence of interaction between carrier (Factor A) and type (Factor B) at level of significance $\alpha = 0.05$. This result agrees with the interaction plot in Exercise 13. **(b)** H_0: There is no carrier (Factor A) effect. That is, the population means do not differ by carrier. H_a: There is a carrier (Factor A) effect. That is, the population means do differ by carrier. Reject H_0 if the p-value ≤ 0.05. The p-value $= 0.020$, which is ≤ 0.05; therefore we reject H_0. There is evidence for a carrier (Factor A) effect. In other words, there is evidence at level of significance $\alpha = 0.05$ that the population means do differ by carrier. **(c)** H_0: There is no type (Factor B) effect. That is, the population means do not differ by type. H_a: There is a type (Factor B) effect. That is, the population means do differ by type. Reject H_0 if the p-value ≤ 0.05. The p-value $= 0.003$, which is ≤ 0.05; therefore we reject H_0. There is evidence for a type (Factor B) effect. In other words, there is evidence at level of significance $\alpha = 0.05$ that the population means do differ by type.

18. (a) H_0: There is no interaction between carrier (Factor A) and type (Factor B). H_a: There is interaction between carrier (Factor A) and type (Factor B). Reject H_0 if the p-value ≤ 0.05. The p-value $= 0.001$, which is ≤ 0.05; therefore we reject H_0. There is evidence of interaction between carrier (Factor A) and type (Factor B) at level of significance $\alpha = 0.05$. This result agrees with the interaction plot in Exercise 14. **(b)** Not appropriate **(c)** Not appropriate

19. (a) H_0: There is no interaction between carrier (Factor A) and type (Factor B). H_a: There is interaction between carrier (Factor A) and type (Factor B). Reject H_0 if the p-value ≤ 0.05. The p-value $= 0.252$, which is not ≤ 0.05; therefore we do not reject H_0. There is insufficient evidence of interaction between carrier (Factor A) and type (Factor B) at level of significance $\alpha = 0.05$. This result agrees with the interaction plot in Exercise 15. **(b)** H_0: There is no carrier (Factor A) effect. That is, the population means do not differ by carrier. H_a: There is a carrier (Factor A) effect. That is, the population means do differ by carrier. Reject H_0 if the p-value ≤ 0.05. The p-value $= 0$, which is ≤ 0.05; therefore we reject H_0. There is evidence for a carrier (Factor A) effect. In other words, there is evidence at level of significance $\alpha = 0.05$ that the population means do differ by carrier. **(c)** H_0: There is no type (Factor B) effect. That is, the population means do not differ by type. H_a: There is a type (Factor B) effect. That is, the population means do differ by type. Reject H_0 if the p-value ≤ 0.05. The p-value $= 0$, which is ≤ 0.05; therefore we reject H_0. There is evidence for a type (Factor B) effect. In other words, there is evidence at level of significance $\alpha = 0.05$ that the population means do differ by type.

20. (a) H_0: There is no interaction between carrier (Factor A) and type (Factor B). H_a: There is interaction between carrier (Factor A) and type (Factor B). Reject H_0 if the p-value ≤ 0.05. The p-value $= 0.332$, which is not ≤ 0.05; therefore we do not reject H_0. There is insufficient evidence of interaction between carrier (Factor A) and type (Factor B) at level of significance $\alpha = 0.05$. This result agrees with the interaction plot in Exercise 16. **(b)** H_0: There is no carrier (Factor A) effect. That is, the population means do not differ by carrier. H_a: There is a carrier (Factor A) effect. That is, the population means do differ by carrier. Reject H_0 if the p-value ≤ 0.05. The p-value $= 0.729$, which is not ≤ 0.05; therefore we do not reject H_0. There is insufficient evidence for a carrier (Factor A) effect. In other words, there is insufficient evidence at level of significance $\alpha = 0.05$ that the population means do differ by carrier. **(c)** H_0: There is no type (Factor B) effect. That is, the population means do not differ by type. H_a: There is a type (Factor B) effect. That is, the population means do differ by type. Reject H_0 if the p-value ≤ 0.05. The p-value $= 0.001$, which is ≤ 0.05; therefore we reject H_0. There is evidence for a type (Factor B) effect. In other words, there is evidence at level of significance $\alpha = 0.05$ that the population means do differ by type.

21. (a) H_0: There is no interaction between hypertension (Factor A) and gender (Factor B). H_a: There is interaction between hypertension (Factor A) and gender (Factor B). Reject H_0 if the p-value ≤ 0.05. The p-value $= 0.443$, which is not ≤ 0.05; therefore we do not reject H_0. There is insufficient evidence of interaction between hypertension (Factor A) and gender (Factor B) at level of significance $\alpha = 0.05$. **(b)** H_0: There is no hypertension (Factor A) effect. That is, the population means do not differ by whether or not the person has hypertension. H_a: There is a hypertension (Factor A) effect. That is, the population means do differ by whether or not the person has hypertension.

Reject H_0 if the p-value ≤ 0.05. The p-value $= 0.014$, which is ≤ 0.05; therefore we reject H_0. There is evidence for a hypertension (Factor A) effect. Thus we can conclude at level of significance $\alpha = 0.05$ that there is a significant difference in mean cardiac output between patients who have hypertension and patients who don't. **(c)** H_0: There is no gender (Factor B) effect. That is, the population means do not differ by gender. H_a: There is a gender (Factor B) effect. That is, the population means do differ by gender. Reject H_0 if the p-value ≤ 0.05. The p-value $= 0.105$, which is not ≤ 0.05; therefore we do not reject H_0. There is insufficient evidence for a gender (Factor B) effect. Thus we can conclude at level of significance $\alpha = 0.05$ that there is no significant difference in mean cardiac output between females and males.

22. (a) H_0: There is no interaction between problem type (Factor A) and course (Factor B). H_a: There is interaction between problem type (Factor A) and course (Factor B). Reject H_0 if the p-value ≤ 0.05. The p-value $= 0.887$, which is not ≤ 0.05; therefore we do not reject H_0. There is insufficient evidence of interaction between problem type (Factor A) and course (Factor B) at level of significance $\alpha = 0.05$. **(b)** H_0: There is no problem type (Factor A) effect. That is, the population means do not differ by problem type. H_a: There is a problem type (Factor A) effect. That is, the population means do differ by problem type. Reject H_0 if the p-value ≤ 0.05. The p-value $= 0.140$, which is not ≤ 0.05; therefore we do not reject H_0. There is insufficient evidence for a problem type (Factor A) effect. Thus we can conclude at level of significance $\alpha = 0.05$ that there is no significant difference in mean test scores between multiple-choice tests, short-answer tests, and word-problem tests. **(c)** H_0: There is no course (Factor B) effect. That is, the population means do not differ by course. H_a: There is a course (Factor B) effect. That is, the population means do differ by course. Reject H_0 if the p-value ≤ 0.05. The p-value $= 0.933$, which is not ≤ 0.05; therefore we do not reject H_0. There is insufficient evidence for a course (Factor B) effect. Thus we can conclude at level of significance $\alpha = 0.05$. that there is no significant difference in mean test scores between statistics tests and algebra tests.

23. H_0: There is no interaction between brand (Factor A) and fiber (Factor B). H_a: There is interaction between brand (Factor A) and fiber (Factor B). Reject H_0 if the p-value ≤ 0.05. The p-value $= 0.888$, which is not ≤ 0.05; therefore we do not reject H_0. There is insufficient evidence of interaction between brand (Factor A) and fiber (Factor B) at level of significance $\alpha = 0.05$. H_0: There is no brand (Factor A) effect. That is, the population means do not differ by brand. H_a: There is a brand (Factor A) effect. That is, the population means do differ by brand. Reject H_0 if the p-value ≤ 0.05. The p-value $= 0.045$, which is ≤ 0.05; therefore we reject H_0. There is evidence for a brand (Factor A) effect. Thus we can conclude at level of significance $\alpha = 0.05$ that there is a significant difference in mean nutritional ratings between Kellogg's cereals and General Mills cereals. H_0: There is no fiber (Factor B) effect. That is, the population means do not differ by whether or not the cereal has fiber. H_a: There is a fiber (Factor B) effect. That is, the population means do differ by whether or not the cereal has fiber. Reject H_0 if the p-value ≤ 0.05. The p-value $= 0.006$, which is ≤ 0.05; therefore we reject H_0. There is evidence for a fiber (Factor B) effect. Thus we can conclude at level of significance $\alpha = 0.05$ that there is a significant difference in mean nutritional ratings between cereals with a high fiber content and cereals with a low fiber content.

24. (a) H_0: There is no interaction between industry (Factor A) and exchange (Factor B). H_a: There is interaction between industry (Factor A) and exchange (Factor B). Reject H_0 if the p-value ≤ 0.01. The p-value $= 0.142$, which is not ≤ 0.01; therefore we do not reject H_0. There is insufficient evidence of interaction between industry (Factor A) and exchange (Factor B) at level of significance $\alpha = 0.01$. **(b)** H_0: There is no industry (Factor A) effect. That is, the population means do not differ by industry. H_a: There is an industry (Factor A) effect. That is, the population means do differ by industry. Reject H_0 if the p-value ≤ 0.01. The p-value $= 0.042$, which is not ≤ 0.01; therefore we do not reject H_0. There is insufficient evidence for an industry (Factor A) effect. Thus we can conclude at level of significance $\alpha = 0.01$ that there is no significant difference in mean stock price between the three industries. **(c)** H_0: There is no exchange (Factor B) effect. That is, the population means do not differ by exchange. H_a: There is an exchange (Factor B) effect. That is, the population means do differ by exchange. Reject H_0 if the p-value ≤ 0.01. The p-value $= 0.002$, which is ≤ 0.01; therefore we reject H_0. There is evidence for an exchange (Factor B) effect. Thus we can conclude at level of significance $\alpha = 0.01$ that there is a significant difference in mean stock prices for the NYSE and the NASDAQ. **(d)** The conclusion in part (b) would change to reject H_0; there is an industry effect. The conclusion in part (c) would remain the same.

Chapter 12 Review

1. **(a)** $df_1 = 3$, $df_2 = 296$ **(b)** 10 **(c)** 10,000 **(d)** 1,157.5 **(e)** 11,157.5 **(f)** 3,333.3333 **(g)** 3.910472973 **(h)** 852.4117985

2.

Source of variation	Sum of squares	Degrees of freedom	Mean square	F-test statistic
Treatment	SSTR = 10,000	$df_1 = 3$	MSTR = 3333.3333	F_{data} = 852.4117985
Error	SSE = 1157.5	$df_2 = 96$	MSE = 3.910472973	
Total	SST = 11,157.5			

3. $H_0 : \mu_1 = \mu_2 = \mu_3$. H_a: Not all the population means are equal. μ_1 = population mean level of satisfaction for Medical Treatment 1. μ_2 = population mean level of satisfaction for Medical Treatment 2. μ_3 = population mean level of satisfaction for Medical Treatment 3. $F_{crit} = 3.55$. Reject H_0 if $F_{data} \geq 3.55$. $F_{data} = 3.19$. Since F_{data} is not ≥ 3.55 we do not reject H_0. There is insufficient evidence that not all of the population means are equal.

```
Source   DF      SS     MS      F      P
Factor    2    4114   2057   3.19   0.065
Error    18   11600    644
Total    20   15714
```

4. $H_0 : \mu_A = \mu_B = \mu_C = \mu_D$. H_a: Not all the population means are equal. μ_A = the population mean customer satisfaction at Store A, μ_B = the population mean customer satisfaction at Store B, μ_C = the population mean customer satisfaction at Store C, and μ_D = the population mean customer satisfaction at Store D. Reject H_0 if the p-value ≤ 0.05. $F_{data} = 25.47$. p-value ≈ 0. Since the p-value ≤ 0.05, we reject H_0. There is evidence that not all the population means are equal.

Source	df	SS	MS	F	P
Factor	3	7321.4	2440.5	25.47	0.000
Error	24	2300.0	95.8		
Total	27	9621.4			

5. $t_{data} = -2.5$
6. $t_{data} = -1.5$
7. $t_{data} = 1$
8. **(a)** Test 1: $H_0 : \mu_1 = \mu_2$ vs. $H_a : \mu_1 \neq \mu_2$; Test 2: $H_0 : \mu_1 = \mu_3$ vs. $H_a : \mu_1 \neq \mu_3$; Test 3: $H_0 : \mu_2 = \mu_3$ vs. $H_a : \mu_2 \neq \mu_3$. For each hypothesis test, reject H_0 if the Bonferroni-adjusted p-value $\leq \alpha = 0.05$. **(b)** Test 1: $t_{data} = -2.5$; Test 2: $t_{data} = -1.5$; Test 3: $t_{data} = 1$ **(c)** Test 1: Bonferroni-adjusted p-value = 0.0441; Test 2: Bonferroni-adjusted p-value = 0.4140; Test 3: Bonferroni-adjusted p-value = 0.9620 **(d)** Test 1: The adjusted p-value = 0.0441, which is ≤ 0.05; therefore we reject H_0. There is evidence at the $\alpha = 0.05$ level of significance that the population mean of Population 1 differs from the population mean of Population 2. Test 2: The adjusted p-value = 0.4140, which is not ≤ 0.05; therefore we do not reject H_0. There is insufficient evidence at the $\alpha = 0.05$ level of significance that the population mean of Population 1 differs from the population mean of Population 3. Test 3: The adjusted p-value = 0.9620, which is not ≤ 0.05; therefore we do not reject H_0. There is insufficient evidence at the $\alpha = 0.05$ level of significance that the population mean of Population 2 differs from the population mean of Population 3.
9. $q_{data} = 2$
10. $q_{data} = 4$
11. $q_{data} = 2$
12. $q_{crit} = 3.356$
13. **(a)** Test 1: $H_0 : \mu_1 = \mu_2$ vs. $H_a : \mu_1 \neq \mu_2$; Test 2: $H_0 : \mu_1 = \mu_3$ vs. $H_a : \mu_1 \neq \mu_3$; Test 3: $H_0 : \mu_2 = \mu_3$ vs. $H_a : \mu_2 \neq \mu_3$. **(b)** $q_{crit} = 3.356$. Reject H_0 if $q_{data} \geq 3.356$. **(c)** Test 1: $q_{data} = 2$; Test 2: $q_{data} = 4$; Test 3: $q_{data} = 2$ **(d)** Test 1: $q_{data} = 2$, which is not ≥ 3.356; therefore we do not reject H_0. There

is insufficient evidence at the $\alpha = 0.05$ level of significance that the population mean of Population 1 differs from the population mean of Population 2. Test 2: $q_{data} = 4$, which is ≥ 3.356; therefore we reject H_0. There is evidence at the $\alpha = 0.05$ level of significance that the population mean of Population 1 differs from the population mean of Population 3. Test 3: $q_{data} = 2$, which is not ≥ 3.356; therefore we do not reject H_0. There is insufficient evidence at the $\alpha = 0.05$ level of significance that the population mean of Population 2 differs from the population mean of Population 3.
14. **(a)** $H_0 : \mu_1 = \mu_2 = \mu_3 = \mu_4 = \mu_5$ vs. H_a: Not all of the population means are equal. Reject H_0 if p-value ≤ 0.05. **(b)** Answers to missing values are in red.

Source	Sum of squares	Degrees of freedom	Mean square	F
Treatments	80	4	20	2
Blocks	420	5	84	
Error	200	20	10	
Total	700	29		

(c) The p-value = 0.1333, which is not ≤ 0.05; therefore we do not reject H_0. There is insufficient evidence at level of significance $\alpha = 0.05$ that the population means are not all equal.
15. $H_0 : \mu_{women} = \mu_{men}$ vs. H_a: Not all of the population means are equal. Reject H_0 if p-value ≤ 0.05. The p-value = 1.000, which is not ≤ 0.05; therefore we do not reject H_0. There is insufficient evidence at level of significance $\alpha = 0.05$ that the population mean number of small businesses owned by women differs from the population mean number of small businesses owned by men.
16. The lines are not parallel, so there is some interaction between Factor A and Factor B.

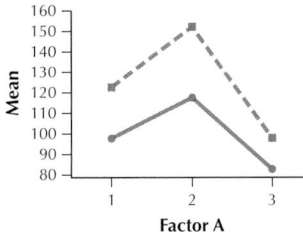

17. There is significant interaction between Factor A and Factor B.

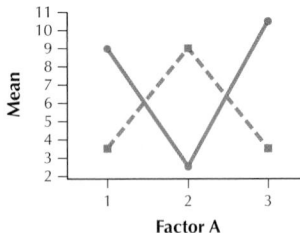

18. **(a)** H_0: There is no interaction between carrier (Factor A) and type (Factor B). H_a: There is interaction between carrier (Factor A) and type (Factor B). Reject H_0 if the p-value ≤ 0.05. The p-value = 0.020, which is ≤ 0.05; therefore we reject H_0. There is evidence of interaction between carrier (Factor A) and type (Factor B) at level of significance $\alpha = 0.05$. This result agrees with the interaction plot in Exercise 16. **(b)** Not appropriate **(c)** Not appropriate
19. **(a)** H_0: There is no interaction between carrier (Factor A) and type (Factor B). H_a: There is interaction between carrier (Factor A) and type (Factor B). Reject H_0 if the p-value ≤ 0.05. The p-value = 0.004, which is ≤ 0.05; therefore we reject H_0. There is evidence of interaction between carrier (Factor A) and type (Factor B) at level of significance $\alpha = 0.05$. This result agrees with the interaction plot in Exercise 17. **(b)** Not appropriate **(c)** Not appropriate

Chapter 12 Quiz

1. False
2. True

3. False
4. Mean square
5. Mean square treatment
6. Mean square error
7. $\bar{\bar{x}}$
8. Mean square equals the sum of squares divided by the degrees of freedom.
9. F_{data}
10. $H_0 : \mu_4 = \mu_6 = \mu_8$. H_a: Not all the population means are equal.
μ_4 = population mean miles per gallon for 4-cylinder cars. μ_6 = population mean miles per gallon for 6-cylinder cars. μ_8 = population mean miles per gallon for 8-cylinder cars. Reject H_0 if the p-value ≤ 0.05. F_{data} = 341.4932. p-value = 0. Since the p-value ≤ 0.05, we reject H_0. There is evidence that not all the population mean gas mileages are equal.
11. $H_0 : \mu_{married} = \mu_{widowed} = \mu_{divorced} = \mu_{separated} = \mu_{nevermarried}$. H_a: Not all the population means are equal. $\mu_{married}$ = population mean number of hours worked by people who are married. $\mu_{widowed}$ = population mean number of hours worked by people who are widowed. $\mu_{divorced}$ = population mean number of hours worked by people who are divorced. $\mu_{separated}$ = population mean number of hours worked by people who are separated. $\mu_{nevermarried}$ = population mean number of hours worked by people who have never been married. Reject H_0 if the p-value ≤ 0.05. F_{data} = 2.5102. p-value = 0.0401. Since the p-value ≤ 0.05, we reject H_0. There is evidence that not all the population mean numbers of hours worked are equal.
12. $H_0 : \mu_{Kelloggs} = \mu_{Quaker} = \mu_{RalstonPurina}$. H_a: Not all the population means are equal. $\mu_{Kelloggs}$ = population mean number of calories per serving in breakfast cereals made by Kellogg's. μ_{Quaker} = population mean number of calories per serving in breakfast cereals made by Quaker. $\mu_{RalstonPurina}$ = population mean number of calories per serving in breakfast cereals made by Ralston Purina. F_{crit} = 3.26. Reject H_0 if $F_{data} \geq 3.26$. F_{data} = 1.5573. Since F_{data} is not ≥ 3.26, we do not reject H_0. There is insufficient evidence that not all the population mean numbers of calories per serving are equal.

Chapter 13

Note to instructors and students: Some answers may vary slightly depending on whether you round at intermediate steps or wait until you get the final answer to round. Also, different software and different forms of technology may give slightly different answers.

Section 13.1

1. The regression equation is calculated from a sample and is valid only for values of x in the range of the sample data. The population regression equation may be used to approximate the relationship between the predictor variable x and the response variable y for the entire population of (x, y) pairs.
2. (1) Zero mean assumption (2) Constant variance assumption (3) Independence assumption (4) Normality assumption
3. We construct a scatterplot of the residuals against the fitted values and a normal probability plot of the residuals. We must make sure that the scatterplot contains no strong evidence of any unhealthy patterns and that the normal probability plot indicates no evidence of departures from normality in residuals.
4. b_0 and b_1 are statistics and β_0 and β_1 are parameters.
5. There is no relationship between x and y.
6. $s = \sqrt{MSE/(n) - 2} = \sqrt{\sum(y - \hat{y})^2/(n) - 2}$ is the standard error of the estimate and $s_x = \sqrt{\sum(x - \bar{x})^2/(n) - 1}$ is the sample standard deviation of x.

7. **(a)** and **(b)**

x	y	Predicted value $\hat{y} = 13.5 + 2.5x$	Residual $(y - \hat{y})$
1	15	16	-1
2	20	18.5	1.5
3	20	21	-1
4	25	23.5	1.5
5	25	26	-1

(c) and **(d)** See Solutions Manual. **(e)** The scatterplot of the residuals contains an unhealthy pattern, so the regression assumptions are not verified.

8. **(a)** and **(b)**

x	y	Predicted value $\hat{y} = 8 + 3.2x$	Residual $(y - \hat{y})$
0	10	8	2
5	20	24	-4
10	45	40	5
15	50	56	-6
20	75	72	3

(c) and **(d)** See Solutions Manual. **(e)** The scatterplot of the residuals contains strong evidence of an unhealthy pattern. Therefore we conclude that the regression assumptions are not verified.

9. **(a)** and **(b)**

x	y	Predicted value $\hat{y} = 21.6 + 4x$	Residual $(y - \hat{y})$
-5	0	1.6	-1.6
-4	8	5.6	2.4
-3	8	9.6	-1.6
-2	16	13.6	2.4
-1	16	17.6	-1.6

(c) and **(d)** See Solutions Manual. **(e)** The scatterplot of the residuals contains an unhealthy pattern, so the regression assumptions are not verified.

10. **(a)** and **(b)**

x	y	Predicted value $\hat{y} = -16 - 3x$	Residual $(y - \hat{y})$
-3	-5	-7	2
-1	-15	-13	-2
1	-20	-19	-1
3	-25	-25	0
5	-30	-31	1

(c)

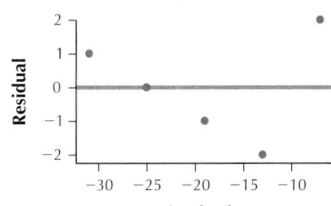

(d)

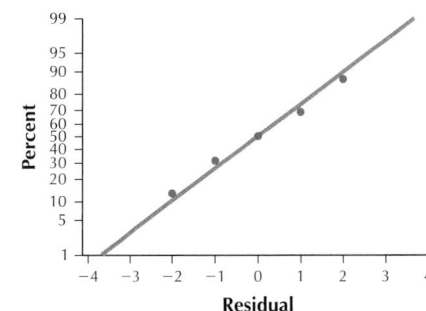

(e) The scatterplot of the residuals contains strong evidence of an unhealthy pattern. Therefore, we conclude that the regression assumptions are not verified.

11. **(a)** and **(b)**

x	y	Predicted value $\hat{y} = 104 - 0.5x$	Residual $(y - \hat{y})$
10	100	99	1
20	95	94	1
30	85	89	-4
40	85	84	1
50	80	79	1

(c) and **(d)** See Solutions Manual. **(e)** The scatterplot of the residuals contains an unhealthy pattern, so the regression assumptions are not verified.

12. (a) and **(b)**

x	y	Predicted value $\hat{y} = 9 + 0.2x$	Residual $(y - \hat{y})$
0	11	9	2
20	11	13	-2
40	16	17	-1
60	21	21	0
80	26	25	1

(c)

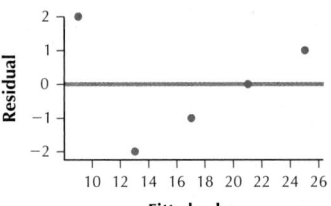

(d)

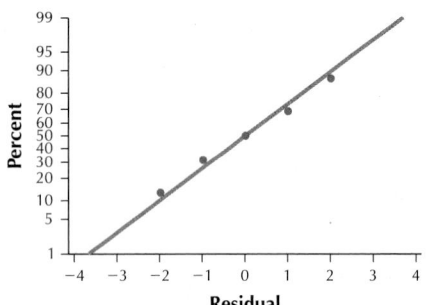

(e) The scatterplot of the residuals contains strong evidence of an unhealthy pattern. Therefore, we conclude that the regression assumptions are not verified.

13. (a) and **(b)**

x	y	$\hat{y} = 0.6x + 0.2$	$y - \hat{y}$
1	1	0.8	0.2
2	1	1.4	-0.4
3	2	2	0
4	3	2.6	0.4
5	3	3.2	-0.2

(c)

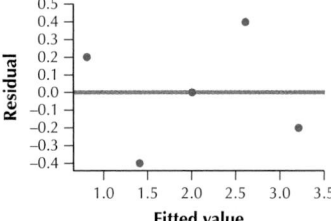

(d)

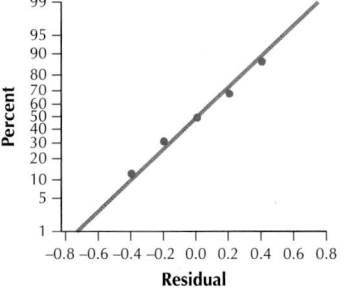

(e) The scatterplot in (c) of the residuals versus fitted values shows no strong evidence of unhealthy patterns. Thus, the independence assumption, the

constant variance assumption, and the zero-mean assumption are verified. Also, the normal probability plot of the residuals in (d) indicates no evidence of departure from normality of the residuals. Therefore we conclude that the regression assumptions are verified.

14. (a) and **(b)**

x	y	$\hat{y} = -2x + 8$	$y - \hat{y}$
1	6	6	0
2	5	4	1
2	4	4	0
2	3	4	-1
3	2	2	0

(c)

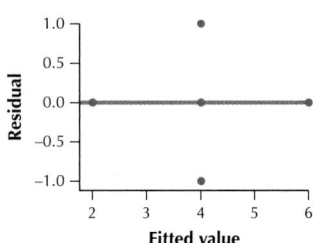

(d)

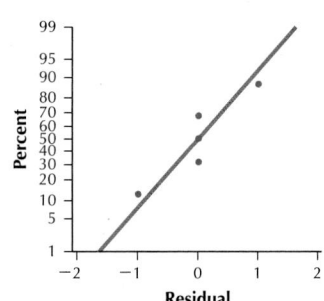

(e) The scatterplot in (c) of the residuals versus fitted values shows no strong evidence of unhealthy patterns. Thus, the independence assumption, the constant variance assumption, and the zero-mean assumption are verified. Also, the normal probability plot of the residuals in (d) indicates no evidence of departure from normality of the residuals. Therefore we conclude that the regression assumptions are verified.

15. (a) $t_{crit} = 3.182$ **(b)** $s = 1.58113883$ **(c)** $\sum(x - \bar{x})^2 = 10$ **(d)** $t_{data} = 5$ **(e)** $H_0 : \beta_1 = 0$: There is no linear relationship between x and y. $H_a : \beta_1 \neq 0$: There is a linear relationship between x and y. Reject H_0 if $t_{data} \geq 3.182$ or $t_{data} \leq -3.182$. Since $t_{data} = 5 \geq 3.182$, we reject H_0. There is evidence at level of significance $\alpha = 0.05$ that $\beta_1 \neq 0$ and that there is a linear relationship between x and y.

16. (a) $t_{crit} = 3.182$ **(b)** $s = 5.477225575$ **(c)** $\sum(x - \bar{x})^2 = 250$ **(d)** $t_{data} = 9.2376$ **(e)** $H_0 : \beta_1 = 0$: There is no linear relationship between x and y. $H_a : \beta_1 \neq 0$: There is a linear relationship between x and y. Reject H_0 if $t_{data} \geq 3.182$ or $t_{data} \leq -3.182$. Since $t_{data} = 9.2376 \geq 3.182$, we reject H_0. There is evidence at level of significance $\alpha = 0.05$ that $\beta_1 \neq 0$ and that there is a linear relationship between x and y.

17. (a) $t_{crit} = 3.182$ **(b)** $s = 2.529822128$ **(c)** $\sum(x - \bar{x})^2 = 10$ **(d)** $t_{data} = 5$ **(e)** $H_0 : \beta_1 = 0$: There is no linear relationship between x and y. $H_a : \beta_1 \neq 0$: There is a linear relationship between x and y. Reject H_0 if $t_{data} \geq 3.182$ or $t_{data} \leq -3.182$. Since $t_{data} = 5 \geq 3.182$, we reject H_0. There is evidence at level of significance $\alpha = 0.05$ that $\beta_1 \neq 0$ and that there is a linear relationship between x and y.

18. (a) $t_{crit} = 3.182$ **(b)** $s = 1.825741858$ **(c)** $\sum(x - \bar{x})^2 = 40$ **(d)** $t_{data} = -10.3923$ **(e)** $H_0 : \beta_1 = 0$: There is no linear relationship between x and y. $H_a : \beta_1 \neq 0$: There is a linear relationship between x and y. Reject H_0 if $t_{data} \geq 3.182$ or $t_{data} \leq -3.182$. Since $t_{data} = -10.3923 \leq -3.182$, we reject H_0. There is evidence at level of significance $\alpha = 0.05$ that $\beta_1 \neq 0$ and that there is a linear relationship between x and y.

19. (a) $s = 2.581988897$ **(b)** $\sum(x - \bar{x})^2 = 1000$ **(c)** $t_{data} = -6.1237$ **(d)** p-value $= 0.0088$ **(e)** $H_0 : \beta_1 = 0$: There is no linear relationship between x and y. $H_a : \beta_1 \neq 0$: There is a linear relationship between x and y. Reject H_0 if p-value ≤ 0.05. Since p-value $= 0.0088 \leq 0.05$, we reject H_0. There is evidence at level of significance $\alpha = 0.05$ that $\beta_1 \neq 0$ and that there is a linear relationship between x and y.

20. (a) $s = 1.825741858$ **(b)** $\sum(x - \bar{x})^2 = 4000$ **(c)** $t_{data} = 6.9282$
(d) p-value $= 0.0062$ **(e)** $H_0 : \beta_1 = 0$: There is no linear relationship between x and y. $H_a : \beta_1 \neq 0$: There is a linear relationship between x and y. Reject H_0 if p-value ≤ 0.05. Since p-value $= 0.0062 \leq 0.05$, we reject H_0. There is evidence at level of significance $\alpha = 0.05$ that $\beta_1 \neq 0$ and that there is a linear relationship between x and y.
21. (a) $s = 0.3651483717$ **(b)** $\sum(x - \bar{x})^2 = 10$ **(c)** $t_{data} = 5.1962$
(d) p-value $= 0.0138$ **(e)** $H_0 : \beta_1 = 0$: There is no linear relationship between x and y. $H_a : \beta_1 \neq 0$: There is a linear relationship between x and y. Reject H_0 if p-value ≤ 0.05. Since p-value $= 0.0138 \leq 0.05$, we reject H_0. There is evidence at level of significance $\alpha = 0.05$ that $\beta_1 \neq 0$ and that there is a linear relationship between x and y.
22. (a) $s = 0.8164965809$ **(b)** $\sum(x - \bar{x})^2 = 2$ **(c)** $t_{data} = -3.4641$
(d) p-value $= 0.0405$ **(e)** $H_0 : \beta_1 = 0$: There is no linear relationship between x and y. $H_a : \beta_1 \neq 0$: There is a linear relationship between x and y. Reject H_0 if p-value ≤ 0.05. Since p-value $= 0.0405 \leq 0.05$, we reject H_0. There is evidence at level of significance $\alpha = 0.05$ that $\beta_1 \neq 0$ and that there is a linear relationship between x and y.
23. (a) $t_{\alpha/2} = 3.182$ **(b)** $E = 1.591$ **(c)** $(0.909, 4.091)$
24. (a) $t_{\alpha/2} = 3.182$ **(b)** $E = 1.1023$ **(c)** $(2.0977, 4.3023)$, TI-83/84: $(2.0976, 4.3024)$
25. (a) $t_{\alpha/2} = 3.182$ **(b)** $E = 2.5456$ **(c)** $(1.4544, 6.5456)$
26. (a) $t_{\alpha/2} = 3.182$ **(b)** $E = 0.9186$ **(c)** $(-3.9186, -2.0814)$
27. (a) $t_{\alpha/2} = 3.182$ **(b)** $E = 0.2598$ **(c)** $(-0.7598, -0.2402)$
28. (a) $t_{\alpha/2} = 3.182$ **(b)** $E = 0.0919$ **(c)** $(0.1081, 0.2919)$.
29. (a) $t_{\alpha/2} = 3.182$ **(b)** $E = 0.3674$ **(c)** $(0.2326, 0.9674)$. TI-83/84: $(0.2325, 0.9675)$
30. (a) $t_{\alpha/2} = 3.182$ **(b)** $E = 1.8371$ **(c)** $(-3.8371, -0.1629)$. TI-83/84: $(-3.8370, -0.1626)$
31. $H_0 : \beta_1 = 0$: There is no linear relationship between x and y. $H_a : \beta_1 \neq 0$: There is a linear relationship between x and y. Since the confidence interval from Exercise 23 (c) does not contain zero, we may conclude that $\beta_1 \neq 0$ and that a linear relationship exists between x and y, at level of significance $\alpha = 0.05$.
32. $H_0 : \beta_1 = 0$: There is no linear relationship between x and y. $H_a : \beta_1 \neq 0$: There is a linear relationship between x and y. Since the confidence interval from Exercise 24 (c) does not contain zero, we may conclude that $\beta_1 \neq 0$ and that a linear relationship exists between x and y, at level of significance $\alpha = 0.05$.
33. $H_0 : \beta_1 = 0$: There is no linear relationship between x and y. $H_a : \beta_1 \neq 0$: There is a linear relationship between x and y. Since the confidence interval from Exercise 25 (c) does not contain zero, we may conclude that $\beta_1 \neq 0$ and that a linear relationship exists between x and y, at level of significance $\alpha = 0.05$.
34. $H_0 : \beta_1 = 0$: There is no linear relationship between x and y. $H_a : \beta_1 \neq 0$: There is a linear relationship between x and y. Since the confidence interval from Exercise 26 (c) does not contain zero, we may conclude that $\beta_1 \neq 0$ and that a linear relationship exists between x and y, at level of significance $\alpha = 0.05$.
35. $H_0 : \beta_1 = 0$: There is no linear relationship between x and y. $H_a : \beta_1 \neq 0$: There is a linear relationship between x and y. Since the confidence interval from Exercise 27 (c) does not contain zero, we may conclude that $\beta_1 \neq 0$ and that a linear relationship exists between x and y, at level of significance $\alpha = 0.05$.
36. $H_0 : \beta_1 = 0$: There is no linear relationship between x and y. $H_a : \beta_1 \neq 0$: There is a linear relationship between x and y. Since the confidence interval from Exercise 28 (c) does not contain zero, we may conclude that $\beta_1 \neq 0$ and that a linear relationship exists between x and y, at level of significance $\alpha = 0.05$.
37. $H_0 : \beta_1 = 0$: There is no linear relationship between x and y. $H_a : \beta_1 \neq 0$: There is a linear relationship between x and y. Since the confidence interval from Exercise 29 (c) does not contain zero, we may conclude that $\beta_1 \neq 0$ and that a linear relationship exists between x and y, at level of significance $\alpha = 0.05$.
38. $H_0 : \beta_1 = 0$: There is no linear relationship between x and y. $H_a : \beta_1 \neq 0$: There is a linear relationship between x and y. Since the confidence interval from Exercise 30 (c) does not contain zero, we may conclude that $\beta_1 \neq 0$ and that a linear relationship exists between x and y, at level of significance $\alpha = 0.05$.
39. $H_0 : \beta_1 = 0$: There is no relationship between volume (x) and weight (y). $H_a : \beta_1 \neq 0$: There is a linear relationship between volume (x) and weight (y). Reject H_0 if the p-value ≤ 0.05. Since the p-value ≈ 0.0006 is ≤ 0.05, we reject H_0. There is evidence for a linear relationship between volume (x) and weight (y).

40. $H_0 : \beta_1 = 0$: There is no relationship between *Family Size* (x) and *Pets* (y). $H_a : \beta_1 \neq 0$: There is a linear relationship between *Family Size* (x) and *Pets* (y). Reject H_0 if $t_{data} \leq -3.182$ or $t_{data} \geq 3.182$. Since $t_{data} = 5$ is ≥ 3.182, we reject H_0. There is evidence for a linear relationship between *Family Size* (x) and *Pets* (y).
41. $H_0 : \beta_1 = 0$: There is no relationship between *Low* (x) and *High* (y). $H_a : \beta_1 \neq 0$: There is a linear relationship between *Low* (x) and *High* (y). Reject H_0 if $t_{data} \geq 2.776$. Since $t_{data} = 12.09$ is ≥ 2.776, we reject H_0. There is evidence for a linear relationship between *Low* (x) and *High* (y).
42. $H_0 : \beta_1 = 0$: No linear relationship exists between weeks and total sales. $H_a : \beta_1 \neq 0$: A linear relationship exists between weeks and total sales. Reject H_0 if the p-value $\leq \alpha = 0.05$. $t_{data} \approx 2.0208$. p-value $= 0.1365542636$. The p-value $= 0.1365542636$ is not $\leq \alpha = 0.05$, so we do not reject H_0. Insufficient evidence exists, at level of significance $\alpha = 0.05$, for a linear relationship between weeks on the top 30 list and total sales.
43. $H_0 : \beta_1 = 0$: No linear relationship exists between DJIA and darts. $H_a : \beta_1 \neq 0$: A linear relationship exists between DJIA and darts. Reject H_0 if the p-value $\leq \alpha = 0.05$. $t_{data} \approx 2.1749$. p-value $= 0.0725703888$. The p-value $= 0.0725703888$ is not $\leq \alpha = 0.05$, so we do not reject H_0. Insufficient evidence exists, at level of significance $\alpha = 0.05$, for a linear relationship between DJIA and darts.
44. $H_0 : \beta_1 = 0$: No linear relationship exists between age and height. $H_a : \beta_1 \neq 0$: A linear relationship exists between age and height. Reject H_0 if the p-value $\leq \alpha = 0.05$. $t_{data} \approx -0.7712$. p-value $= 0.4698711979$. The p-value $= 0.4698711979$ is not $\leq \alpha = 0.05$, so we do not reject H_0. Insufficient evidence exists, at level of significance $\alpha = 0.05$, for a linear relationship between a women's age and a woman's height.
45. $H_0 : \beta_1 = 0$: No linear relationship exists between age and shots. $H_a : \beta_1 \neq 0$: A linear relationship exists between age and shots. Reject H_0 if the p-value $\leq \alpha = 0.05$. $t_{data} \approx 0.1817$. p-value $= 0.8603048707$. The p-value $= 0.8603048707$ is not $\leq \alpha = 0.05$, so we do not reject H_0. Insufficient evidence exists, at level of significance $\alpha = 0.05$, for a linear relationship between a patient's age and the number of shots taken by the patient.
46. $H_0 : \beta_1 = 0$: No linear relationship exists between winning proportion and power rating. $H_a : \beta_1 \neq 0$: A linear relationship exists between winning proportion and power rating. Reject H_0 if the p-value $\leq \alpha = 0.05$. $t_{data} \approx 4.2557$. p-value $= 0.0027775932$. The p-value $= 0.0027775932$ is $\leq \alpha = 0.05$, so we reject H_0. Evidence exists, at level of significance $\alpha = 0.05$, for a linear relationship between a team's winning proportion and the team's power rating.
47. (a) $E = 0.3312$ **(b)** $(1.2688, 1.9312)$ **(c)** We are 95% confident that the interval $(1.2688, 1.9312)$ captures the population slope β_1 of the relationship between volume and weight.
48. (a) $E = 0.3182$ **(b)** $(0.1818, 0.8182)$ **(c)** We are 95% confident that the interval $(0.1818, 0.8182)$ captures the population slope β_1 of the relationship between *Family Size* and *Pets*.
49. (a) $E = 0.2402$ **(b)** $(0.8064, 1.2868)$. TI-83/84: $(0.8063, 1.2868)$ **(c)** We are 95% confident that the interval $(0.8063, 1.2868)$ captures the population slope β_1 of the relationship between *Low* and *High*.
50. (a) 24460.5 game units **(b)** $(-8929, 39992)$ **(c)** We are 95% confident that the interval $(-8929, 39992)$ captures the slope β_1 of the population regression line. That is, we are 95% confident that for each additional week on the top 30 list, the change in the total number of game units sold lies between -8929 units and $39,992$ units.
51. (a) 0.35074 **(b)** $(-0.039, 0.66248)$ **(c)** We are 95% confident that the interval $(-0.039, 0.66248)$ captures the slope β_1 of the population regression line. That is, we are 95% confident that for each additional $1 the stocks in the DJIA gain in one day, the daily change in the stocks in the portfolio predicted by the darts lies between $-\$0.039$ and $\$0.66248$.
52. (a) 0.35327 inch **(b)** $(-0.4646, 0.24194)$ **(c)** We are 95% confident that the interval $(-0.4646, 0.24194)$ captures the slope β_1 of the population regression line. That is, we are 95% confident that for each additional year in a woman's age, the change in her height lies between -0.4646 inch and 0.24194 inch.
53. (a) 0.19824 shot **(b)** $(-0.1826, 0.21388)$ **(c)** We are 95% confident that the interval $(-0.1826, 0.21388)$ captures the slope β_1 of the population regression line. That is, we are 95% confident that for each additional year in a patient's age, the change in the number of shots taken by the patient lies between -0.1826 shot and 0.21388 shot.
54. (a) 11.07125 **(b)** $(9.3605, 31.503)$ **(c)** We are 95% confident that the interval $(9.3605, 31.503)$ captures the slope β_1 of the population regression line.

That is, we are 95% confident that for each additional increase of 1 in the winning proportion, the change in the power rating of a team lies between 9.3605 and 31.503

55. (a) $(-6.647, 49.817)$ **(b)** 0 lies in the interval, so we do not reject H_0.
56. (a) $(-4.715, -1.846)$ **(b)** 0 does not lie in the interval, so we reject H_0.

57.

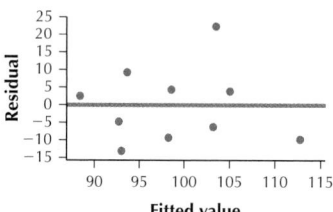

The scatterplot of the residuals versus the fitted values shows no evidence of the unhealthy patterns shown in Figure 4. Thus, the independence assumption, the constant variance assumption, and the zero-mean assumption are verified.

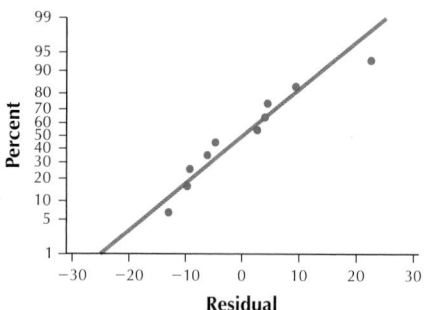

Also, the normal probability plot of the residuals indicates no evidence of departures from normality in the residuals. Therefore, we conclude that the regression assumptions are verified and it is okay to proceed with the regression.

58.

Regression Analysis: Runs scored versus Batting average

Analysis of Variance

Source	DF	Adj SS	Adj MS	F-Value	P-Value
Regression	1	480.1	480.1	3.67	0.092
Batting average	1	480.1	480.1	3.67	0.092
Error	8	1046.8	130.9		
Total	9	1526.9			

Model Summary

S	R-sq	R-sq(adj)	R-sq(pred)
11.4391	31.44%	22.87%	0.00%

Coefficients

Term	Coef	SE Coef	T-Value	P-Value	VIF
Constant	5.5	48.9	0.11	0.913	
Batting average	308	161	1.92	0.092	1.00

Regression Equation

Runs scored = 5.5 + 308 Batting average

Fits and Diagnostics for Unusual Observations

Obs	Runs scored	Fit	Resid	Std Resid
3	126.00	103.49	22.51	2.13 R

R Large residual

$H_0 : \beta_1 = 0$: No linear relationship exists between batting average and runs scored. $H_a : \beta_1 \neq 0$: A linear relationship exists between batting average and runs scored. Reject H_0 if the p-value $\leq \alpha = 0.10$. $t_{data} \approx 1.92$. p-value $= 0.092$. The p-value $= 0.092$ is $\leq \alpha = 0.10$, so we reject H_0. Evidence exists, at level of significance $\alpha = 0.10$, for a linear relationship between a player's batting average and the number of runs a player scores.
59. 22.51 runs; by far the highest number of runs scored but the third highest batting average.
60. $H_0 : \beta_1 = 0$: No linear relationship exists between batting average and runs scored. $H_a : \beta_1 \neq 0$: A linear relationship exists between batting average and runs scored. Reject H_0 if the p-value $\leq \alpha = 0.05$. $t_{data} \approx 1.92$.

p-value $= 0.092$. The p-value $= 0.092$ is not $\leq \alpha = 0.05$, so we do not reject H_0. Insufficient evidence exists, at level of significance $\alpha = 0.05$, for a linear relationship between a player's batting average and the number of runs a player scores.
We reject H_0 when $\alpha = 0.10$ and we do not reject H_0 when $\alpha = 0.05$. Turn to a direct assessment of the strength of evidence against the null hypothesis or obtain more data.
61. Against; the points appear to lie near a line with a positive slope.
62. No, the regression assumptions are not violated.
63. $H_0 : \beta_1 = 0$: There is no linear relationship between SAT Reading score (x) and SAT Math score (y). $H_a : \beta_1 \neq 0$: There is a linear relationship between SAT Reading score (x) and SAT Math score (y). Reject H_0 if $t_{data} \geq 2.353$ or $t_{data} \leq -2.353$. Since $t_{data} = 3.1963 \geq 2.353$, we reject H_0. There is evidence at level of significance $\alpha = 0.10$ that $\beta_1 \neq 0$ and that there is a linear relationship between SAT Reading score (x) and SAT Math score (y).
64. (0.1125, 0.7403) TI-83/84: (0.1125, 0.7404). We are 90% confident that the interval (0.1125, 0.7403) ((0.1125, 0.7404)) captures the slope β_1 of the regression line. That is, we are 90% confident that, for each additional point on the SAT Reading score, the increase in the SAT Math score lies between 0.1125 and 0.7403 (0.7404).
65. Yes. Since the confidence interval from Exercise 64 does not contain zero, we may conclude that $\beta_1 \neq 0$ and that a linear relationship exists between SAT Reading score (x) and SAT Math score (y), at level of significance $\alpha = 0.10$.
66. (a) Increase by 5 **(b)** Remain unchanged **(c)–(d)** Remain unchanged **(e)** Decrease **(f)** Remain unchanged
67. (a) t_{data} increases if b_1 is positive and decreases if b_1 is negative. **(b)** r^2 remains the same. **(c)** s decreases. **(d)** p-value decreases. **(e)** Since we don't know what the new p-value will be, we don't know if the p-value will decrease enough to change the conclusion from "Do not reject H_0" to "Reject H_0."
68. (a) Increase by 10 **(b)** Remain unchanged **(c)–(d)** Increase **(e)** Decrease **(f)** Increase
69. (a) t_{data} increases if b_1 is positive and decreases if b_1 is negative. **(b)** r^2 increases. **(c)** s decreases. **(d)** p-value decreases. **(e)** Unchanged.
70. (a) Unchanged **(b)** Increase **(c)** Unchanged **(d)** Decrease **(e)** Increase **(f)–(g)** Decrease
71. (a–b) Decrease **(c–d)** Increase **(e)** Depends on the new p-value.
72. (a) See the Solutions Manual. The scatterplot of the residuals contains no strong evidence of unhealthy patterns and the normal probability plot indicates no evidence of departures from normality in the residuals. Therefore we conclude that the regression assumptions are verified. **(b)** (1.0203, 1.9577). We are 95% confident that the interval (1.0203, 1.9577) captures the population slope β_1 of the relationship between *Dow Jones Industrial Average* (x) and *pros' performance* (y). **(c)** Since 0 does not lie in the confidence interval, we would expect to reject the null hypothesis that $\beta_1 = 0$. **(d)** $H_0 : \beta_1 = 0$. There is no relationship between *Dow Jones Industrial Average* (x) and *pros' performance* (y). $H_a : \beta_1 \neq 0$. There is a linear relationship between *Dow Jones Industrial Average* (x) and *pros' performance* (y). Reject H_0 if p-value ≤ 0.05. $t_{data} = 6.31$. p-value ≈ 0. Since the p-value ≤ 0.05, we reject H_0. There is evidence for a linear relationship between *Dow Jones Industrial Average* (x) and *pros' performance* (y).
73. (a) The scatterplot of the residuals contains evidence of an unhealthy pattern and the normal probability plot indicates evidence of departures from normality in the residuals. Therefore we conclude that the regression assumptions are not verified. **(b)** (7.821, 8.437). We are 95% confident that the interval (13.5483, 14.6201) captures the population slope β_1 of the relationship between *fat per gram* and *calories per gram*. **(c)** Yes **(d)** $H_0 : \beta_1 = 0$. There is no relationship between *fat per gram* (x) and *calories per gram* (y). $H_a : \beta_1 \neq 0$. There is a linear relationship between *fat per gram* (x) and *calories per gram* (y). Reject H_0 if p-value ≤ 0.05. $t_{data} = 51.84$ p-value ≈ 0. Since the p-value ≤ 0.05, we reject H_0. There is evidence for a linear relationship between *fat per gram* (x) and *calories per gram* (y).
74. (a) See the Solutions Manual. The scatterplot of the residuals contains no strong evidence of unhealthy patterns and the normal probability plot indicates no evidence of departures from normality in the residuals. Therefore we conclude that the regression assumptions are verified. **(b)** (0.0087, 0.0439). We are 95% confident that the interval (0.0087, 0.0439) captures the population slope β_1 of the relationship between *heart rate* and *body temperature*. **(c)** Since 0 does not lie in the confidence interval, we would expect to reject the null hypothesis that $\beta_1 = 0$. **(d)** $H_0 : \beta_1 = 0$. There is no relationship between *heart rate* (x) and *body temperature* (y). $H_a : \beta_1 \neq 0$. There is a linear relationship between *heart rate* (x) and *body temperature* (y). Reject H_0 if p-value ≤ 0.05. $t_{data} = 2.97$.

p-value = 0.004. Since the *p*-value ≤ 0.05, we reject H_0. There is evidence for a linear relationship between *heart rate* (*x*) and *body temperature* (*y*).
75. (a) No (b) Positive relationship (c) Unclear
76. (a) The predicted severity of a head injury from a car crash equals −438.0 plus 27.653 times the severity of a chest injury from the crash. (b) $H_0 : \beta_1 = 0$. There is no relationship between *chest injury severity* (*x*) and *head injury severity* (*y*). $H_0 : \beta_1 \neq 0$. There is a linear relationship between *chest injury severity* (*x*) and *head injury severity* (*y*). Reject H_0 if *p*-value ≤ 0.01. $t_{data} = 12.59$. *p*-value ≈ 0. Since the *p*-value ≤ 0.01, we reject H_0. There is evidence for a linear relationship between *chest injury severity* (*x*) and *head injury severity* (*y*). (c) $b_1 = 27.653$ means that for each 1-point increase in the chest injury severity, the head injury severity increases by 27.653. (d) (21.8863, 33.4197). We are 99% confident that the interval (21.8863, 33.4197) captures the population slope β_1 of the relationship between *chest injury severity* and *head injury severity*. Since 0 does not lie in the confidence interval, we would expect to reject the null hypothesis that $\beta_1 = 0$. In (b), we rejected the null hypothesis.
77. (a) No (b) No apparent relationship between the variables (c) The weight of vehicles is the predictor variable and the severity of the leg injuries should be the response variable.
78. (a) The predicted severity of a left leg injury from a car crash equals 569.4 plus 0.16531 times the weight of the car in pounds. Since the *p*-value ≈ 0 is ≤ 0.05, we reject H_0. There is evidence for a linear relationship between *weight of the car* (*x*) and *severity of left leg injury* (*y*). (c) $b_1 = 0.16531$ means that for every additional pound the car weighs, the estimated severity of the left leg injury increases by 0.16531.

Section 13.2
1. The confidence interval is for the mean value of *y* for a given *x* and the prediction interval is for a randomly selected value of *y* for a given *x*.
2. False
3. Lower bound: 18.7500 (Minitab: 18.7497); Upper bound: 23.5000 (Minitab: 23.2500)
4. Lower bound: 32.2057 (Minitab: 32.2046); Upper bound: 47.7943 (Minitab: 47.7954)
5. Lower bound: 1.1909 (Minitab: 1.19027); Upper bound: 10.0091 (Minitab: 10.0097)
6. Lower bound: −28.182 (Minitab: −28.1824); Upper bound: −21.818 (Minitab: −21.8176)
7. Lower bound: 92.636 (Minitab: 92.6351); Upper bound: 105.364 (Minitab: 105.365)
8. Lower bound: 9.818 (Minitab: 9.81755); Upper bound: 16.182 (Minitab: 16.1824)
9. Lower bound: 15.4886 (Minitab: 15.4878); Upper bound: 26.5114 (Minitab: 26.5122)
10. Lower bound: 20.9080 (Minitab: 20.9053); Upper bound: 59.0920 (Minitab: 59.0947)
11. Lower bound: −3.5783 (Minitab: −3.57958); Upper bound: 14.7783 (Minitab: 14.7796)
12. Lower bound: −31.6239 (Minitab: −31.6248); Upper bound: −18.3761 (Minitab: −18.3752)
13. Lower bound: 88.6076 (Minitab: 88.6062); Upper bound: 109.3924 (Minitab: 109.394)
14. Lower bound: 6.3761 (Minitab: 6.37521); Upper bound: 19.6239 (Minitab: 19.6248)
15. (a) 10.4 kilograms (b) Lower bound: 7.1550 kilograms (Minitab: 7.15453); Upper bound: 13.6450 kilograms (Minitab: 13.6455)
16. (a) 2.7 pets (b) Lower bound: 2.1489 pets (Minitab: 2.14878); Upper bound: 3.2511 pets (Minitab: 3.25122)
17. (a) 43.3443 degrees (b) Lower bound: 34.4488 degrees (Minitab: 34.4486); Upper bound: 52.2398 degrees (Minitab: 52.2400)
18. (a) 739,219 units (b) (−481,043, 1,959,482). Minitab (−481,024, 1,959,463); We are 99% confident that the population mean total sales for all video games that have been on the top 30 list for 20 weeks lies between −481,024 game units and 1,959,463 game units.
19. (a) Lower bound: 5.1009 kilograms (Minitab: 5.1009); Upper bound: 15.6991 kilograms (Minitab: 15.6998) (b) The interval in (a) is wider. Individual values are more variable than their mean. The interval in Exercise 15(b) is a 95% confidence interval for the mean value of *y* for the given value *x* = 4 cubic meters and the interval in (a) is a prediction interval for a randomly chosen value of *y* for *x* = 4 cubic meters.

20. (a) Lower bound: 1.5527 pets (Minitab: 1.5255); Upper bound: 3.8473 pets (Minitab: 3.84745) (b) The interval in (a) is wider. Individual values are more variable than their mean. The interval in Exercise 16(b) is a 95% confidence interval for the mean value of *y* for the given value *x* = 5 and the interval in (a) is a prediction interval for a randomly chosen value of *y* for the given *x* = 5.
21 (a) Lower bound: 22.1568 degrees (Minitab: 22.1564); Upper bound: 64.5317 degrees (Minitab: 64.5322) (b) The range of the low temperatures in the data set is from 7°C to 70°C, inclusive. Therefore, a low temperature of 0°C is outside of the range of our data set, and any predictions or estimates using the regression equation for 0°C represent extrapolation. Extrapolation should be avoided if possible because the relationship between the variables may no longer be linear outside the range of *x*.
22. (a) (−2,046,244.16, 3,524,683.098). Minitab (−2,046,201, 3,524,640). We are 99% confident that the total sales for a randomly selected video game that has been on the top 30 list for 20 weeks lies between −2,046,201 game units and 3,524,640 game units. (b) The range of the weeks on the top 30 list in the data set is from 10 weeks to 78 weeks, inclusive. Therefore, a video game that has been on the top 30 list for 104 weeks is outside of the range of our data set, and any predictions or estimates using the regression equation for 104 weeks represent extrapolation. Extrapolation should be avoided if possible because the relationship between the variables may no longer be linear outside the range of *x*.
23. (a) (98.153, 98.521) (b) (96.891, 99.783)
24. (a) (5.2140, 5.4798) (b) (3.3932, 7.3006)

Section 13.3
1. $\hat{y} = b_0 + b_1 x_1 + b_2 x_2 + b_3 x_3$
2. R_{adj}^2 is preferable to R^2 as a measure of the goodness of a regression equation, because R_{adj}^2 will decrease if an unhelpful *x* variable is added to the regression equation.
3. The *F* test for the overall significance of the multiple regression
4. Yes. If the conclusion of the *F* test is that our multiple regression is useful, then we can conclude that at least one of the β_i's is not equal to zero. This does not indicate that all of the β_i's are not equal to zero.
5. The *F* test is for the overall significance of the multiple regression and the *t* test is for testing whether a particular *x*-variable has a significant relationship with the response variable *y*.
6. *k* (One for each predictor *x* variable).
7. The coefficient of a dummy variable can be interpreted as the estimated increase in *y* for those observations with the value of the dummy variable equal to 1 as compared to those with the value of the dummy variable equal to 0 when all of the other *x* variables are held constant.
8. The *F* test; the *t* tests; verify the assumptions; report and interpret your final model
9. For each increase in one unit of the variable x_1, the estimated value of *y* increases by 5 units when the value of x_2 is held constant.
10. For each increase in one unit of the variable x_2, the estimated value of *y* increases by 8 units when the value of x_1 is held constant.
11. The estimated value of *y* when $x_1 = 0$ and $x_2 = 0$ is $b_0 = 10$. $b_1 = 5$ means that for each increase of one unit of the variable x_1, the estimated value of *y* increases by 5 units when the value of x_2 is held constant. $b_2 = 8$ means that for each increase of one unit of the variable x_2, the estimated value of *y* increases by 8 units when the value of x_1 is held constant.
12. (a) 72 (b) 124
13. For each increase in one unit of the variable x_1, the estimated value of *y* decreases by 0.1 unit when the value of x_2 is held constant.
14. For each increase in one unit of the variable x_2, the estimated value of *y* increases by 0.9 unit when the value of x_1 is held constant.
15. The estimated value of *y* when $x_1 = 0$ and $x_2 = 0$ is $b_0 = 0.5$. $b_1 = -0.1$ means that for each increase of one unit of the variable x_1, the estimated value of *y* decreases by 0.1 unit when the value of x_2 is held constant. $b_2 = 0.9$ means that for each increase of one unit of the variable x_2, the estimated value of *y* increases by 0.9 unit when the value of x_1 is held constant.
16. (a) 5.4 (b) 2.8
17. 50% of the variability in *y* is accounted for by this multiple regression equation.
18. 0.4412
19. 75% of the variability in *y* is accounted for by this multiple regression equation.
20. 0.7394

21. $\hat{y} = -37.8 + 4.50x_1 + 3.37x_2 + 0.306x_3$

22. (a) $H_0 : \beta_1 = \beta_2 = \beta_3 = 0$: There is no linear relationship between y and the set x_1, x_2, and x_3. The overall multiple regression is not significant. H_a: At least one of the β's $\neq 0$: There is a linear relationship between y and the set x_1, x_2, and x_3. The overall multiple regression is significant. Reject H_0 if the p-value $\leq \alpha = 0.05$. **(b)** $F = 14.37$. p-value $= 0.004$ **(c)** The p-value $= 0.004 \leq \alpha = 0.05$, so we reject H_0. There is evidence at level of significance $\alpha = 0.05$ for a linear relationship between y and the set x_1, x_2, and x_3. The overall multiple regression is significant.

23. (a) Test 1: $H_0 : \beta_1 = 0$: There is no linear relationship between y and x_1. $H_a : \beta_1 \neq 0$: There is a linear relationship between y and x_1. Reject H_0 if the p-value $\leq \alpha = 0.05$. Test 2: $H_0 : \beta_2 = 0$: There is no linear relationship between y and x_2. $H_a : \beta_2 \neq 0$: There is a linear relationship between y and x_2. Reject H_0 if the p-value $\leq \alpha = 0.05$. Test 3: $H_0 : \beta_3 = 0$: There is no linear relationship between y and x_3. $H_a : \beta_3 \neq 0$: There is a linear relationship between y and x_3. Reject H_0 if the p-value $\leq \alpha = 0.05$. **(b)** Test 1: $t = 3.72$, with p-value $= 0.010$. Test 2: $t = 2.75$, with p-value $= 0.034$. Test 3: $t = 0.87$, with p-value $= 0.418$. **(c)** Test 1: The p-value $= 0.010$, which is $\leq \alpha = 0.05$. Therefore we reject H_0. There is evidence of a linear relationship between y and x_1. Test 2: The p-value $= 0.034$, which is $\leq \alpha = 0.05$. Therefore we reject H_0. There is evidence of a linear relationship between y and x_2. Test 3: The p-value $= 0.418$, which is not $\leq \alpha = 0.05$. Therefore we do not reject H_0. There is insufficient evidence of a linear relationship between y and x_3.

24. The p-value for x_3 is the only p-value greater than $\alpha = 0.05$, so we eliminate x_3 from the multiple regression equation. The new regression equation is $\hat{y} = -37.3 + 4.46x_1 + 3.31x_2$. **(a)** Test 1: $H_0 : \beta_1 = 0$: There is no linear relationship between y and x_1. $H_a : \beta_1 \neq 0$: There is a linear relationship between y and x_1. Reject H_0 if the p-value $\leq \alpha = 0.05$. Test 2: $H_0 : \beta_2 = 0$: There is no linear relationship between y and x_2. $H_a : \beta_2 \neq 0$: There is a linear relationship between y and x_2. Reject H_0 if the p-value $\leq \alpha = 0.05$. **(b)** Test 1: $t = 3.76$, with p-value $= 0.007$. Test 2: $t = 2.75$, with p-value $= 0.029$. **(c)** Test 1: The p-value $= 0.007$, which is $\leq \alpha = 0.05$. Therefore we reject H_0. There is evidence of a linear relationship between y and x_1. Test 2: The p-value $= 0.029$, which is $\leq \alpha = 0.05$. Therefore we reject H_0. There is evidence of a linear relationship between y and x_2. Since all variables are significant, we have our final multiple regression equation.

25.

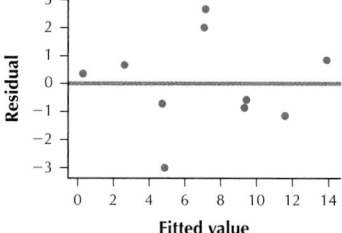

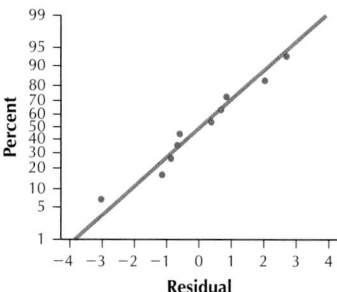

The scatterplot above of the residuals versus fitted values shows no strong evidence of unhealthy patterns. Thus, the independence assumption, the constant variance assumption, and the zero-mean assumption are verified. Also, the normal probability plot of the residuals above indicates no evidence of departure from normality of the residuals. Therefore we conclude that the regression assumptions are verified.

26. (a) $\hat{y} = -37.3 + 4.46x_1 + 3.31x_2$ **(b)** The estimated value of y when $x_1 = 0$ and $x_2 = 0$ is $b_0 = -37.3$. For each increase in one unit of the variable x_1, the estimated value of y increases by 4.46 units when the value of x_2 is held constant. For each increase in one unit of the variable x_2, the estimated value of y increases by 3.31 units when the value of x_1 is held constant. **(c)** Using the multiple regression equation in (a), the size of the typical prediction error

will be about 1.87731. 82.31% of the variability in y is accounted for by this multiple regression equation.

27. The regression equation is $\hat{y} = -2.98 + 1.13x_1 - 0.175x_2 + 3.55x_3$.

28. (a) $H_0 : \beta_1 = \beta_2 = \beta_3 = 0$: There is no linear relationship between y and the set x_1, x_2, and x_3. The overall multiple regression is not significant. H_a: At least one of the β's $\neq 0$: There is a linear relationship between y and the set x_1, x_2, and x_3. The overall multiple regression is significant. Reject H_0 if the p-value $\leq \alpha = 0.01$. **(b)** $F = 185.68$. p-value $= 0.000$ **(c)** The p-value $= 0.000 \leq \alpha = 0.01$, so we reject H_0. There is evidence at level of significance $\alpha = 0.01$ for a linear relationship between y and the set x_1, x_2, and x_3. The overall multiple regression is significant. For $x_3 = 0$, the regression equation is $\hat{y} = -2.98 + 1.13x_1 - 0.175x_2$. For $x_3 = 1$, the regression equation is $\hat{y} = 0.57 + 1.13x_1 - 0.175x_2$.

29. For each increase in one unit of the variable x_3, the estimated value of y increases by 3.55 units when the values of x_1 and x_2 are held constant.

30. (a) Test 1: $H_0 : \beta_1 = 0$: There is no linear relationship between y and x_1. $H_a : \beta_1 \neq 0$: There is a linear relationship between y and x_1. Reject H_0 if the p-value $\leq \alpha = 0.01$. Test 2: $H_0 : \beta_2 = 0$: There is no linear relationship between y and x_2. $H_a : \beta_2 \neq 0$: There is a linear relationship between y and x_2. Reject H_0 if the p-value $\leq \alpha = 0.01$. Test 3: $H_0 : \beta_3 = 0$: There is no linear relationship between y and x_3. $H_a : \beta_3 \neq 0$: There is a linear relationship between y and x_3. Reject H_0 if the p-value $\leq \alpha = 0.01$. **(b)** Test 1: $t = 20.81$, with p-value $= 0.000$; Test 2: $t = -1.03$, with p-value $= 0.344$; Test 3: $t = 5.76$, with p-value $= 0.001$. **(c)** Test 1: The p-value $= 0.000$, which is $\leq \alpha = 0.01$. Therefore we reject H_0. There is evidence of a linear relationship between y and x_1. Test 2: The p-value $= 0.344$, which is not $\leq \alpha = 0.01$. Therefore we do not reject H_0. There is insufficient evidence of a linear relationship between y and x_2. Test 3: The p-value $= 0.001$, which is $\leq \alpha = 0.01$. Therefore we reject H_0. There is evidence of a linear relationship between y and x_3.

31. The p-value for x_2 is the only p-value greater than $\alpha = 0.01$, so we eliminate x_2 from the multiple regression equation. The new regression equation is $\hat{y} = -3.12 + 1.15x_1 + 3.61x_3$. **(a)** Test 1: $H_0 : \beta_1 = 0$: There is no linear relationship between y and x_1. $H_a : \beta_1 \neq 0$: There is a linear relationship between y and x_1. Reject H_0 if the p-value $\leq \alpha = 0.01$. Test 2: $H_0 : \beta_3 = 0$: There is no linear relationship between y and x_3. $H_a : \beta_3 \neq 0$: There is a linear relationship between y and x_3. Reject H_0 if the p-value $\leq \alpha = 0.01$. **(b)** Test 1: $t = 21.38$, with p-value $= 0.000$; Test 2: $t = 5.86$, with p-value $= 0.001$. **(c)** Test 1: The p-value $= 0.000$, which is $\leq \alpha = 0.01$. Therefore we reject H_0. There is evidence of a linear relationship between y and x_1. Test 2: The p-value $= 0.001$, which is $\leq \alpha = 0.01$. Therefore we reject H_0. There is evidence of a linear relationship between y and x_3. Since all of the variables are significant, we have our final multiple regression equation.

32.

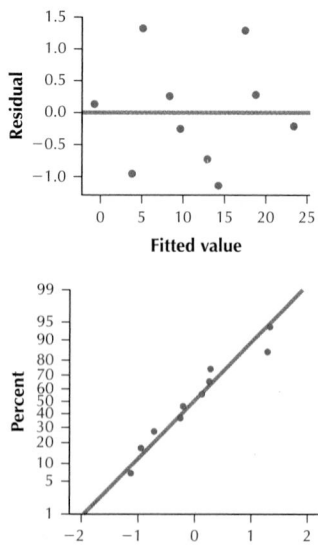

The scatterplot above of the residuals versus fitted values shows no strong evidence of unhealthy patterns. Thus, the independence assumption, the constant variance assumption, and the zero-mean assumption are verified. Also, the normal probability plot of the residuals above indicates no evidence of departure from normality of the residuals. Therefore we conclude that the regression assumptions are verified.

33. (a) The final multiple regression equation is $\hat{y} = -3.12 + 1.15x_1 + 3.61x_3$. For $x_3 = 0$, the regression equation is $\hat{y} = -3.12 + 1.15x_1$. For $x_3 = 1$, the regression equation is $\hat{y} = 0.49 + 1.15x_1$. (b) For each increase in one unit of the variable x_1, the estimated value of y increases by 1.15 units. The estimated increase in y for those observations with $x_3 = 1$, as compared to those with $x_3 = 0$, when x_1 is held constant, is 3.61. (c) Using the multiple regression equation in (a), the size of the typical prediction error will be about 0.959129. 98.4% of the variability in y is accounted for by this multiple regression equation.

34. (a) $H_0: \beta_1 = \beta_2 = \beta_3 = 0$: There is no linear relationship between *Overall Dating Score* and the set *Percentage 18–24 years old, Percentage 18–24 and single*, and *Online Dating Score*. The overall multiple regression is not significant. H_a: At least one of the β's $\neq 0$: There is a linear relationship between *Overall Dating Score* and the set *Percentage 18–24 years old, Percentage 18–24 and single*, and *Online Dating Score*. The overall multiple regression is significant. Reject H_0 if the p-value $\leq \alpha = 0.05$. (b) $F = 3.22$. p-value $= 0.104$ (c) The p-value $= 0.104$, which is not $\leq \alpha = 0.05$, so we do not reject H_0. There is insufficient evidence at level of significance $\alpha = 0.05$ for a linear relationship between *Overall Dating Score* and the set *Percentage 18–24 years old, Percentage 18–24 and single*, and *Online Dating Score*. The overall multiple regression is not significant. (b)–(d) Not appropriate to perform.

35. See Solutions Manual.

36. See Solutions Manual.

37. See Solutions Manual.

38. (a) $H_0: \beta_1 = \beta_2 = \beta_3 = 0$: There is no linear relationship between *Current Ratio* and the set *Price–Earnings Ratio, Assets*, and *Liabilities*. The overall multiple regression is not significant. H_a: At least one of the β's $\neq 0$: There is a linear relationship between *Current Ratio* and the set *Price–Earnings Ratio, Assets*, and *Liabilities*. The overall multiple regression is significant. Reject H_0 if the p-value $\leq \alpha = 0.10$. (b) $F = 0.56$. p-value $= 0.728$ (c) The p-value $= 0.728$, which is not $\leq \alpha = 0.10$, so we do not reject H_0. There is insufficient evidence at level of significance $\alpha = 0.10$ for a linear relationship between *Current Ratio* and the set *Price–Earnings Ratio, Assets*, and *Liabilities*. The overall multiple regression is not significant. (b) There is no linear relationship between *Current Ratio* and the set *Price–Earnings Ratio, Assets*, and *Liabilities*. Therefore there is no final regression model (c) $H_0: \beta_1 = \beta_2 = \beta_3 = 0$: There is no linear relationship between *Current Ratio* and the set *Price–Earnings Ratio, Assets*, and *Liabilities*. The overall multiple regression is not significant. H_a: At least one of the β's $\neq 0$: There is a linear relationship between *Current Ratio* and the set *Price–Earnings Ratio, Assets*, and *Liabilities*. The overall multiple regression is significant. Reject H_0 if the p-value $\leq \alpha = 0.13$. $F = 0.56$. p-value $= 0.728$. The p-value $= 0.728$, which is not $\leq \alpha = 0.13$, so we do not reject H_0. There is insufficient evidence at level of significance $\alpha = 0.13$ for a linear relationship between *Current Ratio* and the set *Price–Earnings Ratio, Assets*, and *Liabilities*. The overall multiple regression is not significant. (d) Not applicable.

39. See Solutions Manual.

40. See Solutions Manual.

41.

```
Regression Equation

CALORIES = 1.880 + 4.720 PROTEIN + 8.7914 FAT + 3.8915 CARBO + 0.02778 CALCIUM
           - 0.04329 PHOSPHOR - 2.190 IRON - 0.02128 POTASS + 0.00360 SODIUM
           + 25.38 THIAMIN


Model Summary

     S    R-sq   R-sq(adj)  R-sq(pred)
16.7233  99.91%    99.91%     99.90%
```

The standard error in the estimate for the final model is $s = 16.7233$. That is, using the multiple regression equation given above, the size of the typical prediction error will be about 16.7233 calories. The adjusted coefficient of variation is $R^2_{adj} = 99.91\%$. In other words, 99.91% of the variation in calories is accounted for by this multiple regression equation.

42. $b_0 = 1.880$ means that the predicted number of calories for a serving of food with 0 grams of protein, fat, carbohydrates, calcium, phosphorous, iron, potassium, sodium, and thiamin is 1.880 calories. $b_1 = 4.720$ means that for each additional gram of protein that a serving of food has with the amount of fat, carbohydrates, calcium, phosphorous, iron, potassium, sodium, and thiamin held constant, the predicted number of calories the serving of food has increases by 4.720. $b_2 = 8.7914$ means that for each additional gram of fat that a serving of food has with the amount of protein, carbohydrates,

calcium, phosphorous, iron, potassium, sodium, and thiamin held constant, the predicted number of calories the serving of food has increases by 8.7914. $b_3 = 3.8915$ means that for each additional gram of carbohydrates that a serving of food has with the amount of protein, fat, calcium, phosphorous, iron, potassium, sodium, and thiamin held constant, the predicted number of calories the serving of food has increases by 3.8915. $b_6 = 0.02778$ means that for each additional gram of calcium that a serving of food has with the amount of protein, fat, carbohydrates, phosphorous, iron, potassium, sodium, and thiamin held constant, the predicted number of calories the serving of food has increases by 0.02778. $b_7 = -0.04329$ means that for each additional gram of phosphorous that a serving of food has with the amount of protein, fat, carbohydrates, calcium, iron, potassium, sodium, and thiamin held constant, the predicted number of calories the serving of food has increases by 0.04329. $b_8 = -2.190$ means that for each additional gram of iron that a serving of food has with the amount of protein, fat, carbohydrates, calcium, phosphorous, potassium, sodium, and thiamin held constant, the predicted number of calories the serving of food has decreases by 2.190. $b_9 = -0.02128$ means that for each additional gram of potassium that a serving of food has with the amount of protein, fat, carbohydrates, calcium, phosphorous, iron, sodium, and thiamin held constant, the predicted number of calories the serving of food has decreases by 0.02128. $b_{10} = 0.00360$ means that for each additional gram of sodium that a serving of food has with the amount of protein, fat, carbohydrates, calcium, phosphorous, iron, and thiamin held constant, the predicted number of calories the serving of food has increases by 0.00360. $b_{11} = 25.38$ means that for each additional gram of thiamin that a serving of food has with the amount of protein, fat, carbohydrates, calcium, phosphorous, iron, potassium, and sodium held constant, the predicted number of calories the serving of food has increases by 25.38. The variables that are most important for predicting the number of calories that a serving of food has are the variables with the largest absolute values of coefficients in the regression model. These are, in decreasing order of absolute value of the coefficients, thiamin, fat, protein, carbohydrates, and iron.

Chapter 13 Review

1. $H_0: \beta_1 = 0$: There is no linear relationship between *Education* (x) and *Annual Earnings* (y). $H_a: \beta_1 \neq 0$: There is a linear relationship between *Education* (x) and *Annual Earnings* (y). Reject H_0 if $t_{data} \geq 2.571$ or $t_{data} \leq -2.571$. Since $t_{data} = 7.542 \geq 2.571$, we reject H_0. There is evidence at level of significance $\alpha = 0.05$ that $\beta_1 \neq 0$ and that there is a linear relationship between *Education* (x) and *Annual Earnings* (y).

2. $H_0: \beta_1 = 0$: There is no linear relationship between *High school GPA* (x) and *First-year college GPA* (y). $H_a: \beta_1 \neq 0$: There is a linear relationship between *High school GPA* (x) and *First-year college GPA* (y). Reject H_0 if $t_{data} \geq 2.306$ or $t_{data} \leq -2.306$. Since $t_{data} = 4.5727 \geq 2.306$, we reject H_0. There is evidence at level of significance $\alpha = 0.05$ that $\beta_1 \neq 0$ and that there is a linear relationship between *High school GPA* (x) and *First-year college GPA* (y).

3. $H_0: \beta_1 = 0$: There is no linear relationship between *Age* (x) and *Price* (y). $H_a: \beta_1 \neq 0$: There is a linear relationship between *Age* (x) and *Price* (y). Reject H_0 if $t_{data} \geq 2.306$ or $t_{data} \leq -2.306$. Since $t_{data} = -15.0124 \leq -2.306$, we reject H_0. There is evidence at level of significance $\alpha = 0.05$ that $\beta_1 \neq 0$ and that there is a linear relationship between *Age* (x) and *Price* (y).

4. (2.946, 5.992). We are 95% confident that the interval (2.946, 5.992) captures the population slope β_1 of the relationship between *Education* and *Annual Earnings*.

5. (0.339, 1.029). We are 95% confident that the interval (0.339, 1.029) captures the population slope β_1 of the relationship between *high school GPA* and *first-year college GPA*.

6. (−1.873, −1.374). We are 95% confident that the interval (−1.873, −1.374) captures the population slope β_1 of the relationship between *Age* and *Price*.

7. (a) 20.92 thousand dollars (b) (14.28, 27.56). We are 95% confident that the mean annual salary for people with 10 years of education lies between 14.28 thousand dollars and 27.56 thousand dollars. (c) (6.55, 35.29). We are 95% confident that the annual salary for a randomly selected person with 10 years of education lies between 6.55 thousand dollars and 35.29 thousand dollars.

8. (a) 2.65 (b) (2.48, 2.81). We are 95% confident that the mean first-year college GPA for students with a high school GPA of 3.0 lies between 2.48 and 2.81. (c) (2.11, 3.19). We are 95% confident that the first-year college GPA for a randomly selected student with a high school GPA of 3.0 lies between 2.11 and 3.19.

9. (a) 6.818 thousand dollars **(b)** (5.805, 7.831). We are 95% confident that the mean price for 8-year-old cars of this make and model lies between 5.805 thousand dollars and 7.831 thousand dollars **(c)** (4.902, 8.733). We are 95% confident that the price for a randomly selected 8-year-old car of this make and model lies between 4.902 thousand dollars and 8.733 thousand dollars.
10. (a) $H_0 : \beta_1 = \beta_2 = \beta_3 = 0$: There is no linear relationship between y and the set x_1, x_2, and x_3. The overall multiple regression is not significant. H_a: At least one of the β's $\neq 0$: There is a linear relationship between y and the set x_1, x_2, and x_3. The overall multiple regression is significant. Reject H_0 if the p-value $\leq \alpha = 0.05$. **(b)** $F = 72.24$. p-value $= 0.000$ **(c)** The p-value $= 0.000 \leq \alpha = 0.05$, so we reject H_0. There is evidence at level of significance $\alpha = 0.05$ for a linear relationship between y and the set x_1, x_2, and x_3. The overall multiple regression is significant.
11. (a) Test 1: $H_0 : \beta_1 = 0$: There is no linear relationship between y and x_1. $H_a : \beta_1 \neq 0$: There is a linear relationship between y and x_1. Reject H_0 if the p-value $\leq \alpha = 0.05$. Test 2: $H_0 : \beta_2 = 0$: There is no linear relationship between y and x_2. $H_a : \beta_2 \neq 0$: There is a linear relationship between y and x_2. Reject H_0 if the p-value $\leq \alpha = 0.05$. Test 3: $H_0 : \beta_3 = 0$: There is no linear relationship between y and x_3. $H_a : \beta_3 \neq 0$: There is a linear relationship between y and x_3. Reject H_0 if the p-value $\leq \alpha = 0.05$. **(b)** Test 1: $t = 1.86$, with p-value $= 0.112$; Test 2: $t = -0.27$, with p-value $= 0.798$; Test 3: $t = -1.07$, with p-value $= 0.326$. **(c)** Test 1: The p-value $= 0.112$, which is not $\leq \alpha = 0.05$. Therefore we do not reject H_0. There is insufficient evidence of a linear relationship between y and x_1. Test 2: The p-value $= 0.798$, which is not $\leq \alpha = 0.05$. Therefore we do not reject H_0. There is insufficient evidence of a linear relationship between y and x_2. Test 3: The p-value $= 0.326$, which is not $\leq \alpha = 0.05$. Therefore we do not reject H_0. There is insufficient evidence of a linear relationship between y and x_3.
12. See Solutions Manual.
13.

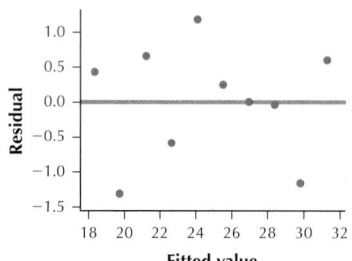

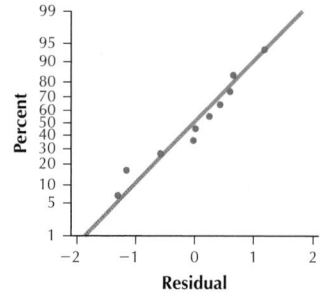

The preceding scatterplot of the residuals versus fitted values shows no strong evidence of unhealthy patterns. Thus, the independence assumption, the constant variance assumption, and the zero-mean assumption are verified. Also, the normal probability plot of the residuals indicates no evidence of departure from normality of the residuals. Therefore we conclude that the regression assumptions are verified.

Chapter 13 Quiz
1. False
2. False
3. True
4. Estimated, predicted y
5. β_1, b_1
6. Dummy
7. Zero-mean assumption $E(\varepsilon) = 0$, constant variance assumption, independence assumption, normality assumption
8. The t test for the slope β_1
9. R^2_{adj} is preferable to R^2 as a measure of the goodness of a regression equation, because R^2_{adj} will decrease if an unhelpful x variable is added to the regression equation.
10. $H_0 : \beta_1 = 0$: There is no linear relationship between *Height* (x) and *Weight* (y). $H_a : \beta_1 \neq 0$: There is a linear relationship between *Height* (x) and *Weight* (y). Reject H_0 if p-value $\leq \alpha = 0.01$. $t_{data} = 8.47$. p-value $= 0.000$. Since p-value $= 0.000 \leq \alpha = 0.01$, we reject H_0. There is evidence at level of significance $\alpha = 0.01$ that $\beta_1 \neq 0$ and that there is a linear relationship between *Height* (x) and *Weight* (y).
11. (a) No. Since H_0 was rejected at level of significance $\alpha = 0.01$ in the hypothesis test in Exercise 10, the conclusion is that $\beta_1 \neq 0$ **(b)** (4.3998, 10.17). We are 99% confident that the interval (4.3998, 10.17) captures the slope β_1 of the regression line. That is, we are 99% confident that, for each additional inch in height of a male student, the increase in the weight lies between 4.3998 and 10.17.
12. (a) 168.1314 pounds **(b)** (163.1156, 173.1472) Minitab: (163.11, 173.14). We are 95% confident that the mean weight for male college students 70 inches tall lies between 163.11 pounds and 173.14 pounds. **(c)** (151.9267, 184.3361). Minitab: (151.92, 184.33). We are 95% confident that the weight for a randomly selected male college student with a height of 70 inches lies between 151.92 pounds and 184.33 pounds.
13. (a) The regression equation is $\hat{y} = 16.8 + 0.718x_1$. **(b)** The estimated value of y when $x_1 = 0$ is 16.8. For each increase in one unit of the variable x_1, the estimated value of y increases by 0.718. **(c)** Using the multiple regression equation above, the size of the typical prediction error will be about 0.854028. 96.3% of the variability in y is accounted for by this multiple regression equation.

Notes and Data Sources

Chapter 1

1. T. J. Scanlon et al., "Is Friday the 13th bad for your health?" *British Medical Journal* 307 (6919, 1993): 1584–86.
2. U.S. Census Bureau, *The Population Profile of the United States: 2000*, http://www.census.gov/population/pop-profile/2000/slideshow/.
3. Amy Branum, Lauren Rossen, and Kenneth Schoendorf, "Trends in caffeine intake among U.S. children and adolescents," *Pediatrics* 133 (3, 2014): 386–93. Our sample size is estimated from the reported standard error.
4. https://studentaid.ed.gov/sa/about/data-center/student/title-iv.
5. www.ers.usda.gov/data-products/adoption-of-genetically-engineered-crops-in-the-us.aspx.
6. http://edition.cnn.com/2013/01/30/us/deadliest-tornadoes/.
7. Michel de Lorgeril, et al., "Mediterranean diet, traditional risk factors, and the rate of cardiovascular complications after myocardial infarction, final report of the Lyon Diet Heart Study," *Circulation* 99 (6, 1999): 779–85. The American Heart Association (www.heart.org/HEARTORG/) identifies the following characteristics as common to most Mediterranean diets. There is a "high consumption of fruits, vegetables, bread and other cereals, potatoes, beans, nuts and seeds. Olive oil is an important monounsaturated fat source. Dairy products, fish and poultry are consumed in low to moderate amounts, and little red meat is eaten."
8. U.S. Department of Health and Human Services, *The Health Consequences of Involuntary Exposure to Tobacco Smoke: A Report of the Surgeon General—Executive Summary* (U.S. Department of Health and Human Services, Centers for Disease Control and Prevention, Coordinating Center for Health Promotion, National Center for Chronic Disease Prevention and Health Promotion, Office on Smoking and Health, 2006).
9. Iain McGregor and Wayne Hall, "MDMA (Ecstasy) neurotoxicity: assessing and communicating the risks," *Lancet* 355 (9217, 2000): 1818–21.
10. R. L. Bratton et al., "Effect of 'ionized' wrist bracelets on musculoskeletal pain: a randomized, double-blind, placebo-controlled trial," *Mayo Clinic Proceedings* 77 (11, 2002):1164–68.

Chapter 2

1. Concetta A. Depaolo and David F. Robinson (Indiana State University), "Café data," *Journal of Statistics Education* 19 (1, 2011).
2. Lars Engebretsen et al., "Sports injuries and illnesses during the Winter Olympic Games 2010," *British Journal of Sports Medicine* 44 (11, 2010): 772–80.
3. Amanda Lenhart, *Cell Phones and American Adults*, Pew Internet and American Life Project, September 2, 2010.
4. See Note 3.
5. Candace A. Howell, Nicolas G. Nelson, and Lara B. McKenzie, "Pediatric and adolescent sledding-related injuries treated in U.S. emergency departments in 1997–2007," *Pediatrics* 126 (3, 2010): 517–24.
6. Eileen M. Burd, "Human Papillomavirus and Cervical Cancer," *Clinical Microbiology Reviews* 16 (1, January 2003): 1–17.
7. Christopher E. Barat, Courtney Wright, and Betty Chou, "Examining potential predictors for completion of the Gardasil vaccine sequence based on data gathered at clinics of Johns Hopkins Medical Institutions," *Journal of Statistics Education* 19 (1, 2011).
8. M. A. Chase and G. M. Dummer, "The role of sports as a social determinant for children," *Research Quarterly for Exercise and Sport* 63 (4, 1992): 418–24.
9. U.S. Bureau of Labor Statistics.
10. World Health Organization, Global Health Indicators, www.who.int/whosis/whostat/EN_WHS10_Part2.pdf.

Chapter 3

1. www.pewsocialtrends.org/2014/02/11/the-rising-cost-of-not-going-to-college/, 2014.
2. P. A. Mackowiak, S. S. Wasserman, and M. M. Levine, "A critical appraisal of 98.6 degrees F, the upper limit of the normal body temperature, and other legacies of Carl Reinhold August Wunderlich, *Journal of the American Medical Association* 268 (12, 1992): 1578–80.
3. Centers for Disease Control and Prevention: www.cdc.gov/nchhstp/stateprofiles/usmap.htm.
4. Dr. Peter Nonacs, "Foraging habits of thatch ants," Department of Statistics, University of California at Los Angeles and the Sierra Nevada Aquatic Research Laboratory.
5. National Center for Education Statistics, 2005.
6. http://radar.oreilly.com/2013/12/tweets-loud-and-quiet.html, December 2013.

Chapter 4

1. Grete Heinz et al., "Exploring relationships in body dimensions," *Journal of Statistics Education* 11 (2, 2003).
2. Data set excerpted from: T. Allisonand D.V. Cicchetti, "Sleep in mammals: ecological and constitutional correlates," *Science* 194 (4266, 1976): 732–34.
3. A. Johnson, *Results from Analyzing Metals in 1999 Spokane River Fish and Crayfish Samples*, Washington State Department of Ecology Report, 2000. https://fortress.wa.gov/ecy/publications/summarypages/0003017.html.
4. See Note 2.
5. *Global Digital Communication: Texting, Social Networking Popular Worldwide*, Pew Research Center Global Attitudes Project, December 2011. www.pewglobal.org/files/2011/12/Pew-Global-Attitudes-Technology-Report-FINAL-December-20-2011.pdf.

Chapter 5

1. L. E. Markowitz et al., "Quadrivalent human papillomavirus vaccine: recommendations of the Advisory Committee on Immunization Practices (ACIP)." *Morbidity and Mortality Weekly Report* 56 (RR-2, 2007): 1–24.
2. Christopher E. Barat, Courtney Wright, and Betty Chou, "Examining potential predictors for completion of the Gardasil vaccine sequence based on data gathered at clinics of Johns Hopkins Medical Institutions," *Journal of Statistics Education* 19 (1, 2011). www.amstat.org/publications/jse/v19n1/barat.pdf.
3. U.S. Census Bureau, 2004 American Community Survey.
4. Gallup.com. *Three in 10 in U.S. Own an Array of Consumer Electronics*. www.gallup.com/poll/166760/three-own-array-consumer-electronics.aspx, 2014.
5. Pew Research Internet Project, *U.S. Views of Technology and the Future*. www.pewinternet.org/2014/04/17/us-views-of-technology-and-the-future/.
6. www.amstat.org/publications/jse/datasets/babyboom.txt.
7. Pew Research Internet Project. *Couples, the Internet, and Social Media*. www.pewinternet.org/2014/02/11/main-report-30/.
8. See Note 4.
9. Kristen Purcell, Roger Enner, and Nicole Henderson, *The Rise of Apps Culture*, Pew Research Center's Internet and American Life Project. www.pewinternet.org, 2010.
10. Andrew Rocco Tresolini Fiore, "Romantic regressions: an analysis of behavior in online dating systems," Master's thesis, Program in Media Arts and Sciences, Massachusetts Institute of Technology, 2004.
11. See Note 9.
12. See Note 2.
13. www.businessinsider.com/encrypted-gmail-data-2014-6, 2014.
14. Frank N. Magid Associates, Inc., 2014.
15. *Profile of Hired Farmworkers, a 2008 Update/ERR-60*, Economic Research Service/USDA.

Chapter 6

1. U.S. National Center for Education Statistics. The category "5 or more" has been changed to "5" for this exercise.

2. Gunter Hitsch, Ali Hortacsu, and Dan Ariely, *What Makes You Click: An Empirical Analysis of Online Dating*. www.aeaweb.org/annual_mtg _papers/2006/0106_0800_0502.pdf.

3. *Women in Management: Analysis of Female Managers' Representation, Characteristics, and Pay*, Government Accountability Office publication GAO-10-892R, September 20, 2010.

4. E. Skogvoll and B. H. Lindqvist, "Modeling the occurrence of cardiac arrest as a Poisson process," *Annals of Emergency Medicine* 33 (4, 1999): 409–17.

5. "British Airways joins RBS and Ethoca in the collaborative fight against fraud," press release, *Marketwire*, February 19, 2009.

6. *Herger-Feinstein Quincy Library Group Forest Recovery Act Implementation on the Eagle Lake Ranger District, Lassen NF*. www.qlg.org /pub/act/lnf/lnf1.htm.

7. Zdravko Markov and Daniel T. Larose, *Data Mining the Web: Uncovering Patterns in Web Content, Structure, and Usage* (John Wiley and Sons, 2007).

8. F. Giannelli, T. Anagnostopolous, and P. M. Green, "Mutation rates in humans. II. Sporadic mutation-specific rates and rate of detrimental human mutations inferred from hemophilia B," *American Journal of Human Genetics* 65 (6, 1999): 1580–87.

9. Edwin R. Van Teijlingen, George W. Lowis, Maureen Porter, and Peter McCaffery, eds., *Midwifery and the Medicalization of Childbirth: Comparative Perspectives* (Nova Publishers, 2004).

10. Department of Health and Human Services. www.mass.gov.

11. The Campus Security Data Analysis Cutting Tool. http://ope.ed.gov /security/.

12. Frederick Mueller and Robert Cantu, *Annual Survey of Football Injury Research*, National Center for Catastrophic Sport Injury Research, 2008.

13. Amanda Lenhart et al., *Teens, Video Games, and Civics*, Pew Internet and American Life Project, September 2008. www.pewinternet.org /2008/09/16/teens-video-games-and-civics/.

14. D. L. Olds et al., "Improving the delivery of prenatal care and outcomes of pregnancy: a randomized trial of nurse home visitation," *Pediatrics* 77 (1, 1986): 16–28.

15. Allen J. Wilcox, National Institutes of Health, "The analysis of birth weight and infant mortality," *International Journal of Epidemiology* (December 2001).

16. Lynn Unruh and Myron Fottler, "Patient turnover and nursing staff adequacy," *Health Services Research*, April 2006.

17. Lauren Dutra and Stanton Glantz, "Electronic cigarettes and conventional cigarette use among U.S. adolescents: a cross-sectional study," *Journal of the American Medical Association—Pediatrics* 168 (7, 2014): 610–17.

18. www.fivethirtyeight.com/datalab/how-americans-like-their-steak/, May 16, 2014.

19. Harvard School of Public Health, Survey of 5046 Adults in Hurricane High-Risk Areas, June–July 2007.

20. The Associated Press/Ipsos Poll actually contacted 1000 adults in June 2007.

21. P. Muntner et al., "Trends in blood pressure among children and adolescents," *Journal of the American Medical Association* 291 (17, 2004): 2107–13.

22. Phillida Bunkle and John Lepper, *Women's Participation in Gambling: Whose Reality? A Public Health Issue*. Paper presented to the European Association for the Study of Gambling Conference, Barcelona, Spain, October 2002.

Chapter 7

1. Bureau of Labor Statistics. www.bls.gov/tus/charts/.

2. B. S. Glenn et al., "Changes in systolic blood pressure associated with lead in blood and bone," *Epidemiology* 17 (5, 2006): 538–44.

3. The Project on Student Debt. http://ticas.org/posd/home.

4. Gallup "State of the States." www.gallup.com/poll/125066/state-states .aspx.

5. George Miller, "The magical number seven, plus or minus two: some limits on our capacity for processing information," *Psychological Review* 63 (1956): 81–97.

6. http://www.iop.harvard.edu

7. Ibid.

8. Sloan Burke et al., "Using technology to control intimate partners: an exploratory study of college undergraduates," *Computers in Human Behavior* 27 (3, May 2011): 1162–67.

9. Murray Mittleman et al., "Determinants of myocardial onset study," *Circulation* June 1999. http://circ.ahajournals.org/content/99/21/2737.full.pdf

Chapter 8

1. Fuel Economy Guide: Model Year 2014. www.fueleconomy.gov/feg/pdfs /guides/FEG2014.pdf.

2. Kevin Crowley et al., "Parents explain more often to boys than girls during shared scientific thinking," *Psychological Science* 12 (3, 2001): 258–61.

3. U.S. Energy Information Administration, 2005.

4. Florida Department of Financial Services, 2011.

5. National Weather Service, 2011.

6. George Miller, "The magical number seven, plus or minus two: some limits on our capacity for processing information. 1956.," *Psychological Review* 101 (2, 1994): 343–52.

7. Mary H. Ward et al., "Proximity to crops and residential exposure to agricultural herbicides in Iowa," *Environmental Health Perspectives* 114 (6, 2006): 893–97.

8. Irene Yen et al., "Perceived neighborhood problems and quality of life, physical functioning, and depressive symptoms among adults with asthma," *American Journal of Public Health* 96 (5, 2006): 873–79.

9. Gary Bennett et al., "Television viewing and pedometer-determined physical activity among multiethnic residents of low-income housing," *American Journal of Public Health* 96 (9, 2006), 1681–85.

10. Adapted from A. Johnson, "Results from analyzing metals in 1999 Spokane River fish and crayfish samples," Quantitative Environmental Learning Project, Washington State Department of Ecology report 00-03-017. www.seattlecentral.edu/qelp/sets/021/021.html.

11. www.vgchartz.com, April 1, 2011.

12. See Note 3.

13. See Note 4.

14. See Note 11.

15. See Note 5.

16. http://blogs.wsj.com/corporate-intelligence/2013/07/26/starbucks-talks -about-its-future-more-food-more-digital/.

17. Christopher Reynolds, "Prey tell," *American Demographics* 25 (8, 2003): 48.

18. www.gallup.com/poll/146885/Positivity-Optimism-Norm-Thriving -States.aspx.

19. *Couples, the Internet, and Social Media*. www.pewinternet.org/2014 /02/11/couples-the-internet-and-social-media, February 2014

20. Mildred Cho and Lisa Bero, "The quality of drug studies published in symposium proceedings," *Annals of Internal Medicine*, 124 (5, 1996): 485–89.

21. http://biomassmagazine.com/plants/listplants/biomass/US.

22. See Note 3.

23. See Note 4.

24. See Note 11.

25. See Note 21.

Chapter 9

1. "Consumers report eating at home more in the wake of high gas prices," press release, NPD Group, Inc., Port Washington, NY, August 23, 2007.

2. "When it comes to height, Americans no longer stand tallest," *Research News*, The Ohio State University. http://researchnews.osu.edu/archive /taller.htm

3. K. Marien, A. Conseur, and M. Sanderson, "The effect of fish consumption on DDT and DDE levels in breast milk among Hispanic immigrants," *Journal of Human Lactation* 14 (3, 1998): 237–42.

4. C. J. Earle, L. B. Brubaker, and G. Segura, International Tree Ring Data Base, NOAA/NGDC Paleoclimatology Program, Boulder, CO.

5. See Note 2.

6. Caroline Davis, Elizabeth Blackmore, Deborah Katzman, and John Fox, "Anorexia nervosa case study," paper presented at Statistical Society of Canada Annual Conference, Montreal, 2004. We have reversed the research question from that of the original case study.

7. Courtesy American Heritage Center, University of Wyoming, Laramie, WY.

8. Data courtesy of OzDASL (Australian Data and Story Library) at www.statsci.org. The original source is Cara Dubois, ed., *Lowie's Selected Papers in Anthropology* (University of California Press, 1960).
9. https://gigaom.com/2010/09/09/419-average-number-of-apps-downloaded -to-iphone-40-android-25/
10. Florida Department of Financial Services, 2011.
11. www.vgchartz.com, April 1, 2011.
12. National Weather Service, 2011.
13. Mary Madden and Amanda Lenhart, *Online Dating*, Pew Internet and American Life Project, 2006.
14. "Patterns and trends in nonmedical prescription pain reliever use: 2002 to 2005," *NSDUH Report*, Substance Abuse and Mental Health Services Administration, April 6, 2007.
15. The National Survey on Environmental Management of Asthma and Children's Exposure to Environmental Tobacco Smoke (NSEMA/CEE) (U.S. Environmental Protection Agency, 2004).
16. U.S. Energy Information Administration, 2014.
17. Ibid.
18. Florida Department of Financial Services, 2011.
19. Steve Strand, Ian Deary, and Pauline Smith, "Sex differences in cognitive abilities test scores: a UK national picture," *British Journal of Educational Psychology* 76 (Pt 3, 2006): 463–80.
20. Siobhan Banks and David Dinges, "Behavioral and physiological consequences of sleep restriction," *Journal of Clinical Sleep Medicine* 3 (5, 2007): 519–28.
21. Michael Smith, Ilona Croy, and Kerstin Persson Waye. "Human sleep and cortical reactivity are influenced by lunar phase," *Current Biology* 24 (12, 2014): R551–22.
22. *A Nation Online: Entering the Broadband Age*, Economics and Statistics Administration, U.S. Department of Commerce, 2004.
23. U.S. Energy Information Administration, 2006.

Chapter 10

1. Daniel Larose and Chantal Larose, *Data Mining and Predictive Analytics* (John Wiley and Sons, 2015). See also Daniel Larose and Chantal Larose, *Discovering Knowledge in Data: An Introduction to Data Mining* (John Wiley and Sons, 2014).
2. K. J. Thomas et al., "Randomised controlled trial of a short course of traditional acupuncture compared with usual care for persistent non-specific low back pain," *British Medical Journal* 333 (7569, 2006).
3. Karin Olson and John Hanson, "Using Reiki to manage pain," *Cancer Prevention and Control* 1 (2, 1997): 108–13.
4. T. J. Scanlon et al., "Is Friday the 13th bad for your health?" *British Medical Journal* 307 (6919, 1993): 1584–86.
5. George W. Snedecor and William G. Cochran, *Statistical Methods*, 8th Ed. (Iowa State University Press, 1989).
6. P. A. Mackowiak, S. S. Wasserman, and M. M. Levine, "A critical appraisal of 98.6 degrees F, the upper limit of the normal body temperature, and other legacies of Carl Reinhold August Wunderlich," *Journal of the American Medical Association* 268 (12, 1992): 1578–80.
7. See note 5.
8. See Barry K. Moser and Gary R. Stevens, "Homogeneity of variance in the two-sample means test," *American Statistician* 46 (1, 1992): 19–21.
9. Steven Reinberg, "U.S. kids using media almost 8 hours a day: survey finds few parents set rules on use of 'smart' phones, computers soars," *Bloomberg Business Week: Executive Health*, January 20, 2010.
10. D. L. Olds et al., "Improving the delivery of prenatal care and outcomes of pregnancy: a randomized trial of nurse home visitation," *Pediatrics* 77 (1, 1986): 16–28.
11. Michael Brett and Charles Goldman, "A meta-analysis of the freshwater trophic cascade," *Proceedings of the National Academy of Sciences of the United States of America* 93 (15, 1996): 7723–26.
12. Faria Sana, Tina Weston, and Nicholas Cepeda, "Laptop multitasking hinders classroom learning for both users and nearby peers," *Computers and Education* 62 (March 2013): 24–31.
13. H. H. Kelley, "The warm-cold variable in first impression of persons," *Journal of Personality* 18 (4, 1950): 431–39.
14. A. Towler and R. L. Dipboye, "The effect of instructor reputation and need for cognition on student behavior," poster presented at American Psychological Society conference, May 1998, San Francisco, CA.

15. www.pewinternet.org/2014/04/17/us-views-of-technology-and-the-future /pi_2014-04-16_techfuture_better_or_worse/.
16. V. K. Gadi et al., "Case-control study of fetal microchimerism and breast cancer," *PLoS One* 3 (3, 2008): e1706.
17. R. L. Bratton et al., "Effect of 'ionized' wrist bracelets on musculoskeletal pain: a randomized, double-blind, placebo-controlled trial," *Mayo Clinic Proceedings* 77 (11, 2002): 1164–68.
18. D. B. Allison et al., "Weight loss increases and fat loss decreases all-cause mortality rate: results from two independent cohort studies," *International Journal of Obesity* 23 (6, 1999): 603–11.
19. Siddharth Chandra et al., "Daily smoking patterns, their determinants, and implications for quitting," *Experimental and Clinical Psychopharmacology* 15 (1, 2007): 67–80.
20. Eric Zander and Sven Olof Dahlgren, "WISC-III index score profiles of 520 Swedish children with pervasive developmental disorders," *Psychological Assessment* 22 (2, 2010): 213–22.
21. G. Danaei et al., "The promise of prevention: the effects of four preventable risk factors and national life expectancy disparities by race and county in the United States," *PLoS Medicine* 7 (3, 2010): 1–14.
22. Letitia Williams et al., *Surveillance for Selected Maternal Behaviors and Experiences Before, During, and After Pregnancy*, Centers for Disease Control and Prevention, November 14, 2003. www.cdc.gov/mmwr/preview /mmwrhtml/ss5211a1.htm#tab5.

Chapter 11

1. Mary Madden and Amanda Lenhart, *Online Dating*, Pew Internet and American Life Project, 2005.
2. www.disastercenter.com/.
3. U.S. Department of Education, National Center for Education Statistics, Adult Education Survey of the 2005 National Household Education Surveys Program.
4. Derek M. Burnett et al., "Impact of minority status following traumatic spinal cord injury," *NeuroRehabilitation* 17 (3, 2002): 187–94.
5. Pew Research Center for the People and the Press, *How Young People View Their Lives, Futures, and Politics: A Portrait of "Generation Next"* (2007).
6. Andrew Rocco Tresolini Fiore, "Romantic regressions: an analysis of behavior in online dating systems," Master's thesis, Program in Media Arts and Sciences, Massachusetts Institute of Technology, 2004.
7. See Note 1.
8. M. A. Chase and G. M. Dummer, "The role of sports as a social determinant for children," *Research Quarterly for Exercise and Sport* 63 (4, 1992): 418–24.
9. J. R. Knight et al., "Alcohol abuse and dependence among U.S. college students," *Journal of Studies on Alcohol* 63 (3, 2002): 263–70.
10. Donald Garrow and Leonard Egede, "National patterns and correlates of complementary and alternative medicine use in adults with diabetes," *Journal of Alternative and Complementary Medicine* 12 (9, 2006): 895–902.
11. J. E. Anderson and S. Sansom, "HIV testing in a national sample of pregnant U.S. women: who is not getting tested?" *AIDS Care* 19 (3, 2007): 375–80.
12. National Agricultural Statistics Service, *Agricultural Statistics*, www.nass.usda.gov, 2006.

Chapter 12

1. Joseph Maze, Richard Murphy, and Cheri Simonds, "I'll see you on Facebook: the effects of computer-mediated teacher self-disclosure on student motivation, affective learning, and classroom climate," *Communication Edition* 56 (1, 2007): 1–17.
2. S. Blackman and D. Catalina, "The moon and the emergency room," *Perceptual and Motor Skills* 37 (2, 1973): 624–26.
3. William S. Cleveland, *Visualizing Data* (Hobart Press, 1993).
4. The data set is adapted from the **Cereals** data set from the Data and Story Library. http://ib.stat.cmu.edu/DASL.

Chapter 13

1. Manufacturer's Web site for Scrabble. www.hasbro.com/scrabble/en_US/.

Index

Note: Page numbers in **boldface** indicates a definition; *italics* indicates a figure; *t* indicates a table.